IDENTITIES AND FORMULAS

Basic Identities [5.1]

	Basic Identities	Common Equivalent Forms
Reciprocal	$\csc\theta = \dfrac{1}{\sin\theta}$	$\sin\theta = \dfrac{1}{\csc\theta}$
	$\sec\theta = \dfrac{1}{\cos\theta}$	$\cos\theta = \dfrac{1}{\sec\theta}$
	$\cot\theta = \dfrac{1}{\tan\theta}$	$\tan\theta = \dfrac{1}{\cot\theta}$
Ratio	$\tan\theta = \dfrac{\sin\theta}{\cos\theta}$	
	$\cot\theta = \dfrac{\cos\theta}{\sin\theta}$	
Pythagorean	$\cos^2\theta + \sin^2\theta = 1$	$\sin^2\theta = 1 - \cos^2\theta$
		$\sin\theta = \pm\sqrt{1 - \cos^2\theta}$
		$\cos^2\theta = 1 - \sin^2\theta$
		$\cos\theta = \pm\sqrt{1 - \sin^2\theta}$
	$1 + \tan^2\theta = \sec^2\theta$	
	$1 + \cot^2\theta = \csc^2\theta$	

Sum and Difference Formulas [5.2]

$\sin(A + B) = \sin A\cos B + \cos A\sin B$

$\sin(A - B) = \sin A\cos B - \cos A\sin B$

$\cos(A + B) = \cos A\cos B - \sin A\sin B$

$\cos(A - B) = \cos A\cos B + \sin A\sin B$

$$\tan(A + B) = \frac{\tan A + \tan B}{1 - \tan A\tan B}$$

$$\tan(A - B) = \frac{\tan A - \tan B}{1 + \tan A\tan B}$$

Double-Angle Formulas [5.3]

$\sin 2A = 2\sin A\cos A$

$\cos 2A = \cos^2 A - \sin^2 A$ First form

$= 2\cos^2 A - 1$ Second form

$= 1 - 2\sin^2 A$ Third form

$$\tan 2A = \frac{2\tan A}{1 - \tan^2 A}$$

Cofunction Theorem [2.1]

$\sin x = \cos(90° - x)$

$\cos x = \sin(90° - x)$

$\tan x = \cot(90° - x)$

Half-Angle Formulas [5.4]

$$\sin\frac{A}{2} = \pm\sqrt{\frac{1 - \cos A}{2}}$$

$$\cos\frac{A}{2} = \pm\sqrt{\frac{1 + \cos A}{2}}$$

$$\tan\frac{A}{2} = \frac{1 - \cos A}{\sin A} = \frac{\sin A}{1 + \cos A}$$

Even/Odd Functions [4.1]

$\cos(-\theta) = \cos\theta$ Even

$\sin(-\theta) = -\sin\theta$, $\tan(-\theta) = -\tan\theta$ Odd

Sum to Product Formulas [5.5]

$$\sin\alpha + \sin\beta = 2\sin\frac{\alpha + \beta}{2}\cos\frac{\alpha - \beta}{2}$$

$$\sin\alpha - \sin\beta = 2\cos\frac{\alpha + \beta}{2}\sin\frac{\alpha - \beta}{2}$$

$$\cos\alpha + \cos\beta = 2\cos\frac{\alpha + \beta}{2}\cos\frac{\alpha - \beta}{2}$$

$$\cos\alpha - \cos\beta = -2\sin\frac{\alpha + \beta}{2}\sin\frac{\alpha - \beta}{2}$$

Product to Sum Formulas [5.5]

$$\sin A\cos B = \frac{1}{2}[\sin(A + B) + \sin(A - B)]$$

$$\cos A\sin B = \frac{1}{2}[\sin(A + B) - \sin(A - B)]$$

$$\cos A\cos B = \frac{1}{2}[\cos(A + B) + \cos(A - B)]$$

$$\sin A\sin B = \frac{1}{2}[\cos(A - B) - \cos(A + B)]$$

Pythagorean Theorem [1.1]

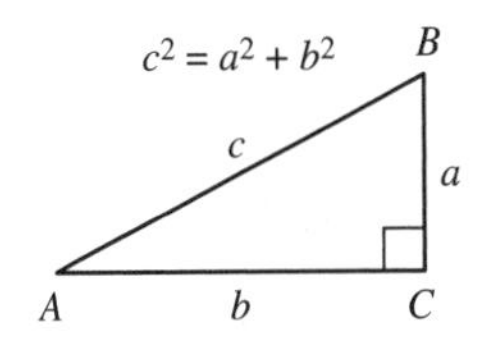

The Law of Sines [7.1]

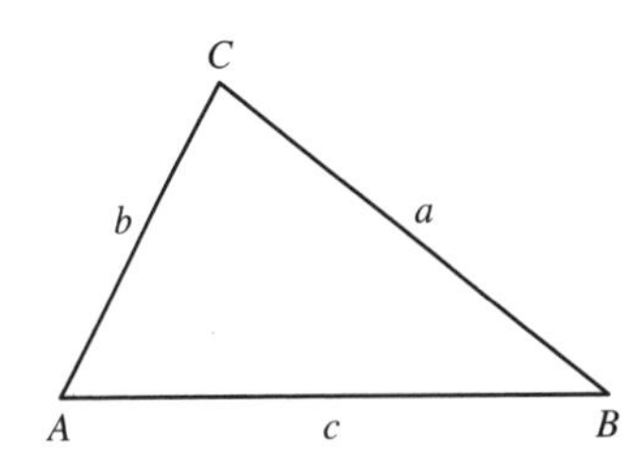

$$\frac{\sin A}{a} = \frac{\sin B}{b} = \frac{\sin C}{c}$$

or, equivalently,

$$\frac{a}{\sin A} = \frac{b}{\sin B} = \frac{c}{\sin C}$$

The Law of Cosines [7.2]

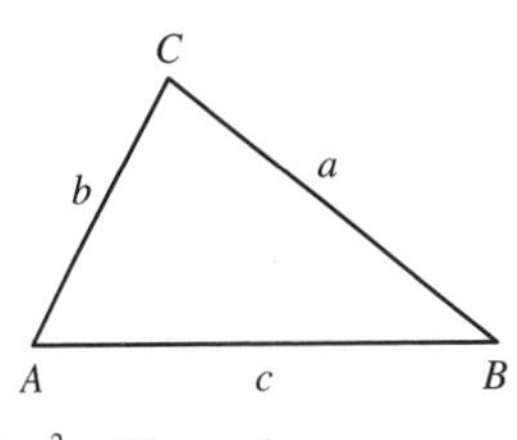

$a^2 = b^2 + c^2 - 2bc\cos A$

$b^2 = a^2 + c^2 - 2ac\cos B$

$c^2 = a^2 + b^2 - 2ab\cos C$

or, equivalently,

$$\cos A = \frac{b^2 + c^2 - a^2}{2bc}$$

$$\cos B = \frac{a^2 + c^2 - b^2}{2ac}$$

$$\cos C = \frac{a^2 + b^2 - c^2}{2ab}$$

TRIGONOMETRY 7E

Charles P. McKeague
Cuesta College

Mark D. Turner
Cuesta College

BROOKS/COLE
CENGAGE Learning™

Australia • Brazil • Japan • Korea • Mexico • Singapore • Spain • United Kingdom • United States

***Trigonometry*, Seventh Edition**
Charles P. McKeague and Mark D. Turner

Acquisitions Editor: Gary Whalen

Developmental Editor: Stacy Green

Assistant Editor: Cynthia Ashton

Editorial Assistant: Sabrina Black

Media Editor: Lynh Pham

Senior Marketing Manager: Danae April

Marketing Coordinator: Shannon Maier

Marketing Communications Manager: Mary Anne Payumo

Content Project Manager: Jennifer Risden

Design Director: Rob Hugel

Art Director: Vernon Boes

Print Buyer: Becky Cross

Rights Acquisitions Specialist: Roberta Broyer

Production Service: MPS Limited, a Macmillan Company

Text Designer: Terri Wright

Text Design Images: J. A. Kraulis/Mastefile, Masterfile Royalty-Free/Masterfile, Aaron Graubert/Getty Images, Fotosearch/Photolibrary

Photo Researcher: Bill Smith Group

Text Researcher: Sue Howard

Copy Editor: Martha Williams

Illustrator: Lori Heckelman; MPS Limited, a Macmillan Company

Cover Designer: Larry Didona

Cover Image: Ferris wheel–Masterfile Royalty-free; Compass–Fotosearch Value/Photolibrary

Compositor: MPS Limited, a Macmillan Company

Library of Congress Control Number: 2011933694

ISBN-13: 978-1-111-82685-7
ISBN-10: 1-111-82685-4

Brooks/Cole
20 Davis Drive
Belmont, CA 94002-3098
USA

Cengage Learning is a leading provider of customized learning solutions with office locations around the globe, including Singapore, the United Kingdom, Australia, Mexico, Brazil, and Japan. Locate your local office at **www.cengage.com/global.**

Cengage Learning products are represented in Canada by Nelson Education, Ltd.

To learn more about Brooks/Cole, visit **www.cengage.com/brookscole.**

Purchase any of our products at your local college store or at our preferred online store **www.cengagebrain.com.**

Printed in China
3 4 5 6 17 16 15 14

BRIEF CONTENTS

BRIEF CONTENTS

CONTENTS

CHAPTER 4 Graphing and Inverse Functions 178

CHAPTER 5 Identities and Formulas 270

CHAPTER 6 Equations 318

CHAPTER 7 Triangles 360

CHAPTER 8 Complex Numbers and Polar Coordinates 418

APPENDIX A Review of Functions 478

PREFACE

This seventh edition of *Trigonometry* preserves the popular format and style of the previous editions. It is a standard right triangle approach to trigonometry. Nearly every section is written so it can be discussed in a typical 50-minute class session. The clean layout and conversational style encourage students to read the text.

The focus of the textbook is on understanding the definitions and principles of trigonometry and their applications to problem solving. Exact values of the trigonometric functions are emphasized throughout the textbook. In addition, this seventh edition emphasizes student learning objectives and assessment.

The text covers all the material usually taught in trigonometry. There is also an appendix on functions and inverse functions. The appendix sections can be used as a review of topics that students may already be familiar with, or they can be used to provide instruction for students encountering these concepts for the first time.

Numerous calculator notes are placed throughout the text to help students calculate values when appropriate. As there are many different models of graphing calculators, and each model has its own set of commands, we have tried to avoid an overuse of specific key icons or command names.

NEW TO THIS EDITION

Content Changes

The following list describes the major content changes you will see in this seventh edition.

- **Section 2.5:** General formulas for the horizontal and vertical vector components of a vector are now given.
- **Section 4.3:** To be consistent with the terminology used in other disciplines, we have eliminated any reference to "phase shift" when describing the horizontal translation and simply refer to this quantity as the horizontal shift. Also, the concept of phase is introduced as the fraction of a period by which one graph lags or leads another.
- **Chapter 7:** The law of cosines is now presented in Section 7.2 prior to a discussion of the ambiguous case for oblique triangles, which follows in Section 7.3. The material on the ambiguous case has been expanded so that these problems are now solved using both the law of sines and the law of cosines. Definitions for heading and true course remain in Section 7.2.
- **Section 8.2:** The complex plane is more formally presented and some of the history behind its development is given.
- **Appendix B:** The content on exponential and logarithmic functions has been moved to CengageBrain.com.

New Features

STUDENT LEARNING OBJECTIVES Each section begins with a list of student learning objectives which describe the specific, measurable knowledge and skills that students are expected to achieve. Learning objectives help the student identify and focus on the important concepts in each section and increase the likelihood of their success by having established and clear goals. For instructors, learning objectives can help in organizing class lessons and learning activities and in creating student assessments.

LEARNING OBJECTIVES ASSESSMENTS Multiple-choice questions have been added at the end of every problem set and are designed to be used in class or outside of class to assess student learning. Each question directly corresponds to one of the student learning objectives for that section. Answers to these questions are not available to students but are provided for instructors in the Instructor's Edition and the Instructor's Solutions Manual. These problems can be especially useful for schools and institutions required to provide documentation and data relating to assessment of student learning outcomes.

MATCHED PRACTICE PROBLEMS In every section of this book, each example is now paired with a matched practice problem that is similar to the example. These problems give students an opportunity to practice what they have just learned before moving on to the next example. Instructors may want to use them as in-class examples or to provide guided practice activities in class. Answers are given in the answers section in the back of the book.

CONCEPTS AND VOCABULARY Each problem set begins with a new set of questions that focus on grasping the main ideas and understanding the vocabulary and terminology presented in that particular section. Most of these questions are short-answer, but in some cases also include matching or other formats.

CUMULATIVE TESTS To help students review previous learning and retain information better, three cumulative tests have been added to the book. These are similar to the chapter tests, except that the questions pertain to all the sections in the book up to that point. The Cumulative Tests are good resources for students studying for a midterm or final exam. Answers to both odd and even problems for cumulative tests are given in the back of the book.

NEW EXERCISES AND APPLICATIONS New exercises and applications have been added in some sections to help students gain a better grasp of key concepts and to help motivate students and stimulate their interest in trigonometry.

ENHANCED WEB ASSIGN This revision is accompanied by a significant increase in the number of exercises that are included within Enhanced Web Assign. All the odd problems in each problem set, and in some cases a number of even problems, are now available in electronic form.

CONTINUING FEATURES

CHAPTER INTRODUCTIONS Each chapter opens with an introduction in which a real-world application, historical example, or a link between topics is used to stimulate interest in the chapter. Many of these introductions are expanded in the chapter and then carried through to topics found later in the book. Many sections open in a similar fashion.

STUDY SKILLS *Study Skills* sections are found in the first six chapter openings and help students become organized and efficient with their time.

USING TECHNOLOGY *Using Technology* sections throughout the book show how graphing calculator technology can be used to enhance the topics covered. All graphing calculator material is optional, but even if you are not using graphing calculator technology in your classroom, these segments can provide additional insight into the standard trigonometric procedures and problem solving found in the section.

THREE DEFINITIONS All three definitions for the trigonometric functions are contained in the text. The point-on-the-terminal-side definition is contained in Section 1.3. The right triangle definition is in Section 2.1. Circular functions are given in Section 3.3.

GETTING READY FOR CLASS Located before each problem set, *Getting Ready for Class* sections require written responses from students and can be answered by reading the preceding section. They are to be done before the students come to class.

THEMES There are a number of themes that run throughout the textbook. We have clearly marked these themes in the problem sets with appropriate icons. Here is a list of the icons and corresponding themes.

 Ferris Wheels

 Human Cannonball

 Navigation

 Sports

APPLICATIONS Application problems are titled according to subject. We have found that students are more likely to put some time and effort into trying application problems if they do not have to work an overwhelming number of them at one time, and if they work on them every day. For this reason, a few application problems are included toward the end of almost every problem set in the book.

GRAPHING CALCULATOR EXERCISES For those of you who are using graphing calculators in your classes, exercises that require graphing calculators are included in some of the problem sets. These exercises are clearly marked with a special icon and may easily be omitted if you are not using this technology in your classroom. Some of these exercises are investigative in nature and help prepare students for concepts that are introduced in following sections.

REVIEW PROBLEMS Each problem set, beginning with Chapter 2, contains a few review problems. Where appropriate, the review problems cover material that will be needed in the next section. Otherwise, they cover material from the previous chapter. Continual review will help students retain what they have learned in previous chapters and reinforce main ideas.

EXTENDING THE CONCEPTS Scattered throughout the text, *Extending the Concepts* problems involve students in researching, writing, and group work on topics such as English, history, religion, and map making. These problems address the goals of increased writing and group activities recommended by various professional mathematics organizations. Some also require research on the Internet.

GROUP PROJECTS Each chapter concludes with a group project involving some interesting problem or application that relates to or extends the ideas introduced in the chapter. Many of the projects emphasize the connection of mathematics with other disciplines or illustrate real-life situations in which trigonometry is used. The projects are designed to be used in class with groups of three or four students each, but the problems could also be given as individual assignments for students wanting an additional challenge.

RESEARCH PROJECTS At least one research project is also offered at the end of each chapter. The research projects ask students to investigate a historical topic or personage that is in some way connected to the material in the chapter, and are intended to promote an appreciation for the rich history behind trigonometry. Students may find the local library or the Internet to be helpful resources in doing their research.

CHAPTER SUMMARIES Each chapter summary lists the main properties and definitions found in the chapter. The margins in the chapter summaries contain examples that illustrate the topics being reviewed.

CHAPTER TESTS Every chapter ends with a chapter test that contains a representative sample of the problems covered in the chapter. The chapter tests were designed to be short enough so that a student may work all the problems in a reasonable amount of time. If you want to reduce the number of problems even further, you can just assign the odd problems or the even problems. Answers to both odd and even problems for chapter tests are given in the back of the book.

SUPPLEMENTS AND ADDITIONAL RESOURCES

For the Instructor

INSTRUCTOR'S EDITION The instructor's version of the complete student text contains answers to all the problems in the problem sets, chapter tests, and cumulative tests.

ENHANCED WEBASSIGN Exclusively from Cengage Learning, Enhanced WebAssign offers an extensive online program for *Trigonometry* to encourage the practice that is so critical for concept mastery. The meticulously crafted pedagogy and exercises in this text become even more effective in Enhanced WebAssign, supplemented by multimedia tutorial support and immediate feedback as students complete their assignments. Algorithmic problems allow you to assign unique versions to each student. The Practice Another Version feature (activated at your discretion) allows students to attempt the questions with new sets of values until they feel confident enough to work the original problem. Students benefit from Personal Study Plans (based on diagnostic quizzing) that identify chapter topics they still need to master and links to video solutions, interactive tutorials, and even online help.

TEST BANK The Test Bank includes multiple tests per chapter as well as final exams. The tests are made up of a combination of multiple-choice, free response, and fill-in-the-blank questions.

INSTRUCTOR'S SOLUTIONS MANUAL The Instructor's Solutions Manual provides worked-out solutions to all the even problems in the text and answers to all Learning Objectives Assessment questions.

EXAMVIEW COMPUTERIZED TESTING ExamView testing software allows you to quickly create, deliver, and customize tests for class in print or online formats and features automatic grading. It includes a test bank with hundreds of questions customized directly to the text. ExamView is available within the PowerLecture CD-ROM.

SOLUTION BUILDER This online instructor database offers complete worked solutions to all exercises in the text, allowing you to create customized, secure solutions printouts (in PDF format) matched exactly to the problems you assign in class. For more information, see *www.cengage.com/solutionbuilder*.

POWERLECTURE WITH EXAMVIEW CD-ROM This CD-ROM provides you with dynamic media tools for teaching. You can create, deliver, and customize tests (both print and online) in minutes with ExamView Computerized Testing, as well as easily build solution sets for homework or exams using Solution Builder's online solution manual. Microsoft® PowerPoint® lecture slides and figures from the book are also included on this CD-ROM.

For the Student

STUDENT SOLUTIONS MANUAL The Student Solutions Manual provides worked-out solutions to the odd-numbered problems from the text's problem sets and all problems from the chapter tests and cumulative tests.

ENHANCED WEBASSIGN Exclusively from Cengage Learning, Enhanced WebAssign offers an extensive online program for *Trigonometry* to encourage the practice that's so critical for concept mastery. You'll receive multimedia tutorial support as you complete your assignments. You'll also benefit from Personal Study Plans (based on diagnostic quizzing) that identify chapter topics you still need to master and links to video solutions, interactive tutorials, and even live online help.

CENGAGEBRAIN.COM Visit *www.cengagebrain.com* to access additional course materials and companion resources. At the CengageBrain.com homepage, search for the ISBN of your title (from the back cover of your book) using the search box at the top of the page. This will take you to the product page where free companion resources can be found.

TEXT-SPECIFIC DVDs These text-specific DVDs feature 10- to 20-minute problem-solving lessons that cover each section of every chapter with worked-out solutions to selected problems of the text.

ACKNOWLEDGMENTS

First and foremost, we are thankful to our many loyal users who have helped make this book one of the most popular trigonometry texts on the market. We sincerely appreciate your comments and words of encouragement offered at conferences and by e-mail.

In producing this seventh edition, we relied on the help of many hardworking people. Gary Whalen, Stacy Green, Cynthia Ashton, and Lynh Pham, our editors at Cengage Learning, were instrumental in initiating the revision, helped us form a central vision, and took care of the many supplements. Vemon Boes provided us with the stunning cover and design. Jennifer Risden guided us through the production process. Charu Khanna and Martha Williams at MPS Limited did an outstanding job with the composition and copyediting. Our accuracy checkers, Jeffrey Saikali, San Diego Miramar College, and Ann Ostberg, Grace University, logged many tedious hours looking for errors. Judy Barclay, formerly at Cuesta College, and Ross Rueger, College of the Sequoias, continued in their roles as authors of the solutions manuals. Our thanks go to all these people; this book would not have been possible without them.

Special thanks go to Diane McKeague and Anne Turner for their continued encouragement and support, and to Allison Turner and Kaitlin Turner for giving up some "daddy" time while this revision was in production.

Finally, a number of people provided us with suggestions and helpful comments on this revision, including some who performed a review or were asked to give feedback on specific topics. We are grateful to all of the following people for their help with this revision.

Thomas Beatty, *Florida Gulf Coast University*
Taylor Braz, Student, *Fullerton College*
Laura Crow, *Lone Star College — Montgomery*
Roger Davidson, *Yuba College*
David Dudley, *Scottsdale Community College*
Joan Van Glabek, *Edison State College*
Shari Harris, *John Wood Community College*
Susan Howell, *University of Southern Mississippi*
Lisa Johnson, *Collin College*
Mikal McDowell, *Cedar Valley College*
Manouchehr Misaghian, *Prairie View A&M University*
Joseph N. O'Brien, *Sam Houston State University*
Alan Roebuck, *Chaffey College*
Jeffrey Saikali, *San Diego Miramar College*
Marcia Carol Siderow, *California State University—Northridge*
Patricia B. Simeon, *Louisiana Tech University*
Mark E. Williams, *University of Maryland Eastern Shore*

Charles P. McKeague
Mark D. Turner
September 2011

A SPECIAL NOTE to the Student

Trigonometry can be a very enjoyable subject to study. You will find that there are many interesting and useful problems that trigonometry can be used to solve. However, many trigonometry students are apprehensive at first because they are worried they will not understand the topics we cover. When we present a new topic that they do not grasp completely, they think something is wrong with them for not understanding it. On the other hand, some students are excited about the course from the beginning. They are not worried about understanding trigonometry and, in fact, expect to find some topics difficult.

What is the difference between these two types of students?

Those who are excited about the course know from experience (as you do) that a certain amount of confusion is associated with most new topics in mathematics. They don't worry about it because they also know that the confusion gives way to understanding in the process of reading the textbook, working problems, and getting their questions answered. If they find a topic difficult, they work as many problems as necessary to grasp the subject. They don't wait for the understanding to come to them; they go out and get it by working lots of problems. In contrast, the students who lack confidence tend to give up when they become confused. Instead of working more problems, they sometimes stop working problems altogether, and that, of course, guarantees that they will remain confused.

If you are worried about this course because you lack confidence in your ability to understand trigonometry, and you want to change the way you feel about mathematics, then look forward to the first topic that causes you some confusion. As soon as that topic comes along, make it your goal to master it, in spite of your apprehension. You will see that each and every topic covered in this course is one you can eventually master, even if your initial introduction to it is accompanied by some confusion. As long as you have passed a college-level intermediate algebra course (or its equivalent), you are ready to take this course.

It also helps a great deal if you make a solid commitment to your trigonometry course. If you are not completely committed to a class, then you will tend to give less than your full effort. Consider this quote from Johann Wolfgang Von Goethe's *Faust*:

> *Until one is committed, there is hesitancy, the chance to draw back, always ineffectiveness. Concerning all acts of initiative and creation, there is one elementary truth the ignorance of which kills countless ideas and splendid plans: that the moment one definitely commits oneself, then providence moves too. All sorts of things occur to help one that would never otherwise have occurred. A whole stream of events issues from the decision, raising in one's favor all manner of unforeseen incidents, meetings and material assistance which no man could have dreamed would have come his way. Whatever you can do or dream you can, begin it. Boldness has genius, power and magic in it.*

If you are committed to doing well in trigonometry, the following suggestions will be important to you.

How to Be Successful in Trigonometry

1. **Attend all class sessions on time.** You cannot know exactly what goes on in class unless you are there. Missing class and then expecting to find out what went on from someone else is not the same as being there yourself.

2. **Read the book.** This textbook was written for you! It is best to read the section that will be covered in class beforehand. Reading in advance, even if you do not understand everything you read, helps prepare a foundation for what you will see in class. Also, your instructor may not have time to discuss everything you need to know from a section, so you may need to pick up some things on your own.

3. **Work problems every day and check your answers.** The key to success in mathematics is working problems. The more problems you work, the better you will become at working them. The answers to the odd-numbered problems are given in the back of the book. When you have finished an assignment, be sure to compare your answers with those in the book. If you have made a mistake, find out what it is, and correct it.

4. **Do it on your own.** Don't be misled into thinking someone else's work is your own. Having someone else show you how to work a problem is not the same as working that problem yourself. It is okay to get help when you are stuck. As a matter of fact, it is a good idea. Just be sure you do the work yourself and that you can work the entire problem correctly on your own later on.

5. **Review every day.** After you have finished the problems your instructor has assigned, take another 15 minutes and review a section you have already completed. The more you review, the longer you will retain the material you have learned.

6. **Don't expect to understand every new topic the first time you see it.** Sometimes you will understand everything you are doing, and sometimes you won't. That's just the way things are in mathematics. Expecting to understand each new topic the first time you see it can lead to disappointment and frustration. The process of understanding trigonometry takes time. It requires you to read the book, work problems, and get your questions answered.

7. **Spend as much time as it takes for you to master the material.** No set formula exists for the exact amount of time you need to spend on trigonometry to master it. You will find out as you go along what is or isn't enough time for you. If you end up spending two or more hours on each section in order to master the material there, then that's how much time it takes; trying to get by with less will not work.

8. **Relax.** It's probably not as difficult as you think.

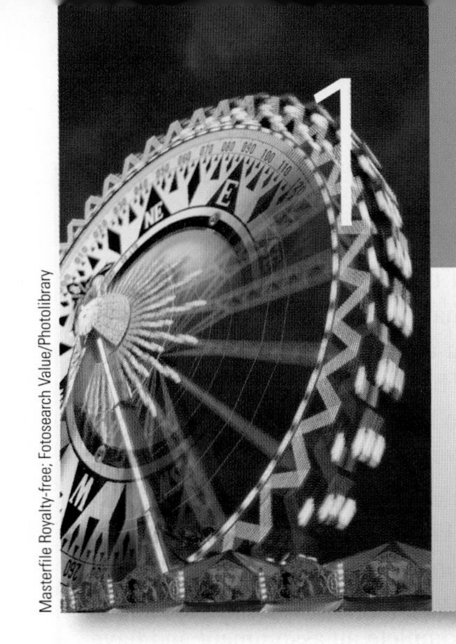

1 The Six Trigonometric Functions

Without Thales there would not have been a Pythagoras—or such a Pythagoras; and without Pythagoras there would not have been a Plato—or such a Plato.

➤ *D. E. Smith*

INTRODUCTION

The history of mathematics is a spiral of knowledge passed down from one generation to another. Each person in the history of mathematics is connected to the others along this spiral. In Italy, around 500 B.C., the Pythagoreans discovered a relationship between the sides of any right triangle. That discovery, known as the Pythagorean Theorem, is the foundation on which the Spiral of Roots shown in Figure 1 is built. The Spiral of Roots gives us a way to visualize square roots of positive integers.

Figure 1

In Problem Set 1.1, you will have a chance to construct the Spiral of Roots yourself.

STUDY SKILLS 1

At the beginning of the first few chapters of this book you will find a Study Skills section in which we list the skills that are necessary for success in trigonometry. If you have just completed an algebra class successfully, you have acquired most of these skills. If it has been some time since you have taken a math class, you must pay attention to the sections on study skills.

Here is a list of things you can do to develop effective study skills.

1. **Put Yourself on a Schedule** The general rule is that you spend two hours on homework for every hour you are in class. Make a schedule for yourself, setting aside at least six hours a week to work on trigonometry. Once you make the schedule, stick to it. Don't just complete your assignments and then stop. Use all the time you have set aside. If you complete an assignment and have time left over, read the next section in the book and work more problems. As the course progresses you may find that six hours a week is not enough time for you to master the material in this course. If it takes you longer than that to reach your goals for this course, then that's how much time it takes. Trying to get by with less will not work.
2. **Find Your Mistakes and Correct Them** There is more to studying trigonometry than just working problems. You must always check your answers with those in the back of the book. When you have made a mistake, find out what it is and correct it. Making mistakes is part of the process of learning mathematics. The key to discovering what you do not understand can be found by correcting your mistakes.
3. **Imitate Success** Your work should look like the work you see in this book and the work your instructor shows. The steps shown in solving problems in this book were written by someone who has been successful in mathematics. The same is true of your instructor. Your work should imitate the work of people who have been successful in mathematics.
4. **Memorize Definitions and Identities** You may think that memorization is not necessary if you understand a topic you are studying. In trigonometry, memorization is especially important. In this first chapter, you will be presented with the definition of the six trigonometric functions that you will use throughout the rest of the course. We have seen many bright students struggle with trigonometry simply because they did not memorize the definitions and identities when they were first presented. ▲

SECTION 1.1 Angles, Degrees, and Special Triangles

LEARNING OBJECTIVES

1. Compute the complement and supplement of an angle.
2. Use the Pythagorean Theorem to find the third side of a right triangle.
3. Find the other two sides of a $30°–60°–90°$ or $45°–45°–90°$ triangle given one side.
4. Solve a real-life problem using the special triangle relationships.

Introduction

Table 1 is taken from the trail map given to skiers at Northstar at Tahoe Ski Resort in Lake Tahoe, California. The table gives the length of each chair lift at Northstar, along with the change in elevation from the beginning of the lift to the end of the lift.

Right triangles are good mathematical models for chair lifts. In this section we review some important items from geometry, including right triangles. Let's begin by looking at some of the terminology associated with angles.

TABLE 1
From the Trail Map for Northstar at Tahoe Ski Resort

Lift Information		
Lift	**Vertical Rise (ft)**	**Length (ft)**
Big Springs Gondola	480	4,100
Bear Paw Double	120	790
Echo Triple	710	4,890
Aspen Express Quad	900	5,100
Forest Double	1,170	5,750
Lookout Double	960	4,330
Comstock Express Quad	1,250	5,900
Rendezvous Triple	650	2,900
Schaffer Camp Triple	1,860	6,150
Chipmunk Tow Lift	28	280
Bear Cub Tow Lift	120	750

Angles in General

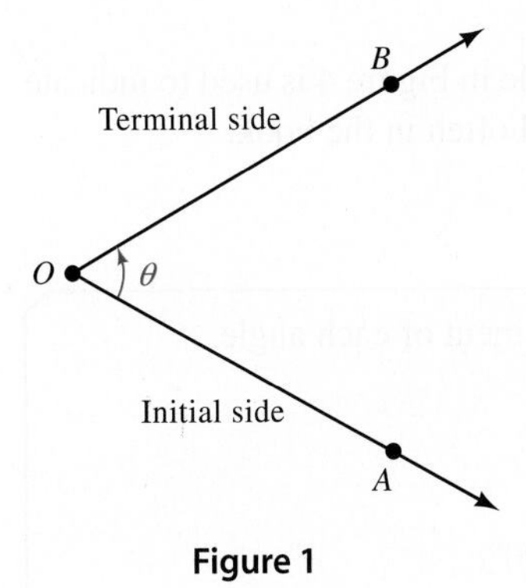

Figure 1

An angle is formed by two rays with the same end point. The common end point is called the *vertex* of the angle, and the rays are called the *sides* of the angle.

In Figure 1 the vertex of angle θ (theta) is labeled O, and A and B are points on each side of θ. Angle θ can also be denoted by AOB, where the letter associated with the vertex is written between the letters associated with the points on each side.

We can think of θ as having been formed by rotating side OA about the vertex to side OB. In this case, we call side OA the *initial side* of θ and side OB the *terminal side* of θ.

When the rotation from the initial side to the terminal side takes place in a counterclockwise direction, the angle formed is considered a *positive angle.* If the rotation is in a clockwise direction, the angle formed is a *negative angle* (Figure 2).

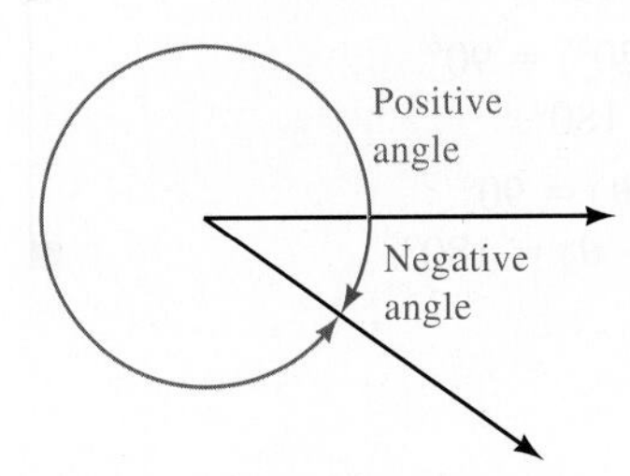

Figure 2

Degree Measure

One way to measure the size of an angle is with degree measure. The angle formed by rotating a ray through one complete revolution has a measure of 360 degrees, written 360° (Figure 3).

One degree (1°), then, is 1/360 of a full rotation. Likewise, 180° is one-half of a full rotation, and 90° is half of that (or a quarter of a rotation). Angles that measure 90° are called *right angles,* while angles that measure 180° are called *straight angles.* Angles that measure between 0° and 90° are called *acute angles,* while angles that measure between 90° and 180° are called *obtuse angles* (see Figure 4).

If two angles have a sum of 90°, then they are called *complementary angles,* and we say each is the *complement* of the other. Two angles with a sum of 180° are called *supplementary angles.*

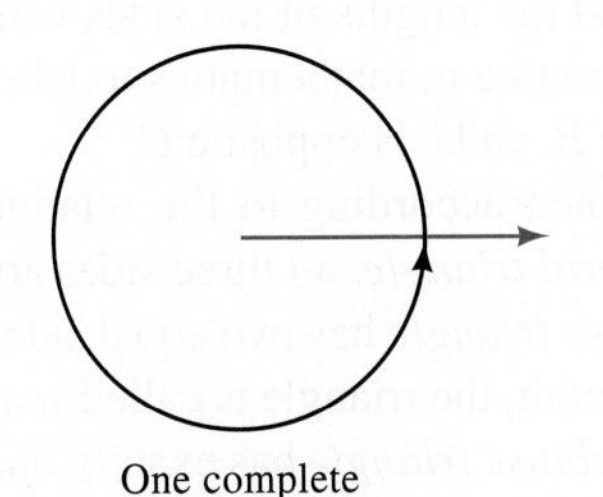
One complete revolution = 360°
Figure 3

NOTE To be precise, we should say "two angles, the sum of the measures of which is 180°, are called supplementary angles" because there is a difference between an angle and its measure. However, in this book, we will not always draw the distinction

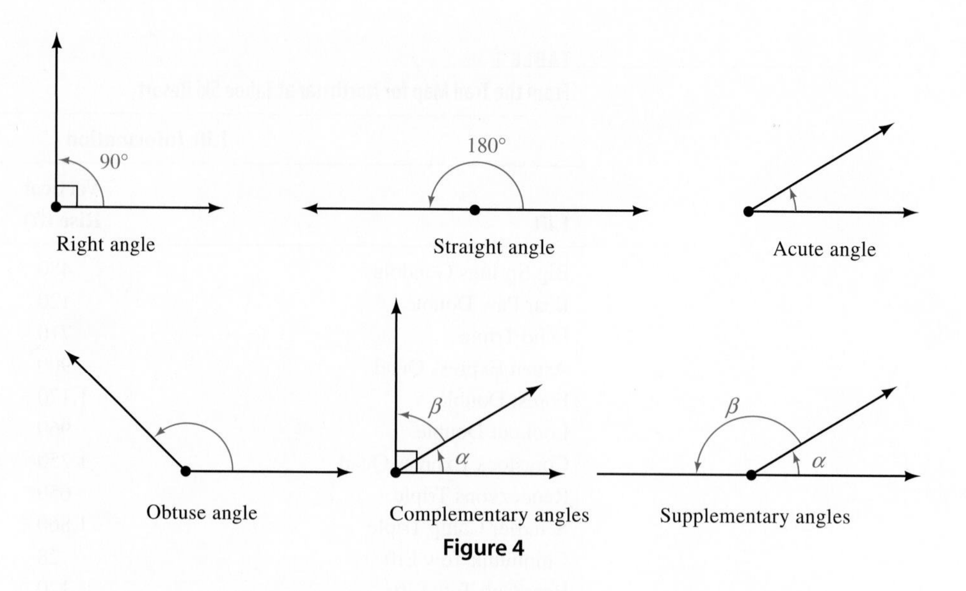

Figure 4

between an angle and its measure. Many times we will refer to "angle θ" when we actually mean "the measure of angle θ."

NOTE The little square by the vertex of the right angle in Figure 4 is used to indicate that the angle is a right angle. You will see this symbol often in the book.

PROBLEM 1
Give the complement and supplement of each angle.
a. 25°
b. 118°
c. β

EXAMPLE 1 Give the complement and the supplement of each angle.

a. 40° **b.** 110° **c.** θ

SOLUTION

a. The complement of 40° is 50° since $40° + 50° = 90°$.
The supplement of 40° is 140° since $40° + 140° = 180°$.

b. The complement of 110° is $-20°$ since $110° + (-20°) = 90°$.
The supplement of 110° is 70° since $110° + 70° = 180°$.

c. The complement of θ is $90° - \theta$ since $\theta + (90° - \theta) = 90°$.
The supplement of θ is $180° - \theta$ since $\theta + (180° - \theta) = 180°$. ■

Triangles

A triangle is a three-sided polygon. Every triangle has three sides and three angles. We denote the angles (or vertices) with uppercase letters and the lengths of the sides with lowercase letters, as shown in Figure 5. It is standard practice in mathematics to label the sides and angles so that a is opposite A, b is opposite B, and c is opposite C.

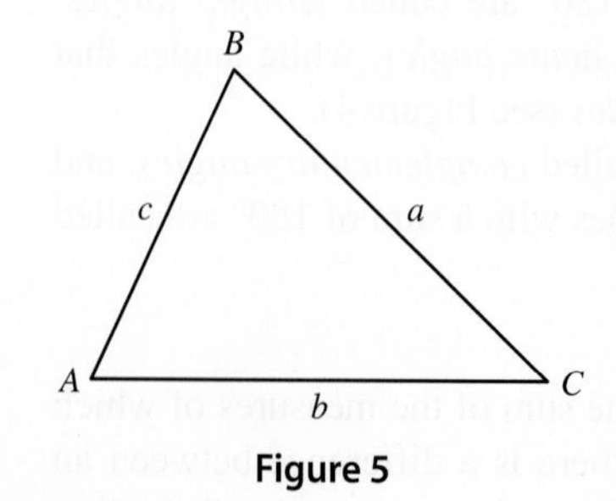

Figure 5

There are different types of triangles that are named according to the relative lengths of their sides or angles (Figure 6). In an *equilateral triangle,* all three sides are of equal length and all three angles are equal. An *isosceles triangle* has two equal sides and two equal angles. If all the sides and angles are different, the triangle is called *scalene*. In an *acute triangle,* all three angles are acute. An *obtuse triangle* has exactly one obtuse angle, and a *right triangle* has one right angle.

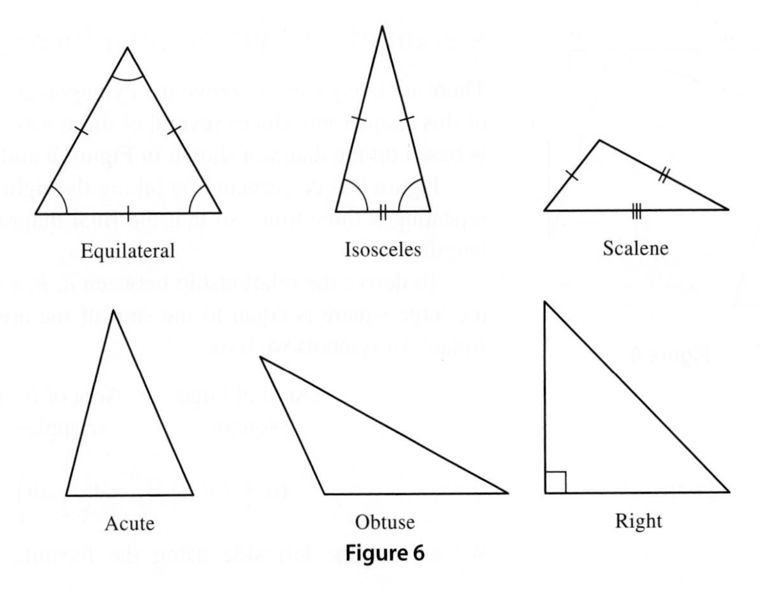

Figure 6

Special Triangles

As we will see throughout this text, right triangles are very important to the study of trigonometry. In every right triangle, the longest side is called the *hypotenuse,* and it is always opposite the right angle. The other two sides are called the *legs* of the right triangle. Because the sum of the angles in any triangle is 180°, the other two angles in a right triangle must be complementary, acute angles. The Pythagorean Theorem that we mentioned in the introduction to this chapter gives us the relationship that exists among the sides of a right triangle. First we state the theorem.

PYTHAGOREAN ■ THEOREM

In any right triangle, the square of the length of the longest side (called the hypotenuse) is equal to the sum of the squares of the lengths of the other two sides (called legs).

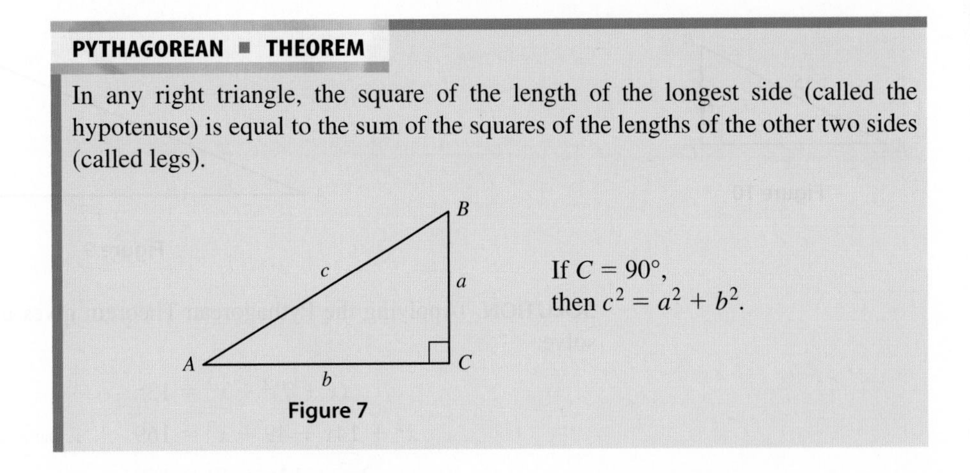

Figure 7

Next we will prove the Pythagorean Theorem. Part of the proof involves finding the area of a triangle. In any triangle, the area is given by the formula

$$\text{Area} = \frac{1}{2}(\text{base})(\text{height})$$

For the right triangle shown in Figure 7, the base is b, and the height is a. Therefore the area is $A = \frac{1}{2}ab$.

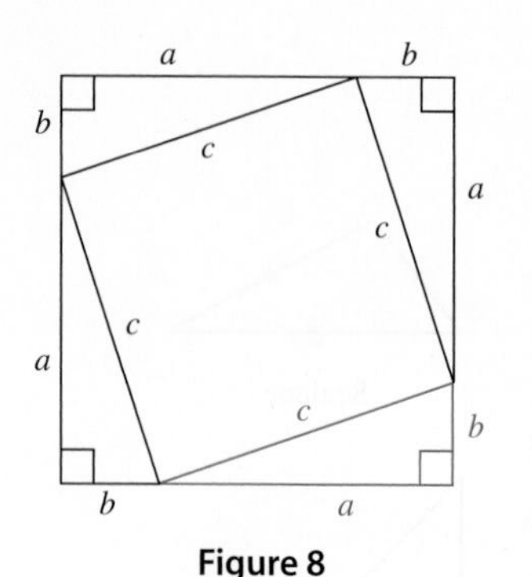

Figure 8

A Proof of the Pythagorean Theorem

There are many ways to prove the Pythagorean Theorem. The Group Project at the end of this chapter introduces several of these ways. The method that we are offering here is based on the diagram shown in Figure 8 and the formula for the area of a triangle.

Figure 8 is constructed by taking the right triangle in the lower right corner and repeating it three times so that the final diagram is a square in which each side has length $a + b$.

To derive the relationship between a, b, and c, we simply notice that the area of the large square is equal to the sum of the areas of the four triangles and the inner square. In symbols we have

Area of large square		Area of four triangles		Area of inner square
$(a + b)^2$	$=$	$4\left(\frac{1}{2}ab\right)$	$+$	c^2

We expand the left side using the formula for the square of a binomial, from algebra. We simplify the right side by multiplying 4 by $\frac{1}{2}$.

$$a^2 + 2ab + b^2 = 2ab + c^2$$

Adding $-2ab$ to each side, we have the relationship we are after:

$$a^2 + b^2 = c^2$$

PROBLEM 2
Solve for x in Figure 10.

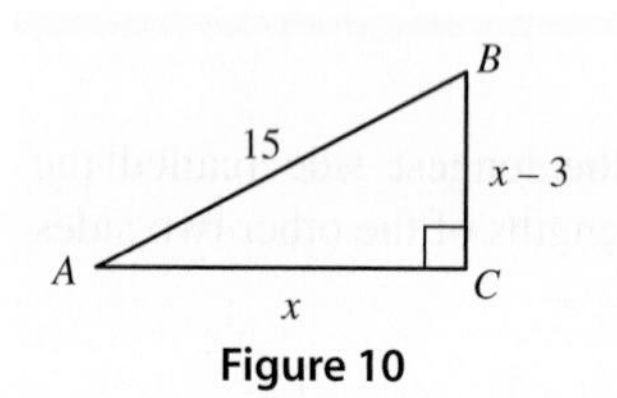

Figure 10

EXAMPLE 2 Solve for x in the right triangle in Figure 9.

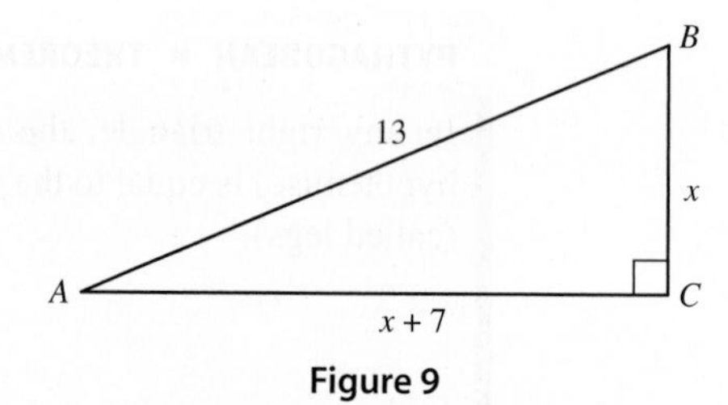

Figure 9

SOLUTION Applying the Pythagorean Theorem gives us a quadratic equation to solve.

$$
\begin{aligned}
(x + 7)^2 + x^2 &= 13^2 & & \\
x^2 + 14x + 49 + x^2 &= 169 & & \text{Expand } (x+7)^2 \text{ and } 13^2 \\
2x^2 + 14x + 49 &= 169 & & \text{Combine similar terms} \\
2x^2 + 14x - 120 &= 0 & & \text{Add } -169 \text{ to both sides} \\
x^2 + 7x - 60 &= 0 & & \text{Divide both sides by 2} \\
(x - 5)(x + 12) &= 0 & & \text{Factor the left side}
\end{aligned}
$$

$$x - 5 = 0 \quad \text{or} \quad x + 12 = 0 \qquad \text{Set each factor to 0}$$

$$x = 5 \quad \text{or} \quad x = -12$$

Our only solution is $x = 5$. We cannot use $x = -12$ because x is the length of a side of triangle ABC and therefore cannot be negative. ■

NOTE The lengths of the sides of the triangle in Example 2 are 5, 12, and 13. Whenever the three sides in a right triangle are natural numbers, those three numbers are called a *Pythagorean triple.*

PROBLEM 3
Repeat Example 3 for the Schaffer Camp Triple chair lift.

EXAMPLE 3 Table 1 in the introduction to this section gives the vertical rise of the Forest Double chair lift (Figure 11) as 1,170 feet and the length of the chair lift as 5,750 feet. To the nearest foot, find the horizontal distance covered by a person riding this lift.

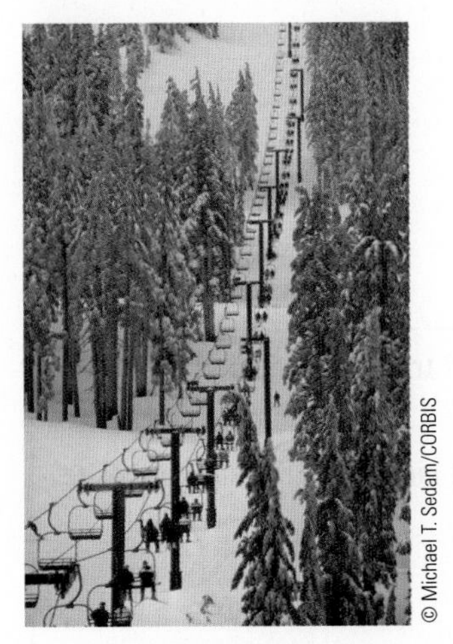

Figure 11

SOLUTION Figure 12 is a model of the Forest Double chair lift. A rider gets on the lift at point A and exits at point B. The length of the lift is AB.

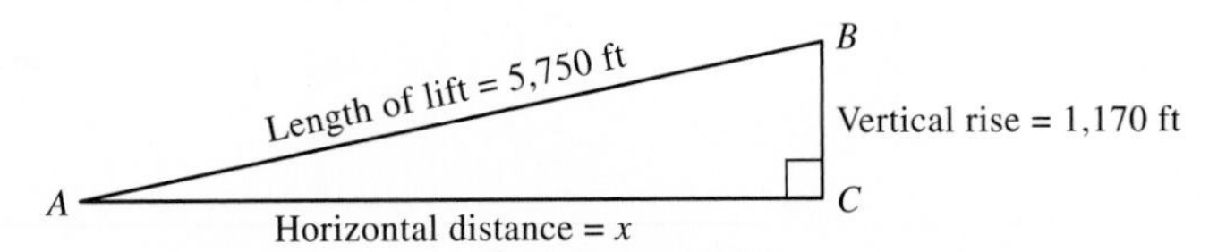

Figure 12

To find the horizontal distance covered by a person riding the chair lift, we use the Pythagorean Theorem:

$$5{,}750^2 = x^2 + 1{,}170^2 \quad \text{Pythagorean Theorem}$$

$$33{,}062{,}500 = x^2 + 1{,}368{,}900 \quad \text{Simplify squares}$$

$$x^2 = 33{,}062{,}500 - 1{,}368{,}900 \quad \text{Solve for } x^2$$

$$x^2 = 31{,}693{,}600 \quad \text{Simplify the right side}$$

$$x = \sqrt{31{,}693{,}600}$$

$$x = 5{,}630 \text{ ft} \quad \text{To the nearest foot}$$

A rider getting on the lift at point A and riding to point B will cover a horizontal distance of approximately 5,630 feet.

Before leaving the Pythagorean Theorem, we should mention something about Pythagoras and his followers, the Pythagoreans. They established themselves as a secret society around the year 540 B.C. The Pythagoreans kept no written record of their work; everything was handed down by spoken word. Their influence was not only in mathematics, but also in religion, science, medicine, and music. Among other things, they discovered the correlation between musical notes and the reciprocals of counting numbers, $\frac{1}{2}$, $\frac{1}{3}$, $\frac{1}{4}$, and so on. In their daily lives they followed strict dietary and moral rules to achieve a higher rank in future lives. The British philosopher Bertrand Russell has referred to Pythagoras as "intellectually one of the most important men that ever lived."

30°
2t
$t\sqrt{3}$
60°
t

30°–60°–90°

Figure 13

THE 30°–60°–90° TRIANGLE

In any right triangle in which the two acute angles are 30° and 60°, the longest side (the hypotenuse) is always twice the shortest side (the side opposite the 30° angle), and the side of medium length (the side opposite the 60° angle) is always $\sqrt{3}$ times the shortest side (Figure 13).

NOTE The shortest side t is opposite the smallest angle 30°. The longest side $2t$ is opposite the largest angle 90°.

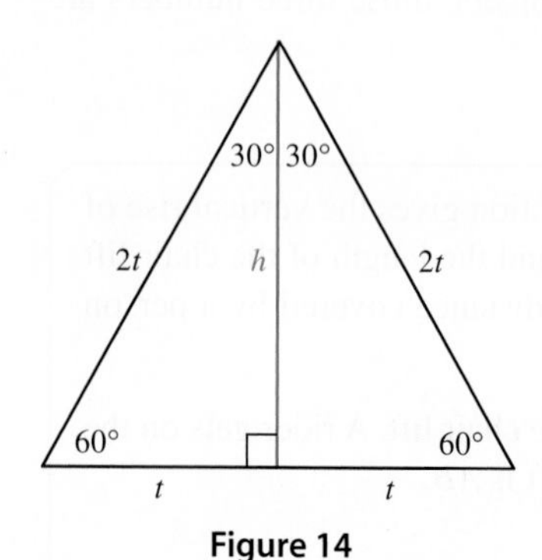

Figure 14

To verify the relationship between the sides in this triangle, we draw an equilateral triangle (one in which all three sides are equal) and label half the base with t (Figure 14).

The altitude h (the colored line) bisects the base. We have two 30°–60°–90° triangles. The longest side in each is $2t$. We find that h is $t\sqrt{3}$ by applying the Pythagorean Theorem.

$$\begin{aligned} t^2 + h^2 &= (2t)^2 \\ h &= \sqrt{4t^2 - t^2} \\ &= \sqrt{3t^2} \\ &= t\sqrt{3} \end{aligned}$$

PROBLEM 4
If the longest side of a 30°–60°–90° triangle is 14, find the lengths of the other two sides.

EXAMPLE 4 If the shortest side of a 30°–60°–90° triangle is 5, find the other two sides.

SOLUTION The longest side is 10 (twice the shortest side), and the side opposite the 60° angle is $5\sqrt{3}$ (Figure 15).

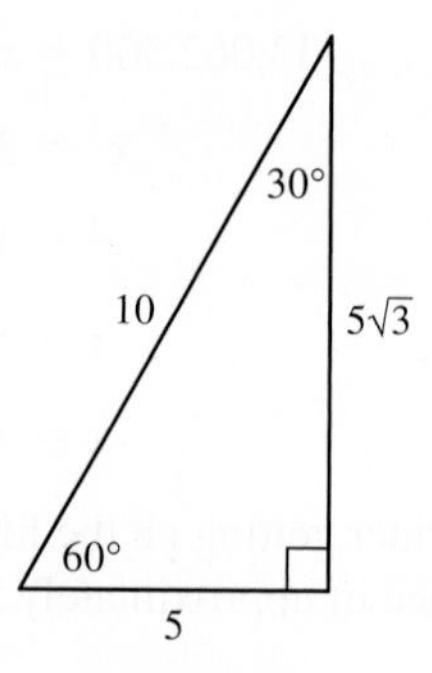

Figure 15

PROBLEM 5
A ladder is leaning against a wall. The bottom of the ladder makes an angle of 60° with the ground and is 3 feet from the base of the wall. How long is the ladder and how high up the wall does it reach?

EXAMPLE 5 A ladder is leaning against a wall. The top of the ladder is 4 feet above the ground and the bottom of the ladder makes an angle of 60° with the ground (Figure 16). How long is the ladder, and how far from the wall is the bottom of the ladder?

SOLUTION The triangle formed by the ladder, the wall, and the ground is a 30°–60°–90° triangle. If we let x represent the distance from the bottom of the ladder to the wall, then the length of the ladder can be represented by $2x$. The distance from the top of the ladder to the ground is $x\sqrt{3}$, since it is opposite the 60° angle (Figure 17). It is also given as 4 feet. Therefore,

$$\begin{aligned} x\sqrt{3} &= 4 \\ x &= \frac{4}{\sqrt{3}} \\ &= \frac{4\sqrt{3}}{3} \end{aligned}$$

Rationalize the denominator by multiplying the numerator and denominator by $\sqrt{3}$.

Figure 16

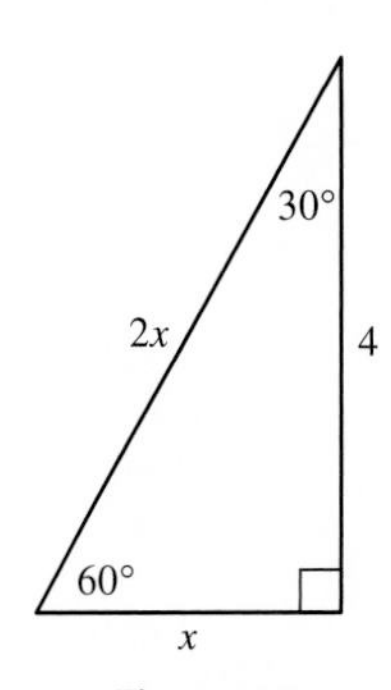

Figure 17

The distance from the bottom of the ladder to the wall, x, is $4\sqrt{3}/3$ feet, so the length of the ladder, $2x$, must be $8\sqrt{3}/3$ feet. Note that these lengths are given in exact values. If we want a decimal approximation for them, we can replace $\sqrt{3}$ with 1.732 to obtain

$$\frac{4\sqrt{3}}{3} \approx \frac{4(1.732)}{3} = 2.309 \text{ ft}$$

$$\frac{8\sqrt{3}}{3} \approx \frac{8(1.732)}{3} = 4.619 \text{ ft}$$

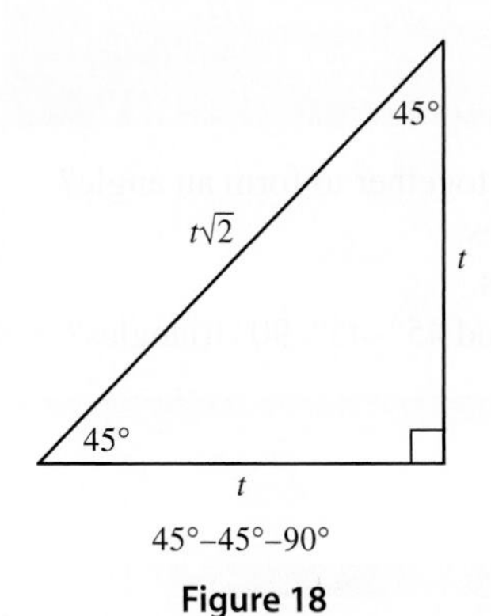

45°–45°–90°

Figure 18

THE 45°–45°–90° TRIANGLE

If the two acute angles in a right triangle are both 45°, then the two shorter sides (the legs) are equal and the longest side (the hypotenuse) is $\sqrt{2}$ times as long as the shorter sides. That is, if the shorter sides are of length t, then the longest side has length $t\sqrt{2}$ (Figure 18).

To verify this relationship, we simply note that if the two acute angles are equal, then the sides opposite them are also equal. We apply the Pythagorean Theorem to find the length of the hypotenuse.

$$\begin{aligned}\text{hypotenuse} &= \sqrt{t^2 + t^2}\\ &= \sqrt{2t^2}\\ &= t\sqrt{2}\end{aligned}$$

PROBLEM 6
Repeat Example 6 if an 8-foot rope is used.

EXAMPLE 6 A 10-foot rope connects the top of a tent pole to the ground. If the rope makes an angle of 45° with the ground, find the length of the tent pole (Figure 19).

SOLUTION Assuming that the tent pole forms an angle of 90° with the ground, the triangle formed by the rope, tent pole, and the ground is a 45°–45°–90° triangle (Figure 20).

If we let x represent the length of the tent pole, then the length of the rope, in terms of x, is $x\sqrt{2}$. It is also given as 10 feet. Therefore,

$$x\sqrt{2} = 10$$

$$x = \frac{10}{\sqrt{2}} = 5\sqrt{2}$$

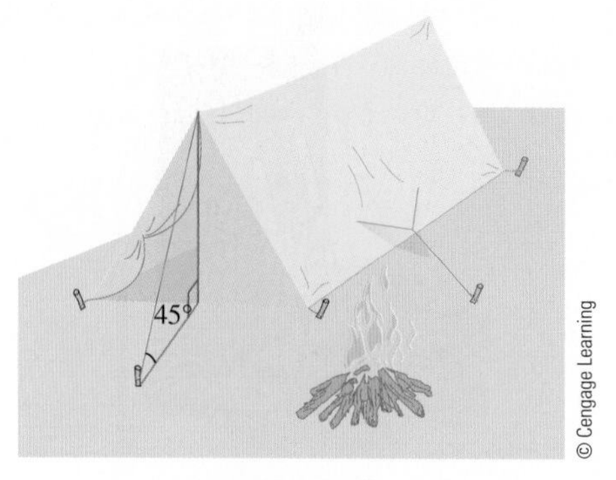

Figure 19

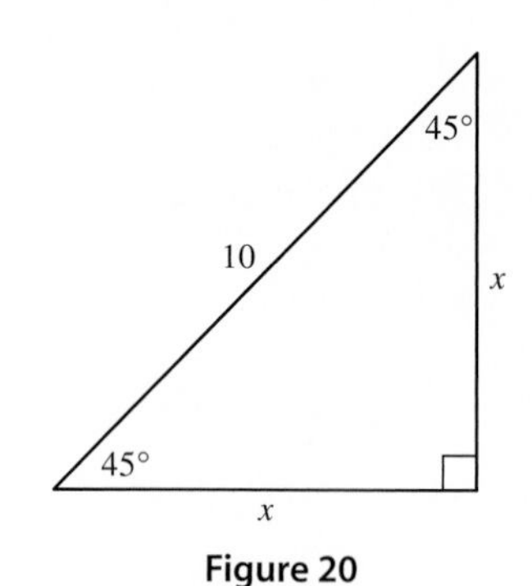

Figure 20

The length of the tent pole is $5\sqrt{2}$ feet. Again, $5\sqrt{2}$ is the exact value of the length of the tent pole. To find a decimal approximation, we replace $\sqrt{2}$ with 1.414 to obtain

$$5\sqrt{2} \approx 5(1.414) = 7.07 \text{ ft}$$

Getting Ready for Class

After reading through the preceding section, respond in your own words and in complete sentences.

a. What do we call the point where two rays come together to form an angle?
b. In your own words, define complementary angles.
c. In your own words, define supplementary angles.
d. Why is it important to recognize 30°–60°–90° and 45°–45°–90° triangles?

1.1 PROBLEM SET

➤ CONCEPTS AND VOCABULARY

For Questions 1 through 7, fill in each blank with the appropriate word or number.

1. For a positive angle, the rotation from the initial side to the terminal side takes place in a __________ direction. For a negative angle, the rotation takes place in a __________ direction.
2. Two angles with a sum of 90° are called __________ angles, and when the sum is 180° they are called __________ angles.
3. In any triangle, the sum of the three interior angles is always _______.
4. In a right triangle, the longest side opposite the right angle is called the _________ and the other two sides are called _______.
5. The Pythagorean Theorem states that the square of the ________ is equal to the ______ of the squares of the ________.

6. In a 30°–60°–90° triangle, the hypotenuse is always ________ the shortest side, and the side opposite the 60° angle is always ________ times the shortest side.

7. In a 45°–45°–90° triangle, the legs are always ________ and the hypotenuse is always ________ times either leg.

8. Match each term with the appropriate angle measure.

a. Right **i.** $0° < \theta < 90°$
b. Straight **ii.** $\theta = 90°$
c. Acute **iii.** $90° < \theta < 180°$
d. Obtuse **iv.** $\theta = 180°$

➤ EXERCISES

Indicate which of the angles below are acute angles and which are obtuse angles. Then give the complement and the supplement of each angle.

9. 10° **10.** 50° **11.** 45° **12.** 90°
13. 120° **14.** 160° **15.** x **16.** y

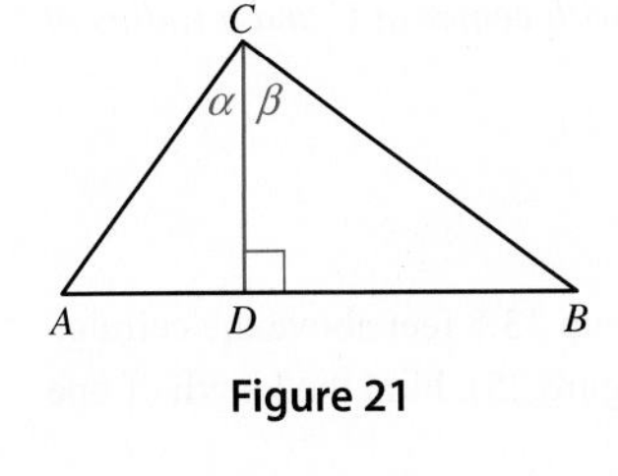

Figure 21

Problems 17 through 22 refer to Figure 21. (Remember: *The sum of the three angles in any triangle is always 180°.*)

17. Find α if $A = 30°$. **18.** Find B if $\beta = 45°$.

19. Find α if $A = 2\alpha$. **20.** Find α if $A = \alpha$.

21. Find A if $B = 30°$ and $\alpha + \beta = 100°$.

22. Find B if $\alpha + \beta = 80°$ and $A = 80°$.

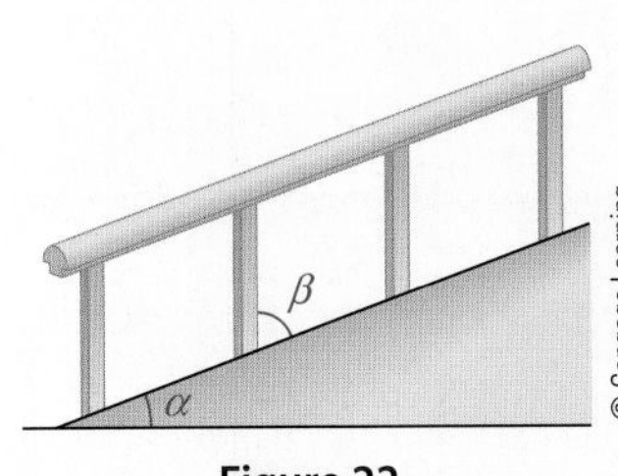

Figure 22

Figure 22 shows a walkway with a handrail. Angle α is the angle between the walkway and the horizontal, while angle β is the angle between the vertical posts of the handrail and the walkway. Use Figure 22 to work Problems 23 through 26. (Assume that the vertical posts are perpendicular to the horizontal.)

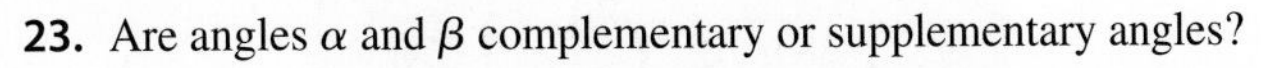

23. Are angles α and β complementary or supplementary angles?

24. If we did not know that the vertical posts were perpendicular to the horizontal, could we answer Problem 23?

25. Find α if $\beta = 52°$.

26. Find β if $\alpha = 25°$.

27. Rotating Light A searchlight rotates through one complete revolution every 4 seconds. How long does it take the light to rotate through 90°?

28. Clock Through how many degrees does the hour hand of a clock move in 4 hours?

29. Geometry An isosceles triangle is a triangle in which two sides are equal in length. The angle between the two equal sides is called the vertex angle, while the other two angles are called the base angles. If the vertex angle is 40°, what is the measure of the base angles?

30. Geometry An equilateral triangle is a triangle in which all three sides are equal. What is the measure of each angle in an equilateral triangle?

Problems 31 through 36 refer to right triangle ABC with C = 90°.

31. If $a = 4$ and $b = 3$, find c. **32.** If $a = 6$ and $b = 8$, find c.

33. If $a = 8$ and $c = 17$, find b. **34.** If $a = 2$ and $c = 6$, find b.

35. If $b = 12$ and $c = 13$, find a. **36.** If $b = 10$ and $c = 26$, find a.

Solve for x in each of the following right triangles:

37.

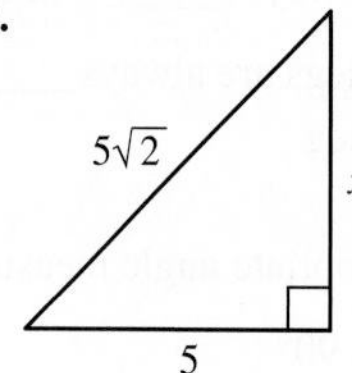

38.

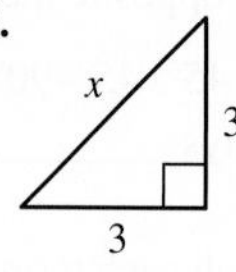

39.

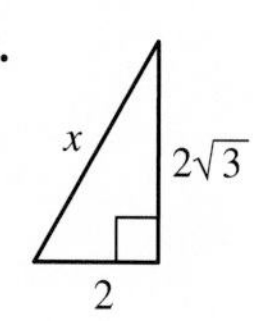

40.

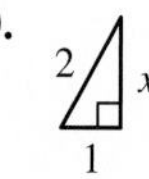

41.

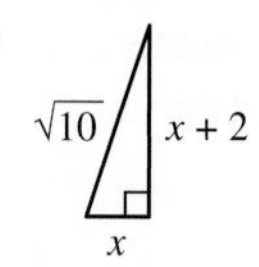

42.

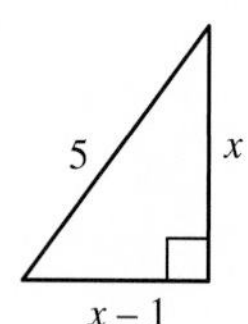

Figure 23

Problems 43 and 44 refer to Figure 23.

43. Find AB if $BC = 4$, $BD = 5$, and $AD = 2$.

44. Find BD if $BC = 5$, $AB = 13$, and $AD = 4$.

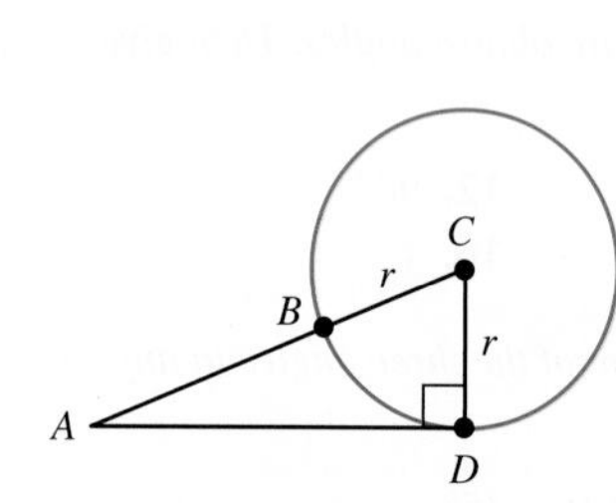

Figure 24

Problems 45 and 46 refer to Figure 24, which shows a circle with center at C and a radius of r, and right triangle ADC.

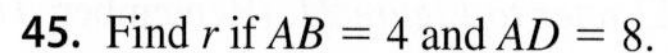

45. Find r if $AB = 4$ and $AD = 8$.

46. Find r if $AB = 8$ and $AD = 12$.

47. Pythagorean Theorem The roof of a house is to extend up 13.5 feet above the ceiling, which is 36 feet across, forming an isosceles triangle (Figure 25). Find the length of one side of the roof.

Figure 25

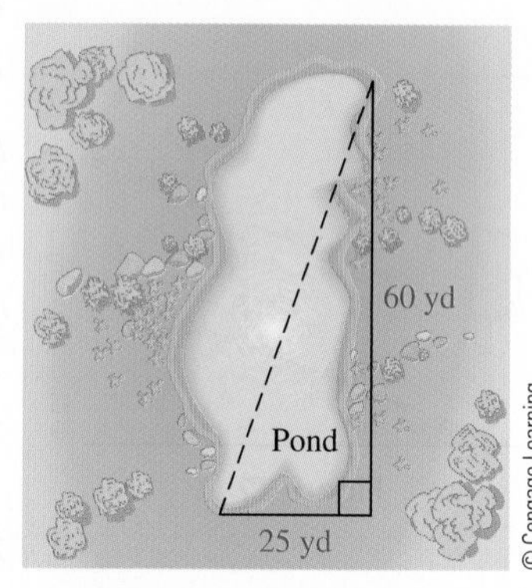

Figure 26

48. Surveying A surveyor is attempting to find the distance across a pond. From a point on one side of the pond he walks 25 yards to the end of the pond and then makes a 90° turn and walks another 60 yards before coming to a point directly across the pond from the point at which he started. What is the distance across the pond? (See Figure 26.)

Find the remaining sides of a 30°–60°–90° triangle if

49. the shortest side is 1

50. the shortest side is 3

51. the longest side is 8

52. the longest side is 5

53. the side opposite 60° is 6

54. the side opposite 60° is 4

55. Escalator An escalator in a department store is to carry people a vertical distance of 20 feet between floors. How long is the escalator if it makes an angle of 30° with the ground?

56. Escalator What is the length of the escalator in Problem 55 if it makes an angle of 60° with the ground?

57. **Tent Design** A two-person tent is to be made so that the height at the center is 4 feet. If the sides of the tent are to meet the ground at an angle of 60°, and the tent is to be 6 feet in length, how many square feet of material will be needed to make the tent? (Figure 27; assume that the tent has a floor and is closed at both ends, and give your answer in exact form and approximate to the nearest tenth of a square foot.)

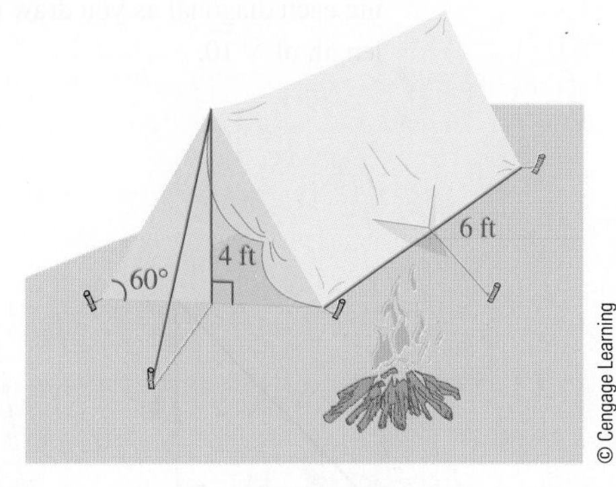

Figure 27

58. **Tent Design** If the height at the center of the tent in Problem 57 is to be 3 feet, how many square feet of material will be needed to make the tent?

Find the remaining sides of a 45°–45°–90° triangle if

59. the shorter sides are each $\frac{4}{5}$

60. the shorter sides are each $\frac{1}{2}$

61. the longest side is $8\sqrt{2}$

62. the longest side is $5\sqrt{2}$

63. the longest side is 4

64. the longest side is 12

65. **Distance a Bullet Travels** A bullet is fired into the air at an angle of 45°. How far does it travel before it is 1,000 feet above the ground? (Assume that the bullet travels in a straight line; neglect the forces of gravity, and give your answer to the nearest foot.)

66. **Time a Bullet Travels** If the bullet in Problem 65 is traveling at 2,828 feet per second, how long does it take for the bullet to reach a height of 1,000 feet?

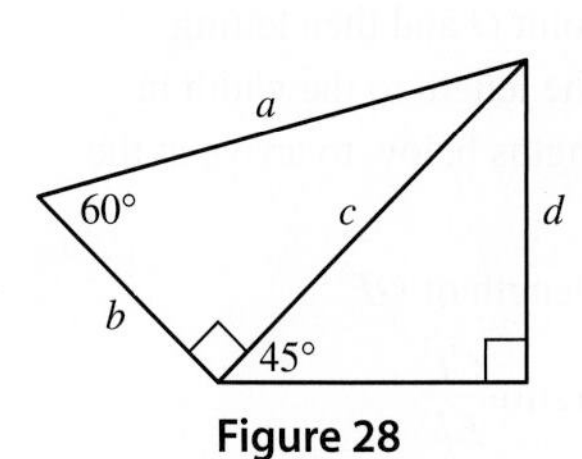

Figure 28

Problems 67 and 68 refer to Figure 28.

67. Find the lengths of sides a, b, and d if $c = 3$.

68. Find the lengths of sides b, c, and d if $a = 4$.

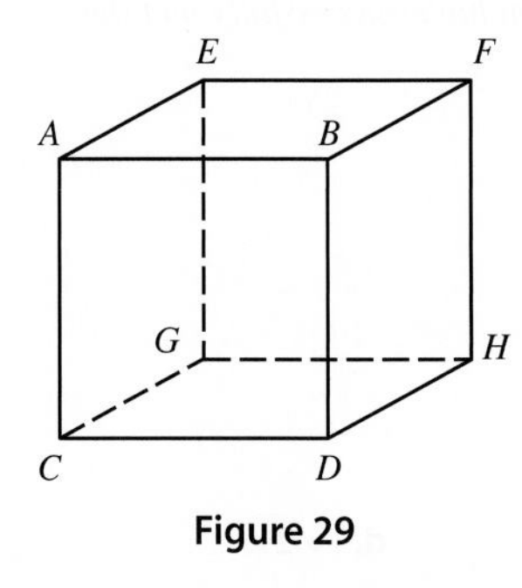

Figure 29

Geometry: Characteristics of a Cube *The object shown in Figure 29 is a cube (all edges are equal in length). Use this diagram for Problems 69 through 72.*

69. If the length of each edge of the cube shown in Figure 29 is 1 inch, find
    a. the length of diagonal CH
    b. the length of diagonal CF

70. If the length of each edge of the cube shown in Figure 29 is 5 centimeters, find
    a. the length of diagonal GD
    b. the length of diagonal GB

71. If the length of each edge of the cube shown in Figure 29 is unknown, we can represent it with the variable x. Then we can write formulas for the lengths of any of the diagonals. Finish each of the following statements:
    a. If the length of each edge of a cube is x, then the length of the diagonal of any face of the cube will be ______.
    b. If the length of each edge of a cube is x, then the length of any diagonal that passes through the center of the cube will be ______.

72. What is the measure of $\angle GDH$ in Figure 29?

➤ EXTENDING THE CONCEPTS

73. The Spiral of Roots The introduction to this chapter shows the Spiral of Roots. The following three figures (Figures 30, 31, and 32) show the first three stages in the construction of the Spiral of Roots. Using graph paper and a ruler, construct the Spiral of Roots, labeling each diagonal as you draw it, to the point where you can see a line segment with a length of $\sqrt{10}$.

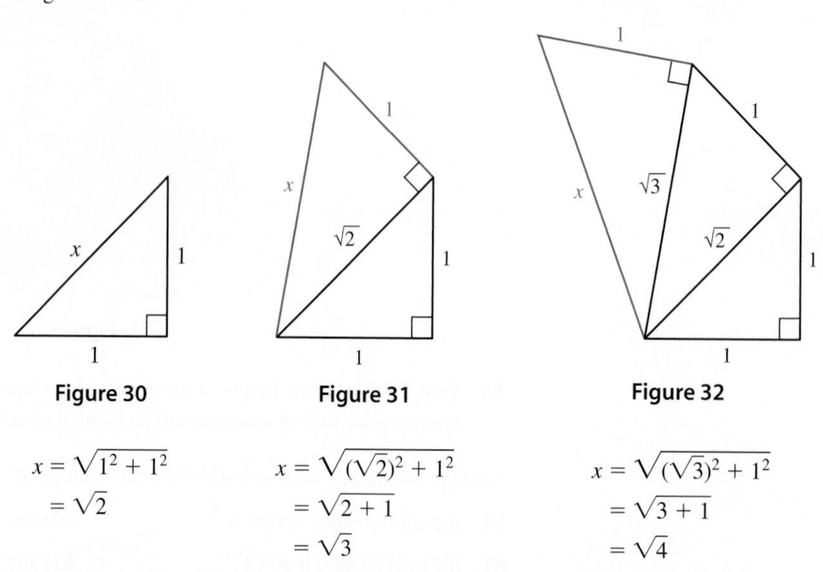

Figure 30 Figure 31 Figure 32

$$x = \sqrt{1^2 + 1^2} = \sqrt{2}$$

$$x = \sqrt{(\sqrt{2})^2 + 1^2} = \sqrt{2 + 1} = \sqrt{3}$$

$$x = \sqrt{(\sqrt{3})^2 + 1^2} = \sqrt{3 + 1} = \sqrt{4}$$

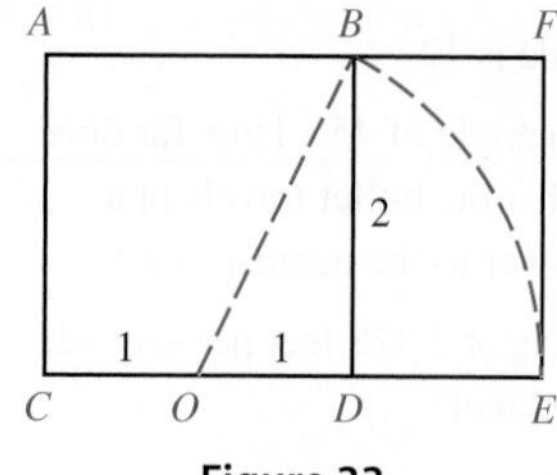

Figure 33

74. The Golden Ratio Rectangle $ACEF$ (Figure 33) is a golden rectangle. It is constructed from square $ACDB$ by holding line segment OB fixed at point O and then letting point B drop down until OB aligns with CD. The ratio of the length to the width in the golden rectangle is called the *golden ratio.* Find the lengths below to arrive at the golden ratio.

a. Find the length of OB. **b.** Find the length of OE.

c. Find the length of CE. **d.** Find the ratio $\dfrac{CE}{EF}$.

➤ LEARNING OBJECTIVES ASSESSMENT

These questions are available for instructors to help assess if you have successfully met the learning objectives for this section.

75. Compute the complement and supplement of 61°.

a. Complement = −61°, supplement = 119°
b. Complement = −61°, supplement = 299°
c. Complement = 119°, supplement = 29°
d. Complement = 29°, supplement = 119°

76. In right triangle ABC, find b if $a = 2$, $c = 5$, and $C = 90°$.

a. 7 **b.** 3 **c.** $\sqrt{21}$ **d.** $\sqrt{29}$

77. Find the remaining sides of a 30°–60°–90° triangle if the longest side is 6.

a. $6, 6\sqrt{2}$ **b.** $3\sqrt{2}, 3\sqrt{2}$ **c.** $3, 3\sqrt{3}$ **d.** $12, 6\sqrt{3}$

78. An escalator in a department store makes an angle of 45° with the ground. How long is the escalator if it carries people a vertical distance of 24 feet?

a. $12\sqrt{2}$ ft **b.** $24\sqrt{2}$ ft **c.** $8\sqrt{3}$ ft **d.** 48 ft

SECTION 1.2 The Rectangular Coordinate System

LEARNING OBJECTIVES

1. Verify a point lies on the graph of the unit circle.
2. Find the distance between two points.
3. Draw an angle in standard position.
4. Find an angle that is coterminal with a given angle.

René Descartes

The book *The Closing of the American Mind* by Allan Bloom was published in 1987 and spent many weeks on the bestseller list. In the book, Mr. Bloom recalls being in a restaurant in France and overhearing a waiter call another waiter a "Cartesian." He goes on to say that French people today define themselves in terms of the philosophy of either René Descartes (1595–1650) or Blaise Pascal (1623–1662). Followers of Descartes are sometimes referred to as *Cartesians.* As a philosopher, Descartes is responsible for the statement "I think, therefore I am." In mathematics, Descartes is credited with, among other things, the invention of the rectangular coordinate system, which we sometimes call the *Cartesian coordinate system.* Until Descartes invented his coordinate system in 1637, algebra and geometry were treated as separate subjects. The rectangular coordinate system allows us to connect algebra and geometry by associating geometric shapes with algebraic equations. For example, every nonvertical straight line (a geometric concept) can be paired with an equation of the form $y = mx + b$ (an algebraic concept), where m and b are real numbers, and x and y are variables that we associate with the axes of a coordinate system. In this section we will review some of the concepts developed around the rectangular coordinate system and graphing in two dimensions.

The rectangular (or Cartesian) coordinate system is shown in Figure 1. The axes divide the plane into four *quadrants* that are numbered I through IV in a counterclockwise direction. Looking at Figure 1, we see that any point in quadrant I will have both coordinates positive; that is, $(+, +)$. In quadrant II, the form is $(-, +)$. In quadrant III, the form is $(-, -)$, and in quadrant IV it is $(+, -)$. Also, any point on the x-axis will have a y-coordinate of 0 (it has no vertical displacement), and any point on the y-axis will have an x-coordinate of 0 (no horizontal displacement).

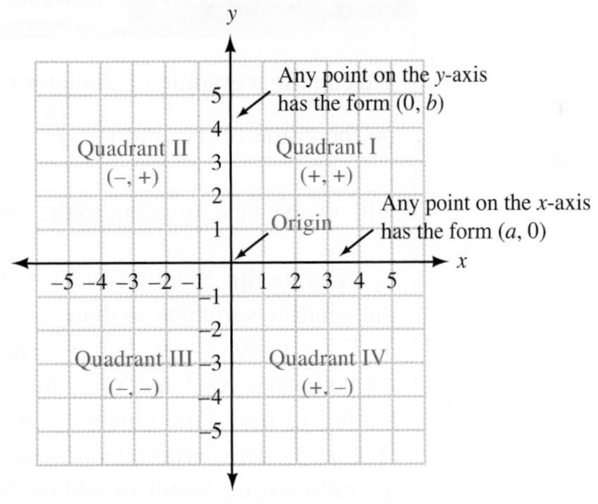

Figure 1

Graphing Lines

PROBLEM 1

Graph $y = -\frac{1}{2}x$.

EXAMPLE 1 Graph the line $y = \frac{3}{2}x$.

SOLUTION Because the equation of the line is written in slope-intercept form, we see that the slope of the line is $\frac{3}{2} = 1.5$ and the y-intercept is 0. To graph the line, we begin at the origin and use the slope to locate a second point. For every unit we traverse to the

NOTE Example 1 illustrates the connection between algebra and geometry that we mentioned in the introduction to this section. The rectangular coordinate system allows us to associate the equation

$$y = \frac{3}{2}x$$

(an algebraic concept) with a specific straight line (a geometric concept). The study of the relationship between equations in algebra and their associated geometric figures is called *analytic geometry* and is based on the coordinate system credited to Descartes.

right, the line will rise 1.5 units. If we traverse 2 units to the right, the line will rise 3 units, giving us the point (2, 3). Or, if we traverse 3 units to the right, the line will rise 4.5 units yielding the point (3, 4.5). The graph of the line is shown in Figure 2.

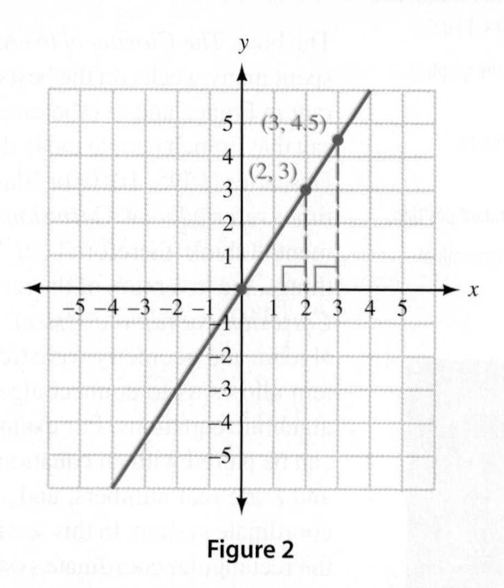

Figure 2

Notice in Example 1 that the points (2, 3) and (3, 4.5) create two similar right triangles whose corresponding sides are in proportion. That is,

$$\frac{4.5}{3} = \frac{3}{2}$$

Using Technology Verifying Slope

We can use a graphing calculator to verify that for any point (other than the origin) on the graph of the line $y = \frac{3}{2}x$, the ratio of the y-coordinate to the x-coordinate will always be equivalent to the slope of $\frac{3}{2}$, or 1.5 as a decimal.

Define this function as $Y_1 = 3x/2$. To match the graph shown in Figure 3, set the window variables so that $-6 \le x \le 6$ and $-6 \le y \le 6$. (By this, we mean that $X_{min} = -6$, $X_{max} = 6$, $Y_{min} = -6$, and $Y_{max} = 6$. We will assume that the scales for both axes, Xscl and Yscl, are set to 1 unless noted otherwise.) Use the TRACE feature to move the cursor to any point on the line other than the origin itself (Figure 3). The current coordinates are stored in the variables x and y. If we check the ratio y/x, the result should be the slope of 1.5 as shown in Figure 4. Try this for several different points to see that the ratio is always the same.

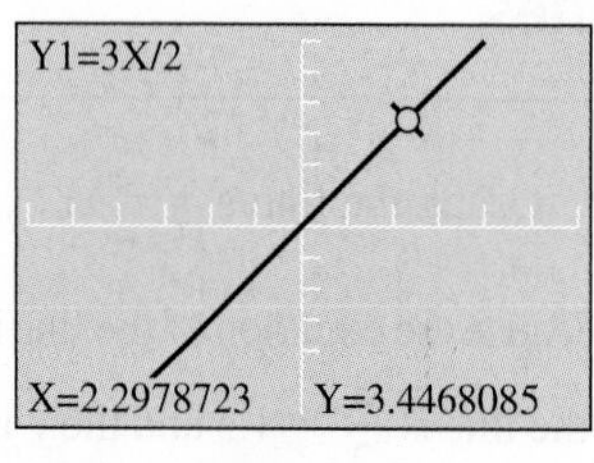

Figure 3

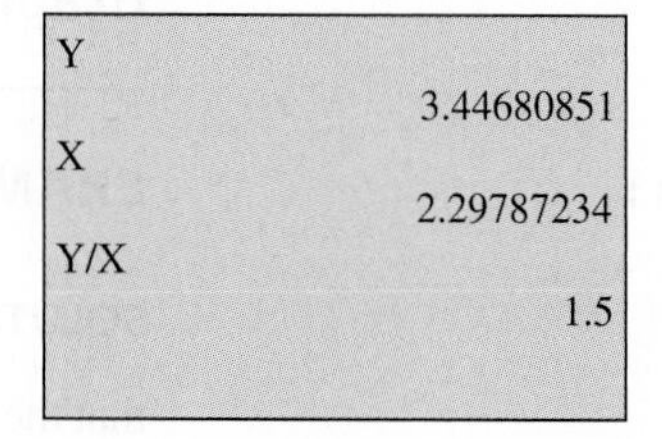

Figure 4

Graphing Parabolas

Recall from your algebra classes that any parabola that opens up or down can be described by an equation of the form

$$y = a(x - h)^2 + k$$

Likewise, any equation of this form will have a graph that is a parabola. The highest or lowest point on the parabola is called the vertex. The coordinates of the vertex are (h, k). The value of a determines how wide or narrow the parabola will be and whether it opens upward or downward.

PROBLEM 2
Repeat Example 2 if the net had been placed 140 feet from the cannon.

NOTE Although quadratic functions are not pertinent to a study of trigonometry, this example introduces the Human Cannonball theme that runs throughout the text and lays the foundation for problems that will follow in later sections, where trigonometric concepts are used.

EXAMPLE 2 At the 1997 Washington County Fair in Oregon, David Smith, Jr., The Bullet, was shot from a cannon. As a human cannonball, he reached a height of 70 feet before landing in a net 160 feet from the cannon. Sketch the graph of his path, and then find the equation of the graph.

SOLUTION We assume that the path taken by the human cannonball is a parabola. If the origin of the coordinate system is at the opening of the cannon, then the net that catches him will be at 160 on the x-axis. Figure 5 shows a graph of this path.

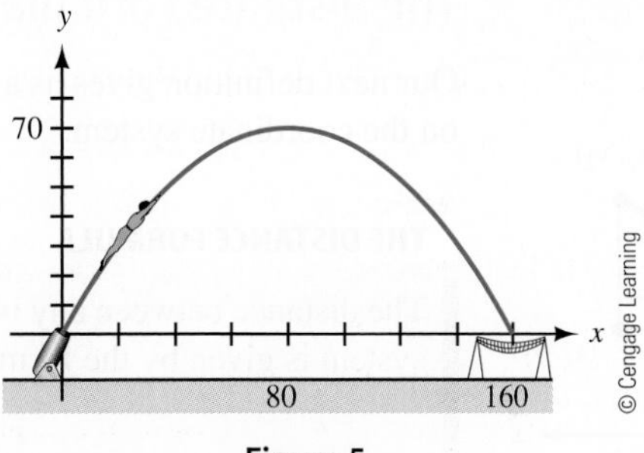

Figure 5

Because the curve is a parabola, we know that the equation will have the form

$$y = a(x - h)^2 + k$$

Because the vertex of the parabola is at (80, 70), we can fill in two of the three constants in our equation, giving us

$$y = a(x - 80)^2 + 70$$

To find a we note that the landing point will be (160, 0). Substituting the coordinates of this point into the equation, we solve for a.

$$0 = a(160 - 80)^2 + 70$$

$$0 = a(80)^2 + 70$$

$$0 = 6400a + 70$$

$$a = -\frac{70}{6400} = -\frac{7}{640}$$

The equation that describes the path of the human cannonball is

$$y = -\frac{7}{640}(x - 80)^2 + 70 \text{ for } 0 \le x \le 160$$

■

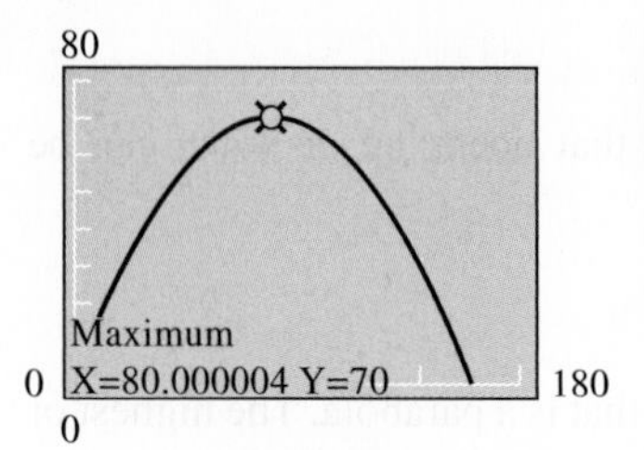

Figure 6

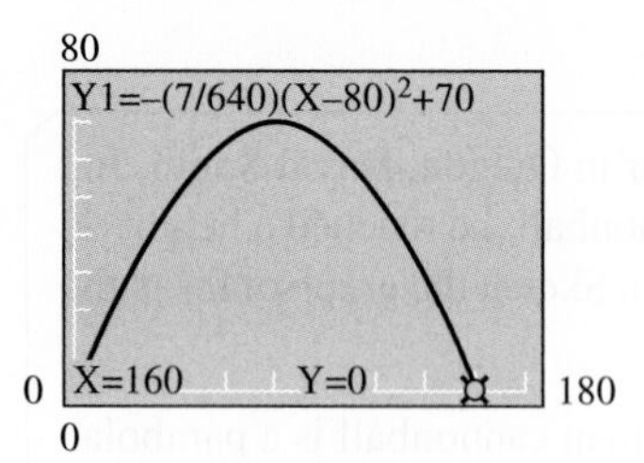

Figure 7

Using Technology

To verify that the equation from Example 2 is correct, we can graph the parabola and check the vertex and the x-intercepts. Graph the equation using the following window settings:

$$0 \leq x \leq 180, \text{ scale} = 20; 0 \leq y \leq 80, \text{ scale} = 10$$

Use the appropriate command on your calculator to find the maximum point on the graph (Figure 6), which is the vertex. Then evaluate the function at $x = 0$ and again at $x = 160$ to verify the x-intercepts (Figure 7).

NOTE There are many different models of graphing calculators, and each model has its own set of commands. For example, to perform the previous steps on a TI-84 we would press [2nd] [CALC] and use the **maximum** and **value** commands. Because we have no way of knowing which model of calculator you are working with, we will generally avoid providing specific key icons or command names throughout the remainder of this book. Check your calculator manual to find the appropriate command for your particular model.

The Distance Formula

Our next definition gives us a formula for finding the distance between any two points on the coordinate system.

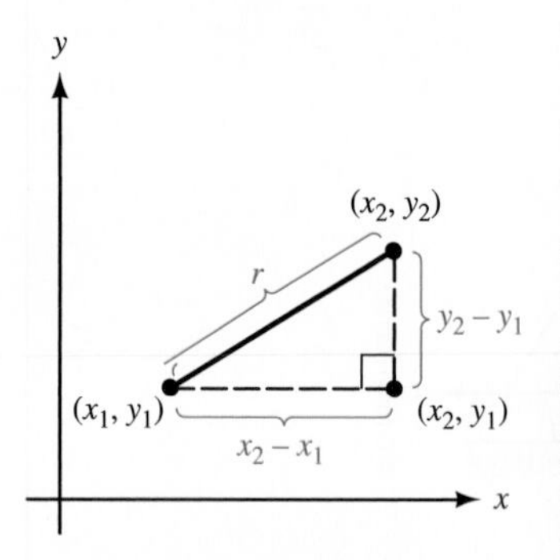

$r^2 = (x_2 - x_1)^2 + (y_2 - y_1)^2$

$r = \sqrt{(x_2 - x_1)^2 + (y_2 - y_1)^2}$

Figure 8

THE DISTANCE FORMULA

The distance between any two points (x_1, y_1) and (x_2, y_2) in a rectangular coordinate system is given by the formula

$$r = \sqrt{(x_2 - x_1)^2 + (y_2 - y_1)^2}$$

The distance formula can be derived by applying the Pythagorean Theorem to the right triangle in Figure 8. Because r is a distance, $r \geq 0$.

PROBLEM 3
Find the distance between (0, 3) and (4, 0).

EXAMPLE 3 Find the distance between the points $(-1, 5)$ and $(2, 1)$.

SOLUTION It makes no difference which of the points we call (x_1, y_1) and which we call (x_2, y_2) because this distance will be the same between the two points regardless (Figure 9).

$$\begin{aligned} r &= \sqrt{(2 - (-1))^2 + (1 - 5)^2} \\ &= \sqrt{3^2 + (-4)^2} \\ &= \sqrt{9 + 16} \\ &= \sqrt{25} \\ &= 5 \end{aligned}$$

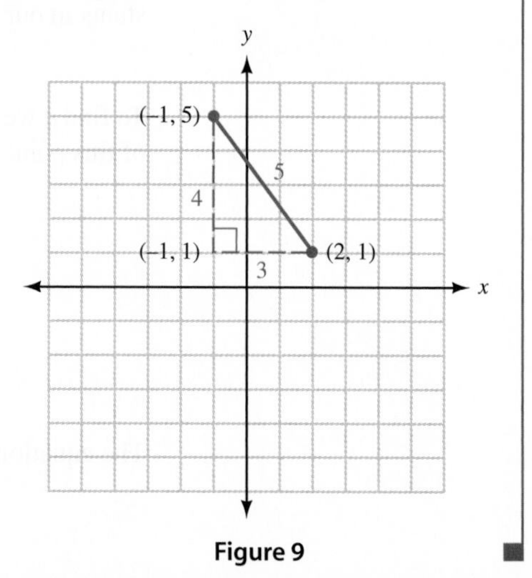

Figure 9

PROBLEM 4
Find the distance from the origin to the point $(-5, 12)$.

EXAMPLE 4 Find the distance from the origin to the point (x, y).

SOLUTION The coordinates of the origin are $(0, 0)$. As shown in Figure 10, applying the distance formula, we have

$$r = \sqrt{(x-0)^2 + (y-0)^2} = \sqrt{x^2 + y^2}$$

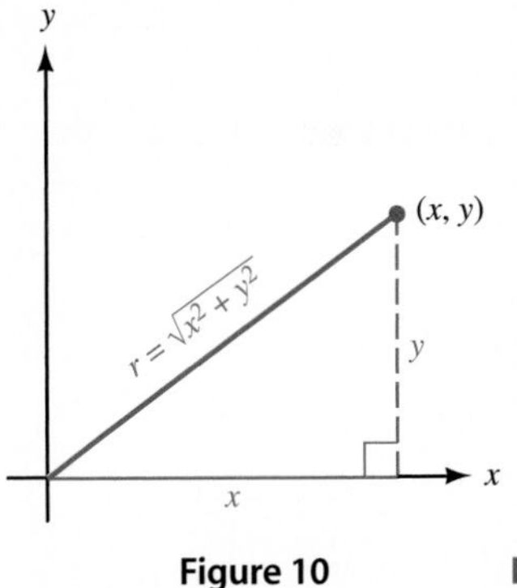

Figure 10

NOTE Because of their symmetry, circles have been used for thousands of years in many disciplines. For example, Stonehenge is based on a circular plan that is thought to have both religious and astronomical significance.

Circles

A *circle* is defined as the set of all points in the plane that are a fixed distance from a given fixed point. The fixed distance is the *radius* of the circle, and the fixed point is called the *center*. If we let $r > 0$ be the radius, (h, k) the center, and (x, y) represent any point on the circle, then (x, y) is r units from (h, k) as Figure 11 illustrates. Applying the distance formula, we have

$$\sqrt{(x-h)^2 + (y-k)^2} = r$$

Squaring both sides of this equation gives the formula for a circle.

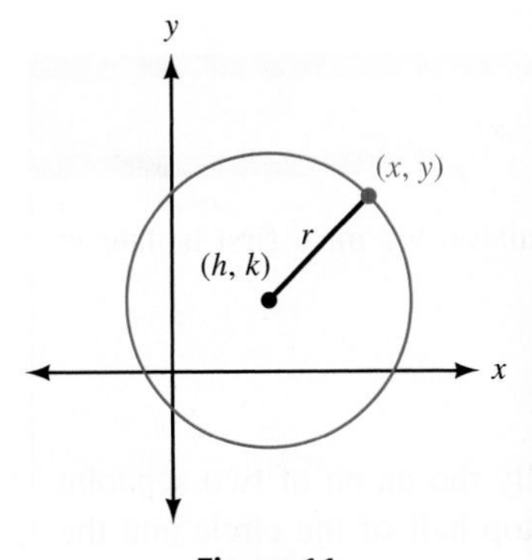

Figure 11

EQUATION OF A CIRCLE

The equation of a circle with center (h, k) and radius $r > 0$ is given by the formula

$$(x-h)^2 + (y-k)^2 = r^2$$

If the center is at the origin so that $(h, k) = (0, 0)$, this simplifies to

$$x^2 + y^2 = r^2$$

PROBLEM 5
Verify that the point $\left(\frac{\sqrt{2}}{3}, -\frac{\sqrt{7}}{3}\right)$ lies on the unit circle.

EXAMPLE 5 Verify that the points $\left(\frac{\sqrt{2}}{2}, \frac{\sqrt{2}}{2}\right)$ and $\left(-\frac{\sqrt{3}}{2}, -\frac{1}{2}\right)$ both lie on a circle of radius 1 centered at the origin.

SOLUTION Because $r = 1$, the equation of the circle is $x^2 + y^2 = 1$. We check each point by showing that the coordinates satisfy the equation.

If $x = \frac{\sqrt{2}}{2}$ and $y = \frac{\sqrt{2}}{2}$

then
$$x^2 + y^2 = \left(\frac{\sqrt{2}}{2}\right)^2 + \left(\frac{\sqrt{2}}{2}\right)^2 = \frac{1}{2} + \frac{1}{2} = 1$$

If $x = -\frac{\sqrt{3}}{2}$ and $y = -\frac{1}{2}$

then
$$x^2 + y^2 = \left(-\frac{\sqrt{3}}{2}\right)^2 + \left(-\frac{1}{2}\right)^2 = \frac{3}{4} + \frac{1}{4} = 1$$

The graph of the circle and the two points are shown in Figure 12.

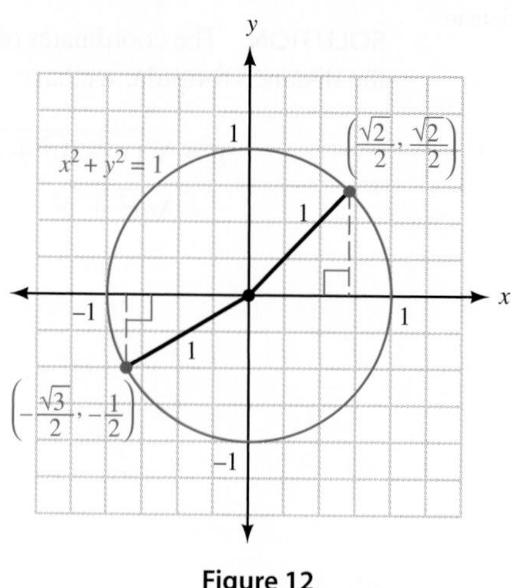

Figure 12

The circle $x^2 + y^2 = 1$ from Example 5 is called the *unit circle* because its radius is 1. As you will see, it will be an important part of one of the definitions that we will give in Chapter 3.

Using Technology

Plot1 Plot2 Plot3
\Y1=√(1−X²)
\Y2=−√(1−X²)
\Y3=
\Y4=
\Y5=
\Y6=
\Y7=

Figure 13

To graph the circle $x^2 + y^2 = 1$ on a graphing calculator, we must first isolate y:

$$y^2 = 1 - x^2$$
$$y = \pm\sqrt{1 - x^2}$$

Because of the ± sign, we see that a circle is really the union of two separate functions. The positive square root represents the top half of the circle and the negative square root represents the bottom half. Define these functions separately (Figure 13).

When you graph both functions, you will probably see a circle that is not quite round and has gaps near the x-axis (Figure 14). Getting rid of the gaps can be tricky, and it requires a good understanding of how your calculator works. However, we can easily make the circle appear round using the zoom-square command (Figure 15).

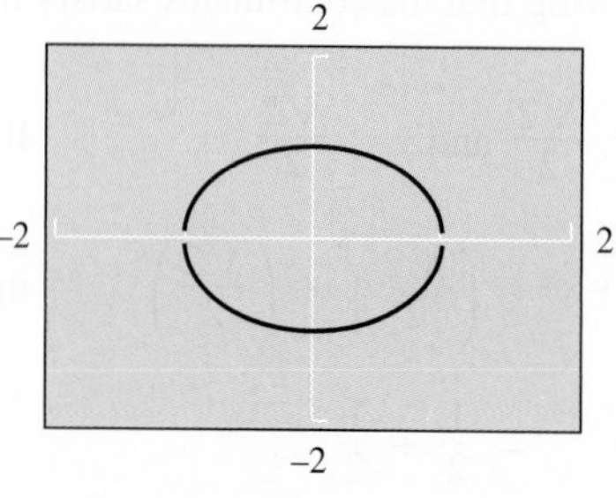

Figure 14

Figure 15

Angles in Standard Position

DEFINITION

An angle is said to be in *standard position* if its initial side is along the positive x-axis and its vertex is at the origin.

PROBLEM 6
Draw 30° in standard position and find a point on the terminal side.

EXAMPLE 6 Draw an angle of 45° in standard position and find a point on the terminal side.

SOLUTION If we draw 45° in standard position, we see that the terminal side is along the line $y = x$ in quadrant I (Figure 16). Because the terminal side of 45° lies along the line $y = x$ in the first quadrant, any point on the terminal side will have positive coordinates that satisfy the equation $y = x$. Here are some of the points that do just that.

$$(1, 1) \qquad (2, 2) \qquad (3, 3) \qquad (\sqrt{2}, \sqrt{2}) \qquad \left(\frac{1}{2}, \frac{1}{2}\right) \qquad \left(\frac{7}{8}, \frac{7}{8}\right)$$

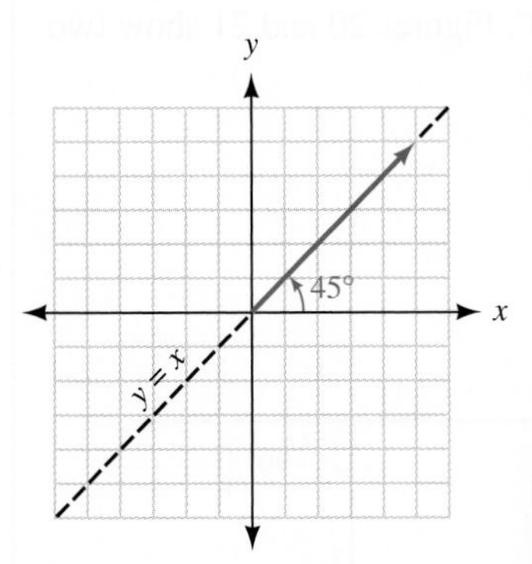

Figure 16

VOCABULARY

If angle θ is in standard position and the terminal side of θ lies in quadrant I, then we say θ lies in quadrant I and we abbreviate it like this:

$$\theta \in \text{QI}$$

Likewise, $\theta \in$ QII means θ is in standard position with its terminal side in quadrant II.

If the terminal side of an angle in standard position lies along one of the axes, then that angle is called a *quadrantal angle.* For example, an angle of 90° drawn in standard position would be a quadrantal angle, because the terminal side would lie along the positive y-axis. Likewise, 270° in standard position is a quadrantal angle because the terminal side would lie along the negative y-axis (Figure 17).

Two angles in standard position with the same terminal side are called *coterminal angles.* Figure 18 shows that 60° and $-300°$ are coterminal angles when they are in standard position. Notice that these two angles differ by 360°. That is, $60° - (-300°) = 360°$. Coterminal angles always differ from each other by some multiple of 360°.

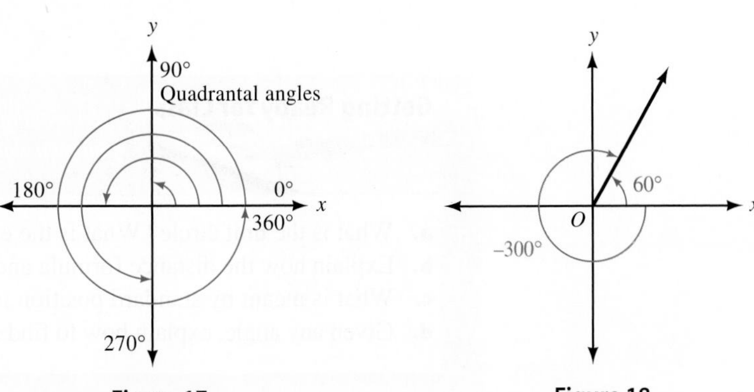

Figure 17

Figure 18

PROBLEM 7
Draw $-45°$ in standard position and name two positive angles coterminal with it.

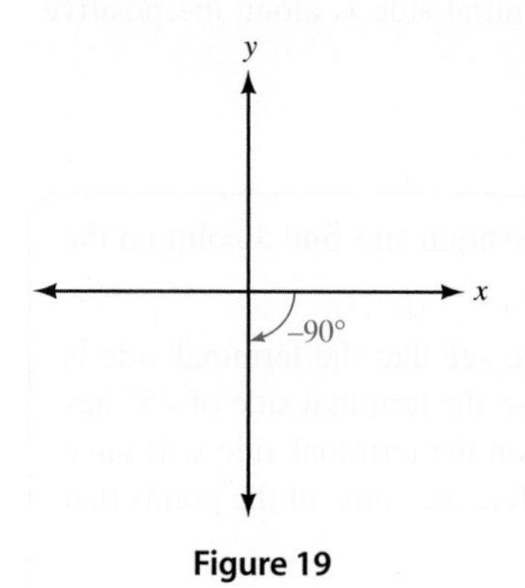

Figure 19

EXAMPLE 7 Draw $-90°$ in standard position and find two positive angles and two negative angles that are coterminal with $-90°$.

SOLUTION Figure 19 shows $-90°$ in standard position. To find a coterminal angle, we must traverse a full revolution in the positive direction or the negative direction.

One revolution in the positive direction:	$-90° + 360° = 270°$
A second revolution in the positive direction:	$270° + 360° = 630°$
One revolution in the negative direction:	$-90° - 360° = -450°$
A second revolution in the negative direction:	$-450° - 360° = -810°$

Thus, 270° and 630° are two positive angles coterminal with $-90°$ and $-450°$ and $-810°$ are two negative angles coterminal with $-90°$. Figures 20 and 21 show two of these angles.

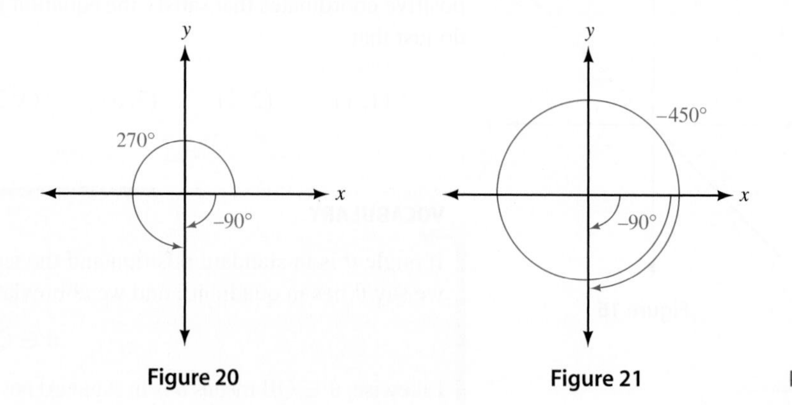

Figure 20

Figure 21

NOTE There are actually an infinite number of angles that are coterminal with $-90°$. We can find a coterminal angle by adding or subtracting 360° any number of times. If we let k be any integer, then the angle $-90° + 360°k$ will be coterminal with $-90°$. For example, if $k = 2$, then

$$-90° + 360°(2) = -90° + 720° = 630°$$

is a coterminal angle as shown previously in Example 7.

PROBLEM 8
Find all angles coterminal with 90°.

EXAMPLE 8 Find all angles that are coterminal with 120°.

SOLUTION For any integer k, $120° + 360°k$ will be coterminal with 120°.

Getting Ready for Class

After reading through the preceding section, respond in your own words and in complete sentences.

a. What is the unit circle? What is the equation for it?
b. Explain how the distance formula and the Pythagorean Theorem are related.
c. What is meant by standard position for an angle?
d. Given any angle, explain how to find another angle that is coterminal with it.

1.2 PROBLEM SET

➤ CONCEPTS AND VOCABULARY

For Questions 1 through 6, fill in each blank with the appropriate word or expression.

1. The Cartesian plane is divided into four regions, or ________, numbered ________ through ________ in a ____________ direction.

2. The unit circle has center ________ and a radius of ________.

3. An angle is in standard position if its vertex is at the ________ and its initial side is along the ________ ________.

4. The notation $\theta \in \text{QIII}$ means that θ is in standard position and its ________ side lies in ________ ________.

5. When the terminal side of an angle in standard position lies along one of the axes, it is called a ____________ angle.

6. Coterminal angles are two angles in standard position having the ________ ________ side. We can find a coterminal angle by adding or subtracting any multiple of ________.

7. State the formula for the distance between (x_1, y_1) and (x_2, y_2).

8. State the formula for a circle with center (h, k) and radius r.

➤ EXERCISES

Determine which quadrant contains each of the following points.

9. $(2, -4)$ **10.** $(-4, -2)$ **11.** $(-\sqrt{3}, 1)$ **12.** $(1, \sqrt{3})$

Graph each of the following lines.

13. $y = x$ **14.** $y = -x$ **15.** $y = \frac{1}{2}x$ **16.** $y = -2x$

17. In what two quadrants do all the points have negative x-coordinates?

18. In what two quadrants do all the points have negative y-coordinates?

19. For points (x, y) in quadrant I, the ratio x/y is always positive because x and y are always positive. In what other quadrant is the ratio x/y always positive?

20. For points (x, y) in quadrant II, the ratio x/y is always negative because x is negative and y is positive in quadrant II. In what other quadrant is the ratio x/y always negative?

Graph each of the following parabolas.

21. $y = x^2 - 4$ **22.** $y = (x - 2)^2$

23. $y = (x + 2)^2 + 4$ **24.** $y = \frac{1}{4}(x + 2)^2 + 4$

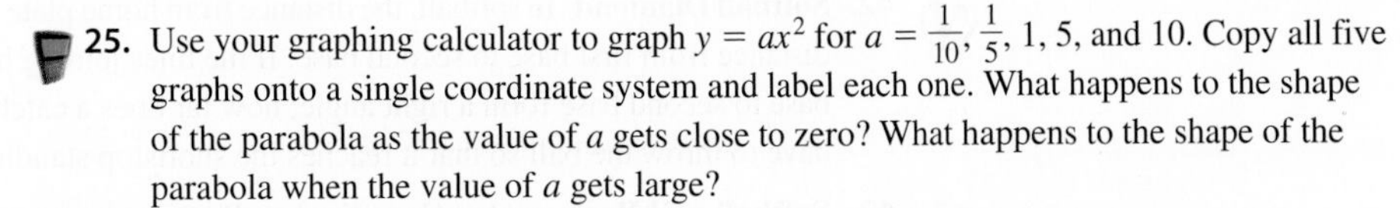

25. Use your graphing calculator to graph $y = ax^2$ for $a = \frac{1}{10}, \frac{1}{5}$, 1, 5, and 10. Copy all five graphs onto a single coordinate system and label each one. What happens to the shape of the parabola as the value of a gets close to zero? What happens to the shape of the parabola when the value of a gets large?

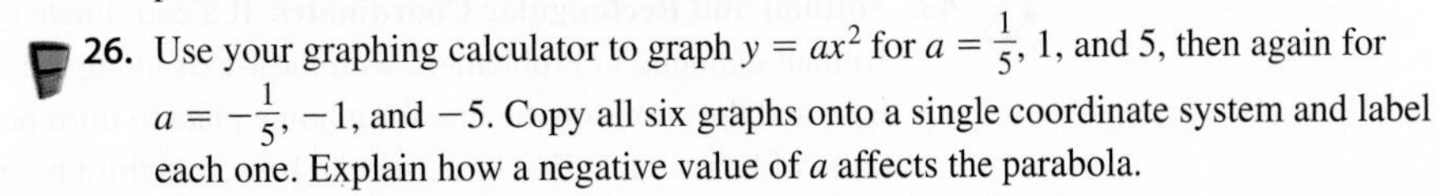

26. Use your graphing calculator to graph $y = ax^2$ for $a = \frac{1}{5}$, 1, and 5, then again for $a = -\frac{1}{5}$, -1, and -5. Copy all six graphs onto a single coordinate system and label each one. Explain how a negative value of a affects the parabola.

27. Use your graphing calculator to graph $y = (x - h)^2$ for $h = -3$, 0, and 3. Copy all three graphs onto a single coordinate system, and label each one. What happens to the position of the parabola when $h < 0$? What if $h > 0$?

28. Use your graphing calculator to graph $y = x^2 + k$ for $k = -3$, 0, and 3. Copy all three graphs onto a single coordinate system, and label each one. What happens to the position of the parabola when $k < 0$? What if $k > 0$?

29. Human Cannonball A human cannonball is shot from a cannon at the county fair. He reaches a height of 60 feet before landing in a net 160 feet from the cannon. Sketch the graph of his path, and then find the equation of the graph. Verify that your equation is correct using your graphing calculator.

30. Human Cannonball Referring to Problem 29, find the height reached by the human cannonball after he has traveled 30 feet horizontally, and after he has traveled 150 feet horizontally. Verify that your answers are correct using your graphing calculator.

Find the distance between the following pairs of points.

31. (3, 7), (6, 3)

32. (4, 7), (8, 1)

33. (0, 12), (5, 0)

34. (−3, 0), (0, 4)

35. (−1, −2), (−10, 5)

36. (−3, 8), (−1, 6)

37. Find the distance from the origin out to the point (3, −4).

38. Find the distance from the origin out to the point (12, −5).

39. Find x so the distance between $(x, 2)$ and (1, 5) is $\sqrt{13}$.

40. Find y so the distance between $(7, y)$ and (3, 3) is 5.

41. Pythagorean Theorem An airplane is approaching Los Angeles International Airport at an altitude of 2,640 feet. If the horizontal distance from the plane to the runway is 1.2 miles, use the Pythagorean Theorem to find the diagonal distance from the plane to the runway (Figure 22). (5,280 feet equals 1 mile.)

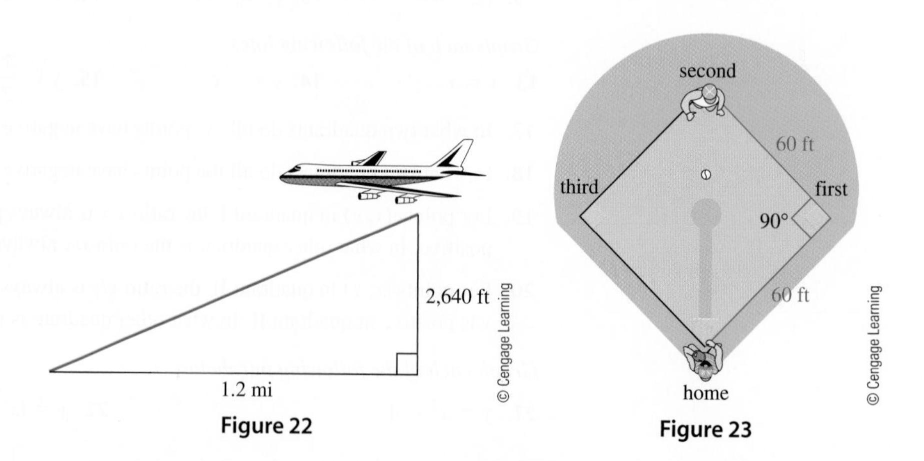

Figure 22

Figure 23

42. Softball Diamond In softball, the distance from home plate to first base is 60 feet, as is the distance from first base to second base. If the lines joining home plate to first base and first base to second base form a right angle, how far does a catcher standing on home plate have to throw the ball so that it reaches the shortstop standing on second base (Figure 23)?

43. Softball and Rectangular Coordinates If a coordinate system is superimposed on the softball diamond in Problem 42 with the x-axis along the line from home plate to first base and the y-axis on the line from home plate to third base, what would be the coordinates of home plate, first base, second base, and third base?

44. Softball and Rectangular Coordinates If a coordinate system is superimposed on the softball diamond in Problem 42 with the origin on home plate and the positive x-axis along the line joining home plate to second base, what would be the coordinates of first base and third base?

Verify that each point lies on the graph of the unit circle.

45. $(0, -1)$

46. $(-1, 0)$

47. $\left(\frac{1}{2}, \frac{\sqrt{3}}{2}\right)$

48. $\left(\frac{\sqrt{5}}{3}, -\frac{2}{3}\right)$

Graph each of the following circles.

49. $x^2 + y^2 = 25$

50. $x^2 + y^2 = 36$

Graph the circle $x^2 + y^2 = 1$ with your graphing calculator. Use the feature on your calculator that allows you to evaluate a function from the graph to find the coordinates of all points on the circle that have the given x-coordinate. Write your answers as ordered pairs and round to four places past the decimal point when necessary.

51. $x = \frac{1}{2}$

52. $x = -\frac{1}{2}$

53. $x = \frac{\sqrt{2}}{2}$

54. $x = -\frac{\sqrt{2}}{2}$

55. $x = -\frac{\sqrt{3}}{2}$

56. $x = \frac{\sqrt{3}}{2}$

57. Use the graph of Problem 49 to name the points at which the line $x + y = 5$ will intersect the circle $x^2 + y^2 = 25$.

58. Use the graph of Problem 50 to name the points at which the line $x - y = 6$ will intersect the circle $x^2 + y^2 = 36$.

59. At what points will the line $y = x$ intersect the unit circle $x^2 + y^2 = 1$?

60. At what points will the line $y = -x$ intersect the unit circle $x^2 + y^2 = 1$?

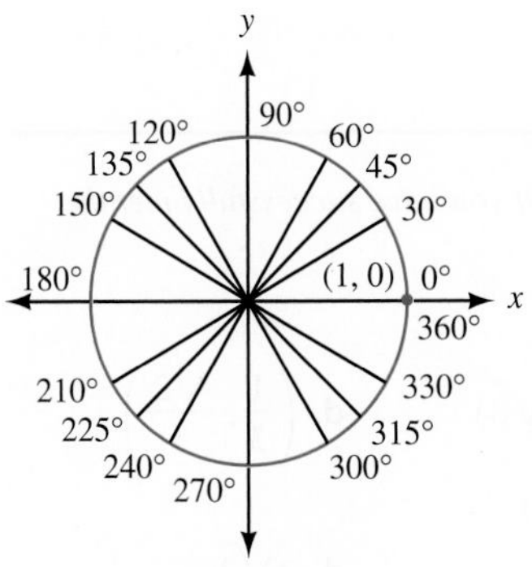

Figure 24 shows some of the more common positive angles. Each angle is in standard position. Next to the terminal side of each angle is the degree measure of the angle. To simplify the diagram, we have made each of the terminal sides one unit long by drawing them out to the unit circle only. Use this figure for Problems 61 through 72.

Figure 24

Find the complement of each of the following angles.

61. 45° **62.** 30° **63.** 60° **64.** 0°

Find the supplement of each of the following angles.

65. 120° **66.** 150° **67.** 90° **68.** 45°

Name an angle between 0° and 360° that is coterminal with each of the following angles.

69. −135° **70.** −60° **71.** −210° **72.** −300°

Draw each of the following angles in standard position, and find one positive angle and one negative angle that is coterminal with the given angle.

73. 300° **74.** 225° **75.** −150° **76.** −330°

Draw each of the following angles in standard position and then do the following:

a. *Name a point on the terminal side of the angle.*

b. *Find the distance from the origin to that point.*

c. *Name another angle that is coterminal with the angle you have drawn.*

77. 135° **78.** 45° **79.** 225° **80.** 315°

81. 90° **82.** 360° **83.** $-45°$ **84.** $-90°$

Find all angles that are coterminal with the given angle.

85. 30° **86.** 150° **87.** $-135°$ **88.** $-45°$

89. Draw 30° in standard position. Then find a if the point $(a, 1)$ is on the terminal side of 30°.

90. Draw 60° in standard position. Then find b if the point $(2, b)$ is on the terminal side of 60°.

91. Draw an angle in standard position whose terminal side contains the point $(3, -2)$. Find the distance from the origin to this point.

92. Draw an angle in standard position whose terminal side contains the point $(2, -3)$. Find the distance from the origin to this point.

For Problems 93 and 94, use the converse of the Pythagorean Theorem, which states that if $c^2 = a^2 + b^2$, then the triangle must be a right triangle.

93. Plot the points (0, 0), (5, 0), and (5, 12) and show that, when connected, they are the vertices of a right triangle.

94. Plot the points $(0, 2)$, $(-3, 2)$, and $(-3, -2)$ and show that they form the vertices of a right triangle.

➤ EXTENDING THE CONCEPTS

95. Descartes and Pascal In the introduction to this section we mentioned two French philosophers, Descartes and Pascal. Many people see the philosophies of the two men as being opposites. Why is this?

96. Pascal's Triangle Pascal has a triangular array of numbers named after him, Pascal's triangle. What part does Pascal's triangle play in the expansion of $(a + b)^n$, where n is a positive integer?

➤ LEARNING OBJECTIVES ASSESSMENT

These questions are available for instructors to help assess if you have successfully met the learning objectives for this section.

97. Which point lies on the graph of the unit circle?

a. $\left(\frac{3}{4}, \frac{\sqrt{7}}{4}\right)$ **b.** $(\sqrt{2}, \sqrt{2})$ **c.** $(1, \sqrt{3})$ **d.** $\left(\frac{1}{3}, \frac{\sqrt{2}}{3}\right)$

98. Find the distance between the points $(-2, 8)$ and $(4, 5)$.

a. 45 **b.** $3\sqrt{5}$ **c.** 9 **d.** $3\sqrt{3}$

99. To draw 140° in standard position, place the vertex at the origin and draw the terminal side 140° ______ from the ______.

a. clockwise, positive x-axis

b. counterclockwise, positive x-axis

c. counterclockwise, positive y-axis

d. clockwise, positive y-axis

100. Which angle is coterminal with 160°?

a. 500° **b.** $-70°$ **c.** 20° **d.** $-200°$

SECTION 1.3 Definition I: Trigonometric Functions

LEARNING OBJECTIVES

1. Find the value of a trigonometric function of an angle given a point on the terminal side.
2. Use Definition I to answer a conceptual question about a trigonometric function.
3. Determine the quadrants an angle could terminate in.
4. Find the value of a trigonometric function given one of the other values.

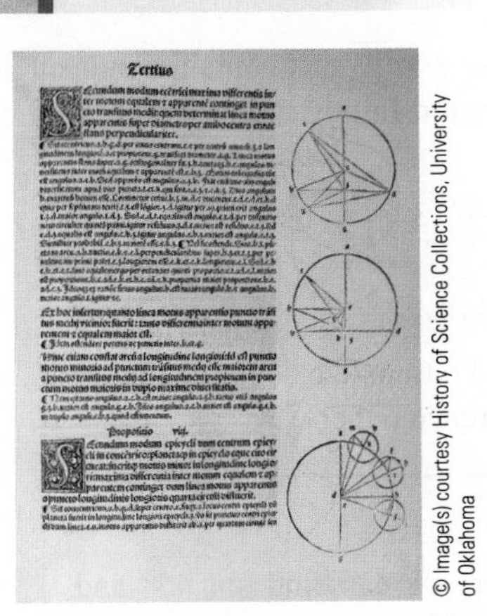

In this section we begin our work with trigonometry. The formal study of trigonometry dates back to the Greeks, when it was used mainly in the design of clocks and calendars and in navigation. The trigonometry of that period was spherical in nature, as it was based on measurement of arcs and chords associated with spheres (see image at left). Unlike the trigonometry of the Greeks, our introduction to trigonometry takes place on a rectangular coordinate system. It concerns itself with angles, line segments, and points in the plane.

The definition of the trigonometric functions that begins this section is one of three definitions we will use. For us, it is the most important definition in the book. What should you do with it? Memorize it. Remember, in mathematics, definitions are simply accepted. That is, unlike theorems, there is no proof associated with a definition; we simply accept them exactly as they are written, memorize them, and then use them. When you are finished with this section, be sure that you have memorized this first definition. It is the most valuable thing you can do for yourself at this point in your study of trigonometry.

DEFINITION I

If θ is an angle in standard position, and the point (x, y) is any point on the terminal side of θ other than the origin, then the six trigonometric functions of angle θ are defined as follows:

Function		Abbreviation		Definition
The sine of θ	=	$\sin\theta$	=	$\dfrac{y}{r}$
The cosine of θ	=	$\cos\theta$	=	$\dfrac{x}{r}$
The tangent of θ	=	$\tan\theta$	=	$\dfrac{y}{x}$ $(x \neq 0)$
The cotangent of θ	=	$\cot\theta$	=	$\dfrac{x}{y}$ $(y \neq 0)$
The secant of θ	=	$\sec\theta$	=	$\dfrac{r}{x}$ $(x \neq 0)$
The cosecant of θ	=	$\csc\theta$	=	$\dfrac{r}{y}$ $(y \neq 0)$

where $x^2 + y^2 = r^2$, or $r = \sqrt{x^2 + y^2}$. That is, r is the distance from the origin to (x, y).

Figure 1

As you can see, the six trigonometric functions are simply names given to the six possible ratios that can be made from the numbers x, y, and r as shown in Figure 1. In particular, notice that $\tan\theta$ can be interpreted as the slope of the line corresponding to the terminal side of θ. Both $\tan\theta$ and $\sec\theta$ will be undefined when $x = 0$, which will occur any time the terminal side of θ coincides with the y-axis. Likewise, both $\cot\theta$ and $\csc\theta$ will be undefined when $y = 0$, which will occur any time the terminal side of θ coincides with the x-axis.

PROBLEM 1
Find the six trigonometric functions of θ if θ is in standard position and the point $(1, -4)$ is on the terminal side.

EXAMPLE 1 Find the six trigonometric functions of θ if θ is in standard position and the point $(-2, 3)$ is on the terminal side of θ.

SOLUTION We begin by making a diagram showing θ, $(-2, 3)$, and the distance r from the origin to $(-2, 3)$, as shown in Figure 2.

Applying the definition for the six trigonometric functions using the values $x = -2$, $y = 3$, and $r = \sqrt{13}$, we have

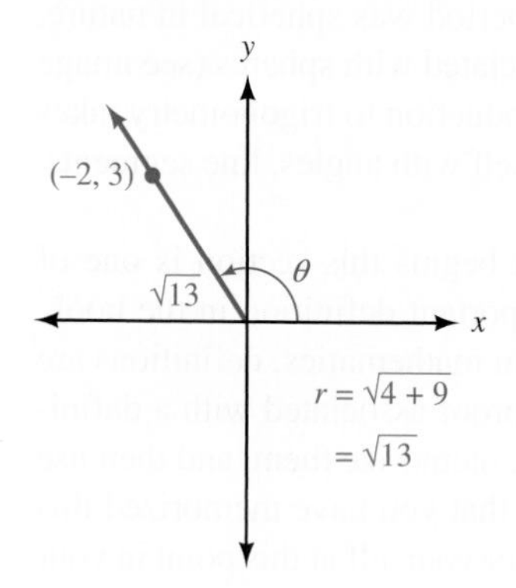

Figure 2

$$\sin\theta = \frac{y}{r} = \frac{3}{\sqrt{13}} = \frac{3\sqrt{13}}{13} \qquad \csc\theta = \frac{r}{y} = \frac{\sqrt{13}}{3}$$

$$\cos\theta = \frac{x}{r} = -\frac{2}{\sqrt{13}} = -\frac{2\sqrt{13}}{13} \qquad \sec\theta = \frac{r}{x} = -\frac{\sqrt{13}}{2}$$

$$\tan\theta = \frac{y}{x} = -\frac{3}{2} \qquad \cot\theta = \frac{x}{y} = -\frac{2}{3}$$

NOTE In algebra, when we encounter expressions like $3/\sqrt{13}$ that contain a radical in the denominator, we usually rationalize the denominator; in this case, by multiplying the numerator and the denominator by $\sqrt{13}$.

$$\frac{3}{\sqrt{13}} = \frac{3}{\sqrt{13}} \cdot \frac{\sqrt{13}}{\sqrt{13}} = \frac{3\sqrt{13}}{13}$$

In trigonometry, it is sometimes convenient to use $3\sqrt{13}/13$, and at other times it is easier to use $3/\sqrt{13}$. In most cases we will go ahead and rationalize denominators, but you should check and see if your instructor has a preference either way.

PROBLEM 2
Find the sine and cosine of 60°.

EXAMPLE 2 Find the sine and cosine of 45°.

SOLUTION According to the definition given earlier, we can find sin 45° and cos 45° if we know a point (x, y) on the terminal side of 45°, when 45° is in standard position. Figure 3 is a diagram of 45° in standard position. Because the terminal side of 45° lies along the line $y = x$, any point on the terminal side will have equal coordinates. A convenient point to use is the point $(1, 1)$. (We say it is a convenient point because the coordinates are easy to work with.)

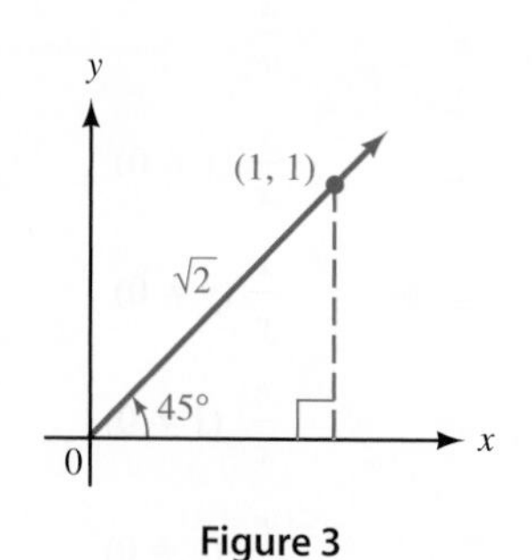

Figure 3

Because $x = 1$ and $y = 1$ and $r = \sqrt{x^2 + y^2}$, we have

$$r = \sqrt{1^2 + 1^2}$$
$$= \sqrt{2}$$

Substituting these values for x, y, and r into our definition for sine and cosine, we have

$$\sin 45^\circ = \frac{y}{r} = \frac{1}{\sqrt{2}} = \frac{\sqrt{2}}{2} \quad \text{and} \quad \cos 45^\circ = \frac{x}{r} = \frac{1}{\sqrt{2}} = \frac{\sqrt{2}}{2}$$

PROBLEM 3
Find the six trigonometric functions of 180°.

EXAMPLE 3 Find the six trigonometric functions of 270°.

SOLUTION Again, we need to find a point on the terminal side of 270°. From Figure 4, we see that the terminal side of 270° lies along the negative y-axis.

A convenient point on the terminal side of 270° is $(0, -1)$. Therefore,

$$r = \sqrt{x^2 + y^2}$$
$$= \sqrt{0^2 + (-1)^2}$$
$$= \sqrt{1}$$
$$= 1$$

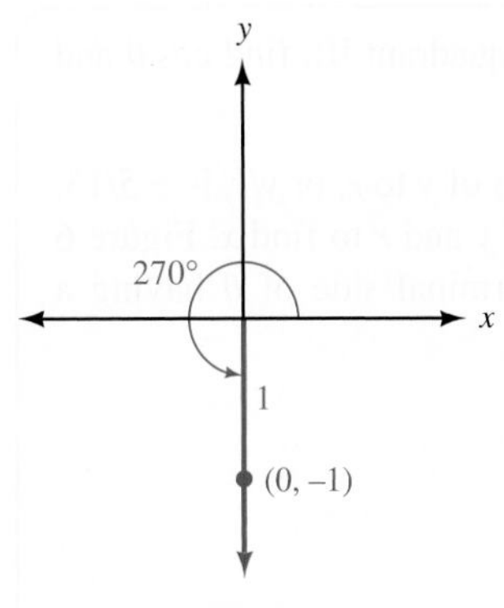

Figure 4

We have $x = 0$, $y = -1$, and $r = 1$. Here are the six trigonometric ratios for $\theta = 270°$.

$$\sin 270° = \frac{y}{r} = \frac{-1}{1} = -1 \qquad \csc 270° = \frac{r}{y} = \frac{1}{-1} = -1$$

$$\cos 270° = \frac{x}{r} = \frac{0}{1} = 0 \qquad \sec 270° = \frac{r}{x} = \frac{1}{0} = \text{undefined}$$

$$\tan 270° = \frac{y}{x} = \frac{-1}{0} = \text{undefined} \qquad \cot 270° = \frac{x}{y} = \frac{0}{-1} = 0$$

Note that tan 270° and sec 270° are undefined since division by 0 is undefined. ■

We can use Figure 1 to get some idea of how large or small each of the six trigonometric ratios might be based on the relative sizes of x, y, and r, as illustrated in the next example.

PROBLEM 4
Which will be greater, cot 30° or cot 60°?

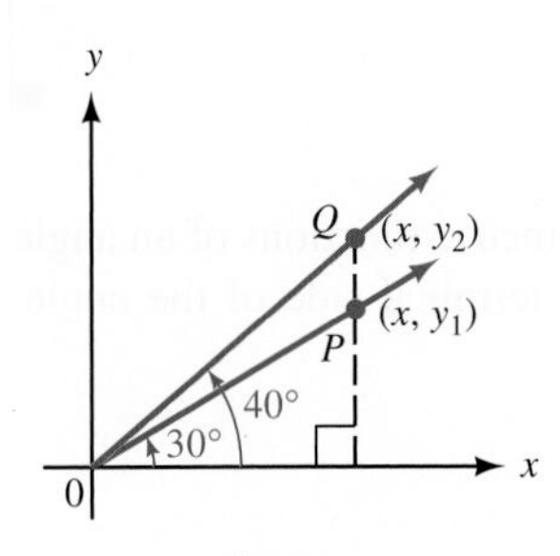

Figure 5

EXAMPLE 4 Which will be greater, tan 30° or tan 40°? How large could tan θ be?

SOLUTION In Figure 5 we have chosen points P and Q on the terminal sides of 30° and 40° so that the x-coordinate is the same for both points. Because 40° > 30°, we can see that $y_2 > y_1$. Therefore, the ratio y_2/x must be greater than the ratio y_1/x, and so tan 40° > tan 30°.

As θ continues to increase, the terminal side of θ will get steeper and steeper. Because tan θ can be interpreted as the slope of the terminal side of θ, tan θ will become larger and larger. As θ nears 90°, the terminal side of θ will be almost vertical, and its slope will become exceedingly large. Theoretically, there is no limit as to how large tan θ can be. ■

Algebraic Signs of Trigonometric Functions

The algebraic sign, + or −, of each of the six trigonometric functions will depend on the quadrant in which θ terminates. For example, in quadrant I all six trigonometric functions are positive because x, y, and r are all positive. In quadrant II, only sin θ and csc θ are positive because y and r are positive and x is negative. Table 1 shows the signs of all the ratios in each of the four quadrants.

TABLE 1

For θ in	QI	QII	QIII	QIV
$\sin\theta = \frac{y}{r}$ and $\csc\theta = \frac{r}{y}$	+	+	−	−
$\cos\theta = \frac{x}{r}$ and $\sec\theta = \frac{r}{x}$	+	−	−	+
$\tan\theta = \frac{y}{x}$ and $\cot\theta = \frac{x}{y}$	+	−	+	−

PROBLEM 5
If $\tan \theta = 4/3$ and θ terminates in quadrant III, find $\sin \theta$ and $\cos \theta$.

EXAMPLE 5 If $\sin \theta = -5/13$, and θ terminates in quadrant III, find $\cos \theta$ and $\tan \theta$.

SOLUTION Because $\sin \theta = -5/13$, we know the ratio of y to r, or y/r, is $-5/13$. We can let y be -5 and r be 13 and use these values of y and r to find x. Figure 6 shows θ in standard position with the point on the terminal side of θ having a y-coordinate of -5.

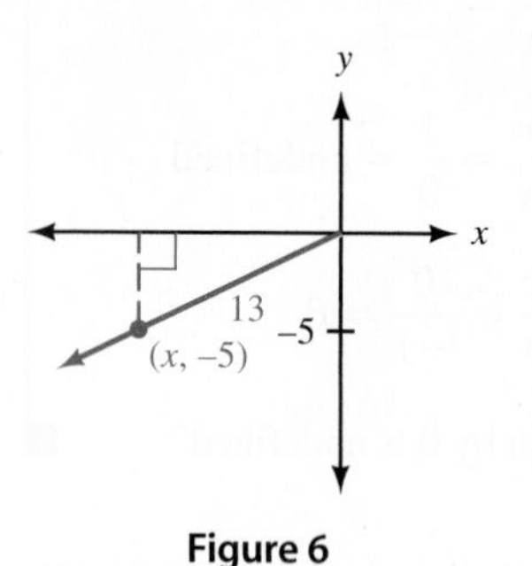

Figure 6

To find x, we use the fact that $x^2 + y^2 = r^2$.

$$x^2 + y^2 = r^2$$
$$x^2 + (-5)^2 = 13^2$$
$$x^2 + 25 = 169$$
$$x^2 = 144$$
$$x = \pm 12$$

Is x the number 12 or -12?

Because θ terminates in quadrant III, we know any point on its terminal side will have a negative x-coordinate; therefore,

$$x = -12$$

Using $x = -12$, $y = -5$, and $r = 13$ in our original definition, we have

$$\cos \theta = \frac{x}{r} = \frac{-12}{13} = -\frac{12}{13}$$

and

$$\tan \theta = \frac{y}{x} = \frac{-5}{-12} = \frac{5}{12}$$

NOTE In Example 5 we are not saying that if $y/r = -5/13$, then y *must* be -5 and r *must* be 13. There are many pairs of numbers whose ratio is $-5/13$, not just -5 and 13. Our definition for sine and cosine indicates we can choose *any* point on the terminal side of θ. However, because r is always positive, we *must* associate the negative sign with y.

As a final note, we should emphasize that the trigonometric functions of an angle are independent of the choice of the point (x, y) on the terminal side of the angle. Figure 7 shows an angle θ in standard position.

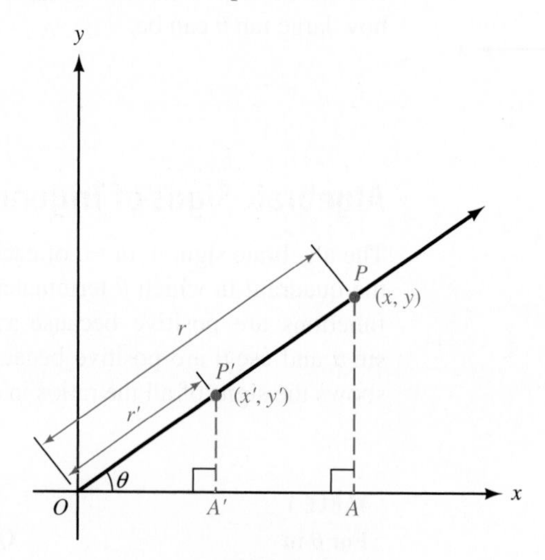

Figure 7

Points $P(x, y)$ and $P'(x', y')$ are both points on the terminal side of θ. Because triangles $P'OA'$ and POA are similar triangles, their corresponding sides are proportional. That is,

$$\sin \theta = \frac{y'}{r'} = \frac{y}{r} \qquad \cos \theta = \frac{x'}{r'} = \frac{x}{r} \qquad \tan \theta = \frac{y'}{x'} = \frac{y}{x}$$

Getting Ready for Class *After reading through the preceding section, respond in your own words and in complete sentences.*

a. Find the six trigonometric functions of θ, if θ is an angle in standard position and the point (x, y) is a point on the terminal side of θ.
b. If r is the distance from the origin to the point (x, y), state the six ratios, or definitions, corresponding to the six trigonometric functions above.
c. Find the sine and cosine of 45°.
d. Find the sine, cosine, and tangent of 270°.

1.3 PROBLEM SET

➤ CONCEPTS AND VOCABULARY

For Questions 1 and 2, fill in each blank with the appropriate word or number.

1. In Definition I, (x, y) is any point on the __________ side of θ when in standard position, and r is the __________ from the _______ to (x, y).

2. In quadrant I, _____ of the six trigonometric functions are positive. In each of the other three quadrants, only _____ of the six functions are positive.

3. Which of the six trigonometric functions are undefined when $x = 0$? Which of the six trigonometric functions are undefined when $y = 0$?

4. Which of the six trigonometric functions do not depend on the value of r?

➤ EXERCISES

Find all six trigonometric functions of θ if the given point is on the terminal side of θ. (In Problem 15, assume that a is a positive number.)

5. $(3, 4)$ **6.** $(-3, -4)$ **7.** $(-5, 12)$ **8.** $(12, -5)$

9. $(-1, -2)$ **10.** $(\sqrt{2}, \sqrt{2})$ **11.** $(\sqrt{3}, -1)$ **12.** $(-3, \sqrt{7})$

13. $(0, -5)$ **14.** $(-3, 0)$ **15.** $(-9a, -12a)$ **16.** (m, n)

In the following diagrams, angle θ is in standard position. In each case, find $\sin\theta$, $\cos\theta$, and $\tan\theta$.

17.

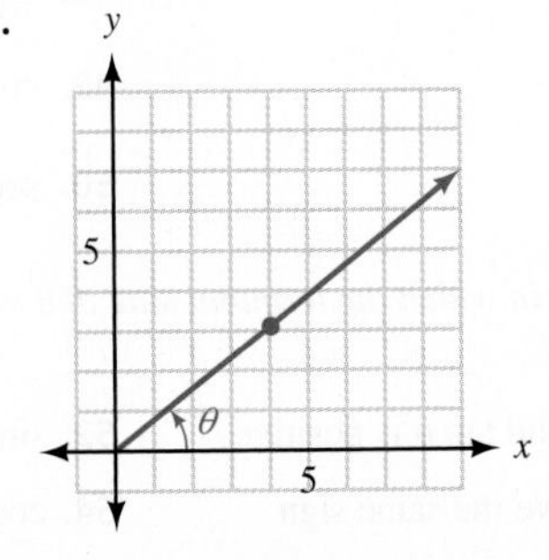

18.

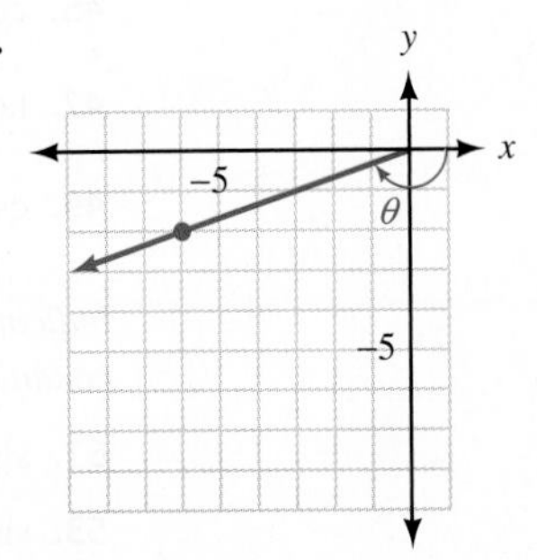

19.

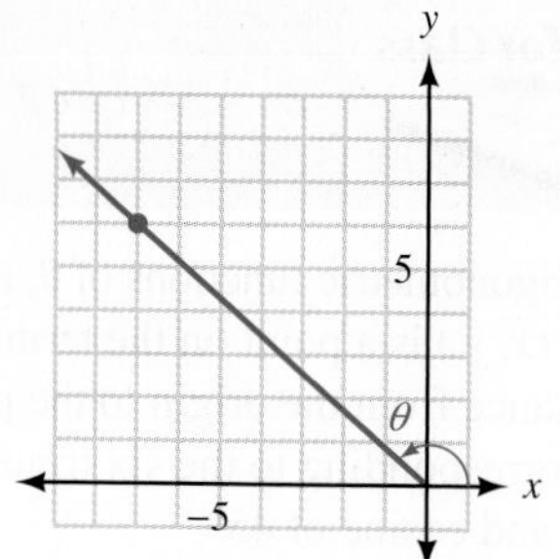

20.

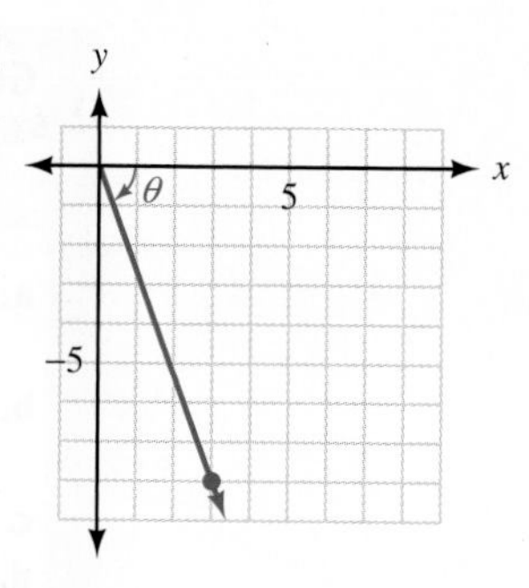

21. Use your calculator to find $\sin \theta$ and $\cos \theta$ if the point (9.36, 7.02) is on the terminal side of θ.

22. Use your calculator to find $\sin \theta$ and $\cos \theta$ if the point (6.36, 2.65) is on the terminal side of θ.

Draw each of the following angles in standard position, find a point on the terminal side, and then find the sine, cosine, and tangent of each angle:

23. $135°$ **24.** $225°$ **25.** $90°$ **26.** $180°$

27. $-45°$ **28.** $-90°$ **29.** $0°$ **30.** $-135°$

Determine whether each statement is true or false.

31. $\cos 35° < \cos 45°$ **32.** $\sin 35° < \sin 45°$

33. $\sec 60° < \sec 75°$ **34.** $\cot 60° < \cot 75°$

Use Definition I and Figure 1 to answer the following.

35. Explain why there is no angle θ such that $\sin \theta = 2$.

36. Explain why there is no angle θ such that $\sec \theta = \frac{1}{2}$.

37. Why is $|\csc \theta| \geq 1$ for any angle θ in standard position?

38. Why is $\cos \theta \leq 1$ for any angle θ in standard position?

39. As θ increases from $0°$ to $90°$, the value of $\sin \theta$ tends toward what number?

40. As θ increases from $0°$ to $90°$, the value of $\cos \theta$ tends toward what number?

41. As θ increases from $0°$ to $90°$, the value of $\tan \theta$ tends toward ______________.

42. As θ increases from $0°$ to $90°$, the value of $\sec \theta$ tends toward ______________.

Indicate the two quadrants θ could terminate in given the value of the trigonometric function.

43. $\sin \theta = \dfrac{3}{5}$ **44.** $\cos \theta = \dfrac{1}{2}$

45. $\cos \theta = -0.45$ **46.** $\sin \theta = -3/\sqrt{10}$

47. $\tan \theta = \dfrac{7}{24}$ **48.** $\cot \theta = -\dfrac{21}{20}$

49. $\csc \theta = -2.45$ **50.** $\sec \theta = 2$

Indicate the quadrants in which the terminal side of θ must lie under each of the following conditions.

51. $\sin \theta$ is negative and $\tan \theta$ is positive **52.** $\sin \theta$ is positive and $\cos \theta$ is negative

53. $\sin \theta$ and $\tan \theta$ have the same sign **54.** $\cos \theta$ and $\cot \theta$ have the same sign

For Problems 55 through 68, find the remaining trigonometric functions of θ based on the given information.

55. $\sin\theta = \frac{12}{13}$ and θ terminates in QI

56. $\cos\theta = \frac{24}{25}$ and θ terminates in QIV

57. $\cos\theta = -\frac{20}{29}$ and θ terminates in QII

58. $\sin\theta = -\frac{20}{29}$ and θ terminates in QIII

59. $\cos\theta = \frac{\sqrt{3}}{2}$ and θ terminates in QIV

60. $\cos\theta = \frac{\sqrt{2}}{2}$ and θ terminates in QI

61. $\tan\theta = \frac{3}{4}$ and θ terminates in QIII

62. $\cot\theta = -2$ and θ terminates in QII

63. $\csc\theta = \frac{13}{5}$ and $\cos\theta < 0$

64. $\sec\theta = \frac{13}{5}$ and $\sin\theta < 0$

65. $\cot\theta = \frac{1}{2}$ and $\cos\theta > 0$

66. $\tan\theta = -\frac{1}{2}$ and $\sin\theta > 0$

67. $\tan\theta = \frac{a}{b}$ where a and b are both positive

68. $\cot\theta = \frac{m}{n}$ where m and n are both positive

69. Find $\sin\theta$ and $\cos\theta$ if the terminal side of θ lies along the line $y = 2x$ in QI.

70. Find $\sin\theta$ and $\cos\theta$ if the terminal side of θ lies along the line $y = 2x$ in QIII.

71. Find $\sin\theta$ and $\tan\theta$ if the terminal side of θ lies along the line $y = -3x$ in QII.

72. Find $\sin\theta$ and $\tan\theta$ if the terminal side of θ lies along the line $y = -3x$ in QIV.

73. Draw 45° and −45° in standard position and then show that $\cos(-45°) = \cos 45°$.

74. Draw 45° and −45° in standard position and then show that $\sin(-45°) = -\sin 45°$.

75. Find y if the point $(5, y)$ is on the terminal side of θ and $\cos\theta = 5/13$.

76. Find x if the point $(x, -3)$ is on the terminal side of θ and $\sin\theta = -3/5$.

➤ LEARNING OBJECTIVES ASSESSMENT

These questions are available for instructors to help assess if you have successfully met the learning objectives for this section.

77. Find $\cos\theta$ if $(1, -3)$ is a point on the terminal side of θ.

a. $2\sqrt{2}$ **b.** $-\frac{1}{3}$ **c.** $-\frac{3\sqrt{10}}{10}$ **d.** $\frac{\sqrt{10}}{10}$

78. As θ increases from 0° to 90°, the value of $\cos\theta$ tends toward which of the following?

a. 0 **b.** ∞ **c.** 1 **d.** $-\infty$

79. Which quadrants could θ terminate in if $\cos\theta$ is negative?

a. QII, QIII **b.** QII, QIV **c.** QI, QIV **d.** QIII, QIV

80. Find $\tan\theta$ if $\sin\theta = \frac{4}{5}$ and θ terminates in QII.

a. $-\frac{4}{3}$ **b.** $\frac{3}{5}$ **c.** $-\frac{5}{3}$ **d.** $\frac{3}{4}$

SECTION 1.4 Introduction to Identities

LEARNING OBJECTIVES

1. Find the value of a trigonometric function using a reciprocal identity.
2. Find the value of a trigonometric function using a ratio identity.
3. Evaluate a trigonometric function raised to an exponent.
4. Use a Pythagorean identity to find the value of a trigonometric function.

You may recall from the work you have done in algebra that an expression such as $\sqrt{x^2 + 9}$ cannot be simplified further because the square root of a sum is not equal to the sum of the square roots. (In other words, it would be a mistake to write $\sqrt{x^2 + 9}$ as $x + 3$.) However, expressions such as $\sqrt{x^2 + 9}$ occur frequently enough in mathematics that we would like to find expressions equivalent to them that do not contain square roots. As it turns out, the relationships that we develop in this section and the next are the key to rewriting expressions such as $\sqrt{x^2 + 9}$ in a more convenient form, which we will show in Section 1.5.

Before we begin our introduction to identities, we need to review some concepts from arithmetic and algebra.

In algebra, statements such as $2x = x + x$, $x^3 = x \cdot x \cdot x$, and $\frac{x}{4x} = \frac{1}{4}$ are called identities. They are identities because they are true for all replacements of the variable for which they are defined.

NOTE $\frac{x}{4x}$ is not equal to $\frac{1}{4}$ when x is 0. The statement $\frac{x}{4x} = \frac{1}{4}$ is still an identity, however, since it is true for all values of x for which $\frac{x}{4x}$ is defined.

The eight basic trigonometric identities we will work with in this section are all derived from our definition of the trigonometric functions. Because many trigonometric identities have more than one form, we will list the basic identity first and then give the most common equivalent forms of that identity.

Reciprocal Identities

Our definition for the sine and cosecant functions indicate that they are reciprocals; that is,

$$\csc\theta = \frac{1}{\sin\theta} \qquad \text{because} \qquad \frac{1}{\sin\theta} = \frac{1}{y/r} = \frac{r}{y} = \csc\theta$$

NOTE We can also write this same relationship between $\sin\theta$ and $\csc\theta$ in another form as

$$\sin\theta = \frac{1}{\csc\theta} \qquad \text{because} \qquad \frac{1}{\csc\theta} = \frac{1}{r/y} = \frac{y}{r} = \sin\theta$$

The first identity we wrote, $\csc\theta = 1/\sin\theta$, is the basic identity. The second one, $\sin\theta = 1/\csc\theta$, is an equivalent form of the first.

From the preceding discussion and from the definition of $\cos\theta$, $\sec\theta$, $\tan\theta$, and $\cot\theta$, it is apparent that $\sec\theta$ is the reciprocal of $\cos\theta$, and $\cot\theta$ is the reciprocal of $\tan\theta$. Table 1 lists three basic reciprocal identities and their common equivalent forms.

TABLE 1
(Memorize)

Reciprocal Identities	Equivalent Forms
$\csc\theta = \frac{1}{\sin\theta}$	$\sin\theta = \frac{1}{\csc\theta}$
$\sec\theta = \frac{1}{\cos\theta}$	$\cos\theta = \frac{1}{\sec\theta}$
$\cot\theta = \frac{1}{\tan\theta}$	$\tan\theta = \frac{1}{\cot\theta}$

The examples that follow show some of the ways in which we use these reciprocal identities.

PROBLEMS

1. If $\sin\theta = -2/3$, find $\csc\theta$.
2. If $\cos\theta = 3/5$, find $\sec\theta$.
3. If $\tan\theta = 1/8$, find $\cot\theta$.
4. If $\csc\theta = 4/3$, find $\sin\theta$.
5. If $\sec\theta = 7/3$, find $\cos\theta$.
6. If $\cot\theta = -7/5$, find $\tan\theta$.

EXAMPLES

1. If $\sin\theta = \dfrac{3}{5}$, then $\csc\theta = \dfrac{5}{3}$, because

$$\csc\theta = \frac{1}{\sin\theta} = \frac{1}{3/5} = \frac{5}{3}$$

2. If $\cos\theta = -\dfrac{\sqrt{3}}{2}$, then $\sec\theta = -\dfrac{2}{\sqrt{3}} = -\dfrac{2\sqrt{3}}{3}$.

(*Remember:* Reciprocals always have the same algebraic sign.)

3. If $\tan\theta = 2$, then $\cot\theta = \dfrac{1}{2}$.

4. If $\csc\theta = a$, then $\sin\theta = \dfrac{1}{a}$.

5. If $\sec\theta = 1$, then $\cos\theta = 1$ (1 is its own reciprocal).

6. If $\cot\theta = -1$, then $\tan\theta = -1$.

Ratio Identities

There are two ratio identities, one for $\tan\theta$ and one for $\cot\theta$ (see Table 2).

TABLE 2
(Memorize)

Ratio Identities		
$\tan\theta = \dfrac{\sin\theta}{\cos\theta}$	because	$\dfrac{\sin\theta}{\cos\theta} = \dfrac{y/r}{x/r} = \dfrac{y}{x} = \tan\theta$
$\cot\theta = \dfrac{\cos\theta}{\sin\theta}$	because	$\dfrac{\cos\theta}{\sin\theta} = \dfrac{x/r}{y/r} = \dfrac{x}{y} = \cot\theta$

PROBLEM 7
If $\cos\theta = -3/5$ and $\sin\theta = 4/5$, find $\tan\theta$ and $\cot\theta$.

EXAMPLE 7 If $\sin\theta = -3/5$ and $\cos\theta = 4/5$, find $\tan\theta$ and $\cot\theta$.

SOLUTION Using the ratio identities, we have

$$\tan\theta = \frac{\sin\theta}{\cos\theta} = \frac{-3/5}{4/5} = -\frac{3}{5}\cdot\frac{5}{4} = -\frac{3}{4}$$

$$\cot\theta = \frac{\cos\theta}{\sin\theta} = \frac{4/5}{-3/5} = \frac{4}{5}\cdot\left(-\frac{5}{3}\right) = -\frac{4}{3}$$

NOTE Once we found $\tan\theta$, we could have used a reciprocal identity to find $\cot\theta$.

$$\cot\theta = \frac{1}{\tan\theta} = \frac{1}{-3/4} = -\frac{4}{3}$$

NOTATION

The notation $\sin^2\theta$ is a shorthand notation for $(\sin\theta)^2$. It indicates we are to square the number that is the sine of θ.

PROBLEMS	EXAMPLES
8. If $\cos\theta = -\frac{1}{4}$, find $\cos^2\theta$.	**8.** If $\sin\theta = \frac{3}{5}$, then $\sin^2\theta = \left(\frac{3}{5}\right)^2 = \frac{9}{25}$.
9. If $\tan\theta = \frac{4}{3}$, find $\tan^3\theta$.	**9.** If $\cos\theta = -\frac{1}{2}$, then $\cos^3\theta = \left(-\frac{1}{2}\right)^3 = -\frac{1}{8}$.

Pythagorean Identities

To derive our first Pythagorean identity, we start with the relationship among x, y, and r as given in the definition of $\sin\theta$ and $\cos\theta$.

$$x^2 + y^2 = r^2$$

$$\frac{x^2}{r^2} + \frac{y^2}{r^2} = 1 \qquad \text{Divide through by } r^2$$

$$\left(\frac{x}{r}\right)^2 + \left(\frac{y}{r}\right)^2 = 1 \qquad \text{Property of exponents}$$

$$(\cos\theta)^2 + (\sin\theta)^2 = 1 \qquad \text{Definition of } \sin\theta \text{ and } \cos\theta$$

$$\cos^2\theta + \sin^2\theta = 1 \qquad \text{Notation}$$

This last line is our first Pythagorean identity. We will use it many times throughout the book. It states that, for any angle θ, the sum of the squares of $\sin\theta$ and $\cos\theta$ is *always* 1.

There are two very useful equivalent forms of the first Pythagorean identity. One form occurs when we solve $\cos^2\theta + \sin^2\theta = 1$ for $\cos\theta$, and the other form is the result of solving for $\sin\theta$.

Solving for $\cos\theta$ we have

$$\cos^2\theta + \sin^2\theta = 1$$

$$\cos^2\theta = 1 - \sin^2\theta \qquad \text{Add } -\sin^2\theta \text{ to both sides}$$

$$\cos\theta = \pm\sqrt{1 - \sin^2\theta} \qquad \text{Take the square root of both sides}$$

Similarly, solving for $\sin\theta$ gives us

$$\sin^2\theta = 1 - \cos^2\theta$$

$$\sin\theta = \pm\sqrt{1 - \cos^2\theta}$$

Our next Pythagorean identity is derived from the first Pythagorean identity, $\cos^2\theta + \sin^2\theta = 1$, by dividing both sides by $\cos^2\theta$. Here is that derivation:

$$\cos^2\theta + \sin^2\theta = 1 \qquad \text{First Pythagorean identity}$$

$$\frac{\cos^2\theta + \sin^2\theta}{\cos^2\theta} = \frac{1}{\cos^2\theta} \qquad \text{Divide each side by } \cos^2\theta$$

$$\frac{\cos^2\theta}{\cos^2\theta} + \frac{\sin^2\theta}{\cos^2\theta} = \frac{1}{\cos^2\theta} \qquad \text{Write the left side as two fractions}$$

$$\left(\frac{\cos\theta}{\cos\theta}\right)^2 + \left(\frac{\sin\theta}{\cos\theta}\right)^2 = \left(\frac{1}{\cos\theta}\right)^2 \qquad \text{Property of exponents}$$

$$1 + \tan^2\theta = \sec^2\theta \qquad \text{Ratio and reciprocal identities}$$

This last expression, $1 + \tan^2\theta = \sec^2\theta$, is our second Pythagorean identity. To arrive at our third, and last, Pythagorean identity, we proceed as we have earlier, but instead of dividing each side by $\cos^2 \theta$, we divide by $\sin^2 \theta$. Without showing the work involved in doing so, the result is

$$1 + \cot^2 \theta = \csc^2 \theta$$

We summarize our derivations in Table 3.

TABLE 3
(Memorize)

Pythagorean Identities	Equivalent Forms
$\cos^2 \theta + \sin^2 \theta = 1$	$\cos \theta = \pm\sqrt{1 - \sin^2 \theta}$ $\sin \theta = \pm\sqrt{1 - \cos^2 \theta}$
$1 + \tan^2 \theta = \sec^2 \theta$	
$1 + \cot^2 \theta = \csc^2 \theta$	

Notice the $\pm$ sign in the equivalent forms. It occurs as part of the process of taking the square root of both sides of the preceding equation. (*Remember:* We would obtain a similar result in algebra if we solved the equation $x^2 = 9$ to get $x = \pm 3$.) In Example 10 we will see how to deal with the $\pm$ sign.

PROBLEM 10
If $\sin \theta = -1/3$ and θ terminates in QIII, find $\cos \theta$ and $\tan \theta$.

EXAMPLE 10 If $\sin \theta = 3/5$ and θ terminates in QII, find $\cos \theta$ and $\tan \theta$.

SOLUTION We begin by using the identity $\cos \theta = \pm\sqrt{1 - \sin^2 \theta}$. First, we must decide whether $\cos \theta$ will be positive or negative (it can't be both). We are told that θ terminates in QII, so we know that $\cos \theta$ must be negative. Therefore

$$\cos \theta = -\sqrt{1 - \sin^2 \theta}$$

Now we can substitute the given value of $\sin \theta$.

If $\quad \sin \theta = \frac{3}{5}$

the identity $\quad \cos \theta = -\sqrt{1 - \sin^2 \theta}$

becomes $\quad \cos \theta = -\sqrt{1 - \left(\frac{3}{5}\right)^2}$

$$= -\sqrt{1 - \frac{9}{25}}$$

$$= -\sqrt{\frac{16}{25}}$$

$$= -\frac{4}{5}$$

To find $\tan \theta$, we use a ratio identity.

$$\tan \theta = \frac{\sin \theta}{\cos \theta}$$

$$= \frac{3/5}{-4/5}$$

$$= -\frac{3}{4}$$

■

PROBLEM 11
If $\cos\theta = -5/13$ and θ terminates in QII, find the remaining trigonometric ratios for θ.

EXAMPLE 11 If $\cos\theta = 1/2$ and θ terminates in QIV, find the remaining trigonometric ratios for θ.

SOLUTION The first, and easiest, ratio to find is $\sec\theta$ because it is the reciprocal of $\cos\theta$.

$$\sec\theta = \frac{1}{\cos\theta} = \frac{1}{1/2} = 2$$

Next we find $\sin\theta$. Using one of the equivalent forms of our Pythagorean identity, we have

$$\sin\theta = \pm\sqrt{1 - \cos^2\theta}$$

Because θ terminates in QIV, $\sin\theta$ will be negative. This gives us

$$\begin{aligned}
\sin\theta &= -\sqrt{1-\cos^2\theta} && \text{Negative sign because } \theta \in \text{QIV} \\
&= -\sqrt{1-\left(\frac{1}{2}\right)^2} && \text{Substitute } \tfrac{1}{2} \text{ for } \cos\theta \\
&= -\sqrt{1-\frac{1}{4}} && \text{Square } \tfrac{1}{2} \text{ to get } \tfrac{1}{4} \\
&= -\sqrt{\frac{3}{4}} && \text{Subtract} \\
&= -\frac{\sqrt{3}}{2} && \text{Take the square root of the numerator and denominator separately}
\end{aligned}$$

Now that we have $\sin\theta$ and $\cos\theta$, we can find $\tan\theta$ by using a ratio identity.

$$\tan\theta = \frac{\sin\theta}{\cos\theta} = \frac{-\sqrt{3}/2}{1/2} = -\frac{\sqrt{3}}{2}\cdot\frac{2}{1} = -\sqrt{3}$$

Cot θ and csc θ are the reciprocals of $\tan\theta$ and $\sin\theta$, respectively. Therefore,

$$\cot\theta = \frac{1}{\tan\theta} = -\frac{1}{\sqrt{3}} = -\frac{\sqrt{3}}{3}$$

$$\csc\theta = \frac{1}{\sin\theta} = -\frac{2}{\sqrt{3}} = -\frac{2\sqrt{3}}{3}$$

Here are all six ratios together:

$$\sin\theta = -\frac{\sqrt{3}}{2} \qquad \csc\theta = -\frac{2\sqrt{3}}{3}$$

$$\cos\theta = \frac{1}{2} \qquad \sec\theta = 2$$

$$\tan\theta = -\sqrt{3} \qquad \cot\theta = -\frac{\sqrt{3}}{3}$$

As a final note, we should mention that the eight basic identities we have derived here, along with their equivalent forms, are very important in the study of trigonometry. It is essential that you memorize them. It may be a good idea to practice writing them from memory until you can write each of the eight, and their equivalent forms, perfectly. As time goes by, we will increase our list of identities, so you will want to keep up with them as we go along.

Getting Ready for Class

After reading through the preceding section, respond in your own words and in complete sentences.

a. State the reciprocal identities for csc θ, sec θ, and cot θ.
b. State the equivalent forms of the reciprocal identities for sin θ, cos θ, and tan θ.
c. State the ratio identities for tan θ and cot θ.
d. State the three Pythagorean identities.

1.4 PROBLEM SET

➤ CONCEPTS AND VOCABULARY

For Questions 1 and 2, fill in each blank with the appropriate word or expression.

1. An identity is an equation that is ______ for all replacements of the variable for which it is __________.

2. The notation $\cos^2\theta$ is a shorthand for (______ $)^2$.

3. Match each trigonometric function with its reciprocal.

a. sine	**i.** cotangent
b. cosine	**ii.** cosecant
c. tangent	**iii.** secant

4. Match the trigonometric functions that are related through a Pythagorean identity.

a. sine	**i.** secant
b. tangent	**ii.** cosecant
c. cotangent	**iii.** cosine

➤ EXERCISES

Give the reciprocal of each number.

5. 7 **6.** 4 **7.** $-2/3$ **8.** $-5/13$

9. $-1/\sqrt{2}$ **10.** $-\sqrt{3}/2$ **11.** x **12.** $1/a$

Use the reciprocal identities for the following problems.

13. If $\sin\theta = 4/5$, find $\csc\theta$.

14. If $\cos\theta = \sqrt{3}/2$, find $\sec\theta$.

15. If $\sec\theta = -2$, find $\cos\theta$.

16. If $\csc\theta = -13/12$, find $\sin\theta$.

17. If $\tan\theta = a$ $(a \neq 0)$, find $\cot\theta$.

18. If $\cot\theta = -b$ $(b \neq 0)$, find $\tan\theta$.

Use a ratio identity to find tan θ given the following values.

19. $\sin\theta = \dfrac{2\sqrt{5}}{5}$ and $\cos\theta = \dfrac{\sqrt{5}}{5}$

20. $\sin\theta = \dfrac{3}{5}$ and $\cos\theta = -\dfrac{4}{5}$

Use a ratio identity to find cot θ given the following values.

21. $\sin\theta = -\dfrac{5}{13}$ and $\cos\theta = -\dfrac{12}{13}$

22. $\sin\theta = \dfrac{2\sqrt{13}}{13}$ and $\cos\theta = \dfrac{3\sqrt{13}}{13}$

For Problems 23 through 26, recall that $\sin^2\theta$ means $(\sin\theta)^2$.

23. If $\sin\theta = \dfrac{\sqrt{2}}{2}$, find $\sin^2\theta$.

24. If $\cos\theta = \dfrac{1}{3}$, find $\cos^2\theta$.

25. If $\tan\theta = 2$, find $\tan^3\theta$.

26. If $\sec\theta = -3$, find $\sec^3\theta$.

For Problems 27 through 30, let $\sin\theta = -12/13$, and $\cos\theta = -5/13$, and find the indicated value.

27. $\tan\theta$

28. $\cot\theta$

29. $\sec\theta$

30. $\csc\theta$

Use the equivalent forms of the first Pythagorean identity on Problems 31 through 38.

31. Find $\sin\theta$ if $\cos\theta = \dfrac{3}{5}$ and θ terminates in QI.

32. Find $\sin\theta$ if $\cos\theta = \dfrac{5}{13}$ and θ terminates in QI.

33. Find $\cos\theta$ if $\sin\theta = \dfrac{\sqrt{3}}{2}$ and θ terminates in QII.

34. Find $\cos\theta$ if $\sin\theta = \dfrac{1}{3}$ and θ terminates in QII.

35. If $\sin\theta = -4/5$ and θ terminates in QIII, find $\cos\theta$.

36. If $\sin\theta = -4/5$ and θ terminates in QIV, find $\cos\theta$.

37. If $\cos\theta = \sqrt{3}/2$ and θ terminates in QI, find $\sin\theta$.

38. If $\cos\theta = -1/2$ and θ terminates in QII, find $\sin\theta$.

39. Find $\tan\theta$ if $\sin\theta = 1/3$ and θ terminates in QI.

40. Find $\cot\theta$ if $\sin\theta = 2/3$ and θ terminates in QII.

41. Find $\sec\theta$ if $\tan\theta = 8/15$ and θ terminates in QIII.

42. Find $\csc\theta$ if $\cot\theta = -24/7$ and θ terminates in QIV.

43. Find $\csc\theta$ if $\cot\theta = -21/20$ and $\sin\theta > 0$.

44. Find $\sec\theta$ if $\tan\theta = 7/24$ and $\cos\theta < 0$.

Find the remaining trigonometric ratios of θ based on the given information.

45. $\cos\theta = 12/13$ and θ terminates in QI

46. $\sin\theta = 12/13$ and θ terminates in QI

47. $\sin\theta = -1/2$ and θ is not in QIII

48. $\cos\theta = -1/3$ and θ is not in QII

49. $\csc\theta = 2$ and $\cos\theta$ is negative

50. $\sec\theta = 2$ and $\sin\theta$ is positive

51. $\cos\theta = \dfrac{2\sqrt{13}}{13}$ and $\theta \in$ QIV

52. $\sin\theta = \dfrac{3\sqrt{10}}{10}$ and $\theta \in$ QII

53. $\csc\theta = a$ and θ terminates in QI

54. $\sec\theta = b$ and θ terminates in QI

Using your calculator and rounding your answers to the nearest hundredth, find the remaining trigonometric ratios of θ based on the given information.

55. $\sin\theta = 0.23$ and $\theta \in$ QI

56. $\cos\theta = 0.51$ and $\theta \in$ QI

57. $\sec\theta = -1.24$ and $\theta \in$ QII

58. $\csc\theta = -2.45$ and $\theta \in$ QIII

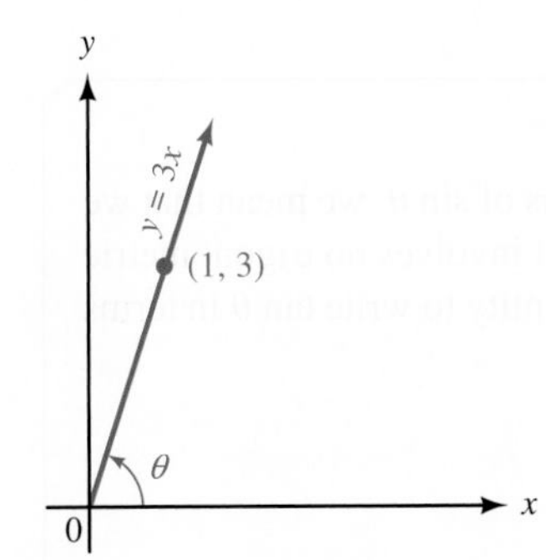

Figure 1

Recall from algebra that the slope of the line through (x_1, y_1) and (x_2, y_2) is

$$m = \frac{y_2 - y_1}{x_2 - x_1}$$

It is the change in the y-coordinates divided by the change in the x-coordinates.

59. The line $y = 3x$ passes through the points (0, 0) and (1, 3). Find its slope.

60. Find the slope of the line $y = mx$. [It passes through the origin and the point $(1, m)$.]

61. Suppose the angle formed by the line $y = 3x$ and the positive x-axis is θ. Find the tangent of θ (Figure 1).

62. Find $\tan\theta$ if θ is the angle formed by the line $y = mx$ and the positive x-axis (Figure 2).

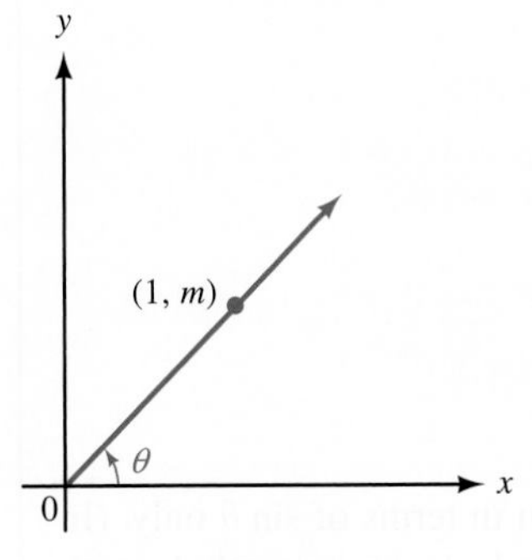

Figure 2

➤ LEARNING OBJECTIVES ASSESSMENT

These questions are available for instructors to help assess if you have successfully met the learning objectives for this section.

63. Find $\sec\theta$ if $\cos\theta = \dfrac{1}{3}$.

a. $2\sqrt{2}$ **b.** $\dfrac{2\sqrt{2}}{3}$ **c.** 3 **d.** $\dfrac{3\sqrt{2}}{4}$

64. Use a ratio identity to find $\tan\theta$ if $\cos\theta = \dfrac{2}{3}$ and $\sin\theta = \dfrac{\sqrt{5}}{3}$.

a. $\dfrac{2\sqrt{5}}{5}$ **b.** $\dfrac{\sqrt{5}}{2}$ **c.** $\dfrac{2\sqrt{5}}{9}$ **d.** $\dfrac{9\sqrt{5}}{10}$

65. If $\cos\theta = \dfrac{1}{4}$, find $\cos^2\theta$.

a. $\dfrac{1}{16}$ **b.** $\cos^2 \dfrac{1}{16}$ **c.** $\dfrac{1}{2}$ **d.** 16

66. Which answer correctly uses a Pythagorean identity to find $\sin\theta$ if $\cos\theta = \dfrac{1}{4}$ and θ terminates in QIV?

a. $\sqrt{1 + \left(\frac{1}{4}\right)^2}$ **b.** $\sqrt{1 - \left(\frac{1}{4}\right)^2}$ **c.** $-\sqrt{1 + \left(\frac{1}{4}\right)^2}$ **d.** $-\sqrt{1 - \left(\frac{1}{4}\right)^2}$

SECTION 1.5 More on Identities

LEARNING OBJECTIVES

1. Write an expression in terms of sines and cosines.
2. Simplify an expression containing trigonometric functions.
3. Use a trigonometric substitution to simplify a radical expression.
4. Verify an equation is an identity.

The topics we will cover in this section are an extension of the work we did with identities in Section 1.4. We can use the eight basic identities we introduced in the previous section to rewrite or simplify certain expressions. It will be helpful at this point if you have already begun to memorize these identities.

The first topic involves writing any of the six trigonometric functions in terms of any of the others. Let's look at an example.

PROBLEM 1
Write $\sec \theta$ in terms of $\sin \theta$.

EXAMPLE 1 Write $\tan \theta$ in terms of $\sin \theta$.

SOLUTION When we say we want $\tan \theta$ written in terms of $\sin \theta$, we mean that we want to write an expression that is equivalent to $\tan \theta$ but involves no trigonometric function other than $\sin \theta$. Let's begin by using a ratio identity to write $\tan \theta$ in terms of $\sin \theta$ and $\cos \theta$.

$$\tan \theta = \frac{\sin \theta}{\cos \theta}$$

Now we need to replace $\cos \theta$ with an expression involving only $\sin \theta$. Since $\cos \theta = \pm\sqrt{1 - \sin^2 \theta}$

$$\begin{aligned} \tan \theta &= \frac{\sin \theta}{\cos \theta} \\ &= \frac{\sin \theta}{\pm\sqrt{1 - \sin^2 \theta}} \\ &= \pm\frac{\sin \theta}{\sqrt{1 - \sin^2 \theta}} \end{aligned}$$

This last expression is equivalent to $\tan \theta$ and is written in terms of $\sin \theta$ only. (In a problem like this, it is okay to include numbers and algebraic symbols with $\sin \theta$.) ■

Here is another example.

PROBLEM 2
Write $\tan \theta \csc \theta$ in terms of $\sin \theta$ and $\cos \theta$ and then simplify.

EXAMPLE 2 Write $\sec \theta \tan \theta$ in terms of $\sin \theta$ and $\cos \theta$, and then simplify.

SOLUTION Since $\sec \theta = 1/\cos \theta$ and $\tan \theta = \sin \theta/\cos \theta$, we have

$$\begin{aligned} \sec \theta \tan \theta &= \frac{1}{\cos \theta} \cdot \frac{\sin \theta}{\cos \theta} \\ &= \frac{\sin \theta}{\cos^2 \theta} \end{aligned}$$

■

The next examples show how we manipulate trigonometric expressions using algebraic techniques.

PROBLEM 3

Add $\sin\theta + \frac{1}{\cos\theta}$.

EXAMPLE 3 Add $\frac{1}{\sin\theta} + \frac{1}{\cos\theta}$.

SOLUTION We can add these two expressions in the same way we would add $\frac{1}{3}$ and $\frac{1}{4}$—by first finding a least common denominator (LCD), and then writing each expression again with the LCD for its denominator.

$$\frac{1}{\sin\theta} + \frac{1}{\cos\theta} = \frac{1}{\sin\theta}\cdot\frac{\mathbf{cos}\ \boldsymbol{\theta}}{\mathbf{cos}\ \boldsymbol{\theta}} + \frac{1}{\cos\theta}\cdot\frac{\mathbf{sin}\ \boldsymbol{\theta}}{\mathbf{sin}\ \boldsymbol{\theta}} \qquad \text{The LCD is } \sin\theta\cos\theta$$

$$= \frac{\cos\theta}{\sin\theta\cos\theta} + \frac{\sin\theta}{\cos\theta\sin\theta}$$

$$= \frac{\cos\theta + \sin\theta}{\sin\theta\cos\theta}$$

PROBLEM 4

Multiply $(2\cos\theta + 1)(3\cos\theta - 2)$.

EXAMPLE 4 Multiply $(\sin\theta + 2)(\sin\theta - 5)$.

SOLUTION We multiply these two expressions in the same way we would multiply $(x + 2)(x - 5)$. (In some algebra books, this kind of multiplication is accomplished using the FOIL method.)

$$(\sin\theta + 2)(\sin\theta - 5) = \sin\theta\sin\theta - 5\sin\theta + 2\sin\theta - 10$$

$$= \sin^2\theta - 3\sin\theta - 10$$

In the introduction to Section 1.4, we mentioned that trigonometric identities are the key to writing the expression $\sqrt{x^2 + 9}$ without the square root symbol. Our next example shows how we do this using a trigonometric substitution.

PROBLEM 5

Simplify the expression $\sqrt{16 - x^2}$ as much as possible after substituting $4\cos\theta$ for x.

EXAMPLE 5 Simplify the expression $\sqrt{x^2 + 9}$ as much as possible after substituting $3\tan\theta$ for x.

SOLUTION Our goal is to write the expression $\sqrt{x^2 + 9}$ without a square root by first making the substitution $x = 3\tan\theta$.

If $\qquad x = 3\tan\theta$

then the expression $\qquad \sqrt{x^2 + 9}$

becomes $\qquad \sqrt{(3\tan\theta)^2 + 9} = \sqrt{9\tan^2\theta + 9}$

$$= \sqrt{9(\tan^2\theta + 1)}$$

$$= \sqrt{9\sec^2\theta}$$

$$= 3\,|\sec\theta|$$

NOTE 1 We must use the absolute value symbol unless we know that $\sec\theta$ is positive. Remember, in algebra, $\sqrt{a^2} = a$ only when a is positive or zero. If it is possible that a is negative, then $\sqrt{a^2} = |a|$. In Section 4.7, we will see how to simplify our answer even further by removing the absolute value symbol.

NOTE 2 After reading through Example 5, you may be wondering if it is mathematically correct to make the substitution $x = 3\tan\theta$. After all, x can be any real number because $x^2 + 9$ will always be positive. (*Remember:* We want to avoid taking the square root of a negative number.) How do we know that any real number can be written as

$3 \tan \theta$? We will take care of this in Section 4.4 by showing that for any real number x, there is a value of θ between $-90°$ and $90°$ for which $x = 3 \tan \theta$.

In the examples that follow, we want to use the eight basic identities we developed in Section 1.4, along with some techniques from algebra, to show that some more complicated identities are true.

PROBLEM 6
Show that $\sin \theta + \cos \theta \cot \theta = \csc \theta$ is true by transforming the left side into the right side.

EXAMPLE 6 Show that the following statement is true by transforming the left side into the right side.

$$\cos \theta \tan \theta = \sin \theta$$

SOLUTION We begin by writing the left side in terms of $\sin \theta$ and $\cos \theta$.

$$\cos \theta \tan \theta = \cos \theta \cdot \frac{\sin \theta}{\cos \theta}$$

$$= \frac{\cos \theta \sin \theta}{\cos \theta}$$

$$= \sin \theta \quad \text{Divide out the } \cos \theta \text{ common to the numerator and denominator}$$

Because we have succeeded in transforming the left side into the right side, we have shown that the statement $\cos \theta \tan \theta = \sin \theta$ is an identity. ■

PROBLEM 7
Prove the identity $(\cos \theta + \sin \theta)^2 = 2 \cos \theta \sin \theta + 1$.

EXAMPLE 7 Prove the identity $(\sin \theta + \cos \theta)^2 = 1 + 2 \sin \theta \cos \theta$.

SOLUTION Let's agree to prove the identities in this section, and the problem set that follows, by transforming the left side into the right side. In this case, we begin by expanding $(\sin \theta + \cos \theta)^2$. (Remember from algebra, $(a + b)^2 = a^2 + 2ab + b^2$.)

$$(\sin \theta + \cos \theta)^2 = \sin^2 \theta + 2 \sin \theta \cos \theta + \cos^2 \theta$$

$$= (\sin^2 \theta + \cos^2 \theta) + 2 \sin \theta \cos \theta \quad \text{Rearrange terms}$$

$$= 1 + 2 \sin \theta \cos \theta \quad \text{Pythagorean identity}$$

■

We should mention that the ability to prove identities in trigonometry is not always obtained immediately. It usually requires a lot of practice. The more you work at it, the better you will become at it. In the meantime, if you are having trouble, check first to see that you have memorized the eight basic identities—reciprocal, ratio, Pythagorean, and their equivalent forms—as given in Section 1.4.

Getting Ready for Class

After reading through the preceding section, respond in your own words and in complete sentences.

a. What do we mean when we want $\tan \theta$ written in terms of $\sin \theta$?
b. Write $\tan \theta$ in terms of $\sin \theta$.
c. What is the least common denominator for the expression $\frac{1}{\sin \theta} + \frac{1}{\cos \theta}$?
d. How would you prove the identity $\cos \theta \tan \theta = \sin \theta$?

1.5 PROBLEM SET

➤ CONCEPTS AND VOCABULARY

For Questions 1 and 2, fill in each blank with the appropriate word.

1. One way to simplify an expression containing trigonometric functions is to first write the expression in terms of ______ and ________ only.

2. To prove, or verify, an identity, we can start with the _____ ______ of the equation and __________ it until it is identical to the ______ ______ of the equation.

➤ EXERCISES

Write each of the following in terms of sin θ only.

3. $\cos\theta$ **4.** $\csc\theta$ **5.** $\cot\theta$ **6.** $\sec\theta$

Write each of the following in terms of cos θ only.

7. $\sec\theta$ **8.** $\sin\theta$ **9.** $\csc\theta$ **10.** $\tan\theta$

Simplify.

11. a. $\dfrac{\frac{1}{a}}{\frac{1}{b}}$ **b.** $\dfrac{\frac{1}{\cos\theta}}{\frac{1}{\sin\theta}}$ **12. a.** $\dfrac{\frac{a}{b}}{\frac{1}{b}}$ **b.** $\dfrac{\frac{\sin\theta}{\cos\theta}}{\frac{1}{\cos\theta}}$

Write each of the following in terms of sin θ and cos θ; then simplify if possible.

13. $\csc\theta\cot\theta$ **14.** $\sec\theta\cot\theta$ **15.** $\csc\theta\tan\theta$ **16.** $\sec\theta\tan\theta\csc\theta$

17. $\dfrac{\sec\theta}{\csc\theta}$ **18.** $\dfrac{\csc\theta}{\sec\theta}$ **19.** $\dfrac{\sec\theta}{\tan\theta}$ **20.** $\dfrac{\csc\theta}{\cot\theta}$

21. $\dfrac{\tan\theta}{\cot\theta}$ **22.** $\dfrac{\cot\theta}{\tan\theta}$ **23.** $\dfrac{\sin\theta}{\csc\theta}$ **24.** $\dfrac{\cos\theta}{\sec\theta}$

25. $\tan\theta + \sec\theta$ **26.** $\cot\theta - \csc\theta$

27. $\sin\theta\cot\theta + \cos\theta$ **28.** $\cos\theta\tan\theta + \sin\theta$

29. $\sec\theta - \tan\theta\sin\theta$ **30.** $\csc\theta - \cot\theta\cos\theta$

Add or subtract as indicated, then simplify if possible. For part (b), leave your answer in terms of sin θ and/or cos θ.

31. a. $\dfrac{1}{a} - \dfrac{1}{b}$ **32. a.** $\dfrac{1}{a} - a$

b. $\dfrac{1}{\sin\theta} - \dfrac{1}{\cos\theta}$ **b.** $\dfrac{1}{\sin\theta} - \sin\theta$

33. a. $\dfrac{b}{a} + \dfrac{1}{b}$ **34. a.** $\dfrac{a}{b} - \dfrac{b}{a}$

b. $\dfrac{\cos\theta}{\sin\theta} + \dfrac{1}{\cos\theta}$ **b.** $\dfrac{\sin\theta}{\cos\theta} - \dfrac{\cos\theta}{\sin\theta}$

Add and subtract as indicated. Then simplify your answers if possible. Leave all answers in terms of sin θ and/or cos θ.

35. $\sin\theta + \dfrac{1}{\cos\theta}$

36. $\cos\theta + \dfrac{1}{\sin\theta}$

37. $\dfrac{1}{\cos\theta} - \cos\theta$

38. $\dfrac{1}{\cos\theta} - \dfrac{1}{\sin\theta}$

39. $\dfrac{\sin\theta}{\cos\theta} + \dfrac{1}{\sin\theta}$

40. $\dfrac{\cos\theta}{\sin\theta} + \dfrac{\sin\theta}{\cos\theta}$

Multiply.

41. **a.** $(a - b)^2$
b. $(\cos\theta - \sin\theta)^2$

42. **a.** $(a + 1)(a - 1)$
b. $(\sec\theta + 1)(\sec\theta - 1)$

43. **a.** $(a - 2)(a - 3)$
b. $(\sin\theta - 2)(\sin\theta - 3)$

44. **a.** $(a + 1)(a - 4)$
b. $(\cos\theta + 1)(\cos\theta - 4)$

45. $(\sin\theta + 4)(\sin\theta + 3)$

46. $(\cos\theta + 2)(\cos\theta - 5)$

47. $(2\cos\theta + 3)(4\cos\theta - 5)$

48. $(3\sin\theta - 2)(5\cos\theta - 4)$

49. $(1 - \sin\theta)(1 + \sin\theta)$

50. $(1 - \cos\theta)(1 + \cos\theta)$

51. $(1 - \tan\theta)(1 + \tan\theta)$

52. $(1 - \cot\theta)(1 + \cot\theta)$

53. $(\sin\theta - \cos\theta)^2$

54. $(\cos\theta + \sin\theta)^2$

55. $(\sin\theta - 4)^2$

56. $(\cos\theta - 2)^2$

57. Simplify the expression $\sqrt{x^2 + 4}$ as much as possible after substituting $2\tan\theta$ for x.

58. Simplify the expression $\sqrt{x^2 + 1}$ as much as possible after substituting $\tan\theta$ for x.

59. Simplify the expression $\sqrt{9 - x^2}$ as much as possible after substituting $3\sin\theta$ for x.

60. Simplify the expression $\sqrt{25 - x^2}$ as much as possible after substituting $5\sin\theta$ for x.

61. Simplify the expression $\sqrt{x^2 - 36}$ as much as possible after substituting $6\sec\theta$ for x.

62. Simplify the expression $\sqrt{x^2 - 64}$ as much as possible after substituting $8\sec\theta$ for x.

63. Simplify the expression $\sqrt{64 - 4x^2}$ as much as possible after substituting $4\sin\theta$ for x.

64. Simplify the expression $\sqrt{4x^2 + 100}$ as much as possible after substituting $5\tan\theta$ for x.

Show that each of the following statements is an identity by transforming the left side of each one into the right side.

65. $\cos\theta\tan\theta = \sin\theta$

66. $\sin\theta\cot\theta = \cos\theta$

67. $\sin\theta\sec\theta\cot\theta = 1$

68. $\cos\theta\csc\theta\tan\theta = 1$

69. $\dfrac{\sin\theta}{\csc\theta} = \sin^2\theta$

70. $\dfrac{\cos\theta}{\sec\theta} = \cos^2\theta$

71. $\dfrac{\csc\theta}{\cot\theta} = \sec\theta$

72. $\dfrac{\sec\theta}{\tan\theta} = \csc\theta$

73. $\dfrac{\csc\theta}{\sec\theta} = \cot\theta$

74. $\dfrac{\sec\theta}{\csc\theta} = \tan\theta$

75. $\dfrac{\sec\theta\cot\theta}{\csc\theta} = 1$

76. $\dfrac{\csc\theta\tan\theta}{\sec\theta} = 1$

77. $\sin\theta\tan\theta + \cos\theta = \sec\theta$

78. $\cos\theta\cot\theta + \sin\theta = \csc\theta$

79. $\tan\theta + \cot\theta = \sec\theta\csc\theta$

80. $\tan^2\theta + 1 = \sec^2\theta$

81. $\csc\theta - \sin\theta = \dfrac{\cos^2\theta}{\sin\theta}$

82. $\sec\theta - \cos\theta = \dfrac{\sin^2\theta}{\cos\theta}$

83. $\csc\theta\tan\theta - \cos\theta = \dfrac{\sin^2\theta}{\cos\theta}$

84. $\sec\theta\cot\theta - \sin\theta = \dfrac{\cos^2\theta}{\sin\theta}$

85. $(1 - \cos\theta)(1 + \cos\theta) = \sin^2\theta$

86. $(1 + \sin\theta)(1 - \sin\theta) = \cos^2\theta$

87. $(\sin\theta + 1)(\sin\theta - 1) = -\cos^2\theta$

88. $(\cos\theta + 1)(\cos\theta - 1) = -\sin^2\theta$

89. $\dfrac{\cos\theta}{\sec\theta} + \dfrac{\sin\theta}{\csc\theta} = 1$

90. $1 - \dfrac{\sin\theta}{\csc\theta} = \cos^2\theta$

91. $(\cos\theta + \sin\theta)^2 - 1 = 2\sin\theta\cos\theta$

92. $(\sin\theta - \cos\theta)^2 - 1 = -2\sin\theta\cos\theta$

93. $\sin\theta(\sec\theta + \csc\theta) = \tan\theta + 1$

94. $\cos\theta(\csc\theta + \tan\theta) = \cot\theta + \sin\theta$

95. $\sin\theta(\csc\theta - \sin\theta) = \cos^2\theta$

96. $\cos\theta(\sec\theta - \cos\theta) = \sin^2\theta$

➤ LEARNING OBJECTIVES ASSESSMENT

These questions are available for instructors to help assess if you have successfully met the learning objectives for this section.

97. Express $\tan\theta$ in terms of $\sin\theta$ only.

a. $\pm\dfrac{\sin\theta}{\sqrt{\sin^2\theta - 1}}$

b. $\pm\dfrac{\sin\theta}{\sqrt{1 - \sin^2\theta}}$

c. $\pm\dfrac{\sqrt{1 - \sin^2\theta}}{\sin\theta}$

d. $\pm\dfrac{\sqrt{\sin^2\theta - 1}}{\sin\theta}$

98. Simplify $\dfrac{\cot\theta}{\csc\theta}$.

a. $\cos\theta$ b. $\sin\theta$ c. $\sec\theta$ d. $\tan\theta$

99. Simplify $\sqrt{x^2 + 16}$ as much as possible after substituting $4\tan\theta$ for x.

a. $4|\cos\theta|$ b. $4|\tan\theta|$ c. $4|\sec\theta|$ d. $4|\cot\theta|$

100. Which of the following is a valid step in proving the identity $\dfrac{1}{\sin\theta} - \sin\theta = \dfrac{\cos^2\theta}{\sec\theta}$?

a. Multiply both sides of the equation by $\sin\theta$.

b. Add $\sin\theta$ to both sides of the equation.

c. Multiply both sides of the equation by $\cos^2\theta$.

d. Write the left side as $\dfrac{1}{\sin\theta} - \dfrac{\sin^2\theta}{\sin\theta}$.

CHAPTER 1 SUMMARY

EXAMPLES

We will use the margin to give examples of the topics being reviewed, whenever it is appropriate.

The number in brackets next to each heading indicates the section in which that topic is discussed.

1

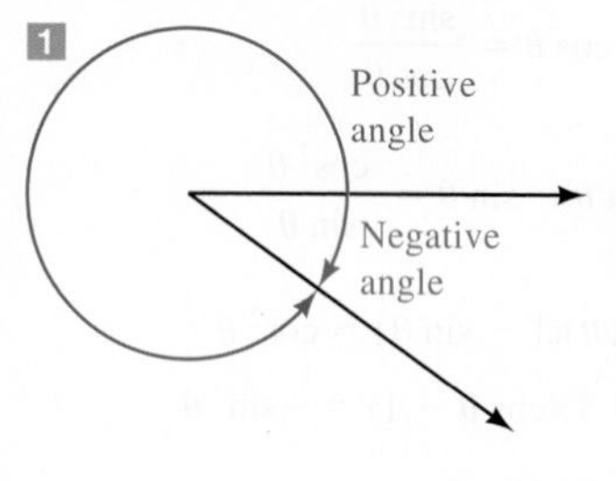

ANGLES [1.1]

An angle is formed by two rays with a common end point. The common end point is called the *vertex,* and the rays are called *sides,* of the angle. If we think of an angle as being formed by rotating the initial side about the vertex to the terminal side, then a counterclockwise rotation gives a *positive angle,* and a clockwise rotation gives a *negative angle.*

2

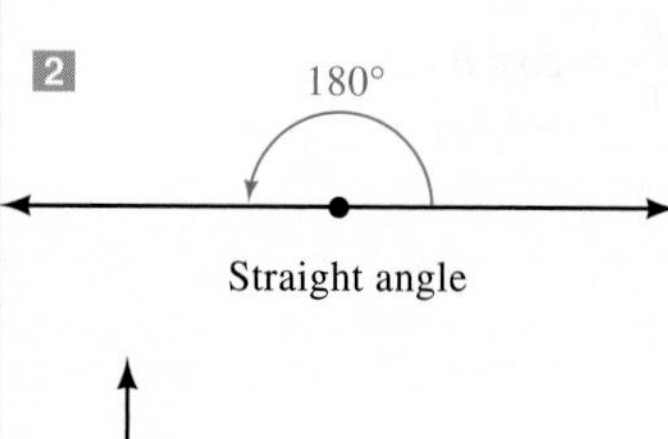

Straight angle

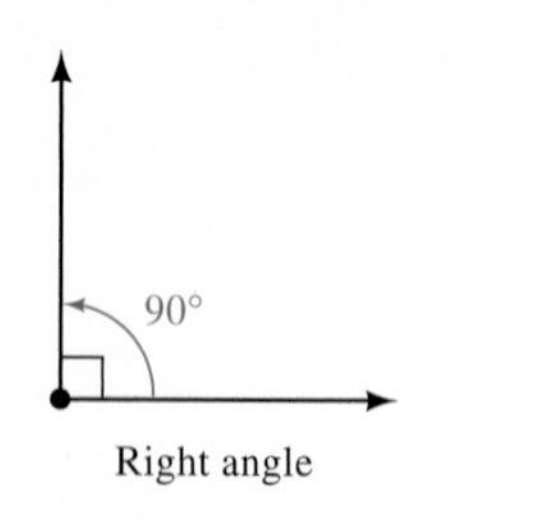

Right angle

DEGREE MEASURE [1.1]

There are 360° in a full rotation. This means that 1° is $\frac{1}{360}$ of a full rotation.

An angle that measures 90° is a *right angle.* An angle that measures 180° is a *straight angle.* Angles that measure between 0° and 90° are called *acute angles,* and angles that measure between 90° and 180° are called *obtuse angles. Complementary angles* have a sum of 90°, and *supplementary angles* have a sum of 180°.

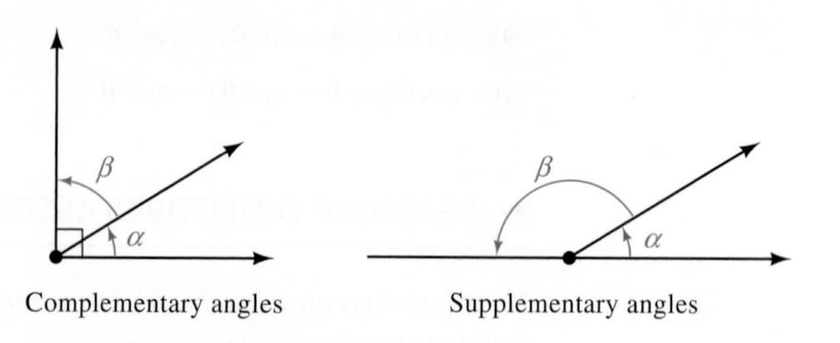

Complementary angles Supplementary angles

3 If ABC is a right triangle with $C = 90°$, and if $a = 4$ and $c = 5$, then

$$4^2 + b^2 = 5^2$$
$$16 + b^2 = 25$$
$$b^2 = 9$$
$$b = 3$$

PYTHAGOREAN THEOREM [1.1]

In any right triangle, the square of the length of the longest side (the *hypotenuse*) is equal to the sum of the squares of the lengths of the other two sides (*legs*).

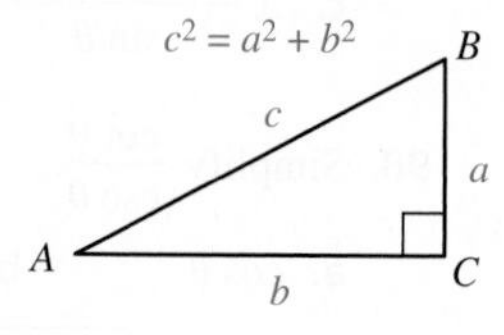

4 If the shortest leg of a 30°–60°–90° triangle is 5, then the hypotenuse is $2(5) = 10$ and the longest leg is $5\sqrt{3}$.

If one leg of a 45°–45°–90° triangle is 5, then the other leg is also 5 and the hypotenuse is $5\sqrt{2}$.

SPECIAL TRIANGLES [1.1]

In any right triangle in which the two acute angles are 30° and 60°, the longest side (the hypotenuse) is always twice the shortest side (the side opposite the 30° angle), and the side of medium length (the side opposite the 60° angle) is always $\sqrt{3}$ times the shortest side.

If the two acute angles in a right triangle are both 45°, then the two shorter sides (the legs) are equal and the longest side (the hypotenuse) is $\sqrt{2}$ times as long as the shorter sides.

DISTANCE FORMULA [1.2]

The distance r between the points (x_1, y_1) and (x_2, y_2) is given by the formula

$$r = \sqrt{(x_2 - x_1)^2 + (y_2 - y_1)^2}$$

5 The distance between (2, 7) and (−1, 3) is

$$r = \sqrt{(2 + 1)^2 + (7 - 3)^2} = \sqrt{9 + 16} = \sqrt{25} = 5$$

CIRCLES [1.2]

A *circle* is defined as the set of all points in the plane that are a fixed distance (the radius) from a given fixed point (the center). If we let $r > 0$ be the radius and (h, k) the center, then the equation for a circle is given by the formula

$$(x - h)^2 + (y - k)^2 = r^2$$

6

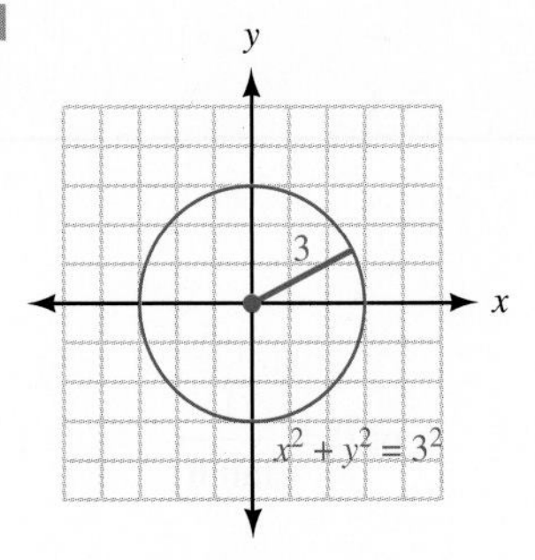

STANDARD POSITION FOR ANGLES [1.2]

An angle is said to be in standard position if its vertex is at the origin and its initial side is along the positive x-axis.

7 135° in standard position is

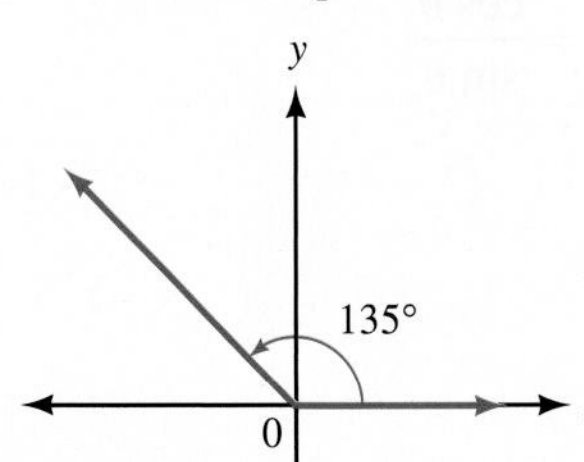

COTERMINAL ANGLES [1.2]

Two angles in standard position with the same terminal side are called *coterminal angles.* Coterminal angles always differ from each other by some multiple of 360°.

8 60° and −300° are coterminal angles.

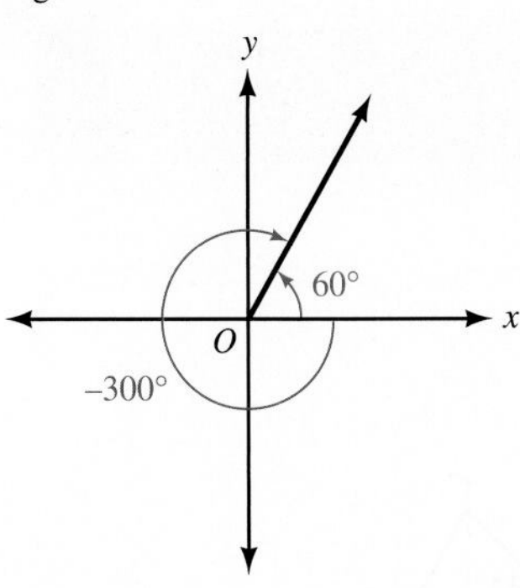

TRIGONOMETRIC FUNCTIONS (DEFINITION I) [1.3]

If θ is an angle in standard position and (x, y) is any point on the terminal side of θ (other than the origin), then

$$\sin\theta = \frac{y}{r} \qquad \csc\theta = \frac{r}{y}$$

$$\cos\theta = \frac{x}{r} \qquad \sec\theta = \frac{r}{x}$$

$$\tan\theta = \frac{y}{x} \qquad \cot\theta = \frac{x}{y}$$

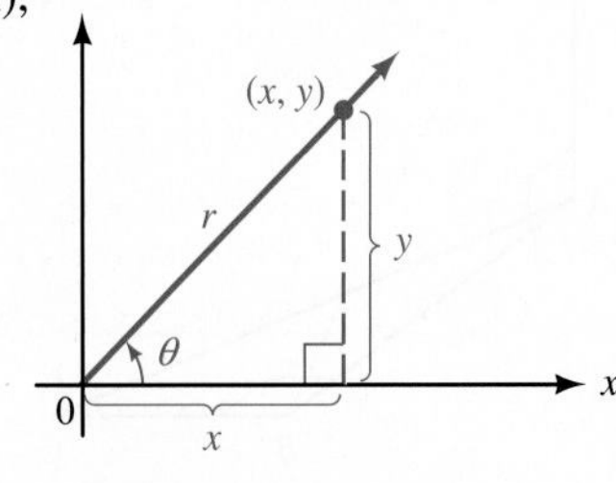

where $x^2 + y^2 = r^2$, or $r = \sqrt{x^2 + y^2}$. That is, r is the distance from the origin to (x, y).

9 If (−3, 4) is on the terminal side of θ, then

$$r = \sqrt{9 + 16} = 5$$

and

$$\sin\theta = \frac{4}{5} \qquad \csc\theta = \frac{5}{4}$$

$$\cos\theta = -\frac{3}{5} \qquad \sec\theta = -\frac{5}{3}$$

$$\tan\theta = -\frac{4}{3} \qquad \cot\theta = -\frac{3}{4}$$

10 If $\sin\theta > 0$, and $\cos\theta > 0$, then θ must terminate in QI.

If $\sin\theta > 0$, and $\cos\theta < 0$, then θ must terminate in QII.

If $\sin\theta < 0$, and $\cos\theta < 0$, then θ must terminate in QIII.

If $\sin\theta < 0$, and $\cos\theta > 0$, then θ must terminate in QIV.

SIGNS OF THE TRIGONOMETRIC FUNCTIONS [1.3]

The algebraic signs, + or −, of the six trigonometric functions depend on the quadrant in which θ terminates.

For θ in	QI	QII	QIII	QIV
$\sin\theta$ and $\csc\theta$	+	+	−	−
$\cos\theta$ and $\sec\theta$	+	−	−	+
$\tan\theta$ and $\cot\theta$	+	−	+	−

11 If $\sin\theta = 1/2$ with θ in QI, then

$$\cos\theta = \sqrt{1-\sin^2\theta} = \sqrt{1-1/4} = \frac{\sqrt{3}}{2}$$

$$\tan\theta = \frac{\sin\theta}{\cos\theta} = \frac{1/2}{\sqrt{3}/2} = \frac{\sqrt{3}}{3}$$

$$\cot\theta = \frac{1}{\tan\theta} = \sqrt{3}$$

$$\sec\theta = \frac{1}{\cos\theta} = \frac{2}{\sqrt{3}} = \frac{2\sqrt{3}}{3}$$

$$\csc\theta = \frac{1}{\sin\theta} = 2$$

BASIC IDENTITIES [1.4]

Reciprocal

$$\csc\theta = \frac{1}{\sin\theta} \qquad \sec\theta = \frac{1}{\cos\theta} \qquad \cot\theta = \frac{1}{\tan\theta}$$

Ratio

$$\tan\theta = \frac{\sin\theta}{\cos\theta} \qquad \cot\theta = \frac{\cos\theta}{\sin\theta}$$

Pythagorean

$$\cos^2\theta + \sin^2\theta = 1$$
$$1 + \tan^2\theta = \sec^2\theta$$
$$1 + \cot^2\theta = \csc^2\theta$$

CHAPTER 1 TEST

1. Find the complement and the supplement of 70°.
2. Solve for x in the right triangle shown in Figure 1.

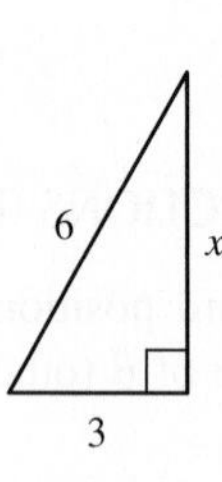

Figure 1

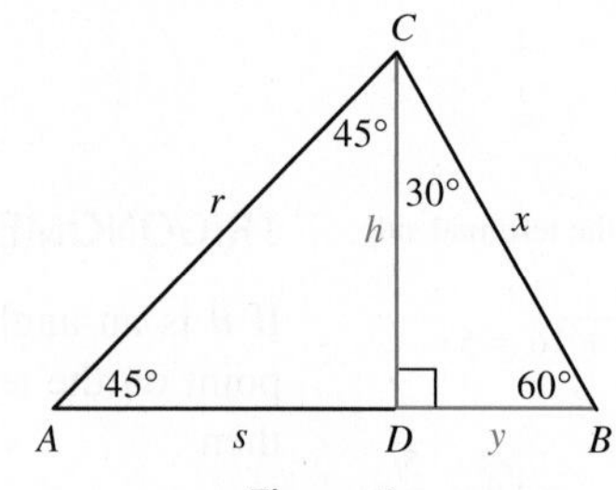

Figure 2

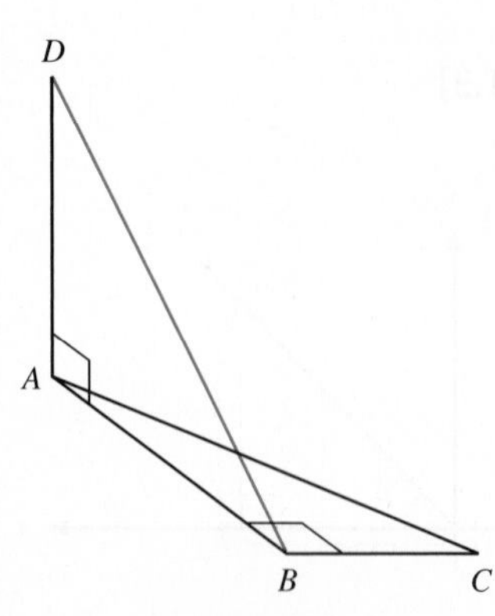

Figure 3

3. Referring to Figure 2, find (in order) h, r, y, and x if $s = 5\sqrt{3}$. (*Note:* s is the distance from A to D, and y is the distance from D to B.)
4. Figure 3 shows two right triangles drawn at 90° to one another. Find the length of DB if $DA = 6$, $AC = 5$, and $BC = 3$.

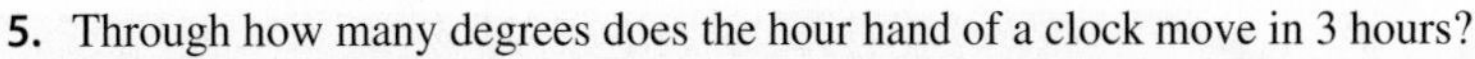

5. Through how many degrees does the hour hand of a clock move in 3 hours?
6. Find the remaining sides of a 30°–60°–90° triangle if the longest side is 5.
7. **Escalator** An escalator in a department store is to carry people a vertical distance of 15 feet between floors. How long is the escalator if it makes an angle of 45° with the ground?

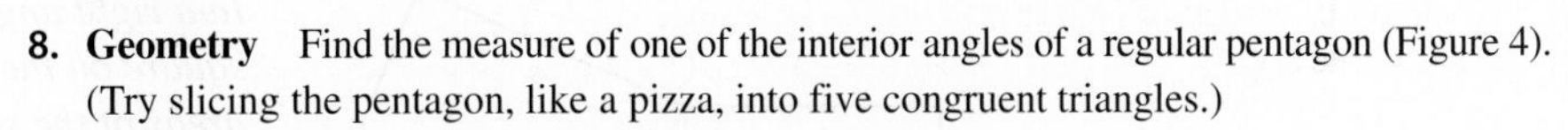

8. **Geometry** Find the measure of one of the interior angles of a regular pentagon (Figure 4). (Try slicing the pentagon, like a pizza, into five congruent triangles.)

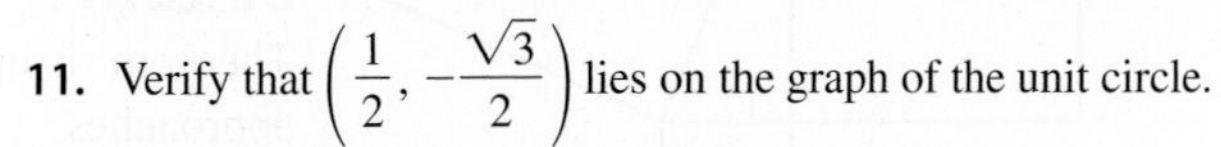

9. Find the distance between the points $(4, -2)$ and $(-1, 10)$.
10. Find x so that the distance between $(-2, 3)$ and $(x, 1)$ is $\sqrt{13}$.

Figure 4

11. Verify that $\left(\frac{1}{2}, -\frac{\sqrt{3}}{2}\right)$ lies on the graph of the unit circle.
12. Find all angles that are coterminal with 225°.
13. **Human Cannonball** A human cannonball is shot from a cannon at the county fair. She reaches a height of 50 feet before landing in a net 140 feet from the cannon. Sketch the graph of her path, and then find the equation of the graph. Use your graphing calculator to verify that your equation is correct.

Find $\sin\theta$, $\cos\theta$, and $\tan\theta$ for each of the following values of θ.

14. 90°
15. −45°
16. Indicate the two quadrants θ could terminate in if $\cos\theta = -\frac{1}{2}$.
17. In which quadrant will θ lie if $\csc\theta > 0$ and $\cos\theta < 0$?
18. Find all six trigonometric functions for θ if the point $(-3, -1)$ lies on the terminal side of θ in standard position.
19. Why is $\sin\theta \leq 1$ for any angle θ in standard position?
20. Find the remaining trigonometric functions of θ if $\sin\theta = \frac{1}{2}$ and θ terminates in QII.
21. Find $\sin\theta$ and $\cos\theta$ if the terminal side of θ lies along the line $y = -2x$ in quadrant IV.
22. If $\sin\theta = -\frac{3}{4}$, find $\csc\theta$.
23. If $\sin\theta = \frac{1}{3}$, find $\sin^3\theta$.
24. If $\sec\theta = 3$ with θ in QIV, find $\cos\theta$, $\sin\theta$, and $\tan\theta$.
25. Expand and simplify $(\cos\theta - \sin\theta)^2$.
26. Subtract $\frac{1}{\sin\theta} - \sin\theta$.
27. Simplify the expression $\sqrt{4 - x^2}$ as much as possible after substituting $2\sin\theta$ for x.

Show that each of the following statements is an identity by transforming the left side of each one into the right side.

28. $\frac{\cot\theta}{\csc\theta} = \cos\theta$
29. $\cot\theta + \tan\theta = \csc\theta\sec\theta$
30. $(1 - \sin\theta)(1 + \sin\theta) = \cos^2\theta$

The Pythagorean Theorem

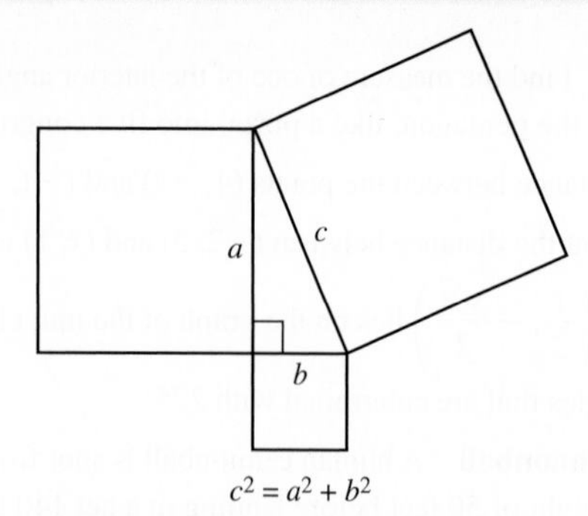

$c^2 = a^2 + b^2$

In a right angled triangle, the area of the square on the hypotenuse is the sum of the areas of the squares on the other two sides.

Theorem of Pythagoras

OBJECTIVE: To prove the Pythagorean Theorem using several different geometric approaches.

Although the Babylonians appear to have had an understanding of this relationship a thousand years before Pythagoras, his name is associated with the theorem because he was the first to provide a proof. In this project you will provide the details for several proofs of the theorem.

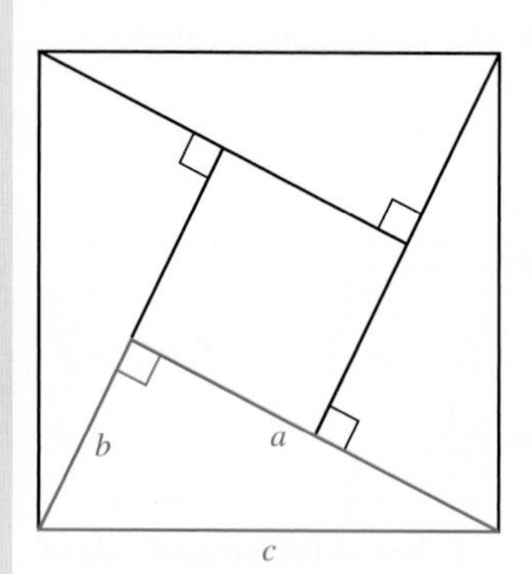

Figure 1

1 The diagram shown in Figure 1 was used by the Hindu mathematician Bhaskara to prove the theorem in the 12th century. His proof consisted only of the diagram and the word "Behold!" Use this figure to derive the Pythagorean Theorem. The right triangle with sides a, b, and c has been repeated three times. The area of the large square is equal to the sum of the areas of the four triangles and the area of the small square in the center. The key to this derivation is in finding the length of a side of the small square. Once you have the equation, simplify the right side to obtain the theorem.

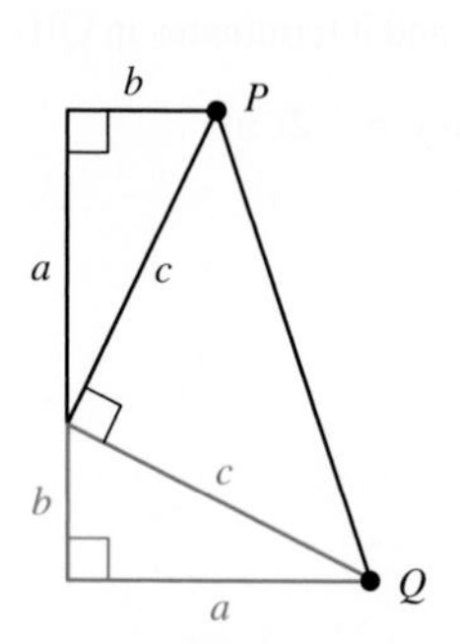

Figure 2

2 The second proof, credited to former president James Garfield, is based on the diagram shown in Figure 2. Use this figure to derive the Pythagorean Theorem. The right triangle with sides a, b, and c has been repeated and a third triangle created by placing a line segment between points P and Q. The area of the entire figure (a trapezoid) is equal to the sum of the areas of the three triangles. Once you have the equation, simplify both sides and then isolate c^2 to obtain the theorem.

3 For the third "proof," consider the two squares shown in Figure 3. Both squares are of the same size, so their areas must be equal. Write a paragraph explaining how this diagram serves as a "visual proof" of the Pythagorean Theorem. The key to this argument is to make use of what the two squares have in common.

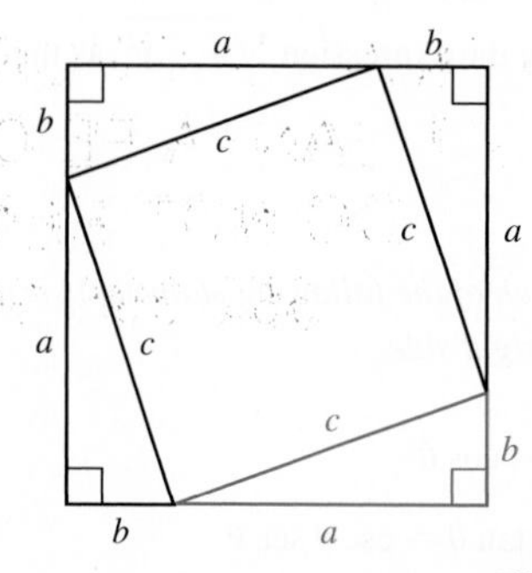

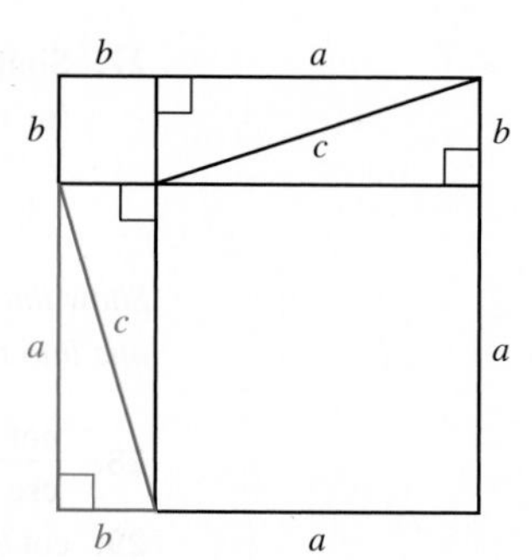

Figure 3

RESEARCH PROJECT Pythagoras and Shakespeare

© World History Archive / Alamy

Pythagoras

Although Pythagoras preceded William Shakespeare by 2,000 years, the philosophy of the Pythagoreans is mentioned in Shakespeare's *The Merchant of Venice.* Here is a quote from that play:

> *Thou almost mak'st me waver in my faith,*
> *To hold opinion with Pythagoras,*
> *That souls of animals infuse themselves*
> *Into the trunks of men.*

Research the Pythagoreans. What were the main beliefs held by their society? What part of the philosophy of the Pythagoreans was Shakespeare referring to specifically with this quote? What present-day religions share a similar belief? Write a paragraph or two about your findings.

2 Right Triangle Trigonometry

We are what we repeatedly do. Excellence, then, is not an act, but a habit.
➤ *Aristotle*

INTRODUCTION

© Mike Agliolo/Corbis

NOTE Many of the chapter introductions touch on one of the themes running throughout the text. For example, the image of the compass rose indicates that this particular introduction relates to the navigation theme.

Most serious hikers are familiar with maps like the one shown in Figure 1. It is a *topographic map*. This particular map is of Bishop's Peak in San Luis Obispo, California.

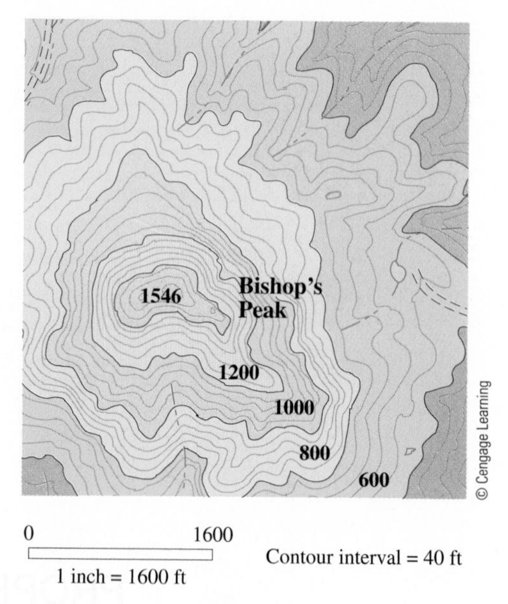

Figure 1

The curved lines on the map are called *contour lines;* they are used to show the changes in elevation of the land shown on the map. On this map, the change in elevation between any two contour lines is 40 feet, meaning that, if you were standing on one contour line and you were to hike to a position two contour lines away from your starting point, your elevation would change by 80 feet. In general, the closer the contour lines are together, the faster the land rises or falls. In this chapter, we use right triangle trigonometry to solve a variety of problems, some of which will involve topographic maps.

STUDY SKILLS 2

The quote at the beginning of this chapter reinforces the idea that success in mathematics comes from practice. We practice trigonometry by working problems. Working problems is probably the single most important thing you can do to achieve excellence in trigonometry.

If you have successfully completed Chapter 1, then you have made a good start at developing the study skills necessary to succeed in all math classes. Here is the list of study skills for this chapter. Some are a continuation of the skills from Chapter 1, while others are new to this chapter.

1. **Continue to Set and Keep a Schedule** Sometimes we find students do well in Chapter 1 and then become overconfident. They will spend less time with their homework. Don't do it. Keep to the same schedule.

2. **Continue to Memorize Definitions and Important Facts** The important definitions in this chapter are those for trigonometric ratios in right triangles. We will point these out as we progress through the chapter.

3. **List Difficult Problems** Begin to make lists of problems that give you the most difficulty. These are the problems that you are repeatedly making mistakes with. Try to spend a few minutes each day reworking some of the problems from your list. Once you are able to work these problems successfully, you will have much more confidence going into the next exam.

4. **Begin to Develop Confidence with Word Problems** It seems that the main difference between people who are good at working word problems and those who are not, is confidence. People with confidence know that no matter how long it takes them, they will eventually be able to solve the problem they are working on. Those without confidence begin by saying to themselves, "I'll never be able to work this problem." If you are in the second category, then instead of telling yourself that you can't do word problems, that you don't like them, or that they're not good for anything anyway, decide to do whatever it takes to master them. As Aristotle said, it is practice that produces excellence. For us, practice means working problems, lots of problems. ▲

SECTION 2.1 Definition II: Right Triangle Trigonometry

LEARNING OBJECTIVES

1. Find the value of a trigonometric function for an angle in a right triangle.
2. Use the Cofunction Theorem to find the value of a trigonometric function.
3. Find the exact value of a trigonometric function for a special angle.
4. Use exact values to simplify an expression involving trigonometric functions.

The word *trigonometry* is derived from two Greek words: *tri'gonon,* which translates as *triangle,* and *met'ron,* which means *measure.* Trigonometry, then, is triangle measure. In this section, we will give a second definition for the trigonometric functions that is, in fact, based on "triangle measure." We will define the trigonometric functions as the ratios of sides in right triangles. As you will see, this new definition does not conflict with the definition from Chapter 1 for the trigonometric functions.

DEFINITION II

If triangle ABC is a right triangle with $C = 90°$ (Figure 1), then the six trigonometric functions for A are defined as follows:

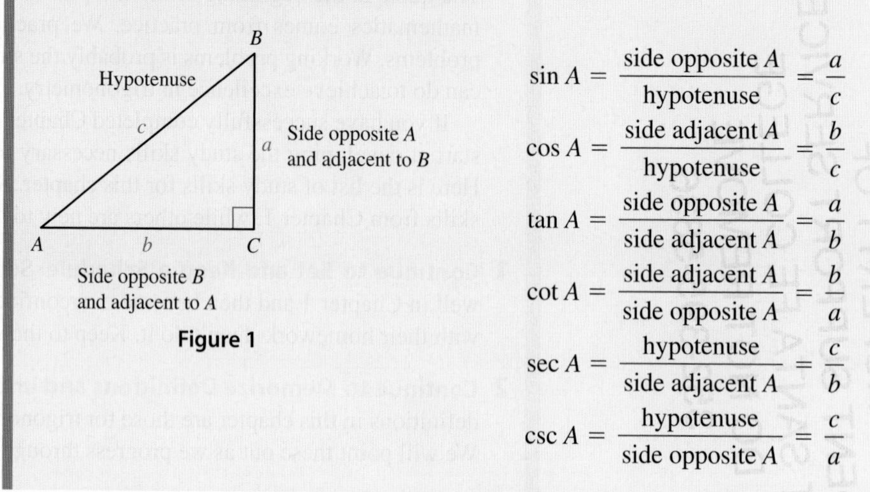

Figure 1

$$\sin A = \frac{\text{side opposite } A}{\text{hypotenuse}} = \frac{a}{c}$$

$$\cos A = \frac{\text{side adjacent } A}{\text{hypotenuse}} = \frac{b}{c}$$

$$\tan A = \frac{\text{side opposite } A}{\text{side adjacent } A} = \frac{a}{b}$$

$$\cot A = \frac{\text{side adjacent } A}{\text{side opposite } A} = \frac{b}{a}$$

$$\sec A = \frac{\text{hypotenuse}}{\text{side adjacent } A} = \frac{c}{b}$$

$$\csc A = \frac{\text{hypotenuse}}{\text{side opposite } A} = \frac{c}{a}$$

PROBLEM 1
Triangle ABC is a right triangle with $C = 90°$. If $a = 12$ and $c = 15$, find the six trigonometric functions of A.

EXAMPLE 1 Triangle ABC is a right triangle with $C = 90°$. If $a = 6$ and $c = 10$, find the six trigonometric functions of A.

SOLUTION We begin by making a diagram of ABC (Figure 2) and then use the given information and the Pythagorean Theorem to solve for b.

$$\begin{aligned} b &= \sqrt{c^2 - a^2} \\ &= \sqrt{100 - 36} \\ &= \sqrt{64} \\ &= 8 \end{aligned}$$

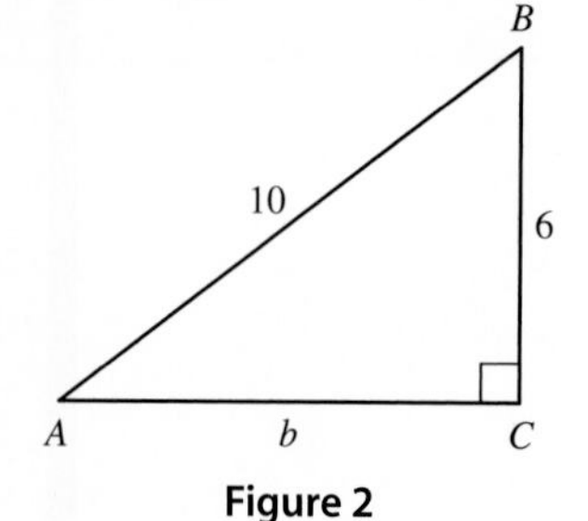

Figure 2

Now we write the six trigonometric functions of A using $a = 6$, $b = 8$, and $c = 10$.

$$\sin A = \frac{a}{c} = \frac{6}{10} = \frac{3}{5} \qquad \csc A = \frac{c}{a} = \frac{5}{3}$$

$$\cos A = \frac{b}{c} = \frac{8}{10} = \frac{4}{5} \qquad \sec A = \frac{c}{b} = \frac{5}{4}$$

$$\tan A = \frac{a}{b} = \frac{6}{8} = \frac{3}{4} \qquad \cot A = \frac{b}{a} = \frac{4}{3}$$

■

PROBLEM 2
Use Definition II to explain why, for any acute angle θ, it is impossible for $\cos \theta = 3/2$.

EXAMPLE 2 Use Definition II to explain why, for any acute angle θ, it is impossible for $\sin \theta = 2$.

SOLUTION Since θ is acute, we can imagine θ as the interior angle of some right triangle. Then we would have

$$\sin \theta = \frac{\text{side opposite } \theta}{\text{hypotenuse}}$$

Because the hypotenuse is always the longest side of a right triangle, the denominator of this ratio is larger than the numerator, forcing the ratio to have a value less than 1. Therefore, it is not possible that this ratio could be equal to 2. ■

Now that we have done a couple of examples using our new definition, let's see how our new definition compares with Definition I from the previous chapter. We can place right triangle ABC on a rectangular coordinate system so that A is in standard position (Figure 3). We then note that a point on the terminal side of A is (b, a).

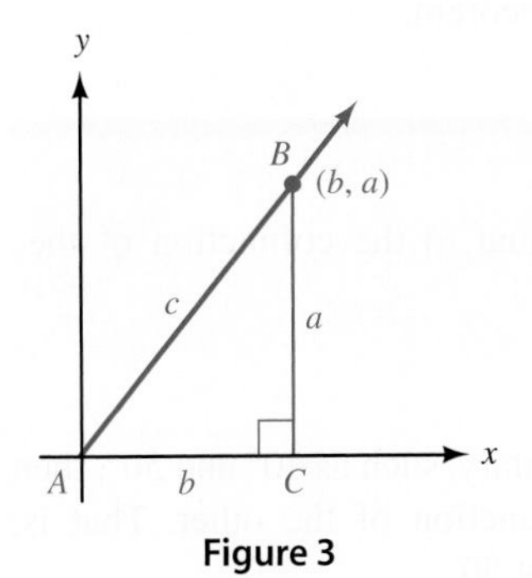

Figure 3

From Definition I in Chapter 1, we have	From Definition II in this chapter, we have
$\sin A = \dfrac{a}{c}$	$\sin A = \dfrac{a}{c}$
$\cos A = \dfrac{b}{c}$	$\cos A = \dfrac{b}{c}$
$\tan A = \dfrac{a}{b}$	$\tan A = \dfrac{a}{b}$

The two definitions agree as long as A is an acute angle. If A is not an acute angle, then Definition II does not apply, because in right triangle ABC, A must be an acute angle.

Here is another definition that we will need before we can take Definition II any further.

DEFINITION

Sine and *co*sine are *co*functions, as are tangent and *co*tangent, and secant and *co*secant. We say sine is the cofunction of cosine, and cosine is the cofunction of sine.

Now let's see what happens when we apply Definition II to B in right triangle ABC.

$$\sin B = \frac{\text{side opposite } B}{\text{hypotenuse}} = \frac{b}{c} = \cos A$$

$$\cos B = \frac{\text{side adjacent } B}{\text{hypotenuse}} = \frac{a}{c} = \sin A$$

$$\tan B = \frac{\text{side opposite } B}{\text{side adjacent } B} = \frac{b}{a} = \cot A$$

$$\cot B = \frac{\text{side adjacent } B}{\text{side opposite } B} = \frac{a}{b} = \tan A$$

$$\sec B = \frac{\text{hypotenuse}}{\text{side adjacent } B} = \frac{c}{a} = \csc A$$

$$\csc B = \frac{\text{hypotenuse}}{\text{side opposite } B} = \frac{c}{b} = \sec A$$

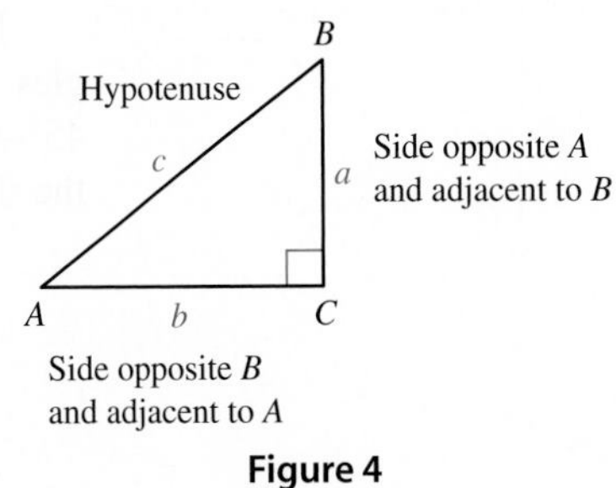

Figure 4

NOTE The prefix *co-* in *co*sine, *co*secant, and *co*tangent is a reference to the complement. Around the year 1463, Regiomontanus used the term *sinus rectus complementi*, presumably referring to the cosine as the sine of the complementary angle. In 1620, Edmund Gunter shortened this to *co.sinus*, which was further abbreviated as *cosinus* by John Newton in 1658.

As you can see in Figure 4, every trigonometric function of A is equal to the cofunction of B. That is, $\sin A = \cos B$, $\sec A = \csc B$, and $\tan A = \cot B$, to name a few. Because A and B are the acute angles in a right triangle, they are always complementary angles; that is, their sum is always 90°. What we actually have here is another property of trigonometric functions: The sine of an angle is the cosine of its complement, the secant of an angle is the cosecant of its complement, and the tangent of an angle is the cotangent of its complement. Or, in symbols,

$$\text{if } A + B = 90°, \text{ then } \begin{cases} \sin A = \cos B \\ \sec A = \csc B \\ \tan A = \cot B \end{cases}$$

and so on.

We generalize this discussion with the following theorem.

COFUNCTION ■ THEOREM

A trigonometric function of an angle is always equal to the cofunction of the complement of the angle.

To clarify this further, if two angles are complementary, such as 40° and 50°, then a trigonometric function of one is equal to the cofunction of the other. That is, $\sin 40° = \cos 50°$, $\sec 40° = \csc 50°$, and $\tan 40° = \cot 50°$.

PROBLEM 3
Fill in the blanks so that each expression becomes a true statement.

a. sin ______ = cos 60°

b. tan 65° = cot ______

c. sec x = csc ______

EXAMPLE 3 Fill in the blanks so that each expression becomes a true statement.

a. sin ______ = cos 30° **b.** tan y = cot ______ **c.** sec 75° = csc ______

SOLUTION Using the theorem on cofunctions of complementary angles, we fill in the blanks as follows:

a. $\sin \underline{\ 60°\ } = \cos 30°$ Because sine and cosine are cofunctions and $60° + 30° = 90°$

b. $\tan y = \cot \underline{(90° - y)}$ Because tangent and cotangent are cofunctions and $y + (90° - y) = 90°$

c. $\sec 75° = \csc \underline{\ 15°\ }$ Because secant and cosecant are cofunctions and $75° + 15° = 90°$

For our next application of Definition II, we need to recall the two special triangles we introduced in Chapter 1. They are the 30°–60°–90° triangle and the 45°–45°–90° triangle. Figure 5 shows both of these triangles for the case in which the shortest side is 1 ($t = 1$).

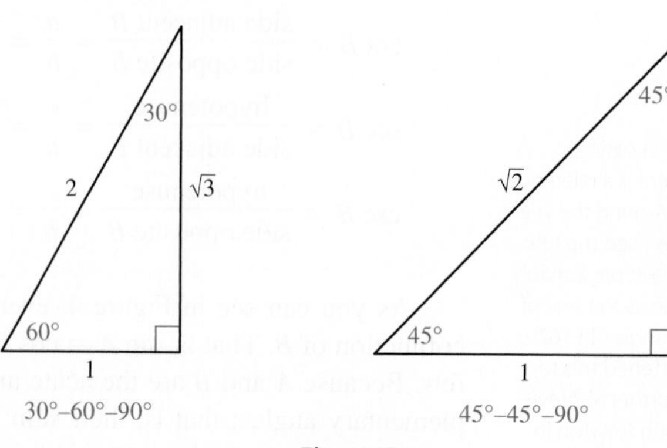

Figure 5

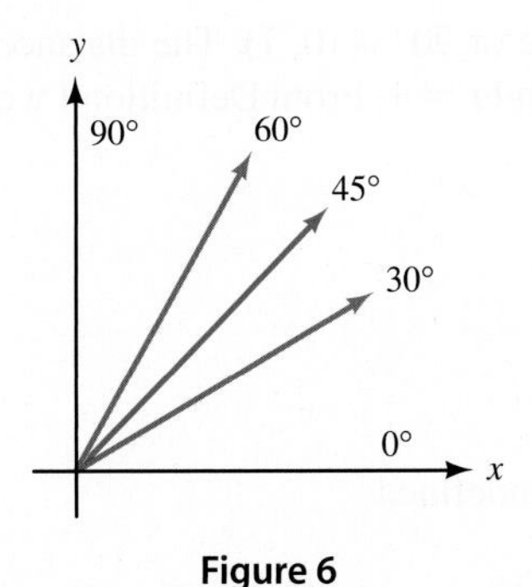

Figure 6

Using these two special triangles and Definition II, we can find the trigonometric functions of 30°, 45°, and 60°. For example,

$$\sin 60° = \frac{\text{side opposite } 60°}{\text{hypotenuse}} = \frac{\sqrt{3}}{2}$$

$$\cos 45° = \frac{\text{side adjacent } 45°}{\text{hypotenuse}} = \frac{1}{\sqrt{2}} = \frac{\sqrt{2}}{2}$$

If we were to continue finding the sine, cosine, and tangent for these special angles (Figure 6), we would obtain the results summarized in Table 1. Because we will be using these values so frequently, you should either memorize the special triangles in Figure 5 (previous page) or the information in Table 1.

TABLE 1
Exact Values

θ	$\sin\theta$	$\cos\theta$	$\tan\theta$
30°	$\frac{1}{2}$	$\frac{\sqrt{3}}{2}$	$\frac{1}{\sqrt{3}} = \frac{\sqrt{3}}{3}$
45°	$\frac{1}{\sqrt{2}} = \frac{\sqrt{2}}{2}$	$\frac{1}{\sqrt{2}} = \frac{\sqrt{2}}{2}$	1
60°	$\frac{\sqrt{3}}{2}$	$\frac{1}{2}$	$\sqrt{3}$

Table 1 is called a table of exact values to distinguish it from a table of approximate values. Later in this chapter we will work with a calculator to obtain tables of approximate values.

PROBLEM 4
Use exact values to show that the following are true.

a. $\sin^2 60° + \cos^2 60° = 1$
b. $\sin^2 45° - \cos^2 45° = 0$

EXAMPLE 4 Use the exact values from either Figure 5 or Table 1 to show that the following are true.

a. $\cos^2 30° + \sin^2 30° = 1$ **b.** $\cos^2 45° + \sin^2 45° = 1$

SOLUTION

a. $\cos^2 30° + \sin^2 30° = \left(\frac{\sqrt{3}}{2}\right)^2 + \left(\frac{1}{2}\right)^2 = \frac{3}{4} + \frac{1}{4} = 1$

b. $\cos^2 45° + \sin^2 45° = \left(\frac{\sqrt{2}}{2}\right)^2 + \left(\frac{\sqrt{2}}{2}\right)^2 = \frac{2}{4} + \frac{2}{4} = 1$ ■

As you will see as we progress through the book, there are times when it is appropriate to use Definition II for the trigonometric functions and times when it is more appropriate to use Definition I. To illustrate, suppose we wanted to find $\sin\theta$, $\cos\theta$, and $\tan\theta$ for $\theta = 90°$. We would not be able to use Definition II because we can't draw a right triangle in which one of the *acute* angles is 90°. Instead, we would draw 90° in standard position, locate a point on the terminal side, and use Definition I to find sin 90°, cos 90°, and tan 90°.

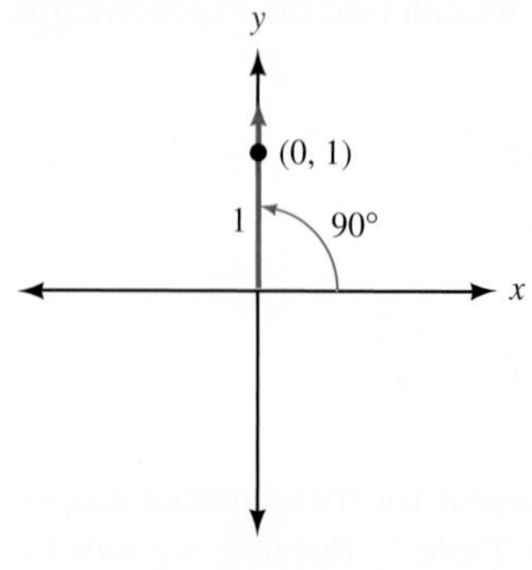

Figure 7

As shown in Figure 7, a point on the terminal side of 90° is (0, 1). The distance from the origin to (0, 1) is 1. Therefore, $x = 0$, $y = 1$, and $r = 1$. From Definition I we have

$$\sin 90° = \frac{y}{r} = \frac{1}{1} = 1$$

$$\cos 90° = \frac{x}{r} = \frac{0}{1} = 0$$

$$\tan 90° = \frac{y}{x} = \frac{1}{0} \text{ which is undefined}$$

In a similar manner, using the point (1, 0) we can show that $\sin 0° = 0$, $\cos 0° = 1$, and $\tan 0° = 0$.

PROBLEM 5
Let $x = 60°$ and $y = 45°$ in each of the following expressions, and then simplify each expression as much as possible.
a. $3 \sin x$
b. $\cos 2y$
c. $5 \sin (3x - 90°)$

EXAMPLE 5 Let $x = 30°$ and $y = 45°$ in each of the expressions that follow, and then simplify each expression as much as possible.

a. $2 \sin x$ **b.** $\sin 2y$ **c.** $4 \sin (3x - 90°)$

SOLUTION

a. $2 \sin x = 2 \sin 30° = 2(1/2) = 1$
b. $\sin 2y = \sin 2(45°) = \sin 90° = 1$
c. $4 \sin (3x - 90°) = 4 \sin [3(30°) - 90°] = 4 \sin 0° = 4(0) = 0$

To conclude this section, we take the information previously obtained for 0° and 90°, along with the exact values in Table 1, and summarize them in Table 2. To make the information in Table 2 a little easier to memorize, we have written some of the exact values differently than we usually do. For example, in Table 2 we have written 2 as $\sqrt{4}$, 0 as $\sqrt{0}$, and 1 as $\sqrt{1}$.

TABLE 2

θ	0°	30°	45°	60°	90°
$\sin \theta$	$\frac{\sqrt{0}}{2}$	$\frac{\sqrt{1}}{2}$	$\frac{\sqrt{2}}{2}$	$\frac{\sqrt{3}}{2}$	$\frac{\sqrt{4}}{2}$
$\cos \theta$	$\frac{\sqrt{4}}{2}$	$\frac{\sqrt{3}}{2}$	$\frac{\sqrt{2}}{2}$	$\frac{\sqrt{1}}{2}$	$\frac{\sqrt{0}}{2}$
$\tan \theta$	0	$\frac{\sqrt{3}}{3}$	1	$\sqrt{3}$	undefined

Getting Ready for Class

After reading through the preceding section, respond in your own words and in complete sentences.

a. In a right triangle, which side is opposite the right angle?
b. If A is an acute angle in a right triangle, how do you define $\sin A$, $\cos A$, and $\tan A$?
c. State the Cofunction Theorem.
d. How are $\sin 30°$ and $\cos 60°$ related?

2.1 PROBLEM SET

➤ CONCEPTS AND VOCABULARY

For Questions 1 through 3, fill in each blank with the appropriate word.

1. Based on its Greek origins, the word "trigonometry" can be translated as __________ __________.

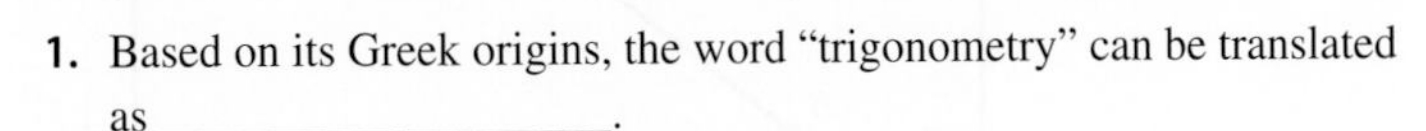

Figure 8

2. Using Definition II and Figure 8, we would refer to a as the side __________ A, b as the side __________ to A, and c as the __________.

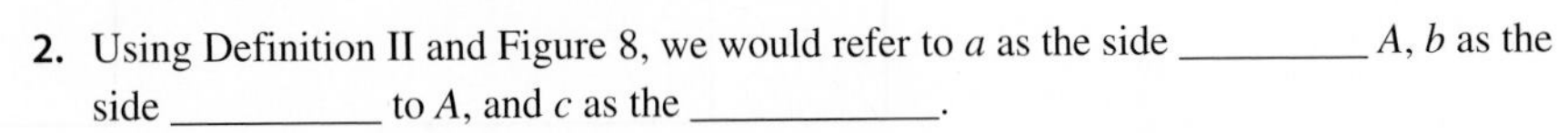

3. A trigonometric function of an angle is always equal to the cofunction of the __________ of the angle.

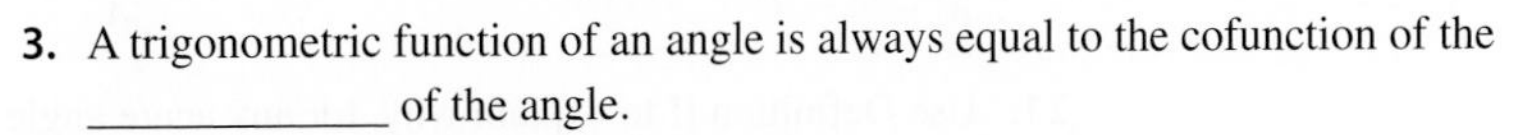

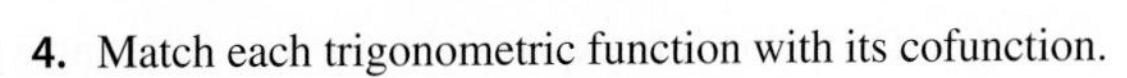

4. Match each trigonometric function with its cofunction.

a. sine **i.** cotangent
b. secant **ii.** cosine
c. tangent **iii.** cosecant

➤ EXERCISES

Problems 5 through 10 refer to right triangle ABC with C = 90°. In each case, use the given information to find the six trigonometric functions of A.

5. $b = 3, c = 5$

6. $b = 5, c = 13$

7. $a = 2, b = 1$

8. $a = 3, b = 2$

9. $a = 2, b = \sqrt{5}$

10. $a = 3, b = \sqrt{7}$

In each of the following right triangles, find sin A, cos A, tan A, and sin B, cos B, tan B.

11.

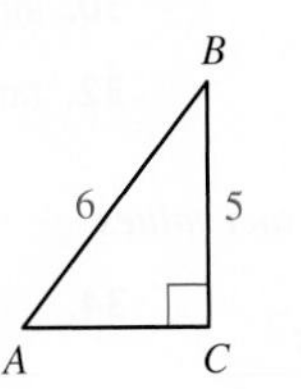

12.

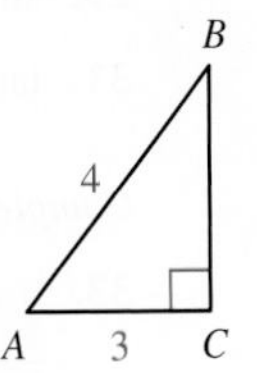

13.

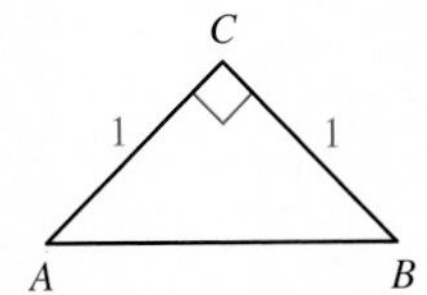

14.

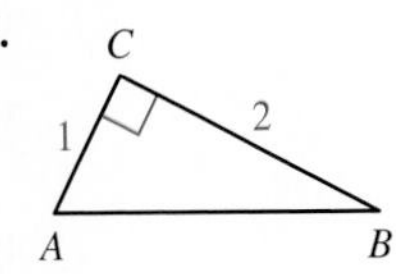

15.

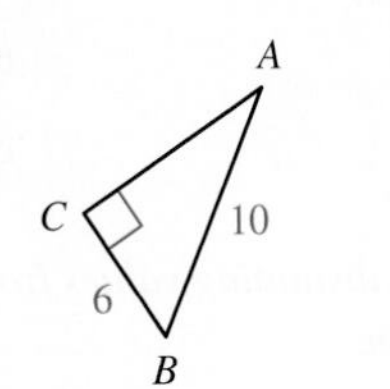

16.

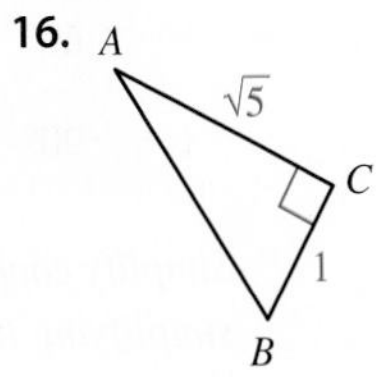

17.

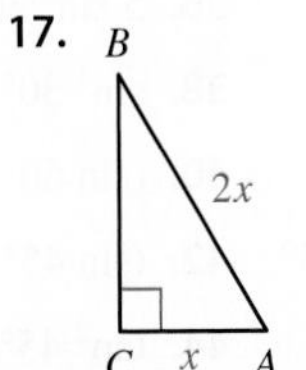

18.

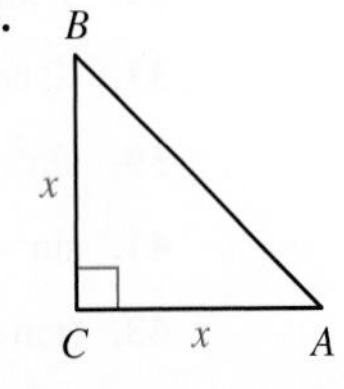

In each of the following diagrams, angle A is in standard position. In each case, find the coordinates of point B and then find sin A, cos A, and tan A.

19.

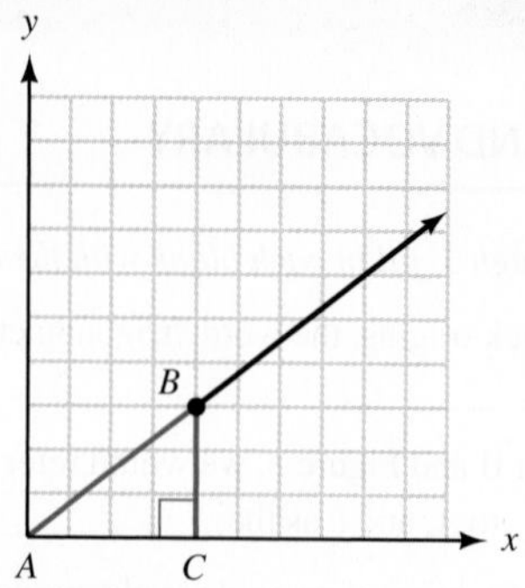

20.

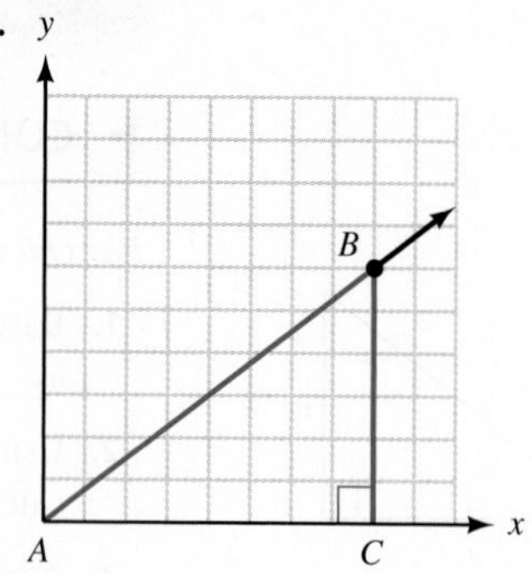

21. Use Definition II to explain why, for any acute angle θ, it is impossible for $\cos\theta = 3$.

22. Use Definition II to explain why, for any acute angle θ, it is impossible for $\sec\theta = \frac{1}{2}$.

23. Use Definition II to explain why it is possible to find an angle θ that will make $\tan\theta$ as large as we wish.

24. Use Definition II to explain why it is possible to find an angle θ that will make $\csc\theta$ as large as we wish.

Use the Cofunction Theorem to fill in the blanks so that each expression becomes a true statement.

25. $\sin 10° = \cos$ ______

26. $\cos 40° = \sin$ ______

27. $\tan 8° = \cot$ ______

28. $\cot 12° = \tan$ ______

29. $\sin x = \cos$ ______

30. $\sin y = \cos$ ______

31. $\tan(90° - x) = \cot$ ______

32. $\tan(90° - y) = \cot$ ______

Complete the following tables using exact values.

33.

x	$\sin x$	$\csc x$
0°	0	
30°	$\frac{1}{2}$	
45°	$\frac{\sqrt{2}}{2}$	
60°	$\frac{\sqrt{3}}{2}$	
90°	1	

34.

x	$\cos x$	$\sec x$
0°	1	
30°	$\frac{\sqrt{3}}{2}$	
45°	$\frac{\sqrt{2}}{2}$	
60°	$\frac{1}{2}$	
90°	0	

Simplify each expression by first substituting values from the table of exact values and then simplifying the resulting expression.

35. $4\sin 30°$

36. $5\sin^2 30°$

37. $(2\cos 30°)^2$

38. $\sin^3 30°$

39. $\sin^2 60° + \cos^2 60°$

40. $(\sin 60° + \cos 60°)^2$

41. $\sin^2 45° - 2\sin 45°\cos 45° + \cos^2 45°$

42. $(\sin 45° - \cos 45°)^2$

43. $(\tan 45° + \tan 60°)^2$

44. $\tan^2 45° + \tan^2 60°$

For each expression that follows, replace x with 30°, y with 45°, and z with 60°, and then simplify as much as possible.

45. $2 \sin x$ **46.** $4 \cos y$

47. $4 \cos(z - 30°)$ **48.** $-2 \sin(y + 45°)$

49. $-3 \sin 2x$ **50.** $3 \sin 2y$

51. $2 \cos(3x - 45°)$ **52.** $2 \sin(90° - z)$

Find exact values for each of the following.

53. sec 30° **54.** csc 30° **55.** csc 60° **56.** sec 60°

57. cot 45° **58.** cot 30° **59.** sec 45° **60.** csc 45°

61. cot 60° **62.** sec 0° **63.** csc 90° **64.** cot 90°

Problems 65 through 68 refer to right triangle ABC with C = 90°. In each case, use a calculator to find sin A, cos A, sin B, and cos B. Round your answers to the nearest hundredth.

65. $b = 8.88$, $c = 9.62$ **66.** $a = 3.42$, $c = 5.70$

67. $a = 19.44$, $b = 5.67$ **68.** $a = 11.28$, $b = 8.46$

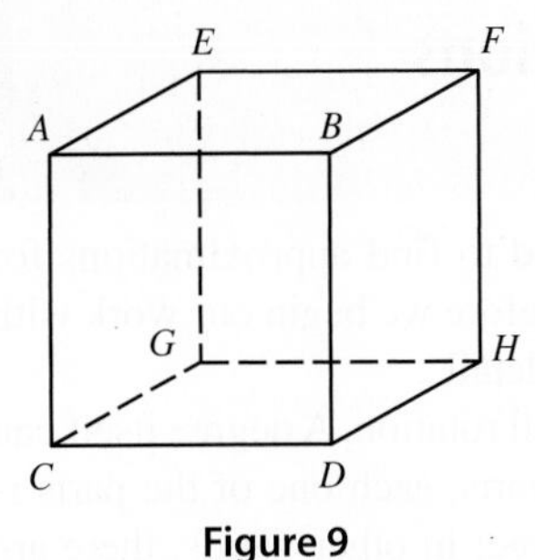

Figure 9

69. Suppose each edge of the cube shown in Figure 9 is 5 inches long. Find the sine and cosine of the angle formed by diagonals *CF* and *CH*.

70. Suppose each edge of the cube shown in Figure 9 is 3 inches long. Find the sine and cosine of the angle formed by diagonals *DE* and *DG*.

71. Suppose each edge of the cube shown in Figure 9 is x inches long. Find the sine and cosine of the angle formed by diagonals *CF* and *CH*.

72. Suppose each edge of the cube shown in Figure 9 is y inches long. Find the sine and cosine of the angle formed by diagonals *DE* and *DG*.

➤ REVIEW PROBLEMS

From here on, each Problem Set will end with a series of review problems. In mathematics, it is very important to review. The more you review, the better you will understand the topics we cover and the longer you will remember them. Also, there will be times when material that seemed confusing earlier will be less confusing the second time around. The problems that follow review material we covered in Section 1.2.

73. Find the distance between the points $(3, -2)$ and $(-1, -4)$.

74. Find x so that the distance between $(x, 2)$ and $(1, 5)$ is $\sqrt{13}$.

Draw each angle in standard position and name a point on the terminal side.

75. 135° **76.** 45°

For each given angle, name a coterminal angle between 0° and 360°.

77. −135° **78.** −210°

➤ LEARNING OBJECTIVES ASSESSMENT

These questions are available for instructors to help assess if you have successfully met the learning objectives for this section.

79. Triangle *ABC* is a right triangle with $C = 90°$. If $a = 16$ and $c = 20$, what is sin *A*?

a. $\frac{3}{5}$ **b.** $\frac{5}{3}$ **c.** $\frac{4}{5}$ **d.** $\frac{5}{4}$

80. According to the Cofunction Theorem, which value is equal to sin 35°?

a. csc 25° **b.** csc 55° **c.** cos 35° **d.** cos 55°

81. Which of the following statements is false?

a. $\sin 30° = \dfrac{\sqrt{3}}{2}$ **b.** $\sin 0° = 0$

c. $\cos 45° = \dfrac{\sqrt{2}}{2}$ **d.** $\cos 90° = 0$

82. Use exact values to simplify $4\cos^2 30° + 2\sin 30°$.

a. $1 + \sqrt{3}$ **b.** 3 **c.** 4 **d.** $3\sqrt{2}$

SECTION 2.2 Calculators and Trigonometric Functions of an Acute Angle

LEARNING OBJECTIVES

1. Add and subtract angles expressed in degrees and minutes.
2. Convert angles from degrees and minutes to decimal degrees or vice-versa.
3. Use a calculator to approximate the value of a trigonometric function.
4. Use a calculator to approximate an acute angle given the value of a trigonometric function.

In this section, we will see how calculators can be used to find approximations for trigonometric functions of angles between 0° and 90°. Before we begin our work with calculators, we need to look at degree measure in more detail.

We previously defined 1 degree (1°) to be $\frac{1}{360}$ of a full rotation. A degree itself can be broken down further. If we divide 1° into 60 equal parts, each one of the parts is called 1 minute, denoted 1′. One minute is $\frac{1}{60}$ of a degree; in other words, there are 60 minutes in every degree. The next smaller unit of angle measure is a second. One second, 1″, is $\frac{1}{60}$ of a minute. There are 60 seconds in every minute.

$$1° = 60' \quad \text{or} \quad 1' = \left(\frac{1}{60}\right)^{\circ}$$

$$1' = 60'' \quad \text{or} \quad 1'' = \left(\frac{1}{60}\right)'$$

Table 1 shows how to read angles written in degree measure.

TABLE 1

The Expression	Is Read
52° 10′	52 degrees, 10 minutes
5° 27′ 30″	5 degrees, 27 minutes, 30 seconds
13° 24′ 15″	13 degrees, 24 minutes, 15 seconds

© Photo courtesy of Gravity Probe B, Stanford University

The ability to measure angles to the nearest second provides enough accuracy for most circumstances, although in some applications, such as surveying and astronomy, even finer measurements may be used. For example, Gravity Probe B, a relativity gyroscope experiment being carried out by the National Aeronautic and Space Administration and Stanford University, will attempt to verify Einstein's general theory of relativity by measuring how space and time are warped by the presence of Earth. To do so will require measuring a rotation of 0.5 milliarcsecond (1 milliarcsecond = 1/1000 second, which is equivalent to the width of a human hair as seen from 10 miles away).[1]

PROBLEM 1
Add 39° 27′ and 65° 36′.

EXAMPLE 1 Add 48° 49′ and 72° 26′.

SOLUTION We can add in columns with degrees in the first column and minutes in the second column.

$$\begin{array}{r} 48^\circ\ 49' \\ +\ 72^\circ\ 26' \\ \hline 120^\circ\ 75' \end{array}$$

Because 60 minutes is equal to 1 degree, we can carry 1 degree from the minutes column to the degrees column.

$$120^\circ\ 75' = 121^\circ\ 15'$$

PROBLEM 2
Subtract 39° 14′ from 87°.

EXAMPLE 2 Subtract 24° 14′ from 90°.

SOLUTION To subtract 24° 14′ from 90°, we will have to "borrow" 1° and write that 1° as 60′.

$$\begin{array}{r} 90^\circ \\ -\ 24^\circ\ 14' \\ \hline \end{array} \quad = \quad \begin{array}{r} 89^\circ\ 60' \text{ (Still } 90^\circ) \\ -\ 24^\circ\ 14' \\ \hline 65^\circ\ 46' \end{array}$$

Decimal Degrees

An alternative to using minutes and seconds to break down degrees into smaller units is decimal degrees. For example, 30.5°, 101.75°, and 62.831° are measures of angles written in decimal degrees.

To convert from decimal degrees to degrees and minutes, we simply multiply the fractional part of the angle (the part to the right of the decimal point) by 60 to convert it to minutes.

PROBLEM 3
Change 18.75° to degrees and minutes.

EXAMPLE 3 Change 27.25° to degrees and minutes.

SOLUTION Multiplying 0.25 by 60, we have the number of minutes equivalent to 0.25°.

$$\begin{aligned} 27.25^\circ &= 27^\circ + 0.25^\circ \\ &= 27^\circ + 0.25(60') \\ &= 27^\circ + 15' \\ &= 27^\circ\ 15' \end{aligned}$$

Of course in actual practice, we would not show all these steps. They are shown here simply to indicate why we multiply only the decimal part of the decimal degree by 60 to change to degrees and minutes.

1. www.einstein.stanford.edu

PROBLEM 4
Change 46° 15′ to decimal degrees.

EXAMPLE 4 Change 10° 45′ to decimal degrees.

SOLUTION We have to reverse the process we used in Example 3. To change 45′ to a decimal, we must divide by 60.

$$\begin{aligned} 10^\circ\, 45' &= 10^\circ + 45' \\ &= 10^\circ + \left(\frac{45}{60}\right)^\circ \\ &= 10^\circ + 0.75^\circ \\ &= 10.75^\circ \end{aligned}$$

48°49′+72°26′►DMS	
	121°15′0″
90–24°14′►DMS	
	65°46′0″
27.25►DMS	
	27°15′0″
10°45′	
	10.75

Figure 1

CALCULATOR NOTE Most scientific and graphing calculators have keys that let you enter angles in degrees, minutes, and seconds format (DMS) or in decimal degrees (DD). You may also have keys or commands that let you convert from one format to the other. Figure 1 shows how Examples 1 through 4 might look when solved on a TI-84 graphing calculator. Consult your calculator's manual to see how this is done with your particular model.

The process of converting back and forth between decimal degrees and degrees and minutes can become more complicated when we use decimal numbers with more digits or when we convert to degrees, minutes, and seconds. In this book, most of the angles written in decimal degrees will be written to the nearest tenth or, at most, the nearest hundredth. The angles written in degrees, minutes, and seconds will rarely go beyond the minutes column.

Table 2 lists the most common conversions between decimal degrees and minutes.

TABLE 2

Decimal Degree	Minutes
0.1°	6′
0.2°	12′
0.3°	18′
0.4°	24′
0.5°	30′
0.6°	36′
0.7°	42′
0.8°	48′
0.9°	54′
1.0°	60′

Trigonometric Functions and Acute Angles

Until now, we have been able to determine trigonometric functions only for angles for which we could find a point on the terminal side or angles that were part of special triangles. We can find decimal approximations for trigonometric functions of any acute angle by using a calculator with keys for sine, cosine, and tangent.

First, there are a couple of things you should know. Just as distance can be measured in feet and also in meters, angles can be measured in degrees and also in radians. We will cover radian measure in Chapter 3. For now, you simply need to be sure that your calculator is set to work in degrees. We will refer to this setting as being in *degree mode.* The most common mistake students make when finding values of trigonometric functions on a calculator is working in the wrong mode. As a rule, you should *always* check the mode settings on your calculator before evaluating a trigonometric function.

PROBLEM 5
Find sin 29.4°.

EXAMPLE 5 Use a calculator to find cos 37.8°.

SOLUTION First, be sure your calculator is set to degree mode. Then, depending on the type of calculator you have, press the indicated keys.

Scientific Calculator	Graphing Calculator
37.8 [cos]	[cos] [(] 37.8 [)] [ENTER]

Your calculator will display a number that rounds to 0.7902. The number 0.7902 is just an approximation of cos 37.8°, which is actually an irrational number, as are the trigonometric functions of most angles.

NOTE We will give answers accurate to four places past the decimal point. You can set your calculator to four-place fixed-point mode, and it will show you the same results without having to round your answers mentally.

CALCULATOR NOTE As we mentioned in Section 1.1, some graphing calculators use parentheses with certain functions. For example, the TI-84 will automatically insert a left parenthesis when the [cos] key is pressed, so TI-84 users can skip the [(] key. Other models do not require them at all. For the sake of clarity, we will often include parentheses throughout this book. You may be able to omit one or both parentheses with your model. Just be sure that you are able to obtain the same results shown for each example.

PROBLEM 6
Find tan 28.6°.

EXAMPLE 6 Find tan 58.75°.

SOLUTION This time, we use the [tan] key:

Scientific Calculator	Graphing Calculator
58.75 [tan]	[tan] [(] 58.75 [)] [ENTER]

Rounding to four places past the decimal point, we have

$$\tan 58.75° = 1.6479$$

PROBLEM 7
Find $\cos^2 25°$.

EXAMPLE 7 Find $\sin^2 14°$.

SOLUTION Because $\sin^2 14° = (\sin 14°)^2$, the calculator sequence is

Scientific Calculator	Graphing Calculator
14 [sin] [x^2]	[(] [sin] [(] 14 [)] [)] [x^2] [ENTER]

Rounding to four digits past the decimal point, we have

$$\sin^2 14° = 0.0585$$

Most calculators do not have additional keys for the secant, cosecant, or cotangent functions. To find a value for one of these functions, we will use the appropriate reciprocal identity.

PROBLEM 8
Find csc 61.2°.

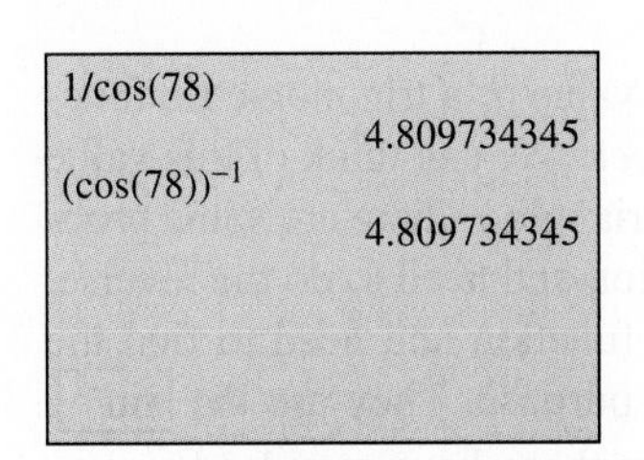
1/cos(78)
4.809734345
$(\cos(78))^{-1}$
4.809734345

Figure 2

EXAMPLE 8 Find sec 78°.

SOLUTION Because $\sec 78° = \dfrac{1}{\cos 78°}$, the calculator sequence is

Scientific Calculator	Graphing Calculator (Figure 2)
78 [cos] [$1/x$]	1 [÷] [cos] [(] 78 [)] [ENTER]
or	or
1 [÷] 78 [cos] [=]	[(] [cos] [(] 78 [)] [)] [x^{-1}] [ENTER]

Rounding to four digits past the decimal point, we have

$$\sec 78° = 4.8097$$

PROBLEM 9
Use a calculator to show that $\sin 52° = \cos 38°$.

EXAMPLE 9 To further justify the Cofunction Theorem introduced in the previous section, use a calculator to find $\sin 24.3°$ and $\cos 65.7°$.

SOLUTION Note that the sum of 24.3° and 65.7° is $24.3° + 65.7° = 90°$; the two angles are complementary. Using a calculator, and rounding our answers as we have previously, we find that the sine of 24.3° is the cosine of its complement 65.7°.

$$\sin 24.3° = 0.4115 \quad \text{and} \quad \cos 65.7° = 0.4115$$

■

Using Technology Working with Tables

Most graphing calculators have the ability to display several values of one or more functions simultaneously in a table format. We can use the table feature to display approximate values for the sine and cosine functions. Then we can easily compare the different trigonometric function values for a given angle, or compare values of a single trigonometric function for different angles.

Create a table for the two functions $Y_1 = \sin x$ and $Y_2 = \cos x$ (Figure 3). Be sure your calculator is set to degree mode. Set up your table so that you can input the values of x yourself. On some calculators, this is done by setting the independent variable to Ask (Figure 4). Display the table, and input the following values for x:

$$x = 0, 30, 45, 60, 90$$

Once you have entered all five x-values, your table should look something like the one shown in Figure 5. Compare the approximate values from your calculator with the exact values we found in Section 2.1, shown here in Table 3.

```
Plot1 Plot2 Plot3
\Y1=sin(X)
\Y2=cos(X)
\Y3=
\Y4=
\Y5=
\Y6=
\Y7=
```

Figure 3

```
TABLE SETUP
  TblStart=0
  ΔTbl=1
Indpnt: Auto Ask
Depend: Auto Ask
```

Figure 4

X	Y1	Y2
0	0	1
30	.5	.86603
45	.70711	.70711
60	.86603	.5
90	1	0
X=		

Figure 5

TABLE 3
Exact Values

x	$\sin x$	$\cos x$
0°	0	1
30°	$\frac{1}{2}$	$\frac{\sqrt{3}}{2}$
45°	$\frac{\sqrt{2}}{2}$	$\frac{\sqrt{2}}{2}$
60°	$\frac{\sqrt{3}}{2}$	$\frac{1}{2}$
90°	1	0

The [sin], [cos], and [tan] keys allow us to find the value of a trigonometric function when we know the measure of the angle. (*Remember:* We can think of this value as being the ratio of the lengths of two sides of a right triangle.) There are some problems, however, where we may be in the opposite situation and need to do the reverse. That is, we may know the value of the trigonometric function and need to find the angle. The calculator has another set of keys for this purpose. They are the [$\sin^{-1}$], [$\cos^{-1}$], and [$\tan^{-1}$] keys. At first glance, the notation on these keys may lead you to believe that they will give us the reciprocals of the trigonometric functions. Instead, this notation is used to denote an inverse function, something we will cover in more detail in Chapter 4. For the time being, just remember that these keys are used to find the angle given the value of one of the trigonometric functions of the angle.

CALCULATOR NOTE Some calculators do not have a key labeled as [$\sin^{-1}$]. You may need to press a combination of keys, such as [INV] [sin], [ARC] [sin], or [2nd] [sin]. If you are not sure which key to press, look in the index of your calculator manual under inverse trigonometric function.

PROBLEM 10
Find the acute angle θ for which $\tan\theta = 2.5129$ to the nearest tenth of a degree.

EXAMPLE 10 Find the acute angle θ for which $\tan\theta = 3.152$. Round your answer to the nearest tenth of a degree.

SOLUTION We are looking for the angle whose tangent is 3.152. We must use the [$\tan^{-1}$] key. First, be sure your calculator is set to degree mode. Then, press the indicated keys.

Scientific Calculator	Graphing Calculator
3.152 [$\tan^{-1}$]	[$\tan^{-1}$] [(] 3.152 [)] [ENTER]

To the nearest tenth of a degree the answer is 72.4°. That is, if $\tan\theta = 3.152$, then $\theta = 72.4°$. ■

PROBLEM 11
Find the acute angle A for which $\sin A = 0.6032$ to the nearest tenth of a degree.

EXAMPLE 11 Find the acute angle A for which $\sin A = 0.3733$. Round your answer to the nearest tenth of a degree.

SOLUTION The sequences are

Scientific Calculator	Graphing Calculator
0.3733 [$\sin^{-1}$]	[$\sin^{-1}$] [(] 0.3733 [)] [ENTER]

The result is $A = 21.9°$. ■

PROBLEM 12
Find the acute angle B for which $\csc B = 1.0338$ to the nearest hundredth of a degree.

EXAMPLE 12 To the nearest hundredth of a degree, find the acute angle B for which $\sec B = 1.0768$.

SOLUTION We do not have a secant key on the calculator, so we must first use a reciprocal to convert this problem into a problem involving $\cos B$ (as in Example 8).

If $\qquad \sec B = 1.0768$

then $\qquad \dfrac{1}{\sec B} = \dfrac{1}{1.0768}$ $\qquad$ Take the reciprocal of each side

$\qquad \cos B = \dfrac{1}{1.0768}$ $\qquad$ Because the cosine is the reciprocal of the secant

From this last line we see that the keys to press are

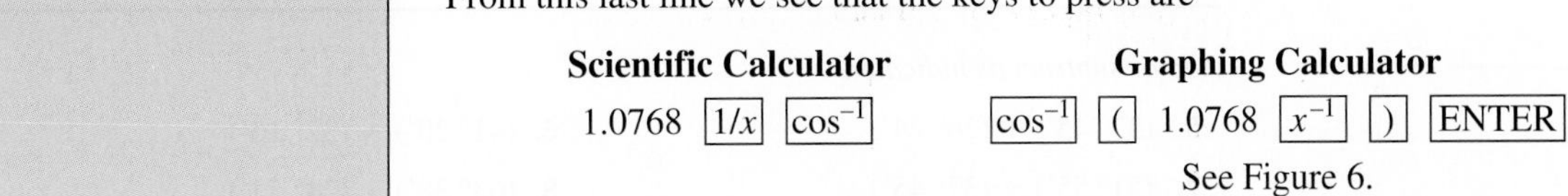

Scientific Calculator	Graphing Calculator
1.0768 [$1/x$] [$\cos^{-1}$]	[$\cos^{-1}$] [(] 1.0768 [x^{-1}] [)] [ENTER]

See Figure 6.

$\cos^{-1}(1.0768^{-1})$
21.77039922

Figure 6

To the nearest hundredth of a degree our answer is $B = 21.77°$. ■

PROBLEM 13
Find the acute angle C for which $\cot C = 1.5224$ to the nearest degree.

EXAMPLE 13 Find the acute angle C for which $\cot C = 0.0975$. Round to the nearest degree.

SOLUTION First, we rewrite the problem in terms of $\tan C$.

If $\cot C = 0.0975$

then $\dfrac{1}{\cot C} = \dfrac{1}{0.0975}$ Take the reciprocal of each side

$\tan C = \dfrac{1}{0.0975}$ Because the tangent is the reciprocal of the cotangent

From this last line we see that the keys to press are

Scientific Calculator	Graphing Calculator
0.0975 [1/x] [$\tan^{-1}$]	[$\tan^{-1}$] [(] 0.0975 [x^{-1}] [)] [ENTER]

To the nearest degree our answer is $C = 84°$.

Getting Ready for Class

After reading through the preceding section, respond in your own words and in complete sentences.

a. What is 1 minute of angle measure?
b. How do you convert from decimal degrees to degrees and minutes?
c. How do you find sin 58.75° on your calculator?
d. If $\tan \theta = 3.152$, how do you use a calculator to find θ?

2.2 PROBLEM SET

➤ CONCEPTS AND VOCABULARY

For Questions 1 through 4, fill in each blank with the appropriate word.

1. One degree is equal to 60 ________ or 3,600 ________.

2. If $\theta = 7.25°$ in decimal degrees, then the digit 7 represents the number of ________, 2 represents the number of ________ of a degree, and 5 represents the number of ________ of a degree.

3. On a calculator, the SIN, COS, and TAN keys allow us to find the ________ of a trigonometric function when we know the ________.

4. On a calculator, the SIN^{-1}, COS^{-1}, and TAN^{-1} keys allow us to find an ________ given the ________ of a trigonometric function.

➤ EXERCISES

Add or subtract as indicated.

5. $(37° 45') + (26° 24')$
6. $(41° 20') + (32° 16')$
7. $(51° 55') + (37° 45')$
8. $(63° 38') + (24° 52')$
9. $(61° 33') + (45° 16')$
10. $(77° 21') + (23° 16')$

11. $90° - (34° 12')$
12. $90° - (62° 25')$
13. $180° - (120° 17')$
14. $180° - (112° 19')$
15. $(76° 24') - (22° 34')$
16. $(89° 38') - (28° 58')$

Convert each of the following to degrees and minutes.

17. 35.4°
18. 63.2°
19. 16.25°
20. 18.75°
21. 92.55°
22. 34.45°
23. 19.9°
24. 18.8°

Change each of the following to decimal degrees. If rounding is necessary, round to the nearest hundredth of a degree.

25. 45° 12′
26. 74° 18′
27. 62° 36′
28. 21° 48′
29. 17° 20′
30. 29° 40′
31. 48° 27′
32. 78° 21′

Use a calculator to find each of the following. Round all answers to four places past the decimal point.

33. sin 27.2°
34. cos 82.9°
35. cos 18°
36. sin 42°
37. tan 87.32°
38. tan 81.43°
39. cot 31°
40. cot 24°
41. sec 48.2°
42. sec 71.8°
43. csc 14.15°
44. csc 12.21°

Use a calculator to find each of the following. Round all answers to four places past the decimal point.

45. cos 24° 30′
46. sin 35° 10′
47. tan 42° 15′
48. tan 19° 45′
49. sin 56° 40′
50. cos 66° 40′
51. sec 45° 54′
52. sec 84° 48′

Use a calculator to complete the following tables. (Be sure your calculator is in degree mode.) Round all answers to four digits past the decimal point. If you have a graphing calculator with table-building capabilities, use it to construct the tables.

53.

x	$\sin x$	$\csc x$
0°		
30°		
45°		
60°		
90°		

54.

x	$\cos x$	$\sec x$
0°		
30°		
45°		
60°		
90°		

Find θ if θ is between 0° and 90°. Round your answers to the nearest tenth of a degree.

55. $\cos\theta = 0.9770$
56. $\sin\theta = 0.3971$
57. $\tan\theta = 0.6873$
58. $\cos\theta = 0.5490$
59. $\sin\theta = 0.9813$
60. $\tan\theta = 0.6273$
61. $\sec\theta = 1.0191$
62. $\sec\theta = 1.0801$
63. $\csc\theta = 1.8214$
64. $\csc\theta = 1.4293$
65. $\cot\theta = 0.6873$
66. $\cot\theta = 0.4327$

Use a calculator to find a value of θ between 0° and 90° that satisfies each statement. Write your answer in degrees and minutes rounded to the nearest minute.

67. $\cos\theta = 0.4112$
68. $\sin\theta = 0.9954$
69. $\cot\theta = 5.5764$
70. $\cot\theta = 4.6252$
71. $\csc\theta = 7.0683$
72. $\sec\theta = 1.0129$

To further justify the Cofunction Theorem, use your calculator to find a value for the given pair of trigonometric functions. In each case, the trigonometric functions are cofunctions of one another, and the angles are complementary angles. Round your answers to four places past the decimal point.

73. $\sin 23°, \cos 67°$

74. $\sin 33°, \cos 57°$

75. $\sec 34.5°, \csc 55.5°$

76. $\sec 56.7°, \csc 33.3°$

77. $\tan 4° 30', \cot 85° 30'$

78. $\tan 10° 30', \cot 79° 30'$

Work each of the following problems on your calculator. Do not write down or round off any intermediate answers.

79. $\cos^2 37° + \sin^2 37°$

80. $\cos^2 85° + \sin^2 85°$

81. What happens when you try to find A for $\sin A = 1.234$ on your calculator? Why does it happen?

82. What happens when you try to find B for $\sin B = 4.321$ on your calculator? Why does this happen?

83. What happens when you try to find $\tan 90°$ on your calculator? Why does this happen?

84. What happens when you try to find $\cot 0°$ on your calculator? Why does this happen?

Complete each of the following tables. Round all answers to the nearest tenth.

85. a.

x	$\tan x$
87°	
87.5°	
88°	
88.5°	
89°	
89.5°	
90°	

b.

x	$\tan x$
89.4°	
89.5°	
89.6°	
89.7°	
89.8°	
89.9°	
90°	

86. a.

x	$\cot x$
3°	
2.5°	
2°	
1.5°	
1°	
0.5°	
0°	

b.

x	$\cot x$
0.6°	
0.5°	
0.4°	
0.3°	
0.2°	
0.1°	
0°	

Sundials *The Moorish sundial is designed so that the shadow of the gnomon (the vertical triangular piece) is consistent at each hour from one day to the next. This allows the hours to be marked on the sundial using a single scale that is not affected by the changes in the Sun's path during the year. If the sundial is positioned so that the gnomon is aligned along a longitudinal line from north to south, then at exactly noon the gnomon will cast a shadow due north.*

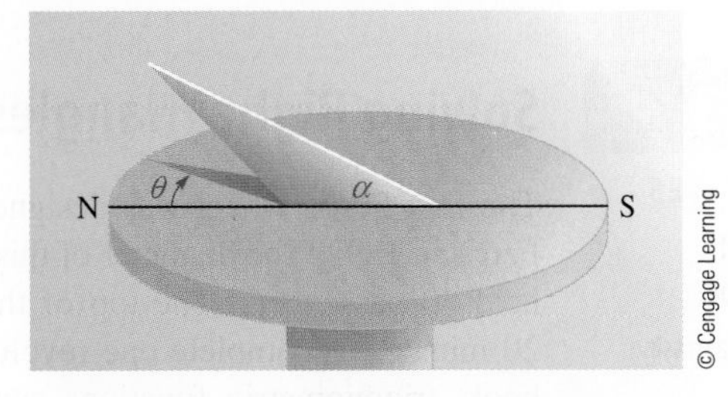

Figure 7

As the sun moves, the shadow will sweep out an angle toward the east, labeled θ in Figure 7, called the shadow angle. *This angle can be calculated using the formula*

$$\tan \theta = \sin \alpha \tan (h \cdot 15°)$$

where α is the latitude of the position of the sundial and h is a number of hours from noon.

87. The latitude of San Luis Obispo, California, is 35.282°. Find the shadow angle for a sundial in San Luis Obispo at 2:00 p.m. Round to the nearest tenth of a degree.

88. The latitude of Fargo, North Dakota, is 46.877°. Find the shadow angle for a sundial in Fargo at 5:00 p.m. Round to the nearest tenth of a degree.

➤ REVIEW PROBLEMS

The problems that follow review material we covered in Section 1.3. Find sin θ, cos θ, and tan θ if the given point is on the terminal side of θ.

89. $(3, -2)$

90. $(-\sqrt{3}, 1)$

Find sin θ, cos θ, and tan θ for each value of θ. (Do not use calculators.)

91. 90°

92. 135°

Find the remaining trigonometric functions of θ based on the given information.

93. $\cos \theta = -5/13$ and θ terminates in QIII

94. $\tan \theta = -3/4$ and θ terminates in QII

In which quadrant must the terminal side of θ lie under the given conditions?

95. $\sin \theta > 0$ and $\cos \theta < 0$

96. $\tan \theta > 0$ and $\sec \theta < 0$

➤ LEARNING OBJECTIVES ASSESSMENT

These questions are available for instructors to help assess if you have successfully met the learning objectives for this section.

97. Subtract (67° 22′) − (34° 30′).

a. 32° 02′ **b.** 33° 02′ **c.** 33° 08′ **d.** 32° 52′

98. Convert 76° 36′ to decimal degrees.

a. 76.36° **b.** 76.6° **c.** 76.4° **d.** 76.54°

99. Use a calculator to approximate sec 31.7°.

a. 0.8508 **b.** 0.5255 **c.** 1.9031 **d.** 1.1753

100. Given $\cot \theta = x$ for some value x, which of the following correctly shows how to use a calculator to approximate θ?

a. $\tan^{-1}(1/x)$ **b.** $1/\tan^{-1}(x)$ **c.** $\cos(x)/\sin(x)$ **d.** $\cos^{-1}(x)/\sin^{-1}(x)$

SECTION 2.3 Solving Right Triangles

LEARNING OBJECTIVES

1. Determine the number of significant digits in a value.
2. Solve a right triangle for a missing side.
3. Solve a right triangle for a missing angle.
4. Solve a real-life problem using right triangle trigonometry.

The first Ferris wheel was designed and built by American engineer George W. G. Ferris in 1893. The diameter of this wheel was 250 feet. It had 36 cars, each of which held 40 passengers. The top of the wheel was 264 feet above the ground. It took 20 minutes to complete one revolution. As you will see as we progress through the book, trigonometric functions can be used to model the motion of a rider on a Ferris wheel. The model can be used to give information about the position of the rider at any time during a ride. For instance, in the last example in this section, we will use Definition II for the trigonometric functions to find the height a rider is above the ground at certain positions on a Ferris wheel.

In this section, we will use Definition II for trigonometric functions of an acute angle, along with our calculators, to find the missing parts to some right triangles. Before we begin, however, we need to talk about significant digits.

DEFINITION

The number of *significant digits* (or figures) in a number is found by counting all the digits from left to right beginning with the first nonzero digit on the left. When no decimal point is present, trailing zeros are not considered significant.

© Bettmann/CORBIS

According to this definition,

0.042 has two significant digits
0.005 has one significant digit
20.5 has three significant digits
6.000 has four significant digits
9,200. has four significant digits
700 has one significant digit

NOTE In actual practice it is not always possible to tell how many significant digits an integer like 700 has. For instance, if exactly 700 people signed up to take history at your school, then 700 has three significant digits. On the other hand, if 700 is the result of a calculation in which the answer, 700, has been rounded to the nearest ten, then it has two significant digits. There are ways to write integers like 700 so that the number of significant digits can be determined exactly. One way is with scientific notation. However, to simplify things, in this book we will assume that measurements (other than angles) written without decimal points have the *least* possible number of significant digits. In the case of 700, that number is one. For angles, we will assume that all angle measures written as integers are accurate to the nearest degree.

The relationship between the accuracy of the sides of a triangle and the accuracy of the angles in the same triangle is shown in Table 1.

We are now ready to use Definition II to solve right triangles. We solve a right triangle by using the information given about it to find all of the missing sides and angles. In all the examples and in the Problem Set that follows, we will assume that C is the right angle in all of our right triangles, unless otherwise noted.

TABLE 1

Accuracy of Sides	Accuracy of Angles
Two significant digits	Nearest degree
Three significant digits	Nearest 10 minutes or tenth of a degree
Four significant digits	Nearest minute or hundredth of a degree

Unless stated otherwise, we round our answers so that the number of significant digits in our answers matches the number of significant digits in the least significant number given in the original problem. Also, we round our answers only and not any of the numbers in the intermediate steps. Finally, we are showing the values of the trigonometric functions to four significant digits simply to avoid cluttering the page with long decimal numbers. This does not mean that you should stop halfway through a problem and round the values of trigonometric functions to four significant digits before continuing.

PROBLEM 1
In right triangle ABC, $A = 50°$ and $c = 14$ centimeters. Find a, b, and B.

EXAMPLE 1 In right triangle ABC, $A = 40°$ and $c = 12$ centimeters. Find a, b, and B.

SOLUTION We begin by making a diagram of the situation (Figure 1). The diagram is very important because it lets us visualize the relationship between the given information and the information we are asked to find.

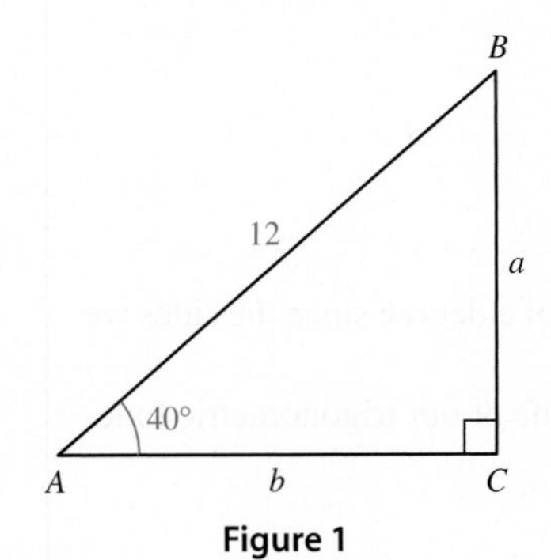

Figure 1

To find B, we use the fact that the sum of the two acute angles in any right triangle is 90°.

$$B = 90° - A$$
$$= 90° - 40°$$
$$B = 50°$$

To find a, we can use the formula for sin A.

$$\sin A = \frac{a}{c}$$

Multiplying both sides of this formula by c and then substituting in our given values of A and c we have

$$a = c \sin A$$
$$= 12 \sin 40°$$
$$a = 12(0.6428) \quad \sin 40° = 0.6428$$
$$a = 7.7 \text{ cm} \quad \text{Answer rounded to two significant digits}$$

There is more than one way to find b.

Using $\cos A = \dfrac{b}{c}$, we have

$$b = c \cos A$$
$$= 12 \cos 40°$$
$$= 12(0.7660)$$
$$b = 9.2 \text{ cm}$$

Using the Pythagorean Theorem, we have

$$c^2 = a^2 + b^2$$
$$b = \sqrt{c^2 - a^2}$$
$$= \sqrt{12^2 - (7.7)^2}$$
$$= \sqrt{144 - 59.29}$$
$$= \sqrt{84.71}$$
$$b = 9.2 \text{ cm}$$

■

In Example 2, we are given two sides and asked to find the remaining parts of a right triangle.

PROBLEM 2
In right triangle ABC, $a = 3.54$ and $b = 5.12$. Find the remaining side and angles.

EXAMPLE 2 In right triangle ABC, $a = 2.73$ and $b = 3.41$. Find the remaining side and angles.

SOLUTION Figure 2 is a diagram of the triangle. We can find A by using the formula for $\tan A$.

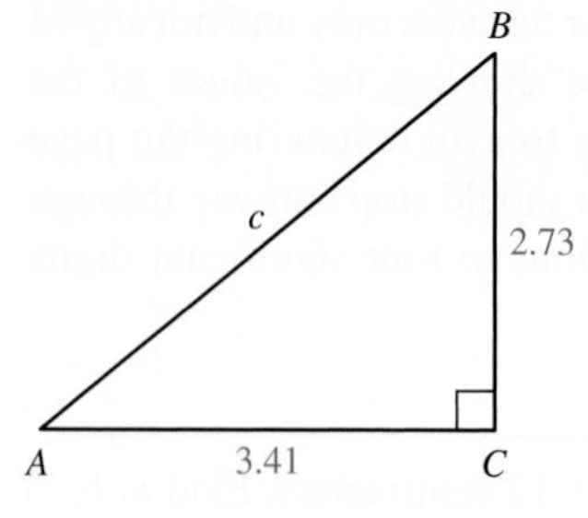

Figure 2

$$\tan A = \frac{a}{b}$$

$$= \frac{2.73}{3.41}$$

$$\tan A = 0.8006$$

Now, to find A, we use a calculator.

$$A = \tan^{-1}(0.8006) = 38.7°$$

Next we find B.

$$B = 90.0° - A$$

$$= 90.0° - 38.7°$$

$$B = 51.3°$$

Notice we are rounding each angle to the nearest tenth of a degree since the sides we were originally given have three significant digits.

We can find c using the Pythagorean Theorem or one of our trigonometric functions. Let's start with a trigonometric function.

If $\quad \sin A = \frac{a}{c}$

then $\quad c = \frac{a}{\sin A}$ $\quad$ Multiply each side by c, then divide each side by $\sin A$

$$= \frac{2.73}{\sin 38.7°}$$

$$= \frac{2.73}{0.6252}$$

$= 4.37$ $\quad$ To three significant digits

Using the Pythagorean Theorem, we obtain the same result.

If $\quad c^2 = a^2 + b^2$

then $\quad c = \sqrt{a^2 + b^2}$

$$= \sqrt{(2.73)^2 + (3.41)^2}$$

$$= \sqrt{19.081}$$

$$= 4.37$$

■

PROBLEM 3
In Figure 3, find x if $A = 25°$ and the radius of the circle is 14 inches.

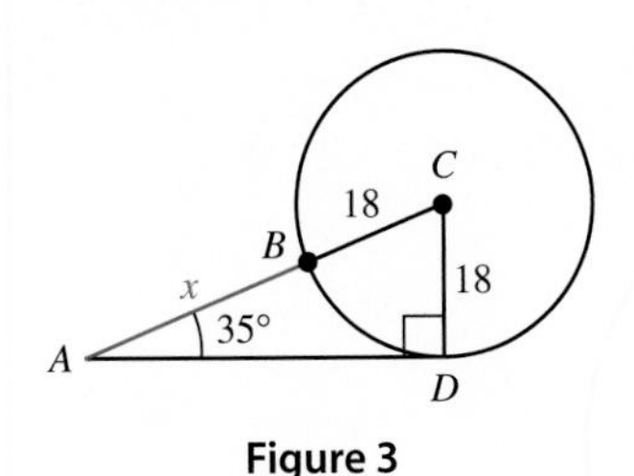

Figure 3

EXAMPLE 3 The circle in Figure 3 has its center at C and a radius of 18 inches. If triangle ADC is a right triangle and A is 35°, find x, the distance from A to B.

SOLUTION In triangle ADC, the side opposite A is 18 and the hypotenuse is $x + 18$. We can use $\sin A$ to write an equation that will allow us to solve for x.

$$\sin 35° = \frac{18}{x + 18}$$

$$(x + 18) \sin 35° = 18 \qquad \text{Multiply each side by } x + 18$$

$$x + 18 = \frac{18}{\sin 35°} \qquad \text{Divide each side by } \sin 35°$$

$$x = \frac{18}{\sin 35°} - 18 \qquad \text{Subtract 18 from each side}$$

$$= \frac{18}{0.5736} - 18$$

$$= 13 \text{ inches} \qquad \text{To two significant digits}$$

PROBLEM 4
In Figure 4, $A = 22°$, $AD = 15$, and $\angle BDC = 48°$. Find DC.

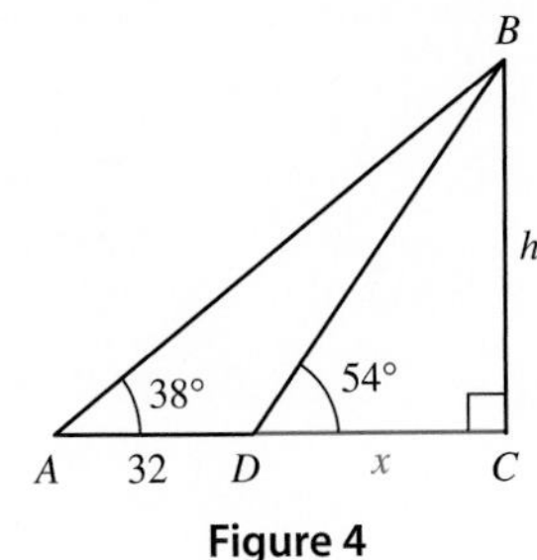

Figure 4

EXAMPLE 4 In Figure 4, the distance from A to D is 32 feet. Use the information in Figure 4 to solve for x, the distance between D and C.

SOLUTION To find x, we write two equations, each of which contains the variables x and h. Then we solve each equation for h and set the two expressions for h equal to each other.

Two equations involving both x and h:

$$\begin{cases} \tan 54° = \dfrac{h}{x} & \Rightarrow h = x \tan 54° \\ \tan 38° = \dfrac{h}{x + 32} & \Rightarrow h = (x + 32) \tan 38° \end{cases} \qquad \text{Solve each equation for } h$$

Setting the two expressions for h equal to each other gives us an equation that involves only x.

$$h = h$$

Therefore

$$x \tan 54° = (x + 32) \tan 38°$$

$$x \tan 54° = x \tan 38° + 32 \tan 38° \qquad \text{Distributive property}$$

$$x \tan 54° - x \tan 38° = 32 \tan 38° \qquad \text{Subtract } x \tan 38° \text{ from each side}$$

$$x(\tan 54° - \tan 38°) = 32 \tan 38° \qquad \text{Factor } x \text{ from each term on the left side}$$

$$x = \frac{32 \tan 38°}{\tan 54° - \tan 38°} \qquad \text{Divide each side by the coefficient of } x$$

$$= \frac{32(0.7813)}{1.3764 - 0.7813}$$

$$= 42 \text{ ft} \qquad \text{To two significant digits}$$

PROBLEM 5
Rework Example 5 if $\theta = 75°$.

EXAMPLE 5 In the introduction to this section, we gave some of the facts associated with the first Ferris wheel. Figure 5 is a simplified model of that Ferris wheel. If θ is the central angle formed as a rider moves from position P_0 to position P_1, find the rider's height above the ground h when θ is 45°.

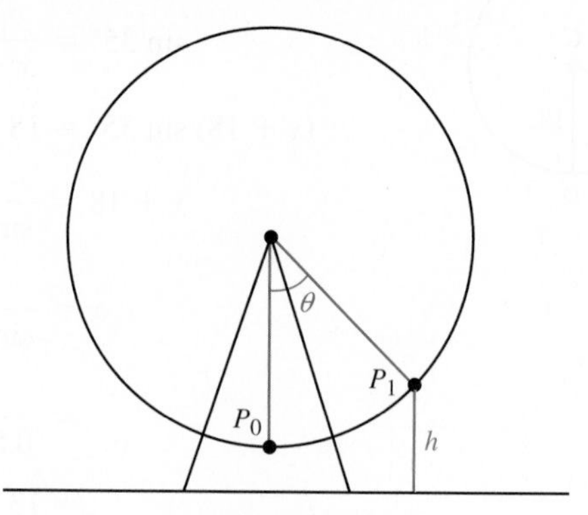

Figure 5

SOLUTION We know from the introduction to this section that the diameter of the first Ferris wheel was 250 feet, which means the radius was 125 feet. Because the top of the wheel was 264 feet above the ground, the distance from the ground to the bottom of the wheel was 14 feet (the distance to the top minus the diameter of the wheel). To form a right triangle, we draw a horizontal line from P_1 to the vertical line connecting the center of the wheel O with P_0. This information is shown in Figure 6.

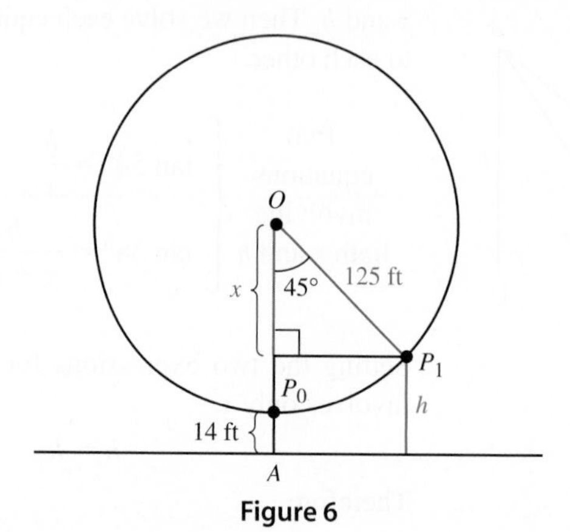

Figure 6

The key to solving this problem is recognizing that x is the difference between OA (the distance from the center of the wheel to the ground) and h. Because OA is 139 feet (the radius of the wheel plus the distance between the bottom of the wheel and the ground: $125 + 14 = 139$), we have

$$x = 139 - h$$

We use a cosine ratio to write an equation that contains h.

$$\cos 45° = \frac{x}{125}$$

$$= \frac{139 - h}{125}$$

Solving for h we have

$$125 \cos 45° = 139 - h$$
$$h = 139 - 125 \cos 45°$$
$$= 139 - 125(0.7071)$$
$$= 139 - 88.4$$
$$= 51 \text{ ft} \quad \text{To two significant digits}$$

If $\theta = 45°$, a rider at position P_1 is $\frac{1}{8}$ of the way around the wheel. At that point, the rider is approximately 51 feet above the ground.

Getting Ready for Class

After reading through the preceding section, respond in your own words and in complete sentences.

a. How do you determine the number of significant digits in a number?
b. Explain the relationship between the accuracy of the sides and the accuracy of the angles in a triangle.
c. In right triangle ABC, angle A is 40°. How do you find the measure of angle B?
d. In right triangle ABC, angle A is 40° and side c is 12 centimeters. How do you find the length of side a?

2.3 PROBLEM SET

➤ CONCEPTS AND VOCABULARY

For Questions 1 through 4, fill in each blank with the appropriate word.

1. To count the number of significant digits in a number, count all the digits from _____ to _______ beginning with the ______ ________ digit on the left. If no decimal point is present, trailing zeros are ____ counted.

2. If the sides of a triangle are accurate to three significant digits, then angles should be measured to the nearest ______ of a degree, or the nearest _____ minutes.

3. To *solve* a right triangle means to find all of the missing _______ and ______.

4. In general, round answers so that the number of significant digits in your answer matches the number of significant digits in the ________ significant number given in the problem.

➤ EXERCISES

For Problems 5 through 8, determine the number of significant digits in each value.

5. a. 65,000	**b.** 6.50	**c.** 650	**d.** 0.0065
6. a. 374	**b.** 0.0374	**c.** 3.7400	**d.** 374,000
7. a. 1,234	**b.** 1,234.00	**c.** 0.01234	**d.** 12.34
8. a. 0.14009	**b.** 140,090	**c.** 1.4009	**d.** 140.0900

Problems 9 through 22 refer to right triangle ABC with $C = 90°$. Begin each problem by drawing a picture of the triangle with both the given and asked for information labeled appropriately. Also, write your answers for angles in decimal degrees.

9. If $A = 42°$ and $c = 89$ cm, find b.

10. If $A = 42°$ and $c = 15$ ft, find a.

11. If $A = 34°$ and $a = 22$ m, find c.

12. If $A = 34°$ and $b = 55$ m, find c.

13. If $B = 16.9°$ and $c = 7.55$ cm, find b.

14. If $B = 24.5°$ and $c = 2.34$ ft, find a.

15. If $B = 55.33°$ and $b = 12.34$ yd, find a.

16. If $B = 77.66°$ and $a = 43.21$ inches, find b.

17. If $a = 42.3$ inches and $b = 32.4$ inches, find B.

18. If $a = 16$ cm and $b = 26$ cm, find A.

19. If $b = 9.8$ mm and $c = 12$ mm, find B.

20. If $b = 6.7$ m and $c = 7.7$ m, find A.

21. If $c = 45.54$ ft and $a = 23.32$ ft, find B.

22. If $c = 5.678$ ft and $a = 4.567$ ft, find A.

Problems 23 through 38 refer to right triangle ABC with $C = 90°$. In each case, solve for all the missing parts using the given information. (In Problems 35 through 38, write your angles in decimal degrees.)

23. $A = 25°$, $c = 24$ m

24. $A = 41°$, $c = 36$ m

25. $A = 32.6°$, $a = 43.4$ inches

26. $A = 48.3°$, $a = 3.48$ inches

27. $A = 10° 42'$, $b = 5.932$ cm

28. $A = 66° 54'$, $b = 28.28$ cm

29. $B = 76°$, $c = 5.8$ ft

30. $B = 21°$, $c = 4.2$ ft

31. $B = 26° 30'$, $b = 324$ mm

32. $B = 53° 30'$, $b = 725$ mm

33. $B = 23.45°$, $a = 5.432$ mi

34. $B = 44.44°$, $a = 5.555$ mi

35. $a = 37$ ft, $b = 87$ ft

36. $a = 91$ ft, $b = 85$ ft

37. $b = 377.3$ inches, $c = 588.5$ inches

38. $a = 62.3$ cm, $c = 73.6$ cm

In Problems 39 and 40, use the information given in the diagram to find A to the nearest degree.

39.

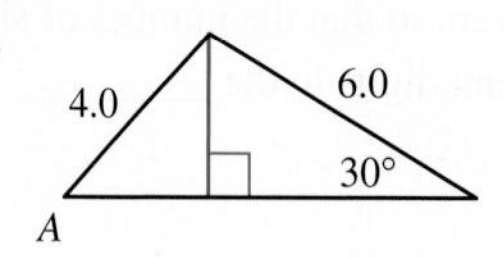

40.

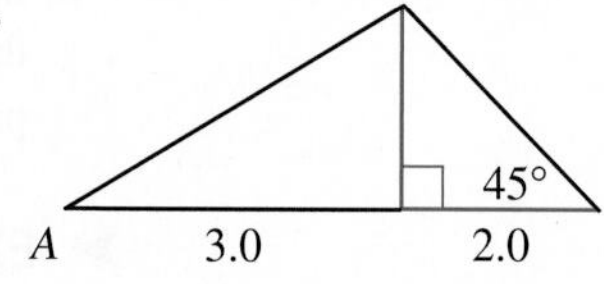

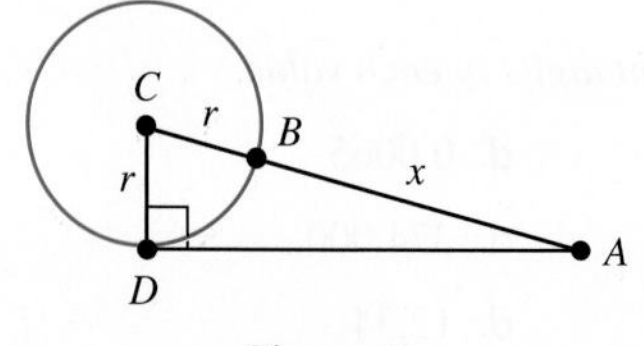

Figure 7

The circle in Figure 7 has a radius of r and center at C. The distance from A to B is x. For Problems 41 through 44, redraw Figure 7, label it as indicated in each problem, and then solve the problem.

41. If $A = 31°$ and $r = 12$, find x.

42. If $C = 26°$ and $r = 19$, find x.

43. If $C = 65°$ and $x = 22$, find r.

44. If $A = 45°$ and $x = 15$, find r.

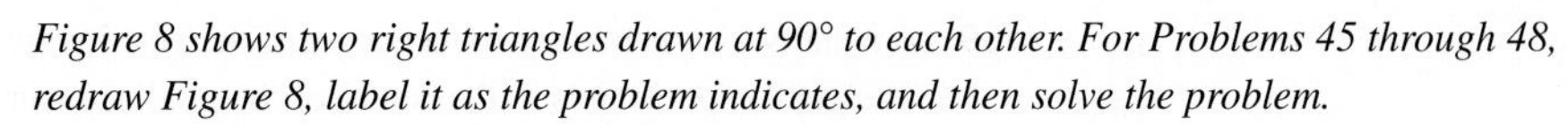

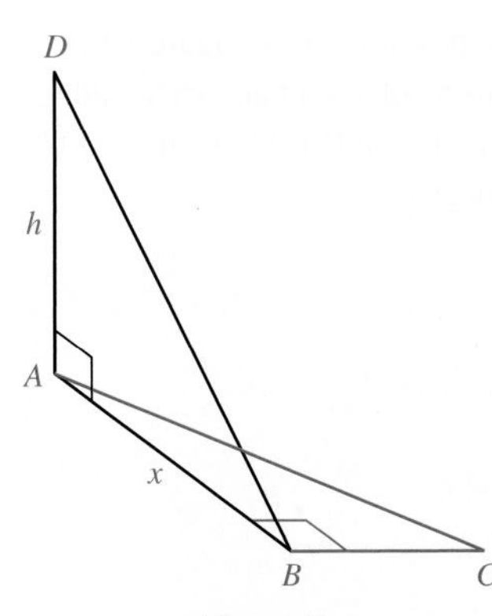

Figure 8

Figure 8 shows two right triangles drawn at 90° to each other. For Problems 45 through 48, redraw Figure 8, label it as the problem indicates, and then solve the problem.

45. If $\angle ABD = 27°$, $C = 62°$, and $BC = 42$, find x and then find h.

46. If $\angle ABD = 53°$, $C = 48°$, and $BC = 42$, find x and then find h.

47. If $AC = 32$, $h = 19$, and $C = 41°$, find $\angle ABD$.

48. If $AC = 19$, $h = 32$, and $C = 49°$, find $\angle ABD$.

In Figure 9, the distance from A to D is y, the distance from D to C is x, and the distance from C to B is h. Use Figure 9 to solve Problems 49 through 54.

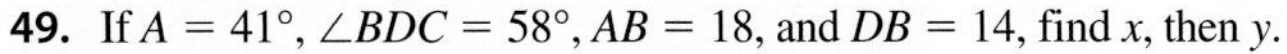

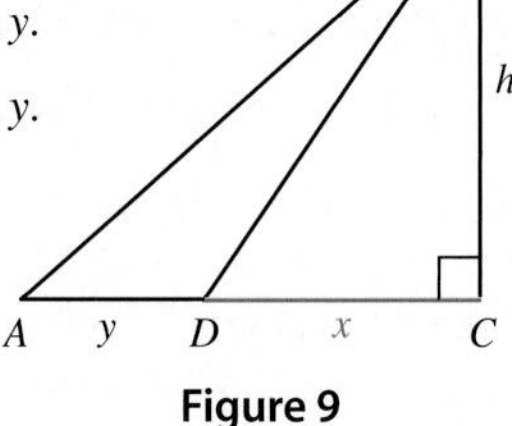

Figure 9

49. If $A = 41°$, $\angle BDC = 58°$, $AB = 18$, and $DB = 14$, find x, then y.

50. If $A = 32°$, $\angle BDC = 48°$, $AB = 17$, and $DB = 12$, find x, then y.

51. If $A = 41°$, $\angle BDC = 58°$, and $AB = 28$, find h, then x.

52. If $A = 32°$, $\angle BDC = 48°$, and $AB = 56$, find h, then x.

53. If $A = 43°$, $\angle BDC = 57°$, and $y = 11$, find x.

54. If $A = 32°$, $\angle BDC = 41°$, and $y = 14$, find x.

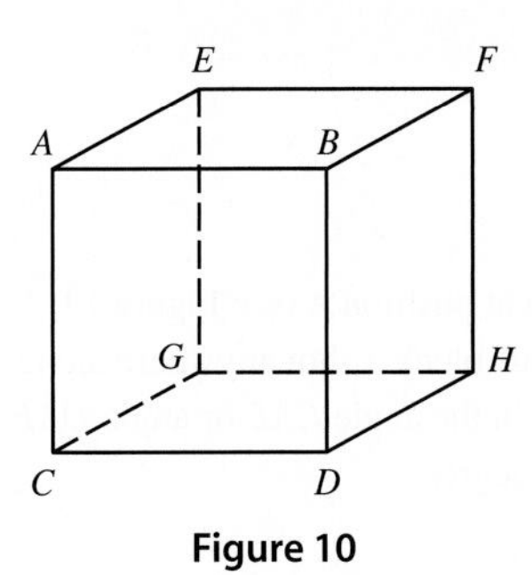

Figure 10

55. Suppose each edge of the cube shown in Figure 10 is 5.00 inches long. Find the measure of the angle formed by diagonals CF and CH. Round your answer to the nearest tenth of a degree.

56. Suppose each edge of the cube shown in Figure 10 is 3.00 inches long. Find the measure of the angle formed by diagonals DE and DG. Round your answer to the nearest tenth of a degree.

57. Suppose each edge of the cube shown is x inches long. Find the measure of the angle formed by diagonals CF and CH in Figure 10. Round your answer to the nearest tenth of a degree.

Soccer *A regulation soccer field has a rectangular penalty area that measures 132 feet by 54 feet. The goal is 24 feet wide and centered along the back of the penalty area. Assume the goalkeeper can block a shot 6 feet to either side of their position for a total coverage of 12 feet. (Source: Fédération Internationale de Football Association)*

58. A penalty kick is taken from a corner of the penalty area at position A (see Figure 11). The goalkeeper stands 6 feet from the goalpost nearest the shooter and can thus block a shot anywhere between the middle of the goal and the nearest goalpost (segment CD). To score, the shooter must kick the ball within the angle CAE. Find the measure of this angle to the nearest tenth of a degree.

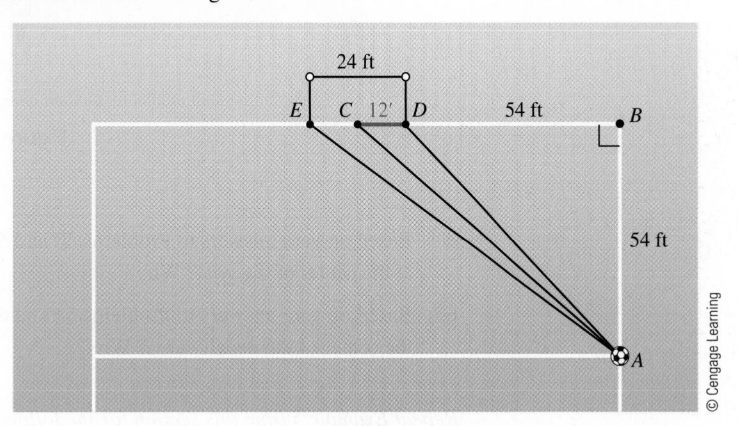

Figure 11

59. A penalty kick is taken from a corner of the penalty area at position A (see Figure 12). The goalkeeper stands in the center of the goal and can thus block a shot anywhere along segment CD. To score, the shooter must kick the ball within the angle CAE or angle DAF. Find the sum of these two angles to the nearest tenth of a degree.

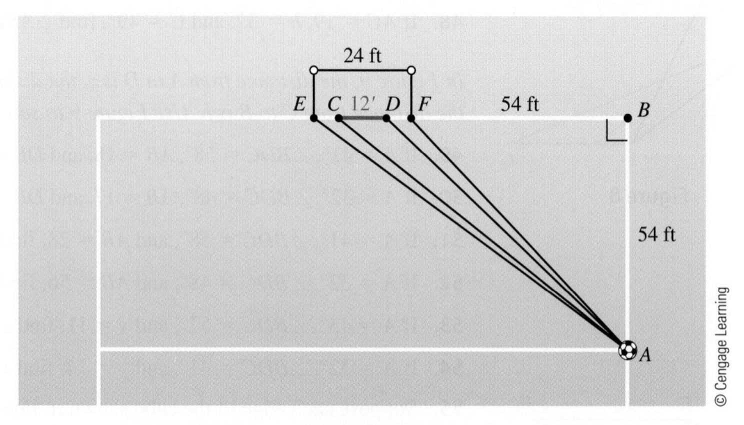

Figure 12

60. A penalty kick is taken from the center of the penalty area at position A (see Figure 13). The goalkeeper stands in the center of the goal and can thus block a shot anywhere along segment CD. To score, the shooter must kick the ball within the angle CAE or angle DAF. Find the sum of these two angles to the nearest tenth of a degree.

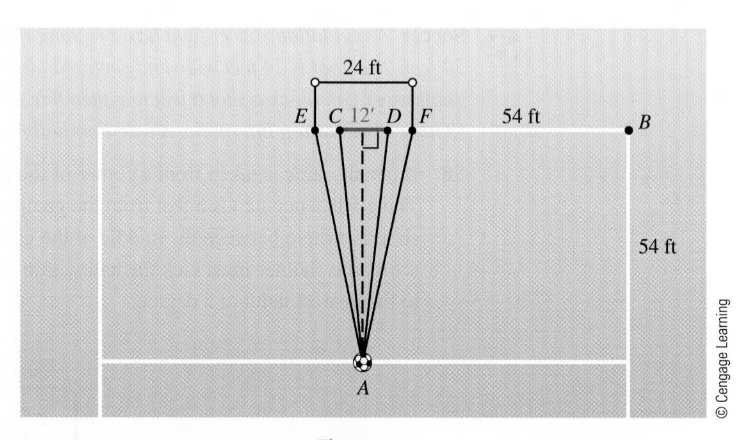

Figure 13

61. Based on your answers to Problems 58 and 59, should the goalkeeper stand to the side or at the center of the goal? Why?

62. Based on your answers to Problems 58 and 60, should the shooter kick from the corner or the center of the penalty area? Why?

Repeat Example 5 from this section for the following values of θ.

63. $\theta = 120°$

64. $\theta = 135°$

65. Ferris Wheel In 1897, a Ferris wheel was built in Vienna that still stands today. It is named the Riesenrad, which translates to the *Great Wheel.* The diameter of the Riesenrad is 197 feet. The top of the wheel stands 209 feet above the ground. Figure 14 is a model of the Riesenrad with angle θ the central angle that is formed as a rider moves from the initial position P_0 to position P_1. The rider is h feet above the ground at position P_1.

a. Find h if θ is 120.0°.
b. Find h if θ is 210.0°.
c. Find h if θ is 315.0°.

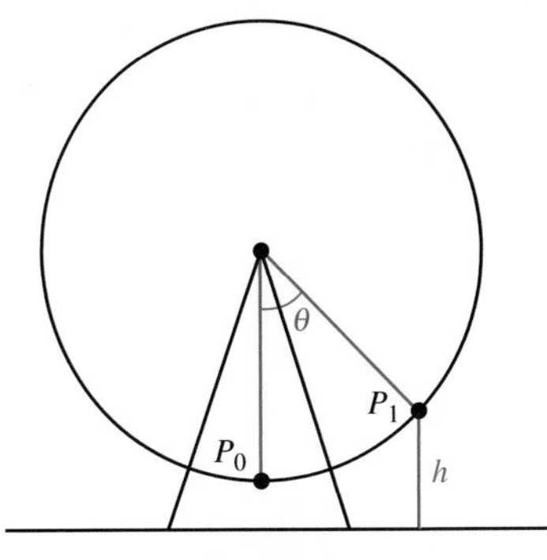

Figure 14

66. Ferris Wheel A Ferris wheel with a diameter of 165 feet was built in St. Louis in 1986. It is called Colossus. The top of the wheel stands 174 feet above the ground. Use the diagram in Figure 14 as a model of Colossus.

a. Find h if θ is 150.0°.
b. Find h if θ is 240.0°.
c. Find h if θ is 315.0°.

67. Observation Wheel The London Eye has a diameter of 135 meters. A rider boards the London Eye at ground level. Through what angle has the wheel rotated when the rider is 44.5 meters above ground for the first time?

68. Observation Wheel The Singapore Flyer is currently the largest observation wheel in the world with a diameter of 150 meters. The top of the wheel stands 165 meters above the ground. Find the height of a rider after the wheel has rotated through an angle of 110.5°. Assume the rider boards at the bottom of the wheel.

➤ REVIEW PROBLEMS

The following problems review material that we covered in Section 1.4.

69. If $\sec B = 2$, find $\cos^2 B$.

70. If $\csc B = 3$, find $\sin^2 B$.

71. If $\cos \theta = -\dfrac{2}{3}$ and θ terminates in QIII, find $\sin \theta$.

72. If $\sin A = \dfrac{1}{4}$ with A in QII, find $\cos A$.

Find the remaining trigonometric ratios for θ based on the given information.

73. $\sin \theta = \dfrac{\sqrt{3}}{2}$ with θ in QII

74. $\cos \theta = \dfrac{1}{\sqrt{5}}$ with θ in QIV

75. $\sec \theta = -2$ with θ in QIII

76. $\csc \theta = -2$ with θ in QIII

➤ EXTENDING THE CONCEPTS

77. Human Cannonball In Example 2 of Section 1.2, we found the equation of the path of the human cannonball. At the 1997 Washington County Fair in Oregon, David Smith, Jr., The Bullet, was shot from a cannon. As a human cannonball, he reached a height of 70 feet before landing in a net 160 feet from the cannon. In that example we found the equation that describes his path is

$$y = -\frac{7}{640}(x - 80)^2 + 70 \quad \text{for } 0 \le x \le 160$$

Graph this equation using the window

$$0 \le x \le 180, \text{ scale} = 20; 0 \le y \le 80, \text{ scale} = 10$$

Then zoom in on the curve near the origin until the graph has the appearance of a straight line. Use TRACE to find the coordinates of any point on the graph. This point defines an approximate right triangle (Figure 15). Use the triangle to find the angle between the cannon and the horizontal.

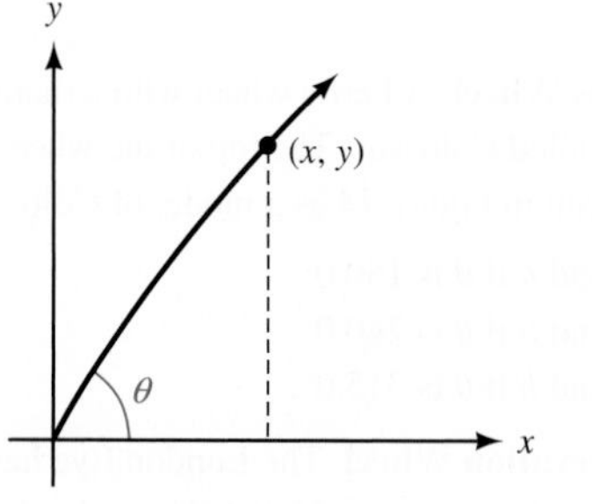

Figure 15

Plot1 Plot2 Plot3
\Y1=(−7/640)(X−80)²+70
\Y2=tan⁻¹(Y1/X)
\Y3=
\Y4=
\Y5=
\Y6=

Figure 16

78. Human Cannonball To get a better estimate of the angle described in Problem 77, we can use the table feature of your calculator. From Figure 15, we see that for any point (x, y) on the curve, $\theta \approx \tan^{-1}(y/x)$. Define functions Y1 and Y2 as shown in Figure 16. Then set up your table so that you can input the following values of x:

$$x = 10, 5, 1, 0.5, 0.1, 0.01$$

Based on the results, what is the angle between the cannon and the horizontal?

➤ LEARNING OBJECTIVES ASSESSMENT

These questions are available for instructors to help assess if you have successfully met the learning objectives for this section.

79. Which number contains three significant digits?

a. 0.0240 **b.** 240 **c.** 0.24 **d.** 24,000

80. Given right triangle ABC with $C = 90°$, if $A = 58°$ and $c = 15$ ft, find b.

a. 28 ft **b.** 13 ft **c.** 7.9 ft **d.** 18 ft

81. Given right triangle ABC with $C = 90°$, if $a = 58$ cm and $b = 35$ cm, find B.

a. 59° **b.** 31° **c.** 53° **d.** 37°

82. A Ferris wheel has a radius of 45 feet and the bottom of the wheel stands 6.5 feet above the ground. Find the height of a rider if the wheel has rotated 140° after the rider was seated.

a. 64 feet **b.** 95 feet **c.** 86 feet **d.** 80 feet

SECTION 2.4 Applications

LEARNING OBJECTIVES

1. Correctly interpret a bearing.
2. Solve a real-life problem involving an angle of elevation or depression.
3. Solve a real-life problem involving bearing.
4. Solve an applied problem using right triangle trigonometry.

As mentioned in the introduction to this chapter, we can use right triangle trigonometry to solve a variety of problems, such as problems involving topographic maps. In this section we will see how this is done by looking at a number of applications of right triangle trigonometry.

PROBLEM 1
The two equal sides of an isosceles triangle are each 14 centimeters. If each of the equal angles measures 48°, find the length of the base and the length of the altitude.

EXAMPLE 1 The two equal sides of an isosceles triangle are each 24 centimeters. If each of the two equal angles measures 52°, find the length of the base and the altitude.

SOLUTION An isosceles triangle is any triangle with two equal sides. The angles opposite the two equal sides are called the base angles, and they are always equal. Figure 1 shows a picture of our isosceles triangle.

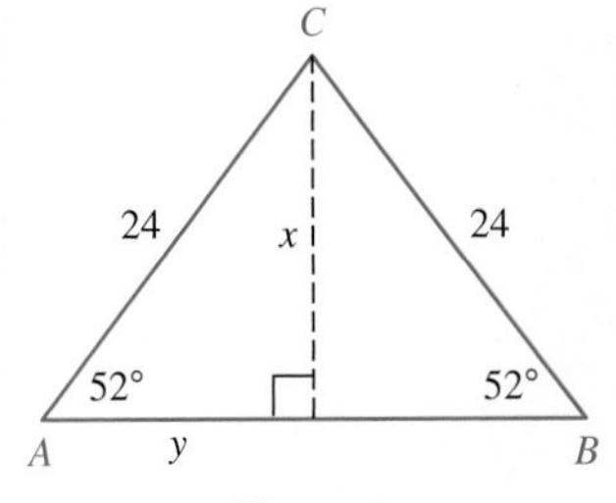

Figure 1

We have labeled the altitude x. We can solve for x using a sine ratio.

$$\begin{aligned} \text{If} \quad \sin 52° &= \frac{x}{24} \\ \text{then} \quad x &= 24 \sin 52° \\ &= 24(0.7880) \\ &= 19 \text{ cm} \quad \text{Rounded to two significant digits} \end{aligned}$$

We have labeled half the base with y. To solve for y, we can use a cosine ratio.

$$\begin{aligned} \text{If} \quad \cos 52° &= \frac{y}{24} \\ \text{then} \quad y &= 24 \cos 52° \\ &= 24(0.6157) \\ &= 15 \text{ cm} \quad \text{To two significant digits} \end{aligned}$$

The base is $2y = 2(15) = 30$ cm.

For our next applications, we need the following definition.

DEFINITION

An angle measured from the horizontal up is called an *angle of elevation*. An angle measured from the horizontal down is called an *angle of depression* (Figure 2).

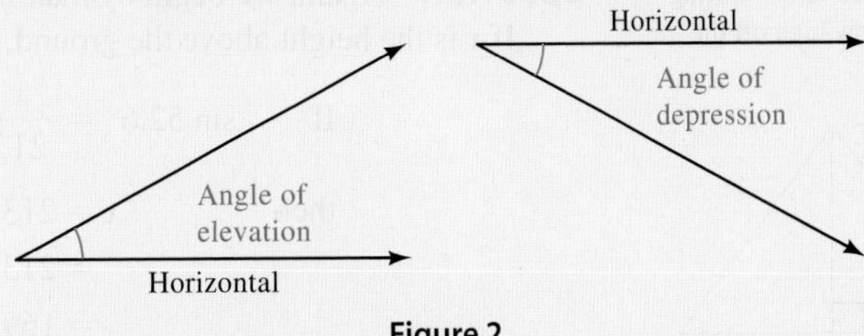

Figure 2

These angles of elevation and depression are always considered positive angles. Also, if an observer positioned at the vertex of the angle views an object in the direction of the nonhorizontal side of the angle, then this side is sometimes called the *line of sight* of the observer.

PROBLEM 2
If a 65.0-foot flagpole casts a shadow of 37.0 feet long, what is the angle of elevation of the sun from the tip of the shadow?

EXAMPLE 2 If a 75.0-foot flagpole casts a shadow 43.0 feet long, to the nearest 10 minutes what is the angle of elevation of the sun from the tip of the shadow?

SOLUTION We begin by making a diagram of the situation (Figure 3).

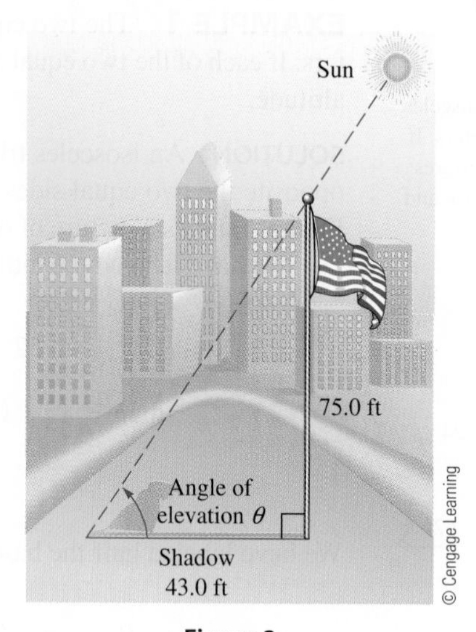

Figure 3

If we let θ = the angle of elevation of the sun, then

$$\tan \theta = \frac{75.0}{43.0}$$

$$\tan \theta = 1.7442$$

which means $\theta = \tan^{-1}(1.7442) = 60° \, 10'$ to the nearest 10 minutes.

PROBLEM 3
A man climbs 235 meters up the side of a pyramid and finds that the angle of depression to his starting point is 56.2°. How high off the ground is he?

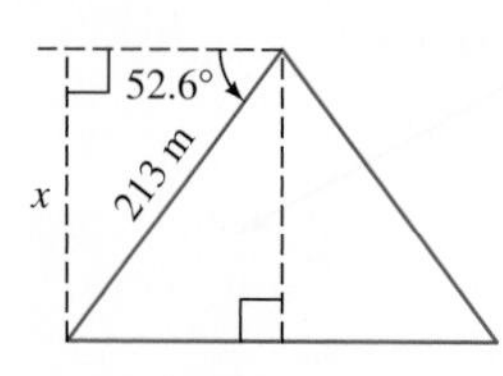

Figure 4

EXAMPLE 3 A man climbs 213 meters up the side of a pyramid and finds that the angle of depression to his starting point is 52.6°. How high off the ground is he?

SOLUTION Again, we begin by making a diagram of the situation (Figure 4).

If x is the height above the ground, we can solve for x using a sine ratio.

$$\text{If} \quad \sin 52.6° = \frac{x}{213}$$

$$\text{then} \quad x = 213 \sin 52.6°$$

$$= 213(0.7944)$$

$$= 169 \text{ m} \qquad \text{To three significant digits}$$

The man is 169 meters above the ground.

PROBLEM 4
Rework Example 4 if the distance between Amy and Stacey is 0.5 inches on the map and there are five contour intervals between them.

EXAMPLE 4 Figure 5 shows the topographic map we mentioned in the introduction to this chapter. Suppose Stacey and Amy are climbing Bishop's Peak. Stacey is at position S, and Amy is at position A. Find the angle of elevation from Amy to Stacey.

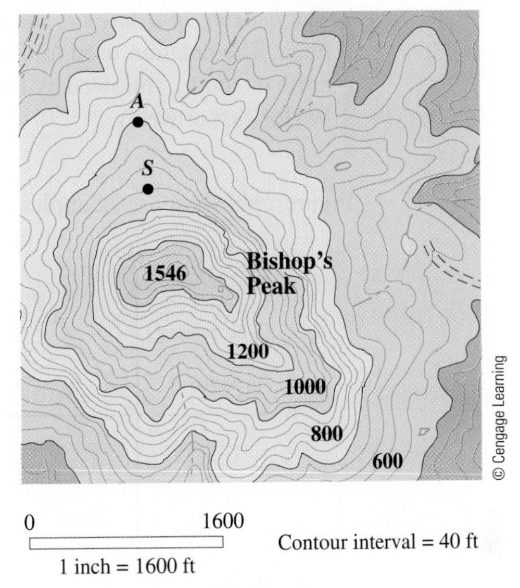

0 1600
1 inch = 1600 ft
Contour interval = 40 ft

Figure 5

NOTE On a map, all digits shown are considered to be significant. That is, 1600 means 1600., which contains four significant figures.

SOLUTION To solve this problem, we have to use two pieces of information from the legend on the map. First, we need to find the horizontal distance between the two people. The legend indicates that 1 inch on the map corresponds to an actual horizontal distance of 1,600 feet. If we measure the distance from Amy to Stacey with a ruler, we find it is $\frac{3}{8}$ inch. Multiplying this by 1,600, we have

$$\frac{3}{8} \cdot 1{,}600. = 600. \text{ ft}$$

which is the actual horizontal distance from Amy to Stacey.

Next, we need the vertical distance Stacey is above Amy. We find this by counting the number of contour intervals between them. There are three. From the legend on the map we know that the elevation changes by 40 feet between any two contour lines. Therefore, Stacey is 120 feet above Amy. Figure 6 shows a triangle that models the information we have so far.

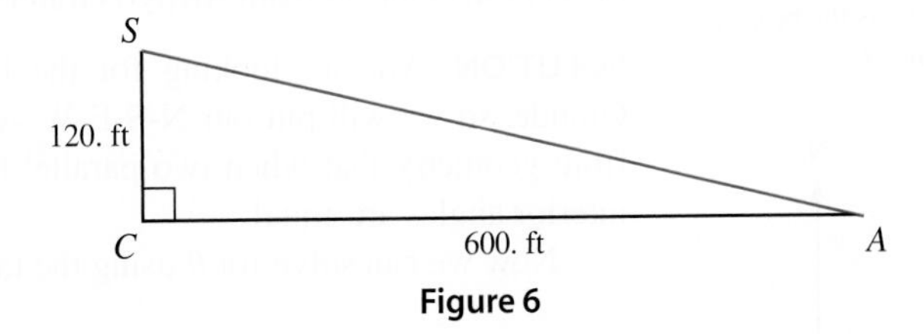

Figure 6

The angle of elevation from Amy to Stacey is angle A. To find A, we use the tangent ratio.

$$\tan A = \frac{120.}{600.} = 0.2$$

$$A = \tan^{-1}(0.2) = 11.3° \quad \text{To the nearest tenth of a degree}$$

Amy must look up at 11.3° from straight ahead to see Stacey. ■

Our next applications are concerned with what is called the *bearing of a line.* It is used in navigation and surveying.

DEFINITION

The *bearing of a line l* is the acute angle formed by the north–south line and the line *l*. The notation used to designate the bearing of a line begins with N or S (for north or south), followed by the number of degrees in the angle, and ends with E or W (for east or west).

Figure 7 shows some examples.

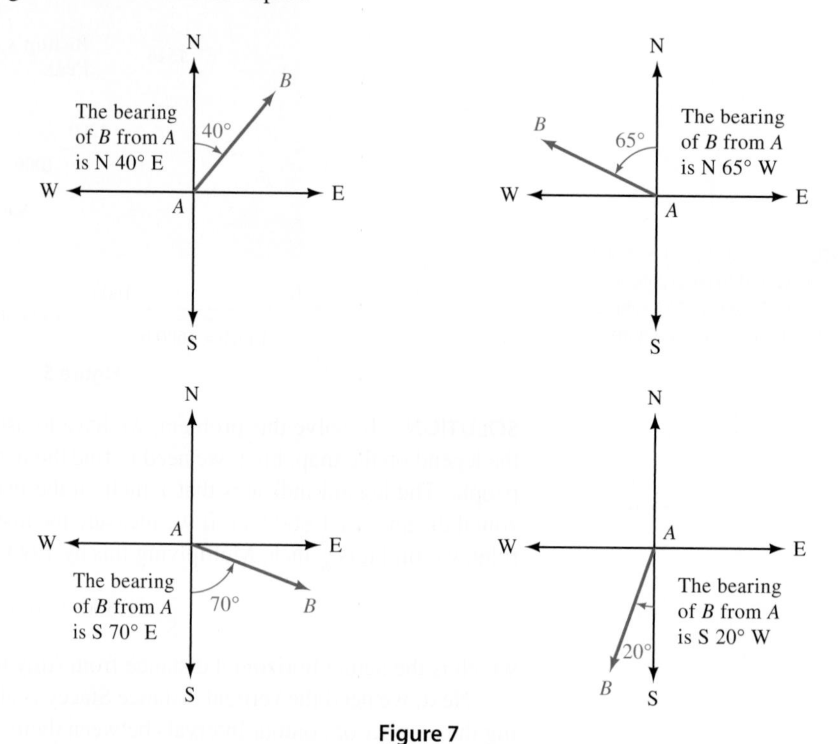

Figure 7

PROBLEM 5
Town A is 14 miles due north of town B. If town C is 3.5 miles due east of town B, what is the bearing from town A to town C?

EXAMPLE 5 San Luis Obispo, California, is 12 miles due north of Grover Beach. If Arroyo Grande is 4.6 miles due east of Grover Beach, what is the bearing of San Luis Obispo from Arroyo Grande?

SOLUTION We are looking for the bearing of San Luis Obispo *from* Arroyo Grande, so we will put our N-S-E-W system on Arroyo Grande (Figure 8). Recall from geometry that when two parallel lines are crossed by a transversal, alternate interior angles are equal.

Now we can solve for θ using the tangent ratio.

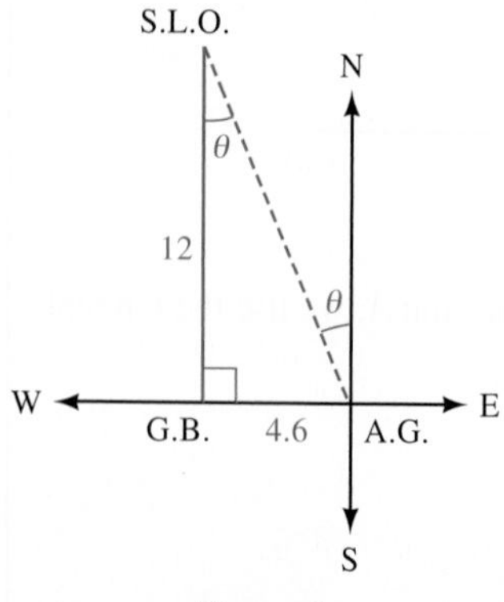

Figure 8

$$\tan\theta = \frac{4.6}{12}$$

$$\tan\theta = 0.3833$$

$$\theta = \tan^{-1}(0.3833) = 21° \quad \text{To the nearest degree}$$

The bearing of San Luis Obispo from Arroyo Grande is N 21° W. ■

PROBLEM 6
A boat travels on a course of bearing N 46° 30′ E for a distance of 248 miles. How many miles north and how many miles east has the boat traveled?

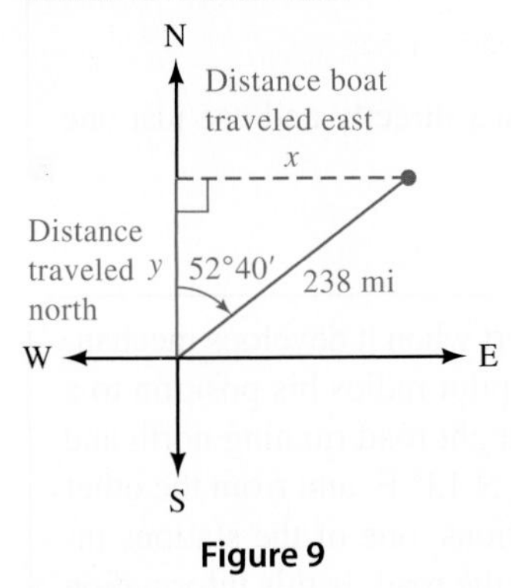

Figure 9

EXAMPLE 6 A boat travels on a course of bearing N 52° 40′ E for a distance of 238 miles. How many miles north and how many miles east has the boat traveled?

SOLUTION In the diagram of the situation, we put our N-S-E-W system at the boat's starting point (Figure 9).

Solving for x with a sine ratio and y with a cosine ratio and rounding our answers to three significant digits, we have

If $\sin 52°\,40' = \dfrac{x}{238}$ If $\cos 52°\,40' = \dfrac{y}{238}$

then $x = 238(0.7951)$ then $y = 238(0.6065)$

$= 189$ mi $= 144$ mi

Traveling 238 miles on a line with bearing N 52° 40′ E will get you to the same place as traveling 144 miles north and then 189 miles east. ■

PROBLEM 7
Rework Example 7 if both angles A and B are 60°.

EXAMPLE 7 Figure 10 is a diagram that shows how Diane estimates the height of a flagpole. She can't measure the distance between herself and the flagpole directly because there is a fence in the way. So she stands at point A facing the pole and finds the angle of elevation from point A to the top of the pole to be 61.7°. Then she turns 90° and walks 25.0 feet to point B, where she measures the angle between her path and a line from B to the base of the pole. She finds that angle is 54.5°. Use this information to find the height of the pole.

Figure 10

SOLUTION First we find x in right triangle ABC with a tangent ratio.

$$\tan 54.5° = \frac{x}{25.0}$$

$$x = 25.0 \tan 54.5°$$

$$= 25.0(1.4019)$$

$$= 35.0487 \text{ ft}$$

Without rounding x, we use it to find h in right triangle ACD using another tangent ratio.

$$\tan 61.7° = \frac{h}{35.0487}$$

$$h = 35.0487(1.8572)$$

$$= 65.1 \text{ ft} \quad \text{To three significant digits}$$

Note that if it weren't for the fence, she could measure x directly and use just one triangle to find the height of the flagpole.

PROBLEM 8
A helicopter is hovering over the desert when it develops mechanical problems and is forced to land. After landing, the pilot radios his position to a pair of radar stations located 27 miles apart along a straight road running north and south. The bearing of the helicopter from one station is N 15° E, and from the other it is S 21° E. How far east is the pilot from the road that connects the two radar stations?

EXAMPLE 8 A helicopter is hovering over the desert when it develops mechanical problems and is forced to land. After landing, the pilot radios his position to a pair of radar stations located 25 miles apart along a straight road running north and south. The bearing of the helicopter from one station is N 13° E, and from the other it is S 19° E. After doing a few trigonometric calculations, one of the stations instructs the pilot to walk due west for 3.5 miles to reach the road. Is this information correct?

SOLUTION Figure 11 is a three-dimensional diagram of the situation. The helicopter is hovering at point D and lands at point C. The radar stations are at A and B, respectively. Because the road runs north and south, the shortest distance from C to the road is due west of C toward point F. To see if the pilot has the correct information, we must find y, the distance from C to F.

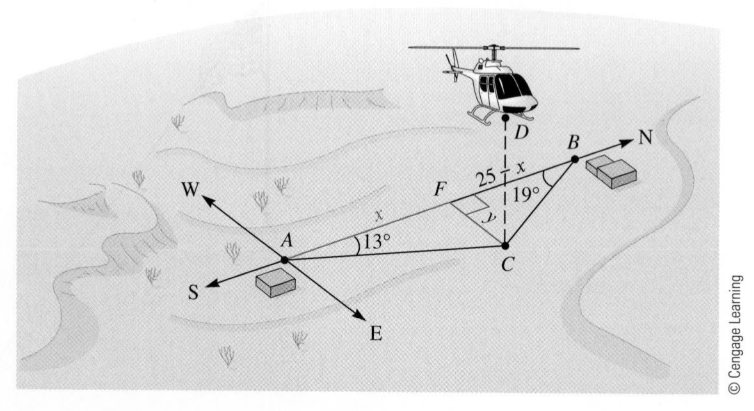

Figure 11

The radar stations are 25 miles apart, thus $AB = 25$. If we let $AF = x$, then $FB = 25 - x$. If we use cotangent ratios in triangles AFC and BFC, we will save ourselves some work.

In triangle AFC $\quad \cot 13° = \dfrac{x}{y}$

so $\quad x = y \cot 13°$

In triangle BFC $\quad \cot 19° = \dfrac{25 - x}{y}$

so $\quad 25 - x = y \cot 19°$

Solving this equation for x we have

$$-x = -25 + y \cot 19° \quad \text{Add } -25 \text{ to each side}$$
$$x = 25 - y \cot 19° \quad \text{Multiply each side by } -1$$

Next, we set our two values of x equal to each other.

$$x = x$$
$$y \cot 13° = 25 - y \cot 19°$$
$$y \cot 13° + y \cot 19° = 25 \quad \text{Add } y \cot 19° \text{ to each side}$$
$$y(\cot 13° + \cot 19°) = 25 \quad \text{Factor } y \text{ from each term}$$
$$y = \frac{25}{\cot 13° + \cot 19°} \quad \text{Divide by the coefficient of } y$$
$$= \frac{25}{4.3315 + 2.9042}$$
$$= \frac{25}{7.2357}$$
$$= 3.5 \text{ mi} \quad \text{To two significant digits}$$

The information given to the pilot is correct.

Getting Ready for Class

After reading through the preceding section, respond in your own words and in complete sentences.

a. What is an angle of elevation?
b. What is an angle of depression?
c. How do you define the bearing of line l?
d. Draw a diagram that shows that point B is N 40° E from point A.

2.4 PROBLEM SET

➤ CONCEPTS AND VOCABULARY

For Questions 1 through 4, fill in each blank with the appropriate word.

1. An angle measured upward from a horizontal line is called an angle of ____________, and an angle measured downward from a horizontal line is called an angle of ____________.
2. If an observer positioned at the vertex of an angle views an object in the direction of the nonhorizontal side of the angle, then this side is called the ________ ___ _________ of the observer.
3. The bearing of a line is the acute angle between the line and a ________-________ line.
4. The bearing of a line is always measured as an angle from the ________ or ________ rotating toward the _____ or ______.

➤ EXERCISES

Problems 5 through 8 refer to Figure 12, which is an illustration of a surveyor standing on top of a building. Use the figure to sketch and label each angle.

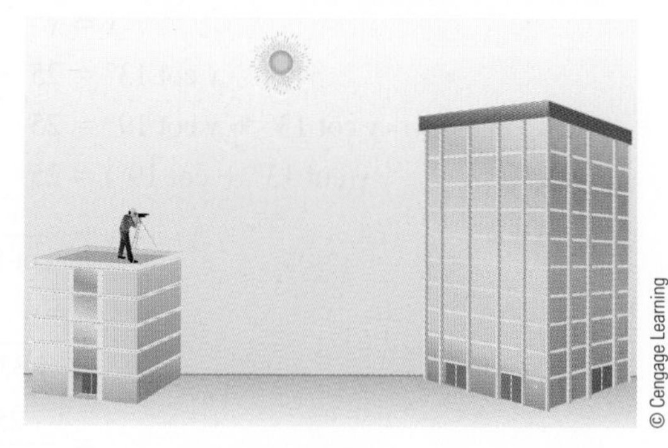

Figure 12

5. The angle of elevation from the eye of the surveyor to the top of the opposite building is 20°.
6. The angle of elevation from the eye of the surveyor to the sun is 50°.
7. The angle of depression from the eye of the surveyor to the base of the opposite building is 30°.
8. The angle of depression from the top of the opposite building to the base of the building the surveyor is on is 40°.

For Problems 9 through 12, draw a diagram that illustrates the given bearing.

9. The bearing of B from A is N 55° E.
10. The bearing of B from A is N 75° W.
11. The bearing of B from A is S 35° W.
12. The bearing of B from A is S 25° E.

Solve each of the following problems. In each case, be sure to make a diagram of the situation with all the given information labeled.

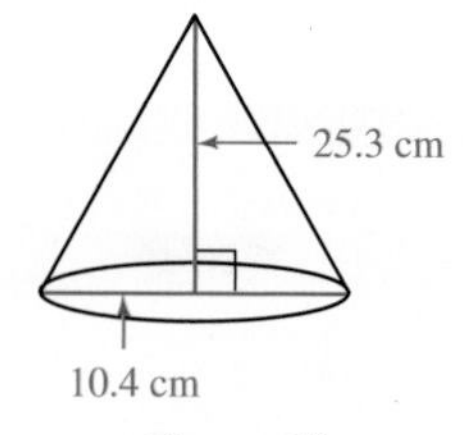

Figure 13

13. **Geometry** The two equal sides of an isosceles triangle are each 42 centimeters. If the base measures 32 centimeters, find the height and the measure of the two equal angles.
14. **Geometry** An equilateral triangle (one with all sides the same length) has an altitude of 4.3 inches. Find the length of the sides.
15. **Geometry** The height of a right circular cone is 25.3 centimeters. If the diameter of the base is 10.4 centimeters, what angle does the side of the cone make with the base (Figure 13)?
16. **Geometry** The diagonal of a rectangle is 348 millimeters, while the longer side is 278 millimeters. Find the shorter side of the rectangle and the angles the diagonal makes with the sides.
17. **Length of an Escalator** How long should an escalator be if it is to make an angle of 33° with the floor and carry people a vertical distance of 21 feet between floors?
18. **Height of a Hill** A road up a hill makes an angle of 5.1° with the horizontal. If the road from the bottom of the hill to the top of the hill is 2.5 miles long, how high is the hill?
19. **Length of a Rope** A 72.5-foot rope from the top of a circus tent pole is anchored to the ground 43.2 feet from the bottom of the pole. What angle does the rope make with the pole? (Assume the pole is perpendicular to the ground.)

20. Angle of a Ladder A ladder is leaning against the top of a 7.0-foot wall. If the bottom of the ladder is 4.5 feet from the wall, what is the angle between the ladder and the wall?

21. Angle of Elevation If a 73.0-foot flagpole casts a shadow 51.0 feet long, what is the angle of elevation of the sun (to the nearest tenth of a degree)?

22. Angle of Elevation If the angle of elevation of the sun is 63.4° when a building casts a shadow of 37.5 feet, what is the height of the building?

23. Angle of Depression A person standing 150 centimeters from a mirror notices that the angle of depression from his eyes to the bottom of the mirror is 12°, while the angle of elevation to the top of the mirror is 11°. Find the vertical dimension of the mirror (Figure 14).

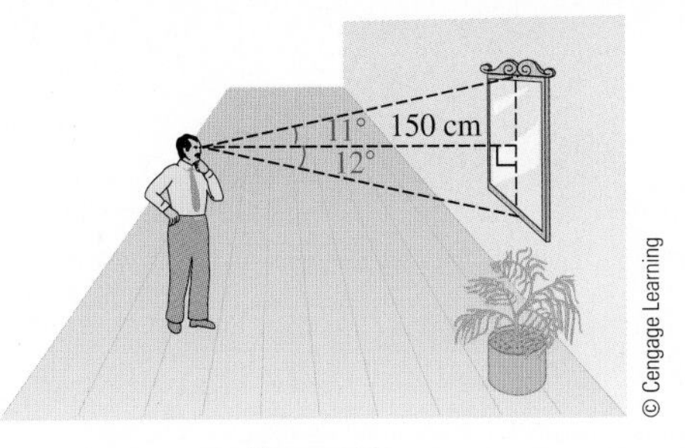

Figure 14

24. Width of a Sand Pile A person standing on top of a 15-foot high sand pile wishes to estimate the width of the pile. He visually locates two rocks on the ground below at the base of the sand pile. The rocks are on opposite sides of the sand pile, and he and the two rocks are in the same vertical plane. If the angles of depression from the top of the sand pile to each of the rocks are 27° and 19°, how far apart are the rocks?

Figure 15 shows the topographic map we used in Example 4 of this section. Recall that Stacey is at position S and Amy is at position A. In Figure 15, Travis, a third hiker, is at position T.

25. Topographic Map Reading If the distance between A and T on the map in Figure 15 is 0.50 inch, find each of the following:

a. the horizontal distance between Amy and Travis

b. the difference in elevation between Amy and Travis

c. the angle of elevation from Travis to Amy

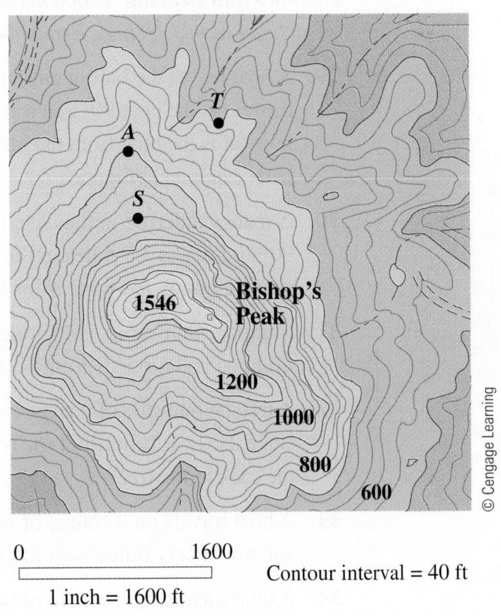

Figure 15

26. Topographic Map Reading If the distance between S and T on the map in Figure 15 is $\frac{5}{8}$ inch, find each of the following:

a. the horizontal distance between Stacey and Travis

b. the difference in elevation between Stacey and Travis

c. the angle of elevation from Travis to Stacey

27. Height of a Door From a point on the floor the angle of elevation to the top of a door is 47°, while the angle of elevation to the ceiling above the door is 59°. If the ceiling is 9.8 feet above the floor, what is the vertical dimension of the door (Figure 16)?

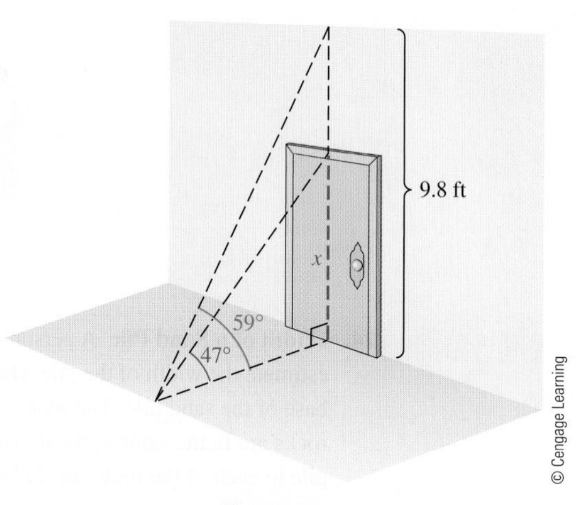

Figure 16

28. Height of a Building A man standing on the roof of a building 60.0 feet high looks down to the building next door. He finds the angle of depression to the roof of that building from the roof of his building to be 34.5°, while the angle of depression from the roof of his building to the bottom of the building next door is 63.2°. How tall is the building next door?

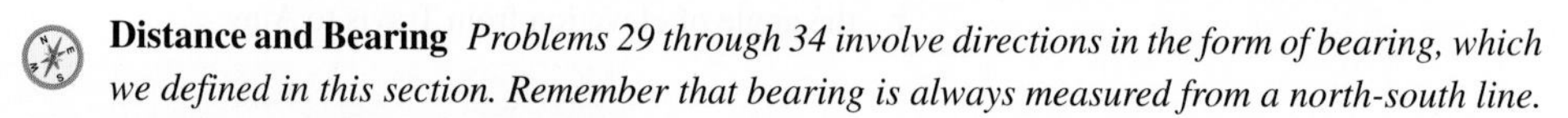

Distance and Bearing *Problems 29 through 34 involve directions in the form of bearing, which we defined in this section. Remember that bearing is always measured from a north-south line.*

29. A boat leaves the harbor entrance and travels 25 miles in the direction N 42° E. The captain then turns the boat 90° and travels another 18 miles in the direction S 48° E. At that time, how far is the boat from the harbor entrance, and what is the bearing of the boat from the harbor entrance (Figure 17)?

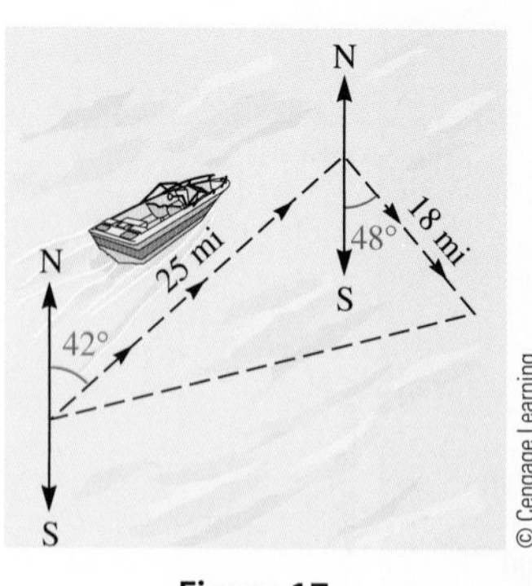

Figure 17

30. A man wandering in the desert walks 2.3 miles in the direction S 31° W. He then turns 90° and walks 3.5 miles in the direction N 59° W. At that time, how far is he from his starting point, and what is his bearing from his starting point?

31. Lompoc, California, is 18 miles due south of Nipomo. Buellton, California, is due east of Lompoc and S 65° E from Nipomo. How far is Lompoc from Buellton?

32. A tree on one side of a river is due west of a rock on the other side of the river. From a stake 21.0 yards north of the rock, the bearing of the tree is S 18.2° W. How far is it from the rock to the tree?

33. A boat travels on a course of bearing N 37° 10′ W for 79.5 miles. How many miles north and how many miles west has the boat traveled?

34. A boat travels on a course of bearing S 63° 50′ E for 101 miles. How many miles south and how many miles east has the boat traveled?

35. Distance In Figure 18, a person standing at point A notices that the angle of elevation to the top of the antenna is $47° 30'$. A second person standing 33.0 feet farther from the antenna than the person at A finds the angle of elevation to the top of the antenna to be $42° 10'$. How far is the person at A from the base of the antenna?

36. Height of an Obelisk Two people decide to find the height of an obelisk. They position themselves 25 feet apart in line with, and on the same side of, the obelisk. If they find that the angles of elevation from the ground where they are standing to the top of the obelisk are 65° and 44°, how tall is the obelisk?

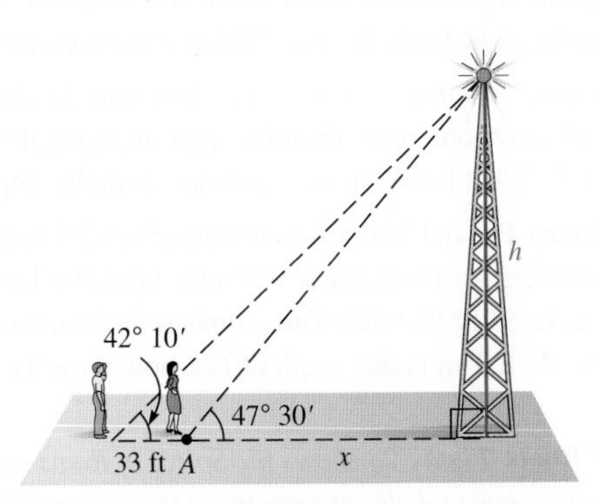

Figure 18

Figure 19

37. Height of a Tree An ecologist wishes to find the height of a redwood tree that is on the other side of a creek, as shown in Figure 19. From point A he finds that the angle of elevation to the top of the tree is 10.7°. He then walks 24.8 feet at a right angle from point A to point B. There he finds that the angle between AB and a line extending from B to the tree is 86.6°. What is the height of the tree?

38. Rescue A helicopter makes a forced landing at sea. The last radio signal received at station C gives the bearing of the helicopter from C as N 57.5° E at an altitude of 426 feet. An observer at C sights the helicopter and gives $\angle DCB$ as 12.3°. How far will a rescue boat at A have to travel to reach any survivors at B (Figure 20)?

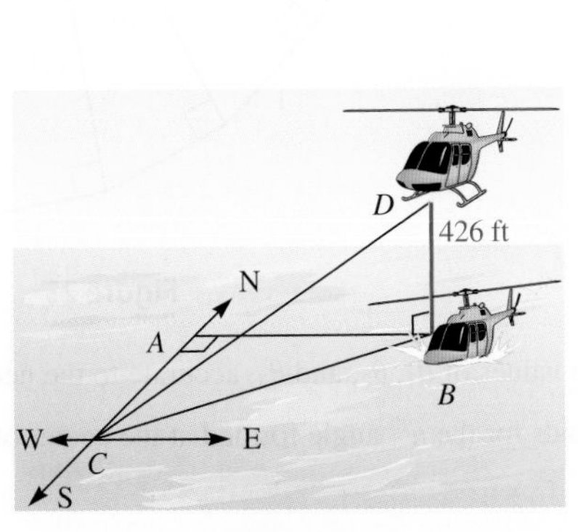

Figure 20

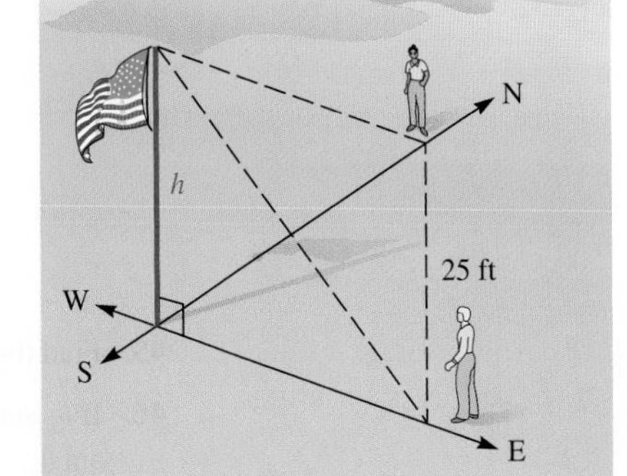

Figure 21

39. Height of a Flagpole Two people decide to estimate the height of a flagpole. One person positions himself due north of the pole and the other person stands due east of the pole. If the two people are the same distance from the pole and 25 feet from each other, find the height of the pole if the angle of elevation from the ground to the top of the pole at each person's position is 56° (Figure 21).

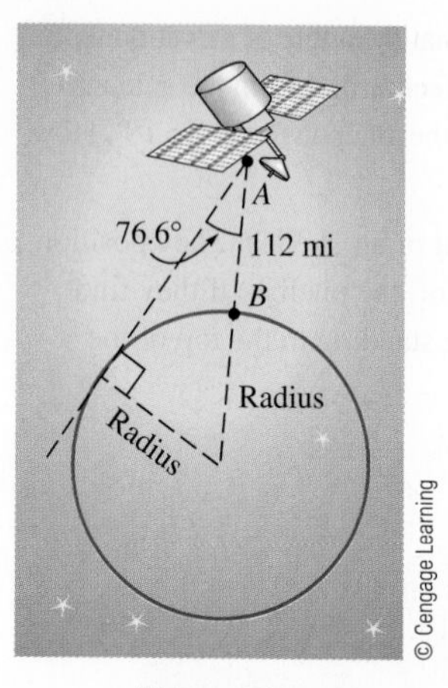

Figure 22

40. Height of a Tree To estimate the height of a tree, one person positions himself due south of the tree, while another person stands due east of the tree. If the two people are the same distance from the tree and 35 feet from each other, what is the height of the tree if the angle of elevation from the ground at each person's position to the top of the tree is 48°?

41. Radius of Earth A satellite is circling 112 miles above Earth, as shown in Figure 22. When the satellite is directly above point *B*, angle *A* is found to be 76.6°. Use this information to find the radius of Earth.

42. Distance Suppose Figure 22 is an exaggerated diagram of a plane flying above Earth. If the plane is 4.55 miles above Earth and the radius of Earth is 3,960 miles, how far is it from the plane to the horizon? What is the measure of angle *A*?

43. Distance A ship is anchored off a long straight shoreline that runs north and south. From two observation points 15 miles apart on shore, the bearings of the ship are N 31° E and S 53° E. What is the shortest distance from the ship to the shore?

44. Distance Pat and Tim position themselves 2.5 miles apart to watch a missile launch from Vandenberg Air Force Base. When the missile is launched, Pat estimates its bearing from him to be S 75° W, while Tim estimates the bearing of the missile from his position to be N 65° W. If Tim is due south of Pat, how far is Tim from the missile when it is launched?

Spiral of Roots *Figure 23 shows the Spiral of Roots we mentioned in the previous chapter. Notice that we have labeled the angles at the center of the spiral with* θ_1, θ_2, θ_3, *and so on.*

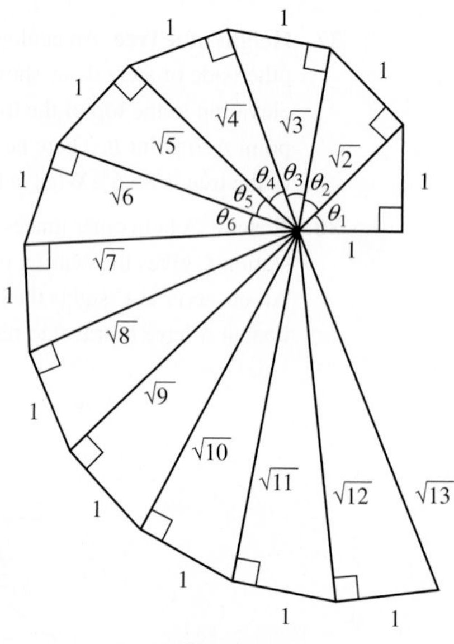

Figure 23

45. Find the values of θ_1, θ_2, and θ_3, accurate to the nearest hundredth of a degree.

46. If θ_n stands for the nth angle formed at the center of the Spiral of Roots, find a formula for $\sin \theta_n$.

➤ REVIEW PROBLEMS

The following problems review material we covered in Section 1.5.

47. Expand and simplify: $(\sin \theta - \cos \theta)^2$

48. Subtract: $\dfrac{1}{\cos \theta} - \cos \theta$

Show that each of the following statements is true by transforming the left side of each one into the right side.

49. $\sin\theta\cot\theta = \cos\theta$

50. $\cos\theta\csc\theta\tan\theta = 1$

51. $\dfrac{\sin\theta}{\tan\theta} = \csc\theta$

52. $(1 - \cos\theta)(1 + \cos\theta) = \sin^2\theta$

53. $\sec\theta - \cos\theta = \dfrac{\sin^2\theta}{\cos\theta}$

54. $1 - \dfrac{\cos\theta}{\sec\theta} = \sin^2\theta$

➤ EXTENDING THE CONCEPTS

55. One of the items we discussed in this section was topographic maps. The process of making one of these maps is an interesting one. It involves aerial photography and different colored projections of the resulting photographs. Research the process used to draw the contour lines on a topographic map, and then give a detailed explanation of that process.

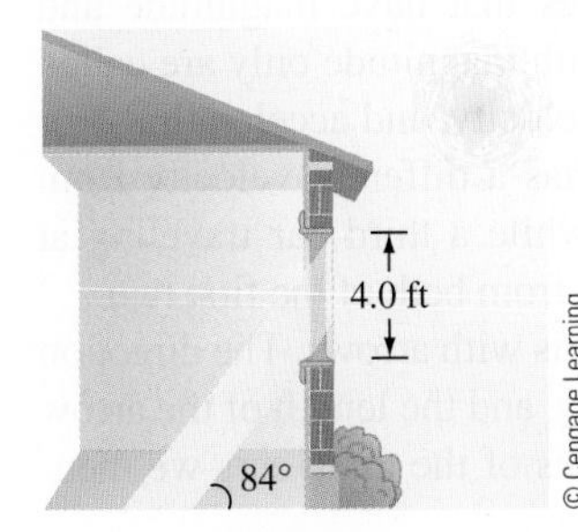

Figure 24

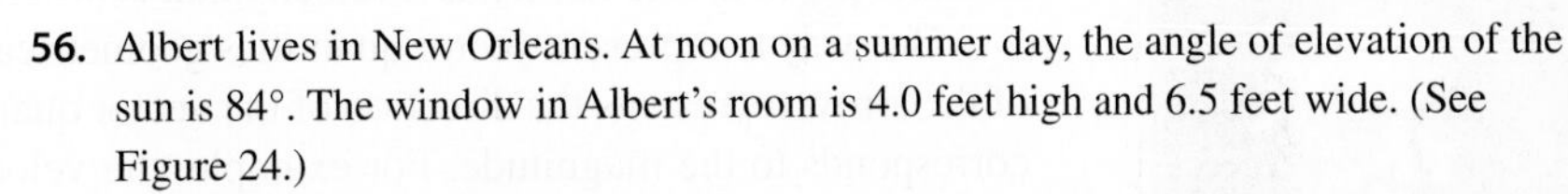

56. Albert lives in New Orleans. At noon on a summer day, the angle of elevation of the sun is 84°. The window in Albert's room is 4.0 feet high and 6.5 feet wide. (See Figure 24.)

a. Calculate the area of the floor surface in Albert's room that is illuminated by the sun when the angle of elevation of the sun is 84°.

b. One winter day, the angle of elevation of the sun outside Albert's window is 37°. Will the illuminated area of the floor in Albert's room be greater on the summer day, or on the winter day?

➤ LEARNING OBJECTIVES ASSESSMENT

These questions are available for instructors to help assess if you have successfully met the learning objectives for this section.

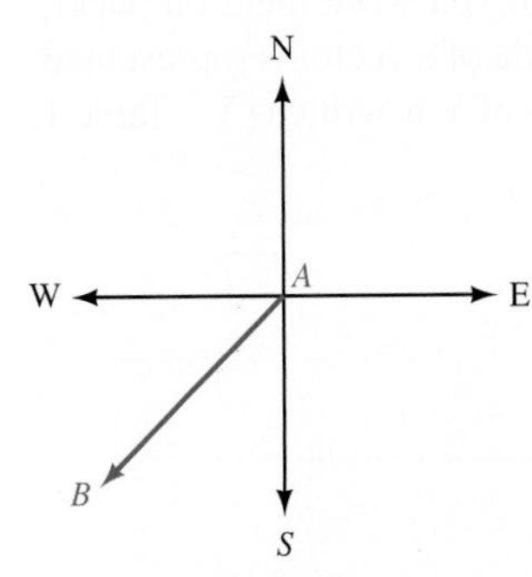

Figure 25

57. What is the bearing of B from A in Figure 25?

a. N 135° W **b.** W 45° S **c.** S 45° W **d.** N 225° E

58. If the angle of elevation to the sun is 74.3° when a flagpole casts a shadow of 22.5 feet, what is the height of the flagpole?

a. 63.2 feet **b.** 79.5 feet **c.** 83.1 feet **d.** 80.0 feet

59. A ship is anchored off a long straight shoreline that runs north and south. From two observation points 4.5 miles apart on shore, the bearings of the ship are S 73° W and N 17° W. What is the distance from the ship to the northernmost observation point?

a. 4.3 mi **b.** 14.7 mi **c.** 4.7 mi **d.** 1.3 mi

60. To estimate the height of a tree, one person stands due north of the tree and a second person stands due east of the tree. If the two people are the same distance from the tree and 35 feet from each other, find the height of the tree if the angle of elevation from the ground to the top of the tree at each person's position is 65°.

a. 53 ft **b.** 110 ft **c.** 75 ft **d.** 130 ft

SECTION 2.5 Vectors: A Geometric Approach

LEARNING OBJECTIVES

1. Find the magnitude of the horizontal and vertical vector components for a vector.
2. Find the magnitude of a vector and the angle it makes with the positive x-axis.
3. Solve an applied problem using vectors.
4. Compute the work done by a force with a given distance.

By most accounts, the study of vectors is based on the work of Irish mathematician Sir William Hamilton (1805–1865). Hamilton was actually studying complex numbers (a topic we will cover in Chapter 8) when he made the discoveries that led to what we now call vectors.

Today, vectors are treated both algebraically and geometrically. In this section, we will focus our attention on the geometric representation of vectors. We will cover the algebraic approach later in the book in Section 7.5. We begin with a discussion of vector quantities and scalar quantities.

Many of the quantities that describe the world around us have both magnitude and direction, while others have only magnitude. Quantities that have magnitude and direction are called *vector quantities,* while quantities with magnitude only are called *scalars.* Some examples of vector quantities are force, velocity, and acceleration. For example, a car traveling 50 miles per hour due south has a different velocity from another car traveling due north at 50 miles per hour, while a third car traveling at 25 miles per hour due north has a velocity that is different from both of the first two.

One way to represent vector quantities geometrically is with arrows. The direction of the arrow represents the direction of the vector quantity, and the length of the arrow corresponds to the magnitude. For example, the velocities of the three cars we mentioned above could be represented as in Figure 1.

William Rowan Hamilton

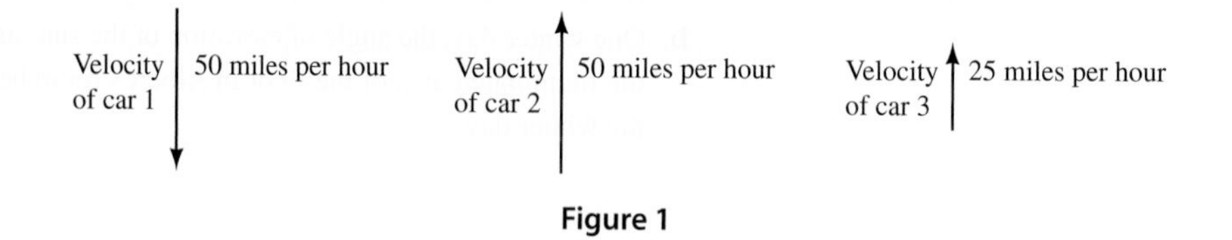

Figure 1

NOTATION

To distinguish between vectors and scalars, we will write the letters used to represent vectors with boldface type, such as **U** or **V**. (When you write them on paper, put an arrow above them like this: $\vec{U}$ or $\vec{V}$.) The magnitude of a vector is represented with absolute value symbols. For example, the magnitude of **V** is written $|\mathbf{V}|$. Table 1 illustrates further.

TABLE 1

Notation	The quantity is
$\mathbf{V}$	a vector
$\vec{V}$	a vector
$\overrightarrow{AB}$	a vector
x	a scalar
$\lvert\mathbf{V}\rvert$	the magnitude of vector **V**, a scalar

Zero Vector

A vector having a magnitude of zero is called a *zero vector* and is denoted by **0**. A zero vector has no defined direction.

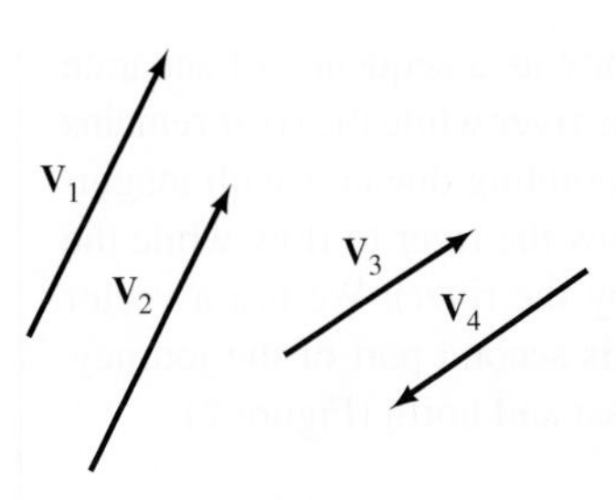

Figure 2

Equality for Vectors

The position of a vector in space is unimportant. Two vectors are equivalent if they have the same magnitude and direction.

In Figure 2, $\mathbf{V}_1 = \mathbf{V}_2 \neq \mathbf{V}_3$. The vectors $\mathbf{V}_1$ and $\mathbf{V}_2$ are equivalent because they have the same magnitude and the same direction. Notice also that $\mathbf{V}_3$ and $\mathbf{V}_4$ have the same magnitude but opposite directions. This means that $\mathbf{V}_4$ is the opposite of $\mathbf{V}_3$, or $\mathbf{V}_4 = -\mathbf{V}_3$.

Addition and Subtraction of Vectors

The sum of the vectors **U** and **V**, written $\mathbf{U} + \mathbf{V}$, is called the *resultant vector.* It is the vector that extends from the tail of **U** to the tip of **V** when the tail of **V** is placed at the tip of **U**, as illustrated in Figure 3. Note that this diagram shows the resultant vector to be a diagonal in the parallelogram that has **U** and **V** as adjacent sides. This being the case, we could also add the vectors by putting the tails of **U** and **V** together to form adjacent sides of that same parallelogram, as shown in Figure 4. In either case, the resultant vector is the diagonal that starts at the tail of **U**.

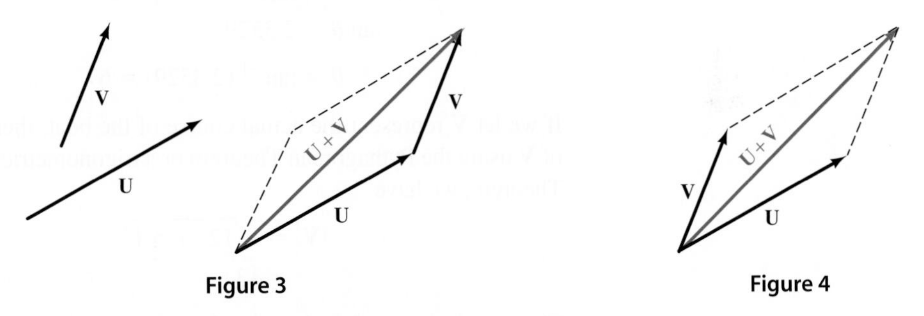

Figure 3

Figure 4

To subtract one vector from another, we can add its opposite. That is,

$$\mathbf{U} - \mathbf{V} = \mathbf{U} + (-\mathbf{V})$$

If **U** and **V** are the vectors shown in Figure 3, then their difference, $\mathbf{U} - \mathbf{V}$, is shown in Figure 5. Another way to find $\mathbf{U} - \mathbf{V}$ is to put the tails of **U** and **V** together and then draw a vector from the tip of **V** to the tip of **U**, completing a triangle as shown in Figure 6.

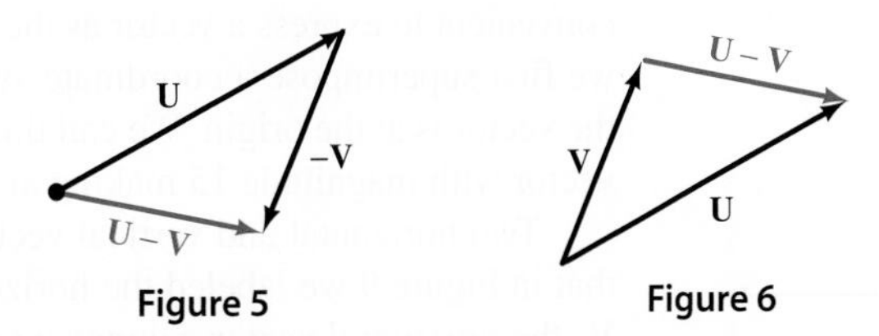

Figure 5

Figure 6

EXAMPLE 1 A boat is crossing a river that runs due north. The boat is pointed due east and is moving through the water at 12 miles per hour. If the current of the river is a constant 5.1 miles per hour, find the actual course of the boat through the water to two significant digits.

PROBLEM 1
A boat is crossing a river that runs due north. The boat is pointed due east and is moving through the water at 15 miles per hour. If the current of the river is a constant 4.8 miles per hour, find the actual course of the boat through the water to two significant digits.

SOLUTION Problems like this are a little difficult to read the first time they are encountered. Even though the boat is "headed" due east, as it travels through the water the water itself is moving northward, so it is actually on a course that will take it east and a little north.

It may help to imagine the first hour of the journey as a sequence of separate events. First, the boat travels 12 miles directly across the river while the river remains still. We can represent this part of the trip by a vector pointing due east with magnitude 12. Then the boat turns off its engine, and we allow the river to flow while the boat remains still (although the boat will be carried by the river). We use a vector pointing due north with magnitude 5.1 to represent this second part of the journey. The end result is that each hour the boat travels both east and north (Figure 7).

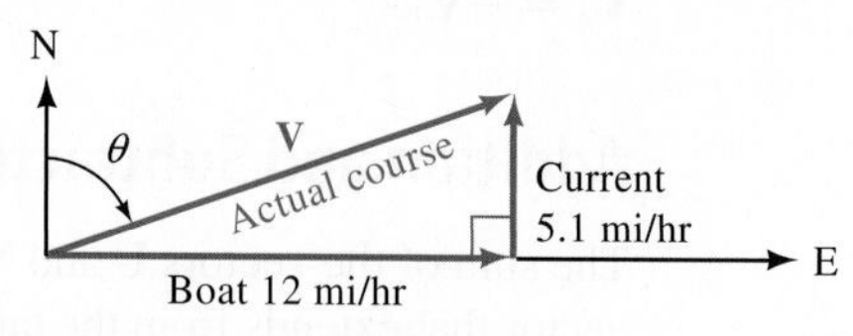

Figure 7

We find θ using a tangent ratio. Note that the angle between the vector representing the actual course and the current vector is also θ.

$$\tan \theta = \frac{12}{5.1}$$

$$\tan \theta = 2.3529$$

$$\theta = \tan^{-1}(2.3529) = 67° \quad \text{To the nearest degree}$$

If we let $\mathbf{V}$ represent the actual course of the boat, then we can find the magnitude of $\mathbf{V}$ using the Pythagorean Theorem or a trigonometric ratio. Using the Pythagorean Theorem, we have

$$|\mathbf{V}| = \sqrt{12^2 + 5.1^2}$$

$$= 13 \quad \text{To two significant digits}$$

The actual course of the boat is 13 miles per hour at N 67° E. That is, the vector $\mathbf{V}$, which represents the motion of the boat with respect to the banks of the river, has a magnitude of 13 miles per hour and a direction of N 67° E.

Horizontal and Vertical Vector Components

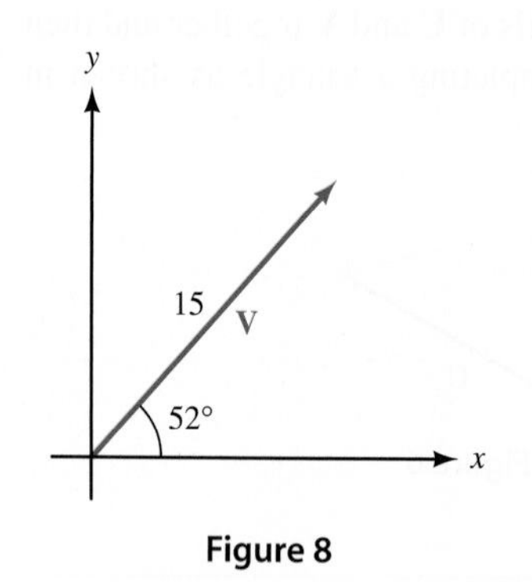

Figure 8

In Example 1, we saw that $\mathbf{V}$ was the sum of a vector pointing east (in a horizontal direction) and a second vector pointing north (in a vertical direction). Many times it is convenient to express a vector as the sum of a horizontal and vertical vector. To do so, we first superimpose a coordinate system on the vector in question so that the tail of the vector is at the origin. We call this *standard position* for a vector. Figure 8 shows a vector with magnitude 15 making an angle of 52° with the horizontal.

Two horizontal and vertical vectors whose sum is $\mathbf{V}$ are shown in Figure 9. Note that in Figure 9 we labeled the horizontal vector as $\mathbf{V}_x$ and the vertical as $\mathbf{V}_y$. We call $\mathbf{V}_x$ the *horizontal vector component* of $\mathbf{V}$ and $\mathbf{V}_y$ the *vertical vector component* of $\mathbf{V}$. We can find the magnitudes of these vectors by using sine and cosine ratios.

$$|\mathbf{V}_x| = |\mathbf{V}| \cos 52°$$

$$= 15(0.6157)$$

$$= 9.2 \text{ to two significant digits}$$

$$|\mathbf{V}_y| = |\mathbf{V}| \sin 52°$$

$$= 15(0.7880)$$

$$= 12 \text{ to two significant digits}$$

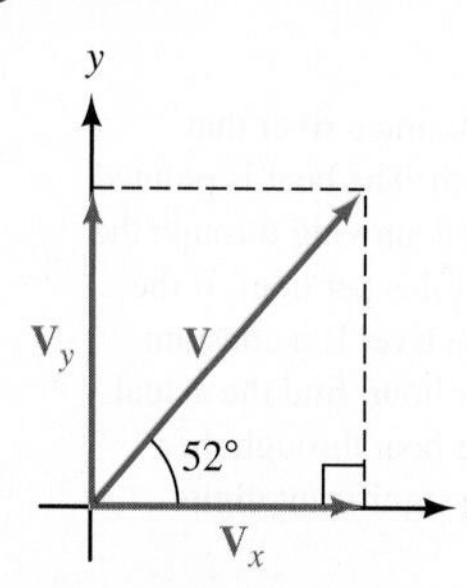

Figure 9

This leads us to the following general result.

VECTOR COMPONENTS

If **V** is a vector in standard position and θ is the angle measured from the positive x-axis to **V**, then the magnitudes of the horizontal and vertical vector components of **V** are given by

$$|\mathbf{V}_x| = |\mathbf{V}| \cdot |\cos\theta| \quad \text{and} \quad |\mathbf{V}_y| = |\mathbf{V}| \cdot |\sin\theta|$$

PROBLEM 2
The human cannonball is shot from a cannon with an initial velocity of 48 miles per hour at an angle of 56° from the horizontal. Find the magnitudes of the horizontal and vertical vector components of the velocity vector.

EXAMPLE 2 The human cannonball is shot from a cannon with an initial velocity of 53 miles per hour at an angle of 60° from the horizontal. Find the magnitudes of the horizontal and vertical vector components of the velocity vector.

SOLUTION Figure 10 is a diagram of the situation.

The magnitudes of $\mathbf{V}_x$ and $\mathbf{V}_y$ from Figure 10 to two significant digits are as follows:

$$|\mathbf{V}_x| = 53\cos 60°$$
$$= 27 \text{ mi/hr}$$
$$|\mathbf{V}_y| = 53\sin 60°$$
$$= 46 \text{ mi/hr}$$

Figure 10

The human cannonball has a horizontal velocity of 27 miles per hour and an initial vertical velocity of 46 miles per hour. ■

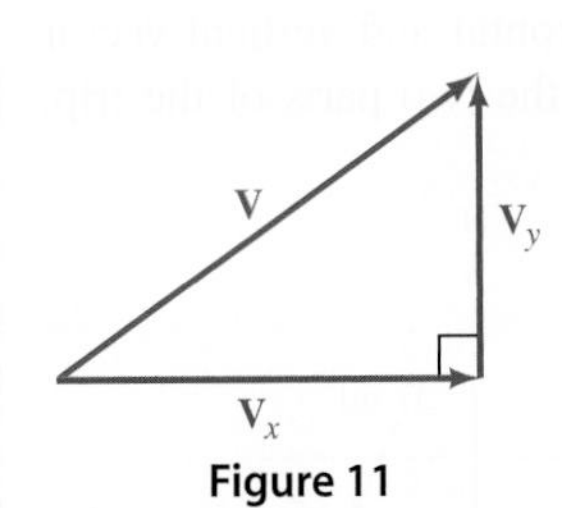

Figure 11

The magnitude of a vector can be written in terms of the magnitude of its horizontal and vertical vector components (Figure 11).

By the Pythagorean Theorem we have

$$|\mathbf{V}| = \sqrt{|\mathbf{V}_x|^2 + |\mathbf{V}_y|^2}$$

PROBLEM 3
An arrow is shot into the air so that its horizontal velocity is 21 feet per second and its vertical velocity is 18 feet per second. Find the velocity of the arrow.

EXAMPLE 3 An arrow is shot into the air so that its horizontal velocity is 25 feet per second and its vertical velocity is 15 feet per second. Find the velocity of the arrow.

SOLUTION Figure 12 shows the velocity vector along with the angle of elevation of the velocity vector.

The magnitude of the velocity is given by

$$|\mathbf{V}| = \sqrt{25^2 + 15^2}$$
$$= 29 \text{ ft/sec} \qquad \text{To the nearest whole number}$$

We can find the angle of elevation using a tangent ratio.

$$\tan\theta = \frac{|\mathbf{V}_y|}{|\mathbf{V}_x|} = \frac{15}{25} = 0.6$$
$$\theta = \tan^{-1}(0.6) = 31° \qquad \text{To the nearest degree}$$

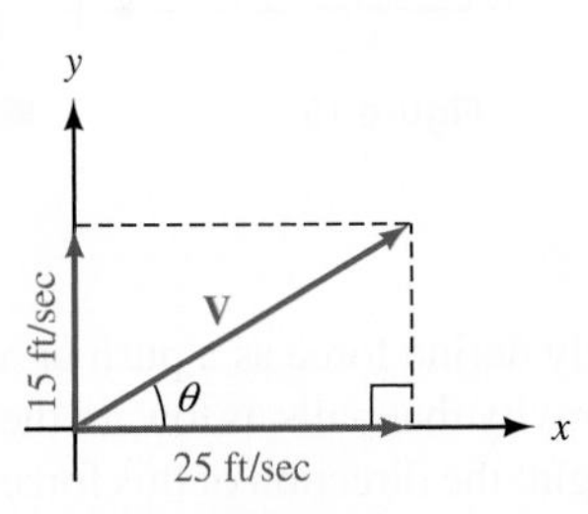

Figure 12

The arrow was shot into the air at 29 feet per second at an angle of elevation of 31°. ■

Using Technology Vector Components

Most graphing calculators have commands that can be used to find the magnitude, angle, or components of a vector. Figure 13 shows how the **P►Rx** and **P►Ry** commands from the ANGLE menu are used on the TI-84 to find the horizontal and vertical vector components for the vector from Example 2. Make sure the calculator is set to degree mode.

In Figure 14 we found the magnitude and direction for the vector from Example 3 using the **R►Pr** and **R►Pθ** commands. Depending on the calculator, you may have a single command that allows you to find both values at once.

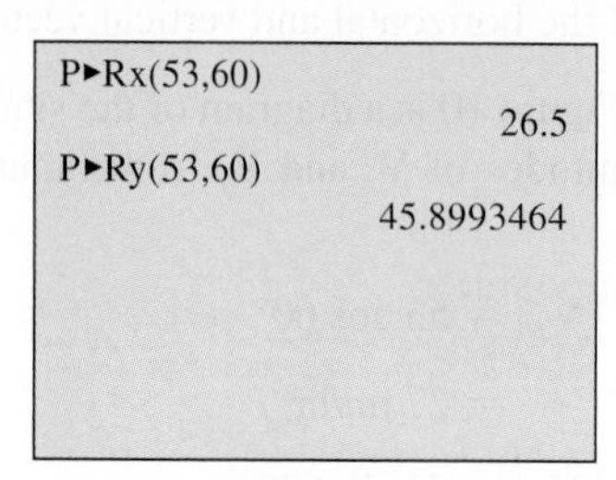

Figure 13

R►Pr(25,15)
29.15475947
R►Pθ(25,15)
30.96375653

Figure 14

PROBLEM 4
A boat travels 68 miles on a course of bearing N 25° E and then changes its course to travel 32 miles at N 60° E. How far north and how far east has the boat traveled on this 100-mile trip?

EXAMPLE 4 A boat travels 72 miles on a course of bearing N 27° E and then changes its course to travel 37 miles at N 55° E. How far north and how far east has the boat traveled on this 109-mile trip?

SOLUTION We can solve this problem by representing each part of the trip with a vector and then writing each vector in terms of its horizontal and vertical vector components. Figure 15 shows the vectors that represent the two parts of the trip. As Figure 15 indicates, the total distance traveled east is given by the sum of the horizontal components, while the total distance traveled north is given by the sum of the vertical components.

Total distance traveled east

$= |\mathbf{U}_x| + |\mathbf{V}_x|$
$= 72 \cos 63° + 37 \cos 35°$
$= 63 \text{ mi}$

Total distance traveled north

$= |\mathbf{U}_y| + |\mathbf{V}_y|$
$= 72 \sin 63° + 37 \sin 35°$
$= 85 \text{ mi}$

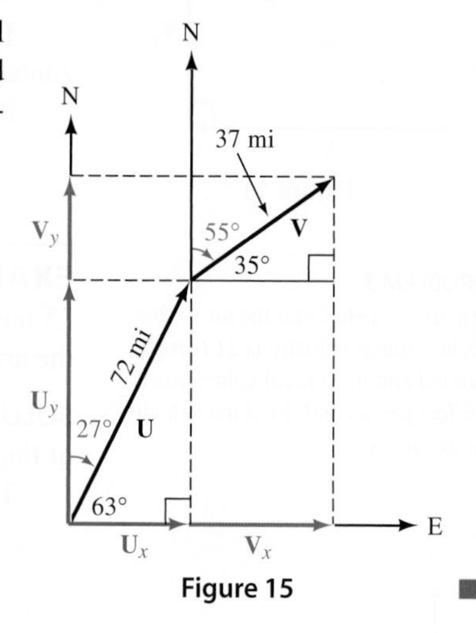

Figure 15

Force

Another important vector quantity is *force*. We can loosely define force as a push or a pull. The most intuitive force in our lives is the force of gravity that pulls us toward the center of the earth. The magnitude of this force is our weight; the direction of this force is always straight down toward the center of the earth.

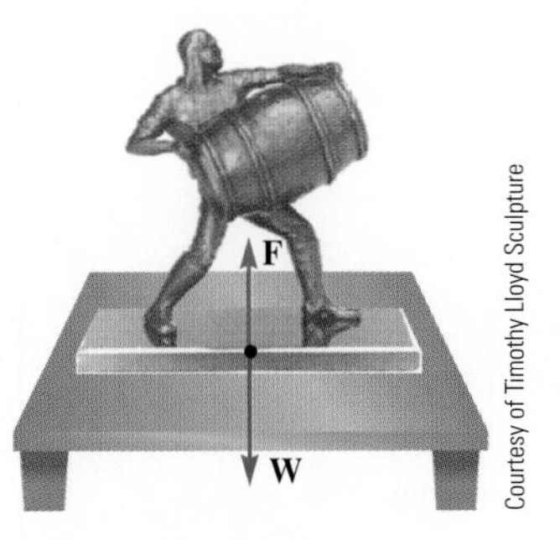

Courtesy of Timothy Lloyd Sculpture

Figure 16

Imagine a 10-pound bronze sculpture sitting on a coffee table. The force of gravity pulls the sculpture downward with a force of magnitude 10 pounds. At the same time, the table pushes the sculpture upward with a force of magnitude 10 pounds. The net result is that the sculpture remains motionless; the two forces, represented by vectors, add to the zero vector (Figure 16).

Although there may be many forces acting on an object at the same time, if the object is stationary, the sum of the forces must be **0**. This leads us to our next definition.

DEFINITION ■ STATIC EQUILIBRIUM

When an object is stationary (at rest), we say it is in a state of *static equilibrium.* When an object is in this state, the sum of the forces acting on the object must be equal to the zero vector **0**.

PROBLEM 5

Jacob is 4 years old and weighs 32.0 pounds. He is sitting on a swing when his brother Aaron pulls him and the swing back horizontally through an angle of 35.0° and then stops. Find the tension in the ropes of the swing and the magnitude of the force exerted by Aaron.

EXAMPLE 5 Danny is 5 years old and weighs 42.0 pounds. He is sitting on a swing when his sister Stacey pulls him and the swing back horizontally through an angle of 30.0° and then stops. Find the tension in the ropes of the swing and the magnitude of the force exerted by Stacey. (Figure 17 is a diagram of the situation.)

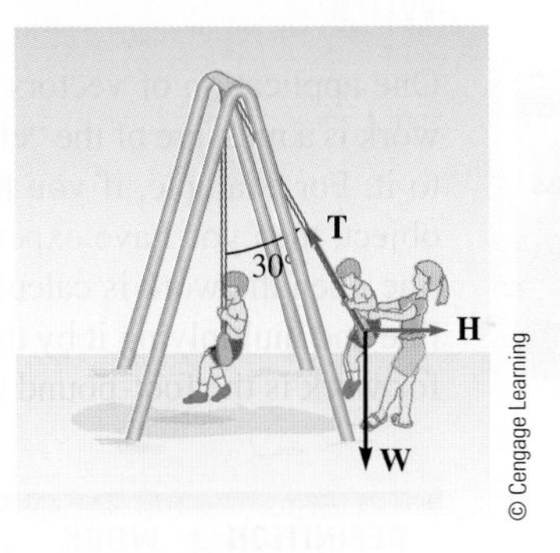

© Cengage Learning

Figure 17

SOLUTION As you can see from Figure 17, there are three forces acting on Danny (and the swing), which we have labeled **W**, **H**, and **T**. The vector **W** is due to the force of gravity, pulling him toward the center of the earth. Its magnitude is $|\mathbf{W}| = 42.0$ pounds, and its direction is straight down. The vector **H** represents the force with which Stacey is pulling Danny horizontally, and **T** is the force acting on Danny in the direction of the ropes. We call this force the *tension* in the ropes.

If we rearrange the vectors from the diagram in Figure 17, we can get a better picture of the situation. Since Stacey is holding Danny in the position shown in Figure 17, he is in a state of static equilibrium. Therefore, the sum of the forces acting on him is **0**. We add the three vectors **W**, **H**, and **T** using the tip-to-tail rule as described on page 99. The resultant vector, extending from the tail of **W** to the tip of **T**, must have a length of 0. This means that the tip of **T** has to coincide with the tail of **W**, forming a right triangle, as shown in Figure 18.

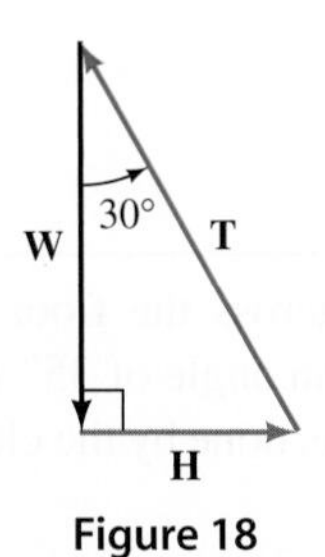

Figure 18

The lengths of the sides of the right triangle shown in Figure 18 are given by the magnitudes of the vectors. We use right triangle trigonometry to find the magnitude of **T**.

$$\cos 30.0° = \frac{|\mathbf{W}|}{|\mathbf{T}|} \qquad \text{Definition of cosine}$$

$$|\mathbf{T}| = \frac{|\mathbf{W}|}{\cos 30.0°} \qquad \text{Solve for } |\mathbf{T}|$$

$$= \frac{42.0}{0.8660} \qquad \text{The magnitude of } \mathbf{W} \text{ is } 42.0$$

$$= 48.5 \text{ lb} \qquad \text{To three significant digits}$$

Next, let's find the magnitude of the force with which Stacey pulls on Danny to keep him in static equilibrium.

$$\tan 30.0° = \frac{|\mathbf{H}|}{|\mathbf{W}|} \qquad \text{Definition of tangent}$$

$$|\mathbf{H}| = |\mathbf{W}| \tan 30.0° \qquad \text{Solve for } |\mathbf{H}|$$

$$= 42.0(0.5774)$$

$$= 24.2 \text{ lb} \qquad \text{To three significant digits}$$

Stacey must pull horizontally with a force of magnitude 24.2 pounds to hold Danny at an angle of 30.0° from vertical.

Work

© Chris Mcpherson/Getty Images

One application of vectors that is related to the concept of force is *work.* Intuitively, work is a measure of the "effort" expended when moving an object by applying a force to it. For example, if you have ever had to push a stalled automobile or lift a heavy object, then you have experienced work. Assuming the object moves along a straight line, then the work is calculated by finding the component of the force parallel to this line and multiplying it by the distance the object is moved. A common unit of measure for work is the foot-pound (ft-lb).

DEFINITION ■ WORK

If a constant force **F** is applied to an object and moves the object in a straight line a distance d, then the *work* W performed by the force is

$$W = (|\mathbf{F}| \cos \theta) \cdot d$$

where θ is the angle between the force **F** and the line of motion of the object.

Our next example illustrates this situation.

PROBLEM 6
A shipping clerk pushes a heavy package across the floor. He applies a force of 55 pounds in a downward direction, making an angle of 38° with the horizontal. If the package is moved 22 feet, how much work is done by the clerk?

EXAMPLE 6 A shipping clerk pushes a heavy package across the floor. He applies a force of 64 pounds in a downward direction, making an angle of 35° with the horizontal. If the package is moved 25 feet, how much work is done by the clerk?

SOLUTION A diagram of the problem is shown in Figure 19.

Because the package moves in a horizontal direction, and not in the direction of the force, we must first find the amount of force that is directed horizontally. This will be the magnitude of the horizontal vector component of **F**.

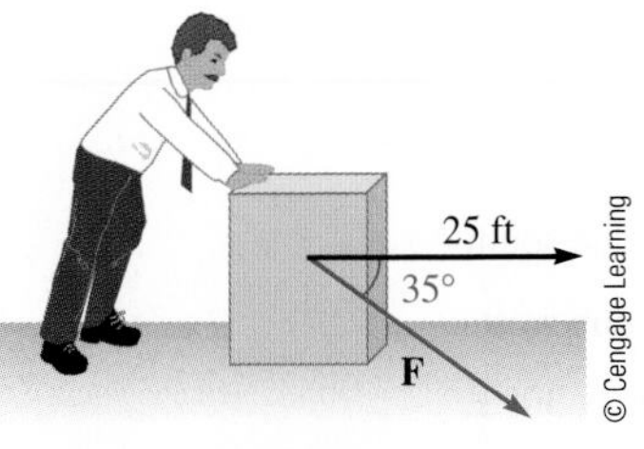

Figure 19

$$|\mathbf{F}_x| = |\mathbf{F}|\cos 35° = 64 \cos 35° \text{ lb}$$

To find the work, we multiply this value by the distance the package moves.

$$\text{Work} = (64 \cos 35°)(25)$$
$$= 1{,}300 \text{ ft-lb} \quad \text{To two significant digits}$$

1,300 foot-pounds of work are performed by the shipping clerk in moving the package.

Getting Ready for Class

After reading through the preceding section, respond in your own words and in complete sentences.

a. What is a vector?
b. How are a vector and a scalar different?
c. How do you add two vectors geometrically?
d. Explain what is meant by static equilibrium.

2.5 PROBLEM SET

➤ CONCEPTS AND VOCABULARY

For Questions 1 through 8, fill in each blank with the appropriate word or expression.

1. A quantity having only a magnitude is called a __________, and a quantity having both a magnitude and direction is called a __________.

2. Two vectors are equivalent if they have the same _______ and _________.

3. The sum of two vectors is called the ___________, and it can be represented geometrically as the __________ of a parallelogram having the two original vectors as adjacent sides.

4. A vector is in standard position if the _____ of the vector is placed at the ______ of a rectangular coordinate system.

5. Every vector $\mathbf{V}$ can be expressed as a sum, $\mathbf{V} = \mathbf{V}_x + \mathbf{V}_y$, where $\mathbf{V}_x$ is called the _________ vector ________ of $\mathbf{V}$ and $\mathbf{V}_y$ is called the _______ vector ________ of $\mathbf{V}$.

6. If $\mathbf{V}$ makes an angle θ with the positive x-axis when in standard position, then $|\mathbf{V}_x| =$ ____________ and $|\mathbf{V}_y| =$ ____________.

7. If an object is stationary (at rest), then the sum of the forces acting on the object equal the _____ vector, and we say the object is in a state of ________________.

8. If a constant force $\mathbf{F}$ is applied to an object and moves the object in a straight line a distance d at an angle θ with the force, then the ______ performed by the force is found by ___________ $|\mathbf{F}|\cos\theta$ and d.

➤ EXERCISES

Draw vectors representing the following velocities:

9. 30 mi/hr due north

10. 30 mi/hr due south

11. 30 mi/hr due east

12. 30 mi/hr due west

13. 50 cm/sec N 30° W

14. 50 cm/sec N 30° E

15. 20 ft/min S 60° E

16. 20 ft/min S 60° W

17. Bearing and Distance A person is riding in a hot air balloon. For the first hour the wind current is a constant 9.50 miles per hour in the direction N 37.5° E. Then the wind current changes to 8.00 miles per hour and heads the balloon in the direction S 52.5° E. If this continues for another 1.5 hours, how far is the balloon from its starting point? What is the bearing of the balloon from its starting point (Figure 20)?

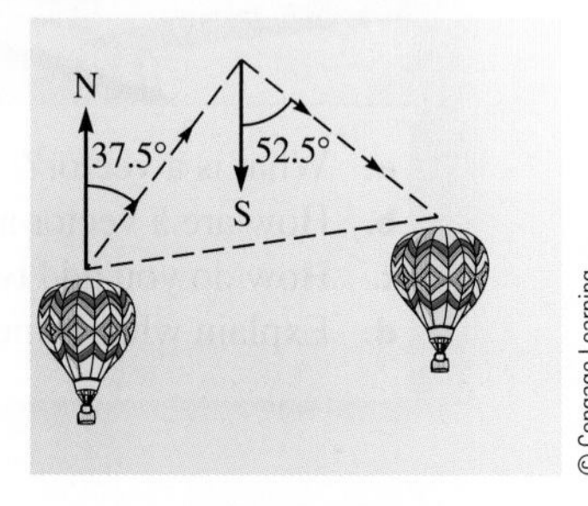

Figure 20

18. Bearing and Distance Two planes take off at the same time from an airport. The first plane is flying at 255 miles per hour on a bearing of S 45.0° E. The second plane is flying in the direction S 45.0° W at 275 miles per hour. If there are no wind currents blowing, how far apart are they after 2 hours? What is the bearing of the second plane from the first after 2 hours?

Each problem below refers to a vector **V** *with magnitude* $|\mathbf{V}|$ *that forms an angle* θ *with the positive x-axis. In each case, give the magnitudes of the horizontal and vertical vector components of* **V**, *namely* $\mathbf{V}_x$ *and* $\mathbf{V}_y$, *respectively.*

19. $|\mathbf{V}| = 13.8, \theta = 24.2°$

20. $|\mathbf{V}| = 17.6, \theta = 67.2°$

21. $|\mathbf{V}| = 425, \theta = 36°\ 10'$

22. $|\mathbf{V}| = 383, \theta = 16°\ 40'$

23. $|\mathbf{V}| = 64, \theta = 0°$

24. $|\mathbf{V}| = 48, \theta = 90°$

For each problem below, the magnitudes of the horizontal and vertical vector components, $\mathbf{V}_x$ *and* $\mathbf{V}_y$, *of vector* **V** *are given. In each case find the magnitude of* **V**.

25. $|\mathbf{V}_x| = 35.0, |\mathbf{V}_y| = 26.0$

26. $|\mathbf{V}_x| = 45.0, |\mathbf{V}_y| = 15.0$

27. $|\mathbf{V}_x| = 4.5, |\mathbf{V}_y| = 3.8$

28. $|\mathbf{V}_x| = 2.2, |\mathbf{V}_y| = 5.8$

29. Navigation A ship is 2.8° off course. If the ship is traveling at 14.0 miles per hour, how far off course will it be after 2 hours?

30. Navigation If a navigation error puts a plane 3.0° off course, how far off course is the plane after flying 135 miles?

31. Velocity of a Bullet A bullet is fired into the air with an initial velocity of 1,200 feet per second at an angle of 45° from the horizontal. Find the magnitude of the horizontal and vertical vector components of the velocity vector.

32. Velocity of a Bullet A bullet is fired into the air with an initial velocity of 1,800 feet per second at an angle of 60° from the horizontal. Find the magnitudes of the horizontal and vertical vector components of the velocity vector.

33. **Distance Traveled by a Bullet** Use the results of Problem 31 to find the horizontal distance traveled by the bullet in 3 seconds. (Neglect the resistance of air on the bullet.)

34. **Distance Traveled by a Bullet** Use the results of Problem 32 to find the horizontal distance traveled by the bullet in 2 seconds.

35. **Distance** A ship travels 130 kilometers on a bearing of S 42° E. How far east and how far south has it traveled?

36. **Distance** A plane flies for 3 hours at 230 kilometers per hour in the direction S 35° W. How far west and how far south does it travel in the 3 hours?

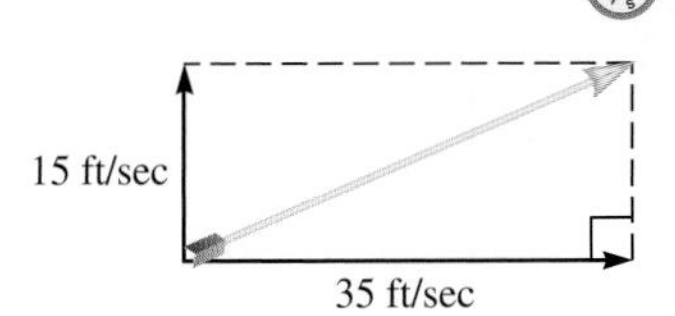

Figure 21

37. **Velocity of an Arrow** An arrow is shot into the air so that its horizontal velocity is 35.0 feet per second and its vertical velocity is 15.0 feet per second (Figure 21). Find the velocity of the arrow.

38. **Velocity of an Arrow** The horizontal and vertical components of the velocity of an arrow shot into the air are 15.0 feet per second and 25.0 feet per second, respectively. Find the velocity of the arrow.

39. **Distance** A plane travels 170 miles on a bearing of N 18° E and then changes its course to N 49° E and travels another 120 miles. Find the total distance traveled north and the total distance traveled east.

40. **Distance** A ship travels in the direction S 12° E for 68 miles and then changes its course to S 60° E and travels another 110 miles. Find the total distance south and the total distance east that the ship traveled.

41. **Static Equilibrium** Repeat the swing problem shown in Example 5 if Stacey pulls Danny through an angle of 45.0° and then holds him at static equilibrium. Find the magnitudes of both **H** and **T**.

42. **Static Equilibrium** The diagram in Figure 18 of this section would change if Stacey were to push Danny forward through an angle of 30° and then hold him in that position. Draw the diagram that corresponds to this new situation.

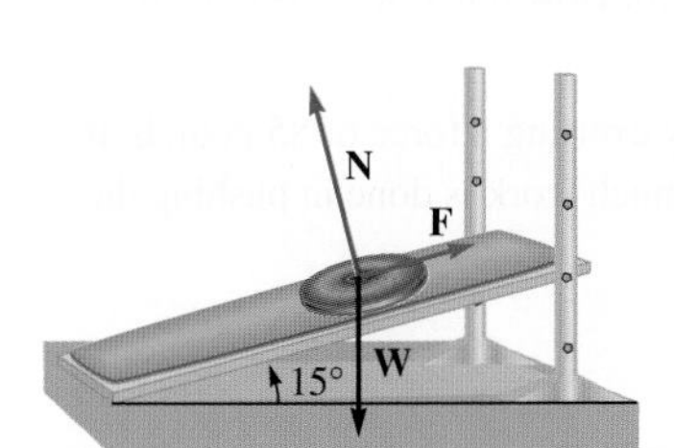

Figure 22

43. **Force** An 8.0-pound weight is lying on a sit-up bench at the gym. If the bench is inclined at an angle of 15°, there are three forces acting on the weight, as shown in Figure 22. **N** is called the normal force and it acts in the direction perpendicular to the bench. **F** is the force due to friction that holds the weight on the bench. If the weight does not move, then the sum of these three forces is **0**. Find the magnitude of **N** and the magnitude of **F**.

44. **Force** Repeat Problem 43 for a 25.0-pound weight and a bench inclined at 10.0°.

45. **Force** Danny and Stacey have gone from the swing (Example 5) to the slide at the park. The slide is inclined at an angle of 52.0°. Danny weighs 42.0 pounds. He is sitting in a cardboard box with a piece of wax paper on the bottom. Stacey is at the top of the slide holding on to the cardboard box (Figure 23). Find the magnitude of the force Stacey must pull with, in order to keep Danny from sliding down the slide. (We are assuming that the wax paper makes the slide into a frictionless surface, so that the only force keeping Danny from sliding is the force with which Stacey pulls.)

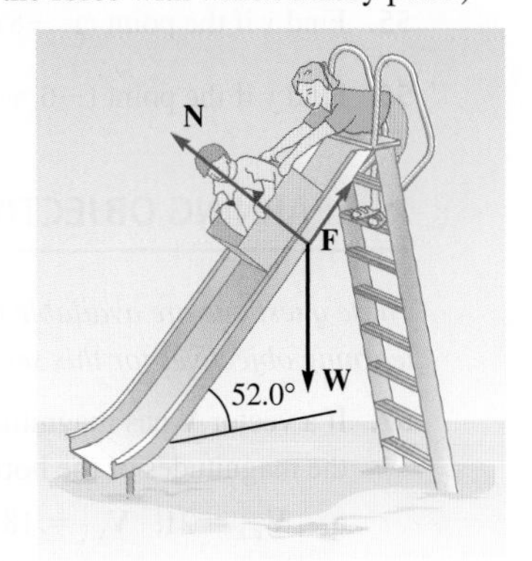

Figure 23

46. Wrecking Ball A 2,200-pound wrecking ball is held in static equilibrium by two cables, one horizontal and one at an angle of 40° from vertical (Figure 24). Find the magnitudes of the tension vectors **T** and **H**.

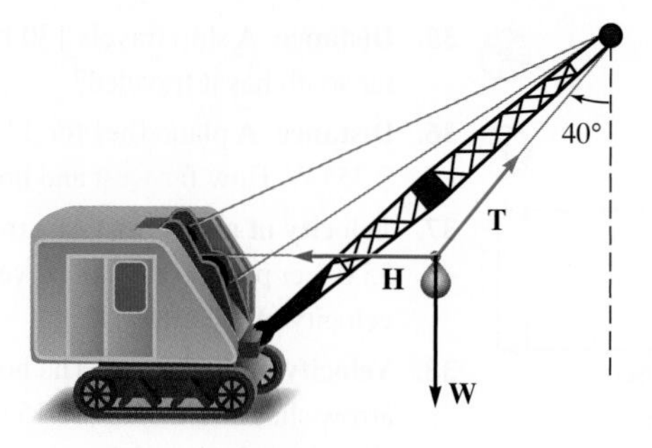

Figure 24

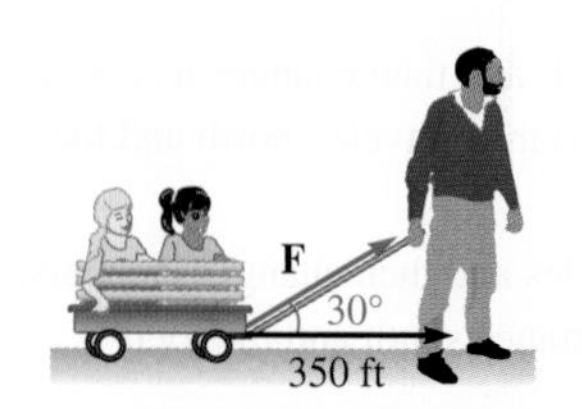

Figure 25

Figure 26

47. Work A package is pushed across a floor a distance of 75 feet by exerting a force of 41 pounds downward at an angle of 20° with the horizontal. How much work is done?

48. Work A package is pushed across a floor a distance of 52 feet by exerting a force of 15 pounds downward at an angle of 25° with the horizontal. How much work is done?

49. Work Mark pulls Allison and Mattie in a wagon by exerting a force of 25 pounds on the handle at an angle of 30° with the horizontal (Figure 25). How much work is done by Mark in pulling the wagon 350 feet?

50. Work An automobile is pushed down a level street by exerting a force of 85 pounds at an angle of 15° with the horizontal (Figure 26). How much work is done in pushing the car 110 feet?

➤ REVIEW PROBLEMS

The problems that follow review material we covered in Section 1.3.

51. Draw 135° in standard position, locate a convenient point on the terminal side, and then find sin 135°, cos 135°, and tan 135°.

52. Draw −270° in standard position, locate a convenient point on the terminal side, and then find sine, cosine, and tangent of −270°.

53. Find $\sin\theta$ and $\cos\theta$ if the terminal side of θ lies along the line $y = 2x$ in quadrant I.

54. Find $\sin\theta$ and $\cos\theta$ if the terminal side of θ lies along the line $y = -x$ in quadrant II.

55. Find x if the point $(x, -8)$ is on the terminal side of θ and $\sin\theta = -\frac{4}{5}$.

56. Find y if the point $(-6, y)$ is on the terminal side of θ and $\cos\theta = -\frac{3}{5}$.

➤ LEARNING OBJECTIVES ASSESSMENT

These questions are available for instructors to help assess if you have successfully met the learning objectives for this section.

57. If a vector **V** has magnitude 28 and makes an angle of 41° with the positive x-axis, find the magnitudes of the horizontal and vertical vector components of **V**.

a. $|\mathbf{V}_x| = 21, |\mathbf{V}_y| = 18$ **b.** $|\mathbf{V}_x| = 18, |\mathbf{V}_y| = 21$

c. $|\mathbf{V}_x| = 16, |\mathbf{V}_y| = 23$ **d.** $|\mathbf{V}_x| = 23, |\mathbf{V}_y| = 16$

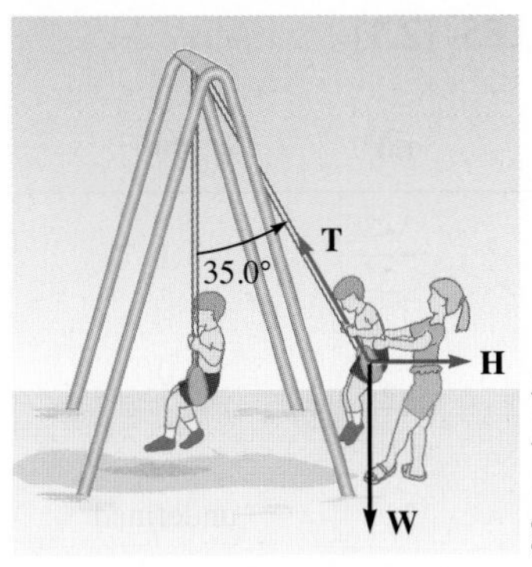

Figure 27

58. If a vector **V** has horizontal and vertical vector components with magnitudes $|\mathbf{V}_x| = 9.6$ and $|\mathbf{V}_y| = 2.3$, find the magnitude of **V** and the angle it makes with the positive x-axis.

a. $|\mathbf{V}| = 11.5, \theta = 13°$ **b.** $|\mathbf{V}| = 11.5, \theta = 13°$
c. $|\mathbf{V}| = 9.9, \theta = 13°$ **d.** $|\mathbf{V}| = 9.9, \theta = 77°$

59. Jadon is 5 years old and weighs 45.0 pounds. He is sitting on a swing when his cousin Allison pulls him and the swing back horizontally through an angle of 35.0° and then stops. Find the magnitude of the force exerted by Allison (Figure 27).

a. 31.5 lb **b.** 64.3 lb **c.** 25.8 lb **d.** 36.9 lb

60. A package is pushed across a floor a distance of 150 feet by exerting a force of 28 pounds downward at an angle of 35°. How much work is done?

a. 2,800 ft-lb **b.** 3,400 ft-lb **c.** 4,200 ft-lb **d.** 3,700 ft-lb

CHAPTER 2 SUMMARY

EXAMPLES

1

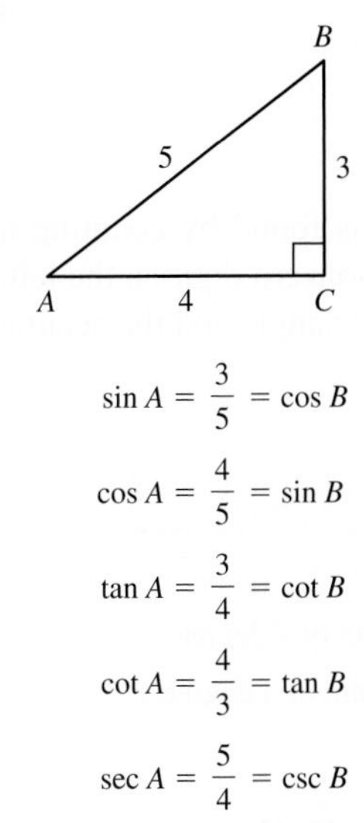

$\sin A = \frac{3}{5} = \cos B$

$\cos A = \frac{4}{5} = \sin B$

$\tan A = \frac{3}{4} = \cot B$

$\cot A = \frac{4}{3} = \tan B$

$\sec A = \frac{5}{4} = \csc B$

$\csc A = \frac{5}{3} = \sec B$

TRIGONOMETRIC FUNCTIONS (DEFINITION II) [2.1]

If triangle ABC is a right triangle with $C = 90°$, then the six trigonometric functions for angle A are

$$\sin A = \frac{\text{side opposite } A}{\text{hypotenuse}} = \frac{a}{c}$$

$$\cos A = \frac{\text{side adjacent } A}{\text{hypotenuse}} = \frac{b}{c}$$

$$\tan A = \frac{\text{side opposite } A}{\text{side adjacent } A} = \frac{a}{b}$$

$$\cot A = \frac{\text{side adjacent } A}{\text{side opposite } A} = \frac{b}{a}$$

$$\sec A = \frac{\text{hypotenuse}}{\text{side adjacent } A} = \frac{c}{b}$$

$$\csc A = \frac{\text{hypotenuse}}{\text{side opposite } A} = \frac{c}{a}$$

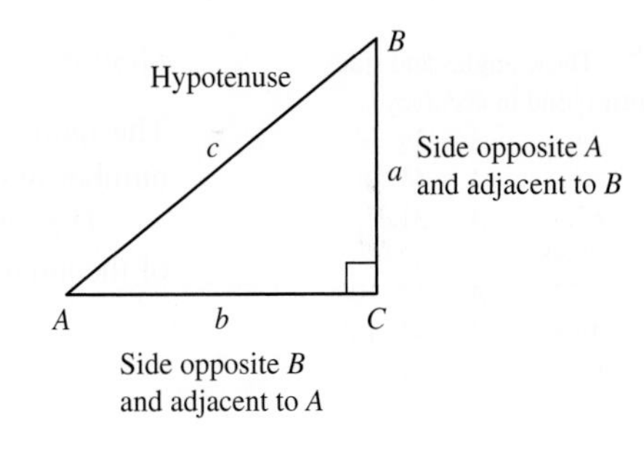

2 $\sin 3° = \cos 87°$
$\cos 10° = \sin 80°$
$\tan 15° = \cot 75°$
$\cot A = \tan (90° - A)$
$\sec 30° = \csc 60°$
$\csc 45° = \sec 45°$

COFUNCTION THEOREM [2.1]

A trigonometric function of an angle is always equal to the cofunction of its complement. In symbols, since the complement of x is $90° - x$, we have

$$\sin x = \cos (90° - x)$$
$$\sec x = \csc (90° - x)$$
$$\tan x = \cot (90° - x)$$

3 The values given in the table are called *exact values* because they are not decimal approximations as you would find on a calculator.

TRIGONOMETRIC FUNCTIONS OF SPECIAL ANGLES [2.1]

θ	0°	30°	45°	60°	90°
$\sin\theta$	0	$\frac{1}{2}$	$\frac{1}{\sqrt{2}}$ or $\frac{\sqrt{2}}{2}$	$\frac{\sqrt{3}}{2}$	1
$\cos\theta$	1	$\frac{\sqrt{3}}{2}$	$\frac{1}{\sqrt{2}}$ or $\frac{\sqrt{2}}{2}$	$\frac{1}{2}$	0
$\tan\theta$	0	$\frac{1}{\sqrt{3}}$ or $\frac{\sqrt{3}}{3}$	1	$\sqrt{3}$	undefined

4
$$\begin{array}{r} 47°\ 30' \\ +\ 23°\ 50' \\ \hline 70°\ 80' = 71°\ 20' \end{array}$$

DEGREES, MINUTES, AND SECONDS [2.2]

There are 360° (degrees) in one complete rotation, 60′ (minutes) in one degree, and 60″ (seconds) in one minute. This is equivalent to saying 1 minute is $\frac{1}{60}$ of a degree, and 1 second is $\frac{1}{60}$ of a minute.

5
$$\begin{aligned} 74.3° &= 74° + 0.3° \\ &= 74° + 0.3(60') \\ &= 74° + 18' \\ &= 74°\ 18' \end{aligned}$$
$$\begin{aligned} 42°\ 48' &= 42° + \left(\frac{48}{60}\right)^\circ \\ &= 42° + 0.8° \\ &= 42.8° \end{aligned}$$

CONVERTING TO AND FROM DECIMAL DEGREES [2.2]

To convert from decimal degrees to degrees and minutes, multiply the fractional part of the angle (that which follows the decimal point) by 60 to get minutes.

To convert from degrees and minutes to decimal degrees, divide minutes by 60 to get the fractional part of the angle.

6 These angles and sides correspond in accuracy.

$a = 24$ $A = 39°$
$a = 5.8$ $A = 45°$
$a = 62.3$ $A = 31.3°$
$a = 0.498$ $A = 42.9°$
$a = 2.77$ $A = 37°\ 10'$
$a = 49.87$ $A = 43°\ 18'$
$a = 6.932$ $A = 24.81°$

SIGNIFICANT DIGITS [2.3]

The number of *significant digits* (or figures) in a number is found by counting the number of digits from left to right, beginning with the first nonzero digit on the left.

The relationship between the accuracy of the sides in a triangle and the accuracy of the angles in the same triangle is given below.

Accuracy of Sides	Accuracy of Angles
Two significant digits	Nearest degree
Three significant digits	Nearest 10 minutes or tenth of a degree
Four significant digits	Nearest minute or hundredth of a degree

7

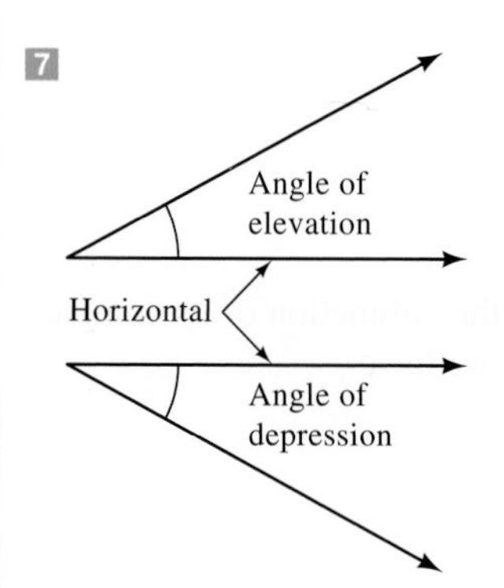

ANGLE OF ELEVATION AND ANGLE OF DEPRESSION [2.4]

An angle measured from the horizontal up is called an *angle of elevation.* An angle measured from the horizontal down is called an *angle of depression.* If an observer positioned at the vertex of the angle views an object in the direction of the nonhorizontal side of the angle, then this side is sometimes called the *line of sight* of the observer.

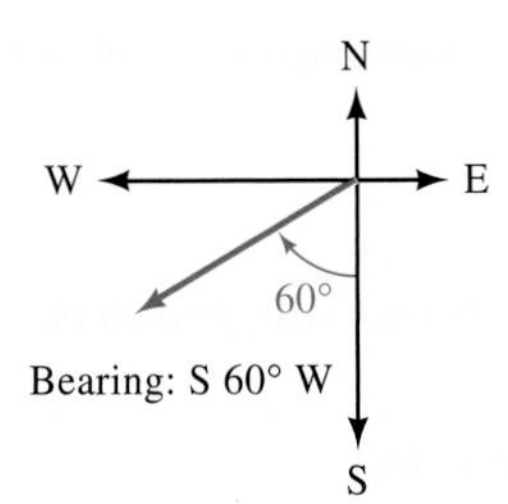

DIRECTION [2.4]

One way to specify the direction of a line or vector is called *bearing*. The notation used to designate bearing begins with N or S, followed by the number of degrees in the angle, and ends with E or W, as in S 60° W.

9 If a car is traveling at 50 miles per hour due south, then its velocity can be represented with a vector.

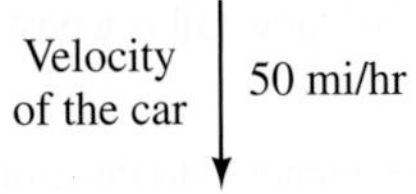

VECTORS [2.5]

Quantities that have both magnitude and direction are called *vector quantities,* while quantities that have only magnitude are called *scalar quantities.* We represent vectors graphically by using arrows. The length of the arrow corresponds to the magnitude of the vector, and the direction of the arrow corresponds to the direction of the vector. In symbols, we denote the magnitude of vector **V** with $|\mathbf{V}|$.

10

HORIZONTAL AND VERTICAL VECTOR COMPONENTS [2.5]

The horizontal and vertical vector components of vector **V** are the horizontal and vertical vectors whose sum is **V**. The horizontal vector component is denoted by $\mathbf{V}_x$, and the vertical vector component is denoted by $\mathbf{V}_y$.

CHAPTER 2 TEST

Find sin A, cos A, tan A, and sin B, cos B, and tan B in right triangle ABC, with C = 90° given the following information.

1. $a = 1$ and $b = 2$

2. $b = 3$ and $c = 6$

3. $a = 3$ and $c = 5$

4. Use Definition II to explain why, for any acute angle θ, it is impossible for $\sin \theta = 2$.

5. Fill in the blank to make the statement true: $\sin 14° = \cos$ ______.

Simplify each expression as much as possible.

6. $\sin^2 45° + \cos^2 30°$

7. $\tan 45° + \cot 45°$

8. $\sin^2 60° - \cos^2 30°$

9. $\dfrac{1}{\csc 30°}$

10. Add 48° 18′ and 24° 52′.

11. Convert 73.2° to degrees and minutes.

12. Convert 2° 48′ to decimal degrees.

Use a calculator to find the following:

13. $\sin 24° 20'$

14. $\cos 37.8°$

15. $\cot 71° 20'$

Use a calculator to find θ to the nearest tenth of a degree if θ is an acute angle and satisfies the given statement.

16. $\sin \theta = 0.9465$

17. $\sec \theta = 1.923$

The following problems refer to right triangle ABC with $C = 90°$. In each case, find all the missing parts.

18. $a = 68.0$ and $b = 104$

19. $a = 24.3$ and $c = 48.1$

20. $b = 305$ and $B = 24.9°$

21. $c = 0.462$ and $A = 35° 30'$

22. Geometry If the altitude of an isosceles triangle is 25 centimeters and each of the two equal angles measures 17°, how long are the two equal sides?

23. Angle of Elevation If the angle of elevation of the sun is 75° 30′, how tall is a post that casts a shadow 1.5 feet long?

24. Distance Two guy wires from the top of a 35-foot tent pole are anchored to the ground below by two stakes. The tent pole is perpendicular to the ground and between the two stakes. If the angles of depression from the top of the pole to each of the stakes are 47° and 43°, how far apart are the stakes?

25. Vector Angle Vector **V** has a horizontal vector component with magnitude 11 and a vertical vector component with magnitude 31. What is the acute angle formed by **V** and the positive x-axis?

26. Velocity A bullet is fired into the air with an initial velocity of 800 feet per second at an angle of 62° from the horizontal. Find the magnitudes of the horizontal and vertical vector components of the velocity vector.

27. Distance and Bearing A ship travels 120 miles on a bearing of S 60° E. How far east and how far south has the ship traveled?

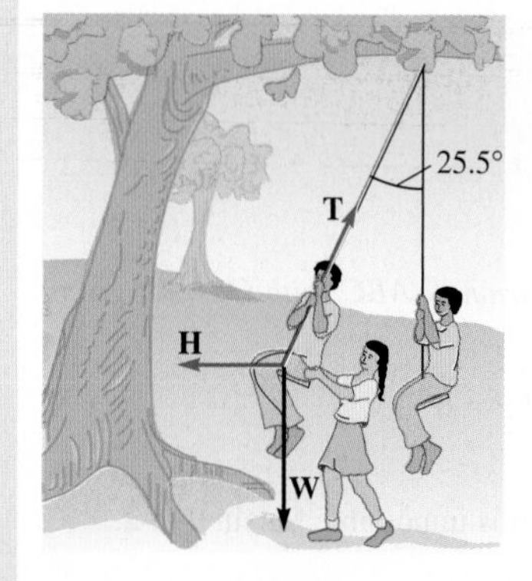

Figure 1

28. Force Tyler and his cousin Kelly have attached a rope to the branch of a tree and tied a board to the other end to form a swing. Tyler sits on the board while his cousin pushes him through an angle of 25.5° and holds him there. If Tyler weighs 95.5 pounds, find the magnitude of the force Kelly must push with horizontally to keep Tyler in static equilibrium. See Figure 1.

29. Force Tyler and Kelly decide to rollerskate. They come to a hill that is inclined at 8.5°. Tyler pushes Kelly halfway up the hill and then holds her there (Figure 2). If Kelly weighs 58.0 pounds, find the magnitude of the force Tyler must push with to keep Kelly from rolling down the hill. (Assume that the only force keeping Kelly from rolling backwards is the force Tyler is pushing with.)

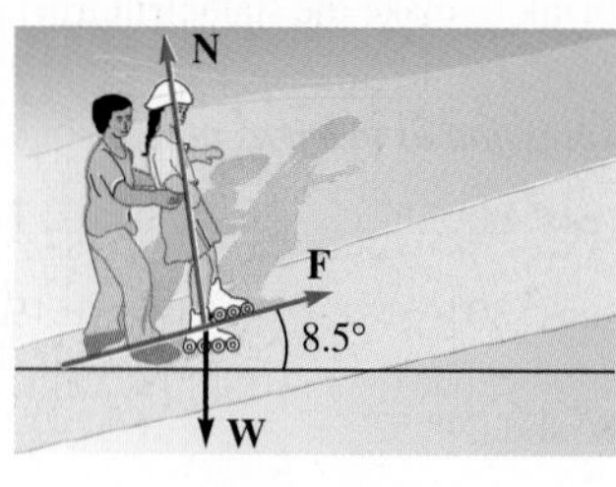

Figure 2

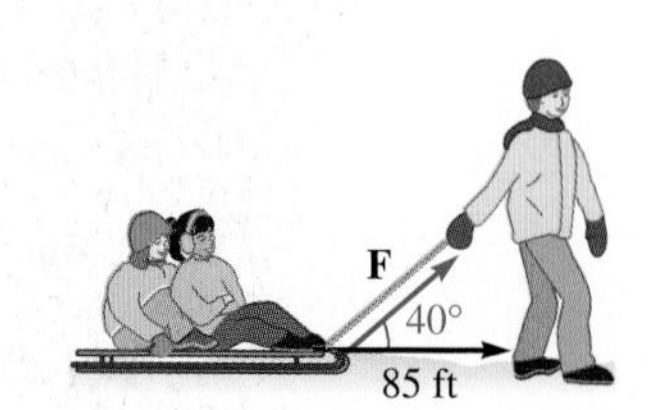

Figure 3

30. Work William pulls his cousins Hannah and Serah on a sled by exerting a force of 35 pounds on a rope at an angle of elevation of 40° (Figure 3). How much work is done by William in pulling the sled 85 feet?

The 15°–75°–90° Triangle

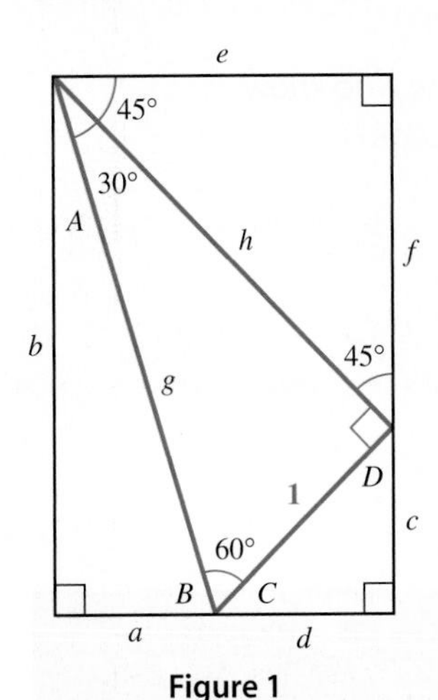

Figure 1

OBJECTIVE: To find exact values of the six trigonometric functions for 15° and 75° angles.

In Section 2.1 we saw how to obtain exact values for the trigonometric functions using two special triangles, the 30°–60°–90° triangle and the 45°–45°–90° triangle. In this project you will use both of these triangles to create a 15°–75°–90° triangle. Once your triangle is complete, you will be able to use it to find additional exact values of the trigonometric functions.

The diagram shown in Figure 1 is called an Ailles rectangle. It is named after a high school teacher, Doug Ailles, who first discovered it. Notice that the rectangle is constructed from four triangles. The triangle in the middle is a 30°–60°–90° triangle, and the triangle above it in the upper right corner is a 45°–45°–90° triangle.

1. Find the measures of the remaining four angles labeled A, B, C, and D.
2. Use your knowledge of a 30°–60°–90° triangle to find sides g and h.
3. Now use your knowledge of a 45°–45°–90° triangle to find sides e and f.
4. Find c and d, and then find sides a and b using your values for sides c, d, e, and f.
5. Now that the diagram is complete, you should have a 15°–75°–90° triangle with all three sides labeled. Use this triangle to fill in the table below with exact values for each trigonometric function.

θ	$\sin\theta$	$\cos\theta$	$\tan\theta$	$\csc\theta$	$\sec\theta$	$\cot\theta$
15°						
75°						

6. Use your calculator to check each of the values in your table. For example, find sin 15° on your calculator and make sure it agrees with the value in your table. (You will need to approximate the value in your table with your calculator also.)

Shadowy Origins

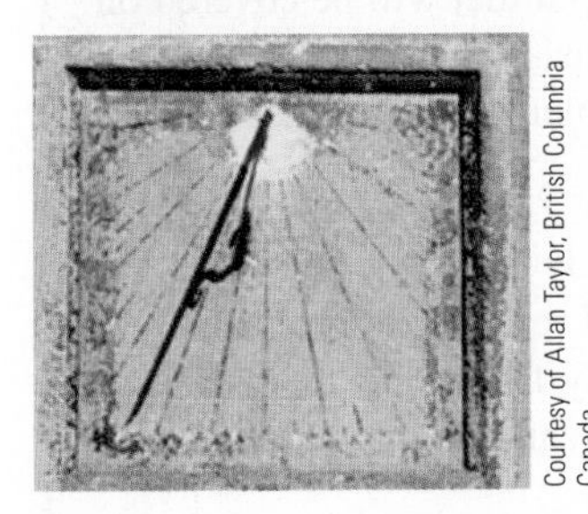

Courtesy of Allan Taylor, British Columbia Canada

The origins of the sine function are found in the tables of chords for a circle constructed by the Greek astronomers/mathematicians Hipparchus and Ptolemy. However, the origins of the tangent and cotangent functions lie primarily with Arabic and Islamic astronomers. Called the *umbra recta* and *umbra versa,* their connection was not to the chord of a circle but to the gnomon of a sundial.

Research the origins of the tangent and cotangent functions. What was the connection to sundials? What other contributions did Arabic astronomers make to trigonometry? Write a paragraph or two about your findings.

3 Radian Measure

The great book of nature can be read only by those who know the language in which it is written. And this language is mathematics.

➤ Galileo

INTRODUCTION

The first cable car was built in San Francisco, California, by Andrew Smith Hallidie. It was completed on August 2, 1873. Today, three cable car lines remain in operation in San Francisco. The cars are propelled by long steel cables. Inside the powerhouse, these cables are driven by large 14-foot drive wheels, called *sheaves,* that turn at a rate of 19 revolutions per minute (Figure 1).

Figure 1

In this chapter we introduce a second type of angle measure that allows us to define rotation in terms of an arc along the circumference of a circle. With radian measure, we can solve problems that involve angular speeds such as revolutions per minute. For example, we will be able to use this information to determine the linear speed at which the cable car travels.

STUDY SKILLS 3

The study skills for this chapter focus on getting ready to take an exam.

1 Getting Ready to Take an Exam Try to arrange your daily study habits so that you have very little studying to do the night before your next exam. The next two goals will help you achieve goal number 1.

2 Review with the Exam in Mind Review material that will be covered on the next exam every day. Your review should consist of working problems. Preferably, the problems you work should be problems from the list of difficult problems you have been recording.

3 Continue to List Difficult Problems This study skill was started in the previous chapter. You should continue to list and rework the problems that give you the most difficulty. Use this list to study for the next exam. Your goal is to go into the next exam knowing that you can successfully work any problem from your list of hard problems.

4 Pay Attention to Instructions Taking a test is different from doing homework. When you take a test, the problems will be mixed up. When you do your homework, you usually work a number of similar problems. Sometimes students do very well on their homework but become confused when they see the same problems on a test because they have not paid attention to the instructions on their homework. ▲

SECTION 3.1 Reference Angle

LEARNING OBJECTIVES

1. Identify the reference angle for a given angle in standard position.
2. Use a reference angle to find the exact value of a trigonometric function.
3. Use a calculator to approximate the value of a trigonometric function.
4. Find an angle given the quadrant and the value of a trigonometric function.

In the previous chapter we found exact values for trigonometric functions of certain angles between 0° and 90°. By using what are called *reference angles,* we can find exact values for trigonometric functions of angles outside the interval 0° to 90°.

DEFINITION

The *reference angle* (sometimes called *related angle*) for any angle θ in standard position is the positive acute angle between the terminal side of θ and the x-axis. In this book, we will denote the reference angle for θ by $\hat{\theta}$.

Note that, for this definition, $\hat{\theta}$ is always positive and always between 0° and 90°. That is, a reference angle is always an acute angle.

PROBLEM 1
Name the reference angle for each of the following angles.

a. 40°
b. 160°
c. 215°
d. 325°
e. −25
f. −13

EXAMPLE 1 Name the reference angle for each of the following angles.

a. 30° **b.** 135° **c.** 240° **d.** 330° **e.** −210° **f.** −140°

SOLUTION We draw each angle in standard position. The reference angle is the positive acute angle formed by the terminal side of the angle in question and the x-axis (Figure 1).

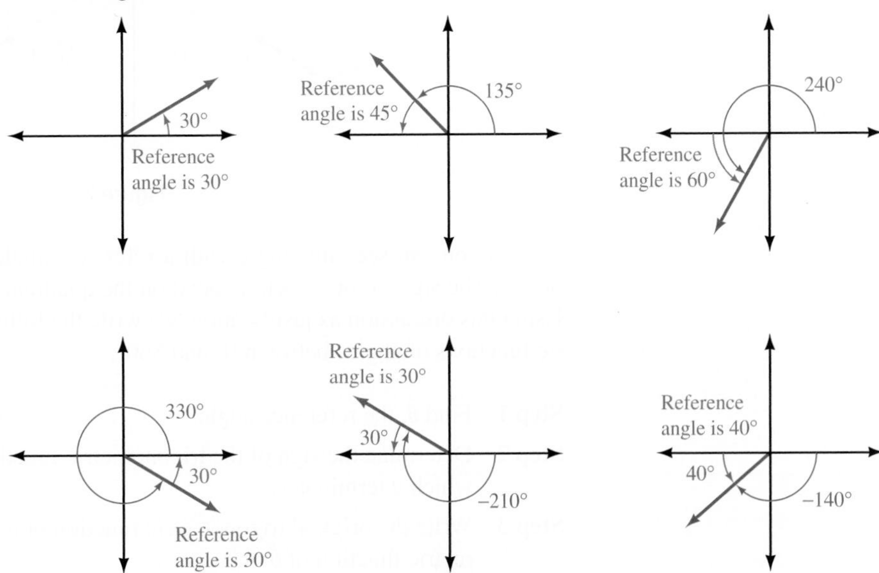

Figure 1

We can generalize the results of Example 1 as follows: If θ is a positive angle between 0° and 360°, and

$$\begin{aligned} &\text{if } \theta \in \text{QI}, && \text{then } \hat{\theta} = \theta \\ &\text{if } \theta \in \text{QII}, && \text{then } \hat{\theta} = 180° - \theta \\ &\text{if } \theta \in \text{QIII}, && \text{then } \hat{\theta} = \theta - 180° \\ &\text{if } \theta \in \text{QIV}, && \text{then } \hat{\theta} = 360° - \theta \end{aligned}$$

We can use our information on reference angles and the signs of the trigonometric functions to write the following theorem.

REFERENCE ANGLE THEOREM

A trigonometric function of an angle and its reference angle are the same, except, perhaps, for a difference in sign.

We will not give a detailed proof of this theorem, but rather, justify it by example. Let's look at the sines of all the angles between 0° and 360° that have a reference angle of 30°. These angles are 30°, 150°, 210°, and 330° (Figure 2).

$$\sin 150° = \sin 30° = \frac{1}{2}$$

They differ in sign only

$$\sin 210° = \sin 330° = -\frac{1}{2}$$

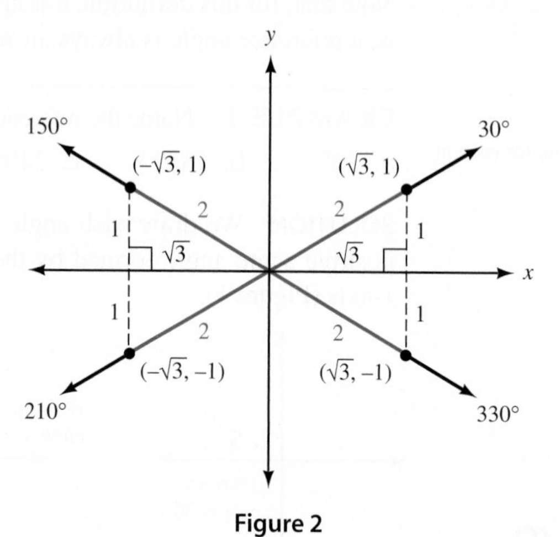

Figure 2

As you can see, any angle with a reference angle of 30° will have a sine of $\frac{1}{2}$ or $-\frac{1}{2}$. The sign, + or −, will depend on the quadrant in which the angle terminates. Using this discussion as justification, we write the following steps to find trigonometric functions of angles between 0° and 360°.

Step 1 Find $\hat{\theta}$, the reference angle.

Step 2 Determine the sign of the trigonometric function based on the quadrant in which θ terminates.

Step 3 Write the original trigonometric function of θ in terms of the same trigonometric function of $\hat{\theta}$.

Step 4 Find the trigonometric function of $\hat{\theta}$.

PROBLEM 2
Find the exact value of sin 210°.

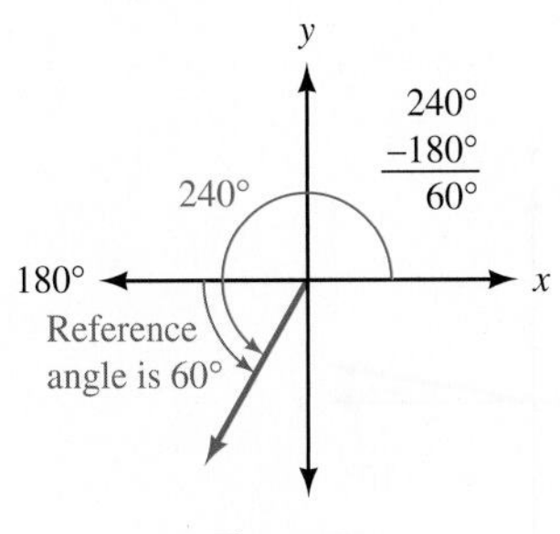

Figure 3

EXAMPLE 2 Find the exact value of sin 240°.

SOLUTION For this example, we will list the steps just given as we use them. Figure 3 is a diagram of the situation.

Step 1 We find $\hat{\theta}$ by subtracting 180° from θ.

$$\hat{\theta} = 240° - 180° = 60°$$

Step 2 Since θ terminates in quadrant III, and the sine function is negative in quadrant III, our answer will be negative. That is, $\sin \theta = -\sin \hat{\theta}$.

Step 3 Using the results of Steps 1 and 2, we write

$$\sin 240° = -\sin 60°$$

Step 4 We finish by finding sin 60°.

$$\sin 240° = -\sin 60° \qquad \text{Sine is negative in QIII}$$

$$= -\left(\frac{\sqrt{3}}{2}\right) \qquad \sin 60° = \frac{\sqrt{3}}{2}$$

$$= -\frac{\sqrt{3}}{2}$$

PROBLEM 3
Find the exact value of tan 225°.

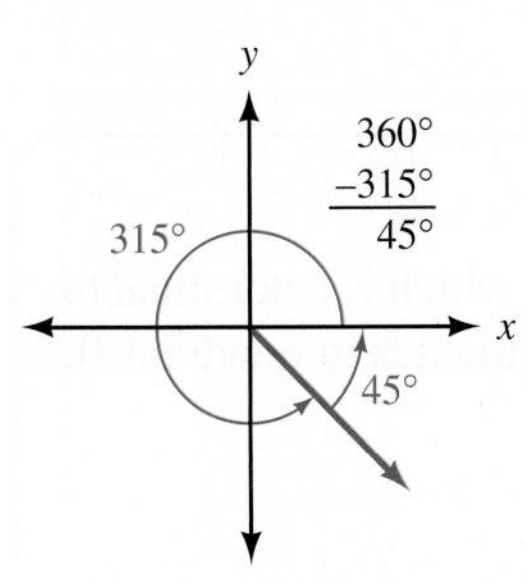

Figure 4

EXAMPLE 3 Find the exact value of tan 315°.

SOLUTION The reference angle is 360° − 315° = 45°. Since 315° terminates in quadrant IV, its tangent will be negative (Figure 4).

$$\tan 315° = -\tan 45° \qquad \text{Because tangent is negative in QIV}$$

$$= -1$$

PROBLEM 4
Find the exact value of sec 300°.

EXAMPLE 4 Find the exact value of csc 300°.

SOLUTION The reference angle is 360° − 300° = 60° (Figure 5). To find the exact value of csc 60°, we use the fact that cosecant and sine are reciprocals.

$$\csc 300° = -\csc 60° \qquad \text{Because cosecant is negative in QIV}$$

$$= -\frac{1}{\sin 60°} \qquad \text{Reciprocal identity}$$

$$= -\frac{1}{\sqrt{3}/2} \qquad \sin 60° = \frac{\sqrt{3}}{2}$$

$$= -\frac{2}{\sqrt{3}}$$

$$= -\frac{2\sqrt{3}}{3}$$

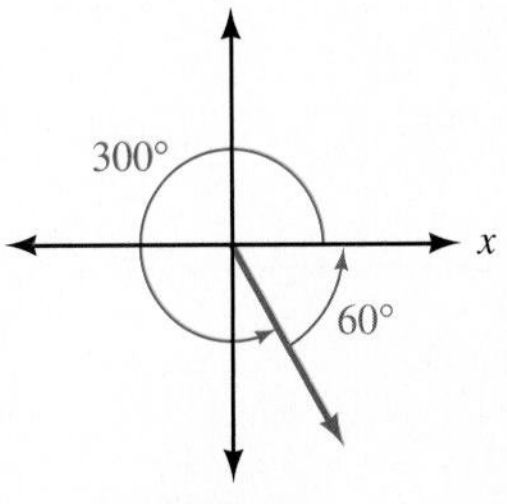

Figure 5

Recall from Section 1.2 that coterminal angles always differ from each other by multiples of 360°. For example, −45° and 315° are coterminal, as are 10° and 370° (Figure 6).

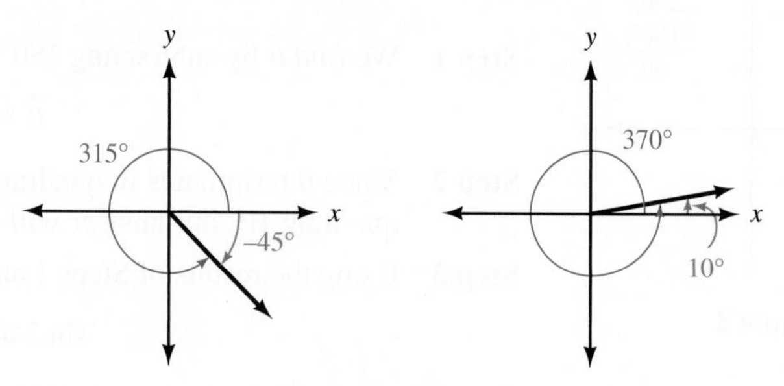

Figure 6

The trigonometric functions of an angle and any angle coterminal to it are always equal. For sine and cosine, we can write this in symbols as follows:

for any integer k,

$$\sin(\theta + 360°k) = \sin\theta \quad \text{and} \quad \cos(\theta + 360°k) = \cos\theta$$

To find values of trigonometric functions for an angle larger than 360° or smaller than 0°, we simply find an angle between 0° and 360° that is coterminal to it and then use the steps outlined in Examples 2 through 4.

PROBLEM 5
Find the exact value of $\sin 405°$.

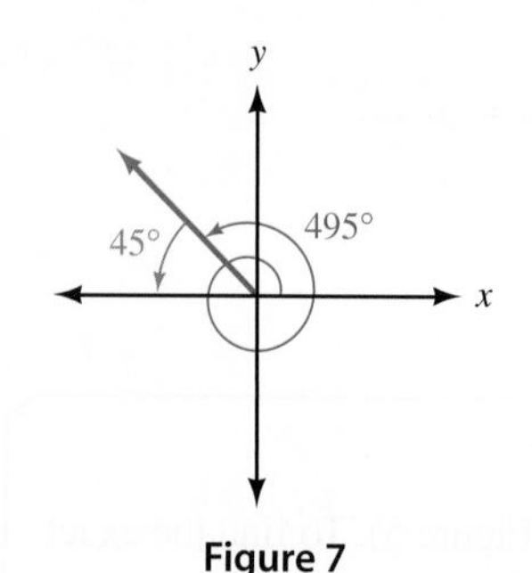

Figure 7

EXAMPLE 5 Find the exact value of $\cos 495°$.

SOLUTION By subtracting 360° from 495°, we obtain 135°, which is coterminal to 495°. The reference angle for 135° is 45°. Because 495° terminates in quadrant II, its cosine is negative (Figure 7).

$$\cos 495° = \cos 135° \quad \text{495° and 135° are coterminal}$$

$$= -\cos 45° \quad \text{In QII } \cos\theta = -\cos\hat{\theta}$$

$$= -\frac{\sqrt{2}}{2} \quad \text{Exact value}$$

Approximations

To find trigonometric functions of angles that do not lend themselves to exact values, we use a calculator. To find an approximation for $\sin\theta$, $\cos\theta$, or $\tan\theta$, we simply enter the angle and press the appropriate key on the calculator. Check to see that you can obtain the following values for sine, cosine, and tangent of 250° and −160° on your calculator. (These answers are rounded to the nearest ten-thousandth.) Make sure your calculator is set to degree mode.

$$\sin 250° = -0.9397 \qquad \sin(-160°) = -0.3420$$

$$\cos 250° = -0.3420 \qquad \cos(-160°) = -0.9397$$

$$\tan 250° = 2.7475 \qquad \tan(-160°) = 0.3640$$

To find csc 250°, sec 250°, and cot 250°, we must use the reciprocals of sin 250°, cos 250°, and tan 250°.

	Scientific Calculator	Graphing Calculator
$\csc 250° = \dfrac{1}{\sin 250°}$	250 [sin] [1/x]	1 [÷] [sin] [(] 250 [)] [ENTER]
$= -1.0642$		
$\sec 250° = \dfrac{1}{\cos 250°}$	250 [cos] [1/x]	1 [÷] [cos] [(] 250 [)] [ENTER]
$= -2.9238$		
$\cot 250° = \dfrac{1}{\tan 250°}$	250 [tan] [1/x]	1 [÷] [tan] [(] 250 [)] [ENTER]
$= 0.3640$		

Next we use a calculator to find an approximation for θ, given one of the trigonometric functions of θ and the quadrant in which θ terminates. To do this we will need to use the [$\sin^{-1}$], [$\cos^{-1}$], and [$\tan^{-1}$] keys, which we introduced earlier in Section 2.2. In using one of these keys to find a reference angle, always enter a *positive* value for the trigonometric function. (*Remember:* A reference angle is between 0° and 90°, so all six trigonometric functions of that angle will be positive.)

CALCULATOR NOTE Some calculators do not have a key labeled as [$\sin^{-1}$]. You may need to press a combination of keys, such as [inv] [sin], [arc] [sin], or [2nd] [sin].

PROBLEM 6
Find θ to the nearest degree if $\cos \theta = -0.8910$ and θ terminates in QII with $0° \le \theta < 360°$.

EXAMPLE 6 Find θ to the nearest degree if $\sin \theta = -0.5592$ and θ terminates in QIII with $0° \le \theta < 360°$.

SOLUTION First we find the reference angle using the [$\sin^{-1}$] key with the positive value 0.5592. From this, we get $\hat{\theta} = 34°$. As shown in Figure 8, the desired angle in QIII whose reference angle is 34° is

$$\begin{aligned}\theta &= 180° + \hat{\theta}\\ &= 180° + 34°\\ &= 214°\end{aligned}$$

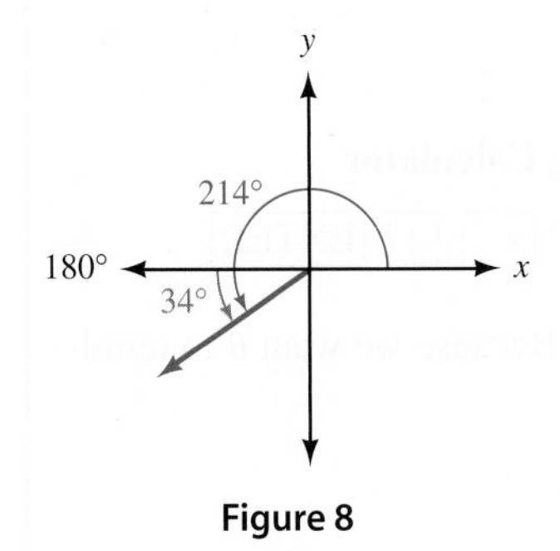

Figure 8

If we wanted to list *all* the angles that terminate in QIII and have a sine of -0.5592, we would write

$$\theta = 214° + 360°k \quad \text{where } k \text{ is any integer}$$

This gives us all angles coterminal with 214°. ■

CALCULATOR NOTE There is a big difference between the [sin] key and the [$\sin^{-1}$] key. The [sin] key can be used to find the value of the sine function for *any* angle. However, the reverse is not true with the [$\sin^{-1}$] key. For example, if you were to try Example 6 by entering -0.5592 and using the [$\sin^{-1}$] key, you would not obtain the angle 214°. To see why this happens you will have to wait until we cover inverse trigonometric functions in Chapter 4. In the meantime, to use a calculator on this kind of problem, use it to find the reference angle and then proceed as we did in Example 6.

PROBLEM 7
Find θ to the nearest tenth of a degree if $\tan \theta = -1.2437$ and θ terminates in QIV with $0° \le \theta < 360°$.

EXAMPLE 7 Find θ to the nearest tenth of a degree if $\tan \theta = -0.8541$ and θ terminates in QIV with $0° \le \theta < 360°$.

SOLUTION Using the [tan⁻¹] key with the positive value 0.8541 gives the reference angle $\hat{\theta} = 40.5°$. The desired angle in QIV with a reference angle of 40.5° is

$$\theta = 360° - 40.5° = 319.5°$$

Again, if we want to list *all* angles in QIV with a tangent of -0.8541, we write

$$\theta = 319.5° + 360°k \qquad \text{where } k \text{ is any integer}$$

to include not only 319.5° but all angles coterminal with it. ■

PROBLEM 8
Find θ if $\sin \theta = -\sqrt{3}/2$ and θ terminates in QIII with $0° \le \theta < 360°$.

EXAMPLE 8 Find θ if $\sin \theta = -\frac{1}{2}$ and θ terminates in QIII with $0° \le \theta < 360°$.

SOLUTION Using our calculator, we find the reference angle to be 30°. The desired angle in QIII with a reference angle of 30° is $180° + 30° = 210°$. ■

PROBLEM 9
Find θ to the nearest degree if $\sec \theta = 1.3054$ and θ terminates in QIV with $0° \le \theta < 360°$.

EXAMPLE 9 Find θ to the nearest degree if $\sec \theta = 3.8637$ and θ terminates in QIV with $0° \le \theta < 360°$.

SOLUTION To find the reference angle on a calculator, we must use the fact that $\sec \theta$ is the reciprocal of $\cos \theta$. That is,

$$\text{If } \sec \theta = 3.8637, \quad \text{then } \cos \theta = \frac{1}{3.8637}$$

From this last line we see that the keys to press are

Scientific Calculator	Graphing Calculator
3.8637 [1/x] [cos⁻¹]	[cos⁻¹] [(] 3.8637 [x⁻¹] [)] [ENTER]

To the nearest degree, the reference angle is $\hat{\theta} = 75°$. Because we want θ to terminate in QIV, we subtract 75° from 360° to get

$$\theta = 360° - 75° = 285°$$

We can check our result on a calculator by entering 285°, finding its cosine, and then finding the reciprocal of the result.

Scientific Calculator	Graphing Calculator
285 [cos] [1/x]	1 [÷] [cos] [(] 285 [)] [ENTER]

The calculator gives a result of approximately 3.8637. ■

PROBLEM 10
Find θ to the nearest degree if $\cot\theta = -1.8040$ and θ terminates in QII with $0° \le \theta < 360°$.

EXAMPLE 10 Find θ to the nearest degree if $\cot\theta = -1.6003$ and θ terminates in QII, with $0° \le \theta < 360°$.

SOLUTION To find the reference angle on a calculator, we ignore the negative sign in -1.6003 and use the fact that $\cot\theta$ is the reciprocal of $\tan\theta$.

$$\text{If } \cot\theta = 1.6003, \quad \text{then } \tan\theta = \frac{1}{1.6003}$$

From this last line we see that the keys to press are

Scientific Calculator	Graphing Calculator
1.6003 [1/x] [tan⁻¹]	[tan⁻¹] [(] 1.6003 [x⁻¹] [)] [ENTER]

To the nearest degree, the reference angle is $\hat{\theta} = 32°$. Because we want θ to terminate in QII, we subtract 32° from 180° to get $\theta = 148°$.

Again, we can check our result on a calculator by entering 148°, finding its tangent, and then finding the reciprocal of the result.

Scientific Calculator	Graphing Calculator
148 [tan] [1/x]	1 [÷] [tan] [(] 148 [)] [ENTER]

The calculator gives a result of approximately -1.6003. ■

Getting Ready for Class *After reading through the preceding section, respond in your own words and in complete sentences.*

a. Define *reference angle.*
b. State the reference angle theorem.
c. What is the first step in finding the exact value of $\cos 495°$?
d. Explain how to find θ to the nearest tenth of a degree, if $\tan\theta = -0.8541$ and θ terminates in QIV with $0° \le \theta < 360°$.

3.1 PROBLEM SET

➤ CONCEPTS AND VOCABULARY

For Questions 1 through 3, fill in each blank with the appropriate word or symbol.

1. For an angle θ in standard position, the ___________ angle is the positive acute angle between the terminal side of θ and the ___-axis.

2. The only possible difference between a trigonometric function of an angle and its reference angle will be the _____ of the value.

3. To find a reference angle using the SIN^{-1}, COS^{-1}, or TAN^{-1} keys on a calculator, always enter a __________ value.

4. Complete each statement regarding an angle θ and its reference angle $\hat{\theta}$.

a. If $\theta \in \text{QI}$, then $\hat{\theta} =$ _________.
b. If $\theta \in \text{QII}$, then $\hat{\theta} =$ _________.
c. If $\theta \in \text{QIII}$, then $\hat{\theta} =$ _________.
d. If $\theta \in \text{QIV}$, then $\hat{\theta} =$ _________.

➤ EXERCISES

Draw each of the following angles in standard position and then name the reference angle.

5. $150°$ **6.** $210°$ **7.** $253.8°$ **8.** $143.4°$
9. $311.7°$ **10.** $93.2°$ **11.** $195°\ 10'$ **12.** $171°\ 40'$
13. $-300°$ **14.** $-330°$ **15.** $-120°$ **16.** $-150°$

Find the exact value of each of the following.

17. $\cos 135°$ **18.** $\cos 225°$ **19.** $\sin 210°$ **20.** $\sin 120°$
21. $\tan 135°$ **22.** $\tan 315°$ **23.** $\cos(-240°)$ **24.** $\cos(-150°)$
25. $\sec(-330°)$ **26.** $\csc(-330°)$ **27.** $\csc 300°$ **28.** $\sec 300°$
29. $\sin 390°$ **30.** $\cos 420°$ **31.** $\cot 480°$ **32.** $\cot 510°$

Use a calculator to find the following.

33. $\cos 347°$ **34.** $\cos 238°$ **35.** $\sec 101.8°$ **36.** $\csc 166.7°$
37. $\tan 143.4°$ **38.** $\tan 253.8°$ **39.** $\sec 311.7°$ **40.** $\csc 93.2°$
41. $\cot 390°$ **42.** $\cot 420°$ **43.** $\csc 575.4°$ **44.** $\sec 590.9°$
45. $\sin(-225°)$ **46.** $\cos(-315°)$ **47.** $\tan 195°\ 10'$ **48.** $\tan 171°\ 40'$
49. $\csc 670°\ 20'$ **50.** $\sec 314°\ 40'$ **51.** $\sin(-120°)$ **52.** $\cos(-150°)$

Use the given information and a calculator to find θ to the nearest tenth of a degree if $0° \le \theta < 360°$.

53. $\sin\theta = -0.3090$ with θ in QIII
54. $\sin\theta = -0.3090$ with θ in QIV
55. $\cos\theta = -0.7660$ with θ in QII
56. $\cos\theta = -0.7660$ with θ in QIII
57. $\tan\theta = 0.5890$ with θ in QIII
58. $\tan\theta = 0.5890$ with θ in QI
59. $\cos\theta = 0.2644$ with θ in QI
60. $\cos\theta = 0.2644$ with θ in QIV
61. $\sin\theta = 0.9652$ with θ in QII
62. $\sin\theta = 0.9652$ with θ in QI
63. $\sec\theta = 1.4325$ with θ in QIV
64. $\csc\theta = 1.4325$ with θ in QII
65. $\csc\theta = 2.4957$ with θ in QII
66. $\sec\theta = -3.4159$ with θ in QII
67. $\cot\theta = -0.7366$ with θ in QII
68. $\cot\theta = -0.1234$ with θ in QIV
69. $\sec\theta = -1.7876$ with θ in QIII
70. $\csc\theta = -1.7876$ with θ in QIII

Find θ, $0° \le \theta < 360°$, given the following information.

71. $\sin\theta = -\dfrac{\sqrt{3}}{2}$ and θ in QIII
72. $\sin\theta = -\dfrac{\sqrt{2}}{2}$ and θ in QIII
73. $\cos\theta = -\dfrac{\sqrt{2}}{2}$ and θ in QII
74. $\cos\theta = -\dfrac{\sqrt{3}}{2}$ and θ in QIII
75. $\sin\theta = -\dfrac{\sqrt{3}}{2}$ and θ in QIV
76. $\sin\theta = \dfrac{\sqrt{2}}{2}$ and θ in QII
77. $\tan\theta = \sqrt{3}$ and θ in QIII
78. $\tan\theta = \dfrac{\sqrt{3}}{3}$ and θ in QIII
79. $\sec\theta = -2$ with θ in QII
80. $\csc\theta = 2$ with θ in QII
81. $\csc\theta = \sqrt{2}$ with θ in QII
82. $\sec\theta = \sqrt{2}$ with θ in QIV
83. $\cot\theta = -1$ with θ in QIV
84. $\cot\theta = \sqrt{3}$ with θ in QIII

➤ REVIEW PROBLEMS

The problems that follow review material we covered in Sections 1.1 and 2.1. Give the complement and supplement of each angle.

85. $70°$ **86.** $120°$ **87.** x **88.** $90° - y$

89. If the longest side in a 30°–60°–90° triangle is 10, find the length of the other two sides.

90. If the two shorter sides of a 45°–45°–90° triangle are both $\frac{3}{4}$, find the length of the hypotenuse.

Simplify each expression by substituting values from the table of exact values and then simplifying the resulting expression.

91. $\sin 30° \cos 60°$ **92.** $4 \sin 60° - 2 \cos 30°$

93. $\sin^2 45° + \cos^2 45°$ **94.** $(\sin 45° + \cos 45°)^2$

➤ LEARNING OBJECTIVES ASSESSMENT

These questions are available to help instructors assess if you have successfully met the learning objectives for this section.

95. Give the reference angle for 153°.

a. $27°$ **b.** $63°$ **c.** $-27°$ **d.** $-63°$

96. Use a reference angle to find the exact value of $\cos 210°$.

a. $-\frac{1}{2}$ **b.** $-\frac{\sqrt{3}}{2}$ **c.** $-\frac{\sqrt{2}}{2}$ **d.** $\frac{1}{2}$

97. Use a calculator to approximate $\csc(-304°)$.

a. 1.7883 **b.** 1.2062 **c.** 0.8290 **d.** 1.4826

98. Approximate θ if $0° \le \theta < 360°$ and $\cos \theta = -0.3256$ with θ in QIII.

a. $109°$ **b.** $199°$ **c.** $251°$ **d.** $213°$

SECTION 3.2 Radians and Degrees

LEARNING OBJECTIVES

1. Find the radian measure of a central angle given the radius and arc length.
2. Convert an angle from degrees to radians or vice versa.
3. Evaluate a trigonometric function using radians.
4. Identify the reference angle for a given angle measured in radians.

If you think back to the work you have done with functions of the form $y = f(x)$ in your algebra class, you will see that the variables x and y were always real numbers. The trigonometric functions we have worked with so far have had the form $y = f(\theta)$, where θ is measured in degrees. To apply the knowledge we have about functions from algebra to our trigonometric functions, we need to write our angles as real numbers, not degrees. The key to doing this is called *radian measure.*

Radian measure is a relatively new concept in the history of mathematics. The first printed use of the term *radian* was by physicist James T. Thomson in examination questions in 1873. It is believed that the concept of radian measure was originally proposed by Roger Cotes (1682–1716), who was also the first to calculate 1 radian in degrees. The introduction of radian measure will allow us to do a number of useful things. For instance, in Chapter 4 we will graph the function $y = \sin x$ on a rectangular coordinate system, where the units on the x- and y-axes are given by real numbers, just as they would be if we were graphing $y = 2x + 3$ or $y = x^2$.

To understand the definition for radian measure, we have to recall from geometry that a central angle in a circle is an angle with its vertex at the center of the circle. Here is the definition for an angle with a measure of 1 radian.

DEFINITION

In a circle, a central angle that cuts off an arc equal in length to the radius of the circle has a measure of 1 *radian* (rad). Figure 1 illustrates this.

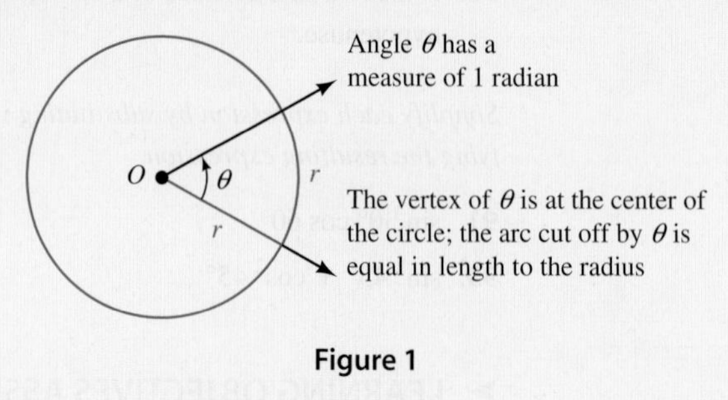

Figure 1

To find the radian measure of *any* central angle, we must find how many radii are in the arc it cuts off. To do so, we divide the arc length by the radius. If the radius is 2 centimeters and the arc cut off by central angle θ is 6 centimeters, then the radian measure of θ is $\frac{6}{2} = 3$ rad. Here is the formal definition:

DEFINITION ■ RADIAN MEASURE

If a central angle θ, in a circle of radius r, cuts off an arc of length s, then the measure of θ, in radians, is given by s/r (Figure 2).

$$\theta \text{ (in radians)} = \frac{s}{r}$$

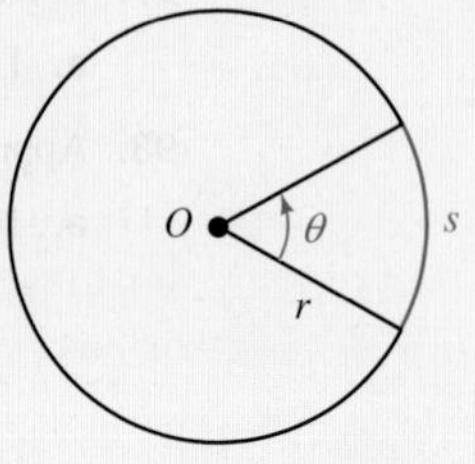

Figure 2

As you will see later in this section, one radian is equal to approximately 57.3°.

PROBLEM 1
A central angle θ in a circle of radius 5 centimeters cuts off an arc of length 15 centimeters. What is the radian measure of θ?

EXAMPLE 1 A central angle θ in a circle of radius 3 centimeters cuts off an arc of length 6 centimeters. What is the radian measure of θ?

SOLUTION We have $r = 3$ cm and $s = 6$ cm (Figure 3); therefore,

$$\begin{aligned} \theta \text{ (in radians)} &= \frac{s}{r} \\ &= \frac{6 \text{ cm}}{3 \text{ cm}} \\ &= 2 \end{aligned}$$

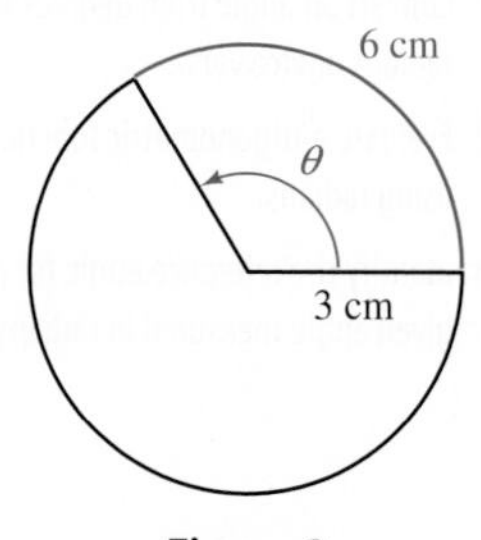

Figure 3

We say the radian measure of θ is 2, or $\theta = 2$ rad. ■

NOTE Because radian measure is defined as a ratio of two lengths, s/r, technically it is a unitless measure. That is, radians are just real numbers. To see why, look again at Example 1. Notice how the units of centimeters divide out, leaving the number 2 without any units. For this reason, it is common practice to omit the word *radian* when using radian measure. (To avoid confusion, we will sometimes use the term *rad* in this book as if it were a unit.) If no units are showing, an angle is understood to be measured in radians; with degree measure, the degree symbol ° must be written.

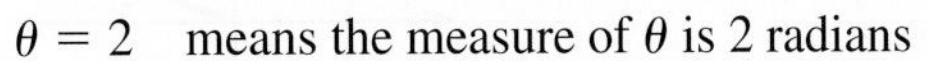

$\theta = 2$ means the measure of θ is 2 radians

$\theta = 2°$ means the measure of θ is 2 degrees

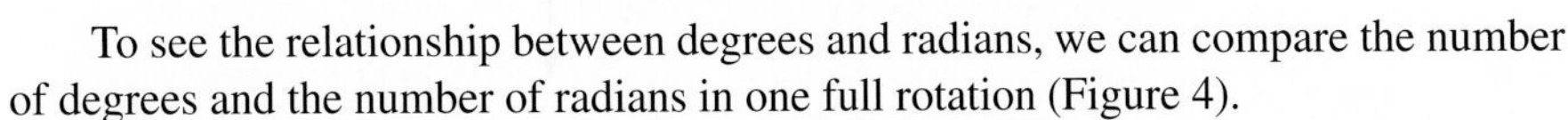

To see the relationship between degrees and radians, we can compare the number of degrees and the number of radians in one full rotation (Figure 4).

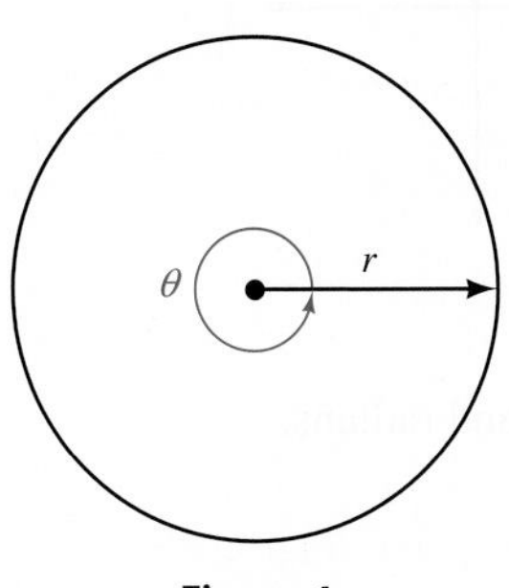

Figure 4

The angle formed by one full rotation about the center of a circle of radius r will cut off an arc equal to the circumference of the circle. Since the circumference of a circle of radius r is $2\pi r$, we have

$$\theta = \frac{2\pi r}{r} = 2\pi$$

θ measures one full rotation — The measure of θ in radians is 2π

Because one full rotation in degrees is 360°, we have the following relationship between radians and degrees.

$$360° = 2\pi \text{ rad}$$

Dividing both sides by 2 we have

$$180° = \pi \text{ rad}$$

NOTE We could also find this conversion factor using the definition of π, which is the ratio of the circumference to the diameter for any circle. Because

$$\pi = \frac{C}{2r} = \frac{\frac{1}{2}C}{r}$$

by our definition of radian measure π, is the angle in radians corresponding to an arc length of half the circumference of the circle. Therefore, π rad = 180°.

To obtain conversion factors that will allow us to change back and forth between degrees and radians, we divide both sides of this last equation alternately by 180 and by π.

$$180° = \pi \text{ rad}$$

Divide both sides by 180: $1° = \frac{\pi}{180} \text{ rad}$

Divide both sides by π: $\left(\frac{180}{\pi}\right)° = 1 \text{ rad}$

To gain some insight into the relationship between degrees and radians, we can approximate π with 3.14 to obtain the approximate number of degrees in 1 radian.

$$\begin{aligned} 1 \text{ rad} &= 1\left(\frac{180}{\pi}\right)° \\ &\approx 1\left(\frac{180}{3.14}\right)° \\ &= 57.3° \end{aligned}$$

To the nearest tenth

We see that 1 radian is approximately 57°. A radian is much larger than a degree. Figure 5 illustrates the relationship between 20° and 20 radians.

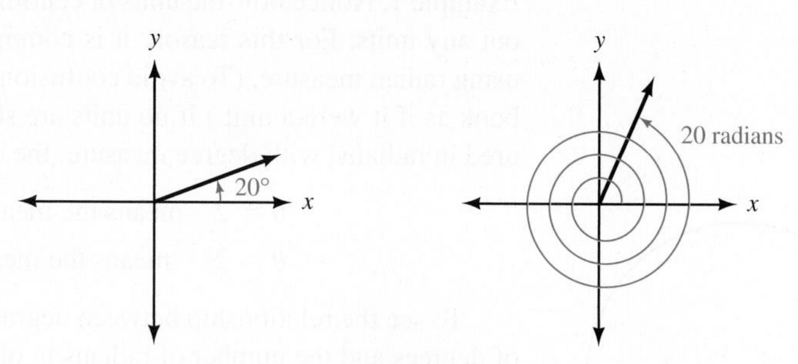

Figure 5

Here are some further conversions between degrees and radians.

Converting from Degrees to Radians

PROBLEM 2
Convert 50° to radians.

EXAMPLE 2 Convert 45° to radians.

SOLUTION Because $1° = \dfrac{\pi}{180}$ radians, and 45° is the same as 45(1°), we have

$$45° = 45\left(\frac{\pi}{180}\right) \text{ rad} = \frac{\pi}{4} \text{ rad}$$

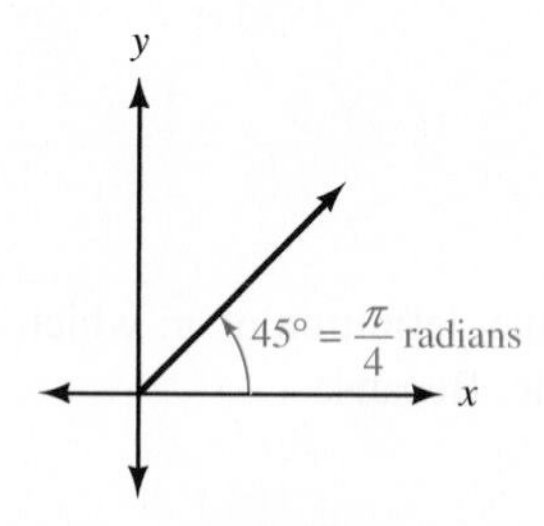

Figure 6

as illustrated in Figure 6. When we have our answer in terms of π, as in $\pi/4$, we are writing an exact value. If we wanted a decimal approximation, we would substitute 3.14 for π.

Exact value $\dfrac{\pi}{4} \approx \dfrac{3.14}{4} = 0.785$ Approximate value

PROBLEM 3
Convert 430° to radians.

EXAMPLE 3 Convert 450° to radians.

SOLUTION As illustrated in Figure 7, multiplying by $\pi/180$ we have

$$450° = 450\left(\frac{\pi}{180}\right) \text{ rad}$$

$$= \frac{5\pi}{2} \text{ rad}$$

$$\approx 7.85$$

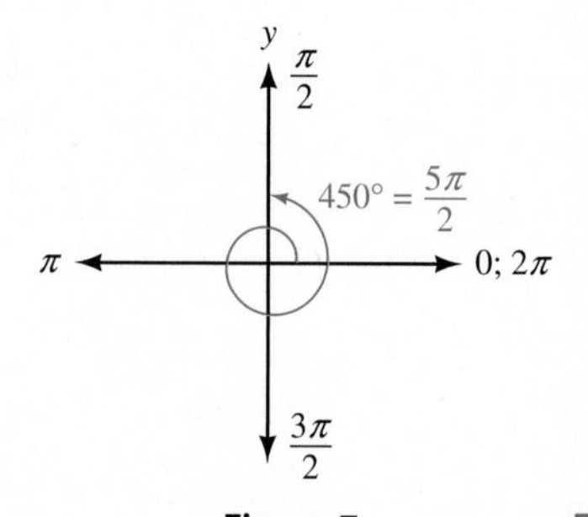

Figure 7

PROBLEM 4
Convert −47.5° to radians. Round to the nearest hundredth.

EXAMPLE 4 Convert −78.4° to radians. Round to the nearest hundredth.

SOLUTION We multiply by $\pi/180$ and then approximate using a calculator.

$$-78.4° = -78.4\left(\frac{\pi}{180}\right) \text{ rad} \approx -1.37 \text{ rad}$$

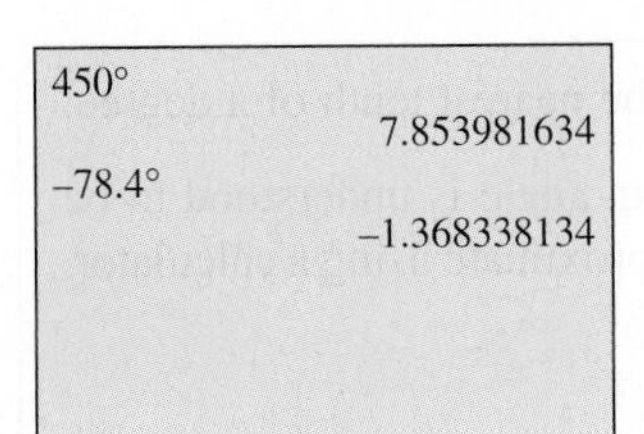

Figure 8

CALCULATOR NOTE Some calculators have the ability to convert angles from degree measure to radian measure, and vice versa. For example, Figure 8 shows how Examples 3 and 4 could be done on a graphing calculator that is set to radian mode. Consult your calculator manual to see if your model is able to perform angle conversions.

Converting from Radians to Degrees

PROBLEM 5

Convert $\frac{\pi}{3}$ to degrees.

EXAMPLE 5 Convert $\pi/6$ to degrees.

SOLUTION To convert from radians to degrees, we multiply by $180/\pi$.

$$\frac{\pi}{6}\text{ (rad)} = \frac{\pi}{6}\left(\frac{180}{\pi}\right)^{\circ} = 30^{\circ}$$

Note that 60° is twice 30°, so $2(\pi/6) = \pi/3$ must be the radian equivalent of 60°. Figure 9 illustrates.

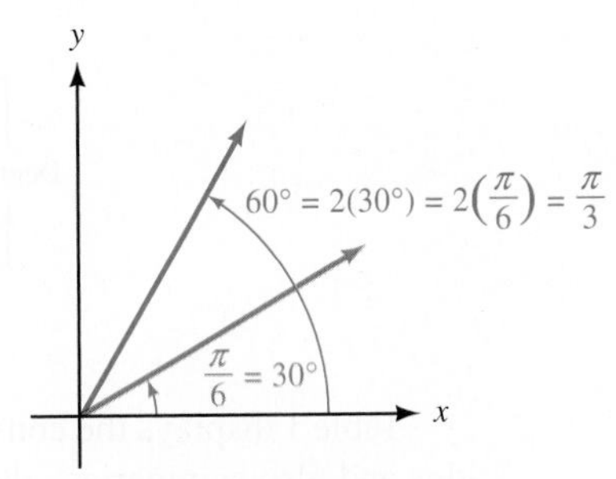

Figure 9

PROBLEM 6

Convert $\frac{7\pi}{3}$ to degrees.

EXAMPLE 6 Convert $4\pi/3$ to degrees.

SOLUTION Multiplying by $180/\pi$ we have

$$\frac{4\pi}{3}\text{ (rad)} = \frac{4\pi}{3}\left(\frac{180}{\pi}\right)^{\circ} = 240^{\circ}$$

Or, knowing that $\pi/3 = 60^{\circ}$, we could write

$$\frac{4\pi}{3} = 4 \cdot \frac{\pi}{3} = 4(60^{\circ}) = 240^{\circ}$$

Note that the reference angle for the angle shown in Figure 10 can be given in either degrees or radians.

$$\text{In degrees: } \hat{\theta} = 240^{\circ} - 180^{\circ}$$
$$= 60^{\circ}$$
$$\text{In radians: } \hat{\theta} = \frac{4\pi}{3} - \pi$$
$$= \frac{4\pi}{3} - \frac{3\pi}{3}$$
$$= \frac{\pi}{3}$$

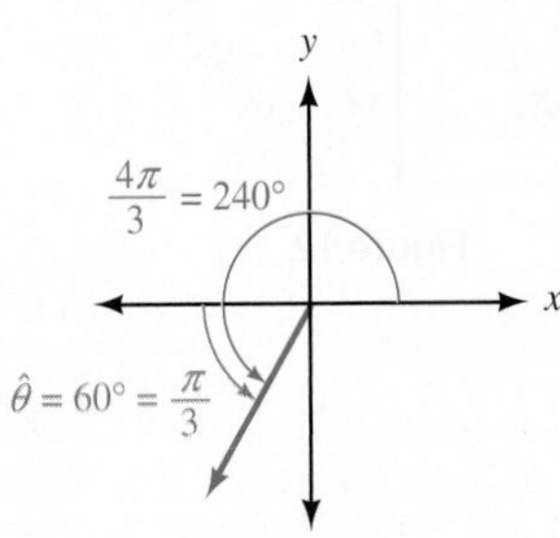

Figure 10

PROBLEM 7
Convert -3.4 to degrees. Round to the nearest tenth of a degree.

EXAMPLE 7 Convert -4.5 to degrees. Round to the nearest tenth of a degree.

SOLUTION Remember that if no units are showing, an angle is understood to be measured in radians. We multiply by $180/\pi$ and then approximate using a calculator.

$$-4.5 = -4.5\left(\frac{180}{\pi}\right)^{\circ}$$

$$\approx -257.8^{\circ}$$

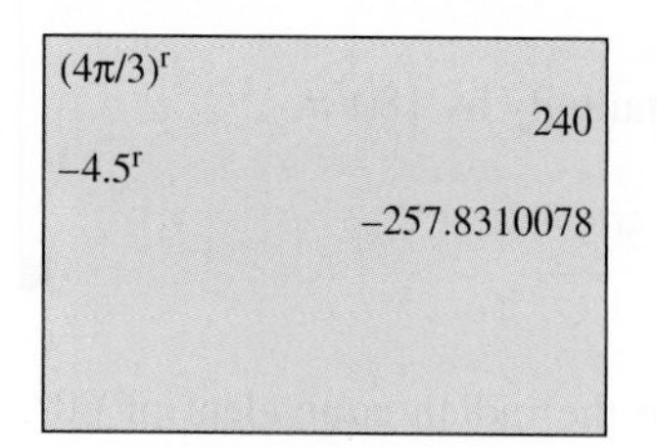

Figure 11

CALCULATOR NOTE Figure 11 shows how Examples 6 and 7 could be done on a graphing calculator that is set to degree mode. If your model is able to perform angle conversions, try these examples and make sure you can get the same values.

As is apparent from the preceding examples, changing from degrees to radians and radians to degrees is simply a matter of multiplying by the appropriate conversion factors.

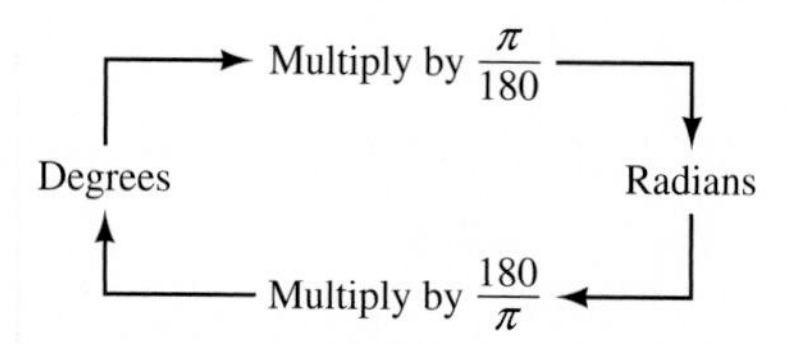

Table 1 displays the conversions between degrees and radians for the special angles and also summarizes the exact values of the sine, cosine, and tangent of these angles for your convenience. In Figure 12 we show each of the special angles in both degrees and radians for all four quadrants.

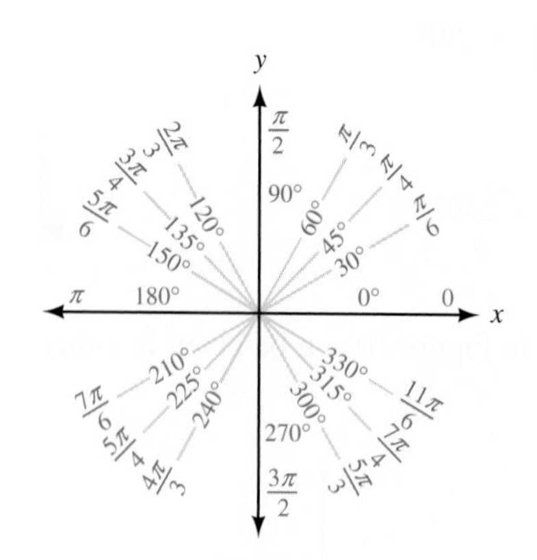

Figure 12

TABLE 1

θ		Value of Trigonometric Function		
Degrees	*Radians*	*sin θ*	*cos θ*	*tan θ*
0°	0	0	1	0
30°	$\frac{\pi}{6}$	$\frac{1}{2}$	$\frac{\sqrt{3}}{2}$	$\frac{\sqrt{3}}{3}$
45°	$\frac{\pi}{4}$	$\frac{\sqrt{2}}{2}$	$\frac{\sqrt{2}}{2}$	1
60°	$\frac{\pi}{3}$	$\frac{\sqrt{3}}{2}$	$\frac{1}{2}$	$\sqrt{3}$
90°	$\frac{\pi}{2}$	1	0	undefined
180°	π	0	-1	0
270°	$\frac{3\pi}{2}$	-1	0	undefined
360°	2π	0	1	0

PROBLEM 8

Find $\sin \frac{2\pi}{3}$.

EXAMPLE 8 Find $\sin \frac{\pi}{6}$.

SOLUTION Because $\pi/6$ and $30°$ are equivalent, so are their sines.

$$\sin \frac{\pi}{6} = \sin 30° = \frac{1}{2}$$

CALCULATOR NOTE To work this problem on a calculator, we must first set the calculator to radian mode. (Consult the manual that came with your calculator to see how to do this.) If your calculator does not have a key labeled π, use 3.1416. Here is the sequence to key in your calculator to work the problem given in Example 8.

Scientific Calculator	Graphing Calculator
3.1416 [÷] 6 [=] [sin]	[sin] [(] [π] [÷] 6 [)] [ENTER]

PROBLEM 9

Find $5 \cos \frac{\pi}{4}$.

EXAMPLE 9 Find $4 \sin \frac{7\pi}{6}$.

SOLUTION Because $7\pi/6$ is just slightly larger than $\pi = 6\pi/6$, we know that $7\pi/6$ terminates in QIII and therefore its sine will be negative (Figure 13).

The reference angle is

$$\frac{7\pi}{6} - \pi = \frac{7\pi}{6} - \frac{6\pi}{6} = \frac{\pi}{6}$$

Then

$$4 \sin \frac{7\pi}{6} = 4\left(-\sin \frac{\pi}{6}\right) = 4\left(-\frac{1}{2}\right) = -2$$

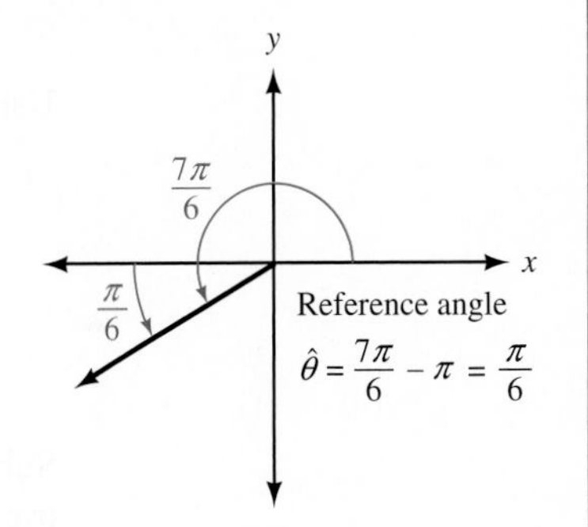

Figure 13

PROBLEM 10

Evaluate $3 \cos (2x + \pi)$ when $x = \pi/6$.

EXAMPLE 10 Evaluate $4 \sin (2x + \pi)$ when $x = \pi/6$.

SOLUTION Substituting $\pi/6$ for x and simplifying, we have

$$\begin{aligned} 4 \sin\left(2 \cdot \frac{\pi}{6} + \pi\right) &= 4 \sin\left(\frac{\pi}{3} + \pi\right) \\ &= 4 \sin \frac{4\pi}{3} \\ &= 4\left(-\sin \frac{\pi}{3}\right) \\ &= 4\left(-\frac{\sqrt{3}}{2}\right) \\ &= -2\sqrt{3} \end{aligned}$$

PROBLEM 11
Repeat Example 11, but use coordinates P_1(N 36° 43.403′, E 25° 16.930′) and P_2(N 36° 23.169′, E 25° 25.819′) corresponding to Ios and Santorini.

EXAMPLE 11 In navigation, distance is not usually measured along a straight line, but along a great circle because the earth is round (Figure 14). The formula to determine the great circle distance between two points $P_1(LT_1, LN_1)$ and $P_2(LT_2, LN_2)$ whose coordinates are given as latitudes and longitudes involves the expression

$$\sin (LT_1) \sin (LT_2) + \cos (LT_1) \cos (LT_2) \cos (LN_1 - LN_2)$$

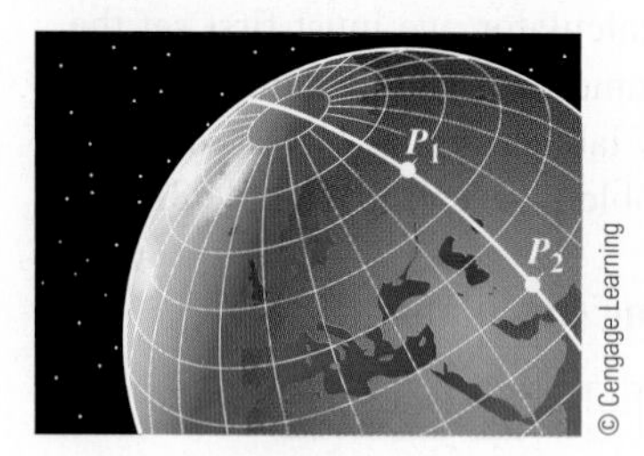

Figure 14

To use this formula, the latitudes and longitudes must be entered as angles in radians. However, most GPS (global positioning system) units give these coordinates in degrees and minutes. To use the formula thus requires converting from degrees to radians.

Evaluate this expression for the pair of coordinates P_1(N 32° 22.108′, W 64° 41.178′) and P_2(N 13° 04.809′, W 59° 29.263′) corresponding to Bermuda and Barbados, respectively.

SOLUTION First, we must convert each angle to radians.

$$\begin{aligned} LT_1 = 32° \; 22.108' &= 32° + \left(\frac{22.108}{60}\right)^{\circ} \\ &\approx 32.3685° \\ &= 32.3685\left(\frac{\pi}{180}\right) \text{ rad} \\ &\approx 0.565 \text{ rad} \end{aligned}$$

Using similar steps for the other three angles, we have

$$LN_1 = 64° \; 41.178' \approx 1.13 \text{ rad}$$

$$LT_2 = 13° \; 04.809' \approx 0.228 \text{ rad}$$

$$LN_2 = 59° \; 29.263' \approx 1.04 \text{ rad}$$

Substituting the angles into the expression and using a calculator to evaluate each trigonometric function, we obtain

$$\begin{aligned} &\sin (LT_1) \sin (LT_2) + \cos (LT_1) \cos (LT_2) \cos (LN_1 - LN_2) \\ &= \sin (0.565) \sin (0.228) + \cos (0.565) \cos (0.228) \cos (1.13 - 1.04) \\ &\approx (0.5354)(0.2260) + (0.8446)(0.9741)(0.9960) \\ &= 0.9404 \end{aligned}$$

Getting Ready for Class

After reading through the preceding section, respond in your own words and in complete sentences.

a. What is a radian?
b. How is radian measure different from degree measure?
c. Explain how to convert an angle from radians to degrees.
d. Explain how to convert an angle from degrees to radians.

3.2 PROBLEM SET

➤ CONCEPTS AND VOCABULARY

For Questions 1 and 2, fill in each blank with the appropriate word or number.

1. A radian is the measure of a central angle in a circle that cuts off an arc equal in length to the ________ of the circle.

2. When converting between degree measure and radian measure, use the fact that ______ degrees equals _______ radians.

3. Complete each statement regarding an angle θ and its reference angle $\hat{\theta}$ using *radian* measure.

a. If $\theta \in$ QI, then $\hat{\theta} =$ _________. **c.** If $\theta \in$ QIII, then $\hat{\theta} =$ _________.

b. If $\theta \in$ QII, then $\hat{\theta} =$ _________. **d.** If $\theta \in$ QIV, then $\hat{\theta} =$ _________.

4. Match each angle (a-d) with its corresponding radian measure (i-iv).

a. 30° **b.** 45° **c.** 60° **d.** 90°

i. $\frac{\pi}{2}$ **ii.** $\frac{\pi}{3}$ **iii.** $\frac{\pi}{6}$ **iv.** $\frac{\pi}{4}$

➤ EXERCISES

Find the radian measure of angle θ, if θ is a central angle in a circle of radius r, and θ cuts off an arc of length s.

5. $r = 3$ cm, $s = 9$ cm

6. $r = 10$ inches, $s = 5$ inches

7. $r = 12$ inches, $s = 3\pi$ inches

8. $r = 4$ inches, $s = 12\pi$ inches

9. $r = \frac{1}{4}$ cm, $s = \frac{1}{2}$ cm

10. $r = \frac{1}{4}$ cm, $s = \frac{1}{8}$ cm

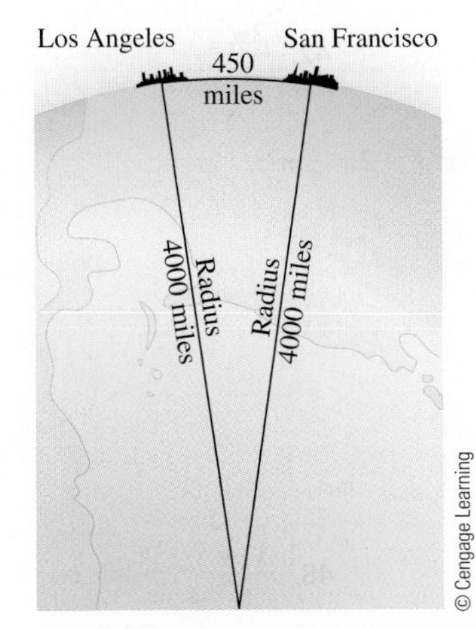

Figure 15

11. Angle Between Cities Los Angeles and San Francisco are approximately 450 miles apart on the surface of the earth. Assuming that the radius of the earth is 4,000 miles, find the radian measure of the central angle with its vertex at the center of the earth that has Los Angeles on one side and San Francisco on the other side (Figure 15).

12. Angle Between Cities Los Angeles and New York City are approximately 2,500 miles apart on the surface of the earth. Assuming that the radius of the earth is 4,000 miles, find the radian measure of the central angle with its vertex at the center of the earth that has Los Angeles on one side and New York City on the other side.

For each of the following angles,

a. *draw the angle in standard position.*

b. *convert to radian measure using exact values.*

c. *name the reference angle in both degrees and radians.*

13. 30° **14.** 135° **15.** 260° **16.** 340°

17. −150° **18.** −120° **19.** 420° **20.** 390°

For Problems 21 through 24, use 3.1416 for π unless your calculator has a key marked π.

21. Use a calculator to convert 120° 40′ to radians. Round your answer to the nearest hundredth. (First convert to decimal degrees, then multiply by the appropriate conversion factor to convert to radians.)

22. Use a calculator to convert 256° 20′ to radians to the nearest hundredth of a radian.

23. Use a calculator to convert 1′ (1 minute) to radians to three significant digits.

24. Use a calculator to convert 1° to radians to three significant digits.

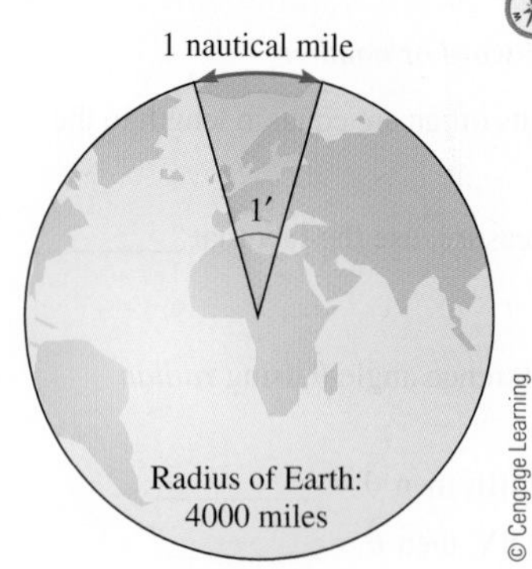

Figure 16

© Cengage Learning

Nautical Miles *If a central angle with its vertex at the center of the earth has a measure of 1′, then the arc on the surface of the earth that is cut off by this angle (known as the* great circle distance*) has a measure of 1 nautical mile (Figure 16).*

25. Find the number of regular (statute) miles in 1 nautical mile to the nearest hundredth of a mile. (Use 4,000 miles for the radius of the earth.)

26. If two ships are 20 nautical miles apart on the ocean, how many statute miles apart are they? (Use the result of Problem 25 to do the calculations.)

27. If two ships are 70 nautical miles apart on the ocean, what is the great circle distance between them in radians?

28. If the great circle distance between two ships on the ocean is 0.1 radian, what is this distance in nautical miles?

29. Clock Through how many radians does the minute hand of a clock turn during a 5-minute period?

30. Clock Through how many radians does the minute hand of a clock turn during a 25-minute period?

Simplify each expression.

31. $\pi - \frac{\pi}{3}$ **32.** $\pi + \frac{\pi}{6}$ **33.** $2\pi - \frac{\pi}{4}$ **34.** $2\pi - \frac{\pi}{3}$

35. $\frac{\pi}{6} + \frac{\pi}{2}$ **36.** $\frac{\pi}{6} - \frac{\pi}{3}$ **37.** $\frac{5\pi}{4} - \frac{\pi}{2}$ **38.** $\frac{3\pi}{4} + \frac{\pi}{2}$

Write each angle as a sum or difference involving π. For example, $5\pi/6 = \pi - \pi/6$.

39. $\frac{2\pi}{3}$ **40.** $\frac{4\pi}{3}$ **41.** $\frac{7\pi}{6}$ **42.** $\frac{3\pi}{4}$

Write each angle as a difference involving 2π. For example, $5\pi/3 = 2\pi - \pi/3$.

43. $\frac{7\pi}{4}$ **44.** $\frac{11\pi}{6}$

For each of the following angles,

a. *draw the angle in standard position.*

b. *convert to degree measure.*

c. *label the reference angle in both degrees and radians.*

45. $\frac{\pi}{3}$ **46.** $\frac{3\pi}{4}$ **47.** $\frac{4\pi}{3}$ **48.** $\frac{11\pi}{6}$

49. $-\frac{7\pi}{6}$ **50.** $-\frac{5\pi}{3}$ **51.** $\frac{11\pi}{4}$ **52.** $\frac{7\pi}{3}$

Use a calculator to convert each of the following to degree measure to the nearest tenth of a degree.

53. 1 **54.** 2.4 **55.** 0.25 **56.** 5

Give the exact value of each of the following:

57. $\sin \frac{4\pi}{3}$ **58.** $\cos \frac{5\pi}{3}$ **59.** $\tan \frac{\pi}{6}$ **60.** $\cot \frac{\pi}{3}$

61. $\sec \frac{2\pi}{3}$ **62.** $\csc \frac{3\pi}{2}$ **63.** $\csc \frac{5\pi}{6}$ **64.** $\sec \frac{5\pi}{6}$

65. $-\sin \frac{\pi}{6}$ **66.** $-\cos \frac{\pi}{6}$ **67.** $4 \cos \left(-\frac{\pi}{4}\right)$ **68.** $4 \sin \left(-\frac{\pi}{4}\right)$

Evaluate each of the following expressions when x is π/6. In each case, use exact values.

69. $\sin 2x$ **70.** $\cos 3x$ **71.** $\frac{1}{2} \cos 2x$ **72.** $6 \sin 3x$

73. $3 - \sin x$ **74.** $2 + \cos x$ **75.** $\sin \left(x + \frac{\pi}{2}\right)$ **76.** $\sin \left(x - \frac{\pi}{3}\right)$

77. $4 \cos \left(2x + \frac{\pi}{3}\right)$ **78.** $-4 \sin \left(3x + \frac{\pi}{6}\right)$

79. $4 - \frac{2}{3} \sin (3x - \pi)$ **80.** $-1 + \frac{3}{4} \cos \left(2x - \frac{\pi}{2}\right)$

For the following expressions, find the value of y that corresponds to each value of x, then write your results as ordered pairs (x, y).

81. $y = \sin x$ for $x = 0, \frac{\pi}{4}, \frac{\pi}{2}, \frac{3\pi}{4}, \pi$

82. $y = \cos x$ for $x = 0, \frac{\pi}{4}, \frac{\pi}{2}, \frac{3\pi}{4}, \pi$

83. $y = -\cos x$ for $x = 0, \frac{\pi}{2}, \pi, \frac{3\pi}{2}, 2\pi$

84. $y = -\sin x$ for $x = 0, \frac{\pi}{2}, \pi, \frac{3\pi}{2}, 2\pi$

85. $y = \frac{1}{2} \cos x$ for $x = 0, \frac{\pi}{2}, \pi, \frac{3\pi}{2}, 2\pi$

86. $y = 2 \sin x$ for $x = 0, \frac{\pi}{2}, \pi, \frac{3\pi}{2}, 2\pi$

87. $y = \sin 2x$ for $x = 0, \frac{\pi}{4}, \frac{\pi}{2}, \frac{3\pi}{4}, \pi$

88. $y = \cos 3x$ for $x = 0, \frac{\pi}{6}, \frac{\pi}{3}, \frac{\pi}{2}, \frac{2\pi}{3}$

89. $y = \cos \left(x - \frac{\pi}{6}\right)$ for $x = \frac{\pi}{6}, \frac{\pi}{3}, \frac{2\pi}{3}, \pi, \frac{7\pi}{6}$

90. $y = \sin \left(x + \frac{\pi}{2}\right)$ for $x = -\frac{\pi}{2}, 0, \frac{\pi}{2}, \pi, \frac{3\pi}{2}$

91. $y = 2 + \cos x$ for $x = 0, \frac{\pi}{2}, \pi, \frac{3\pi}{2}, 2\pi$

92. $y = -3 + \sin x \quad$ for $x = 0, \frac{\pi}{2}, \pi, \frac{3\pi}{2}, 2\pi$

93. $y = 3 \sin\left(2x + \frac{\pi}{2}\right) \quad$ for $x = -\frac{\pi}{4}, 0, \frac{\pi}{4}, \frac{\pi}{2}, \frac{3\pi}{4}$

94. $y = 5 \cos\left(2x - \frac{\pi}{3}\right) \quad$ for $x = \frac{\pi}{6}, \frac{\pi}{3}, \frac{2\pi}{3}, \pi, \frac{7\pi}{6}$

95. Cycling The Campagnolo Hyperon carbon wheel has 22 spokes evenly distributed around the rim of the wheel. What is the measure, in radians, of the central angle formed by adjacent pairs of spokes (Figure 17)?

Figure 17

Figure 18

96. Cycling The Reynolds Stratus DV carbon wheel has 16 spokes evenly distributed around the rim of the wheel. What is the measure, in radians, of the central angle formed by adjacent spokes (Figure 18)?

Navigation *The formula to determine the great circle distance between two points $P_1(LT_1, LN_1)$ and $P_2(LT_2, LN_2)$ whose coordinates are given as latitudes and longitudes involves the expression*

$$\sin(LT_1)\sin(LT_2) + \cos(LT_1)\cos(LT_2)\cos(LN_1 - LN_2)$$

To use this formula, the latitudes and longitudes must be entered as angles in radians (see Example 11). Find the value of this expression for each given pair of coordinates.

97. P_1(N 21° 53.896′, W 159° 36.336′) Kauai, and

P_2(N 19° 43.219′, W 155° 02.930′) Hawaii

98. P_1(N 35° 51.449′, E 14° 28.668′) Malta, and

P_2(N 36° 24.155′, E 25° 28.776′) Santorini

➤ REVIEW PROBLEMS

The problems that follow review material we covered in Section 1.3.
Find all six trigonometric functions of θ, if the given point is on the terminal side of θ.

99. $(1, -3)$ **100.** $(-1, 3)$

101. Find the remaining trigonometric functions of θ, if $\sin\theta = \frac{1}{2}$ and θ terminates in QII.

102. Find the remaining trigonometric functions of θ, if $\cos\theta = -\sqrt{2}/2$ and θ terminates in QII.

103. Find the six trigonometric functions of θ, if the terminal side of θ lies along the line $y = 2x$ in QI.

104. Find the six trigonometric functions of θ, if the terminal side of θ lies along the line $y = 2x$ in QIII.

➤ LEARNING OBJECTIVES ASSESSMENT

These questions are available for instructors to help assess if you have successfully met the learning objectives for this section.

105. Find the radian measure of θ if θ is a central angle in a circle of radius 8 cm that cuts off an arc of length 20 cm.

a. 0.4 **b.** 2.5 **c.** 160 **d.** 143

106. Convert 80° to radians.

a. $\dfrac{7\pi}{18}$ **b.** $\dfrac{4\pi}{9}$ **c.** $\dfrac{5\pi}{11}$ **d.** $\dfrac{9\pi}{20}$

107. Which of the following statements is false?

a. $\cos \pi = 0$ **b.** $\tan \dfrac{\pi}{3} = \sqrt{3}$ **c.** $\sin \dfrac{\pi}{2} = 1$ **d.** $\sec \dfrac{\pi}{4} = \sqrt{2}$

108. Give the reference angle for $\dfrac{11\pi}{6}$.

a. π **b.** $\dfrac{\pi}{2}$ **c.** $\dfrac{\pi}{3}$ **d.** $\dfrac{\pi}{6}$

SECTION 3.3 Definition III: Circular Functions

LEARNING OBJECTIVES

1. Evaluate a trigonometric function using the unit circle.
2. Find the value of a trigonometric function given a point on the unit circle.
3. Use a calculator to approximate the value of a trigonometric function for an angle in radians.
4. Use the unit circle to answer a conceptual question about a trigonometric function.

The origins of the trigonometric functions are actually found in astronomy and the need to find the length of the chord subtended by the central angle of a circle. The Greek mathematician Hipparchus is believed to have been the first to produce a table of chords in 140 B.C., making him the founder of trigonometry in the eyes of many. This table is essentially a table of values of the sine function, because the sine of a central angle on the unit circle is half the chord of twice the angle (Figure 1). In modern notation,

$$\text{chord}(\theta) = AC = 2AB = 2\sin\left(\frac{\theta}{2}\right)$$

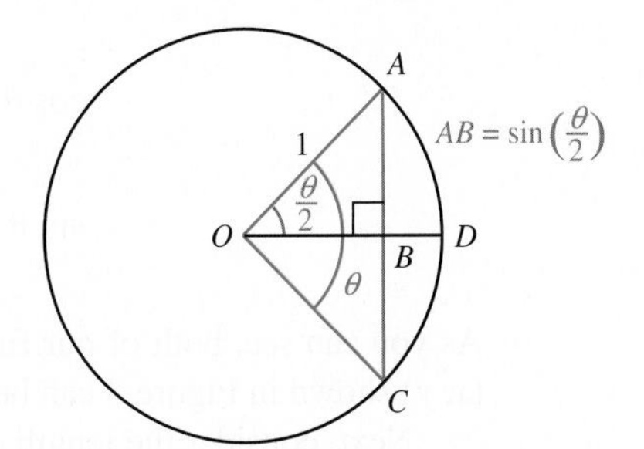

Figure 1

The first explicit use of the sine as a function of an angle occurs in a table of half-chords by the Hindu mathematician and astronomer Aryabhata around the year 510 in his work the *Aryabhatiya*. The word *ardha-jya* was used for the half-chord, which was eventually shortened to simply *jya* or *jiva*. When the Arabs translated the *Aryabhatiya,* they replaced *jya* with the term *jiba,* a meaningless word that sounded the same. The word *jiba* evolved into *jaib,* meaning fold, bosom, or bay. This was then later translated into Latin as *sinus,* which later became *sine* in English.

In this section, we will give a third and final definition for the trigonometric functions. Rather than give the new definition first and then show that it does not conflict with our previous definitions, we will do the reverse and show that either of our first two definitions leads to conclusions that imply a third definition.

To begin, recall that the unit circle (Figure 2) is the circle with center at the origin and radius 1. The equation of the unit circle is $x^2 + y^2 = 1$.

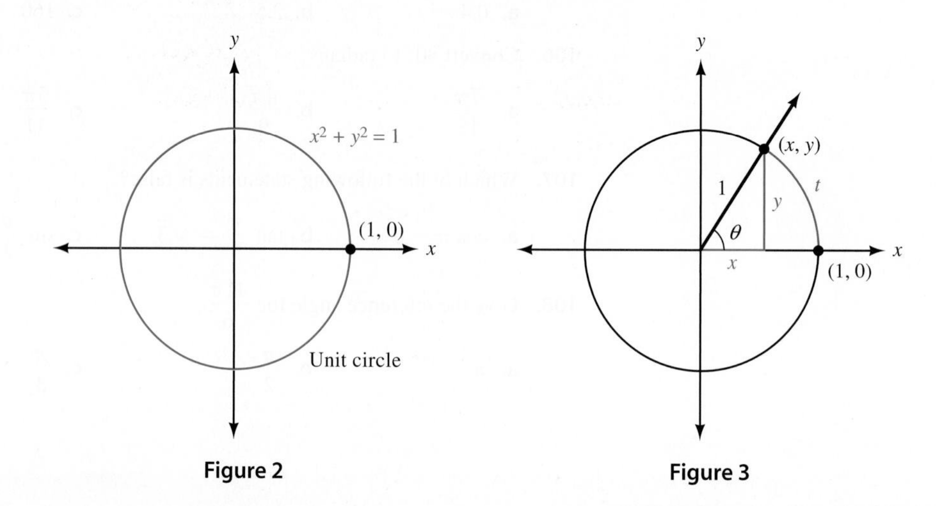

Figure 2

Figure 3

Suppose the terminal side of angle θ, in standard position, intersects the unit circle at point (x, y) as shown in Figure 3. Because the radius of the unit circle is 1, the distance from the origin to the point (x, y) is 1. We can use our first definition for the trigonometric functions from Section 1.3 to write

$$\cos\theta = \frac{x}{r} = \frac{x}{1} = x \quad \text{and} \quad \sin\theta = \frac{y}{r} = \frac{y}{1} = y$$

In other words, the ordered pair (x, y) shown in Figure 3 can also be written as $(\cos\theta, \sin\theta)$. We can arrive at this same result by applying our second definition for the trigonometric functions (from Section 2.1). Because θ is an acute angle in a right triangle, by Definition II we have

$$\cos\theta = \frac{\text{side adjacent } \theta}{\text{hypotenuse}} = \frac{x}{1} = x$$

$$\sin\theta = \frac{\text{side opposite } \theta}{\text{hypotenuse}} = \frac{y}{1} = y$$

As you can see, both of our first two definitions lead to the conclusion that the point (x, y) shown in Figure 3 can be written as $(\cos\theta, \sin\theta)$.

Next, consider the length of the arc from $(1, 0)$ to (x, y), which we have labeled t in Figure 3. If θ is measured in radians, then by definition

$$\theta = \frac{t}{r} = \frac{t}{1} = t$$

The length of the arc from $(1, 0)$ to (x, y) is exactly the same as the radian measure of angle θ. Therefore, we can write

$$\cos\theta = \cos t = x \quad \text{and} \quad \sin\theta = \sin t = y$$

These results give rise to a third definition for the trigonometric functions.

DEFINITION III ■ CIRCULAR FUNCTIONS

If (x, y) is any point on the unit circle, and t is the distance from $(1, 0)$ to (x, y) along the circumference of the unit circle (Figure 4), then,

$$\cos t = x$$
$$\sin t = y$$
$$\tan t = \frac{y}{x} \quad (x \neq 0)$$
$$\cot t = \frac{x}{y} \quad (y \neq 0)$$
$$\csc t = \frac{1}{y} \quad (y \neq 0)$$
$$\sec t = \frac{1}{x} \quad (x \neq 0)$$

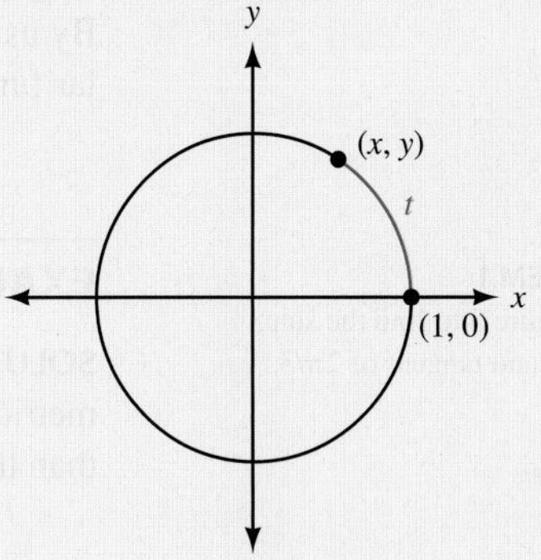

Figure 4

As we travel around the unit circle starting at $(1, 0)$, the points we come across all have coordinates $(\cos t, \sin t)$, where t is the distance we have traveled. (Note that t will be positive if we travel in the counterclockwise direction but negative if we travel in the clockwise direction.) When we define the trigonometric functions this way, we call them *circular functions* because of their relationship to the unit circle.

Figure 5 shows an enlarged version of the unit circle with multiples of $\pi/6$ and $\pi/4$ marked off. Each angle is given in both degrees and radians. The radian measure of each angle is the same as the distance from $(1, 0)$ to the point on the terminal side of the angle, as measured along the circumference of the circle in a counterclockwise direction. The x- and y-coordinate of each point shown are the cosine and sine, respectively, of the associated angle or distance.

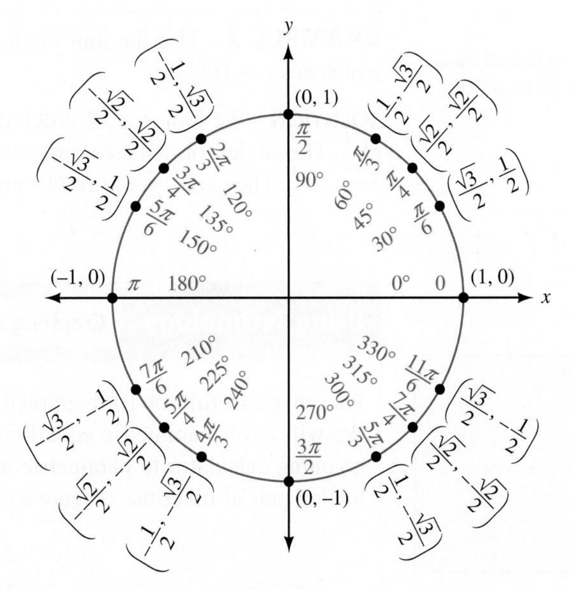

Figure 5

Figure 5 is helpful in visualizing the relationships among the angles shown and the trigonometric functions of those angles. You may want to make a larger copy of this diagram yourself. In the process of doing so you will become more familiar with the relationship between degrees and radians and the exact values of the angles in the diagram. Keep in mind, though, that you do not need to memorize this entire figure. By using reference angles, it is really only necessary to know the values of the circular functions for the quadrantal angles and angles that terminate in QI.

PROBLEM 1
Use Figure 5 to find the sine, cosine, and tangent of $2\pi/3$.

EXAMPLE 1 Use Figure 5 to find the six trigonometric functions of $5\pi/6$.

SOLUTION We obtain cosine and sine directly from Figure 5. The other trigonometric functions of $5\pi/6$ are found by using the ratio and reciprocal identities, rather than the new definition.

$$\sin\frac{5\pi}{6} = y = \frac{1}{2}$$

$$\cos\frac{5\pi}{6} = x = -\frac{\sqrt{3}}{2}$$

$$\tan\frac{5\pi}{6} = \frac{\sin(5\pi/6)}{\cos(5\pi/6)} = \frac{1/2}{-\sqrt{3}/2} = -\frac{1}{\sqrt{3}} = -\frac{\sqrt{3}}{3}$$

$$\cot\frac{5\pi}{6} = \frac{1}{\tan(5\pi/6)} = \frac{1}{-1/\sqrt{3}} = -\sqrt{3}$$

$$\sec\frac{5\pi}{6} = \frac{1}{\cos(5\pi/6)} = \frac{1}{-\sqrt{3}/2} = -\frac{2}{\sqrt{3}} = -\frac{2\sqrt{3}}{3}$$

$$\csc\frac{5\pi}{6} = \frac{1}{\sin(5\pi/6)} = \frac{1}{1/2} = 2$$

PROBLEM 2
Use the unit circle to find all values of t between 0 and 2π for which $\sin t = 1/2$.

EXAMPLE 2 Use the unit circle to find all values of t between 0 and 2π for which $\cos t = 1/2$.

SOLUTION We look for all ordered pairs on the unit circle with an x-coordinate of $\frac{1}{2}$. The angles, or distances, associated with these points are the angles for which $\cos t = \frac{1}{2}$. They are $t = \pi/3$ or 60° and $t = 5\pi/3$ or 300°.

Using Technology Graphing the Unit Circle

We can use Definition III to graph the unit circle with a set of parametric equations. We will cover parametric equations in more detail in Section 6.4. First, set your graphing calculator to parametric mode and to radian mode. Then define the following pair of functions (Figure 6).

$$X_{1T} = \cos t$$
$$Y_{1T} = \sin t$$

```
Plot1 Plot2 Plot3
\X1T=cos(T)
 Y1T=sin(T)
\X2T=
 Y2T=
\X3T=
 Y3T=
\X4T=
```

Figure 6

Set the window variables so that

$$0 \le t \le 2\pi, \text{ scale} = \pi/12; \ -1.5 \le x \le 1.5; \ -1.5 \le y \le 1.5$$

Graph the equations using the zoom-square command. This will make the circle appear round and not like an ellipse. Now you can trace the circle in increments of $\pi/12$. For example, to find the sine and cosine of $\pi/3$ radians, we would trace through four increments (Figure 7).

$$4\left(\frac{\pi}{12}\right) = \frac{4\pi}{12}$$
$$= \frac{\pi}{3}$$
$$\approx 1.0472$$

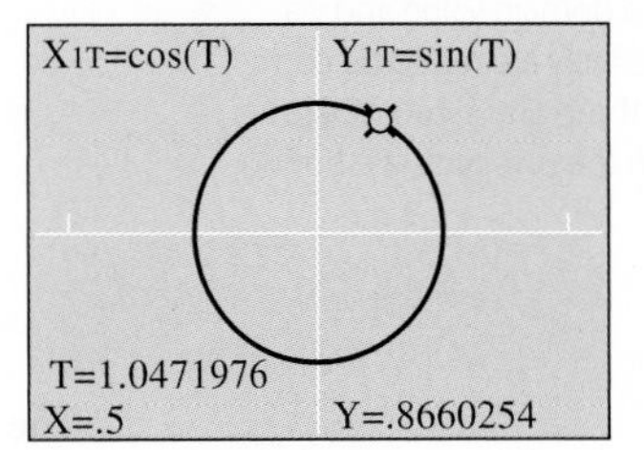

Figure 7

We see that $\cos(\pi/3) = 0.5$ and $\sin(\pi/3) = 0.8660$ to the nearest ten-thousandth.

Because it is easier for a calculator to display angles exactly in degrees rather than in radians, it is sometimes helpful to graph the unit circle using degrees. Set your graphing calculator to degree mode. Change the window settings so that $0 \le t \le 360$ with scale $= 5$. Graph the circle once again. Now when you trace the graph, each increment represents 5°.

We can use this graph to verify our results in Example 2. Trace around the unit circle until the x-coordinate is 0.5. The calculator shows us there are two such points, at $t = 60°$ and at $t = 300°$ (Figure 8).

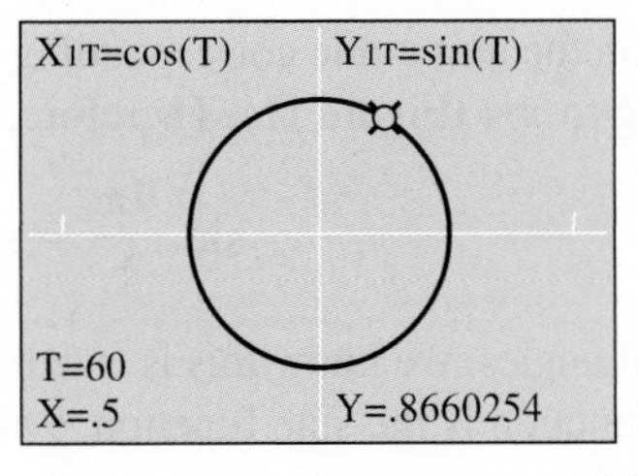

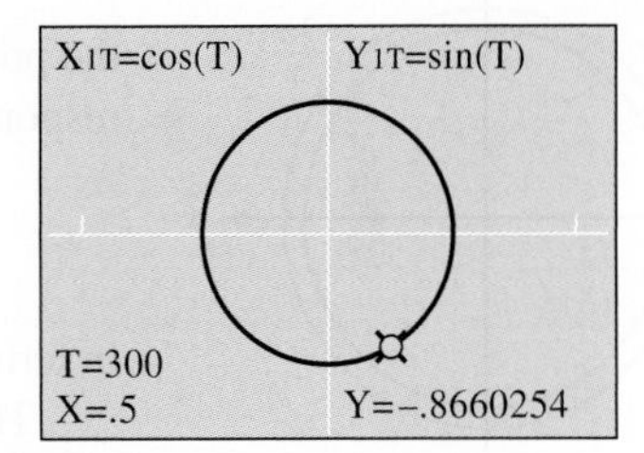

Figure 8

PROBLEM 3
Find tan t if t corresponds to the point (0.8290, 0.5592) on the unit circle.

EXAMPLE 3 Find tan t if t corresponds to the point $(-0.737, 0.675)$ on the unit circle (Figure 9).

SOLUTION Using Definition III we have

$$\tan t = \frac{y}{x}$$
$$= \frac{0.675}{-0.737}$$
$$\approx -0.916$$

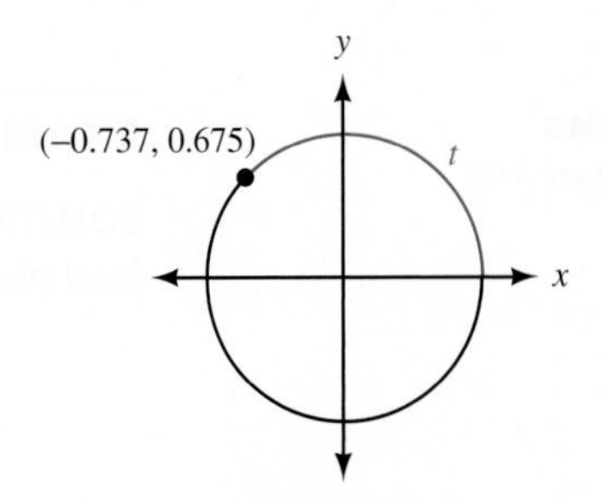

Figure 9

Remember that a function is a rule that pairs each element of the domain with exactly one element from the range. When we see the statement $y = \sin x$, it is identical to the notation $y = f(x)$. In fact, if we wanted to be precise, we would write $y = \sin(x)$. In visual terms, we can picture the sine function as a machine that assigns a single output value to every input value (Figure 10).

NOTE Our use of x and y here follows the traditional use of these variables with functions, where x represents a domain value and y a range value. They are not to be confused with the x- and y-coordinates of points on the unit circle as shown in Figure 4.

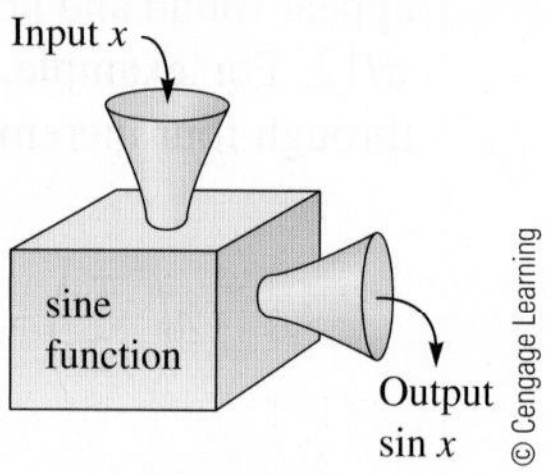

Input x

sine function

Output $\sin x$

Figure 10

The input x is a real number, which can be interpreted as a distance along the circumference of the unit circle or an angle in radians. The input is formally referred to as the *argument* of the function.

PROBLEM 4
Evaluate $\cos(7\pi/3)$. Identify the function, the argument, and the value of the function.

EXAMPLE 4 Evaluate $\sin \dfrac{9\pi}{4}$. Identify the function, the argument of the function, and the value of the function.

SOLUTION Because

$$\frac{9\pi}{4} = \frac{\pi}{4} + \frac{8\pi}{4} = \frac{\pi}{4} + 2\pi$$

the point on the unit circle corresponding to $9\pi/4$ will be the same as the point corresponding to $\pi/4$ (Figure 11). Therefore,

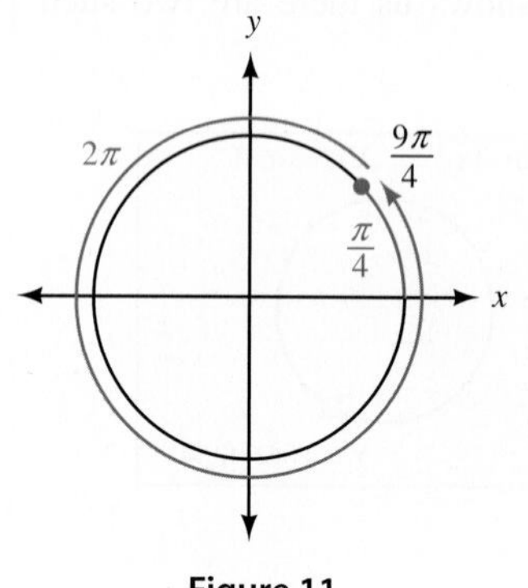

Figure 11

$$\sin \frac{9\pi}{4} = \sin \frac{\pi}{4} = \frac{\sqrt{2}}{2}$$

In terms of angles, we know this is true because $9\pi/4$ and $\pi/4$ are coterminal.

The function is the sine function, $9\pi/4$ is the argument, and $\sqrt{2}/2$ is the value of the function. ■

For angles other than the special angles shown in Figure 5, we can use a calculator to evaluate the circular functions. It is important to remember that the circular functions are functions of *real* numbers, so we must *always have the calculator in radian mode when evaluating these functions.*

PROBLEM 5
Evaluate sec 3.47.

EXAMPLE 5 Evaluate cot 2.37.

SOLUTION Make sure your calculator is set to radian mode. Rounding to four decimal places, we have

$$\cot 2.37 = \frac{1}{\tan 2.37}$$

$$\approx -1.0280$$ ■

Domain and Range

Using Definition III we can find the domain for each of the circular functions. Because any value of t determines a point (x, y) on the unit circle, the sine and cosine functions are always defined and therefore have a domain of all real numbers.

Because $\tan t = y/x$ and $\sec t = 1/x$, the tangent and secant functions will be undefined when $x = 0$, which will occur at the points $(0, 1)$ and $(0, -1)$. Looking back at Figure 5, we see these points correspond to $t = \pi/2$ and $t = 3\pi/2$. However, if we travel around the unit circle multiple times, in either a clockwise or counterclockwise direction, we will encounter these points again and again, such as when $t = 5\pi/2$ and $t = 7\pi/2$, or $t = -\pi/2$ and $t = -3\pi/2$. All these values differ from each other by some multiple of π because $(0, 1)$ and $(0, -1)$ are on opposite sides of the unit circle a half-revolution apart. Therefore, the tangent and secant functions are defined for all real numbers t except $t = \pi/2 + k\pi$, where k is any integer.

In a similar manner, the cotangent and cosecant functions will be undefined when $y = 0$, corresponding to the points $(1, 0)$ or $(-1, 0)$. To avoid either of these two positions, we must avoid the values $t = k\pi$ for any integer k. We summarize these results here:

DOMAINS OF THE CIRCULAR FUNCTIONS

$\sin t$, $\cos t$	All real numbers, or $(-\infty, \infty)$
$\tan t$, $\sec t$	All real numbers except $t = \pi/2 + k\pi$ for any integer k
$\cot t$, $\csc t$	All real numbers except $t = k\pi$ for any integer k

In Chapter 4, we will examine the ranges of the circular functions in detail. For now, it will suffice to simply state the ranges so that we may begin becoming familiar with them.

RANGES OF THE CIRCULAR FUNCTIONS

$\sin t$, $\cos t$	$[-1, 1]$
$\tan t$, $\cot t$	All real numbers, or $(-\infty, \infty)$
$\sec t$, $\csc t$	$(-\infty, -1] \cup [1, \infty)$

PROBLEM 6

Determine which statements are possible for some real number z:

a. $\sin z = 1.5$

b. $\csc z = -345$

c. $\tan \dfrac{3\pi}{2} = z$

EXAMPLE 6 Determine which statements are possible for some real number z.

a. $\cos z = 2$ **b.** $\csc \pi = z$ **c.** $\tan z = 1000$

SOLUTION

a. This statement is not possible because 2 is not within the range of the cosine function. The largest value $\cos z$ can assume is 1.

b. This statement is also not possible, because $\csc \pi$ is undefined and therefore not equal to any real number.

c. This statement is possible because the range of the tangent function is all real numbers, which certainly includes 1,000. ■

Geometric Representations

Based on the circular definitions, we can represent the values of the six trigonometric functions geometrically as indicated in Figure 12. The diagram shows a point $P(x, y)$ that is t units from the point (1, 0) on the circumference of the unit circle. Therefore, $\cos t = x$ and $\sin t = y$. Because triangle BOR is similar to triangle AOP, we have

$$\frac{BR}{OB} = \frac{AP}{OA} = \frac{\sin t}{\cos t} = \tan t$$

Because $OB = 1$, BR is equal to $\tan t$. Notice that this will be the slope of OP. Using a similar argument, it can be shown that $CQ = \cot t$, $OQ = \csc t$, and $OR = \sec t$.

NOTE There are many concepts that can be visualized from Figure 12. One of the more important is the variations that occur in sin *t*, cos *t*, and tan *t* as *P* travels around the unit circle. We illustrate this in Example 7.

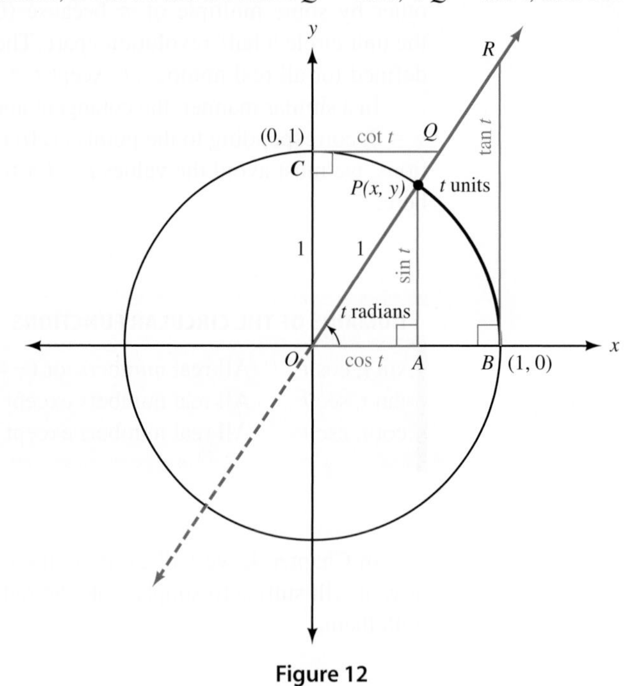

Figure 12

PROBLEM 7
Describe how tan t varies as t increases from 0 to $\pi/2$.

EXAMPLE 7 Describe how sec t varies as t increases from 0 to $\pi/2$.

SOLUTION When $t = 0$, $OR = 1$ so that sec t will begin at a value of 1. As t increases, sec t grows larger and larger. Eventually, when $t = \pi/2$, OP will be vertical so $\sec t = OR$ will no longer be defined.

Getting Ready for Class

After reading through the preceding section, respond in your own words and in complete sentences.

a. If (x, y) is any point on the unit circle, and t is the distance from (1, 0) to (x, y) along the circumference of the unit circle, what are the six trigonometric functions of t?

b. What is the first step in finding all values of t between 0 and 2π on the unit circle for which $\cos t = 1/2$?

c. When we evaluate sin 2, how do we know whether to set the calculator to degree or radian mode?

d. Why are the sine and cosine functions defined for all real numbers?

3.3 PROBLEM SET

➤ CONCEPTS AND VOCABULARY

For Questions 1 through 6, fill in each blank with the appropriate word or number.

1. If (x, y) is a point on the unit circle and t is the distance from $(1, 0)$ to the point along the circumference of the circle, then x is the _______ of t and y is the _______ of t.
2. On the unit circle, the _______ measure of a central angle and the length of _______ it cuts off are exactly the same value.
3. The input to a trigonometric function is formally called the _______ of the function.
4. The circular functions are functions of _______ numbers, so we must always use a calculator set to _______ mode when evaluating them.
5. The two trigonometric functions having a domain of all real numbers are _______ and _______, and the two functions having a range of all real numbers are _______ and _______.
6. The largest possible value for the sine or cosine function is _______ and the smallest possible value is _______.

➤ EXERCISES

Use the unit circle to evaluate each function.

7. $\sin 30°$
8. $\cos 225°$
9. $\cot 90°$
10. $\tan 300°$
11. $\sec 120°$
12. $\csc 210°$
13. $\cos \dfrac{\pi}{3}$
14. $\sin \dfrac{\pi}{2}$
15. $\tan \dfrac{5\pi}{6}$
16. $\sec \pi$
17. $\csc \dfrac{7\pi}{4}$
18. $\cot \dfrac{4\pi}{3}$

Use the unit circle to find the six trigonometric functions of each angle.

19. $150°$
20. $90°$
21. $5\pi/3$
22. $11\pi/6$
23. $180°$
24. $270°$
25. $3\pi/4$
26. $5\pi/4$

Use the unit circle to find all values of θ between 0 and 2π for which the given statement is true.

27. $\sin \theta = 1/2$
28. $\sin \theta = -1/2$
29. $\cos \theta = -\sqrt{3}/2$
30. $\cos \theta = 0$
31. $\tan \theta = -\sqrt{3}$
32. $\cot \theta = \sqrt{3}$

Graph the unit circle using parametric equations with your calculator set to radian mode. Use a scale of $\pi/12$. Trace the circle to find the sine and cosine of each angle to the nearest ten-thousandth.

33. $\dfrac{4\pi}{3}$
34. $\dfrac{\pi}{4}$
35. $\dfrac{7\pi}{6}$
36. $\dfrac{5\pi}{12}$

Graph the unit circle using parametric equations with your calculator set to radian mode. Use a scale of $\pi/12$. Trace the circle to find all values of t between 0 and 2π satisfying each of the following statements. Round your answers to the nearest ten-thousandth.

37. $\cos t = -\dfrac{1}{2}$
38. $\sin t = \dfrac{1}{2}$
39. $\sin t = 1$
40. $\cos t = -1$

Graph the unit circle using parametric equations with your calculator set to degree mode. Use a scale of 5. Trace the circle to find the sine and cosine of each angle to the nearest ten-thousandth.

41. 120° **42.** 75° **43.** 225° **44.** 310°

Graph the unit circle using parametric equations with your calculator set to degree mode. Use a scale of 5. Trace the circle to find all values of t between 0° and 360° satisfying each of the following statements.

45. $\sin t = -\frac{1}{2}$ **46.** $\cos t = 0$ **47.** $\sin t = \cos t$ **48.** $\sin t = -\cos t$

49. If angle θ is in standard position and the terminal side of θ intersects the unit circle at the point $(1/\sqrt{5}, -2/\sqrt{5})$, find $\sin\theta$, $\cos\theta$, and $\tan\theta$.

50. If angle θ is in standard position and the terminal side of θ intersects the unit circle at the point $(-1/\sqrt{10}, -3/\sqrt{10})$, find $\csc\theta$, $\sec\theta$, and $\cot\theta$.

51. If t is the distance from (1, 0) to (0.5403, 0.8415) along the circumference of the unit circle, find $\sin t$, $\cos t$, and $\tan t$.

52. If t is the distance from (1, 0) to (−0.9422, 0.3350) along the circumference of the unit circle, find $\csc t$, $\sec t$, and $\cot t$.

53. Find the coordinates of the point on the unit circle measured 4 units clockwise along the circumference of the circle from (1, 0). Round your answer to four decimal places.

54. Find the coordinates of the point on the unit circle measured 5 units counterclockwise along the circumference of the circle from (1, 0). Round your answer to four decimal places.

For Problems 55 through 62, use Figure 13, which shows the unit circle with increments of 0.1 units marked off along the circumference in a counterclockwise direction starting from (1, 0).

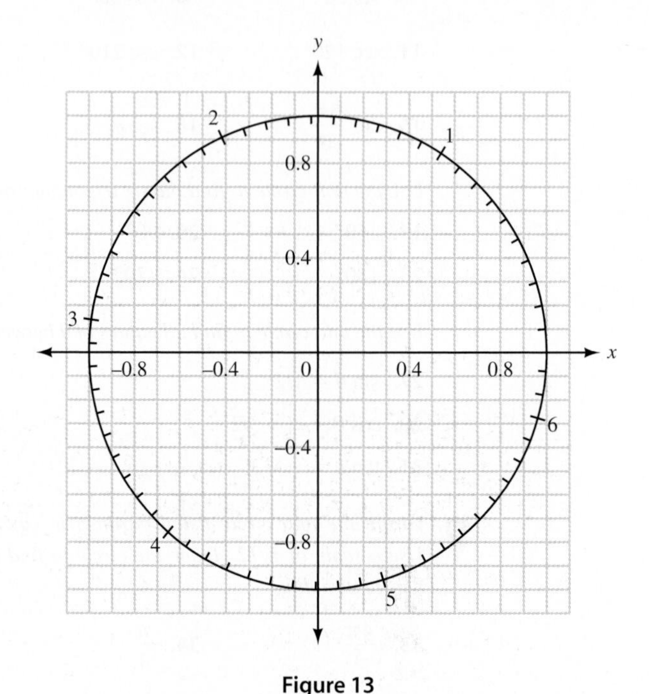

Figure 13

55. Estimate sin 2.7. **56.** Estimate cos 2.7.

57. Estimate sec 4.6. **58.** Estimate csc 0.2.

59. Estimate tan 3.9.

60. Estimate cot 2.2.

61. Estimate θ if $\cos \theta = -0.8$ and $0 \le \theta < 2\pi$.

62. Estimate θ if $\sin \theta = -0.7$ and $0 \le \theta < 2\pi$.

Identify the argument of each function.

63. $\tan 5$

64. $\cos \theta$

65. $\sin 2A$

66. $\cos (A + B)$

67. $\sin \left(x + \frac{\pi}{2}\right)$

68. $\tan \left(\frac{x}{2} + \frac{\pi}{8}\right)$

69. Evaluate $\cos \frac{2\pi}{3}$. Identify the function, the argument of the function, and the function value.

70. Evaluate $\sin \frac{13\pi}{6}$. Identify the function, the argument of the function, and the function value.

Use a calculator to approximate each value to four decimal places.

71. $\sin 4$

72. $\cos (-1.5)$

73. $\tan (-45)$

74. $\cot 30$

75. $\sec 0.8$

76. $\csc \frac{3}{2}$

If we start at the point (1, 0) and travel once around the unit circle, we travel a distance of 2π units and arrive back where we started. If we continue around the unit circle a second time, we will repeat all the values of x and y that occurred during our first trip around. Use this discussion to evaluate the following expressions:

77. $\sin \left(2\pi + \frac{\pi}{2}\right)$

78. $\cos \left(2\pi + \frac{\pi}{2}\right)$

79. $\sin \left(2\pi + \frac{\pi}{6}\right)$

80. $\cos \left(2\pi + \frac{\pi}{3}\right)$

81. Which of the six trigonometric functions are not defined at 0?

82. Which of the six trigonometric functions are not defined at $\pi/2$?

For Problems 83 through 94, determine if the statement is possible for some real number z.

83. $\sec \frac{\pi}{2} = z$

84. $\csc 0 = z$

85. $\cos \pi = z$

86. $\sin 0 = z$

87. $\tan \frac{\pi}{2} = z$

88. $\cot \pi = z$

89. $\sin z = 1.2$

90. $\cos z = \pi$

91. $\tan z = \frac{3\pi}{2}$

92. $\cot z = 0$

93. $\sec z = \frac{1}{2}$

94. $\csc z = 0$

Use the diagram shown in Figure 12 for the following exercises.

95. Describe how csc t varies as t increases from 0 to $\pi/2$.

96. Describe how cot t varies as t increases from 0 to $\pi/2$.

97. Describe how sin t varies as t increases from $\pi/2$ to π.

98. Describe how cos t varies as t increases from $\pi/2$ to π.

The diagram of the unit circle shown in Figure 14 can be found on the Internet encyclopedia, Wikipedia. Note that triangles **FAO** *and* **OCA** *are similar to triangle* **OAE**. *Use the diagram to answer Problems 99 through 106.*

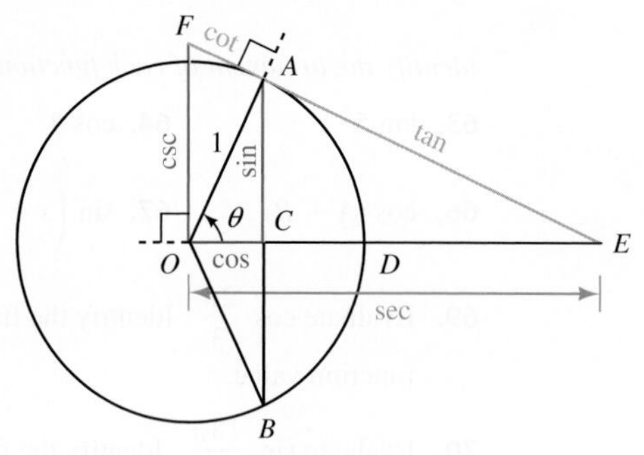

Figure 14

99. Show why **AE** = tan θ.

100. Show why **OF** = csc θ.

101. If θ is close to 0, determine whether the value of each trigonometric function is close to 0, close to 1, or becoming infinitely large.

a. sin θ **b.** cos θ **c.** tan θ

d. csc θ **e.** sec θ **f.** cot θ

102. If θ is close to $\pi/2$, determine whether the value of each trigonometric function is close to 0, close to 1, or becoming infinitely large.

a. sin θ **b.** cos θ **c.** tan θ

d. csc θ **e.** sec θ **f.** cot θ

103. Why is $|\sec \theta| \geq 1$?

104. Why is $|\cos \theta| \leq 1$?

105. Determine whether the value of each trigonometric function is greater when $\theta = \pi/6$ or $\theta = \pi/3$.

a. sin θ **b.** cos θ **c.** tan θ

106. Determine whether the value of each trigonometric function is greater when $\theta = \pi/6$ or $\theta = \pi/4$.

a. csc θ **b.** sec θ **c.** cot θ

➤ REVIEW PROBLEMS

The problems that follow review material we covered in Section 2.3.

Problems 107 through 112 refer to right triangle ABC in which C = 90°. Solve each triangle.

107. $A = 42°, c = 36$ **108.** $A = 58°, c = 17$

109. $B = 22°, b = 320$ **110.** $a = 16.3, b = 20.8$

111. $a = 4.37, c = 6.21$ **112.** $a = 7.12, c = 8.44$

➤ LEARNING OBJECTIVES ASSESSMENT

These questions are available for instructors to help assess if you have successfully met the learning objectives for this section.

113. Use the unit circle in Figure 13 to approximate cos 2.

a. 0.9 **b.** 0.6 **c.** -0.2 **d.** -0.4

114. If angle θ is in standard position and the terminal side of θ intersects the unit circle at the point $\left(-\frac{\sqrt{17}}{17}, \frac{4\sqrt{17}}{17}\right)$, find $\tan\theta$.

a. -4 **b.** $-\frac{1}{4}$ **c.** $-\frac{\sqrt{17}}{17}$ **d.** $\frac{4\sqrt{17}}{17}$

115. Use a calculator to approximate sec 4.

a. -1.3213 **b.** 14.3356 **c.** -1.5299 **d.** 1.0024

116. Which statement is possible for some real number z?

a. $\sin z = 2$ **b.** $\sec z = 2$ **c.** $\cot 0 = z$ **d.** $\csc \pi = z$

SECTION 3.4 Arc Length and Area of a Sector

LEARNING OBJECTIVES

1. Calculate the arc length for a central angle.
2. Calculate the area of a sector formed by a central angle.
3. Solve a real-life problem involving arc length.
4. Solve a real-life problem involving sector area.

In Chapter 2, we discussed some of the aspects of the first Ferris wheel, which was built by George Ferris in 1893. The first topic we will cover in this section is *arc length.* Our study of arc length will allow us to find the distance traveled by a rider on a Ferris wheel at any point during the ride.

In Section 3.2, we found that if a central angle θ, measured in radians, in a circle of radius r cuts off an arc of length s, then the relationship between s, r, and θ can be written as $\theta = \frac{s}{r}$. Figure 1 illustrates this.

If we multiply both sides of this equation by r, we will obtain the equation that gives arc length s in terms of r and θ.

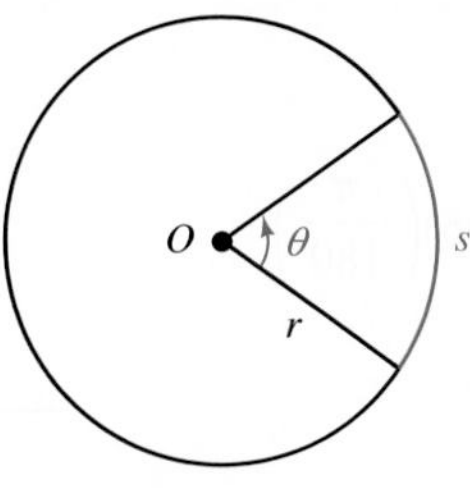

Figure 1

$$\theta = \frac{s}{r} \qquad \text{Definition of radian measure}$$

$$r \cdot \theta = r \cdot \frac{s}{r} \qquad \text{Multiply both sides by } r$$

$$r\theta = s$$

ARC LENGTH

If θ (in radians) is a central angle in a circle with radius r, then the length of the arc cut off by θ is given by

$$s = r\theta \qquad (\theta \text{ in radians})$$

Because r would be constant for a given circle, this formula tells us that the arc length is proportional to the central angle. An angle twice as large would cut off an arc twice as long.

PROBLEM 1
Give the length of the arc cut off by a central angle of 3.1 radians in a circle of radius 5.4 inches.

EXAMPLE 1 Give the length of the arc cut off by a central angle of 2 radians in a circle of radius 4.3 inches.

SOLUTION We have $\theta = 2$ and $r = 4.3$ inches. Applying the formula $s = r\theta$ gives us

$$\begin{aligned} s &= r\theta \\ &= 4.3(2) \\ &= 8.6 \text{ inches} \end{aligned}$$

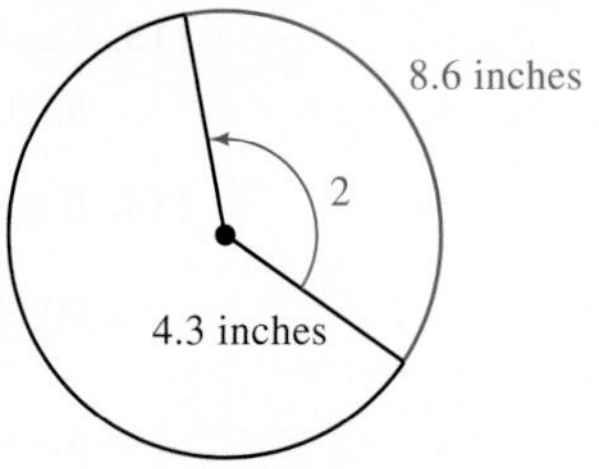

Figure 2 illustrates this example.

8.6 inches
2
4.3 inches

Figure 2

PROBLEM 2
Repeat Example 2, but this time find the distance traveled by the rider when $\theta = 90°$ and when $\theta = 180°$.

EXAMPLE 2 Figure 3 is a model of George Ferris's Ferris wheel. Recall that the diameter of the wheel is 250 feet, and θ is the central angle formed as a rider travels from his or her initial position P_0 to position P_1. Find the distance traveled by the rider if $\theta = 45°$ and if $\theta = 105°$.

SOLUTION The formula for arc length, $s = r\theta$, requires θ to be given in radians. Since θ is given in degrees, we must multiply it by $\pi/180$ to convert to radians. Also, since the diameter of the wheel is 250 feet, the radius is 125 feet.

For $\theta = 45°$:

$$\begin{aligned} s &= r\theta \\ &= 125(45)\left(\frac{\pi}{180}\right) \\ &= \frac{125\pi}{4} \\ &\approx 98 \text{ ft} \end{aligned}$$

For $\theta = 105°$:

$$\begin{aligned} s &= r\theta \\ &= 125(105)\left(\frac{\pi}{180}\right) \\ &= \frac{875\pi}{12} \\ &\approx 230 \text{ ft} \end{aligned}$$

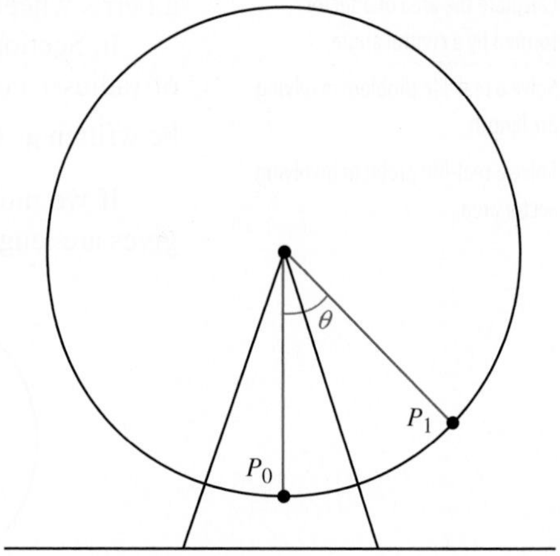

θ
P_1
P_0

Figure 3

PROBLEM 3
The minute hand of a clock is 2.4 centimeters long. To two significant digits, how far does the tip of the minute hand move in 15 minutes?

EXAMPLE 3 The minute hand of a clock is 1.2 centimeters long. To two significant digits, how far does the tip of the minute hand move in 20 minutes?

SOLUTION We have $r = 1.2$ centimeters. Since we are looking for s, we need to find θ. We can use a proportion to find θ. One complete rotation is 60 minutes and 2π radians, so we say θ is to 2π as 20 minutes is to 60 minutes, or

$$\text{If} \quad \frac{\theta}{2\pi} = \frac{20}{60} \quad \text{then} \quad \theta = \frac{2\pi}{3}$$

Now we can find s.

$$\begin{aligned} s &= r\theta \\ &= 1.2\left(\frac{2\pi}{3}\right) \\ &= \frac{2.4\pi}{3} \\ &\approx 2.5 \text{ cm} \end{aligned}$$

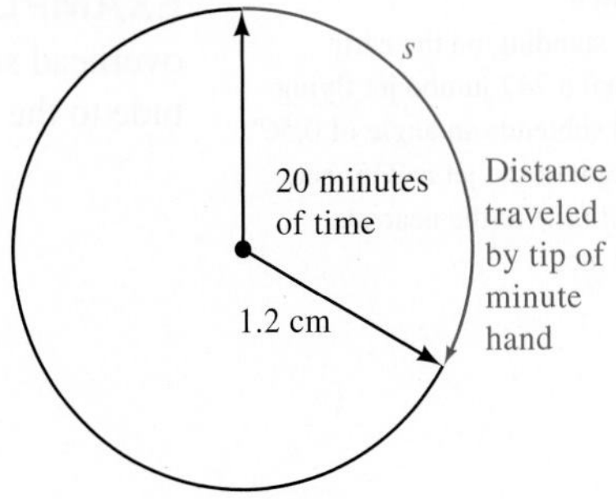

Figure 4

Figure 4 illustrates this example.

The tip of the minute hand will travel approximately 2.5 centimeters every 20 minutes.

If we are working with relatively small central angles in circles with large radii, we can use the length of the intercepted arc to approximate the length of the associated chord. For example, Figure 5 shows a central angle of 1° in a circle of radius 1,800 feet, along with the arc and chord cut off by 1°. (Figure 5 is not drawn to scale.)

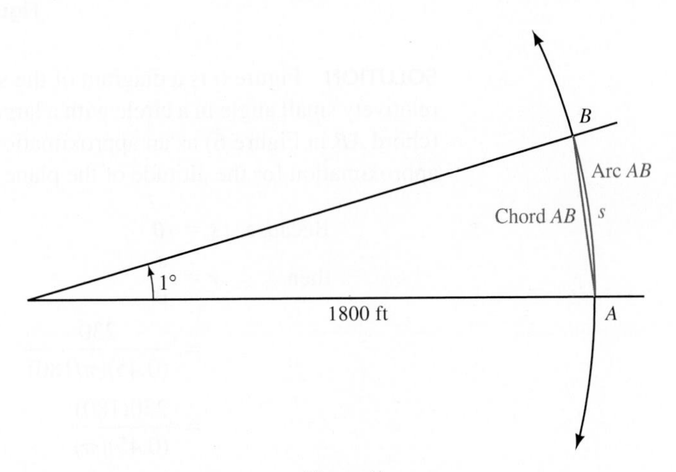

Figure 5

To find the length of arc AB, we convert θ to radians by multiplying by $\pi/180$. Then we apply the formula $s = r\theta$.

$$s = r\theta = 1{,}800(1)\left(\frac{\pi}{180}\right) = 10\pi \approx 31 \text{ ft}$$

If we had carried out the calculation of arc AB to six significant digits, we would have obtained $s = 31.4159$. The length of the chord AB is 31.4155 to six significant digits (found by using the law of sines, which we will cover in Chapter 7). As you can see, the first five digits in each number are the same. It seems reasonable then to approximate the length of chord AB with the length of arc AB.

As our next example illustrates, we can also use the procedure just outlined in the reverse order to find the radius of a circle by approximating arc length with the length of the associated chord.

PROBLEM 4
A person standing on the earth notices that a 747 jumbo jet flying overhead subtends an angle of 0.56°. If the length of the jet is 235 feet, find its altitude to the nearest thousand feet.

EXAMPLE 4 A person standing on the earth notices that a 747 Jumbo Jet flying overhead subtends an angle of 0.45°. If the length of the jet is 230 feet, find its altitude to the nearest thousand feet.

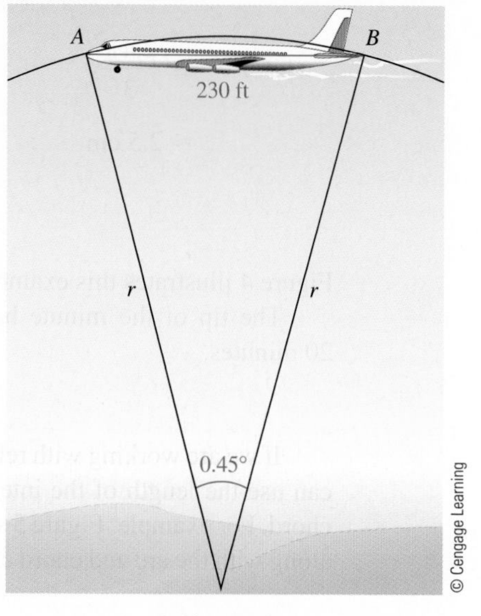

Figure 6

SOLUTION Figure 6 is a diagram of the situation. Because we are working with a relatively small angle in a circle with a large radius, we use the length of the airplane (chord *AB* in Figure 6) as an approximation of the length of the arc *AB*, and *r* as an approximation for the altitude of the plane.

Because $s = r\theta$

then $r = \dfrac{s}{\theta}$

$= \dfrac{230}{(0.45)(\pi/180)}$ We multiply 0.45° by $\pi/180$ to change to radian measure

$= \dfrac{230(180)}{(0.45)(\pi)}$

$= 29{,}000$ ft To the nearest thousand feet

Area of a Sector

Next we want to derive the formula for the area of the sector formed by a central angle θ (Figure 7).

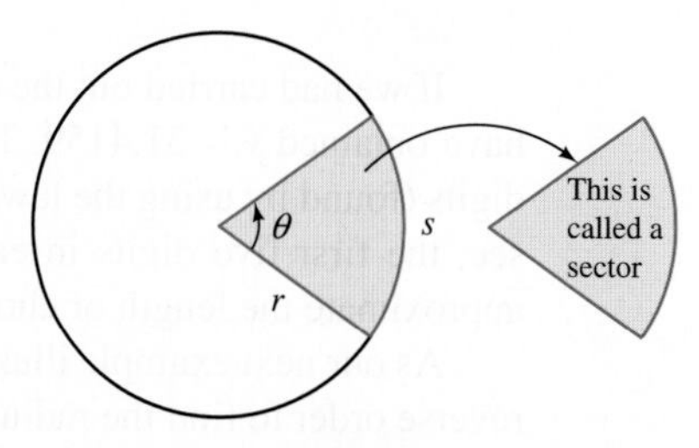

Figure 7

If we let A represent the area of the sector formed by central angle θ, we can find A by setting up a proportion as follows: We say the area A of the sector is to the area of the circle as θ is to one full rotation. That is,

$$\text{Area of sector} \longrightarrow \frac{A}{\pi r^2} = \frac{\theta}{2\pi} \longleftarrow \text{Central angle } \theta$$
$$\text{Area of circle} \longrightarrow \qquad\qquad \longleftarrow \text{One full rotation}$$

We solve for A by multiplying both sides of the proportion by πr^2.

$$\pi r^2 \cdot \frac{A}{\pi r^2} = \frac{\theta}{2\pi} \cdot \pi r^2$$

$$A = \frac{1}{2} r^2 \theta$$

AREA OF A SECTOR

If θ (in radians) is a central angle in a circle with radius r, then the area of the sector (Figure 8) formed by angle θ is given by

$$A = \frac{1}{2} r^2 \theta \qquad (\theta \text{ in radians})$$

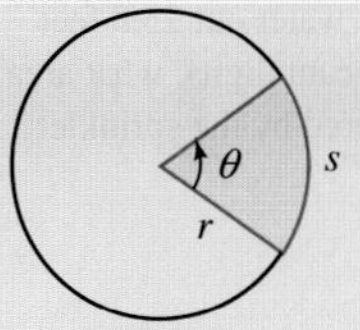

Figure 8

PROBLEM 5
Find the area of the sector formed by a central angle of 1.6 radians in a circle of radius 3.5 meters.

EXAMPLE 5 Find the area of the sector formed by a central angle of 1.4 radians in a circle of radius 2.1 meters.

SOLUTION We have $r = 2.1$ meters and $\theta = 1.4$. Applying the formula for A gives us

$$A = \frac{1}{2} r^2 \theta$$

$$= \frac{1}{2}(2.1)^2(1.4)$$

$$= 3.1 \text{ m}^2 \qquad \text{To the nearest tenth}$$

REMEMBER Area is measured in square units. When $r = 2.1$ m, $r^2 = (2.1 \text{ m})^2 = 4.41 \text{ m}^2$.

PROBLEM 6
If the sector formed by a central angle of 30° has an area of $\frac{2\pi}{3}$ square centimeters, find the radius of the circle.

EXAMPLE 6 If the sector formed by a central angle of 15° has an area of $\pi/3$ square centimeters, find the radius of the circle.

SOLUTION We first convert 15° to radians.

$$\theta = 15\left(\frac{\pi}{180}\right) = \frac{\pi}{12}$$

Then we substitute $\theta = \pi/12$ and $A = \pi/3$ into the formula for A and solve for r.

$$A = \frac{1}{2}r^2\theta$$

$$\frac{\pi}{3} = \frac{1}{2}r^2\frac{\pi}{12} = \frac{\pi}{24}r^2$$

$$r^2 = \frac{\pi}{3} \cdot \frac{24}{\pi}$$

$$r^2 = 8$$

$$r = 2\sqrt{2} \text{ cm}$$

Note that we need only use the positive square root of 8, since we know our radius must be measured with positive units.

PROBLEM 7
A lawn sprinkler located at the corner of a yard is set to rotate through 60° and project water out 25.5 feet. To three significant digits, what area of lawn is watered by the sprinkler?

EXAMPLE 7 A lawn sprinkler located at the corner of a yard is set to rotate through 90° and project water out 30.0 feet. To three significant digits, what area of lawn is watered by the sprinkler?

SOLUTION We have $\theta = 90° = \frac{\pi}{2} \approx 1.57$ radians and $r = 30.0$ feet. Figure 9 illustrates this example.

$$A = \frac{1}{2}r^2\theta$$

$$\approx \frac{1}{2}(30.0)^2(1.57)$$

$$= 707 \text{ ft}^2$$

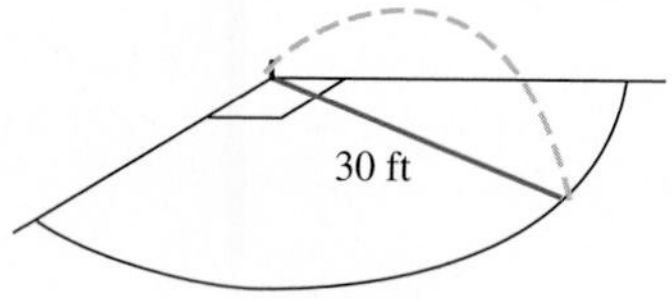

30 ft

Lawn

Figure 9

Getting Ready for Class

After reading through the preceding section, respond in your own words and in complete sentences.

a. What is the definition and formula for arc length?
b. Which is longer, the chord AB or the corresponding arc AB?
c. What is a sector?
d. What is the definition and formula for the area of a sector?

3.4 PROBLEM SET

➤ CONCEPTS AND VOCABULARY

For Questions 1 through 4, fill in each blank with the appropriate word.

1. To find the arc length for a central angle, multiply the ______ by the __________.
2. Arc length is _______________ to the central angle. An angle three times as large would cut off an arc ________ times as long.

3. The region interior to a central angle is called a _________.

4. When calculating arc length or sector area, the angle must be measured in ________.

➤ EXERCISES

Unless otherwise stated, all answers in this Problem Set that need to be rounded should be rounded to three significant digits.

For each of the following problems, θ is a central angle in a circle of radius r. In each case, find the length of arc s cut off by θ.

5. $\theta = 2$, $r = 3$ inches

6. $\theta = 3$, $r = 2$ inches

7. $\theta = 1.5$, $r = 1.5$ ft

8. $\theta = 2.4$, $r = 1.8$ ft

9. $\theta = \pi/6$, $r = 12$ cm

10. $\theta = \pi/3$, $r = 12$ cm

11. $\theta = 30°$, $r = 4$ mm

12. $\theta = 60°$, $r = 4$ mm

13. $\theta = 315°$, $r = 5$ inches

14. $\theta = 240°$, $r = 10$ inches

15. Arc Length The minute hand of a clock is 1.2 centimeters long. How far does the tip of the minute hand travel in 40 minutes?

16. Arc Length The minute hand of a clock is 2.4 centimeters long. How far does the tip of the minute hand travel in 20 minutes?

17. Arc Length A space shuttle 200 miles above the earth is orbiting the earth once every 6 hours. How far does the shuttle travel in 1 hour? (Assume the radius of the earth is 4,000 miles.) Give your answer as both an exact value and an approximation to three significant digits (Figure 10).

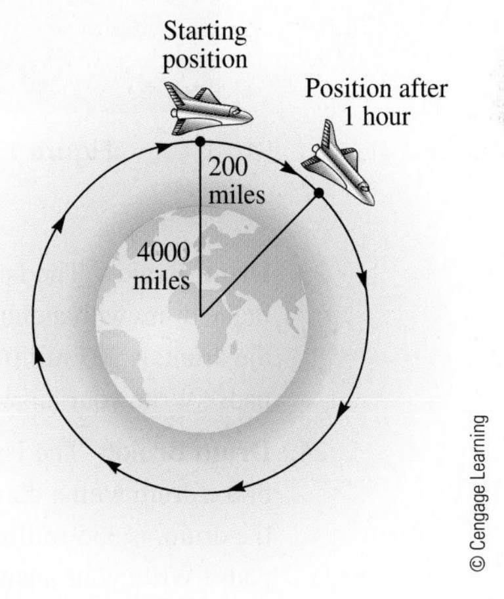

Figure 10

18. Arc Length How long, in hours, does it take the space shuttle in Problem 17 to travel 8,400 miles? Give both the exact value and an approximate value for your answer.

19. Arc Length The pendulum on a grandfather clock swings from side to side once every second. If the length of the pendulum is 4 feet and the angle through which it swings is 20°, how far does the tip of the pendulum travel in 1 second?

20. **Arc Length** Find the total distance traveled in 1 minute by the tip of the pendulum on the grandfather clock in Problem 19.

21. **Cable Car Drive System** The current San Francisco cable railway is driven by two large 14-foot-diameter drive wheels, called *sheaves.* Because of the figure-eight system used, the cable subtends a central angle of 270° on each sheave. Find the length of cable riding on one of the drive sheaves at any given time (Figure 11).

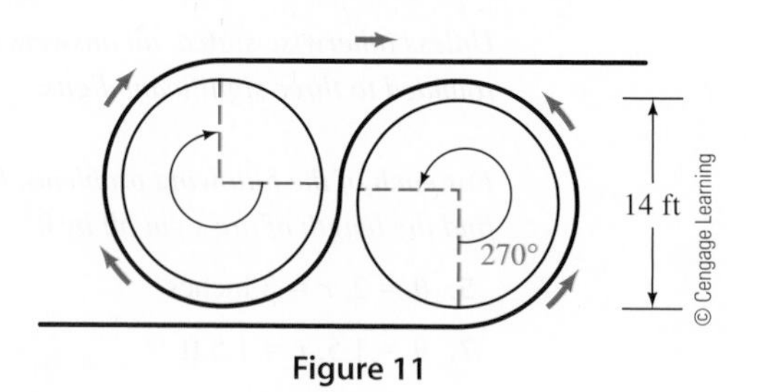

Figure 11

22. **Cable Car Drive System** The first cable railway to make use of the figure-eight drive system was the Sutter Street Railway in San Francisco in 1883 (Figure 12). Each drive sheave was 12 feet in diameter. Find the length of cable riding on one of the drive sheaves. (See Problem 21.)

Figure 12

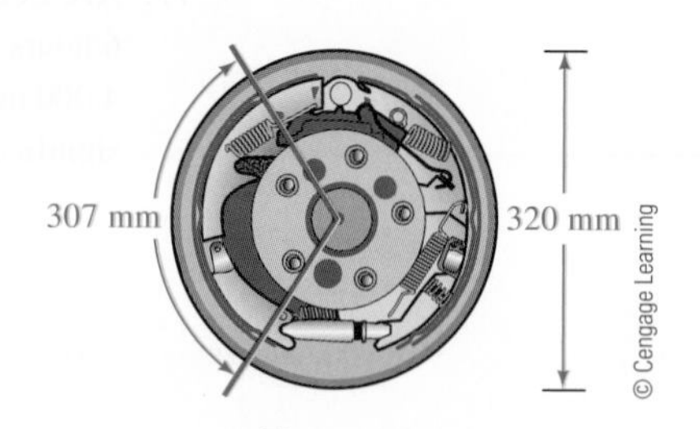

Figure 13

23. **Drum Brakes** The Isuzu NPR 250 light truck with manual transmission has a circular brake drum with a diameter of 320 millimeters. Each brake pad, which presses against the drum, is 307 millimeters long. What central angle is subtended by one of the brake pads? Write your answer in both radians and degrees (Figure 13).

24. **Drum Brakes** The Isuzu NPR 250 truck with automatic transmission has a circular brake drum with a diameter of 320 millimeters. Each brake pad, which presses against the drum, is 335 millimeters long. What central angle is subtended by one of the brake pads? Write your answer in both radians and degrees.

25. **Diameter of the Moon** From the earth, the moon subtends an angle of approximately 0.5°. If the distance to the moon is approximately 240,000 miles, find an approximation for the diameter of the moon accurate to the nearest hundred miles. (See Example 4 and the discussion that precedes it.)

26. **Diameter of the Sun** If the distance to the sun is approximately 93 million miles, and, from the earth, the sun subtends an angle of approximately 0.5°, estimate the diameter of the sun to the nearest 10,000 miles.

Repeat Example 2 from this section for the following values of θ.

27. $\theta = 30°$ **28.** $\theta = 60°$

29. $\theta = 220°$ **30.** $\theta = 315°$

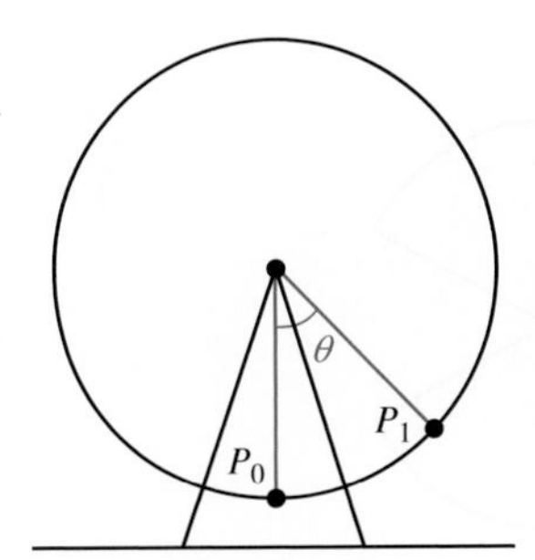

Figure 14

31. Ferris Wheel In Problem Set 2.3, we mentioned a Ferris wheel built in Vienna in 1897 known as the Great Wheel. The diameter of this wheel is 197 feet. Use Figure 14 as a model of the Great Wheel. For each value of θ, find the distance traveled by a rider in going from initial position P_0 to position P_1.

a. 60° **b.** 210° **c.** 285°

32. Ferris Wheel A Ferris Wheel called Colossus that we mentioned in Problem Set 2.3 has a diameter of 165 feet. Using Figure 14 as a model, for each value of θ find the distance traveled by someone starting at initial position P_0 and moving to position P_1.

a. 150° **b.** 240° **c.** 345°

In each of the following problems, θ is a central angle that cuts off an arc of length s. In each case, find the radius of the circle.

33. $\theta = 6$, $s = 3$ ft **34.** $\theta = 1$, $s = 2$ ft

35. $\theta = 1.4$, $s = 4.2$ inches **36.** $\theta = 5.1$, $s = 10.2$ inches

37. $\theta = \pi/4$, $s = \pi$ cm **38.** $\theta = 3\pi/4$, $s = \pi$ cm

39. $\theta = 90°$, $s = \pi/2$ m **40.** $\theta = 180°$, $s = \pi/2$ m

41. $\theta = 150°$, $s = 5$ km **42.** $\theta = 225°$, $s = 4$ km

Find the area of the sector formed by the given central angle θ in a circle of radius r.

43. $\theta = 2$, $r = 3$ cm **44.** $\theta = 3$, $r = 2$ cm

45. $\theta = 2.4$, $r = 4$ inches **46.** $\theta = 1.8$, $r = 2$ inches

47. $\theta = 2\pi/5$, $r = 3$ m **48.** $\theta = \pi/5$, $r = 3$ m

49. $\theta = 15°$, $r = 5$ m **50.** $\theta = 15°$, $r = 10$ m

51. Area of a Sector An arc of length 3 feet is cut off by a central angle of $\pi/4$ radians. Find the area of the sector formed.

52. Area of a Sector A central angle of 2 radians cuts off an arc of length 4 inches. Find the area of the sector formed.

53. Radius of a Circle If the sector formed by a central angle of 30° has an area of $\pi/3$ square centimeters, find the radius of the circle.

54. Arc Length What is the length of the arc cut off by angle θ in Problem 53?

55. Radius of a Circle A sector of area $2\pi/3$ square inches is formed by a central angle of 45°. What is the radius of the circle?

56. Radius of a Circle A sector of area 25 square inches is formed by a central angle of 4 radians. Find the radius of the circle.

57. Lawn Sprinkler A lawn sprinkler is located at the corner of a yard. The sprinkler is set to rotate through 90° and project water out 60 feet. What is the area of the yard watered by the sprinkler?

58. Windshield Wiper An automobile windshield wiper 10 inches long rotates through an angle of 60°. If the rubber part of the blade covers only the last 9 inches of the wiper, find the area of the windshield cleaned by the windshield wiper.

Figure 15

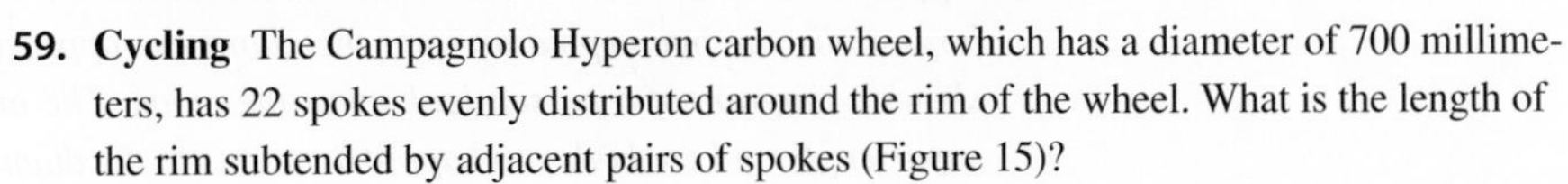

59. Cycling The Campagnolo Hyperon carbon wheel, which has a diameter of 700 millimeters, has 22 spokes evenly distributed around the rim of the wheel. What is the length of the rim subtended by adjacent pairs of spokes (Figure 15)?

60. Cycling The Reynolds Stratus DV carbon wheel, which has a diameter of 700 millimeters, has 16 spokes evenly distributed around the rim of the wheel. What is the length of the rim subtended by adjacent pairs of spokes (Figure 16)?

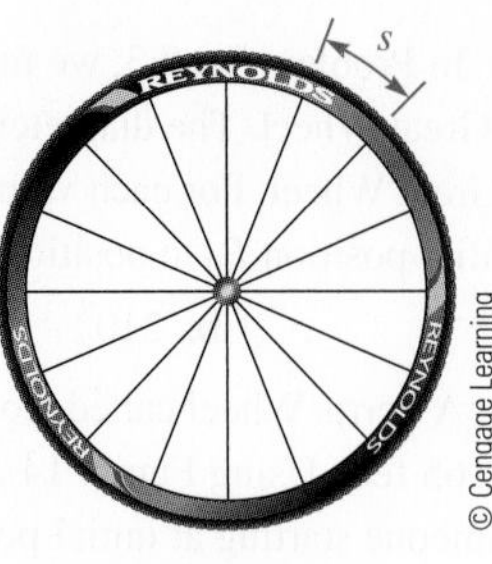

Figure 16

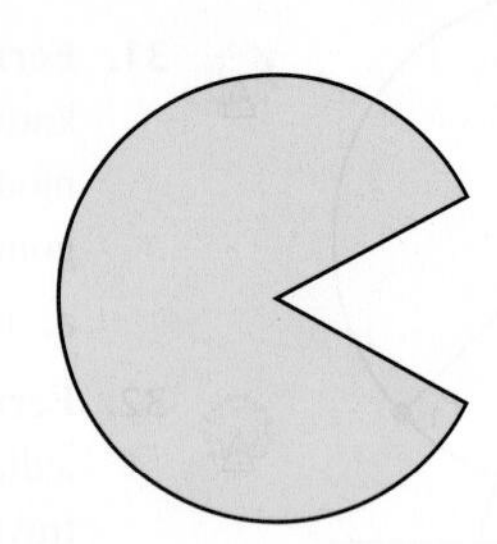

Figure 17

61. Pac-Man Figure 17 shows an image of Pac-Man from the classic video game. The "mouth" of Pac-Man forms a central angle of 55.0° in a circle with radius 4.00 millimeters.

a. Find the length of the perimeter of Pac-Man (including the sides of the mouth).

b. Find the area enclosed by Pac-Man.

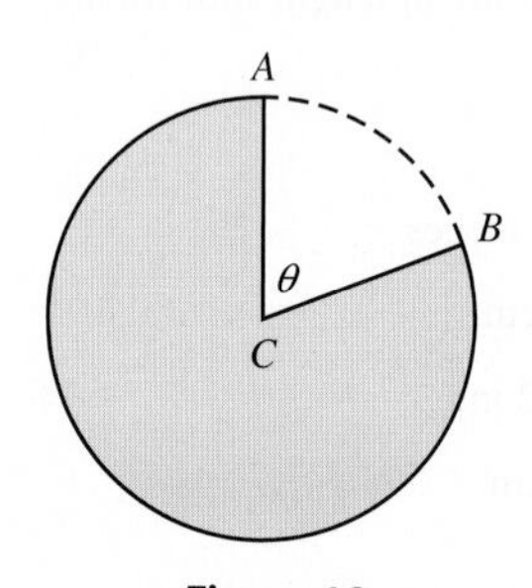

Figure 18

62. Coffee Filter A conical coffee filter is made from a circular piece of paper of diameter 10.0 inches by cutting out a sector and joining the edges CA and CB (Figure 18). The arc length of the sector that is removed is 3.50 inches.

a. Find the central angle, θ, of the sector that was removed, in degrees.

b. Find the area that remains once the sector has been removed.

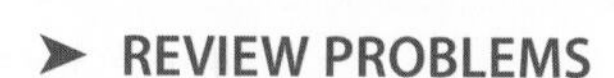

➤ REVIEW PROBLEMS

The problems that follow review material we covered in Section 2.4.

63. Angle of Elevation If a 75-foot flagpole casts a shadow 43 feet long, what is the angle of elevation of the sun from the tip of the shadow?

64. Height of a Hill A road up a hill makes an angle of 5° with the horizontal. If the road from the bottom of the hill to the top of the hill is 2.5 miles long, how high is the hill?

65. Angle of Depression A person standing 5.2 feet from a mirror notices that the angle of depression from his eyes to the bottom of the mirror is 13°, while the angle of elevation to the top of the mirror is 12°. Find the vertical dimension of the mirror.

66. Distance and Bearing A boat travels on a course of bearing S 63° 50′ E for 114 miles. How many miles south and how many miles east has the boat traveled?

67. Geometry The height of a right circular cone is 35.8 centimeters. If the diameter of the base is 20.5 centimeters, what angle does the side of the cone make with the base?

68. Height of a Tree Two people decide to find the height of a tree. They position themselves 35 feet apart in line with, and on the same side of, the tree. If they find the angles of elevation from the ground where they are standing to the top of the tree are 65° and 44°, how tall is the tree?

➤ EXTENDING THE CONCEPTS

Apparent Diameter *As we mentioned in this section, for small central angles in circles with large radii, the intercepted arc and the chord are approximately the same length. Figure 19 shows a diagram of a person looking at the moon. The arc and the chord are essentially the same length and are both good approximations to the diameter.*

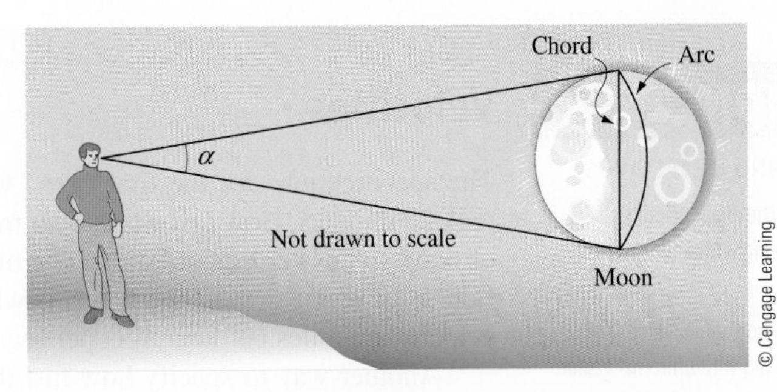

Figure 19

In astronomy, the angle α at which an object is seen by the eye is called the apparent diameter *of the object. Apparent diameter depends on the size of the object and its distance from the observer. To three significant figures, the diameter of the moon is 3,470 kilometers and its average distance to the earth is 384,000 kilometers. The diameter of the sun is 1,390,000 kilometers and its mean distance to the earth is 150,000,000 kilometers.*

69. Calculate the apparent diameter of the moon as seen from the earth. Give your answer in both radians and degrees. Round each to the nearest thousandth.

70. Suppose you have a friend who is 6 feet tall. At what distance must your friend stand so that he or she has the same apparent diameter as the moon?

71. The moon and the sun are, at first sight, about equal in size. How can you explain this apparent similarity in size even though the diameter of the sun is about 400 times greater than the diameter of the moon?

72. The distance between the earth and the moon varies as the moon travels its orbit around the earth. Accordingly, the apparent diameter of the moon must also vary. The smallest distance between the earth and the moon is 356,340 kilometers, while the greatest distance is 406,630 kilometers. Find the largest and smallest apparent diameter of the moon. Round your answers to the nearest hundredth.

➤ LEARNING OBJECTIVES ASSESSMENT

These questions are available for instructors to help assess if you have successfully met the learning objectives for this section.

73. If $\theta = 40°$ is a central angle in a circle of radius 9 feet, find the arc length cut off by θ.

a. 1620 ft **b.** 9π ft **c.** 2π ft **d.** 360 ft

74. If $\theta = 40°$ is a central angle in a circle of radius 9 feet, find the area of the sector formed by θ.

a. 1620 ft^2 **b.** 9π ft^2 **c.** 2π ft^2 **d.** 360 ft^2

75. A curve along a highway is an arc of a circle with a 250-meter radius. If the curve corresponds to a central angle of 1.5 radians, find the length of the highway along the curve.

a. 6.5 m **b.** 50π m **c.** 167 m **d.** 375 m

76. A lawn sprinkler has been set to rotate through an angle of 110° and project water out to 12 feet. What is the approximate area of the yard watered by the sprinkler?

a. 11.5 ft^2 **b.** 7,920 ft^2 **c.** 660 ft^2 **d.** 138 ft^2

SECTION 3.5 Velocities

LEARNING OBJECTIVES

1. Calculate the linear velocity of a point moving with uniform circular motion.
2. Calculate the angular velocity of a point moving with uniform circular motion.
3. Convert from linear velocity to angular velocity and vice versa.
4. Solve real-life problems involving linear and angular velocity.

The specifications for the first Ferris wheel indicate that one trip around the wheel took 20 minutes. How fast was a rider traveling around the wheel? There are a number of ways to answer this question. The most intuitive measure of the rate at which the rider is traveling around the wheel is what we call *linear velocity.* The units of linear velocity are miles per hour, feet per second, and so forth.

Another way to specify how fast the rider is traveling around the wheel is with what we call *angular velocity.* Angular velocity is given as the amount of central angle through which the rider travels over a given amount of time. The central angle swept out by a rider traveling once around the wheel is 360°, or 2π radians. If one trip around the wheel takes 20 minutes, then the angular velocity of a rider is

$$\frac{2\pi \text{ rad}}{20 \text{ min}} = \frac{\pi}{10} \text{ rad/min}$$

In this section, we will learn more about angular velocity and linear velocity and the relationship between them. Let's start with the formal definition for the linear velocity of a point moving on the circumference of a circle.

DEFINITION

If P is a point on a circle of radius r, and P moves a distance s on the circumference of the circle in an amount of time t, then the *linear velocity,* v, of P is given by the formula

$$v = \frac{s}{t}$$

To calculate the linear velocity, we simply divide distance traveled by time. It does not matter whether the object moves on a curve or in a straight line.

EXAMPLE 1 A point on a circle travels 5 centimeters in 2 seconds. Find the linear velocity of the point.

SOLUTION Substituting $s = 5$ and $t = 2$ into the equation $v = s/t$ gives us

$$v = \frac{5 \text{ cm}}{2 \text{ sec}}$$

$$= 2.5 \text{ cm/sec}$$

PROBLEM 1
A point on a circle travels 8 centimeters in 5 seconds. Find the linear velocity of the point.

NOTE In all the examples and problems in this section, we are assuming that the point on the circle moves with uniform circular motion. That is, the velocity of the point is constant.

DEFINITION

If P is a point moving with uniform circular motion on a circle of radius r, and the line from the center of the circle through P sweeps out a central angle θ in an amount of time t, then the *angular velocity,* ω (omega), of P is given by the formula

$$\omega = \frac{\theta}{t} \quad \text{where } \theta \text{ is measured in radians}$$

PROBLEM 2
A point P on a circle rotates through $\frac{2\pi}{3}$ radians in 4 seconds. Give the angular velocity of P.

EXAMPLE 2 A point P on a circle rotates through $3\pi/4$ radians in 3 seconds. Give the angular velocity of P.

SOLUTION Substituting $\theta = 3\pi/4$ and $t = 3$ into the equation $\omega = \theta/t$ gives us

$$\omega = \frac{3\pi/4 \text{ rad}}{3 \text{ sec}}$$

$$= \frac{\pi}{4} \text{ rad/sec}$$

NOTE There are a number of equivalent ways to express the units of velocity. For example, the answer to Example 2 can be expressed in each of the following ways; they are all equivalent.

$$\frac{\pi}{4} \text{ radians per second} = \frac{\pi}{4} \text{ rad/sec} = \frac{\pi \text{ radians}}{4 \text{ seconds}} = \frac{\pi \text{ rad}}{4 \text{ sec}}$$

Likewise, you can express the answer to Example 1 in any of the following ways:

$$2.5 \text{ centimeters per second} = 2.5 \text{ cm/sec} = \frac{2.5 \text{ cm}}{1 \text{ sec}}$$

PROBLEM 3
A bicycle wheel with a radius of 15.0 inches turns with an angular velocity of 5 radians per second. Find the distance traveled by a point on the bicycle tire in 1 minute.

EXAMPLE 3 A bicycle wheel with a radius of 13.0 inches turns with an angular velocity of 3 radians per second. Find the distance traveled by a point on the bicycle tire in 1 minute.

SOLUTION We have $\omega = 3$ radians per second, $r = 13.0$ inches, and $t = 60$ seconds. First we find θ using $\omega = \theta/t$.

$$\text{If} \quad \omega = \frac{\theta}{t}$$

$$\text{then} \quad \theta = \omega t$$

$$= (3 \text{ rad/sec})(60 \text{ sec})$$

$$= 180 \text{ rad}$$

To find the distance traveled by the point in 60 seconds, we use the formula $s = r\theta$ from Section 3.4, with $r = 13.0$ inches and $\theta = 180$.

$$s = 13.0(180)$$

$$= 2{,}340 \text{ inches}$$

If we want this result expressed in feet, we divide by 12.

$$s = \frac{2{,}340}{12}$$

$$= 195 \text{ ft}$$

A point on the tire of the bicycle will travel 195 feet in 1 minute. If the bicycle were being ridden under these conditions, the rider would travel 195 feet in 1 minute.

PROBLEM 4
Repeat Example 4, but assume the light is 12 feet from the wall and rotates through two complete revolutions every second.

EXAMPLE 4 Figure 1 shows a fire truck parked on the shoulder of a freeway next to a long block wall. The red light on the top of the truck is 10 feet from the wall and rotates through one complete revolution every 2 seconds. Find the equations that give the lengths d and l in terms of time t.

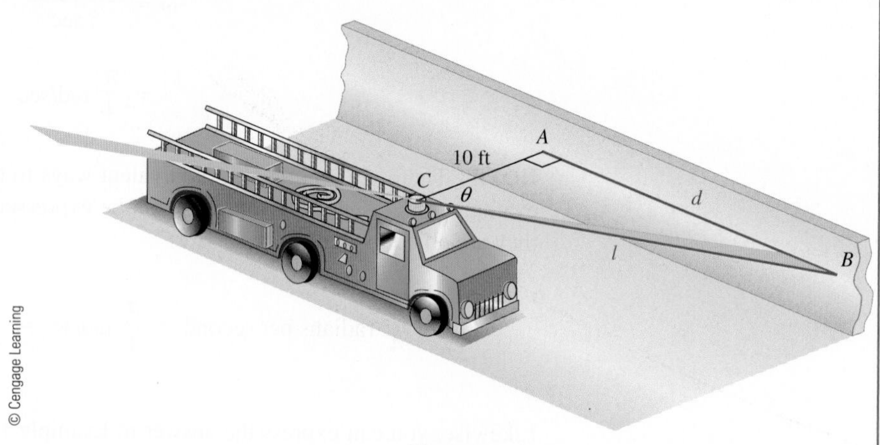

Figure 1

SOLUTION The angular velocity of the rotating red light is

$$\omega = \frac{\theta}{t} = \frac{2\pi \text{ rad}}{2 \text{ sec}} = \pi \text{ rad/sec}$$

From right triangle ABC, we have the following relationships:

$$\tan\theta = \frac{d}{10} \quad \text{and} \quad \sec\theta = \frac{l}{10}$$

$$d = 10\tan\theta \qquad\qquad l = 10\sec\theta$$

Now, these equations give us d and l in terms of θ. To write d and l in terms of t, we solve $\omega = \theta/t$ for θ to obtain $\theta = \omega t = \pi t$. Substituting this for θ in each equation, we have d and l expressed in terms of t.

$$d = 10\tan\pi t \quad \text{and} \quad l = 10\sec\pi t$$

The Relationship Between the Two Velocities

To find the relationship between the two kinds of velocities we have developed so far, we can take the equation that relates arc length and central angle measure, $s = r\theta$, and divide both sides by time, t.

$$\text{If} \quad s = r\theta$$

$$\text{then} \quad \frac{s}{t} = \frac{r\theta}{t}$$

$$\frac{s}{t} = r\frac{\theta}{t}$$

$$v = r\omega$$

Linear velocity is the product of the radius and the angular velocity. Just as with arc length, this relationship implies that the linear velocity is proportional to the angular velocity because r would be constant for a given circle.

LINEAR AND ANGULAR VELOCITY

If a point is moving with uniform circular motion on a circle of radius r, then the linear velocity v and angular velocity ω of the point are related by the formula

$$v = r\omega$$

As suggested by William McClure of Orange Coast Community College, a helpful analogy to visualize the relationship between linear and angular velocity is to imagine a group of ice skaters linked arm to arm on the ice in a straight line, with half the skaters facing in one direction and the other half facing in the opposite direction. As the line of skaters rotates in a circular motion, every skater is moving with the same angular velocity. However, in terms of linear velocity, the skaters near the center of the line are barely moving while the skaters near the ends of the line are moving very rapidly because they have to skate a much larger circle in the same amount of time.

NOTE When using this formula to relate linear velocity to angular velocity, keep in mind that the angular velocity ω must be expressed in radians per unit of time.

PROBLEM 5
A phonograph record is turning at 33 revolutions per minute (rpm). If the distance from the center of the record to a point on the edge of the record is 4 inches, find the angular velocity and the linear velocity of the point (in feet per minute).

EXAMPLE 5 A phonograph record is turning at 45 revolutions per minute (rpm). If the distance from the center of the record to a point on the edge of the record is 3 inches, find the angular velocity and the linear velocity of the point in feet per minute.

SOLUTION The quantity 45 revolutions per minute is another way of expressing the rate at which the point on the record is moving. We can obtain the angular velocity from it by remembering that one complete revolution is equivalent to 2π radians. Therefore,

$$\omega = 45 \text{ rpm} = \frac{45 \text{ rev}}{1 \text{ min}} \cdot \frac{2\pi \text{ rad}}{1 \text{ rev}}$$

$$= 90\pi \text{ rad/min}$$

Because one revolution is equivalent to 2π radians, the fraction

$$\frac{2\pi \text{ rad}}{1 \text{ rev}}$$

is just another form of the number 1. We call this kind of fraction a *conversion factor.* Notice how the units of revolutions divide out in much the same way that common factors divide out when we reduce fractions to lowest terms. The conversion factor allows us to convert from revolutions to radians by dividing out revolutions.

To find the linear velocity, we multiply ω by the radius.

$$
\begin{aligned}
v &= r\omega \\
&= (3 \text{ in.})(90\pi \text{ rad/min}) \\
&= 270\pi \text{ in./min} && \text{Exact value} \\
&\approx 270(3.14) && \text{Approximate value} \\
&= 848 \text{ in./min} && \text{To three significant digits}
\end{aligned}
$$

To convert 848 inches per minute to feet per minute, we use another conversion factor relating feet to inches. Here is our work:

$$848 \text{ in./min} = \frac{848 \text{ in.}}{1 \text{ min}} \cdot \frac{1 \text{ ft}}{12 \text{ in.}} = \frac{848}{12} \text{ ft/min} \approx 70.7 \text{ ft/min}$$

NOTE In light of our previous discussion about units, you may be wondering how we went from (inches)(radians/minute) to just inches/minute in Example 5. Remember that radians are just real numbers and technically have no units. We sometimes write them as if they did for our own convenience, but one of the advantages of radian measure is that it does not introduce a new unit into calculations.

PROBLEM 6
Repeat Example 6, but assume the diameter is 200 feet and one complete revolution takes 10 minutes.

EXAMPLE 6 The Ferris wheel shown in Figure 2 is a model of the one we mentioned in the introduction to this section. If the diameter of the wheel is 250 feet, the distance from the ground to the bottom of the wheel is 14 feet, and one complete revolution takes 20 minutes, find the following.

a. The linear velocity, in miles per hour, of a person riding on the wheel.

b. The height of the rider in terms of the time t, where t is measured in minutes.

SOLUTION

a. We found the angular velocity in the introduction to this section. It is

$$\omega = \frac{\pi}{10} \text{ rad/min}$$

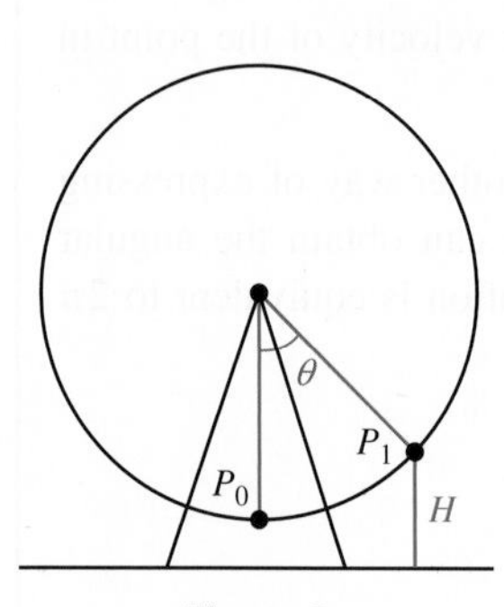

Figure 2

Next, we use the formula $v = r\omega$ to find the linear velocity. That is, we multiply the angular velocity by the radius to find the linear velocity.

$$
\begin{aligned}
v &= r\omega \\
&= (125 \text{ ft})\left(\frac{\pi}{10} \text{ rad/min}\right) \\
&\approx 39.27 \text{ ft/min (intermediate answer)}
\end{aligned}
$$

To convert to miles per hour, we use the facts that there are 60 minutes in 1 hour and 5,280 feet in 1 mile.

$$
\begin{aligned}
39.27 \text{ ft/min} &= \frac{39.27 \text{ ft}}{1 \text{ min}} \cdot \frac{60 \text{ min}}{1 \text{ hr}} \cdot \frac{1 \text{ mi}}{5{,}280 \text{ ft}} \\
&= \frac{(39.27)(60) \text{ mi}}{5{,}280 \text{ hr}} \\
&\approx 0.45 \text{ mi/hr}
\end{aligned}
$$

b. Suppose the person riding on the wheel is at position P_1 as shown in Figure 2. Based on our work in Example 5 of Section 2.3, the height, H, of the rider can be found from the central angle θ using the equation

$$H = 139 - 125 \cos \theta$$

Because the angular velocity of the wheel is $\omega = \pi/10$ radians per minute, then assuming θ is measured in radians we have

$$\omega = \frac{\theta}{t} \quad \text{Definition of angular velocity}$$

$$\theta = \omega t \quad \text{Multiply both sides by } t$$

$$\theta = \frac{\pi}{10}t \quad \text{Substitute } \omega = \pi/10$$

So

$$H = 139 - 125 \cos\left(\frac{\pi}{10}t\right)$$

This equation gives us the height of the rider at any time t, where t is measured in minutes from the beginning of the ride. Later, in Section 4.5, we will find how to obtain this same equation using a graphical approach. ■

To gain a more intuitive understanding of the relationship between the radius of a circle and the linear velocity of a point on the circle, imagine that a bird is sitting on one of the wires that connects the center of the Ferris wheel in Example 6 to the wheel itself. Imagine further that the bird is sitting exactly halfway between the center of the wheel and a rider on the wheel. How does the linear velocity of the bird compare with the linear velocity of the rider? The angular velocity of the bird and the rider are equal (both sweep out the same amount of central angle in a given amount of time), and linear velocity is the product of the radius and the angular velocity, so we can simply multiply the linear velocity of the rider by $\frac{1}{2}$ to obtain the linear velocity of the bird. Therefore, the bird is traveling at

$$\frac{1}{2} \cdot 0.45 \text{ mi/hr} = 0.225 \text{ mi/hr}$$

In one revolution around the wheel, the rider will travel a greater distance than the bird in the same amount of time, so the rider's velocity is greater than that of the bird. Twice as great, to be exact.

Getting Ready for Class

After reading through the preceding section, respond in your own words and in complete sentences.

a. Define linear velocity and give its formula.
b. Define angular velocity and give its formula.
c. What is the relationship between linear and angular velocity?
d. Explain the difference in linear velocities between a rider on a Ferris wheel and a bird positioned between the rider and the center of the wheel. Why is their angular velocity the same?

3.5 PROBLEM SET

➤ CONCEPTS AND VOCABULARY

For Questions 1 through 4, fill in each blank with the appropriate word or equation.

1. If a point travels along a circle with constant velocity, then the motion of the point is called __________ motion.

2. For a point moving with uniform circular motion, the distance traveled per unit time by the point is called __________ velocity, and the amount of rotation per unit time is called _________ velocity.

3. Linear velocity is ____________ to angular velocity. A point rotating twice as fast will travel _________ the distance.

4. The formula relating linear and angular velocity is _________, where the angular velocity must be measured in __________ per unit time.

➤ EXERCISES

In this Problem Set, round answers as necessary to three significant digits.

Find the linear velocity of a point moving with uniform circular motion, if the point covers a distance s in the given amount of time t.

5. $s = 3$ ft and $t = 2$ min

6. $s = 10$ ft and $t = 2$ min

7. $s = 12$ cm and $t = 4$ sec

8. $s = 12$ cm and $t = 2$ sec

9. $s = 30$ mi and $t = 2$ hr

10. $s = 100$ mi and $t = 4$ hr

Find the distance s covered by a point moving with linear velocity v for the given time t.

11. $v = 20$ ft/sec and $t = 4$ sec

12. $v = 10$ ft/sec and $t = 4$ sec

13. $v = 45$ mi/hr and $t = \frac{1}{2}$ hr

14. $v = 55$ mi/hr and $t = \frac{1}{2}$ hr

15. $v = 21$ mi/hr and $t = 20$ min

16. $v = 63$ mi/hr and $t = 10$ sec

Point P sweeps out central angle θ as it rotates on a circle of radius r as given below. In each case, find the angular velocity of point P.

17. $\theta = 24$, $t = 6$ min

18. $\theta = 12$, $t = 3$ min

19. $\theta = 8\pi$, $t = 3\pi$ sec

20. $\theta = 12\pi$, $t = 5\pi$ sec

21. $\theta = 45\pi$, $t = 1.2$ hr

22. $\theta = 24\pi$, $t = 1.8$ hr

23. Rotating Light Figure 3 shows a lighthouse that is 100 feet from a long straight wall on the beach. The light in the lighthouse rotates through one complete rotation once every 4 seconds. Find an equation that gives the distance d in terms of time t, then find d when t is $\frac{1}{2}$ second and $\frac{3}{2}$ seconds. What happens when you try $t = 1$ second in the equation? How do you interpret this?

24. Rotating Light Using the diagram in Figure 3, find an equation that expresses l in terms of time t. Find l when t is 0.5 second, 1.0 second, and 1.5 seconds. (Assume the light goes through one rotation every 4 seconds.)

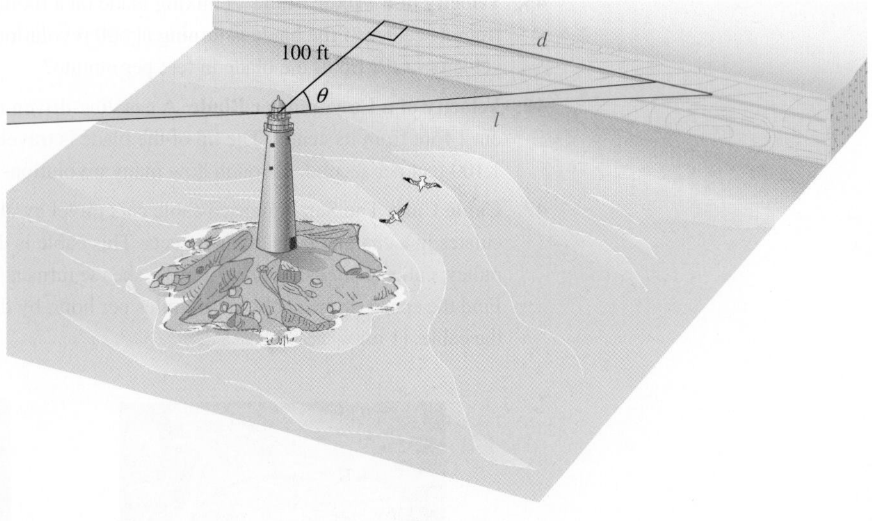

Figure 3

In the problems that follow, point P moves with angular velocity ω on a circle of radius r. In each case, find the distance s traveled by the point in time t.

25. $\omega = 4$ rad/sec, $r = 2$ inches, $t = 5$ sec

26. $\omega = 2$ rad/sec, $r = 4$ inches, $t = 5$ sec

27. $\omega = 3\pi/2$ rad/sec, $r = 4$ m, $t = 30$ sec

28. $\omega = 4\pi/3$ rad/sec, $r = 8$ m, $t = 20$ sec

29. $\omega = 10$ rad/sec, $r = 6$ ft, $t = 2$ min

30. $\omega = 15$ rad/sec, $r = 5$ ft, $t = 1$ min

For each of the following problems, find the angular velocity, in radians per minute, associated with the given revolutions per minute (rpm).

31. 10 rpm **32.** 20 rpm **33.** $33\frac{1}{3}$ rpm **34.** $16\frac{2}{3}$ rpm

35. 5.8 rpm **36.** 7.2 rpm

For each of the following problems, a point is rotating with uniform circular motion on a circle of radius r.

37. Find v if $r = 2$ inches and $\omega = 5$ rad/sec.

38. Find v if $r = 8$ inches and $\omega = 4$ rad/sec.

39. Find ω if $r = 6$ cm and $v = 3$ cm/sec.

40. Find ω if $r = 3$ cm and $v = 8$ cm/sec.

41. Find v if $r = 4$ ft and the point rotates at 10 rpm.

42. Find v if $r = 1$ ft and the point rotates at 20 rpm.

43. Velocity at the Equator The earth rotates through one complete revolution every 24 hours. Since the axis of rotation is perpendicular to the equator, you can think of a person standing on the equator as standing on the edge of a disc that is rotating through one complete revolution every 24 hours. Find the angular velocity of a person standing on the equator.

44. Velocity at the Equator Assuming the radius of the earth is 4,000 miles, use the information from Problem 43 to find the linear velocity of a person standing on the equator.

45. **Velocity of a Mixer Blade** A mixing blade on a food processor extends out 3 inches from its center. If the blade is turning at 600 revolutions per minute, what is the linear velocity of the tip of the blade in feet per minute?

46. **Velocity of a Lawnmower Blade** A gasoline-driven lawnmower has a blade that extends out 1 foot from its center. The tip of the blade is traveling at the speed of sound, which is 1,100 feet per second. Through how many revolutions per minute is the blade turning?

47. **Cable Cars** The San Francisco cable cars travel by clamping onto a steel cable that circulates in a channel beneath the streets. This cable is driven by a large 14-foot-diameter pulley, called a *sheave* (Figure 4). The sheave turns at a rate of 19 revolutions per minute. Find the speed of the cable car, in miles per hour, by determining the linear velocity of the cable. (1 mi = 5,280 ft)

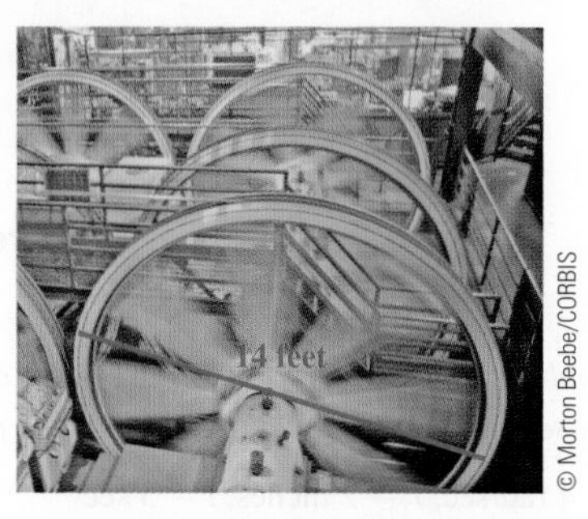

© Morton Beebe/CORBIS

Figure 4

© CORBIS

Figure 5

48. **Cable Cars** The Los Angeles Cable Railway was driven by a 13-foot-diameter drum that turned at a rate of 18 revolutions per minute. Find the speed of the cable car, in miles per hour, by determining the linear velocity of the cable.

49. **Cable Cars** The Cleveland City Cable Railway had a 14-foot-diameter pulley to drive the cable. In order to keep the cable cars moving at a linear velocity of 12 miles per hour, how fast would the pulley need to turn (in revolutions per minute)?

50. **Cable Cars** The old Sutter Street cable car line in San Francisco (Figure 5) used a 12-foot-diameter sheave to drive the cable. In order to keep the cable cars moving at a linear velocity of 10 miles per hour, how fast would the sheave need to turn (in revolutions per minute)?

51. **Ski Lift** A ski lift operates by driving a wire rope, from which chairs are suspended, around a bullwheel (Figure 6). If the bullwheel is 12 feet in diameter and turns at a rate of 9 revolutions per minute, what is the linear velocity, in feet per second, of someone riding the lift?

Courtesy of Doppelmayr CTEC, Inc.

Figure 6

52. **Ski Lift** An engineering firm is designing a ski lift. The wire rope needs to travel with a linear velocity of 2.0 meters per second, and the angular velocity of the bullwheel will be 10 revolutions per minute. What diameter bullwheel should be used to drive the wire rope?

Figure 7

 **53. Velocity of a Ferris Wheel** Figure 7 is a model of the Ferris wheel known as the Riesenrad, or Great Wheel, that was built in Vienna in 1897. The diameter of the wheel is 197 feet, and one complete revolution takes 15 minutes. Find the linear velocity of a person riding on the wheel. Give your answer in miles per hour and round to the nearest hundredth.

 **54. Velocity of a Ferris Wheel** Use Figure 7 as a model of the Ferris wheel called Colossus that was built in St. Louis in 1986. The diameter of the wheel is 165 feet. A brochure that gives some statistics associated with Colossus indicates that it rotates at 1.5 revolutions per minute and also indicates that a rider on the wheel is traveling at 10 miles per hour. Explain why these two numbers, 1.5 revolutions per minute and 10 miles per hour, cannot both be correct.

 **55. Ferris Wheel** For the Ferris wheel described in Problem 53, find the height of the rider, h, in terms of the time, t, where t is measured in minutes from the beginning of the ride. The distance from the ground to the bottom of the wheel is 12.0 feet (see Section 2.3, Problem 65).

 **56. Ferris Wheel** For the Ferris wheel described in Problem 54, find the height of the rider, h, in terms of the time, t, where t is measured in minutes from the beginning of the ride. The distance from the ground to the bottom of the wheel is 9.00 feet (see Section 2.3, Problem 66). Assume the wheel rotates at a rate of 1.5 revolutions per minute.

 **57. Velocity of a Bike Wheel** A woman rides a bicycle for 1 hour and travels 16 kilometers (about 10 miles). Find the angular velocity of the wheel if the radius is 30 centimeters.

 **58. Velocity of a Bike Wheel** Find the number of revolutions per minute for the wheel in Problem 57.

Gear Trains *Figure 8 shows a single-stage* gear train. *Gear trains are used in many products, such as clocks and automotive transmissions, to reduce or increase the angular velocity of a component. The size of each gear is measured by the number of teeth rather than the radius. Suppose the first gear has n_1 teeth and the second gear has n_2 teeth.*

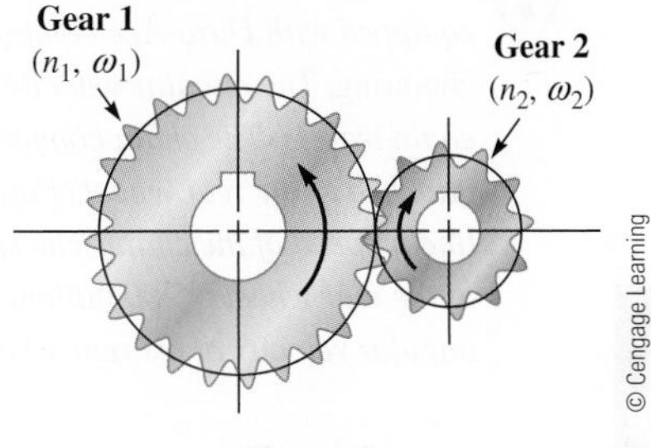

Figure 8

Because the spacing of the teeth is the same for both gears, the ratio of their radii will be equivalent to the corresponding ratio of the number of teeth. When two gears are meshed together, they share the same linear velocity. If ω_1 and ω_2 are the angular velocities of the first and second gears, respectively, then

$$v_2 = v_1$$

$$r_2\omega_2 = r_1\omega_1$$

$$\omega_2 = \frac{r_1}{r_2}\omega_1$$

$$\omega_2 = \frac{n_1}{n_2}\omega_1$$

59. The first gear in a single-stage gear train has 42 teeth and an angular velocity of 2 revolutions per second. The second gear has 7 teeth. Find the angular velocity of the second gear.

60. The second gear in a single-stage gear train has 6 teeth and an angular velocity of 90 revolutions per minute. The first gear has 54 teeth. Find the angular velocity of the first gear.

61. A gear train consists of three gears meshed together (Figure 9). The middle gear is known as an idler. Show that the angular velocity of the third gear does not depend on the number of teeth of the idler gear (Gear 2).

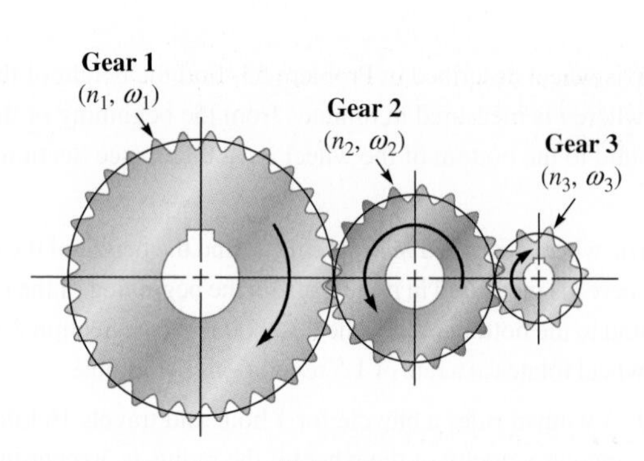

Figure 9

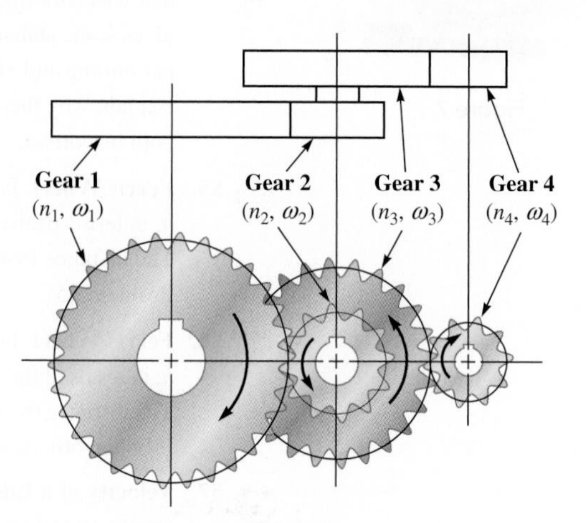

Figure 10

62. A two-stage gear train consists of four gears meshed together (Figure 10). The second and third gears are attached, so that they share the same angular velocity ($\omega_2 = \omega_3$). Find a formula giving the angular velocity of the fourth gear, ω_4, in terms of ω_1 and the values of n_1, n_2, n_3, and n_4.

Cycling *Lance Armstrong, seven-time winner of the Tour de France, rides a Trek bicycle equipped with Dura-Ace components (Figure 11). When Lance pedals, he turns a gear, called a* chainring. *The angular velocity of the chainring will determine the linear speed at which the chain travels. The chain connects the chainring to a smaller gear, called a* sprocket, *which is attached to the rear wheel (Figure 12). The angular velocity of the sprocket depends upon the linear speed of the chain. The sprocket and rear wheel rotate at the same rate, and the diameter of the rear wheel is 700 millimeters. The speed at which Lance travels is determined by the angular velocity of his rear wheel. Use this information to answer Problems 63 through 68.*

Figure 11

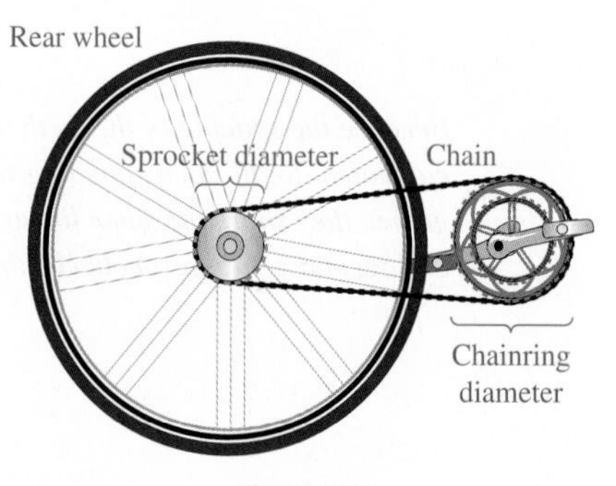

Figure 12

63. When Lance Armstrong blazed up Mount Ventoux in the 2002 Tour, he was equipped with a 150-millimeter-diameter chainring and a 95-millimeter-diameter sprocket. Lance is known for maintaining a very high *cadence,* or pedal rate. If he was pedaling at a rate of 90 revolutions per minute, find his speed in kilometers per hour. (1 km = 1,000,000 mm)

64. On level ground, Lance would use a larger chainring and a smaller sprocket. If he shifted to a 210-millimeter-diameter chainring and a 40-millimeter-diameter sprocket, how fast would he be traveling in kilometers per hour if he pedaled at a rate of 80 revolutions per minute?

65. If Lance was using his 210-millimeter-diameter chainring and pedaling at a rate of 85 revolutions per minute, what diameter sprocket would he need in order to maintain a speed of 45 kilometers per hour?

66. If Lance was using his 150-millimeter-diameter chainring and pedaling at a rate of 95 revolutions per minute, what diameter sprocket would he need in order to maintain a speed of 24 kilometers per hour?

67. Suppose Lance was using a 150-millimeter-diameter chainring and an 80-millimeter-diameter sprocket. How fast would he need to pedal, in revolutions per minute, in order to maintain a speed of 20 kilometers per hour?

68. Suppose Lance was using a 210-millimeter-diameter chainring and a 40-millimeter-diameter sprocket. How fast would he need to pedal, in revolutions per minute, in order to maintain a speed of 40 kilometers per hour?

➤ REVIEW PROBLEMS

The problems that follow review material we covered in Section 2.5.

69. **Magnitude of a Vector** Find the magnitudes of the horizontal and vertical vector components of a velocity vector of 68 feet per second with angle of elevation 37°.

70. **Magnitude of a Vector** The magnitude of the horizontal component of a vector is 75, while the magnitude of its vertical component is 45. What is the magnitude of the vector?

71. **Distance and Bearing** A ship sails for 85.5 miles on a bearing of S 57.3° W. How far west and how far south has the boat traveled?

72. **Distance and Bearing** A plane flying with a constant speed of 285.5 miles per hour flies for 2 hours on a course with bearing N 48.7° W. How far north and how far west does the plane fly?

➤ LEARNING OBJECTIVES ASSESSMENT

These questions are available for instructors to help assess if you have successfully met the learning objectives for this section.

73. Find the linear velocity of a point moving with uniform circular motion if the point covers a distance of 45 meters in 15 seconds.

 a. 6π m/sec b. 3 m/sec c. 180π m/sec d. 675 m/sec

74. Find the angular velocity of a point moving with uniform circular motion if the point rotates through an angle of 24π radians in 3 minutes.

 a. 8π rad/min b. 72π rad/min c. 0.4 rad/min d. 36 rad/min

75. Find the linear velocity of a point rotating at 30 revolutions per second on a circle of radius 6 centimeters.

 a. 180 cm/sec b. 10π cm/sec c. $\frac{90}{\pi}$ cm/sec d. 360π cm/sec

76. A pulley is driven by a belt moving at a speed of 15.7 feet per second. If the pulley is 6 inches in diameter, approximate the angular velocity of the pulley in revolutions per second.

 a. 0.83 rev/sec b. 7.5 rev/sec c. 10 rev/sec d. 63 rev/sec

CHAPTER 3 SUMMARY

REFERENCE ANGLE [3.1]

The *reference angle* $\hat{\theta}$ for any angle θ in standard position is the positive acute angle between the terminal side of θ and the x-axis.

A trigonometric function of an angle and its reference angle differ at most in sign.

We find trigonometric functions for angles between 0° and 360° by first finding the reference angle. We then find the value of the trigonometric function of the reference angle and use the quadrant in which the angle terminates to assign the correct sign.

$$\sin 150° = \sin 30° = \frac{1}{2} \qquad \sin 330° = -\sin 30° = -\frac{1}{2}$$

EXAMPLES

1 30° is the reference angle for 30°, 150°, 210°, and 330°.

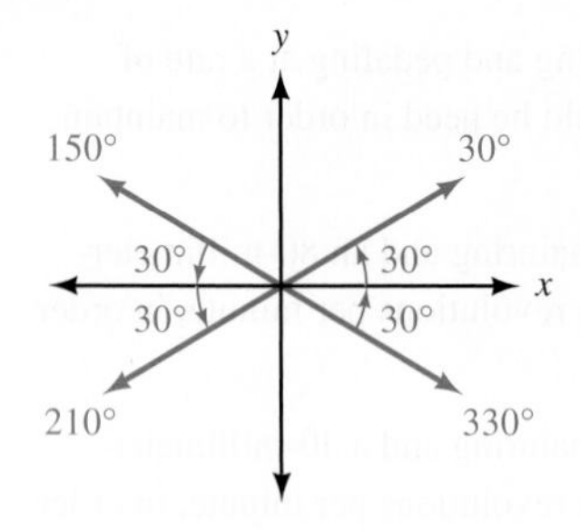

RADIAN MEASURE [3.2]

In a circle with radius r, if central angle θ cuts off an arc of length s, then the radian measure of θ is given by

$$\theta = \frac{s}{r}$$

2 If $r = 3$ cm and $s = 6$ cm, then

$$\theta = \frac{6}{3} = 2 \text{ radians}$$

RADIANS AND DEGREES [3.2]

Changing from degrees to radians and radians to degrees is simply a matter of multiplying by the appropriate conversion factor.

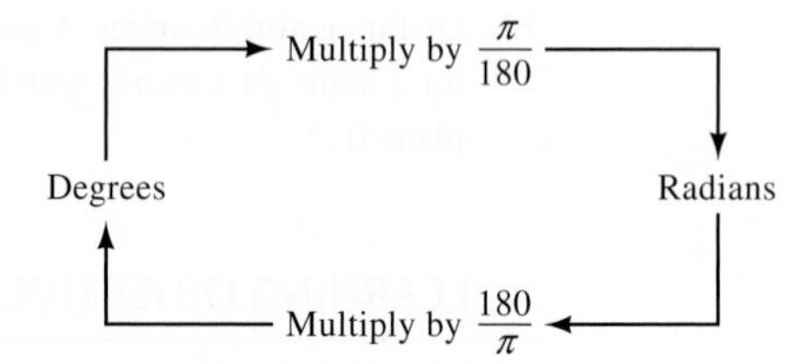

3 Radians to degrees

$$\frac{4\pi}{3} \text{ rad} = \frac{4\pi}{3}\left(\frac{180}{\pi}\right)^{\circ} = 240°$$

Degrees to radians

$$450° = 450\left(\frac{\pi}{180}\right) = \frac{5\pi}{2} \text{ rad}$$

CIRCULAR FUNCTIONS (DEFINITION III) [3.3]

If (x, y) is any point on the unit circle, and t is the distance from $(1, 0)$ to (x, y) along the circumference of the unit circle, then

$$\cos t = x$$

$$\sin t = y$$

$$\tan t = \frac{y}{x} \quad (x \neq 0)$$

$$\cot t = \frac{x}{y} \quad (y \neq 0)$$

$$\csc t = \frac{1}{y} \quad (y \neq 0)$$

$$\sec t = \frac{1}{x} \quad (x \neq 0)$$

4 If t corresponds to the point $(-0.1288, 0.9917)$ on the unit circle, then

$$\sin t = 0.9917$$

$$\cos t = -0.1288$$

$$\tan t = \frac{0.9917}{-0.1288} = -7.6995$$

5 There is no value t for which $\cos t = 4$ since the range of the cosine function is $[-1, 1]$.

DOMAIN AND RANGE OF THE CIRCULAR FUNCTIONS [3.3]

	Domain	Range
$\cos t$	$(-\infty, \infty)$	$[-1, 1]$
$\sin t$	$(-\infty, \infty)$	$[-1, 1]$
$\tan t$	$t \neq \frac{\pi}{2} + k\pi$	$(-\infty, \infty)$
$\cot t$	$t \neq k\pi$	$(-\infty, \infty)$
$\csc t$	$t \neq k\pi$	$(-\infty, -1] \cup [1, \infty)$
$\sec t$	$t \neq \frac{\pi}{2} + k\pi$	$(-\infty, -1] \cup [1, \infty)$

6 The arc cut off by 2.5 radians in a circle of radius 4 inches is

$$s = 4(2.5) = 10 \text{ inches}$$

ARC LENGTH [3.4]

If s is an arc cut off by a central angle θ, measured in radians, in a circle of radius r, then

$$s = r\theta$$

7 The area of the sector formed by a central angle of 2.5 radians in a circle of radius 4 inches is

$$A = \frac{1}{2}(4)^2(2.5) = 20 \text{ inches}^2$$

AREA OF A SECTOR [3.4]

The area of the sector formed by a central angle θ in a circle of radius r is

$$A = \frac{1}{2}r^2\theta$$

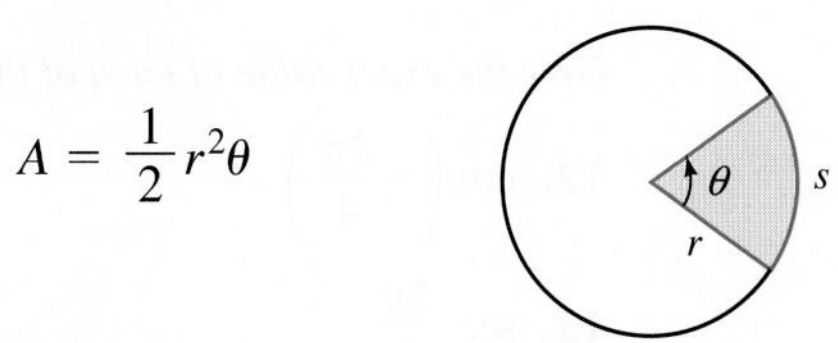

where θ is measured in radians.

8 If a point moving at a uniform speed on a circle travels 12 centimeters every 3 seconds, then the linear velocity of the point is

$$v = \frac{12 \text{ cm}}{3 \text{ sec}} = 4 \text{ cm/sec}$$

LINEAR VELOCITY [3.5]

If P is a point on a circle of radius r, and P moves a distance s on the circumference of the circle in an amount of time t, then the *linear velocity*, v, of P is given by the formula

$$v = \frac{s}{t}$$

9 If a point moving at uniform speed on a circle of radius 4 inches rotates through $3\pi/4$ radians every 3 seconds, then the angular velocity of the point is

$$\omega = \frac{3\pi/4 \text{ rad}}{3 \text{ sec}} = \frac{\pi}{4} \text{ rad/sec}$$

The linear velocity of the same point is given by

$$v = 4\left(\frac{\pi}{4}\right) = \pi \text{ inches/sec}$$

ANGULAR VELOCITY [3.5]

If P is a point moving with uniform circular motion on a circle of radius r, and the line from the center of the circle through P sweeps out a central angle θ in an amount of time t, then the *angular velocity*, ω, of P is given by the formula

$$\omega = \frac{\theta}{t} \quad \text{where } \theta \text{ is measured in radians}$$

The relationship between linear velocity and angular velocity is given by the formula

$$v = r\omega$$

CHAPTER 3 TEST

Draw each of the following angles in standard position and then name the reference angle:

1. $235°$

2. $-225°$

Use a calculator to find each of the following (round to four decimal places):

3. $\cot 320°$

4. $\csc(-236.7°)$

5. $\sec 140° 20'$

Use a calculator to find θ, to the nearest tenth of a degree, if θ is between 0° and 360° and

6. $\sin\theta = 0.1045$ with θ in QII

7. $\cot\theta = 0.9659$ with θ in QIII

Give the exact value of each of the following:

8. $\sin 225°$

9. $\tan 330°$

10. Convert 250° to radian measure. Write your answer as an exact value.

11. Convert $7\pi/12$ to degree measure.

Give the exact value of each of the following:

12. $\cos\left(-\dfrac{3\pi}{4}\right)$

13. $\sec\dfrac{5\pi}{6}$

14. If t is the distance from $(1, 0)$ to $\left(\dfrac{2\sqrt{13}}{13}, -\dfrac{3\sqrt{13}}{13}\right)$ along the circumference of the unit circle, find $\sin t$, $\cos t$, and $\tan t$.

15. Identify the argument of $\cos 4x$.

16. Use a calculator to approximate $\cos 5$ to four decimal places.

17. Determine if the statement $\sin z = 2$ is possible for some real number z.

18. Evaluate $2\cos\left(3x - \dfrac{\pi}{2}\right)$ when x is $\dfrac{\pi}{3}$.

19. If $\theta = 60°$ is a central angle in a circle of radius 6 feet, find the length of the arc cut off by θ to the nearest hundredth.

20. If $\theta = \pi/4$ is a central angle that cuts off an arc length of π centimeters, find the radius of the circle.

21. Find the area of the sector formed by central angle $\theta = 2.4$ in a circle of radius 3 centimeters.

22. **Distance** A boy is twirling a model airplane on a string 5 feet long. If he twirls the plane at 0.5 revolutions per minute, how far does the plane travel in 2 minutes? Round to the nearest tenth.

23. **Area of a Sector** A central angle of 4 radians cuts off an arc of length 8 inches. Find the area of the sector formed.

24. Point P moves with angular velocity $\omega = 4$ radians per second on a circle of radius 3 inches. Find the distance s traveled by the point in 6 seconds.

For each of the following problems, a point is rotating with uniform circular motion on a circle of radius r. Give your answer in exact form.

25. Find ω if $r = 10$ cm and $v = 5$ cm/sec.

26. Find v if $r = 2$ ft and the point rotates at 20 rpm.

27. **Angular Velocity** A belt connects a pulley of radius 8 centimeters to a pulley of radius 6 centimeters. Each point on the belt is traveling at 24 centimeters per second. Find the angular velocity of each pulley (Figure 1).

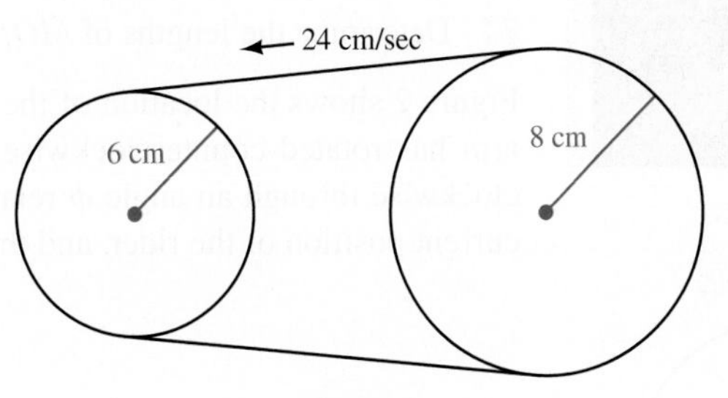

Figure 1

28. **Linear Velocity** A propeller with radius 1.50 feet is rotating at 900 revolutions per minute. Find the linear velocity of the tip of the propeller. Give the exact value and an approximation to three significant digits.

29. **Cable Cars** If a 14-foot-diameter sheave is used to drive a cable car, at what angular velocity must the sheave turn in order for the cable car to travel 11 miles per hour? Give your answer in revolutions per minute. (1 mi = 5,280 ft)

30. **Cycling** Alberto Contador is riding in the Vuelta a España. He is using a 210-millimeter-diameter chainring and a 50-millimeter-diameter sprocket with a 700-millimeter-diameter rear wheel (Figure 2). Find his linear velocity if he is pedaling at a rate of 75 revolutions per minute. Give your answer in kilometers per hour to the nearest tenth. (1 km = 1,000,000 mm)

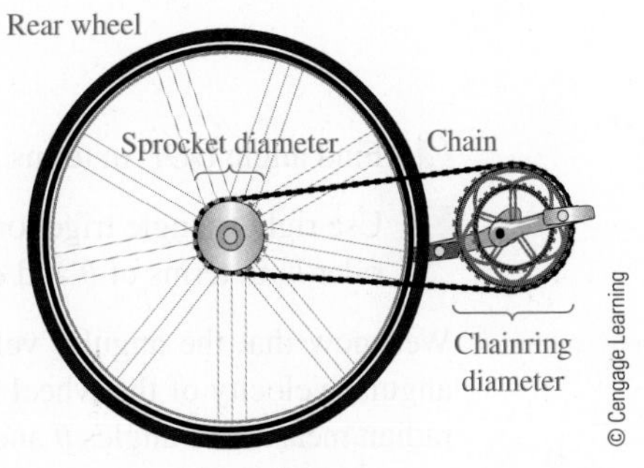

Figure 2

GROUP PROJECT Modeling a Double Ferris Wheel

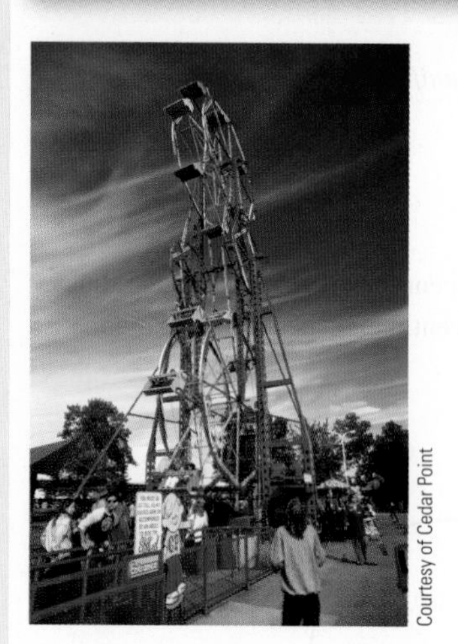
Courtesy of Cedar Point

OBJECTIVE: To find a model for the height of a rider on a double Ferris wheel.

In 1939, John Courtney invented the first double Ferris wheel, called a Sky Wheel, consisting of two smaller wheels spinning at the ends of a rotating arm.

For this project, we will model a double Ferris wheel with a 50-foot arm that is spinning at a rate of 3 revolutions per minute in a counterclockwise direction. The center of the arm is 44 feet above the ground. The diameter of each wheel is 32 feet, and the wheels turn at a rate of 5 revolutions per minute in a clockwise direction. A diagram of the situation is shown in Figure 1. M is the midpoint of the arm, and O is the center of the lower wheel. Assume the rider is initially at point P on the wheel.

1 Determine the lengths of MO, OP, and MG.

Figure 2 shows the location of the rider after a short amount of time has passed. The arm has rotated counterclockwise through an angle θ, while the wheel has rotated clockwise through an angle ϕ relative to the direction of the arm. Point P shows the current position of the rider, and the height of the rider is h.

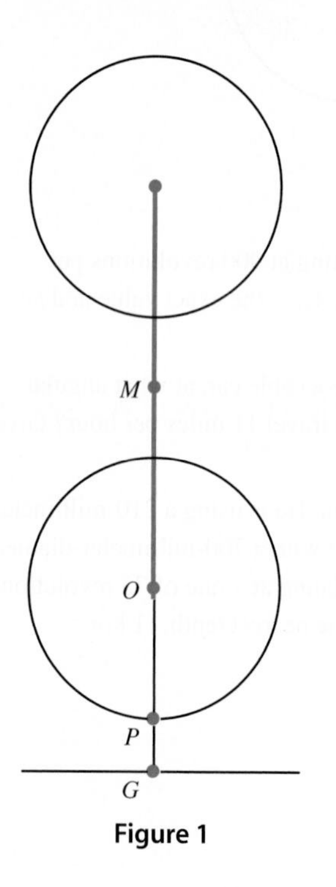

Figure 1

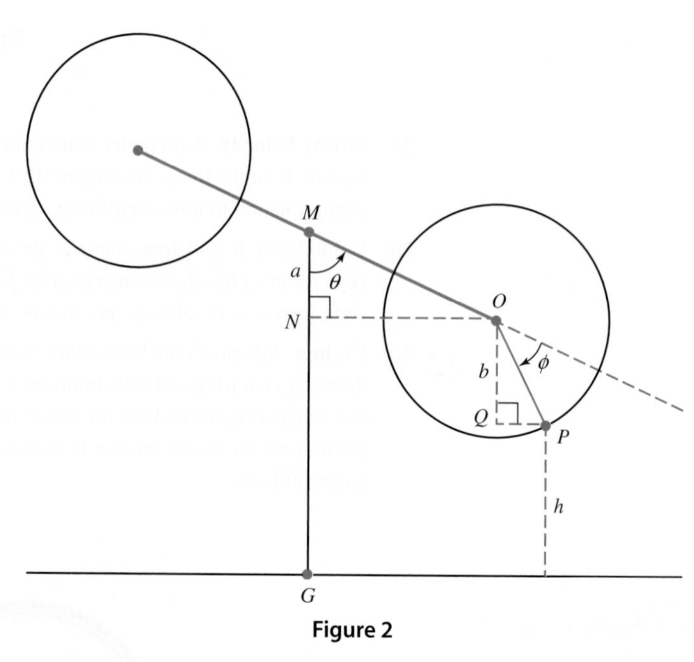

Figure 2

2 Find angle QOP in terms of θ and ϕ.

3 Use right triangle trigonometry to find lengths a and b, and then the height of the rider h, in terms of θ and ϕ.

We know that the angular velocity of the arm is 3 revolutions per minute, and the angular velocity of the wheel is 5 revolutions per minute. The last step is to find the radian measure of angles θ and ϕ.

4 Let t be the number of minutes that have passed since the ride began. Use the angular velocities of the arm and wheel to find θ and ϕ in radians in terms of t. Replace θ and ϕ in your answer from Step 3 to obtain the height h as a function of time t (in minutes).

5 Use this function to find the following:

a. The height of the rider after 30 seconds have passed.

b. The height of the rider after the arm has completed its first revolution.

c. The height of the rider after the wheel has completed two revolutions.

6 Graph this function with your graphing calculator. Make sure your calculator is set to radian mode. Use the graph to find the following:

a. The maximum height of the rider.

b. The minimum number of minutes required for the rider to return to their original position (point P in Figure 1).

The Third Man

The Riesenrad in Vienna

We mentioned in Chapter 2 that the Ferris wheel called the *Riesenrad,* built in Vienna in 1897, is still in operation today. A brochure that gives some statistics associated with the Riesenrad indicates that passengers riding it travel at 2.5 feet per second. The Orson Welles movie *The Third Man* contains a scene in which Orson Welles rides the Riesenrad through one complete revolution. Watch *The Third Man* so you can view the Riesenrad in operation. Then devise a method of using the movie to estimate the angular velocity of the wheel. Give a detailed account of the procedure you use to arrive at your estimate. Finally, use your results either to prove or to disprove the claim that passengers travel at 2.5 feet per second on the Riesenrad.

CUMULATIVE TEST 1–3

1. Use the information given in Figure 1 to find x, h, s, and r if $y = 3$. (*Note:* s is the distance from A to D, and y is the distance from D to B.)

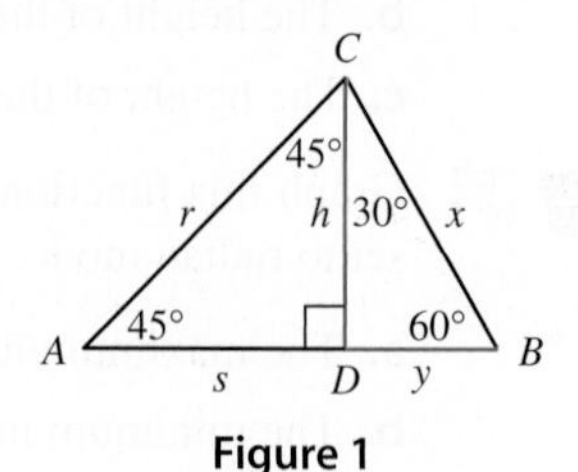

Figure 1

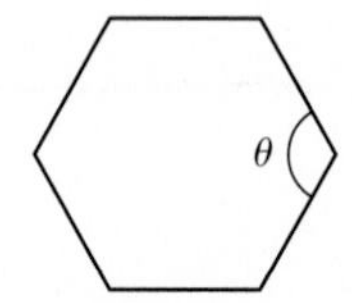

Figure 2

2. **Geometry** Find the measure of one of the interior angles of a regular hexagon (see Figure 2).
3. Find the distance from the origin to the point (a, b).
4. Find all angles that are coterminal with $120°$.
5. In which quadrant will θ lie if $\sin\theta < 0$ and $\cos\theta > 0$?
6. Find all six trigonometric functions for θ if the point $(-6, 8)$ lies on the terminal side of θ in standard position.
7. Find the remaining trigonometric functions of θ if $\tan\theta = \frac{12}{5}$ and θ terminates in QIII.
8. If $\sec\theta = -2$, find $\cos\theta$.
9. Multiply $(\sin\theta + 3)(\sin\theta - 7)$.
10. Prove that $\sin\theta(\csc\theta + \cot\theta) = 1 + \cos\theta$ is an identity by transforming the left side into the right side.
11. Find $\sin A$, $\cos A$, $\tan A$, and $\sin B$, $\cos B$, and $\tan B$ in right triangle ABC, with $C = 90°$, if $a = 5$ and $b = 12$.
12. Fill in the blank to make the statement true: sec _____ = csc 73°.
13. Subtract $15°32'$ from $25°15'$.
14. Use a calculator to find θ to the nearest tenth of a degree if θ is an acute angle and $\tan\theta = 0.0816$.
15. Give the number of significant digits in each number.

    a. 0.00028 **b.** 280 **c.** 2,800.

16. If triangle ABC is a right triangle with $B = 48°$, $C = 90°$, and $b = 270$, solve the triangle by finding the remaining sides and angles.

17. **Distance and Bearing** A ship leaves the harbor entrance and travels 35 miles in the direction N 42° E. The captain then turns the ship 90° and travels another 24 miles in the direction S 48° E. At that time, how far is the ship from the harbor entrance, and what is the bearing of the ship from the harbor entrance?
18. **Angle of Depression** A man standing on the roof of a building 86.0 feet above the ground looks down to the building next door. He finds the angle of depression to the roof of that building from the roof of his building to be 14.5°, while the angle of depression from the roof of his building to the bottom of the building next door is 43.2°. How tall is the building next door?
19. **Vector Magnitude** If vector **V** has magnitude 5.0 and makes an angle of 30° with the positive x-axis, find the magnitudes of the horizontal and vertical vector components of **V**.

20. **Static Equilibrium** David and his sister Jessica are playing on a tree swing. David sits on the swing while Jessica pushes him horizontally through an angle of 28.5° and holds him there (Figure 3). If David weighs 80.5 pounds, find the magnitude of the tension in the rope and the magnitude of the force Jessica must push with to keep David in static equilibrium.

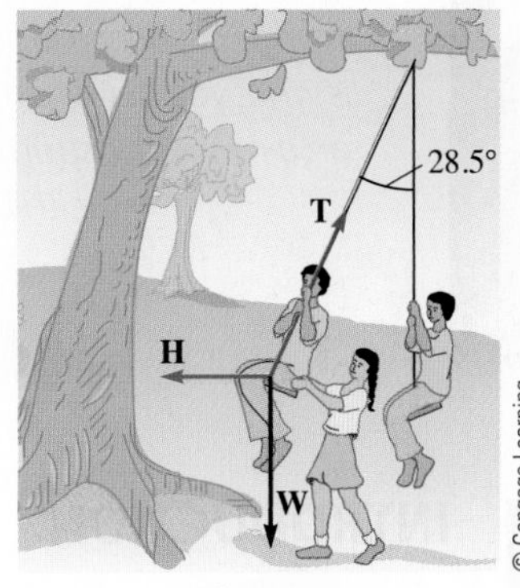

Figure 3

21. Draw 117.8° in standard position and then name the reference angle.
22. Use a calculator to find θ, to the nearest tenth of a degree, if $0° \le \theta < 360°$ and $\cos\theta = -0.4772$ with θ in QIII.
23. Convert $-390°$ to radian measure. Write your answer as an exact value.
24. Give the exact value of $\sin \frac{2\pi}{3}$.
25. Use the unit circle to find all values of θ between 0 and 2π for which $\cos\theta = -\frac{1}{2}$.
26. Use the unit circle to explain how the value of $\tan t$ varies as t increases from $\frac{\pi}{2}$ to π.
27. If $\theta = \frac{\pi}{6}$ is a central angle in a circle of radius 12 meters, find the length of the arc cut off by θ to the nearest hundredth.
28. Find the area of the sector formed by central angle $\theta = 90°$ in a circle of radius 4 inches.
29. Point P moves with angular velocity $\frac{3\pi}{4}$ radians per second on a circle of radius 8 feet. Find the distance s traveled by the point in 20 seconds.
30. **Velocity of a DVD** A DVD, when placed in a DVD player, rotates at 2,941 revolutions per minute. Find the linear velocity, in miles per hour, of a point 1.5 inches from the center of the DVD. Round your answer to the nearest tenth. (1 mile = 5,280 feet)

Graphing and Inverse Functions

A page of sheet music represents a piece of music; the music itself is what you get when the notes on the page are sung or performed on a musical instrument.

➤ Keith Devlin

© Flat Earth Royalty Free Photograph/ Fotosearch

INTRODUCTION

Trigonometric functions are periodic functions because they repeat all their range values at regular intervals. The riders on the Ferris wheels we have been studying repeat their positions around the wheel periodically as well. That is why trigonometric functions are good models for the motion of a rider on a Ferris wheel. When we use a rectangular coordinate system to plot the distance between the ground and a rider during a ride, we find that the shape of the graph matches exactly the shape of the graph of one of the trigonometric functions.

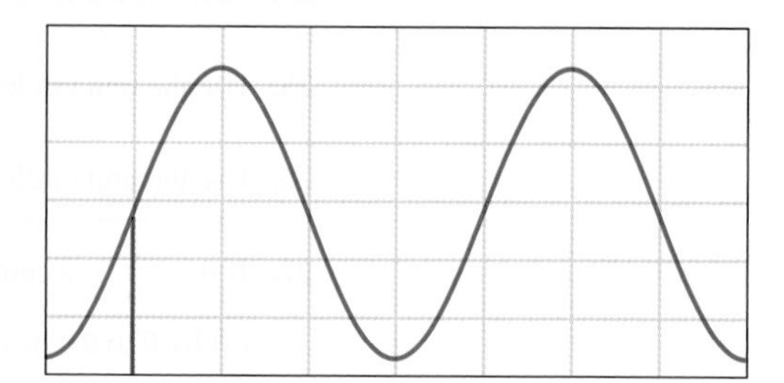

Once we have modeled the motion of the rider with a trigonometric function, we have a new set of mathematical tools with which to investigate the motion of the rider on the wheel.

STUDY SKILLS 4

The study skills for this chapter are about attitude. They are points of view that point toward success.

1 Be Focused, Not Distracted We have students who begin their assignments by asking themselves, "Why am I taking this class?" If you are asking yourself similar questions, you are distracting yourself from doing the things that will produce the results you want in this course. Don't dwell on questions and evaluations of the class that can be used as excuses for not doing well. If you want to succeed in this course, focus your energy and efforts toward success, rather than distracting yourself from your goals.

2 Be Resilient Don't let setbacks keep you from your goals. You want to put yourself on the road to becoming a person who can succeed in this class, or any college class. Failing a test or quiz, or having a difficult time on some topics, is normal. No one goes through college without some setbacks. Don't let a temporary disappointment keep you from succeeding in this course. A low grade on a test or quiz is simply a signal that some reevaluation of your study habits needs to take place.

3 Intend to Succeed We always have a few students who simply go through the motions of studying without intending on mastering the material. It is more important to them to look as if they are studying than to actually study. You need to study with the intention of being successful in the course. Intend to master the material, no matter what it takes. ▲

SECTION 4.1 Basic Graphs

LEARNING OBJECTIVES

1. Sketch the graph of a basic trigonometric function.
2. Analyze the graph of a trigonometric function.
3. Evaluate a trigonometric function using the even and odd function relationships.
4. Prove an equation is an identity.

In Section 3.3 we introduced the circular functions, which define the six trigonometric functions as functions of real numbers, and investigated their domains. Now we will sketch the graphs of these functions and determine their ranges.

The Sine Graph

To graph the function $y = \sin x$, we begin by making a table of values of x and y that satisfy the equation (Table 1), and then use the information in the table to sketch the graph. To make it easy on ourselves, we will let x take on values that are multiples of $\pi/4$. As an aid in sketching the graphs, we will approximate $\sqrt{2}/2$ with 0.7.

Graphing each ordered pair and then connecting them with a smooth curve, we obtain the graph in Figure 1:

TABLE 1

x	$y = \sin x$
0	$\sin 0 = 0$
$\frac{\pi}{4}$	$\sin \frac{\pi}{4} = \frac{\sqrt{2}}{2}$
$\frac{\pi}{2}$	$\sin \frac{\pi}{2} = 1$
$\frac{3\pi}{4}$	$\sin \frac{3\pi}{4} = \frac{\sqrt{2}}{2}$
π	$\sin \pi = 0$
$\frac{5\pi}{4}$	$\sin \frac{5\pi}{4} = -\frac{\sqrt{2}}{2}$
$\frac{3\pi}{2}$	$\sin \frac{3\pi}{2} = -1$
$\frac{7\pi}{4}$	$\sin \frac{7\pi}{4} = -\frac{\sqrt{2}}{2}$
2π	$\sin 2\pi = 0$

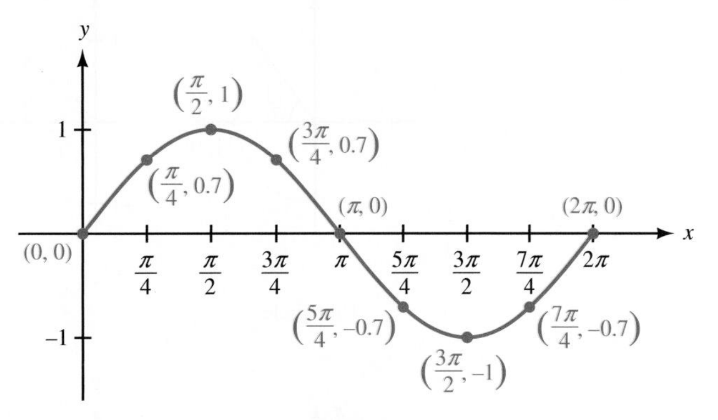

Figure 1

Graphing $y = \sin x$ Using the Unit Circle

We can also obtain the graph of the sine function by using the unit circle definition (Definition III). Figure 2 shows a diagram of the unit circle we introduced earlier in Section 3.3. If the point (x, y) is t units from $(1, 0)$ along the circumference of the unit circle, then $\sin t = y$. Therefore, if we start at the point $(1, 0)$ and travel once around the unit circle (a distance of 2π units), we can find the value of y in the equation $y = \sin t$ by simply keeping track of the y-coordinates of the points that are t units from $(1, 0)$.

NOTE In Figure 2 we are using x differently than we are in Figure 1. Because in Figure 2 we need x to represent the coordinate of a point on the unit circle, we let t be the input value to the trigonometric function. The variable x in Figure 1 represents the same quantity as the variable t in Figure 2.

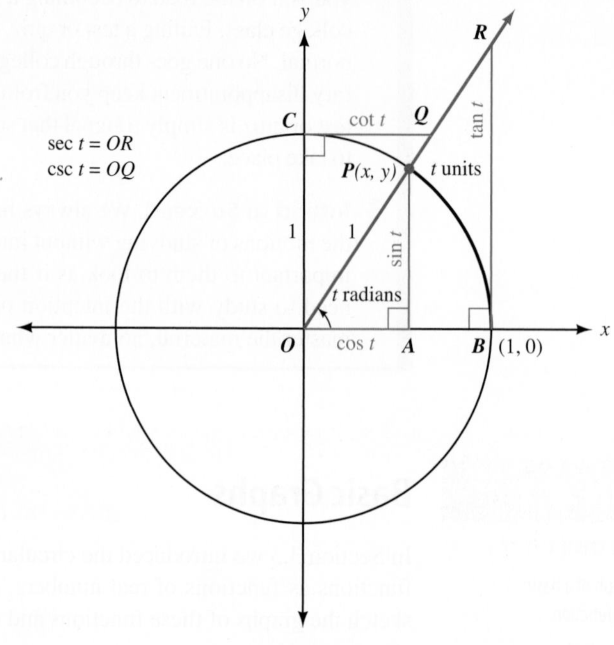

Figure 2

As t increases from 0 to $\pi/2$, meaning P travels from $(1, 0)$ to $(0, 1)$, $y = \sin t$ increases from 0 to 1. As t continues in QII from $\pi/2$ to π, y decreases from 1 back to 0. In QIII the length of segment AP increases from 0 to 1, but because it is located below the x-axis the y-coordinate is negative. So, as t increases from π to $3\pi/2$, y decreases from 0 to -1. Finally, as t increases from $3\pi/2$ to 2π in QIV, bringing P back to $(1, 0)$, y increases from -1 back to 0. Figure 3 illustrates how the y-coordinate of P (or AP) is used to construct the graph of the sine function as t increases.

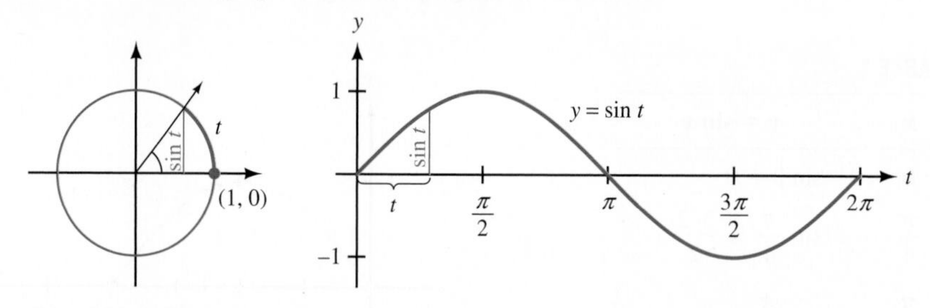

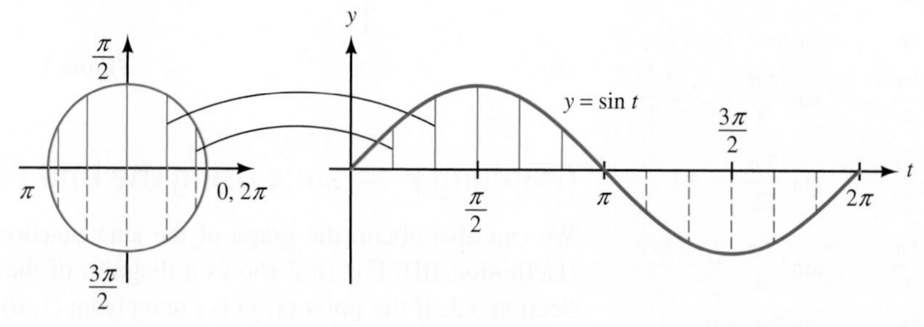

Figure 3

Extending the Sine Graph

Figures 1 and 3 each show one complete cycle of $y = \sin x$. (In Figure 3 we have used t in place of x, but you can see the graphs are the same.) We can extend the graph of $y = \sin x$ to the right of $x = 2\pi$ by realizing that, once we go past $x = 2\pi$, we will begin to name angles that are coterminal with the angles between 0 and 2π. Because of this, we will start to repeat the values of $\sin x$. Likewise, if we let x take on values to the left of $x = 0$, we will simply get the values of $\sin x$ between 0 and 2π in the reverse order. Figure 4 shows the graph of $y = \sin x$ extended beyond the interval from $x = 0$ to $x = 2\pi$.

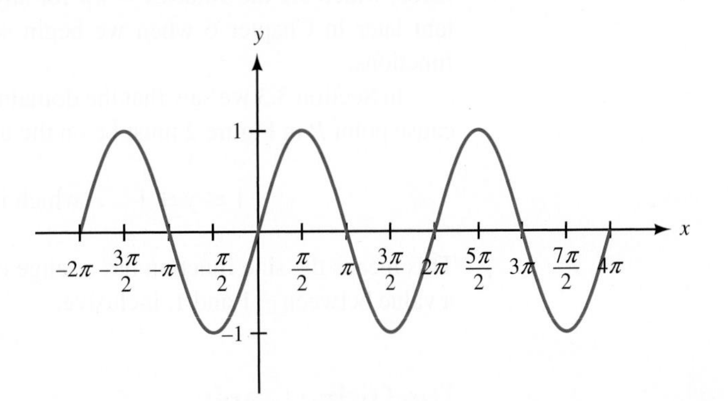

Figure 4

The graph of $y = \sin x$ never goes above 1 or below -1, repeats itself every 2π units on the x-axis, and crosses the x-axis at multiples of π. This gives rise to the following three definitions.

DEFINITION ■ PERIOD

For any function $y = f(x)$, the smallest positive number p for which

$$f(x + p) = f(x)$$

for all x in the domain of f is called the *period* of $f(x)$.

In the case of $y = \sin x$, the period is 2π because $p = 2\pi$ is the smallest positive number for which $\sin(x + p) = \sin x$ for all x.

DEFINITION ■ AMPLITUDE

If the greatest value of y is M and the least value of y is m, then the *amplitude* of the graph of y is defined to be

$$A = \frac{1}{2}|M - m|$$

In the case of $y = \sin x$, the amplitude is 1 because

$$\frac{1}{2}|1 - (-1)| = \frac{1}{2}(2) = 1$$

DEFINITION ■ ZERO

A *zero* of a function $y = f(x)$ is any domain value $x = c$ for which $f(c) = 0$. If c is a real number, then $x = c$ will be an x-intercept of the graph of $y = f(x)$.

From the graph of $y = \sin x$, we see that the sine function has an infinite number of zeros, which are the values $x = k\pi$ for any integer k. These values will be very important later in Chapter 6 when we begin solving equations that involve trigonometric functions.

In Section 3.3 we saw that the domain for the sine function is all real numbers. Because point P in Figure 2 must be on the unit circle, we have

$$-1 \le y \le 1 \quad \text{which implies} \quad -1 \le \sin t \le 1$$

This means the sine function has a range of $[-1, 1]$. The sine of any angle can only be a value between -1 and 1, inclusive.

The Cosine Graph

The graph of $y = \cos x$ has the same general shape as the graph of $y = \sin x$.

PROBLEM 1
Graph $y = \sin x$.

EXAMPLE 1 Sketch the graph of $y = \cos x$.

SOLUTION We can arrive at the graph by making a table of convenient values of x and y (Table 2). Plotting points, we obtain the graph shown in Figure 5.

TABLE 2

x	$y = \cos x$
0	$\cos 0 = 1$
$\frac{\pi}{4}$	$\cos \frac{\pi}{4} = \frac{\sqrt{2}}{2}$
$\frac{\pi}{2}$	$\cos \frac{\pi}{2} = 0$
$\frac{3\pi}{4}$	$\cos \frac{3\pi}{4} = -\frac{\sqrt{2}}{2}$
π	$\cos \pi = -1$
$\frac{5\pi}{4}$	$\cos \frac{5\pi}{4} = -\frac{\sqrt{2}}{2}$
$\frac{3\pi}{2}$	$\cos \frac{3\pi}{2} = 0$
$\frac{7\pi}{4}$	$\cos \frac{7\pi}{4} = \frac{\sqrt{2}}{2}$
2π	$\cos 2\pi = 1$

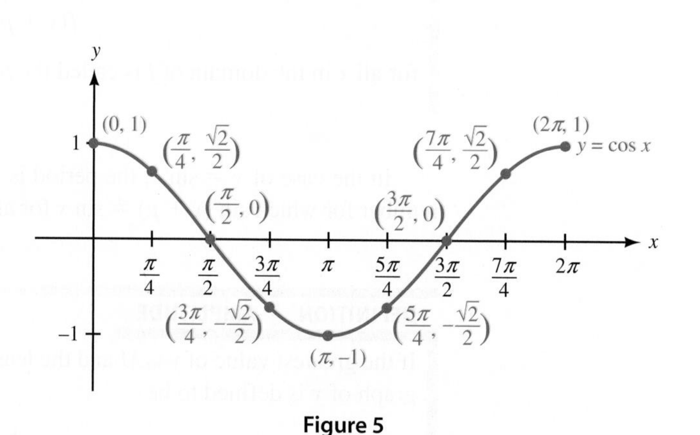

Figure 5

We can generate the graph of the cosine function using the unit circle just as we did for the sine function. By Definition III, if the point (x, y) is t units from $(1, 0)$ along the circumference of the unit circle, then $\cos t = x$. We start at the point $(1, 0)$ and travel once around the unit circle, keeping track of the x-coordinates of the points that are t units from $(1, 0)$. To help visualize how the x-coordinates generate the cosine graph, we have rotated the unit circle 90° counterclockwise so that we may represent the x-coordinates as vertical line segments (Figure 6).

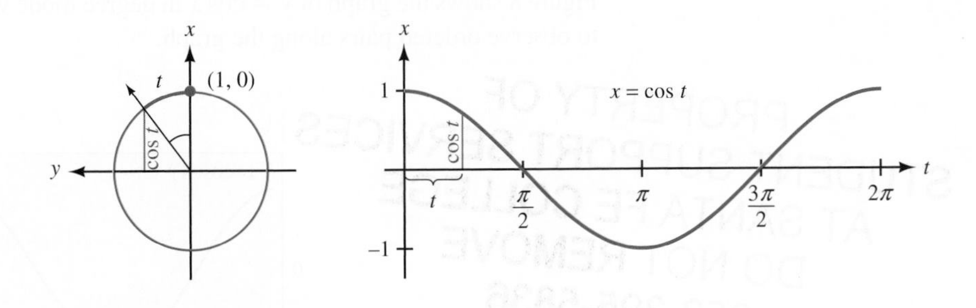

Unit circle (rotated)

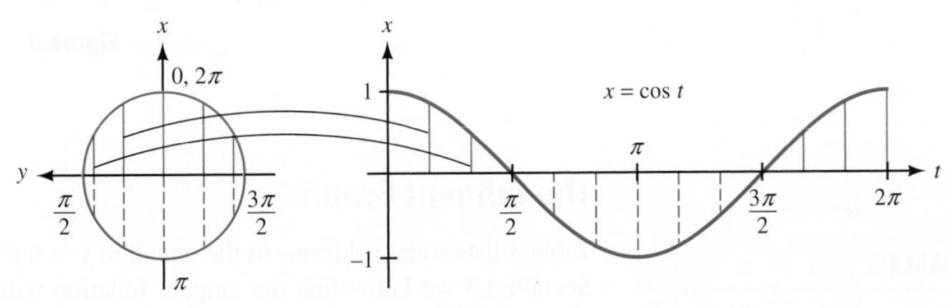

Figure 6

Extending this graph to the right of 2π and to the left of 0, we obtain the graph shown in Figure 7. As this figure indicates, the period, amplitude, and range of the cosine function are the same as for the sine function. The zeros, or x-intercepts, of $y = \cos x$ are the values $x = \frac{\pi}{2} + k\pi$ for any integer k.

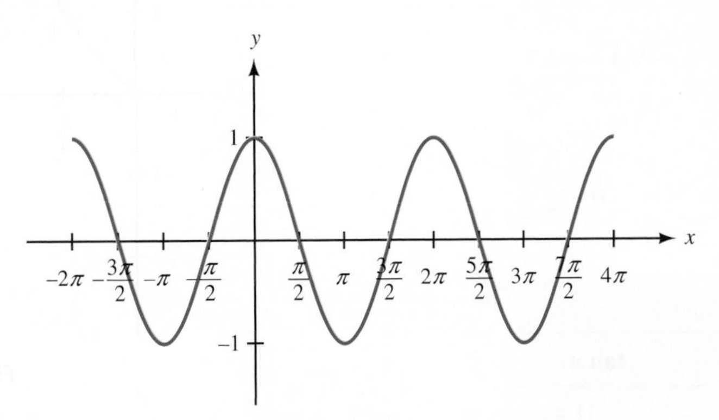

Figure 7

CALCULATOR NOTE To graph one cycle of the sine or cosine function using your graphing calculator in radian mode, define Y1 = sin(x) or Y1 = cos (x) and set your window variables so that

$$0 \le x \le 2\pi,\ \text{scale} = \pi/2;\ -1.5 \le y \le 1.5$$

To graph either function in degree mode, set your window variables to

$$0 \le x \le 360, \text{ scale} = 90; \ -1.5 \le y \le 1.5$$

Figure 8 shows the graph of $y = \cos x$ in degree mode with the trace feature being used to observe ordered pairs along the graph.

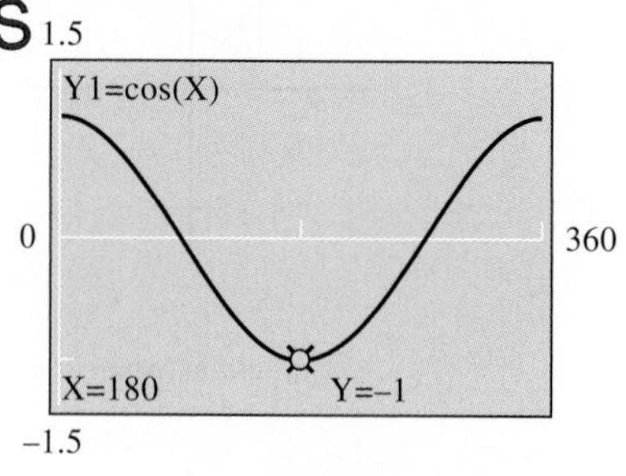

Figure 8

The Tangent Graph

Table 3 lists some solutions to the equation $y = \tan x$ between $x = 0$ and $x = \pi$. From Section 3.3 we know that the tangent function will be undefined at $x = \pi/2$ because of the division by zero. Figure 9 shows the graph based on the information from Table 3.

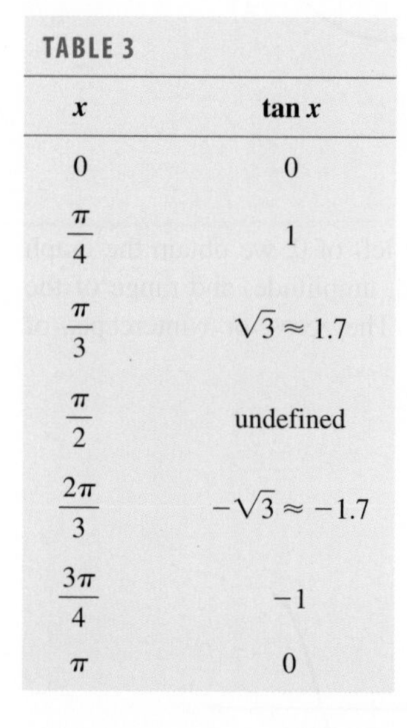

TABLE 3

x	$\tan x$
0	0
$\frac{\pi}{4}$	1
$\frac{\pi}{3}$	$\sqrt{3} \approx 1.7$
$\frac{\pi}{2}$	undefined
$\frac{2\pi}{3}$	$-\sqrt{3} \approx -1.7$
$\frac{3\pi}{4}$	-1
π	0

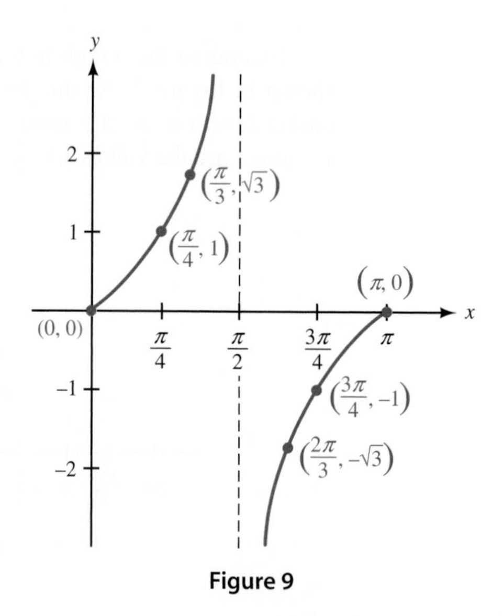

Figure 9

TABLE 4

x	$\tan x$
85°	11.4
89°	57.3
89.9°	573.0
89.99°	5729.6
90.01°	−5729.6
90.1°	−573.0
91°	−57.3
95°	−11.4

Because $y = \tan x$ is undefined at $x = \pi/2$, there is no point on the graph with an x-coordinate of $\pi/2$. To help us remember this, we have drawn a dotted vertical line through $x = \pi/2$. This vertical line is called an *asymptote*. The graph will never cross or touch this line. If we were to calculate values of $\tan x$ when x is very close to $\pi/2$ (or very close to 90° in degree mode), we would find that $\tan x$ would become very large for values of x just to the left of the asymptote and very large in the negative direction for values of x just to the right of the asymptote, as shown in Table 4.

In terms of the unit circle (Definition III), visualize segment BR in Figure 10a as t increases from 0 to $\pi/2$. When $t = 0$, OP is horizontal and so $BR = 0$. As t increases, BR grows in length, getting very large as t nears $\pi/2$. At $t = \pi/2$, OP is vertical and BR is no longer defined. The exact reverse happens as t increases from $\pi/2$ to π, except that now $\tan t$ is negative because BR will be located below the x-axis (Figure 10b). As t continues through QIII, the values of $\tan t$ will be a repeat of the values we saw in QI. The same can be said of QIV and QII.

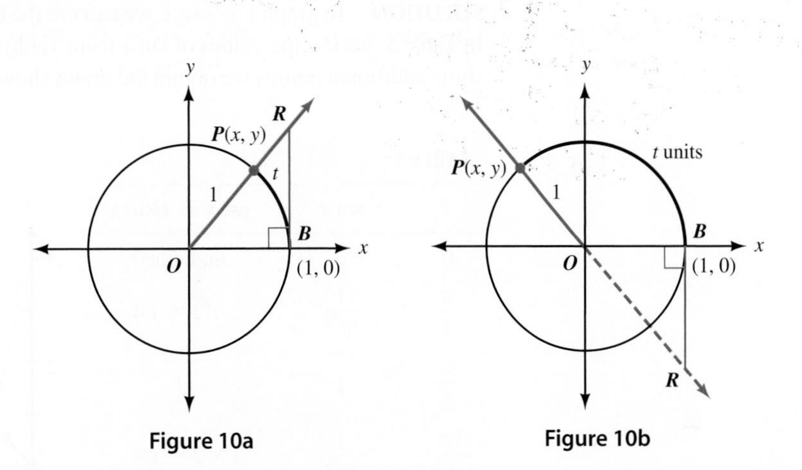

Figure 10a

Figure 10b

We can also visualize $\tan t$ as the slope of OP. At $t = 0$, OP is horizontal and the slope is zero. As P travels around the unit circle through QI, you can see how OP gets very steep when t nears $\pi/2$. At $t = \pi/2$, OP is vertical and the slope is not defined. As P travels through QII the slopes are all negative, reaching a zero slope once again at $t = \pi$. For P in QIII and QIV, the slope of OP will simply repeat the values seen in QI and QII.

Extending the graph in Figure 9 to the right of π and to the left of 0, we obtain the graph shown in Figure 11. As this figure indicates, the period of $y = \tan x$ is π. The tangent function has no amplitude because there is no largest or smallest value of y on the graph of $y = \tan x$. For this same reason, the range of the tangent function is all real numbers. Because $\tan x = \sin x/\cos x$, the zeros for the tangent function are the same as for the sine; that is, $x = k\pi$ for any integer k. The vertical asymptotes correspond to the zeros of the cosine function, which are $x = \pi/2 + k\pi$ for any integer k.

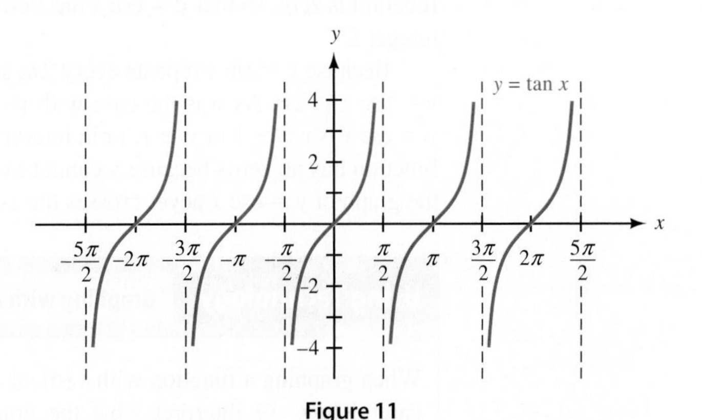

Figure 11

The Cosecant Graph

Now that we have the graph of the sine, cosine, and tangent functions, we can use the reciprocal identities to sketch the graph of the remaining three trigonometric functions.

PROBLEM 2
Graph $y = \sec x$.

EXAMPLE 2 Sketch the graph of $y = \csc x$.

SOLUTION To graph $y = \csc x$, we can use the fact that $\csc x$ is the reciprocal of $\sin x$. In Table 5, we use the values of $\sin x$ from Table 1 and take reciprocals. Filling in with some additional points, we obtain the graph shown in Figure 12.

TABLE 5

x	$\sin x$	$\csc x = 1/\sin x$
0	0	undefined
$\frac{\pi}{4}$	$\frac{1}{\sqrt{2}}$	$\sqrt{2} \approx 1.4$
$\frac{\pi}{2}$	1	1
$\frac{3\pi}{4}$	$\frac{1}{\sqrt{2}}$	$\sqrt{2} \approx 1.4$
π	0	undefined
$\frac{5\pi}{4}$	$-\frac{1}{\sqrt{2}}$	$-\sqrt{2} \approx -1.4$
$\frac{3\pi}{2}$	-1	-1
$\frac{7\pi}{4}$	$-\frac{1}{\sqrt{2}}$	$-\sqrt{2} \approx -1.4$
2π	0	undefined

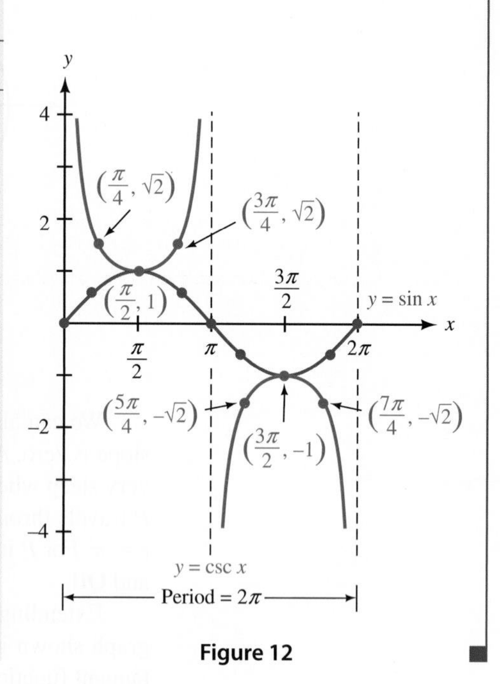

Figure 12

As you can see in Figure 12, the reciprocals of 1 and -1 are themselves, so these points are common to both the graphs of the sine and cosecant functions. When $\sin x$ is close to 1 or -1, so is $\csc x$. When $\sin x$ is close to zero, $\csc x$ will be a very large positive or negative number. The cosecant function will be undefined whenever the sine function is zero, so that $y = \csc x$ has vertical asymptotes at the values $x = k\pi$ for any integer k.

Because $y = \sin x$ repeats every 2π, so do the reciprocals of $\sin x$, so the period of $y = \csc x$ is 2π. As was the case with $y = \tan x$, there is no amplitude. The range of $y = \csc x$ is $y \le -1$ or $y \ge 1$, or in interval notation, $(-\infty, -1] \cup [1, \infty)$. The cosecant function has no zeros because y cannot ever be equal to zero. Notice in Figure 12 that the graph of $y = \csc x$ never crosses the x-axis.

Using Technology **Graphing with Asymptotes**

When graphing a function with vertical asymptotes, such as $y = \csc x$, we must be careful how we interpret what the graphing calculator shows us. For example,

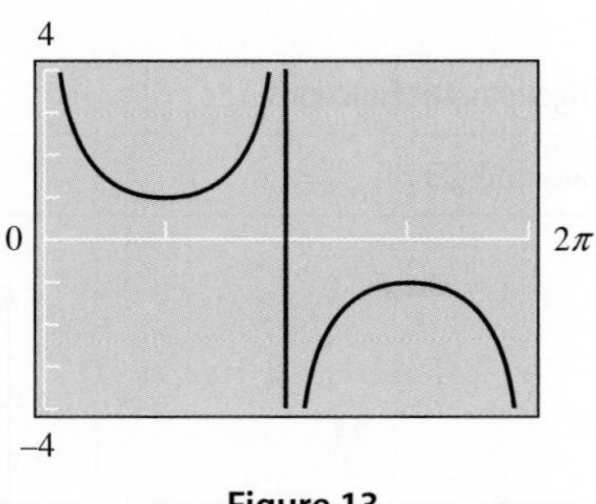

Figure 13

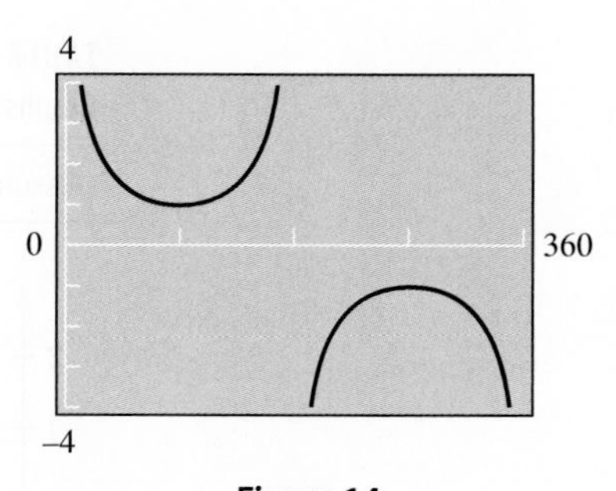

Figure 14

having defined $Y_1 = 1/\sin(x)$, Figure 13 shows the graph of one cycle of this function in radian mode with the window variables set so that

$$0 \le x \le 2\pi, \text{ scale} = \pi/2; \ -4 \le y \le 4$$

and Figure 14 shows the same graph in degree mode with window settings

$$0 \le x \le 360, \text{ scale} = 90; \ -4 \le y \le 4$$

Because the calculator graphs a function by plotting points and connecting them, the vertical line seen in Figure 13 will sometimes appear where an asymptote exists. We just need to remember that this line is not part of the graph of the cosecant function, as Figure 14 indicates. To make sure that these artificial lines do not appear in the graph, we can graph the function in dot mode or with the dot style, as illustrated in Figure 15.

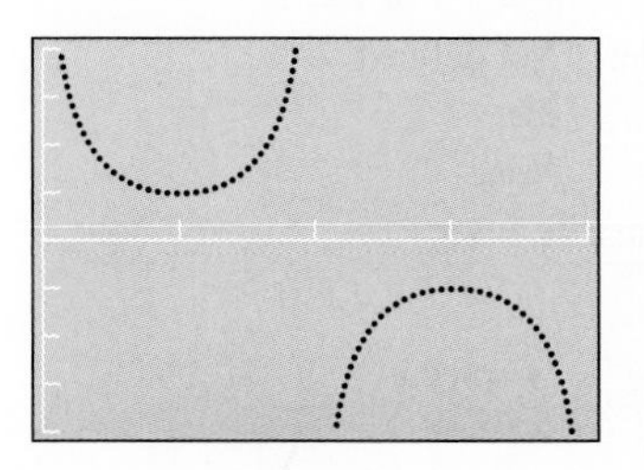

Figure 15

The Cotangent and Secant Graphs

In Problem Set 4.1, you will be asked to graph $y = \cot x$ and $y = \sec x$. These graphs are shown in Figures 16 and 17 for reference.

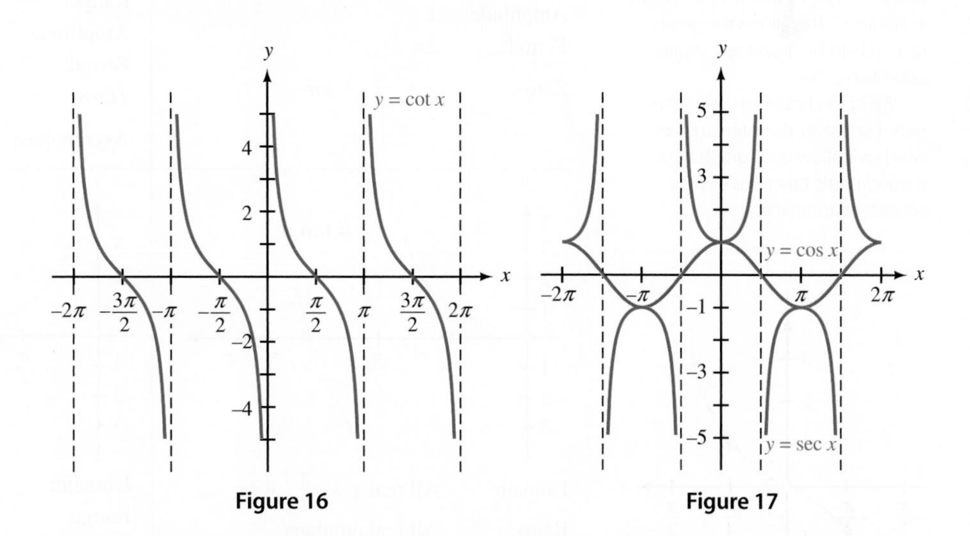

Figure 16

Figure 17

Table 6 is a summary of the important facts associated with the graphs of our trigonometric functions. Each graph shows one cycle for the corresponding function, which we will refer to as the *basic cycle.* Keep in mind that all these graphs repeat indefinitely to the left and to the right.

TABLE 6
Graphs of the Trigonometric Functions

(Assume k is any integer)

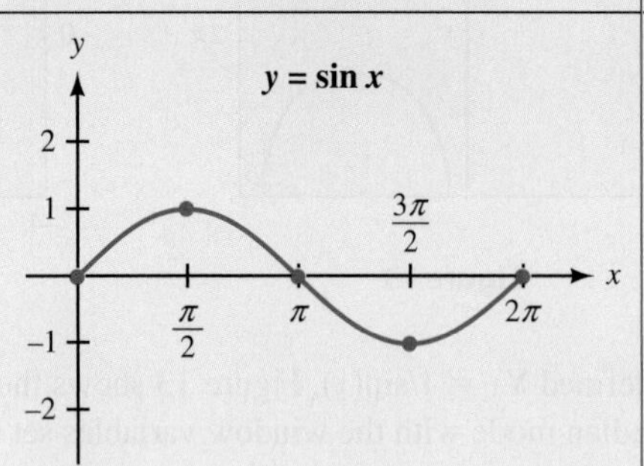

Domain:	All real numbers
Range:	$-1 \le y \le 1$
Amplitude:	1
Period:	2π
Zeros:	$x = k\pi$

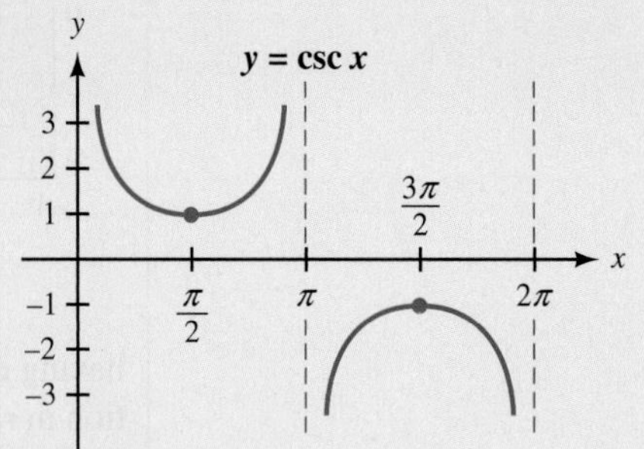

Domain:	All real $x \neq k\pi$
Range:	$y \le -1$ or $y \ge 1$
Amplitude:	Not defined
Period:	2π
Zeros:	None
Asymptotes:	$x = k\pi$

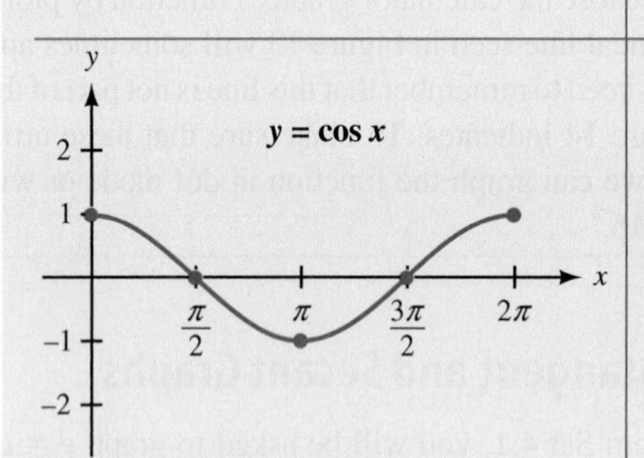

Domain:	All real numbers
Range:	$-1 \le y \le 1$
Amplitude:	1
Period:	2π
Zeros:	$x = \frac{\pi}{2} + k\pi$

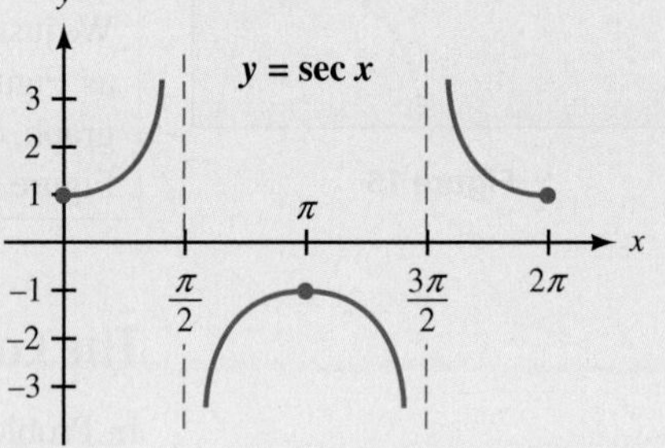

Domain:	All real $x \neq \frac{\pi}{2} + k\pi$
Range:	$y \le -1$ or $y \ge 1$
Amplitude:	Not defined
Period:	2π
Zeros:	None
Asymptotes:	$x = \frac{\pi}{2} + k\pi$

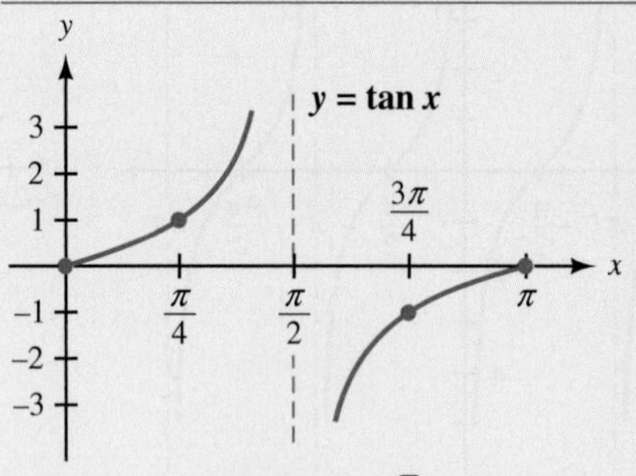

Domain:	All real $x \neq \frac{\pi}{2} + k\pi$
Range:	All real numbers
Amplitude:	Not defined
Period:	π
Zeros:	$x = k\pi$
Asymptotes:	$x = \frac{\pi}{2} + k\pi$

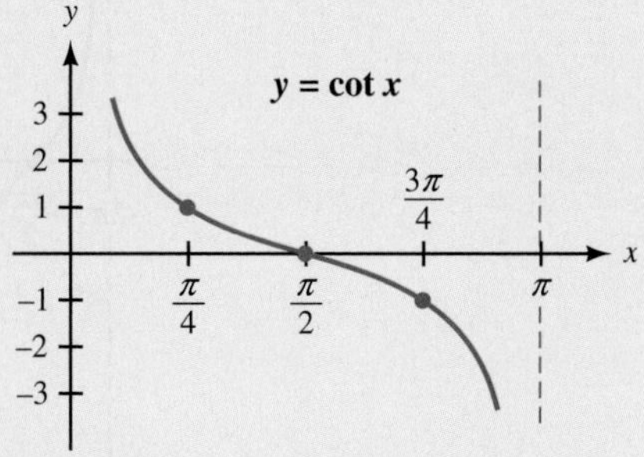

Domain:	All real $x \neq k\pi$
Range:	All real numbers
Amplitude:	Not defined
Period:	π
Zeros:	$x = \frac{\pi}{2} + k\pi$
Asymptotes:	$x = k\pi$

NOTE Some instructors prefer to define the basic cycle for the tangent function to be $(-\pi/2, \pi/2)$ as shown in Figure 18. This allows the graph of a cycle to be drawn as a single, unbroken curve.

We have chosen to define the basic cycle over the interval $[0, \pi]$, which will allow us to graph all six trigonometric functions using a consistent approach.

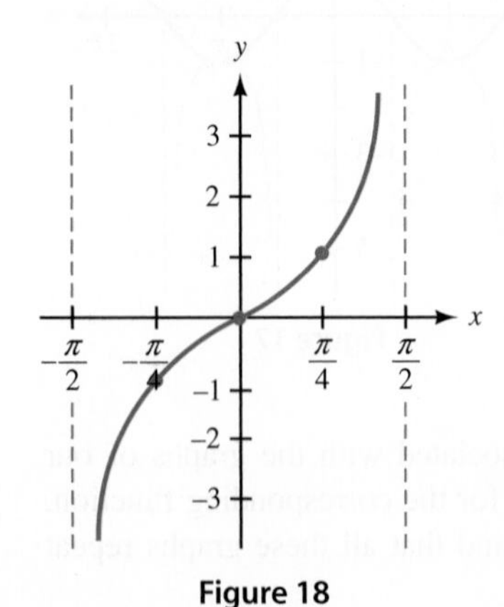

Figure 18

Even and Odd Functions

Recall from algebra the definitions of even and odd functions.

DEFINITION

An *even function* is a function for which

$$f(-x) = f(x) \text{ for all } x \text{ in the domain of } f$$

The graph of an even function is symmetric about the y-axis.

An even function is a function for which replacing x with $-x$ leaves the expression that defines the function unchanged. If a function is even, then every time the point (x, y) is on the graph, so is the point $(-x, y)$. The function $f(x) = x^2 + 3$ is an even function because

$$\begin{aligned} f(-x) &= (-x)^2 + 3 \\ &= x^2 + 3 \\ &= f(x) \end{aligned}$$

DEFINITION

An *odd function* is a function for which

$$f(-x) = -f(x) \text{ for all } x \text{ in the domain of } f$$

The graph of an odd function is symmetric about the origin.

An odd function is a function for which replacing x with $-x$ changes the sign of the expression that defines the function. If a function is odd, then every time the point (x, y) is on the graph, so is the point $(-x, -y)$. The function $f(x) = x^3 - x$ is an odd function because

$$\begin{aligned} f(-x) &= (-x)^3 - (-x) \\ &= -x^3 + x \\ &= -(x^3 - x) \\ &= -f(x) \end{aligned}$$

From the unit circle it is apparent that sine is an odd function and cosine is an even function. To begin to see that this is true, we locate $\pi/6$ and $-\pi/6$ ($-\pi/6$ is coterminal with $11\pi/6$) on the unit circle and notice that

$$\cos\left(-\frac{\pi}{6}\right) = \frac{\sqrt{3}}{2} = \cos\frac{\pi}{6}$$

and

$$\sin\left(-\frac{\pi}{6}\right) = -\frac{1}{2} = -\sin\frac{\pi}{6}$$

We can generalize this result by drawing an angle θ and its opposite $-\theta$ in standard position and then labeling the points where their terminal sides intersect the unit circle with (x, y) and $(x, -y)$, respectively. (Can you see from Figure 19 why we label these two points in this way? That is, does it make sense that if (x, y) is on the terminal side of θ, then $(x, -y)$ must be on the terminal side of $-\theta$?)

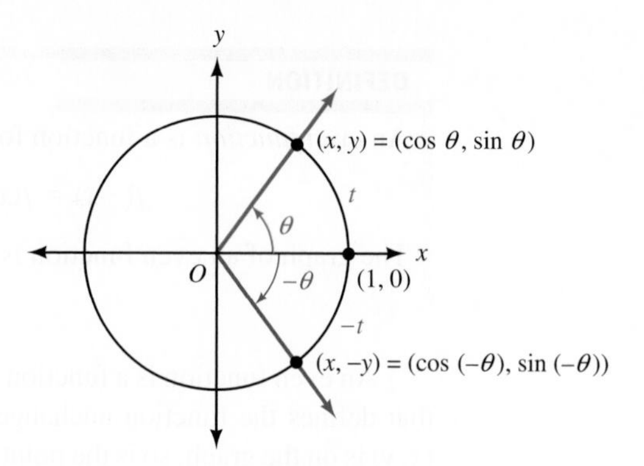

Figure 19

On the unit circle, $\cos\theta = x$ and $\sin\theta = y$, so we have

$$\cos(-\theta) = x = \cos\theta$$

indicating that cosine is an even function and

$$\sin(-\theta) = -y = -\sin\theta$$

indicating that sine is an odd function.

Now that we have established that sine is an odd function and cosine is an even function, we can use our ratio and reciprocal identities to find which of the other trigonometric functions are even and which are odd. Example 3 shows how this is done for the cosecant function.

PROBLEM 3
Show that secant is an even function.

EXAMPLE 3 Show that cosecant is an odd function.

SOLUTION We must prove that $\csc(-\theta) = -\csc\theta$. That is, we must turn $\csc(-\theta)$ into $-\csc\theta$. Here is how it goes:

$$\csc(-\theta) = \frac{1}{\sin(-\theta)} \qquad \text{Reciprocal identity}$$

$$= \frac{1}{-\sin\theta} \qquad \text{Sine is an odd function}$$

$$= -\frac{1}{\sin\theta} \qquad \text{Algebra}$$

$$= -\csc\theta \qquad \text{Reciprocal identity}$$

Because the sine function is odd, we can see in Figure 4 that the graph of $y = \sin x$ is symmetric about the origin. On the other hand, the cosine function is even, so the graph of $y = \cos x$ is symmetric about the y-axis as can be seen in Figure 7. We summarize the nature of all six trigonometric functions for easy reference.

Even Functions	Odd Functions
$y = \cos x$, $y = \sec x$	$y = \sin x$, $y = \csc x$
	$y = \tan x$, $y = \cot x$
Graphs are symmetric about the y-axis	Graphs are symmetric about the origin

PROBLEM 4

Find exact values for each of the following.

a. $\sin\left(-\frac{5\pi}{6}\right)$

b. $\sec(-150°)$

EXAMPLE 4 Use the even and odd function relationships to find exact values for each of the following.

a. $\cos\left(-\frac{2\pi}{3}\right)$ **b.** $\csc(-225°)$

SOLUTION

a. $\cos\left(-\frac{2\pi}{3}\right) = \cos\left(\frac{2\pi}{3}\right)$ Cosine is an even function

$= -\frac{1}{2}$ Unit circle

b. $\csc(-225°) = \frac{1}{\sin(-225°)}$ Reciprocal identity

$= \frac{1}{-\sin 225°}$ Sine is an odd function

$= \frac{1}{-(-1/\sqrt{2})}$ Unit circle

$= \sqrt{2}$

Getting Ready for Class *After reading through the preceding section, respond in your own words and in complete sentences.*

a. How do we use the unit circle to graph the function $y = \sin x$?
b. What is the period of a function?
c. What is the definition of an even function?
d. What type of symmetry will the graph of an odd function have?

4.1 PROBLEM SET

➤ CONCEPTS AND VOCABULARY

For Questions 1 through 6, fill in each blank with the appropriate word or expression.

1. The graph of the sine function illustrates how the ___________ of a point on the unit circle varies with ____ __________.

2. For a periodic function f, the period p is the ___________ positive number for which $f($______$) = f(x)$ for all x in the domain of f.

3. To calculate the amplitude for a function, take half the __________ of the ___________ value and the __________ value of the function.

4. If c is a domain value for a function f and $f(c) = 0$, then $x = c$ is called a _____ of the function. If c is a real number, then $x = c$ will appear as an ___________ for the graph of f.

5. A function is even if the opposite input results in an ________ output. A function is odd if the opposite input results in the __________ output.

6. The graph of an even function is symmetric about the ________ and the graph of an odd function is symmetric about the _________.

7. Which trigonometric functions have a defined amplitude?

8. Which trigonometric functions have asymptotes?

9. Which trigonometric functions do not have real zeros?

10. Which trigonometric functions have a period of π?

11. Which trigonometric functions have a period of 2π?

12. Which trigonometric functions are even?

➤ EXERCISES

Make a table of values for Problems 13 through 18 using multiples of $\pi/4$ for x. Then use the entries in the table to sketch the graph of each function for x between 0 and 2π.

13. $y = \cos x$
14. $y = \cot x$
15. $y = \csc x$
16. $y = \sin x$
17. $y = \tan x$
18. $y = \sec x$

Sketch the graphs of each of the following between $x = -4\pi$ and $x = 4\pi$ by extending the graphs you made in Problems 13 through 18:

19. $y = \cos x$
20. $y = \cot x$
21. $y = \csc x$
22. $y = \sin x$
23. $y = \tan x$
24. $y = \sec x$

Use the graphs you made in Problems 13 through 18 to find all values of x, $0 \le x \le 2\pi$, for which the following are true:

25. $\sin x = 0$
26. $\cos x = 0$
27. $\sin x = 1$
28. $\cos x = 1$
29. $\tan x = 0$
30. $\cot x = 0$
31. $\sec x = 1$
32. $\csc x = 1$
33. $\tan x$ is undefined
34. $\cot x$ is undefined
35. $\csc x$ is undefined
36. $\sec x$ is undefined

Use the unit circle and the fact that cosine is an even function to find each of the following:

37. $\cos(-60°)$
38. $\cos(-120°)$
39. $\cos\left(-\frac{5\pi}{6}\right)$
40. $\cos\left(-\frac{4\pi}{3}\right)$

Use the unit circle and the fact that sine is an odd function to find each of the following:

41. $\sin(-30°)$
42. $\sin(-90°)$
43. $\sin\left(-\frac{3\pi}{4}\right)$
44. $\sin\left(-\frac{7\pi}{4}\right)$

45. If $\sin\theta = -1/3$, find $\sin(-\theta)$.

46. If $\cos\theta = -1/3$, find $\cos(-\theta)$.

Make a diagram of the unit circle with an angle θ in QI and its supplement $180° - \theta$ in QII. Label the point on the terminal side of θ and the unit circle with (x, y) and the point on the terminal side of $180° - \theta$ and the unit circle with $(-x, y)$. Use the diagram to show the following.

47. $\sin(180° - \theta) = \sin\theta$

48. $\cos(180° - \theta) = -\cos\theta$

49. Show that tangent is an odd function.

50. Show that cotangent is an odd function.

Prove each identity.

51. $\sin(-\theta)\cot(-\theta) = \cos\theta$

52. $\cos(-\theta)\tan\theta = \sin\theta$

53. $\sin(-\theta)\sec(-\theta)\cot(-\theta) = 1$

54. $\cos(-\theta)\csc(-\theta)\tan(-\theta) = 1$

55. $\csc\theta + \sin(-\theta) = \frac{\cos^2\theta}{\sin\theta}$

56. $\sec\theta - \cos(-\theta) = \frac{\sin^2\theta}{\cos\theta}$

➤ REVIEW PROBLEMS

The problems that follow review material we covered in Sections 1.5 and 3.2.

Prove the following identities.

57. $\cos\theta\tan\theta = \sin\theta$

58. $\sin\theta\tan\theta + \cos\theta = \sec\theta$

59. $(1 + \sin\theta)(1 - \sin\theta) = \cos^2\theta$

60. $(\sin\theta + \cos\theta)^2 = 1 + 2\sin\theta\cos\theta$

Write each of the following in degrees.

61. $\dfrac{2\pi}{3}$

62. $\dfrac{5\pi}{4}$

63. $\dfrac{11\pi}{6}$

64. $\dfrac{\pi}{2}$

➤ EXTENDING THE CONCEPTS

Problems 65 through 74 will help prepare you for the next section.

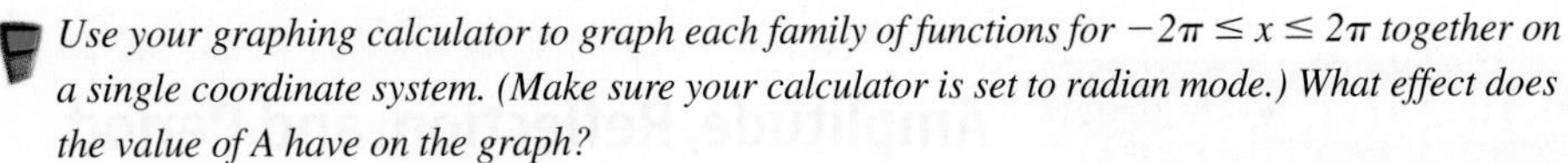

Use your graphing calculator to graph each family of functions for $-2\pi \le x \le 2\pi$ together on a single coordinate system. (Make sure your calculator is set to radian mode.) What effect does the value of A have on the graph?

65. $y = A\sin x$ for $A = 1, 2, 3$

66. $y = A\sin x$ for $A = 1, \frac{1}{2}, \frac{1}{3}$

67. $y = A\cos x$ for $A = 1, 0.6, 0.2$

68. $y = A\cos x$ for $A = 1, 3, 5$

Use your graphing calculator to graph each pair of functions for $-2\pi \le x \le 2\pi$ together on a single coordinate system. (Make sure your calculator is set to radian mode.) What effect does the negative sign have on the graph?

69. $y = 3\sin x, \quad y = -3\sin x$

70. $y = 4\cos x, \quad y = -4\cos x$

Use your graphing calculator to graph each pair of functions for $0 \le x \le 4\pi$. (Make sure your calculator is set to radian mode.) What effect does the value of B have on the graph?

71. $y = \sin Bx$ for $B = 1, 2$

72. $y = \sin Bx$ for $B = 1, 4$

73. $y = \cos Bx$ for $B = 1, \frac{1}{2}$

74. $y = \cos Bx$ for $B = 1, \frac{1}{3}$

➤ LEARNING OBJECTIVES ASSESSMENT

These questions are available for instructors to help assess if you have successfully met the learning objectives for this section.

75. Sketch the graph of $y = \cos x$. Which of the following matches your graph?

a.

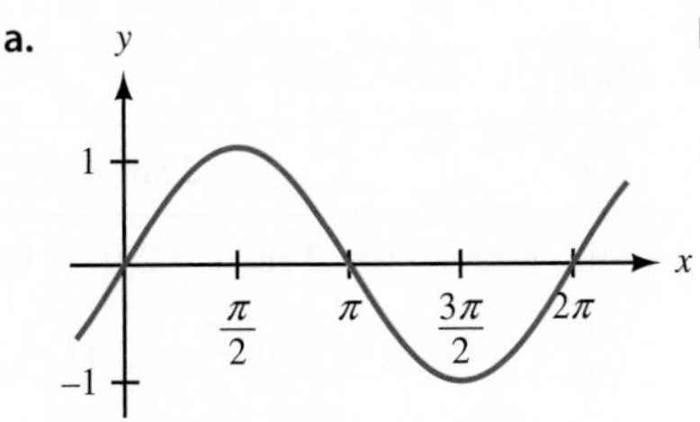

b.

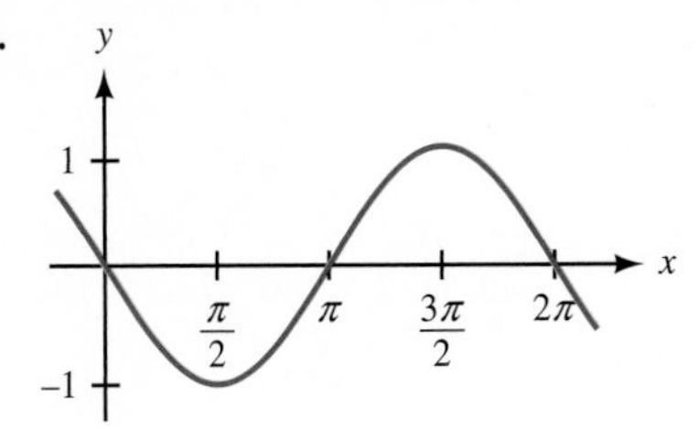

c.

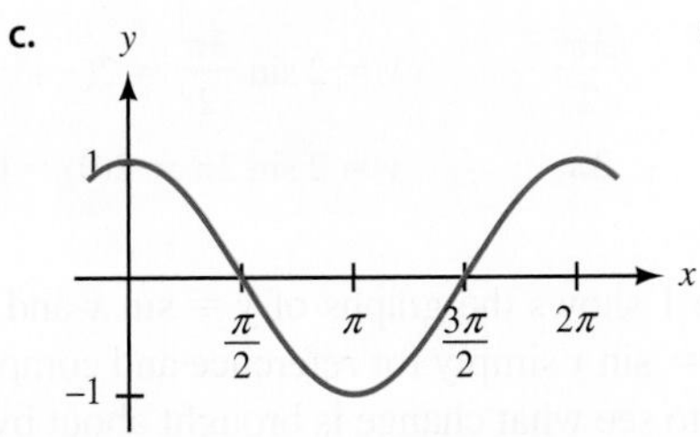

d.

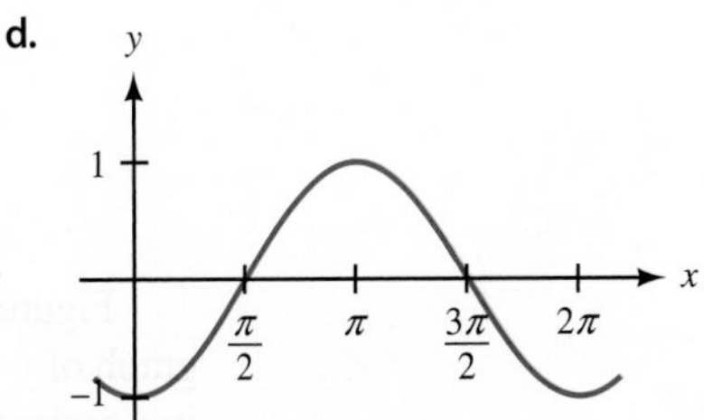

76. Sketch the graph of $y = \sec x$. For which values of x, $0 \le x \le 2\pi$, is $\sec x$ undefined?

a. $\frac{\pi}{4}, \frac{5\pi}{4}$ b. $0, \pi, 2\pi$ c. $\frac{\pi}{2}, \frac{3\pi}{2}$ d. $\frac{3\pi}{4}, \frac{7\pi}{4}$

77. If $\tan \theta = -3$, find $\tan(-\theta)$.

a. 3 b. -3 c. $\frac{1}{3}$ d. $-\frac{1}{3}$

78. To prove $\cos(-\theta)\csc(-\theta)\tan(-\theta) = 1$, which of the following is correct as a first step?

a. $\cos(-\theta)\csc(-\theta)\tan(-\theta) = \cos\theta(-\csc\theta)(-\tan\theta)$

b. $\cos(-\theta)\csc(-\theta)\tan(-\theta) = -\cos\theta(\csc\theta)(-\tan\theta)$

c. $\cos(-\theta)\csc(-\theta)\tan(-\theta) = \cos\theta(\csc\theta)(\tan\theta)$

d. $\cos(-\theta)\csc(-\theta)\tan(-\theta) = \cos\theta(-\csc\theta)(\tan\theta)$

SECTION 4.2 Amplitude, Reflection, and Period

LEARNING OBJECTIVES

1. Find the amplitude of a sine or cosine function.
2. Find the period of a sine or cosine function.
3. Graph a sine or cosine function having a different amplitude and period.
4. Solve a real-life problem involving a trigonometric function as a model.

In Section 4.1, the graphs of $y = \sin x$ and $y = \cos x$ were shown to have an amplitude of 1 and a period of 2π. In this section, we will extend our work with these two functions by considering what happens to the graph when we allow for a coefficient with a trigonometric function.

Amplitude

First, we will consider the effect that multiplying a trigonometric function by a numerical factor has on the graph.

PROBLEM 1
Graph $y = 4\cos x$ for $0 \le x \le 2\pi$.

EXAMPLE 1 Sketch the graph of $y = 2\sin x$ for $0 \le x \le 2\pi$.

SOLUTION The coefficient 2 on the right side of the equation will simply multiply each value of $\sin x$ by a factor of 2. Therefore, the values of y in $y = 2\sin x$ should all be twice the corresponding values of y in $y = \sin x$. Table 1 contains some values for $y = 2\sin x$.

TABLE 1

x	$y = 2\sin x$	(x, y)
0	$y = 2\sin 0 = 2(0) = 0$	$(0, 0)$
$\frac{\pi}{2}$	$y = 2\sin\frac{\pi}{2} = 2(1) = 2$	$\left(\frac{\pi}{2}, 2\right)$
π	$y = 2\sin\pi = 2(0) = 0$	$(\pi, 0)$
$\frac{3\pi}{2}$	$y = 2\sin\frac{3\pi}{2} = 2(-1) = -2$	$\left(\frac{3\pi}{2}, -2\right)$
2π	$y = 2\sin 2\pi = 2(0) = 0$	$(2\pi, 0)$

Figure 1 shows the graphs of $y = \sin x$ and $y = 2\sin x$. (We are including the graph of $y = \sin x$ simply for reference and comparison. With both graphs to look at, it is easier to see what change is brought about by the coefficient 2.)

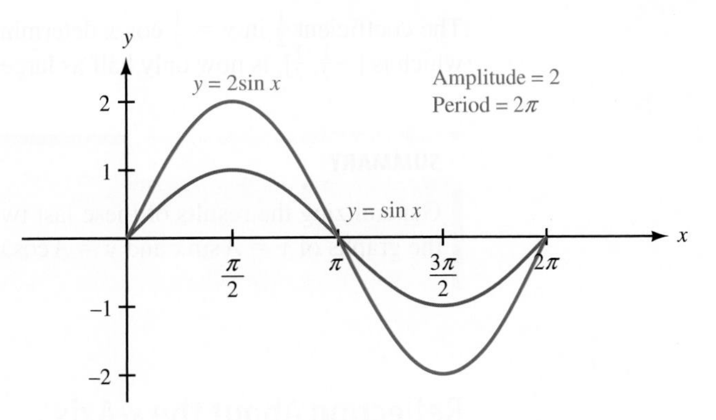

Figure 1

The coefficient 2 in $y = 2 \sin x$ changes the amplitude from 1 to 2 but does not affect the period. That is, we can think of the graph of $y = 2 \sin x$ as if it were the graph of $y = \sin x$ with the amplitude extended to 2 instead of 1. Observe that the range has doubled from $[-1, 1]$ to $[-2, 2]$.

PROBLEM 2

Graph one cycle of $y = \frac{1}{3} \sin x$.

EXAMPLE 2 Sketch one complete cycle of the graph of $y = \frac{1}{2} \cos x$.

SOLUTION Table 2 gives us some points on the curve $y = \frac{1}{2} \cos x$. Figure 2 shows the graphs of both $y = \frac{1}{2} \cos x$ and $y = \cos x$ on the same set of axes, from $x = 0$ to $x = 2\pi$.

TABLE 2

x	$y = \frac{1}{2} \cos x$	(x, y)
0	$y = \frac{1}{2} \cos 0 = \frac{1}{2}(1) = \frac{1}{2}$	$\left(0, \frac{1}{2}\right)$
$\frac{\pi}{2}$	$y = \frac{1}{2} \cos \frac{\pi}{2} = \frac{1}{2}(0) = 0$	$\left(\frac{\pi}{2}, 0\right)$
π	$y = \frac{1}{2} \cos \pi = \frac{1}{2}(-1) = -\frac{1}{2}$	$\left(\pi, -\frac{1}{2}\right)$
$\frac{3\pi}{2}$	$y = \frac{1}{2} \cos \frac{3\pi}{2} = \frac{1}{2}(0) = 0$	$\left(\frac{3\pi}{2}, 0\right)$
2π	$y = \frac{1}{2} \cos 2\pi = \frac{1}{2}(1) = \frac{1}{2}$	$\left(2\pi, \frac{1}{2}\right)$

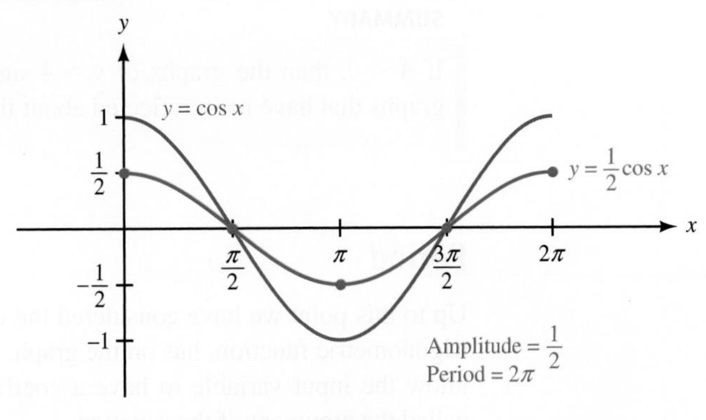

Figure 2

The coefficient $\frac{1}{2}$ in $y = \frac{1}{2}\cos x$ determines the amplitude of the graph. The range, which is $[-\frac{1}{2}, \frac{1}{2}]$, is now only half as large.

SUMMARY

Generalizing the results of these last two examples, we can say that if $A > 0$, then the graphs of $y = A\sin x$ and $y = A\cos x$ will have amplitude A and range $[-A, A]$.

Reflecting About the *x*-Axis

In the previous examples we only considered changes to the graph if the coefficient A was a positive number. To see how a negative value of A affects the graph, we will consider the function $y = -2\cos x$.

PROBLEM 3
Graph $y = -3\sin x$ for $-2\pi \le x \le 4\pi$.

EXAMPLE 3 Graph $y = -2\cos x$, from $x = -2\pi$ to $x = 4\pi$.

SOLUTION Each value of y on the graph of $y = -2\cos x$ will be the opposite of the corresponding value of y on the graph of $y = 2\cos x$. The result is that the graph of $y = -2\cos x$ is the reflection of the graph of $y = 2\cos x$ about the x-axis. Figure 3 shows the extension of one complete cycle of $y = -2\cos x$ to the interval $-2\pi \le x \le 4\pi$.

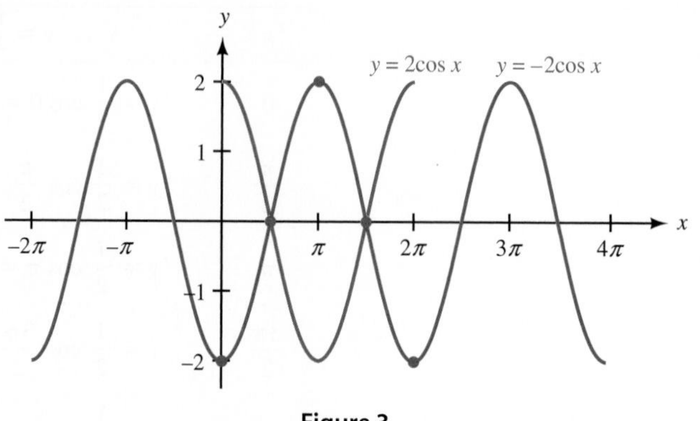

Figure 3

SUMMARY

If $A < 0$, then the graphs of $y = A\sin x$ and $y = A\cos x$ will be sine and cosine graphs that have been reflected about the x-axis. The amplitude will be $|A|$.

Period

Up to this point we have considered the effect that a coefficient, which multiplies the trigonometric function, has on the graph. Now we will investigate what happens if we allow the input variable to have a coefficient. Remember that the input is formally called the argument of the function.

PROBLEM 4
Graph $y = \cos 3x$ for $0 \le x \le 2\pi$.

NOTE Remember that sin 2*x* really means sin (2*x*). The product 2*x* is the argument to the sine function.

EXAMPLE 4 Graph $y = \sin 2x$ for $0 \le x \le 2\pi$.

SOLUTION To see how the coefficient 2 in $y = \sin 2x$ affects the graph, we can make a table in which the values of x are multiples of $\pi/4$. (Multiples of $\pi/4$ are convenient because the coefficient 2 divides the 4 in $\pi/4$ exactly.) Table 3 shows the values of x and y, while Figure 4 contains the graphs of $y = \sin x$ and $y = \sin 2x$.

TABLE 3

x	$y = \sin 2x$	(x, y)
0	$y = \sin 2 \cdot 0 = \sin 0 = 0$	$(0, 0)$
$\frac{\pi}{4}$	$y = \sin 2 \cdot \frac{\pi}{4} = \sin \frac{\pi}{2} = 1$	$\left(\frac{\pi}{4}, 1\right)$
$\frac{\pi}{2}$	$y = \sin 2 \cdot \frac{\pi}{2} = \sin \pi = 0$	$\left(\frac{\pi}{2}, 0\right)$
$\frac{3\pi}{4}$	$y = \sin 2 \cdot \frac{3\pi}{4} = \sin \frac{3\pi}{2} = -1$	$\left(\frac{3\pi}{4}, -1\right)$
π	$y = \sin 2 \cdot \pi = \sin 2\pi = 0$	$(\pi, 0)$
$\frac{5\pi}{4}$	$y = \sin 2 \cdot \frac{5\pi}{4} = \sin \frac{5\pi}{2} = 1$	$\left(\frac{5\pi}{4}, 1\right)$
$\frac{3\pi}{2}$	$y = \sin 2 \cdot \frac{3\pi}{2} = \sin 3\pi = 0$	$\left(\frac{3\pi}{2}, 0\right)$
$\frac{7\pi}{4}$	$y = \sin 2 \cdot \frac{7\pi}{4} = \sin \frac{7\pi}{2} = -1$	$\left(\frac{7\pi}{4}, -1\right)$
2π	$y = \sin 2 \cdot 2\pi = \sin 4\pi = 0$	$(2\pi, 0)$

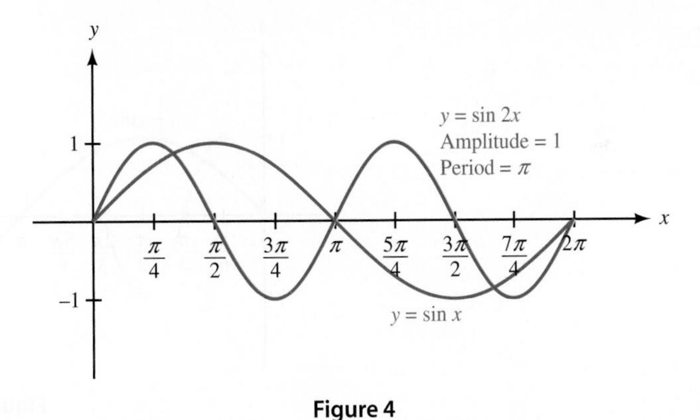

Figure 4

The graph of $y = \sin 2x$ has a period of π. It goes through two complete cycles in 2π units on the x-axis. Notice that doubling the argument to the function has the reverse effect of halving the period. This may be surprising at first, but we can see the reason for it by looking at a basic cycle. We know that the sine function completes one cycle when the input value, or argument, varies between 0 and 2π.

One cycle: $0 \le \text{argument} \le 2\pi$

$0 \le 2x \le 2\pi$ — The argument is $2x$

$0 \le x \le \pi$ — Divide by 2 to isolate x

Because of the factor of 2, the variable x only needs to reach π to complete one cycle, thus shortening the period. ■

PROBLEM 5
Graph $y = \sin 4x$ for $0 \le x \le 2\pi$.

EXAMPLE 5 Graph $y = \sin 3x$ for $0 \le x \le 2\pi$.

SOLUTION We begin by investigating the effect that the coefficient of 3 in the argument of the sine function will have on one cycle.

One cycle: $0 \le \text{argument} \le 2\pi$

$$0 \le 3x \le 2\pi \qquad \text{The argument is } 3x$$

$$0 \le x \le \frac{2\pi}{3} \qquad \text{Divide by 3 to isolate } x$$

The period of the sine function will be one-third as long. To aid in sketching the graph, we divide the length of the period, which is $2\pi/3$, into 4 intervals of equal width.

$$\frac{2\pi/3}{4} = \frac{1}{4} \cdot \frac{2\pi}{3} = \frac{\pi}{6}$$

Starting with the beginning of the cycle at $x = 0$, we mark off the x-axis every $\pi/6$ units. The coordinates of these points will be

$$0, \quad 1 \cdot \frac{\pi}{6} = \frac{\pi}{6}, \quad 2 \cdot \frac{\pi}{6} = \frac{\pi}{3}, \quad 3 \cdot \frac{\pi}{6} = \frac{\pi}{2}, \quad 4 \cdot \frac{\pi}{6} = \frac{2\pi}{3}$$

Because we already know the cycle begins at 0 and ends at $2\pi/3$, we really only need to compute the middle three values.

Knowing that for the basic sine graph, a cycle begins and ends on the x-axis, and crosses the x-axis halfway through, we sketch the graph of $y = \sin 3x$ as shown in Figure 5. The graph of $y = \sin x$ is also shown for comparison. Notice that $y = \sin 3x$ completes three cycles in 2π units.

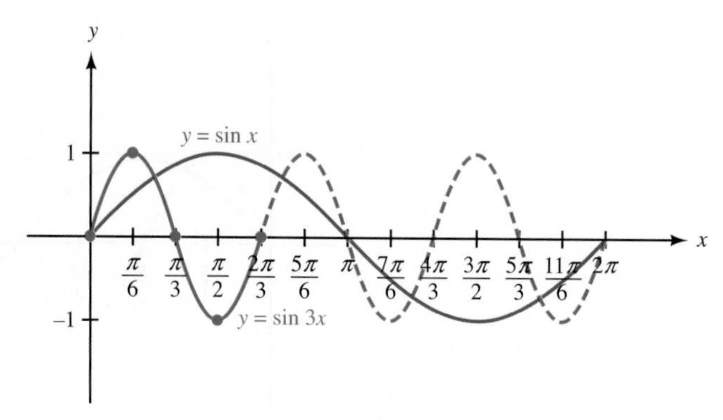

Figure 5

PROBLEM 6
Graph one cycle of $y = \sin \frac{1}{4}x$.

EXAMPLE 6 Graph one complete cycle of $y = \cos \frac{1}{2}x$.

SOLUTION As before, we begin by determining the effect that the coefficient of $\frac{1}{2}$ in the argument of the cosine function will have on one cycle.

One cycle: $0 \le \text{argument} \le 2\pi$

$$0 \le \frac{1}{2}x \le 2\pi \qquad \text{The argument is } \frac{1}{2}x$$

$$0 \le x \le 4\pi \qquad \text{Multiply by 2 to isolate } x$$

The period of the cosine function will be twice as long. To aid in sketching the graph, we divide the length of the period, 4π, into 4 intervals of equal width.

$$\frac{4\pi}{4} = \pi$$

Starting at $x = 0$, we mark off the x-axis every π units. The coordinates of these points will be $0, \pi, 2\pi, 3\pi$, and 4π. Figure 6 shows the graph, along with the graph of $y = \cos x$ for comparison. Notice that $y = \cos \frac{1}{2}x$ completes one-half cycle in 2π units.

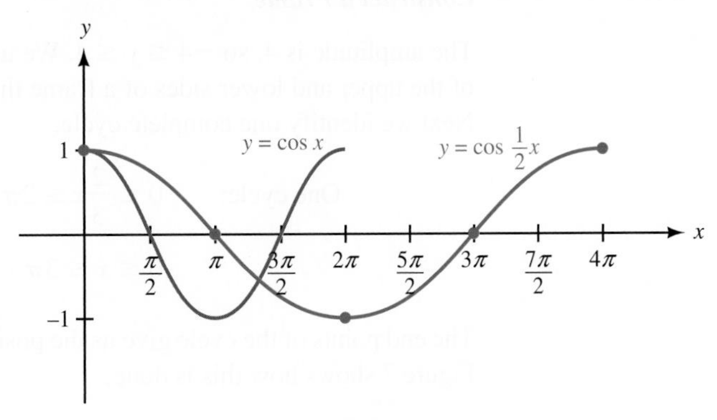

Figure 6

In general, for $y = \sin Bx$ or $y = \cos Bx$ to complete one cycle, the product Bx must vary from 0 to 2π. Therefore

$$0 \le Bx \le 2\pi \quad \text{if} \quad 0 \le x \le \frac{2\pi}{B}$$

The period will be $2\pi/B$, and the graph will complete B cycles in 2π units. We summarize all of the information gathered from the previous examples as follows.

AMPLITUDE AND PERIOD FOR SINE AND COSINE

If A is any real number and $B > 0$, then the graphs of $y = A \sin Bx$ and $y = A \cos Bx$ will have

$$\text{Amplitude} = |A| \quad \text{and} \quad \text{Period} = \frac{2\pi}{B}$$

NOTE We are not just "moving" the negative out of the argument. The properties of even and odd functions allow us to write the functions differently when a negative is involved. For the same reason, we cannot just "move" the factor of 2 out from the argument. That is, $\sin 2x$ is not the same as $2 \sin x$.

In the situation where $B < 0$, we can use the properties of even and odd functions to rewrite the function so that B is positive. For example,

$y = 3 \sin(-2x)$ is equivalent to $y = -3 \sin(2x)$ because sine is an odd function

$y = 3 \cos(-2x)$ is equivalent to $y = 3 \cos(2x)$ because cosine is an even function

In the next two examples, we use this information about amplitude and period to graph one complete cycle of a sine and cosine curve and then extend these graphs to cover more than the one cycle. We also take this opportunity to introduce a method of drawing the graph by constructing a "frame" for the basic cycle.

PROBLEM 7
Graph $y = 2\sin\left(-\frac{3}{4}x\right)$ for $-\frac{8\pi}{3} \le x \le \frac{8\pi}{3}$.

EXAMPLE 7 Graph $y = 4\cos\left(-\frac{2}{3}x\right)$ for $-\frac{15\pi}{4} \le x \le \frac{15\pi}{4}$.

SOLUTION Because cosine is an even function,

$$y = 4\cos\left(-\frac{2}{3}x\right) = 4\cos\left(\frac{2}{3}x\right)$$

Construct a Frame

The amplitude is 4, so $-4 \le y \le 4$. We use the amplitude to determine the position of the upper and lower sides of a frame that will act as a boundary for a basic cycle. Next we identify one complete cycle.

One cycle: $0 \le \frac{2}{3}x \le 2\pi$

$0 \le x \le 3\pi$ Multiply by $\frac{3}{2}$ to isolate x

The end points of the cycle give us the position of the left and right sides of the frame. Figure 7 shows how this is done.

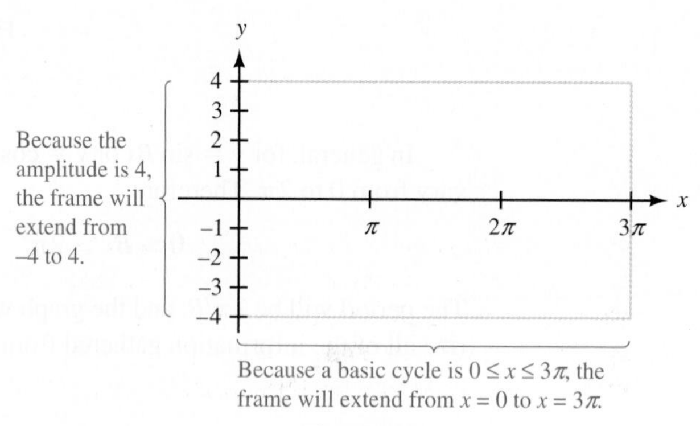

Figure 7

Subdivide the Frame

NOTE We could also find the period by dividing 2π by $B = \frac{2}{3}$:

$$\frac{2\pi}{B} = \frac{2\pi}{\frac{2}{3}} = 3\pi$$

The advantage of using a cycle will become more apparent in the next section.

The period is 3π. Dividing by 4 gives us $3\pi/4$, so we will mark the x-axis in increments of $3\pi/4$. We already know where the cycle begins and ends, so we compute the three middle values:

$$1 \cdot \frac{3\pi}{4} = \frac{3\pi}{4}, \quad 2 \cdot \frac{3\pi}{4} = \frac{3\pi}{2}, \quad 3 \cdot \frac{3\pi}{4} = \frac{9\pi}{4}$$

We divide our frame in Figure 7 into four equal sections, marking the x-axis accordingly. Figure 8 shows the result.

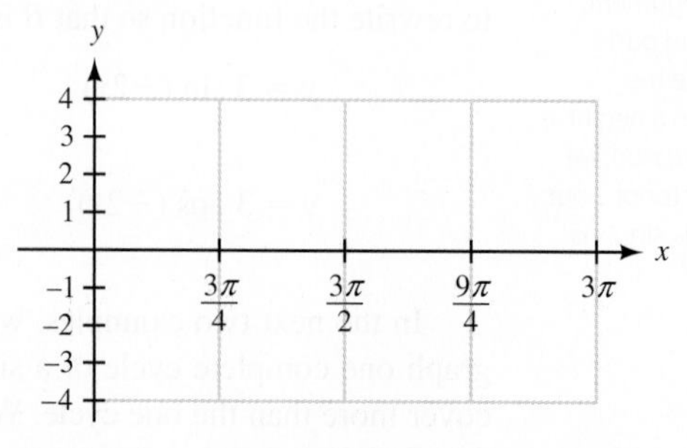

Figure 8

Graph One Cycle

Now we use the frame to plot the key points that will define the shape of one complete cycle of the graph and then draw the graph itself (Figure 9).

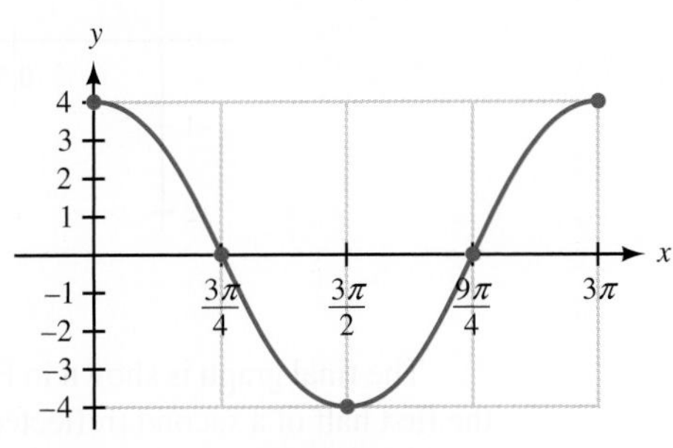

Figure 9

Extend the Graph, if Necessary

The original problem asked for the graph on the interval $-\frac{15\pi}{4} \le x \le \frac{15\pi}{4}$. We extend the graph to the right by adding the first quarter of a second cycle. On the left, we add another complete cycle (which takes the graph to -3π) and then add the last quarter of an additional cycle to reach $-15\pi/4$. The final graph is shown in Figure 10.

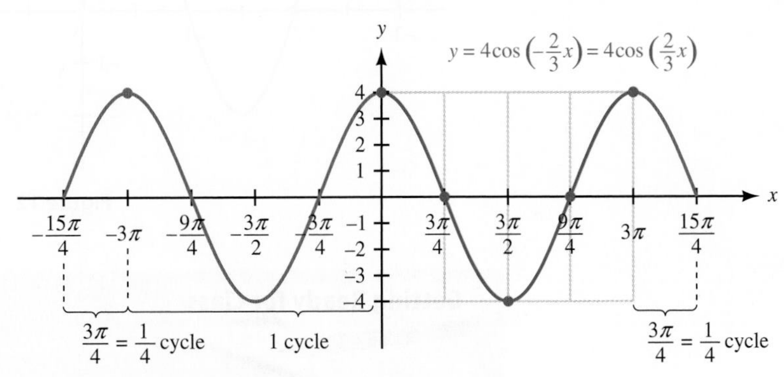

Figure 10

PROBLEM 8

Graph $y = 3\cos\left(-\frac{\pi}{9}x\right)$ for $-18 \le x \le 18$.

EXAMPLE 8 Graph $y = 2\sin(-\pi x)$ for $-3 \le x \le 3$.

SOLUTION Because sine is an odd function,

$$y = 2\sin(-\pi x) = -2\sin(\pi x)$$

The amplitude is $|-2| = 2$. The range will be $-2 \le y \le 2$. Next we identify one complete cycle.

$$\text{One cycle:} \quad 0 \le \pi x \le 2\pi$$

$$0 \le x \le 2 \qquad \text{Divide by } \pi \text{ to isolate } x$$

The period is 2. Dividing this by 4, we will mark the x-axis every $\frac{1}{2}$ unit. Figure 11 shows the frame we have constructed for one cycle and the key points. Because of the negative sign, we must remember that the graph will be reflected about the x-axis.

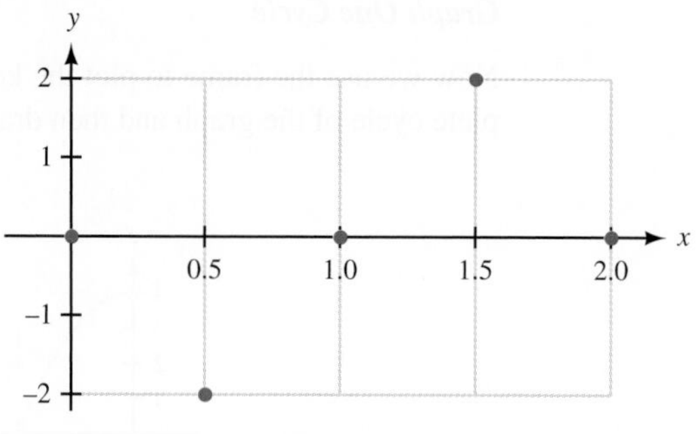

Figure 11

The final graph is shown in Figure 12. We extend the graph to the right by adding the first half of a second (reflected) cycle. On the left, we add another complete cycle (which takes the graph to -2), and then add the last half of an additional cycle to reach -3.

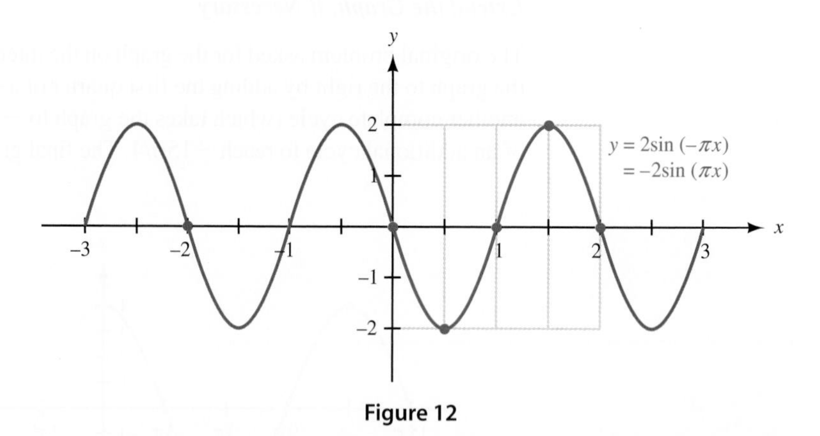

Figure 12

Getting Ready for Class

After reading through the preceding section, respond in your own words and in complete sentences.

a. How does the graph of $y = -2\sin x$ differ from the graph of $y = \sin x$?
b. How does the graph of $y = \sin 2x$ differ from the graph of $y = \sin x$?
c. What is the amplitude of $y = A \cos Bx$?
d. What is the period of $y = A \sin Bx$?

4.2 PROBLEM SET

➤ CONCEPTS AND VOCABULARY

For Questions 1 through 6, fill in each blank with the appropriate word or expression.

1. The graph of $y = A \sin x$ will have an amplitude of _______ and a range of _______.

2. If A is negative, then the graph of $y = A \sin x$ will be ___________ about the _______.

3. The graph of $y = \sin Bx$ will have a period of ________.

4. If $0 < B < 1$, then the period of $y = \sin Bx$ will be __________ than 2π, and if $B > 1$ the period will be _________ than 2π.

5. To graph $y = \sin(-Bx)$, first write the function as $y =$ _________ because sine is an ______ function.

6. To graph $y = \cos(-Bx)$, first write the function as $y =$ _________ because cosine is an ______ function.

➤ EXERCISES

7. Sketch the graph of $y = 2 \sin x$ from $x = 0$ to $x = 2\pi$ by making a table using multiples of $\pi/2$ for x. What is the amplitude of the graph you obtain?

8. Sketch the graph of $y = \frac{1}{2} \cos x$ from $x = 0$ to $x = 2\pi$ by making a table using multiples of $\pi/2$ for x. What is the amplitude of the graph you obtain?

Identify the amplitude for each of the following. Do not sketch the graph.

9. $y = 5 \sin x$

10. $y = 2.5 \cos x$

11. $y = -\dfrac{1}{4} \cos x$

12. $y = -\dfrac{2}{5} \sin x$

Graph one complete cycle of each of the following. In each case, label the axes accurately and identify the amplitude for each graph.

13. $y = 6 \sin x$

14. $y = 6 \cos x$

15. $y = \dfrac{1}{2} \cos x$

16. $y = \dfrac{1}{3} \sin x$

17. $y = -3 \cos x$

18. $y = -4 \sin x$

19. Make a table using multiples of $\pi/4$ for x to sketch the graph of $y = \sin 2x$ from $x = 0$ to $x = 2\pi$. After you have obtained the graph, state the number of complete cycles your graph goes through between 0 and 2π.

20. Make a table using multiples of $\pi/6$ for x to sketch the graph of $y = \sin 3x$ from $x = 0$ to $x = 2\pi$. After you have obtained the graph, state the number of complete cycles your graph goes through between 0 and 2π.

Identify the period for each of the following. Do not sketch the graph.

21. $y = \cos 4x$

22. $y = \cos 6x$

23. $y = \sin \dfrac{1}{6} x$

24. $y = \sin \dfrac{1}{4} x$

25. $y = \cos 2\pi x$

26. $y = \cos \dfrac{\pi}{3} x$

Graph one complete cycle of each of the following. In each case, label the axes accurately and identify the period for each graph.

27. $y = \sin 2x$

28. $y = \sin \dfrac{1}{2} x$

29. $y = \cos \dfrac{1}{3} x$

30. $y = \cos 3x$

31. $y = \sin \pi x$

32. $y = \cos \pi x$

33. $y = \sin \dfrac{\pi}{2} x$

34. $y = \cos \dfrac{\pi}{2} x$

Give the amplitude and period of each of the following graphs:

35.

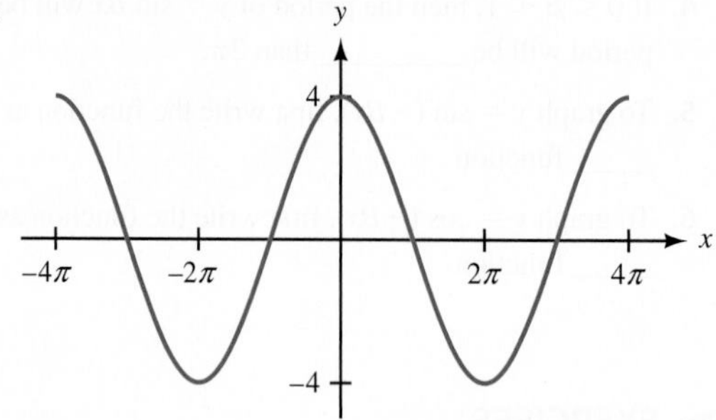

36.

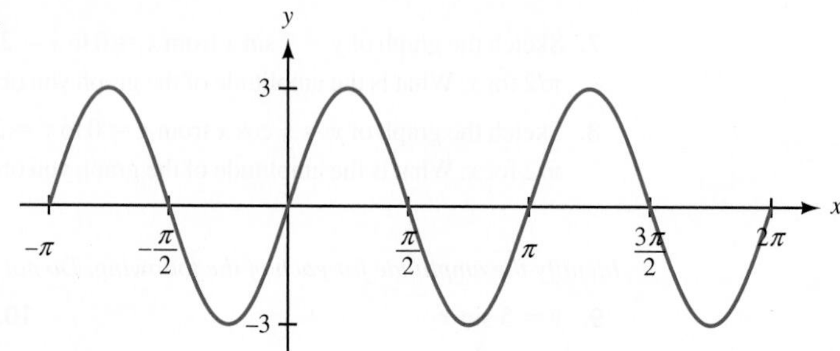

37.

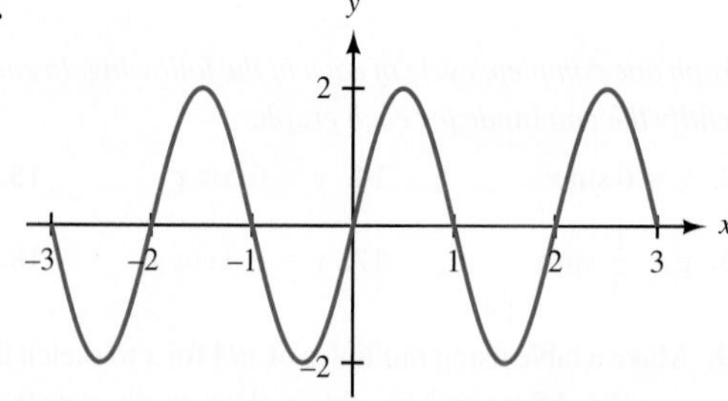

38.

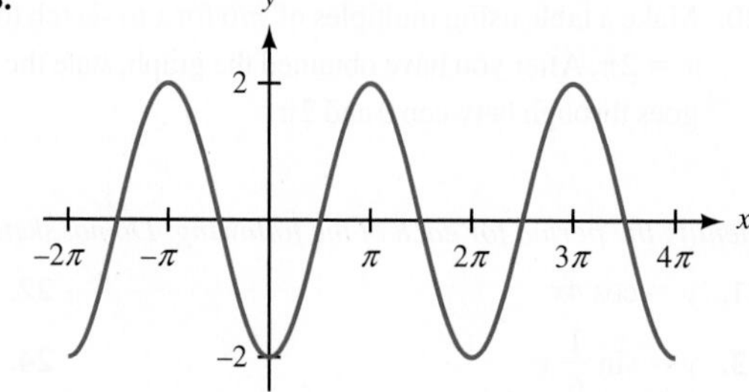

39.

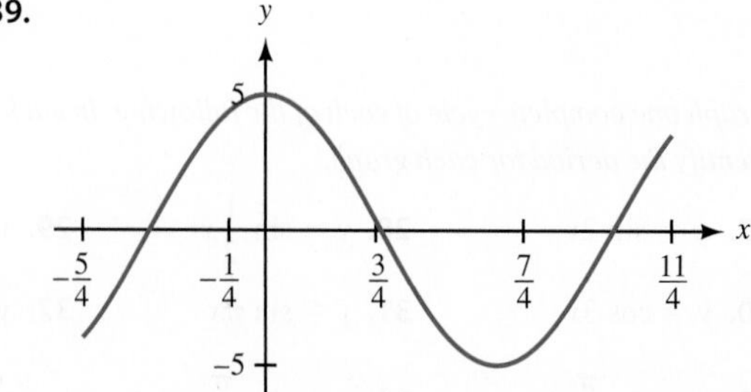

40.

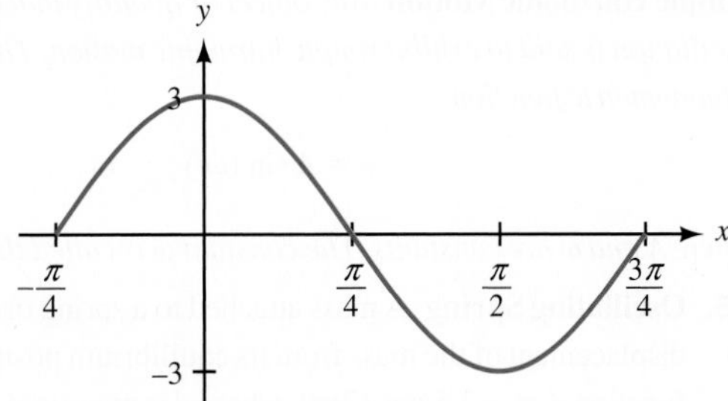

Identify the amplitude and period for each of the following. Do not sketch the graph.

41. $y = \frac{1}{3} \sin 3x$

42. $y = \frac{1}{4} \sin 6x$

43. $y = -10 \cos \frac{x}{10}$

44. $y = -8 \cos \frac{\pi}{8} x$

Graph one complete cycle for each of the following. In each case, label the axes so that the amplitude and period are easy to read.

45. $y = 4 \sin 2x$

46. $y = 2 \sin 4x$

47. $y = 3 \sin \frac{1}{2} x$

48. $y = 2 \sin \frac{1}{3} x$

49. $y = \frac{1}{2} \cos 3x$

50. $y = \frac{1}{2} \sin 3x$

51. $y = -\frac{1}{2} \sin \frac{\pi}{2} x$

52. $y = -2 \sin \frac{\pi}{2} x$

Graph each of the following over the given interval. Label the axes so that the amplitude and period are easy to read.

53. $y = 3 \cos \pi x, -2 \le x \le 4$

54. $y = 2 \sin \pi x, -4 \le x \le 4$

55. $y = 3 \sin 2x, -\pi \le x \le 2\pi$

56. $y = -3 \sin 2x, -2\pi \le x \le 2\pi$

57. $y = -3 \cos \frac{1}{2} x, -2\pi \le x \le 6\pi$

58. $y = 3 \cos \frac{1}{2} x, -4\pi \le x \le 4\pi$

59. $y = -2 \sin (-3x), 0 \le x \le 2\pi$

60. $y = -2 \cos (-3x), 0 \le x \le 2\pi$

61. Electric Current The current in an alternating circuit varies in intensity with time. If I represents the intensity of the current and t represents time, then the relationship between I and t is given by

$$I = 20 \sin (120\pi t)$$

where I is measured in amperes and t is measured in seconds. Find the maximum value of I and the time it takes for I to go through one complete cycle.

62. Maximum Velocity and Distance A weight is hung from a spring and set in motion so that it moves up and down continuously. The velocity v of the weight at any time t is given by the equation

$$v = 3.5 \cos (2\pi t)$$

where v is measured in meters per second and t is measured in seconds. Find the maximum velocity of the weight and the amount of time it takes for the weight to move from its lowest position to its highest position.

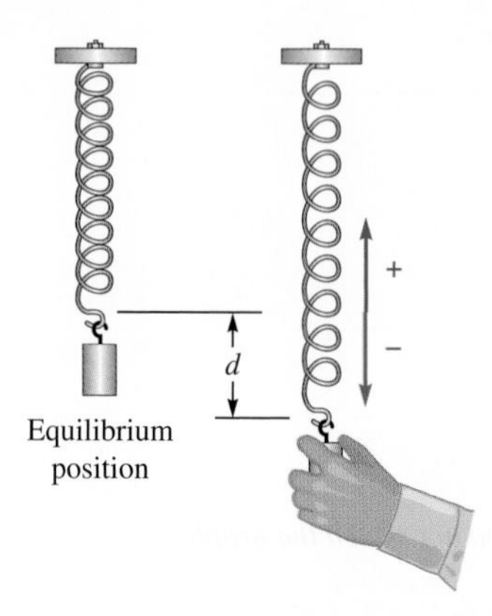

Figure 13

Simple Harmonic Motion *Any object or quantity that is moving with a periodic sinusoidal oscillation is said to exhibit simple harmonic motion. This motion can be modeled by the trigonometric function*

$$y = A \sin(\omega t) \qquad \text{or} \qquad y = A \cos(\omega t)$$

where A and ω are constants. The constant ω is called the angular frequency.

63. **Oscillating Spring** A mass attached to a spring oscillates upward and downward. The displacement of the mass from its equilibrium position after t seconds is given by the function $d = -3.5 \cos(2\pi t)$, where d is measured in centimeters (Figure 13).
    a. Sketch the graph of this function for $0 \le t \le 5$.
    b. What is the furthest distance of the mass from its equilibrium position?
    c. How long does it take for the mass to complete one oscillation?

64. **Pendulum** A pendulum swings back and forth. The angular displacement θ of the pendulum from its rest position after t seconds is given by the function $\theta = 20 \cos(3\pi t)$, where θ is measured in degrees (Figure 14).
    a. Sketch the graph of this function for $0 \le t \le 6$.
    b. What is the maximum angular displacement?
    c. How long does it take for the pendulum to complete one oscillation?

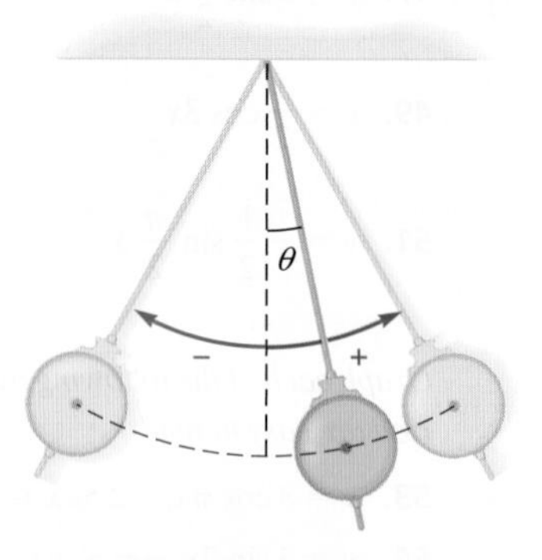

Figure 14

65. **Alternating Current** In North America, the voltage of the alternating current coming through an electrical outlet can be modeled by the function $V = 163 \sin(120\pi t)$, where t is measured in seconds and V in volts. Sketch the graph of this function for $0 \le t \le 0.1$.

66. **Sound Wave** The oscillations in air pressure representing the sound wave for a tone at the standard pitch of A can be modeled by the equation $y = 0.02 \sin(880\pi t)$, where y is the sound pressure in pascals after t seconds. Sketch the graph of this function for $0 \le t \le 0.01$.

Frequency *With simple harmonic motion, the reciprocal of the period is called the frequency. The frequency, given by*

$$f = 1/\text{period}$$

represents the number of cycles (or oscillations) that are completed per unit time. The units used to describe frequency are Hertz, where 1 Hz = 1 *cycle per second.*

67. **Alternating Current** In Europe, the voltage of the alternating current coming through an electrical outlet can be modeled by the function $V = 230 \sin(100\pi t)$, where t is measured in seconds and V in volts. What is the frequency of the voltage?

68. **Sound Wave** The oscillations in air pressure representing the sound wave for a particular musical tone can be modeled by the equation $y = 0.3 \sin(600\pi t)$, where y is the sound pressure in pascals after t seconds. What is the frequency of the tone?

➤ REVIEW PROBLEMS

The problems that follow review material we covered in Section 3.2. Reviewing these problems will help you with the next section.

Evaluate each of the following if x is π/2 and y is π/6.

69. $\sin\left(x + \frac{\pi}{2}\right)$

70. $\sin\left(x - \frac{\pi}{2}\right)$

71. $\cos\left(y - \frac{\pi}{6}\right)$

72. $\cos\left(y + \frac{\pi}{6}\right)$

73. $\sin(x + y)$

74. $\cos(x + y)$

75. $\sin x + \sin y$

76. $\cos x + \cos y$

Convert each of the following to radians without using a calculator.

77. 150°

78. 90°

79. 225°

80. 300°

➤ EXTENDING THE CONCEPTS

Problems 81 through 88 will help prepare you for the next section.

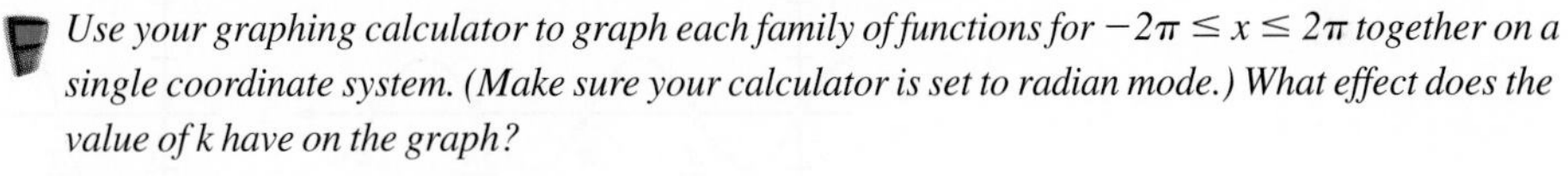

Use your graphing calculator to graph each family of functions for $-2\pi \le x \le 2\pi$ together on a single coordinate system. (Make sure your calculator is set to radian mode.) What effect does the value of k have on the graph?

81. $y = k + \sin x \quad$ for $k = 0, 2, 4$

82. $y = k + \cos x \quad$ for $k = 0, \frac{1}{2}, -\frac{1}{2}$

83. $y = k + \sin x \quad$ for $k = 0, -2, -4$

84. $y = k + \cos x \quad$ for $k = 0, 1, -1$

Use your graphing calculator to graph each family of functions for $-2\pi \le x \le 2\pi$ together on a single coordinate system. (Make sure your calculator is set to radian mode.) What effect does the value of h have on the graph?

85. $y = \sin(x - h) \quad$ for $h = 0, \frac{\pi}{4}, \frac{\pi}{2}$

86. $y = \cos(x - h) \quad$ for $h = 0, \frac{\pi}{3}, -\frac{\pi}{3}$

87. $y = \sin(x - h) \quad$ for $h = 0, -\frac{\pi}{4}, -\frac{\pi}{2}$

88. $y = \cos(x - h) \quad$ for $h = 0, \frac{\pi}{6}, -\frac{\pi}{6}$

➤ LEARNING OBJECTIVES ASSESSMENT

These questions are available for instructors to help assess if you have successfully met the learning objectives for this section.

89. Find the amplitude of $y = -4\cos(2x)$.

a. -2 **b.** 2 **c.** -4 **d.** 4

90. Find the period of $y = -4 \cos (2x)$.

a. 4π **b.** $\frac{\pi}{2}$ **c.** 2π **d.** π

91. Sketch the graph of $y = -\frac{1}{2} \sin (\pi x)$. Which of the following matches your graph?

a.

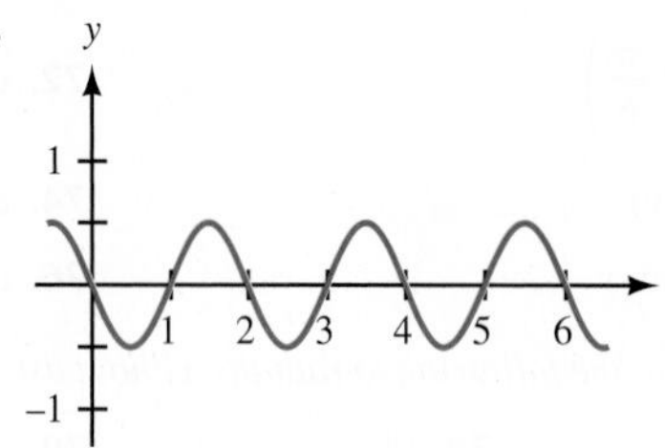

b.

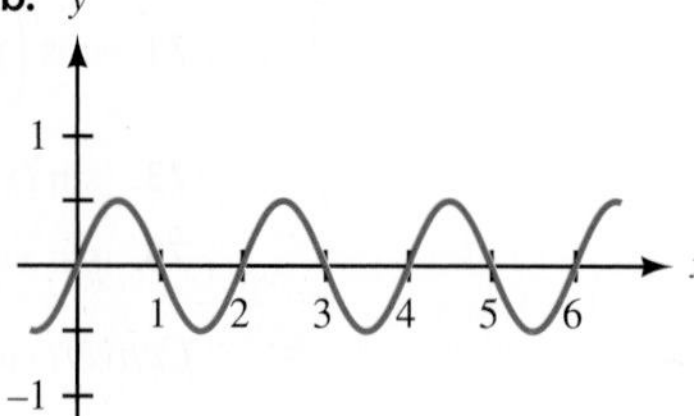

c.

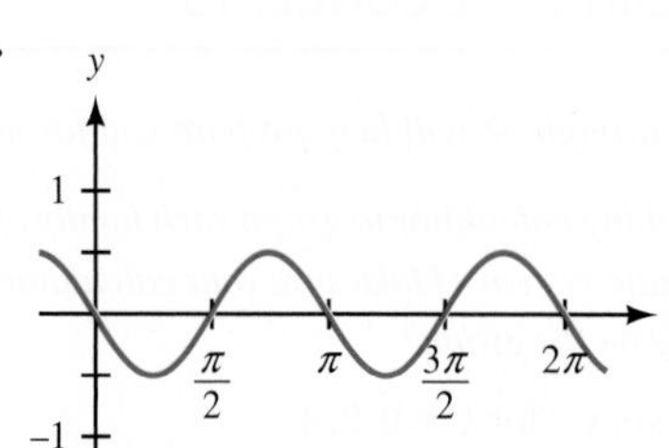

d.

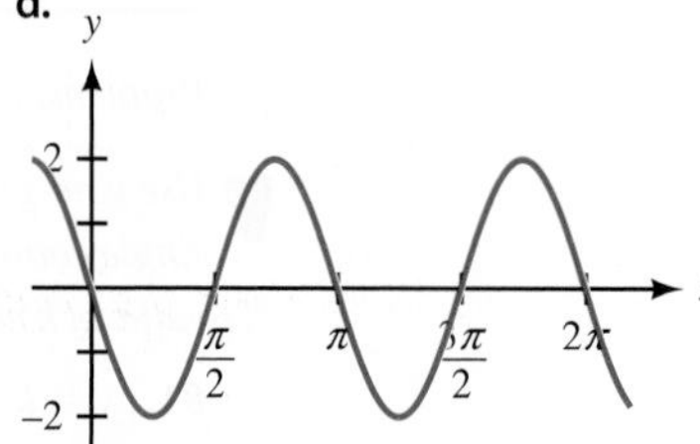

92. A mass attached to a spring oscillates upward and downward. The displacement of the mass from its equilibrium position after t seconds is given by $d = -4 \cos (2\pi t)$. How long does it take for the mass to travel from it lowest position to its highest position?

a. 1 sec **b.** 0.5 sec **c.** 4 sec **d.** 2 sec

SECTION 4.3 Vertical and Horizontal Translations

LEARNING OBJECTIVES

1. Find the vertical translation of a sine or cosine function.
2. Find the horizontal translation of a sine or cosine function.
3. Identify the phase for a sine or cosine function.
4. Graph a sine or cosine function having a horizontal and vertical translation.

In the previous section, we considered what happens to the graph when we allow for a coefficient (a multiplier) with a sine or cosine function. We will conclude our study of the graphs of these two functions by investigating the effect caused by inserting a term (that is, adding or subtracting a number) in the equation of the function. As we will see, the addition of a term creates a translation, in which the position, but not shape, of the graph is changed.

Vertical Translations

Recall from algebra the relationship between the graphs of $y = x^2$ and $y = x^2 - 3$. Figures 1 and 2 show the graphs. The graph of $y = x^2 - 3$ has the same shape as the graph of $y = x^2$ but with its vertex (and all other points) moved down three units. If we were to graph $y = x^2 + 2$, the graph would have the same shape as the graph of $y = x^2$, but it would be shifted up two units. In general, the graph of $y = f(x) + k$ is the graph of $y = f(x)$ translated k units vertically. If k is a positive number, the translation is up. If k is a negative number, the translation is down.

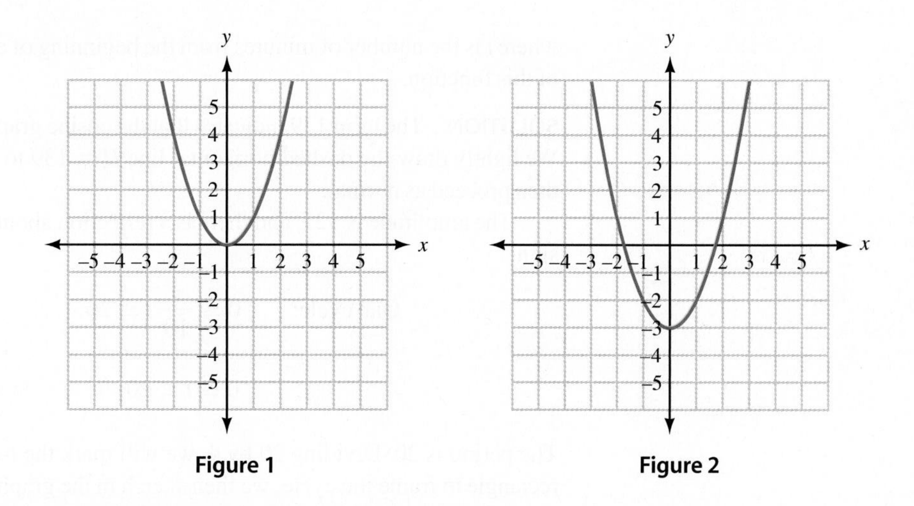

Figure 1 **Figure 2**

PROBLEM 1
Graph $y = 4 + 3 \cos \pi x$.

EXAMPLE 1 Sketch the graph of $y = -3 - 2 \sin \pi x$.

SOLUTION By rewriting the function slightly

$$y = -3 - 2 \sin \pi x = -2 \sin \pi x - 3$$

we see that $y = -3 - 2 \sin \pi x$ will have a graph that is identical to that of $y = -2 \sin \pi x$, except that all points on the graph will be shifted downward three units. Using the result of Example 8 in the previous section, we move every point 3 units downward to obtain the graph shown in Figure 3.

NOTE When we write $y = -2 \sin \pi x - 3$, the argument of the sine function is still πx, not $\pi x - 3$. For the argument to be $\pi x - 3$ we would need to use parentheses and write the function as $y = -2 \sin (\pi x - 3)$.

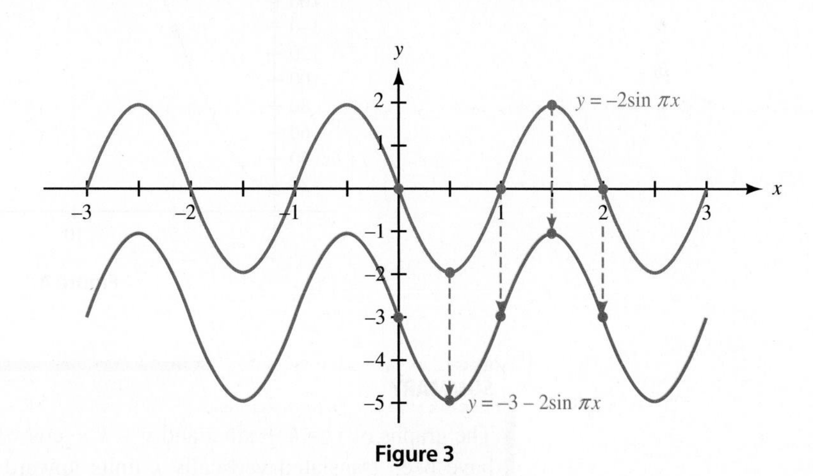

Figure 3

Notice how the vertical translation has changed the range as well. By subtracting 3 units, the original range $[-2, 2]$ becomes $[-5, -1]$.

In our next example we show how to graph a trigonometric function involving a vertical translation directly. Because the sine and cosine functions are centered about the x-axis, we can shift the axis first and then sketch the graph as usual.

PROBLEM 2
Graph $y = 120 - 110\cos\left(\frac{\pi}{12}t\right)$.

EXAMPLE 2 In Example 6 of Section 3.5, we found that the height of a rider on a Ferris wheel was given by the function

$$H = 139 - 125\cos\left(\frac{\pi}{10}t\right)$$

where t is the number of minutes from the beginning of a ride. Graph a complete cycle of this function.

SOLUTION The term 139 indicates that the cosine graph is shifted 139 units upward. We lightly draw the dashed horizontal line $H = 139$ to act in place of the t-axis, and then proceed as normal.

The amplitude is 125, and there is a reflection about the t-axis due to the negative sign.

$$\text{One cycle:} \quad 0 \le \frac{\pi}{10}t \le 2\pi$$

$$0 \le t \le 20 \qquad \text{Multiply by } \frac{10}{\pi}$$

The period is 20. Dividing 20 by 4, we will mark the t-axis at intervals of 5. Using a rectangle to frame the cycle, we then sketch in the graph as shown in Figure 4. Notice that we measure 125 units above and below the line $H = 139$ to create the frame and that we must remember to plot points for a cycle that is reflected.

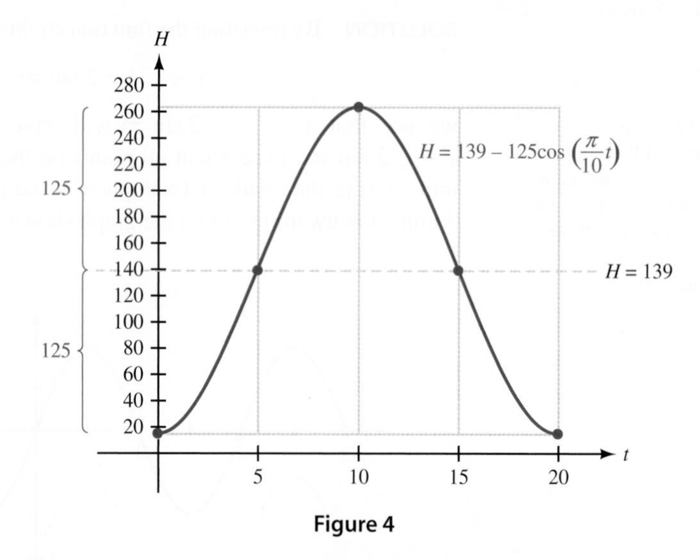

Figure 4

SUMMARY

The graphs of $y = k + \sin x$ and $y = k + \cos x$ will be sine and cosine graphs that have been translated vertically k units upward if $k > 0$, or k units downward if $k < 0$.

Horizontal Translations

If we add a term to the argument of the function, the graph will be translated in a horizontal direction instead of a vertical direction as demonstrated in the next example.

PROBLEM 3

Graph $y = \cos\left(x - \frac{\pi}{3}\right)$ for $\frac{\pi}{3} \le x \le \frac{7\pi}{3}$.

EXAMPLE 3 Graph $y = \sin\left(x + \frac{\pi}{2}\right)$, if $-\frac{\pi}{2} \le x \le \frac{3\pi}{2}$.

SOLUTION Because we have not graphed an equation of this form before, it is a good idea to begin by making a table (Table 1). In this case, multiples of $\pi/2$ will be the most convenient replacements for x in the table. Also, if we start with $x = -\pi/2$, our first value of y will be 0.

TABLE 1

x	$y = \sin\left(x + \frac{\pi}{2}\right)$	(x, y)
$-\frac{\pi}{2}$	$y = \sin\left(-\frac{\pi}{2} + \frac{\pi}{2}\right) = \sin 0 = 0$	$\left(-\frac{\pi}{2}, 0\right)$
0	$y = \sin\left(0 + \frac{\pi}{2}\right) = \sin\frac{\pi}{2} = 1$	$(0, 1)$
$\frac{\pi}{2}$	$y = \sin\left(\frac{\pi}{2} + \frac{\pi}{2}\right) = \sin \pi = 0$	$\left(\frac{\pi}{2}, 0\right)$
π	$y = \sin\left(\pi + \frac{\pi}{2}\right) = \sin\frac{3\pi}{2} = -1$	$(\pi, -1)$
$\frac{3\pi}{2}$	$y = \sin\left(\frac{3\pi}{2} + \frac{\pi}{2}\right) = \sin 2\pi = 0$	$\left(\frac{3\pi}{2}, 0\right)$

Graphing these points and then drawing the sine curve that connects them gives us the graph of $y = \sin(x + \pi/2)$, as shown in Figure 5.

NOTE Figure 5 also includes the graph of $y = \sin x$ for reference; we are trying to discover how the graphs of $y = \sin(x + \pi/2)$ and $y = \sin x$ differ.

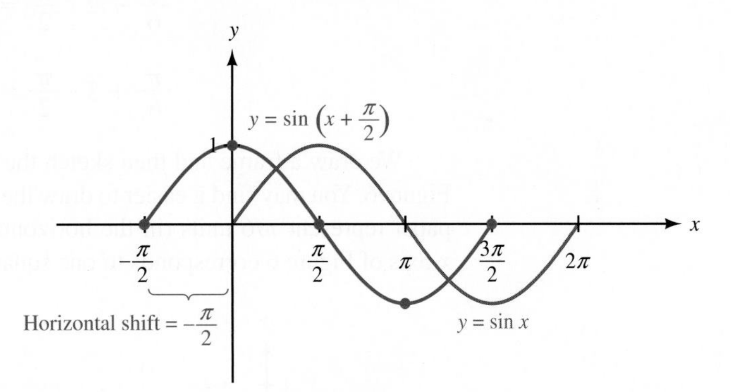

Figure 5

It seems that the graph of $y = \sin(x + \pi/2)$ is shifted $\pi/2$ units to the left of the graph of $y = \sin x$. We say the graph of $y = \sin(x + \pi/2)$ has a *horizontal translation,* or *horizontal shift*, of $-\pi/2$, where the negative sign indicates the shift is to the left (in the negative direction).

We can see why the graph was shifted to the left by looking at how the extra term affects a basic cycle of the sine function. We know that the sine function completes one cycle when the input value, or argument, varies between 0 and 2π.

One cycle: $0 \le \text{argument} \le 2\pi$

$$0 \le x + \frac{\pi}{2} \le 2\pi \qquad \text{The argument is } x + \frac{\pi}{2}$$

$$-\frac{\pi}{2} \le x \le \frac{3\pi}{2} \qquad \text{Subtract } \frac{\pi}{2} \text{ to isolate } x$$

Notice that a cycle will now begin at $x = -\pi/2$ instead of at zero, and will end at $3\pi/2$ instead of 2π, which agrees with the graph in Figure 5. The graph has simply shifted $\pi/2$ units to the left. The horizontal shift is the value of x at which the basic cycle begins, which will always be the left value in the above inequality *after x has been isolated.*

PROBLEM 4
Graph one cycle of
$y = \sin\left(x + \frac{2\pi}{3}\right)$.

EXAMPLE 4 Graph one complete cycle of $y = \cos\left(x - \frac{\pi}{6}\right)$.

SOLUTION There are no additional coefficients present, so the amplitude is 1 and the period is 2π. The term in the argument will cause a horizontal translation. We consider a basic cycle:

$$\text{One cycle:} \quad 0 \le x - \frac{\pi}{6} \le 2\pi$$

$$\frac{\pi}{6} \le x \le \frac{13\pi}{6} \qquad \text{Add } \frac{\pi}{6} \text{ to isolate } x$$

A cycle will begin at $x = \pi/6$ and end at $x = 13\pi/6$. Notice that the period has not changed, because

$$\frac{13\pi}{6} - \frac{\pi}{6} = \frac{12\pi}{6} = 2\pi$$

Dividing 2π by 4, we get $\pi/2$. To mark the x-axis, we begin at $x = \pi/6$ and add increments of $\pi/2$ as follows:

$$\frac{\pi}{6} + \frac{\pi}{2} = \frac{4\pi}{6} = \frac{2\pi}{3}$$

$$\frac{\pi}{6} + 2\cdot\frac{\pi}{2} = \frac{\pi}{6} + \pi = \frac{7\pi}{6}$$

$$\frac{\pi}{6} + 3\cdot\frac{\pi}{2} = \frac{10\pi}{6} = \frac{5\pi}{3}$$

We draw a frame and then sketch the graph of one complete cycle as shown in Figure 6. You may find it easier to draw the graph if you let each square on your graph paper represent $\pi/6$ units (in the horizontal direction). Then each tick mark on the x-axis of Figure 6 corresponds to one square on your paper.

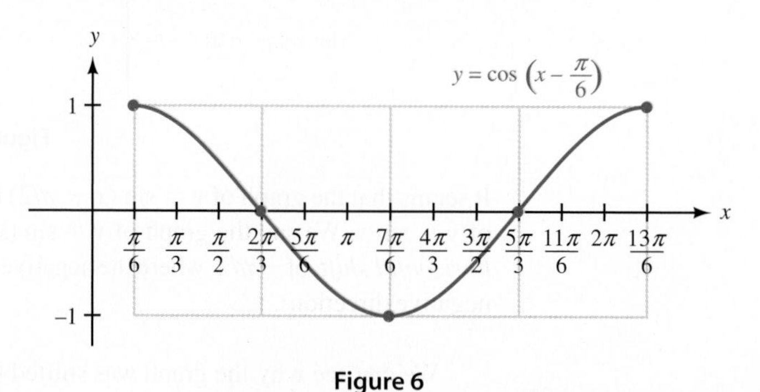

Figure 6

SUMMARY

The graphs of $y = \sin(x - h)$ and $y = \cos(x - h)$ will be sine and cosine graphs that have been translated horizontally h units to the right if $h > 0$, or h units to the left if $h < 0$.

Next we look at an example that involves a combination of a period change and a horizontal shift.

PROBLEM 5

Graph $y = 4\sin\left(3x + \frac{\pi}{2}\right)$ for $0 \le x \le 2\pi$.

EXAMPLE 5 Graph $y = 4\cos\left(2x - \frac{3\pi}{2}\right)$ for $0 \le x \le 2\pi$.

SOLUTION The amplitude is 4. There is no vertical translation because no number has been added to or subtracted from the cosine function. We determine the period and horizontal translation from a basic cycle.

$$\text{One cycle:} \quad 0 \le 2x - \frac{3\pi}{2} \le 2\pi$$

$$\frac{3\pi}{2} \le 2x \le \frac{7\pi}{2} \qquad \text{Add } \frac{3\pi}{2} \text{ first}$$

$$\frac{3\pi}{4} \le x \le \frac{7\pi}{4} \qquad \text{Divide by 2}$$

A cycle will begin at $x = 3\pi/4$, so the horizontal shift is $3\pi/4$. To find the period, we can subtract the left and right endpoints of the cycle

$$\text{Period} = \frac{7\pi}{4} - \frac{3\pi}{4} = \frac{4\pi}{4} = \pi$$

or, equivalently, divide 2π by $B = 2$

$$\text{Period} = \frac{2\pi}{B} = \frac{2\pi}{2} = \pi$$

Dividing the period by 4 gives us $\pi/4$. To mark the x-axis, we begin at $x = 3\pi/4$ and add increments of $\pi/4$ as follows:

$$\frac{3\pi}{4} + \frac{\pi}{4} = \frac{4\pi}{4} = \pi \qquad \frac{3\pi}{4} + 2 \cdot \frac{\pi}{4} = \frac{5\pi}{4}$$

$$\frac{3\pi}{4} + 3 \cdot \frac{\pi}{4} = \frac{6\pi}{4} = \frac{3\pi}{2}$$

We draw a frame and then sketch the graph of one complete cycle as shown in Figure 7.

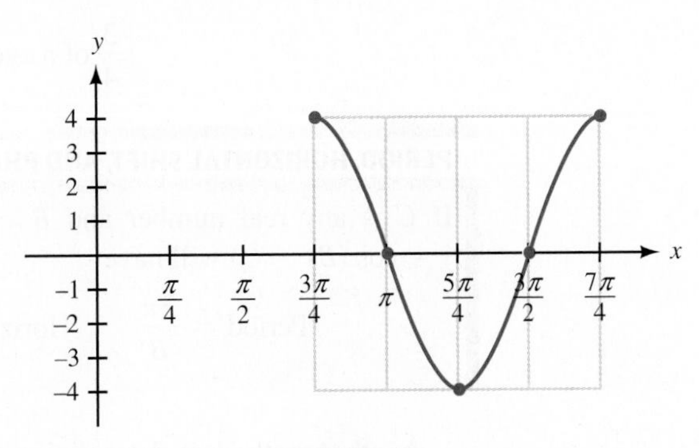

Figure 7

Because the original problem asked for the graph on the interval $0 \le x \le 2\pi$, we extend the graph to the right by adding the first quarter of a second cycle. On the left, we add the last three quarters of an additional cycle to reach 0. The final graph is shown in Figure 8.

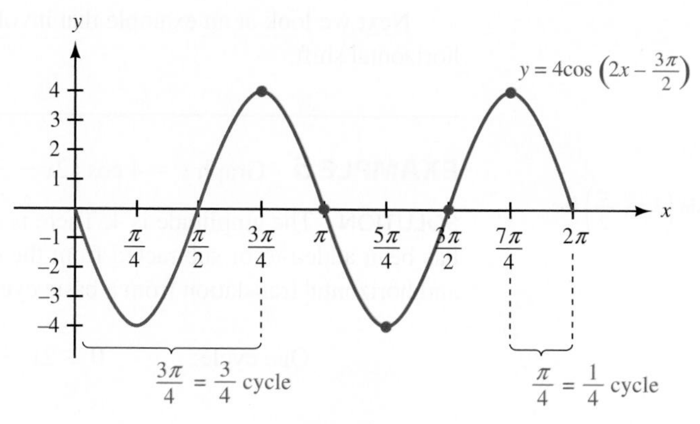

Figure 8

In general, for $y = \sin (Bx + C)$ or $y = \cos (Bx + C)$ to complete one cycle, the quantity $Bx + C$ must vary from 0 to 2π. Therefore, assuming $B > 0$,

$$0 \le Bx + C \le 2\pi \quad \text{if} \quad -\frac{C}{B} \le x \le \frac{2\pi - C}{B}$$

The horizontal shift will be the left end point of the cycle, or $-C/B$. If you find the difference between the end points, you will see that the period is $2\pi/B$ as before.

The constant C in $y = \sin (Bx + C)$ or $y = \cos (Bx + C)$ is called the *phase*. Phase is important in situations, such as when working with alternating currents, where two sinusoidal curves are being compared to one another. If x represents time, then the phase is the fraction of a standard period of 2π that a point on the graph of $y = \sin (Bx + C)$ lags or leads a corresponding point on the graph of $y = \sin Bx$.

For instance, the phase of the equation in Example 3 is $\pi/2$, which indicates that the graph of $y = \sin (x + \pi/2)$ leads the graph of $y = \sin x$ by 1/4 of a complete cycle. In Figure 5 you can see how all the points on the red graph would occur $\pi/2$ units of time *before* the corresponding points on the blue graph, and

$$\frac{1}{4} \text{ of a cycle} = \frac{1}{4} \cdot 2\pi = \frac{\pi}{2}$$

Likewise, in Example 5 the phase is $-3\pi/2$. As you can see in Figure 8, the graph of $y = 4 \cos (2x - 3\pi/2)$ lags behind the graph of $y = 4 \cos 2x$ by 3/4 of a cycle, and

$$\frac{3}{4} \text{ of a cycle} = \frac{3}{4} \cdot 2\pi = \frac{3\pi}{4}$$

PERIOD, HORIZONTAL SHIFT, AND PHASE FOR SINE AND COSINE

If C is any real number and $B > 0$, then the graphs of $y = \sin (Bx + C)$ and $y = \cos (Bx + C)$ will have

$$\text{Period} = \frac{2\pi}{B} \qquad \text{Horizontal shift} = -\frac{C}{B} \qquad \text{Phase} = C$$

Another method of determining the period and horizontal shift is to rewrite the function so that the argument looks like $B(x - h)$ instead of combined as $Bx + C$. For instance, using the function from Example 5 we have

$$y = 4 \cos \left(2x - \frac{3\pi}{2}\right) = 4 \cos \left(2\left(x - \frac{3\pi}{4}\right)\right)$$

which is accomplished by factoring out the coefficient of 2. We can now easily identify that $B = 2$ and $h = 3\pi/4$.

Before working a final example that ties everything together, we summarize all the information we have covered about the graphs of the sine and cosine functions.

GRAPHING THE SINE AND COSINE FUNCTIONS

The graphs of $y = k + A\sin(B(x - h))$ and $y = k + A\cos(B(x - h))$, where $B > 0$, will have the following characteristics:

$$\text{Amplitude} = |A| \qquad \text{Period} = \frac{2\pi}{B}$$

$$\text{Horizontal translation} = h \qquad \text{Vertical translation} = k$$

In addition, if $A < 0$ the graph will be reflected about the x-axis.

PROBLEM 6
Graph one cycle of
$y = 2 - 4\cos\left(2\pi x - \frac{\pi}{3}\right)$.

EXAMPLE 6 Graph one complete cycle of $y = 3 - 5\sin\left(\pi x + \frac{\pi}{4}\right)$.

SOLUTION First, we rewrite the function by factoring out the coefficient of π.

$$y = 3 - 5\sin\left(\pi x + \frac{\pi}{4}\right) = 3 - 5\sin\left(\pi\left(x + \frac{1}{4}\right)\right)$$

In this case, the values are $A = -5$, $B = \pi$, $h = -1/4$, and $k = 3$. This gives us

$$\text{Amplitude} = |-5| = 5 \qquad \text{Period} = \frac{2\pi}{\pi} = 2$$

$$\text{Horizontal shift} = -\frac{1}{4} \qquad \text{Vertical shift} = 3$$

To verify the period and horizontal shift, and to help sketch the graph, we examine one cycle.

One cycle: $0 \le \pi x + \frac{\pi}{4} \le 2\pi$

$-\frac{\pi}{4} \le \pi x \le \frac{7\pi}{4}$ Subtract $\frac{\pi}{4}$ first

$-\frac{1}{4} \le x \le \frac{7}{4}$ Divide by π

Dividing the period by 4 gives $\frac{1}{2}$, so we will mark the x-axis in increments of $\frac{1}{2}$ starting with $x = -\frac{1}{4}$. Notice that our frame for the cycle (Figure 9) has been shifted upward 3 units, and we have plotted the key points to account for the x-axis reflection. The graph is shown in Figure 10.

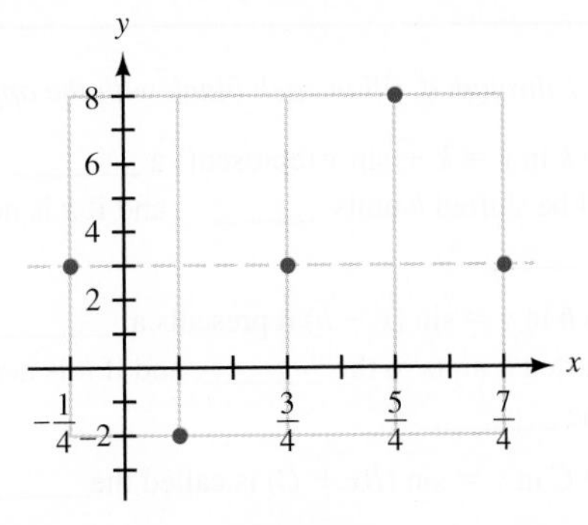

Figure 9

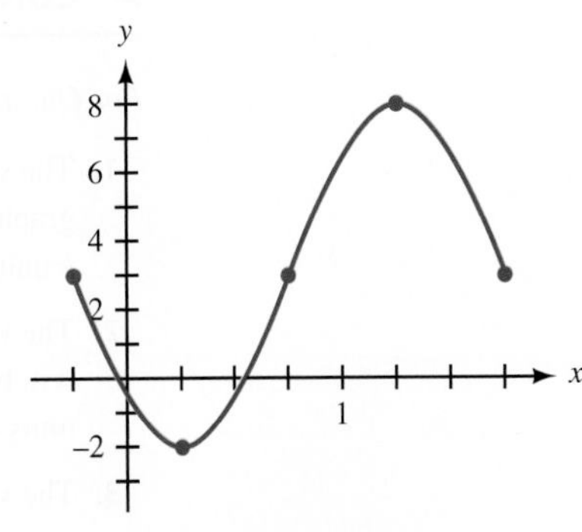

Figure 10

Using Technology — Verifying a Graph

To verify the graph in Figure 10, set your calculator to radian mode, define

$$Y_1 = 3 - 5 \sin (\pi x + \pi/4)$$

and set the window variables to match the characteristics for the cycle that was drawn (set the viewing rectangle to agree with the rectangle shown in the figure):

$$-0.25 \le x \le 1.75, \text{ scale} = 0.25; \ -2 \le y \le 8$$

If our drawing is correct, we should see one cycle of the sine function that exactly fits the screen and matches our graph perfectly. Figure 11 shows the result.

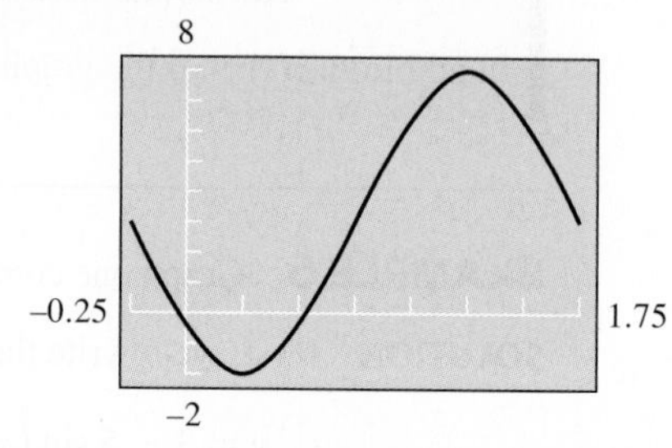

Figure 11

Getting Ready for Class

After reading through the preceding section, respond in your own words and in complete sentences.

a. How does the term $\frac{\pi}{2}$ affect the graph of $y = \sin\left(x + \frac{\pi}{2}\right)$?

b. How does the term $\frac{\pi}{2}$ affect the graph of $y = \sin\left(x - \frac{\pi}{2}\right)$?

c. How do you find the period for the graph of $y = A \sin (Bx + C)$?

d. How do you find the horizontal translation for the graph of $y = A \sin (Bx + C)$?

4.3 PROBLEM SET

➤ CONCEPTS AND VOCABULARY

For Questions 1 through 4, fill in each blank with the appropriate word or expression.

1. The value k in $y = k + \sin x$ represents a __________ translation. If k is positive, the graph will be shifted k units _________, and if k is negative, the graph will be shifted k units _________.

2. The value h in $y = \sin (x - h)$ represents a _________ translation. If h is positive, the graph will be shifted h units to the ________, and if h is negative, the graph will be shifted h units to the _________.

3. The value C in $y = \sin (Bx + C)$ is called the _________.

4. The graph of $y = \sin (Bx + C)$ will have a horizontal shift of _________.

➤ EXERCISES

Identify the vertical translation for each equation. Do not sketch the graph.

5. $y = 5 + \sin x$
6. $y = 2 + \cos x$
7. $y = -\dfrac{1}{4} + \cos x$
8. $y = -\dfrac{1}{2} - \sin x$

Graph one complete cycle of each of the following. In each case, label the axes accurately and identify the vertical translation for each graph.

9. $y = 2 + \sin x$
10. $y = 4 + \sin x$
11. $y = -5 + \cos x$
12. $y = -1 + \cos x$
13. $y = 3 - \sin x$
14. $y = 6 - \sin x$
15. $y = \dfrac{1}{2} - \cos x$
16. $y = -\dfrac{3}{2} - \cos x$

Graph one complete cycle of each of the following. In each case, label the axes accurately and identify the amplitude, period, and vertical translation for each graph.

17. $y = 4 + 4 \sin 2x$
18. $y = -2 + 2 \sin 4x$
19. $y = -1 + \dfrac{1}{2} \cos 3x$
20. $y = 1 + \dfrac{1}{2} \sin 3x$

Identify the horizontal translation for each equation. Do not sketch the graph.

21. $y = \cos (x + \pi)$
22. $y = \cos (x - \pi)$
23. $y = \sin \left(x - \dfrac{2\pi}{3}\right)$
24. $y = \sin \left(x + \dfrac{3\pi}{4}\right)$

Graph one complete cycle of each of the following. In each case, label the axes accurately and identify the horizontal translation for each graph.

25. $y = \sin \left(x + \dfrac{\pi}{4}\right)$
26. $y = \sin \left(x + \dfrac{\pi}{6}\right)$
27. $y = \sin \left(x - \dfrac{\pi}{4}\right)$
28. $y = \sin \left(x - \dfrac{\pi}{6}\right)$
29. $y = \sin \left(x + \dfrac{\pi}{3}\right)$
30. $y = \sin \left(x - \dfrac{\pi}{3}\right)$
31. $y = \cos \left(x - \dfrac{\pi}{2}\right)$
32. $y = \cos \left(x + \dfrac{\pi}{2}\right)$
33. $y = \cos \left(x + \dfrac{\pi}{3}\right)$
34. $y = \cos \left(x - \dfrac{\pi}{4}\right)$

For each equation, identify the period, horizontal shift, and phase. Do not sketch the graph.

35. $y = \cos \left(\dfrac{\pi}{3}x + \dfrac{\pi}{3}\right)$
36. $y = \cos \left(\dfrac{\pi}{2}x - \dfrac{\pi}{2}\right)$
37. $y = \sin (6x - \pi)$
38. $y = \sin (4x + \pi)$
39. $y = 3 - \sin \left(\dfrac{1}{2}x + \dfrac{\pi}{6}\right)$
40. $y = 2 + \cos \left(\dfrac{1}{3}x - \dfrac{\pi}{6}\right)$

For each equation, identify the amplitude, period, horizontal shift, and phase. Then label the axes accordingly and sketch one complete cycle of the curve.

41. $y = \sin (2x - \pi)$
42. $y = \sin (2x + \pi)$
43. $y = \sin \left(\pi x + \dfrac{\pi}{2}\right)$
44. $y = \sin \left(\pi x - \dfrac{\pi}{2}\right)$

45. $y = -\cos\left(2x + \dfrac{\pi}{2}\right)$

46. $y = -\cos\left(2x - \dfrac{\pi}{2}\right)$

47. $y = 2\sin\left(\dfrac{1}{2}x + \dfrac{\pi}{2}\right)$

48. $y = 3\cos\left(\dfrac{1}{2}x + \dfrac{\pi}{3}\right)$

49. $y = \dfrac{1}{2}\cos\left(3x - \dfrac{\pi}{2}\right)$

50. $y = \dfrac{4}{3}\cos\left(3x + \dfrac{\pi}{2}\right)$

51. $y = 3\sin\left(\dfrac{\pi}{3}x - \dfrac{\pi}{3}\right)$

52. $y = 3\cos\left(\dfrac{\pi}{3}x - \dfrac{\pi}{3}\right)$

Use your answers for Problems 41 through 46 for reference, and graph one complete cycle of each of the following equations.

53. $y = 1 + \sin(2x - \pi)$

54. $y = -1 + \sin(2x + \pi)$

55. $y = -3 + \sin\left(\pi x + \dfrac{\pi}{2}\right)$

56. $y = 3 + \sin\left(\pi x - \dfrac{\pi}{2}\right)$

57. $y = 2 - \cos\left(2x + \dfrac{\pi}{2}\right)$

58. $y = -2 - \cos\left(2x - \dfrac{\pi}{2}\right)$

Graph one complete cycle of each of the following. In each case, label the axes accurately and identify the amplitude, period, vertical and horizontal translation, and phase for each graph.

59. $y = -2 + 3\cos\left(\dfrac{1}{2}x - \dfrac{\pi}{3}\right)$

60. $y = 3 + 2\sin\left(\dfrac{1}{2}x - \dfrac{\pi}{2}\right)$

61. $y = \dfrac{3}{2} - \dfrac{1}{2}\sin(3x + \pi)$

62. $y = \dfrac{2}{3} - \dfrac{4}{3}\cos(3x - \pi)$

Graph each of the following equations over the given interval. In each case, be sure to label the axes so that the amplitude, period, vertical translation, and horizontal translation are easy to read.

63. $y = 4\cos\left(2x - \dfrac{\pi}{2}\right),\ -\dfrac{\pi}{4} \le x \le \dfrac{3\pi}{2}$

64. $y = -\dfrac{2}{3}\sin\left(3x + \dfrac{\pi}{2}\right),\ -\pi \le x \le \pi$

65. $y = \dfrac{5}{2} - 3\cos\left(\pi x - \dfrac{\pi}{4}\right),\ -2 \le x \le 2$

66. $y = 2 - \dfrac{1}{3}\cos\left(\pi x + \dfrac{3\pi}{2}\right),\ -\dfrac{1}{4} \le x \le \dfrac{15}{4}$

67. Oscillating Spring A mass attached to a spring oscillates upward and downward. The length L of the spring after t seconds is given by the function $L = 15 - 3.5\cos(2\pi t)$, where L is measured in centimeters (Figure 12).

a. Sketch the graph of this function for $0 \le t \le 5$.

b. What is the length the spring when it is at equilibrium?

c. What is the length the spring when it is shortest?

d. What is the length the spring when it is longest?

68. Oscillating Spring A mass attached to a spring oscillates upward and downward. The length L of the spring after t seconds is given by the function $L = 8 - 2\cos(4\pi t)$, where L is measured in inches (Figure 12).

a. Sketch the graph of this function for $0 \le t \le 3$.

b. What is the length the spring when it is at equilibrium?

c. What is the length the spring when it is shortest?

d. What is the length the spring when it is longest?

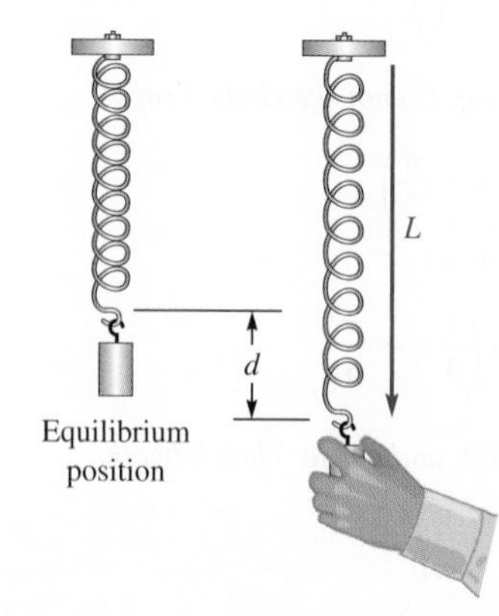

Figure 12

Simple Harmonic Motion *In the Section 4.2 problem set, we introduced the concept of simple harmonic motion. A more general model for this type of motion is given by*

$$y = A \sin(\omega t + \phi) \qquad \text{or} \qquad y = A \cos(\omega t + \phi)$$

where ϕ is the phase (sometimes called the phase angle).

69. **Sound Wave** The oscillations in air pressure representing the sound wave for a musical tone can be modeled by the equation $y = 0.05 \sin(500\pi t + 10\pi)$, where y is the sound pressure in pascals after t seconds.
    a. Sketch the graph of one complete cycle of the sound wave.
    b. What is the phase?

70. **RLC Circuit** The electric current in an RLC circuit can be modeled by the equation $y = 2 \cos(990t - 0.64)$, where y is the current in milliamps after t seconds.
    a. Sketch the graph of one complete cycle of the current.
    b. What is the phase?

➤ REVIEW PROBLEMS

The following problems review material we covered in Section 3.4.

71. **Arc Length** Find the length of arc cut off by a central angle of $\pi/6$ radians in a circle of radius 10 centimeters.

72. **Arc Length** How long is the arc cut off by a central angle of 90° in a circle with radius 22 centimeters?

73. **Radius of a Circle** Find the radius of a circle if a central angle of 6 radians cuts off an arc of length 4 feet.

74. **Radius of a Circle** In a circle, a central angle of 135° cuts off an arc of length 75 meters. Find the radius of the circle.

➤ LEARNING OBJECTIVES ASSESSMENT

These questions are available for instructors to help assess if you have successfully met the learning objectives for this section.

75. Sketch the graph of $y = 2 + \cos\left(\dfrac{\pi}{2}x - \dfrac{\pi}{2}\right)$. Which of the following matches your graph?

a.

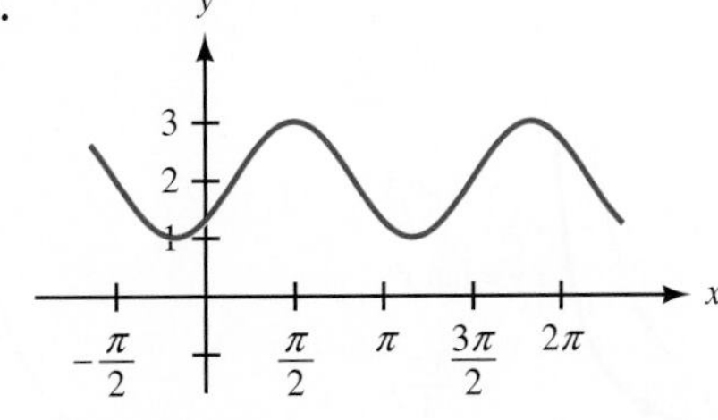

b.

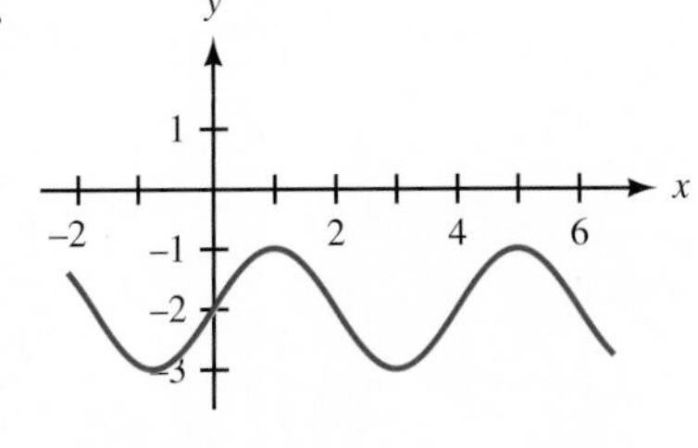

c.

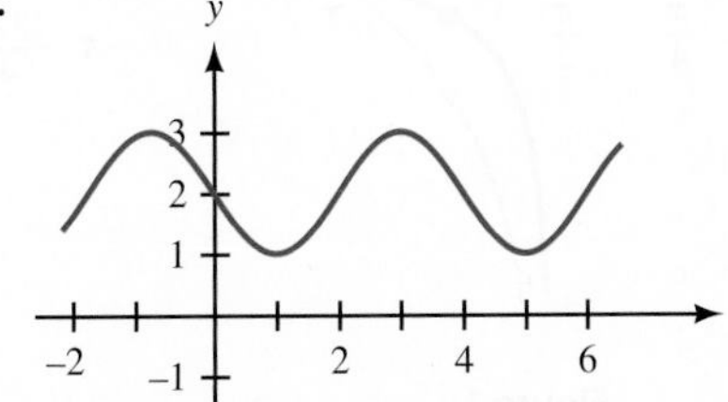

d.

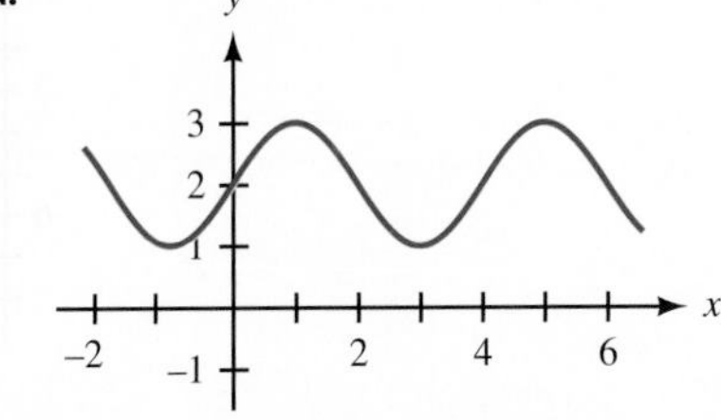

76. Identify the vertical translation for the graph of $y = 1 + 4\cos\left(3x + \frac{\pi}{2}\right)$.

a. $\frac{\pi}{2}$ **b.** $-\frac{\pi}{2}$ **c.** 1 **d.** 4

77. Identify the horizontal translation for the graph of $y = 1 + 4\cos\left(3x + \frac{\pi}{2}\right)$.

a. $-\frac{\pi}{2}$ **b.** $-\frac{\pi}{6}$ **c.** $\frac{\pi}{2}$ **d.** 1

78. Identify the phase for the graph of $y = 1 + 4\cos\left(3x + \frac{\pi}{2}\right)$.

a. $-\frac{\pi}{2}$ **b.** $-\frac{\pi}{6}$ **c.** $\frac{\pi}{2}$ **d.** 1

SECTION 4.4 The Other Trigonometric Functions

LEARNING OBJECTIVES

1. Find the period and sketch the graph of a tangent function.
2. Find the period and sketch the graph of a cotangent function.
3. Find the period and sketch the graph of a secant function.
4. Find the period and sketch the graph of a cosecant function.

The same techniques that we used to graph the sine and cosine functions can be used with the other four trigonometric functions.

Tangent and Cotangent

EXAMPLE 1 Graph $y = 3\tan x$ for $-\pi \le x \le \pi$.

SOLUTION Although the tangent does not have a defined amplitude, we know from our work in the previous sections that the factor of 3 will triple all of the y-coordinates. That is, for the same x, the value of y in $y = 3\tan x$ will be three times the corresponding value of y in $y = \tan x$.

To sketch the graph of one cycle, remember that a cycle begins with an x-intercept, has the vertical asymptote in the middle, and ends with an x-intercept. At $x = \pi/4$, the normal y-value of 1 must be tripled, so we plot a point at $(\pi/4, 3)$. For the same reason we plot a point at $(3\pi/4, -3)$. Figure 1 shows a complete cycle for $y = 3\tan x$ (we have included the graph of $y = \tan x$ for comparison).

The original problem asked for the graph on the interval $-\pi \le x \le \pi$. We extend the graph to the left by adding a second complete cycle. The final graph is shown in Figure 2.

PROBLEM 1

Graph $y = \frac{1}{3}\tan x$ for $-\pi \le x \le \pi$.

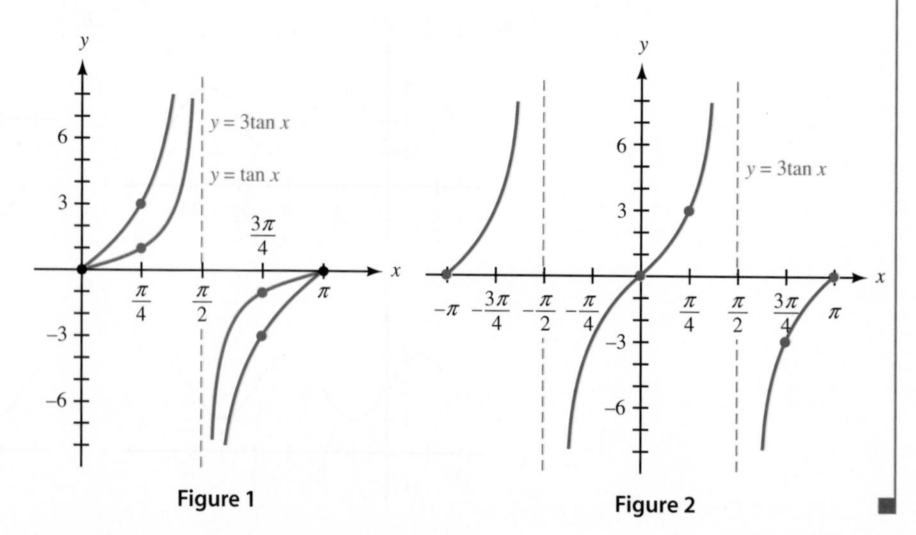

Figure 1 Figure 2

In Section 1.5, we found that the expression $\sqrt{x^2+9}$ could be rewritten without a square root by making the substitution $x = 3 \tan \theta$. Then we noted that the substitution itself was questionable because we did not know at that time if every real number x could be written as $3 \tan \theta$, for some value of θ. As you can see from the graph in Figure 2, the range for $y = 3 \tan \theta$ is all real numbers. This means that any real number can indeed be written as $3 \tan \theta$, which confirms the validity of the substitution.

PROBLEM 2

Graph one cycle of $y = 2 \cot \frac{1}{3} x$.

EXAMPLE 2 Graph one complete cycle of $y = \frac{1}{2} \cot(-2x)$.

SOLUTION Because the cotangent is an odd function,

$$y = \frac{1}{2} \cot(-2x) = -\frac{1}{2} \cot(2x)$$

The factor of $-\frac{1}{2}$ will halve all of the y-coordinates of $y = \cot(2x)$ and cause an x-axis reflection. In addition, there is a coefficient of 2 in the argument. To see how this will affect the period, we identify a complete cycle. Remember that the period of the cotangent function is π, not 2π.

One cycle:	$0 \le \text{argument} \le \pi$	
	$0 \le 2x \le \pi$	The argument is $2x$
	$0 \le x \le \frac{\pi}{2}$	Divide by 2 to isolate x

The period is $\pi/2$. Dividing this by 4 gives us $\pi/8$, so we will mark the x-axis in increments of $\pi/8$ starting at $x = 0$.

$$0, \quad 1 \cdot \frac{\pi}{8} = \frac{\pi}{8}, \quad 2 \cdot \frac{\pi}{8} = \frac{\pi}{4}, \quad 3 \cdot \frac{\pi}{8} = \frac{3\pi}{8}, \quad 4 \cdot \frac{\pi}{8} = \frac{\pi}{2}$$

The basic cycle of a cotangent graph begins with a vertical asymptote, has an x-intercept in the middle, and ends with another vertical asymptote (see Table 6, Section 4.1). We can sketch a frame to help plot the key points and draw the asymptotes, much as we did with the sine and cosine functions. The difference is that the upper and lower sides of the frame do not indicate maximum and minimum values of the graph but are only used to define the position of two key points. Figure 3 shows the result.

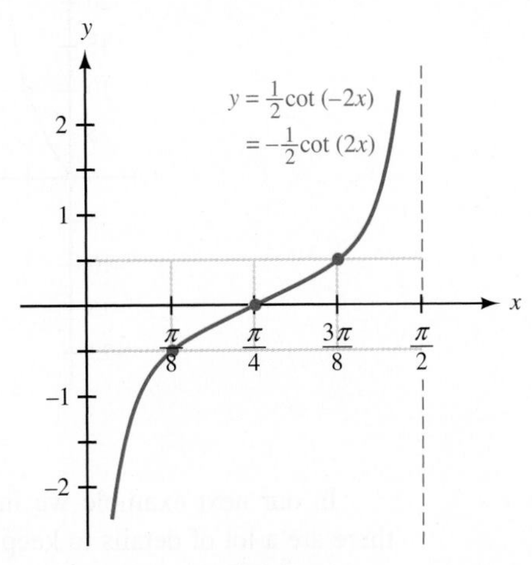

Figure 3

For our next example, we look at the graph of one of the equations we found in Example 4 of Section 3.5.

PROBLEM 3
Graph $y = 10 \tan 2\pi t$ for $0 \le t \le 1$.

EXAMPLE 3 Figure 4 shows a fire truck parked on the shoulder of a freeway next to a long block wall. The red light on the top of the truck is 10 feet from the wall and rotates through one complete revolution every 2 seconds. Graph the function that gives the length d in terms of time t from $t = 0$ to $t = 2$.

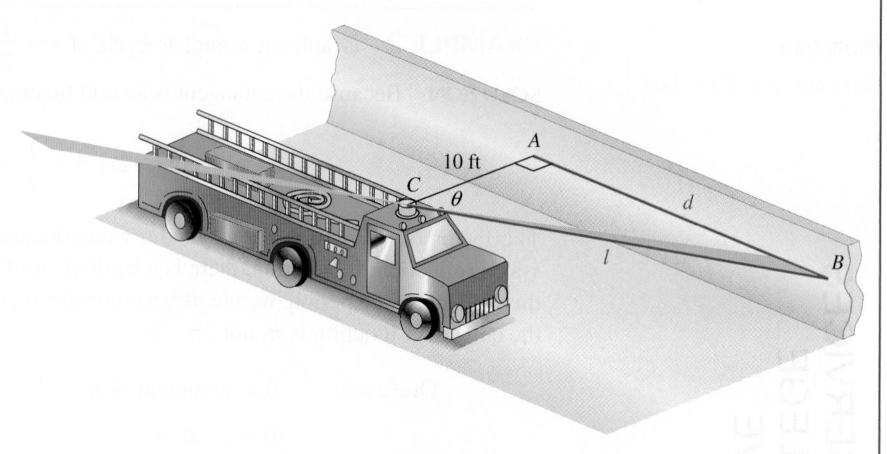

Figure 4

SOLUTION From Example 4 of Section 3.5, we know that $d = 10 \tan \pi t$. We must multiply all the y-values of the basic tangent function by 10. Also, the coefficient of π will change the period.

One cycle: $0 \le \pi t \le \pi$

$0 \le t \le 1$ Divide by π to isolate t

The period is 1. Dividing by 4 gives us $\frac{1}{4}$, so we mark the t-axis in increments of $\frac{1}{4}$. The graph is shown in Figure 5. Because the original problem asks for the graph from $t = 0$ to $t = 2$, we extended the graph by adding an additional cycle on the right.

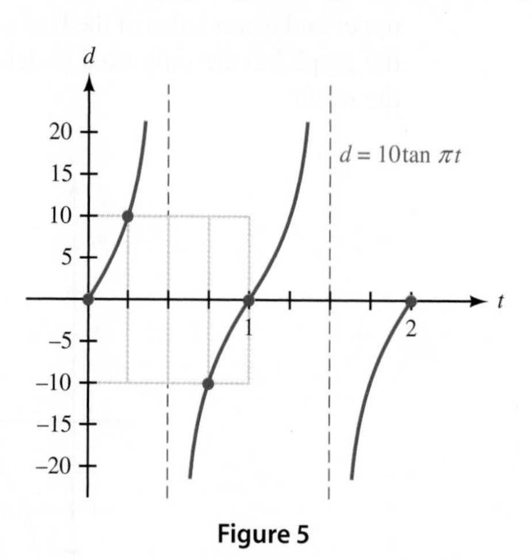

Figure 5

In our next example we include a vertical and horizontal translation. Although there are a lot of details to keep track of, the steps are nearly identical to the ones we presented in the previous section.

PROBLEM 4
Graph one cycle of
$y = 2 + 4\tan\left(\frac{x}{2} - \frac{\pi}{6}\right)$.

EXAMPLE 4 Graph one complete cycle of $y = 3 + 2\tan\left(\frac{x}{2} + \frac{\pi}{8}\right)$.

SOLUTION The first term of 3 will shift the graph of $y = \tan x$ vertically 3 units upward. The factor of 2 will double the y-coordinates. To determine the horizontal shift and period, we check one cycle.

$$\text{One cycle:} \quad 0 \le \frac{x}{2} + \frac{\pi}{8} \le \pi$$

$$-\frac{\pi}{8} \le \frac{x}{2} \le \frac{7\pi}{8} \qquad \text{Subtract } \frac{\pi}{8}$$

$$-\frac{\pi}{4} \le x \le \frac{7\pi}{4} \qquad \text{Multiply by 2}$$

A cycle will begin at $-\pi/4$, so the graph is shifted $\pi/4$ units to the left. To find the period, we compute the difference between endpoints for the cycle:

$$\text{Period} = \frac{7\pi}{4} - \left(-\frac{\pi}{4}\right) = \frac{8\pi}{4} = 2\pi$$

Dividing the period by 4 gives us $\pi/2$, so we mark the x-axis at intervals of $\pi/2$ beginning with $x = -\pi/4$ as follows (we only need to find the three middle points).

$$-\frac{\pi}{4} + \frac{\pi}{2} = \frac{\pi}{4} \qquad -\frac{\pi}{4} + 2 \cdot \frac{\pi}{2} = \frac{3\pi}{4} \qquad -\frac{\pi}{4} + 3 \cdot \frac{\pi}{2} = \frac{5\pi}{4}$$

A cycle will begin and end with x-intercepts at $x = -\pi/4$ and $x = 7\pi/4$ (on the shifted axis). There will be a vertical asymptote at $x = 3\pi/4$. Figure 6 shows the key points, and the final graph is given in Figure 7.

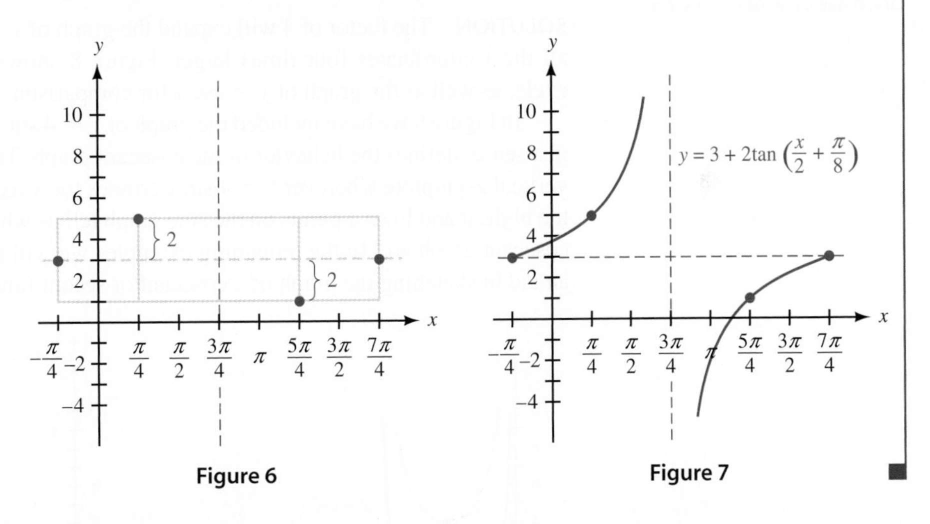

Figure 6 **Figure 7**

We summarize the characteristics for the tangent and cotangent as we did for the sine and cosine functions. The derivation of these formulas is similar, except that a period of π is used.

PERIOD AND HORIZONTAL SHIFT FOR TANGENT AND COTANGENT

If C is any real number and $B > 0$, then the graphs of $y = \tan(Bx + C)$ and $y = \cot(Bx + C)$ will have

$$\text{Period} = \frac{\pi}{B} \quad \text{and} \quad \text{Horizontal shift} = -\frac{C}{B}$$

We can also rewrite the function by factoring so that the argument looks like $B(x - h)$ instead of combined as $Bx + C$. For instance, using the function from Example 4 we have

$$y = 3 + 2\tan\left(\frac{x}{2} + \frac{\pi}{8}\right) = 3 + 2\tan\left(\frac{1}{2}\left(x + \frac{\pi}{4}\right)\right)$$

Now it is apparent that $B = \frac{1}{2}$ and $h = -\pi/4$.

GRAPHING THE TANGENT AND COTANGENT FUNCTIONS

The graphs of $y = k + A\tan(B(x - h))$ and $y = k + A\cot(B(x - h))$, where $B > 0$, will have the following characteristics:

Period $= \dfrac{\pi}{B}$ Horizontal translation $= h$ Vertical translation $= k$

In addition, $|A|$ is the factor by which the basic graphs are expanded or contracted vertically. If $A < 0$ the graph will be reflected about the x-axis.

Secant and Cosecant

Because the secant and cosecant are reciprocals of the cosine and sine, respectively, there is a natural relationship between their graphs. We will take advantage of this relationship to graph the secant and cosecant functions by first graphing a corresponding cosine or sine function.

PROBLEM 5
Graph one cycle of $y = 3\sec x$.

EXAMPLE 5 Graph one complete cycle of $y = 4\csc x$.

SOLUTION The factor of 4 will expand the graph of $y = \csc x$ vertically by making all the y-coordinates four times larger. Figure 8 shows the resulting graph of one cycle, as well as the graph of $y = \csc x$ for comparison.

In Figure 9 we have included the graph of $y = 4\sin x$. Notice how the sine graph, in a sense, defines the behavior of the cosecant graph. The graph of $y = 4\csc x$ has a vertical asymptote wherever $y = 4\sin x$ crosses the x-axis (has a zero). Furthermore, the highest and lowest points on the sine graph tell us where the two key points on the cosecant graph are. In the remaining examples we will use a sine or cosine graph as an aid in sketching the graph of a cosecant or secant function, respectively.

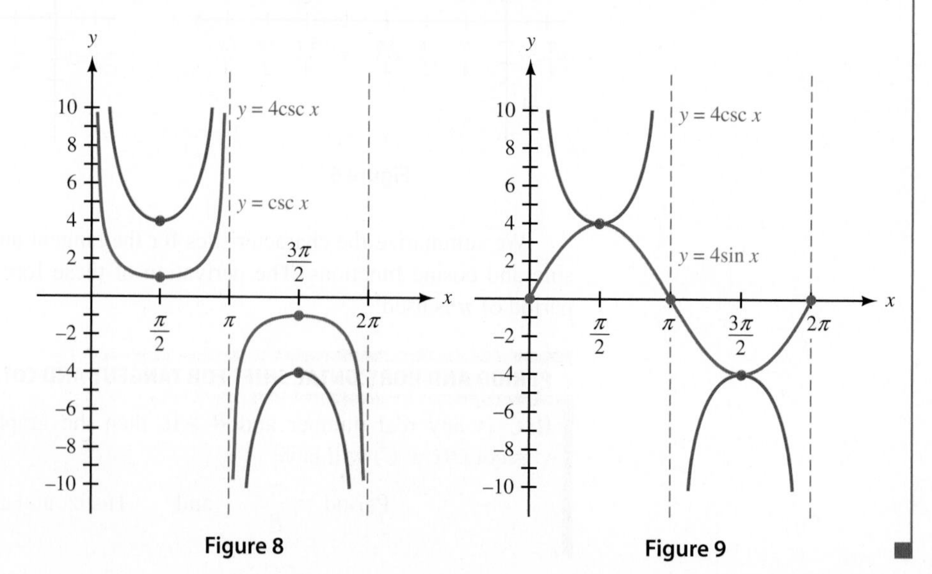

Figure 8 Figure 9

PROBLEM 6

Graph $y = \frac{1}{4}\csc 3x$ for $0 \le x \le 2\pi$.

EXAMPLE 6 Graph $y = \frac{1}{3}\sec 2x$ for $-\frac{3\pi}{2} \le x \le \frac{3\pi}{2}$.

SOLUTION We begin with the graph of $y = \frac{1}{3}\cos 2x$, which will have an amplitude of $\frac{1}{3}$. We check one cycle:

$$\text{One cycle:} \quad 0 \le 2x \le 2\pi$$
$$0 \le x \le \pi$$

The period is π. We can sketch two complete cycles between $-\pi$ and π, and then use two half-cycles to extend the graph out to $-3\pi/2$ and $3\pi/2$. Figure 10 shows the resulting graph.

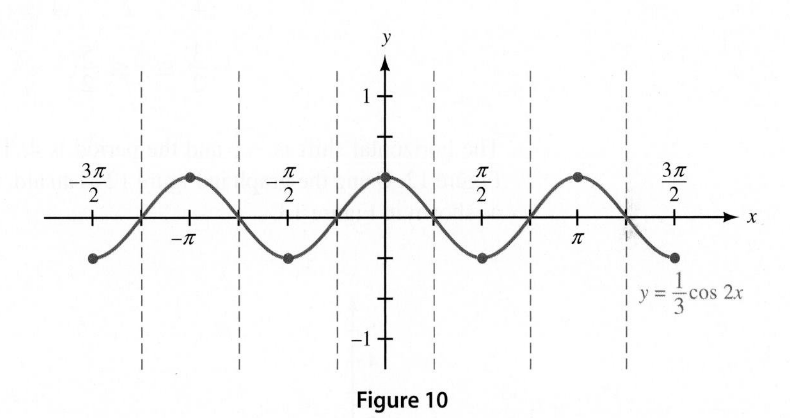

Figure 10

To sketch the graph of the secant function, we note that the zeros of the cosine graph correspond to the vertical asymptotes of the secant graph, and the peaks and valleys of the cosine graph correspond to the valleys and peaks of the secant graph, respectively.

The graph of $y = \frac{1}{3}\cos 2x$ is shown in Figure 11. Notice the range of the function is $y \le -\frac{1}{3}$ or $y \ge \frac{1}{3}$.

NOTE In sketching the graph of the cosecant function in Example 5 and the secant function in Example 6, we are not actually taking reciprocals of the y-coordinates of points on the sine and cosine graphs. Because of the change in amplitude, we would not obtain the correct graphs if we did so. We are simply using the shape of a sine or cosine graph with the proper amplitude to guide us in drawing the correct graph of the cosecant or secant function.

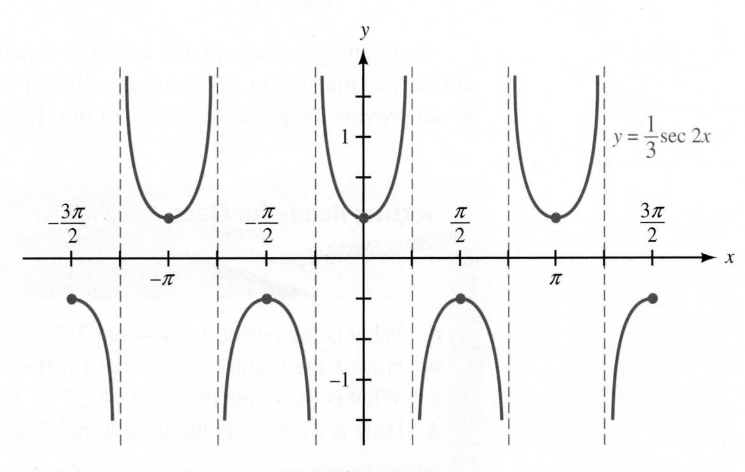

Figure 11

PROBLEM 7
Graph one cycle of
$y = -2 - 4\csc\left(\frac{\pi x}{3} - \frac{\pi}{6}\right)$.

EXAMPLE 7 Graph one cycle of $y = -1 - 3\csc\left(\frac{\pi x}{2} + \frac{3\pi}{4}\right)$.

SOLUTION First, we sketch the graph of

$$y = -1 - 3\sin\left(\frac{\pi x}{2} + \frac{3\pi}{4}\right)$$

There is a vertical translation of the graph of $y = \sin x$ one unit downward. The amplitude is 3 and there is a reflection about the x-axis. We check one cycle:

One cycle: $0 \le \frac{\pi x}{2} + \frac{3\pi}{4} \le 2\pi$

$-\frac{3\pi}{4} \le \frac{\pi x}{2} \le \frac{5\pi}{4}$ Subtract $\frac{3\pi}{4}$

$-\frac{3}{2} \le x \le \frac{5}{2}$ Multiply by $\frac{2}{\pi}$

The horizontal shift is $-\frac{3}{2}$ and the period is 4. The graph of one cycle is shown in Figure 12. Using the graph in Figure 12 as an aid, we sketch the graph of the cosecant as shown in Figure 13.

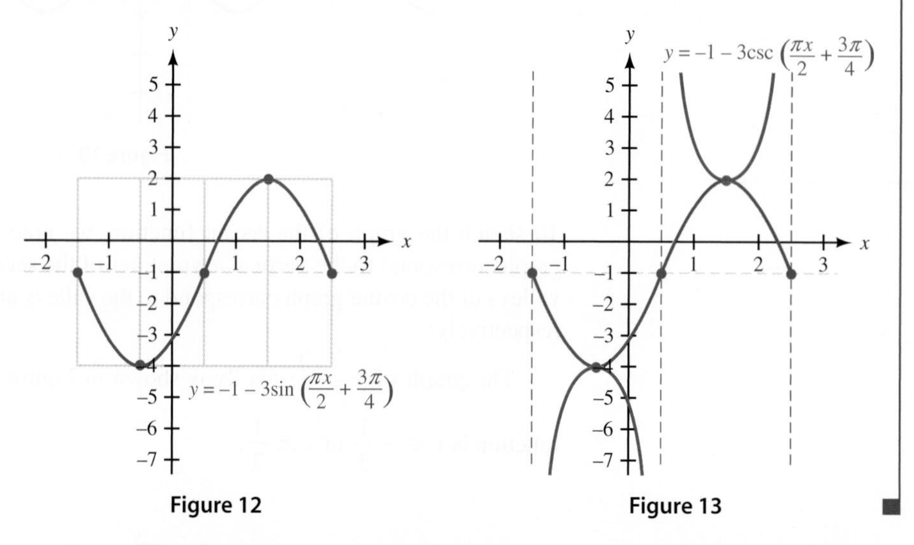

Figure 12

Figure 13

In graphing the secant and cosecant functions, the period, horizontal translation, and vertical translation are identical to those for the sine and cosine. For this reason, we do not provide a separate summary of the characteristics for these two functions.

Getting Ready for Class

After reading through the preceding section, respond in your own words and in complete sentences.

a. What is the period for $y = \tan 2x$?
b. How is the graph of $y = 2\tan x$ different from the graph of $y = \tan x$?
c. What is the range of $y = 4\csc x$?
d. How is a cosine graph used to help sketch the graph of a secant function?

4.4 PROBLEM SET

➤ CONCEPTS AND VOCABULARY

For Questions 1 through 4, fill in each blank with the appropriate word or expression.

1. In $y = A \tan x$, $|A|$ is the factor by which the basic tangent graph will be __________ or __________ vertically. To sketch the graph, __________ the y-coordinate of each point on the basic tangent graph by A.
2. The number 3 in $y = 3 \sec x$ will ________ the graph of $y = \sec x$ vertically by a factor of ____. The number $\frac{1}{3}$ in $y = \frac{1}{3} \sec x$ will ________ the graph of $y = \sec x$ vertically by a factor of ____.
3. The period of $y = \tan Bx$ is _____.
4. To graph $y = k + A \csc (Bx + C)$, first sketch the graph of the corresponding _____ function to use as a guide.

➤ EXERCISES

Graph one complete cycle of each of the following. In each case, label the axes accurately.

5. $y = 4 \tan x$
6. $y = 3 \cot x$
7. $y = \frac{1}{2} \csc x$
8. $y = \frac{1}{2} \sec x$
9. $y = -\frac{1}{3} \cot x$
10. $y = -\frac{1}{4} \tan x$
11. $y = -5 \sec x$
12. $y = -4 \csc x$

Graph one complete cycle of each of the following. In each case, label the axes accurately and identify the period for each graph.

13. $y = \csc 3x$
14. $y = \csc 4x$
15. $y = \sec \frac{\pi}{4} x$
16. $y = \sec \frac{1}{4} x$
17. $y = -\tan \frac{1}{2} x$
18. $y = \tan \frac{1}{3} x$
19. $y = \cot 4x$
20. $y = -\cot \pi x$

Graph one complete cycle for each of the following. In each case, label the axes accurately and state the period for each graph.

21. $y = \frac{1}{2} \csc 3x$
22. $y = \frac{1}{2} \sec 3x$
23. $y = 3 \sec \frac{1}{2} x$
24. $y = 3 \csc \frac{1}{2} x$
25. $y = 2 \tan 3x$
26. $y = 3 \tan 2x$
27. $y = \frac{1}{2} \cot \frac{\pi}{2} x$
28. $y = \frac{1}{3} \cot \frac{1}{2} x$

Graph each of the following over the given interval. In each case, label the axes accurately and state the period for each graph.

29. $y = -\cot 2x, 0 \leq x \leq \pi$

30. $y = -\tan 4x, 0 \leq x \leq \pi$

31. $y = -2 \csc 3x, 0 \leq x \leq 2\pi$

32. $y = -2 \sec 3x, 0 \leq x \leq 2\pi$

Use your graphing calculator to graph each pair of functions together for $-2\pi \leq x \leq 2\pi$. (Make sure your calculator is set to radian mode.)

33. **a.** $y = \tan x, y = 2 + \tan x$
b. $y = \tan x, y = -2 + \tan x$
c. $y = \tan x, y = -\tan x$

34. **a.** $y = \cot x, y = 5 + \cot x$
b. $y = \cot x, y = -5 + \cot x$
c. $y = \cot x, y = -\cot x$

35. **a.** $y = \sec x, y = 1 + \sec x$
b. $y = \sec x, y = -1 + \sec x$
c. $y = \sec x, y = -\sec x$

36. **a.** $y = \csc x, y = 3 + \csc x$
b. $y = \csc x, y = -3 + \csc x$
c. $y = \csc x, y = -\csc x$

Use your graphing calculator to graph each pair of functions together for $-2\pi \leq x \leq 2\pi$. (Make sure your calculator is set to radian mode.) How does the value of C affect the graph in each case?

37. **a.** $y = \tan x, y = \tan (x + C)$ for $C = \frac{\pi}{6}$
b. $y = \tan x, y = \tan (x + C)$ for $C = -\frac{\pi}{6}$

38. **a.** $y = \csc x, y = \csc (x + C)$ for $C = \frac{\pi}{4}$
b. $y = \csc x, y = \csc (x + C)$ for $C = -\frac{\pi}{4}$

Use your answers for Problems 23 through 27 for reference, and graph one complete cycle of each of the following equations.

39. $y = -2 + 3 \sec \frac{1}{2}x$

40. $y = 3 + 3 \csc \frac{1}{2} x$

41. $y = 3 + \frac{1}{2} \cot \frac{\pi}{2} x$

42. $y = -4 + 3 \tan 2x$

Graph one complete cycle for each of the following. In each case, label the axes accurately and state the period and horizontal shift for each graph.

43. $y = \tan \left(x + \frac{\pi}{4}\right)$

44. $y = \tan \left(x - \frac{\pi}{4}\right)$

45. $y = \cot \left(x - \frac{\pi}{4}\right)$

46. $y = \cot \left(x + \frac{\pi}{4}\right)$

47. $y = \tan \left(2x - \frac{\pi}{2}\right)$

48. $y = \tan \left(2x + \frac{\pi}{2}\right)$

Sketch one complete cycle of each of the following by first graphing the appropriate sine or cosine curve and then using the reciprocal relationships.

49. $y = \csc \left(x + \frac{\pi}{4}\right)$

50. $y = \sec \left(x + \frac{\pi}{4}\right)$

51. $y = 2 \sec \left(2x - \frac{\pi}{2}\right)$

52. $y = 2 \csc \left(2x - \frac{\pi}{2}\right)$

53. $y = -3 \csc \left(2x + \frac{\pi}{3}\right)$

54. $y = -3 \sec \left(2x - \frac{\pi}{3}\right)$

Graph one complete cycle for each of the following. In each case, label the axes accurately and state the period, vertical translation, and horizontal translation for each graph.

55. $y = -1 - \tan\left(\frac{1}{2}x + \frac{\pi}{4}\right)$

56. $y = 1 + \tan\left(2x - \frac{\pi}{4}\right)$

57. $y = \frac{3}{2} - \frac{1}{2}\cot\left(\frac{\pi}{2}x - \frac{3\pi}{2}\right)$

58. $y = -\frac{1}{2} - 3\cot\left(\pi x + \frac{\pi}{2}\right)$

59. $y = \frac{1}{3} + \csc\left(3x - \frac{\pi}{2}\right)$

60. $y = -2 + \sec\left(\frac{1}{2}x - \frac{\pi}{3}\right)$

61. $y = -3 - 2\sec\left(\pi x + \frac{\pi}{3}\right)$

62. $y = 1 - \frac{1}{2}\csc\left(\pi x + \frac{\pi}{4}\right)$

63. Rotating Light Figure 14 shows a lighthouse that is 100 feet from a long straight wall on the beach. The light in the lighthouse rotates through one complete rotation once every 4 seconds. In Problem 23 of Problem Set 3.5, you found the equation that gives d in terms of t to be $d = 100 \tan \frac{\pi}{2}t$. Graph this equation by making a table in which t assumes all multiples of $\frac{1}{2}$ from $t = 0$ to $t = 4$.

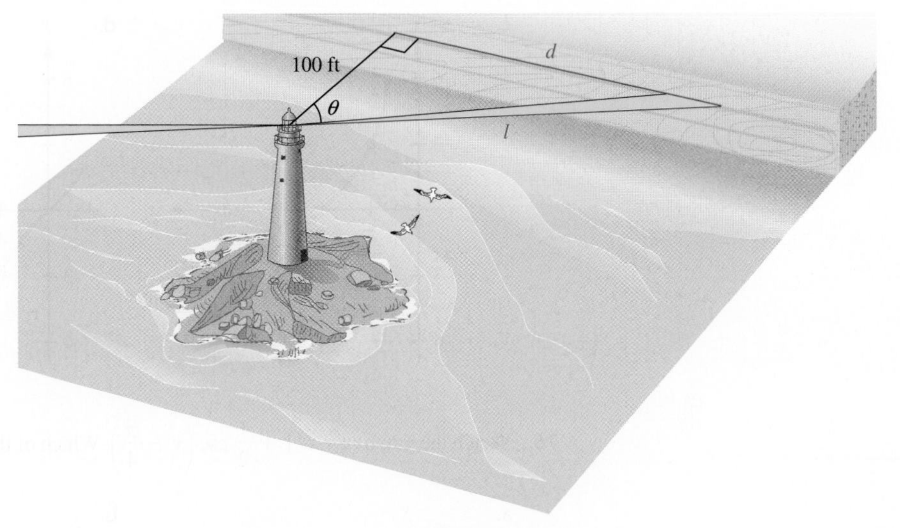

Figure 14

64. Rotating Light In Figure 14, the equation that gives l in terms of time t is $l = 100 \sec \frac{\pi}{2}t$. Graph this equation from $t = 0$ to $t = 4$.

➤ REVIEW PROBLEMS

The problems that follow review material we covered in Sections 2.2 and 3.3. Identify the argument of each function.

65. $\cos 4\theta$

66. $\sin (A - B)$

67. $\cot (2x + \pi)$

68. $\sec\left(\frac{x}{2} + \frac{\pi}{4}\right)$

Use a calculator to approximate each value to four decimal places.

69. $\cos 10$

70. $\cos 10°$

71. $\tan (-25°)$

72. $\tan (-25)$

73. $\csc 16.3$

74. $\csc 16.3°$

➤ LEARNING OBJECTIVES ASSESSMENT

These questions are available for instructors to help assess if you have successfully met the learning objectives for this section.

75. Sketch the graph of $y = -3\tan\left(\frac{1}{2}x\right)$. Which of the following matches your graph?

a.

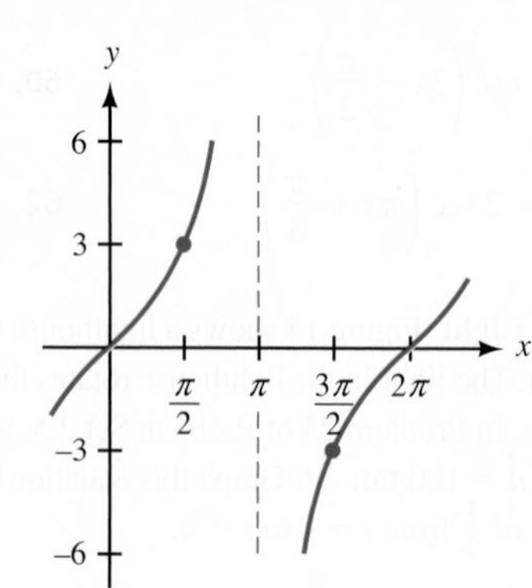

b.

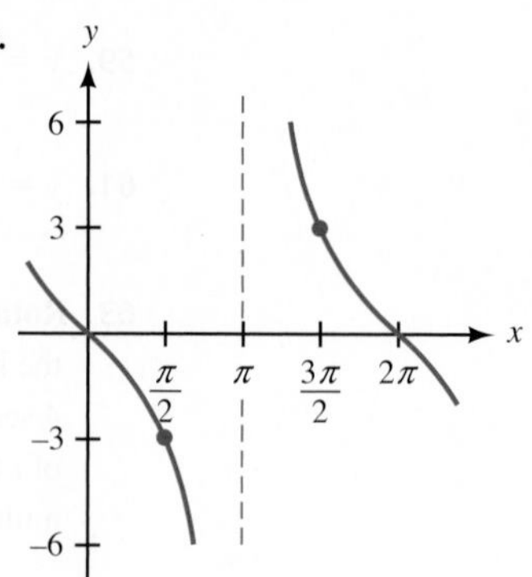

c.

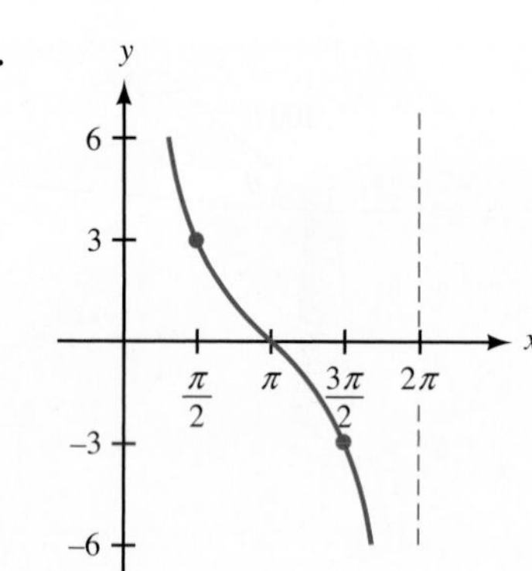

d.

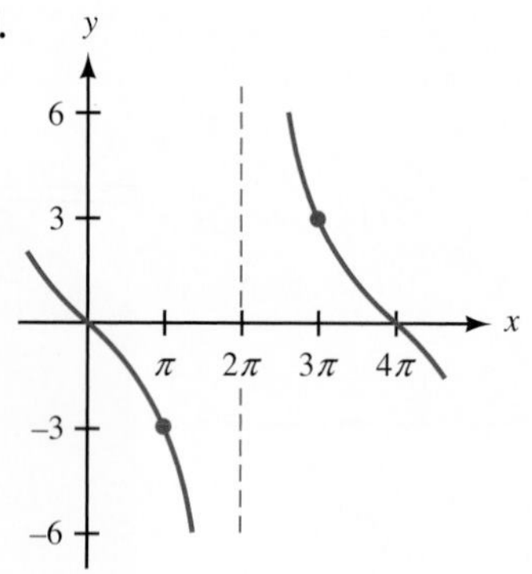

76. Sketch the graph of $y = 1 + \frac{1}{2}\csc\left(x - \frac{\pi}{4}\right)$. Which of the following matches your graph?

a.

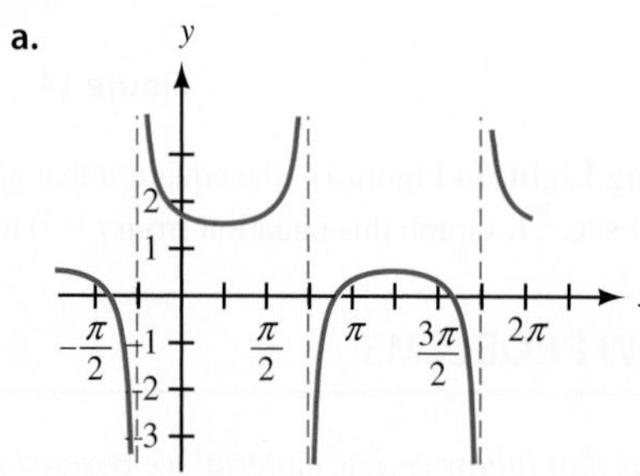

b.

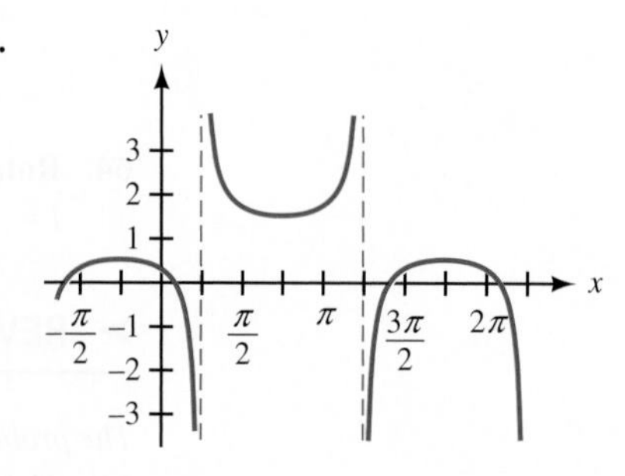

c.

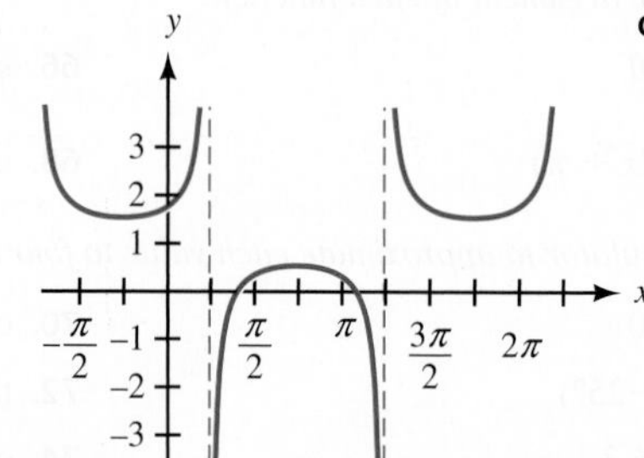

d.

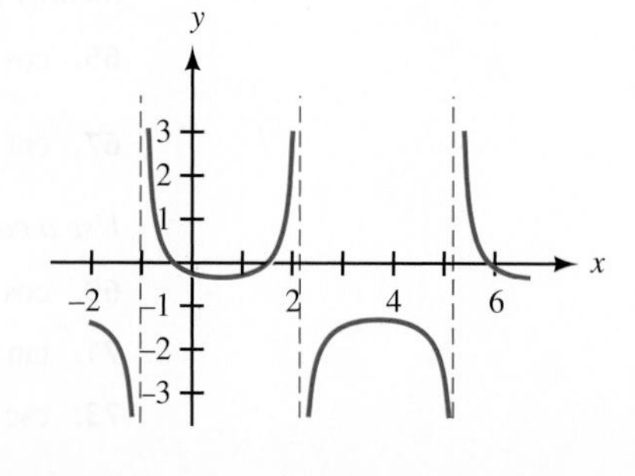

77. Find the period of $y = 3\cot\left(\frac{x}{4}\right)$.

a. π **b.** 4π **c.** 8π **d.** $\frac{\pi}{4}$

78. Find the period of $y = \frac{1}{3}\sec\left(\frac{\pi x}{2} - \frac{\pi}{2}\right)$.

a. 4 **b.** 2 **c.** 6π **d.** $\frac{\pi}{2}$

SECTION 4.5 Finding an Equation from Its Graph

LEARNING OBJECTIVES

1. Find the equation of a line given its graph.
2. Find an equation of a sine or cosine function for a given graph.
3. Find a sinusoidal model for a real-life problem.

In this section, we will reverse what we have done in the previous sections of this chapter and produce an equation that describes a graph, rather than a graph that describes an equation. Let's start with an example from algebra.

EXAMPLE 1 Find the equation of the line shown in Figure 1.

PROBLEM 1
Find the equation of the line shown in Figure 2.

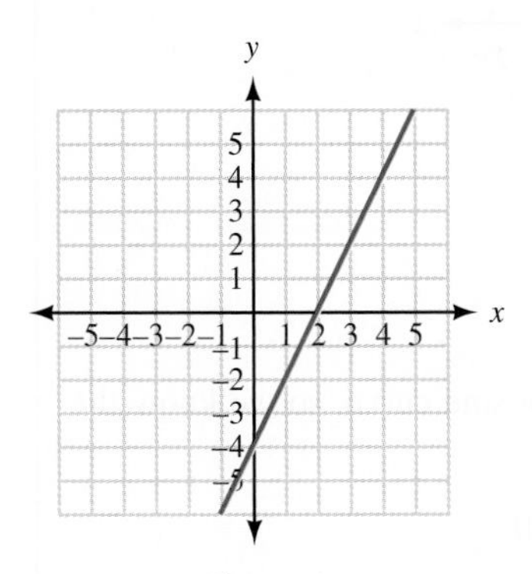

Figure 2

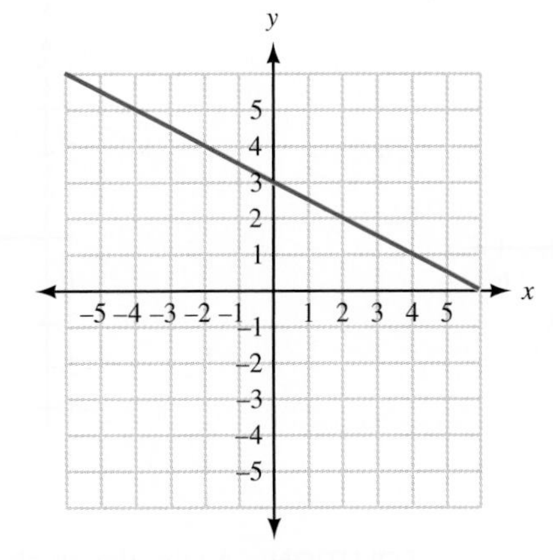

Figure 1

SOLUTION From algebra we know that the equation of any straight line (except a vertical one) can be written in slope-intercept form as

$$y = mx + b$$

where m is the slope of the line, and b is its y-intercept.

Because the line in Figure 1 crosses the y-axis at 3, we know the y-intercept b is 3. To find the slope of the line, we find the ratio of the vertical change to the horizontal change between any two points on the line (sometimes called rise/run). From Figure 1 we see that this ratio is $-1/2$. Therefore, $m = -1/2$. The equation of our line must be

$$y = -\frac{1}{2}x + 3$$

■

PROBLEM 2
Find an equation to match the graph shown in Figure 3.

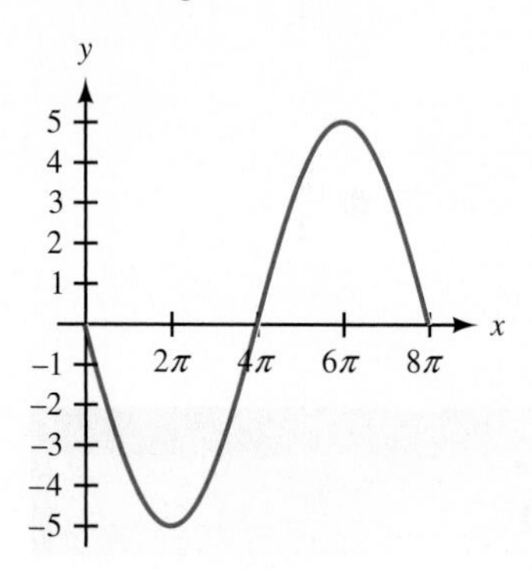

Figure 3

EXAMPLE 2 One cycle of the graph of a trigonometric function is shown in Figure 4. Find an equation to match the graph.

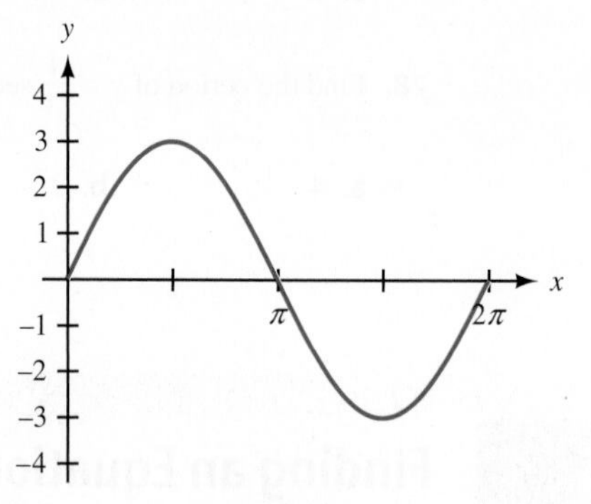

Figure 4

SOLUTION The graph most closely resembles a sine curve with an amplitude of 3, period 2π, and no horizontal shift. The equation is

$$y = 3 \sin x \qquad 0 \leq x \leq 2\pi$$

■

PROBLEM 3
Find an equation to match the graph shown in Figure 5.

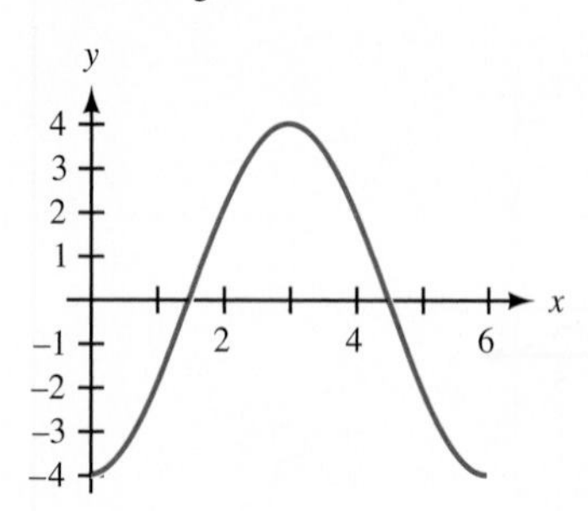

Figure 5

EXAMPLE 3 Find an equation of the graph shown in Figure 6.

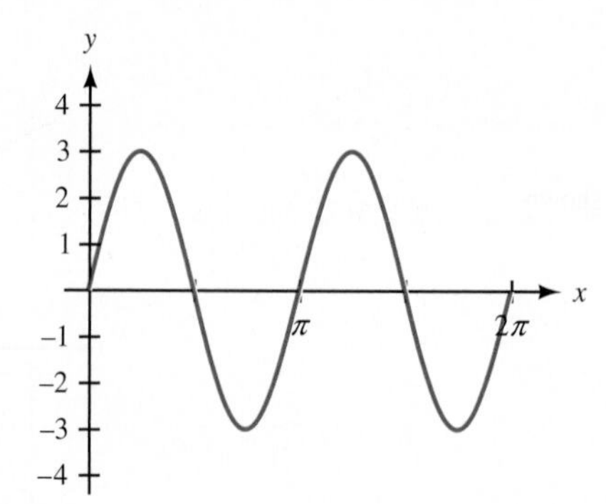

Figure 6

SOLUTION Again, the graph most closely matches a sine curve, so we know the equation will have the form

$$y = k + A \sin (B(x - h))$$

From Figure 6 we see that the amplitude is 3, which means that $A = 3$. There is no horizontal shift, nor is there any vertical translation of the graph. Therefore, both h and k are 0.

To find B, we notice that the period is π. Because the formula for the period is $2\pi/B$, we have

$$\pi = \frac{2\pi}{B}$$

which means that B is 2. Our equation must be

$$y = 0 + 3 \sin (2(x - 0))$$

which simplifies to

$$y = 3 \sin 2x \qquad \text{for} \qquad 0 \leq x \leq 2\pi$$

■

PROBLEM 4
Find an equation to match the graph shown in Figure 7.

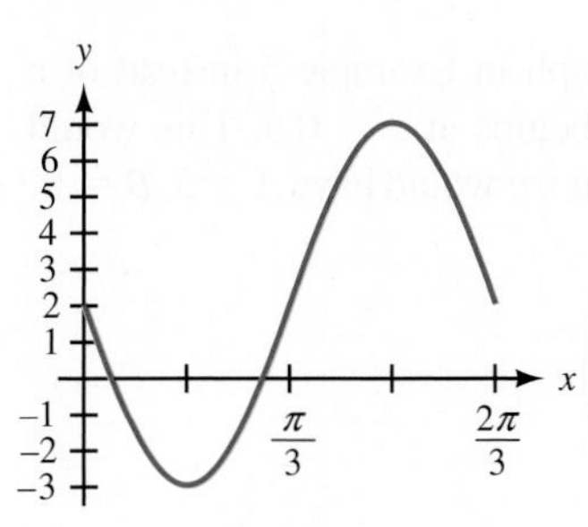

Figure 7

EXAMPLE 4 Find an equation of the graph shown in Figure 8.

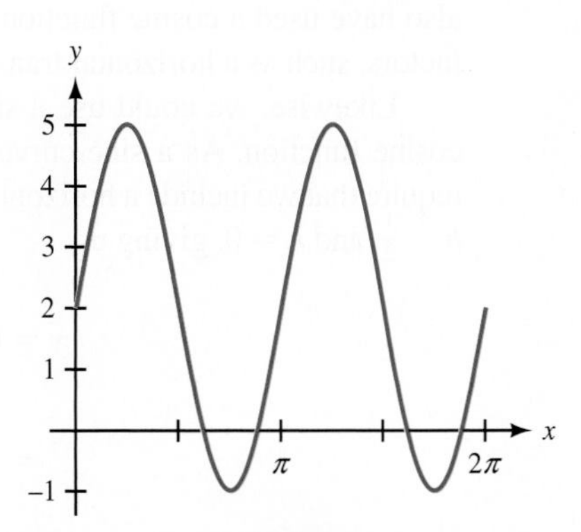

Figure 8

SOLUTION The graph in Figure 8 has the same shape (amplitude, period, and horizontal shift) as the graph shown in Figure 6. In addition, it has undergone a vertical shift up of two units; therefore, the equation is

$$y = 2 + 3 \sin 2x \qquad \text{for} \qquad 0 \leq x \leq 2\pi$$

■

PROBLEM 5
Find an equation to match the graph shown in Figure 9.

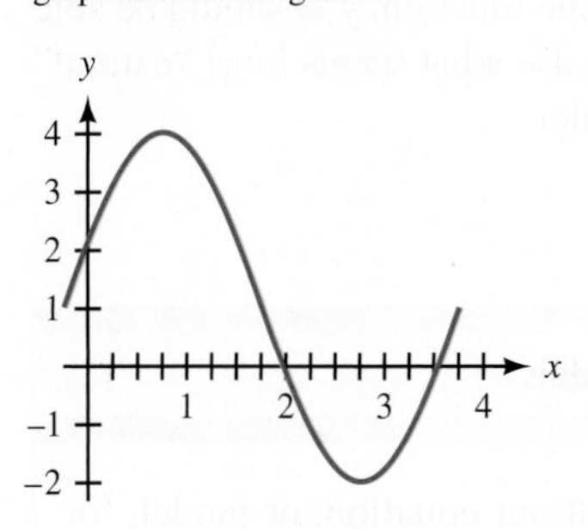

Figure 9

EXAMPLE 5 Find an equation of the graph shown in Figure 10.

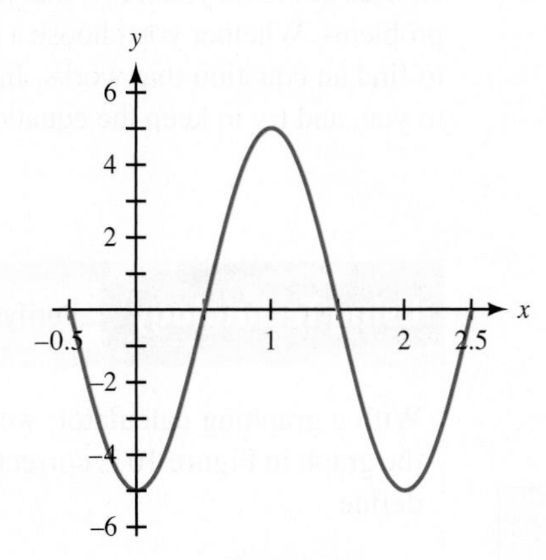

Figure 10

SOLUTION If we look at the graph from $x = 0$ to $x = 2$, it looks like a cosine curve that has been reflected about the x-axis. The general form for a cosine curve is

$$y = k + A \cos (B(x - h))$$

From Figure 10 we see that the amplitude is 5. Because the graph has been reflected about the x-axis, $A = -5$. The period is 2, giving us an equation to solve for B:

$$\text{Period} = \frac{2\pi}{B} = 2 \Rightarrow B = \pi$$

There is no horizontal or vertical translation of the curve (if we assume it is a cosine curve), so h and k are both 0. An equation that describes this graph is

$$y = -5 \cos \pi x \qquad \text{for} \qquad -0.5 \leq x \leq 2.5$$

■

In Examples 2 through 4, a sine function was the most natural choice given the appearance of the graph, and this resulted in the simplest equation possible. We could also have used a cosine function, but it would have required us to consider additional factors, such as a horizontal translation.

Likewise, we could use a sine function for the graph in Example 5 instead of a cosine function. As a sine curve, we see that a cycle begins at $x = 0.5$. This would require that we include a horizontal shift of $0.5 = \frac{1}{2}$. Then we would have $A = 5$, $B = \pi$, $h = \frac{1}{2}$, and $k = 0$, giving us

$$\begin{aligned} y &= 0 + 5 \sin\left(\pi\left(x - \frac{1}{2}\right)\right) \\ &= 5 \sin\left(\pi x - \frac{\pi}{2}\right) \end{aligned}$$

If we use a reflected sine function, then $A = -5$ and a cycle begins at -0.5. This would give us a horizontal shift of $-\frac{1}{2}$, so

$$\begin{aligned} y &= 0 - 5 \sin\left(\pi\left(x - \left(-\frac{1}{2}\right)\right)\right) \\ &= -5 \sin\left(\pi x + \frac{\pi}{2}\right) \end{aligned}$$

As you can see, there are many possible equations we could find to represent the graph of a trigonometric function. The point is not to overwhelm you with too many choices but to let you know that there is more than one correct answer for each of these problems. Whether you choose a sine function or a cosine function, you should be able to find an equation that works. In making your choice, use what seems most "natural" to you, and try to keep the equation as simple as possible.

Using Technology — Verifying Trigonometric Models

With a graphing calculator, we can easily verify that our equation, or model, for the graph in Figure 10 is correct. First, set your calculator to radian mode, and then define

$$Y_1 = -5 \cos(\pi x)$$

Set the window variables to match the given graph:

$$-0.5 \le x \le 2.5,\ \text{scale} = 0.5;\ -6 \le y \le 6$$

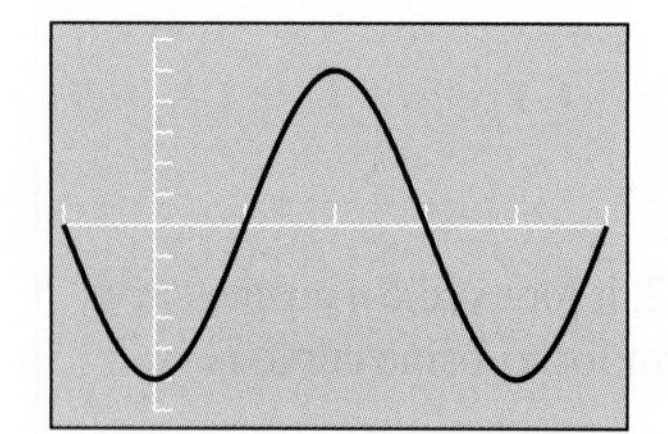

Figure 11

If our equation is correct, the graph drawn by our calculator should be identical to the graph provided. Because the graph shown in Figure 11 is the same as the graph in Figure 10, our equation must be correct.

In our next example, we use a table of values for two variables to obtain a graph. From the graph, we find the equation.

PROBLEM 6
A Ferris wheel with diameter 220 feet completes one revolution every 24 minutes. The bottom of the wheel stands 8 feet above the ground. Find an equation that gives the height of a rider, H, as a function of the time, t, in minutes.

EXAMPLE 6 The Ferris wheel built by George Ferris that we encountered in Chapters 2 and 3 is shown in Figure 12. Recall that the diameter is 250 feet and it rotates through one complete revolution every 20 minutes.

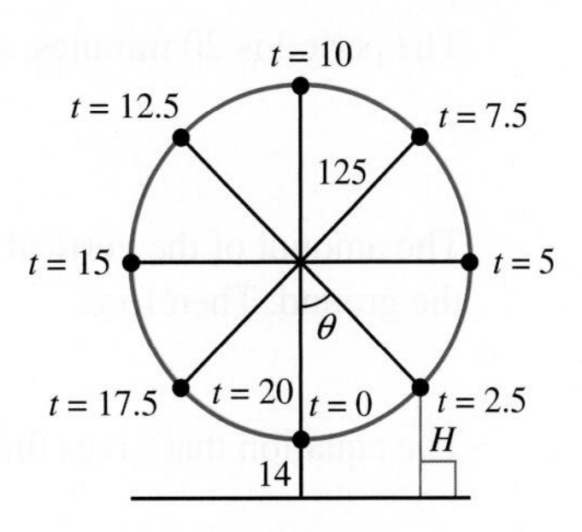

Figure 12

First, make a table that shows the rider's height H above the ground for each of the nine values of t shown in Figure 12. Then graph the ordered pairs indicated by the table. Finally, use the graph to find an equation that gives H as a function of t.

SOLUTION Here is the reasoning we use to find the value of H for each of the given values of t: When $t = 0$ and $t = 20$, the rider is at the bottom of the wheel, 14 feet above the ground. When $t = 10$, the rider is at the top of the wheel, which is 264 feet above the ground. At times $t = 5$ and $t = 15$, the rider is even with the center of the wheel, which is 139 feet above the ground. The other four values of H are found using right triangle trigonometry, as we did in Section 2.3. Table 1 shows our corresponding values of t and H.

TABLE 1

t	H
0 min	14 ft
2.5 min	51 ft
5 min	139 ft
7.5 min	227 ft
10 min	264 ft
12.5 min	227 ft
15 min	139 ft
17.5 min	51 ft
20 min	14 ft

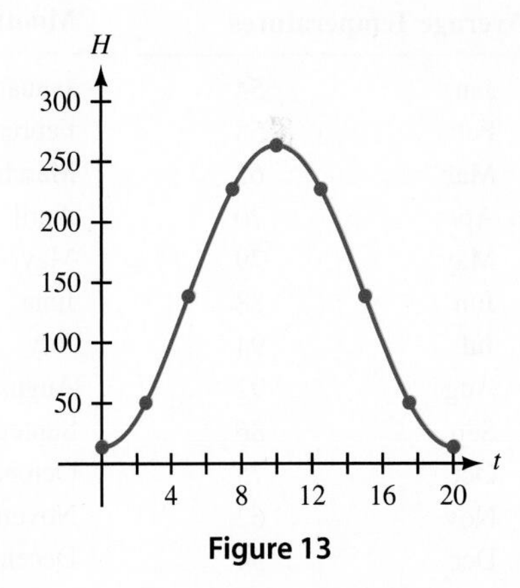

Figure 13

If we graph the points (t, H) on a rectangular coordinate system and then connect them with a smooth curve, we produce the diagram shown in Figure 13. The curve is a cosine curve that has been reflected about the t-axis and then shifted up vertically.

The curve is a cosine curve, so the equation will have the form

$$H = k + A\cos(B(x - h))$$

To find the equation for the curve in Figure 13, we must find values for k, A, B, and h. We begin by finding h. Because the curve starts at $t = 0$, the horizontal shift is 0, which gives us

$$h = 0$$

The amplitude, half the difference between the highest point and the lowest point on the graph, must be 125 (also equal to the radius of the wheel). However, because the graph is a *reflected* cosine curve, we have

$$A = -125$$

The period is 20 minutes, so we find B with the equation

$$20 = \frac{2\pi}{B} \Rightarrow B = \frac{\pi}{10}$$

The amount of the vertical translation, k, is the distance the center of the wheel is off the ground. Therefore,

$$k = 139$$

The equation that gives the height of the rider at any time t during the ride is

$$H = 139 - 125 \cos\left(\frac{\pi}{10}t\right)$$

Notice that this equation is identical to the one we obtained in Example 6 of Section 3.5. ■

PROBLEM 7
Table 2 shows the average monthly temperatures, in degrees Fahrenheit, for Phoenix, Arizona. Find a trigonometric function that models the data, assuming the months are numbered from 1 to 12. (*Source:* CityRating.com)

TABLE 2

Phoenix, AZ Average Temperatures	
Jan	54
Feb	58
Mar	62
Apr	70
May	79
Jun	88
Jul	94
Aug	92
Sep	86
Oct	75
Nov	62
Dec	54

EXAMPLE 7 Table 3 shows the average monthly attendance at Lake Nacimiento in California. Find the equation of a trigonometric function to use as a model for this data. (*Source:* Nacimiento Water Supply Project: Report on Recreational Use at Lake Nacimiento)

SOLUTION A graph of the data is shown in Figure 14, where y is the average attendance and x is the month, with $x = 1$ corresponding to January.

TABLE 3

Month	Attendance
January	6,500
February	6,600
March	15,800
April	26,000
May	38,000
June	36,000
July	31,300
August	23,500
September	12,000
October	4,000
November	900
December	2,100

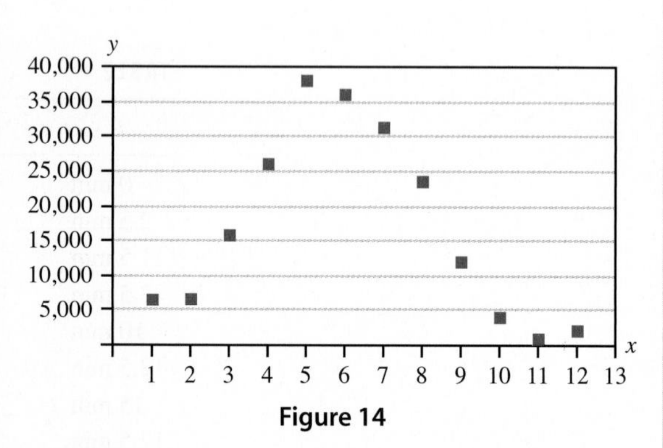

Figure 14

We will use a cosine function to model the data. The general form is

$$y = k + A \cos(B(x - h))$$

To find the amplitude, we calculate half the difference between the maximum and minimum values.

$$A = \frac{1}{2}(38{,}000 - 900) = 18{,}550$$

For the vertical translation k, we average the maximum and minimum values.

$$k = \frac{1}{2}(38{,}000 + 900) = 19{,}450$$

The horizontal distance between the maximum and minimum values is half the period. The maximum value occurs at $x = 5$ and the minimum value occurs at $x = 11$, so

$$\text{Period} = 2(11 - 5) = 12$$

Now we can find B. Because the formula for the period is $2\pi/B$, we have

$$12 = \frac{2\pi}{B} \Rightarrow B = \frac{2\pi}{12} = \frac{\pi}{6}$$

Because a cosine cycle begins at its maximum value, the horizontal shift will be the corresponding x-coordinate of this point. The maximum occurs at $x = 5$, so we have $h = 5$. Substituting the values for A, B, k, and h into the general form gives us

$$y = 19{,}450 + 18{,}550 \cos\left(\frac{\pi}{6}(x - 5)\right)$$

$$= 19{,}450 + 18{,}550 \cos\left(\frac{\pi}{6}x - \frac{5\pi}{6}\right), \qquad 0 \le x \le 12$$

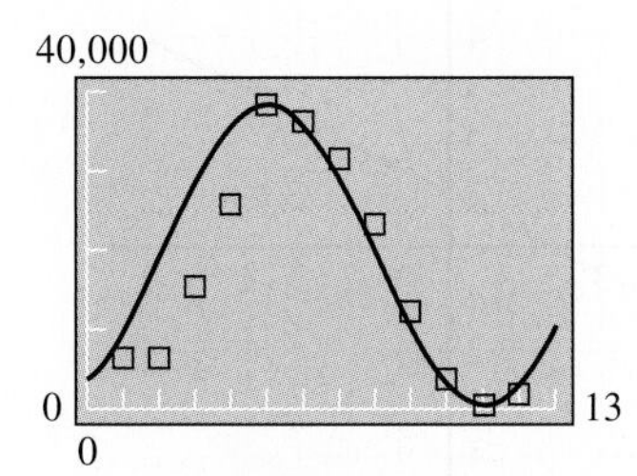

Figure 15

In Figure 15 we have used a graphing calculator to graph this function against the data. As you can see, the model is a good fit for the second half of the year but does not fit the data as well during the spring months.

Getting Ready for Class

After reading through the preceding section, respond in your own words and in complete sentences.

a. How do you find the equation of a line from its graph?
b. How do you find the coefficient A by looking at the graph of $y = A \sin (Bx + C)$?
c. How do you find the period for a sine function by looking at its graph?
d. How do you find the number k by looking at the graph of $y = k + A \sin (Bx + C)$?

4.5 PROBLEM SET

➤ CONCEPTS AND VOCABULARY

For Questions 1 through 4, fill in each blank with the appropriate word.

1. Given a sine or cosine graph, the horizontal distance between a maximum point and the following minimum point is equal to _____ the _________.

2. Given a sine or cosine graph, the vertical distance between a maximum point and a minimum point is equal to _____ the _________.

3. Given a sine or cosine graph, the vertical translation is equal to the _________ of the maximum and minimum values.

4. To find the horizontal shift for a cosine graph, find a _________ point and use the __-coordinate of that point. To find the horizontal shift for a reflected cosine graph, find a _________ point and use the __-coordinate of that point.

➤ EXERCISES

Find the equation of each of the following lines. Write your answers in slope-intercept form, $y = mx + b$.

5.

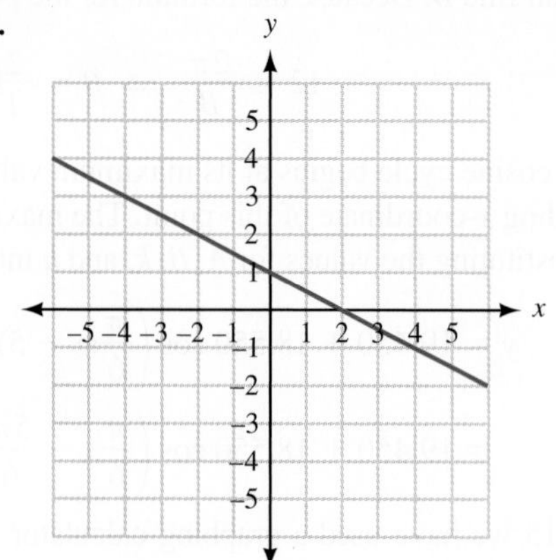

6.

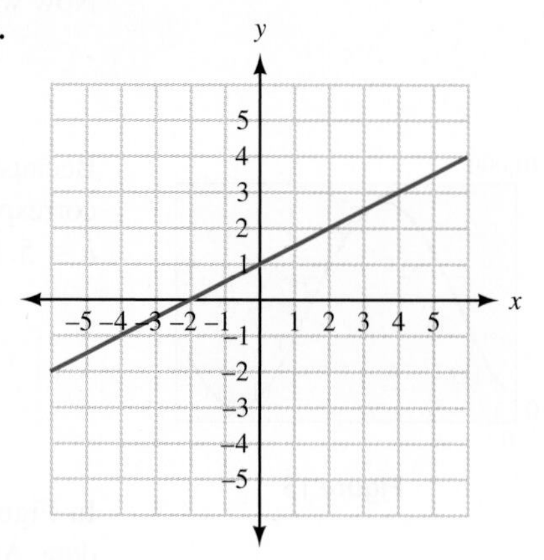

7.

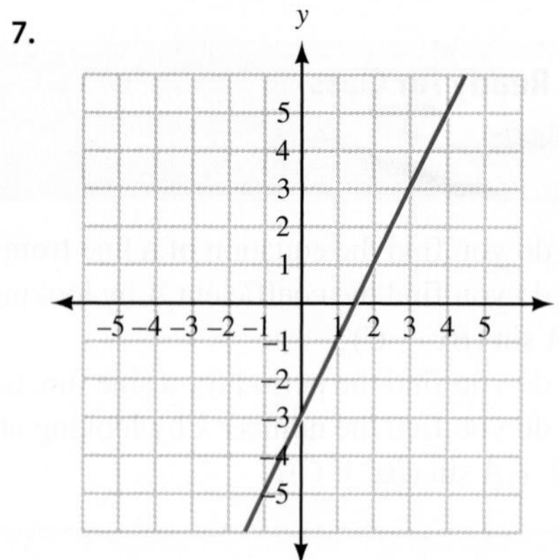

8.

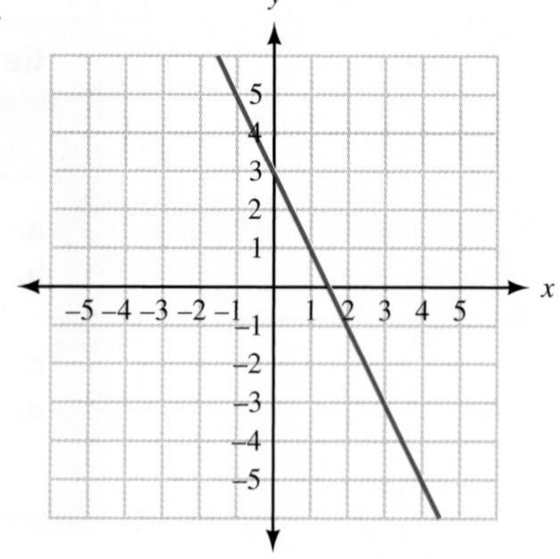

Each of the following graphs shows at least one complete cycle of the graph of an equation containing a trigonometric function. In each case, find an equation to match the graph. If you are using a graphing calculator, graph your equation to verify that it is correct.

9.

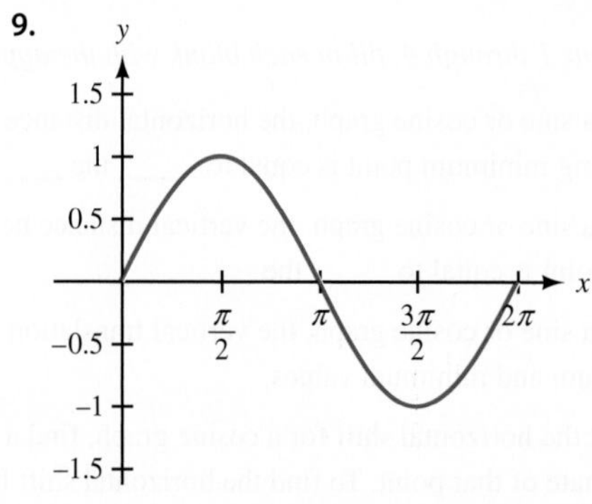

10.

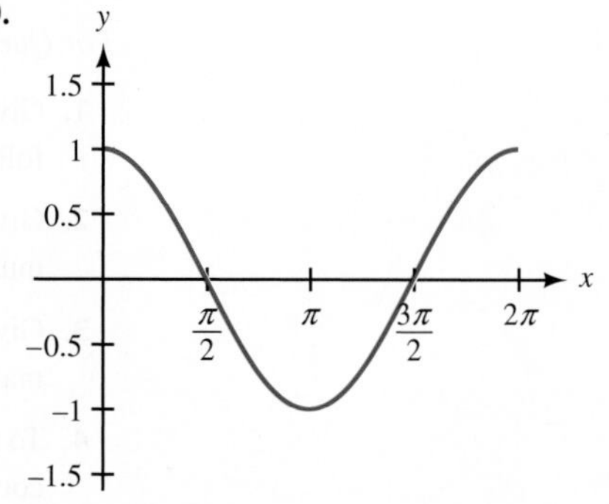

11.

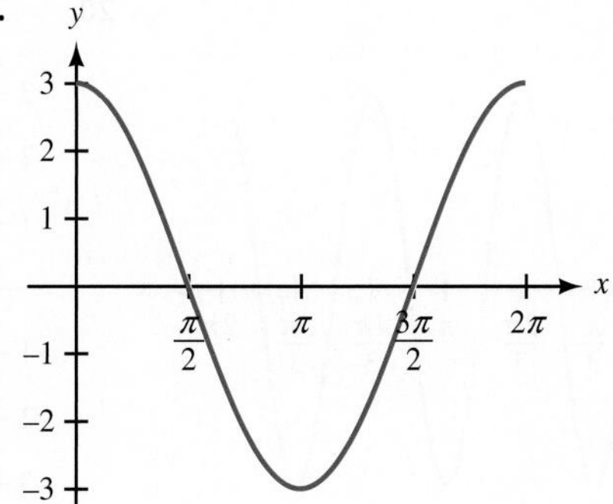

12.

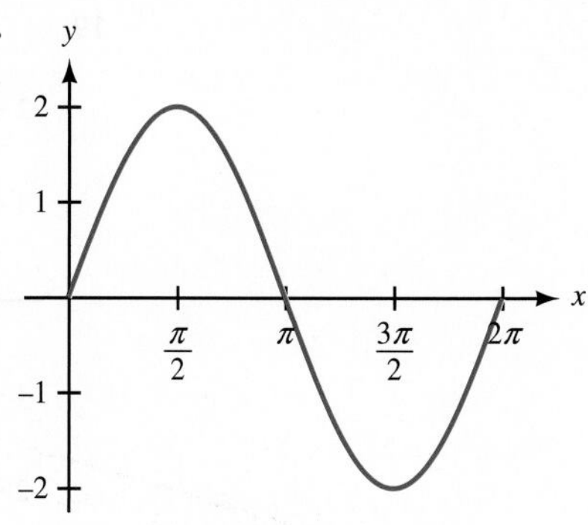

13.

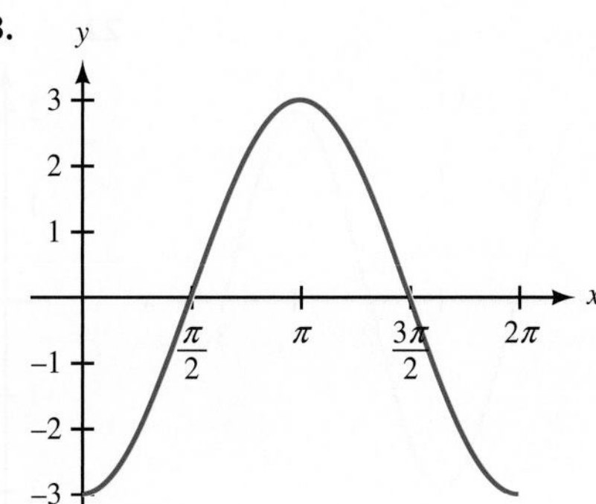

14.

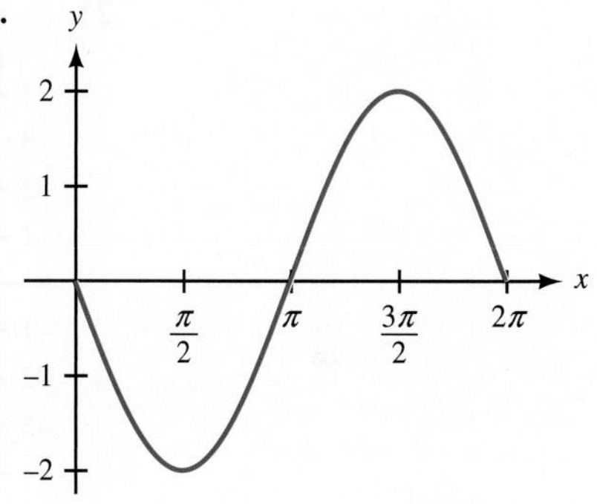

15.

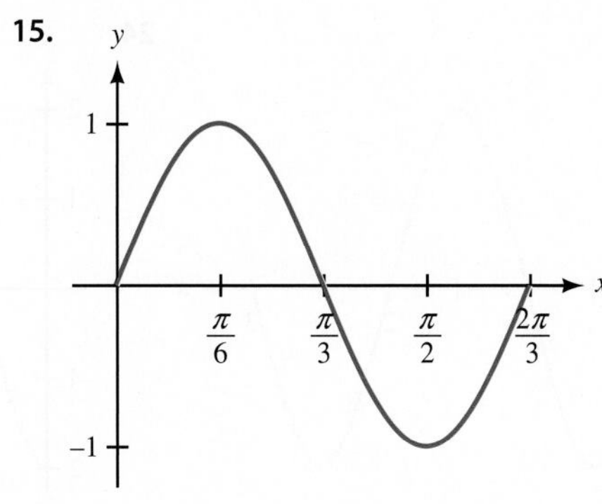

16.

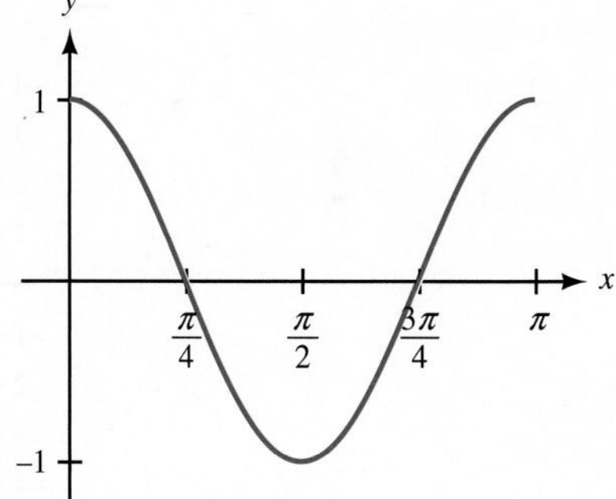

17.

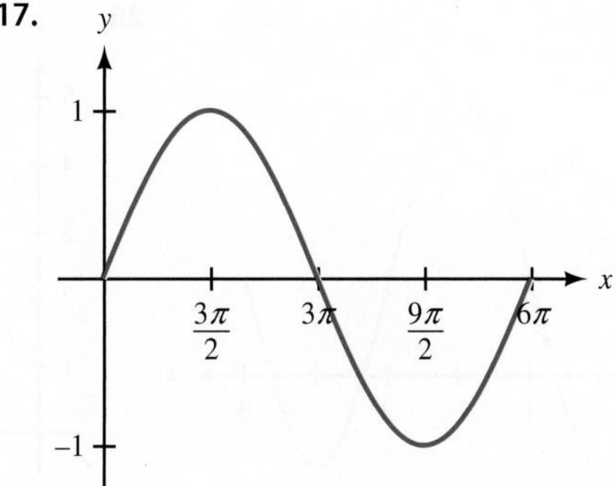

18.

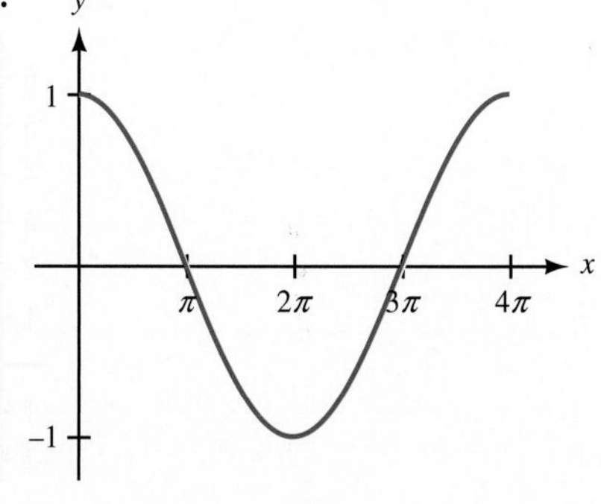

19.

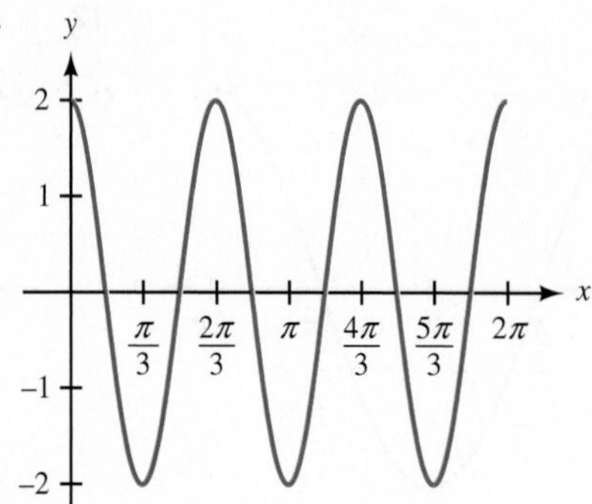

20.

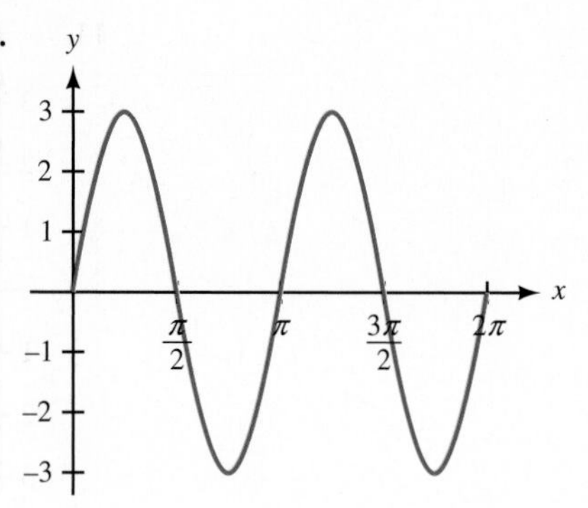

21.

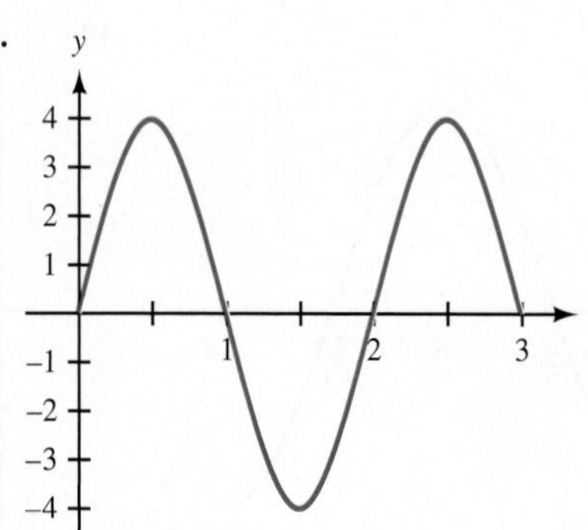

22.

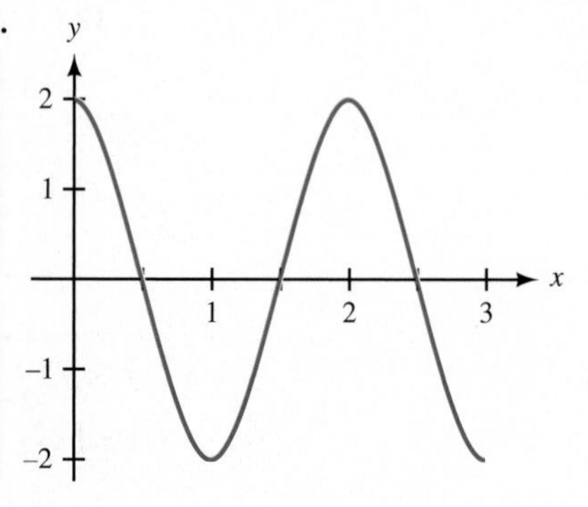

23.

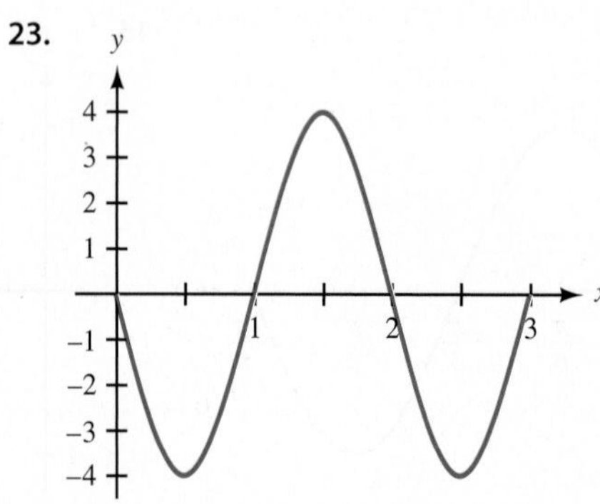

24.

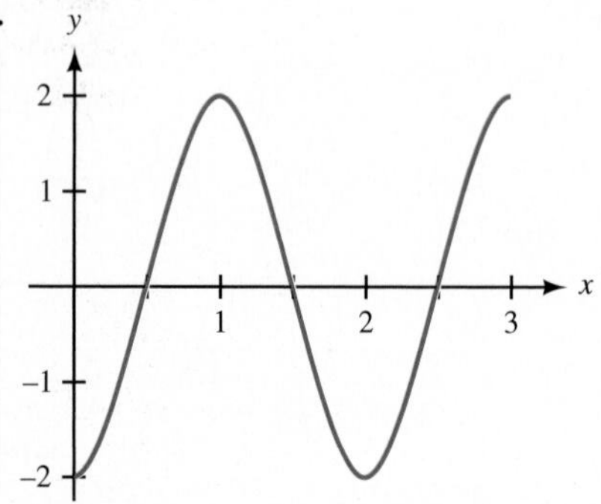

25.

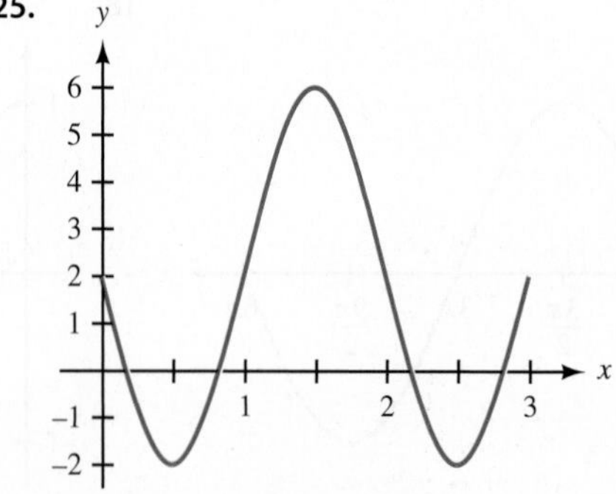

26.

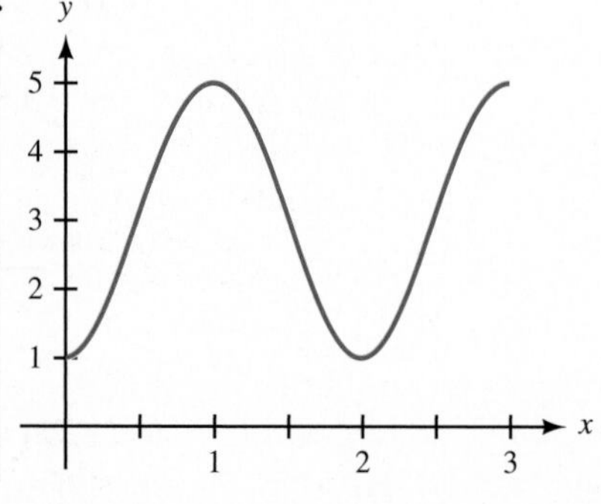

27.

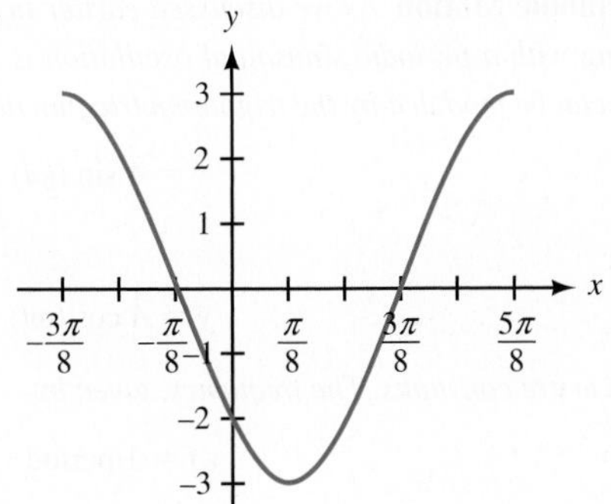

28.

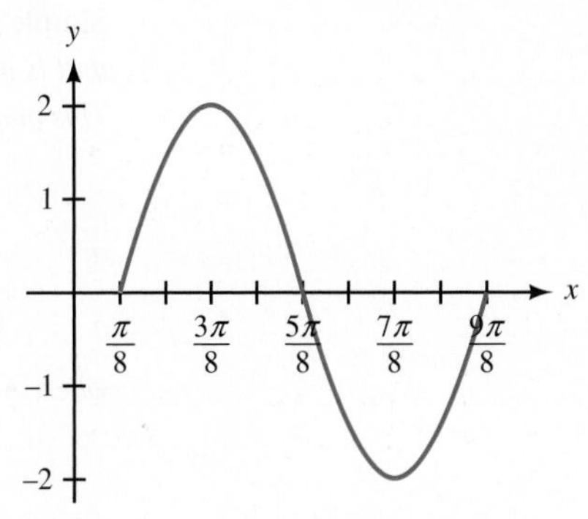

29.

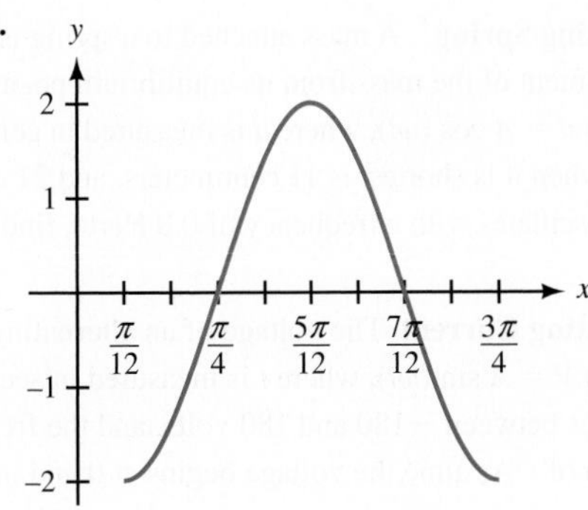

30.

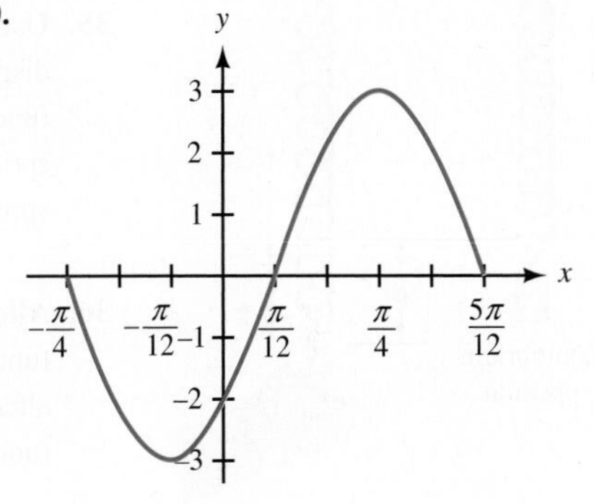

31.

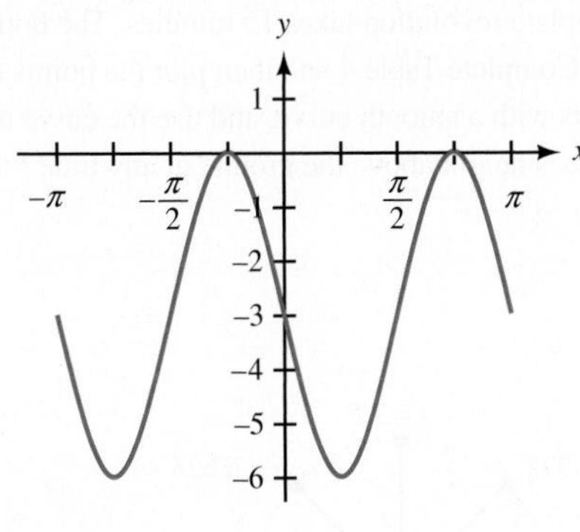

32.

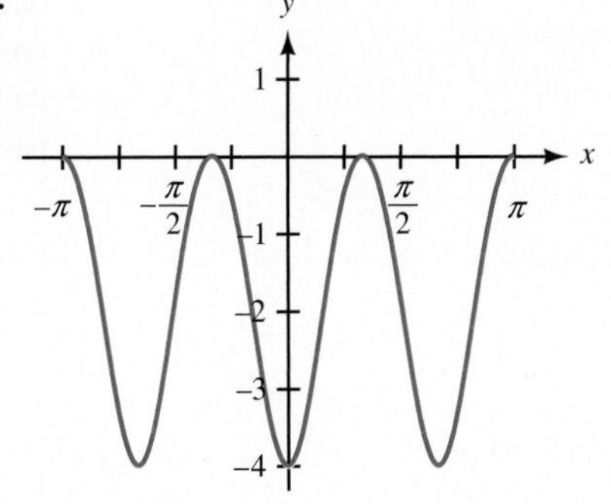

33.

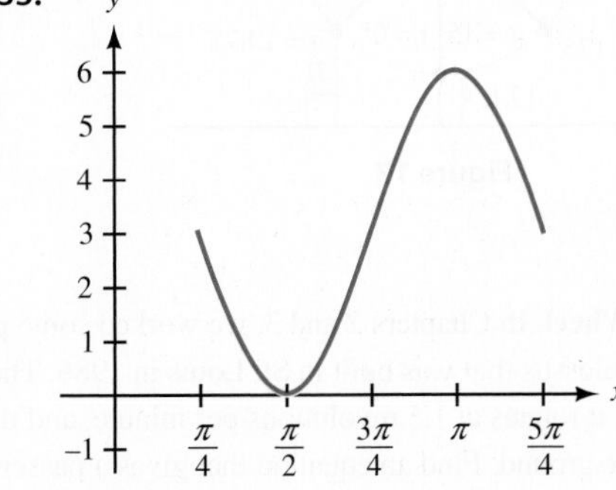

34.

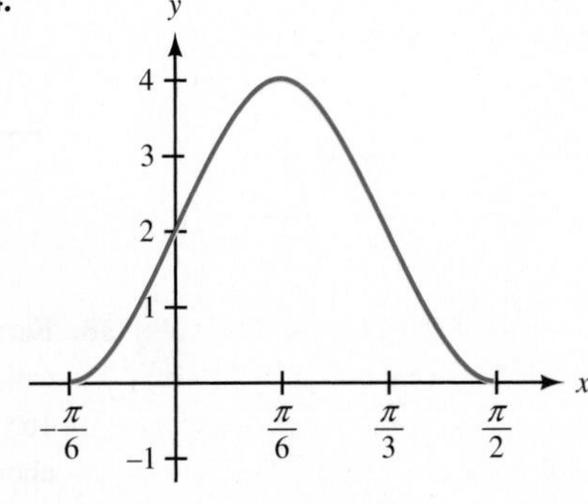

Simple Harmonic Motion *As we discussed earlier in Problem Set 4.2, any object or quantity that is moving with a periodic sinusoidal oscillation is said to exhibit simple harmonic motion. This motion can be modeled by the trigonometric function*

$$y = A \sin(\omega t)$$

or

$$y = A \cos(\omega t)$$

where A and ω are constants. The frequency, given by

$$f = 1/\text{period}$$

represents the number of cycles (or oscillations) that are completed per unit time. The unit used to describe frequency is the Hertz, where 1 Hz = 1 cycle per second.

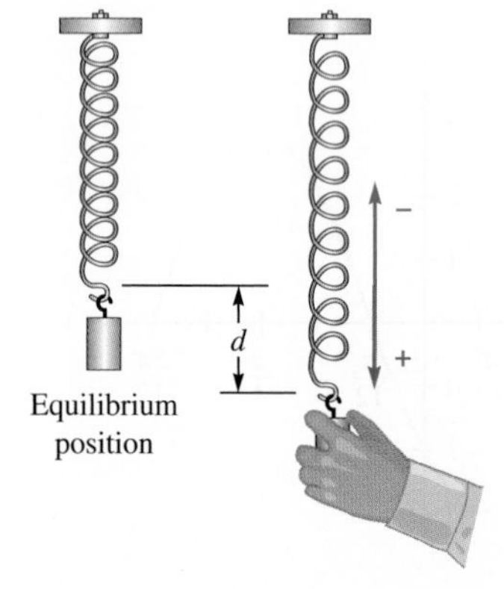

Figure 16

35. Oscillating Spring A mass attached to a spring is pulled downward and released. The displacement of the mass from its equilibrium position after t seconds is given by the function $d = A \cos(\omega t)$, where d is measured in centimeters (Figure 16). The length of the spring when it is shortest is 11 centimeters, and 21 centimeters when it is longest. If the spring oscillates with a frequency of 0.8 Hertz, find d as a function of t.

36. Alternating Current The voltage of an alternating current can be modeled by the function $V = A \sin(\omega t)$, where t is measured in seconds and V in volts. If the voltage alternates between −180 and 180 volts, and the frequency is 70 Hertz, find V as a function of t. Assume the voltage begins at 0 and increases at first.

37. Ferris Wheel Figure 17 is a model of the Ferris wheel known as the Riesenrad that we encountered in Chapters 2 and 3. Recall that the diameter of the wheel is 197 feet, and one complete revolution takes 15 minutes. The bottom of the wheel is 12 feet above the ground. Complete Table 4 and then plot the points (t, H) from the table. Finally, connect the points with a smooth curve, and use the curve to find an equation that will give a passenger's height above the ground at any time t during the ride.

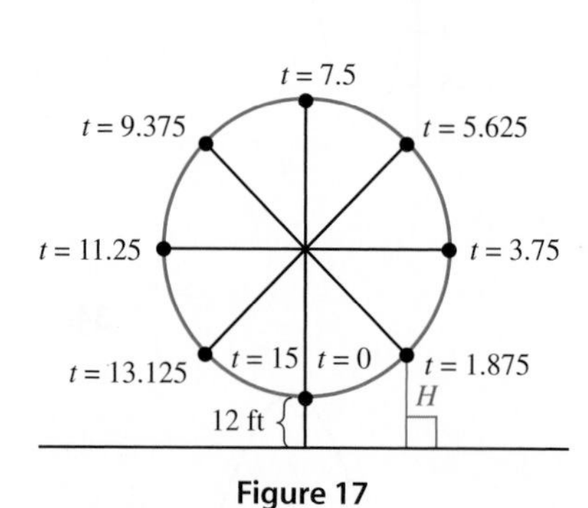

Figure 17

TABLE 4

t	H
0 min	
1.875 min	
3.75 min	
5.625 min	
7.5 min	
9.375 min	
11.25 min	
13.125 min	
15 min	

38. Ferris Wheel In Chapters 2 and 3, we worked some problems involving the Ferris wheel called Colossus that was built in St. Louis in 1986. The diameter of the wheel is 165 feet, it rotates at 1.5 revolutions per minute, and the bottom of the wheel is 9 feet above the ground. Find an equation that gives a passenger's height above the ground at any time t during the ride. Assume the passenger starts the ride at the bottom of the wheel.

TABLE 5

Year	Tornadoes
1986	765
1987	656
1988	702
1989	856
1990	1133
1991	1132
1992	1297
1993	1173
1994	1082

39. Tornadoes Table 5 shows the total number of tornadoes in the United States for various past years. A graph of the data is given in Figure 18, where we have used t to represent the year and n the number of tornadoes. Find the equation of a trigonometric function to use as a model for these data. (*Source*: NOAA Storm Prediction Center)

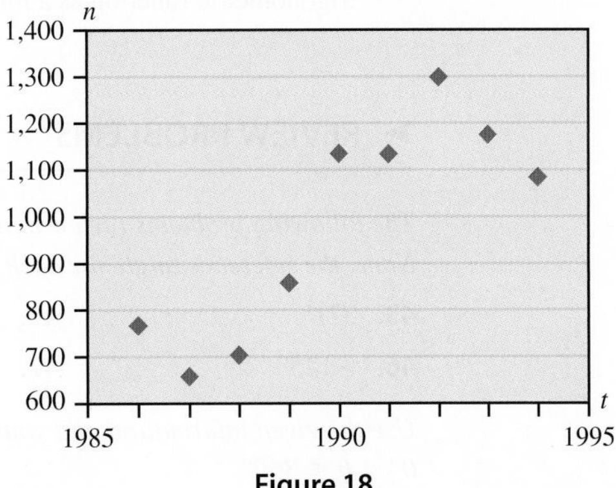

Figure 18

TABLE 6

Day	Hours Daylight
0	9.81
30	10.39
60	11.39
90	12.50
120	13.55
150	14.33
180	14.52
210	14.02
240	13.10
270	12.01
300	10.94
330	10.08
360	9.78

40. Hours of Daylight Table 6 shows the number of hours of daylight for the city of San Luis Obispo, California, at a certain number of days into the year. A graph of the data is given in Figure 19, where we have used d to represent the day (with $d = 1$ corresponding to January 1) and h the number of hours of daylight. Find the equation of a trigonometric function to use as a model for these data.

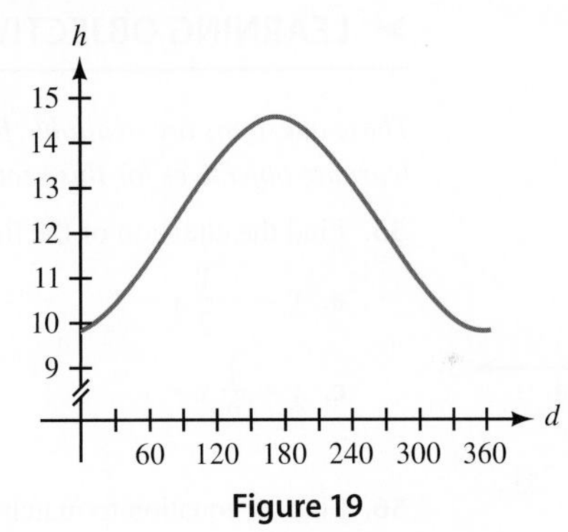

Figure 19

TABLE 7

Month	Average Maximum Temperature (°F)
January	28.6
February	34.0
March	39.6
April	49.4
May	60.4
June	70.0
July	79.6
August	78.3
September	67.8
October	55.7
November	38.7
December	30.5

41. Maximum Temperature Table 7 shows the average maximum temperature in Yellowstone National Park throughout the year. A graph of the data is given in Figure 20, where we have used m to represent the month and T the average maximum temperature. Find the equation of a trigonometric function to use as a model for these data. Assume $m = 1$ corresponds to January.

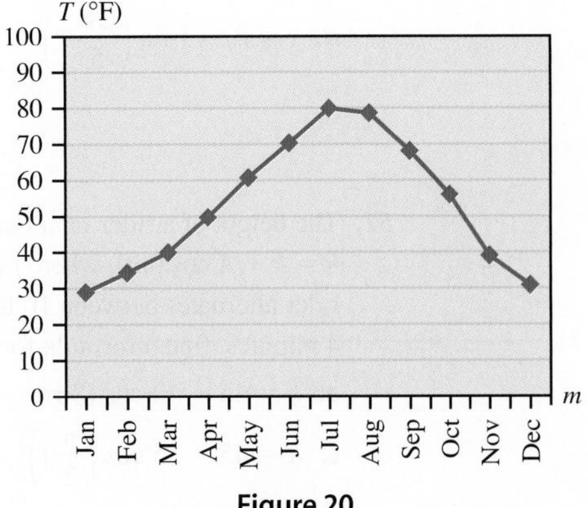

Figure 20

42. The water temperature T at a particular time of the day in March for a lake follows a periodic cycle. The temperature varies between 38°F and 45°F. At 10 a.m. the lake reaches its average temperature and continues to warm until late afternoon. Let t represent the number of hours after midnight, and assume the temperature cycle repeats each day. Using a trigonometric function as a model, write T as a function of t.

➤ REVIEW PROBLEMS

The following problems review material we covered in Section 3.1.
Name the reference angle for each angle below.

43. 321° **44.** 148° **45.** 236°

46. −125° **47.** −276° **48.** 460°

Use the given information and your calculator to find θ to the nearest tenth of a degree if $0° < \theta < 360°$.

49. $\sin\theta = 0.7455$ with θ in QII

50. $\cos\theta = 0.7455$ with θ in QIV

51. $\csc\theta = -2.3228$ with θ in QIII

52. $\sec\theta = -3.1416$ with θ in QIII

53. $\cot\theta = -0.2089$ with θ in QIV

54. $\tan\theta = 0.8156$ with θ in QIII

➤ LEARNING OBJECTIVES ASSESSMENT

These questions are available for instructors to help assess if you have successfully met the learning objectives for this section.

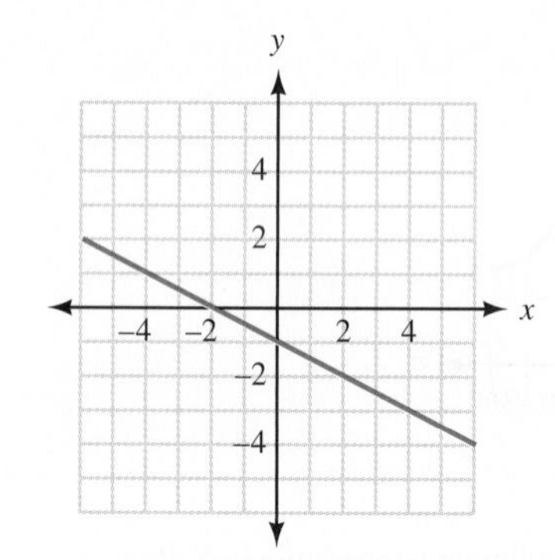

Figure 21

55. Find the equation of the line whose graph is shown in Figure 21.

a. $y = -\frac{1}{2}x - 1$ **b.** $y = -2x - 1$

c. $y = \frac{1}{2}x - 1$ **d.** $y = -2x - 2$

56. Find an equation to match the graph shown in Figure 22.

a. $y = 2 + 3\cos\left(5x + \frac{3}{2}\right)$

b. $y = -\frac{5}{2} + 3\sin\left(\frac{\pi}{5}x + 2\right)$

c. $y = 2 - 3\cos\left(x - \frac{3}{2}\right)$

d. $y = 2 - 3\sin\left(\frac{2\pi}{5}x\right)$

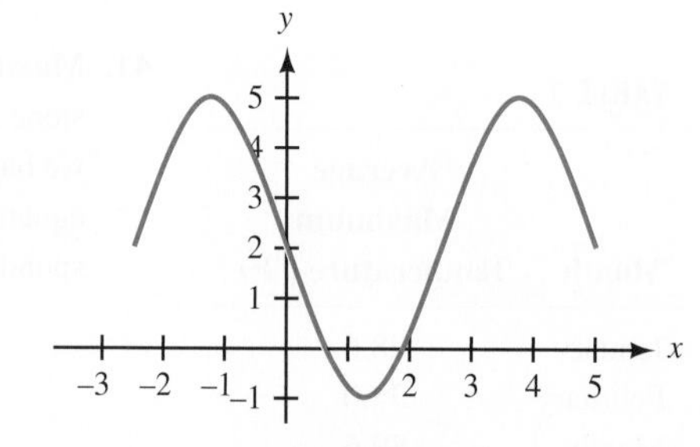

Figure 22

57. The height of a rider on a Ferris wheel can be modeled by the function $h = k - A\cos(\omega t)$, where t is measured in minutes and h in feet. If the height of the rider alternates between 10 feet and 80 feet and the rider completes one revolution every 4 minutes, find a formula for h.

a. $h = 45 - 40\cos(8\pi t)$ **b.** $h = 40 - 45\cos(4t)$

c. $h = 45 - 35\cos\left(\frac{\pi}{2}t\right)$ **d.** $h = 35 - 45\cos\left(\frac{\pi}{2}t\right)$

SECTION 4.6 Graphing Combinations of Functions

LEARNING OBJECTIVES

1. Use addition of y-coordinates to evaluate a function.
2. Use addition of y-coordinates to graph a function.

In this section, we will graph equations of the form $y = y_1 + y_2$, where y_1 and y_2 are algebraic or trigonometric functions of x. For instance, the equation $y = 1 + \sin x$ can be thought of as the sum of the two functions $y_1 = 1$ and $y_2 = \sin x$. That is,

$$\text{if} \quad y_1 = 1 \quad \text{and} \quad y_2 = \sin x$$
$$\text{then} \quad y = y_1 + y_2$$

Using this kind of reasoning, the graph of $y = 1 + \sin x$ is obtained by adding each value of y_2 in $y_2 = \sin x$ to the corresponding value of y_1 in $y_1 = 1$. Graphically, we can show this by adding the values of y from the graph of y_2 to the corresponding values of y from the graph of y_1 (Figure 1). If $y_2 > 0$, then $y_1 + y_2$ will be *above* y_1 by a distance equal to y_2. If $y_2 < 0$, then $y_1 + y_2$ will be *below* y_1 by a distance equal to the absolute value of y_2.

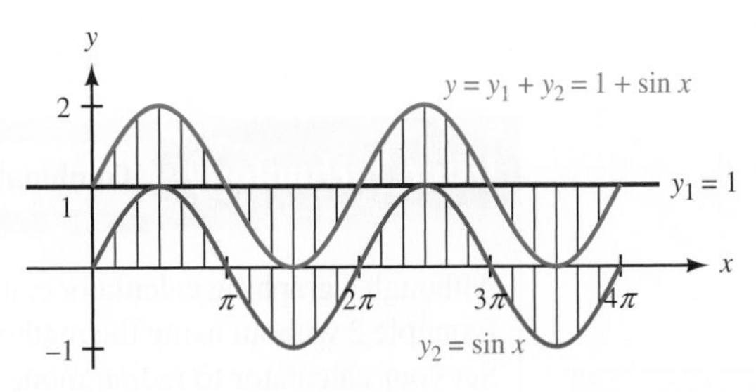

Figure 1

Although in actual practice you may not draw in the little vertical lines we have shown here, they do serve the purpose of allowing us to visualize the idea of adding the y-coordinates on one graph to the corresponding y-coordinates on another graph.

PROBLEM 1

Graph $y = \frac{3}{2}x + \sin x$ for $-2\pi \le x \le 2\pi$.

EXAMPLE 1 Graph $y = \dfrac{1}{3}x - \sin x$ between $x = 0$ and $x = 4\pi$.

SOLUTION We can think of the equation $y = \frac{1}{3}x - \sin x$ as the sum of the equations $y_1 = \frac{1}{3}x$ and $y_2 = -\sin x$. Graphing each of these two equations on the same set of axes and then adding the values of y_2 to the corresponding values of y_1, we have the graph shown in Figure 2.

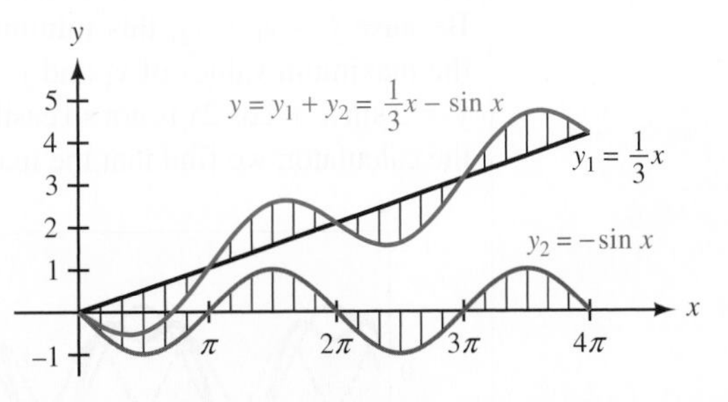

Figure 2

For the rest of the examples in this section, we will not show the vertical lines used to visualize the process of adding y-coordinates. Sometimes the graphs become too confusing to read when the vertical lines are included. It is the idea behind the vertical lines that is important, not the lines themselves.

PROBLEM 2
Graph $y = 2 \sin x - \cos 2x$ for $0 \le x \le 4\pi$.

EXAMPLE 2 Graph $y = 2 \sin x + \cos 2x$ for x between 0 and 4π.

SOLUTION We can think of y as the sum of y_1 and y_2, where

$$y_1 = 2 \sin x \qquad \text{and} \qquad y_2 = \cos 2x$$

The graphs of y_1, y_2, and $y = y_1 + y_2$ are shown in Figure 3.

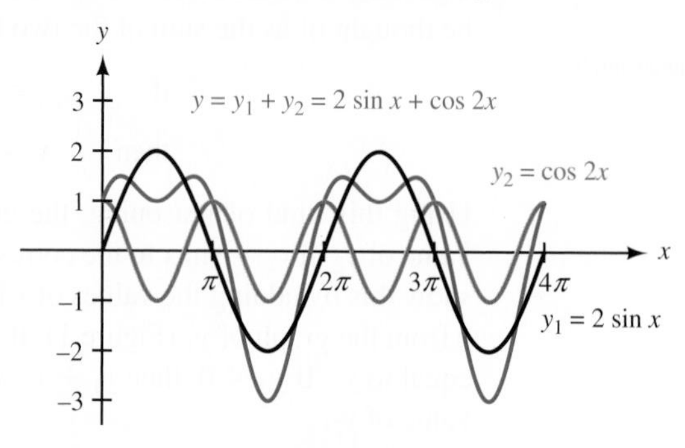

Figure 3

Using Technology Combinations of Functions

Although a graphing calculator can easily graph the function $y = 2 \sin x + \cos 2x$ in Example 2 without using the methods in this section, it can also help reinforce them. Set your calculator to radian mode and enter the following functions:

$$\text{Y}_1 = 2 \sin(x), \text{Y}_2 = \cos(2x), \qquad \text{and} \qquad \text{Y}_3 = \text{Y}_1 + \text{Y}_2$$

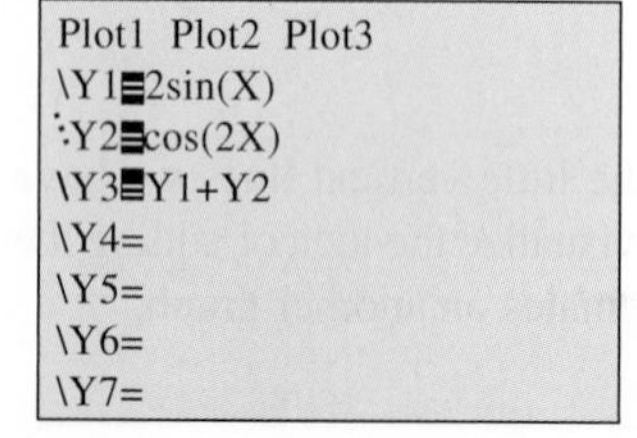

Figure 4

If your calculator has the ability to set different graphing styles, set the first function Y_1 to graph in the normal style, the second function Y_2 in the dot style, and the third function Y_3 in the bold style. Figure 4 shows how the functions would be defined on a TI-84. Now set your window variables so that

$$0 \le x \le 4\pi, \text{ scale} = \pi; \ -4 \le y \le 4$$

When you graph the functions, your calculator screen should look similar to Figure 5. For any x, the dotted graph (Y_2) indicates the distance we should travel above or below the normal graph (Y_1) in order to locate the bold graph (Y_3).

Notice that the period of the function $y = 2 \sin x + \cos 2x$ is 2π. We can also see that the function is at a minimum when the individual component functions $y_1 = 2 \sin x$ and $y_2 = \cos 2x$ are both at their minimum values of -2 and -1. Because $y = y_1 + y_2$, this minimum value will be $(-2) + (-1) = -3$. However, the maximum values of y_1 and y_2 do not occur simultaneously, so the maximum of $y = 2 \sin x + \cos 2x$ is not so easily determined. Using the appropriate command on the calculator, we find that the maximum value is 1.5, as shown in Figure 6.

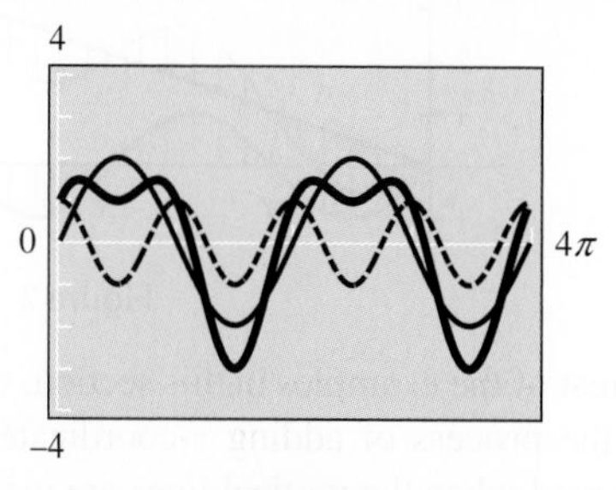

Figure 5

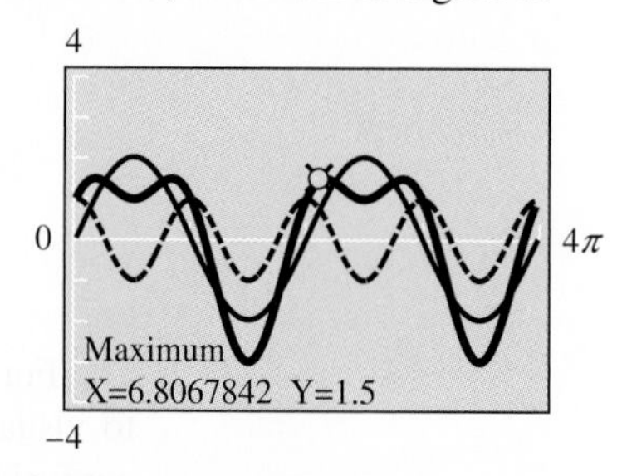

Figure 6

PROBLEM 3
Graph $y = \sin x + \sin 2x$ for $-2\pi \le x \le 2\pi$.

EXAMPLE 3 Graph $y = \cos x + \cos 2x$ for $0 \le x \le 4\pi$.

SOLUTION We let $y = y_1 + y_2$, where

$$y_1 = \cos x \qquad \text{(amplitude 1, period } 2\pi)$$

and

$$y_2 = \cos 2x \qquad \text{(amplitude 1, period } \pi)$$

Figure 7 illustrates the graphs of y_1, y_2, and $y = y_1 + y_2$.

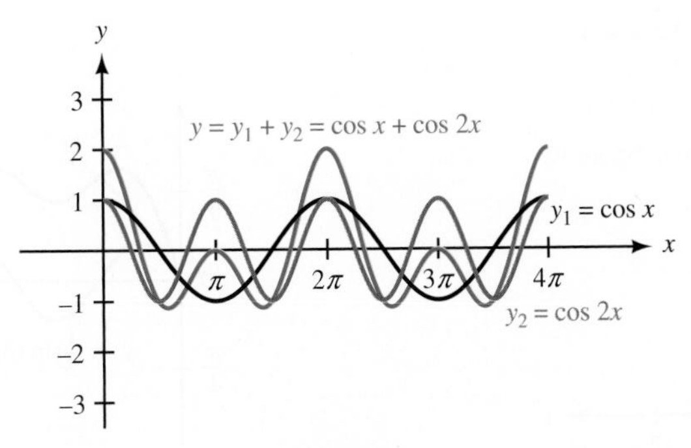

Figure 7

PROBLEM 4
Graph $y = \sin x - \cos x$ for $0 \le x \le 4\pi$.

EXAMPLE 4 Graph $y = \sin x + \cos x$ for x between 0 and 4π.

SOLUTION We let $y_1 = \sin x$ and $y_2 = \cos x$ and graph y_1, y_2, and $y = y_1 + y_2$. Figure 8 illustrates these graphs.

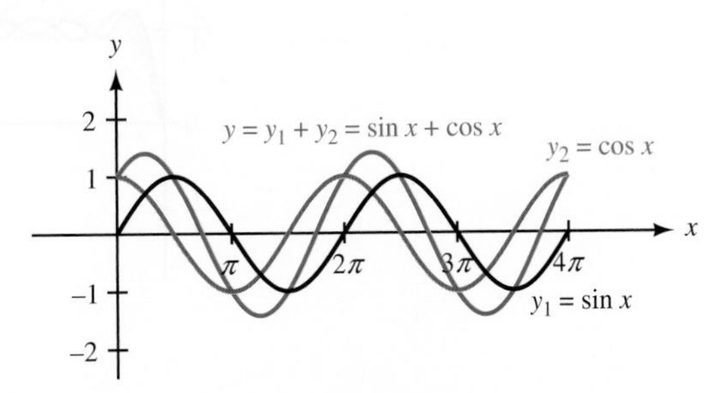

Figure 8

The graph of $y = \sin x + \cos x$ has amplitude $\sqrt{2}$. If we were to extend the graph to the left, we would find it crossed the x-axis at $-\pi/4$. It would then be apparent that the graph of $y = \sin x + \cos x$ is the same as the graph of $y = \sqrt{2}\sin(x + \pi/4)$; both are sine curves with amplitude $\sqrt{2}$ and horizontal shift $-\pi/4$.

One application of combining trigonometric functions can be seen with Fourier series. Fourier series are used in physics and engineering to represent certain waveforms as an infinite sum of sine and/or cosine functions.

PROBLEM 5

Graph $y = \cos \pi x + \frac{1}{3}\cos 3\pi x$ for $-2 \le x \le 2$.

EXAMPLE 5 The following function, which consists of an infinite number of terms, is called a Fourier series.

$$f(x) = \sin(\pi x) + \frac{\sin(3\pi x)}{3} + \frac{\sin(5\pi x)}{5} + \frac{\sin(7\pi x)}{7} + \cdots, 0 \le x < 2$$

The second partial sum of this series only includes the first two terms. Graph the second partial sum.

NOTE The Fourier series in Example 5 is a series representation of the square wave function

$$f(x) = \begin{cases} \frac{\pi}{4} & 0 \le x < 1 \\ -\frac{\pi}{4} & 1 \le x < 2 \end{cases}$$

whose graph is shown in Figure 10.

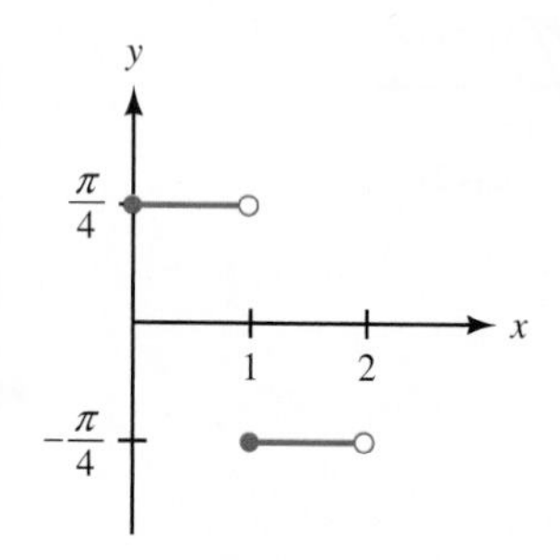

Figure 10

SOLUTION We let $y_1 = \sin(\pi x)$ and $y_2 = \frac{\sin(3\pi x)}{3}$ and graph y_1, y_2, and $y = y_1 + y_2$. Figure 9 illustrates these graphs.

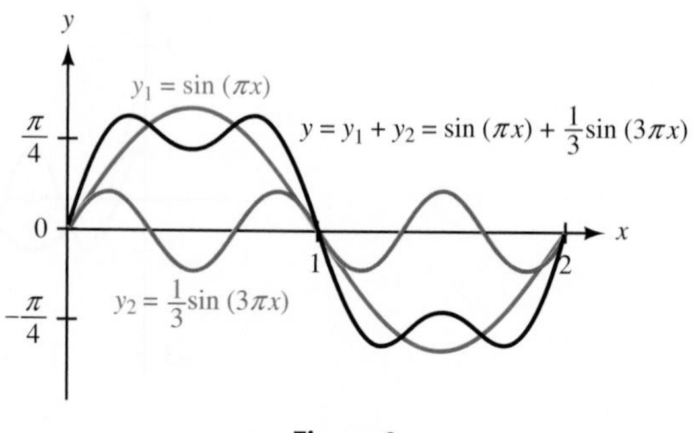

Figure 9

In Figure 11 we have graphed the first, second, third, and fourth partial sums for the Fourier series in Example 5. You can see that as we include more terms from the series, the resulting curve gives a better approximation of the square wave graph shown in Figure 10.

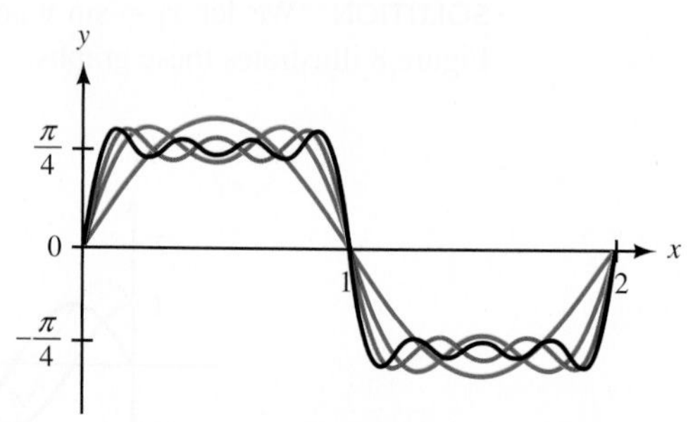

Figure 11

Getting Ready for Class

After reading through the preceding section, respond in your own words and in complete sentences.

a. We can think of the function $y = 1 + \sin x$ as the sum of what two functions?
b. How do you graph the function $y = 1 + \sin x$?
c. How do you graph the function $y = 2\sin x + \cos 2x$?
d. What is the period of the function $y = 2\sin x + \cos 2x$?

4.6 PROBLEM SET

➤ CONCEPTS AND VOCABULARY

For Questions 1 through 4, fill in each blank with the appropriate word or symbol.

1. To graph the sum of two functions $y = y_1 + y_2$, at each point add the ___-coordinate from the point on the y_2 graph to the ____-coordinate of the point on the y_1 graph.

2. If y_2 is positive, then $y_1 + y_2$ will lie _______ y_1 by a distance equal to ____.

3. If y_2 is negative, then $y_1 + y_2$ will lie _______ y_1 by a distance equal to ____.

4. A __________ series can be used to represent certain waveforms as an infinite sum of sine and/or cosine functions.

➤ EXERCISES

For Problems 5 through 8, A and B are points on the graphs of functions y_1 and y_2, respectively. Find the corresponding point on the graph of $y = y_1 + y_2$.

5. $A = (1, 1), B = (1, 2)$

6. $A = (0, 1), B = (0, -2)$

7. $A = (\pi, -0.5), B = (\pi, -1)$

8. $A = \left(\frac{\pi}{4}, -\frac{\sqrt{2}}{2}\right), B = \left(\frac{\pi}{4}, \frac{\sqrt{2}}{2}\right)$

Use addition of y-coordinates to sketch the graph of each of the following between $x = 0$ and $x = 4\pi$.

9. $y = 1 + \sin x$

10. $y = 1 + \cos x$

11. $y = 2 - \cos x$

12. $y = 2 - \sin x$

13. $y = 4 + 2 \sin x$

14. $y = 4 + 2 \cos x$

15. $y = \frac{1}{2}x - \sin x$

16. $y = \frac{1}{3}x - \cos x$

17. $y = \frac{1}{3}x + \cos x$

18. $y = \frac{1}{2}x - \cos x$

Sketch the graph of each equation from $x = 0$ to $x = 8$.

19. $y = x + \sin \pi x$

20. $y = x + \cos \pi x$

Sketch the graph from $x = 0$ to $x = 4\pi$.

21. $y = 3 \sin x + \cos 2x$

22. $y = 3 \cos x + \sin 2x$

23. $y = 2 \cos x - \sin 2x$

24. $y = 2 \sin x - \cos 2x$

25. $y = \sin x + \sin \frac{x}{2}$

26. $y = \cos x + \cos \frac{x}{2}$

27. $y = \cos x + \cos 2x$

28. $y = \sin x + \sin 2x$

29. $y = \sin x + \frac{1}{2} \cos 2x$

30. $y = \cos x + \frac{1}{2} \sin 2x$

31. $y = \sin x - \cos x$

32. $y = \cos x - \sin x$

33. Make a table using multiples of $\pi/2$ for x between 0 and 4π to help sketch the graph of $y = x \sin x$.

34. Sketch the graph of $y = x \cos x$.

Use your graphing calculator to graph each of the following between $x = 0$ and $x = 4\pi$. In each case, show the graph of y_1, y_2, and $y = y_1 + y_2$. (Make sure your calculator is set to radian mode.)

35. $y = 2 + \cos x$

36. $y = 2 + \sin x$

37. $y = x + \cos x$

38. $y = x - \sin x$

39. $y = \sin x - 2\cos x$

40. $y = \cos x + 2\sin x$

41. $y = \sin 2x - 2\sin 3x$

42. $y = \cos 2x + 2\cos 3x$

43. In Example 5 we stated that a square wave can be represented by the Fourier series

$$f(x) = \sin(\pi x) + \frac{\sin(3\pi x)}{3} + \frac{\sin(5\pi x)}{5} + \frac{\sin(7\pi x)}{7} + \cdots, 0 \le x < 2$$

Use your graphing calculator to graph the following partial sums of this series.

a. $f(x) = \sin(\pi x) + \dfrac{\sin(3\pi x)}{3}$

b. $f(x) = \sin(\pi x) + \dfrac{\sin(3\pi x)}{3} + \dfrac{\sin(5\pi x)}{5}$

c. $f(x) = \sin(\pi x) + \dfrac{\sin(3\pi x)}{3} + \dfrac{\sin(5\pi x)}{5} + \dfrac{\sin(7\pi x)}{7}$

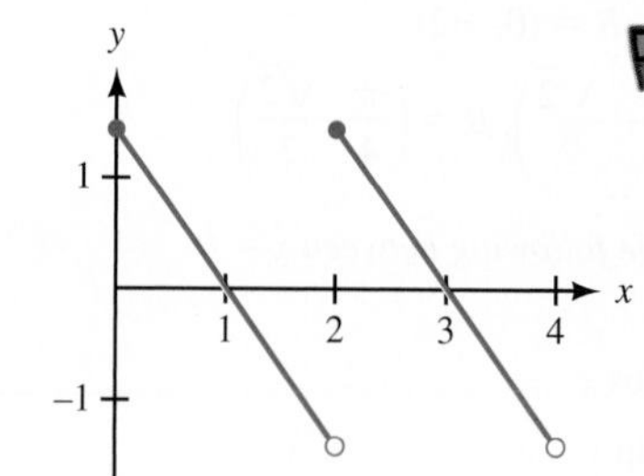

Figure 12

44. The waveform shown in Figure 12, called a *sawtooth wave,* can be represented by the Fourier series

$$f(x) = \sin(\pi x) + \frac{\sin(2\pi x)}{2} + \frac{\sin(3\pi x)}{3} + \frac{\sin(4\pi x)}{4} + \cdots, 0 \le x < 4$$

Use your graphing calculator to graph the following partial sums of this series.

a. $f(x) = \sin(\pi x) + \dfrac{\sin(2\pi x)}{2}$

b. $f(x) = \sin(\pi x) + \dfrac{\sin(2\pi x)}{2} + \dfrac{\sin(3\pi x)}{3}$

c. $f(x) = \sin(\pi x) + \dfrac{\sin(2\pi x)}{2} + \dfrac{\sin(3\pi x)}{3} + \dfrac{\sin(4\pi x)}{4}$

➤ REVIEW PROBLEMS

The following problems review material we covered in Section 3.5.

45. Linear Velocity A point moving on the circumference of a circle covers 5 feet every 20 seconds. Find the linear velocity of the point.

46. Arc Length A point moves at 65 meters per second on the circumference of a circle. How far does the point travel in 1 minute?

47. Arc Length A point is moving with an angular velocity of 3 radians per second on a circle of radius 6 meters. How far does the point travel in 10 seconds?

48. Angular Velocity Convert 30 revolutions per minute (rpm) to angular velocity in radians per second.

49. Linear Velocity A point is rotating at 5 revolutions per minute on a circle of radius 6 inches. What is the linear velocity of the point?

50. Arc Length How far does the tip of a 10-centimeter minute hand on a clock travel in 2 hours?

➤ LEARNING OBJECTIVES ASSESSMENT

These questions are available for instructors to help assess if you have successfully met the learning objectives for this section.

51. If (3, 5) is a point on the graph of a function y_1, and $(3, -1)$ is a point on the graph of a function y_2, then what is the value of $y = y_1 + y_2$ at $x = 3$?

a. -4 **b.** 4 **c.** 6 **d.** -6

52. Given the graphs of y_1 and y_2 shown in Figure 13, use addition of y-coordinates to graph the function $y = y_1 + y_2$.

Figure 13

a.

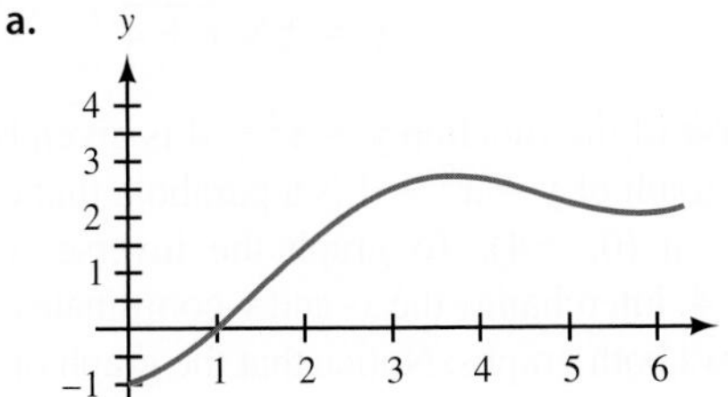

b.

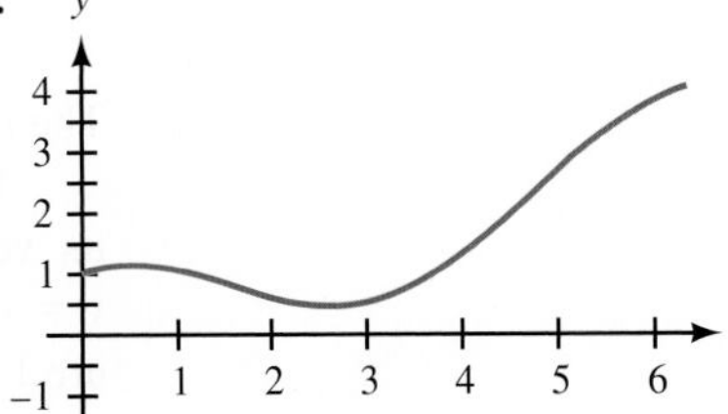

c.

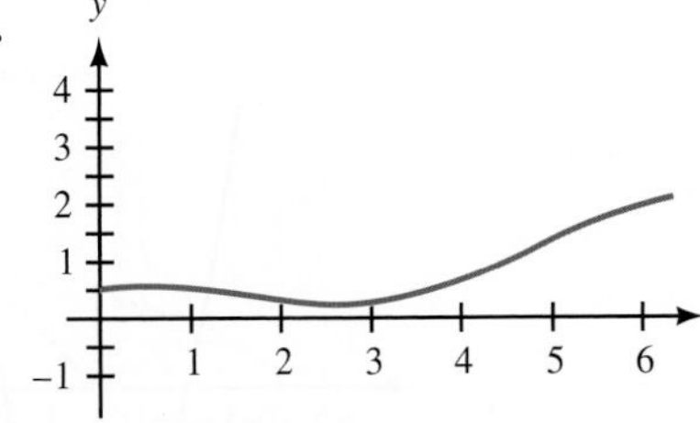

d.

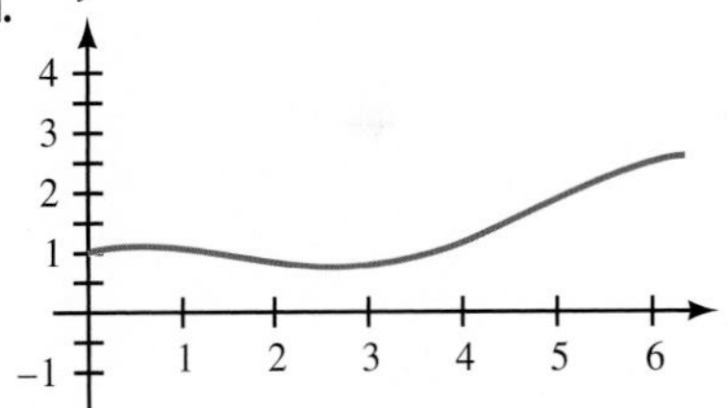

SECTION 4.7 Inverse Trigonometric Functions

LEARNING OBJECTIVES

1. Find the exact value of an inverse trigonometric function.
2. Use a calculator to approximate the value of an inverse trigonometric function.
3. Evaluate a composition involving a trigonometric function and its inverse.
4. Simplify a composition involving a trigonometric and inverse trigonometric function.

In Chapter 2, we encountered situations in which we needed to find an angle given a value of one of the trigonometric functions. This is the reverse of what a trigonometric function is designed to do. When we try to use a function in the reverse direction, we are really using what is called the *inverse* of the function. We begin this section with a brief review of the inverse of a function. If this is a new topic for you, a more thorough presentation of functions and inverse functions is provided in Appendix A.

First, let us review the definition of a function and its inverse.

DEFINITION

A *function* is a rule or correspondence that pairs each element of the domain with exactly one element from the range. That is, a function is a set of ordered pairs in which no two different ordered pairs have the same first coordinate.

The *inverse* of a function is found by interchanging the coordinates in each ordered pair that is an element of the function.

With the inverse, the domain and range of the function have switched roles. To find the equation of the inverse of a function, we simply exchange x and y in the equation and then solve for y. For example, to find the inverse of the function $y = x^2 - 4$, we would proceed like this:

The inverse of	$y = x^2 - 4$	
is	$x = y^2 - 4$	Exchange x and y
or	$y^2 - 4 = x$	
	$y^2 = x + 4$	Add 4 to both sides
	$y = \pm\sqrt{x + 4}$	Take the square root of both sides

The inverse of the function $y = x^2 - 4$ is given by the equation $y = \pm\sqrt{x + 4}$.

The graph of $y = x^2 - 4$ is a parabola that crosses the x-axis at -2 and 2 and has its vertex at $(0, -4)$. To graph the inverse, we take each point on the graph of $y = x^2 - 4$, interchange the x- and y-coordinates, and then plot the resulting point. Figure 1 shows both graphs. Notice that the graph of the inverse is a reflection of the graph of the original function about the line $y = x$.

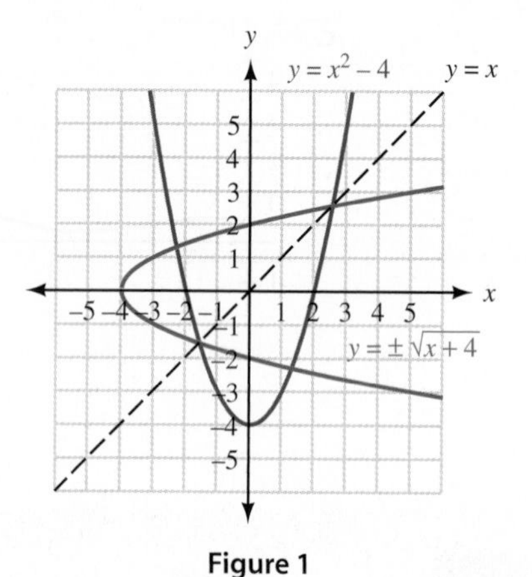

Figure 1

From Figure 1 we see that the inverse of $y = x^2 - 4$ is not a function because the graph of $y = \pm\sqrt{x + 4}$ does not pass the vertical line test. For the inverse to also be a function, the graph of the original function must pass the *horizontal* line test. Functions with this property are called *one-to-one* functions. If a function is one-to-one, then we know its inverse will be a function as well.

INVERSE FUNCTION NOTATION

If $y = f(x)$ is a one-to-one function, then the inverse of f is also a function and can be denoted by $y = f^{-1}(x)$.

Because the graphs of all six trigonometric functions do not pass the horizontal line test, the inverse relations for these functions will not be functions themselves. However, we will see that it is possible to define an inverse that is a function if we restrict the original trigonometric function to certain angles. In this section, we will limit our discussion of inverse trigonometric functions to the inverses of the three major functions: sine, cosine, and tangent. The other three inverse trigonometric functions can be handled with the use of reciprocal identities.

The Inverse Sine Relation

To find the inverse of $y = \sin x$, we interchange x and y to obtain

$$x = \sin y$$

This is the equation of the inverse sine relation.

To graph $x = \sin y$, we simply reflect the graph of $y = \sin x$ about the line $y = x$, as shown in Figure 2.

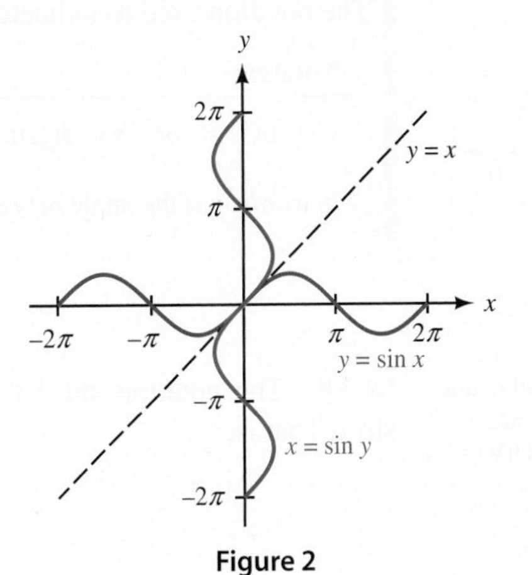

Figure 2

As you can see from the graph, $x = \sin y$ is a relation but not a function. For every value of x in the domain, there are many values of y. The graph of $x = \sin y$ fails the vertical line test.

The Inverse Sine Function

If the function $y = \sin x$ is to have an inverse that is also a function, it is necessary to restrict the values that x can assume so that we may satisfy the horizontal line test. The interval we restrict it to is $-\pi/2 \le x \le \pi/2$. Figure 3 displays the graph of $y = \sin x$ with the restricted interval showing. Notice that this segment of the sine graph passes the horizontal line test, and it maintains the full range of the function $-1 \le y \le 1$. Figure 4 shows the graph of the inverse relation $x = \sin y$ with the restricted interval after the sine curve has been reflected about the line $y = x$.

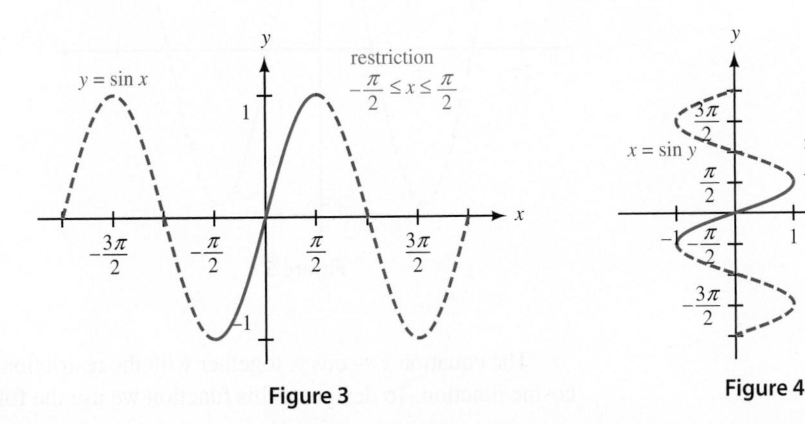

Figure 3

Figure 4

It is apparent from Figure 4 that if $x = \sin y$ is restricted to the interval $-\pi/2 \le y \le \pi/2$, then each value of x between -1 and 1 is associated with exactly one value of y, and we have a function rather than just a relation. The equation $x = \sin y$, together with the restriction $-\pi/2 \le y \le \pi/2$, forms the inverse sine function. To designate this function, we use the following notation.

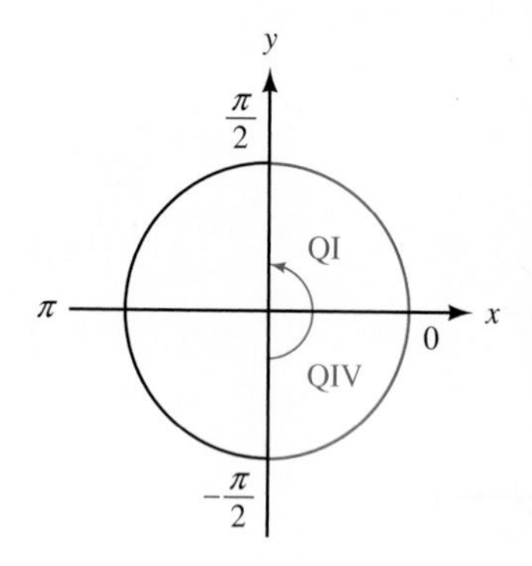

The inverse sine function will return an angle between $-\pi/2$ and $\pi/2$, inclusive, corresponding to QIV or QI.

NOTATION

The notation used to indicate the inverse sine function is as follows:

Notation	Meaning
$y = \sin^{-1} x$ or $y = \arcsin x$	$x = \sin y$ and $-\frac{\pi}{2} \le y \le \frac{\pi}{2}$

In words: y is the angle between $-\pi/2$ and $\pi/2$, inclusive, whose sine is x.

NOTE The notation $\sin^{-1} x$ is not to be interpreted as meaning the reciprocal of $\sin x$. That is,

$$\sin^{-1} x \ne \frac{1}{\sin x}$$

If we want the reciprocal of $\sin x$, we use $\csc x$ or $(\sin x)^{-1}$, but never $\sin^{-1} x$.

The Inverse Cosine Function

Just as we did for the sine function, we must restrict the values that x can assume with the cosine function in order to satisfy the horizontal line test. The interval we restrict it to is $0 \le x \le \pi$. Figure 5 shows the graph of $y = \cos x$ with the restricted interval. Figure 6 shows the graph of the inverse relation $x = \cos y$ with the restricted interval after the cosine curve has been reflected about the line $y = x$.

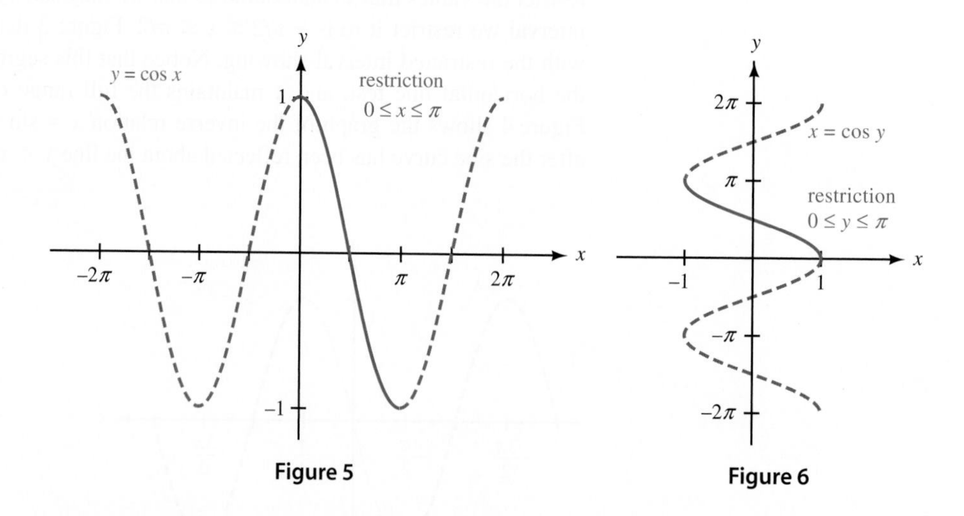

Figure 5

Figure 6

The equation $x = \cos y$, together with the restriction $0 \le y \le \pi$, forms the inverse cosine function. To designate this function we use the following notation.

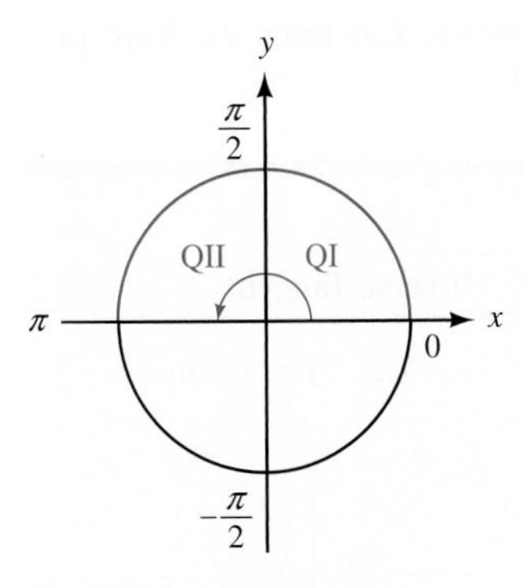

The inverse cosine function will return an angle between 0 and π, inclusive, corresponding to QI or QII.

NOTATION

The notation used to indicate the inverse cosine function is as follows:

Notation	Meaning
$y = \cos^{-1} x$ or $y = \arccos x$	$x = \cos y$ and $0 \le y \le \pi$

In words: y is the angle between 0 and π, inclusive, whose cosine is x.

The Inverse Tangent Function

For the tangent function, we restrict the values that x can assume to the interval $-\pi/2 < x < \pi/2$. Figure 7 shows the graph of $y = \tan x$ with the restricted interval. Figure 8 shows the graph of the inverse relation $x = \tan y$ with the restricted interval after it has been reflected about the line $y = x$.

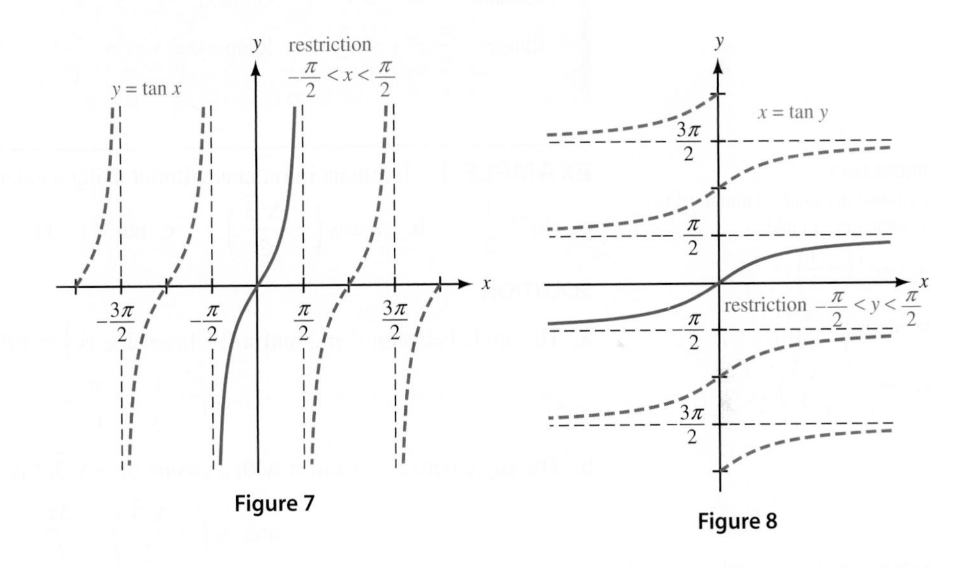

Figure 7

Figure 8

The equation $x = \tan y$, together with the restriction $-\pi/2 < y < \pi/2$, forms the inverse tangent function. To designate this function we use the following notation.

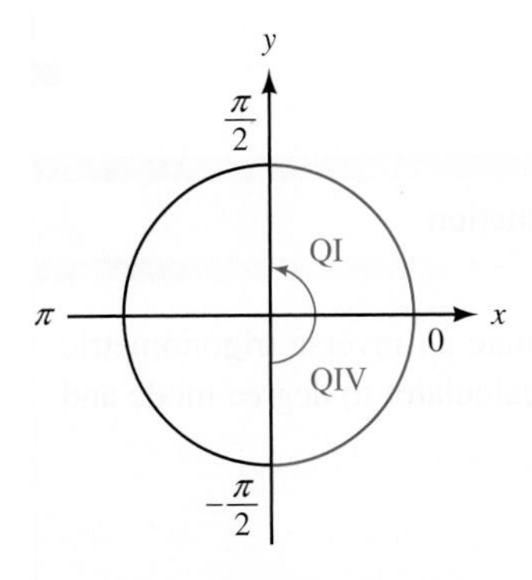

The inverse tangent function will return an angle between $-\pi/2$ and $\pi/2$, corresponding to QIV or QI.

NOTATION

The notation used to indicate the inverse tangent function is as follows:

Notation	Meaning
$y = \tan^{-1} x$ or $y = \arctan x$	$x = \tan y$ and $-\frac{\pi}{2} < y < \frac{\pi}{2}$

In words: y is the angle between $-\pi/2$ and $\pi/2$ whose tangent is x.

To summarize, here are the three inverse trigonometric functions we have presented, along with the domain, range, and graph for each.

INVERSE TRIGONOMETRIC FUNCTIONS

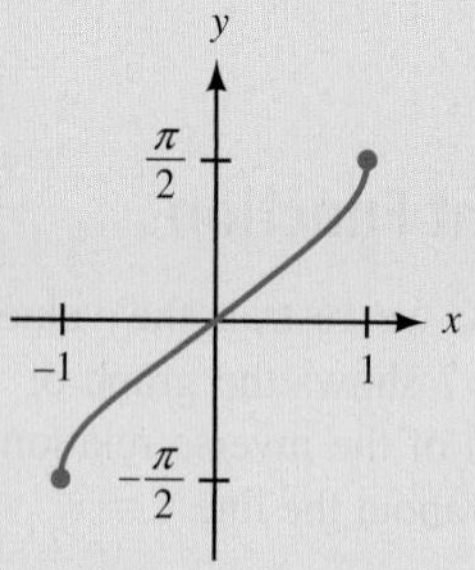

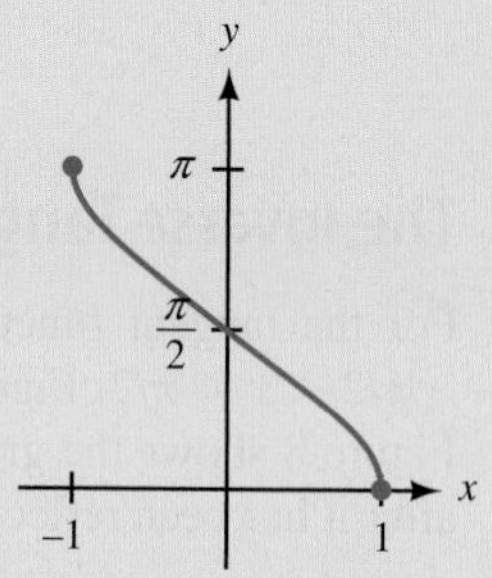

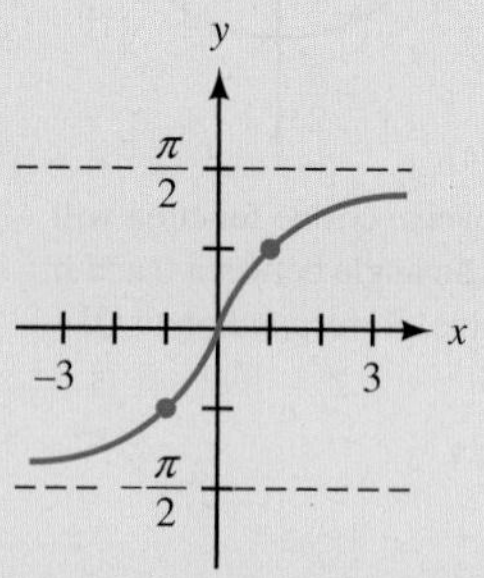

Inverse Sine	Inverse Cosine	Inverse Tangent
$y = \sin^{-1} x = \arcsin x$	$y = \cos^{-1} x = \arccos x$	$y = \tan^{-1} x = \arctan x$
Domain: $-1 \le x \le 1$	Domain: $-1 \le x \le 1$	Domain: all real numbers
Range: $-\frac{\pi}{2} \le y \le \frac{\pi}{2}$	Range: $0 \le y \le \pi$	Range: $-\frac{\pi}{2} < y < \frac{\pi}{2}$

PROBLEM 1
Evaluate in radians without using a calculator or table.

a. $\cos^{-1}\left(-\frac{1}{2}\right)$

b. $\arcsin\left(-\frac{\sqrt{3}}{2}\right)$

c. $\tan^{-1}\left(-\frac{\sqrt{3}}{3}\right)$

EXAMPLE 1 Evaluate in radians without using a calculator or tables.

a. $\sin^{-1} \frac{1}{2}$ **b.** $\arccos\left(-\frac{\sqrt{3}}{2}\right)$ **c.** $\tan^{-1}(-1)$

SOLUTION

a. The angle between $-\pi/2$ and $\pi/2$ whose sine is $\frac{1}{2}$ is $\pi/6$.

$$\sin^{-1} \frac{1}{2} = \frac{\pi}{6}$$

b. The angle between 0 and π with a cosine of $-\sqrt{3}/2$ is $5\pi/6$.

$$\arccos\left(-\frac{\sqrt{3}}{2}\right) = \frac{5\pi}{6}$$

c. The angle between $-\pi/2$ and $\pi/2$ the tangent of which is -1 is $-\pi/4$.

$$\tan^{-1}(-1) = -\frac{\pi}{4}$$

NOTE In part c of Example 1, it would be incorrect to give the answer as $7\pi/4$. It is true that $\tan 7\pi/4 = -1$, but $7\pi/4$ is not between $-\pi/2$ and $\pi/2$. There is a difference.

Using Technology **Graphing the Inverse Sine Function**

A graphing calculator can be used to graph and evaluate an inverse trigonometric function. To graph the inverse sine function, set your calculator to degree mode and enter the function

$$Y_1 = \sin^{-1} x$$

Set your window variables so that

$$-1.5 \le x \le 1.5;\ -120 \le y \le 120,\ \text{scale} = 45$$

The graph of the inverse sine function is shown in Figure 9.

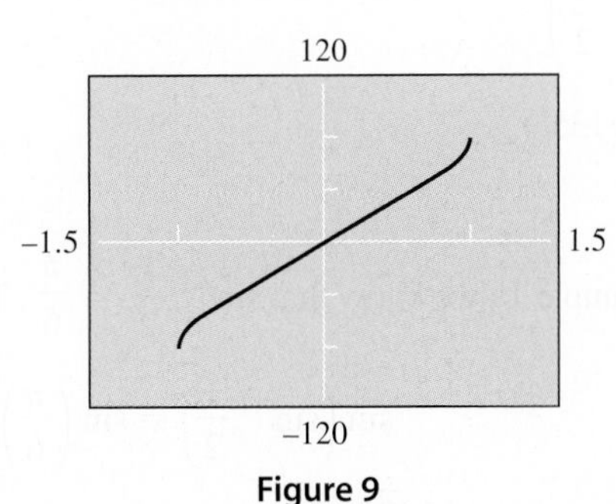

Figure 9

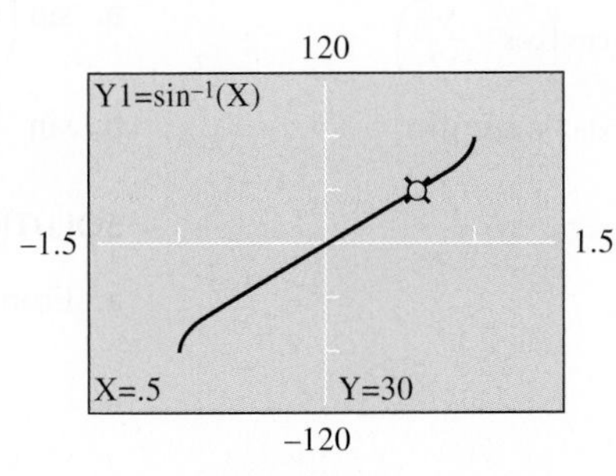

Figure 10

To find $\sin^{-1}\frac{1}{2}$ in Example 1a, use the appropriate command to evaluate the function for $x = 0.5$. As Figure 10 illustrates, the result is $y = 30°$. The angle between $-90°$ and $90°$ having a sine of 1/2 is 30°, or $\pi/6$ in radians.

PROBLEM 2
Use a calculator to find each value to the nearest tenth of a degree.
a. arcsin (0.4328)
b. $\sin^{-1}(-0.4328)$
c. arcos (-0.7219)
d. $\cos^{-1}(0.7219)$
e. $\tan^{-1}(3.825)$
f. arctan (-3.825)

EXAMPLE 2 Use a calculator to evaluate each expression to the nearest tenth of a degree.

a. arcsin (0.5075) **b.** arcsin (-0.5075) **c.** $\cos^{-1}(0.6428)$
d. $\cos^{-1}(-0.6428)$ **e.** arctan (4.474) **f.** arctan (-4.474)

SOLUTION Make sure the calculator is set to degree mode, and then enter the number and press the appropriate key. Scientific and graphing calculators are programmed so that the restrictions on the inverse trigonometric functions are automatic.

a. arcsin (0.5075) = 30.5°
b. arcsin $(-0.5075) = -30.5°$ } Reference angle 30.5°

c. $\cos^{-1}(0.6428) = 50.0°$
d. $\cos^{-1}(-0.6428) = 130.0°$ } Reference angle 50°

e. arctan (4.474) = 77.4°
f. arctan $(-4.474) = -77.4°$ } Reference angle 77.4°

In Example 5 of Section 1.5, we simplified the expression $\sqrt{x^2 + 9}$ using the trigonometric substitution $x = 3 \tan \theta$ to eliminate the square root. We will now see how to simplify the result of that example further by removing the absolute value symbol.

PROBLEM 3
Simplify $2\,|\cos \theta|$ if $\theta = \sin^{-1}\frac{x}{2}$ for some real number x.

EXAMPLE 3 Simplify $3\,|\sec \theta|$ if $\theta = \tan^{-1}\frac{x}{3}$ for some real number x.

SOLUTION Because $\theta = \tan^{-1}\frac{x}{3}$, we know from the definition of the inverse tangent function that $-\frac{\pi}{2} < \theta < \frac{\pi}{2}$. For any angle θ within this interval, $\sec \theta$ will be a positive value. Therefore, $|\sec \theta| = \sec \theta$ and we can simplify the expression further as

$$3\,|\sec \theta| = 3 \sec \theta$$

PROBLEM 4
Evaluate each expression.

a. $\cos\left(\cos^{-1}\dfrac{\sqrt{3}}{2}\right)$

b. $\sin^{-1}(\sin 210°)$

EXAMPLE 4 Evaluate each expression.

a. $\sin\left(\sin^{-1}\dfrac{1}{2}\right)$

b. $\sin^{-1}(\sin 135°)$

SOLUTION

a. From Example 1a we know that $\sin^{-1}\dfrac{1}{2} = \dfrac{\pi}{6}$. Therefore,

$$\sin\left(\sin^{-1}\frac{1}{2}\right) = \sin\left(\frac{\pi}{6}\right) = \frac{1}{2}$$

b. Because $\sin 135° = \dfrac{\sqrt{2}}{2}$, $\sin^{-1}(\sin 135°) = \sin^{-1}\left(\dfrac{\sqrt{2}}{2}\right)$ will be the angle y, $-90° \le y \le 90°$, for which $\sin y = \dfrac{\sqrt{2}}{2}$. The angle satisfying this requirement is $y = 45°$. So,

$$\sin^{-1}(\sin 135°) = \sin^{-1}\left(\frac{\sqrt{2}}{2}\right) = 45°$$

NOTE Notice in Example 4a that the result of 1/2 is the same as the value that appeared in the original expression. Because $y = \sin x$ and $y = \sin^{-1} x$ are inverse functions, the one function will "undo" the action performed by the other. (For more details on this property of inverse functions, see Appendix A.) The reason this same process did not occur in Example 4b is that the original angle 135° is not within the restricted domain of the sine function and is therefore outside the range of the inverse sine function. In a sense, the functions in Example 4b are not really inverses because we did not choose an input within the agreed-upon interval for the sine function.

PROBLEM 5
Simplify $\cos^{-1}(\cos x)$ if $0 \le x \le \pi$.

EXAMPLE 5 Simplify $\tan^{-1}(\tan x)$ if $-\dfrac{\pi}{2} < x < \dfrac{\pi}{2}$.

SOLUTION Because x is within the restricted domain for the tangent function, the two functions are inverses. Whatever value the tangent function assigns to x, the inverse tangent function will reverse this association and return the original value of x. So, by the property of inverse functions,

$$\tan^{-1}(\tan x) = x$$

PROBLEM 6
Evaluate $\sin\left(\cos^{-1}\dfrac{12}{13}\right)$ without using a calculator.

EXAMPLE 6 Evaluate $\sin\left(\tan^{-1}\frac{3}{4}\right)$ without using a calculator.

SOLUTION We begin by letting $\theta = \tan^{-1}\frac{3}{4}$. (Remember, $\tan^{-1} x$ is the angle whose tangent is x.) Then we have

$$\text{If } \theta = \tan^{-1}\frac{3}{4}, \text{ then } \tan\theta = \frac{3}{4} \text{ and } 0° < \theta < 90°$$

We can draw a triangle in which one of the acute angles is θ (Figure 11). Because $\tan\theta = \frac{3}{4}$, we label the side opposite θ with 3 and the side adjacent to θ with 4. The hypotenuse is found by applying the Pythagorean Theorem.

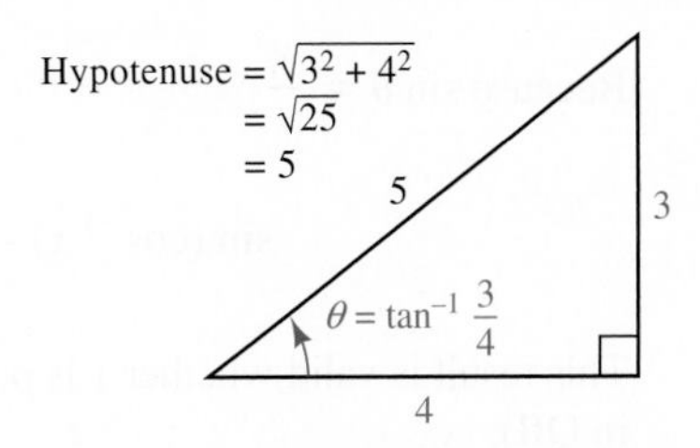

Figure 11

From Figure 11 we find $\sin \theta$ using the ratio of the side opposite θ to the hypotenuse.

$$\sin\left(\tan^{-1}\frac{3}{4}\right) = \sin \theta = \frac{3}{5}$$

CALCULATOR NOTE If we were to do the same problem with the aid of a calculator, the sequence would look like this:

Scientific Calculator

3 [÷] 4 [=] [tan⁻¹] [sin]

Graphing Calculator

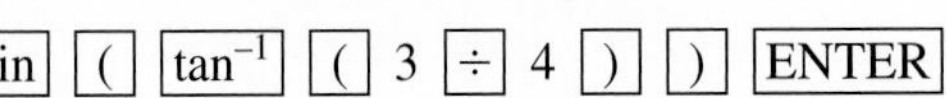

The display would read 0.6, which is $\frac{3}{5}$.

Although it is a lot easier to use a calculator on problems like the one in Example 6, solving it without a calculator will be of more use to you in the future.

PROBLEM 7
Write $\cos\left(\sin^{-1}\frac{x}{2}\right)$ as an equivalent algebraic expression in x only.

EXAMPLE 7 Write the expression $\sin(\cos^{-1} x)$ as an equivalent algebraic expression in x only.

SOLUTION We let $\theta = \cos^{-1} x$. Then

$$\cos \theta = x = \frac{x}{1} \quad \text{and} \quad 0 \le \theta \le \pi$$

We can visualize the problem by drawing θ in standard position with terminal side in either QI or QII (Figure 12). Let $P = (x, y)$ be a point on the terminal side of θ. By Definition I, $\cos \theta = \frac{x}{r}$, so r must be equal to 1. We can find y by applying the Pythagorean Theorem. Notice that y will be a positive value in either quadrant.

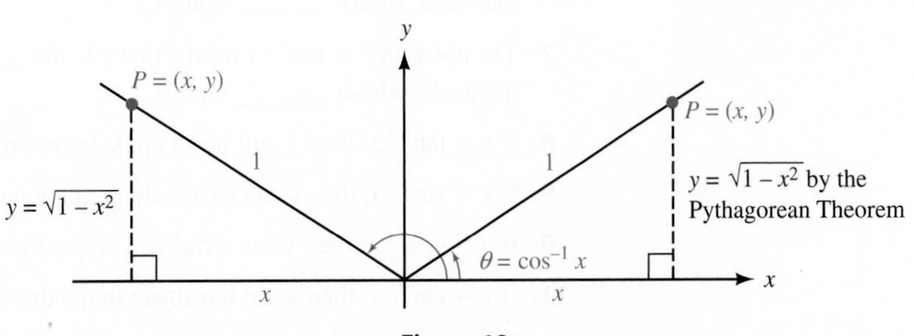

Figure 12

Because $\sin \theta = \dfrac{y}{r}$,

$$\sin(\cos^{-1} x) = \sin \theta = \frac{\sqrt{1 - x^2}}{1} = \sqrt{1 - x^2}$$

This result is valid whether x is positive (θ terminates in QI) or negative (θ terminates in QII).

Getting Ready for Class

After reading through the preceding section, respond in your own words and in complete sentences.

a. Why must the graph of $y = \sin x$ be restricted to $-\dfrac{\pi}{2} \le x \le \dfrac{\pi}{2}$ in order to have an inverse function?

b. What restriction is made on the values of x for $y = \cos x$ so that it will have an inverse that is a function?

c. What is the meaning of the notation $y = \tan^{-1} x$?

d. If $y = \arctan x$, then what restriction is placed on the value of y?

4.7 PROBLEM SET

➤ CONCEPTS AND VOCABULARY

For Questions 1 through 12, fill in each blank with the appropriate word, number, or equation.

1. To find the inverse of a function, ____________ the coordinates in each ordered pair for the function.

2. Only functions whose graphs pass the ____________ ______ test will have an inverse that is also a function. We call this type of function a _____ ___ ______ function.

3. The graph of the inverse of a function is a ____________ of the graph of the function about the line ________.

4. Because all six trigonometric functions are not one-to-one, we must __________ the domain in order to define an inverse function.

5. The two notations for the inverse of $y = \sin x$ are ________ and __________.

6. The notation $y = \arccos x$ means that y is the __________ between _____ and _____, inclusive, whose ________ equals x.

7. The notation $y = \sin^{-1} x$ means that y is the __________ between _____ and _____, inclusive, whose ________ equals x.

8. If $y = \tan^{-1} x$, then y will be an angle between _____ and _____.

9. If $y = \sin^{-1} x$, then y can terminate in quadrant _____ or _____.

10. If $y = \cos^{-1} x$, then y can terminate in quadrant _____ or _____.

11. If $y = \tan^{-1} x$, then y can terminate in quadrant _____ or _____.

12. In order for $\sin^{-1}(\sin x) = x$, x must be a value between _____ and _____.

➤ EXERCISES

13. Graph $y = \cos x$ for x between -2π and 2π, and then reflect the graph about the line $y = x$ to obtain the graph of $x = \cos y$.

14. Graph $y = \sin x$ for x between $-\pi/2$ and $\pi/2$, and then reflect the graph about the line $y = x$ to obtain the graph of $y = \sin^{-1} x$ between $-\pi/2$ and $\pi/2$.

15. Graph $y = \tan x$ for x between $-\pi/2$ and $\pi/2$, and then reflect the graph about the line $y = x$ to obtain the graph of $y = \tan^{-1} x$.

16. Graph $y = \cot x$ for x between 0 and π, and then reflect the graph about the line $y = x$ to obtain the graph of $y = \cot^{-1} x$.

Evaluate each expression without using a calculator, and write your answers in radians.

17. $\sin^{-1}\left(\frac{\sqrt{3}}{2}\right)$

18 $\cos^{-1}\left(\frac{1}{2}\right)$

19. $\cos^{-1}(-1)$

20. $\cos^{-1}(0)$

21. $\tan^{-1}(1)$

22. $\tan^{-1}(0)$

23. $\arccos\left(-\frac{\sqrt{2}}{2}\right)$

24. $\arccos(1)$

25. $\sin^{-1}\left(-\frac{1}{2}\right)$

26. $\sin^{-1}\left(\frac{\sqrt{2}}{2}\right)$

27. $\arctan(\sqrt{3})$

28. $\arctan\left(\frac{\sqrt{3}}{3}\right)$

29. $\arcsin(0)$

30. $\arcsin\left(-\frac{\sqrt{3}}{2}\right)$

31. $\tan^{-1}\left(-\frac{\sqrt{3}}{3}\right)$

32. $\tan^{-1}(-\sqrt{3})$

33. $\sin^{-1}(1)$

34. $\cos^{-1}\left(-\frac{1}{2}\right)$

35. $\arccos\left(\frac{\sqrt{3}}{2}\right)$

36. $\arcsin(-1)$

Use a calculator to evaluate each expression to the nearest tenth of a degree.

37. $\sin^{-1}(0.1702)$

38. $\sin^{-1}(-0.1702)$

39. $\arccos(0.8425)$

40. $\arccos(0.9627)$

41. $\tan^{-1}(0.3799)$

42. $\tan^{-1}(-0.3799)$

43. $\cos^{-1}(-0.4664)$

44. $\sin^{-1}(-0.4664)$

45. $\arctan(-2.748)$

46. $\arctan(-0.3640)$

47. $\sin^{-1}(-0.7660)$

48. $\cos^{-1}(-0.7660)$

49. Use your graphing calculator to graph $y = \cos^{-1} x$ in degree mode. Use the graph with the appropriate command to evaluate each expression.

a. $\cos^{-1}\left(\frac{1}{2}\right)$ **b.** $\cos^{-1}\left(-\frac{\sqrt{3}}{2}\right)$ **c.** $\arccos\left(\frac{\sqrt{2}}{2}\right)$

50. Use your graphing calculator to graph $y = \sin^{-1} x$ in degree mode. Use the graph with the appropriate command to evaluate each expression.

a. $\sin^{-1}\left(-\frac{1}{2}\right)$ **b.** $\sin^{-1}\left(\frac{\sqrt{3}}{2}\right)$ **c.** $\arcsin\left(-\frac{\sqrt{2}}{2}\right)$

51. Use your graphing calculator to graph $y = \tan^{-1} x$ in degree mode. Use the graph with the appropriate command to evaluate each expression.

a. $\tan^{-1}(-1)$ **b.** $\tan^{-1}(\sqrt{3})$ **c.** $\arctan\left(-\frac{\sqrt{3}}{3}\right)$

52. Simplify $2\,|\sin\theta|$ if $\theta = \cos^{-1}\frac{x}{2}$ for some real number x.

53. Simplify $4\,|\cos\theta|$ if $\theta = \sin^{-1}\frac{x}{4}$ for some real number x.

54. Simplify $5\,|\sec\theta|$ if $\theta = \tan^{-1}\frac{x}{5}$ for some real number x.

Evaluate without using a calculator.

55. $\sin\left(\sin^{-1}\frac{3}{5}\right)$

56. $\cos\left(\cos^{-1}\frac{3}{5}\right)$

57. $\cos\left(\cos^{-1}\frac{1}{2}\right)$

58. $\sin\left(\sin^{-1}\frac{\sqrt{2}}{2}\right)$

59. $\tan\left(\tan^{-1}\frac{1}{2}\right)$

60. $\tan\left(\tan^{-1}\frac{3}{4}\right)$

61. $\sin^{-1}(\sin 225°)$

62. $\sin^{-1}(\sin 330°)$

63. $\sin^{-1}\left(\sin\frac{\pi}{3}\right)$

64. $\sin^{-1}\left(\sin\frac{\pi}{4}\right)$

65. $\cos^{-1}(\cos 120°)$

66. $\cos^{-1}(\cos 45°)$

67. $\cos^{-1}\left(\cos\frac{7\pi}{4}\right)$

68. $\cos^{-1}\left(\cos\frac{7\pi}{6}\right)$

69. $\tan^{-1}(\tan 45°)$

70. $\tan^{-1}(\tan 60°)$

71. $\tan^{-1}\left(\tan\frac{5\pi}{6}\right)$

72. $\tan^{-1}\left(\tan\frac{2\pi}{3}\right)$

Evaluate without using a calculator.

73. $\tan\left(\sin^{-1}\frac{3}{5}\right)$

74. $\csc\left(\tan^{-1}\frac{3}{4}\right)$

75. $\sec\left(\cos^{-1}\frac{1}{\sqrt{5}}\right)$

76. $\tan\left(\cos^{-1}\frac{3}{5}\right)$

77. $\sin\left(\cos^{-1}\frac{1}{2}\right)$

78. $\cos\left(\tan^{-1}\frac{3}{4}\right)$

79. $\cot\left(\tan^{-1}\frac{1}{2}\right)$

80. $\cos\left(\sin^{-1}\frac{1}{2}\right)$

81. Simplify $\sin^{-1}(\sin x)$ if $-\pi/2 \le x \le \pi/2$.

82. Simplify $\cos^{-1}(\cos x)$ if $0 \le x \le \pi$.

For each expression below, write an equivalent algebraic expression that involves x only. (For Problems 89 through 92, assume x is positive.)

83. $\cos(\cos^{-1} x)$

84. $\sin(\sin^{-1} x)$

85. $\cos(\sin^{-1} x)$

86. $\tan(\cos^{-1} x)$

87. $\sin(\tan^{-1} x)$

88. $\cos(\tan^{-1} x)$

89. $\sin\left(\cos^{-1}\frac{1}{x}\right)$

90. $\cos\left(\sin^{-1}\frac{1}{x}\right)$

91. $\sec\left(\cos^{-1}\frac{1}{x}\right)$

92. $\csc\left(\sin^{-1}\frac{1}{x}\right)$

Navigation *The great circle distance between two points $P_1(LT_1, LN_1)$ and $P_2(LT_2, LN_2)$, whose coordinates are given as latitudes and longitudes, is given by the formula*

$$d = \cos^{-1}(\sin(LT_1)\sin(LT_2) + \cos(LT_1)\cos(LT_2)\cos(LN_1 - LN_2))$$

where d is the distance along a great circle of the earth measured in radians (Figure 13). To use this formula, the latitudes and longitudes must be entered as angles in radians. Find the great circle distance, in miles, between the given points assuming the radius of Earth is 3,960 miles.

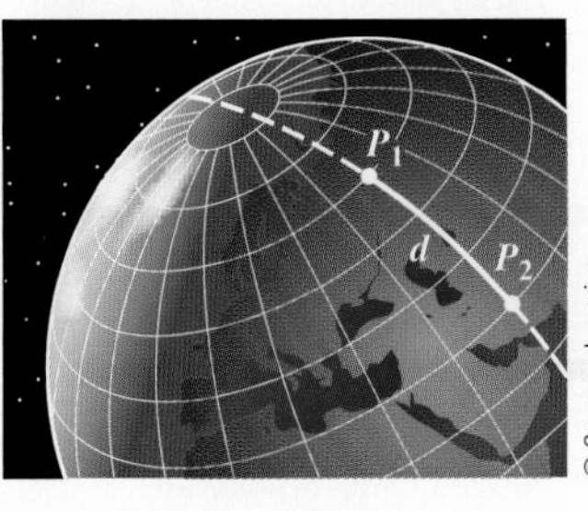

Figure 13

93. P_1(N 32° 22.108′, W 64° 41.178′) Bermuda, and
P_2(N 13° 04.809′, W 59° 29.263′) Barbados

94. P_1(N 21° 53.896′, W 159° 36.336′) Kauai, and
P_2(N 19° 43.219′, W 155° 02.930′) Hawaii

➤ REVIEW PROBLEMS

The problems that follow review material we covered in Sections 4.2 and 4.3.

Graph each of the following equations over the indicated interval. Be sure to label the x- and y-axes so that the amplitude and period are easy to see.

95. $y = 2\sin\pi x,\ -4 \le x \le 4$

96. $y = 3\cos\pi x,\ -2 \le x \le 4$

97. $y = -3\cos\frac{1}{2}x,\ -2\pi \le x \le 6\pi$

98. $y = -3\sin 2x,\ -2\pi \le x \le 2\pi$

Graph one complete cycle of each of the following equations. Be sure to label the x- and y-axes so that the amplitude, period, and horizontal shift for each graph are easy to see.

99. $y = \sin\left(x - \frac{\pi}{4}\right)$

100. $y = \sin\left(x + \frac{\pi}{6}\right)$

101. $y = 3\sin\left(2x - \frac{\pi}{3}\right)$

102. $y = 3\cos\left(2x - \frac{\pi}{3}\right)$

➤ LEARNING OBJECTIVES ASSESSMENT

These questions are available for instructors to help assess if you have successfully met the learning objectives for this section.

103. Which of the following is a false statement?

a. $\tan^{-1}(-1) = -45°$

b. $\arcsin(1) = \frac{\pi}{2}$

c. $\cos^{-1}\left(-\frac{1}{2}\right) = -\frac{\pi}{3}$

d. $\sin^{-1}\left(\frac{\sqrt{3}}{2}\right) = 60°$

104. Use a calculator to approximate arcsin (−0.8855) to the nearest tenth of a degree.

a. −62.3° **b.** −1.01° **c.** 1.1° **d.** 62.3°

105. Find the exact value of $\cos^{-1}\left(\cos\dfrac{11\pi}{12}\right)$.

a. $\dfrac{11\pi}{12}$ **b.** $-\dfrac{11\pi}{12}$ **c.** $-\dfrac{\pi}{12}$ **d.** $\dfrac{\pi}{12}$

106. Simplify $\tan\left(\cos^{-1}\dfrac{1}{\sqrt{10}}\right)$.

a. $\dfrac{1}{3}$ **b.** $\dfrac{3\sqrt{10}}{10}$ **c.** $\dfrac{\sqrt{10}}{3}$ **d.** 3

CHAPTER 4 SUMMARY

PERIODIC FUNCTIONS [4.1]

A function $y = f(x)$ is said to be periodic with period p if p is the smallest positive number such that $f(x + p) = f(x)$ for all x in the domain of f.

EXAMPLES

1 Because

$$\sin(x + 2\pi) = \sin x$$

the function $y = \sin x$ is periodic with period 2π. Likewise, because

$$\tan(x + \pi) = \tan x$$

the function $y = \tan x$ is periodic with period π.

AMPLITUDE [4.1, 4.2]

The *amplitude* A of a curve is half the absolute value of the difference between the largest value of y, denoted by M, and the smallest value of y, denoted by m.

$$A = \frac{1}{2}|M - m|$$

2

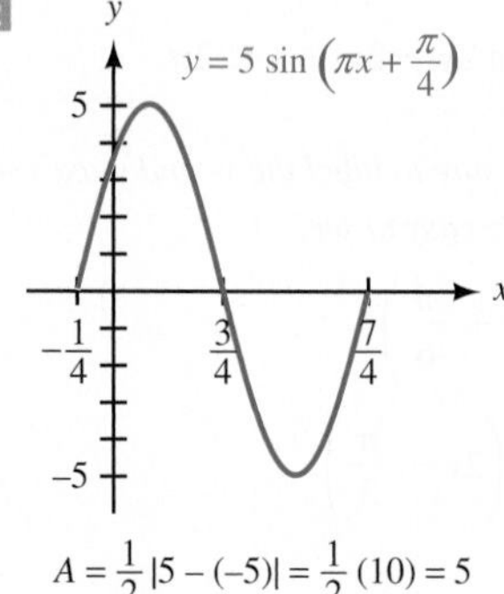

$A = \frac{1}{2}|5 - (-5)| = \frac{1}{2}(10) = 5$

BASIC GRAPHS [4.1]

The graphs of $y = \sin x$ and $y = \cos x$ are both periodic with period 2π. The amplitude of each graph is 1. The sine curve passes through 0 on the y-axis, while the cosine curve passes through 1 on the y-axis.

The graphs of $y = \csc x$ and $y = \sec x$ are also periodic with period 2π. We graph them by using the fact that they are reciprocals of sine and cosine. There is no largest or smallest value of y, so we say the secant and cosecant curves have no amplitude.

The graphs of $y = \tan x$ and $y = \cot x$ are periodic with period π. The tangent curve passes through the origin, while the cotangent is undefined when x is 0. There is no amplitude for either graph.

3 **a.**

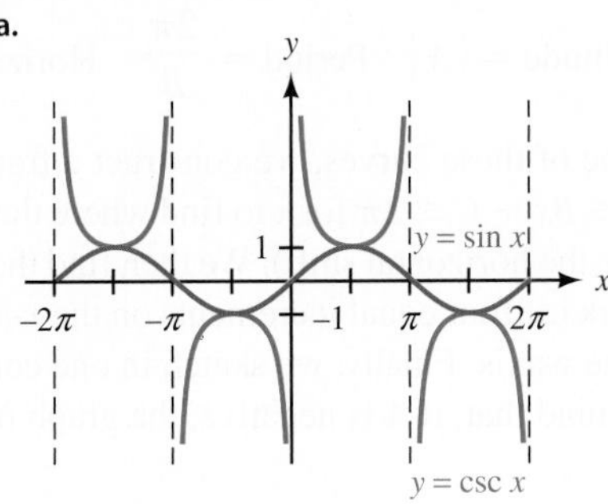

b.

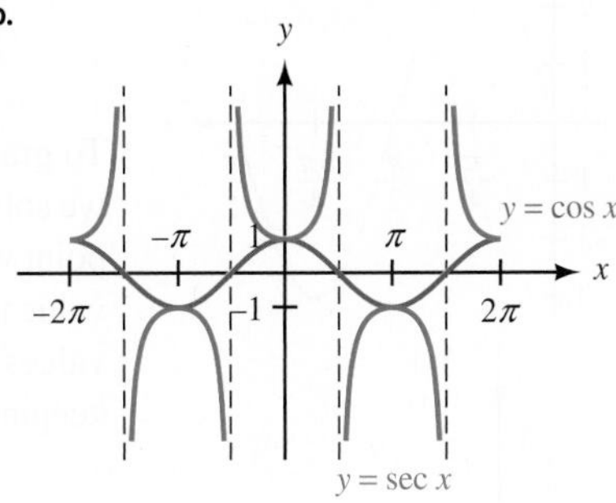

c.

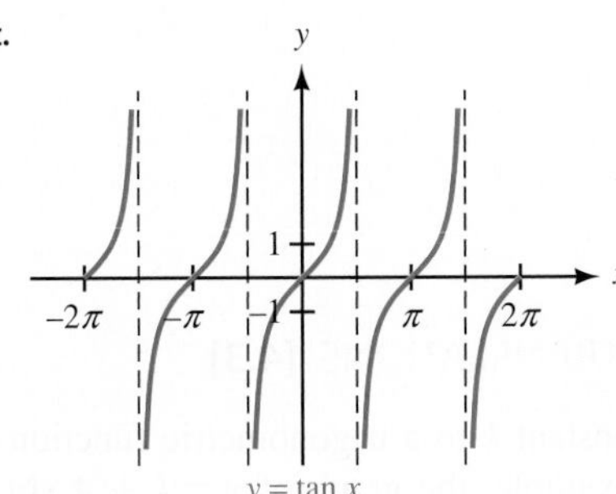

d.

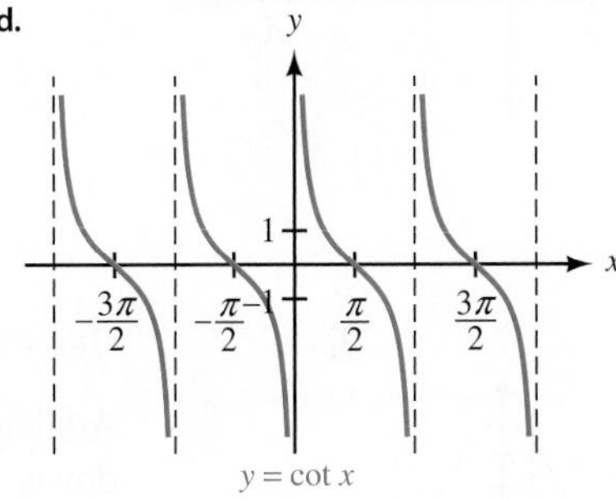

4 Tan θ is an odd function.

$$\tan(-\theta) = \frac{\sin(-\theta)}{\cos(-\theta)} = \frac{-\sin\theta}{\cos\theta} = -\frac{\sin\theta}{\cos\theta} = -\tan\theta$$

EVEN AND ODD FUNCTIONS [4.1]

An *even function* is a function for which

$$f(-x) = f(x) \text{ for all } x \text{ in the domain of } f$$

and an *odd function* is a function for which

$$f(-x) = -f(x) \text{ for all } x \text{ in the domain of } f$$

Cosine is an even function, and sine is an odd function. That is,

$$\cos(-\theta) = \cos\theta \qquad \text{Cosine is an even function}$$

and

$$\sin(-\theta) = -\sin\theta \qquad \text{Sine is an odd function}$$

The graph of an even function is symmetric about the y-axis, and the graph of an odd function is symmetric about the origin.

5 The horizontal shift for the graph in Example 2 is $-\frac{1}{4}$.

HORIZONTAL TRANSLATIONS [4.3]

The *horizontal translation* for a sine or cosine curve is the distance the curve has moved right or left from the curve $y = \sin x$ or $y = \cos x$. For example, we usually think of the graph of $y = \sin x$ as starting at the origin. If we graph another sine curve that starts at $\pi/4$, then we say this curve has a horizontal translation, or horizontal shift, of $\pi/4$.

6

a.

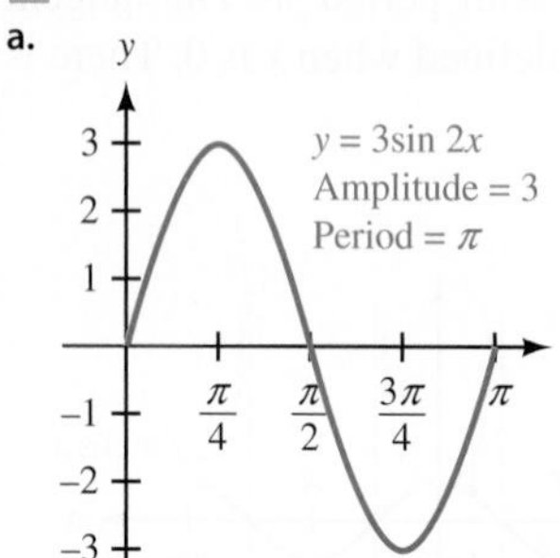

b.

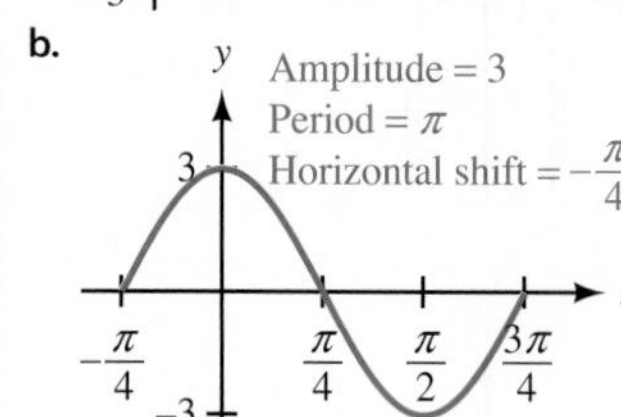

GRAPHING SINE AND COSINE CURVES [4.2, 4.3]

The graphs of $y = A \sin (Bx + C)$ and $y = A \cos (Bx + C)$, where $B > 0$, will have the following characteristics:

$$\text{Amplitude} = |A| \quad \text{Period} = \frac{2\pi}{B} \quad \text{Horizontal shift} = -\frac{C}{B} \quad \text{Phase} = C$$

To graph one of these curves, we construct a frame for one cycle. To mark the x-axis, we solve $0 \le Bx + C \le 2\pi$ for x to find where the cycle will begin and end (the left end point will be the horizontal shift). We then find the period and divide it by 4 and use this value to mark off four equal increments on the x-axis. We use the amplitude to mark off values on the y-axis. Finally, we sketch in one complete cycle of the curve in question, keeping in mind that, if A is negative, the graph must be reflected about the x-axis.

7

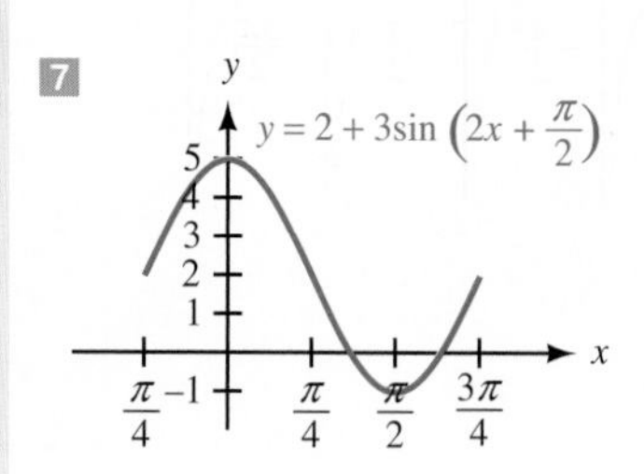

VERTICAL TRANSLATIONS [4.3]

Adding a constant k to a trigonometric function translates the graph vertically up or down. For example, the graph of $y = k + A \sin (Bx + C)$ will have the same shape (amplitude, period, horizontal shift, and reflection, if indicated) as $y = A \sin (Bx + C)$ but will be translated k units vertically from the graph of $y = A \sin (Bx + C)$.

8

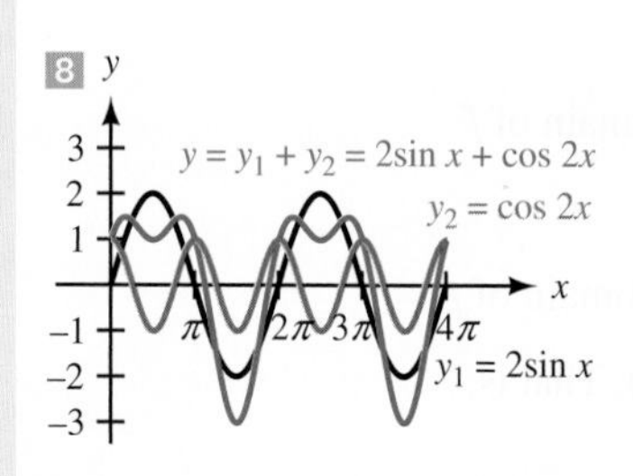

GRAPHING BY ADDITION OF Y-COORDINATES [4.6]

To graph equations of the form $y = y_1 + y_2$, where y_1 and y_2 are algebraic or trigonometric functions of x, we graph y_1 and y_2 separately on the same coordinate system and then add the two graphs to obtain the graph of y.

9 Evaluate in radians without using a calculator.

a. $\sin^{-1} \frac{1}{2}$

The angle between $-\pi/2$ and $\pi/2$ whose sine is $\frac{1}{2}$ is $\pi/6$.

$$\sin^{-1} \frac{1}{2} = \frac{\pi}{6}$$

b. $\arccos \left(-\frac{\sqrt{3}}{2}\right)$

The angle between 0 and π with a cosine of $-\sqrt{3}/2$ is $5\pi/6$.

$$\arccos \left(-\frac{\sqrt{3}}{2}\right) = \frac{5\pi}{6}$$

INVERSE TRIGONOMETRIC FUNCTIONS [4.7]

Inverse Function	Meaning
$y = \sin^{-1} x$ or $y = \arcsin x$	$x = \sin y$ and $-\frac{\pi}{2} \le y \le \frac{\pi}{2}$
In words: y is the angle between $-\pi/2$ and $\pi/2$, inclusive, whose sine is x.	
$y = \cos^{-1} x$ or $y = \arccos x$	$x = \cos y$ and $0 \le y \le \pi$
In words: y is the angle between 0 and π, inclusive, whose cosine is x.	
$y = \tan^{-1} x$ or $y = \arctan x$	$x = \tan y$ and $-\frac{\pi}{2} < y < \frac{\pi}{2}$
In words: y is the angle between $-\pi/2$ and $\pi/2$ whose tangent is x.	

CHAPTER 4 TEST

Graph each of the following between $x = -4\pi$ and $x = 4\pi$.

1. $y = \sin x$

2. $y = \cos x$

3. $y = \tan x$

4. $y = \sec x$

5. Show that cotangent is an odd function.

6. Prove the identity $\sin(-\theta)\sec(-\theta)\cot(-\theta) = 1$.

For each equation below, first identify the amplitude and period and then use this information to sketch one complete cycle of the graph.

7. $y = \cos \pi x$

8. $y = -3\cos x$

Graph each of the following on the given interval.

9. $y = 2 + 3\sin 2x,\ -\pi \le x \le 2\pi$

10. $y = 2\sin \pi x,\ -4 \le x \le 4$

For each of the following functions, identify the amplitude, period, and horizontal shift and then use this information to sketch one complete cycle of the graph.

11. $y = \sin\left(x + \dfrac{\pi}{4}\right)$

12. $y = 3\sin\left(2x - \dfrac{\pi}{3}\right)$

13. $y = -3 + 3\sin\left(\dfrac{\pi}{3}x - \dfrac{\pi}{3}\right)$

14. $y = 1 + \dfrac{1}{2}\csc\left(x + \dfrac{\pi}{4}\right)$

15. $y = -3\tan\left(2x - \dfrac{\pi}{2}\right)$

Graph each of the following on the given interval.

16. $y = 2\sin(3x - \pi),\ -\dfrac{\pi}{3} \le x \le \dfrac{5\pi}{3}$

17. $y = \dfrac{1}{2} - 2\sin\left(\dfrac{\pi}{2}x - \dfrac{\pi}{4}\right),\ -\dfrac{1}{2} \le x \le \dfrac{13}{2}$

Find an equation for each of the following graphs.

18.

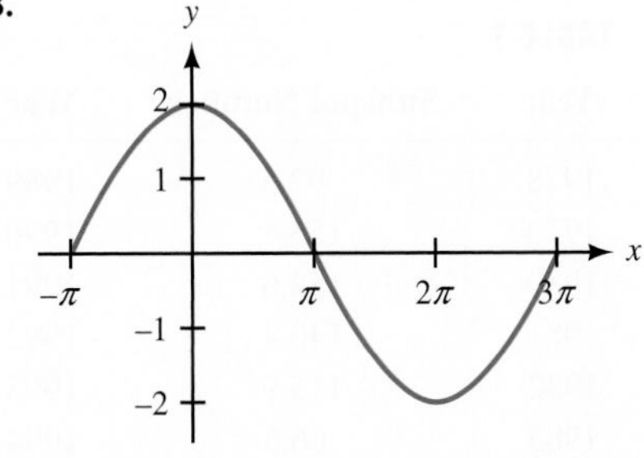

19.

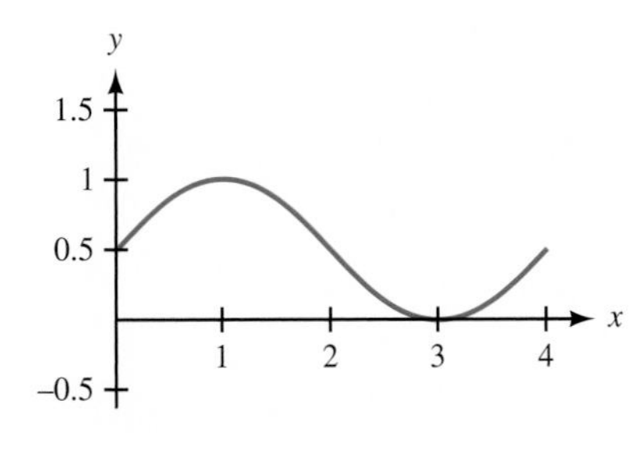

Sketch the following between $x = 0$ and $x = 4\pi$.

20. $y = \dfrac{1}{2}x - \sin x$

21. $y = \sin x + \cos 2x$

Graph each of the following.

22. $y = \cos^{-1} x$

23. $y = \arcsin x$

Evaluate each expression without using a calculator and write your answer in radians.

24. $\sin^{-1}\left(\frac{1}{2}\right)$

25. $\arctan(-1)$

Use a calculator to evaluate each expression to the nearest tenth of a degree.

26. $\arcsin(0.5934)$

27. $\arctan(-0.8302)$

Evaluate without using a calculator.

28. $\tan\left(\cos^{-1}\frac{2}{3}\right)$

29. $\tan^{-1}\left(\tan\frac{7\pi}{6}\right)$

30. Write an equivalent algebraic expression for $\sin(\cos^{-1}x)$ that involves x only.

GROUP PROJECT Modeling the Sunspot Cycle

NASA/Getty Images

From my earlier observations, which I have reported every year in this journal, it appears that there is a certain periodicity in the appearance of sunspots and this theory seems more and more probable from the results of this year . . .

Heinrich Schwabe, 1843

OBJECTIVE: To find a sinusoidal model for the average annual sunspot number using recent data.

Sunspots have been observed and recorded for thousands of years. Some of the earliest observations were made by Chinese astronomers. Further observations were made following the invention of the telescope in the 17th century by a number of astronomers, including Galileo and William Herschel. However, it was the German astronomer Heinrich Schwabe who, in searching for evidence of other planets between Mercury and the Sun, first observed an apparent periodic cycle in sunspot activity.

Table 1 provides the average annual sunspot number for each year between 1978 and 1998.

TABLE 1

Year	Sunspot Number	Year	Sunspot Number
1978	92.5	1989	157.6
1979	155.4	1990	142.6
1980	154.6	1991	145.7
1981	140.4	1992	94.3
1982	115.9	1993	54.6
1983	66.6	1994	29.9
1984	45.9	1995	17.5
1985	17.9	1996	8.6
1986	13.4	1997	21.5
1987	29.4	1998	64.3
1988	100.2		

1. Sketch a graph of the data, listing the year along the horizontal axis and the sunspot number along the vertical axis.

2. Using the data and your graph from Question 1, find an equation of the form $y = k + A \sin(B(x - h))$ or $y = k + A \cos(B(x - h))$ to match the graph, where x is the year and y is the sunspot number for that year. Explain how you determined the values of k, A, B, and h, and show supporting work.

3. According to your model, what is the length of the sunspot cycle (in years)? In 1848, Rudolph Wolf determined the length of the sunspot cycle to be 11.1 years. How does your value compare with his?

4. Use your model to predict the sunspot number for 2001. How does your prediction compare with the actual value of 111?

5. Using your graphing calculator, make a scatter plot of the data from the table. Then graph your model from Question 2 along with the data. How well does your model fit the data? What could you do to try to improve your model?

6. If your graphing calculator is capable of computing a least-squares sinusoidal regression model, use it to find a second model for the data. Graph this new equation along with your first model. How do they compare?

RESEARCH PROJECT The Sunspot Cycle

ETH-Bibliothek Zurich Image Archive

Johann Rudolph Wolf

Although many astronomers made regular observations of sunspots, it was the Swiss astronomer Rudolph Wolf who devised the first universal method for counting sunspots. The daily sunspot number, sometimes called the Wolf number, follows a periodic cycle. Wolf calculated the length of this cycle to be about 11.1 years.

Research the astronomer Rudolph Wolf and the sunspot cycle. Why do modern astronomers consider the length of the sunspot cycle to be 22 years? What are some of the effects that the sunspot cycle causes here on Earth? Write a paragraph or two about your findings.

5 Identities and Formulas

Mathematics, rightly viewed, possesses not only truth, but supreme beauty.

➤ Bertrand Russell

INTRODUCTION

Although it doesn't look like it, Figure 1 shows the graphs of two functions, namely

$$y = \cos^2 x \qquad \text{and} \qquad y = \frac{1 - \sin^4 x}{1 + \sin^2 x}$$

Although these two functions look quite different from one another, they are in fact the same function. This means that, for all values of x,

$$\cos^2 x = \frac{1 - \sin^4 x}{1 + \sin^2 x}$$

This last expression is an *identity,* and identities are one of the topics we will study in this chapter.

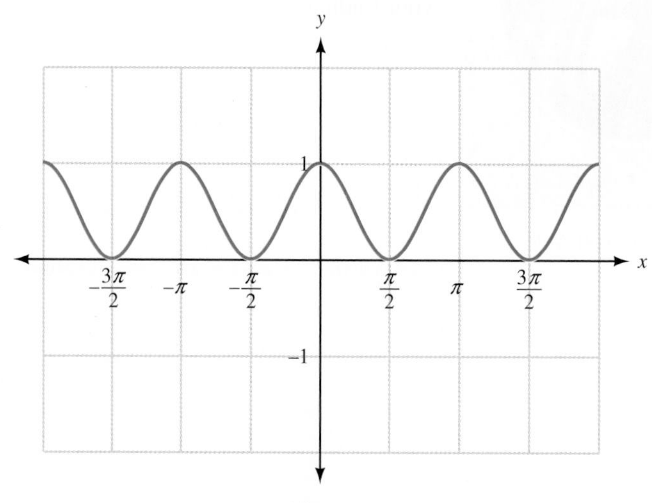

Figure 1

STUDY SKILLS 5

The study skills for this chapter focus on the way you approach new situations in mathematics. The first study skill is a point of view you hold about your natural instincts for what does and doesn't work in mathematics. The second study skill gives you a way of testing your instincts.

1 Don't Let Your Intuition Fool You As you become more experienced and more successful in mathematics, you will be able to trust your mathematical intuition. For now, though, it can get in the way of your success. For example, if you ask a beginning algebra student to expand $(a + b)^2$, many will write $a^2 + b^2$, which is incorrect. In trigonometry, at first glance it may seem that a statement such as $\sin (A + B) = \sin A + \sin B$ is true. However, it too is false, as you will see in this chapter.

2 Test Properties You Are Unsure Of From time to time you will be in a situation in which you would like to apply a property or rule, but you are not sure it is true. You can always test a property or statement by substituting numbers for variables. For instance, we always have students that rewrite $(x + 3)^2$ as $x^2 + 9$, thinking that the two expressions are equivalent. The fact that the two expressions are not equivalent becomes obvious when we substitute 10 for x in each one.

When $x = 10$, the expression $(x + 3)^2$ is $(10 + 3)^2 = 13^2 = 169$
When $x = 10$, the expression $x^2 + 9 = 10^2 + 9 = 100 + 9 = 109$

Similarly, there may come a time when you are wondering if $\sin 2A$ is the same as $2 \sin A$. If you try $A = 30°$ in each expression, you will find out quickly that the two expressions are not the same.

When $A = 30°$, the expression $\sin 2A = \sin 2(30°) = \sin 60° = \dfrac{\sqrt{3}}{2}$

When $A = 30°$, the expression $2 \sin A = 2 \sin 30° = 2 \cdot \dfrac{1}{2} = 1$

When you test the equivalence of expressions by substituting numbers for the variable, make it easy on yourself by choosing numbers that are easy to work with, as we did above.

It is not good practice to trust your intuition or instincts in every new situation in mathematics. If you have any doubt about generalizations you are making, test them by replacing variables with numbers and simplifying. ▲

SECTION 5.1 Proving Identities

LEARNING OBJECTIVES

1. Prove an equation is an identity.
2. Use a counterexample to prove an equation is not an identity.
3. Use a graphing calculator to determine if an equation appears to be an identity.

We began proving identities in Chapter 1. In this section, we will extend the work we did in that chapter to include proving more complicated identities. For review, Table 1 lists the basic identities and some of their more important equivalent forms.

TABLE 1

	Basic Identities	Common Equivalent Forms
Reciprocal	$\csc\theta = \dfrac{1}{\sin\theta}$	$\sin\theta = \dfrac{1}{\csc\theta}$
	$\sec\theta = \dfrac{1}{\cos\theta}$	$\cos\theta = \dfrac{1}{\sec\theta}$
	$\cot\theta = \dfrac{1}{\tan\theta}$	$\tan\theta = \dfrac{1}{\cot\theta}$
Ratio	$\tan\theta = \dfrac{\sin\theta}{\cos\theta}$	
	$\cot\theta = \dfrac{\cos\theta}{\sin\theta}$	
Pythagorean	$\cos^2\theta + \sin^2\theta = 1$	$\sin^2\theta = 1 - \cos^2\theta$ $\sin\theta = \pm\sqrt{1-\cos^2\theta}$ $\cos^2\theta = 1 - \sin^2\theta$ $\cos\theta = \pm\sqrt{1-\sin^2\theta}$
	$1 + \tan^2\theta = \sec^2\theta$ $1 + \cot^2\theta = \csc^2\theta$	

NOTE The last two Pythagorean identities can be derived from $\cos^2\theta + \sin^2\theta = 1$ by dividing each side by $\cos^2\theta$ and $\sin^2\theta$, respectively. For example, if we divide each side of $\cos^2\theta + \sin^2\theta = 1$ by $\cos^2\theta$, we have

$$\cos^2\theta + \sin^2\theta = 1$$
$$\frac{\cos^2\theta + \sin^2\theta}{\cos^2\theta} = \frac{1}{\cos^2\theta}$$
$$\frac{\cos^2\theta}{\cos^2\theta} + \frac{\sin^2\theta}{\cos^2\theta} = \frac{1}{\cos^2\theta}$$
$$1 + \tan^2\theta = \sec^2\theta$$

To derive the last Pythagorean identity, we would need to divide both sides of $\cos^2\theta + \sin^2\theta = 1$ by $\sin^2\theta$ to obtain $1 + \cot^2\theta = \csc^2\theta$.

The rest of this section is concerned with using the basic identities (or their equivalent forms) just listed, along with our knowledge of algebra, to prove other identities.

Recall that an identity in trigonometry is a statement that two expressions are equal for all replacements of the variable for which each expression is defined. To prove (or verify) a trigonometric identity, we use trigonometric substitutions and algebraic manipulations to either

1. transform the right side of the identity into the left side, or
2. transform the left side of the identity into the right side.

The main thing to remember in proving identities is to work on each side of the identity separately. We do not want to use properties from algebra that involve both sides of the identity—like the addition property of equality. To do so would be to assume that the two sides *are* equal, which is what we are trying to establish. We are not allowed to treat the problem as an equation.

We prove identities to develop the ability to transform one trigonometric expression into another. When we encounter problems in other courses that require the techniques used to verify identities, we usually find that the solution to these problems hinges on transforming an expression containing trigonometric functions into less complicated expressions. In these cases, we do not usually have an equal sign to work with.

PROBLEM 1
Prove $\dfrac{\tan\theta}{\sec\theta} = \sin\theta$.

EXAMPLE 1 Prove $\sin\theta\cot\theta = \cos\theta$.

PROOF To prove this identity, we transform the left side into the right side.

$$\begin{aligned}\sin\theta\cot\theta &= \sin\theta\cdot\frac{\cos\theta}{\sin\theta} && \text{Ratio identity}\\ &= \frac{\sin\theta\cos\theta}{\sin\theta} && \text{Multiply}\\ &= \cos\theta && \text{Divide out common factor } \sin\theta\end{aligned}$$

In this example, we have transformed the left side into the right side. Remember, we verify identities by transforming one expression into another. ■

PROBLEM 2
Prove $\cot x + 1 = \cos x(\csc x + \sec x)$.

EXAMPLE 2 Prove $\tan x + \cos x = \sin x(\sec x + \cot x)$.

PROOF We can begin by applying the distributive property to the right side to multiply through by $\sin x$. Then we can change the right side to an equivalent expression involving only $\sin x$ and $\cos x$.

$$\begin{aligned}\sin x(\sec x + \cot x) &= \sin x\sec x + \sin x\cot x && \text{Multiply}\\ &= \sin x\cdot\frac{1}{\cos x} + \sin x\cdot\frac{\cos x}{\sin x} && \text{Reciprocal and ratio identities}\\ &= \frac{\sin x}{\cos x} + \cos x && \text{Multiply}\\ &= \tan x + \cos x && \text{Ratio identity}\end{aligned}$$

In this case, we transformed the right side into the left side. ■

Before we go on to the next example, let's list some guidelines that may be useful in learning how to prove identities.

GUIDELINES FOR PROVING IDENTITIES

1. It is usually best to work on the more complicated side first.
2. Look for trigonometric substitutions involving the basic identities that may help simplify things.
3. Look for algebraic operations, such as adding fractions, the distributive property, or factoring, that may simplify the side you are working with or that will at least lead to an expression that will be easier to simplify.
4. If you cannot think of anything else to do, change everything to sines and cosines and see if that helps.
5. Always keep an eye on the side you are not working with to be sure you are working toward it. There is a certain sense of direction that accompanies a successful proof.

Probably the best advice is to remember that these are simply guidelines. The best way to become proficient at proving trigonometric identities is to practice. The more identities you prove, the more you will be able to prove and the more confident you will become. *Don't be afraid to stop and start over if you don't seem to be getting anywhere.* With most identities, there are a number of different proofs that will lead to the same result. Some of the proofs will be longer than others.

PROBLEM 3

Prove $\dfrac{\csc^2 x - 1}{\csc^2 x - \csc x} = 1 + \sin x$.

EXAMPLE 3 Prove $\dfrac{\cos^4 t - \sin^4 t}{\cos^2 t} = 1 - \tan^2 t$.

PROOF In this example, factoring the numerator on the left side will reduce the exponents there from 4 to 2.

$$\begin{aligned} \frac{\cos^4 t - \sin^4 t}{\cos^2 t} &= \frac{(\cos^2 t + \sin^2 t)(\cos^2 t - \sin^2 t)}{\cos^2 t} && \text{Factor} \\ &= \frac{1\,(\cos^2 t - \sin^2 t)}{\cos^2 t} && \text{Pythagorean identity} \\ &= \frac{\cos^2 t}{\cos^2 t} - \frac{\sin^2 t}{\cos^2 t} && \text{Separate into two fractions} \\ &= 1 - \tan^2 t && \text{Ratio identity} \end{aligned}$$

■

PROBLEM 4

Prove $1 - \cos\theta = \dfrac{\sin^2\theta}{1 + \cos\theta}$.

EXAMPLE 4 Prove $1 + \cos\theta = \dfrac{\sin^2\theta}{1 - \cos\theta}$.

PROOF We begin this proof by applying an alternate form of the Pythagorean identity to the right side to write $\sin^2\theta$ as $1 - \cos^2\theta$. Then we factor $1 - \cos^2\theta$ as the difference of two squares and reduce to lowest terms.

$$\begin{aligned} \frac{\sin^2\theta}{1 - \cos\theta} &= \frac{1 - \cos^2\theta}{1 - \cos\theta} && \text{Pythagorean identity} \\ &= \frac{(1 - \cos\theta)(1 + \cos\theta)}{1 - \cos\theta} && \text{Factor} \\ &= 1 + \cos\theta && \text{Reduce} \end{aligned}$$

■

Using Technology Verifying Identities

You can use your graphing calculator to decide if an equation is an identity or not. If the two expressions are indeed equal for all defined values of the variable, then they should produce identical graphs. Although this does not constitute a proof, it does give strong evidence that the identity is true.

We can verify the identity in Example 4 by defining the expression on the left as a function Y_1 and the expression on the right as a second function Y_2. If your calculator is equipped with different graphing styles, set the style of Y_2 so that you will be able to distinguish the second graph from the first. (In Figure 1, we have used the *path* style on a TI-84 for the second function.) Also, be sure your calculator is set to radian mode. Set your window variables so that

$$-2\pi \le x \le 2\pi,\ \text{scale} = \pi/2;\ -4 \le y \le 4,\ \text{scale} = 1$$

```
Plot1 Plot2 Plot3
\Y1=1+cos(X)
-0Y2=(sin(X))²/(1–cos(X))
\Y3=
\Y4=
\Y5=
\Y6=
\Y7=
```

Figure 1

When you graph the functions, your calculator screen should look similar to Figure 2 (the small circle is a result of the path style in action). Observe that the two graphs are identical.

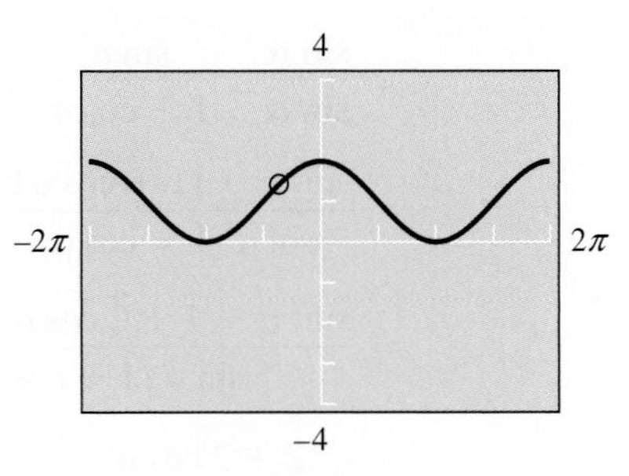

Figure 2

If your calculator is not equipped with different graphing styles, it may be difficult to tell if the second graph really coincides with the first. In this case you can trace the graph, and switch between the two functions at several points to convince yourself that the two graphs are indeed the same.

PROBLEM 5

Prove $\sin\theta\cos\theta = \dfrac{\sin\theta - \cos\theta}{\sec\theta - \csc\theta}$.

EXAMPLE 5 Prove $\tan x + \cot x = \sec x \csc x$.

PROOF We begin this proof by writing the left side in terms of $\sin x$ and $\cos x$. Then we simplify the left side by finding a common denominator in order to add the resulting fractions.

$$\tan x + \cot x = \frac{\sin x}{\cos x} + \frac{\cos x}{\sin x} \qquad \text{Changes to sines and cosines}$$

$$= \frac{\sin x}{\cos x} \cdot \frac{\mathbf{\sin x}}{\mathbf{\sin x}} + \frac{\cos x}{\sin x} \cdot \frac{\mathbf{\cos x}}{\mathbf{\cos x}} \qquad \text{LCD}$$

$$= \frac{\sin^2 x + \cos^2 x}{\cos x \sin x} \qquad \text{Add fractions}$$

$$= \frac{1}{\cos x \sin x} \qquad \text{Pythagorean identity}$$

$$= \frac{1}{\cos x} \cdot \frac{1}{\sin x} \qquad \text{Write as separate fractions}$$

$$= \sec x \csc x \qquad \text{Reciprocal identities}$$

■

PROBLEM 6

Prove

$\dfrac{1}{1 - \sin x} + \dfrac{1 - \sin x}{\cos^2 x} = 2\sec^2 x$.

EXAMPLE 6 Prove $\dfrac{\sin\alpha}{1 + \cos\alpha} + \dfrac{1 + \cos\alpha}{\sin\alpha} = 2\csc\alpha$.

PROOF The common denominator for the left side of the equation is $\sin\alpha\,(1 + \cos\alpha)$. We multiply the first fraction by $(\sin\alpha)/(\sin\alpha)$ and the second fraction by $(1 + \cos \alpha)/(1 + \cos \alpha)$ to produce two equivalent fractions with the same denominator.

$$\frac{\sin \alpha}{1 + \cos \alpha} + \frac{1 + \cos \alpha}{\sin \alpha}$$

$$= \frac{\mathbf{sin}\ \boldsymbol{\alpha}}{\mathbf{sin}\ \boldsymbol{\alpha}} \cdot \frac{\sin \alpha}{1 + \cos \alpha} + \frac{1 + \cos \alpha}{\sin \alpha} \cdot \frac{\mathbf{1 + cos}\ \boldsymbol{\alpha}}{\mathbf{1 + cos}\ \boldsymbol{\alpha}} \qquad \text{LCD}$$

$$= \frac{\sin^2 \alpha + (1 + \cos \alpha)^2}{\sin \alpha\,(1 + \cos \alpha)} \qquad \text{Add numerators}$$

$$= \frac{\sin^2 \alpha + 1 + 2\cos \alpha + \cos^2 \alpha}{\sin \alpha\,(1 + \cos \alpha)} \qquad \text{Expand } (1 + \cos \alpha)^2$$

$$= \frac{2 + 2\cos \alpha}{\sin \alpha\,(1 + \cos \alpha)} \qquad \text{Pythagorean identity}$$

$$= \frac{2(1 + \cos \alpha)}{\sin \alpha\,(1 + \cos \alpha)} \qquad \text{Factor out a 2}$$

$$= \frac{2}{\sin \alpha} \qquad \text{Reduce}$$

$$= 2 \csc \alpha \qquad \text{Reciprocal identity}$$

PROBLEM 7

Prove $\dfrac{1 + \cos t}{\sin t} = \dfrac{\sin t}{1 - \cos t}$.

EXAMPLE 7 Prove $\dfrac{1 + \sin t}{\cos t} = \dfrac{\cos t}{1 - \sin t}$.

PROOF The key to proving this identity requires that we multiply the numerator and denominator on the right side by $1 + \sin t$. (This is similar to rationalizing the denominator.)

$$\frac{\cos t}{1 - \sin t} = \frac{\cos t}{1 - \sin t} \cdot \frac{\mathbf{1 + sin}\ \boldsymbol{t}}{\mathbf{1 + sin}\ \boldsymbol{t}} \qquad \text{Multiply numerator and denominator by } 1 + \sin t$$

$$= \frac{\cos t\,(1 + \sin t)}{1 - \sin^2 t} \qquad \text{Multiply out the denominator}$$

$$= \frac{\cos t\,(1 + \sin t)}{\cos^2 t} \qquad \text{Pythagorean identity}$$

$$= \frac{1 + \sin t}{\cos t} \qquad \text{Reduce}$$

Note that it would have been just as easy for us to verify this identity by multiplying the numerator and denominator on the left side by $1 - \sin t$.

In the previous examples, we have concentrated on methods for proving that a statement is an identity, meaning that the two expressions are equal for all replacements of the variable for which each expression is defined. To show that a statement is *not* an identity is usually much simpler. All we must do is find a single value of the variable for which each expression is defined, but which makes the statement false. This is known as finding a *counterexample*.

PROBLEM 8
Show $\cos x = \sqrt{1 - \sin^2 x}$ is not an identity by finding a counterexample.

EXAMPLE 8 Show that $\cot^2 \theta + \cos^2 \theta = \cot^2 \theta \cos^2 \theta$ is not an identity by finding a counterexample.

SOLUTION Because $\cot \theta$ is undefined for $\theta = k\pi$, where k is any integer, we must choose some other value of θ as a counterexample. Using $\theta = \pi/4$, we find

$$\cot^2 \frac{\pi}{4} + \cos^2 \frac{\pi}{4} \stackrel{?}{=} \cot^2 \frac{\pi}{4} \cos^2 \frac{\pi}{4}$$

$$\left(\cot \frac{\pi}{4}\right)^2 + \left(\cos \frac{\pi}{4}\right)^2 \stackrel{?}{=} \left(\cot \frac{\pi}{4}\right)^2 \left(\cos \frac{\pi}{4}\right)^2$$

$$(1)^2 + \left(\frac{\sqrt{2}}{2}\right)^2 \stackrel{?}{=} (1)^2 \left(\frac{\sqrt{2}}{2}\right)^2$$

$$1 + \frac{1}{2} \stackrel{?}{=} 1 \cdot \frac{1}{2}$$

$$\frac{3}{2} \neq \frac{1}{2}$$

Therefore, $\cot^2 \theta + \cos^2 \theta \neq \cot^2 \theta \cos^2 \theta$ when $\theta = \pi/4$, so the statement is not an identity. ■

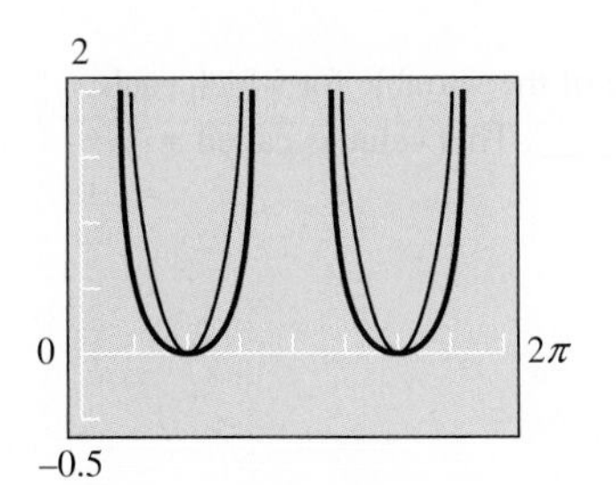

Figure 3

Using Technology Finding Counterexamples

In the previous Using Technology section, we saw that a graphing calculator can be used to determine whether an equation appears to be an identity or not. Using this approach with Example 8, we define

$$\text{Y}_1 = (\tan(x))^{-2} + (\cos(x))^2$$

$$\text{and} \quad \text{Y}_2 = (\tan(x))^{-2}(\cos(x))^2 \qquad \text{(Bold style)}$$

As shown in Figure 3, the two graphs are not identical so we suspect that this equation is not an identity. To find a counterexample, simply choose any value of the variable where the two graphs do not overlap. In Figure 4 we have used the calculator to evaluate each expression at $\pi/4$. You can see how the expressions give different results.

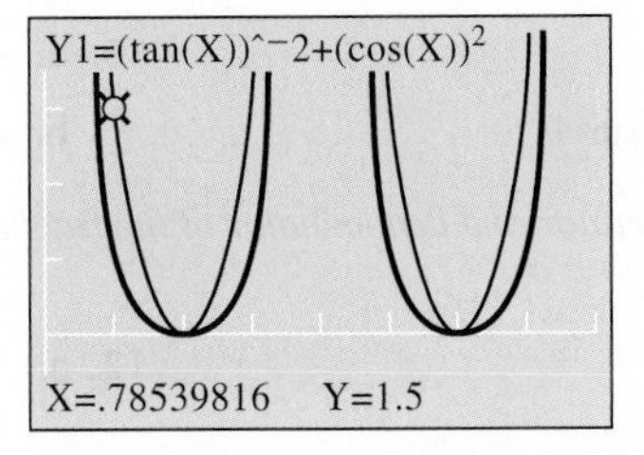

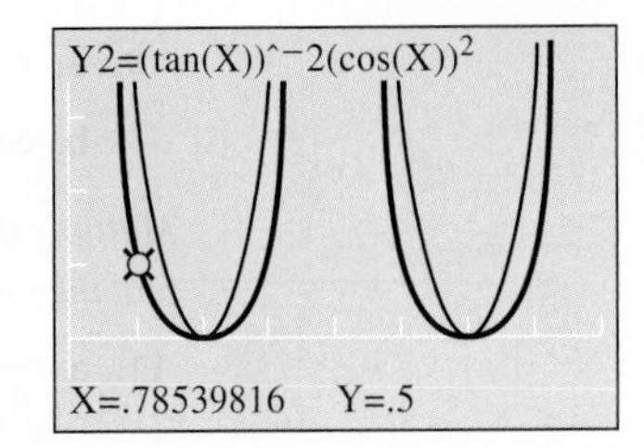

Figure 4

Getting Ready for Class

After reading through the preceding section, respond in your own words and in complete sentences.

a. What is an identity?
b. In trigonometry, how do we prove an identity?
c. How do we prove an equation is not an identity?
d. What is a counterexample?

5.1 PROBLEM SET

➤ CONCEPTS AND VOCABULARY

For Questions 1 through 6, fill in each blank with the appropriate word.

1. An identity is a statement that two expressions are ______ for all replacements of the variable for which each expression is _________.

2. To prove, or verify, an identity, we start with ______ ______ of the equation and __________ it until it is identical to the ______ ______ of the equation.

3. To prove an identity, it is usually best to start with the more ___________ side first.

4. If nothing else comes to mind, try changing everything in the expression into ______ and ______ and then simplify.

5. To investigate if an equation is an identity, graph the left side and the right side separately and see if the two graphs are _________.

6. To prove that an equation is not an identity, find a value of the variable for which each expression is ______, but which makes the statement ______. This value is called a ______________.

➤ EXERCISES

Factor each expression completely.

7. a. $x^2 - xy$
b. $\sin^2 \theta - \sin \theta \cos \theta$

8. a. $1 - y^2$
b. $1 - \sin^2 \theta$

9. a. $x^2 - y^2$
b. $\cos^2 \theta - \sin^2 \theta$

10. a. $x^4 - y^4$
b. $\sin^4 \theta - \cos^4 \theta$

Multiply the numerator and denominator of the fraction by the conjugate of the denominator, and then simplify.

11. a. $\dfrac{1}{1 + \sqrt{3}}$
b. $\dfrac{1}{1 + \cos x}$

12. a. $\dfrac{2}{1 - \sqrt{3}}$
b. $\dfrac{\cos x}{1 - \sin x}$

13. **a.** $\dfrac{1 - \sqrt{2}}{1 + \sqrt{2}}$

b. $\dfrac{1 - \sin x}{1 + \sin x}$

14. **a.** $\dfrac{\sqrt{3} + 1}{\sqrt{3} - 1}$

b. $\dfrac{\csc x + 1}{\csc x - 1}$

Prove that each of the following identities is true.

15. $\csc \theta \tan \theta = \sec \theta$

16. $\sec \theta \cot \theta = \csc \theta$

17. $\dfrac{\tan A}{\sec A} = \sin A$

18. $\dfrac{\cot A}{\csc A} = \cos A$

19. $\cos x (\csc x + \tan x) = \cot x + \sin x$

20. $\sin x (\sec x + \csc x) = \tan x + 1$

21. $\cot x - 1 = \cos x (\csc x - \sec x)$

22. $\tan x (\cos x + \cot x) = \sin x + 1$

23. $\sin^2 x (\cot^2 x + 1) = 1$

24. $\cos^2 x (1 + \tan^2 x) = 1$

25. $\dfrac{\cos^4 t - \sin^4 t}{\sin^2 t} = \cot^2 t - 1$

26. $\dfrac{\sin^4 t - \cos^4 t}{\sin^2 t \cos^2 t} = \sec^2 t - \csc^2 t$

27. $1 + \sin \theta = \dfrac{\cos^2 \theta}{1 - \sin \theta}$

28. $1 - \sin \theta = \dfrac{\cos^2 \theta}{1 + \sin \theta}$

29. $\dfrac{1 - \sin^4 \theta}{1 + \sin^2 \theta} = \cos^2 \theta$

30. $\dfrac{1 - \cos^4 \theta}{1 + \cos^2 \theta} = \sin^2 \theta$

31. $\sec^2 \theta - \tan^2 \theta = 1$

32. $\csc^2 \theta - \cot^2 \theta = 1$

33. $\sec^4 \theta - \tan^4 \theta = \dfrac{1 + \sin^2 \theta}{\cos^2 \theta}$

34. $\csc^4 \theta - \cot^4 \theta = \dfrac{1 + \cos^2 \theta}{\sin^2 \theta}$

35. $\tan \theta - \cot \theta = \dfrac{\sin^2 \theta - \cos^2 \theta}{\sin \theta \cos \theta}$

36. $\sec \theta - \csc \theta = \dfrac{\sin \theta - \cos \theta}{\sin \theta \cos \theta}$

37. $\csc B - \sin B = \cot B \cos B$

38. $\sec B - \cos B = \tan B \sin B$

39. $\cot \theta \cos \theta + \sin \theta = \csc \theta$

40. $\tan \theta \sin \theta + \cos \theta = \sec \theta$

41. $\dfrac{\cos x}{1 + \sin x} + \dfrac{1 + \sin x}{\cos x} = 2 \sec x$

42. $\dfrac{\cos x}{1 + \sin x} - \dfrac{1 - \sin x}{\cos x} = 0$

43. $\dfrac{1}{1 + \cos x} + \dfrac{1}{1 - \cos x} = 2 \csc^2 x$

44. $\dfrac{1}{1 - \sin x} + \dfrac{1}{1 + \sin x} = 2 \sec^2 x$

45. $\dfrac{1 - \sec x}{1 + \sec x} = \dfrac{\cos x - 1}{\cos x + 1}$

46. $\dfrac{\csc x - 1}{\csc x + 1} = \dfrac{1 - \sin x}{1 + \sin x}$

47. $\dfrac{\sin t}{1 + \cos t} = \dfrac{1 - \cos t}{\sin t}$

48. $\dfrac{\cos t}{1 + \sin t} = \dfrac{1 - \sin t}{\cos t}$

49. $\dfrac{(1 - \sin t)^2}{\cos^2 t} = \dfrac{1 - \sin t}{1 + \sin t}$

50. $\dfrac{\sin^2 t}{(1 - \cos t)^2} = \dfrac{1 + \cos t}{1 - \cos t}$

51. $\dfrac{\sec \theta + 1}{\tan \theta} = \dfrac{\tan \theta}{\sec \theta - 1}$

52. $\dfrac{\csc \theta - 1}{\cot \theta} = \dfrac{\cot \theta}{\csc \theta + 1}$

53. $\dfrac{1 + \cos x}{1 - \cos x} = (\csc x + \cot x)^2$

54. $\dfrac{1 - \sin x}{1 + \sin x} = (\sec x - \tan x)^2$

55. $\sec x + \tan x = \dfrac{1}{\sec x - \tan x}$

56. $\dfrac{1}{\csc x - \cot x} = \csc x + \cot x$

57. $\dfrac{\sin x + 1}{\cos x + \cot x} = \tan x$

58. $\dfrac{\cos x + 1}{\cot x} = \sin x + \tan x$

59. $\sin^4 A - \cos^4 A = 1 - 2 \cos^2 A$

60. $\cos^4 A - \sin^4 A = 1 - 2 \sin^2 A$

61. $\dfrac{\cot^2 B - \cos^2 B}{\csc^2 B - 1} = \cos^2 B$

62. $\dfrac{\sin^2 B - \tan^2 B}{1 - \sec^2 B} = \sin^2 B$

63. $\dfrac{\sec^4 y - \tan^4 y}{\sec^2 y + \tan^2 y} = 1$

64. $\dfrac{\csc^2 y + \cot^2 y}{\csc^4 y - \cot^4 y} = 1$

65. $\dfrac{\sin^3 A - 8}{\sin A - 2} = \sin^2 A + 2\sin A + 4$

66. $\dfrac{1 - \cos^3 A}{1 - \cos A} = \cos^2 A + \cos A + 1$

67. $\dfrac{1 + \cot^3 t}{1 + \cot t} = \csc^2 t - \cot t$

68. $\dfrac{1 - \tan^3 t}{1 - \tan t} = \sec^2 t + \tan t$

Prove that each of the following statements is not an identity by finding a counterexample.

69. $\sin\theta = \sqrt{1 - \cos^2\theta}$

70. $\sin\theta + \cos\theta = 1$

71. $\sin\theta = \dfrac{1}{\cos\theta}$

72. $\tan^2\theta + \cot^2\theta = 1$

73. $\sqrt{\sin^2\theta + \cos^2\theta} = \sin\theta + \cos\theta$

74. $\sin\theta\cos\theta = 1$

Use your graphing calculator to determine if each equation appears to be an identity or not by graphing the left expression and right expression together. If so, verify the identity. If not, find a counterexample.

75. $(\sec B - 1)(\sec B + 1) = \tan^2 B$

76. $\dfrac{1 - \sec\theta}{\cos\theta} = \dfrac{\cos\theta}{1 + \sec\theta}$

77. $\sec x + \cos x = \tan x \sin x$

78. $\dfrac{\tan t}{\sec t + 1} = \dfrac{\sec t - 1}{\tan t}$

79. $\sec A - \csc A = \dfrac{\cos A - \sin A}{\cos A \sin A}$

80. $\cos^4\theta - \sin^4\theta = 2\cos^2\theta - 1$

81. $\dfrac{1}{1 - \sin x} + \dfrac{1}{1 + \sin x} = 2\sec^2 x$

82. $\cot^4 t - \tan^4 t = \dfrac{\sin^2 t + 1}{\cos^2 t}$

83. Show that $\sin(A + B)$ is not, in general, equal to $\sin A + \sin B$ by substituting 30° for A and 60° for B in both expressions and simplifying.

84. Show that $\sin 2x \neq 2\sin x$ by substituting 30° for x and then simplifying both sides.

➤ REVIEW PROBLEMS

The problems that follow review material we covered in Sections 1.4, 3.2, and 4.1. Reviewing these problems will help you with some of the material in the next section.

85. If $\sin A = 3/5$ and A terminates in quadrant I, find $\cos A$ and $\tan A$.

86. If $\cos B = -5/13$ with B in quadrant III, find $\sin B$ and $\tan B$.

Give the exact value of each of the following.

87. $\sin\dfrac{\pi}{3}$

88. $\cos\dfrac{\pi}{3}$

89. $\cos\dfrac{\pi}{6}$

90. $\sin\dfrac{\pi}{6}$

Convert to degrees.

91. $\dfrac{\pi}{12}$

92. $\dfrac{5\pi}{12}$

Prove each identity.

93. $\csc\theta + \sin(-\theta) = \dfrac{\cos^2\theta}{\sin\theta}$

94. $\sec\theta - \cos(-\theta) = \dfrac{\sin^2\theta}{\cos\theta}$

➤ LEARNING OBJECTIVES ASSESSMENT

These questions are available for instructors to help assess if you have successfully met the learning objectives for this section.

95. Which of the following is a valid step in proving the identity $\dfrac{\cos x}{1 + \sin x} = \dfrac{1 - \sin x}{\cos x}$?

a. $\cos^2 x = (1 - \sin x)(1 + \sin x)$

b. $\dfrac{\cos x}{1 + \sin x} = \dfrac{\cos x}{1 + \sin x} \cdot \dfrac{1 - \sin x}{1 - \sin x}$

c. $\dfrac{\cos x}{1 + \sin x} = \dfrac{\cos x}{1} + \dfrac{\cos x}{\sin x}$

d. $\dfrac{\cos x}{1 + \sin x} = \dfrac{\cos^2 x}{(1 + \sin x)^2}$

96. Which value is a counterexample to prove that $\dfrac{1 + \cos x}{\cot x} = \sin x - \tan x$ is not an identity?

a. $x = \dfrac{3\pi}{4}$ **b.** $x = 0$ **c.** $x = \pi$ **d.** $x = \dfrac{\pi}{2}$

97. Use your graphing calculator to identify which of the following equations is most likely an identity.

a. $\dfrac{\sin x + 1}{\cos x + \cot x} = \sec x$ **b.** $\dfrac{1}{\csc x - \cot x} = \csc x + \cot x$

c. $\dfrac{1 + \cos^2 x}{\sin^2 x} = \csc^4 x + \cot^4 x$ **d.** $\dfrac{1 + \cos x}{1 - \cos x} = \csc^2 x + \cot^2 x$

SECTION 5.2 Sum and Difference Formulas

LEARNING OBJECTIVES

1. Use a sum or difference formula to find the exact value of a trigonometric function.
2. Prove an equation is an identity using a sum or difference formula.
3. Simplify an expression using a sum or difference formula.
4. Use a sum or difference formula to graph an equation.

The expressions $\sin(A + B)$ and $\cos(A + B)$ occur frequently enough in mathematics that it is necessary to find expressions equivalent to them that involve sines and cosines of single angles. The most obvious question to begin with is

$$\text{Is } \sin(A + B) = \sin A + \sin B?$$

The answer is no. Substituting almost any pair of numbers for A and B in the formula will yield a false statement. As a counterexample, we can let $A = 30°$ and $B = 60°$ in the preceding formula and then simplify each side.

$$\sin(30° + 60°) \stackrel{?}{=} \sin 30° + \sin 60°$$

$$\sin 90° \stackrel{?}{=} \frac{1}{2} + \frac{\sqrt{3}}{2}$$

$$1 \neq \frac{1 + \sqrt{3}}{2}$$

The formula just doesn't work. The next question is, what are the formulas for $\sin(A + B)$ and $\cos(A + B)$? The answer to that question is what this section is all about. Let's start by deriving the formula for $\cos(A + B)$.

We begin by drawing angle A in standard position and then adding B to it. Next we draw $-B$ in standard position. Figure 1 shows these angles in relation to the unit circle. Note that the points on the unit circle through which the terminal sides of the angles A, $A + B$, and $-B$ pass have been labeled with the sines and cosines of those angles.

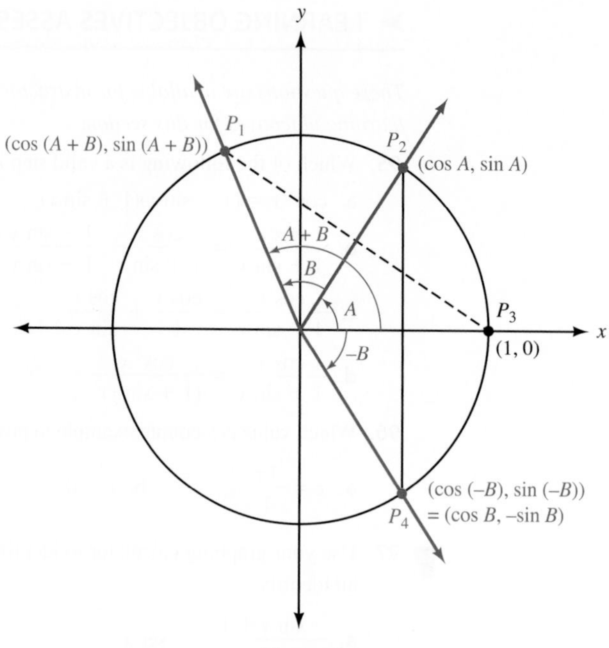

Figure 1

To derive the formula for $\cos(A + B)$, we simply have to see that length $P_1 P_3$ is equal to length $P_2 P_4$. (From geometry, they are chords cut off by equal central angles.)

$$P_1 P_3 = P_2 P_4$$

Squaring both sides gives us

$$(P_1 P_3)^2 = (P_2 P_4)^2$$

Now, applying the distance formula, we have

$$[\cos(A + B) - 1]^2 + [\sin(A + B) - 0]^2 = (\cos A - \cos B)^2 + (\sin A + \sin B)^2$$

Let's call this Equation 1. Taking the left side of Equation 1, expanding it, and then simplifying by using the first Pythagorean identity gives us

Left Side of Equation 1

$$\cos^2(A + B) - 2\cos(A + B) + 1 + \sin^2(A + B) \quad \text{Expand squares}$$
$$= -2\cos(A + B) + 2 \quad \text{Pythagorean identity}$$

Applying the same two steps to the right side of Equation 1 gives us

Right Side of Equation 1

$$\cos^2 A - 2\cos A\cos B + \cos^2 B + \sin^2 A + 2\sin A\sin B + \sin^2 B$$
$$= -2\cos A\cos B + 2\sin A\sin B + 2$$

Equating the simplified versions of the left and right sides of Equation 1 we have

$$-2\cos(A + B) + 2 = -2\cos A\cos B + 2\sin A\sin B + 2$$

Adding -2 to both sides and then dividing both sides by -2 gives us the formula we are after:

$$\boxed{\cos(A + B) = \cos A\cos B - \sin A\sin B}$$

This is the first formula in a series of formulas for trigonometric functions of the sum or difference of two angles. It must be memorized. Before we derive the others, let's look at some of the ways we can use our first formula.

PROBLEM 1
Find the exact value of $\cos 15°$.

EXAMPLE 1 Find the exact value for $\cos 75°$.

SOLUTION We write 75° as 45° + 30° and then apply the formula for $\cos(A + B)$.

$$\begin{aligned}\cos 75° &= \cos(45° + 30°)\\ &= \cos 45° \cos 30° - \sin 45° \sin 30°\\ &= \frac{\sqrt{2}}{2}\cdot\frac{\sqrt{3}}{2} - \frac{\sqrt{2}}{2}\cdot\frac{1}{2}\\ &= \frac{\sqrt{6}-\sqrt{2}}{4}\end{aligned}$$

NOTE If you completed the Chapter 2 Group Project, compare your value of $\cos 75°$ from the project with our result in Example 1. Convince yourself that the two values are the same.

PROBLEM 2
Prove $\sin\left(x + \frac{\pi}{2}\right) = \cos x$.

EXAMPLE 2 Show that $\cos(x + 2\pi) = \cos x$.

SOLUTION Applying the formula for $\cos(A + B)$, we have

$$\begin{aligned}\cos(x + 2\pi) &= \cos x \cos 2\pi - \sin x \sin 2\pi\\ &= \cos x \cdot 1 - \sin x \cdot 0\\ &= \cos x\end{aligned}$$

Notice that this is not a new relationship. We already know that if two angles are coterminal, then their cosines are equal—and $x + 2\pi$ and x are coterminal. What we have done here is shown this to be true with a formula instead of the definition of cosine.

PROBLEM 3
Write $\sin 5x \cos x + \cos 5x \sin x$ as a single sine.

EXAMPLE 3 Write $\cos 3x \cos 2x - \sin 3x \sin 2x$ as a single cosine.

SOLUTION We apply the formula for $\cos(A + B)$ in the reverse direction from the way we applied it in the first two examples.

$$\begin{aligned}\cos 3x \cos 2x - \sin 3x \sin 2x &= \cos(3x + 2x)\\ &= \cos 5x\end{aligned}$$

Here is the derivation of the formula for $\cos(A - B)$. It involves the formula for $\cos(A + B)$ and the formulas for even and odd functions.

$$\begin{aligned}\cos(A - B) &= \cos[A + (-B)] && \text{Write } A - B \text{ as a sum}\\ &= \cos A \cos(-B) - \sin A \sin(-B) && \text{Sum formula}\\ &= \cos A \cos B - \sin A(-\sin B) && \text{Cosine is an even function, sine is odd}\\ &= \cos A \cos B + \sin A \sin B\end{aligned}$$

The only difference in the formulas for the expansion of $\cos(A + B)$ and $\cos(A - B)$ is the sign between the two terms. Here are both formulas.

$$\cos(A + B) = \cos A \cos B - \sin A \sin B$$
$$\cos(A - B) = \cos A \cos B + \sin A \sin B$$

Again, both formulas are important and should be memorized.

PROBLEM 4
Show that $\sin (90° - B) = \cos B$.

EXAMPLE 4 Show that $\cos (90° - A) = \sin A$.

SOLUTION We will need this formula when we derive the formula for $\sin (A + B)$.

$$\begin{aligned} \cos (90° - A) &= \cos 90° \cos A + \sin 90° \sin A \\ &= 0 \cdot \cos A + 1 \cdot \sin A \\ &= \sin A \end{aligned}$$

Note that the formula we just derived is not a new formula. The angles $90° - A$ and A are complementary angles, and we already know by the Cofunction Theorem from Chapter 2 that the sine of an angle is always equal to the cosine of its complement. We could also state it this way:

$$\sin (90° - A) = \cos A$$

We can use this information to derive the formula for $\sin (A + B)$. To understand this derivation, you must recognize that $A + B$ and $90° - (A + B)$ are complementary angles.

$$\sin (A + B) = \cos [90° - (A + B)] \quad \text{The sine of an angle is the cosine of its complement}$$

$$= \cos [90° - A - B] \quad \text{Remove parentheses}$$

$$= \cos [(90° - A) - B] \quad \text{Regroup within brackets}$$

Now we expand using the formula for the cosine of a difference.

$$\begin{aligned} &= \cos (90° - A) \cos B + \sin (90° - A) \sin B \\ &= \sin A \cos B + \cos A \sin B \end{aligned}$$

This gives us an expansion formula for $\sin (A + B)$.

$$\boxed{\sin (A + B) = \sin A \cos B + \cos A \sin B}$$

This is the formula for the sine of a sum. To find the formula for $\sin (A - B)$, we write $A - B$ as $A + (-B)$ and proceed as follows:

$$\begin{aligned} \sin (A - B) &= \sin (A + (-B)) \\ &= \sin A \cos (-B) + \cos A \sin (-B) \\ &= \sin A \cos B - \cos A \sin B \end{aligned}$$

This gives us the formula for the sine of a difference.

$$\boxed{\sin (A - B) = \sin A \cos B - \cos A \sin B}$$

PROBLEM 5
Graph $y = 5 (\cos 6x \cos 4x + \sin 6x \sin 4x)$ for $0 \le x \le 2\pi$.

EXAMPLE 5 Graph $y = 4 \sin 5x \cos 3x - 4 \cos 5x \sin 3x$ from $x = 0$ to $x = 2\pi$.

SOLUTION To write the equation in the form $y = A \sin Bx$, we factor 4 from each term on the right and then apply the formula for $\sin (A - B)$ to the remaining expression to write it as a single trigonometric function.

$$\begin{aligned} y &= 4 \sin 5x \cos 3x - 4 \cos 5x \sin 3x \\ &= 4 (\sin 5x \cos 3x - \cos 5x \sin 3x) \\ &= 4 \sin (5x - 3x) \\ &= 4 \sin 2x \end{aligned}$$

The graph of $y = 4 \sin 2x$ will have an amplitude of 4 and a period of $2\pi/2 = \pi$. The graph is shown in Figure 2.

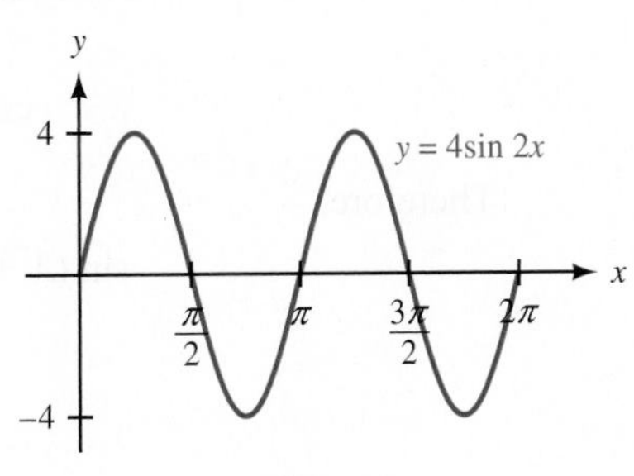

Figure 2

PROBLEM 6

Find the exact value of $\cos \dfrac{5\pi}{12}$.

EXAMPLE 6 Find the exact value of $\sin \dfrac{\pi}{12}$.

SOLUTION We have to write $\pi/12$ in terms of two numbers for which the exact values are known. The numbers $\pi/3$ and $\pi/4$ will work because their difference is $\pi/12$.

$$\begin{aligned} \sin \frac{\pi}{12} &= \sin\left(\frac{\pi}{3} - \frac{\pi}{4}\right) \\ &= \sin \frac{\pi}{3} \cos \frac{\pi}{4} - \cos \frac{\pi}{3} \sin \frac{\pi}{4} \\ &= \frac{\sqrt{3}}{2} \cdot \frac{\sqrt{2}}{2} - \frac{1}{2} \cdot \frac{\sqrt{2}}{2} \\ &= \frac{\sqrt{6} - \sqrt{2}}{4} \end{aligned}$$

This is the same answer we obtained in Example 1 when we found the exact value of cos 75°. It should be, though, because $\pi/12 = 15°$, which is the complement of 75°, and the cosine of an angle is equal to the sine of its complement.

PROBLEM 7

If $\cos A = -7/25$ with A in QIII and $\sin B = 40/41$ with B in QI, find $\sin (A - B)$, $\cos (A - B)$, and $\tan (A - B)$.

EXAMPLE 7 If $\sin A = \frac{3}{5}$ with A in QI and $\cos B = -\frac{5}{13}$ with B in QIII, find $\sin (A + B)$, $\cos (A + B)$, and $\tan (A + B)$.

SOLUTION We have $\sin A$ and $\cos B$. We need to find $\cos A$ and $\sin B$ before we can apply any of our formulas. Some equivalent forms of our first Pythagorean identity will help here.

If $\sin A = \frac{3}{5}$ with A in QI, then

$$\begin{aligned} \cos A &= \pm\sqrt{1 - \sin^2 A} \\ &= +\sqrt{1 - \left(\frac{3}{5}\right)^2} && A \in \text{QI, so } \cos A \text{ is positive} \\ &= \frac{4}{5} \end{aligned}$$

If $\cos B = -\frac{5}{13}$ with B in QIII, then

$$\begin{aligned} \sin B &= \pm\sqrt{1 - \cos^2 B} \\ &= -\sqrt{1 - \left(-\frac{5}{13}\right)^2} && B \in \text{QIII, so } \sin B \text{ is negative} \\ &= -\frac{12}{13} \end{aligned}$$

We have

$$\sin A = \frac{3}{5} \qquad \sin B = -\frac{12}{13}$$

$$\cos A = \frac{4}{5} \qquad \cos B = -\frac{5}{13}$$

Therefore,

$$\begin{aligned}\sin (A + B) &= \sin A \cos B + \cos A \sin B \\ &= \frac{3}{5}\left(-\frac{5}{13}\right) + \frac{4}{5}\left(-\frac{12}{13}\right) \\ &= -\frac{63}{65}\end{aligned}$$

$$\begin{aligned}\cos (A + B) &= \cos A \cos B - \sin A \sin B \\ &= \frac{4}{5}\left(-\frac{5}{13}\right) - \frac{3}{5}\left(-\frac{12}{13}\right) \\ &= \frac{16}{65}\end{aligned}$$

$$\begin{aligned}\tan (A + B) &= \frac{\sin (A + B)}{\cos (A + B)} \\ &= \frac{-63/65}{16/65} \\ &= -\frac{63}{16}\end{aligned}$$

Notice also that $A + B$ must terminate in QIV because

$$\sin (A + B) < 0 \qquad \text{and} \qquad \cos (A + B) > 0$$

While working through the last part of Example 7, you may have wondered if there is a separate formula for $\tan (A + B)$. (More likely, you are hoping there isn't.) There is, and it is derived from the formulas we already have.

$$\begin{aligned}\tan (A + B) &= \frac{\sin (A + B)}{\cos (A + B)} \\ &= \frac{\sin A \cos B + \cos A \sin B}{\cos A \cos B - \sin A \sin B}\end{aligned}$$

To be able to write this last line in terms of tangents only, we must divide numerator and denominator by $\cos A \cos B$.

$$\begin{aligned}&= \frac{\dfrac{\sin A \cos B}{\cos A \cos B} + \dfrac{\cos A \sin B}{\cos A \cos B}}{\dfrac{\cos A \cos B}{\cos A \cos B} - \dfrac{\sin A \sin B}{\cos A \cos B}} \\ &= \frac{\tan A + \tan B}{1 - \tan A \tan B}\end{aligned}$$

The formula for $\tan (A + B)$ is

$$\boxed{\tan (A + B) = \frac{\tan A + \tan B}{1 - \tan A \tan B}}$$

Because tangent is an odd function, the formula for $\tan(A - B)$ will look like this:

$$\tan(A - B) = \frac{\tan A - \tan B}{1 + \tan A \tan B}$$

PROBLEM 8
If $\sin A = -8/17$ with A in QIII and $\cos B = 7/25$ with B in QI, find $\tan(A - B)$ using the formula $\tan(A - B) = \dfrac{\tan A - \tan B}{1 + \tan A \tan B}$.

EXAMPLE 8 If $\sin A = \frac{3}{5}$ with A in QI and $\cos B = -\frac{5}{13}$ with B in QIII, find $\tan(A + B)$ by using the formula

$$\tan(A + B) = \frac{\tan A + \tan B}{1 - \tan A \tan B}$$

SOLUTION The angles A and B as given here are the same ones used previously in Example 7. Looking over Example 7 again, we find that

$$\tan A = \frac{\sin A}{\cos A} \quad \text{and} \quad \tan B = \frac{\sin B}{\cos B}$$

$$= \frac{\frac{3}{5}}{\frac{4}{5}} \qquad\qquad = \frac{-\frac{12}{13}}{-\frac{5}{13}}$$

$$= \frac{3}{4} \qquad\qquad = \frac{12}{5}$$

Therefore,

$$\tan(A + B) = \frac{\tan A + \tan B}{1 - \tan A \tan B}$$

$$= \frac{\frac{3}{4} + \frac{12}{5}}{1 - \frac{3}{4} \cdot \frac{12}{5}}$$

$$= \frac{\frac{15}{20} + \frac{48}{20}}{1 - \frac{9}{5}}$$

$$= \frac{\frac{63}{20}}{-\frac{4}{5}}$$

$$= -\frac{63}{16}$$

which is the same result we obtained previously. ■

Getting Ready for Class

After reading through the preceding section, respond in your own words and in complete sentences.

a. Why is it necessary to have sum and difference formulas for sine, cosine, and tangent?
b. Write both the sum and the difference formulas for cosine.
c. Write both the sum and the difference formulas for sine.
d. Write both the sum and the difference formulas for tangent.

5.2 PROBLEM SET

➤ CONCEPTS AND VOCABULARY

For Questions 1 through 6, complete each sum or difference identity.

1. $\sin(x + y) =$ ________
2. $\sin(x - y) =$ ________
3. $\cos(\theta + \phi) =$ ________
4. $\cos(\theta - \phi) =$ ________
5. $\tan(C + D) =$ ________
6. $\tan(C - D) =$ ________

For Questions 7 and 8, determine if the statement is true or false.

7. The sine of the sum of two angles equals the sum of the sines of those angles.

8. The cosine of the difference of two angles equals the difference of the cosines of those angles.

➤ EXERCISES

Find exact values for each of the following.

9. $\sin 15°$
10. $\sin 75°$
11. $\tan 15°$
12. $\tan 75°$
13. $\cos \dfrac{7\pi}{12}$
14. $\sin \dfrac{7\pi}{12}$
15. $\cos 105°$
16. $\sin 105°$

Show that each of the following is true.

17. $\sin(x + 2\pi) = \sin x$
18. $\cos(x - 2\pi) = \cos x$
19. $\cos\left(x - \dfrac{\pi}{2}\right) = \sin x$
20. $\sin\left(x - \dfrac{\pi}{2}\right) = -\cos x$
21. $\cos(180° - \theta) = -\cos\theta$
22. $\sin(180° - \theta) = \sin\theta$
23. $\sin(90° + \theta) = \cos\theta$
24. $\cos(90° + \theta) = -\sin\theta$
25. $\tan\left(x + \dfrac{\pi}{4}\right) = \dfrac{1 + \tan x}{1 - \tan x}$
26. $\tan\left(x - \dfrac{\pi}{4}\right) = \dfrac{\tan x - 1}{\tan x + 1}$
27. $\sin\left(\dfrac{3\pi}{2} - x\right) = -\cos x$
28. $\cos\left(x - \dfrac{3\pi}{2}\right) = -\sin x$

Write each expression as a single trigonometric function.

29. $\sin 3x \cos 2x + \cos 3x \sin 2x$

30. $\cos 3x \cos 2x + \sin 3x \sin 2x$

31. $\cos 5x \cos x - \sin 5x \sin x$

32. $\sin 8x \cos x - \cos 8x \sin x$

33. $\cos 15° \cos 75° - \sin 15° \sin 75°$

34. $\cos 15° \cos 75° + \sin 15° \sin 75°$

Graph each of the following from $x = 0$ to $x = 2\pi$.

35. $y = \sin 5x \cos 3x - \cos 5x \sin 3x$

36. $y = \sin x \cos 2x + \cos x \sin 2x$

37. $y = 3 \cos 7x \cos 5x + 3 \sin 7x \sin 5x$

38. $y = 2 \cos 4x \cos x + 2 \sin 4x \sin x$

39. Graph one complete cycle of $y = \sin x \cos \frac{\pi}{6} - \cos x \sin \frac{\pi}{6}$ by first rewriting the right side in the form $\sin (A - B)$.

40. Graph one complete cycle of $y = \sin x \cos \frac{\pi}{4} + \cos x \sin \frac{\pi}{4}$ by first rewriting the right side in the form $\sin (A + B)$.

41. Graph one complete cycle of $y = 2 (\sin x \cos \frac{\pi}{3} + \cos x \sin \frac{\pi}{3})$ by first rewriting the right side in the form $2 \sin (A + B)$.

42. Graph one complete cycle of $y = 2 (\sin x \cos \frac{\pi}{3} - \cos x \sin \frac{\pi}{3})$ by first rewriting the right side in the form $2 \sin (A - B)$.

43. Let $\sin A = \frac{3}{5}$ with A in QII and $\sin B = -\frac{5}{13}$ with B in QIII. Find $\sin (A + B)$, $\cos (A + B)$, and $\tan (A + B)$. In what quadrant does $A + B$ terminate?

44. Let $\cos A = -\frac{5}{13}$ with A in QII and $\sin B = \frac{3}{5}$ with B in QI. Find $\sin (A - B)$, $\cos (A - B)$, and $\tan (A - B)$. In what quadrant does $A - B$ terminate?

45. If $\sin A = \sqrt{5}/5$ with A in QI and $\tan B = \frac{3}{4}$ with B in QI, find $\tan (A + B)$ and $\cot (A + B)$. In what quadrant does $A + B$ terminate?

46. If $\sec A = \sqrt{5}$ with A in QI and $\sec B = \sqrt{10}$ with B in QI, find $\sec (A + B)$. [First find $\cos (A + B)$.]

47. If $\tan (A + B) = 3$ and $\tan B = \frac{1}{2}$, find $\tan A$.

48. If $\tan (A + B) = 2$ and $\tan B = \frac{1}{3}$, find $\tan A$.

49. Write a formula for $\cos 2x$ by writing $\cos 2x$ as $\cos (x + x)$ and using the formula for the cosine of a sum.

50. Write a formula for $\sin 2x$ by writing $\sin 2x$ as $\sin (x + x)$ and using the formula for the sine of a sum.

Prove each identity.

51. $\sin (90° + x) + \sin (90° - x) = 2 \cos x$

52. $\sin (90° + x) - \sin (90° - x) = 0$

53. $\cos (x - 90°) - \cos (x + 90°) = 2 \sin x$

54. $\cos (x + 90°) + \cos (x - 90°) = 0$

55. $\sin \left(\frac{\pi}{6} + x\right) + \sin \left(\frac{\pi}{6} - x\right) = \cos x$

56. $\cos \left(\frac{\pi}{3} + x\right) + \cos \left(\frac{\pi}{3} - x\right) = \cos x$

57. $\cos \left(x + \frac{\pi}{4}\right) + \cos \left(x - \frac{\pi}{4}\right) = \sqrt{2} \cos x$

58. $\sin\left(\frac{\pi}{4}+x\right)+\sin\left(\frac{\pi}{4}-x\right)=\sqrt{2}\cos x$

59. $\cos\left(x+\frac{3\pi}{2}\right)+\cos\left(x-\frac{3\pi}{2}\right)=0$

60. $\sin\left(\frac{3\pi}{2}+x\right)+\sin\left(\frac{3\pi}{2}-x\right)=-2\cos x$

61. $\sin(A+B)+\sin(A-B)=2\sin A\cos B$

62. $\cos(A+B)+\cos(A-B)=2\cos A\cos B$

63. $\dfrac{\sin(A-B)}{\cos A\cos B}=\tan A-\tan B$

64. $\dfrac{\cos(A+B)}{\sin A\cos B}=\cot A-\tan B$

65. $\sec(A+B)=\dfrac{\cos(A-B)}{\cos^2 A-\sin^2 B}$

66. $\sec(A-B)=\dfrac{\cos(A+B)}{\cos^2 A-\sin^2 B}$

Use your graphing calculator to determine if each equation appears to be an identity by graphing the left expression and right expression together. If so, prove the identity. If not, find a counterexample.

67. $\sin x=\cos\left(\frac{\pi}{2}-x\right)$

68. $\cos x=\sin\left(\frac{\pi}{2}-x\right)$

69. $-\cos x=\sin\left(\frac{\pi}{2}+x\right)$

70. $\sin x=\cos\left(\frac{\pi}{2}+x\right)$

71. $-\sin x=\cos\left(\frac{\pi}{2}+x\right)$

72. $-\cos x=\cos(\pi-x)$

➤ REVIEW PROBLEMS

The problems that follow review material we covered in Section 4.2. Graph one complete cycle of each of the following.

73. $y=3\sin\frac{1}{2}x$

74. $y=2\sin 4x$

75. $y=\csc 3x$

76. $y=\sec 3x$

77. $y=\frac{1}{2}\cos 3x$

78. $y=2\sin\frac{\pi}{2}x$

➤ LEARNING OBJECTIVES ASSESSMENT

These questions are available for instructors to help assess if you have successfully met the learning objectives for this section.

79. Find the exact value of cos 165° by writing 165° as 120° + 45° and using a sum identity.

a. $-\dfrac{\sqrt{2}+\sqrt{6}}{4}$ **b.** $\dfrac{\sqrt{6}-\sqrt{2}}{4}$ **c.** $\dfrac{-1+\sqrt{2}}{2}$ **d.** $\dfrac{1-\sqrt{2}}{2}$

80. Which of the following is an identity?

a. $\sin(45° + x) + \sin(45° - x) = \sqrt{2} + 2\sin x$

b. $\sin(45° + x) + \sin(45° - x) = \sqrt{2}$

c. $\sin(45° + x) + \sin(45° - x) = \sqrt{2}\sin x$

d. $\sin(45° + x) + \sin(45° - x) = \sqrt{2}\cos x$

81. Simplify $\cos(270° - x)$.

a. $\cos x$ **b.** $-\cos x$ **c.** $-\sin x$ **d.** $\sin x$

82. Graph the equation $y = \sin 3x \cos x - \cos 3x \sin x$.

a.

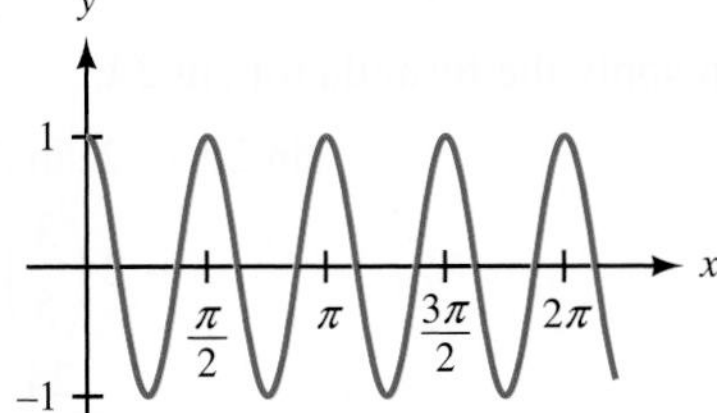

b.

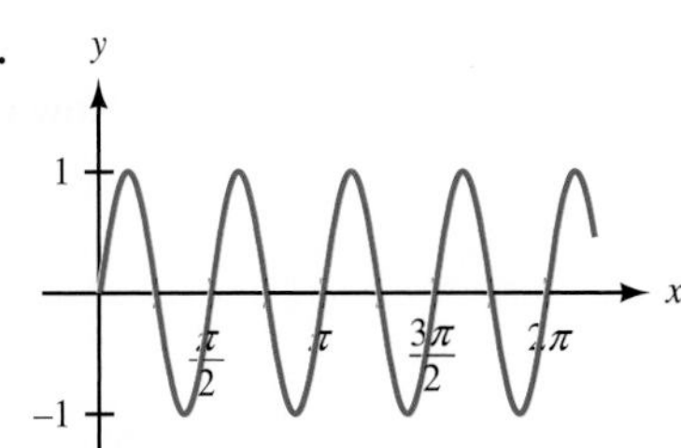

c.

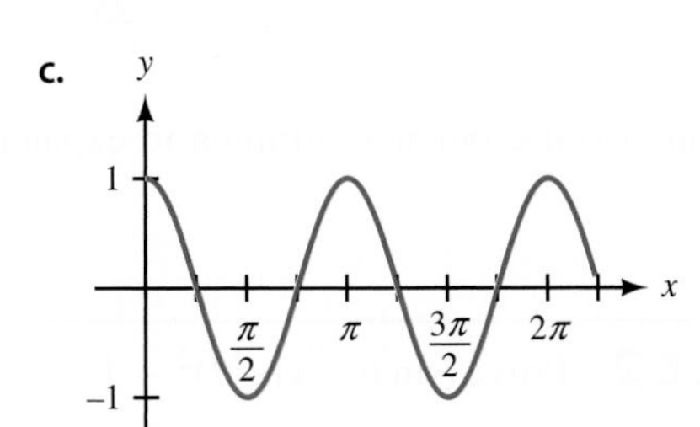

d.

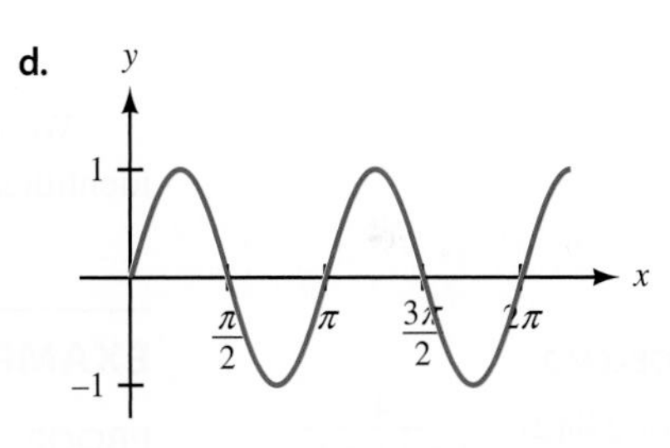

SECTION 5.3 Double-Angle Formulas

LEARNING OBJECTIVES

1. Use a double-angle formula to find the exact value of a trigonometric function.
2. Simplify an expression using a double-angle formula.
3. Use a double-angle formula to graph an equation.
4. Prove an equation is an identity using a double-angle formula.

We will begin this section by deriving the formulas for $\sin 2A$ and $\cos 2A$ using the formulas for $\sin(A + B)$ and $\cos(A + B)$. The formulas we derive for $\sin 2A$ and $\cos 2A$ are called *double-angle* formulas. Here is the derivation of the formula for $\sin 2A$.

$$\begin{aligned} \sin 2A &= \sin(A + A) && \text{Write } 2A \text{ as } A + A \\ &= \sin A \cos A + \cos A \sin A && \text{Sum formula} \\ &= \sin A \cos A + \sin A \cos A && \text{Commutative property} \\ &= 2 \sin A \cos A \end{aligned}$$

The last line gives us our first double-angle formula.

$$\boxed{\sin 2A = 2 \sin A \cos A}$$

The first thing to notice about this formula is that it indicates the 2 in $\sin 2A$ *cannot* be factored out and written as a coefficient. That is,

$$\sin 2A \neq 2 \sin A$$

For example, if $A = 30°$, $\sin 2 \cdot 30° = \sin 60° = \sqrt{3}/2$, which is not the same as $2 \sin 30° = 2\left(\frac{1}{2}\right) = 1$.

PROBLEM 1
If $\cos A = -3/8$ with A in QIII, find $\sin 2A$.

EXAMPLE 1 If $\sin A = \frac{3}{5}$ with A in QII, find $\sin 2A$.

SOLUTION To apply the formula for $\sin 2A$, we must first find $\cos A$. Because A terminates in QII, $\cos A$ is negative.

$$\begin{aligned} \cos A &= \pm\sqrt{1 - \sin^2 A} \\ &= -\sqrt{1 - \left(\frac{3}{5}\right)^2} && A \in \text{QII, so } \cos A \text{ is negative} \\ &= -\frac{4}{5} \end{aligned}$$

Now we can apply the formula for $\sin 2A$.

$$\begin{aligned} \sin 2A &= 2 \sin A \cos A \\ &= 2\left(\frac{3}{5}\right)\left(-\frac{4}{5}\right) \\ &= -\frac{24}{25} \end{aligned}$$

We can also use our new formula to expand the work we did previously with identities.

PROBLEM 2
Prove $2 \sin 2x = \dfrac{4}{\tan x + \cot x}$.

EXAMPLE 2 Prove $(\sin \theta + \cos \theta)^2 = 1 + \sin 2\theta$.

PROOF

$$\begin{aligned} (\sin \theta + \cos \theta)^2 &= \sin^2 \theta + 2 \sin \theta \cos \theta + \cos^2 \theta && \text{Expand} \\ &= 1 + 2 \sin \theta \cos \theta && \text{Pythagorean identity} \\ &= 1 + \sin 2\theta && \text{Double-angle identity} \end{aligned}$$

PROBLEM 3
Prove $\sin 2\theta = \dfrac{2 \tan \theta}{1 + \tan^2 \theta}$.

EXAMPLE 3 Prove $\sin 2x = \dfrac{2 \cot x}{1 + \cot^2 x}$.

PROOF

$$\begin{aligned} \frac{2 \cot x}{1 + \cot^2 x} &= \frac{2 \cdot \frac{\cos x}{\sin x}}{1 + \frac{\cos^2 x}{\sin^2 x}} && \text{Ratio identity} \\ &= \frac{2 \sin x \cos x}{\sin^2 x + \cos^2 x} && \text{Multiply numerator and denominator by } \sin^2 x \\ &= 2 \sin x \cos x && \text{Pythagorean identity} \\ &= \sin 2x && \text{Double-angle identity} \end{aligned}$$

There are three forms of the double-angle formula for $\cos 2A$. The first involves both $\sin A$ and $\cos A$, the second involves only $\cos A$, and the third involves only $\sin A$. Here is how we obtain the three formulas.

$$\begin{aligned} \cos 2A &= \cos (A + A) && \text{Write } 2A \text{ as } A + A \\ &= \cos A \cos A - \sin A \sin A && \text{Sum formula} \\ &= \cos^2 A - \sin^2 A \end{aligned}$$

To write this last formula in terms of $\cos A$ only, we substitute $1 - \cos^2 A$ for $\sin^2 A$.

$$\begin{aligned} \cos 2A &= \cos^2 A - (1 - \cos^2 A) \\ &= \cos^2 A - 1 + \cos^2 A \\ &= 2\cos^2 A - 1 \end{aligned}$$

To write the formula in terms of $\sin A$ only, we substitute $1 - \sin^2 A$ for $\cos^2 A$ in the last line above.

$$\begin{aligned} \cos 2A &= 2\cos^2 A - 1 \\ &= 2(1 - \sin^2 A) - 1 \\ &= 2 - 2\sin^2 A - 1 \\ &= 1 - 2\sin^2 A \end{aligned}$$

Here are the three forms of the double-angle formula for $\cos 2A$.

$\cos 2A = \cos^2 A - \sin^2 A$	First form
$= 2\cos^2 A - 1$	Second form
$= 1 - 2\sin^2 A$	Third form

Which form we choose will depend on the circumstances of the problem, as the next three examples illustrate.

PROBLEM 4

If $\cos x = -\frac{3}{4}$, find $\cos 2x$.

EXAMPLE 4 If $\sin A = \frac{1}{\sqrt{5}}$, find $\cos 2A$.

SOLUTION In this case, because we are given $\sin A$, applying the third form of the formula for $\cos 2A$ will give us the answer more quickly than applying either of the other two forms.

$$\begin{aligned} \cos 2A &= 1 - 2\sin^2 A \\ &= 1 - 2\left(\frac{1}{\sqrt{5}}\right)^2 \\ &= 1 - \frac{2}{5} \\ &= \frac{3}{5} \end{aligned}$$

PROBLEM 5

Prove $\sin 2x - 4\cos x \sin^3 x = \sin 2x \cos 2x$.

EXAMPLE 5 Prove $\cos 4x = 8\cos^4 x - 8\cos^2 x + 1$.

PROOF We can write $\cos 4x$ as $\cos(2 \cdot 2x)$ and apply our double-angle formula. Because the right side is written in terms of $\cos x$ only, we will choose the second form of our double-angle formula for $\cos 2A$.

$$\begin{aligned} \cos 4x &= \cos(2 \cdot 2x) \\ &= 2\cos^2 2x - 1 && \text{Double-angle formula} \\ &= 2(2\cos^2 x - 1)^2 - 1 && \text{Double-angle formula} \\ &= 2(4\cos^4 x - 4\cos^2 x + 1) - 1 && \text{Square} \\ &= 8\cos^4 x - 8\cos^2 x + 2 - 1 && \text{Distribute} \\ &= 8\cos^4 x - 8\cos^2 x + 1 && \text{Simplify} \end{aligned}$$

PROBLEM 6
Graph $y = 8\cos^2 x - 4$ when $0 \le x \le 2\pi$.

EXAMPLE 6 Graph $y = 3 - 6\sin^2 x$ from $x = 0$ to $x = 2\pi$.

SOLUTION To write the equation in the form $y = A\cos Bx$, we factor 3 from each term on the right side and then apply the formula for $\cos 2A$ to the remaining expression to write it as a single trigonometric function.

$$\begin{aligned} y &= 3 - 6\sin^2 x \\ &= 3(1 - 2\sin^2 x) && \text{Factor 3 from each term} \\ &= 3\cos 2x && \text{Double-angle formula} \end{aligned}$$

The graph of $y = 3\cos 2x$ will have an amplitude of 3 and a period of $2\pi/2 = \pi$. The graph is shown in Figure 1.

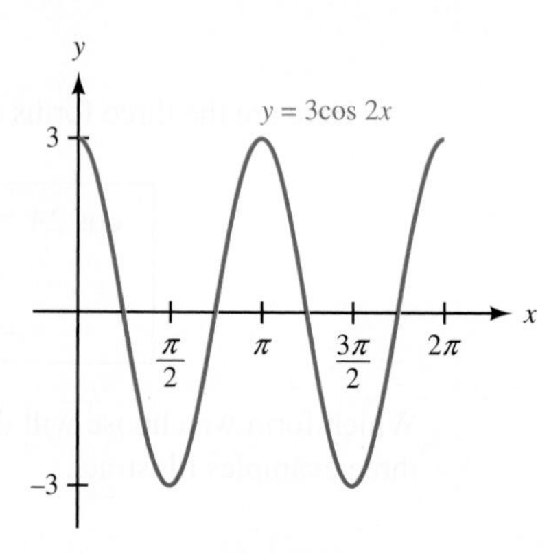

Figure 1

PROBLEM 7
Prove $\cot\theta = \dfrac{1 + \cos 2\theta}{\sin 2\theta}$.

EXAMPLE 7 Prove $\tan\theta = \dfrac{1 - \cos 2\theta}{\sin 2\theta}$.

PROOF

$$\begin{aligned} \frac{1 - \cos 2\theta}{\sin 2\theta} &= \frac{1 - (1 - 2\sin^2\theta)}{2\sin\theta\cos\theta} && \text{Double-angle formulas} \\ &= \frac{2\sin^2\theta}{2\sin\theta\cos\theta} && \text{Simplify numerator} \\ &= \frac{\sin\theta}{\cos\theta} && \text{Divide out common factor } 2\sin\theta \\ &= \tan\theta && \text{Ratio identity} \end{aligned}$$

We end this section by deriving the formula for $\tan 2A$.

$$\begin{aligned} \tan 2A &= \tan(A + A) \\ &= \frac{\tan A + \tan A}{1 - \tan A\tan A} \\ &= \frac{2\tan A}{1 - \tan^2 A} \end{aligned}$$

Our double-angle formula for $\tan 2A$ is

$$\boxed{\tan 2A = \frac{2\tan A}{1 - \tan^2 A}}$$

NOTE Of course, if you already know the value of $\sin 2A$ and $\cos 2A$, then you can find $\tan 2A$ using the ratio identity

$$\tan 2A = \frac{\sin 2A}{\cos 2A}$$

PROBLEM 8
Simplify $2\cos^2 22.5° - 1$.

EXAMPLE 8 Simplify $\dfrac{2\tan 15°}{1 - \tan^2 15°}$.

SOLUTION The expression has the same form as the right side of our double-angle formula for tan 2*A*. Therefore,

$$\begin{aligned}\frac{2\tan 15°}{1 - \tan^2 15°} &= \tan(2 \cdot 15°)\\ &= \tan 30°\\ &= \frac{\sqrt{3}}{3}\end{aligned}$$

PROBLEM 9
If $x = 2\cos\theta$, write $\theta - \dfrac{\sin 2\theta}{2}$ in terms of x.

EXAMPLE 9 If $x = 3\tan\theta$, write the following expression in terms of just x.

$$\frac{\theta}{2} + \frac{\sin 2\theta}{4}$$

SOLUTION To substitute for the first term above, we need to write θ in terms of x. To do so, we solve the equation $x = 3\tan\theta$ for θ.

$$\begin{aligned}&\text{If} && 3\tan\theta = x\\ &\text{then} && \tan\theta = \frac{x}{3}\\ &\text{and} && \theta = \tan^{-1}\frac{x}{3}\end{aligned}$$

Next, because the inverse tangent function can take on values only between $-\pi/2$ and $\pi/2$, we can visualize θ by drawing a right triangle in which θ is one of the acute angles. Because $\tan\theta = x/3$, we label the side opposite θ with x and the side adjacent to θ with 3.

By the Pythagorean Theorem, the hypotenuse of the triangle in Figure 2 must be $\sqrt{x^2 + 9}$, which means $\sin\theta = x/\sqrt{x^2 + 9}$ and $\cos\theta = 3/\sqrt{x^2 + 9}$. Now we are ready to simplify and substitute to solve our problem.

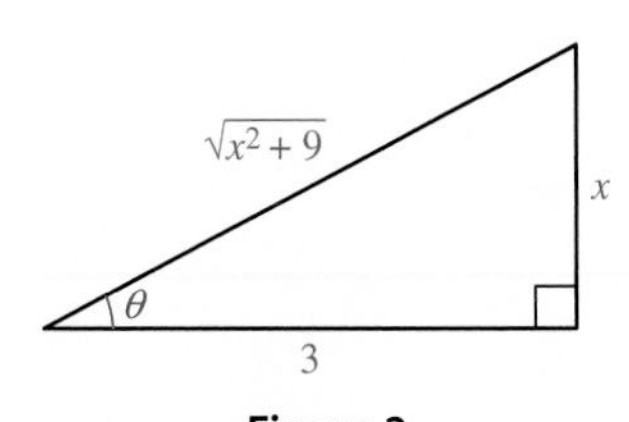

Figure 2

$$\begin{aligned}\frac{\theta}{2} + \frac{\sin 2\theta}{4} &= \frac{\theta}{2} + \frac{2\sin\theta\cos\theta}{4}\\ &= \frac{\theta}{2} + \frac{\sin\theta\cos\theta}{2}\\ &= \frac{1}{2}(\theta + \sin\theta\cos\theta)\\ &= \frac{1}{2}\left(\tan^{-1}\frac{x}{3} + \frac{x}{\sqrt{x^2+9}} \cdot \frac{3}{\sqrt{x^2+9}}\right)\\ &= \frac{1}{2}\left(\tan^{-1}\frac{x}{3} + \frac{3x}{x^2+9}\right)\end{aligned}$$

Note that we do not have to use absolute value symbols when we multiply and simplify the square roots in the second to the last line above because we know that $x^2 + 9$ is always positive.

Getting Ready for Class

After reading through the preceding section, respond in your own words and in complete sentences.

a. What is the formula for $\sin 2A$?
b. What are the three formulas for $\cos 2A$?
c. What is the formula for $\tan 2A$?
d. Explain why, in general, $\sin 2A \neq 2 \sin A$.

5.3 PROBLEM SET

➤ CONCEPTS AND VOCABULARY

For Questions 1 through 3, complete each double-angle identity.

1. $\sin 2x =$ ________

2. $\cos 2\theta =$ ________

3. $\tan 2y =$ ________

For Questions 4 through 6, determine if the statement is true or false.

4. The sine of twice an angle equals twice the sine of the angle.

5. $\cos 2A = 2 \cos A$

6. $\dfrac{\tan 2A}{2} = \tan A$

➤ EXERCISES

Let $\sin A = -\dfrac{3}{5}$ *with A in QIII and find the following.*

7. $\sin 2A$

8. $\cos 2A$

9. $\tan 2A$

10. $\cot 2A$

Let $\cos x = \dfrac{1}{\sqrt{10}}$ *with x in QIV and find the following.*

11. $\cos 2x$

12. $\sin 2x$

13. $\cot 2x$

14. $\tan 2x$

Let $\tan \theta = \dfrac{5}{12}$ *with θ in QI and find the following.*

15. $\sin 2\theta$

16. $\cos 2\theta$

17. $\sec 2\theta$

18. $\csc 2\theta$

Let $\csc t = \sqrt{5}$ *with t in QII and find the following.*

19. $\cos 2t$

20. $\sin 2t$

21. $\sec 2t$

22. $\csc 2t$

Graph each of the following from $x = 0$ to $x = 2\pi$.

23. $y = 2 - 4\sin^2 x$

24. $y = 4 - 8\sin^2 x$

25. $y = 6\cos^2 x - 3$

26. $y = 4\cos^2 x - 2$

27. $y = 1 - 2\sin^2 2x$

28. $y = 2\cos^2 2x - 1$

Use exact values to show that each of the following is true.

29. $\sin 60° = 2\sin 30° \cos 30°$

30. $\cos 60° = 1 - 2\sin^2 30°$

31. $\cos 120° = \cos^2 60° - \sin^2 60°$

32. $\sin 90° = 2\sin 45° \cos 45°$

33. If $\tan A = -\sqrt{3}$, find $\tan 2A$.

34. If $\tan A = \frac{3}{4}$, find $\tan 2A$.

Simplify each of the following.

35. $2\sin 15° \cos 15°$

36. $\cos^2 15° - \sin^2 15°$

37. $1 - 2\sin^2 75°$

38. $2\cos^2 105° - 1$

39. $\sin \frac{\pi}{12} \cos \frac{\pi}{12}$

40. $\sin \frac{\pi}{8} \cos \frac{\pi}{8}$

41. $\dfrac{\tan 22.5°}{1 - \tan^2 22.5°}$

42. $\dfrac{\tan \frac{3\pi}{8}}{1 - \tan^2 \frac{3\pi}{8}}$

Prove each of the following identities.

43. $(\sin x - \cos x)^2 = 1 - \sin 2x$

44. $(\cos x - \sin x)(\cos x + \sin x) = \cos 2x$

45. $\cos^2 \theta = \dfrac{1 + \cos 2\theta}{2}$

46. $\sin^2 \theta = \dfrac{1 - \cos 2\theta}{2}$

47. $\cot \theta = \dfrac{\sin 2\theta}{1 - \cos 2\theta}$

48. $\cos 2\theta = \dfrac{1 - \tan^2 \theta}{1 + \tan^2 \theta}$

49. $2\csc 2x = \tan x + \cot x$

50. $2\cot 2x = \cot x - \tan x$

51. $\sin 3\theta = 3\sin \theta - 4\sin^3 \theta$

52. $\cos 3\theta = 4\cos^3 \theta - 3\cos \theta$

53. $\cos^4 x - \sin^4 x = \cos 2x$

54. $2\sin^4 x + 2\sin^2 x \cos^2 x = 1 - \cos 2x$

55. $\cot \theta - \tan \theta = \dfrac{\cos 2\theta}{\sin \theta \cos \theta}$

56. $\csc \theta - 2\sin \theta = \dfrac{\cos 2\theta}{\sin \theta}$

57. $\sin 4A = 4\sin A \cos^3 A - 4\sin^3 A \cos A$

58. $\cos 4A = \cos^4 A - 6\cos^2 A \sin^2 A + \sin^4 A$

59. $\dfrac{1 - \tan x}{1 + \tan x} = \dfrac{1 - \sin 2x}{\cos 2x}$

60. $\dfrac{2 - 2\cos 2x}{\sin 2x} = \sec x \csc x - \cot x + \tan x$

Use your graphing calculator to determine if each equation appears to be an identity by graphing the left expression and right expression together. If so, prove the identity. If not, find a counterexample.

61. $\cot 2x = \dfrac{\cos x - \sin x \tan x}{\sec x \sin 2x}$

62. $\tan 2x = \dfrac{2 \cot x}{\csc^2 x - 2}$

63. $\sec 2x = \dfrac{\sec^2 x \csc^2 x}{\csc^2 x + \sec^2 x}$

64. $\csc 2x = \dfrac{\sec x + \csc x}{2 \sin x - 2 \cos x}$

65. If $x = 5 \tan \theta$, write the expression $\dfrac{\theta}{2} - \dfrac{\sin 2\theta}{4}$ in terms of just x.

66. If $x = 4 \sin \theta$, write the expression $\dfrac{\theta}{2} - \dfrac{\sin 2\theta}{4}$ in terms of just x.

67. If $x = 3 \sin \theta$, write the expression $\dfrac{\theta}{2} - \dfrac{\sin 2\theta}{4}$ in terms of just x.

68. If $x = 2 \sin \theta$, write the expression $2\theta - \tan 2\theta$ in terms of just x.

➤ REVIEW PROBLEMS

The problems that follow review material we covered in Sections 4.3 and 4.6. Graph each of the following from $x = 0$ to $x = 4\pi$.

69. $y = 2 - 2 \cos x$

70. $y = 3 + 3 \cos x$

71. $y = \cos x + \dfrac{1}{2} \sin 2x$

72. $y = \sin x + \dfrac{1}{2} \cos 2x$

Graph each of the following from $x = 0$ to $x = 8$.

73. $y = \dfrac{1}{2}x + \sin \pi x$

74. $y = x + \sin \dfrac{\pi}{2}x$

➤ LEARNING OBJECTIVES ASSESSMENT

These questions are available for instructors to help assess if you have successfully met the learning objectives for this section.

75. Find the exact value of $\sin 2\theta$ if $\tan \theta = -\dfrac{1}{2}$ with θ in QII.

a. $\dfrac{3}{5}$ **b.** $\dfrac{2\sqrt{5}}{5}$ **c.** $-\dfrac{4}{5}$ **d.** $-\dfrac{4\sqrt{5}}{5}$

76. Simplify $\cos^2 \dfrac{\pi}{12} - \sin^2 \dfrac{\pi}{12}$.

a. $\dfrac{\sqrt{3}}{2}$ **b.** $\dfrac{1}{2}$ **c.** 1 **d.** $\dfrac{1 - \sqrt{3}}{2}$

77. Graph the equation $y = 3 - 6\sin^2 x$.

a.

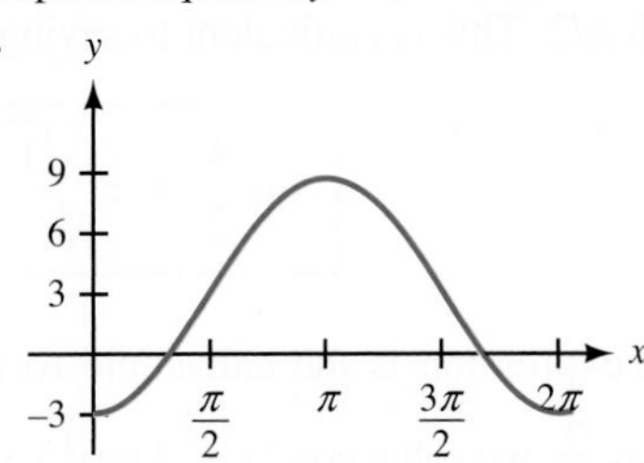

b.

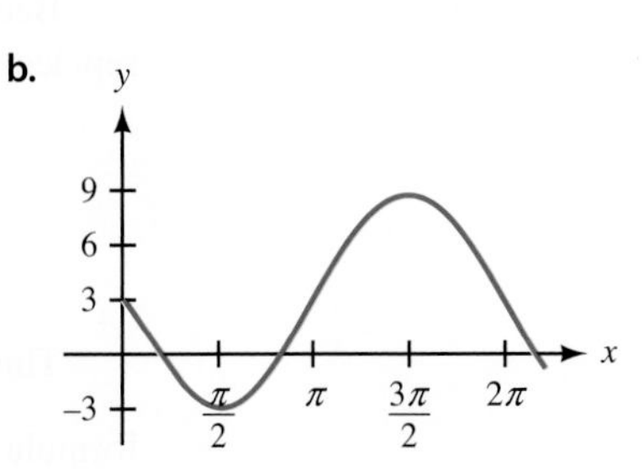

c.

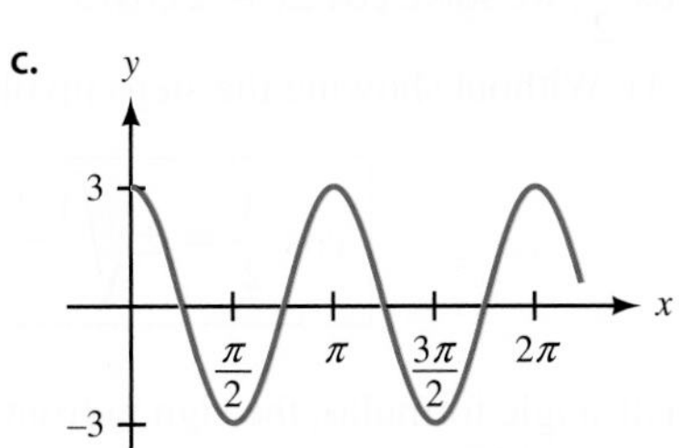

d.

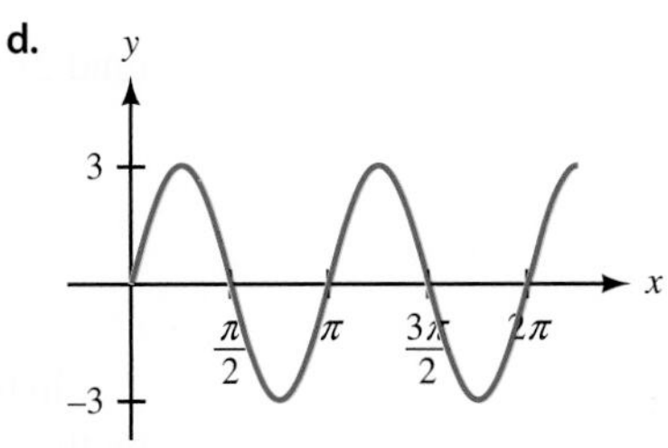

78. Which of the following is an identity?

a. $\dfrac{2\tan x}{1+\tan^2 x} = \sec 2x$ **b.** $\dfrac{2\tan x}{1+\tan^2 x} = \csc 2x$

c. $\dfrac{2\tan x}{1+\tan^2 x} = \cos 2x$ **d.** $\dfrac{2\tan x}{1+\tan^2 x} = \sin 2x$

SECTION 5.4 Half-Angle Formulas

LEARNING OBJECTIVES

1. Determine the quadrant in which a half-angle terminates.
2. Use a half-angle formula to find the exact value of a trigonometric function.
3. Use a half-angle formula to graph an equation.
4. Prove an equation is an identity using a half-angle formula.

In this section, we will derive formulas for $\sin\dfrac{A}{2}$ and $\cos\dfrac{A}{2}$. These formulas are called *half-angle* formulas and are derived from the double-angle formulas for $\cos 2A$.

In Section 5.3, we developed three ways to write the formula for $\cos 2A$, two of which were

$$\cos 2A = 1 - 2\sin^2 A \quad \text{and} \quad \cos 2A = 2\cos^2 A - 1$$

The choice of the letter we use to denote the angles in these formulas is arbitrary, so we can use x instead of A.

$$\cos 2x = 1 - 2\sin^2 x \quad \text{and} \quad \cos 2x = 2\cos^2 x - 1$$

Let us exchange sides in the first formula and solve for $\sin x$.

$$1 - 2\sin^2 x = \cos 2x \qquad \text{Exchange sides}$$

$$-2\sin^2 x = -1 + \cos 2x \qquad \text{Add } -1 \text{ to both sides}$$

$$\sin^2 x = \frac{1 - \cos 2x}{2} \qquad \text{Divide both sides by } -2$$

$$\sin x = \pm\sqrt{\frac{1 - \cos 2x}{2}} \qquad \text{Take the square root of both sides}$$

Because every value of x can be written as $\frac{1}{2}$ of some other number A, we can replace x with $A/2$. This is equivalent to saying $2x = A$.

$$\sin \frac{A}{2} = \pm\sqrt{\frac{1 - \cos A}{2}}$$

This last expression is the half-angle formula for $\sin \frac{A}{2}$. To find the half-angle formula for $\cos \frac{A}{2}$, we solve $\cos 2x = 2\cos^2 x - 1$ for $\cos x$ and then replace x with $A/2$ (and $2x$ with A). Without showing the steps involved in this process, here is the result:

$$\cos \frac{A}{2} = \pm\sqrt{\frac{1 + \cos A}{2}}$$

In both half-angle formulas, the sign in front of the radical, + or −, is determined by the quadrant in which $A/2$ terminates.

PROBLEM 1
If $\sin A = -8/17$ with $180° \le A \le 270°$, find $\sin \frac{A}{2}$, $\cos \frac{A}{2}$, and $\tan \frac{A}{2}$.

EXAMPLE 1 If $\cos A = \frac{3}{5}$ with $270° < A < 360°$, find $\sin \frac{A}{2}$, $\cos \frac{A}{2}$, and $\tan \frac{A}{2}$.

SOLUTION First of all, we determine the quadrant in which $A/2$ terminates.

$$\text{If} \quad 270° < A < 360°$$

$$\text{then} \quad \frac{270°}{2} < \frac{A}{2} < \frac{360°}{2}$$

$$135° < \frac{A}{2} < 180° \quad \text{so } \frac{A}{2} \in \text{QII}$$

In QII, sine is positive, and cosine is negative. Using the half-angle formulas, we have

$$\sin \frac{A}{2} = \sqrt{\frac{1 - \cos A}{2}} = \sqrt{\frac{1 - 3/5}{2}} = \sqrt{\frac{1}{5}} = \frac{1}{\sqrt{5}} = \frac{\sqrt{5}}{5}$$

$$\cos \frac{A}{2} = -\sqrt{\frac{1 + \cos A}{2}} = -\sqrt{\frac{1 + 3/5}{2}} = -\sqrt{\frac{4}{5}} = -\frac{2}{\sqrt{5}} = -\frac{2\sqrt{5}}{5}$$

Now we can use a ratio identity to find $\tan \frac{A}{2}$.

$$\tan \frac{A}{2} = \frac{\sin \frac{A}{2}}{\cos \frac{A}{2}} = \frac{\frac{\sqrt{5}}{5}}{-\frac{2\sqrt{5}}{5}} = -\frac{1}{2}$$

■

PROBLEM 2
If $\sin A = -7/25$ with $270° \le A \le 360°$, find the six trigonometric functions of $\frac{A}{2}$.

EXAMPLE 2 If $\sin A = -\frac{12}{13}$ with $180° < A < 270°$, find the six trigonometric functions of $A/2$.

SOLUTION To use the half-angle formulas, we need to find $\cos A$. Because $A \in \text{QIII}$, $\cos A$ is negative, so

$$\begin{aligned}\cos A &= \pm\sqrt{1 - \sin^2 A}\\ &= -\sqrt{1 - \left(-\frac{12}{13}\right)^2} \quad \text{We use the negative square root because cos A is negative}\\ &= -\sqrt{\frac{25}{169}}\\ &= -\frac{5}{13}\end{aligned}$$

Also, $A/2$ terminates in QII because

$$\text{if} \quad 180° < A < 270°$$

$$\text{then} \quad \frac{180°}{2} < \frac{A}{2} < \frac{270°}{2}$$

$$90° < \frac{A}{2} < 135° \qquad \text{So } \frac{A}{2} \in \text{QII}$$

In QII, sine is positive and cosine is negative. The half-angle formulas give us

$$\begin{aligned}\sin\frac{A}{2} &= \pm\sqrt{\frac{1 - \cos A}{2}} & \cos\frac{A}{2} &= \pm\sqrt{\frac{1 + \cos A}{2}}\\ &= +\sqrt{\frac{1 - (-5/13)}{2}} & &= -\sqrt{\frac{1 + (-5/13)}{2}}\\ &= \sqrt{\frac{9}{13}} & &= -\sqrt{\frac{4}{13}}\\ &= \frac{3}{\sqrt{13}} & &= -\frac{2}{\sqrt{13}}\\ &= \frac{3\sqrt{13}}{13} & &= -\frac{2\sqrt{13}}{13}\end{aligned}$$

Now that we have sine and cosine of $A/2$, we can apply the ratio identity for tangent to find $\tan\frac{A}{2}$.

$$\tan\frac{A}{2} = \frac{\sin\frac{A}{2}}{\cos\frac{A}{2}} = \frac{\frac{3\sqrt{13}}{13}}{-\frac{2\sqrt{13}}{13}} = -\frac{3}{2}$$

Next, we apply our reciprocal identities to find cosecant, secant, and cotangent of $A/2$.

$$\begin{aligned}\csc\frac{A}{2} &= \frac{1}{\sin\frac{A}{2}} & \sec\frac{A}{2} &= \frac{1}{\cos\frac{A}{2}} & \cot\frac{A}{2} &= \frac{1}{\tan\frac{A}{2}}\\ &= \frac{1}{\frac{3}{\sqrt{13}}} & &= \frac{1}{-\frac{2}{\sqrt{13}}} & &= \frac{1}{-\frac{3}{2}}\\ &= \frac{\sqrt{13}}{3} & &= -\frac{\sqrt{13}}{2} & &= -\frac{2}{3}\end{aligned}$$

In the previous two examples, we found $\tan \frac{A}{2}$ by using the ratio of $\sin \frac{A}{2}$ to $\cos \frac{A}{2}$. There are formulas that allow us to find $\tan \frac{A}{2}$ directly from $\sin A$ and $\cos A$. In Example 7 of Section 5.3, we proved the following identity:

$$\tan \theta = \frac{1 - \cos 2\theta}{\sin 2\theta}$$

If we let $\theta = \frac{A}{2}$ in this identity, we obtain a formula for $\tan \frac{A}{2}$ that involves only $\sin A$ and $\cos A$. Here it is.

$$\tan \frac{A}{2} = \frac{1 - \cos A}{\sin A}$$

If we multiply the numerator and denominator of the right side of this formula by $1 + \cos A$ and simplify the result, we have a second formula for $\tan \frac{A}{2}$.

$$\tan \frac{A}{2} = \frac{\sin A}{1 + \cos A}$$

PROBLEM 3
Find $\cos 112.5°$.

EXAMPLE 3 Find $\tan 15°$.

SOLUTION Because $15° = 30°/2$, we can use a half-angle formula to find $\tan 15°$.

$$\begin{aligned} \tan 15° &= \tan \frac{30°}{2} \\ &= \frac{1 - \cos 30°}{\sin 30°} \\ &= \frac{1 - \frac{\sqrt{3}}{2}}{\frac{1}{2}} \\ &= 2 - \sqrt{3} \end{aligned}$$

PROBLEM 4
Graph $y = 6 \sin^2 \frac{x}{2}$ for $0 \le x \le 4\pi$.

EXAMPLE 4 Graph $y = 4 \cos^2 \frac{x}{2}$ from $x = 0$ to $x = 4\pi$.

SOLUTION Applying our half-angle formula for $\cos \frac{x}{2}$ to the right side, we have

$$\begin{aligned} y &= 4 \cos^2 \frac{x}{2} \\ &= 4\left(\pm\sqrt{\frac{1 + \cos x}{2}}\right)^2 \\ &= 4\left(\frac{1 + \cos x}{2}\right) \\ &= 2 + 2 \cos x \end{aligned}$$

The graph of $y = 2 + 2\cos x$ has an amplitude of 2 and a vertical translation 2 units upward. The graph is shown in Figure 1.

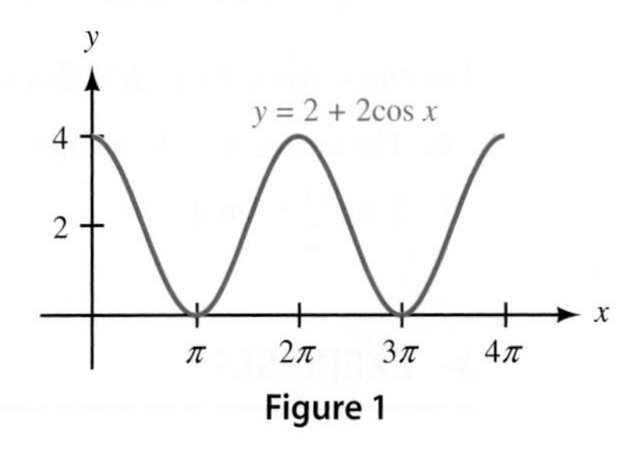

Figure 1

PROBLEM 5
Prove $2\tan\theta\cos^2\dfrac{\theta}{2} = \sin\theta + \tan\theta$.

EXAMPLE 5 Prove $\sin^2\dfrac{x}{2} = \dfrac{\tan x - \sin x}{2\tan x}$.

PROOF We can use a half-angle formula on the left side. In this case, because we have $\sin^2(x/2)$, we write the half-angle formula without the square root sign. After that, we multiply the numerator and denominator on the left side by $\tan x$ because the right side has $\tan x$ in both the numerator and the denominator.

$$\begin{aligned}
\sin^2\frac{x}{2} &= \frac{1-\cos x}{2} && \text{Square of half-angle formula} \\
&= \frac{\tan x}{\tan x}\cdot\frac{1-\cos x}{2} && \text{Multiply numerator and denominator by } \tan x \\
&= \frac{\tan x - \tan x\cos x}{2\tan x} && \text{Distributive property} \\
&= \frac{\tan x - \sin x}{2\tan x} && \tan x\cos x \text{ is } \sin x
\end{aligned}$$

Getting Ready for Class

After reading through the preceding section, respond in your own words and in complete sentences.

a. From what other formulas are half-angle formulas derived?
b. What is the formula for $\sin(A/2)$?
c. What is the formula for $\cos(A/2)$?
d. What are the two formulas for $\tan(A/2)$?

5.4 PROBLEM SET

➤ CONCEPTS AND VOCABULARY

For Questions 1 and 2, fill in the blank with an appropriate word or expression.

1. When using the half-angle formula for $\sin(A/2)$ or $\cos(A/2)$, the sign in front of the radical is determined by the quadrant in which _______ terminates.

2. The half-angle formula for $\cos(A/2)$ allows us to find the cosine of ______ the angle if we know the ________ of the angle.

For Questions 3 through 5, complete each half-angle identity.

3. $\sin \frac{x}{2} =$ ________ **4.** $\cos \frac{\theta}{2} =$ ________ **5.** $\tan \frac{y}{2} =$ ________

For Questions 6 through 8, determine if the statement is true or false.

6. The cosine of half an angle equals half the cosine of the angle.

7. $2 \sin \frac{A}{2} = \sin A$ **8.** $\tan \frac{A}{2} = \frac{\tan A}{2}$

➤ EXERCISES

9. If $0° < A < 90°$, then $A/2$ terminates in which quadrant?

10. If $90° < A < 180°$, then $A/2$ terminates in which quadrant?

11. If $180° < A < 270°$, then $A/2$ terminates in which quadrant?

12. If $270° < A < 360°$, then $A/2$ terminates in which quadrant?

13. If $270° < A < 360°$, then is $\cos (A/2)$ positive or negative?

14. If $180° < A < 270°$, then is $\sin (A/2)$ positive or negative?

15. True or false: If $\sin A$ is positive, then $\sin (A/2)$ is positive as well.

16. True or false: If $\cos A$ is negative, then $\cos (A/2)$ is negative as well.

NOTE *For the following problems, assume that all the given angles are in simplest form, so that if A is in QIV you may assume that $270° < A < 360°$.*

If $\cos A = \frac{1}{2}$ with A in QIV, find the following.

17. $\sin \frac{A}{2}$ **18.** $\cos \frac{A}{2}$

19. $\sec \frac{A}{2}$ **20.** $\csc \frac{A}{2}$

If $\sin A = -\frac{3}{5}$ with A in QIII, find the following.

21. $\cos \frac{A}{2}$ **22.** $\sin \frac{A}{2}$

23. $\sec \frac{A}{2}$ **24.** $\csc \frac{A}{2}$

If $\sin B = -\frac{1}{3}$ with B in QIII, find the following.

25. $\sin \frac{B}{2}$ **26.** $\csc \frac{B}{2}$

27. $\cos \frac{B}{2}$ **28.** $\sec \frac{B}{2}$

29. $\cot \frac{B}{2}$ **30.** $\tan \frac{B}{2}$

If $\sin A = \frac{4}{5}$ with A in QII, and $\sin B = \frac{3}{5}$ with B in QI, find the following.

31. $\cos \frac{A}{2}$ **32.** $\sin \frac{A}{2}$

33. $\sin \frac{B}{2}$ **34.** $\cos \frac{B}{2}$

Graph each of the following from $x = 0$ to $x = 4\pi$.

35. $y = 4\sin^2 \frac{x}{2}$

36. $y = 6\cos^2 \frac{x}{2}$

37. $y = 2\cos^2 \frac{x}{2}$

38. $y = 2\sin^2 \frac{x}{2}$

Use half-angle formulas to find exact values for each of the following.

39. $\cos 15°$

40. $\tan 15°$

41. $\sin 75°$

42. $\cos 75°$

43. $\cos 105°$

44. $\sin 105°$

Prove the following identities.

45. $\sin^2 \frac{\theta}{2} = \frac{\csc\theta - \cot\theta}{2\csc\theta}$

46. $2\cos^2 \frac{\theta}{2} = \frac{\sin^2\theta}{1 - \cos\theta}$

47. $\sec^2 \frac{A}{2} = \frac{2\sec A}{\sec A + 1}$

48. $\csc^2 \frac{A}{2} = \frac{2\sec A}{\sec A - 1}$

49. $\tan \frac{B}{2} = \csc B - \cot B$

50. $\tan \frac{B}{2} = \frac{\sec B}{\sec B \csc B + \csc B}$

51. $\tan \frac{x}{2} + \cot \frac{x}{2} = 2\csc x$

52. $\tan \frac{x}{2} - \cot \frac{x}{2} = -2\cot x$

53. $\cos^2 \frac{\theta}{2} = \frac{\tan\theta + \sin\theta}{2\tan\theta}$

54. $2\sin^2 \frac{\theta}{2} = \frac{\sin^2\theta}{1 + \cos\theta}$

55. $\cos^4\theta = \frac{1}{4} + \frac{\cos 2\theta}{2} + \frac{\cos^2 2\theta}{4}$

56. $4\sin^4\theta = 1 - 2\cos 2\theta + \cos^2 2\theta$

➤ REVIEW PROBLEMS

The following problems review material we covered in Section 4.7. Reviewing these problems will help you with the next section.

Evaluate without using a calculator or tables.

57. $\sin\left(\arcsin \frac{3}{5}\right)$

58. $\cos\left(\arcsin \frac{3}{5}\right)$

59. $\cos(\arctan 2)$

60. $\sin(\arctan 2)$

Write an equivalent algebraic expression that involves x only.

61. $\sin(\tan^{-1} x)$

62. $\cos(\tan^{-1} x)$

63. $\tan(\sin^{-1} x)$

64. $\tan(\cos^{-1} x)$

➤ LEARNING OBJECTIVES ASSESSMENT

These questions are available for instructors to help assess if you have successfully met the learning objectives for this section.

65. If $270° < \theta < 360°$, then which quadrant does $\frac{\theta}{2}$ terminate in?

a. QI **b.** QII **c.** QIII **d.** QIV

66. Find $\cos \frac{A}{2}$ if $\cos A = \frac{2}{5}$ with $270° < A < 360°$.

a. $\frac{3}{10}$ **b.** $\frac{7}{10}$ **c.** $-\frac{\sqrt{70}}{10}$ **d.** $\frac{\sqrt{30}}{10}$

67. Graph the equation $y = 4\cos^2 \frac{x}{2}$.

a.

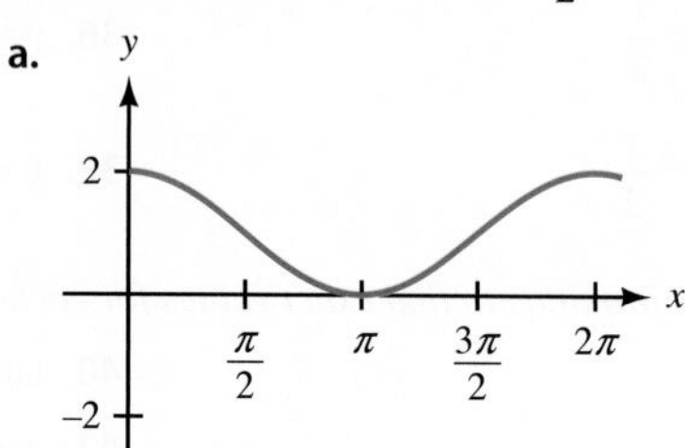

b.

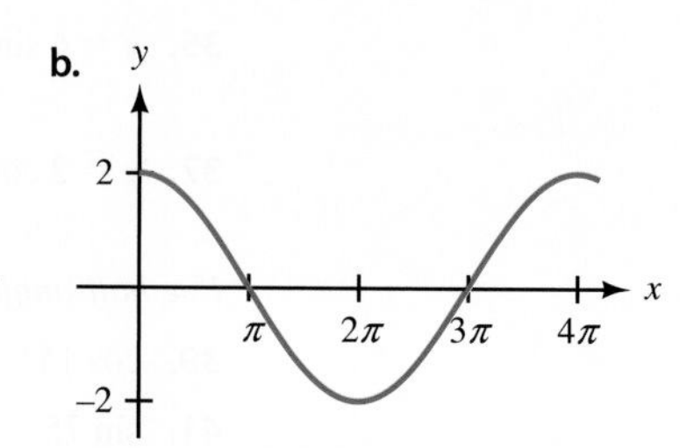

c.

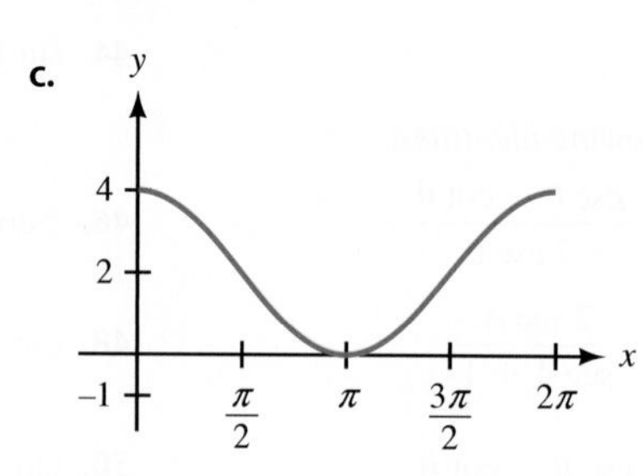

d.

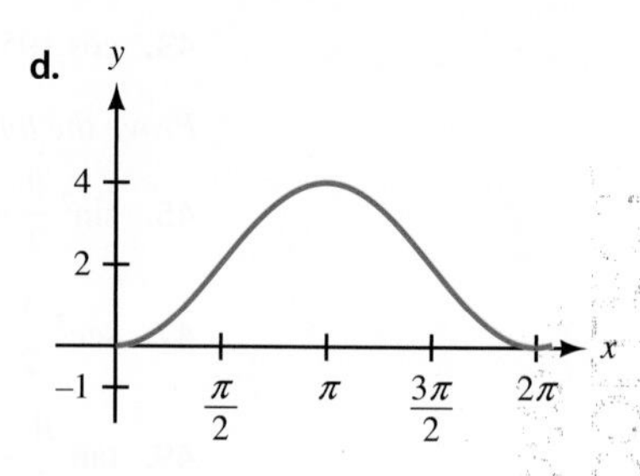

68. Which of the following is a valid step in proving the identity $\cos^2 \frac{\theta}{2} = \frac{\tan\theta + \sin\theta}{2\tan\theta}$?

a. $\cos^2 \frac{\theta}{2} = \frac{1}{2}\cos^2\theta$

b. $\cos^2\theta = \frac{\tan\theta + \sin\theta}{\tan\theta}$

c. $\cos^2 \frac{\theta}{2} = \left(\pm\sqrt{\frac{1 - \cos\theta}{2}}\right)^2$

d. $\cos^2 \frac{\theta}{2} = \left(\pm\sqrt{\frac{1 + \cos\theta}{2}}\right)^2$

SECTION 5.5 Additional Identities

LEARNING OBJECTIVES

1. Use an identity to simplify a composition of trigonometric and inverse trigonometric functions.
2. Express a product of two trigonometric functions as a sum or difference.
3. Express a sum or difference of two trigonometric functions as a product.
4. Use the sum to product and product to sum formulas to prove an identity.

There are two main parts to this section, both of which rely on the work we have done previously with identities and formulas. In the first part of this section, we will extend our work on identities to include problems that involve inverse trigonometric functions. In the second part, we will use the formulas we obtained for the sine and cosine of a sum or difference to write some new formulas involving sums and products.

Identities and Formulas Involving Inverse Functions

The solutions to our first two examples require that we combine a knowledge of inverse trigonometric functions along with formulas from previous sections of this chapter.

PROBLEM 1
Evaluate
$\cos\left(\arcsin\frac{4}{5} + \text{arcsec}(-2)\right)$.

EXAMPLE 1 Evaluate $\sin\left(\arcsin\frac{3}{5} + \arctan 2\right)$ without using a calculator.

SOLUTION We can simplify things somewhat if we let $\alpha = \arcsin\frac{3}{5}$ and $\beta = \arctan 2$.

$$\sin\left(\arcsin\frac{3}{5} + \arctan 2\right) = \sin(\alpha + \beta)$$
$$= \sin\alpha\cos\beta + \cos\alpha\sin\beta$$

Drawing and labeling a triangle for α and another for β, we have

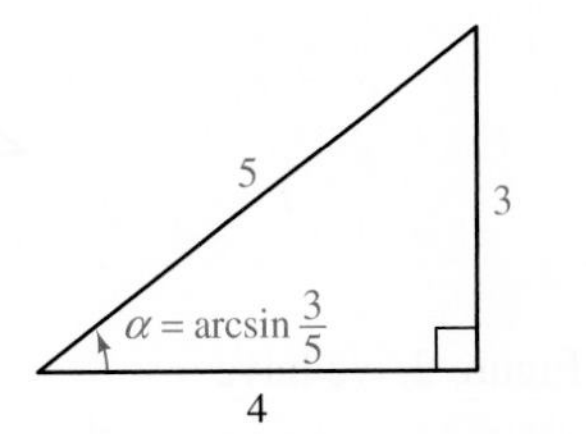

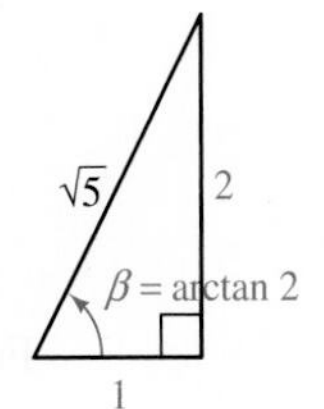

Figure 1

From the triangles in Figure 1, we have

$$\sin\alpha = \frac{3}{5} \qquad \sin\beta = \frac{2}{\sqrt{5}} = \frac{2\sqrt{5}}{5}$$
$$\cos\alpha = \frac{4}{5} \qquad \cos\beta = \frac{1}{\sqrt{5}} = \frac{\sqrt{5}}{5}$$

Substituting these numbers into

$$\sin\alpha\cos\beta + \cos\alpha\sin\beta$$

gives us

$$\frac{3}{5}\cdot\frac{\sqrt{5}}{5} + \frac{4}{5}\cdot\frac{2\sqrt{5}}{5} = \frac{11\sqrt{5}}{25}$$

■

CALCULATOR NOTE To check our answer for Example 1 using a calculator, we would use the following sequence:

Scientific Calculator

3 [÷] 5 [=] [sin⁻¹] [+] 2 [tan⁻¹] [=] [sin]

Graphing Calculator

[sin] [(] [sin⁻¹] [(] 3 [÷] 5 [)] [+] [tan⁻¹] [(] 2 [)] [)] [ENTER]

The display would show 0.9839 to four decimal places, which is the decimal approximation of $\frac{11\sqrt{5}}{25}$. It is appropriate to check your work on problems like this by using your calculator. The concepts are best understood, however, by working through the problems without using a calculator.

PROBLEM 2
Write $\cos(2\tan^{-1} x)$ as an expression involving x only.

EXAMPLE 2 Write $\sin(2\tan^{-1} x)$ as an equivalent algebraic expression involving only x. (Assume x is positive.)

SOLUTION We begin by letting $\theta = \tan^{-1} x$, which means

$$\tan\theta = x = \frac{x}{1}$$

Draw a right triangle with an acute angle of θ and label the opposite side with x and the adjacent side with 1. Then the ratio of the side opposite θ to the side adjacent to θ is $x/1 = x$.

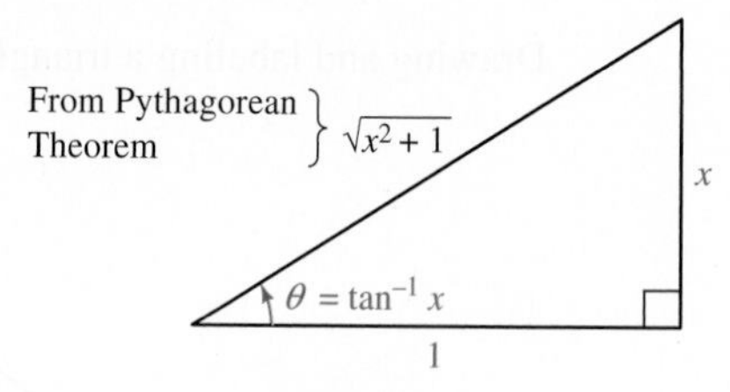

Figure 2

From Figure 2, we have

$$\sin\theta = \frac{x}{\sqrt{x^2+1}} \quad \text{and} \quad \cos\theta = \frac{1}{\sqrt{x^2+1}}$$

Therefore,

$$\begin{aligned}
\sin(2\tan^{-1} x) &= \sin 2\theta && \text{Substitute } \theta \text{ for } \tan^{-1} x \\
&= 2\sin\theta\cos\theta && \text{Double-angle identity} \\
&= 2 \cdot \frac{x}{\sqrt{x^2+1}} \cdot \frac{1}{\sqrt{x^2+1}} && \text{From Figure 2} \\
&= \frac{2x}{x^2+1} && \text{Multiplication}
\end{aligned}$$

Product to Sum Formulas

If we add the formula for $\sin(A - B)$ to the formula for $\sin(A + B)$, we will eventually arrive at a formula for the product $\sin A \cos B$.

$$\begin{aligned}
\sin A\cos B + \cos A\sin B &= \sin(A + B) \\
\sin A\cos B - \cos A\sin B &= \sin(A - B) \\
\hline
2\sin A\cos B \qquad\qquad &= \sin(A + B) + \sin(A - B)
\end{aligned}$$

Dividing both sides of this result by 2 gives us

$$\sin A\cos B = \frac{1}{2}[\sin(A + B) + \sin(A - B)] \qquad (1)$$

By similar methods, we can derive the formulas that follow.

NOTE These four product formulas are of use in calculus. The reason they are useful is that they indicate how we can convert a product into a sum. In calculus, it is sometimes much easier to work with sums of trigonometric functions than it is to work with products.

$$\cos A\sin B = \frac{1}{2}[\sin(A + B) - \sin(A - B)] \qquad (2)$$

$$\cos A\cos B = \frac{1}{2}[\cos(A + B) + \cos(A - B)] \qquad (3)$$

$$\sin A\sin B = \frac{1}{2}[\cos(A - B) - \cos(A + B)] \qquad (4)$$

PROBLEM 3
Verify the product formula for $\cos A \sin B$ if $A = 240°$ and $B = 30°$.

EXAMPLE 3 Verify product formula (3) for $A = 30°$ and $B = 120°$.

SOLUTION Substituting $A = 30°$ and $B = 120°$ into

$$\cos A \cos B = \frac{1}{2}[\cos (A + B) + \cos (A - B)]$$

we have

$$\cos 30° \cos 120° = \frac{1}{2}[\cos 150° + \cos (-90°)]$$

$$\frac{\sqrt{3}}{2} \cdot \left(-\frac{1}{2}\right) = \frac{1}{2}\left(-\frac{\sqrt{3}}{2} + 0\right)$$

$$-\frac{\sqrt{3}}{4} = -\frac{\sqrt{3}}{4} \quad \text{A true statement}$$

PROBLEM 4
Write $6 \cos 7x \cos 3x$ as a sum or difference.

EXAMPLE 4 Write $10 \cos 5x \sin 3x$ as a sum or difference.

SOLUTION Product formula (2) is appropriate for an expression of the form $\cos A \sin B$. If we substitute $A = 5x$ and $B = 3x$, then

$$10 \cos 5x \sin 3x = 10 \cdot \frac{1}{2}[\sin (5x + 3x) - \sin (5x - 3x)]$$

$$= 5(\sin 8x - \sin 2x)$$

Sum to Product Formulas

By some simple manipulations we can change our product formulas into sum formulas. If we take the formula for $\sin A \cos B$, exchange sides, and then multiply through by 2 we have

$$\sin (A + B) + \sin (A - B) = 2 \sin A \cos B$$

If we let $\alpha = A + B$ and $\beta = A - B$, then we can solve for A by adding the left sides and the right sides.

$$\begin{aligned} A + B &= \alpha \\ A - B &= \beta \\ \hline 2A \quad &= \alpha + \beta \\ A \quad &= \frac{\alpha + \beta}{2} \end{aligned}$$

By subtracting the expression for β from the expression for α, we have

$$B = \frac{\alpha - \beta}{2}$$

Writing the equation $\sin (A + B) + \sin (A - B) = 2 \sin A \cos B$ in terms of α and β gives us our first sum formula:

$$\boxed{\sin \alpha + \sin \beta = 2 \sin \frac{\alpha + \beta}{2} \cos \frac{\alpha - \beta}{2}} \quad (5)$$

Similarly, the following sum formulas can be derived from the other product formulas:

$$\sin\alpha - \sin\beta = 2\cos\frac{\alpha+\beta}{2}\sin\frac{\alpha-\beta}{2} \quad (6)$$

$$\cos\alpha + \cos\beta = 2\cos\frac{\alpha+\beta}{2}\cos\frac{\alpha-\beta}{2} \quad (7)$$

$$\cos\alpha - \cos\beta = -2\sin\frac{\alpha+\beta}{2}\sin\frac{\alpha-\beta}{2} \quad (8)$$

PROBLEM 5
Verify formula (5) for $\alpha = 240°$ and $\beta = 60°$.

EXAMPLE 5 Verify sum formula (7) for $\alpha = 30°$ and $\beta = 90°$.

SOLUTION We substitute $\alpha = 30°$ and $\beta = 90°$ into sum formula (7) and simplify each side of the resulting equation.

$$\cos 30° + \cos 90° = 2\cos\frac{30° + 90°}{2}\cos\frac{30° - 90°}{2}$$

$$\cos 30° + \cos 90° = 2\cos 60°\cos(-30°)$$

$$\frac{\sqrt{3}}{2} + 0 = 2\left(\frac{1}{2}\right)\left(\frac{\sqrt{3}}{2}\right)$$

$$\frac{\sqrt{3}}{2} = \frac{\sqrt{3}}{2} \qquad \text{A true statement}$$

PROBLEM 6
Prove $\cot x = \dfrac{\sin 5x - \sin 7x}{\cos 5x - \cos 7x}$.

EXAMPLE 6 Verify the identity $-\tan x = \dfrac{\cos 3x - \cos x}{\sin 3x + \sin x}$.

PROOF Applying the formulas for $\cos\alpha - \cos\beta$ and $\sin\alpha + \sin\beta$ to the right side and then simplifying, we arrive at $-\tan x$.

$$\frac{\cos 3x - \cos x}{\sin 3x + \sin x} = \frac{-2\sin\frac{3x+x}{2}\sin\frac{3x-x}{2}}{2\sin\frac{3x+x}{2}\cos\frac{3x-x}{2}} \qquad \text{Sum to product formulas}$$

$$= \frac{-2\sin 2x\sin x}{2\sin 2x\cos x} \qquad \text{Simplify}$$

$$= -\frac{\sin x}{\cos x} \qquad \text{Divide out common factors}$$

$$= -\tan x \qquad \text{Ratio identity}$$

Getting Ready for Class

After reading through the preceding section, respond in your own words and in complete sentences.

a. How would you describe arcsin (3/5) to a classmate?
b. How do we arrive at a formula for the product $\sin A\cos B$?
c. Explain the difference between product to sum formulas and sum to product formulas.
d. Write $\sin\alpha + \sin\beta$ as a product using only sine and cosine.

5.5 PROBLEM SET

➤ CONCEPTS AND VOCABULARY

For Questions 1 and 2, fill in the blank with an appropriate word.

1. When simplifying expressions that involve inverse trigonometric functions, a good first step is to assign a new variable to represent the value of each inverse function, which will be an _______ of some kind.

2. After assigning a new variable to each inverse trigonometric function, draw a _________ triangle and use right triangle trigonometry to label two of the three ______.

For Questions 3 and 4, complete each product to sum formula.

3. $\cos x \sin y =$ ___________

4. $\sin x \sin y =$ ___________

For Questions 5 and 6, complete each sum to product formula.

5. $\sin x - \sin y =$ ___________

6. $\cos x + \cos y =$ ___________

➤ EXERCISES

Evaluate each expression below without using a calculator. (Assume any variables represent positive numbers.)

7. $\cos\left(\arcsin\frac{3}{5} - \arctan 2\right)$

8. $\sin\left(\arcsin\frac{3}{5} - \arctan 2\right)$

9. $\cos\left(\tan^{-1}\frac{1}{2} + \sin^{-1}\frac{1}{2}\right)$

10. $\sin\left(\tan^{-1}\frac{1}{2} - \sin^{-1}\frac{1}{2}\right)$

11. $\sin\left(2\cos^{-1}\frac{\sqrt{5}}{5}\right)$

12. $\sin\left(2\tan^{-1}\frac{3}{4}\right)$

13. $\cos\left(2\tan^{-1}\frac{3}{2}\right)$

14. $\cos\left(2\sin^{-1}\frac{1}{3}\right)$

Write each expression as an equivalent algebraic expression involving only x. (Assume x is positive.)

15. $\tan(\sin^{-1} x)$

16. $\tan(\cos^{-1} x)$

17. $\cot\left(\sin^{-1}\frac{x}{3}\right)$

18. $\cos\left(\tan^{-1}\frac{x}{2}\right)$

19. $\sin(2\sin^{-1} x)$

20. $\sin(2\cos^{-1} x)$

21. $\cos(2\cos^{-1} x)$

22. $\cos(2\sin^{-1} x)$

23. $\sin\left(\sec^{-1}\frac{x+1}{3}\right)$

24. $\sec\left(\tan^{-1}\frac{x-2}{2}\right)$

25. Verify product formula (4) for $A = 30°$ and $B = 120°$.

26. Verify product formula (1) for $A = 120°$ and $B = 30°$.

Rewrite each expression as a sum or difference, then simplify if possible.

27. $10 \sin 5x \cos 3x$

28. $10 \sin 5x \sin 3x$

29. $\cos 8x \cos 2x$

30. $\cos 2x \sin 8x$

31. $\cos 90° \cos 180°$

32. $\sin 60° \cos 30°$

33. $\cos 3\pi \sin \pi$

34. $\sin 4\pi \sin 2\pi$

35. Verify sum formula (6) for $\alpha = 30°$ and $\beta = 90°$.

36. Verify sum formula (8) for $\alpha = 90°$ and $\beta = 30°$.

Rewrite each expression as a product. Simplify if possible.

37. $\sin 7x + \sin 3x$

38. $\cos 5x - \cos 3x$

39. $\cos 45° + \cos 15°$

40. $\sin 75° - \sin 15°$

41. $\sin \frac{7\pi}{12} - \sin \frac{\pi}{12}$

42. $\cos \frac{\pi}{12} + \cos \frac{7\pi}{12}$

Verify each identity.

43. $-\cot x = \frac{\sin 3x + \sin x}{\cos 3x - \cos x}$

44. $\cot x = \frac{\cos 3x + \cos x}{\sin 3x - \sin x}$

45. $\cot 2x = \frac{\sin 3x - \sin x}{\cos x - \cos 3x}$

46. $\cot x = \frac{\sin 4x + \sin 6x}{\cos 4x - \cos 6x}$

47. $\tan 4x = \frac{\sin 5x + \sin 3x}{\cos 3x + \cos 5x}$

48. $-\tan 4x = \frac{\cos 3x - \cos 5x}{\sin 3x - \sin 5x}$

➤ REVIEW PROBLEMS

The problems that follow review material we covered in Section 4.3. Graph one complete cycle.

49. $y = \sin\left(x + \frac{\pi}{4}\right)$

50. $y = \cos\left(x - \frac{\pi}{3}\right)$

51. $y = \sin\left(2x - \frac{\pi}{2}\right)$

52. $y = \sin\left(2x - \frac{\pi}{3}\right)$

53. $y = \frac{1}{2}\cos\left(3x - \frac{\pi}{2}\right)$

54. $y = 4\sin\left(2\pi x - \frac{\pi}{2}\right)$

➤ LEARNING OBJECTIVES ASSESSMENT

These questions are available for instructors to help assess if you have successfully met the learning objectives for this section.

55. Write an expression equivalent to $\sin(2\tan^{-1} x)$ involving x only. Assume x is positive.

a. $\frac{x - 1}{\sqrt{x^2 + 1}}$ **b.** $\frac{x^2 - 1}{x^2 + 1}$ **c.** $\frac{2x}{\sqrt{x^2 + 1}}$ **d.** $\frac{2x}{x^2 + 1}$

56. Rewrite $6 \sin 4x \sin 3x$ as a sum or difference and simplify if possible.

a. $3\cos 7x + 3\cos x$ **b.** $3\cos x - 3\cos 7x$

c. $3\sin 7x - 3\sin x$ **d.** $3\sin 7x + 3\sin x$

57. Rewrite $\cos 6x + \cos 4x$ as a product and simplify if possible.

a. $2\cos 5x \cos x$ **b.** $-2\sin 5x \sin x$

c. $2\cos 5x \sin x$ **d.** $2\sin 5x \cos x$

58. Which of the following is an identity?

a. $\frac{\sin 4x + \sin 2x}{\cos 4x - \cos 2x} = -\cot x$ **b.** $\frac{\sin 4x + \sin 2x}{\cos 4x - \cos 2x} = \tan x$

c. $\frac{\sin 4x + \sin 2x}{\cos 4x - \cos 2x} = \cot x$ **d.** $\frac{\sin 4x + \sin 2x}{\cos 4x - \cos 2x} = -\tan x$

CHAPTER 5 SUMMARY

EXAMPLES

BASIC IDENTITIES [5.1]

	Basic Identities	Common Equivalent Forms
Reciprocal	$\csc\theta = \dfrac{1}{\sin\theta}$	$\sin\theta = \dfrac{1}{\csc\theta}$
	$\sec\theta = \dfrac{1}{\cos\theta}$	$\cos\theta = \dfrac{1}{\sec\theta}$
	$\cot\theta = \dfrac{1}{\tan\theta}$	$\tan\theta = \dfrac{1}{\cot\theta}$
Ratio	$\tan\theta = \dfrac{\sin\theta}{\cos\theta}$	
	$\cot\theta = \dfrac{\cos\theta}{\sin\theta}$	
Pythagorean	$\cos^2\theta + \sin^2\theta = 1$	$\sin^2\theta = 1 - \cos^2\theta$
		$\sin\theta = \pm\sqrt{1-\cos^2\theta}$
		$\cos^2\theta = 1 - \sin^2\theta$
		$\cos\theta = \pm\sqrt{1-\sin^2\theta}$
	$1 + \tan^2\theta = \sec^2\theta$	
	$1 + \cot^2\theta = \csc^2\theta$	

1 To prove

$\tan x + \cos x = \sin x\,(\sec x + \cot x)$,

we can multiply through by $\sin x$ on the right side and then change to sines and cosines.

$$\begin{aligned} &\sin x\,(\sec x + \cot x) \\ &= \sin x \sec x + \sin x \cot x \\ &= \sin x \cdot \frac{1}{\cos x} + \sin x \cdot \frac{\cos x}{\sin x} \\ &= \frac{\sin x}{\cos x} + \cos x \\ &= \tan x + \cos x \end{aligned}$$

PROVING IDENTITIES [5.1]

An identity in trigonometry is a statement that two expressions are equal for all replacements of the variable for which each expression is defined. To prove a trigonometric identity, we use trigonometric substitutions and algebraic manipulations to either

1. transform the right side into the left side, or
2. transform the left side into the right side.

Remember to work on each side separately. We do not want to use properties from algebra that involve both sides of the identity—like the addition property of equality.

2 To find the exact value for $\cos 75°$, we write $75°$ as $45° + 30°$ and then apply the formula for $\cos(A + B)$.

$$\begin{aligned} &\cos 75° \\ &= \cos(45° + 30°) \\ &= \cos 45° \cos 30° \\ &\quad - \sin 45° \sin 30° \\ &= \frac{\sqrt{2}}{2} \cdot \frac{\sqrt{3}}{2} - \frac{\sqrt{2}}{2} \cdot \frac{1}{2} \\ &= \frac{\sqrt{6} - \sqrt{2}}{4} \end{aligned}$$

SUM AND DIFFERENCE FORMULAS [5.2]

$$\sin(A + B) = \sin A \cos B + \cos A \sin B$$
$$\sin(A - B) = \sin A \cos B - \cos A \sin B$$
$$\cos(A + B) = \cos A \cos B - \sin A \sin B$$
$$\cos(A - B) = \cos A \cos B + \sin A \sin B$$
$$\tan(A + B) = \frac{\tan A + \tan B}{1 - \tan A \tan B}$$
$$\tan(A - B) = \frac{\tan A - \tan B}{1 + \tan A \tan B}$$

DOUBLE-ANGLE FORMULAS [5.3]

$$\sin 2A = 2\sin A\cos A$$

$$\cos 2A = \cos^2 A - \sin^2 A \quad \text{First form}$$
$$= 2\cos^2 A - 1 \quad \text{Second form}$$
$$= 1 - 2\sin^2 A \quad \text{Third form}$$

$$\tan 2A = \frac{2\tan A}{1 - \tan^2 A}$$

3 If $\sin A = \frac{3}{5}$ with A in QII, then

$$\cos 2A = 1 - 2\sin^2 A = 1 - 2\left(\frac{3}{5}\right)^2 = \frac{7}{25}$$

HALF-ANGLE FORMULAS [5.4]

$$\sin\frac{A}{2} = \pm\sqrt{\frac{1-\cos A}{2}}$$

$$\cos\frac{A}{2} = \pm\sqrt{\frac{1+\cos A}{2}}$$

$$\tan\frac{A}{2} = \frac{1-\cos A}{\sin A} = \frac{\sin A}{1+\cos A}$$

4 We can use a half-angle formula to find the exact value of sin 15° by writing 15° as 30°/2.

$$\sin 15° = \sin\frac{30°}{2} = \sqrt{\frac{1-\cos 30°}{2}} = \sqrt{\frac{1-\sqrt{3}/2}{2}} = \sqrt{\frac{2-\sqrt{3}}{4}} = \frac{\sqrt{2-\sqrt{3}}}{2}$$

PRODUCT TO SUM FORMULAS [5.5]

$$\sin A\cos B = \frac{1}{2}[\sin(A+B) + \sin(A-B)]$$

$$\cos A\sin B = \frac{1}{2}[\sin(A+B) - \sin(A-B)]$$

$$\cos A\cos B = \frac{1}{2}[\cos(A+B) + \cos(A-B)]$$

$$\sin A\sin B = \frac{1}{2}[\cos(A-B) - \cos(A+B)]$$

5 We can write the product

$$10\cos 5x\sin 3x$$

as a difference by applying the second product to sum formula:

$$10\cos 5x\sin 3x = 10\cdot\tfrac{1}{2}[\sin(5x+3x) - \sin(5x-3x)] = 5(\sin 8x - \sin 2x)$$

SUM TO PRODUCT FORMULAS [5.5]

$$\sin\alpha + \sin\beta = 2\sin\frac{\alpha+\beta}{2}\cos\frac{\alpha-\beta}{2}$$

$$\sin\alpha - \sin\beta = 2\cos\frac{\alpha+\beta}{2}\sin\frac{\alpha-\beta}{2}$$

$$\cos\alpha + \cos\beta = 2\cos\frac{\alpha+\beta}{2}\cos\frac{\alpha-\beta}{2}$$

$$\cos\alpha - \cos\beta = -2\sin\frac{\alpha+\beta}{2}\sin\frac{\alpha-\beta}{2}$$

6 Prove

$$-\tan x = \frac{\cos 3x - \cos x}{\sin 3x + \sin x}.$$

Proof

$$\frac{\cos 3x - \cos x}{\sin 3x + \sin x} = \frac{-2\sin\frac{3x+x}{2}\sin\frac{3x-x}{2}}{2\sin\frac{3x+x}{2}\cos\frac{3x-x}{2}} = \frac{-2\sin 2x\sin x}{2\sin 2x\cos x} = -\frac{\sin x}{\cos x} = -\tan x$$

CHAPTER 5 TEST

Prove each identity.

1. $\dfrac{\cot\theta}{\csc\theta} = \cos\theta$
2. $(\sec x - 1)(\sec x + 1) = \tan^2 x$
3. $\sec\theta - \cos\theta = \tan\theta\sin\theta$
4. $\dfrac{\cos t}{1 - \sin t} = \dfrac{1 + \sin t}{\cos t}$
5. $\dfrac{1}{1 - \sin t} + \dfrac{1}{1 + \sin t} = 2\sec^2 t$
6. $\cos\left(\dfrac{\pi}{2} + \theta\right) = -\sin\theta$
7. $\cos^4 A - \sin^4 A = \cos 2A$
8. $\cot A = \dfrac{\sin 2A}{1 - \cos 2A}$
9. $\cot x - \tan x = \dfrac{\cos 2x}{\sin x\cos x}$
10. $\tan\dfrac{x}{2} = \dfrac{\tan x}{\sec x + 1}$

Use your graphing calculator to determine if each equation appears to be an identity by graphing the left expression and right expression together. If so, prove the identity. If not, find a counterexample. (Some of these identities are from the book Plane and Spherical Trigonometry, *written by Leonard M. Passano and published by The Macmillan Company in 1918.)*

11. $\dfrac{\sec^2\alpha}{\sin^2\alpha} = \csc^2\alpha + \sec^2\alpha$
12. $\dfrac{1}{\tan\theta + \cot\theta} = \csc\theta\sec\theta$

Let $\sin A = -\dfrac{3}{5}$ *with* $270° \le A \le 360°$ *and* $\sin B = \dfrac{12}{13}$ *with* $90° \le B \le 180°$ *and find the following.*

13. $\sin(A + B)$
14. $\cos(A - B)$
15. $\cos 2B$
16. $\sin\dfrac{A}{2}$

Find exact values for each of the following.

17. $\sin 75°$
18. $\tan\dfrac{\pi}{12}$

Write each expression as a single trigonometric function.

19. $\cos 4x\cos 5x - \sin 4x\sin 5x$
20. $\sin 15°\cos 75° + \cos 15°\sin 75°$
21. If $\sin A = -\dfrac{\sqrt{5}}{5}$ with $180° \le A \le 270°$, find $\cos 2A$ and $\cos\dfrac{A}{2}$.
22. If $\sec A = \sqrt{10}$ with $0° \le A \le 90°$, find $\sin 2A$ and $\sin\dfrac{A}{2}$.
23. Find $\tan A$ if $\tan B = \dfrac{1}{2}$ and $\tan(A + B) = 3$.
24. Find $\cos x$ if $\cos 2x = \dfrac{1}{2}$.

Evaluate each expression below without using a calculator. (Assume any variables represent positive numbers.)

25. $\cos\left(\arcsin\dfrac{4}{5} - \arctan 2\right)$
26. $\sin\left(\arccos\dfrac{4}{5} + \arctan 2\right)$

Write each expression as an equivalent algebraic expression involving only x. (Assume x is positive.)

27. $\cos(2\sin^{-1} x)$

28. $\sin(2\cos^{-1} x)$

29. Rewrite the product $\sin 6x \sin 4x$ as a sum or difference.

30. Rewrite the sum $\cos 15° + \cos 75°$ as a product and simplify.

Combinations of Functions

Figure 1

OBJECTIVE: To write $y = a \sin Bx + b \cos Bx$ as a single trigonometric function $y = A \sin(Bx + C)$.

The GPS (global positioning system) allows a person to locate their position anywhere in the world. To determine position, a GPS receiver downloads a signal that contains navigation data from a NAVSTAR satellite (Figure 1) and performs a series of calculations using this data. Several of the calculations require writing the sum of a sine and cosine function as a single sine function. In this project you will learn how this is done.

In Example 4 of Section 4.6, we showed how we could graph the function $y = \sin x + \cos x$ by adding y-coordinates. We observed that the resulting graph appeared to have the shape of a sine function.

1 Graph $y = \sin x + \cos x$ with your graphing calculator. Use your calculator to find the coordinates of a high point on the graph and a low point (you can use your TRACE key, or your calculator may have special commands for finding maximum and minimum points). Find an equation of the graph in the form $y = A \sin(Bx + C)$ by using these coordinates to estimate values of A, B, and C. Check your equation by graphing it. The two graphs should be nearly identical.

To find an equation for this graph using an algebraic method, we need to find a way to write $y = a \sin Bx + b \cos Bx$ as $y = A \sin(Bx + C)$. That is, we would like to find values of A, B, and C so that the equation $a \sin Bx + b \cos Bx = A \sin(Bx + C)$ is an identity. Because the value of B will be the same for both functions, we only need to find values for A and C.

2 First, we will assume $a \sin Bx + b \cos Bx = A \sin(Bx + C)$. Express the right side of this equation as a sum using a sum identity. In order for the resulting equation to be an identity, explain why it must be true that $A \cos C = a$ and $A \sin C = b$.

3 Solve for A in terms of a and b using the Pythagorean identity

$$\sin^2 C + \cos^2 C = 1$$

Assume A is positive. To find the value of C, we simply need to find an angle satisfying the criteria:

$$\sin C = \frac{b}{A} \quad \text{and} \quad \cos C = \frac{a}{A}$$

Figure 2 shows a visual way to remember the relationships between a, b, A, and C. You just have to remember to choose angle C so that its terminal side lies in the correct quadrant.

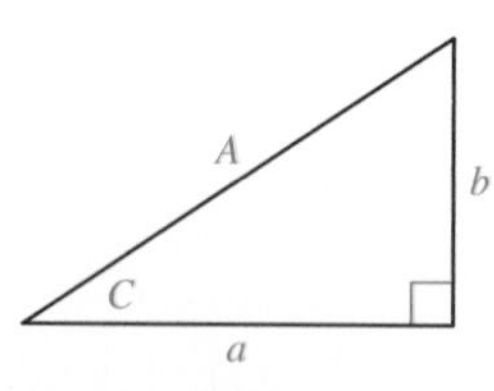

Figure 2

4 Write $y = \sin x + \cos x$ in the form $y = A \sin(Bx + C)$. First identify the values of a, b, and B. Then solve for A and C using the results from Question 3. How does this equation compare with the one you found in Question 1?

5 Graph your equation from Question 4 and compare it with the graph of $y = \sin x + \cos x$. The two graphs should be identical!

RESEARCH PROJECT Ptolemy

© Michael Nicholson/CORBIS

Claudius Ptolemy

Many of the identities presented in this chapter were known to the Greek astronomer and mathematician Claudius Ptolemy (A.D. 85–165). In his work *Almagest,* he was able to find the sine of sums and differences of angles and half-angles.

Research Ptolemy and his use of chords and the chord function. What was the notation that he used, and how is it different from modern notation? What other significant contributions did Ptolemy make to science? Write a paragraph or two about your findings.

6 Equations

Use, use your powers: what now costs you effort will in the end become mechanical.

➤ Georg C. Lichtenberg

© Agripicture Images / Alamy

INTRODUCTION

The human cannonball is launched from the cannon with an initial velocity of 64 feet per second at an angle of θ from the horizontal.

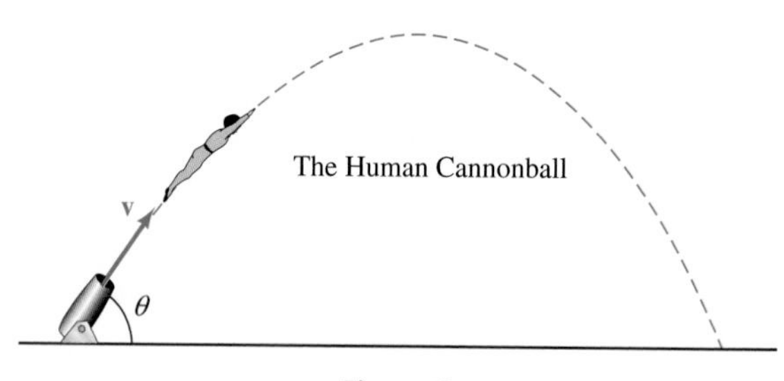

Figure 1

The maximum height attained by the human cannonball is a function of both his initial velocity and the angle that the cannon is inclined from the horizontal. His height t seconds after the cannon is fired is given by the equation

$$h(t) = -16t^2 + 64t \sin \theta$$

which you can see involves a trigonometric function. Equations that contain trigonometric functions are what we will study in this chapter.

STUDY SKILLS 6

This is the last chapter where we will mention study skills. You should know by now what works best for you and what you have to do to achieve your goals for this course. From now on it is simply a matter of sticking with the things that work for you and avoiding the things that do not. It seems simple, but as with anything that takes effort, it is up to you to see that you maintain the skills that get you where you want to be in the course. ▲

SECTION 6.1 Solving Trigonometric Equations

LEARNING OBJECTIVES

1. Solve a simple trigonometric equation.
2. Use factoring to solve a trigonometric equation.
3. Use the quadratic formula to solve a trigonometric equation.
4. Solve a trigonometric equation involving a horizontal translation.

The solution set for an equation is the set of all numbers that, when used in place of the variable, make the equation a true statement. In the previous chapter we worked with identities, which are equations that are true for every replacement of the variable for which it is defined. Most equations, however, are not identities. They are only true for certain values of the variable. For example, the solution set for the equation $4x^2 - 9 = 0$ is $\left\{-\frac{3}{2}, \frac{3}{2}\right\}$ because these are the only two numbers that, when used in place of x, turn the equation into a true statement. We call these types of equations *conditional equations*.

In algebra, the first kind of equations you learned to solve were linear (or first-degree) equations in one variable. Solving these equations was accomplished by applying two important properties: the *addition property of equality* and the *multiplication property of equality*. These two properties were stated as follows:

ADDITION PROPERTY OF EQUALITY

For any three algebraic expressions A, B, and C

$$\text{If} \quad A = B$$
$$\text{then} \quad A + C = B + C$$

In words: Adding the same quantity to both sides of an equation will not change the solution set.

MULTIPLICATION PROPERTY OF EQUALITY

For any three algebraic expressions A, B, and C, with $C \neq 0$,

$$\text{If} \quad A = B$$
$$\text{then} \quad AC = BC$$

In words: Multiplying both sides of an equation by the same nonzero quantity will not change the solution set.

Here is an example that shows how we use these two properties to solve a linear equation in one variable.

PROBLEM 1
Solve $6x - 9 = 2x - 5$.

EXAMPLE 1 Solve $5x + 7 = 2x - 5$ for x.

SOLUTION

$$5x + 7 = 2x - 5$$
$$3x + 7 = -5 \quad \text{Add } -2x \text{ to each side}$$
$$3x = -12 \quad \text{Add } -7 \text{ to each side}$$
$$x = -4 \quad \text{Multiply each side by } \frac{1}{3}$$

Notice in the last step we could just as easily have divided both sides by 3 instead of multiplying both sides by $\frac{1}{3}$. Division by a number and multiplication by its reciprocal are equivalent operations.

The process of solving trigonometric equations is very similar to the process of solving algebraic equations. With trigonometric equations, we look for values of an *angle* that will make the equation into a true statement. We usually begin by solving for a specific trigonometric function of that angle and then use the concepts we have developed earlier to find the angle. Here are some examples that illustrate this procedure.

PROBLEM 2
Solve $2 \cos x - 1 = 0$ if $0° \le x < 360°$.

NOTE Remember that sin *x* is function notation, meaning sin (*x*), where *x* is the input and "sin" is an abbreviation for "sine of."

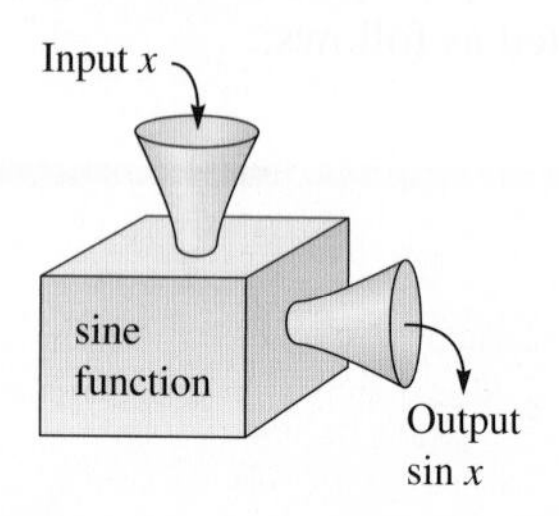

By itself, "sin" has no meaning or value. We cannot, for instance, isolate *x* in Example 2 by dividing both sides by "sin."

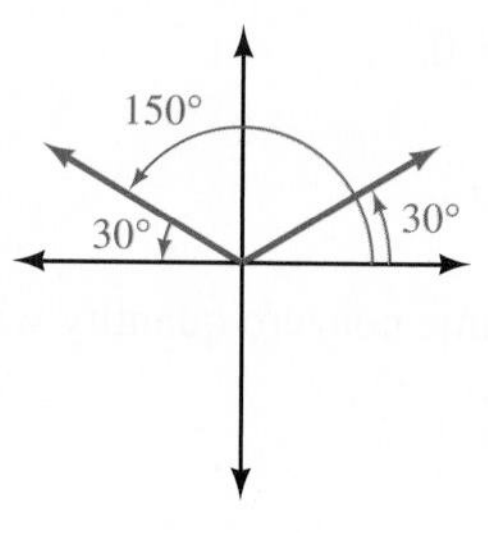

Figure 2

EXAMPLE 2 Solve $2 \sin x - 1 = 0$ for *x*.

SOLUTION We can solve for sin *x* using our methods from algebra. We then use our knowledge of trigonometry to find *x* itself.

$$2 \sin x - 1 = 0$$
$$2 \sin x = 1$$
$$\sin x = \frac{1}{2}$$

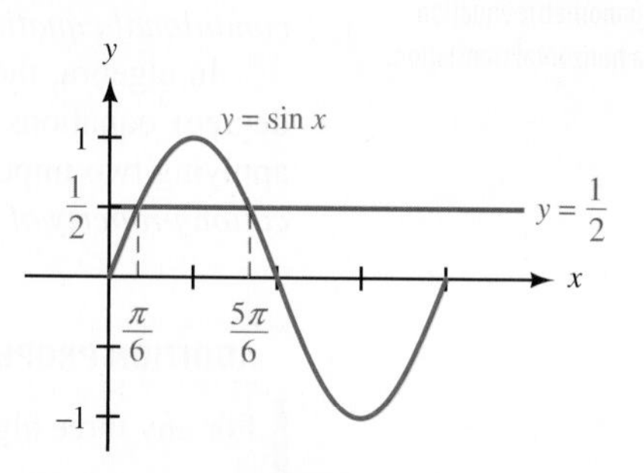

Figure 1

From Figure 1 we can see that if we are looking for radian solutions between 0 and 2π, then *x* is either $\pi/6$ or $5\pi/6$. On the other hand, if we want degree solutions between 0° and 360°, then our solutions will be 30° and 150°. Without the aid of Figure 1, we would reason that, because $\sin x = \frac{1}{2}$, the reference angle for *x* is 30°. Then, because $\frac{1}{2}$ is a positive number and the sine function is positive in QI and QII, *x* must be 30° or 150° (Figure 2).

Solutions Between 0° and 360° or 0 and 2π

In Degrees	In Radians
$x = 30°$ or $x = 150°$	$x = \frac{\pi}{6}$ or $x = \frac{5\pi}{6}$

Because the sine function is periodic with period 2π (or 360°), any angle coterminal with $x = \pi/6$ (or 30°) or $x = 5\pi/6$ (or 150°) will also be a solution of the equation. For any integer *k*, adding $2k\pi$ (or $360°k$) will result in a coterminal angle. Therefore, we can represent *all* solutions to the equation as follows.

All Solutions (*k* Is an Integer)

In Degrees	In Radians
$x = 30° + 360°k$	$x = \frac{\pi}{6} + 2k\pi$
or $x = 150° + 360°k$	or $x = \frac{5\pi}{6} + 2k\pi$

Using Technology Solving Equations: Finding Zeros

We can solve the equation in Example 2 using a graphing calculator. If we define the left side of the equation as function Y1, then the solutions of the equation will be the x-intercepts of this function. We sometimes refer to these values as *zeros*.

Set your calculator to degree mode and define $Y_1 = 2\sin(x) - 1$. Set your window variables so that

$$0 \le x \le 360, \text{ scale} = 90; \ -4 \le y \le 4, \text{ scale} = 1$$

Graph the function and use the appropriate command on your calculator to find both zeros. From Figure 3 we see that the solutions between 0° and 360° are $x = 30°$ and $x = 150°$.

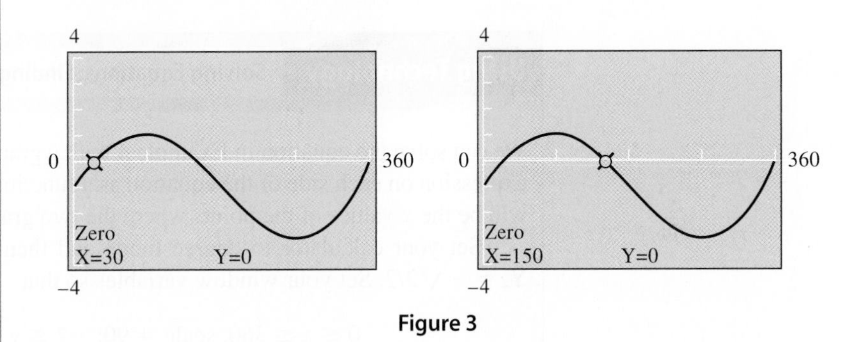

Figure 3

PROBLEM 3
Solve $3\cos x - 5 = 0$ if $0° \le x < 360°$.

EXAMPLE 3 Solve $2\sin\theta - 3 = 0$, if $0° \le \theta < 360°$.

SOLUTION We begin by solving for $\sin\theta$.

$$2\sin\theta - 3 = 0$$

$$2\sin\theta = 3 \quad \text{Add 3 to both sides}$$

$$\sin\theta = \frac{3}{2} \quad \text{Divide both sides by 2}$$

Because $\sin\theta$ is between -1 and 1 for all values of θ, $\sin\theta$ can never be $\frac{3}{2}$. Therefore, there is no solution to our equation.

To justify our conclusion further, we can graph $y = 2\sin x - 3$. The graph is a sine curve with amplitude 2 that has been shifted down 3 units vertically. The graph is shown in Figure 4. The graph does not cross the x-axis, so there is no solution to the equation $2\sin x - 3 = 0$.

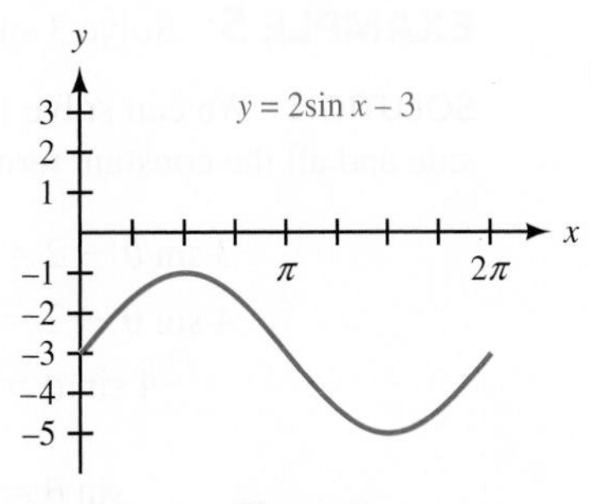

Figure 4 ■

PROBLEM 4
Find all degree solutions to
$\sin(A - 60°) = -\frac{\sqrt{3}}{2}$.

EXAMPLE 4 Find all degree solutions to $\cos(A - 25°) = -\frac{\sqrt{2}}{2}$.

SOLUTION The reference angle is given by $\cos^{-1}(\sqrt{2}/2) = 45°$. Because the cosine function is negative in QII or QIII, the expression $A - 25°$ must be coterminal with 135° or 225°. Therefore,

$$A - 25° = 135° + 360°k \quad \text{or} \quad A - 25° = 225° + 360°k$$

for any integer k. We can now solve for A by adding 25° to both sides.

$$\begin{aligned} A - 25° &= 135° + 360°k & \text{or} \quad A - 25° &= 225° + 360°k \\ A &= 160° + 360°k & A &= 250° + 360°k \end{aligned}$$

Using Technology — Solving Equations: Finding Intersection Points

We can solve the equation in Example 4 with a graphing calculator by defining the expression on each side of the equation as a function. The solutions to the equation will be the x-values of the points where the two graphs intersect.

Set your calculator to degree mode and then define $Y_1 = \cos(x - 25)$ and $Y_2 = -\sqrt{2}/2$. Set your window variables so that

$$0 \le x \le 360, \text{ scale} = 90; \ -2 \le y \le 2, \text{ scale} = 1$$

Graph both functions and use the appropriate command on your calculator to find the coordinates of the two intersection points. From Figure 5 we see that the x-coordinates of these points are $x = 160°$ and $x = 250°$.

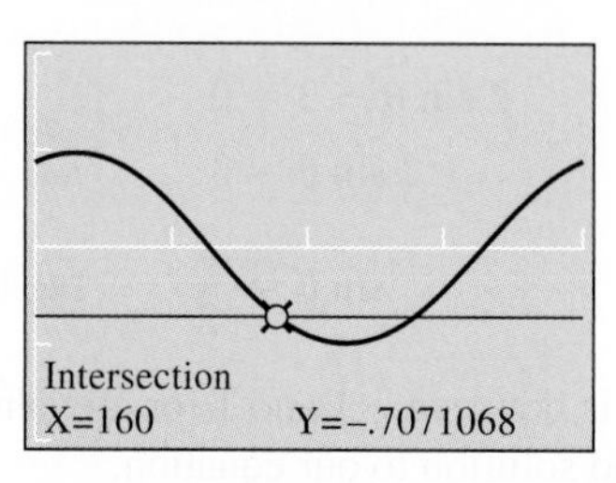

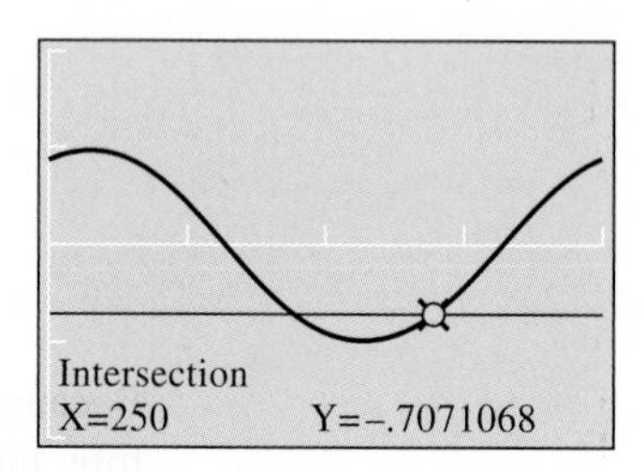

Figure 5

PROBLEM 5
Solve $2\cos\theta - 3 = 5\cos\theta - 2$ if $0° \le \theta < 360°$.

EXAMPLE 5 Solve $3\sin\theta - 2 = 7\sin\theta - 1$, if $0° \le \theta < 360°$.

SOLUTION We can solve for $\sin\theta$ by collecting all the variable terms on the left side and all the constant terms on the right side of the equation.

$$\begin{aligned} 3\sin\theta - 2 &= 7\sin\theta - 1 & & \\ -4\sin\theta - 2 &= -1 & & \text{Add } -7\sin\theta \text{ to each side} \\ -4\sin\theta &= 1 & & \text{Add 2 to each side} \\ \sin\theta &= -\frac{1}{4} & & \text{Divide each side by } -4 \end{aligned}$$

We have not memorized the angle whose sine is $-\frac{1}{4}$, so we must use a calculator to find the reference angle.

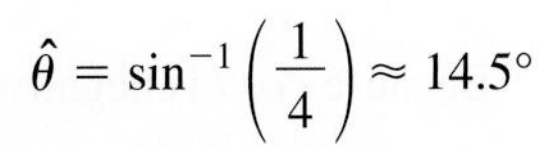

$$\hat{\theta} = \sin^{-1}\left(\frac{1}{4}\right) \approx 14.5°$$

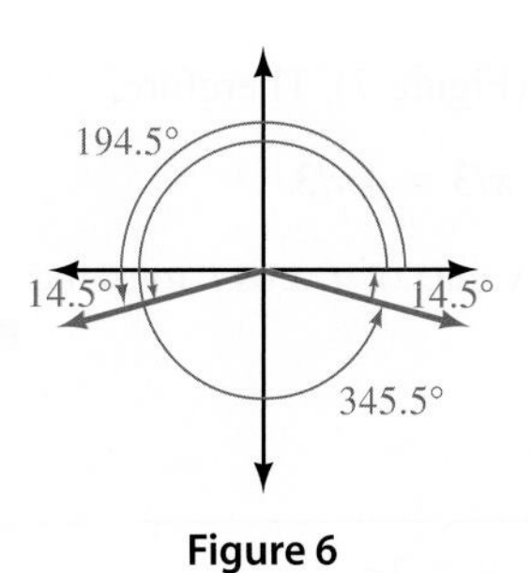

Figure 6

We find that the acute angle whose sine is nearest to $\frac{1}{4}$ is 14.5°. Therefore, the reference angle is 14.5°. Because $\sin \theta$ is negative, θ will terminate in QIII or QIV (Figure 6).

In QIII we have	In QIV we have
$\theta = 180° + 14.5°$	$\theta = 360° - 14.5°$
$= 194.5°$	$= 345.5°$

NOTE Unless directed otherwise, let's agree to write all approximate degree solutions to our equations in decimal degrees to the nearest tenth of a degree and all approximate radian solutions to the nearest hundredth.

CALCULATOR NOTE Remember, because of the restricted values on your calculator, if you use the $\boxed{\sin^{-1}}$ key with $-\frac{1}{4}$, your calculator will display approximately $-14.5°$, which is not within the desired interval for Example 5. The best way to proceed is to first find the reference angle using $\boxed{\sin^{-1}}$ with the positive value $\frac{1}{4}$. Then do the rest of the calculations based on the quadrants that the angle terminates in.

The next kind of trigonometric equation we will solve is quadratic in form. In algebra, the two most common methods of solving quadratic equations are factoring and applying the quadratic formula. Here is an example that reviews the factoring method.

PROBLEM 6
Solve $2x^2 + 3x = -1$.

EXAMPLE 6 Solve $2x^2 - 9x = 5$ for x.

SOLUTION We begin by writing the equation in standard form (0 on the right side—decreasing powers of the variable on the left side). We then factor the left side and set each factor equal to 0.

$$2x^2 - 9x = 5$$
$$2x^2 - 9x - 5 = 0 \quad \text{Standard form}$$
$$(2x + 1)(x - 5) = 0 \quad \text{Factor}$$
$$2x + 1 = 0 \quad \text{or} \quad x - 5 = 0 \quad \text{Set each factor to 0}$$
$$x = -\frac{1}{2} \quad \text{or} \quad x = 5 \quad \text{Solving resulting equations}$$

The two solutions, $x = -\frac{1}{2}$ and $x = 5$, are the only two numbers that satisfy the original equation.

PROBLEM 7
Solve $2\cos^2 t + 3\cos t = -1$ if $0 \le t < 2\pi$.

EXAMPLE 7 Solve $2\cos^2 t - 9\cos t = 5$, if $0 \le t < 2\pi$.

SOLUTION This equation is the equation from Example 6 with $\cos t$ in place of x. The fact that $0 \le t < 2\pi$ indicates we are to write our solutions in radians.

$$2\cos^2 t - 9\cos t = 5$$
$$2\cos^2 t - 9\cos t - 5 = 0 \quad \text{Standard form}$$
$$(2\cos t + 1)(\cos t - 5) = 0 \quad \text{Factor}$$
$$2\cos t + 1 = 0 \quad \text{or} \quad \cos t - 5 = 0 \quad \text{Set each factor to 0}$$
$$\cos t = -\frac{1}{2} \quad \text{or} \quad \cos t = 5 \quad \text{Isolate } \cos t$$

Figure 7

The first result, $\cos t = -\frac{1}{2}$, gives us a reference angle of

$$\hat{\theta} = \cos^{-1}(1/2) = \pi/3$$

Because $\cos t$ is negative, t must terminate in QII or QIII (Figure 7). Therefore,

$$t = \pi - \pi/3 = 2\pi/3 \quad \text{or} \quad t = \pi + \pi/3 = 4\pi/3$$

The second result, $\cos t = 5$, has no solution. For any value of t, $\cos t$ must be between -1 and 1. It can never be 5. ■

PROBLEM 8
Solve $2\sin^2 t - \sin t - 2 = 0$ if $0 \le t < 2\pi$.

EXAMPLE 8 Solve $2\sin^2\theta + 2\sin\theta - 1 = 0$, if $0 \le \theta < 2\pi$.

SOLUTION The equation is already in standard form. If we try to factor the left side, however, we find it does not factor. We must use the quadratic formula. The quadratic formula states that the solutions to the equation

$$ax^2 + bx + c = 0$$

will be

$$x = \frac{-b \pm \sqrt{b^2 - 4ac}}{2a}$$

In our case, the coefficients a, b, and c are

$$a = 2, \quad b = 2, \quad c = -1$$

Using these numbers, we can solve for $\sin\theta$ as follows:

$$\begin{aligned}
\sin\theta &= \frac{-2 \pm \sqrt{4 - 4(2)(-1)}}{2(2)} \\
&= \frac{-2 \pm \sqrt{12}}{4} \\
&= \frac{-2 \pm 2\sqrt{3}}{4} \\
&= \frac{-1 \pm \sqrt{3}}{2}
\end{aligned}$$

Using the approximation $\sqrt{3} = 1.7321$, we arrive at the following decimal approximations for $\sin\theta$:

$$\sin\theta = \frac{-1 + 1.7321}{2} \quad \text{or} \quad \sin\theta = \frac{-1 - 1.7321}{2}$$

$$\sin\theta = 0.3661 \quad \text{or} \quad \sin\theta = -1.3661$$

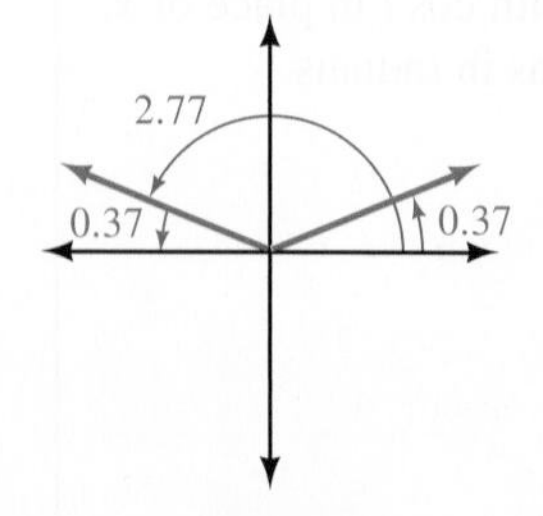

Figure 8

We will not obtain any solutions from the second expression, $\sin\theta = -1.3661$, because $\sin\theta$ must be between -1 and 1. For $\sin\theta = 0.3661$, we use a calculator to find the angle whose sine is nearest to 0.3661. That angle is approximately 0.37 radian, and it is the reference angle for θ. Since $\sin\theta$ is positive, θ must terminate in QI or QII (Figure 8). Therefore,

$$\theta = 0.37 \quad \text{or} \quad \theta = \pi - 0.37 = 2.77$$

■

Getting Ready for Class

After reading through the preceding section, respond in your own words and in complete sentences.

a. State the multiplication property of equality.
b. What is the solution set for an equation?
c. How many solutions between 0° and 360° does the equation $2 \sin x - 1 = 0$ contain?
d. Under what condition is factoring part of the process of solving an equation?

6.1 PROBLEM SET

➤ CONCEPTS AND VOCABULARY

For Questions 1 through 4, fill in the blank with an appropriate word or number.

1. To solve a trigonometric equation means to find all values of an _______ that will make the equation ______.

2. When solving a simple trigonometric equation, first __________ the trigonometric function, and then determine the values for the _______ itself.

3. To find the correct values (in degrees) for the angle, begin by finding the __________ angle, and use it to find all solutions between ____ and _____. Then represent all solutions by adding _______ to each value.

4. Graphically, each solution will appear as either an __________ for a single graph, or a point of ____________ of two graphs.

➤ EXERCISES

For each of the following equations, solve for **(a)** *all degree solutions and* **(b)** θ *if* $0° \le \theta < 360°$. *Do not use a calculator.*

5. $2 \sin \theta = 1$

6. $2 \cos \theta = 1$

7. $2 \cos \theta - \sqrt{3} = 0$

8. $2 \cos \theta + \sqrt{3} = 0$

9. $\sqrt{3} \cot \theta - 1 = 0$

10. $2 \tan \theta + 2 = 0$

For each of the following equations, solve for **(a)** *all radian solutions and* **(b)** t *if* $0 \le t < 2\pi$. *Give all answers as exact values in radians. Do not use a calculator.*

11. $4 \sin t - \sqrt{3} = 2 \sin t$

12. $\sqrt{3} + 5 \sin t = 3 \sin t$

13. $2 \cos t = 6 \cos t - 2\sqrt{3}$

14. $5 \cos t + 2\sqrt{3} = \cos t$

15. $3 \sin t + 5 = -2 \sin t$

16. $3 \sin t + 4 = 4$

For each of the following equations, solve for **(a)** *all degree solutions and* **(b)** θ *if* $0° \le \theta < 360°$. *Use a calculator to approximate all answers to the nearest tenth of a degree.*

17. $4 \sin \theta - 3 = 0$

18. $4 \sin \theta + 3 = 0$

19. $2 \cos \theta - 5 = 3 \cos \theta - 2$

20. $4 \cos \theta - 1 = 3 \cos \theta + 4$

21. $\sin \theta - 3 = 5 \sin \theta$

22. $\sin \theta - 4 = -2 \sin \theta$

Factor each expression completely.

23. a. $x - 2xy$

b. $\sin\theta - 2\sin\theta\cos\theta$

24. a. $y + 2xy$

b. $\tan\theta + 2\cos\theta\tan\theta$

25. a. $2x^2 - 7x + 3$

b. $2\cos^2\theta - 7\cos\theta + 3$

26. a. $2x^2 + 3x + 1$

b. $2\sin^2\theta + 3\sin\theta + 1$

For each of the following equations, solve for **(a)** *all radian solutions and* **(b)** *x if $0 \le x < 2\pi$. Give all answers as exact values in radians. Do not use a calculator.*

27. $(\sin x - 1)(2\sin x - 1) = 0$

28. $\tan x(\tan x + 1) = 0$

29. $\sin x + 2\sin x\cos x = 0$

30. $\cos x - 2\sin x\cos x = 0$

31. $2\sin^2 x - \sin x - 1 = 0$

32. $2\cos^2 x + \cos x - 1 = 0$

For each of the following equations, solve for **(a)** *all degree solutions and* **(b)** *θ if $0° \le \theta < 360°$. Do not use a calculator.*

33. $(2\cos\theta + \sqrt{3})(2\cos\theta + 1) = 0$

34. $(2\sin\theta - \sqrt{3})(2\sin\theta - 1) = 0$

35. $\sqrt{3}\tan\theta - 2\sin\theta\tan\theta = 0$

36. $\tan\theta - 2\cos\theta\tan\theta = 0$

37. $2\cos^2\theta + 11\cos\theta = -5$

38. $2\sin^2\theta - 7\sin\theta = -3$

Use the quadratic formula to find **(a)** *all degree solutions and* **(b)** *θ if $0° \le \theta < 360°$. Use a calculator to approximate all answers to the nearest tenth of a degree.*

39. $2\sin^2\theta - 2\sin\theta - 1 = 0$

40. $2\cos^2\theta + 2\cos\theta - 1 = 0$

41. $\cos^2\theta + \cos\theta - 1 = 0$

42. $\sin^2\theta - \sin\theta - 1 = 0$

43. $2\sin^2\theta + 1 = 4\sin\theta$

44. $1 - 4\cos\theta = -2\cos^2\theta$

Find all degree solutions to the following equations.

45. $\cos(A - 50°) = \dfrac{\sqrt{3}}{2}$

46. $\sin(A + 50°) = \dfrac{\sqrt{3}}{2}$

47. $\sin(A + 30°) = \dfrac{1}{2}$

48. $\cos(A + 30°) = \dfrac{1}{2}$

Find all radian solutions to the following equations.

49. $\cos\left(A - \dfrac{\pi}{9}\right) = -\dfrac{1}{2}$

50. $\sin\left(A - \dfrac{\pi}{9}\right) = -\dfrac{1}{2}$

51. $\sin\left(A + \dfrac{\pi}{12}\right) = -\dfrac{\sqrt{2}}{2}$

52. $\cos\left(A + \dfrac{\pi}{12}\right) = -\dfrac{\sqrt{2}}{2}$

Use your graphing calculator to find the solutions to the equations you solved in the following problems by graphing the function represented by the left side of the equation and then finding its zeros. Make sure your calculator is set to degree mode.

53. Problem 7

54. Problem 8

55. Problem 17

56. Problem 18

57. Problem 35

58. Problem 36

59. Problem 39

60. Problem 40

Use your graphing calculator to find the solutions to the equations you solved in the following problems by defining the left side and right side of the equation as functions and then finding the intersection points of their graphs. Make sure your calculator is set to degree mode.

61. Problem 5

62. Problem 6

63. Problem 21

64. Problem 22

65. Problem 37

66. Problem 38

67. Problem 43

68. Problem 44

69. Problem 45

70. Problem 46

Motion of a Projectile *If a projectile (such as a bullet) is fired into the air with an initial velocity v at an angle of elevation θ (see Figure 9), then the height h of the projectile at time t is given by*

$$h = -16t^2 + vt \sin \theta$$

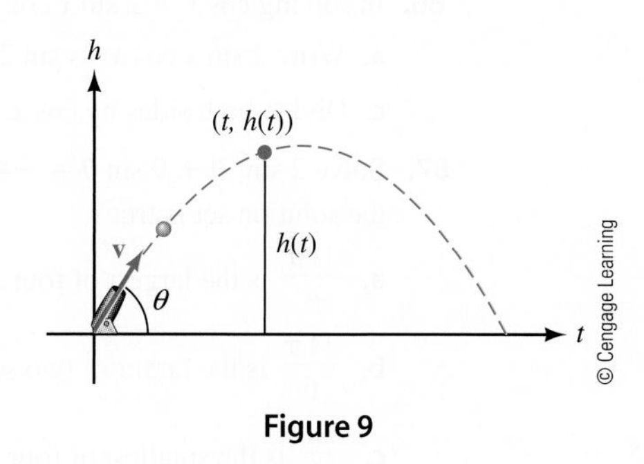

Figure 9

71. Give the equation for the height, if v is 1,500 feet per second and θ is 30°.

72. Give the equation for h, if v is 600 feet per second and θ is 45°. (Leave your answer in exact value form.)

73. Use the equation found in Problem 71 to find the height of the object after 2 seconds.

74. Use the equation found in Problem 72 to find the height of the object after $\sqrt{3}$ seconds to the nearest tenth.

75. Find the angle of elevation θ of a rifle barrel, if a bullet fired at 1,500 feet per second takes 2 seconds to reach a height of 750. Give your answer to the nearest tenth of a degree.

76. Find the angle of elevation of a rifle, if a bullet fired at 1,500 feet per second takes 3 seconds to reach a height of 750 feet. Give your answer to the nearest tenth of a degree.

➤ REVIEW PROBLEMS

The problems that follow review material we covered in Sections 5.2 and 5.3. Reviewing these problems will help you with the next section.

77. Write $\cos 2A$ in terms of $\sin A$ only.

78. Write $\cos 2A$ in terms of $\cos A$ only.

79. Expand $\sin(\theta + 45°)$ and then simplify.

80. Expand $\sin(\theta + 30°)$ and then simplify.

81. Find the exact value of $\sin 75°$.

82. Find the exact value of $\cos 105°$.

83. Prove the identity $\cos 2x = \dfrac{1 - \tan^2 x}{1 + \tan^2 x}$.

84. Prove the identity $\sin 2x = \dfrac{2}{\tan x + \cot x}$.

➤ LEARNING OBJECTIVES ASSESSMENT

These questions are available for instructors to help assess if you have successfully met the learning objectives for this section.

85. Solve $6 \sin x + \sqrt{3} = 4 \sin x$ for $0° \leq x < 360°$. Which of the following statements about the solution set is true?

a. 210° is the smaller of two solutions. **b.** 210° is the larger of two solutions.

c. 240° is the smaller of two solutions. **d.** 240° is the larger of two solutions.

86. In solving $\cos x + 2 \sin x \cos x = 0$, which of the following is the best first step?

a. Write $2 \sin x \cos x$ as $\sin 2x$. **b.** Factor $\cos x$ from the left side.

c. Divide both sides by $\cos x$. **d.** Subtract $\cos x$ from both sides.

87. Solve $2 \sin^2 \theta + 9 \sin \theta = -4$ for $0 \leq \theta < 2\pi$. Which of the following statements about the solution set is true?

a. $\frac{11\pi}{6}$ is the largest of four solutions.

b. $\frac{11\pi}{6}$ is the larger of two solutions.

c. $\frac{7\pi}{6}$ is the smallest of four solutions.

d. $\frac{7\pi}{6}$ is the smallest of three solutions.

88. Solve $2 \cos (A + 40°) = \sqrt{2}$ for all degree solutions. What is the *sum* of the first two positive solutions?

a. 100° **b.** 360° **c.** 440° **d.** 280°

SECTION 6.2 More on Trigonometric Equations

LEARNING OBJECTIVES

1. Use an identity to solve a trigonometric equation.
2. Solve a trigonometric equation by clearing denominators.
3. Solve a trigonometric equation by squaring both sides.
4. Use a graphing calculator to approximate the solutions to a trigonometric equation.

In this section, we will use our knowledge of identities to replace some parts of the equations we are solving with equivalent expressions that will make the equations easier to solve. Here are some examples.

EXAMPLE 1 Solve $2 \cos x - 1 = \sec x$, if $0 \leq x < 2\pi$.

SOLUTION To solve this equation as we have solved the equations in the previous section, we must write each term using the same trigonometric function. To do so, we can use a reciprocal identity to write $\sec x$ in terms of $\cos x$.

$$2 \cos x - 1 = \frac{1}{\cos x}$$

To clear the equation of fractions, we multiply both sides by $\cos x$. (Note that we must assume $\cos x \neq 0$ in order to multiply both sides by it. If we obtain solutions for which $\cos x = 0$, we will have to discard them.)

$$\cos x\,(2 \cos x - 1) = \frac{1}{\cos x} \cdot \cos x$$

$$2 \cos^2 x - \cos x = 1$$

PROBLEM 1
Solve $2 \cos x - 3 \sec x = -1$ if $0 \leq x < 2\pi$.

We are left with a quadratic equation that we write in standard form and then solve.

$$2\cos^2 x - \cos x - 1 = 0 \quad \text{Standard form}$$
$$(2\cos x + 1)(\cos x - 1) = 0 \quad \text{Factor}$$
$$2\cos x + 1 = 0 \quad \text{or} \quad \cos x - 1 = 0 \quad \text{Set each factor to 0}$$
$$\cos x = -\frac{1}{2} \quad \text{or} \quad \cos x = 1 \quad \text{Isolate } \cos x$$
$$x = \frac{2\pi}{3}, \frac{4\pi}{3} \quad \text{or} \quad x = 0$$

The solutions are 0, $2\pi/3$, and $4\pi/3$. ■

PROBLEM 2
Solve $\sin 2\theta - \sin\theta = 0$ if $0° \le \theta < 360°$.

EXAMPLE 2 Solve $\sin 2\theta + \sqrt{2}\cos\theta = 0$, $0° \le \theta < 360°$.

SOLUTION To solve this equation, both trigonometric functions must be functions of the same angle. As the equation stands now, one angle is 2θ, while the other is θ. We can write everything as a function of θ by using the double-angle identity $\sin 2\theta = 2\sin\theta\cos\theta$.

$$\sin 2\theta + \sqrt{2}\cos\theta = 0$$
$$2\sin\theta\cos\theta + \sqrt{2}\cos\theta = 0 \quad \text{Double-angle identity}$$
$$\cos\theta\,(2\sin\theta + \sqrt{2}) = 0 \quad \text{Factor out } \cos\theta$$
$$\cos\theta = 0 \quad \text{or} \quad 2\sin\theta + \sqrt{2} = 0 \quad \text{Set each factor to 0}$$
$$\sin\theta = -\frac{\sqrt{2}}{2}$$
$$\theta = 90°, 270° \quad \text{or} \quad \theta = 225°, 315°$$

■

PROBLEM 3
Solve $\cos 2\theta + \cos\theta = 0$ if $0° \le \theta < 360°$.

EXAMPLE 3 Solve $\cos 2\theta + 3\sin\theta - 2 = 0$, if $0° \le \theta < 360°$.

SOLUTION We have the same problem with this equation that we did with the equation in Example 2. We must rewrite $\cos 2\theta$ in terms of functions of just θ. Recall that there are three forms of the double-angle identity for $\cos 2\theta$. We choose the double-angle identity that involves $\sin\theta$ only, because the middle term of our equation involves $\sin\theta$, and it is best to have all terms involve the same trigonometric function.

$$\cos 2\theta + 3\sin\theta - 2 = 0$$
$$1 - 2\sin^2\theta + 3\sin\theta - 2 = 0 \quad \cos 2\theta = 1 - 2\sin^2\theta$$
$$-2\sin^2\theta + 3\sin\theta - 1 = 0 \quad \text{Simplify}$$
$$2\sin^2\theta - 3\sin\theta + 1 = 0 \quad \text{Multiply each side by } -1$$
$$(2\sin\theta - 1)(\sin\theta - 1) = 0 \quad \text{Factor}$$
$$2\sin\theta - 1 = 0 \quad \text{or} \quad \sin\theta - 1 = 0 \quad \text{Set factors to 0}$$
$$\sin\theta = \frac{1}{2} \qquad \sin\theta = 1$$
$$\theta = 30°, 150° \quad \text{or} \quad \theta = 90°$$

■

PROBLEM 4
Solve $4\cos^2 x + 4\sin x = 5$ if $0 \le x < 2\pi$.

EXAMPLE 4 Solve $4\cos^2 x + 4\sin x - 5 = 0$, $0 \le x < 2\pi$.

SOLUTION We cannot factor and solve this quadratic equation until each term involves the same trigonometric function. If we change the $\cos^2 x$ in the first term to $1 - \sin^2 x$, we will obtain an equation that involves the sine function only.

$$4\cos^2 x + 4\sin x - 5 = 0$$

$$4(1 - \sin^2 x) + 4\sin x - 5 = 0 \qquad \cos^2 x = 1 - \sin^2 x$$

$$4 - 4\sin^2 x + 4\sin x - 5 = 0 \qquad \text{Distributive property}$$

$$-4\sin^2 x + 4\sin x - 1 = 0 \qquad \text{Add 4 and } -5$$

$$4\sin^2 x - 4\sin x + 1 = 0 \qquad \text{Multiply each side by } -1$$

$$(2\sin x - 1)^2 = 0 \qquad \text{Factor}$$

$$2\sin x - 1 = 0 \qquad \text{Set factor to 0}$$

$$\sin x = \frac{1}{2}$$

$$x = \frac{\pi}{6}, \frac{5\pi}{6}$$

PROBLEM 5
Solve $\cos\theta - \sqrt{3}\sin\theta = 0$ if $0° \le \theta < 360°$.

EXAMPLE 5 Solve $\sin\theta - \cos\theta = 1$, if $0 \le \theta < 2\pi$.

SOLUTION If we separate $\sin\theta$ and $\cos\theta$ on opposite sides of the equal sign, and then square both sides of the equation, we will be able to use an identity to write the equation in terms of one trigonometric function only.

$$\sin\theta - \cos\theta = 1$$

$$\sin\theta = 1 + \cos\theta \qquad \text{Add } \cos\theta \text{ to each side}$$

$$\sin^2\theta = (1 + \cos\theta)^2 \qquad \text{Square each side}$$

$$\sin^2\theta = 1 + 2\cos\theta + \cos^2\theta \qquad \text{Expand } (1 + \cos\theta)^2$$

$$1 - \cos^2\theta = 1 + 2\cos\theta + \cos^2\theta \qquad \sin^2\theta = 1 - \cos^2\theta$$

$$0 = 2\cos\theta + 2\cos^2\theta \qquad \text{Standard form}$$

$$0 = 2\cos\theta(1 + \cos\theta) \qquad \text{Factor}$$

$$2\cos\theta = 0 \quad \text{or} \quad 1 + \cos\theta = 0 \qquad \text{Set factors to 0}$$

$$\cos\theta = 0 \qquad \cos\theta = -1$$

$$\theta = \pi/2, 3\pi/2 \quad \text{or} \quad \theta = \pi$$

We have three possible solutions, some of which may be extraneous because we squared both sides of the equation in Step 2. Any time we raise both sides of an equation to an even power, we have the possibility of introducing extraneous solutions. We must check each possible solution in our original equation.

Checking $\theta = \pi/2$	*Checking* $\theta = \pi$	*Checking* $\theta = 3\pi/2$
$\sin\pi/2 - \cos\pi/2 \stackrel{?}{=} 1$	$\sin\pi - \cos\pi \stackrel{?}{=} 1$	$\sin 3\pi/2 - \cos 3\pi/2 \stackrel{?}{=} 1$
$1 - 0 \stackrel{?}{=} 1$	$0 - (-1) \stackrel{?}{=} 1$	$-1 - 0 \stackrel{?}{=} 1$
$1 = 1$	$1 = 1$	$-1 \ne 1$
$\theta = \pi/2$ is a solution	$\theta = \pi$ is a solution	$\theta = 3\pi/2$ is not a solution

All possible solutions, except $\theta = 3\pi/2$, produce true statements when used in place of the variable in the original equation. $\theta = 3\pi/2$ is an extraneous solution produced by squaring both sides of the equation. Our solution set is $\{\pi/2, \pi\}$.

Using Technology Verifying Solutions Graphically

We can verify the solutions in Example 5 and confirm that there are only two solutions between 0 and 2π using a graphing calculator. First, write the equation so that all terms are on one side and a zero is on the other.

$$\sin\theta - \cos\theta = 1$$

$$\sin\theta - \cos\theta - 1 = 0 \quad \text{Add } -1 \text{ to both sides}$$

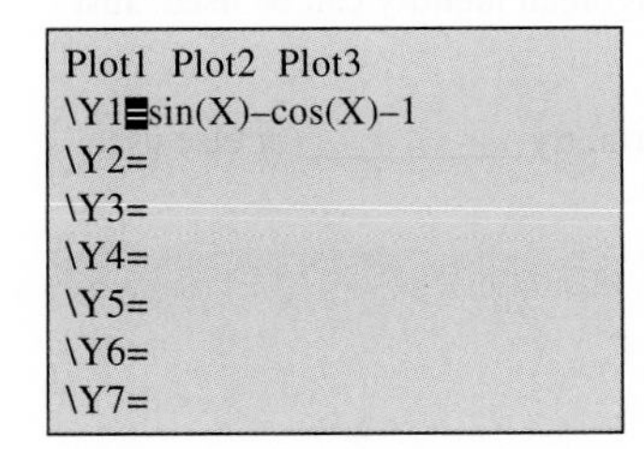

Figure 1

Define the expression on the left side of this equation as function Y_1 (Figure 1). Set your calculator to radian mode and graph the function using the following window settings:

$$0 \le x \le 2\pi, \text{ scale} = \pi/2; \ -3 \le y \le 2, \text{ scale} = 1$$

The solutions of the equation will be the zeros (x-intercepts) of this function. From the graph, we see that there are only two solutions. Use the feature of your calculator that will allow you to evaluate the function from the graph, and verify that $x = \pi/2$ and $x = \pi$ are x-intercepts (Figure 2). It is clear from the graph that $x = 3\pi/2$ is not a solution.

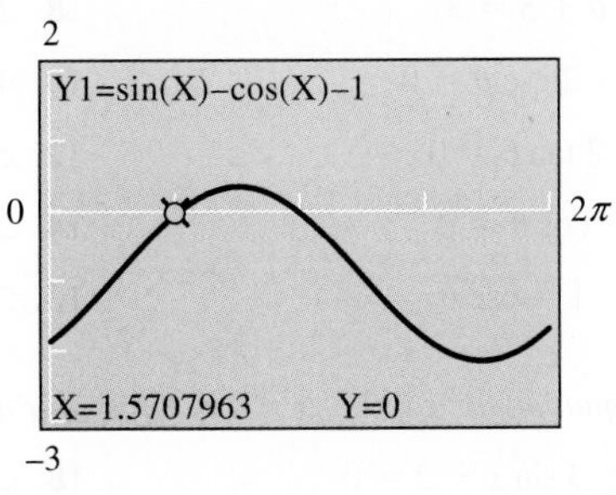

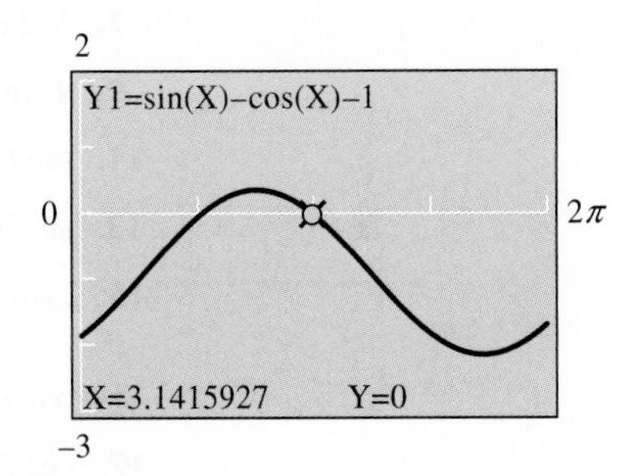

Figure 2

Getting Ready for Class

After reading through the preceding section, respond in your own words and in complete sentences.

a. What is the first step in solving the equation $2\cos x - 1 = \sec x$?

b. Why do we need zero on one side of a quadratic equation to solve the equation?

c. How many solutions between 0 and 2π does the equation $\cos x = 0$ contain?

d. How do you factor the left side of the equation $2\sin\theta\cos\theta + \sqrt{2}\cos\theta = 0$?

6.2 PROBLEM SET

➤ CONCEPTS AND VOCABULARY

For Questions 1 through 4, fill in the blank with an appropriate word.

1. To solve an equation containing secant, cosecant, or cotangent functions, first rewrite each function using a ________ identity and then _______ fractions by ___________ both sides of the equation by the least common denominator.
2. For equations containing trigonometric functions with different arguments, use an appropriate ________ to write all of the functions in terms of the same _______.
3. When solving an equation containing a single sine and cosine, sometimes it is necessary to ________ both sides of the equation so that the Pythagorean identity can be used. Just be sure to check for _____________ solutions.
4. To solve a trigonometric equation that is quadratic in form, try __________ or else use the ________ formula.

➤ EXERCISES

Solve each equation for θ if $0° \le \theta < 360°$.

5. $\sqrt{3}\sec\theta = 2$
6. $\sqrt{2}\csc\theta = 2$
7. $\sqrt{2}\csc\theta + 5 = 3$
8. $2\sqrt{3}\sec\theta + 7 = 3$
9. $4\sin\theta - 2\csc\theta = 0$
10. $4\cos\theta - 3\sec\theta = 0$
11. $\sec\theta - 2\tan\theta = 0$
12. $\csc\theta + 2\cot\theta = 0$
13. $\sin 2\theta - \cos\theta = 0$
14. $2\sin\theta + \sin 2\theta = 0$
15. $2\sin\theta - 1 = \csc\theta$
16. $2\cos\theta + 1 = \sec\theta$

Solve each equation for x if $0 \le x < 2\pi$. Give your answers in radians using exact values only.

17. $\cos 2x - 3\sin x - 2 = 0$
18. $\cos 2x - \cos x - 2 = 0$
19. $\cos x - \cos 2x = 0$
20. $\sin x = -\cos 2x$
21. $2\cos^2 x + \sin x - 1 = 0$
22. $2\sin^2 x - \cos x - 1 = 0$
23. $4\sin^2 x + 4\cos x - 5 = 0$
24. $4\cos^2 x - 4\sin x - 5 = 0$
25. $2\sin x + \cot x - \csc x = 0$
26. $2\cos x + \tan x = \sec x$
27. $\sin x + \cos x = \sqrt{2}$
28. $\sin x - \cos x = \sqrt{2}$

Solve for θ if $0° \le \theta < 360°$.

29. $\sqrt{3}\sin\theta + \cos\theta = \sqrt{3}$
30. $\sin\theta - \sqrt{3}\cos\theta = \sqrt{3}$
31. $\sqrt{3}\sin\theta - \cos\theta = 1$
32. $\sin\theta - \sqrt{3}\cos\theta = 1$
33. $\sin\frac{\theta}{2} - \cos\theta = 0$
34. $\sin\frac{\theta}{2} + \cos\theta = 1$
35. $\cos\frac{\theta}{2} - \cos\theta = 1$
36. $\cos\frac{\theta}{2} - \cos\theta = 0$

For each equation, find all degree solutions in the interval $0° \leq \theta < 360°$. If rounding is necessary, round to the nearest tenth of a degree. Use your graphing calculator to verify each solution graphically.

37. $6 \cos \theta + 7 \tan \theta = \sec \theta$

38. $13 \cot \theta + 11 \csc \theta = 6 \sin \theta$

39. $18 \sec^2 \theta - 17 \tan \theta \sec \theta - 12 = 0$

40. $23 \csc^2 \theta - 22 \cot \theta \csc \theta - 15 = 0$

41. $7 \sin^2 \theta - 9 \cos 2\theta = 0$

42. $16 \cos 2\theta - 18 \sin^2 \theta = 0$

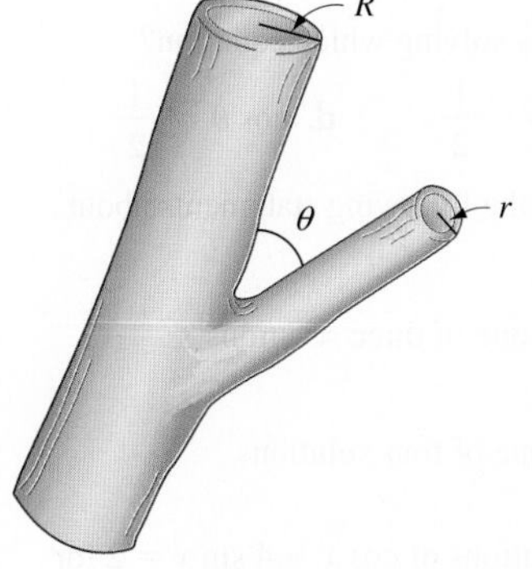

Figure 3

Write expressions that give all solutions to the equations you solved in the following problems.

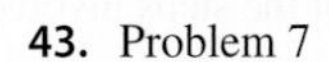

43. Problem 7

44. Problem 8

45. Problem 27

46. Problem 28

47. Problem 35

48. Problem 36

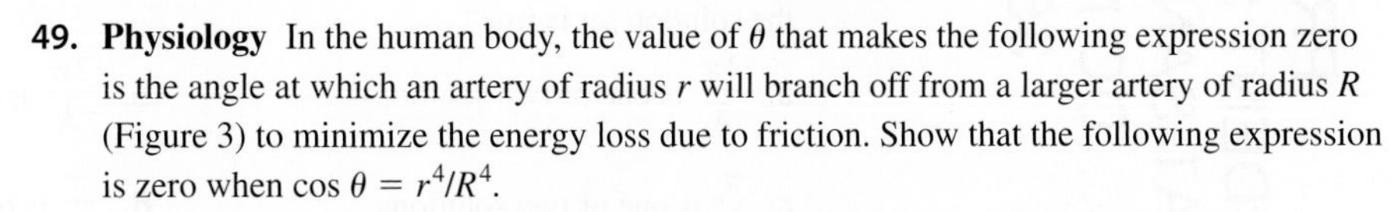

49. Physiology In the human body, the value of θ that makes the following expression zero is the angle at which an artery of radius r will branch off from a larger artery of radius R (Figure 3) to minimize the energy loss due to friction. Show that the following expression is zero when $\cos \theta = r^4/R^4$.

$$r^4 \csc^2 \theta - R^4 \csc \theta \cot \theta$$

50. Physiology Find the value of θ that makes the expression in Problem 49 zero, if $r = 2$ mm and $R = 4$ mm. (Give your answer to the nearest tenth of a degree.)

Solving the following equations will require you to use the quadratic formula. Solve each equation for θ between $0°$ and $360°$, and round your answers to the nearest tenth of a degree.

51. $2 \sin^2 \theta - 2 \cos \theta - 1 = 0$

52. $2 \cos^2 \theta + 2 \sin \theta - 1 = 0$

53. $\cos^2 \theta + \sin \theta = 0$

54. $\sin^2 \theta = \cos \theta$

55. $2 \sin^2 \theta = 3 - 4 \cos \theta$

56. $4 \sin \theta = 3 - 2 \cos^2 \theta$

Use your graphing calculator to find all radian solutions in the interval $0 \leq x < 2\pi$ for each of the following equations. Round your answers to four decimal places.

57. $\cos x + 3 \sin x - 2 = 0$

58. $2 \cos x + \sin x + 1 = 0$

59. $\sin^2 x - 3 \sin x - 1 = 0$

60. $\cos^2 x - 3 \cos x + 1 = 0$

61. $\sec x + 2 = \cot x$

62. $\csc x - 3 = \tan x$

➤ REVIEW PROBLEMS

The problems that follow review material we covered in Section 5.4.

If $\sin A = \frac{2}{3}$ with A in the interval $0° \leq A \leq 90°$, find

63. $\sin \frac{A}{2}$

64. $\cos \frac{A}{2}$

65. $\tan \frac{A}{2}$

66. $\cot \frac{A}{2}$

67. Graph $y = 4 \sin^2 \frac{x}{2}$.

68. Graph $y = 6 \cos^2 \frac{x}{2}$.

69. Use a half-angle formula to find $\sin 22.5°$.

70. Use a half-angle formula to find $\cos 15°$.

➤ LEARNING OBJECTIVES ASSESSMENT

These questions are available for instructors to help assess if you have successfully met the learning objectives for this section.

71. Solve $2 \cos 2\theta - 4 \sin \theta = -1$ for $0° \le \theta < 360°$. Which of the following statements about the solution set is true?

a. $30°$ is the smallest of four solutions. **b.** $150°$ is the smaller of two solutions.

c. $150°$ is the larger of two solutions. **d.** $150°$ is the smallest of four solutions.

72. In solving $\csc \theta - 2 \cot \theta = 0$, one of the steps involves solving which equation?

a. $\sin \theta = 1$ **b.** $\sin \theta = -1$ **c.** $\cos \theta = -\dfrac{1}{2}$ **d.** $\cos \theta = \dfrac{1}{2}$

73. Solve $\sqrt{3} \sin x + \cos x = 1$ for $0 \le x < 2\pi$. Which of the following statements about the solution set is true?

a. $\dfrac{2\pi}{3}$ is one of two solutions. **b.** $\dfrac{2\pi}{3}$ is one of three solutions.

c. $\dfrac{\pi}{6}$ is one of two solutions. **d.** $\dfrac{\pi}{6}$ is one of four solutions.

74. Use a graphing calculator to approximate all radian solutions of $\cos x - 4 \sin x = 2$ for $0 \le x < 2\pi$ to four decimal places.

a. 0.7154, 2.1808 **b.** 0.7514, 2.8801

c. 3.9308, 6.1270 **d.** 3.8930, 6.0217

SECTION 6.3 Trigonometric Equations Involving Multiple Angles

LEARNING OBJECTIVES

1. Solve a simple trigonometric equation involving a multiple angle.
2. Use an identity to solve a trigonometric equation involving a multiple angle.
3. Solve a trigonometric equation involving a multiple angle by factoring or the quadratic formula.
4. Solve a real-life problem using a trigonometric equation.

In this section, we will consider equations that contain multiple angles. We will use most of the same techniques to solve these equations that we have used in the past. We have to be careful at the last step, however, when our equations contain multiple angles. Here is an example.

EXAMPLE 1 Solve $\cos 2\theta = \dfrac{\sqrt{3}}{2}$, if $0° \le \theta < 360°$.

SOLUTION The equation cannot be simplified further. The reference angle is given by $\cos^{-1}(\sqrt{3}/2) = 30°$. Because the cosine function is positive in QI or QIV, the expression 2θ must be coterminal with $30°$ or $330°$. Therefore, for any integer k, all solutions will have the form

$$2\theta = 30° + 360°k \quad \text{or} \quad 2\theta = 330° + 360°k$$

$$\theta = \frac{30°}{2} + \frac{360°k}{2} \qquad \theta = \frac{330°}{2} + \frac{360°k}{2}$$

$$\theta = 15° + 180°k \qquad \theta = 165° + 180°k$$

PROBLEM 1
Solve $\sin 3\theta = \sqrt{3}/2$ if $0° \le \theta < 360°$.

Because we were asked only for values of θ between 0° and 360°, we substitute the appropriate integers in place of k.

If	$k = 0$	If	$k = 1$
then	$\theta = 15° + 180°(0) = 15°$	then	$\theta = 15° + 180°(1) = 195°$
or	$\theta = 165° + 180°(0) = 165°$	or	$\theta = 165° + 180°(1) = 345°$

For all other values of k, θ will fall outside the given interval. ■

Using Technology

To solve the equation in Example 1 graphically, we can use the intersection of graphs method that was introduced in Section 6.1. Set your graphing calculator to degree mode, then define each side of the equation as an individual function.

$$Y_1 = \cos(2x),\ Y_2 = \sqrt{3}/2$$

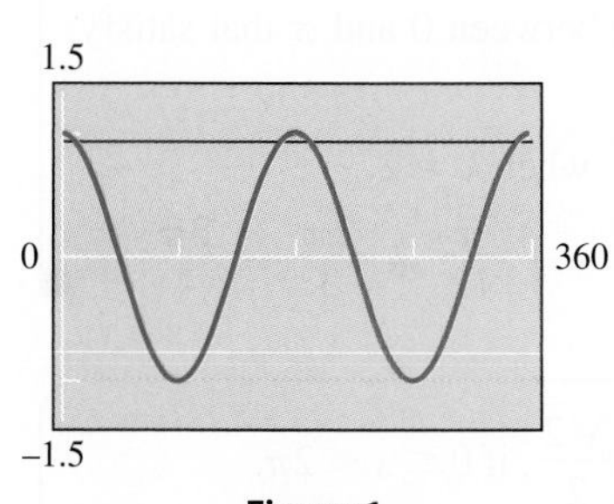

Figure 1

Graph the two functions using the following window settings:

$$0 \le x \le 360, \text{ scale} = 90;\ -1.5 \le y \le 1.5, \text{ scale} = 1$$

Because the period of $y = \cos 2x$ is $360°/2 = 180°$, we see that two cycles of the cosine function occur between 0° and 360° (Figure 1). Within the first cycle there are two intersection points. Using the appropriate command, we find that the x-coordinates of these points are $x = 15°$ and $x = 165°$ (Figure 2).

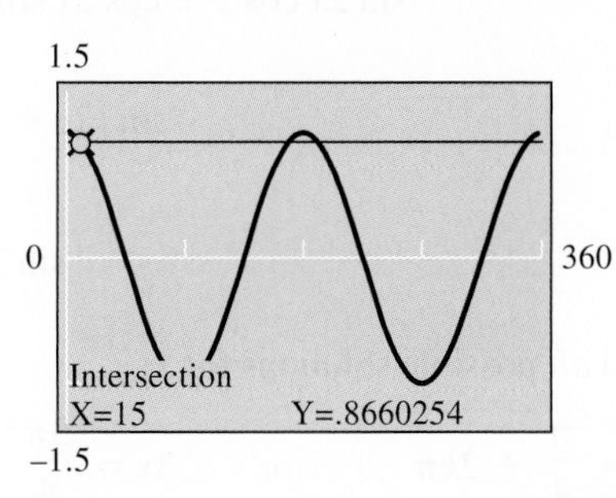

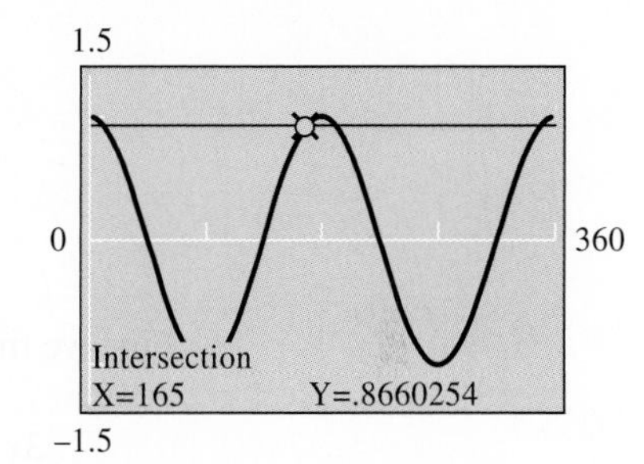

Figure 2

Because the period is 180°, the next pair of intersection points (located on the second cycle) is found by adding 180° to each of the above values (Figure 3). Use the intersection command on your calculator to verify that the next solutions occur at $x = 195°$ and $x = 345°$.

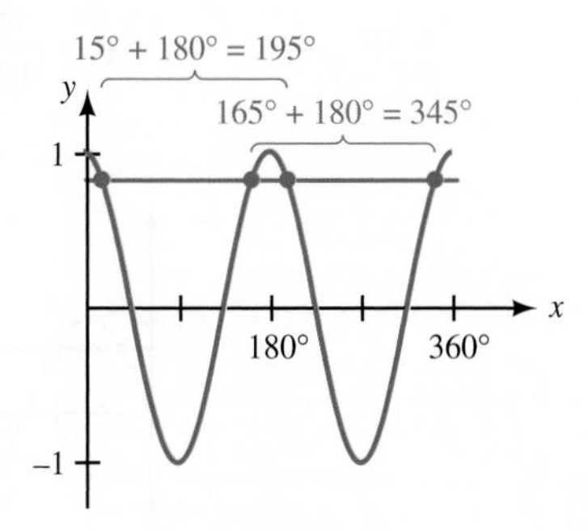

Figure 3

PROBLEM 2
Find all radian solutions for $\tan 2x = -1$.

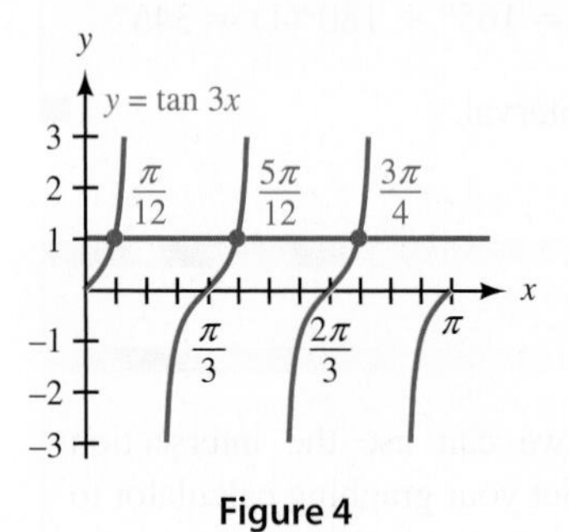

Figure 4

EXAMPLE 2 Find all solutions to $\tan 3x = 1$, if x is measured in radians with exact values.

SOLUTION First we find all values of $3x$ in the interval $0 \le 3x < \pi$ that satisfy $\tan 3x = 1$, and then we add on multiples of π because the period of the tangent function is π. After that, we simply divide by 3 to solve for x.

If $\quad \tan 3x = 1$

then $\quad 3x = \dfrac{\pi}{4} + k\pi \qquad$ k is any integer

$$x = \frac{\pi}{12} + \frac{k\pi}{3} \qquad \text{Divide by 3}$$

Note that $k = 0$, 1, and 2 will give us all values of x between 0 and π that satisfy $\tan 3x = 1$ (Figure 4):

when $k = 0$: $x = \dfrac{\pi}{12}$

when $k = 1$: $x = \dfrac{\pi}{12} + \dfrac{\pi}{3} = \dfrac{5\pi}{12}$

when $k = 2$: $x = \dfrac{\pi}{12} + \dfrac{2\pi}{3} = \dfrac{3\pi}{4}$ ■

PROBLEM 3
Solve $\cos 2x \cos x - \sin 2x \sin x = \dfrac{1}{2}$ if $0 \le x < 2\pi$.

EXAMPLE 3 Solve $\sin 2x \cos x + \cos 2x \sin x = \dfrac{\sqrt{2}}{2}$, if $0 \le x < 2\pi$.

SOLUTION We can simplify the left side by using the formula for $\sin (A + B)$.

$$\sin 2x \cos x + \cos 2x \sin x = \frac{\sqrt{2}}{2}$$

$$\sin (2x + x) = \frac{\sqrt{2}}{2}$$

$$\sin 3x = \frac{\sqrt{2}}{2}$$

First we find *all* possible solutions for x:

$$3x = \frac{\pi}{4} + 2k\pi \quad \text{or} \quad 3x = \frac{3\pi}{4} + 2k\pi \qquad k \text{ is any integer}$$

$$x = \frac{\pi}{12} + \frac{2k\pi}{3} \quad \text{or} \quad x = \frac{\pi}{4} + \frac{2k\pi}{3} \qquad \text{Divide by 3}$$

To find those solutions that lie in the interval $0 \le x < 2\pi$, we let k take on values of 0, 1, and 2. Doing so results in the following solutions (Figure 5):

$$x = \frac{\pi}{12}, \frac{\pi}{4}, \frac{3\pi}{4}, \frac{11\pi}{12}, \frac{17\pi}{12}, \text{ and } \frac{19\pi}{12}$$

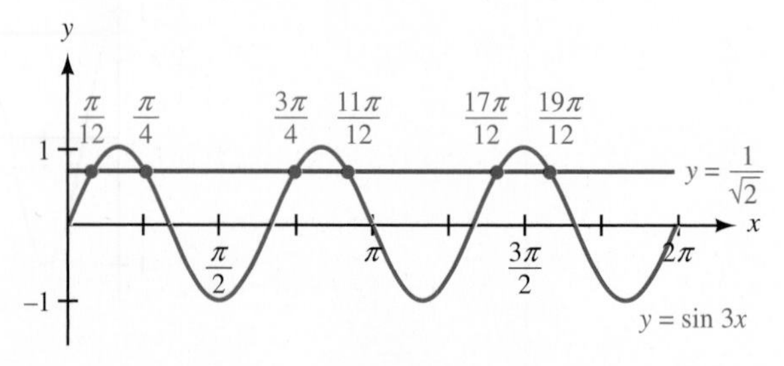

Figure 5 ■

PROBLEM 4
Solve $4 \sin^2 2\theta - 3 = 0$ if $0° \le \theta < 360°$.

EXAMPLE 4 Find all solutions to $2 \sin^2 3\theta - \sin 3\theta - 1 = 0$, if θ is measured in degrees.

SOLUTION We have an equation that is quadratic in $\sin 3\theta$. We factor and solve as usual.

$$2 \sin^2 3\theta - \sin 3\theta - 1 = 0 \quad \text{Standard form}$$
$$(2 \sin 3\theta + 1)(\sin 3\theta - 1) = 0 \quad \text{Factor}$$
$$2 \sin 3\theta + 1 = 0 \quad \text{or} \quad \sin 3\theta - 1 = 0 \quad \text{Set factors to 0}$$
$$\sin 3\theta = -\frac{1}{2} \quad \text{or} \quad \sin 3\theta = 1$$
$$3\theta = 210° + 360°k \quad \text{or} \quad 3\theta = 330° + 360°k \quad \text{or} \quad 3\theta = 90° + 360°k$$
$$\theta = 70° + 120°k \quad \text{or} \quad \theta = 110° + 120°k \quad \text{or} \quad \theta = 30° + 120°k$$

PROBLEM 5
Find all radian solutions for $\tan^2 2x = 3$.

EXAMPLE 5 Find all radian solutions for $\tan^2 3x = 1$.

SOLUTION Taking the square root of both sides we have

$$\tan^2 3x = 1$$
$$\tan 3x = \pm 1 \quad \text{Square root of both sides}$$

The period of the tangent function is π, so we have

$$3x = \frac{\pi}{4} + k\pi \quad \text{or} \quad 3x = \frac{3\pi}{4} + k\pi$$
$$x = \frac{\pi}{12} + \frac{k\pi}{3} \quad \text{or} \quad x = \frac{\pi}{4} + \frac{k\pi}{3}$$

PROBLEM 6
Solve $\sin \theta - \cos \theta = \sqrt{2}$ if $0° \le \theta < 360°$.

EXAMPLE 6 Solve $\sin \theta - \cos \theta = 1$ if $0° \le \theta < 360°$.

SOLUTION We have solved this equation before—in Section 6.2. Here is a second solution.

$$\sin \theta - \cos \theta = 1$$
$$(\sin \theta - \cos \theta)^2 = 1^2 \quad \text{Square both sides}$$
$$\sin^2 \theta - 2 \sin \theta \cos \theta + \cos^2 \theta = 1 \quad \text{Expand left side}$$
$$-2 \sin \theta \cos \theta + 1 = 1 \quad \sin^2 \theta + \cos^2 \theta = 1$$
$$-2 \sin \theta \cos \theta = 0 \quad \text{Add } -1 \text{ to both sides}$$
$$-\sin 2\theta = 0 \quad \text{Double-angle identity}$$
$$\sin 2\theta = 0 \quad \text{Multiply both sides by } -1$$

The sine function is zero for 0° or 180°. Therefore, the expression 2θ must be coterminal with either of these. For any integer k, we can represent all solutions more concisely as

$$2\theta = 0° + 180°k$$
$$2\theta = 180°k$$
$$\theta = 90°k$$

We were asked only for values of θ in the interval $0° \le \theta < 360°$. Choosing $k = 0$, 1, 2, or 3 we have $\theta = 0°, 90°, 180°$, or $270°$. For all other values of k, θ will fall outside the given interval.

Because our first step involved squaring both sides of the equation, we must check all the possible solutions to see if they satisfy the original equation. Doing so gives us solutions $\theta = 90°$ and $\theta = 180°$. The other two values are extraneous, as can be seen in Figure 6.

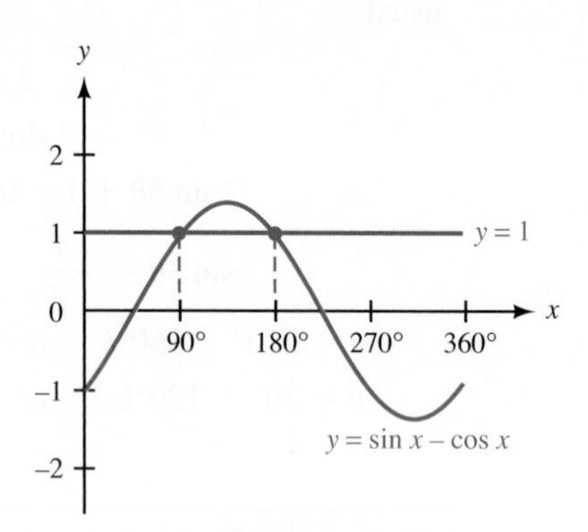

Figure 6

PROBLEM 7
Rework Example 7 using an attendance of at least 15,000.

EXAMPLE 7 In Section 4.5 we found that the average monthly attendance at Lake Nacimiento could be modeled by the function

$$y = 19{,}450 + 18{,}550 \cos\left(\frac{\pi}{6}x - \frac{5\pi}{6}\right), \qquad 0 \le x \le 12$$

where x is the month, with $x = 1$ corresponding to January. Use this model to determine the percentage of the year that the average attendance is at least 25,000.

SOLUTION To begin, we will find the values of x for which y is exactly 25,000. To do so, we must solve the equation

$$19{,}450 + 18{,}550 \cos\left(\frac{\pi}{6}x - \frac{5\pi}{6}\right) = 25{,}000$$

First we isolate the trigonometric function.

$$19{,}450 + 18{,}550 \cos\left(\frac{\pi}{6}x - \frac{5\pi}{6}\right) = 25{,}000$$

$$18{,}550 \cos\left(\frac{\pi}{6}x - \frac{5\pi}{6}\right) = 5{,}550$$

$$\cos\left(\frac{\pi}{6}x - \frac{5\pi}{6}\right) = 0.2992$$

The reference angle is $\cos^{-1}(0.2992) \approx 1.27$ radians. The cosine is positive in QI and QIV, so we have

$$\frac{\pi}{6}x - \frac{5\pi}{6} = 1.27 \qquad \text{or} \qquad \frac{\pi}{6}x - \frac{5\pi}{6} = 2\pi - 1.27$$

$$\frac{\pi}{6}x = 3.89 \qquad\qquad \frac{\pi}{6}x - \frac{5\pi}{6} = 5.01$$

$$x = 7.43 \qquad\qquad \frac{\pi}{6}x = 7.63$$

$$x = 14.57$$

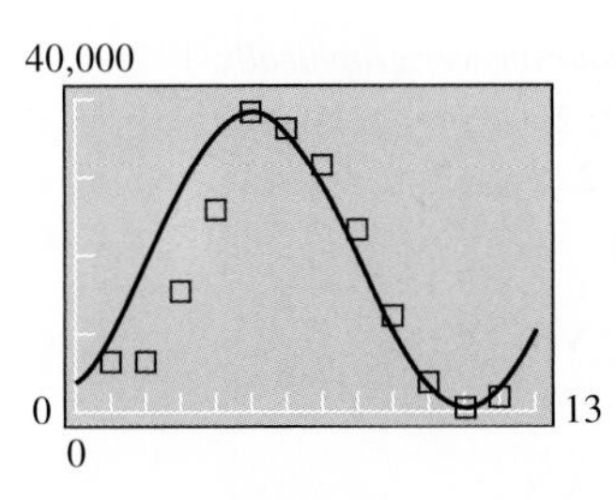

Figure 7

The solution $x = 14.57$ is outside the given interval $0 \le x \le 12$. However, because the period is 12, adding or subtracting any multiple of 12 to 14.57 will give another solution. Thus, $x = 14.57 - 12 = 2.57$ is a valid solution within the given interval.

Looking at Figure 7, we see that y will be at least 25,000 between $x = 2.57$ and $x = 7.43$, inclusive. This gives us

$$7.43 - 2.57 = 4.86 \text{ months}$$

or

$$\frac{4.86}{12} = 0.405 = 40.5\% \text{ of the year}$$

Getting Ready for Class

After reading through the preceding section, respond in your own words and in complete sentences.

a. If θ is between 0° and 360°, then what can you conclude about 2θ?
b. If $2\theta = 30° + 360°k$, then what can you say about θ?
c. What is an extraneous solution to an equation?
d. Why is it necessary to check solutions to equations that occur after squaring both sides of the equation?

6.3 PROBLEM SET

➤ CONCEPTS AND VOCABULARY

For Questions 1 through 3, fill in the blank with an appropriate word or expression.

1. To solve a trigonometric equation involving a multiple angle $n\theta$, first find the possible values of ______, then divide by ____ to find the values of θ.

2. For a trigonometric equation involving $\cos n\theta$ in degrees, add _______ to the values of θ to find all coterminal solutions. If solving in radians, add _______ to the values of θ to find all coterminal solutions.

3. To solve the equation $\sin 3x \cos x + \cos 3x \sin x = 1$, use a _____ identity to rewrite the left side.

4. True or False: To solve the equation $\cos 2x = \sqrt{3}/2$, use a double-angle formula to rewrite the left side.

➤ EXERCISES

Find all solutions if $0° \le \theta < 360°$. Verify your answer graphically.

5. $\sin 2\theta = \dfrac{\sqrt{3}}{2}$

6. $\sin 2\theta = -\dfrac{\sqrt{3}}{2}$

7. $\tan 2\theta = -1$

8. $\cot 2\theta = 1$

9. $\cos 3\theta = -1$

10. $\sin 3\theta = -1$

Find all solutions if $0 \le x < 2\pi$. Use exact values only. Verify your answer graphically.

11. $\cos 2x = \frac{\sqrt{2}}{2}$

12. $\sin 2x = \frac{\sqrt{2}}{2}$

13. $\sec 3x = -1$

14. $\csc 3x = 1$

15. $\tan 2x = \sqrt{3}$

16. $\tan 2x = -\sqrt{3}$

Find all degree solutions for each of the following:

17. $\sin 2\theta = \frac{1}{2}$

18. $\sin 2\theta = -\frac{\sqrt{3}}{2}$

19. $\cos 3\theta = 0$

20. $\cos 3\theta = -1$

21. $\sin 10\theta = \frac{\sqrt{3}}{2}$

22. $\cos 8\theta = \frac{1}{2}$

Use your graphing calculator to find all degree solutions in the interval $0° \le x < 360°$ for each of the following equations.

23. $\sin 2x = -\frac{\sqrt{2}}{2}$

24. $\cos 2x = -\frac{1}{2}$

25. $\cos 3x = \frac{1}{2}$

26. $\sin 3x = \frac{\sqrt{3}}{2}$

27. $\tan 2x = \frac{\sqrt{3}}{3}$

28. $\tan 2x = 1$

Find all solutions if $0 \le x < 2\pi$. Use exact values only.

29. $\sin 2x \cos x + \cos 2x \sin x = \frac{1}{2}$

30. $\sin 2x \cos x + \cos 2x \sin x = -\frac{1}{2}$

31. $\cos 2x \cos x - \sin 2x \sin x = -\frac{\sqrt{3}}{2}$

32. $\cos 2x \cos x - \sin 2x \sin x = \frac{\sqrt{2}}{2}$

Find all solutions in radians using exact values only.

33. $\sin 3x \cos 2x + \cos 3x \sin 2x = 1$

34. $\sin 2x \cos 3x + \cos 2x \sin 3x = -1$

35. $\sin^2 4x = 1$

36. $\cos^2 4x = 1$

37. $\cos^3 5x = -1$

38. $\sin^3 5x = -1$

Find all degree solutions.

39. $2 \sin^2 3\theta + \sin 3\theta - 1 = 0$

40. $2 \sin^2 3\theta + 3 \sin 3\theta + 1 = 0$

41. $2 \cos^2 2\theta + 3 \cos 2\theta + 1 = 0$

42. $2 \cos^2 2\theta - \cos 2\theta - 1 = 0$

43. $\tan^2 3\theta = 3$

44. $\cot^2 3\theta = 1$

Find all solutions if $0° \le \theta < 360°$. When necessary, round your answers to the nearest tenth of a degree.

45. $\cos\theta - \sin\theta = 1$

46. $\sin\theta - \cos\theta = 1$

47. $\sin\theta + \cos\theta = -1$

48. $\cos\theta - \sin\theta = -1$

49. $\sin^2 2\theta - 4 \sin 2\theta - 1 = 0$

50. $\cos^2 3\theta - 6 \cos 3\theta + 4 = 0$

51. $4 \cos^2 3\theta - 8 \cos 3\theta + 1 = 0$

52. $2 \sin^2 2\theta - 6 \sin 2\theta + 3 = 0$

53. $2 \cos^2 4\theta + 2 \sin 4\theta = 1$

54. $2 \sin^2 4\theta - 2 \cos 4\theta = 1$

55. Ferris Wheel In Example 6 of Section 4.5, we found the equation that gives the height h of a passenger on a Ferris wheel at any time t during the ride to be

$$h = 139 - 125 \cos \frac{\pi}{10} t$$

where h is given in feet and t is given in minutes. Use this equation to find the times at which a passenger will be 100 feet above the ground. Round your answers to the nearest tenth of a minute. Use your graphing calculator to graph the function and verify your answers.

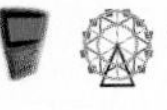

56. Ferris Wheel In Problem 37 of Problem Set 4.5, you found the equation that gives the height h of a passenger on a Ferris wheel at any time t during the ride to be

$$h = 110.5 - 98.5 \cos \frac{2\pi}{15} t$$

where the units for h are feet and the units for t are minutes. Use this equation to find the times at which a passenger will be 100 feet above the ground. Round your answers to the nearest tenth of a minute. Use your graphing calculator to graph the function and verify your answers.

57. Geometry The following formula gives the relationship between the number of sides n, the radius r, and the length of each side l in a regular polygon (Figure 8). Find n, if $l = r$.

$$l = 2r \sin \frac{180°}{n}$$

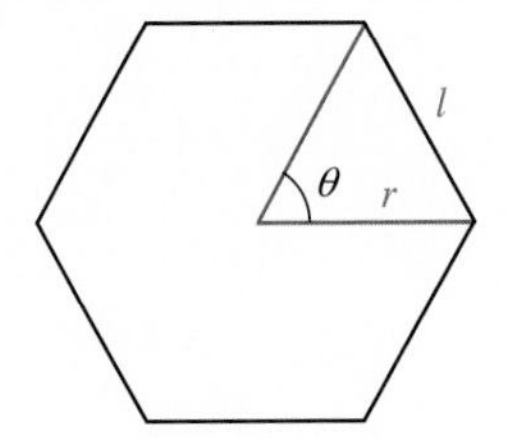

Figure 8

Figure 9

58. Geometry If central angle θ cuts off a chord of length c in a circle of radius r (Figure 9), then the relationship between θ, c, and r is given by

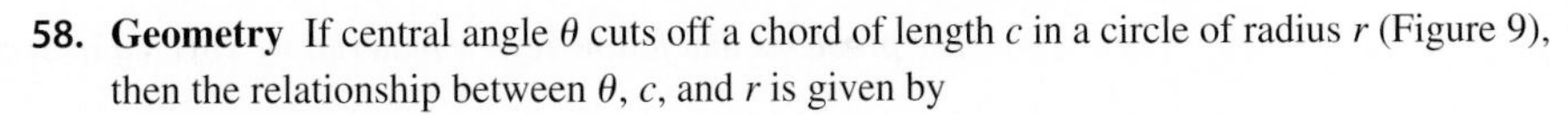

$$2r \sin \frac{\theta}{2} = c$$

Find θ, if $c = \sqrt{3}r$.

59. Rotating Light In Example 4 of Section 3.5, we found the equation that gives d in terms of t in Figure 10 to be $d = 10 \tan \pi t$, where d is measured in feet and t is measured in seconds. If a person is standing against the wall, 10 feet from point A, how long after the light is at point A will it reach the person? (*Hint:* You must find t when d is 10.)

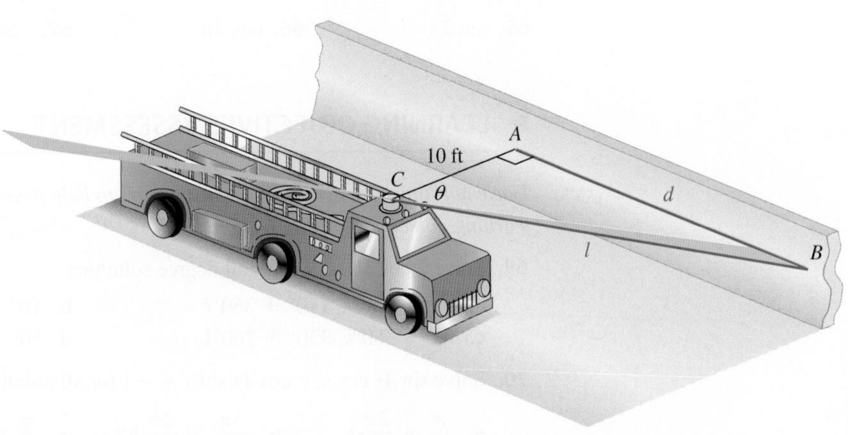

Figure 10

60. Rotating Light In Problem 23 of Problem Set 3.5, you found the equation that gives d in terms of t in Figure 11 to be $d = 100 \tan \frac{1}{2}\pi t$, where d is measured in feet and t is measured in seconds. Two people are sitting on the wall. One of them is directly opposite the lighthouse, while the other person is 100 feet further down the wall. How long after one of them sees the light does the other one see the light? (*Hint:* There are two solutions depending on who sees the light first.)

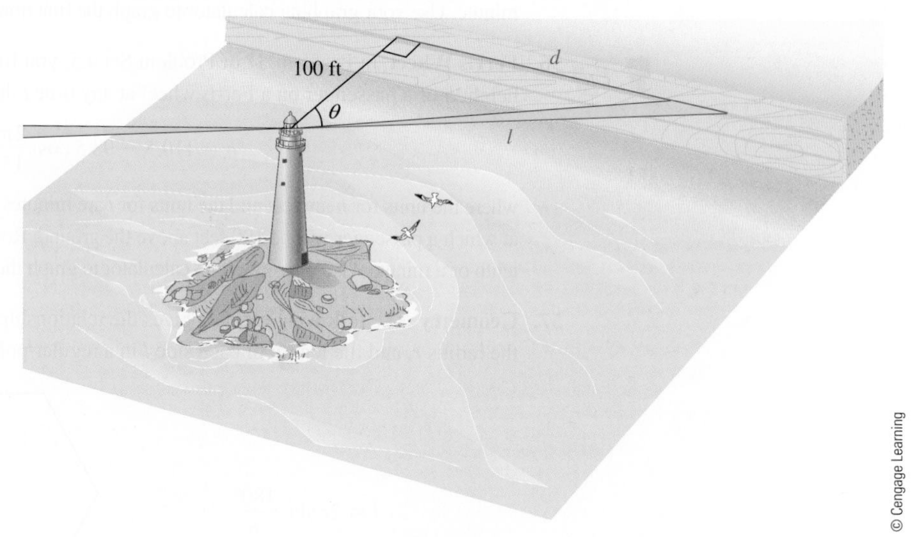

Figure 11

➤ REVIEW PROBLEMS

The problems that follow review material we covered in Sections 5.1 through 5.4.

Prove each identity.

61. $\dfrac{\sin x}{1 + \cos x} = \dfrac{1 - \cos x}{\sin x}$

62. $\dfrac{\sin^2 x}{(1 - \cos x)^2} = \dfrac{1 + \cos x}{1 - \cos x}$

63. $\dfrac{1}{1 + \cos t} + \dfrac{1}{1 - \cos t} = 2 \csc^2 t$

64. $\dfrac{1}{1 - \sin t} + \dfrac{1}{1 + \sin t} = 2 \sec^2 t$

If $\sin A = \frac{1}{3}$ *with* $90° \le A \le 180°$ *and* $\sin B = \frac{3}{5}$ *with* $0° \le B \le 90°$*, find each of the following.*

65. $\sin 2A$

66. $\cos 2B$

67. $\cos \dfrac{A}{2}$

68. $\sin \dfrac{B}{2}$

➤ LEARNING OBJECTIVES ASSESSMENT

These questions are available for instructors to help assess if you have successfully met the learning objectives for this section.

69. Solve $2 \cos 3\theta = \sqrt{3}$ for all degree solutions.

a. $10° + 360°k,\ 110° + 360°k$
b. $10° + 120°k,\ 110° + 120°k$
c. $30° + 360°k,\ 330° + 360°k$
d. $30° + 120°k,\ 330° + 120°k$

70. Solve $\sin 4x \cos x + \cos 4x \sin x = -1$ for all radian solutions.

a. $\dfrac{\pi}{5} + \dfrac{2\pi}{5}k$
b. $\dfrac{3\pi}{10} + \dfrac{2\pi}{5}k$
c. $\dfrac{\pi}{2} + \dfrac{2\pi}{3}k$
d. $\dfrac{\pi}{3} + \dfrac{2\pi}{3}k$

71. Solve $2\cos^2 4\theta + \cos 4\theta - 1 = 0$ for all degree solutions.

a. $15° + 30°k$ **b.** $7.5° + 30°k$

c. $45° + 90°k, 75° + 90°k$ **d.** $7.5° + 90°k, 45° + 90°k, 82.5° + 90°k$

72. The height of a passenger on a Ferris wheel at any time t is given by

$$h = 101 - 95\cos(\pi t),$$

where h is measured in feet and t in minutes. Find the times at which a passenger will be 150 feet above the ground.

a. 2.1 min and 4.2 min **b.** 2.1 min and 5.3 min

c. 0.7 min and 1.3 min **d.** 0.7 min and 1.8 min

SECTION 6.4 Parametric Equations and Further Graphing

LEARNING OBJECTIVES

1. Graph a plane curve by plotting points.
2. Indicate the orientation of a plane curve.
3. Eliminate the parameter from a pair of parametric equations.
4. Use parametric equations as a model in a real-life problem.

Up to this point we have had a number of encounters with the Ferris wheel problem. In Sections 3.5 and 4.5, we were able to find a function that gives the rider's height above the ground at any time t during the ride. A third way to derive this function is with *parametric equations*, which we will study in this section.

Let's begin with a model of the giant wheel built by George Ferris. The diameter of this wheel is 250 feet, and the bottom of the wheel sits 14 feet above the ground. We can superimpose a coordinate system on our model, so that the origin of the coordinate system is at the center of the wheel. This is shown in Figure 1.

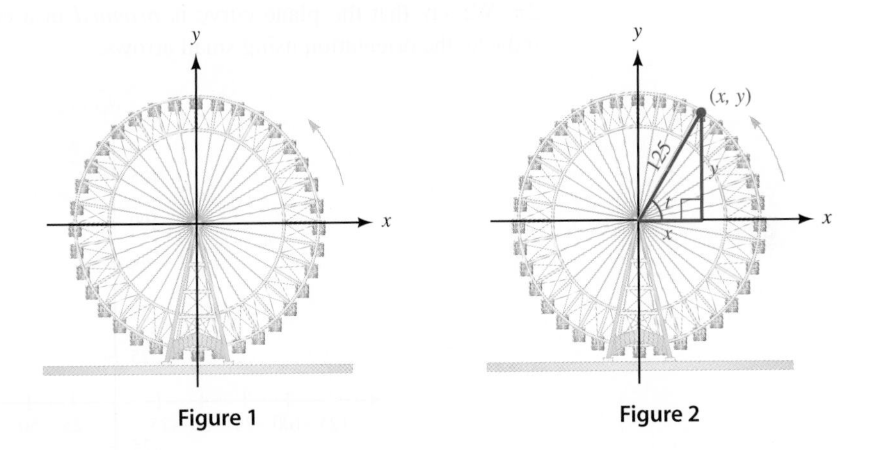

Figure 1 Figure 2

Choosing one of the carriages in the first quadrant as the point (x, y), we draw a line from the origin to the carriage. Then we draw the right triangle shown in Figure 2. From the triangle we have

$$\cos t = \frac{x}{125} \Rightarrow x = 125\cos t$$

$$\sin t = \frac{y}{125} \Rightarrow y = 125\sin t$$

The two equations on the right are called *parametric equations*. They show x and y as functions of a third variable, t, called the *parameter*. Each number we substitute for t gives us an ordered pair (x, y). If we graph each of these ordered pairs on a rectangular coordinate system, the resulting curve is called a *plane curve*. We demonstrate this in the following example.

PROBLEM 1
Graph the plane curve defined by $x = 6 \sin t$ and $y = 6 \cos t$.

EXAMPLE 1 Graph the plane curve defined by the parametric equations $x = 125 \cos t$ and $y = 125 \sin t$.

SOLUTION We can find points on the graph by choosing values of t and using the two equations to find corresponding values of x and y. Table 1 shows a number of ordered pairs, which we obtained by letting t be different multiples of $\pi/4$.

TABLE 1

t	$x = 125 \cos t$	$y = 125 \sin t$	(x, y)
0	125	0	(125, 0)
$\pi/4$	88	88	(88, 88)
$\pi/2$	0	125	(0, 125)
$3\pi/4$	−88	88	(−88, 88)
π	−125	0	(−125, 0)
$5\pi/4$	−88	−88	(−88, −88)
$3\pi/2$	0	−125	(0, −125)
$7\pi/4$	88	−88	(88, −88)
2π	125	0	(125, 0)

The graph of the plane curve is shown in Figure 3. We have labeled each point with its associated value of the parameter t. As expected, the curve is a circle. More importantly, notice in which direction the circle is traversed as t increases from 0 to 2π. We say that the plane curve is *oriented* in a counterclockwise direction and indicate the orientation using small arrows.

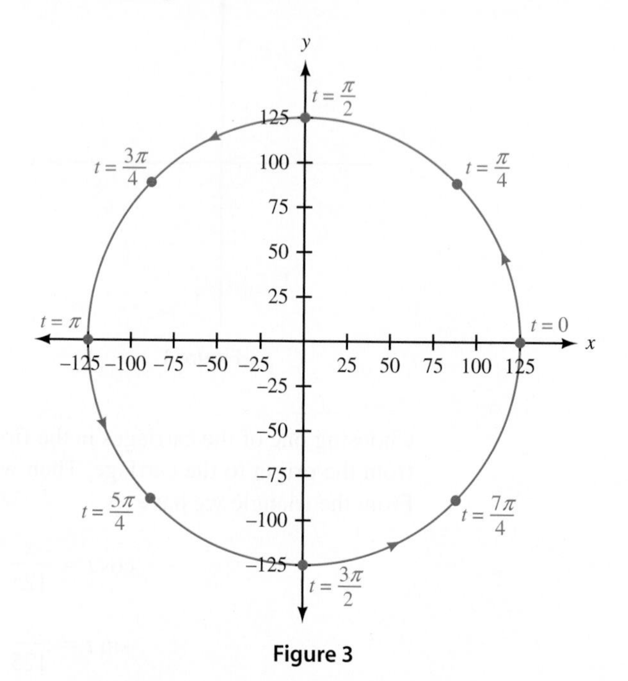

Figure 3

In general, the orientation of a plane curve is the direction the curve is traversed as the parameter increases. The ability of parametric equations to associate points on a curve with values of a parameter makes them especially suited for describing the path of an object in motion.

Using Technology — Graphing Parametric Equations

```
Plot1 Plot2 Plot3
\X1T=125cos(T)
 Y1T=125sin(T)
\X2T=
 Y2T=
\X3T=
 Y3T=
\X4T=
```

Figure 4

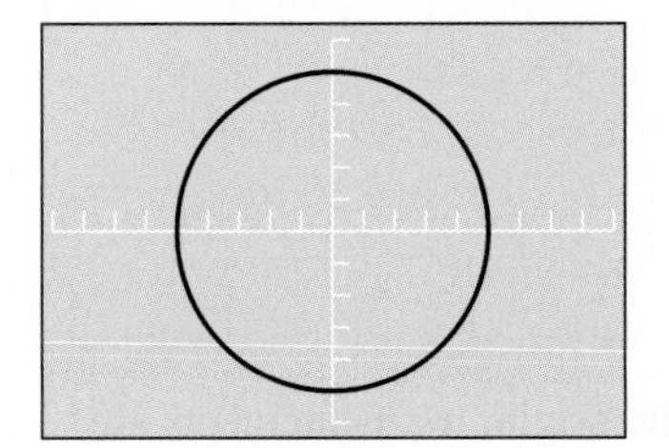

Figure 5

We can use a graphing calculator to graph the plane curve for a set of parametric equations. For instance, to graph the plane curve in Example 1, set your calculator to radian mode and parametric mode. Define the parametric equations as shown in Figure 4, and set your window as follows:

$$0 \le t \le 2\pi, \quad \text{step} = \pi/24$$
$$-150 \le x \le 150, \text{ scale} = 25$$
$$-150 \le y \le 150, \text{ scale} = 25$$

To make the circle look right, graph the equations using the zoom-square command. The result is shown in Figure 5.

To see the orientation, we can trace the circle and follow the cursor as the parameter t increases (Figure 6). We can also evaluate the equations for any given value of the parameter. Figure 7 shows how this might look for $t = 3\pi/4$. Notice that the coordinates match our results in Table 1.

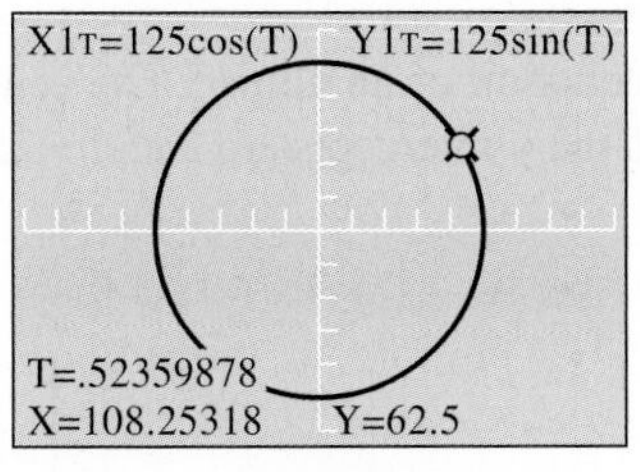

Figure 6

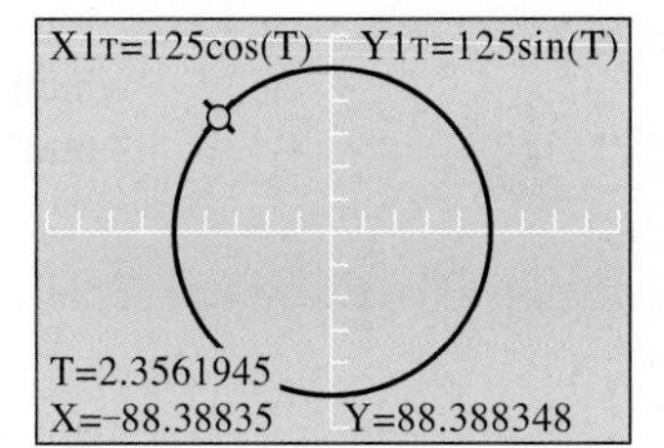

Figure 7

Eliminating the Parameter

Let's go back to our original set of parametric equations and solve for $\cos t$ and $\sin t$:

$$x = 125 \cos t \Rightarrow \cos t = \frac{x}{125}$$

$$y = 125 \sin t \Rightarrow \sin t = \frac{y}{125}$$

Substituting the expressions above for $\cos t$ and $\sin t$ into the Pythagorean identity $\cos^2 t + \sin^2 t = 1$, we have

$$\left(\frac{x}{125}\right)^2 + \left(\frac{y}{125}\right)^2 = 1$$

$$\frac{x^2}{125^2} + \frac{y^2}{125^2} = 1$$

$$x^2 + y^2 = 125^2$$

We recognize this last equation as the equation of a circle with a radius of 125 and center at the origin. What we have done is eliminate the parameter t to obtain an equation in just x and y whose graph we recognize. This process is called *eliminating the parameter.* Note that it gives us further justification that the graph of our set of parametric equations is a circle.

PROBLEM 2
Eliminate the parameter t from $x = 5 \sin t$ and $y = 2 \cos t$.

EXAMPLE 2 Eliminate the parameter t from the parametric equations $x = 3 \cos t$ and $y = 2 \sin t$.

SOLUTION Again, we will use the identity $\cos^2 t + \sin^2 t = 1$. Before we do so, however, we must solve the first equation for $\cos t$ and the second equation for $\sin t$.

$$x = 3 \cos t \Rightarrow \cos t = \frac{x}{3}$$

$$y = 2 \sin t \Rightarrow \sin t = \frac{y}{2}$$

Substituting $x/3$ and $y/2$ for $\cos t$ and $\sin t$ into the first Pythagorean identity gives us

$$\left(\frac{x}{3}\right)^2 + \left(\frac{y}{2}\right)^2 = 1$$

$$\frac{x^2}{9} + \frac{y^2}{4} = 1$$

which is the equation of an ellipse. The center is at the origin, the x-intercepts are 3 and -3, and the y-intercepts are 2 and -2. Figure 8 shows the graph.

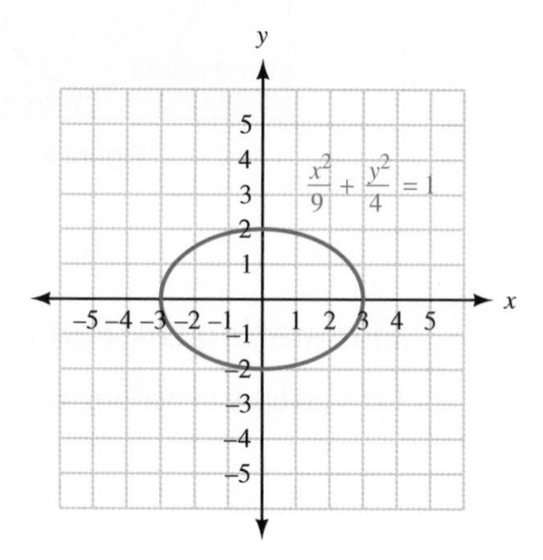

Figure 8

PROBLEM 3
Eliminate the parameter t from $x = \sin t + 5$ and $y = \cos t - 3$.

EXAMPLE 3 Eliminate the parameter t from the parametric equations $x = 3 + \sin t$ and $y = \cos t - 2$.

SOLUTION Solving the first equation for $\sin t$ and the second equation for $\cos t$, we have

$$\sin t = x - 3 \quad \text{and} \quad \cos t = y + 2$$

Substituting these expressions for $\sin t$ and $\cos t$ into the Pythagorean identity gives us

$$(x - 3)^2 + (y + 2)^2 = 1$$

which is the equation of a circle with a radius of 1 and center at $(3, -2)$. The graph of this circle is shown in Figure 9.

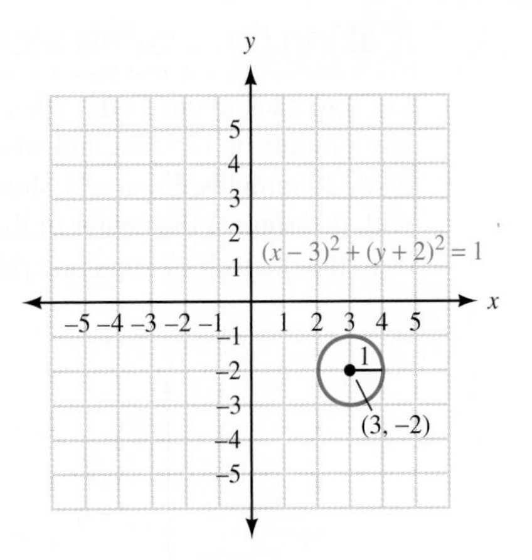

Figure 9

PROBLEM 4
Eliminate the parameter t from $x = 3 \sec t - 4$ and $y = 5 \tan t - 1$.

EXAMPLE 4 Eliminate the parameter t from the parametric equations $x = 3 + 2 \sec t$ and $y = 2 + 4 \tan t$.

SOLUTION In this case, we solve for $\sec t$ and $\tan t$ and then use the identity $1 + \tan^2 t = \sec^2 t$.

$$x = 3 + 2 \sec t \Rightarrow \sec t = \frac{x-3}{2}$$

$$y = 2 + 4 \tan t \Rightarrow \tan t = \frac{y-2}{4}$$

so

$$1 + \left(\frac{y-2}{4}\right)^2 = \left(\frac{x-3}{2}\right)^2$$

or

$$\frac{(x-3)^2}{4} - \frac{(y-2)^2}{16} = 1$$

This is the equation of a hyperbola. Figure 10 shows the graph.

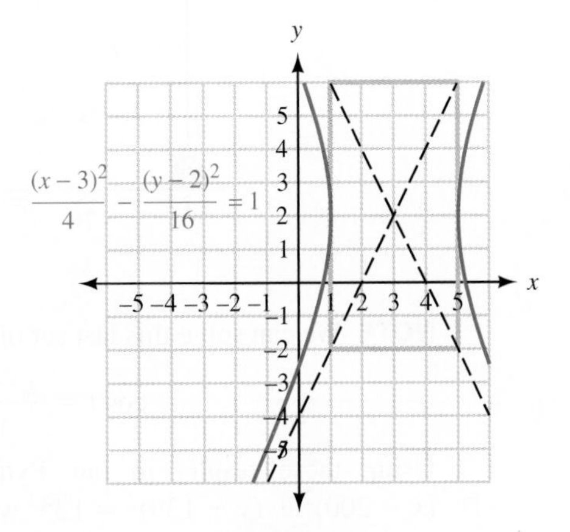

Figure 10

Making Our Models More Realistic

Let's go back to our Ferris wheel model from the beginning of this section. Our wheel has a radius of 125 feet and sits 14 feet above the ground. One trip around the wheel takes 20 minutes. Figure 11 shows this model with a coordinate system superimposed with its origin at the center of the wheel. Also shown are the parametric equations that describe the path of someone riding the wheel.

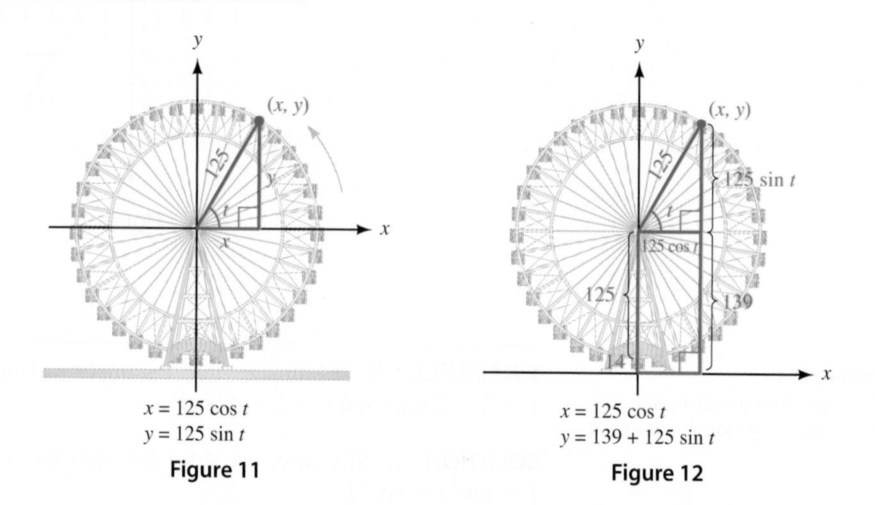

Figure 11 **Figure 12**

Our model would be more realistic if the x-axis was along the ground, below the wheel. We can accomplish this very easily by moving everything up 139 feet (125 feet for the radius and another 14 feet for the wheel's height above the ground). Figure 12 shows our new graph next to a new set of parametric equations.

Next, let's assume there is a ticket booth 200 feet to the left of the wheel. If we want to use the ticket booth as our starting point, we can move our graph to the right 200 feet. Figure 13 shows this model, along with the corresponding set of parametric equations.

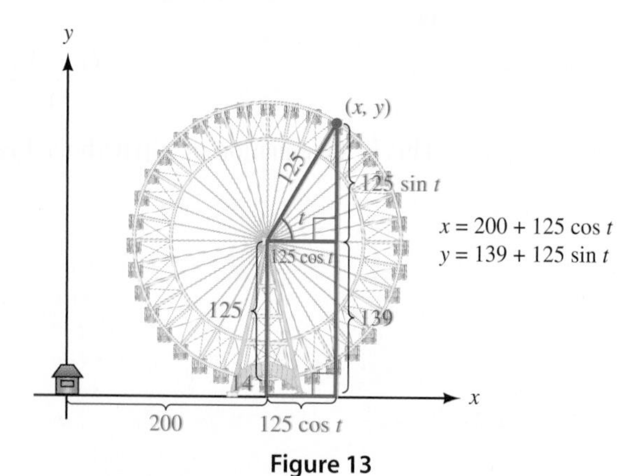

Figure 13

NOTE We can solve this last set of equations for cos t and sin t to get

$$\cos t = \frac{x - 200}{125} \qquad \sin t = \frac{y - 139}{125}$$

Using these results in our Pythagorean identity, $\cos^2 t + \sin^2 t = 1$, we have $(x - 200)^2 + (y - 139)^2 = 125^2$ which we recognize as the equation of a circle with center at (200, 139) and radius 125.

It's just that easy. If we want to translate our graph horizontally, we simply add the translation amount to our expression for x. Likewise, vertical translations are accomplished by adding to our expression for y.

Continuing to improve our model, we would like the rider on the wheel to start their ride at the bottom of the wheel. Assuming t is in radians, we accomplish this by subtracting $\pi/2$ from t giving us

$$x = 200 + 125 \cos\left(t - \frac{\pi}{2}\right)$$

$$y = 139 + 125 \sin\left(t - \frac{\pi}{2}\right)$$

Finally, one trip around the wheel takes 20 minutes, so we can write our equations in terms of time T by using the proportion

$$\frac{t}{2\pi} = \frac{T}{20}$$

Solving for t we have

$$t = \frac{\pi}{10}T$$

Substituting this expression for t into our parametric equations we have

$$x = 200 + 125 \cos\left(\frac{\pi}{10}T - \frac{\pi}{2}\right)$$

$$y = 139 + 125 \sin\left(\frac{\pi}{10}T - \frac{\pi}{2}\right)$$

These last equations give us a very accurate model of the path taken by someone riding on this Ferris wheel. To graph these equations on our graphing calculator, we can use the following window:

Radian mode: $0 \le T \le 20$, step $= 1$

$-50 \le x \le 330$, scale $= 20$

$-50 \le y \le 330$, scale $= 20$

Using the zoom-square command, we have the graph shown in Figure 14. Figure 15 shows a trace of the graph. Tracing around this graph gives us the position of the rider at each minute of the 20-minute ride.

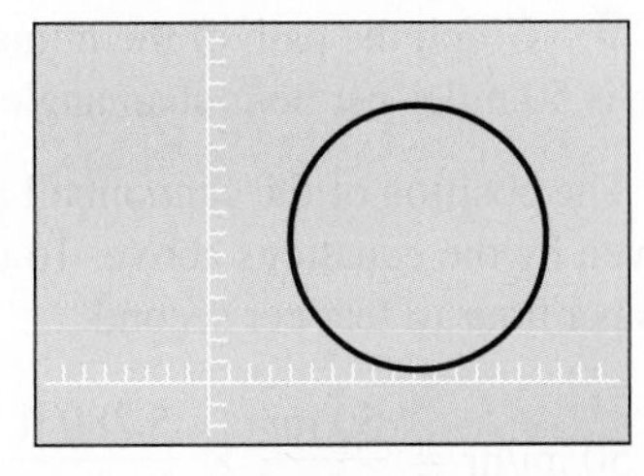

Figure 14

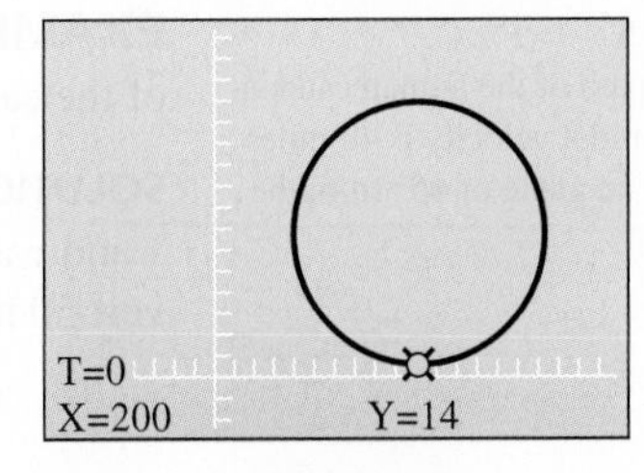

Figure 15

Parametric Equations and the Human Cannonball

In Section 2.5, we found that the human cannonball, when shot from a cannon at 50 miles per hour at 60° from the horizontal, will have a horizontal velocity of 25 miles per hour and an initial vertical velocity of 43 miles per hour (Figure 16). We can generalize this so that a human cannonball, if shot from a cannon at $|\mathbf{V}_0|$ miles per hour at an angle of elevation θ, will have a horizontal speed of $|\mathbf{V}_0| \cos \theta$ miles per hour and an initial vertical speed of $|\mathbf{V}_0| \sin \theta$ miles per hour (Figure 17).

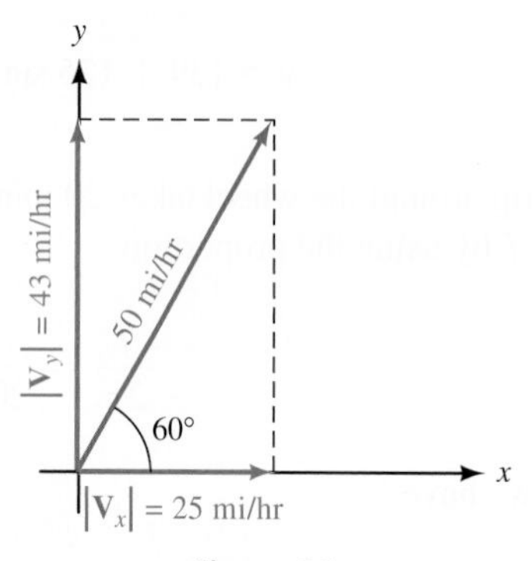

Figure 16

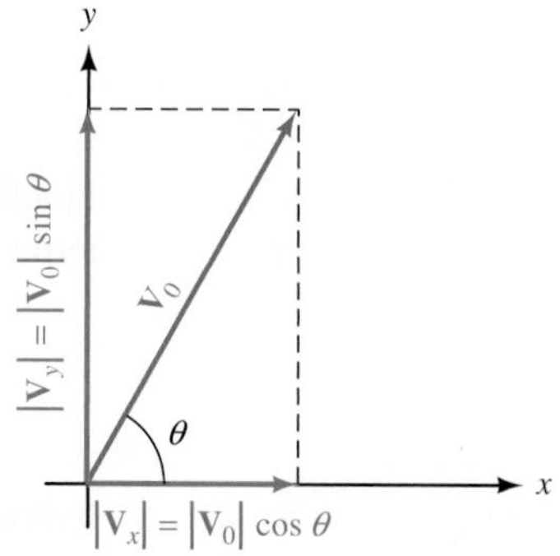

Figure 17

Neglecting the resistance of air, the only force acting on the human cannonball is the force of gravity, which is an acceleration of 32 feet per second squared toward the earth. Because the cannonball's horizontal speed is constant, we can find the distance traveled after t seconds by simply multiplying speed and time. Therefore, the distance traveled horizontally after t seconds is

$$x = (|\mathbf{V}_0| \cos \theta)t$$

To find the cannonball's vertical distance from the cannon after t seconds, we use a formula from physics:

$$y = (|\mathbf{V}_0| \sin \theta)t - \frac{1}{2}gt^2$$ where $g = 32$ ft/sec^2 (the acceleration of gravity on Earth)

This gives us the following set of parametric equations:

$$x = (|\mathbf{V}_0| \cos \theta)t$$
$$y = (|\mathbf{V}_0| \sin \theta)t - 16t^2$$

The equations describe the path of a human cannonball shot from a cannon at a speed of $|\mathbf{V}_0|$, at an angle of θ degrees from horizontal. So that the units will agree, $|\mathbf{V}_0|$ must be in feet per second because t is in seconds.

PROBLEM 5
Graph the path of the human cannonball if the initial velocity is 40 miles per hour at an angle of 65° from the horizontal.

EXAMPLE 5 Graph the path of the human cannonball if the initial velocity out of the cannon is 50 miles per hour at an angle of 60° from the horizontal.

SOLUTION The position of the cannonball at time t is described parametrically by x and y as given by the equations above. To use these equations, we must first convert 50 miles per hour to feet per second.

$$50 \text{ mi/hr} = \frac{50 \text{ mi}}{\text{hr}} \times \frac{5{,}280 \text{ ft}}{1 \text{ mi}} \times \frac{1 \text{ hr}}{3{,}600 \text{ sec}} \approx 73.3 \text{ ft/sec}$$

Substituting 73.3 for $|\mathbf{V}_0|$ and 60° for θ gives us parametric equations that together describe the path of the human cannonball.

$$x = (73.3 \cos 60°)t$$
$$y = (73.3 \sin 60°)t - 16t^2$$

First setting the calculator to degree and parametric modes, we then set up our function list (Figure 18) and window as follows.

$$\text{Degree mode: } 0 \le t \le 6, \quad \text{step} = 0.1$$
$$0 \le x \le 160, \quad \text{scale} = 20$$
$$0 \le y \le 80, \quad \text{scale} = 20$$

The graph is shown in Figure 19.

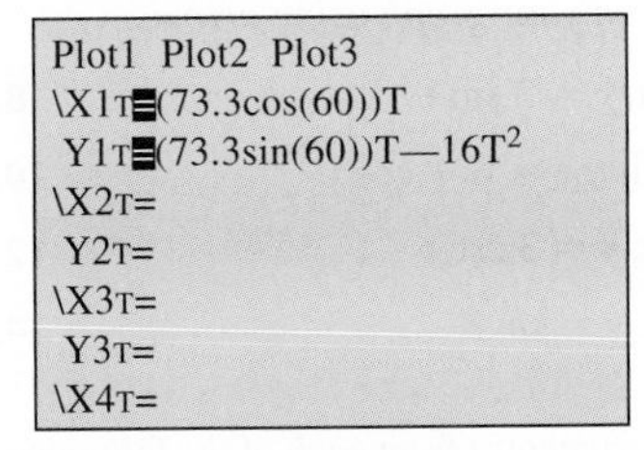

Figure 18

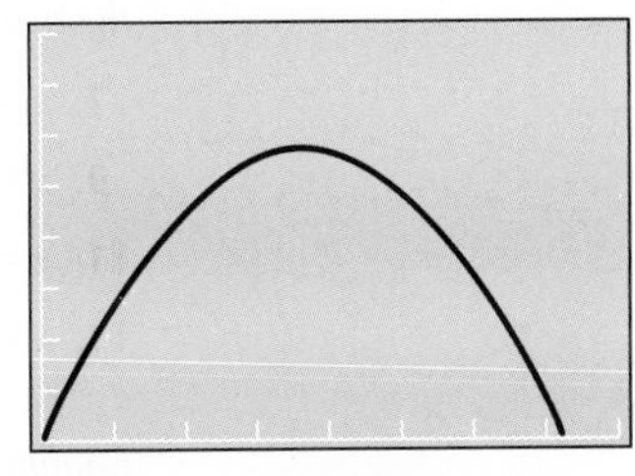

Figure 19

Getting Ready for Class

After reading through the preceding section, respond in your own words and in complete sentences.

a. What are parametric equations?

b. What trigonometric identity do we use to eliminate the parameter t from the equations $x = 3 \cos t$ and $y = 2 \sin t$?

c. What is the first step in eliminating the parameter t from the equations $x = 3 + \sin t$ and $y = \cos t - 2$?

d. How do you convert 50 miles per hour into an equivalent quantity in feet per second?

6.4 PROBLEM SET

➤ CONCEPTS AND VOCABULARY

For Questions 1 through 6, fill in the blank with an appropriate word or expression.

1. The pair of equations $x = 125 \cos t$, $y = 125 \sin t$ are called ____________ equations, and the common variable t is called the ____________.

2. When we sketch the graph for a pair of parametric equations, the resulting curve is called a ________ ________.

3. The ____________ of a plane curve shows the ____________ the curve is traversed as the parameter ____________.

4. *Eliminating the parameter* means algebraically eliminating the variable _______ from a pair of parametric equations to obtain a single equation in the variables _______ and _______ only.

5. In order to eliminate the parameter from the parametric equations $x = 3 + \sin t$, $y = \cos t - 2$, isolate ______ and ______ and then substitute these expressions into the ___________ identity.

6. Given values $|\mathbf{V}_0|$ and θ, the plane curve for the parametric equations $x = (|\mathbf{V}_0| \cos \theta)t$, $y = (|\mathbf{V}_0| \sin \theta)t - 16t^2$ is a ___________ passing through the point _______.

➤ EXERCISES

Graph the plane curve for each pair of parametric equations by plotting points, and indicate the orientation on your graph using arrows.

7. $x = 3 \cos t$, $y = 3 \sin t$
8. $x = 2 \sin t$, $y = 2 \cos t$
9. $x = 2 + \sin t$, $y = 3 + \cos t$
10. $x = 3 \cos t - 3$, $y = 3 \sin t + 1$
11. $x = 3 \cot t$, $y = 3 \csc t$
12. $x = 3 \sec t$, $y = 3 \tan t$
13. $x = \cos 2t$, $y = \sin t$
14. $x = \cos 2t$, $y = \cos t$

Eliminate the parameter t from each of the following and then sketch the graph of the plane curve:

15. $x = \sin t$, $y = \cos t$
16. $x = -\sin t$, $y = \cos t$
17. $x = 3 \cos t$, $y = 3 \sin t$
18. $x = 2 \cos t$, $y = 2 \sin t$
19. $x = 2 \sin t$, $y = 4 \cos t$
20. $x = 3 \sin t$, $y = 4 \cos t$
21. $x = 2 + \sin t$, $y = 3 + \cos t$
22. $x = 3 + \sin t$, $y = 2 + \cos t$
23. $x = \sin t - 2$, $y = \cos t - 3$
24. $x = \cos t - 3$, $y = \sin t + 2$
25. $x = 3 + 2 \sin t$, $y = 1 + 2 \cos t$
26. $x = 2 + 3 \sin t$, $y = 1 + 3 \cos t$
27. $x = 3 \cos t - 3$, $y = 3 \sin t + 1$
28. $x = 4 \sin t - 5$, $y = 4 \cos t - 3$

Eliminate the parameter t in each of the following:

29. $x = \sec t$, $y = \tan t$
30. $x = \tan t$, $y = \sec t$
31. $x = 3 \sec t$, $y = 3 \tan t$
32. $x = 3 \cot t$, $y = 3 \csc t$
33. $x = 2 + 3 \tan t$, $y = 4 + 3 \sec t$
34. $x = 3 + 5 \tan t$, $y = 2 + 5 \sec t$
35. $x = \cos 2t$, $y = \sin t$
36. $x = \cos 2t$, $y = \cos t$
37. $x = \sin t$, $y = \sin t$
38. $x = \cos t$, $y = \cos t$
39. $x = 3 \sin t$, $y = 2 \sin t$
40. $x = 2 \sin t$, $y = 3 \sin t$

For Problems 41 through 44, use parametric equations to model the path of a rider on the wheel. You want to end up with parametric equations that will give you the position of the rider every minute of the ride. Graph your results on a graphing calculator.

41. **Ferris Wheel** The Ferris wheel built in Vienna in 1897 has a diameter of 197 feet and sits 12 feet above the ground. It rotates in a counterclockwise direction, making one complete revolution every 15 minutes. Place your coordinate system so that the origin of the coordinate system is on the ground below the bottom of the wheel.

42. **Observation Wheel** The London Eye has a diameter of 135 meters. A rider boards the London Eye at ground level. It rotates in a counterclockwise direction, making one complete revolution every 30 minutes. Place your coordinate system so that the origin of the coordinate system is on the ground below the bottom of the wheel.

jasonleehl / Shutterstock.com

43. Observation Wheel The Singapore Flyer is currently the largest observation wheel in the world. It has a diameter of 150 meters and sits 15 meters above the ground. It rotates in a counterclockwise direction, making one complete revolution every 32 minutes. Place your coordinate system so that the origin of the coordinate system is on the ground 125 meters to the left of the wheel.

44. Ferris Wheel A Ferris wheel named Colossus was built in St. Louis in 1986. It has a diameter of 165 feet and sits 9 feet above the ground. It rotates in a counterclockwise direction, making one complete revolution every 1.5 minutes. Place your coordinate system so that the origin of the coordinate system is on the ground 100 feet to the left of the wheel.

45. Human Cannonball Graph the parametric equations in Example 5 and then find the maximum height of the cannonball, the maximum distance traveled horizontally, and the time at which the cannonball hits the net. (Assume the barrel of the cannon and the net are the same distance above the ground.)

46. Human Cannonball A human cannonball is fired from a cannon with an initial velocity of 50 miles per hour. On the same screen on your calculator, graph the paths taken by the cannonball if the angle between the cannon and the horizontal is 20°, 30°, 40°, 50°, 60°, 70°, and 80°.

➤ REVIEW PROBLEMS

The problems that follow review material we covered in Sections 4.7 and 5.5. Evaluate each expression.

47. $\cos(\sin^{-1} x)$

48. $\sin(\cos^{-1} x)$

49. $\cos(2 \tan^{-1} x)$

50. $\sin(2 \tan^{-1} x)$

51. Write $8 \sin 3x \cos 2x$ as a sum.

52. Write $\sin 8x + \sin 4x$ as a product.

➤ EXTENDING THE CONCEPTS

53. For an NBA regulation basket court, the hoop is 10 feet above the ground, and the horizontal distance from the center of the hoop to the free throw line is 13.75 feet. In making a free throw, Shaquille O'Neal releases the ball approximately 8.5 feet directly above the free throw line (see Figure 20). The ball reaches the center of the hoop 0.98 second later. The path of the ball can be described by the parametric equations

$$x = (|\mathbf{V}_0| \cos \theta)t$$
$$y = (|\mathbf{V}_0| \sin \theta)t - 16t^2 + 8.5$$

where x is the distance (in feet) traveled horizontally and y is the height (in feet) of the ball above the ground t seconds after the ball is released. Find the initial speed $|\mathbf{V}_0|$ and the angle θ at which the basketball is thrown.

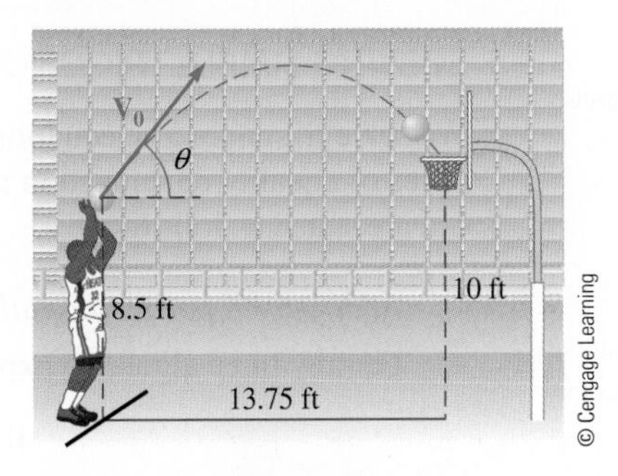

Figure 20

➤ LEARNING OBJECTIVES ASSESSMENT

These questions are available for instructors to help assess if you have successfully met the learning objectives for this section.

54. Graph the plane curve for the parametric equations $x = 2 + 4\cos t$, $y = -1 + 3\sin t$ by plotting points. Use values of t that are multiples of $\pi/4$. Which for the following points lie on the graph?

a. $(6, -1)$ and $(2, 2)$ **b.** $(-2, -1)$ and $(1, -1)$
c. $(2, -4)$ and $(-4, 3)$ **d.** $(-2, -2)$ and $(4, -1)$

55. Which graph shows the correct orientation for the plane curve for $x = \csc t$, $y = \cot t$?

a.

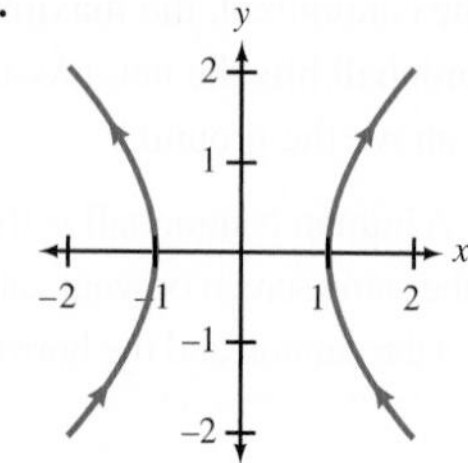

b.

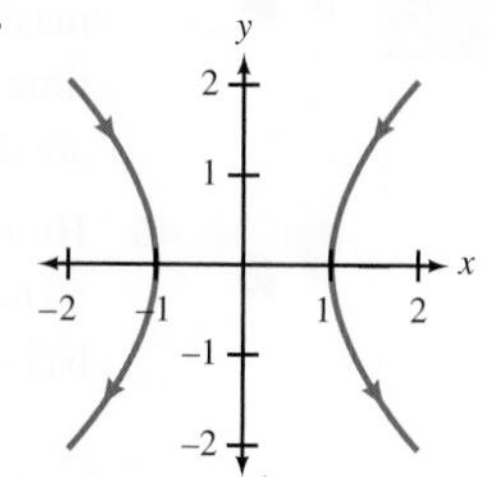

56. Eliminate the parameter from the parametric equations $x = 3\sin t$, $y = 2 + \cos t$. The resulting equation is equivalent to which of the following?

a. $\dfrac{x^2}{9} + (y-2)^2 = 1$ **b.** $\dfrac{x}{3(y-2)} = 1$

c. $\dfrac{x^2}{9} - \dfrac{y^2}{4} = 1$ **d.** $x^2 + y^2 = 13$

57. The path taken by a human cannonball can be modeled by the parametric equations $x = (65\cos 70°)t$, $y = (65\sin 70°)t - 16t^2$, where t is measured in seconds and x and y are measured in feet. Use a graphing calculator to approximate the maximum distance traveled horizontally by the human cannonball (assume the barrel of the cannon and the landing net are the same distance above ground).

a. 58 feet **b.** 85 feet **c.** 71 feet **d.** 104 feet

CHAPTER 6 SUMMARY

EXAMPLES

1 Solve for all radian solutions:

$$2\cos x - \sqrt{3} = 0$$
$$2\cos x = \sqrt{3}$$
$$\cos x = \frac{\sqrt{3}}{2}$$

The reference angle is $\pi/6$, and x must be an angle terminating in QI or QIV.

SOLVING SIMPLE TRIGONOMETRIC EQUATIONS [6.1]

We solve trigonometric equations that are linear in $\sin x$ or $\cos x$ by applying the properties of equality developed in algebra. The two most important properties from algebra are stated as follows:

Addition Property of Equality

For any three algebraic expressions A, B, and C,

If $A = B$
then $A + C = B + C$

Therefore, for any integer k,

$$x = \frac{\pi}{6} + 2k\pi$$

or

$$x = \frac{11\pi}{6} + 2k\pi$$

Multiplication Property of Equality

For any three algebraic expressions A, B, and C, with $C \neq 0$,

If $A = B$

then $AC = BC$

To solve a trigonometric equation that is quadratic in $\sin x$ or $\cos x$, we write it in standard form and then factor it or use the quadratic formula.

2 Solve if $0° \leq \theta < 360°$:

$$\cos 2\theta + 3 \sin \theta - 2 = 0$$
$$1 - 2 \sin^2 \theta + 3 \sin \theta - 2 = 0$$
$$2 \sin^2 \theta - 3 \sin \theta + 1 = 0$$
$$(2 \sin \theta - 1)(\sin \theta - 1) = 0$$
$$2 \sin \theta - 1 = 0 \quad \text{or} \quad \sin \theta - 1 = 0$$
$$\sin \theta = \frac{1}{2} \qquad \sin \theta = 1$$
$$\theta = 30°, 150°, 90°$$

USING IDENTITIES IN TRIGONOMETRIC EQUATIONS [6.2]

Sometimes it is necessary to use identities to make trigonometric substitutions when solving equations. Identities are usually required if the equation contains more than one trigonometric function or if there is more than one angle named in the equation. In the example to the left, we begin by using a double-angle identity to replace $\cos 2\theta$ with $1 - 2 \sin^2 \theta$. Doing so gives us a quadratic equation in $\sin \theta$, which we put in standard form and solve by factoring.

3 Solve:

$$\sin 3x = \frac{\sqrt{2}}{2}$$

For any integer k,

$$3x = \frac{\pi}{4} + 2k\pi \text{ or } 3x = \frac{3\pi}{4} + 2k\pi$$

$$x = \frac{\pi}{12} + \frac{2k\pi}{3} \text{ or } x = \frac{\pi}{4} + \frac{2k\pi}{3}$$

EQUATIONS INVOLVING MULTIPLE ANGLES [6.3]

Sometimes the equations we solve in trigonometry reduce to equations that contain multiple angles. When this occurs, we have to be careful in the last step that we do not leave out any solutions. First we find all solutions as an expression involving k, where k is an integer. Then we choose appropriate values for k to obtain the desired solutions.

4 Eliminate the parameter t from the equations $x = 3 + \sin t$ and $y = \cos t - 2$.

Solving for $\sin t$ and $\cos t$ we have

$$\sin t = x - 3 \quad \text{and} \quad \cos t = y + 2$$

Substituting these expressions into $\sin^2 t + \cos^2 t = 1$, we have

$$(x - 3)^2 + (y + 2)^2 = 1$$

which is the equation of a circle with a radius of 1 and center $(3, -2)$.

PARAMETRIC EQUATIONS [6.4]

When the coordinates of point (x, y) are described separately by two equations of the form $x = f(t)$ and $y = g(t)$, then the two equations are called *parametric equations* and t is called the *parameter.* One way to graph the plane curve for a set of points (x, y) that are given in terms of the parameter t is to make a table of values and plot points. Another way to graph the plane curve is to eliminate the parameter and obtain an equation in just x and y that gives the same set of points (x, y).

CHAPTER 6 TEST

Find all solutions in the interval $0° \leq \theta < 360°$. If rounding is necessary, round to the nearest tenth of a degree.

1. $2 \sin \theta - 1 = 0$

2. $\sqrt{3} \tan \theta + 1 = 0$

3. $\cos \theta - 2 \sin \theta \cos \theta = 0$

4. $\tan \theta - 2 \cos \theta \tan \theta = 0$

5. $4 \cos \theta - 2 \sec \theta = 0$

6. $2 \sin \theta - \csc \theta = 1$

7. $\sin \frac{\theta}{2} + \cos \theta = 0$

8. $\cos \frac{\theta}{2} - \cos \theta = 0$

9. $4\cos 2\theta + 2\sin\theta = 1$

10. $\sin(3\theta - 45°) = -\frac{\sqrt{3}}{2}$

11. $\sin\theta + \cos\theta = 1$

12. $\sin\theta - \cos\theta = 1$

13. $\cos 3\theta = -\frac{1}{2}$

14. $\tan 2\theta = 1$

Find all solutions for the following equations. Write your answers in radians using exact values.

15. $\cos 2x - 3\cos x = -2$

16. $\sqrt{3}\sin x - \cos x = 0$

17. $\sin 2x\cos x + \cos 2x\sin x = -1$

18. $\sin^3 4x = 1$

Find all solutions, to the nearest tenth of a degree, in the interval $0° \le \theta < 360°$.

19. $5\sin^2\theta - 3\sin\theta = 2$

20. $4\cos^2\theta - 4\cos\theta = 2$

Use your graphing calculator to find all radian solutions in the interval $0 \le x < 2\pi$ for each of the following equations. Round your answers to four decimal places.

21. $3\sin x - 2 = 0$

22. $\cos x + 3 = 4\sin x$

23. $\sin^2 x + 3\sin x - 1 = 0$

24. $\sin 2x = \frac{3}{5}$

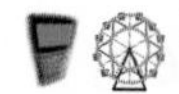

25. Ferris Wheel In Example 6 of Section 4.5, we found the equation that gives the height h of a passenger on a Ferris wheel at any time t during the ride to be

$$h = 139 - 125\cos\frac{\pi}{10}t$$

where h is given in feet and t is given in minutes. Use this equation to find the times at which a passenger will be 150 feet above the ground. Round your answers to the nearest tenth of a minute.

Eliminate the parameter t from each of the following and then sketch the graph.

26. $x = 3\cos t,\ y = 3\sin t$

27. $x = \sec t,\ y = \tan t$

28. $x = 3 + 2\sin t,\ y = 1 + 2\cos t$

29. $x = 3\cos t - 3,\ y = 3\sin t + 1$

30. Ferris Wheel A Ferris wheel has a diameter of 180 feet and sits 8 feet above the ground. It rotates in a counterclockwise direction, making one complete revolution every 3 minutes. Use parametric equations to model the path of a rider on this wheel. Place your coordinate system so that the origin is on the ground below the bottom of the wheel. You want to end up with parametric equations that will give you the position of the rider every minute of the ride. Graph your results on a graphing calculator.

GROUP PROJECT Deriving the Cycloid

OBJECTIVE: To derive parametric equations for the cycloid.

The cycloid is a famous planar curve with a rich history. First studied and named by Galileo, the cycloid is defined as the curve traced by a point on the circumference of a circle that is rolling along a line without slipping (Figure 1).

Figure 1

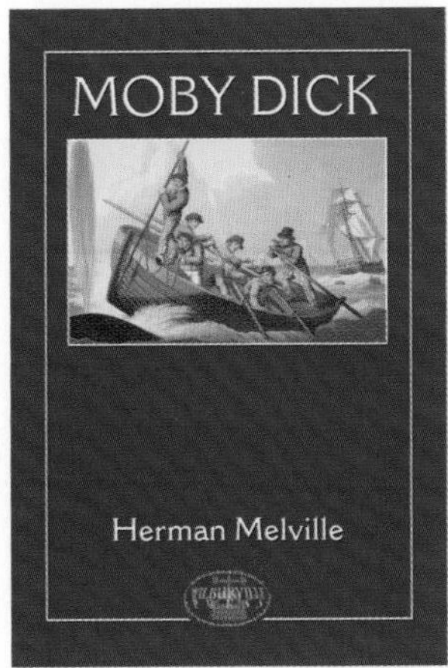

Herman Melville referred to the cycloid in this passage from *Moby-Dick:*

> It was in the left-hand try-pot of the Pequod, with the soapstone diligently circling round me, that I was first indirectly struck by the remarkable fact, that in geometry all bodies gliding along a cycloid, my soapstone, for example, will descend from any point in precisely the same time.

To begin, we will assume the circle has radius r and is positioned on a rectangular coordinate system with its center on the positive y-axis and tangent to the x-axis (Figure 2). The x-axis will serve as the line that the circle will roll along. We will choose point P, initially at the origin, to be the fixed point on the circumference of the circle that will trace the cycloid. Figure 3 shows the position of the circle after it has rolled a short distance d along the x-axis. We will use t to represent the angle (in radians) through which the circle has rotated.

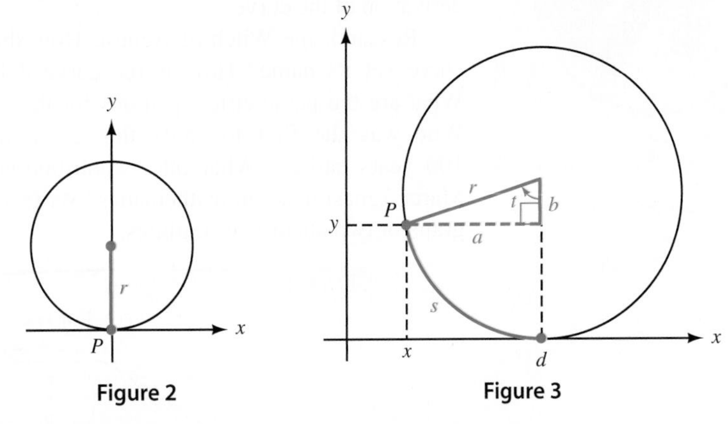

Figure 2 **Figure 3**

1 If d is the distance that the circle has rolled, then what is the length of the arc s? Use this to find a relationship between d and t.

2 Use the lengths a, b, r, and d to find the coordinates of point P. That is, find equations for x and y in terms of these values.

3 Now use right triangle trigonometry to find a and b in terms of r and t.

4 Using your answers to Questions 1 through 3, find equations for x and y in terms of r and t only. Because r is a constant, you now have parametric equations for the cycloid, using the angle t as the parameter!

5 Suppose that $r = 1$. Complete the table by finding x and y for each value of t given. Then use your results to sketch the graph of the cycloid for $0 \le t \le 2\pi$.

t	0	$\pi/4$	$\pi/2$	$3\pi/4$	π	$5\pi/4$	$3\pi/2$	$7\pi/4$	2π
x									
y									

6 Use your graphing calculator to graph the cycloid for $0 \le t \le 6\pi$. First, put your calculator into radian and parametric modes. Then define x and y using your equations from Question 4 (using $r = 1$). Set your window so that $0 \le t \le 6\pi$, $0 \le x \le 20$, and $0 \le y \le 3$.

7 Research the cycloid. What property of the cycloid is Melville referring to? What other famous property does the cycloid have? What was the connection between the cycloid and navigation at sea? Write a paragraph or two about your findings.

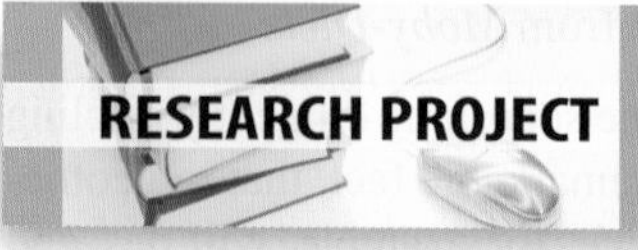

RESEARCH PROJECT The Witch of Agnesi

Maria Gaetana Agnesi

Maria Gaetana Agnesi (1718–1799) was the author of *Instituzioni Analitiche ad uso Della Gioventu Italiana,* a calculus textbook considered to be the best book of its time and the first surviving mathematical work written by a woman. Within this text Maria Agnesi describes a famous curve that has come to be known as the *Witch of Agnesi.* Figure 1 shows a diagram from the book used to illustrate the derivation of the curve.

Research the Witch of Agnesi. How did this curve get its name? How is the curve defined? What are the parametric equations for the curve? Who was the first to study this curve almost 100 years earlier? What other contributions did Maria Agnesi make in mathematics? Write a paragraph or two about your findings.

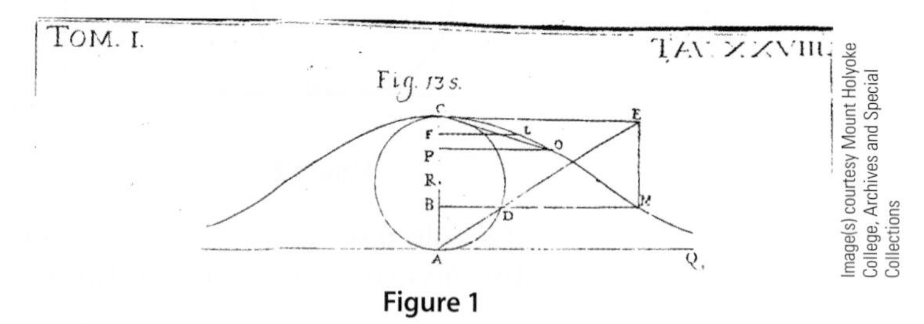

Figure 1

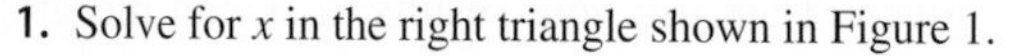

CUMULATIVE TEST 1–6

1. Solve for x in the right triangle shown in Figure 1.

2. Draw 210° in standard position, and find one positive angle and one negative angle that is coterminal.

3. Find the remaining trigonometric functions of θ if $\sin\theta = \dfrac{24}{25}$ and θ terminates in QII.

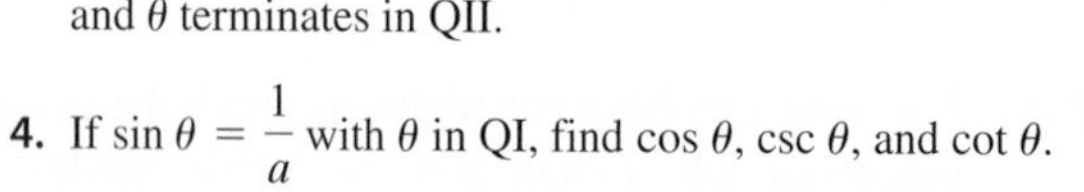

4. If $\sin\theta = \dfrac{1}{a}$ with θ in QI, find $\cos\theta$, $\csc\theta$, and $\cot\theta$.

Figure 1 (right triangle with hypotenuse $x + 2$, vertical leg x, base 4)

5. Prove that the equation $\tan\theta = \sin\theta\sec\theta$ is an identity.

6. Evaluate the expression $\cos^2 30° - \sin^2 30°$ by substituting exact values and simplifying.

7. Convert 16.45° to degrees and minutes.

8. In Figure 2, the distance from A to D is y, the distance from D to C is x, and the distance from C to B is h. If $A = 38°$, $\angle BDC = 55°$, $AB = 41$, and $DB = 29$, find x and y.

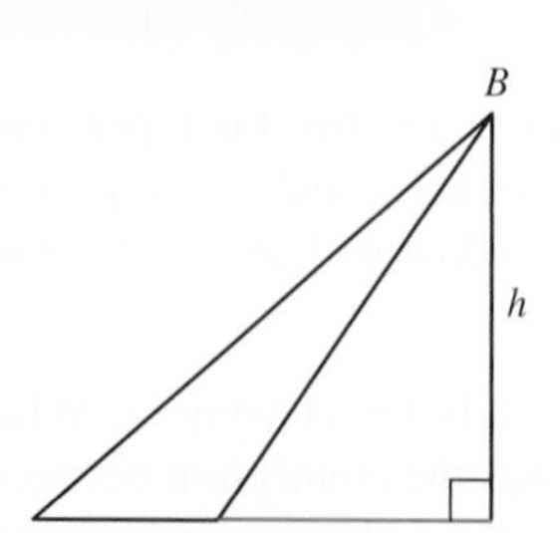

Figure 2

9. Distance A sailboat that has overturned is observed by two people who are at different points along a straight shoreline that runs north and south. From one observer, the bearing of the sailboat is S 75° W, and from the other observer the bearing is N 65° W. If the distance between the two observers is 1.7 miles, what is the shortest distance from the sailboat to the shore?

10. Velocity of an Arrow An arrow is shot with an initial velocity of 48 feet per second at an angle of elevation of 33°. Find the magnitude of the horizontal and vertical components of the velocity vector.

11. Draw 410° 20′ in standard position and then name the reference angle.

12. Evaluate $4\sin\left(2x + \frac{\pi}{4}\right)$ when x is $\frac{\pi}{4}$.

13. If an angle θ is in standard position, and the terminal side of θ intersects the unit circle at the point $(-0.2537, 0.9673)$, find $\sin\theta$, $\cos\theta$, and $\tan\theta$.

14. Arc Length The minute hand of a clock is 2 centimeters long. How far does the tip of the minute hand travel in 30 minutes? Give your answer in exact form.

15. Uniform Circular Motion A point is rotating with uniform circular motion on a circle of radius 3 centimeters. Find ω if $v = 5$ centimeters per second. Give your answer in exact form.

16. Use the graph of $y = \sec x$ to find all values of x between -4π and 4π for which $\sec x = -1$.

17. Graph one complete cycle of $y = \frac{1}{2}\sin 3x$. State the amplitude and period.

18. For the equation $y = 3 - \cos\left(x - \frac{\pi}{4}\right)$, identify the horizontal shift and vertical shift. Then use this information to sketch one complete cycle of the graph.

19. Graph one complete cycle of $y = 3\tan\left(\frac{1}{2}x + \frac{\pi}{4}\right)$. State the period and horizontal shift.

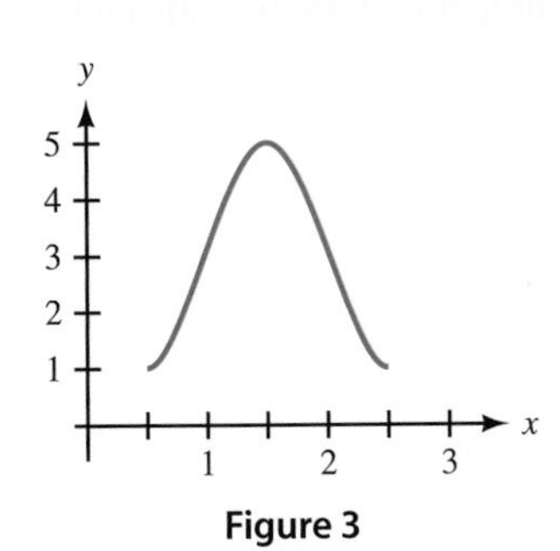

Figure 3

20. The graph in Figure 3 shows one complete cycle of an equation containing a trigonometric function. Find an equation to match the graph.

21. Evaluate $\cos\left(\tan^{-1}\frac{2}{3}\right)$ without using a calculator.

Prove each identity.

22. $(1 + \sec x)(1 - \cos x) = \tan^2 x \cos x$

23. $\sin(\theta - 90°) = -\cos\theta$

Let $\sin A = -\frac{3}{5}$ *with* $270° \le A \le 360°$ *and* $\sin B = \frac{12}{13}$ *with* $90° \le B \le 180°$ *and find the following:*

24. $\sin 2B$

25. $\cos\frac{A}{2}$

26. Rewrite the expression $4\sin 7x \sin 3x$ as a sum or difference, then simplify if possible.

27. Solve $2\cos^2\theta - \cos\theta - 1 = 0$ for θ if $0° \le \theta < 360°$.

28. Solve $\sin 2x - \sin x = 0$ if $0 \le x < 2\pi$.

29. Find all degree solutions for $\cos 4x \cos x + \sin 4x \sin x = -\frac{1}{2}$.

30. Eliminate the parameter t from the parametric equations $x = 3\cos t$, $y = 5\sin t$ and then sketch the graph of the plane curve.

7 Triangles

Mathematics, the nonempirical science par excellence . . . the science of sciences, delivering the key to those laws of nature and the universe which are concealed by appearances.

➤ *Hannah Arendt*

© Joel W. Rogers/CORBIS

INTRODUCTION

The aerial tram in Rio de Janeiro, which carries sightseers up to the top of Sugarloaf Mountain, provides a good example of forces acting in equilibrium. The force representing the weight of the gondola and its passengers must be withstood by the tension in the cable, which can be represented by a pair of forces pointing in the same direction as the cable on either side (Figure 1). Because these three forces are in static equilibrium, they must form a triangle as shown in Figure 2.

© Joel W. Rogers/CORBIS

Figure 1

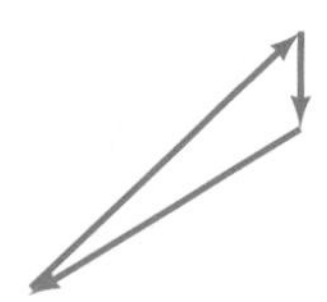

Figure 2

In this chapter we will continue our study of triangles and their usefulness as mathematical models in many situations. Specifically, we will consider triangles like the one shown in Figure 2 that are not right triangles.

SECTION 7.1 The Law of Sines

LEARNING OBJECTIVES

1. Use the law of sines to find a missing side in an oblique triangle.
2. Solve a real-life problem using the law of sines.
3. Use vectors and the law of sines to solve an applied problem.

In this chapter we return to the matter of solving triangles. Up to this point, the triangles we have worked with have always been right triangles. We will now look at situations that involve *oblique triangles,* which are triangles that do not have a right angle.

It is important to remember that our triangle definitions of the six trigonometric functions (Definition II) involve ratios of the sides of a right triangle only. These definitions do not apply to oblique triangles. For example, the triangle shown in Figure 1 is oblique, so we cannot say that $\sin A = a/c$. In this chapter we will see how the sine and cosine functions can be used properly to solve these kinds of triangles.

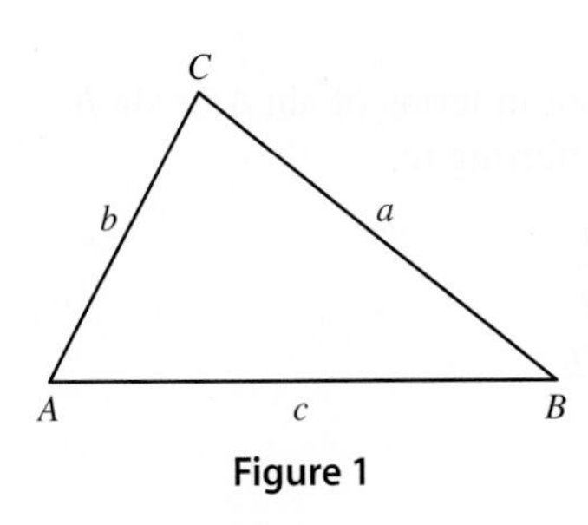

Figure 1

Every triangle has three sides and three angles. To solve any triangle, we must know at least three of these six values. Table 1 summarizes the possible cases and shows which method can be used to solve the triangle in each case.

TABLE 1
Solving Oblique Triangles

Case		Method
AAA	*Angle-angle-angle* This case cannot be solved because knowing all three angles does not determine a unique triangle. There are an infinite number of similar triangles that share the same angles.	None
AAS	*Angle-angle-side* Given two angles and a side opposite one of the angles, a unique triangle is determined.	Law of sines
ASA	*Angle-side-angle* Given two angles and the included side, a unique triangle is determined.	
SAS	*Side-angle-side* Given two sides and the included angle, a unique triangle is determined.	Law of cosines
SSS	*Side-side-side* Given all three sides, a unique triangle is determined.	
SSA	*Side-side-angle* Given two sides and an angle opposite one of the sides, there may be one, two, or no triangles that are possible. This is known as the ambiguous case.	Law of sines or Law of cosines

We will begin with oblique triangles that can be solved using the law of sines. The law of cosines will be introduced in Section 7.2.

The Law of Sines

There are many relationships that exist between the sides and angles in a triangle. One such relationship is called the *law of sines,* which states that the ratio of the sine of an angle to the length of the side opposite that angle is constant in any triangle.

LAW OF SINES

Given triangle ABC shown in Figure 2,

$$\frac{\sin A}{a} = \frac{\sin B}{b} = \frac{\sin C}{c}$$

or, equivalently,

$$\frac{a}{\sin A} = \frac{b}{\sin B} = \frac{c}{\sin C}$$

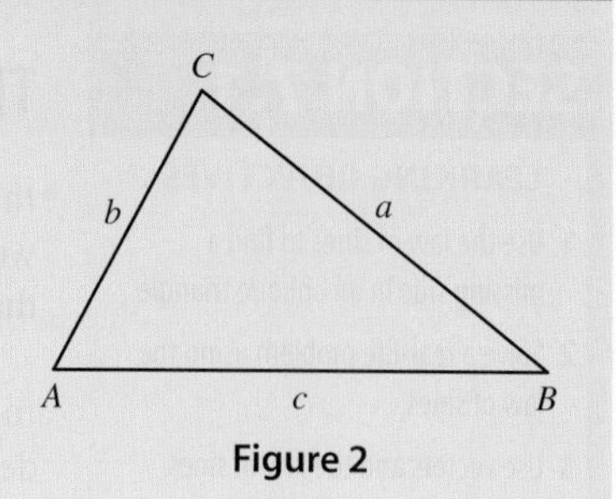

Figure 2

PROOF

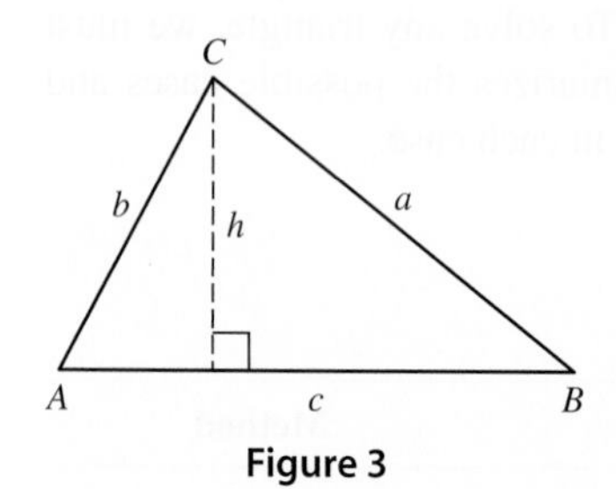

Figure 3

The altitude h of the triangle in Figure 3 can be written in terms of $\sin A$ or $\sin B$ depending on which of the two right triangles we are referring to:

$$\sin A = \frac{h}{b} \qquad \sin B = \frac{h}{a}$$

$$h = b \sin A \qquad h = a \sin B$$

Because h is equal to itself, we have

$$h = h$$

$$b \sin A = a \sin B$$

$$\frac{b \sin A}{ab} = \frac{a \sin B}{ab} \qquad \text{Divide both sides by } ab$$

$$\frac{\sin A}{a} = \frac{\sin B}{b} \qquad \text{Divide out common factors}$$

If we do the same kind of thing with the altitude that extends from A, we will have the third ratio in the law of sines, $\frac{\sin C}{c}$, equal to the two preceding ratios.

Note that the derivation of the law of sines will proceed in the same manner if triangle ABC contains an obtuse angle, as in Figure 4.

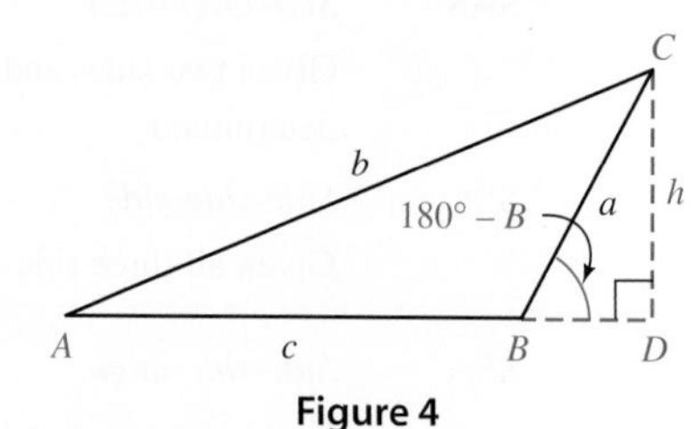

Figure 4

In triangle BDC we have

$$\sin (180° - B) = \frac{h}{a}$$

but,

$$\begin{aligned} \sin (180° - B) &= \sin 180° \cos B - \cos 180° \sin B \\ &= (0) \cos B - (-1) \sin B \\ &= \sin B \end{aligned}$$

So, $\sin B = h/a$, which is the result we obtained previously. Using triangle ADC, we have $\sin A = h/b$. As you can see, these are the same two expressions we began with when deriving the law of sines for the acute triangle in Figure 3. From this point on, the derivation would match our previous derivation. ■

We can use the law of sines to find missing parts of triangles for which we are given two angles and a side.

Two Angles and One Side

In our first example, we are given two angles and the side opposite one of them. (You may recall that in geometry these were the parts we needed equal in two triangles in order to prove them congruent using the AAS Theorem.)

PROBLEM 1
In triangle ABC, $A = 40°$, $C = 60°$, and $a = 7.0$ inches. Find the length of side b.

EXAMPLE 1 In triangle ABC, $A = 30°$, $B = 70°$, and $a = 8.0$ cm. Find the length of side c.

SOLUTION We begin by drawing a picture of triangle ABC (it does not have to be accurate) and labeling it with the information we have been given (Figure 5).

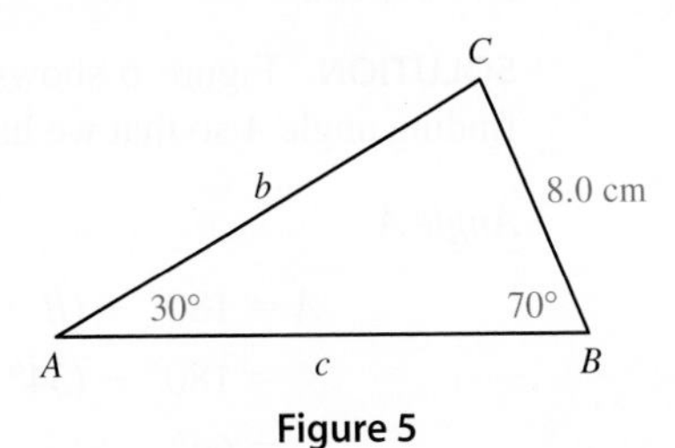

Figure 5

When we use the law of sines, we must have one of the ratios given to us. In this case, since we are given a and A, we have the ratio $\dfrac{a}{\sin A}$. To solve for c, we need to first find angle C. The sum of the angles in any triangle is 180°, so we have

$$\begin{aligned} C &= 180° - (A + B) \\ &= 180° - (30° + 70°) \\ &= 80° \end{aligned}$$

To find side c, we use the following two ratios given in the law of sines.

$$\frac{c}{\sin C} = \frac{a}{\sin A}$$

To solve for c, we multiply both sides by $\sin C$ and then substitute.

$$\begin{aligned} c &= \frac{a \sin C}{\sin A} && \text{Multiply both sides by } \sin C \\ &= \frac{8.0 \sin 80°}{\sin 30°} && \text{Substitute in known values} \\ &= \frac{8.0(0.9848)}{0.5000} && \text{Calculator} \\ &= 16 \text{ cm} && \text{To two significant digits} \end{aligned}$$

■

NOTE 1 The equal sign in the third line should actually be replaced by the *approximately equal to* symbol, ≈, because the decimal 0.9848 is an approximation to sin 80°. (Remember, most values of the trigonometric functions are irrational numbers.) In this chapter, we will use an equal sign in the solutions to all of our examples, even when the ≈ symbol would be more appropriate, to make the examples a little easier to follow.

NOTE 2 As in Chapter 2, we round our answers so that the number of significant digits in our answers matches the number of significant digits in the least significant number given in the original problem. Also, we round our answers only, not any of the numbers in the intermediate steps. We are showing the values of the trigonometric functions to four significant digits simply to avoid cluttering the page with long decimal numbers. This does not mean that you should stop halfway through a problem and round the values of trigonometric functions to four significant digits before continuing.

In our next example, we are given two angles and the side included between them (ASA) and are asked to find all the missing parts.

PROBLEM 2
Find the missing parts of triangle ABC if $A = 46°$, $C = 78°$, and $b = 7.5$ cm.

EXAMPLE 2 Find the missing parts of triangle ABC if $B = 34°$, $C = 82°$, and $a = 5.6$ cm.

SOLUTION Figure 6 shows a diagram with the given information. We begin by finding angle A so that we have one of the ratios in the law of sines completed.

Angle A

$$\begin{aligned} A &= 180° - (B + C) \\ &= 180° - (34° + 82°) \\ &= 64° \end{aligned}$$

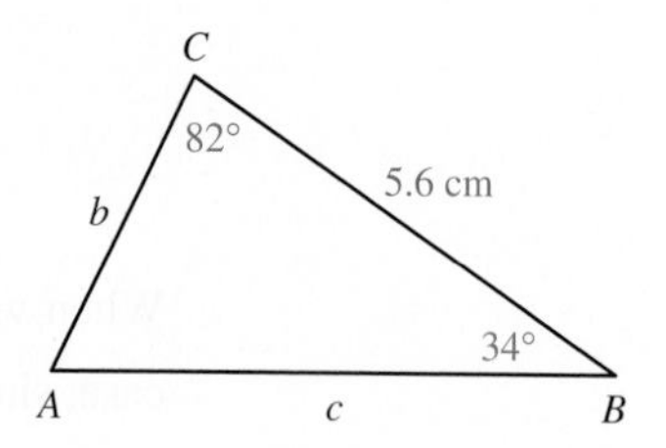

Figure 6

Side b

$$\text{If} \quad \frac{b}{\sin B} = \frac{a}{\sin A}$$

$$\text{then} \quad b = \frac{a \sin B}{\sin A} \quad \text{Multiply both sides by } \sin B$$

$$= \frac{5.6 \sin 34°}{\sin 64°} \quad \text{Substitute in known values}$$

$$= \frac{5.6(0.5592)}{0.8988} \quad \text{Calculator}$$

$$= 3.5 \text{ cm} \quad \text{To two significant digits}$$

Side c

$$\text{If} \quad \frac{c}{\sin C} = \frac{a}{\sin A}$$

$$\text{then} \quad c = \frac{a \sin C}{\sin A} \quad \text{Multiply both sides by } \sin C$$

$$= \frac{5.6 \sin 82°}{\sin 64°} \quad \text{Substitute in known values}$$

$$= \frac{5.6(0.9903)}{0.8988} \quad \text{Calculator}$$

$$= 6.2 \text{ cm} \quad \text{To two significant digits}$$

■

The law of sines, along with some fancy electronic equipment, was used to obtain the results of some of the field events in one of the recent Olympic Games. For instance, Figure 7 is a diagram of a shot put ring. The shot is tossed (put) from the left and lands at A. A small electronic device is then placed at A (there is usually a dent in the ground where the shot lands, so it is easy to find where to place the device). The device at A sends a signal to a booth in the stands that gives the measures of angles A and B. The distance a is found ahead of time. With this information, we have the case AAS. To find the distance x, the law of sines is used.

$$\frac{x}{\sin B} = \frac{a}{\sin A} \Rightarrow x = \frac{a \sin B}{\sin A}$$

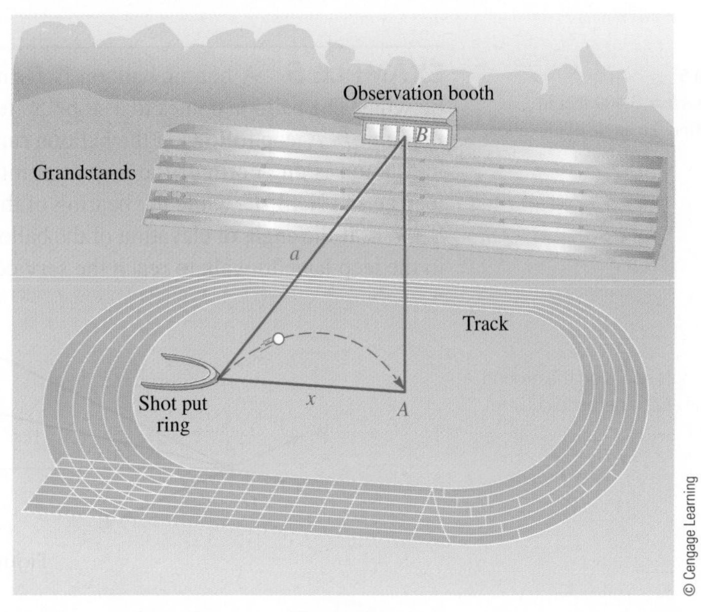

Figure 7

PROBLEM 3
Find x in Figure 7 if $a = 738$ ft, $B = 6.3°$, and $A = 78.4°$.

EXAMPLE 3 Find distance x in Figure 7 if $a = 562$ ft, $B = 5.7°$, and $A = 85.3°$.

SOLUTION

$$x = \frac{a \sin B}{\sin A} = \frac{562 \sin 5.7°}{\sin 85.3°} = 56.0 \text{ ft} \quad \text{To three significant digits}$$

PROBLEM 4
Rework Example 4 if the distance BD on the circumference of the earth is 1,050 miles and angle A is 79.2°.

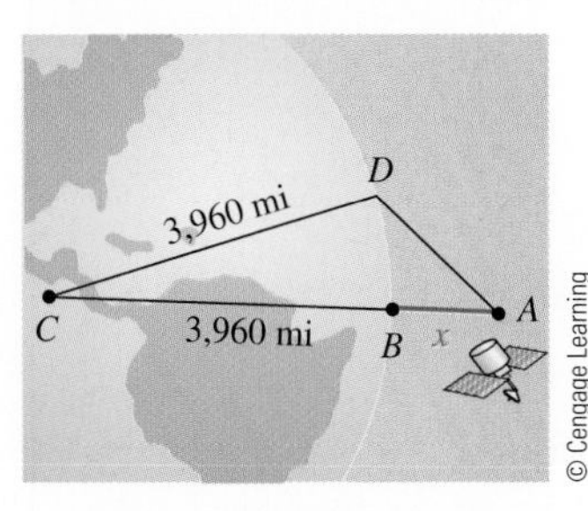

Figure 8

EXAMPLE 4 A satellite is circling above the earth as shown in Figure 8. When the satellite is directly above point B, angle A is 75.4°. If the distance between points B and D on the circumference of the earth is 910 miles and the radius of the earth is 3,960 miles, how far above the earth is the satellite?

SOLUTION First we find the radian measure of central angle C by dividing the arc length BD by the radius of the earth. Multiplying this number by $180/\pi$ will give us the degree measure of angle C.

$$C = \underbrace{\frac{910}{3{,}960}}_{\text{Angle } C \text{ in radians}} \cdot \underbrace{\frac{180}{\pi}}_{\text{Convert to degrees}} = 13.2°$$

Next we find angle CDA.

$$\angle CDA = 180° - (75.4° + 13.2°) = 91.4°$$

We now have the case ASA. To find x, we use the law of sines.

$$\frac{x + 3{,}960}{\sin 91.4°} = \frac{3{,}960}{\sin 75.4°}$$

$$x + 3{,}960 = \frac{3{,}960 \sin 91.4°}{\sin 75.4°}$$

$$x = \frac{3{,}960 \sin 91.4°}{\sin 75.4°} - 3{,}960$$

$$x = 130 \text{ mi} \quad \text{To two significant digits}$$

PROBLEM 5
Assuming $AB = 3{,}400$ feet in Figure 9, find the angle of elevation $\angle CBD$.

EXAMPLE 5 A hot-air balloon is flying over a dry lake when the wind stops blowing. The balloon comes to a stop 450 feet above the ground at point D as shown in Figure 9. A jeep following the balloon runs out of gas at point A. The nearest service station is due north of the jeep at point B. The bearing of the balloon from the jeep at A is N 13° E, while the bearing of the balloon from the service station at B is S 19° E. If the angle of elevation of the balloon from A is 12°, how far will the people in the jeep have to walk to reach the service station at point B?

NOTE Triangle ABC is on the ground, but triangle ACD is perpendicular to the ground.

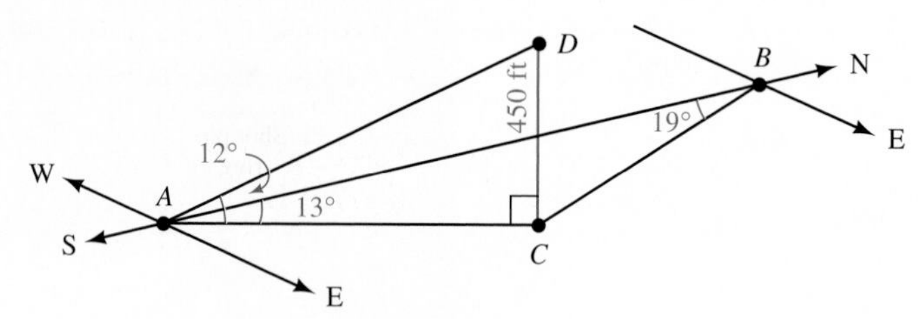

Figure 9

SOLUTION First we find the distance between C and A using right triangle trigonometry. Because this is an intermediate calculation, which we will use again, we keep more than two significant digits for AC.

$$\tan 12° = \frac{450}{AC}$$

$$AC = \frac{450}{\tan 12°}$$

$$= 2{,}117 \text{ ft}$$

Next we find angle ACB. This will give us the case ASA.

$$\angle ACB = 180° - (13° + 19°) = 148°$$

Finally, we find AB using the law of sines.

$$\frac{AB}{\sin 148°} = \frac{2{,}117}{\sin 19°}$$

$$AB = \frac{2{,}117 \sin 148°}{\sin 19°}$$

$$= 3{,}400 \text{ ft} \quad \text{To two significant digits}$$

There are 5,280 feet in a mile, so the people at A will walk approximately $3{,}400/5{,}280 = 0.64$ mile to get to the service station at B.

Our next example involves vectors. It is taken from the text *College Physics* by Miller and Schroeer, published by Saunders College Publishing.

PROBLEM 6
Rework Example 6 if the two angles shown in Figure 10 are instead 35° and 50°.

EXAMPLE 6 A traffic light weighing 22 pounds is suspended by two wires as shown in Figure 10. Find the magnitude of the tension in wire AB, and the magnitude of the tension in wire AC.

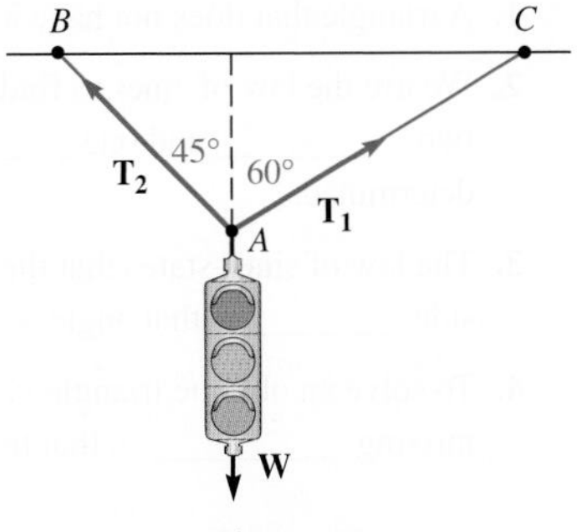

Figure 10

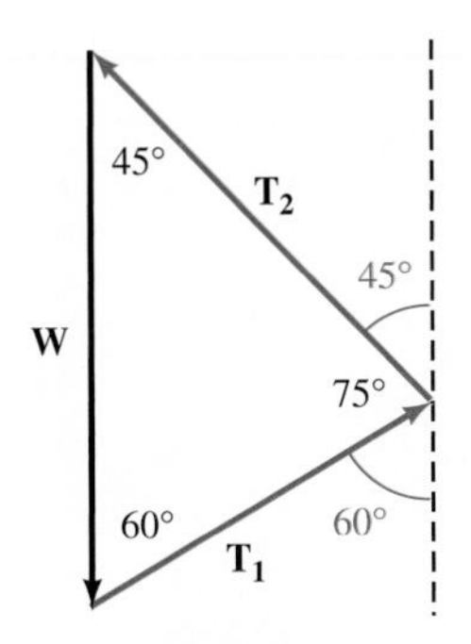

Figure 11

SOLUTION We assume that the traffic light is not moving and is therefore in a state of *static equilibrium.* When an object is in this state, the sum of the forces acting on the object must be 0. It is because of this fact that we can redraw the vectors from Figure 10 and be sure that they form a closed triangle. Figure 11 shows a convenient redrawing of the two tension vectors $\mathbf{T}_1$ and $\mathbf{T}_2$, and the vector $\mathbf{W}$ that is due to gravity. Notice that we have the case ASA.

Using the law of sines we have:

$$\frac{|\mathbf{T}_1|}{\sin 45^\circ} = \frac{22}{\sin 75^\circ}$$

$$|\mathbf{T}_1| = \frac{22 \sin 45^\circ}{\sin 75^\circ}$$

$$= 16 \text{ lb} \qquad \text{To two significant figures}$$

$$\frac{|\mathbf{T}_2|}{\sin 60^\circ} = \frac{22}{\sin 75^\circ}$$

$$|\mathbf{T}_2| = \frac{22 \sin 60^\circ}{\sin 75^\circ}$$

$$= 20 \text{ lb} \qquad \text{To two significant figures}$$

Getting Ready for Class

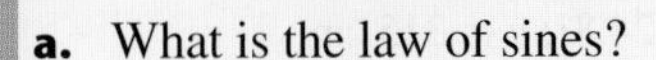

After reading through the preceding section, respond in your own words and in complete sentences.

a. What is the law of sines?
b. For which of the six cases discussed is the law of sines required?
c. Why is it always possible to find the third angle of any triangle when you are given the first two?
d. Why must you always have an angle and the length of the side opposite that angle to use the law of sines?

7.1 PROBLEM SET

➤ CONCEPTS AND VOCABULARY

For Questions 1 through 4, fill in the blank with an appropriate word.

1. A triangle that does not have a right angle is called an __________ triangle.
2. We use the law of sines to find the missing parts of triangles for which we are given two __________ and one __________. In either case, a __________ triangle is determined.
3. The law of sines states that the ratio of the __________ of an angle to the length of the side __________ that angle is __________ in any triangle.
4. To solve an oblique triangle given the case ASA, the first step is to find the missing __________ so that the law of sines can be used.

➤ EXERCISES

Each problem that follows refers to triangle ABC.

5. If $A = 80°$, $B = 30°$, and $b = 14$ cm, find a.
6. If $A = 40°$, $B = 60°$, and $a = 12$ cm, find b.
7. If $B = 120°$, $C = 20°$, and $c = 28$ inches, find b.
8. If $B = 110°$, $C = 40°$, and $b = 18$ inches, find c.
9. If $A = 5°$, $C = 125°$, and $c = 510$ yd, find a.
10. If $A = 10°$, $C = 100°$, and $a = 24$ yd, find c.
11. If $A = 50°$, $B = 60°$, and $a = 36$ km, find C and then find c.
12. If $B = 40°$, $C = 70°$, and $c = 42$ km, find A and then find a.
13. If $A = 52°$, $B = 48°$, and $c = 14$ cm, find C and then find a.
14. If $A = 33°$, $C = 82°$, and $b = 18$ cm, find B and then find c.

The following information refers to triangle ABC. In each case, find all the missing parts.

15. $A = 42.5°$, $B = 71.4°$, $a = 215$ inches
16. $A = 110.4°$, $C = 21.8°$, $c = 246$ inches
17. $B = 57°$, $C = 31°$, $a = 7.3$ m
18. $A = 46°$, $B = 95°$, $c = 6.8$ m
19. $A = 43° 30'$, $C = 120° 30'$, $a = 3.48$ ft
20. $B = 14° 20'$, $C = 75° 40'$, $b = 2.72$ ft
21. $B = 13.4°$, $C = 24.8°$, $a = 315$ cm
22. $A = 105°$, $B = 45°$, $c = 630$ cm
23. In triangle ABC, $A = 30°$, $b = 20$ ft, and $a = 2$ ft. Show that it is impossible to solve this triangle by using the law of sines to find $\sin B$.
24. In triangle ABC, $A = 40°$, $b = 19$ ft, and $a = 18$ ft. Use the law of sines to find $\sin B$ and then give two possible values for B.

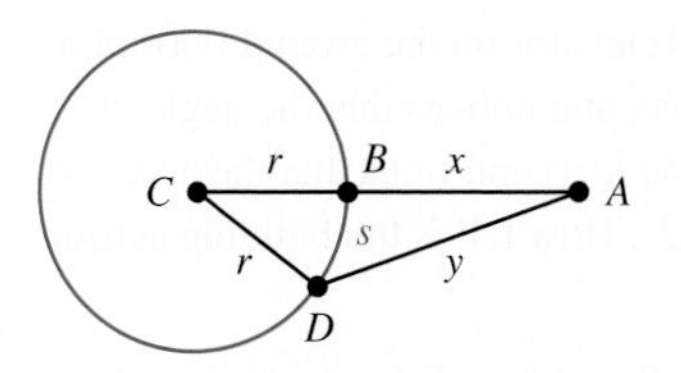

Figure 12

Geometry *The circle in Figure 12 has a radius of r and center at C. The distance from A to B is x, the distance from A to D is y, and the length of arc BD is s. For Problems 25 through 28, redraw Figure 12, label it as indicated in each problem, and then solve the problem.*

25. If $A = 31°$, $s = 11$, and $r = 12$, find x.

26. If $A = 26°$, $s = 22$, and $r = 19$, find x.

27. If $A = 45°$, $s = 18$, and $r = 15$, find y.

28. If $A = 55°$, $s = 21$, and $r = 22$, find y.

29. Angle of Elevation A man standing near a radio station antenna observes that the angle of elevation to the top of the antenna is 64°. He then walks 100 feet further away and observes that the angle of elevation to the top of the antenna is 46° (Figure 13). Find the height of the antenna to the nearest foot. (*Hint:* Find x first.)

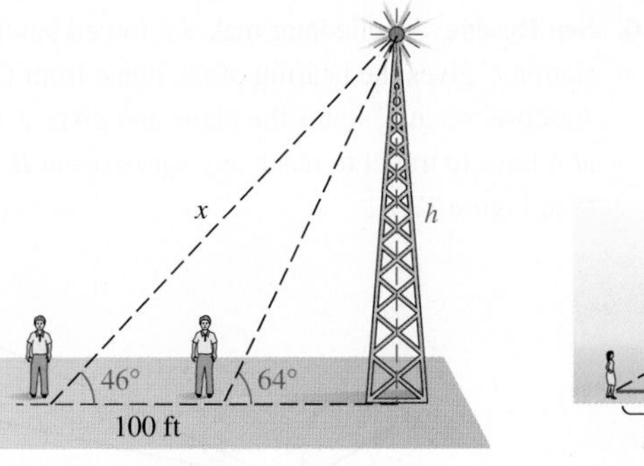

Figure 13

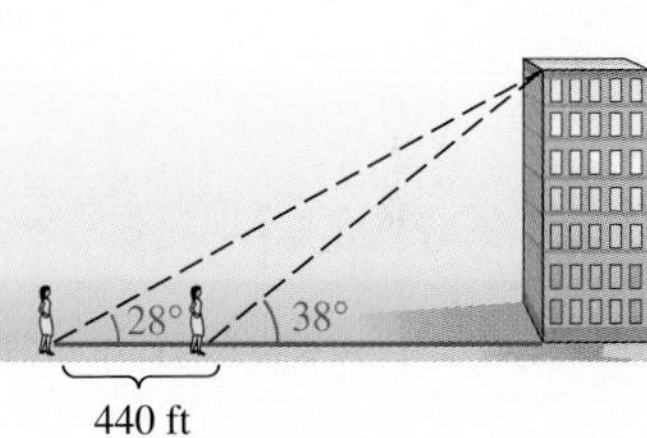

Figure 14

30. Angle of Elevation A person standing on the street looks up to the top of a building and finds that the angle of elevation is 38°. She then walks one block further away (440 feet) and finds that the angle of elevation to the top of the building is now 28°. How far away from the building is she when she makes her second observation? (See Figure 14.)

31. Angle of Depression A man is flying in a hot-air balloon in a straight line at a constant rate of 5 feet per second, while keeping it at a constant altitude. As he approaches the parking lot of a market, he notices that the angle of depression from his balloon to a friend's car in the parking lot is 35°. A minute and a half later, after flying directly over this friend's car, he looks back to see his friend getting into the car and observes the angle of depression to be 36°. At that time, what is the distance between him and his friend? (Round to the nearest foot.)

32. Angle of Elevation From a point on the ground, a person notices that a 110-foot antenna on the top of a hill subtends an angle of 1.5°. If the angle of elevation to the bottom of the antenna is 25°, find the height of the hill. (See Figure 15.)

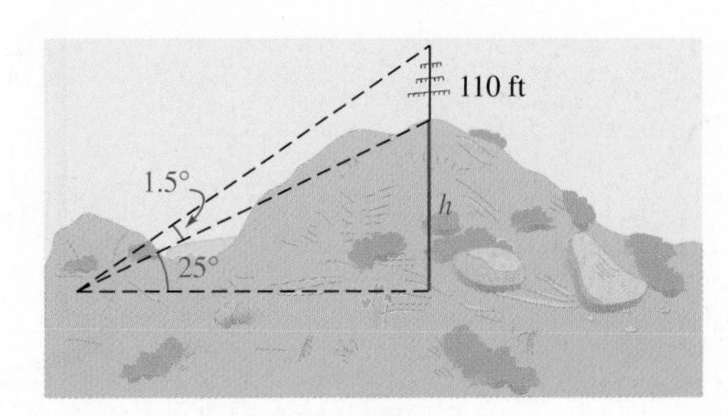

Figure 15

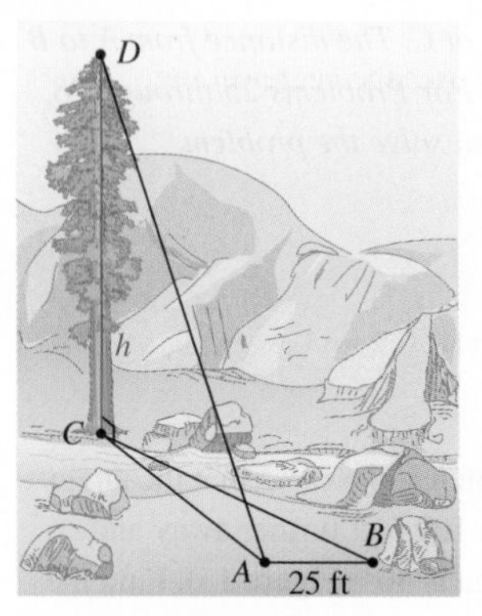

Figure 16

33. Angle of Elevation A woman entering an outside glass elevator on the ground floor of a hotel glances up to the top of the building across the street and notices that the angle of elevation is 48°. She rides the elevator up three floors (60 feet) and finds that the angle of elevation to the top of the building across the street is 32°. How tall is the building across the street? (Round to the nearest foot.)

34. Angle of Elevation A 155-foot antenna is on top of a tall building. From a point on the ground, the angle of elevation to the top of the antenna is 28.5°, while the angle of elevation to the bottom of the antenna from the same point is 23.5°. How tall is the building?

35. Height of a Tree Figure 16 is a diagram that shows how Colleen estimates the height of a tree that is on the other side of a stream. She stands at point *A* facing the tree and finds the angle of elevation from *A* to the top of the tree to be 51°. Then she turns 105° and walks 25 feet to point *B*, where she measures the angle between her path *AB* and the line *BC* extending from her to the base of the tree. She finds that angle to be 44°. Use this information to find the height of the tree.

36. Sea Rescue A helicopter makes a forced landing at sea. The last radio signal received at station *C* gives the bearing of the plane from *C* as N 55.4° E at an altitude of 1,050 feet. An observer at *C* sights the plane and gives $\angle DCB$ as 22.5°. How far will a rescue boat at *A* have to travel to reach any survivors at *B*, if the bearing of *B* from *A* is S 56.4° E? (See Figure 17.)

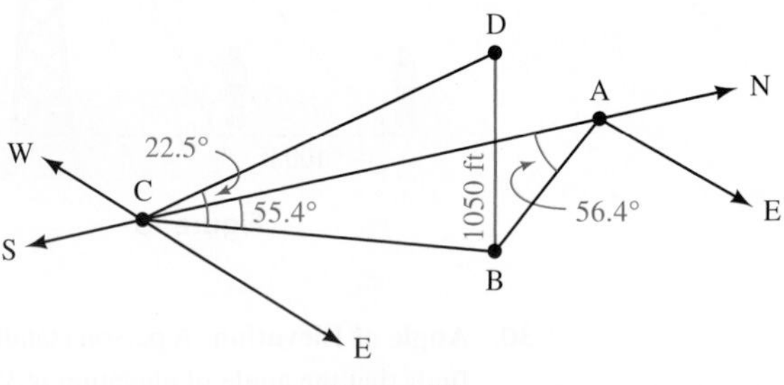

Figure 17

37. Distance to a Ship A ship is anchored off a long straight shoreline that runs north and south. From two observation points 18 miles apart on shore, the bearings of the ship are N 31° E and S 53° E. What is the distance from the ship to each of the observation points?

38. Distance to a Rocket Tom and Fred are 3.5 miles apart watching a rocket being launched from Vandenberg Air Force Base. Tom estimates the bearing of the rocket from his position to be S 75° W, while Fred estimates that the bearing of the rocket from his position is N 65° W. If Fred is due south of Tom, how far is each of them from the rocket?

39. Force A tightrope walker is standing still with one foot on the tightrope as shown in Figure 18. If the tightrope walker weighs 125 pounds, find the magnitudes of the tension in the rope toward each end of the rope.

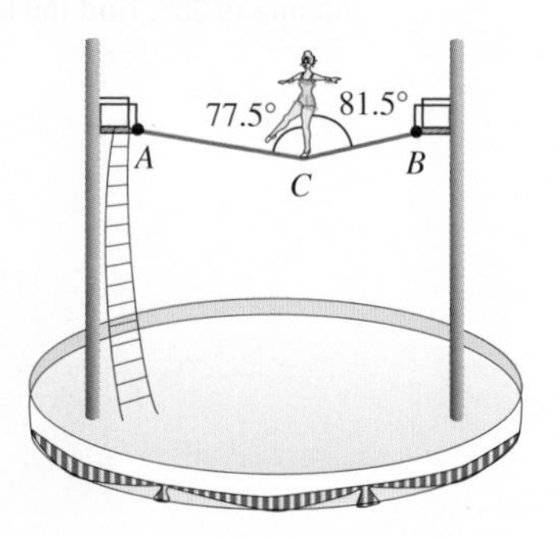

Figure 18

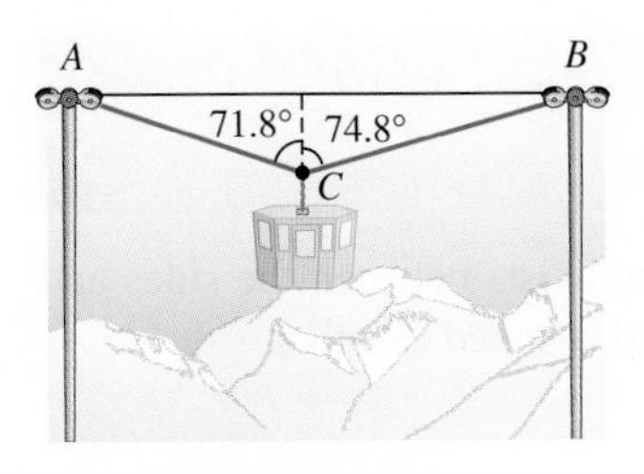

Figure 19

40. **Force** A tightrope walker weighing 145 pounds is standing still at the center of a tightrope that is 46.5 feet long. The weight of the walker causes the center of the tightrope to move down 14.5 inches. Find the magnitude of the tension in the tightrope toward each end of the tightrope.

41. **Force** If you have ever ridden on a chair lift at a ski area and had it stop, you know that the chair will pull down on the cable, dropping you down to a lower height than when the chair is in motion. Figure 19 shows a gondola that is stopped. Find the magnitude of the tension in the cable toward each end of the cable if the total weight of the gondola and its occupants is 1,850 pounds.

42. **Force** A chair lift at a ski resort is stopped halfway between two poles that support the cable to which the chair is attached. The poles are 215 feet apart and the combined weight of the chair and the three people on the chair is 725 pounds. If the weight of the chair and the people riding it causes the chair to move to a position 15.8 feet below the horizontal line that connects the top of the two poles, find the tension in the cable toward each end of the cable.

➤ REVIEW PROBLEMS

The problems that follow review material we covered in Sections 3.1 and 6.1.

Solve each equation for θ if $0° \leq \theta < 360°$. If rounding is necessary, round to the nearest tenth of a degree.

43. $2 \sin \theta - \sqrt{2} = 0$

44. $5 \cos \theta - 3 = 0$

45. $\sin \theta \cos \theta - 2 \cos \theta = 0$

46. $3 \cos \theta - 2 \sin \theta \cos \theta = 0$

47. $2 \sin^2 \theta - 3 \sin \theta = -1$

48. $10 \cos^2 \theta + \cos \theta - 3 = 0$

Find θ to the nearest tenth of a degree if $0 \leq \theta < 360°$, and

49. $\sin \theta = 0.7380$

50. $\sin \theta = 0.2351$

➤ LEARNING OBJECTIVES ASSESSMENT

These questions are available for instructors to help assess if you have successfully met the learning objectives for this section.

51. Find c for triangle ABC if $A = 43°$, $B = 12°$, and $b = 25$ centimeters.
    a. 98 cm **b.** 34 cm **c.** 120 cm **d.** 82 cm

52. Gina is standing near a building and notices that the angle of elevation to the top of the building is 68°. She then walks 72 feet further away from the building and notices that the angle of elevation to the top of the building is now only 51°. Find the height of the building.
    a. 25 ft **b.** 56 ft **c.** 149 ft **d.** 180 ft

53. A 45.0 lb signal light is hanging from a wire attached to two poles, causing the wire to sag as shown in Figure 20. Find the magnitude of the tension in the wire toward the pole labeled A.
    a. 44.6 lb **b.** 143 lb **c.** 73.1 lb **d.** 26.5 lb

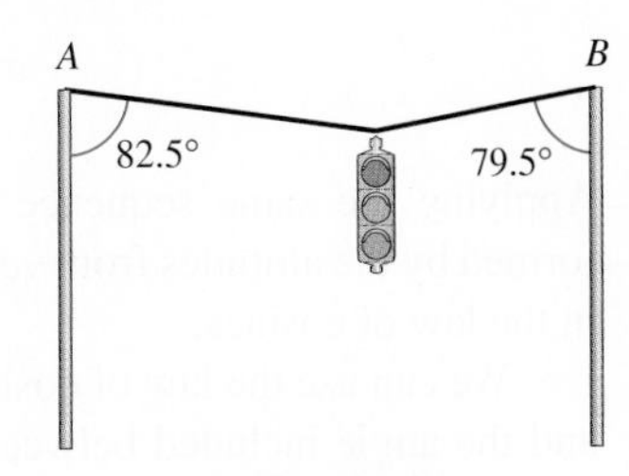

Figure 20

SECTION 7.2 The Law of Cosines

LEARNING OBJECTIVES

1. Use the law of cosines to find a missing side in an oblique triangle.
2. Use the law of cosines to find a missing angle in an oblique triangle.
3. Draw a vector representing a given heading.
4. Use the law of cosines to solve a real-life problem involving heading or true course.

In this section, we will derive another relationship that exists between the sides and angles in any triangle. It is called the *law of cosines* and is stated like this:

LAW OF COSINES (SAS)

Given triangle ABC shown in Figure 1:

$$a^2 = b^2 + c^2 - 2bc \cos A$$
$$b^2 = a^2 + c^2 - 2ac \cos B$$
$$c^2 = a^2 + b^2 - 2ab \cos C$$

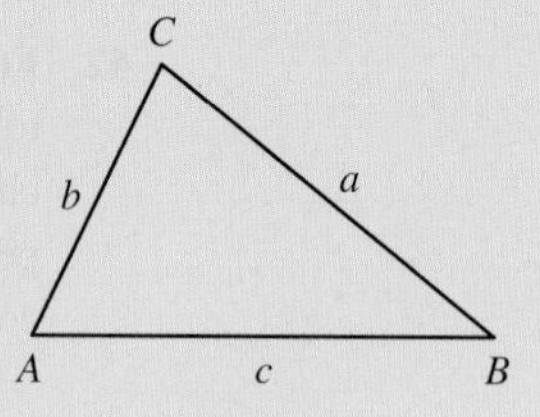

Figure 1

Derivation

To derive the formulas stated in the law of cosines, we apply the Pythagorean Theorem and some of our basic trigonometric identities. Applying the Pythagorean Theorem to right triangle BCD in Figure 2, we have

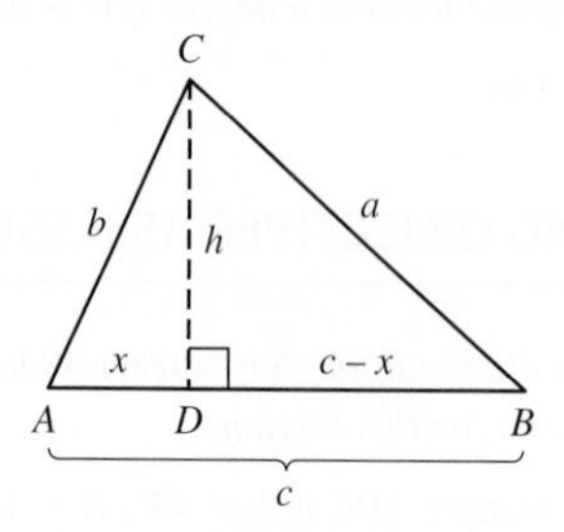

Figure 2

$$\begin{aligned} a^2 &= (c - x)^2 + h^2 \\ &= c^2 - 2cx + x^2 + h^2 \end{aligned}$$

But from right triangle ACD, we have $x^2 + h^2 = b^2$, so

$$\begin{aligned} a^2 &= c^2 - 2cx + b^2 \\ &= b^2 + c^2 - 2cx \end{aligned}$$

Now, since $\cos A = x/b$, we have $x = b \cos A$, or

$$a^2 = b^2 + c^2 - 2bc \cos A$$

Applying the same sequence of substitutions and reasoning to the right triangles formed by the altitudes from vertices A and B will give us the other two formulas listed in the law of cosines.

We can use the law of cosines to solve triangles for which we are given two sides and the angle included between them (SAS) or triangles for which we are given all three sides (SSS).

Two Sides and the Included Angle

PROBLEM 1
Find the missing parts of triangle ABC if $A = 48°$, $b = 15$ inches, and $c = 25$ inches.

EXAMPLE 1 Find the missing parts of triangle ABC if $A = 60°$, $b = 25$ inches, and $c = 32$ inches.

SOLUTION A diagram of the given information is shown in Figure 3.

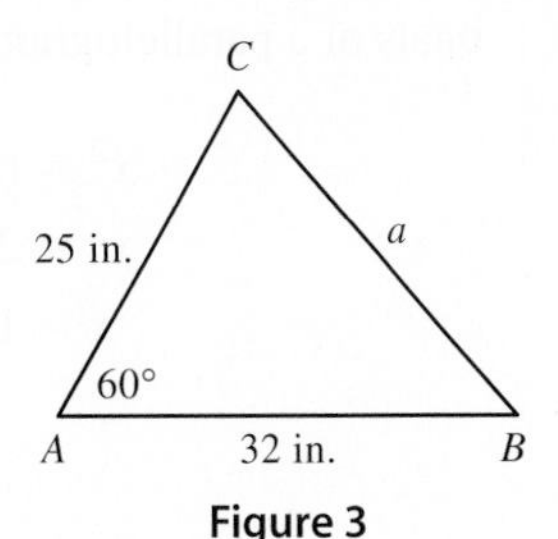

Figure 3

The solution process will include the use of both the law of cosines and the law of sines. We begin by using the law of cosines to find a.

Side a

$$a^2 = b^2 + c^2 - 2bc \cos A \qquad \text{Law of cosines}$$
$$= 25^2 + 32^2 - 2(25)(32) \cos 60° \qquad \text{Substitute in given values}$$
$$= 625 + 1{,}024 - 1{,}600(0.5) \qquad \text{Calculator}$$
$$a^2 = 849$$
$$a = 29 \text{ inches} \qquad \text{To two significant digits}$$

Now that we have a, we can use either the law of sines or the law of cosines to solve for angle B or C.

When we have a choice of angles to solve for, and we are using the law of sines to do so, it is best to solve for the smaller angle. This is because a triangle can have at most one obtuse angle, which, if present, must be opposite the longest side. Since c is the longest side, only angle C might be obtuse. Therefore we solve for angle B first because we know it must be acute. If using the law of cosines, it does not matter which angle we solve for first.

Angle B

Using the law of sines

$$\sin B = \frac{b \sin A}{a}$$
$$= \frac{25 \sin 60°}{29}$$
$$= 0.7466$$

So $B = \sin^{-1}(0.7466)$

$= 48°$ To the nearest degree

Using the law of cosines

$$b^2 = a^2 + c^2 - 2ac \cos B$$
$$25^2 = 29^2 + 32^2 - 2(29)(32) \cos B$$
$$625 = 1{,}865 - 1{,}856 \cos B$$
$$-1{,}240 = -1{,}856 \cos B$$
$$0.6681 = \cos B$$

So $B = \cos^{-1}(0.6681)$

$= 48°$ To the nearest degree

NOTE Because B must be acute, we do not have to check $B' = 180° - 48° = 132°$ when using the law of sines.

Angle C

$$C = 180° - (A + B)$$
$$= 180° - (60° + 48°)$$
$$= 72°$$

■

PROBLEM 2
Repeat Example 2 if the diagonals are 28.2 cm and 36.4 cm, and they intersect at an angle of 60.4°.

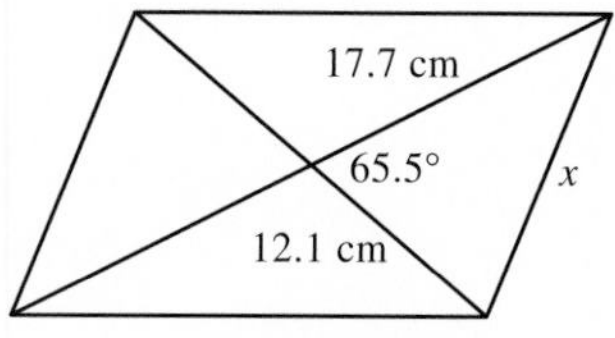

Figure 4

EXAMPLE 2 The diagonals of a parallelogram are 24.2 centimeters and 35.4 centimeters and intersect at an angle of 65.5°. Find the length of the shorter side of the parallelogram.

SOLUTION A diagram of the parallelogram is shown in Figure 4. We used the variable x to represent the length of the shorter side. Note also that we labeled half of each diagonal with its length to give us the sides of a triangle. (Recall that the diagonals of a parallelogram bisect each other.)

$$x^2 = (12.1)^2 + (17.7)^2 - 2(12.1)(17.7)\cos 65.5°$$
$$x^2 = 282.07$$
$$x = 16.8 \text{ cm} \qquad \text{To three significant digits}$$

Three Sides

To use the law of cosines to solve a triangle for which we are given all three sides, it is convenient to rewrite the equations with the cosines isolated on one side.

$$a^2 = b^2 + c^2 - 2bc\cos A$$
$$a^2 + 2bc\cos A = b^2 + c^2 \qquad \text{Add } 2bc\cos A \text{ to both sides}$$
$$2bc\cos A = b^2 + c^2 - a^2 \qquad \text{Add } -a^2 \text{ to both sides}$$
$$\cos A = \frac{b^2 + c^2 - a^2}{2bc} \qquad \text{Divide both sides by } 2bc$$

Here is an equivalent form of the law of cosines. The first formula is the one we just derived.

LAW OF COSINES (SSS)

Given triangle ABC shown in Figure 5,

$$\cos A = \frac{b^2 + c^2 - a^2}{2bc}$$
$$\cos B = \frac{a^2 + c^2 - b^2}{2ac}$$
$$\cos C = \frac{a^2 + b^2 - c^2}{2ab}$$

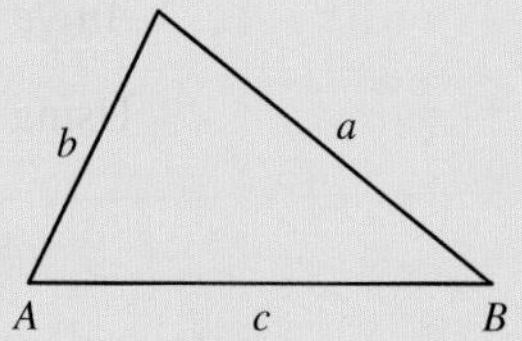

Figure 5

PROBLEM 3
Solve triangle ABC if $a = 26$ km, $b = 19$ km, and $c = 15$ km.

EXAMPLE 3 Solve triangle ABC if $a = 34$ km, $b = 20$ km, and $c = 18$ km.

SOLUTION We will use the law of cosines to solve for one of the angles and then use the law of sines to find one of the remaining angles. Since there is never any confusion as to whether an angle is acute or obtuse if we have its cosine (the cosine of an obtuse angle is negative), it is best to solve for the largest angle first. Since the longest side is a, we solve for A first.

Angle A

$$\cos A = \frac{b^2 + c^2 - a^2}{2bc}$$
$$= \frac{20^2 + 18^2 - 34^2}{(2)(20)(18)}$$
$$= -0.6000$$
$$\text{so} \quad A = \cos^{-1}(-0.6000) = 127° \quad \text{To the nearest degree}$$

There are two ways to find angle C. We can use either the law of sines or the law of cosines.

Angle C

Using the law of sines,

$$\sin C = \frac{c \sin A}{a}$$
$$= \frac{18 \sin 127°}{34}$$
$$\sin C = 0.4228$$
$$\text{so} \quad C = \sin^{-1}(0.4228)$$
$$= 25° \quad \text{To the nearest degree}$$

Using the law of cosines,

$$\cos C = \frac{a^2 + b^2 - c^2}{2ab}$$
$$= \frac{34^2 + 20^2 - 18^2}{2(34)(20)}$$
$$\cos C = 0.9059$$
$$\text{so} \quad C = \cos^{-1}(0.9059)$$
$$= 25° \quad \text{To the nearest degree}$$

NOTE Angle C must be acute, so we do not have to check $C' = 180° - 25° = 155°$ when using the law of sines.

Angle B

$$B = 180° - (A + C) = 180° - (127° + 25°) = 28°$$

Navigation

In Section 2.4, we used bearing to give the position of one point from another and to give the direction of a moving object. Another way to specify the direction of a moving object is to use *heading.*

DEFINITION

The *heading* of an object is the angle, measured clockwise from due north, to the vector representing the *intended* path of the object.

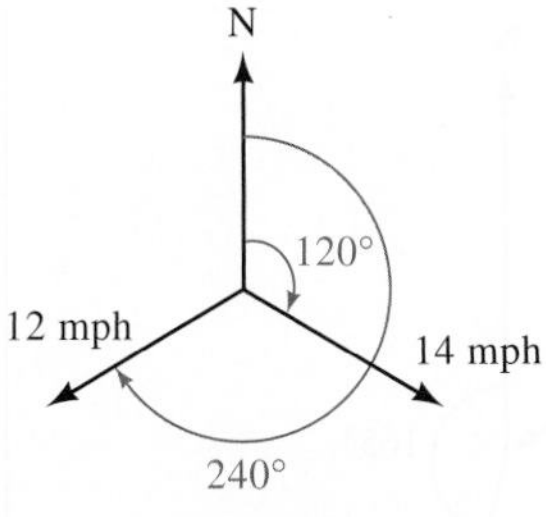

Figure 6

Figure 6 shows two vectors that represent the velocities of two ships—one traveling 14 miles per hour in the direction 120° from due north, and the other traveling 12 miles per hour at 240° from due north. Note that this method of giving the path of a moving object is a little simpler than the bearing method used in Section 2.4. To give the path of the ship traveling 14 miles per hour using the bearing method, we would say it is on a course with bearing S 60° E.

In Example 1 of Section 2.5, we found that although a boat is headed in one direction, its actual path may be in a different direction because of the current of the river. The same type of thing can happen with a plane if the air currents are in a direction different from that of the plane. With heading, the angle is measured from due north to the *intended path* of the object, meaning the path when not influenced by exterior forces such as wind or current. The true course is the angle measured from due north to the *actual path* traveled by the object relative to the ground. If no exterior force is present, then the two paths and the two angles are the same.

DEFINITION

The *true course* of an object is the angle, measured clockwise from due north, to the vector representing the *actual path* of the object.

With the specific case of objects in flight, we use the term *airspeed* for the speed of the object relative to the air and *ground speed* for the speed of the object relative to the ground. The airspeed is the magnitude of the vector representing the velocity of the object in the direction of the heading, and the ground speed is the magnitude of the vector representing the velocity of the object in the direction of the true course. Likewise, the *wind speed* is the magnitude of the vector representing the wind. As illustrated in Figure 7, the ground speed/true course vector is the resultant vector found by adding the airspeed/heading vector and the wind speed/wind direction vector.

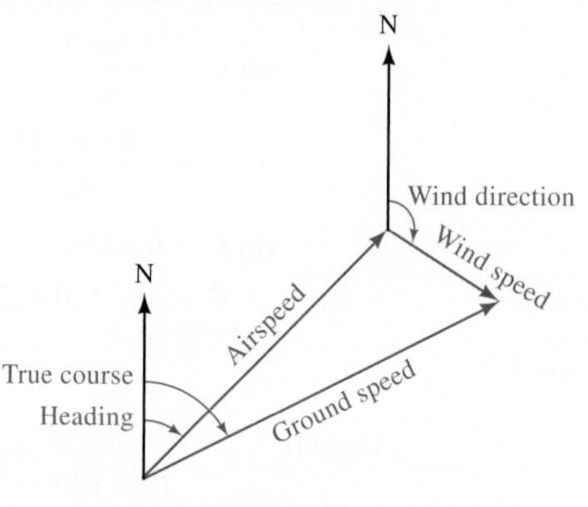

Figure 7

PROBLEM 4
A plane is flying with an airspeed of 205 miles per hour with a heading of 105°. The wind currents are running at a constant 35 miles per hour at 155° clockwise from due north. Find the true course and the ground speed of the plane.

EXAMPLE 4 A plane is flying with an airspeed of 185 miles per hour with heading 120°. The wind currents are running at a constant 32 miles per hour at 165° clockwise from due north. Find the true course and ground speed of the plane.

SOLUTION Figure 8 is a diagram of the situation with the vector **V** representing the airspeed and direction of the plane and **W** representing the speed and direction of the wind currents.

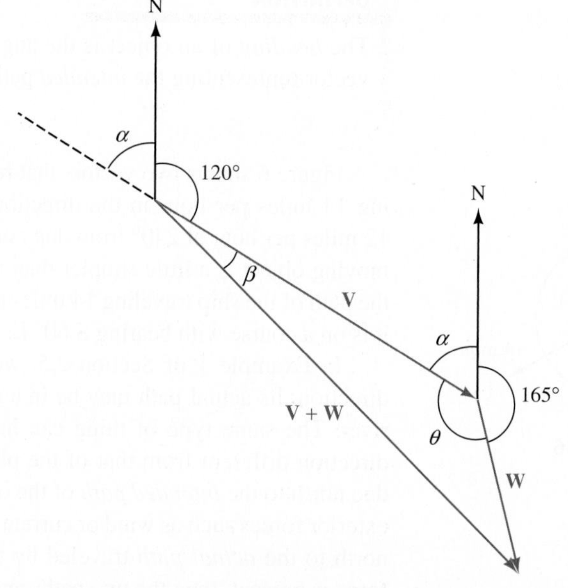

Figure 8

From Figure 8, we see that

$$\alpha = 180° - 120° = 60°$$

and

$$\theta = 360° - (\alpha + 165°) = 135°$$

We now have the case SAS. The magnitude of **V** + **W** can be found from the law of cosines.

$$\begin{aligned} |\mathbf{V} + \mathbf{W}|^2 &= |\mathbf{V}|^2 + |\mathbf{W}|^2 - 2|\mathbf{V}||\mathbf{W}| \cos \theta \\ &= 185^2 + 32^2 - 2(185)(32) \cos 135° \\ &= 43{,}621 \end{aligned}$$

so $|\mathbf{V} + \mathbf{W}| = 210$ mph   To two significant digits

To find the direction of **V** + **W**, we first find β using the law of sines.

$$\frac{\sin \beta}{32} = \frac{\sin \theta}{210}$$

$$\sin \beta = \frac{32 \sin 135°}{210}$$

$$= 0.1077$$

so $\beta = \sin^{-1}(0.1077) = 6°$ To the nearest degree

The true course is $120° + \beta = 120° + 6° = 126°$. The speed of the plane with respect to the ground is 210 mph. ■

Getting Ready for Class

After reading through the preceding section, respond in your own words and in complete sentences.

a. State the law of cosines.
b. What information must we be given in order to use the law of cosines to solve triangles?
c. State the form of the law of cosines used to solve a triangle for which all three sides are given (SSS).
d. Explain the difference between airspeed and ground speed for an object.

7.2 PROBLEM SET

➤ CONCEPTS AND VOCABULARY

For Questions 1 through 8, fill in the blank with an appropriate word.

1. We use the law of cosines to find the missing parts of triangles for which we are given two _______ and the ______ between them, or for which we are given all three _______. In either case, a _________ triangle is determined.

2. An oblique triangle can have at most _____ obtuse angle, which must be opposite the _________ side.
3. To solve an oblique triangle given two sides and the included angle, the first step is to find the missing ______ using the law of __________. Then use the law of ______ to find the __________ of the two remaining angles.
4. Given all three sides, first use the law of _______ to find the ________ angle.
5. The *heading* of an object is the angle, measured _____________ from due _________, to the vector representing the___________ path of the object.
6. The difference between heading and true course is that true course is measured to the vector representing the__________ path of the object.
7. Airspeed is the speed of an object relative to the _____, and is the magnitude of the vector representing the velocity of the object in the direction of the ___________.
8. Ground speed is the speed of an object relative to the ________, and is the magnitude of the vector representing the velocity of the object in the direction of the _____ _______.
9. Complete the formula for the law of cosines: $a^2 =$ _________
10. Complete the formula for the law of cosines: $\cos A =$ _________

➤ EXERCISES

Each of the following problems refers to triangle ABC.

11. If $a = 120$ inches, $b = 66$ inches, and $C = 60°$, find c.
12. If $a = 120$ inches, $b = 66$ inches, and $C = 120°$, find c.
13. If $a = 13$ yd, $b = 14$ yd, and $c = 15$ yd, find the largest angle.
14. If $a = 22$ yd, $b = 24$ yd, and $c = 26$ yd, find the largest angle.
15. If $b = 4.2$ m, $c = 6.8$ m, and $A = 116°$, find a.
16. If $a = 3.7$ m, $c = 6.4$ m, and $B = 23°$, find b.
17. If $a = 38$ cm, $b = 10$ cm, and $c = 31$ cm, find the largest angle.
18. If $a = 51$ cm, $b = 24$ cm, and $c = 31$ cm, find the largest angle.

Solve each of the following triangles.

19. $a = 412$ m, $c = 342$ m, $B = 151.5°$
20. $a = 76.3$ m, $c = 42.8$ m, $B = 16.3°$
21. $a = 48$ yd, $b = 75$ yd, $c = 63$ yd
22. $a = 0.48$ yd, $b = 0.63$ yd, $c = 0.75$ yd
23. $b = 0.923$ km, $c = 0.387$ km, $A = 43° 20'$
24. $b = 63.4$ km, $c = 75.2$ km, $A = 124° 40'$
25. $a = 4.38$ ft, $b = 3.79$ ft, $c = 5.22$ ft
26. $a = 832$ ft, $b = 623$ ft, $c = 345$ ft
27. Use the law of cosines to show that, if $A = 90°$, then $a^2 = b^2 + c^2$.
28. Use the law of cosines to show that, if $a^2 = b^2 + c^2$, then $A = 90°$.
29. **Geometry** The diagonals of a parallelogram are 14 meters and 16 meters and intersect at an angle of 60°. Find the length of the longer side.

30. Geometry The diagonals of a parallelogram are 56 inches and 34 inches and intersect at an angle of 120°. Find the length of the shorter side.

31. Distance Between Two Planes Two planes leave an airport at the same time. Their speeds are 130 miles per hour and 150 miles per hour, and the angle between their courses is 36°. How far apart are they after 1.5 hours?

32. Distance Between Two Ships Two ships leave a harbor entrance at the same time. The first ship is traveling at a constant 18 miles per hour, while the second is traveling at a constant 22 miles per hour. If the angle between their courses is 123°, how far apart are they after 2 hours?

Draw vectors representing the course of a ship that travels

33. 75 miles on a course with heading 330°

34. 75 miles on a course with heading 30°

35. 25 miles on a course with heading 225°

36. 25 miles on a course with heading 135°

For Problems 37 through 42, use your knowledge of bearing, heading, and true course to sketch a diagram that will help you solve each problem.

37. Heading and Distance Two planes take off at the same time from an airport. The first plane is flying at 246 miles per hour on a course of 135.0°. The second plane is flying in the direction 175.0° at 357 miles per hour. Assuming there are no wind currents blowing, how far apart are they after 2 hours?

38. Bearing and Distance Two ships leave the harbor at the same time. One ship is traveling at 14 miles per hour on a course with a bearing of S 13° W, while the other is traveling at 12 miles per hour on a course with a bearing of N 75° E. How far apart are they after 3 hours?

39. True Course and Speed A plane is flying with an airspeed of 160 miles per hour and heading of 150°. The wind currents are running at 35 miles per hour at 165° clockwise from due north. Use vectors to find the true course and ground speed of the plane.

40. True Course and Speed A plane is flying with an airspeed of 244 miles per hour with heading 272.7°. The wind currents are running at a constant 45.7 miles per hour in the direction 262.6°. Find the ground speed and true course of the plane.

41. Speed and Direction A plane has an airspeed of 195 miles per hour and a heading of 30.0°. The ground speed of the plane is 207 miles per hour, and its true course is in the direction of 34.0°. Find the speed and direction of the air currents, assuming they are constants.

42. Speed and Direction The airspeed and heading of a plane are 140 miles per hour and 130°, respectively. If the ground speed of the plane is 135 miles per hour and its true course is 137°, find the speed and direction of the wind currents, assuming they are constants.

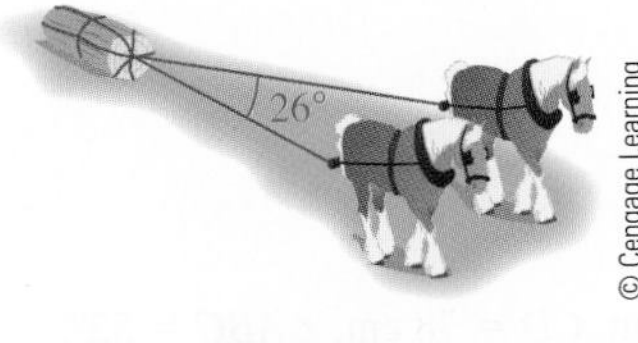

Figure 9

43. Resultant Force A heavy log is dragged across the ground by two horses pulling on ropes (Figure 9). The magnitudes of the tension forces in the direction of the ropes are 58 pounds and 73 pounds. If the angle between the ropes is 26°, find the magnitude of the resultant force.

44. Resultant Force Two trucks are trying to pull an auto out of the mud using chains. The magnitudes of the tension forces in the direction of the chains are 556 pounds and 832 pounds. If the angle between the chains is 38.5°, find the magnitude of the resultant force.

45. Towing a Barge A barge is pulled by two tugboats. The first tugboat is traveling at a speed of 15 knots with heading 130°, and the second tugboat is traveling at a speed of 16 knots with heading 190°. Find the resulting speed and direction of the barge.

46. Pulling a Crate A large crate is pulled across the ice with two ropes. A force of 47 pounds is applied to the first rope in the direction 80°, and a force of 55 pounds is applied to the second rope in the direction 105°. What are the magnitude and direction of the resultant force acting on the crate?

Problems 47 and 48 refer to Figure 10, which is a diagram of the Colnago Dream Plus bike frame. The frame can be approximated as two triangles that have the seat tube as a common side.

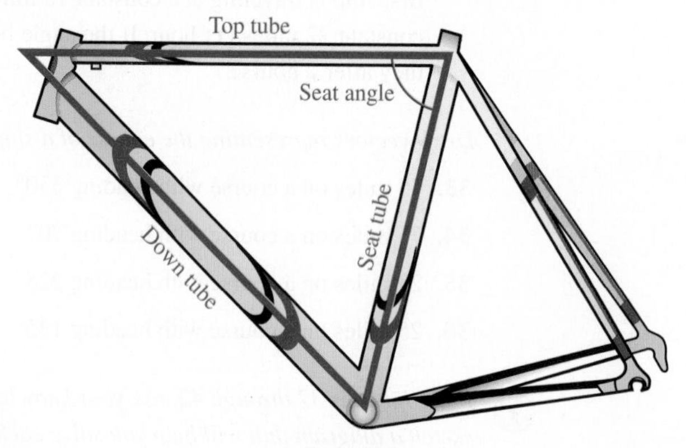

Figure 10

47. Bike Frame Geometry The 50-centimeter Colnago frame has a top tube length of 52.3 centimeters, a seat tube length of 48.0 centimeters, and a seat angle of 75.0°. Find the length of the down tube and the angle between the seat tube and the down tube.

48. Bike Frame Geometry The 59-centimeter Colnago frame has a top tube length of 56.9 centimeters, a seat tube length of 57.0 centimeters, and a seat angle of 73.0°. Find the length of the down tube and the angle between the seat tube and the down tube.

Problems 49 and 50 refer to Figure 11, which shows a diagram for a bike frame.

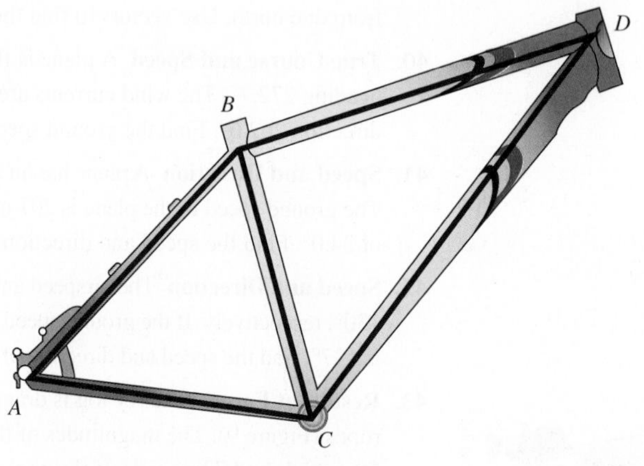

Figure 11

49. Bike Frame Geometry Given $BC = 51$ cm, $BD = 61$ cm, $CD = 78$ cm, $\angle ABC = 52°$, and $\angle ACB = 65°$, find the following.

a. The length of the chainstay, AC **b.** $\angle BCD$

50. Bike Frame Geometry Given $BC = 49$ cm, $BD = 59$ cm, $\angle BAC = 53°$, $\angle ACB = 69°$, and $\angle CBD = 88°$, find the following.

a. The length of the chainstay, AC **b.** $\angle BDC$

51. **Surveying** In performing a geodetic survey, the information shown in Figure 12 is measured. Use this information to find the distance between Cambria and Cayucos.

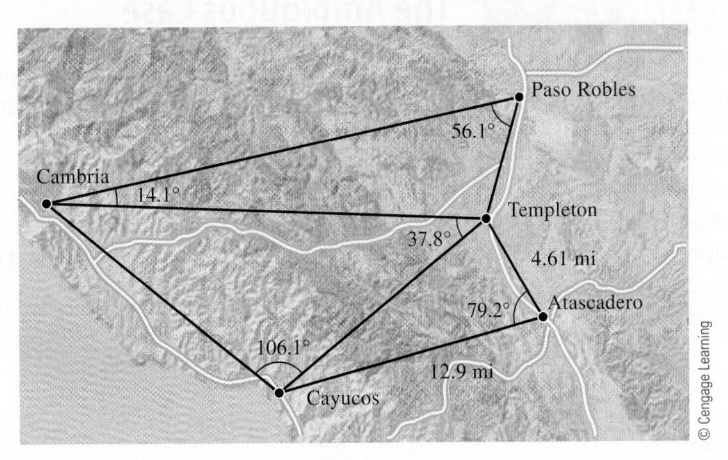

Figure 12

52. **Surveying** Use the information shown in Figure 12 to find the distance between Cambria and Paso Robles.

➤ REVIEW PROBLEMS

The problems that follow review material we covered in Section 6.3.

Find all solutions in radians using exact values only.

53. $\sin 3x = 1/2$

54. $\cos 4x = 0$

55. $\tan^2 3x = 1$

56. $\tan^2 4x = 1$

Find all degree solutions.

57. $2\cos^2 3\theta - 9\cos 3\theta + 4 = 0$

58. $3\sin^2 2\theta - 2\sin 2\theta - 5 = 0$

59. $\sin 4\theta \cos 2\theta + \cos 4\theta \sin 2\theta = -1$

60. $\cos 3\theta \cos 2\theta - \sin 3\theta \sin 2\theta = -1$

Solve each equation for θ if $0° \le \theta < 360°$.

61. $\sin\theta + \cos\theta = 1$

62. $\sin\theta - \cos\theta = 0$

➤ LEARNING OBJECTIVES ASSESSMENT

These questions are available for instructors to help assess if you have successfully met the learning objectives for this section.

63. Find c for triangle ABC if $a = 6.8$ meters, $b = 8.4$ meters, and $C = 48°$.

 a. 5.6 m b. 40 m c. 8.6 m d. 6.4 m

64. Find B for triangle ABC if $a = 13.8$ yards, $b = 22.3$ yards, and $c = 9.50$ yards.

 a. 34.3° b. 145.7° c. 13.8° d. 160°

65. What is the heading of vector $\overrightarrow{AB}$ in Figure 13?

 a. S 45° W b. W 45° S c. 225° d. 135°

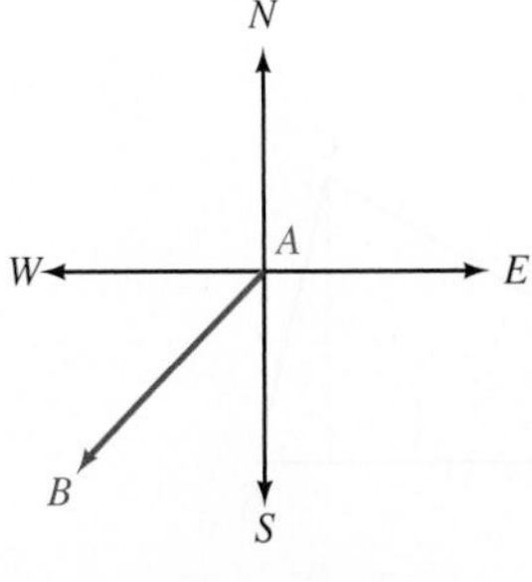

Figure 13

66. A plane with an airspeed of 560 miles per hour and traveling at a heading of 130° encounters a 65 mile per hour wind blowing in the direction N 45° E. Find the resulting ground speed of the plane.

 a. 564 mph b. 569 mph c. 553 mph d. 557 mph

SECTION 7.3 The Ambiguous Case

LEARNING OBJECTIVES

1. Use the law of sines to find all possible solutions for the ambiguous case.
2. Use the law of cosines to find all possible solutions for the ambiguous case.
3. Solve applied problems involving the ambiguous case.

In this section, we will apply the law of sines or the law of cosines to solve triangles for which we are given two sides and the angle opposite one of the given sides.

Using the Law of Sines

EXAMPLE 1 Find angle B in triangle ABC if $a = 2$, $b = 6$, and $A = 30°$.

SOLUTION Applying the law of sines we have

$$\frac{\sin B}{b} = \frac{\sin A}{a}$$

$$\sin B = \frac{b \sin A}{a}$$

$$= \frac{6 \sin 30°}{2}$$

$$= 1.5$$

Figure 1

PROBLEM 1
In triangle ABC, $A = 35°$, $a = 4.0$, and $b = 12$. Find angle B.

Because $\sin B$ can never be larger than 1, no triangle exists for which $a = 2$, $b = 6$, and $A = 30°$. (You may recall from geometry that there was no congruence theorem SSA.) Figure 1 illustrates what went wrong here. No matter how we orient side a, we cannot complete a triangle.

When we are given two sides and an angle opposite one of them (SSA), we have several possibilities for the triangle or triangles that result. As was the case in Example 1, one of the possibilities is that no triangle will fit the given information. If side a in Example 1 had been equal in length to the altitude drawn from vertex C, then we would have had a right triangle that fit the given information, as shown in Figure 2. If side a had been longer than this altitude but shorter than side b, we would have had two triangles that fit the given information, triangle $AB'C$ and triangle ABC as illustrated in Figure 3 and Figure 4. Notice that the two possibilities for side a in Figure 3 create an isosceles triangle $B'CB$.

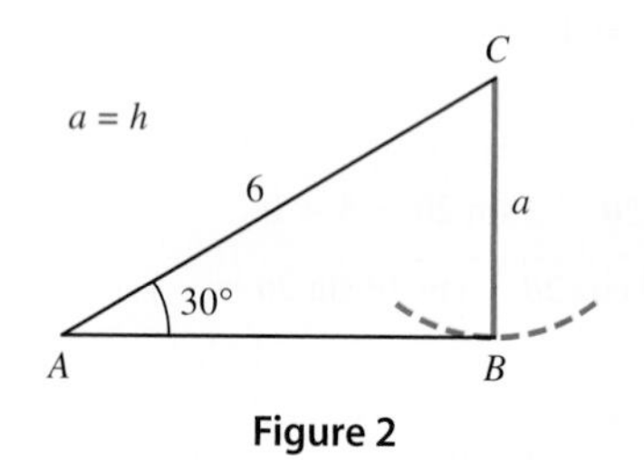

Figure 2

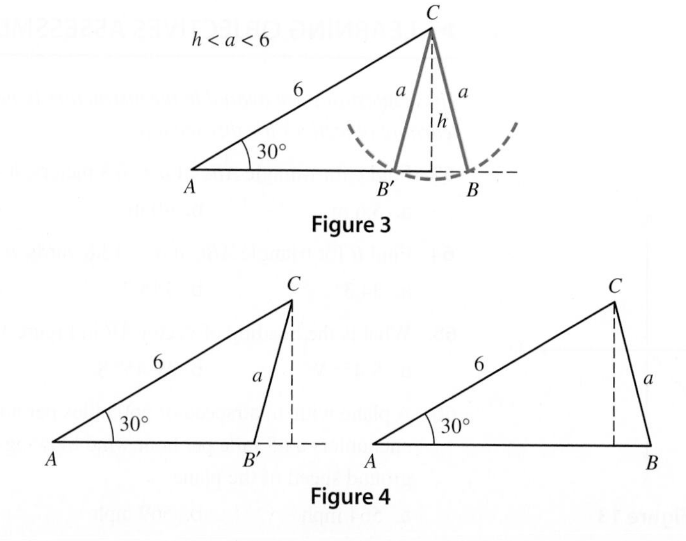

Figure 3

Figure 4

If side a in triangle ABC of Example 1 had been longer than side b, we would have had only one triangle that fit the given information, as shown in Figure 5.

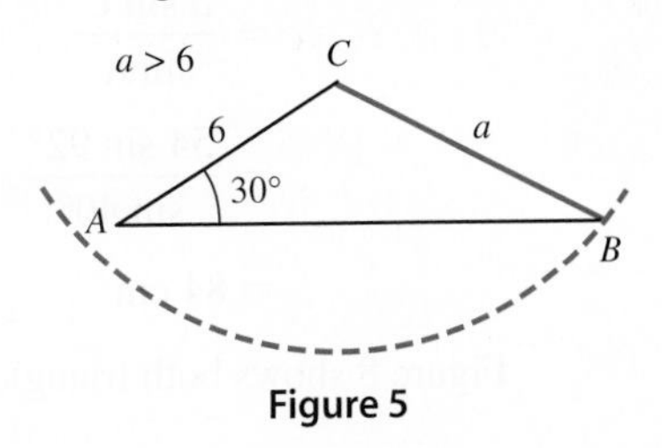

Figure 5

Because of the different possibilities that arise in solving a triangle for which we are given two sides and an angle opposite one of the given sides, we call this situation the *ambiguous case*.

PROBLEM 2
Find the missing parts of triangle ABC if $A = 42°$, $a = 52$ cm, and $b = 64$ cm.

EXAMPLE 2 Find the missing parts in triangle ABC if $a = 54$ cm, $b = 62$ cm, and $A = 40°$.

SOLUTION First we solve for $\sin B$ with the law of sines.

Angle B

$$\frac{\sin B}{b} = \frac{\sin A}{a}$$

$$\sin B = \frac{b \sin A}{a}$$

$$= \frac{62 \sin 40°}{54}$$

$$= 0.7380$$

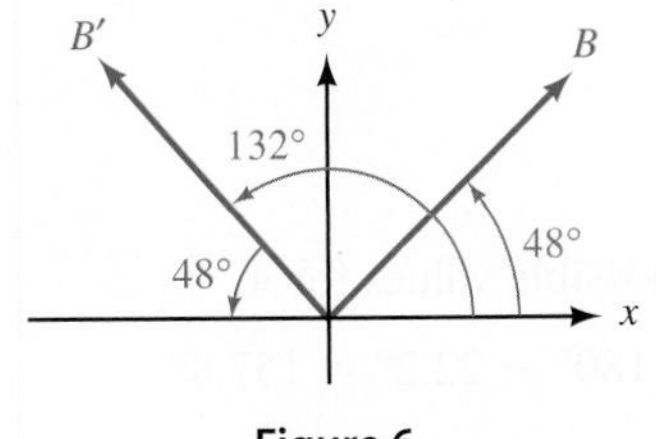

Figure 6

Now, because $\sin B$ is positive for any angle in QI or QII, we have two possibilities (Figure 6). We will call one of them B and the other B'.

$$B = \sin^{-1}(0.7380) = 48° \qquad \text{or} \qquad B' = 180° - 48° = 132°$$

Notice that $B' = 132°$ is the supplement of B. We have two different angles that can be found with $a = 54$ cm, $b = 62$ cm, and $A = 40°$. Figure 7 shows both of them. One is labeled ABC, while the other is labeled $AB'C$. Also, because triangle $B'BC$ is isosceles, we can see that angle $AB'C$ must be supplementary to B.

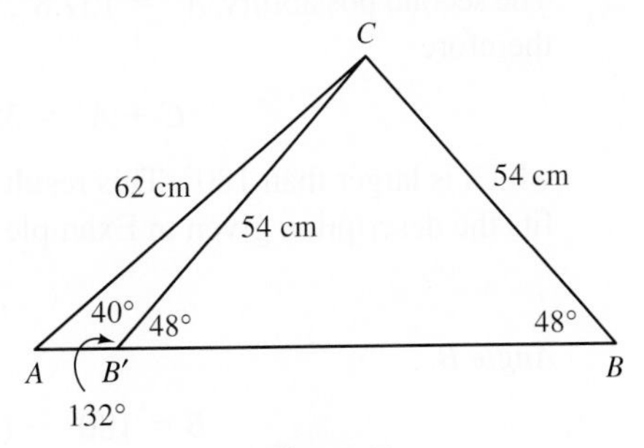

Figure 7

Angles C and C′
There are two values for B, so we have two values for C.

$$C = 180 - (A + B) \qquad \text{and} \qquad C' = 180 - (A + B')$$

$$= 180 - (40° + 48°) \qquad\qquad = 180 - (40° + 132°)$$

$$= 92° \qquad\qquad = 8°$$

Sides c and c′

$$c = \frac{a \sin C}{\sin A} \qquad \text{and} \qquad c' = \frac{a \sin C'}{\sin A}$$

$$= \frac{54 \sin 92°}{\sin 40°} \qquad\qquad = \frac{54 \sin 8°}{\sin 40°}$$

$$= 84 \text{ cm} \qquad\qquad = 12 \text{ cm}$$ To two significant digits

Figure 8 shows both triangles.

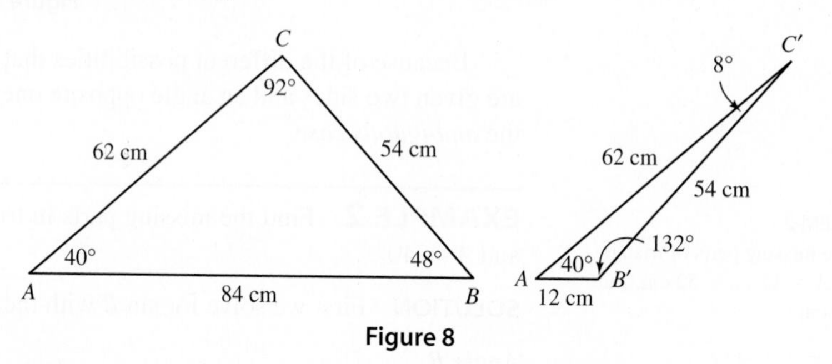

Figure 8

PROBLEM 3
Find the missing parts of triangle ABC if $C = 37.2°$, $a = 105$ ft, and $c = 215$ ft.

EXAMPLE 3 Find the missing parts of triangle ABC if $C = 35.4°$, $a = 205$ ft, and $c = 314$ ft.

SOLUTION Applying the law of sines, we find sin A.

Angle A

$$\sin A = \frac{a \sin C}{c}$$

$$= \frac{205 \sin 35.4°}{314}$$

$$= 0.3782$$

Because sin A is positive in QI and QII, we have two possible values for A.

$$A = \sin^{-1}(0.3782) = 22.2° \qquad \text{and} \qquad A' = 180° - 22.2° = 157.8°$$

The second possibility, $A' = 157.8°$, will not work, however, because C is 35.4° and therefore

$$C + A' = 35.4° + 157.8° = 193.2°$$

which is larger than 180°. This result indicates that there is exactly one triangle that fits the description given in Example 3. In that triangle

$$A = 22.2°$$

Angle B

$$B = 180° - (35.4° + 22.2°) = 122.4°$$

Side b

$$b = \frac{c \sin B}{\sin C}$$

$$b = \frac{314 \sin 122.4°}{\sin 35.4°}$$

$$= 458 \text{ ft}$$ To three significant digits

Figure 9 is a diagram of this triangle.

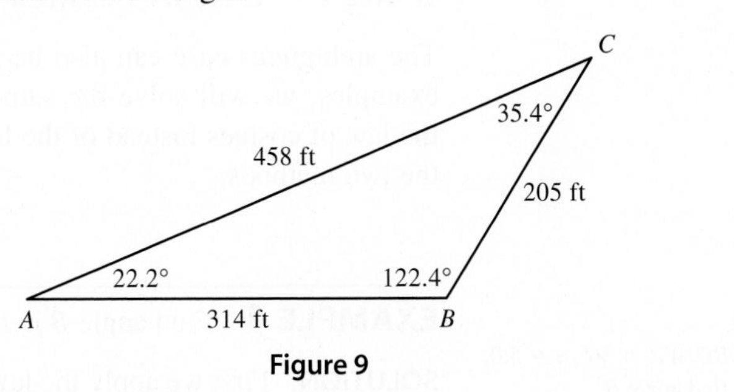

Figure 9

The different cases that can occur when we solve the kinds of triangles given in Examples 1, 2, and 3 become apparent in the process of solving for the missing parts. The following table completes the set of conditions under which we will have 1, 2, or no triangles in the ambiguous case.

In Table 1, we are assuming that we are given angle A and sides a and b in triangle ABC, and that h is the altitude from vertex C.

TABLE 1

Conditions	Number of Triangles	Diagram
$A < 90°$ and $a < h$	0	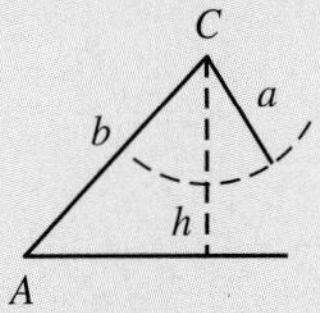
$A > 90°$ and $a < b$	0	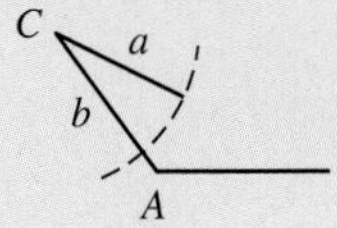
$A < 90°$ and $a = h$	1	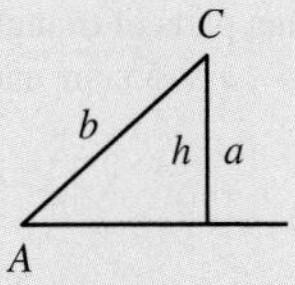
$A < 90°$ and $a \geq b$	1	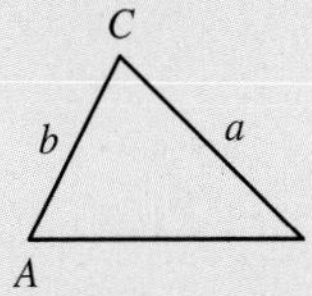
$A > 90°$ and $a > b$	1	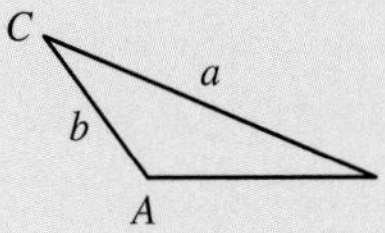
$A < 90°$ and $h < a < b$	2	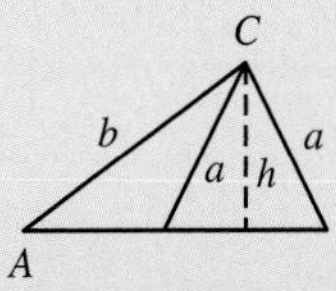

Using the Law of Cosines

The ambiguous case can also be solved using the law of cosines. In the next three examples, we will solve the same triangles from Examples 1, 2, and 3, but using the law of cosines instead of the law of sines. This way you will be able to compare the two methods.

PROBLEM 4
In triangle ABC, if $A = 36°$, $a = 5.0$, and $b = 14$, find angle B.

EXAMPLE 4 Find angle B in triangle ABC if $a = 2$, $b = 6$, and $A = 30°$.

SOLUTION First we apply the law of cosines to find the missing side.

$$\begin{aligned} a^2 &= b^2 + c^2 - 2bc \cos A \\ 2^2 &= 6^2 + c^2 - 2 \cdot 6 \cdot c \cos 30° \\ 4 &= 36 + c^2 - 12c\left(\frac{\sqrt{3}}{2}\right) \\ 0 &= c^2 - 6\sqrt{3}c + 32 \end{aligned}$$

Using the quadratic formula to solve this quadratic equation, we obtain

$$\begin{aligned} c &= \frac{-(-6\sqrt{3}) \pm \sqrt{(-6\sqrt{3})^2 - 4(1)(32)}}{2(1)} \\ &= \frac{6\sqrt{3} \pm \sqrt{-20}}{2} \end{aligned}$$

Because of the negative number in the square root, c is not a real number. Therefore no triangle is possible. ■

PROBLEM 5
Find the missing parts of triangle ABC if $A = 44°$, $a = 55$ cm, and $b = 66$ cm.

EXAMPLE 5 Find the missing parts in triangle ABC if $a = 54$ cm, $b = 62$ cm, and $A = 40°$.

SOLUTION Once again, we begin by using the law of cosines to find side c.

Side c

$$\begin{aligned} a^2 &= b^2 + c^2 - 2bc \cos A \\ 54^2 &= 62^2 + c^2 - 2 \cdot 62 \cdot c \cos 40° \\ 2{,}916 &= 3{,}844 + c^2 - 124c(0.7660) \qquad \text{Approximate } \cos 40° \\ 0 &= c^2 - 94.984c + 928 \end{aligned}$$

Using the quadratic formula to solve for c, we obtain

$$\begin{aligned} c &= \frac{-(-94.984) \pm \sqrt{(-94.984)^2 - 4(1)(928)}}{2(1)} \\ &= \frac{94.984 \pm 72.8695}{2} \end{aligned}$$

which gives us $c = 11$ or $c = 84$, to two significant digits. Since both of these values are positive real numbers, there are two triangles possible. We can use the law of cosines again to find either of the remaining two angles.

Angles C and C′

$$c = 11 \qquad\qquad c' = 84$$

$$\cos C = \frac{a^2 + b^2 - c^2}{2ab} \quad \text{and} \quad \cos C' = \frac{a^2 + b^2 - c^2}{2ab}$$

$$= \frac{54^2 + 62^2 - 11^2}{2(54)(62)} \qquad\qquad = \frac{54^2 + 62^2 - 84^2}{2(54)(62)}$$

$$= 0.9915 \qquad\qquad = -0.0442$$

$$\text{So } C = \cos^{-1}(0.9915) \qquad\qquad \text{So } C' = \cos^{-1}(-0.0442)$$

$$= 7° \qquad\qquad = 93°$$

NOTE Notice that, because of rounding, we obtain slightly different values compared to those in Example 2.

Angles B and B′

$$B = 180° - 40° - 7° \quad \text{and} \quad B' = 180° - 40° - 93°$$

$$= 133° \qquad\qquad = 47°$$

PROBLEM 6
Find the missing parts of triangle ABC if $C = 32.7°$, $a = 115$ ft, and $c = 205$ ft.

EXAMPLE 6 Find the missing parts of triangle ABC if $C = 35.4°$, $a = 205$ ft, and $c = 314$ ft.

SOLUTION First we use the law of cosines to find side b.

Side b

$$c^2 = a^2 + b^2 - 2ab\cos C$$

$$314^2 = 205^2 + b^2 - 2 \cdot 205 \cdot b\cos 35.4°$$

$$98{,}596 = 42{,}025 + b^2 - 410b(0.8151) \qquad \text{Approximate } \cos 35.4°$$

$$0 = b^2 - 334.191b - 56{,}571$$

Using the quadratic formula to solve for b we obtain

$$b = \frac{-(-334.191) \pm \sqrt{(-334.191)^2 - 4(1)(-56{,}571)}}{2(1)}$$

$$= \frac{334.191 \pm 581.3498}{2}$$

which gives us $b = 458$ or $b = -123$, to three significant digits. The side of a triangle cannot be a negative number, so we discard the negative value of b. This means only one triangle is possible. We can use the law of cosines again to find either of the remaining two angles.

Angle B

$$\cos B = \frac{a^2 + c^2 - b^2}{2ac}$$

$$= \frac{205^2 + 314^2 - 458^2}{2(205)(314)}$$

$$= -0.5371$$

$$\text{So } B = \cos^{-1}(-0.5371)$$

$$= 122.5°$$

NOTE Again, we obtain slightly different values for angles A and B compared to those in Example 3 because of rounding.

Angle A

$$A = 180° - 35.4° - 122.5° = 22.1°$$

Applications

As you work the problems in this section, try experimenting with both methods of solving the ambiguous case. We end this section by solving a navigation problem by using the law of sines.

PROBLEM 7
A plane is flying with an airspeed of 165 miles per hour and a heading of 42.3°. Its true course is 52.5° from due north. If the wind currents are a constant 55.0 miles per hour, what are the possibilities for the ground speed of the plane?

EXAMPLE 7 A plane is flying with an airspeed of 170 miles per hour and a heading of 52.5°. Its true course, however, is at 64.1° from due north. If the wind currents are a constant 40.0 miles per hour, what are the possibilities for the ground speed of the plane?

SOLUTION We represent the velocity of the plane and the velocity of the wind with vectors, the resultant of which will be the ground speed and true course of the plane. First, measure an angle of 52.5° from north (clockwise) and draw a vector in that direction with magnitude 170 to represent the heading and airspeed. Then draw a second vector at an angle of 64.1° from north with unknown magnitude to represent the true course and ground speed. Because this second vector must be the resultant, the wind velocity will be some vector originating at the tip of the first vector and ending at the tip of the second vector, and having a length of 40.0. Figure 10 illustrates the situation. From Figure 10, we see that

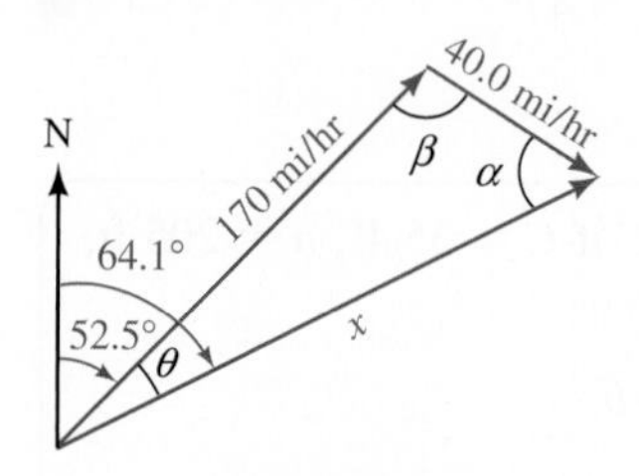

Figure 10

$$\theta = 64.1° - 52.5° = 11.6°$$

First we use the law of sines to find angle α.

$$\frac{\sin \alpha}{170} = \frac{\sin 11.6°}{40.0}$$

$$\sin \alpha = \frac{170 \sin 11.6°}{40.0}$$

$$= 0.8546$$

Because $\sin \alpha$ is positive in quadrants I and II, we have two possible values for α.

$$\alpha = \sin^{-1}(0.8546) = 58.7° \quad \text{and} \quad \alpha' = 180° - 58.7° = 121.3°$$

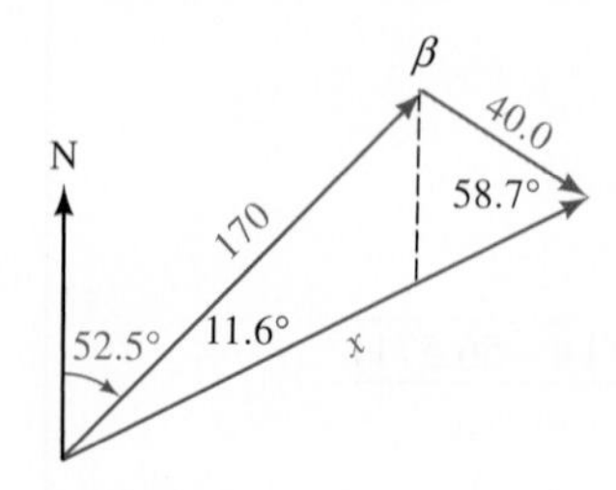

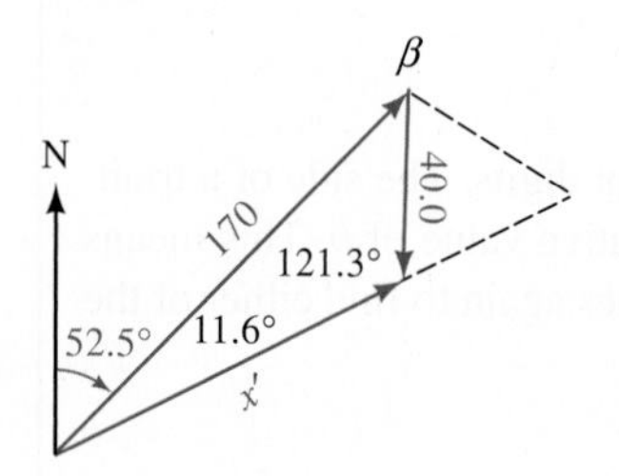

Figure 11

We see there are two different triangles that can be formed with the three vectors. Figure 11 shows both of them. There are two values for α, so there are two values for β.

$$\beta = 180° - (11.6° + 58.7°) = 109.7° \quad \text{and} \quad \beta' = 180° - (11.6° + 121.3°) = 47.1°$$

Now we can use the law of sines again to find the ground speed.

$$\frac{x}{\sin 109.7°} = \frac{40}{\sin 11.6°} \quad \text{and} \quad \frac{x'}{\sin 47.1°} = \frac{40}{\sin 11.6°}$$

$$x = \frac{40 \sin 109.7°}{\sin 11.6°} \qquad x' = \frac{40 \sin 47.1°}{\sin 11.6°}$$

$$= 187 \text{ mi/hr} \qquad = 146 \text{ mi/hr}$$

To three significant digits, the two possibilities for the ground speed of the plane are 187 miles per hour and 146 miles per hour. ■

Getting Ready for Class *After reading through the preceding section, respond in your own words and in complete sentences.*

a. What are the three possibilities for the resulting triangle or triangles when we are given two sides and an angle opposite them?

b. In triangle ABC, $A = 40°$, $B = 48°$, and $B' = 132°$. What are the measures of angles C and C'?

c. How many triangles are possible if angle A is greater than 90° and side a is less than side b?

d. How many triangles are possible if angle A is greater than 90° and side a is greater than side b?

7.3 PROBLEM SET

➤ CONCEPTS AND VOCABULARY

For Questions 1 through 4, fill in the blank with an appropriate word or symbol.

1. The *ambiguous case* refers to an oblique triangle in which we are given ____ sides and an angle __________ one of the given sides.

2. With the ambiguous case, there may be _____, _____, or _____ triangles that are determined.

3. When solving the ambiguous case using the law of sines, the first step is to solve for a missing ______. Use the inverse sine function to find the __________ angle. Then determine the possible values of the angle itself for both quadrants ____ and ____.

4. When solving the ambiguous case using the law of cosines, the first step is to solve for the missing _____. This will result in a _________ equation, which can be solved using the _______ __________.

➤ EXERCISES

For each of the following triangles, solve for B and use the results to explain why the triangle has the given number of solutions.

5. $A = 150°$, $b = 30$ ft, $a = 10$ ft; no solution

6. $A = 30°$, $b = 40$ ft, $a = 10$ ft; no solution

7. $A = 120°$, $b = 20$ cm, $a = 30$ cm; one solution

8. $A = 30°$, $b = 12$ cm, $a = 6$ cm; one solution

9. $A = 60°$, $b = 18$ m, $a = 16$ m; two solutions

10. $A = 20°$, $b = 40$ m, $a = 30$ m; two solutions

Find all solutions to each of the following triangles:

11. $A = 38°$, $a = 41$ ft, $b = 54$ ft

12. $A = 43°$, $a = 31$ ft, $b = 37$ ft

13. $A = 112.2°$, $a = 43.8$ cm, $b = 22.3$ cm

14. $A = 124.3°$, $a = 27.3$ cm, $b = 50.2$ cm

15. $C = 27° 50'$, $c = 347$ m, $b = 425$ m

16. $C = 51° 30'$, $c = 707$ m, $b = 821$ m

17. $B = 62° 40'$, $b = 6.78$ inches, $c = 3.48$ inches

18. $B = 45° 10'$, $b = 1.79$ inches, $c = 1.12$ inches

19. $B = 118°$, $b = 0.68$ cm, $a = 0.92$ cm

20. $B = 30°$, $b = 4.2$ cm, $a = 8.4$ cm

21. $A = 142°$, $b = 2.9$ yd, $a = 1.4$ yd

22. $A = 65°$, $b = 7.6$ yd, $a = 7.1$ yd

23. $C = 26.8°$, $c = 36.8$ km, $b = 36.8$ km

24. $C = 73.4°$, $c = 51.1$ km, $b = 92.4$ km

25. Distance A 51-foot wire running from the top of a tent pole to the ground makes an angle of 58° with the ground. If the length of the tent pole is 44 feet, how far is it from the bottom of the tent pole to the point where the wire is fastened to the ground? (The tent pole is not necessarily perpendicular to the ground.)

26. Distance A hot-air balloon is held at a constant altitude by two ropes that are anchored to the ground. One rope is 120 feet long and makes an angle of 65° with the ground. The other rope is 115 feet long. What is the distance between the points on the ground at which the two ropes are anchored?

27. Current A ship is headed due north at a constant 16 miles per hour. Because of the ocean current, the true course of the ship is 15°. If the currents are a constant 14 miles per hour, in what direction are the currents running?

28. Ground Speed A plane is headed due east with an airspeed of 340 miles per hour. Its true course, however, is at 98° from due north. If the wind currents are a constant 55 miles per hour, what are the possibilities for the ground speed of the plane?

29. Ground Speed A ship headed due east is moving through the water at a constant speed of 12 miles per hour. However, the true course of the ship is 60°. If the currents are a constant 6 miles per hour, what is the ground speed of the ship?

30. True Course A plane headed due east is traveling with an airspeed of 190 miles per hour. The wind currents are moving with constant speed in the direction 240°. If the ground speed of the plane is 95 miles per hour, what is its true course?

31. Leaning Windmill After a wind storm, a farmer notices that his 32-foot windmill may be leaning, but he is not sure. From a point on the ground 31 feet from the base of the windmill, he finds that the angle of elevation to the top of the windmill is 48°. Is the windmill leaning? If so, what is the acute angle the windmill makes with the ground?

32. Distance A boy is riding his motorcycle on a road that runs east and west. He leaves the road at a service station and rides 5.25 miles in the direction N 15.5° E. Then he turns to his right and rides 6.50 miles back to the road, where his motorcycle breaks down. How far will he have to walk to get back to the service station?

33. A sailboat set a course of N 25° E from a small port along a shoreline that runs north and south. Sometime later the boat overturned and the crew sent out a distress call. They estimated that they were 12 miles away from the nearest harbor, which is 28 miles north of the port they had set sail from. If a rescue team leaves from the harbor, find all possible courses the team must follow in order to reach the overturned sailboat.

34. In Problem 33, find the shortest (perpendicular) distance to shore from each possible position of the overturned sailboat. Round your answers to the nearest tenth of a mile.

➤ REVIEW PROBLEMS

The problems that follow review material we covered in Section 6.2.

Find all solutions in the interval $0° \leq \theta < 360°$. If rounding is necessary, round to the nearest tenth of a degree.

35. $4 \sin \theta - \csc \theta = 0$

36. $2 \sin \theta - 1 = \csc \theta$

37. $2 \cos \theta - \sin 2\theta = 0$

38. $\cos 2\theta + 3 \cos \theta - 2 = 0$

39. $18 \sec^2 \theta - 17 \tan \theta \sec \theta - 12 = 0$

40. $7 \sin^2 \theta - 9 \cos 2\theta = 0$

Find all radian solutions using exact values only.

41. $2 \cos x - \sec x + \tan x = 0$

42. $2 \cos^2 x - \sin x = 1$

43. $\sin x + \cos x = 0$

44. $\sin x - \cos x = 1$

➤ LEARNING OBJECTIVES ASSESSMENT

These questions are available for instructors to help assess if you have successfully met the learning objectives for this section.

45. Use the law of sines to find C for triangle ABC if $B = 35°$, $a = 28$ feet, and $b = 19$ feet.

a. 23° or 87°
b. 58°
c. 58° or 87°
d. No triangle is possible

46. Given triangle ABC with $B = 35°$, $a = 28$ feet, and $b = 19$ feet, if the law of cosines is used to solve the triangle, what quadratic equation must first be solved?

a. $c^2 - 45.87c + 423 = 0$
b. $c^2 + 45.87c - 1{,}145 = 0$
c. $c^2 - 31.13c - 423 = 0$
d. $c^2 + 31.3c + 1{,}145 = 0$

47. A plane headed due west is traveling with a constant speed of 214 miles per hour. The wind is blowing at a constant speed in the direction 50.5°. If the ground speed of the plane is 145 miles per hour, what is its true course?

a. 295.5°
b. 25.5° or 334.5°
c. 290.2°
d. 300.3° or 340.7°

SECTION 7.4 The Area of a Triangle

LEARNING OBJECTIVES

1. Calculate the area of a triangle given two sides and the included angle.
2. Calculate the area of a triangle given two angles and one side.
3. Find the semiperimeter of a triangle.
4. Use Heron's formula to find the area of a triangle.

In this section, we will derive three formulas for the area S of a triangle. We will start by deriving the formula used to find the area of a triangle for which two sides and the included angle are given.

Two Sides and the Included Angle

To derive our first formula, we begin with the general formula for the area of a triangle:

$$S = \frac{1}{2}(\text{base})(\text{height})$$

The base of triangle ABC in Figure 1 is c and the height is h. So the formula for S becomes, in this case,

$$S = \frac{1}{2}ch$$

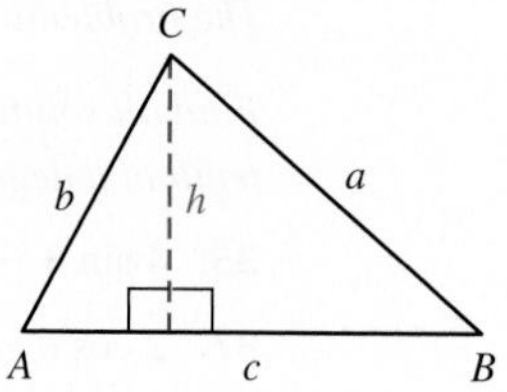

Figure 1

Suppose that, for triangle ABC, we are given the lengths of sides b and c and the measure of angle A. Then we can write $\sin A$ as

$$\sin A = \frac{h}{b}$$

or, by solving for h,

$$h = b \sin A$$

Substituting this expression for h into the formula

$$S = \frac{1}{2}ch$$

we have

$$S = \frac{1}{2}bc \sin A$$

Applying the same kind of reasoning to the heights drawn from A and B, we also have the following.

AREA OF A TRIANGLE (SAS)

$$S = \frac{1}{2}bc \sin A \qquad S = \frac{1}{2}ac \sin B \qquad S = \frac{1}{2}ab \sin C$$

Each of these three formulas indicates that to find the area of a triangle for which we are given two sides and the angle included between them, we multiply half the product of the two sides by the sine of the angle included between them.

PROBLEM 1
Find the area of triangle ABC if $A = 42.3°$, $b = 3.41$ cm, and $c = 4.72$ cm.

EXAMPLE 1 Find the area of triangle ABC if $A = 35.1°$, $b = 2.43$ cm, and $c = 3.57$ cm.

SOLUTION Applying the first formula we derived, we have

$$\begin{aligned} S &= \frac{1}{2}bc \sin A \\ &= \frac{1}{2}(2.43)(3.57) \sin 35.1° \\ &= 2.49 \text{ cm}^2 \quad \text{To three significant digits} \end{aligned}$$

Two Angles and One Side

The next area formula we will derive is used to find the area of triangles for which we are given two angles and one side.

Suppose we were given angles A and B and side a in triangle ABC in Figure 1. We could easily solve for C by subtracting the sum of A and B from 180°.

To find side b, we use the law of sines

$$\frac{b}{\sin B} = \frac{a}{\sin A}$$

Solving this equation for b would give us

$$b = \frac{a \sin B}{\sin A}$$

Substituting this expression for b into the formula

$$S = \frac{1}{2}ab \sin C$$

we have

$$S = \frac{1}{2}a\left(\frac{a \sin B}{\sin A}\right)\sin C$$

$$= \frac{a^2 \sin B \sin C}{2 \sin A}$$

A similar sequence of steps can be used to derive the following. The formula we use depends on the side we are given.

AREA OF A TRIANGLE (AAS)

$$S = \frac{a^2 \sin B \sin C}{2 \sin A} \qquad S = \frac{b^2 \sin A \sin C}{2 \sin B} \qquad S = \frac{c^2 \sin A \sin B}{2 \sin C}$$

PROBLEM 2

Find the area of triangle ABC given $A = 25°15'$, $B = 130°30'$, and $a = 5.42$ ft.

EXAMPLE 2 Find the area of triangle ABC given $A = 24°\,10'$, $B = 120°\,40'$, and $a = 4.25$ ft.

SOLUTION We begin by finding C.

$$C = 180° - (24°\,10' + 120°\,40')$$
$$= 35°\,10'$$

Now, applying the formula

$$S = \frac{a^2 \sin B \sin C}{2 \sin A}$$

with $a = 4.25$, $A = 24°\,10'$, $B = 120°\,40'$, and $C = 35°\,10'$, we have

$$S = \frac{(4.25)^2(\sin 120°\,40')(\sin 35°\,10')}{2 \sin 24°\,10'}$$

$$= \frac{(4.25)^2(0.8601)(0.5760)}{2(0.4094)}$$

$$= 10.9 \text{ ft}^2 \qquad \text{To three significant digits}$$

Heron of Alexandria

Three Sides

The last area formula we will discuss is called *Heron's formula*. It is used to find the area of a triangle in which all three sides are known. Heron's formula is attributed to Heron of Alexandria, a geometer of the first century A.D. Heron proved his formula in Book I of his work, *Metrica*.

AREA OF A TRIANGLE (SSS)

Given a triangle with sides of length a, b, and c, the *semiperimeter* of the triangle is defined as

$$s = \frac{1}{2}(a + b + c)$$

and the area of the triangle is given by

$$S = \sqrt{s(s - a)(s - b)(s - c)}$$

The semiperimeter is simply half the perimeter of the triangle.

PROOF

We begin our proof by squaring both sides of the formula

$$S = \frac{1}{2}ab \sin C$$

to obtain

$$S^2 = \frac{1}{4}a^2b^2 \sin^2 C$$

Next, we multiply both sides of the equation by $4/(a^2b^2)$ to isolate $\sin^2 C$ on the right side.

$$\frac{4S^2}{a^2b^2} = \sin^2 C$$

Replacing $\sin^2 C$ with $1 - \cos^2 C$ and then factoring as the difference of two squares, we have

$$\begin{aligned}\frac{4S^2}{a^2b^2} &= 1 - \cos^2 C \\ &= (1 + \cos C)(1 - \cos C)\end{aligned}$$

From the law of cosines, we know that $\cos C = (a^2 + b^2 - c^2)/(2ab)$.

$$\begin{aligned}&= \left[1 + \frac{a^2 + b^2 - c^2}{2ab}\right]\left[1 - \frac{a^2 + b^2 - c^2}{2ab}\right] \\ &= \left[\frac{2ab + a^2 + b^2 - c^2}{2ab}\right]\left[\frac{2ab - a^2 - b^2 + c^2}{2ab}\right] \\ &= \left[\frac{(a^2 + 2ab + b^2) - c^2}{2ab}\right]\left[\frac{c^2 - (a^2 - 2ab + b^2)}{2ab}\right] \\ &= \left[\frac{(a + b)^2 - c^2}{2ab}\right]\left[\frac{c^2 - (a - b)^2}{2ab}\right]\end{aligned}$$

Now we factor each numerator as the difference of two squares and multiply the denominators.

$$= \frac{[(a + b + c)(a + b - c)][(c + a - b)(c - a + b)]}{4a^2b^2}$$

Now, because $a + b + c = 2s$, it is also true that

$$a + b - c = a + b + c - 2c = 2s - 2c$$
$$c + a - b = a + b + c - 2b = 2s - 2b$$
$$c - a + b = a + b + c - 2a = 2s - 2a$$

Substituting these expressions into our last equation, we have

$$= \frac{2s(2s - 2c)(2s - 2b)(2s - 2a)}{4a^2b^2}$$

Factoring out a 2 from each term in the numerator and showing the left side of our equation along with the right side, we have

$$\frac{4S^2}{a^2b^2} = \frac{16s(s - a)(s - b)(s - c)}{4a^2b^2}$$

Multiplying both sides by $a^2b^2/4$ we have

$$S^2 = s(s - a)(s - b)(s - c)$$

Taking the square root of both sides of the equation, we have Heron's formula.

$$S = \sqrt{s(s - a)(s - b)(s - c)}$$ ■

PROBLEM 3
Find the area of triangle ABC if $a = 15$ m, $b = 13$ m, and $c = 22$ m.

EXAMPLE 3 Find the area of triangle ABC if $a = 12$ m, $b = 14$ m, and $c = 8.0$ m.

SOLUTION We begin by calculating the formula for s, half the perimeter of ABC.

$$s = \frac{1}{2}(12 + 14 + 8) = 17$$

Substituting this value of s into Heron's formula along with the given values of a, b, and c, we have

$$\begin{aligned} S &= \sqrt{17(17 - 12)(17 - 14)(17 - 8)} \\ &= \sqrt{17(5)(3)(9)} \\ &= \sqrt{2{,}295} \\ &= 48 \text{ m}^2 \quad \text{To two significant digits} \end{aligned}$$

■

Getting Ready for Class

After reading through the preceding section, respond in your own words and in complete sentences.

a. State the formula for the area of a triangle given two sides and the included angle.
b. State the formula for the area of a triangle given two angles and one side.
c. State the formula for the area of a triangle given all three sides.
d. What is the name given to this last formula?

7.4 PROBLEM SET

➤ CONCEPTS AND VOCABULARY

For Questions 1 through 4, fill in the blank with an appropriate word or number.

1. To find the area of a triangle given two sides and the angle between them, multiply ______ the product of the ______ by the _______ of the angle between them.
2. To find the area of a triangle given two angles and one side, first find the missing ______ and then use the appropriate formula.
3. The perimeter of a triangle is equal to the ______ of the three sides. To find the semi-perimeter, divide the perimeter by ______.
4. To find the area of a triangle given all three sides, we can use ________ formula. The first step is to find the ____________.

For Questions 5 and 6, assume triangle ABC has sides a, b, and c.

5. State the formula for the semiperimeter of triangle ABC: $s =$ __________
6. State Heron's formula for the area of triangle ABC: $S =$ _________

➤ EXERCISES

Find the semiperimeter of triangle ABC.

7. $a = 3$ ft, $b = 4$ ft, $c = 5$ ft
8. $a = 143$ cm, $b = 175$ cm, $c = 232$ cm
9. $a = 2.1$ m, $b = 2.3$ m, $c = 3.9$ m
10. $a = 33$ yd, $b = 35$ yd, $c = 6$ yd

Each of the following problems refers to triangle ABC. In each case, find the area of the triangle. Round to three significant digits.

11. $a = 50$ cm, $b = 70$ cm, $C = 60°$
12. $a = 10$ cm, $b = 12$ cm, $C = 120°$
13. $a = 41.5$ m, $c = 34.5$ m, $B = 151.5°$
14. $a = 76.3$ m, $c = 42.8$ m, $B = 16.3°$
15. $b = 0.923$ km, $c = 0.387$ km, $A = 43° 20'$
16. $b = 63.4$ km, $c = 75.2$ km, $A = 124° 40'$
17. $A = 46°$, $B = 95°$, $c = 6.8$ m
18. $B = 57°$, $C = 31°$, $a = 7.3$ m
19. $A = 42.5°$, $B = 71.4°$, $a = 210$ in.
20. $A = 110.4°$, $C = 21.8°$, $c = 240$ in.
21. $A = 43° 30'$, $C = 120° 30'$, $a = 3.48$ ft
22. $B = 14° 20'$, $C = 75° 40'$, $b = 2.72$ ft
23. $a = 44$ in., $b = 66$ in., $c = 88$ in.
24. $a = 23$ in., $b = 34$ in., $c = 45$ in.
25. $a = 4.8$ yd, $b = 6.3$ yd, $c = 7.5$ yd
26. $a = 48$ yd, $b = 75$ yd, $c = 63$ yd
27. $a = 4.38$ ft, $b = 3.79$ ft, $c = 5.22$ ft
28. $a = 8.32$ ft, $b = 6.23$ ft, $c = 3.45$ ft
29. **Geometry and Area** Find the area of a parallelogram if the angle between two of the sides is 120° and the two sides are 15 inches and 12 inches in length.
30. **Geometry and Area** Find the area of a parallelogram if the two sides measure 24.1 inches and 31.4 inches and the shorter diagonal is 32.4 inches.
31. **Geometry and Area** The area of a triangle is 40 square centimeters. Find the length of the side included between the angles $A = 30°$ and $B = 50°$.
32. **Geometry and Area** The area of a triangle is 80 square inches. Find the length of the side included between $A = 25°$ and $C = 110°$.

➤ REVIEW PROBLEMS

The problems that follow review material we covered in Section 6.4.

Eliminate the parameter t and graph the resulting equation.

33. $x = \cos t, y = \sin t$

34. $x = \cos t, y = -\sin t$

35. $x = 3 + 2 \sin t, y = 1 + 2 \cos t$

36. $x = \cos t - 3, y = \sin t + 2$

Eliminate the parameter t but do not graph.

37. $x = 2 \tan t, y = 3 \sec t$

38. $x = 4 \cot t, y = 2 \csc t$

39. $x = \sin t, y = \cos 2t$

40. $x = \cos t, y = \cos 2t$

➤ LEARNING OBJECTIVES ASSESSMENT

These questions are available for instructors to help assess if you have successfully met the learning objectives for this section.

41. Find the area of triangle ABC if $a = 73.6$ millimeters, $b = 41.5$ millimeters, and $C = 22.3°$.

a. 1,160 mm² **b.** 1,412 mm² **c.** 902 mm² **d.** 580 mm²

42. Find the area of triangle ABC if $A = 56°$, $B = 71°$, and $c = 21$ inches.

a. 150 in² **b.** 200 in² **c.** 220 in² **d.** 240 in²

43. Find the semiperimeter of triangle ABC with $a = 17$, $b = 41$, and $c = 28$.

a. 86 **b.** 172 **c.** 43 **d.** 29

44. If Heron's formula is used to find the area of triangle ABC having $a = 3$ meters, $b = 5$ meters, and $c = 6$ meters, which of the following shows the correct way to set up the formula?

a. $S = 7\sqrt{(10)(12)(13)}$

b. $S = \sqrt{(4)(2)(1)}$

c. $S = \sqrt{7(3)(5)(6)}$

d. $S = \sqrt{7(4)(2)(1)}$

SECTION 7.5 Vectors: An Algebraic Approach

LEARNING OBJECTIVES

1. Draw a vector in standard position.
2. Express a vector in terms of unit vectors **i** and **j**.
3. Find the magnitude of a vector.
4. Find a sum, difference, or scalar multiple of vectors.

In this section, we will take a second look at vectors, this time from an algebraic point of view. Much of the credit for this treatment of vectors is attributed to both Irish mathematician William Rowan Hamilton (1805–1865) and German mathematician Hermann Grassmann (1809–1877). Grassmann is famous for his book *Ausdehnungslehre (The Calculus of Extension)*, which was first published in 1844. However, it wasn't until 60 years later that this work was fully accepted or the significance realized, when Albert Einstein used it in his theory of relativity.

Hermann Grassmann

Standard Position

As we mentioned in Section 2.5, a vector is in *standard position* when it is placed on a coordinate system so that its tail is located at the origin. If the tip of the vector corresponds to the point (a, b), then the coordinates of this point provide a unique representation for the vector. That is, the point (a, b) determines both the length of the vector and its direction, as shown in Figure 1.

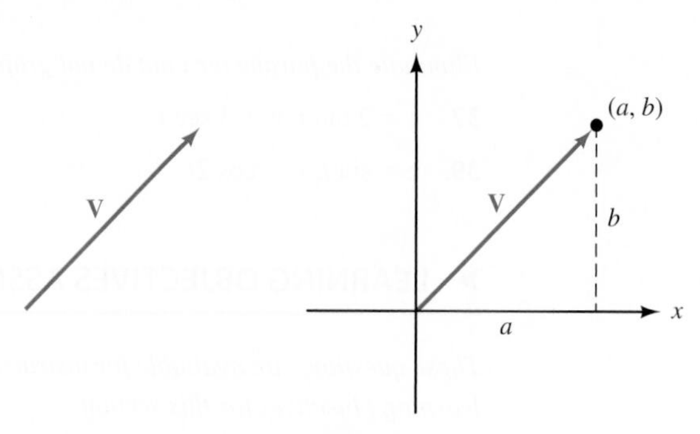

Figure 1

To avoid confusion between the point (a, b) and the vector, which is the ray extending from the origin to the point, we will denote the vector as $\mathbf{V} = \langle a, b \rangle$. We refer to this notation as *component form.* The x-coordinate, a, is called the *horizontal component* of $\mathbf{V}$, and the y-coordinate, b, is called the *vertical component* of $\mathbf{V}$.

Magnitude

As you know from Section 2.5, the magnitude of a vector is its length. Referring to Figure 1, we can find the magnitude of the vector $\mathbf{V} = \langle a, b \rangle$ using the Pythagorean Theorem:

$$|\mathbf{V}| = \sqrt{a^2 + b^2}$$

PROBLEM 1
Draw the vector $\langle 5, -2 \rangle$ in standard position and find its magnitude.

EXAMPLE 1 Draw the vector $\mathbf{V} = \langle 3, -4 \rangle$ in standard position and find its magnitude.

SOLUTION We draw the vector by sketching an arrow from the origin to the point $(3, -4)$, as shown in Figure 2. To find the magnitude of $\mathbf{V}$, we find the positive square root of the sum of the squares of the horizontal and vertical components.

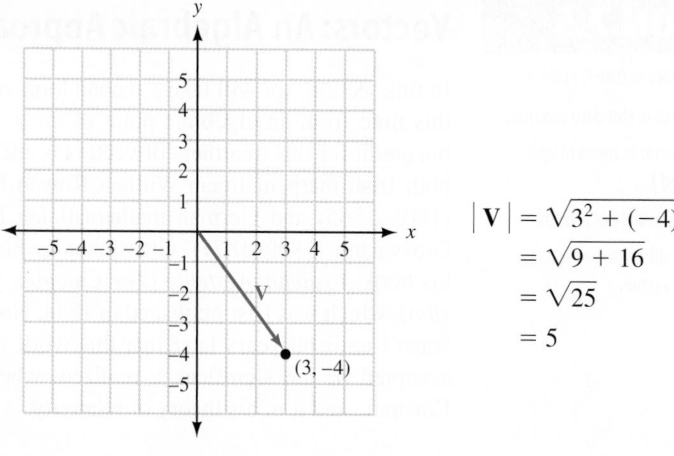

Figure 2

$$\begin{aligned} |\mathbf{V}| &= \sqrt{3^2 + (-4)^2} \\ &= \sqrt{9 + 16} \\ &= \sqrt{25} \\ &= 5 \end{aligned}$$

■

Addition and Subtraction with Algebraic Vectors

Adding and subtracting vectors written in component form is simply a matter of adding (or subtracting) the horizontal components and adding (or subtracting) the vertical components. Figure 3 shows the vector sum of vectors $\mathbf{U} = \langle 6, 2\rangle$ and $\mathbf{V} = \langle -3, 5\rangle$.

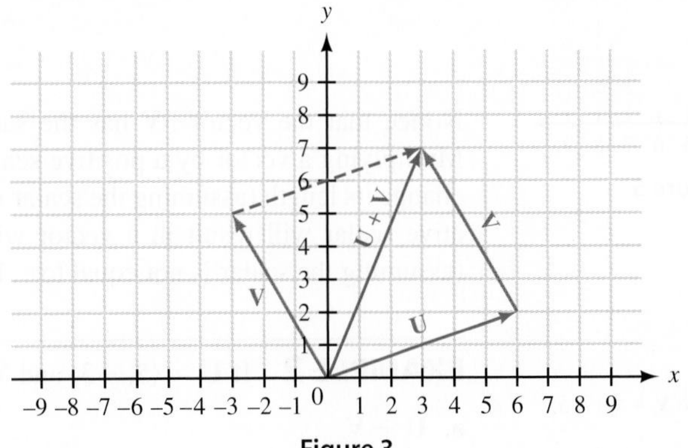

Figure 3

By simply counting squares on the grid you can convince yourself that the sum can be obtained by adding horizontal components and adding vertical components. That is,

$$\begin{aligned}\mathbf{U} + \mathbf{V} &= \langle 6, 2\rangle + \langle -3, 5\rangle \\ &= \langle 6 + (-3), 2 + 5\rangle \\ &= \langle 3, 7\rangle\end{aligned}$$

Figure 4 shows the difference of vectors **U** and **V**. As the diagram in Figure 4 indicates, subtraction of algebraic vectors can be accomplished by subtracting corresponding components. That is,

$$\begin{aligned}\mathbf{U} - \mathbf{V} &= \langle 6, 2\rangle - \langle -3, 5\rangle \\ &= \langle 6 - (-3), 2 - 5\rangle \\ &= \langle 9, -3\rangle\end{aligned}$$

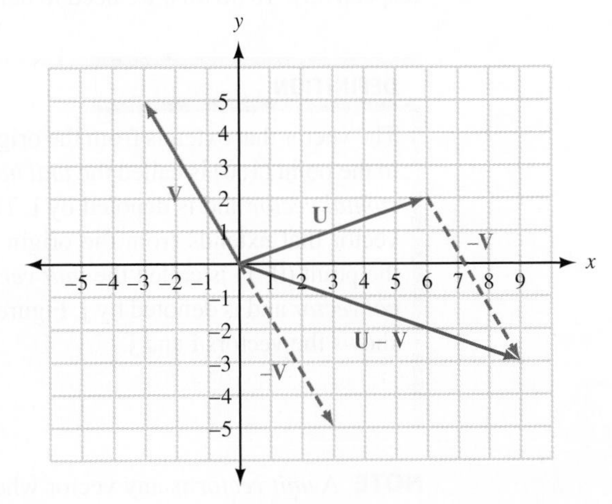

Figure 4

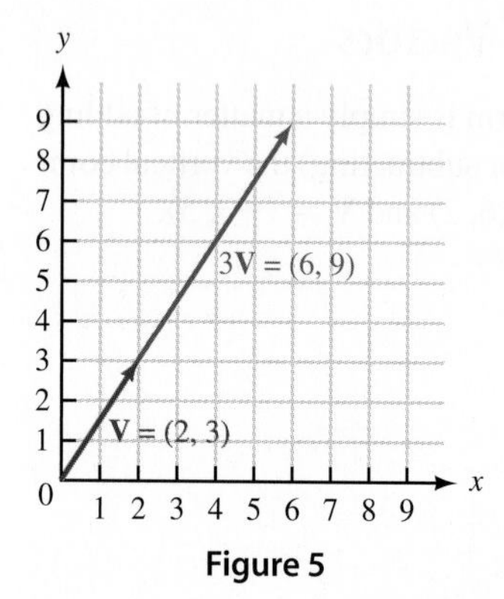

Figure 5

Scalar Multiplication

To multiply a vector in component form by a scalar (real number) we multiply each component of the vector by the scalar. Figure 5 shows the vector $\mathbf{V} = \langle 2, 3\rangle$ and the vector $3\mathbf{V}$. As you can see,

$$\begin{aligned} 3\mathbf{V} &= 3\langle 2, 3\rangle \\ &= \langle 3 \cdot 2, 3 \cdot 3\rangle \\ &= \langle 6, 9\rangle \end{aligned}$$

Notice that the vector $3\mathbf{V}$ has the same direction as $\mathbf{V}$ but is three times as long. Multiplying a vector by a positive scalar will preserve the direction of the vector but change its length (assuming the scalar is not equal to 1). Multiplying a vector by a negative scalar will result in a vector with the opposite direction and different length (assuming the scalar is not equal to -1).

PROBLEM 2
If $\mathbf{U} = \langle 2, -4\rangle$ and $\mathbf{V} = \langle -7, 5\rangle$, find
a. $\mathbf{U} + \mathbf{V}$
b. $3\mathbf{U} - 4\mathbf{V}$

EXAMPLE 2 If $\mathbf{U} = \langle 5, -3\rangle$ and $\mathbf{V} = \langle -6, 4\rangle$, find

a. $\mathbf{U} + \mathbf{V}$

b. $4\mathbf{U} - 5\mathbf{V}$

SOLUTION

a. $$\begin{aligned} \mathbf{U} + \mathbf{V} &= \langle 5, -3\rangle + \langle -6, 4\rangle \\ &= \langle 5 - 6, -3 + 4\rangle \\ &= \langle -1, 1\rangle \end{aligned}$$

b. $$\begin{aligned} 4\mathbf{U} - 5\mathbf{V} &= 4\langle 5, -3\rangle - 5\langle -6, 4\rangle \\ &= \langle 20, -12\rangle + \langle 30, -20\rangle \\ &= \langle 20 + 30, -12 - 20\rangle \\ &= \langle 50, -32\rangle \end{aligned}$$

Component Vector Form

Another way to represent a vector algebraically is to express the vector as the sum of a horizontal vector and a vertical vector. As we mentioned in Section 2.5, any vector $\mathbf{V}$ can be written in terms of its horizontal and vertical component vectors, $\mathbf{V}_x$, and $\mathbf{V}_y$, respectively. To do this, we need to define two special vectors.

DEFINITION

The vector that extends from the origin to the point $(1, 0)$ is called the *unit horizontal vector* and is denoted by $\mathbf{i}$. The vector that extends from the origin to the point $(0, 1)$ is called the *unit vertical vector* and is denoted by $\mathbf{j}$. Figure 6 shows the vectors $\mathbf{i}$ and $\mathbf{j}$.

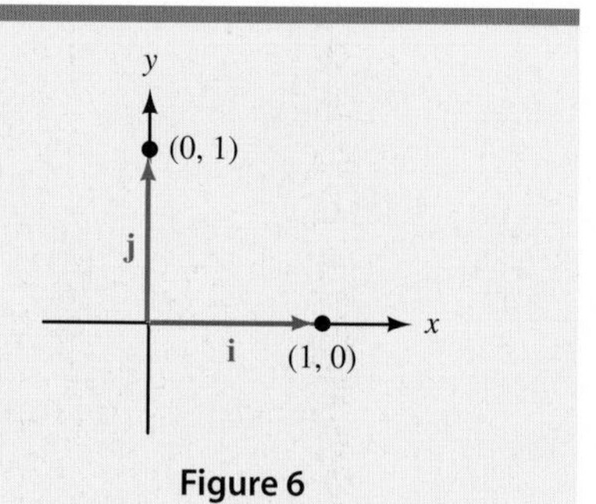

Figure 6

NOTE A *unit vector* is any vector whose magnitude is 1. Using our previous component notation, we would write $\mathbf{i} = \langle 1, 0\rangle$ and $\mathbf{j} = \langle 0, 1\rangle$.

PROBLEM 3
Write $\mathbf{V} = \langle 5, 12 \rangle$ in terms of $\mathbf{i}$ and $\mathbf{j}$.

EXAMPLE 3 Write the vector $\mathbf{V} = \langle 3, 4 \rangle$ in terms of the unit vectors $\mathbf{i}$ and $\mathbf{j}$.

SOLUTION From the origin, we must go three units in the positive x-direction, and then four units in the positive y-direction, to locate the terminal point of $\mathbf{V}$ at $(3, 4)$. Because $\mathbf{i}$ is a vector of length 1 in the positive x-direction, $3\mathbf{i}$ will be a vector of length 3 in that same direction. Likewise, $4\mathbf{j}$ will be a vector of length 4 in the positive y-direction. As shown in Figure 7, $\mathbf{V}$ is the sum of vectors $3\mathbf{i}$ and $4\mathbf{j}$.

Therefore, we can write $\mathbf{V}$ in terms of the unit vectors $\mathbf{i}$ and $\mathbf{j}$ as

$$\mathbf{V} = 3\mathbf{i} + 4\mathbf{j}$$

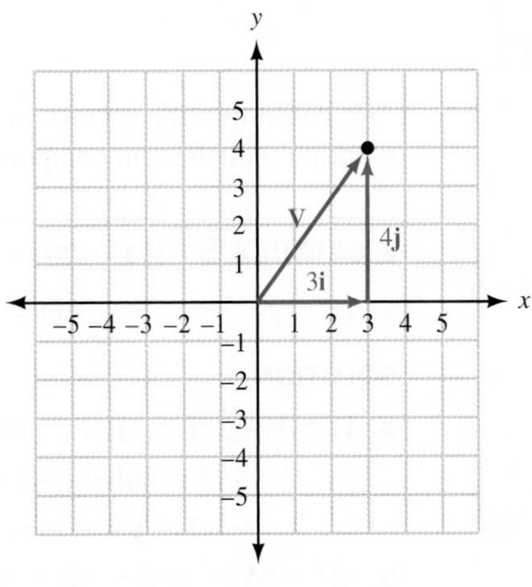

Figure 7

In Example 3, the vector $3\mathbf{i}$ is the *horizontal vector component* of $\mathbf{V}$, which we have previously referred to as $\mathbf{V}_x$. Likewise, $4\mathbf{j}$ is the *vertical vector component* of $\mathbf{V}$, previously referred to as $\mathbf{V}_y$. Notice that the coefficients of these vectors are simply the coordinates of the terminal point of $\mathbf{V}$. That is,

$$\mathbf{V} = \langle 3, 4 \rangle = 3\mathbf{i} + 4\mathbf{j}$$

We refer to the notation $\mathbf{V} = 3\mathbf{i} + 4\mathbf{j}$ as *vector component form.* Every vector $\mathbf{V}$ can be written in terms of horizontal and vertical components and the unit vectors $\mathbf{i}$ and $\mathbf{j}$.

Here is a summary of the information we have developed to this point.

ALGEBRAIC VECTORS

If $\mathbf{i}$ is the unit vector from $(0, 0)$ to $(1, 0)$, and $\mathbf{j}$ is the unit vector from $(0, 0)$ to $(0, 1)$, then any vector $\mathbf{V}$ can be written as

$$\mathbf{V} = a\mathbf{i} + b\mathbf{j} = \langle a, b \rangle$$

where a and b are real numbers (Figure 8).

The magnitude of $\mathbf{V}$ is

$$|\mathbf{V}| = \sqrt{a^2 + b^2}$$

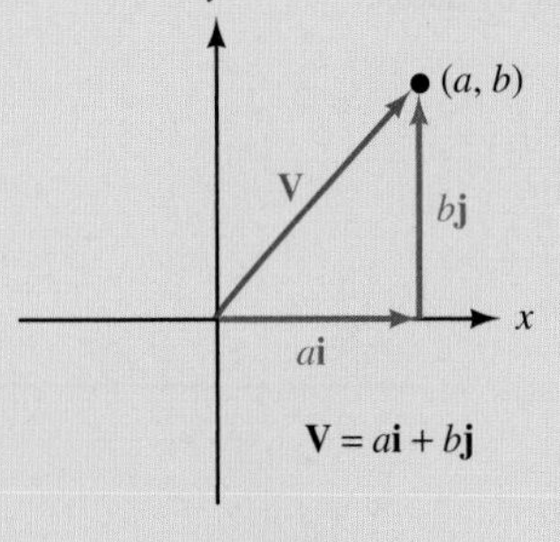

Figure 8

PROBLEM 4
Vector $\mathbf{V}$ has its tail at the origin and makes an angle of 55° with the positive x-axis. Its magnitude is 15. Write $\mathbf{V}$ in terms of $\mathbf{i}$ and $\mathbf{j}$.

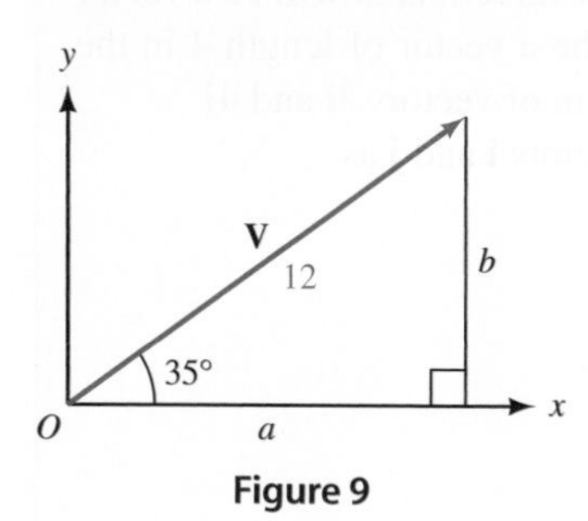

Figure 9

EXAMPLE 4 Vector $\mathbf{V}$ has its tail at the origin, and makes an angle of 35° with the positive x-axis. Its magnitude is 12. Write $\mathbf{V}$ in terms of the unit vectors $\mathbf{i}$ and $\mathbf{j}$.

SOLUTION Figure 9 is a diagram of $\mathbf{V}$. The horizontal and vertical components of $\mathbf{V}$ are a and b, respectively.

We find a and b using right triangle trigonometry.

$$a = 12 \cos 35° = 9.8$$
$$b = 12 \sin 35° = 6.9$$

Writing $\mathbf{V}$ in terms of the unit vectors $\mathbf{i}$ and $\mathbf{j}$ we have

$$\mathbf{V} = 9.8\mathbf{i} + 6.9\mathbf{j}$$

Working with vectors in vector component form $a\mathbf{i} + b\mathbf{j}$ is no different from working with them in component form $\langle a, b\rangle$, as illustrated in the next example.

PROBLEM 5
If $\mathbf{U} = 4\mathbf{i} - 5\mathbf{j}$ and $\mathbf{V} = -3\mathbf{i} + 7\mathbf{j}$, find
a. $\mathbf{U} + \mathbf{V}$
b. $3\mathbf{U} - 4\mathbf{V}$

EXAMPLE 5 If $\mathbf{U} = 5\mathbf{i} - 3\mathbf{j}$ and $\mathbf{V} = -6\mathbf{i} + 4\mathbf{j}$, find

a. $\mathbf{U} + \mathbf{V}$ **b.** $4\mathbf{U} - 5\mathbf{V}$

SOLUTION

a.
$$\begin{aligned}\mathbf{U} + \mathbf{V} &= (5\mathbf{i} - 3\mathbf{j}) + (-6\mathbf{i} + 4\mathbf{j})\\ &= (5 - 6)\mathbf{i} + (-3 + 4)\mathbf{j}\\ &= -\mathbf{i} + \mathbf{j}\end{aligned}$$

b.
$$\begin{aligned}4\mathbf{U} - 5\mathbf{V} &= 4(5\mathbf{i} - 3\mathbf{j}) - 5(-6\mathbf{i} + 4\mathbf{j})\\ &= (20\mathbf{i} - 12\mathbf{j}) + (30\mathbf{i} - 20\mathbf{j})\\ &= (20 + 30)\mathbf{i} + (-12 - 20)\mathbf{j}\\ &= 50\mathbf{i} - 32\mathbf{j}\end{aligned}$$

Notice that these results are equivalent to those of Example 2.

As an application of algebraic vectors, let's return to one of the static equilibrium problems we solved in Section 2.5.

PROBLEM 6
Davey is 6 years old and weighs 61 pounds. He is sitting on a swing when his sister Susan pulls him and the swing back horizontally through an angle of 35° and then stops. Find the tension in the ropes of the swing and the magnitude of the force exerted by Susan.

EXAMPLE 6 Danny is 5 years old and weighs 42 pounds. He is sitting on a swing when his sister Stacey pulls him and the swing back horizontally through an angle of 30° and then stops. Find the tension in the ropes of the swing and the magnitude of the force exerted by Stacey. (Figure 10 is a diagram of the situation.)

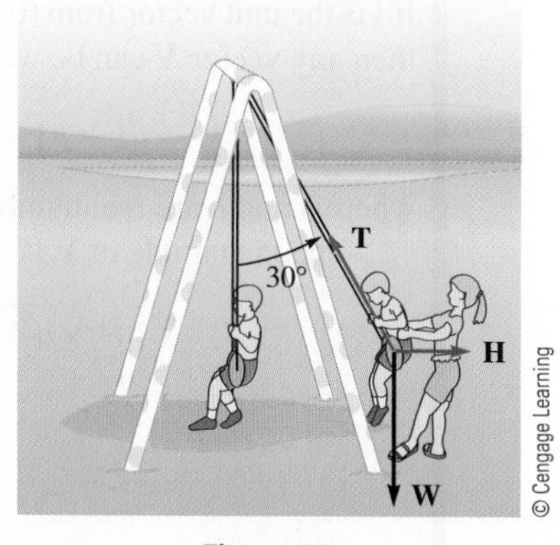

Figure 10

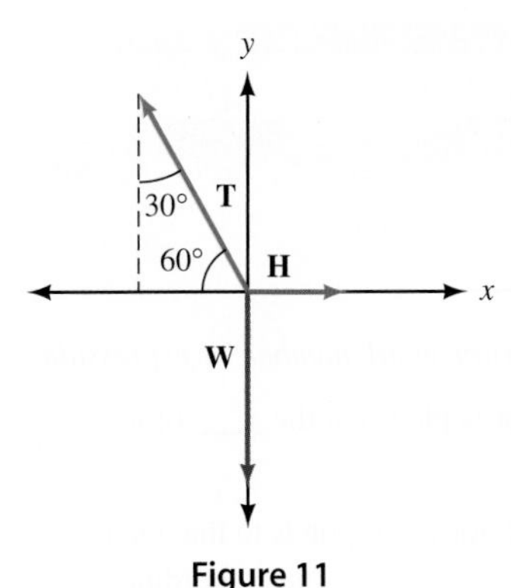

Figure 11

SOLUTION When we solved this problem in Section 2.5, we noted that there are three forces acting on Danny (and the swing), which we have labeled **W**, **H**, and **T**. Recall that the vector **W** is due to the force of gravity: its magnitude is $|\mathbf{W}| = 42$ lb, and its direction is straight down. The vector **H** represents the force with which Stacey is pulling Danny horizontally, and **T** is the force acting on Danny in the direction of the ropes.

If we place the origin of a coordinate system on the point at which the tails of the three vectors intersect, we can write each vector in terms of its magnitude and the unit vectors **i** and **j**. The coordinate system is shown in Figure 11. The direction along which **H** acts is in the positive x direction, so we can write

$$\mathbf{H} = |\mathbf{H}|\mathbf{i}$$

Likewise, the weight vector **W** is straight down, so we can write

$$\mathbf{W} = -|\mathbf{W}|\mathbf{j} = -42\mathbf{j}$$

Using right triangle trigonometry, we write **T** in terms of its horizontal and vertical components:

$$\mathbf{T} = -|\mathbf{T}|\cos 60°\mathbf{i} + |\mathbf{T}|\sin 60°\mathbf{j}$$

Because Danny and the swing are at rest, we have static equilibrium. Therefore, the sum of the vectors is the zero vector.

$$\mathbf{T} + \mathbf{H} + \mathbf{W} = \mathbf{0}$$

$$-|\mathbf{T}|\cos 60°\mathbf{i} + |\mathbf{T}|\sin 60°\mathbf{j} + |\mathbf{H}|\mathbf{i} + (-42\mathbf{j}) = \mathbf{0}$$

Collecting all the **i** components and all the **j** components together, we have

$$(-|\mathbf{T}|\cos 60° + |\mathbf{H}|)\mathbf{i} + (|\mathbf{T}|\sin 60° - 42)\mathbf{j} = \mathbf{0}$$

The only way this can happen is if both components are 0. Setting the coefficient of **j** to 0, we have

$$\begin{aligned} |\mathbf{T}|\sin 60° - 42 &= 0 \\ |\mathbf{T}| &= \frac{42}{\sin 60°} \\ &= 48 \text{ lb} \qquad \text{To two significant digits} \end{aligned}$$

Setting the coefficient of **i** to 0, and then substituting 48 for $|\mathbf{T}|$, we have

$$\begin{aligned} -|\mathbf{T}|\cos 60° + |\mathbf{H}| &= 0 \\ -48\cos 60° + |\mathbf{H}| &= 0 \\ |\mathbf{H}| &= 48\cos 60° \\ &= 24 \text{ lb} \end{aligned}$$

■

Getting Ready for Class

After reading through the preceding section, respond in your own words and in complete sentences.

a. How can we algebraically represent a vector that is in standard position?
b. Explain how to add or subtract two vectors in component form.
c. What is a scalar and how is it used in our study of vectors?
d. State the definitions for the unit horizontal vector and the unit vertical vector.

7.5 PROBLEM SET

➤ CONCEPTS AND VOCABULARY

For Questions 1 through 10, fill in the blank with an appropriate word, number, or expression.

1. A vector is in standard position if the ____ of the vector is placed at the ____ of a rectangular coordinate system.

2. If a vector **V** is in standard position and the tip of the vector corresponds to the point (a, b), then we can write the vector in component form as ________. The x-coordinate, a, is called the __________ __________ of **V**, and the y-coordinate, b, is called the ________ ________ of **V**.

3. To add or subtract two vectors in component form, simply add or subtract the corresponding ____________.

4. A scalar is a ________ number. To multiply a vector in component form by a scalar, simply multiply each ________ by the scalar. This is called ________ multiplication.

5. Multiplying a vector by a positive scalar will change the ________ of the vector but not its ________.

6. The opposite of a vector is a vector with the ______ magnitude and ________ direction. To obtain the opposite of a vector, multiply the vector by ____.

7. A ______ vector is any vector whose magnitude is 1.

8. The unit vector **i** points in the direction of the positive _____ and is called the _____ ________ vector. The unit vector **j** points in the direction of the positive _____ and is called the _____ ________ vector.

9. The notation $\mathbf{V} = a\mathbf{i} + b\mathbf{j}$ is called ______ __________ form, where $a\mathbf{i}$ is the ________ ________ of **V** and $b\mathbf{j}$ is the ________ ________ ________ of **V**.

10. The magnitude of $\mathbf{V} = \langle a, b\rangle = a\mathbf{i} + b\mathbf{j}$ is given by ________.

➤ EXERCISES

Draw the vector **V** *that goes from the origin to the given point. Then write* **V** *in component form* $\langle a, b\rangle$.

11. $(4, 1)$ **12.** $(1, 4)$ **13.** $(-5, 2)$ **14.** $(-2, 5)$

15. $(3, -3)$ **16.** $(5, -5)$ **17.** $(-6, -4)$ **18.** $(-4, -6)$

Draw the vector **V** *that goes from the origin to the given point. Then write* **V** *in terms of the unit vectors* **i** *and* **j**.

19. $(2, 5)$ **20.** $(5, 2)$ **21.** $(-3, 6)$ **22.** $(-6, 3)$

23. $(4, -5)$ **24.** $(5, -4)$ **25.** $(-1, -5)$ **26.** $(-5, -1)$

Find the magnitude of each of the following vectors.

27. $\langle -5, 6\rangle$ **28.** $\langle -8, -3\rangle$ **29.** $\langle 0, 5\rangle$ **30.** $\langle 2, 0\rangle$

31. $\mathbf{U} = 5\mathbf{i} + 12\mathbf{j}$ **32.** $\mathbf{U} = 20\mathbf{i} - 21\mathbf{j}$ **33.** $\mathbf{W} = \mathbf{i} + 2\mathbf{j}$ **34.** $\mathbf{W} = 3\mathbf{i} + \mathbf{j}$

For each vector, find $\frac{1}{2}\mathbf{V}$, $-\mathbf{V}$, *and* $4\mathbf{V}$.

35. $\mathbf{V} = \langle -3, 7\rangle$ **36.** $\mathbf{V} = \langle -2, -5\rangle$ **37.** $\mathbf{V} = 2\mathbf{i} + 4\mathbf{j}$ **38.** $\mathbf{V} = 6\mathbf{i} + 8\mathbf{j}$

For each pair of vectors, find $\mathbf{U} + \mathbf{V}$, $\mathbf{U} - \mathbf{V}$, *and* $2\mathbf{U} - 3\mathbf{V}$.

39. $\mathbf{U} = \langle 4, 4\rangle, \mathbf{V} = \langle 4, -4\rangle$

40. $\mathbf{U} = \langle -4, 4\rangle, \mathbf{V} = \langle 4, 4\rangle$

41. $\mathbf{U} = \langle -5, 0\rangle, \mathbf{V} = \langle 0, 1\rangle$

42. $\mathbf{U} = \langle 2, 0\rangle, \mathbf{V} = \langle 0, -7\rangle$

43. $\mathbf{U} = \langle 4, 1\rangle, \mathbf{V} = \langle -5, 2\rangle$

44. $\mathbf{U} = \langle 1, 4\rangle, \mathbf{V} = \langle -2, 5\rangle$

For each pair of vectors, find $\mathbf{U} + \mathbf{V}$, $\mathbf{U} - \mathbf{V}$, *and* $3\mathbf{U} + 2\mathbf{V}$.

45. $\mathbf{U} = -\mathbf{i} + \mathbf{j}, \mathbf{V} = \mathbf{i} + \mathbf{j}$

46. $\mathbf{U} = \mathbf{i} + \mathbf{j}, \mathbf{V} = \mathbf{i} - \mathbf{j}$

47. $\mathbf{U} = 6\mathbf{i}, \mathbf{V} = -8\mathbf{j}$

48. $\mathbf{U} = -3\mathbf{i}, \mathbf{V} = 5\mathbf{j}$

49. $\mathbf{U} = 2\mathbf{i} + 5\mathbf{j}, \mathbf{V} = 5\mathbf{i} + 2\mathbf{j}$

50. $\mathbf{U} = 5\mathbf{i} + 3\mathbf{j}, \mathbf{V} = 3\mathbf{i} + 5\mathbf{j}$

51. Vector $\mathbf{V}$ is in standard position and makes an angle of 40° with the positive x-axis. Its magnitude is 18. Write $\mathbf{V}$ in component form $\langle a, b\rangle$ and in vector component form $a\mathbf{i} + b\mathbf{j}$.

52. Vector $\mathbf{U}$ is in standard position and makes an angle of 110° with the positive x-axis. Its magnitude is 25. Write $\mathbf{U}$ in component form $\langle a, b\rangle$ and in vector component form $a\mathbf{i} + b\mathbf{j}$.

53. Vector $\mathbf{W}$ is in standard position and makes an angle of 230° with the positive x-axis. Its magnitude is 8. Write $\mathbf{W}$ in component form $\langle a, b\rangle$ and in vector component form $a\mathbf{i} + b\mathbf{j}$.

54. Vector $\mathbf{F}$ is in standard position and makes an angle of 285° with the positive x-axis. Its magnitude is 30. Write $\mathbf{F}$ in component form $\langle a, b\rangle$ and in vector component form $a\mathbf{i} + b\mathbf{j}$.

Find the magnitude of each vector and the angle θ, $0° \le \theta < 360°$, *that the vector makes with the positive x-axis.*

55. $\mathbf{U} = \langle 3, 3\rangle$

56. $\mathbf{V} = \langle 5, -5\rangle$

57. $\mathbf{W} = -\mathbf{i} - \sqrt{3}\mathbf{j}$

58. $\mathbf{F} = -2\sqrt{3}\mathbf{i} + 2\mathbf{j}$

➤ REVIEW PROBLEMS

You may have worked the following problems previously in Chapter 2 or Section 7.1. Solve each problem using the methods shown in Example 6 of this section.

59. Force An 8.0-pound weight is lying on a situp bench at the gym. If the bench is inclined at an angle of 15°, there are three forces acting on the weight, as shown in Figure 12. Find the magnitude of **N** and the magnitude of **F**.

60. Force Repeat Problem 59 for a 25.0-pound weight and a bench inclined at 10.0°.

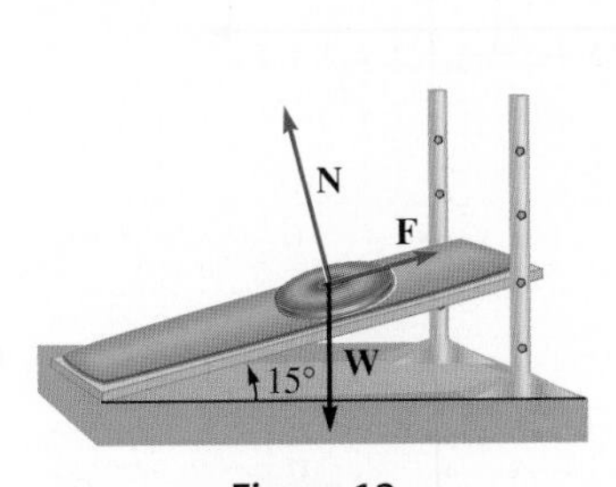

Figure 12

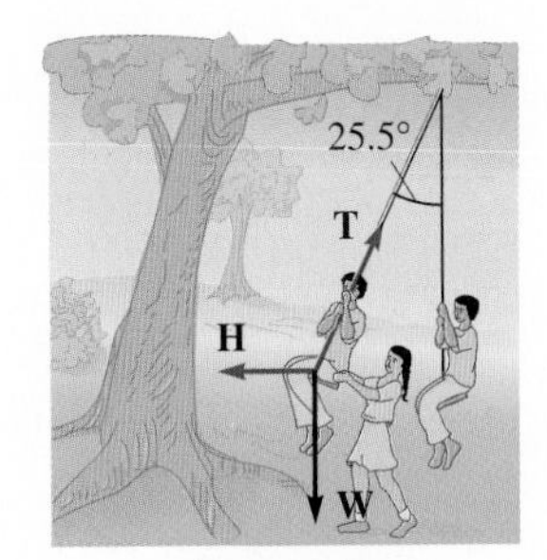

Figure 13

61. Force Tyler and his cousin Kelly have attached a rope to the branch of a tree and tied a board to the other end to form a swing. Tyler stands on the board while his cousin pushes him through an angle of 25.5° and holds him there. If Tyler weighs 95.5 pounds, find the magnitude of the force Kelly must push with horizontally to keep Tyler in static equilibrium. See Figure 13.

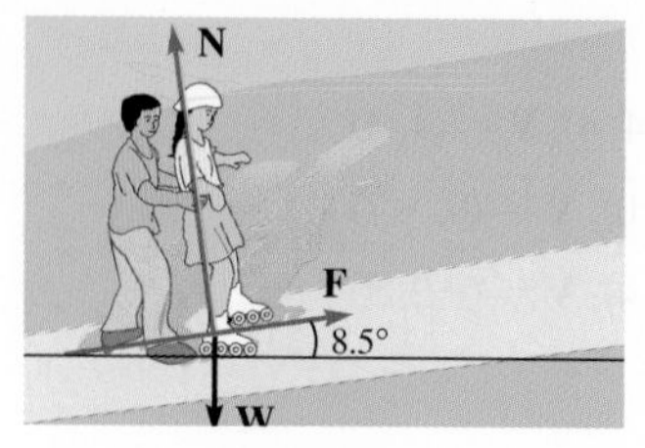

Figure 14

62. Force After they are finished swinging, Tyler and Kelly decide to rollerskate. They come to a hill that is inclined at 8.5°. Tyler pushes Kelly halfway up the hill and then holds her there (Figure 14). If Kelly weighs 58.0 pounds, find the magnitude of the force Tyler must push with to keep Kelly from rolling down the hill. (We are assuming that the rollerskates make the hill into a frictionless surface so that the only force keeping Kelly from rolling backwards down the hill is the force Tyler is pushing with.)

63. Force A traffic light weighing 22 pounds is suspended by two wires as shown in Figure 15. Find the magnitude of the tension in wire AB, and the magnitude of the tension in wire AC.

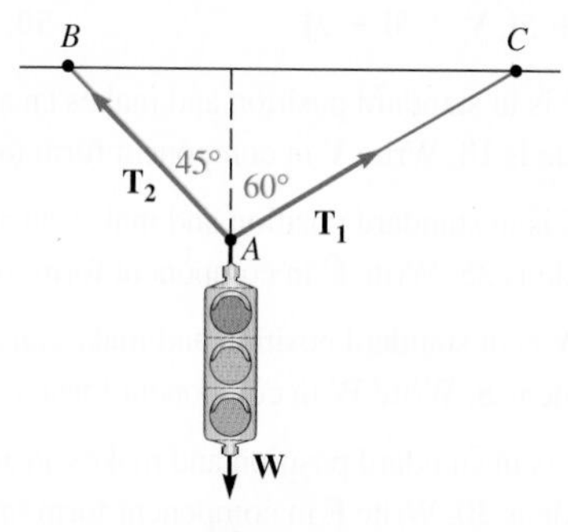

Figure 15

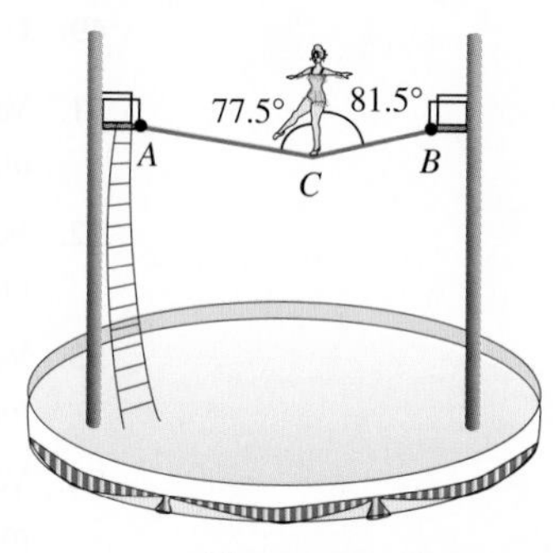

Figure 16

64. Force A tightrope walker is standing still with one foot on the tightrope as shown in Figure 16. If the tightrope walker weighs 125 pounds, find the magnitudes of the tension in the rope toward each end of the rope.

➤ LEARNING OBJECTIVES ASSESSMENT

These questions are available for instructors to help assess if you have successfully met the learning objectives for this section.

65. Draw $\mathbf{V} = \langle 2, 3 \rangle$ in standard position.

a.

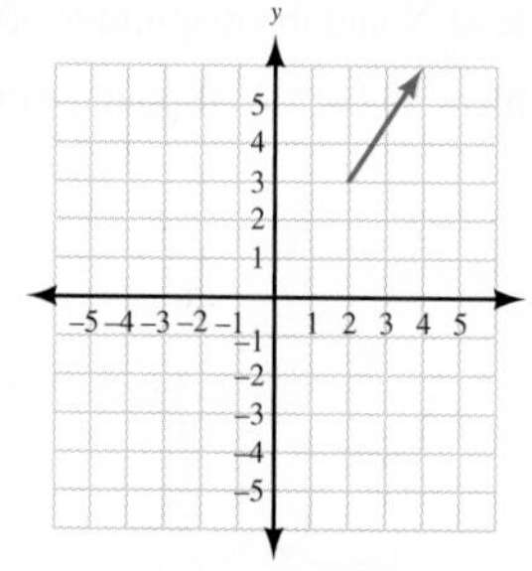

b.

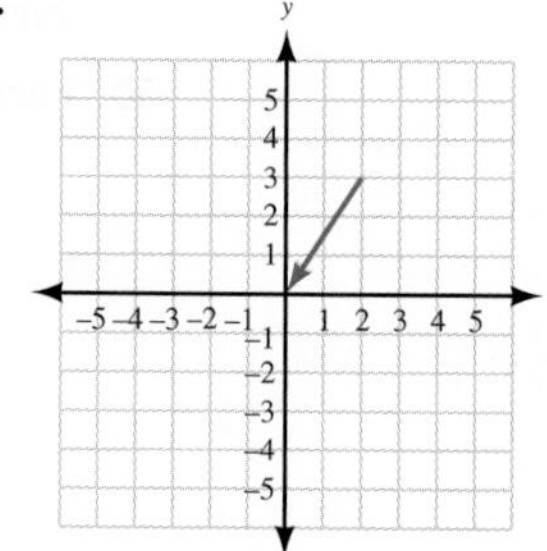

c.

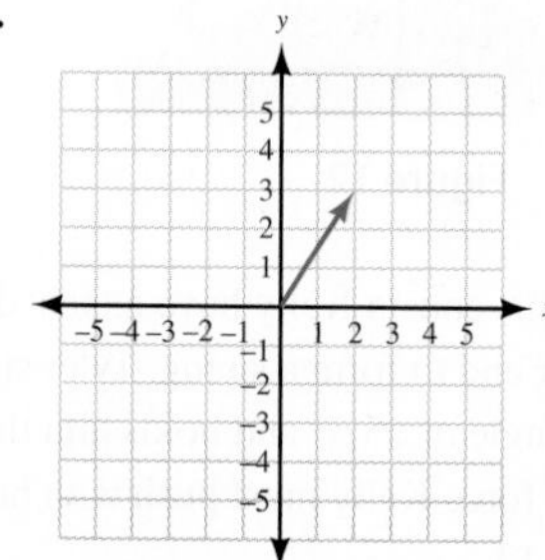

d.

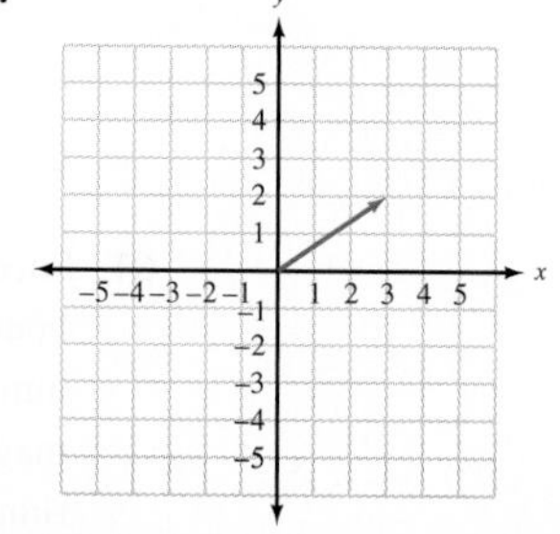

66. If vector **V** is in standard position, has magnitude 12, and makes an angle of 120° with the positive x-axis, write **V** in vector component form.

a. $\mathbf{V} = -6\sqrt{3}\mathbf{i} + 6\mathbf{j}$ b. $\mathbf{V} = -6\mathbf{i} + 6\sqrt{3}\mathbf{j}$
c. $\mathbf{V} = 12\mathbf{i} + 120\mathbf{j}$ d. $\mathbf{V} = 120\mathbf{i} + 12\mathbf{j}$

67. Find the magnitude of vector $\mathbf{V} = 5\mathbf{i} - 2\mathbf{j}$.

a. 10 b. $\sqrt{29}$ c. 7 d. $\sqrt{21}$

68. If $\mathbf{U} = 3\mathbf{i} + 7\mathbf{j}$ and $\mathbf{V} = \mathbf{i} - 4\mathbf{j}$, find $\mathbf{U} + 2\mathbf{V}$.

a. $\mathbf{i} + 15\mathbf{j}$ b. $5\mathbf{i} + 3\mathbf{j}$ c. $10\mathbf{i} - 6\mathbf{j}$ d. $5\mathbf{i} - \mathbf{j}$

SECTION 7.6 Vectors: The Dot Product

LEARNING OBJECTIVES

1. Compute a dot product.
2. Find the angle between two vectors.
3. Determine if two nonzero vectors are perpendicular.
4. Use the dot product to calculate work.

Now that we have a way to represent vectors algebraically, we can define a type of multiplication between two vectors. The *dot product* (also called the *scalar product*) is a form of multiplication that results in a scalar quantity. For our purposes, it will be useful when finding the angle between two vectors or for finding the work done by a force in moving an object. Here is the definition of the dot product of two vectors.

DEFINITION

The *dot product* of two vectors $\mathbf{U} = a\mathbf{i} + b\mathbf{j}$ and $\mathbf{V} = c\mathbf{i} + d\mathbf{j}$ is written $\mathbf{U} \bullet \mathbf{V}$ and is defined as follows:

$$\begin{aligned}\mathbf{U} \bullet \mathbf{V} &= (a\mathbf{i} + b\mathbf{j}) \bullet (c\mathbf{i} + d\mathbf{j}) \\ &= (ac) + (bd)\end{aligned}$$

As you can see, the dot product is a real number (scalar), not a vector.

PROBLEM 1
Find each of the following dot products.
a. $\mathbf{U} \bullet \mathbf{V}$ when $\mathbf{U} = \langle 4, 5\rangle$ and $\mathbf{V} = \langle 1, 7\rangle$
b. $\langle -3, 2\rangle \bullet \langle 5, -8\rangle$
c. $\mathbf{S} \bullet \mathbf{W}$ when $\mathbf{S} = 3\mathbf{i} - 6\mathbf{j}$ and $\mathbf{W} = 4\mathbf{i} + 9\mathbf{j}$

EXAMPLE 1 Find each of the following dot products.

a. $\mathbf{U} \bullet \mathbf{V}$ when $\mathbf{U} = \langle 3, 4\rangle$ and $\mathbf{V} = \langle 2, 5\rangle$
b. $\langle -1, 2\rangle \bullet \langle 3, -5\rangle$
c. $\mathbf{S} \bullet \mathbf{W}$ when $\mathbf{S} = 6\mathbf{i} + 3\mathbf{j}$ and $\mathbf{W} = 2\mathbf{i} - 7\mathbf{j}$

SOLUTION For each problem, we simply multiply the coefficients a and c and add that result to the product of the coefficients b and d.

a. $$\begin{aligned}\mathbf{U} \bullet \mathbf{V} &= 3(2) + 4(5) \\ &= 6 + 20 \\ &= 26\end{aligned}$$

b. $$\begin{aligned}\langle -1, 2\rangle \bullet \langle 3, -5\rangle &= -1(3) + 2(-5) \\ &= -3 + (-10) \\ &= -13\end{aligned}$$

c. $$\begin{aligned}\mathbf{S} \bullet \mathbf{W} &= 6(2) + 3(-7) \\ &= 12 + (-21) \\ &= -9\end{aligned}$$

Finding the Angle Between Two Vectors

One application of the dot product is finding the angle between two vectors. To do this, we will use an alternate form of the dot product, shown in the following theorem.

THEOREM 7.1

The dot product of two vectors is equal to the product of their magnitudes multiplied by the cosine of the angle between them. That is, when θ is the angle between two nonzero vectors **U** and **V**, then

$$\mathbf{U} \cdot \mathbf{V} = |\mathbf{U}||\mathbf{V}| \cos \theta$$

The proof of this theorem is derived from the law of cosines and is left as an exercise.

When we are given two vectors and asked to find the angle between them, we rewrite the formula in Theorem 7.1 by dividing each side by $|\mathbf{U}||\mathbf{V}|$. The result is

$$\cos \theta = \frac{\mathbf{U} \cdot \mathbf{V}}{|\mathbf{U}||\mathbf{V}|}$$

This formula is equivalent to our original formula, but is easier to work with when finding the angle between two nonzero vectors.

PROBLEM 2
Find the angle between vectors **U** and **V**.
a. $\mathbf{U} = \langle 3, 4 \rangle$ and $\mathbf{V} = \langle -1, 2 \rangle$
b. $\mathbf{U} = 3\mathbf{i} - 5\mathbf{j}$ and $\mathbf{V} = 2\mathbf{i} + 5\mathbf{j}$

EXAMPLE 2 Find the angle between the vectors **U** and **V**.

a. $\mathbf{U} = \langle 2, 3 \rangle$ and $\mathbf{V} = \langle -3, 2 \rangle$ **b.** $\mathbf{U} = 6\mathbf{i} - \mathbf{j}$ and $\mathbf{V} = \mathbf{i} + 4\mathbf{j}$

SOLUTION

a.
$$\begin{aligned} \cos \theta &= \frac{\mathbf{U} \cdot \mathbf{V}}{|\mathbf{U}||\mathbf{V}|} \\ &= \frac{2(-3) + 3(2)}{\sqrt{2^2 + 3^2} \cdot \sqrt{(-3)^2 + 2^2}} \\ &= \frac{-6 + 6}{\sqrt{13} \cdot \sqrt{13}} \\ &= \frac{0}{13} \\ \cos \theta &= 0 \\ \theta &= 90^\circ \end{aligned}$$

b.
$$\begin{aligned} \cos \theta &= \frac{\mathbf{U} \cdot \mathbf{V}}{|\mathbf{U}||\mathbf{V}|} \\ &= \frac{6(1) + (-1)4}{\sqrt{6^2 + (-1)^2} \cdot \sqrt{1^2 + 4^2}} \\ &= \frac{6 + (-4)}{\sqrt{37} \cdot \sqrt{17}} \\ &= \frac{2}{25.08} \\ \cos \theta &= 0.0797 \\ \theta &= \cos^{-1}(0.0797) = 85.43^\circ \end{aligned}$$

To the nearest hundredth of a degree ■

Perpendicular Vectors

If two nonzero vectors are perpendicular, then the angle between them is 90°. We sometimes refer to perpendicular vectors as being orthogonal. Because the cosine of 90° is always 0, the dot product of two perpendicular vectors must also be 0. This fact gives rise to the following theorem.

THEOREM 7.2

If **U** and **V** are two nonzero vectors, then

$$\mathbf{U} \cdot \mathbf{V} = 0 \Longleftrightarrow \mathbf{U} \perp \mathbf{V}$$

In Words: Two nonzero vectors are perpendicular, or orthogonal, if and only if their dot product is 0.

PROBLEM 3
Which of the following vectors are perpendicular to each other?
$\mathbf{U} = 2\mathbf{i} + 5\mathbf{j}$, $\mathbf{V} = 4\mathbf{i} + 10\mathbf{j}$, $\mathbf{W} = 5\mathbf{i} - 2\mathbf{j}$

EXAMPLE 3 Given vectors $\mathbf{U} = 8\mathbf{i} + 6\mathbf{j}$, $\mathbf{V} = 3\mathbf{i} - 4\mathbf{j}$, and $\mathbf{W} = 4\mathbf{i} + 3\mathbf{j}$, determine if **U** is perpendicular to either **V** or **W**.

SOLUTION Find **U** • **V** and **U** • **W**. If the dot product is zero, then the two vectors are perpendicular.

$$\begin{aligned} \mathbf{U} \cdot \mathbf{V} &= 8(3) + 6(-4) \\ &= 24 - 24 \\ &= 0 \end{aligned}$$

Therefore, **U** and **V** are perpendicular

$$\begin{aligned} \mathbf{U} \cdot \mathbf{W} &= 8(4) + (6)3 \\ &= 32 + 18 \\ &= 50 \end{aligned}$$

Therefore, **U** and **W** are not perpendicular

Work

In Section 2.5 we introduced the concept of work. Recall that work is performed when a constant force **F** is used to move an object a certain distance. We can represent the movement of the object using a displacement vector, **d**, as shown in Figure 1.

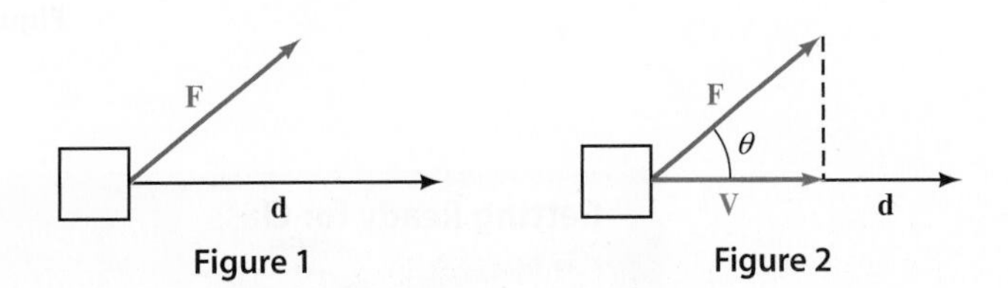

Figure 1 Figure 2

In Figure 2 we let **V** represent the component of **F** that is oriented in the same direction as **d**, since only the amount of the force in the direction of movement can be used in calculating work. **V** is sometimes called the *projection of* **F** *onto* **d**. We can find the magnitude of **V** using right triangle trigonometry:

$$|\mathbf{V}| = |\mathbf{F}| \cos \theta$$

Because $|\mathbf{d}|$ represents the distance the object is moved, the work performed by the force is

$$\begin{aligned} \text{Work} &= |\mathbf{V}||\mathbf{d}| \\ &= (|\mathbf{F}|\cos\theta)\cdot|\mathbf{d}| \\ &= |\mathbf{F}||\mathbf{d}|\cos\theta \\ &= \mathbf{F}\bullet\mathbf{d} \qquad \text{By Theorem 7.1} \end{aligned}$$

We have just established the following theorem.

THEOREM 7.3

If a constant force **F** is applied to an object, and the resulting movement of the object is represented by the displacement vector **d**, then the work performed by the force is

$$\text{Work} = \mathbf{F}\bullet\mathbf{d}$$

PROBLEM 4

A force $\mathbf{F} = 25\mathbf{i} - 16\mathbf{j}$ (in pounds) is used to push an object up a ramp. The resulting movement of the object is represented by the displacement vector $\mathbf{d} = 12\mathbf{i} + 8\mathbf{j}$ (in feet). Find the work done by the force.

EXAMPLE 4 A force $\mathbf{F} = 35\mathbf{i} - 12\mathbf{j}$ (in pounds) is used to push an object up a ramp. The resulting movement of the object is represented by the displacement vector $\mathbf{d} = 15\mathbf{i} + 4\mathbf{j}$ (in feet), as illustrated in Figure 3. Find the work done by the force.

SOLUTION By Theorem 7.3,

$$\begin{aligned} \text{Work} &= \mathbf{F}\bullet\mathbf{d} \\ &= 35(15) + (-12)(4) \\ &= 480 \text{ ft-lb} \qquad \text{To two significant digits} \end{aligned}$$

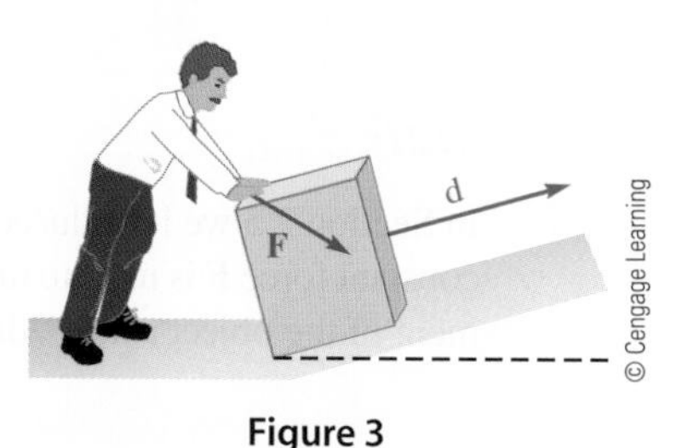

Figure 3

Getting Ready for Class *After reading through the preceding section, respond in your own words and in complete sentences.*

a. Explain how to find the dot product of two vectors.
b. Explain how the dot product is used to find the angle between two vectors.
c. How do we know if two nonzero vectors are perpendicular?
d. Explain how vectors can be used to calculate work.

7.6 PROBLEM SET

➤ CONCEPTS AND VOCABULARY

For Questions 1 through 8, fill in the blank with an appropriate word.

1. To find the dot product of two vectors, _________ corresponding components, and then ____ these two values together.
2. The dot product of two vectors is a ________ quantity. For this reason, it is sometimes called the _______ product.
3. The dot product of two vectors is also equal to the product of their __________ multiplied by the ______ of the angle between them.
4. To find the angle between two nonzero vectors, find the inverse _______ of the quotient of the _____ _______ of the vectors divided by the product of their _________.
5. Perpendicular vectors are also said to be __________.
6. Two nonzero vectors are perpendicular if and only if their _____ _______ is equal to ______.
7. The component of a force **F** that is oriented in the same direction as another vector **d** is called the _________ of **F** ____ **d**.
8. If a constant force **F** is applied to an object, and the resulting movement of the object is represented by the displacement vector **d**, then the work performed by the force is given by the ____ ______ of **F** and **d**.

➤ EXERCISES

Find each of the following dot products.

9. $\langle 3, 4\rangle \cdot \langle 5, 5\rangle$
10. $\langle 6, 6\rangle \cdot \langle 3, 5\rangle$
11. $\langle -23, 4\rangle \cdot \langle 15, -6\rangle$
12. $\langle 11, -8\rangle \cdot \langle 4, -7\rangle$

For each pair of vectors, find $\mathbf{U} \cdot \mathbf{V}$.

13. $\mathbf{U} = \mathbf{i} + \mathbf{j}$, $\mathbf{V} = \mathbf{i} - \mathbf{j}$
14. $\mathbf{U} = -\mathbf{i} + \mathbf{j}$, $\mathbf{V} = \mathbf{i} + \mathbf{j}$
15. $\mathbf{U} = -3\mathbf{i}$, $\mathbf{V} = 5\mathbf{j}$
16. $\mathbf{U} = 6\mathbf{i}$, $\mathbf{V} = -8\mathbf{j}$
17. $\mathbf{U} = 2\mathbf{i} + 5\mathbf{j}$, $\mathbf{V} = 5\mathbf{i} + 2\mathbf{j}$
18. $\mathbf{U} = 5\mathbf{i} + 3\mathbf{j}$, $\mathbf{V} = 3\mathbf{i} + 5\mathbf{j}$
19. $\mathbf{U} = -4\mathbf{i} - 3\mathbf{j}$, $\mathbf{V} = -\mathbf{i} - 2\mathbf{j}$
20. $\mathbf{U} = -2\mathbf{i} - 9\mathbf{j}$, $\mathbf{V} = -3\mathbf{i} - \mathbf{j}$
21. $\mathbf{U} = -11\mathbf{i} + 7\mathbf{j}$, $\mathbf{V} = 9\mathbf{i} - 5\mathbf{j}$
22. $\mathbf{U} = 5\mathbf{i} - 11\mathbf{j}$, $\mathbf{V} = -20\mathbf{i} + 9\mathbf{j}$

Find the angle θ *between the given vectors to the nearest tenth of a degree.*

23. $\mathbf{U} = 13\mathbf{i}$, $\mathbf{V} = -6\mathbf{j}$
24. $\mathbf{U} = -4\mathbf{i}$, $\mathbf{V} = 17\mathbf{j}$
25. $\mathbf{U} = -3\mathbf{i} + 5\mathbf{j}$, $\mathbf{V} = 6\mathbf{i} + 3\mathbf{j}$
26. $\mathbf{U} = 4\mathbf{i} + 5\mathbf{j}$, $\mathbf{V} = 7\mathbf{i} - 4\mathbf{j}$
27. $\mathbf{U} = 13\mathbf{i} - 8\mathbf{j}$, $\mathbf{V} = 2\mathbf{i} + 11\mathbf{j}$
28. $\mathbf{U} = 11\mathbf{i} + 7\mathbf{j}$, $\mathbf{V} = -4\mathbf{i} + 6\mathbf{j}$

Show that each pair of vectors is perpendicular.

29. $\mathbf{i} + \mathbf{j}$ and $\mathbf{i} - \mathbf{j}$
30. $\mathbf{i}$ and $\mathbf{j}$
31. $-\mathbf{i}$ and $\mathbf{j}$
32. $2\mathbf{i} + \mathbf{j}$ and $\mathbf{i} - 2\mathbf{j}$
33. $-4\mathbf{i} - 3\mathbf{j}$ and $6\mathbf{i} - 8\mathbf{j}$
34. $-6\mathbf{i} - 5\mathbf{j}$ and $10\mathbf{i} - 12\mathbf{j}$
35. Find the value of a so that vectors $\mathbf{U} = a\mathbf{i} + 6\mathbf{j}$ and $\mathbf{V} = 9\mathbf{i} + 12\mathbf{j}$ are perpendicular.
36. In general, show that the vectors $\mathbf{V} = a\mathbf{i} + b\mathbf{j}$ and $\mathbf{W} = -b\mathbf{i} + a\mathbf{j}$ are always perpendicular. Assume a and b are not both equal to zero.

Find the work performed when the given force **F** *is applied to an object, whose resulting motion is represented by the displacement vector* **d**. *Assume the force is in pounds and the displacement is measured in feet.*

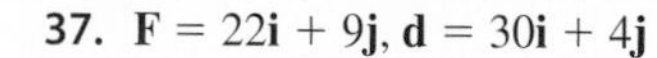

37. $\mathbf{F} = 22\mathbf{i} + 9\mathbf{j}$, $\mathbf{d} = 30\mathbf{i} + 4\mathbf{j}$

38. $\mathbf{F} = 45\mathbf{i} - 12\mathbf{j}$, $\mathbf{d} = 70\mathbf{i} + 15\mathbf{j}$

39. $\mathbf{F} = -6\mathbf{i} + 19\mathbf{j}$, $\mathbf{d} = 8\mathbf{i} + 55\mathbf{j}$

40. $\mathbf{F} = -67\mathbf{i} + 39\mathbf{j}$, $\mathbf{d} = -96\mathbf{i} - 28\mathbf{j}$

41. $\mathbf{F} = 85\mathbf{i}$, $\mathbf{d} = 6\mathbf{i}$

42. $\mathbf{F} = 54\mathbf{i}$, $\mathbf{d} = 20\mathbf{i}$

43. $\mathbf{F} = 13\mathbf{j}$, $\mathbf{d} = 44\mathbf{i}$

44. $\mathbf{F} = 39\mathbf{j}$, $\mathbf{d} = 72\mathbf{i}$

45. Use the diagram shown in Figure 4 along with the law of cosines to prove Theorem 7.1. (Begin by writing $\mathbf{U} = a\mathbf{i} + b\mathbf{j}$ and $\mathbf{V} = c\mathbf{i} + d\mathbf{j}$.)

46. Use Theorem 7.1 to prove Theorem 7.2.

y, x, U, V, U–V, θ

Figure 4

➤ REVIEW PROBLEMS

The problems that follow are problems you may have worked previously in Section 2.5. Solve each problem by first expressing the force and the displacement of the object as vectors in terms of the unit vectors **i** *and* **j**. *Then use Theorem 7.3 to find the work done.*

47. Work A package is pushed across a floor a distance of 75 feet by exerting a force of 41 pounds downward at an angle of 20° with the horizontal. How much work is done?

48. Work A package is pushed across a floor a distance of 52 feet by exerting a force of 15 pounds downward at an angle of 25° with the horizontal. How much work is done?

49. Work An automobile is pushed down a level street by exerting a force of 85 pounds at an angle of 15° with the horizontal (Figure 5). How much work is done in pushing the car 110 feet?

Figure 5

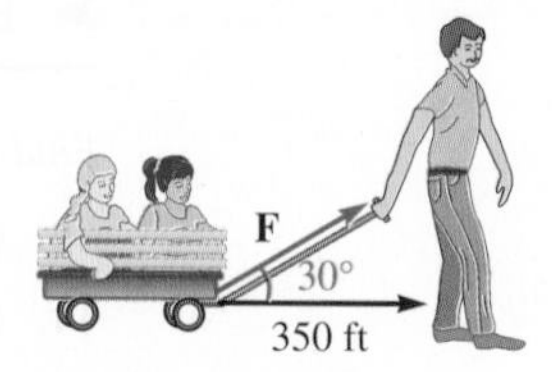

Figure 6

50. Work Mark pulls Allison and Mattie in a wagon by exerting a force of 25 pounds on the handle at an angle of 30° with the horizontal (Figure 6). How much work is done by Mark in pulling the wagon 350 feet?

➤ LEARNING OBJECTIVES ASSESSMENT

These questions are available for instructors to help assess if you have successfully met the learning objectives for this section.

51. If $\mathbf{U} = 7\mathbf{i} + 2\mathbf{j}$ and $\mathbf{V} = 3\mathbf{i} - 4\mathbf{j}$, find $\mathbf{U} \cdot \mathbf{V}$.

a. $21\mathbf{i} - 8\mathbf{j}$ **b.** $21\mathbf{i}^2 - 22\mathbf{ij} - 8\mathbf{j}^2$ **c.** 2 **d.** 13

52. Find the angle between $\mathbf{U} = 3\mathbf{i} + 7\mathbf{j}$ and $\mathbf{V} = \mathbf{i} - 4\mathbf{j}$.

a. 52.8° **b.** 142.8° **c.** 137.2° **d.** 157.9°

53. Which pair of vectors are perpendicular?

a. $3\mathbf{i} + 4\mathbf{j}$ and $8\mathbf{i} - 6\mathbf{j}$ **b.** $3\mathbf{i} + 2\mathbf{j}$ and $2\mathbf{i} + 3\mathbf{j}$ **c.** $\mathbf{i} + 5\mathbf{j}$ and $\mathbf{i} - 5\mathbf{j}$ **d.** $2\mathbf{i} - 5\mathbf{j}$ and $-7\mathbf{i} - 3\mathbf{j}$

54. Find the work performed when a force $\mathbf{F} = 15\mathbf{i} - 9\mathbf{j}$ is applied to an object whose resulting motion is represented by displacement vector $\mathbf{d} = 80\mathbf{i} + 12\mathbf{j}$. Assume the force is measured in pounds and the displacement in feet.

a. 1,415 ft-lb **b.** 552 ft-lb **c.** 1,092 ft-lb **d.** 1,308 ft-lb

CHAPTER 7 SUMMARY

EXAMPLES

1 If $A = 30°$, $B = 70°$, and $a = 8.0$ cm in triangle ABC, then, by the law of sines,

$$b = \frac{a \sin B}{\sin A} = \frac{8 \sin 70°}{\sin 30°}$$

$$= 15 \text{ cm}$$

THE LAW OF SINES [7.1]

For any triangle ABC, the following relationships are always true:

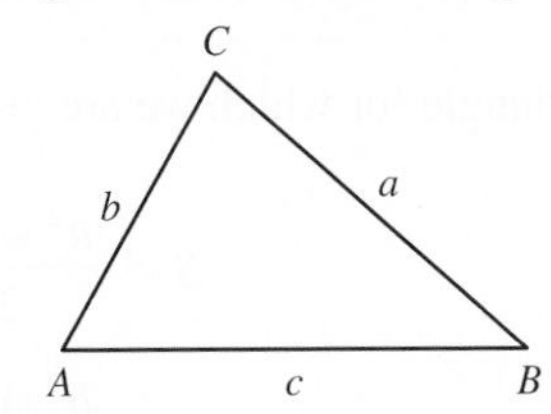

$$\frac{\sin A}{a} = \frac{\sin B}{b} = \frac{\sin C}{c}$$

or, equivalently,

$$\frac{a}{\sin A} = \frac{b}{\sin B} = \frac{c}{\sin C}$$

2 In triangle ABC, if $a = 34$ km, $b = 20$ km, and $c = 18$ km, then we can find A using the law of cosines.

$$\cos A = \frac{b^2 + c^2 - a^2}{2bc}$$

$$= \frac{20^2 + 18^2 - 34^2}{(2)(20)(18)}$$

$$\cos A = -0.6000$$

$$\text{so } A = 127°$$

THE LAW OF COSINES [7.2]

In any triangle ABC, the following relationships are always true:

$$a^2 = b^2 + c^2 - 2bc \cos A \qquad \cos A = \frac{b^2 + c^2 - a^2}{2bc}$$

$$b^2 = a^2 + c^2 - 2ac \cos B \qquad \cos B = \frac{a^2 + c^2 - b^2}{2ac}$$

$$c^2 = a^2 + b^2 - 2ab \cos C \qquad \cos C = \frac{a^2 + b^2 - c^2}{2ab}$$

3

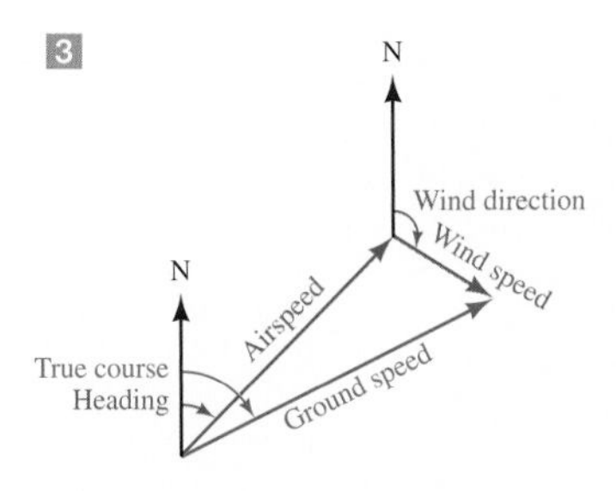

NAVIGATION [7.2]

The *heading* of an object is the angle measured clockwise from due north to the vector representing the path of the object. If the object is subject to wind or currents, then the actual direction of the object is called its *true course.*

4 In triangle ABC, if $a = 54$ cm, $b = 62$ cm, and $A = 40°$, then

$$\sin B = \frac{b \sin A}{a} = \frac{62 \sin 40°}{54}$$

$$= 0.7380$$

Because $\sin B$ is positive for any angle in QI or QII, we have two possibilities for B:

$$B = 48° \quad \text{or} \quad B' = 180° - 48° = 132°$$

THE AMBIGUOUS CASE [7.3]

When we are given two sides and an angle opposite one of them (SSA), we have several possibilities for the triangle or triangles that result. One of the possibilities is that no triangle will fit the given information. Another possibility is that two different triangles can be obtained from the given information, and a third possibility is that exactly one triangle will fit the given information. Because of these different possibilities, we call the situation where we are solving a triangle in which we are given two sides and the angle opposite one of them the *ambiguous case.*

THE AREA OF A TRIANGLE [7.4]

The area of a triangle for which we are given two sides and the included angle is given by

$$S = \frac{1}{2}ab \sin C \qquad S = \frac{1}{2}ac \sin B \qquad S = \frac{1}{2}bc \sin A$$

The area of a triangle for which we are given all three sides is given by the formula

$$S = \sqrt{s(s-a)(s-b)(s-c)}$$

$$\text{where } s = \frac{1}{2}(a+b+c)$$

The area of a triangle for which we are given two angles and a side is given by

$$S = \frac{a^2 \sin B \sin C}{2 \sin A}$$

$$S = \frac{b^2 \sin A \sin C}{2 \sin B}$$

$$S = \frac{c^2 \sin A \sin B}{2 \sin C}$$

5 For triangle ABC,

a. If $a = 12$ cm, $b = 15$ cm, and $C = 20°$, then the area of ABC is

$$S = \frac{1}{2}(12)(15)\sin 20°$$
$$= 30.8 \text{ cm}^2 \text{ to the nearest tenth}$$

b. If $a = 24$ inches, $b = 14$ inches, and $c = 18$ inches, then the area of ABC is

$$S = \sqrt{28(28-24)(28-14)(28-18)}$$
$$= \sqrt{28(4)(14)(10)}$$
$$= \sqrt{15{,}680}$$
$$= 125.2 \text{ inches}^2 \text{ to the nearest tenth}$$

c. If $A = 40°$, $B = 72°$, and $c = 45$ m, then the area of ABC is

$$S = \frac{45^2 \sin 40° \sin 72°}{2 \sin 68°}$$

$$S = \frac{2{,}025(0.6428)(0.9511)}{2(0.9272)}$$
$$= 667.6 \text{ m}^2 \text{ to the nearest tenth}$$

ALGEBRAIC VECTORS [7.5]

The vector that extends from the origin to the point (1, 0) is called the *unit horizontal vector* and is denoted by **i**. The vector that extends from the origin to the point (0, 1) is called the *unit vertical vector* and is denoted by **j**. Any nonzero vector **V** can be written

1. In terms of unit vectors as $\mathbf{V} = a\mathbf{i} + b\mathbf{j}$
2. In component form as $\mathbf{V} = \langle a, b \rangle$

where a and b are real numbers.

6 The vector **V** that extends from the origin to the point $(-3, 4)$ is

$$\mathbf{V} = -3\mathbf{i} + 4\mathbf{j}$$
$$= \langle -3, 4 \rangle$$

MAGNITUDE [7.5]

The magnitude of $\mathbf{V} = a\mathbf{i} + b\mathbf{j} = \langle a, b \rangle$ is

$$|\mathbf{V}| = \sqrt{a^2 + b^2}$$

7 The magnitude of $\mathbf{V} = -3\mathbf{i} + 4\mathbf{j}$ is

$$|\mathbf{V}| = \sqrt{(-3)^2 + 4^2}$$
$$= \sqrt{25}$$
$$= 5$$

ADDITION AND SUBTRACTION WITH ALGEBRAIC VECTORS [7.5]

If $\mathbf{U} = a\mathbf{i} + b\mathbf{j} = \langle a, b \rangle$ and $\mathbf{V} = c\mathbf{i} + d\mathbf{j} = \langle c, d \rangle$, then vector addition and subtraction are defined as follows:

Addition: $\mathbf{U} + \mathbf{V} = (a + c)\mathbf{i} + (b + d)\mathbf{j} = \langle a + c, b + d \rangle$

Subtraction: $\mathbf{U} - \mathbf{V} = (a - c)\mathbf{i} + (b - d)\mathbf{j} = \langle a - c, b - d \rangle$

8 If $\mathbf{U} = 6\mathbf{i} + 2\mathbf{j}$ and $\mathbf{V} = -3\mathbf{i} + 5\mathbf{j}$, then

$$\mathbf{U} + \mathbf{V} = (6\mathbf{i} + 2\mathbf{j}) + (-3\mathbf{i} + 5\mathbf{j})$$
$$= (6 - 3)\mathbf{i} + (2 + 5)\mathbf{j}$$
$$= 3\mathbf{i} + 7\mathbf{j}$$
$$\mathbf{U} - \mathbf{V} = (6\mathbf{i} + 2\mathbf{j}) - (-3\mathbf{i} + 5\mathbf{j})$$
$$= [6 - (-3)]\mathbf{i} + (2 - 5)\mathbf{j}$$
$$= 9\mathbf{i} - 3\mathbf{j}$$

9 If $\mathbf{U} = -3\mathbf{i} + 4\mathbf{j}$ and $\mathbf{V} = 4\mathbf{i} + 3\mathbf{j}$, then

$$\mathbf{U} \cdot \mathbf{V} = -3(4) + 4(3) = 0$$

DOT PRODUCT [7.6]

The *dot product* of two vectors $\mathbf{U} = a\mathbf{i} + b\mathbf{j}$ and $\mathbf{V} = c\mathbf{i} + d\mathbf{j}$ is written $\mathbf{U} \cdot \mathbf{V}$ and is defined as follows:

$$\mathbf{U} \cdot \mathbf{V} = ac + bd$$

If θ is the angle between the two vectors, then it is also true that

$$\mathbf{U} \cdot \mathbf{V} = |\mathbf{U}||\mathbf{V}| \cos \theta \quad \text{or} \quad \cos \theta = \frac{\mathbf{U} \cdot \mathbf{V}}{|\mathbf{U}||\mathbf{V}|}$$

10 The vectors **U** and **V** in Example 9 are perpendicular because their dot product is 0.

PERPENDICULAR VECTORS [7.6]

If **U** and **V** are two nonzero vectors, then

$$\mathbf{U} \cdot \mathbf{V} = 0 \Longleftrightarrow \mathbf{U} \perp \mathbf{V}$$

In Words: Two nonzero vectors are perpendicular, or orthogonal, if and only if their dot product is 0.

11 If $\mathbf{F} = 35\mathbf{i} - 12\mathbf{j}$ and $\mathbf{d} = 15\mathbf{i} + 4\mathbf{j}$, then

$$\text{Work} = \mathbf{F} \cdot \mathbf{d} = 35(15) + (-12)(4) = 477$$

WORK [7.6]

If a constant force **F** is applied to an object, and the resulting movement of the object is represented by the displacement vector **d**, then the work performed by the force is

$$\text{Work} = \mathbf{F} \cdot \mathbf{d}$$

CHAPTER 7 TEST

Problems 1 through 10 refer to triangle ABC, which is not necessarily a right triangle.

1. If $A = 32°$, $B = 70°$, and $a = 3.8$ inches, use the law of sines to find b.
2. If $A = 38.2°$, $B = 63.4°$, and $c = 42.0$ cm, find all the missing parts.
3. If $A = 24.7°$, $C = 106.1°$, and $b = 34.0$ cm, find all the missing parts.
4. If $C = 60°$, $a = 10$ cm, and $b = 12$ cm, use the law of cosines to find c.
5. If $a = 5$ km, $b = 7$ km, and $c = 9$ km, use the law of cosines to find C to the nearest tenth of a degree.
6. Find all the missing parts if $a = 6.4$ m, $b = 2.8$ m, and $C = 119°$.
7. Find all the missing parts if $b = 3.7$ m, $c = 6.2$ m, and $A = 35°$.
8. Use the law of sines to show that no triangle exists for which $A = 60°$, $a = 12$ inches, and $b = 42$ inches.
9. Use the law of sines to show that exactly one triangle exists for which $A = 42°$, $a = 29$ inches, and $b = 21$ inches.
10. Find two triangles for which $A = 51°$, $a = 6.5$ ft, and $b = 7.9$ ft.
11. Find the area of the triangle in Problem 2.
12. Find the area of the triangle in Problem 4.
13. Find the area of the triangle in Problem 5.
14. **Geometry** The two equal sides of an isosceles triangle are each 38 centimeters. If the base measures 48 centimeters, find the measure of the two equal angles.

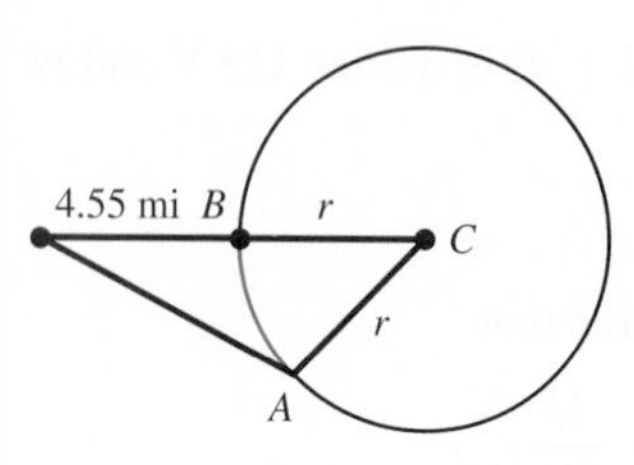

Figure 1

15. **Angle of Elevation** A man standing near a building notices that the angle of elevation to the top of the building is 64°. He then walks 240 feet farther away from the building and finds the angle of elevation to the top to be 43°. How tall is the building?

16. **Geometry** The diagonals of a parallelogram are 26.8 meters and 39.4 meters. If they meet at an angle of 134.5°, find the length of the shorter side of the parallelogram.

17. **Arc Length** Suppose Figure 1 is an exaggerated diagram of a plane flying above Earth. When the plane is 4.55 miles above point B, the pilot finds angle A to be 90.8°. Assuming that the radius of the earth is 3,960 miles, what is the distance from point A to point B along the circumference of the earth? (*Hint:* Find angle C first, then use the formula for arc length.)

18. **Distance and Bearing** A man wandering in the desert walks 3.3 miles in the direction S 44° W. He then turns and walks 2.2 miles in the direction N 55° W. At that time, how far is he from his starting point, and what is his bearing from his starting point?

19. **Distance** Two guy wires from the top of a tent pole are anchored to the ground on each side of the pole by two stakes so that the two stakes and the tent pole lie along the same line. One of the wires is 56 feet long and makes an angle of 47° with the ground. The other wire is 65 feet long and makes an angle of 37° with the ground. How far apart are the stakes that hold the wires to the ground?

20. **Ground Speed** A plane is headed due east with an airspeed of 345 miles per hour. Its true course, however, is at 95.5° from due north. If the wind currents are a constant 55.0 miles per hour, what are the possibilities for the ground speed of the plane?

21. **Height of a Tree** To estimate the height of a tree, two people position themselves 25 feet apart. From the first person, the bearing of the tree is N 48° E and the angle of elevation to the top of the tree is 73°. If the bearing of the tree from the second person is N 38° W, estimate the height of the tree to the nearest foot.

22. **True Course and Speed** A plane flying with an airspeed of 325 miles per hour is headed in the direction 87.6°. The wind currents are running at a constant 65.4 miles per hour at 262.6°. Find the ground speed and true course of the plane.

Let $\mathbf{U} = 5\mathbf{i} + 12\mathbf{j}$, $\mathbf{V} = -4\mathbf{i} + \mathbf{j}$, *and* $\mathbf{W} = \mathbf{i} - 4\mathbf{j}$, *and find:*

23. $|\mathbf{U}|$

24. $3\mathbf{U} + 5\mathbf{V}$

25. $|2\mathbf{V} - \mathbf{W}|$

26. $\mathbf{V} \cdot \mathbf{W}$

27. The angle between **U** and **V** to the nearest tenth of a degree.

28. Show that $\mathbf{V} = 3\mathbf{i} + 6\mathbf{j}$ and $\mathbf{W} = -8\mathbf{i} + 4\mathbf{j}$ are perpendicular.

29. Find the value of b so that vectors $\mathbf{U} = 5\mathbf{i} + 12\mathbf{j}$ and $\mathbf{V} = 4\mathbf{i} + b\mathbf{j}$ are perpendicular.

30. Find the work performed by the force $\mathbf{F} = 33\mathbf{i} - 4\mathbf{j}$ in moving an object, whose resulting motion is represented by the displacement vector $\mathbf{d} = 56\mathbf{i} + 10\mathbf{j}$.

Measuring the Distance to Mars

NASA

OBJECTIVE: To find the distance of Mars from the sun using two Earth-based observations.

The planet Mars orbits the Sun once every 687 days. This is referred to as its sidereal period. When observing Mars from Earth, the angle that Mars makes with the Sun is called the solar elongation (Figure 1). On November 13, 2000, Mars had a solar elongation of 42.8° W. On October 1, 2002, 687 days later, the solar elongation of Mars was 17.1° W. Because the number of days between these observations is equal to the sidereal period of Mars, the planet was in the same position on both occasions. Figure 2 is an illustration of the situation.

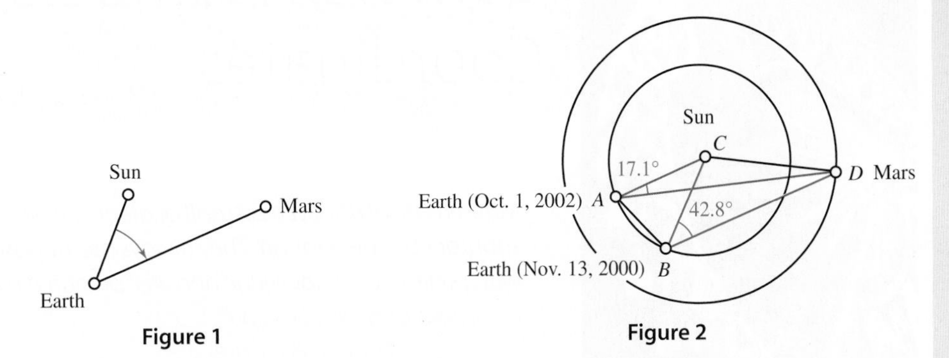

Figure 1 Figure 2

The distance from the Sun to Earth is defined as 1 astronomical unit (1 AU), which is about 93 million miles. Because both AC and BC have a length of 1 AU, triangle ABC is an isosceles triangle.

1. Use the fact that Earth completes one revolution about the Sun every 365.25 days to determine how many degrees the earth travels in one day. On November 13, 2002, Earth would again be at point B. Find the number of days it would take Earth to move from point A to point B. Now, using both of these results, compute the number of degrees for central angle ACB.
2. Find the distance AB using triangle ABC.
3. Because triangle ABC is isosceles, angles CAB and CBA are equal. Find the measure of these two angles. Use this result to find angles DAB and DBA.
4. Use triangle ABD to find the length of AD.
5. Find the length of CD. Give your answer both in terms of AU and in miles.

Quaternions

LECTURES
ON
QUATERNIONS:
CONTAINING A SYSTEMATIC STATEMENT
OF
A New Mathematical Method;
THE ROYAL IRISH ACADEMY;
THE HALLS OF TRINITY COLLEGE, DUBLIN:
BY
SIR WILLIAM ROWAN HAMILTON, LL.D., M.R.I.A.,

DUBLIN:
HODGES AND SMITH, GRAFTON-STREET,
BOOKSELLERS TO THE UNIVERSITY.
LONDON: WHITTAKER & CO., AVE-MARIA LANE.
CAMBRIDGE: MACMILLAN & CO.
1853.

We mentioned in Section 7.5 that our algebraic treatment of vectors could be attributed, in part, to the Irish mathematician William Rowan Hamilton. Hamilton considered his greatest achievement to be the discovery of *quaternions,* which he (incorrectly) predicted would revolutionize physics.

Research the subject of quaternions. What are they? Why did quaternions fail to be as useful for physics as Hamilton predicted? How are quaternions useful, instead, for 3-D computer graphics? Write a paragraph or two about your findings.

8 Complex Numbers and Polar Coordinates

Mathematical discoveries, small or great, are never born of spontaneous generation. They always presuppose a soil seeded with preliminary knowledge and well prepared by labour, both conscious and subconscious.

➤ *Henri Poincaré*

INTRODUCTION

In 1545, Jerome Cardan published the mathematical work *Ars Magna,* the title page of which is shown in Figure 1. The title translates to *The Great Art of Solving Algebraic Equations.* In this book Cardan posed a number of problems, such as solving certain cubic and quartic equations.

HIERONYMI CAR
DANI, PRÆSTANTISSIMI MATHE
MATICI, PHILOSOPHI, AC MEDICI,
ARTIS MAGNÆ,
SIVE DE REGVLIS ALGEBRAICIS,
Lib. unus. Qui & totius operis de Arithmetica, quod
OPVS PERFECTVM
inscripsit, est in ordine Decimus.

Courtesy Biblioteca Nacional Digital, Portugal

Figure 1

In our present-day notation, many of the problems posed by Cardan can be solved with some simple algebra, but the solutions require taking square roots of negative numbers. Square roots of negative numbers are handled with the complex number system, which had not been developed when Cardan wrote his book. The foundation of the complex number system is contained in this chapter, along with the very interesting and useful connection between complex numbers and trigonometric functions.

SECTION 8.1 Complex Numbers

LEARNING OBJECTIVES

1. Express the square root of a negative number as an imaginary number.
2. Add and subtract complex numbers.
3. Simplify powers of *i*.
4. Multiply and divide complex numbers.

One of the problems posed by Cardan in *Ars Magna* is the following:

> If someone says to you, divide 10 into two parts, one of which multiplied into the other shall produce 40, it is evident that this case or question is impossible.

Using our present-day notation, this problem can be solved with a system of equations:

$$x + y = 10$$
$$xy = 40$$

Solving the first equation for y, we have $y = 10 - x$. Substituting this value of y into the second equation, we have

$$x(10 - x) = 40$$

The equation is quadratic. We write it in standard form and apply the quadratic formula.

$$\begin{aligned}
0 &= x^2 - 10x + 40 \\
x &= \frac{10 \pm \sqrt{100 - 4(1)(40)}}{2} \\
&= \frac{10 \pm \sqrt{100 - 160}}{2} \\
&= \frac{10 \pm \sqrt{-60}}{2} \\
&= \frac{10 \pm 2\sqrt{-15}}{2} \\
&= 5 \pm \sqrt{-15}
\end{aligned}$$

This is as far as Cardan could take the problem because he did not know what to do with the square root of a negative number. We handle this situation by using *complex numbers.* Our work with complex numbers is based on the following definition.

DEFINITION

The number i, called the *imaginary unit,* is such that $i^2 = -1$. (That is, i is the number whose square is -1.)

Leonhard Euler

The Swiss mathematician Leonhard Euler was the first to introduce the number i in 1777 in one of his memoirs. The memoir was later formally published in 1794 in his work *Institutionum Calculi Integralis.*

Solutio.

Quoniam mihi quidem alia adhuc via non patet istud praestandi, nisi per imaginaria procedendo, formulam $\sqrt{-1}$ littera i in posterum designabo, ita ut sit $ii = -1$, ideoque $\frac{1}{i} = -i$. Jam ante omnia in numeratore nostrae formulae loco cos. Φ has duas partes substituamus

The number i is not a real number. We can use it to write square roots of negative numbers without a negative sign. To do so, we reason that if $a > 0$, then $\sqrt{-a} = \sqrt{ai^2} = i\sqrt{a}$.

PROBLEM 1
Write each expression in terms of i.
a. $\sqrt{-16}$
b. $\sqrt{-18}$
c. $\sqrt{-7}$

NOTE We usually place i after the coefficient, but in front of the radical, so that we avoid inadvertently writing the i inside the radical.

EXAMPLE 1 Write each expression in terms of i.

a. $\sqrt{-9}$ b. $\sqrt{-12}$ c. $\sqrt{-17}$

SOLUTION

a. $\sqrt{-9} = i\sqrt{9} = 3i$
b. $\sqrt{-12} = i\sqrt{12} = 2i\sqrt{3}$
c. $\sqrt{-17} = i\sqrt{17}$

To simplify expressions that contain square roots of negative numbers by using the properties of radicals developed in algebra, it is necessary to write each square root in terms of i before applying the properties of radicals. For example,

this is correct: $\sqrt{-4}\,\sqrt{-9} = (i\sqrt{4})(i\sqrt{9}) = (2i)(3i) = 6i^2 = -6$

this is incorrect: $\sqrt{-4}\,\sqrt{-9} = \sqrt{-4(-9)} = \sqrt{36} = 6$

Remember, the properties of radicals you developed in algebra hold only for expressions in which the numbers under the radical sign are nonnegative. When the radicals contain negative numbers, you must first write each radical in terms of i and then simplify.

Next, we use i to write a definition for complex numbers.

DEFINITION

A *complex number* is any number that can be written in the form

$$a + bi$$

where a and b are real numbers and $i^2 = -1$. The form $a + bi$ is called *standard form* for complex numbers. The number a is called the *real part* of the complex number. The number b is called the *imaginary part* of the complex number. If $b = 0$, then $a + bi = a$, which is a real number. If $a = 0$ and $b \neq 0$, then $a + bi = bi$, which is called an *imaginary number.*

PROBLEM 2
Find the real part and the imaginary part of each complex number.
a. $4 - 5i$
b. $-9i$
c. 3

EXAMPLE 2

a. The number $3 + 2i$ is a complex number in standard form. The number 3 is the real part, and the number 2 (not $2i$) is the imaginary part.

b. The number $-7i$ is a complex number because it can be written as $0 + (-7)i$. The real part is 0. The imaginary part is -7. The number $-7i$ is also an imaginary number since $a = 0$ and $b \neq 0$.

c. The number 4 is a complex number because it can be written as $4 + 0i$. The real part is 4 and the imaginary part is 0.

From part c in Example 2, it is apparent that real numbers are also complex numbers. The real numbers are a subset of the complex numbers.

Equality for Complex Numbers

DEFINITION

Two complex numbers are equal if and only if their real parts are equal and their imaginary parts are equal. That is, for real numbers a, b, c, and d,

$$a + bi = c + di \quad \text{if and only if} \quad a = c \text{ and } b = d$$

PROBLEM 3
Find x and y if $(4x - 8) + 8i = 20 + (-5y - 2)i$.

EXAMPLE 3 Find x and y if $(-3x - 9) + 4i = 6 + (3y - 2)i$.

SOLUTION The real parts are $-3x - 9$ and 6. The imaginary parts are 4 and $3y - 2$.

$$\begin{aligned} -3x - 9 &= 6 \\ -3x &= 15 \\ x &= -5 \end{aligned} \qquad \text{and} \qquad \begin{aligned} 4 &= 3y - 2 \\ 6 &= 3y \\ y &= 2 \end{aligned}$$

Addition and Subtraction of Complex Numbers

DEFINITION

If $z_1 = a_1 + b_1 i$ and $z_2 = a_2 + b_2 i$ are complex numbers, then the sum and difference of z_1 and z_2 are defined as follows:

$$\begin{aligned} z_1 + z_2 &= (a_1 + b_1 i) + (a_2 + b_2 i) \\ &= (a_1 + a_2) + (b_1 + b_2)i \end{aligned}$$

$$\begin{aligned} z_1 - z_2 &= (a_1 + b_1 i) - (a_2 + b_2 i) \\ &= (a_1 - a_2) + (b_1 - b_2)i \end{aligned}$$

As you can see, we add and subtract complex numbers in the same way we would add and subtract polynomials: by combining similar terms.

PROBLEM 4
If $z_1 = 5 + 3i$ and $z_2 = -7 - 6i$, find $z_1 + z_2$ and $z_1 - z_2$.

EXAMPLE 4 If $z_1 = 3 - 5i$ and $z_2 = -6 - 2i$, find $z_1 + z_2$ and $z_1 - z_2$.

SOLUTION

$$\begin{aligned} z_1 + z_2 &= (3 - 5i) + (-6 - 2i) \\ &= -3 - 7i \\ z_1 - z_2 &= (3 - 5i) - (-6 - 2i) \\ &= 9 - 3i \end{aligned}$$

Powers of *i*

If we assume the properties of exponents hold when the base is i, we can write any integer power of i as i, -1, $-i$, or 1. Using the fact that $i^2 = -1$, we have

$$\begin{aligned} i^1 &= i \\ i^2 &= -1 \\ i^3 &= i^2 \cdot i = -1(i) = -i \\ i^4 &= i^2 \cdot i^2 = -1(-1) = 1 \end{aligned}$$

Because $i^4 = 1$, i^5 will simplify to i and we will begin repeating the sequence $i, -1, -i, 1$ as we increase our exponent by one each time.

$$\begin{aligned} i^5 &= i^4 \cdot i = 1(i) = i \\ i^6 &= i^4 \cdot i^2 = 1(-1) = -1 \\ i^7 &= i^4 \cdot i^3 = 1(-i) = -i \\ i^8 &= i^4 \cdot i^4 = 1(1) = 1 \end{aligned}$$

We can simplify higher powers of i by writing them in terms of i^4 since i^4 is always 1.

PROBLEM 5
Simplify each power of i.
a. i^{24}
b. i^{17}
c. i^{22}

EXAMPLE 5 Simplify each power of i.

a. $i^{20} = (i^4)^5 = 1^5 = 1$

b. $i^{23} = (i^4)^5 \cdot i^3 = 1(-i) = -i$

c. $i^{30} = (i^4)^7 \cdot i^2 = 1(-1) = -1$

Multiplication and Division with Complex Numbers

DEFINITION

If $z_1 = a_1 + b_1 i$ and $z_2 = a_2 + b_2 i$ are complex numbers, then their product is defined as follows:

$$\begin{aligned} z_1 z_2 &= (a_1 + b_1 i)(a_2 + b_2 i) \\ &= (a_1 a_2 - b_1 b_2) + (a_1 b_2 + a_2 b_1)i \end{aligned}$$

This formula is simply a result of binomial multiplication and is much less complicated than it looks. Complex numbers have the form of binomials with i as the variable, so we can multiply two complex numbers using the same methods we use to multiply binomials and not have another formula to memorize.

PROBLEM 6
Multiply $(4 - 2i)(3 + i)$.

EXAMPLE 6 Multiply $(3 - 4i)(2 - 5i)$.

SOLUTION Multiplying as if these were two binomials, we have

$$\begin{aligned} (3 - 4i)(2 - 5i) &= 3 \cdot 2 - 3 \cdot 5i - 2 \cdot 4i + 4i \cdot 5i \\ &= 6 - 15i - 8i + 20i^2 \\ &= 6 - 23i + 20i^2 \end{aligned}$$

Now, because $i^2 = -1$, we can simplify further.

$$\begin{aligned} &= 6 - 23i + 20(-1) \\ &= 6 - 23i - 20 \\ &= -14 - 23i \end{aligned}$$

PROBLEM 7
Multiply $(7 - 2i)(7 + 2i)$.

EXAMPLE 7 Multiply $(4 - 5i)(4 + 5i)$.

SOLUTION This product has the form $(a - b)(a + b)$, which we know results in the difference of two squares, $a^2 - b^2$.

$$\begin{aligned} (4 - 5i)(4 + 5i) &= 4^2 - (5i)^2 \\ &= 16 - 25i^2 \\ &= 16 - 25(-1) \\ &= 16 + 25 \\ &= 41 \end{aligned}$$

The product of the two complex numbers $4 - 5i$ and $4 + 5i$ is the real number 41. This fact is very useful and leads to the following definition.

DEFINITION

The complex numbers $a + bi$ and $a - bi$ are called *complex conjugates.* Their product is the real number $a^2 + b^2$. Here's why:

$$\begin{aligned}(a + bi)(a - bi) &= a^2 - (bi)^2 \\ &= a^2 - b^2i^2 \\ &= a^2 - b^2(-1) \\ &= a^2 + b^2\end{aligned}$$

The fact that the product of two complex conjugates is a real number is the key to division with complex numbers.

PROBLEM 8

Divide $\dfrac{6i}{4 + 2i}$.

EXAMPLE 8 Divide $\dfrac{5i}{2 - 3i}$.

SOLUTION We want to find a complex number in standard form that is equivalent to the quotient $5i/(2 - 3i)$. To do so, we need to replace the denominator with a real number. We can accomplish this by multiplying both the numerator and the denominator by $2 + 3i$, which is the conjugate of $2 - 3i$.

$$\begin{aligned}\frac{5i}{2 - 3i} &= \frac{5i}{2 - 3i} \cdot \frac{\mathbf{(2 + 3}\boldsymbol{i}\mathbf{)}}{\mathbf{(2 + 3}\boldsymbol{i}\mathbf{)}} \\ &= \frac{5i(2 + 3i)}{(2 - 3i)(2 + 3i)} \\ &= \frac{10i + 15i^2}{4 - 9i^2} \\ &= \frac{10i + 15(-1)}{4 - 9(-1)} \\ &= \frac{-15 + 10i}{13} \\ &= -\frac{15}{13} + \frac{10}{13}i\end{aligned}$$

Notice that we have written our answer in standard form. The real part is $-15/13$ and the imaginary part is $10/13$. ■

CALCULATOR NOTE Some graphing calculators have the ability to perform operations with complex numbers. Check to see if your model has a key for entering the number i. Figure 1 shows how Examples 4 and 6 might look when solved on a graphing calculator. Example 8 is shown in Figure 2.

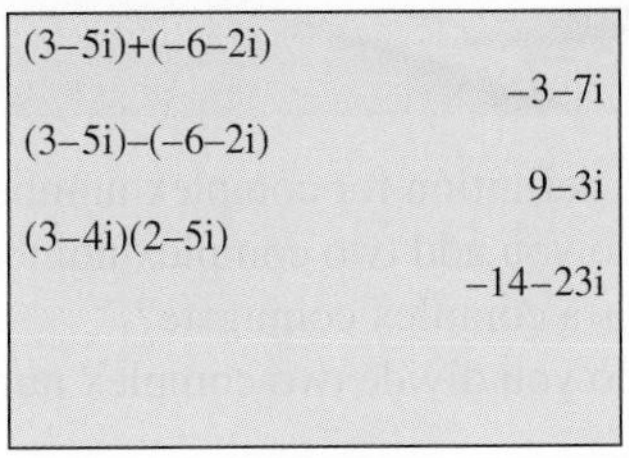

```
(3−5i)+(−6−2i)
                  −3−7i
(3−5i)−(−6−2i)
                   9−3i
(3−4i)(2−5i)
                −14−23i
```

Figure 1

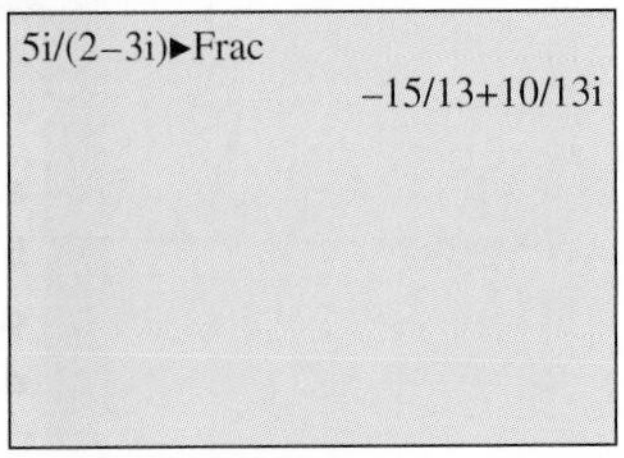

```
5i/(2−3i)►Frac
          −15/13+10/13i
```

Figure 2

PROBLEM 9
Show that $x = 6 + i\sqrt{19}$ and $y = 6 - i\sqrt{19}$ satisfy the system of equations $x + y = 12$, $xy = 55$.

EXAMPLE 9 In the introduction to this section, we discovered that $x = 5 + \sqrt{-15}$ is one part of a solution to the system of equations

$$x + y = 10$$
$$xy = 40$$

Find the value of y that accompanies this value of x. Then show that together they form a solution to the system of equations that describe Cardan's problem.

SOLUTION First we write the number $x = 5 + \sqrt{-15}$ with our complex number notation as $x = 5 + i\sqrt{15}$. Now, to find the value of y that accompanies this value of x, we substitute $5 + i\sqrt{15}$ for x in the equation $x + y = 10$ and solve for y:

$$\begin{aligned} &\text{When} & x &= 5 + i\sqrt{15} \\ &\text{then} & x + y &= 10 \\ &\text{becomes} & 5 + i\sqrt{15} + y &= 10 \\ & & y &= 5 - i\sqrt{15} \end{aligned}$$

All we have left to do is show that these values of x and y satisfy the second equation, $xy = 40$. We do so by substitution:

$$\begin{aligned} &\text{If} & x = 5 + i\sqrt{15} \quad \text{and} \quad y &= 5 - i\sqrt{15} \\ &\text{then the equation} & xy &= 40 \\ &\text{becomes} & (5 + i\sqrt{15})(5 - i\sqrt{15}) &= 40 \\ & & 5^2 - (i\sqrt{15})^2 &= 40 \\ & & 25 - i^2(\sqrt{15})^2 &= 40 \\ & & 25 - (-1)(15) &= 40 \\ & & 25 + 15 &= 40 \\ & & 40 &= 40 \end{aligned}$$

Because our last statement is a true statement, our two expressions for x and y together form a solution to the equation.

Getting Ready for Class

After reading through the preceding section, respond in your own words and in complete sentences.

a. Give a definition for complex numbers.
b. How do you add two complex numbers?
c. What is a complex conjugate?
d. How do you divide two complex numbers?

8.1 PROBLEM SET

➤ CONCEPTS AND VOCABULARY

For Questions 1 through 10, fill in the blank with an appropriate word or number.

1. The number i, called the _________ ______, is defined such that $i^2 =$ ____.
2. When writing the square root of a negative number in terms of i, we usually place i after the ________ but in front of the ______.
3. When simplifying radical expressions, it is necessary to write any square root of a _______ number in terms of i ________ using the properties of radicals.
4. The product of any nonzero real number and i is called an ___________ number.
5. A complex number can be a real number, an imaginary number, or the _____ of a real number and an imaginary number.
6. For any complex number $a + bi$, a is called the _____ part and b is called the ________ part.
7. Two complex numbers are equal if and only if their ______ parts are equal and their ________ parts are equal.
8. To add or subtract two complex numbers, simply combine _______ terms.
9. The number $a - bi$ is called the complex __________ of $a + bi$. The product of $a + bi$ and $a - bi$ will always be a _____ number.
10. To divide complex numbers, multiply the numerator and denominator by the complex ________ of the ________.
11. Match each power of i with its corresponding value when simplified.

a. i^1	i. -1
b. i^2	ii. 1
c. i^3	iii. i
d. i^4	iv. $-i$

For Questions 12 through 14, determine if the statement is true or false.

12. Every real number is also a complex number.
13. Every imaginary number is also a complex number.
14. Every complex number is either a real number or an imaginary number.

➤ EXERCISES

Write each expression in terms of i.

15. $\sqrt{-16}$
16. $\sqrt{-49}$
17. $\sqrt{-121}$
18. $\sqrt{-400}$
19. $\sqrt{-18}$
20. $\sqrt{-45}$
21. $\sqrt{-8}$
22. $\sqrt{-20}$

Write in terms of i and then simplify.

23. $\sqrt{-4} \cdot \sqrt{-9}$
24. $\sqrt{-25} \cdot \sqrt{-1}$
25. $\sqrt{-1} \cdot \sqrt{-9}$
26. $\sqrt{-16} \cdot \sqrt{-4}$

Find x and y so that each of the following equations is true.

27. $4 + 7i = 6x - 14yi$
28. $2 - 5i = -x + 10yi$
29. $(x^2 - 2x) + y^2 i = 8 + (2y - 1)i$
30. $(x^2 - 6) + 9i = x + y^2 i$

Find all x and y ($0 \le x, y < 2\pi$) so that each of the following equations is true.

31. $\cos x + i \sin y = \sin x + i$

32. $\sin x + i \cos y = -\cos x - i$

33. $(\sin^2 x + 1) + i \tan y = 2 \sin x + i$

34. $(\cos^2 x + 1) + i \tan y = 2 \cos x - i$

Combine the following complex numbers.

35. $(7 + 2i) + (3 - 4i)$

36. $(3 - 5i) + (2 + 4i)$

37. $(5 + 2i) - (3 + 6i)$

38. $(6 + 7i) - (4 + i)$

39. $(11 - 6i) - (2 - 4i)$

40. $(7 - 3i) - (4 + 10i)$

41. $(3 \cos x + 4i \sin y) + (2 \cos x - 7i \sin y)$

42. $(2 \cos x - 3i \sin y) + (3 \cos x - 2i \sin y)$

43. $[(3 + 2i) - (6 + i)] + (5 + i)$

44. $[(4 - 5i) - (2 + i)] + (2 + 5i)$

45. $(7 - 4i) - [(-2 + i) - (3 + 7i)]$

46. $(10 - 2i) - [(2 + i) - (3 - i)]$

Simplify each power of i.

47. i^{12}

48. i^{13}

49. i^{15}

50. i^{14}

51. i^{34}

52. i^{32}

53. i^{33}

54. i^{35}

Find the following products.

55. $-6i(3 - 8i)$

56. $6i(3 + 8i)$

57. $(2 + 4i)(3 - i)$

58. $(2 - 4i)(3 + i)$

59. $(3 - 2i)^2$

60. $(3 + 2i)^2$

61. $(4 + 5i)(4 - 5i)$

62. $(5 + 4i)(5 - 4i)$

63. $(7 + 2i)(7 - 2i)$

64. $(2 + 7i)(2 - 7i)$

65. $3i(1 + i)^2$

66. $4i(1 - i)^2$

Find the following quotients. Write all answers in standard form for complex numbers.

67. $\dfrac{2i}{3 + i}$

68. $\dfrac{3i}{2 + i}$

69. $\dfrac{2 + 3i}{2 - 3i}$

70. $\dfrac{3 + 2i}{3 - 2i}$

71. $\dfrac{5 - 2i}{i}$

72. $\dfrac{5 - 2i}{-i}$

73. $\dfrac{5 + 4i}{3 + 6i}$

74. $\dfrac{2 + i}{5 - 6i}$

Let $z_1 = 2 + 3i$, $z_2 = 2 - 3i$, and $z_3 = 4 + 5i$, and find

75. z_1z_2

76. z_2z_1

77. z_1z_3

78. z_3z_1

79. $2z_1 + 3z_2$

80. $3z_1 + 2z_2$

81. $z_3(z_1 + z_2)$

82. $z_3(z_1 - z_2)$

83. Assume x represents a real number and multiply $(x + 3i)(x - 3i)$.

84. Assume x represents a real number and multiply $(x - 4i)(x + 4i)$.

85. Show that $x = 2 + 3i$ is a solution to the equation $x^2 - 4x + 13 = 0$.

86. Show that $x = 3 + 2i$ is a solution to the equation $x^2 - 6x + 13 = 0$.

87. Show that $x = a + bi$ is a solution to the equation $x^2 - 2ax + (a^2 + b^2) = 0$.

88. Show that $x = a - bi$ is a solution to the equation $x^2 - 2ax + (a^2 + b^2) = 0$.

Use the method shown in the introduction to this section to solve each system of equations.

89. $x + y = 8$
$xy = 20$

90. $x - y = 10$
$xy = -40$

91. $2x + y = 4$
$xy = 8$

92. $3x + y = 6$
$xy = 9$

93. If z is a complex number, show that the product of z and its conjugate is a real number.

94. If z is a complex number, show that the sum of z and its conjugate is a real number.

➤ REVIEW PROBLEMS

The problems that follow review material we covered in Sections 1.3, 3.1, and Chapter 7. Reviewing these problems will help you with some of the material in the next section.

Find $\sin\theta$ and $\cos\theta$ if the given point lies on the terminal side of θ.

95. $(3, -4)$

96. $(-5, 12)$

Find θ between 0° and 360° based on the given information.

97. $\sin\theta = \dfrac{\sqrt{2}}{2}$ and $\cos\theta = -\dfrac{\sqrt{2}}{2}$

98. $\sin\theta = \dfrac{1}{2}$ and θ terminates in QII

Solve triangle ABC given the following information.

99. $A = 73.1°$, $b = 243$ cm, and $c = 157$ cm

100. $B = 24.2°$, $C = 63.8°$, and $b = 5.92$ inches

101. $a = 42.1$ m, $b = 56.8$ m, and $c = 63.4$ m

102. $B = 32.8°$, $a = 625$ ft, and $b = 521$ ft

➤ LEARNING OBJECTIVES ASSESSMENT

These questions are available for instructors to help assess if you have successfully met the learning objectives for this section.

103. Write $\sqrt{-50}$ in terms of i.

a. $2i\sqrt{5}$ **b.** $50i$ **c.** $5\sqrt{2i}$ **d.** $5i\sqrt{2}$

104. Simplify $[(8 + 5i) - (4 - 2i)] + (-6 + 3i)$.

a. $12 + 6i$ **b.** $-2 + 6i$ **c.** $12 + 10i$ **d.** $-2 + 10i$

105. Simplify i^{31}.

a. i **b.** $-i$ **c.** 1 **d.** -1

106. Divide $\dfrac{5 + 2i}{4 - 3i}$.

a. $-\dfrac{7}{25} + \dfrac{26}{25}i$ **b.** $\dfrac{5}{4} - \dfrac{2}{3}i$ **c.** $\dfrac{14}{25} + \dfrac{23}{25}i$ **d.** $\dfrac{26}{7} + \dfrac{23}{7}i$

SECTION 8.2 Trigonometric Form for Complex Numbers

LEARNING OBJECTIVES

1. Find the absolute value of a complex number.
2. Find the conjugate of a complex number.
3. Write a complex number in trigonometric form.
4. Convert a complex number from trigonometric form to standard form.

As you know, the quadratic formula, $x = \dfrac{-b \pm \sqrt{b^2 - 4ac}}{2a}$, can be used to solve any quadratic equation in the form $ax^2 + bx + c = 0$. In his book *Ars Magna,* Jerome Cardan gives a similar formula that can be used to solve certain cubic equations. Here it is in our notation:

If $\quad x^3 = ax + b$

then $$x = \sqrt[3]{\frac{b}{2} + \sqrt{\left(\frac{b}{2}\right)^2 - \left(\frac{a}{3}\right)^3}} + \sqrt[3]{\frac{b}{2} - \sqrt{\left(\frac{b}{2}\right)^2 - \left(\frac{a}{3}\right)^3}}$$

Cardan

This formula is known as Cardan's formula. In his book, Cardan attempts to use his formula to solve the equation

$$x^3 = 15x + 4$$

This equation has the form $x^3 = ax + b$, where $a = 15$ and $b = 4$. Substituting these values for a and b in Cardan's formula, we have

$$\begin{aligned} x &= \sqrt[3]{\frac{4}{2} + \sqrt{\left(\frac{4}{2}\right)^2 - \left(\frac{15}{3}\right)^3}} + \sqrt[3]{\frac{4}{2} - \sqrt{\left(\frac{4}{2}\right)^2 - \left(\frac{15}{3}\right)^3}} \\ &= \sqrt[3]{2 + \sqrt{4 - 125}} + \sqrt[3]{2 - \sqrt{4 - 125}} \\ &= \sqrt[3]{2 + \sqrt{-121}} + \sqrt[3]{2 - \sqrt{-121}} \end{aligned}$$

Cardan couldn't go any further than this because he didn't know what to do with $\sqrt{-121}$. In fact, he wrote

> I have sent to enquire after the solution to various problems for which you have given me no answer, one of which concerns the cube equal to an unknown plus a number. I have certainly grasped this rule, but when the cube of one-third of the coefficient of the unknown is greater in value than the square of one-half of the number, then, it appears, I cannot make it fit into the equation.

Notice in his formula that if $(a/3)^3 > (b/2)^2$, then the result will be a negative number inside the square root.

In this section, we will take the first step in finding cube roots of complex numbers by learning how to write complex numbers in *trigonometric form.* Before we do, let's look at a definition that will give us a visual representation for complex numbers.

To graphically represent a complex number $a + bi$, we need a system which allows us to indicate the values of both a and b. To do this, we set up a rectangular coordinate system much like the Cartesian coordinate system, except that we use the horizontal axis to indicate the real part and the vertical axis to indicate the imaginary part. We refer to the horizontal axis as the *real axis,* which represents the real part of a complex number, and the vertical axis as the *imaginary axis,* which represents the imaginary part. The resulting two-dimensional coordinate system is called the *complex plane,* or *Argand plane* (Figure 1).

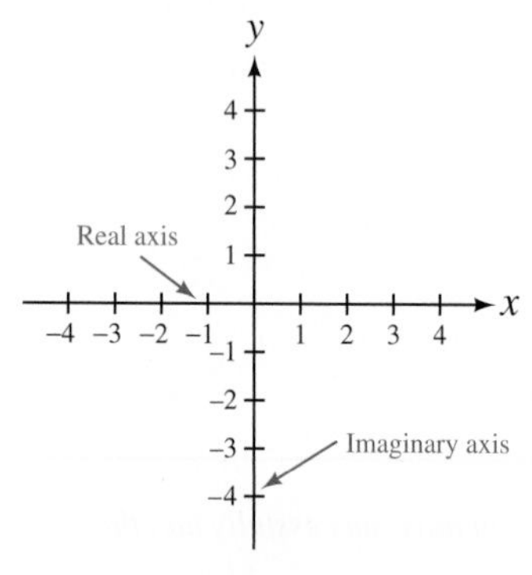

Figure 1

NOTE The complex plane is named after the Frenchman Jean-Robert Argand, who was thought to have provided the first geometrical interpretation of complex numbers in 1806. It was discovered 100 years later that the Norwegian-Danish surveyor Caspar Wessel was actually the first to have done so in a presentation to the Royal Danish Academy in 1797.

DEFINITION

The graph of the complex number $x + yi$ is a vector (arrow) that extends from the origin out to the point (x, y) in the complex plane.

PROBLEM 1
Graph each complex number:
$3 + 4i$, $-3 - 4i$, and $3 - 4i$.

EXAMPLE 1 Graph each complex number: $2 + 4i$, $-2 - 4i$, and $2 - 4i$.

SOLUTION The graphs are shown in Figure 2. Notice how the graphs of $2 + 4i$ and $2 - 4i$, which are conjugates, have symmetry about the real axis, and that the graphs of $2 + 4i$ and $-2 - 4i$, which are opposites, have symmetry about the origin.

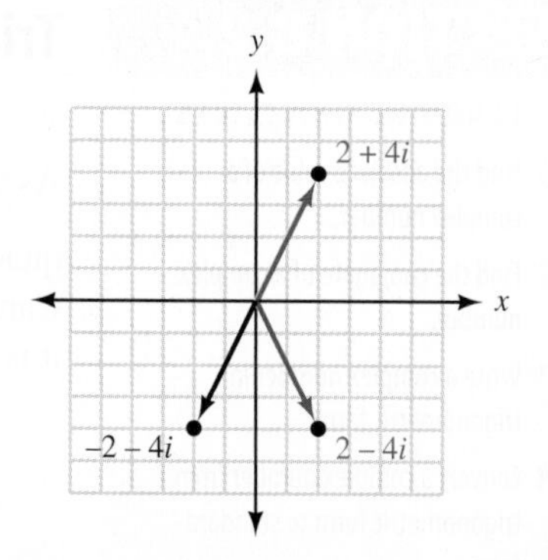

Figure 2

■

PROBLEM 2
Graph the complex numbers 3, −3, $4i$, and $-4i$.

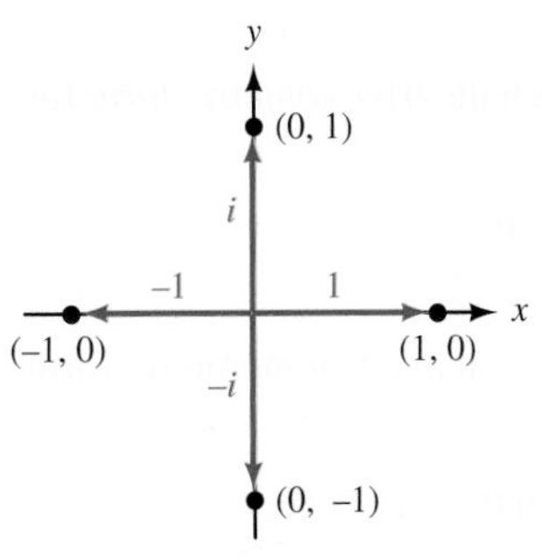

Figure 3

EXAMPLE 2 Graph the complex numbers $1, i, -1$, and $-i$.

SOLUTION Here are the four complex numbers written in standard form.

$$1 = 1 + 0i$$
$$i = 0 + i$$
$$-1 = -1 + 0i$$
$$-i = 0 - i$$

The graph of each is shown in Figure 3. ■

DEFINITION

The *absolute value* or *modulus* of the complex number $z = x + yi$ is the distance from the origin to the point (x, y) in the complex plane. If this distance is denoted by r, then

$$r = |z| = |x + yi| = \sqrt{x^2 + y^2}$$

PROBLEM 3
Find the modulus of each complex number: $4i$, 6, and $1 + 3i$.

EXAMPLE 3 Find the modulus of each of the complex numbers $5i$, 7, and $3 + 4i$.

SOLUTION Writing each number in standard form and then applying the definition of modulus, we have

For $z = 5i = 0 + 5i$, $r = |z| = |0 + 5i| = \sqrt{0^2 + 5^2} = 5$
For $z = 7 = 7 + 0i$, $r = |z| = |7 + 0i| = \sqrt{7^2 + 0^2} = 7$
For $z = 3 + 4i$, $r = |z| = |3 + 4i| = \sqrt{3^2 + 4^2} = 5$ ■

DEFINITION

The *argument* of the complex number $z = x + yi$, denoted $\arg(z)$, is the smallest positive angle θ from the positive real axis to the graph of z.

Figure 4 illustrates the relationships between the complex number $z = x + yi$, its graph, and the modulus r and argument θ of z. From Figure 4 we see that

$$\cos\theta = \frac{x}{r} \quad \text{or} \quad x = r\cos\theta$$

and

$$\sin\theta = \frac{y}{r} \quad \text{or} \quad y = r\sin\theta$$

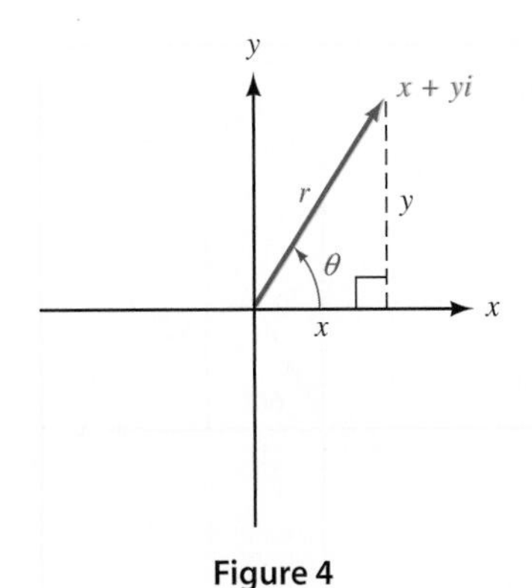

Figure 4

We can use this information to write z in terms of r and θ.

$$\begin{aligned} z &= x + yi \\ &= r\cos\theta + (r\sin\theta)i \\ &= r\cos\theta + ri\sin\theta \\ &= r(\cos\theta + i\sin\theta) \end{aligned}$$

This last expression is called the *trigonometric form* for z, which can be abbreviated as $r \text{ cis } \theta$. The formal definition follows.

DEFINITION

If $z = x + yi$ is a complex number in standard form, then the *trigonometric form* for z is given by

$$z = r(\cos\theta + i\sin\theta) = r \text{ cis } \theta$$

where r is the modulus of z and θ is the argument of z.

We can convert back and forth between standard form and trigonometric form by using the relationships that follow:

$$\text{For} \qquad z = x + yi = r(\cos\theta + i\sin\theta) = r \text{ cis } \theta$$

$$r = \sqrt{x^2 + y^2} \quad \text{and } \theta \text{ is such that}$$

$$\cos\theta = \frac{x}{r}, \quad \sin\theta = \frac{y}{r}, \quad \text{and} \quad \tan\theta = \frac{y}{x}$$

PROBLEM 4
Write $z = 1 + i\sqrt{3}$ in trigonometric form.

EXAMPLE 4 Write $z = -1 + i$ in trigonometric form.

SOLUTION We have $x = -1$ and $y = 1$; therefore,

$$r = \sqrt{(-1)^2 + 1^2} = \sqrt{2}$$

Angle θ is the smallest positive angle for which

$$\cos\theta = \frac{x}{r} = -\frac{1}{\sqrt{2}} = -\frac{\sqrt{2}}{2} \quad \text{and} \quad \sin\theta = \frac{y}{r} = \frac{1}{\sqrt{2}} = \frac{\sqrt{2}}{2}$$

Therefore, θ must be 135°, or $3\pi/4$ radians.

Using these values of r and θ in the formula for trigonometric form, we have

$$\begin{aligned} z &= r(\cos\theta + i\sin\theta) \\ &= \sqrt{2}(\cos 135° + i\sin 135°) \\ &= \sqrt{2} \text{ cis } 135° \end{aligned}$$

In radians, $z = \sqrt{2} \text{ cis } (3\pi/4)$. The graph of z is shown in Figure 5.

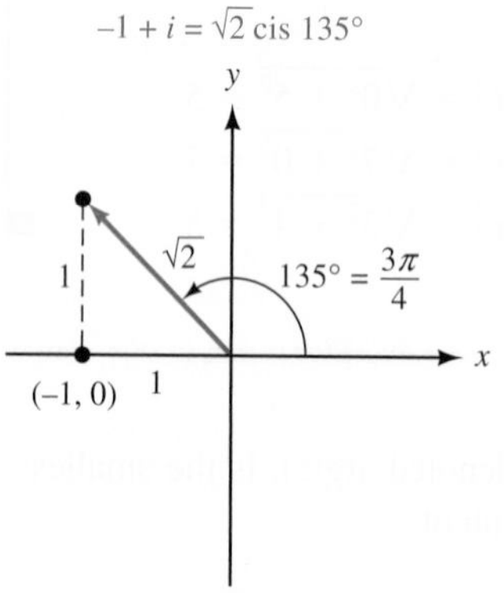

$-1 + i = \sqrt{2} \text{ cis } 135°$

Figure 5

PROBLEM 5
Write $z = 3 \text{ cis } 45°$ in standard form.

EXAMPLE 5 Write $z = 2 \text{ cis } 60°$ in standard form.

SOLUTION Using exact values for cos 60° and sin 60°, we have

$$\begin{aligned} z &= 2 \text{ cis } 60° \\ &= 2(\cos 60° + i\sin 60°) \\ &= 2\left(\frac{1}{2} + i\frac{\sqrt{3}}{2}\right) \\ &= 1 + i\sqrt{3} \end{aligned}$$

The graph of z is shown in Figure 6.

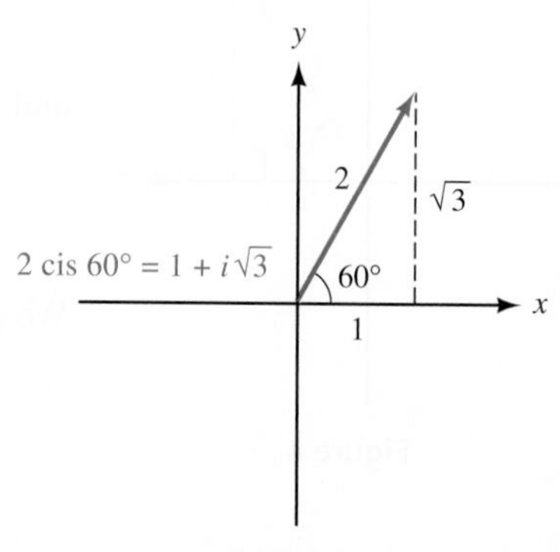

Figure 6

As you can see, converting from trigonometric form to standard form is usually more direct than converting from standard form to trigonometric form.

Using Technology **Converting Between Trigonometric and Rectangular Form**

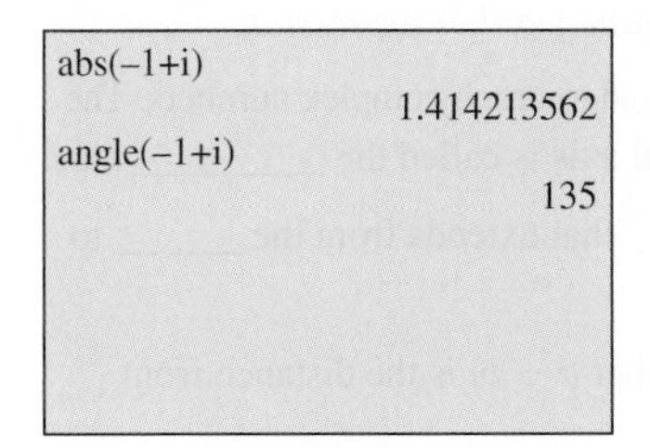

Figure 7

Most graphing calculators are able to convert a complex number between standard form (sometimes called *rectangular form*) and trigonometric form (also called *polar form*). For example, on a TI-84 we could solve Example 4 using the **abs** and **angle** commands found in the MATH CPX menu, as shown in Figure 7. We have set the calculator to degree mode.

Check the manual for your model to see if your calculator is able to convert complex numbers to trigonometric form. You may find the commands listed in the index under polar coordinates (we will explore polar coordinates later in this chapter).

PROBLEM 6
Write $4 + \sqrt{-36}$ and $4 - \sqrt{-36}$ in trigonometric form.

EXAMPLE 6 In the introduction to this section, we mentioned two complex numbers that Jerome Cardan had difficulty with: $2 + \sqrt{-121}$ and $2 - \sqrt{-121}$. Write each of these numbers in trigonometric form.

SOLUTION First we write them as complex numbers

$$2 + \sqrt{-121} = 2 + 11i$$
$$2 - \sqrt{-121} = 2 - 11i$$

The modulus of each is

$$r = \sqrt{2^2 + 11^2} = \sqrt{125} = 5\sqrt{5}$$

For $2 + 11i$, we have $\tan\theta = 11/2$ and $\theta \in$ QI. Using a calculator and rounding to the nearest hundredth of a degree, we find that $\theta = \tan^{-1}(5.5) = 79.70°$. Therefore,

$$\begin{aligned} 2 + 11i &= 5\sqrt{5}(\cos 79.70° + i \sin 79.70°) \\ &= 5\sqrt{5} \text{ cis } 79.70° \end{aligned}$$

For $2 - 11i$, we have $\tan\theta = -11/2$ and $\theta \in$ QIV, giving us $\theta = 360° - 79.70° = 280.30°$ to the nearest hundredth of a degree. Therefore,

$$\begin{aligned} 2 - 11i &= 5\sqrt{5}(\cos 280.30° + i \sin 280.30°) \\ &= 5\sqrt{5} \text{ cis } 280.30° \end{aligned}$$

Getting Ready for Class *After reading through the preceding section, respond in your own words and in complete sentences.*

a. How do you draw the graph of a complex number?
b. Define the absolute value of a complex number.
c. What is the argument of a complex number?
d. What is the first step in writing the complex number $-1 + i$ in trigonometric form?

8.2 PROBLEM SET

➤ CONCEPTS AND VOCABULARY

For Questions 1 through 8, fill in the blank with an appropriate word or expression.

1. The complex plane, also called the ________ plane, is used to graph complex numbers. The horizontal axis is called the _____ axis, and the vertical axis is called the _________ axis.

2. The graph of the complex number $a + bi$ is a ________ that extends from the ______ to the point ______ in the complex plane.

3. The absolute value, or _________, of a complex number $a + bi$ is the distance from the ______ to the point ______ in the complex plane.

4. The argument of a complex number z is the _________ _________ angle from the _______ ______ axis to the graph of z.

5. The notation $a + bi$ is called ________ form for a complex number, and the notation $r(\cos\theta + i\sin\theta)$ is called _________ form.

6. When we write $z = r \text{ cis } \theta$, r is the ________ of z and θ is the ________.

7. To convert a number from trigonometric form into standard form, simply _________ the trigonometric functions and then ________ the value of r.

8. To convert a complex number $z = x + yi$ from standard form into trigonometric form, use the relationships $r =$ _________ and $\tan\theta =$ ______, keeping in mind the quadrant that the graph of z lies in.

➤ EXERCISES

Graph each complex number. In each case, give the absolute value of the number.

9. $1 + i$ **10.** $1 - i$ **11.** $-3 - 4i$ **12.** $-4 + 3i$

13. $-5i$ **14.** $4i$ **15.** -4 **16.** 2

Graph each complex number along with its opposite and conjugate.

17. $2 - i$ **18.** $2 + i$ **19.** $-3i$ **20.** $4i$

21. 5 **22.** -3 **23.** $-5 - 2i$ **24.** $-2 - 5i$

Write each complex number in standard form.

25. $2(\cos 30° + i\sin 30°)$ **26.** $4(\cos 30° + i\sin 30°)$

27. $4\left(\cos\frac{2\pi}{3} + i\sin\frac{2\pi}{3}\right)$ **28.** $8\left(\cos\frac{2\pi}{3} + i\sin\frac{2\pi}{3}\right)$

29. $1 \text{ cis } 210°$ **30.** $1 \text{ cis } 240°$ **31.** $1 \text{ cis } \frac{7\pi}{4}$ **32.** $\sqrt{2} \text{ cis } \frac{7\pi}{4}$

Use a calculator to help write each complex number in standard form. Round the numbers in your answers to the nearest hundredth.

33. $10(\cos 12° + i\sin 12°)$ **34.** $100(\cos 70° + i\sin 70°)$

35. $100(\cos 2.5 + i\sin 2.5)$ **36.** $100(\cos 3 + i\sin 3)$

37. $1 \text{ cis } 205°$ **38.** $1 \text{ cis } 261°$ **39.** $10 \text{ cis } 6$ **40.** $10 \text{ cis } 5.5$

Write each complex number in trigonometric form, once using degrees and once using radians. In each case, begin by sketching the graph to help find the argument θ.

41. $-1 + i$ **42.** $1 + i$ **43.** $1 - i$ **44.** $-1 - i$

45. $3 + 3i$ **46.** $5 + 5i$ **47.** $-8i$ **48.** $8i$

49. -9 **50.** 2 **51.** $-2 + 2i\sqrt{3}$ **52.** $-2\sqrt{3} + 2i$

Write each complex number in trigonometric form. Round all angles to the nearest hundredth of a degree.

53. $3 - 4i$ **54.** $3 + 4i$ **55.** $-20 + 21i$ **56.** $-21 - 20i$

57. $7 - 24i$ **58.** $11 + 2i$ **59.** $-11 + 2i$ **60.** $-8 - 15i$

Use your graphing calculator to convert the complex number to trigonometric form in the following problems.

61. Problem 53 **62.** Problem 54 **63.** Problem 55 **64.** Problem 56

65. Problem 57 **66.** Problem 58 **67.** Problem 59 **68.** Problem 60

69. Show that 2 cis 30° and 2 cis (−30°) are conjugates.

70. Show that 2 cis 60° and 2 cis (−60°) are conjugates.

71. Show that if $z = \cos\theta + i\sin\theta$, then $|z| = 1$.

72. Show that if $z = \cos\theta - i\sin\theta$, then $|z| = 1$.

➤ REVIEW PROBLEMS

The problems that follow review material we covered in Sections 5.2 and 7.3. Reviewing the problems from Section 5.2 will help you understand the next section.

73. Use the formula for $\cos(A + B)$ to find the exact value of cos 75°.

74. Use the formula for $\sin(A + B)$ to find the exact value of sin 75°.

Let $\sin A = 3/5$ with A in QI and $\sin B = 5/13$ with B in QI and find the following.

75. $\sin(A + B)$ **76.** $\cos(A + B)$

Simplify each expression to a single trigonometric function.

77. $\sin 30° \cos 90° + \cos 30° \sin 90°$ **78.** $\cos 30° \cos 90° - \sin 30° \sin 90°$

In triangle ABC, $A = 45.6°$ and $b = 567$ inches. Find B for each given value of a. (You may get one or two values of B, or you may find that no triangle fits the given description.)

79. $a = 234$ inches **80.** $a = 678$ inches

81. $a = 456$ inches **82.** $a = 789$ inches

➤ LEARNING OBJECTIVES ASSESSMENT

These questions are available for instructors to help assess if you have successfully met the learning objectives for this section.

83. Find the absolute value of $3 - 2i$.

a. 5 **b.** 1 **c.** $\sqrt{13}$ **d.** $\sqrt{5}$

84. Write $-4 + 4i$ in trigonometric form.

a. $4\sqrt{2}(\cos 135° + i\sin 135°)$ **b.** $4(\cos 315° + i\sin 315°)$

c. $4\sqrt{2}(\cos 315° + i\sin 315°)$ **d.** $4\sqrt{2}(\sin 135° + i\cos 135°)$

85. Write $6 \text{ cis } \frac{2\pi}{3}$ in standard form.

a. $-3 + 3i\sqrt{3}$ **b.** $-3\sqrt{3} + 3i$ **c.** $3 - 3i\sqrt{3}$ **d.** $3\sqrt{3} - 3i$

86. Graph $4 - 2i$.

a.

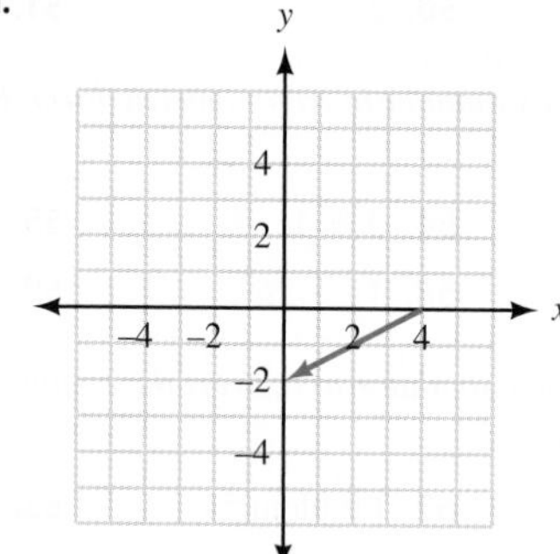

b.

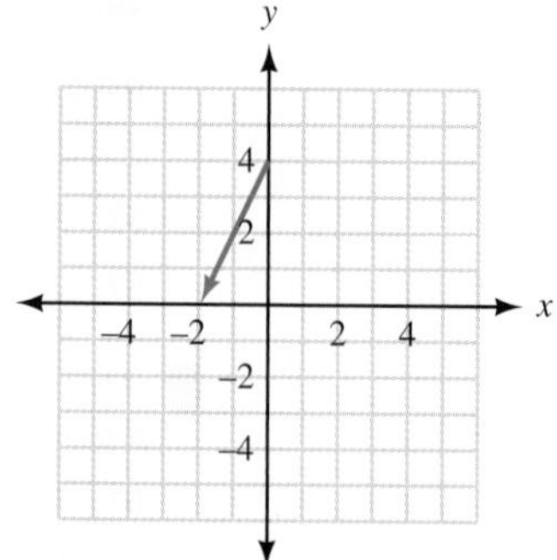

c.

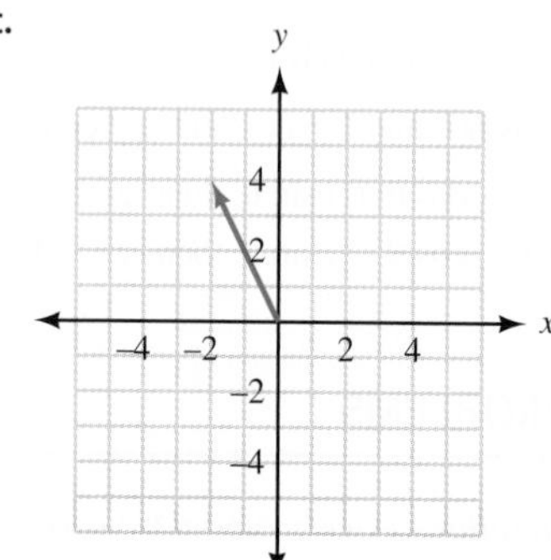

d.

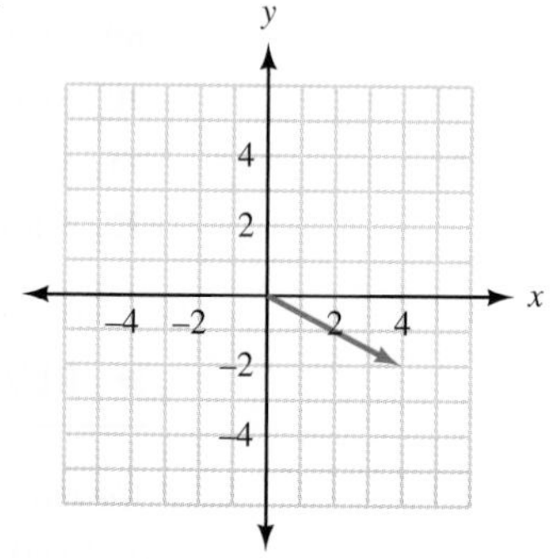

SECTION 8.3 Products and Quotients in Trigonometric Form

LEARNING OBJECTIVES

1. Multiply complex numbers in trigonometric form.
2. Divide complex numbers in trigonometric form.
3. Use de Moivre's Theorem to find powers of complex numbers.
4. Use trigonometric form to simplify expressions involving complex numbers.

Multiplication and division with complex numbers becomes a very simple process when the numbers are written in trigonometric form. Let's state the rule for finding the product of two complex numbers written in trigonometric form as a theorem and then prove the theorem.

THEOREM (MULTIPLICATION)

If

$$z_1 = r_1(\cos\theta_1 + i\sin\theta_1) = r_1 \operatorname{cis}\theta_1$$

and

$$z_2 = r_2(\cos\theta_2 + i\sin\theta_2) = r_2 \operatorname{cis}\theta_2$$

are two complex numbers in trigonometric form, then their product, z_1z_2, is

$$\begin{aligned} z_1z_2 &= [r_1(\cos\theta_1 + i\sin\theta_1)][r_2(\cos\theta_2 + i\sin\theta_2)] \\ &= r_1r_2[\cos(\theta_1 + \theta_2) + i\sin(\theta_1 + \theta_2)] \\ &= r_1r_2 \operatorname{cis}(\theta_1 + \theta_2) \end{aligned}$$

In words: To multiply two complex numbers in trigonometric form, multiply the moduli and add the arguments.

PROOF

We begin by multiplying algebraically. Then we simplify our product by using the sum formulas we introduced in Section 5.2.

$$\begin{aligned} z_1z_2 &= [r_1(\cos\theta_1 + i\sin\theta_1)][r_2(\cos\theta_2 + i\sin\theta_2)] \\ &= r_1r_2(\cos\theta_1 + i\sin\theta_1)(\cos\theta_2 + i\sin\theta_2) \\ &= r_1r_2(\cos\theta_1\cos\theta_2 + i\cos\theta_1\sin\theta_2 + i\sin\theta_1\cos\theta_2 + i^2\sin\theta_1\sin\theta_2) \\ &= r_1r_2[\cos\theta_1\cos\theta_2 + i(\cos\theta_1\sin\theta_2 + \sin\theta_1\cos\theta_2) - \sin\theta_1\sin\theta_2] \\ &= r_1r_2[(\cos\theta_1\cos\theta_2 - \sin\theta_1\sin\theta_2) + i(\sin\theta_1\cos\theta_2 + \cos\theta_1\sin\theta_2)] \\ &= r_1r_2[\cos(\theta_1 + \theta_2) + i\sin(\theta_1 + \theta_2)] \end{aligned}$$

This completes our proof. As you can see, to multiply two complex numbers in trigonometric form, we multiply absolute values, r_1r_2, and add angles, $\theta_1 + \theta_2$. ■

PROBLEM 1
Multiply 4 cis 20° and 3 cis 55°.

EXAMPLE 1 Find the product of 3 cis 40° and 5 cis 10°.

SOLUTION Applying the formula from our theorem on products, we have

$$\begin{aligned} (3\text{ cis }40°)(5\text{ cis }10°) &= 3 \cdot 5\text{ cis }(40° + 10°) \\ &= 15\text{ cis }50° \end{aligned}$$

■

PROBLEM 2
Find the product of $3i$ and $2 + 2i$ in both standard form and trigonometric form.

EXAMPLE 2 Find the product of $z_1 = 1 + i\sqrt{3}$ and $z_2 = -\sqrt{3} + i$ in standard form, and then write z_1 and z_2 in trigonometric form and find their product again.

SOLUTION Leaving each complex number in standard form and multiplying we have

$$\begin{aligned} z_1z_2 &= (1 + i\sqrt{3})(-\sqrt{3} + i) \\ &= -\sqrt{3} + i - 3i + i^2\sqrt{3} \\ &= -2\sqrt{3} - 2i \end{aligned}$$

Changing z_1 and z_2 to trigonometric form and multiplying looks like this:

$$\begin{aligned} z_1 &= 1 + i\sqrt{3} = 2(\cos 60° + i\sin 60°) \\ z_2 &= -\sqrt{3} + i = 2(\cos 150° + i\sin 150°) \\ z_1z_2 &= [2(\cos 60° + i\sin 60°)][2(\cos 150° + i\sin 150°)] \\ &= 4(\cos 210° + i\sin 210°) \end{aligned}$$

To compare our two products, we convert our product in trigonometric form to standard form.

$$\begin{aligned} 4(\cos 210° + i\sin 210°) &= 4\left(-\frac{\sqrt{3}}{2} - \frac{1}{2}i\right) \\ &= -2\sqrt{3} - 2i \end{aligned}$$

As you can see, both methods of multiplying complex numbers produce the same result. ■

The next theorem is an extension of the work we have done so far with multiplication. It is named after Abraham de Moivre, who published the formula in a paper in 1722. De Moivre was born in France but emigrated to London to avoid

The Art Archive / National Gallery London / Eileen Tweedy

Abraham de Moivre

religious persecution. Although he was a competent mathematician, as a foreigner he was unable to attain a post at a university and had to earn a living as a private tutor of mathematics. De Moivre made significant contributions to the development of analytic geometry and the theory of probability. We will not give a formal proof of the theorem.

DE MOIVRE'S THEOREM

If $z = r(\cos\theta + i\sin\theta) = r\operatorname{cis}\theta$ is a complex number in trigonometric form and n is an integer, then

$$\begin{aligned} z^n &= [r(\cos\theta + i\sin\theta)]^n \\ &= r^n(\cos n\theta + i\sin n\theta) \\ &= r^n \operatorname{cis} n\theta \end{aligned}$$

The theorem seems reasonable after the work we have done with multiplication. For example, if n is 2,

$$\begin{aligned} [r(\cos\theta + i\sin\theta)]^2 &= r(\cos\theta + i\sin\theta)\cdot r(\cos\theta + i\sin\theta) \\ &= r\cdot r[\cos(\theta+\theta) + i\sin(\theta+\theta)] \\ &= r^2(\cos 2\theta + i\sin 2\theta) \end{aligned}$$

PROBLEM 3
Find $(2 - 2i)^5$.

EXAMPLE 3 Find $(1 + i)^{10}$.

SOLUTION First we write $1 + i$ in trigonometric form:

$$1 + i = \sqrt{2}\left(\cos\frac{\pi}{4} + i\sin\frac{\pi}{4}\right)$$

Then we use de Moivre's Theorem to raise this expression to the 10th power.

$$\begin{aligned} (1+i)^{10} &= \left[\sqrt{2}\left(\cos\frac{\pi}{4} + i\sin\frac{\pi}{4}\right)\right]^{10} \\ &= (\sqrt{2})^{10}\left(\cos 10\cdot\frac{\pi}{4} + i\sin 10\cdot\frac{\pi}{4}\right) \\ &= 32\left(\cos\frac{5\pi}{2} + i\sin\frac{5\pi}{2}\right) \end{aligned}$$

which we can simplify to

$$= 32\left(\cos\frac{\pi}{2} + i\sin\frac{\pi}{2}\right)$$

because $\pi/2$ and $5\pi/2$ are coterminal. In standard form, our result is

$$\begin{aligned} &= 32(0 + i) \\ &= 32i \end{aligned}$$

That is,

$$(1 + i)^{10} = 32i$$

■

Multiplication with complex numbers in trigonometric form is accomplished by multiplying the moduli and adding the arguments, so we should expect that division is accomplished by dividing the moduli and subtracting the arguments.

THEOREM (DIVISION)

If

$$z_1 = r_1(\cos\theta_1 + i\sin\theta_1) = r_1 \operatorname{cis}\theta_1$$

and

$$z_2 = r_2(\cos\theta_2 + i\sin\theta_2) = r_2 \operatorname{cis}\theta_2$$

are two complex numbers in trigonometric form, then their quotient, z_1/z_2, is

$$\begin{aligned}\frac{z_1}{z_2} &= \frac{r_1(\cos\theta_1 + i\sin\theta_1)}{r_2(\cos\theta_2 + i\sin\theta_2)} \\ &= \frac{r_1}{r_2}[\cos(\theta_1 - \theta_2) + i\sin(\theta_1 - \theta_2)] \\ &= \frac{r_1}{r_2}\operatorname{cis}(\theta_1 - \theta_2)\end{aligned}$$

PROOF

As was the case with division of complex numbers in standard form, the major step in this proof is multiplying the numerator and denominator of our quotient by the conjugate of the denominator.

$$\begin{aligned}&\frac{r_1(\cos\theta_1 + i\sin\theta_1)}{r_2(\cos\theta_2 + i\sin\theta_2)} \\ &\quad= \frac{r_1(\cos\theta_1 + i\sin\theta_1)}{r_2(\cos\theta_2 + i\sin\theta_2)} \cdot \frac{\mathbf{(\cos\theta_2 - \mathit{i}\sin\theta_2)}}{\mathbf{(\cos\theta_2 - \mathit{i}\sin\theta_2)}} \\ &\quad= \frac{r_1(\cos\theta_1 + i\sin\theta_1)(\cos\theta_2 - i\sin\theta_2)}{r_2(\cos^2\theta_2 + \sin^2\theta_2)} \\ &\quad= \frac{r_1}{r_2}(\cos\theta_1\cos\theta_2 - i\cos\theta_1\sin\theta_2 + i\sin\theta_1\cos\theta_2 - i^2\sin\theta_1\sin\theta_2) \\ &\quad= \frac{r_1}{r_2}[(\cos\theta_1\cos\theta_2 + \sin\theta_1\sin\theta_2) + i(\sin\theta_1\cos\theta_2 - \cos\theta_1\sin\theta_2)] \\ &\quad= \frac{r_1}{r_2}[\cos(\theta_1 - \theta_2) + i\sin(\theta_1 - \theta_2)]\end{aligned}$$

■

PROBLEM 4

Divide 40 cis 55° by 5 cis 25°.

EXAMPLE 4 Find the quotient when $20(\cos 75° + i\sin 75°)$ is divided by $4(\cos 40° + i\sin 40°)$.

SOLUTION We divide according to the formula given in our theorem on division.

$$\begin{aligned}\frac{20(\cos 75° + i\sin 75°)}{4(\cos 40° + i\sin 40°)} &= \frac{20}{4}[\cos(75° - 40°) + i\sin(75° - 40°)] \\ &= 5(\cos 35° + i\sin 35°)\end{aligned}$$

■

PROBLEM 5
Find the quotient when $4i$ is divided by $1 + i\sqrt{3}$ in both standard form and trigonometric form.

EXAMPLE 5 Divide $z_1 = 1 + i\sqrt{3}$ by $z_2 = \sqrt{3} + i$ and leave the answer in standard form. Then change each to trigonometric form and divide again.

SOLUTION Dividing in standard form, we have

$$\begin{aligned}\frac{z_1}{z_2} &= \frac{1 + i\sqrt{3}}{\sqrt{3} + i} \\ &= \frac{1 + i\sqrt{3}}{\sqrt{3} + i} \cdot \frac{\sqrt{3} - i}{\sqrt{3} - i} \\ &= \frac{\sqrt{3} - i + 3i - i^2\sqrt{3}}{3 + 1} \\ &= \frac{2\sqrt{3} + 2i}{4} \\ &= \frac{\sqrt{3}}{2} + \frac{1}{2}i\end{aligned}$$

Changing z_1 and z_2 to trigonometric form,

$$z_1 = 1 + i\sqrt{3} = 2 \text{ cis } \frac{\pi}{3}$$

$$z_2 = \sqrt{3} + i = 2 \text{ cis } \frac{\pi}{6}$$

and dividing again, we have

$$\begin{aligned}\frac{z_1}{z_2} &= \frac{2 \text{ cis } (\pi/3)}{2 \text{ cis } (\pi/6)} \\ &= \frac{2}{2} \text{ cis } \left(\frac{\pi}{3} - \frac{\pi}{6}\right) \\ &= 1 \text{ cis } \frac{\pi}{6}\end{aligned}$$

which, in standard form, is

$$\frac{\sqrt{3}}{2} + \frac{1}{2}i$$

Getting Ready for Class

After reading through the preceding section, respond in your own words and in complete sentences.

a. How do you multiply two complex numbers written in trigonometric form?
b. How do you divide two complex numbers written in trigonometric form?
c. What is the argument of the complex number that results when $\sqrt{2}(\cos 45° + i \sin 45°)$ is raised to the 10th power?
d. What is the modulus of the complex number that results when $\sqrt{2}(\cos 45° + i \sin 45°)$ is raised to the 10th power?

8.3 PROBLEM SET

➤ CONCEPTS AND VOCABULARY

For Questions 1 through 4, fill in the blank with an appropriate word or symbol.

1. To multiply two complex numbers in trigonometric form, ________ their moduli and _____ their arguments.
2. To divide two complex numbers in trigonometric form, ________ their moduli and _____ their arguments.
3. We use de Moivre's Theorem to find _______ of complex numbers written in trigonometric form.
4. To find the nth power of a complex number, find the nth power of the _________ and multiply the ___________ by ___.

➤ EXERCISES

Multiply. Leave all answers in trigonometric form.

5. $5(\cos 15° + i \sin 15°) \cdot 2(\cos 25° + i \sin 25°)$
6. $3(\cos 20° + i \sin 20°) \cdot 4(\cos 30° + i \sin 30°)$
7. $9(\cos 115° + i \sin 115°) \cdot 4(\cos 51° + i \sin 51°)$
8. $7(\cos 110° + i \sin 110°) \cdot 8(\cos 47° + i \sin 47°)$
9. $2 \text{ cis } \frac{3\pi}{4} \cdot 2 \text{ cis } \frac{\pi}{4}$
10. $2 \text{ cis } \frac{2\pi}{3} \cdot 4 \text{ cis } \frac{\pi}{6}$

Find the product z_1z_2 in standard form. Then write z_1 and z_2 in trigonometric form and find their product again. Finally, convert the answer that is in trigonometric form to standard form to show that the two products are equal.

11. $z_1 = 1 + i, z_2 = -1 + i$
12. $z_1 = 1 + i, z_2 = 2 + 2i$
13. $z_1 = 1 + i\sqrt{3}, z_2 = -\sqrt{3} + i$
14. $z_1 = -1 + i\sqrt{3}, z_2 = \sqrt{3} + i$
15. $z_1 = 3i, z_2 = -4i$
16. $z_1 = 2i, z_2 = -5i$
17. $z_1 = 1 + i, z_2 = 4i$
18. $z_1 = 1 + i, z_2 = 3i$
19. $z_1 = -5, z_2 = 1 + i\sqrt{3}$
20. $z_1 = -3, z_2 = \sqrt{3} + i$
21. We know that $2i \cdot 3i = 6i^2 = -6$. Change $2i$ and $3i$ to trigonometric form, and then show that their product in trigonometric form is still -6.
22. Change $4i$ and 2 to trigonometric form and then multiply. Show that this product is $8i$.

Use de Moivre's Theorem to find each of the following. Write your answer in standard form.

23. $[2(\cos 10° + i \sin 10°)]^6$
24. $[4(\cos 15° + i \sin 15°)]^3$
25. $\left[3\left(\cos \frac{\pi}{3} + i \sin \frac{\pi}{3}\right)\right]^4$
26. $\left[3\left(\cos \frac{\pi}{6} + i \sin \frac{\pi}{6}\right)\right]^4$
27. $(\text{cis } 12°)^{10}$
28. $(\text{cis } 18°)^{10}$
29. $\left(\sqrt{2} \text{ cis } \frac{7\pi}{18}\right)^6$
30. $\left(\sqrt{2} \text{ cis } \frac{\pi}{4}\right)^{10}$
31. $(1 + i)^4$
32. $(1 + i)^5$
33. $(-\sqrt{3} + i)^4$
34. $(\sqrt{3} + i)^4$
35. $(1 + i)^6$
36. $(-1 + i)^8$
37. $(-2 + 2i)^3$
38. $(-2 - 2i)^3$

Divide. Leave your answers in trigonometric form.

39. $\dfrac{20(\cos 75° + i \sin 75°)}{5(\cos 40° + i \sin 40°)}$

40. $\dfrac{30(\cos 80° + i \sin 80°)}{10(\cos 30° + i \sin 30°)}$

41. $\dfrac{18(\cos 51° + i \sin 51°)}{12(\cos 32° + i \sin 32°)}$

42. $\dfrac{21(\cos 63° + i \sin 63°)}{14(\cos 44° + i \sin 44°)}$

43. $\dfrac{4 \text{ cis } \frac{\pi}{2}}{8 \text{ cis } \frac{\pi}{6}}$

44. $\dfrac{6 \text{ cis } \frac{2\pi}{3}}{8 \text{ cis } \frac{\pi}{2}}$

Find the quotient z_1/z_2 in standard form. Then write z_1 and z_2 in trigonometric form and find their quotient again. Finally, convert the answer that is in trigonometric form to standard form to show that the two quotients are equal.

45. $z_1 = 2 + 2i, z_2 = 1 + i$

46. $z_1 = 2 - 2i, z_2 = 1 - i$

47. $z_1 = \sqrt{3} + i, z_2 = 2i$

48. $z_1 = 1 + i\sqrt{3}, z_2 = 2i$

49. $z_1 = 4 + 4i, z_2 = 2 - 2i$

50. $z_1 = 6 + 6i, z_2 = -3 - 3i$

51. $z_1 = 8, z_2 = -4$

52. $z_1 = -6, z_2 = 3$

Convert all complex numbers to trigonometric form and then simplify each expression. Write all answers in standard form.

53. $\dfrac{(1 + i)^4(2i)^2}{-2 + 2i}$

54. $\dfrac{(\sqrt{3} + i)^4(2i)^5}{(1 + i)^{10}}$

55. $\dfrac{(2 + 2i)^5(-3 + 3i)^3}{(\sqrt{3} + i)^{10}}$

56. $\dfrac{(1 + i\sqrt{3})^4(\sqrt{3} - i)^2}{(1 - i\sqrt{3})^3}$

57. Show that $x = 2(\cos 60° + i \sin 60°)$ is a solution to the quadratic equation $x^2 - 2x + 4 = 0$ by replacing x with $2(\cos 60° + i \sin 60°)$ and simplifying.

58. Show that $x = 2(\cos 300° + i \sin 300°)$ is a solution to the quadratic equation $x^2 - 2x + 4 = 0$.

59. Show that $w = 2(\cos 15° + i \sin 15°)$ is a fourth root of $z = 8 + 8i\sqrt{3}$ by raising w to the fourth power and simplifying to get z. (The number w is a fourth root of z, $w = z^{1/4}$, if the fourth power of w is z, $w^4 = z$.)

60. Show that $x = 1/2 + (\sqrt{3}/2)i$ is a cube root of -1.

De Moivre's Theorem can be used to find reciprocals of complex numbers. Recall from algebra that the reciprocal of x is $1/x$, which can be expressed as x^{-1}. Use this fact, along with de Moivre's Theorem, to find the reciprocal of each number below.

61. $1 + i$

62. $1 - i$

63. $\sqrt{3} - i$

64. $\sqrt{3} + i$

➤ REVIEW PROBLEMS

The problems that follow review material we covered in Sections 5.3, 5.4, 7.1, and 7.2.

If $\cos A = -1/3$ and A is between 90° and 180°, find

65. $\cos 2A$

66. $\sin 2A$

67. $\sin \dfrac{A}{2}$

68. $\cos \dfrac{A}{2}$

69. Distance and Bearing A crew member on a fishing boat traveling due north off the coast of California observes that the bearing of Morro Rock from the boat is N 35° E. After sailing another 9.2 miles, the crew member looks back to find that the bearing of Morro Rock from the ship is S 27° E. At that time, how far is the boat from Morro Rock?

70. **Length** A tent pole is held in place by two guy wires on opposite sides of the pole. One of the guy wires makes an angle of 43.2° with the ground and is 10.1 feet long. How long is the other guy wire if it makes an angle of 34.5° with the ground?

71. **True Course and Speed** A plane is flying with an airspeed of 170 miles per hour with a heading of 112°. The wind currents are a constant 28 miles per hour in the direction of due north. Find the true course and ground speed of the plane.

72. **Geometry** If a parallelogram has sides of 33 centimeters and 22 centimeters that meet at an angle of 111°, how long is the longer diagonal?

➤ LEARNING OBJECTIVES ASSESSMENT

These questions are available for instructors to help assess if you have successfully met the learning objectives for this section.

73. Multiply $5(\cos 10° + i \sin 10°) \cdot 4(\cos 15° + i \sin 15°)$.

a. $9(\cos 150° + i \sin 150°)$
b. $\cos 100° + i \sin 100°$
c. $20(\cos 150° + i \sin 150°)$
d. $20(\cos 25° + i \sin 25°)$

74. Divide $\dfrac{8(\cos 120° + i \sin 120°)}{2(\cos 40° + i \sin 40°)}$.

a. $4(\cos 30° + i \sin 30°)$
b. $4(\cos 80° + i \sin 80°)$
c. $4(\cos 160° + i \sin 160°)$
d. $6(\cos 30° + i \sin 30°)$

75. Find $\left(4 \text{ cis } \dfrac{\pi}{6}\right)^3$.

a. $12 \text{ cis } \dfrac{\pi}{2}$
b. $64 \text{ cis } \dfrac{\pi^3}{216}$
c. $64 \text{ cis } \dfrac{\pi}{2}$
d. $12 \text{ cis } \dfrac{\pi}{18}$

76. Simplify $\dfrac{(1 - i)^4(-\sqrt{3} - i)^2}{(1 + i\sqrt{3})^5}$ by first writing each complex number in trigonometric form. Convert your answer back to standard form.

a. $\dfrac{\sqrt{3}}{4} + \dfrac{1}{4}i$
b. $-\dfrac{1}{2} + \dfrac{\sqrt{3}}{2}i$
c. $-\dfrac{\sqrt{2}}{2} - \dfrac{\sqrt{2}}{2}i$
d. $\dfrac{1}{4} - \dfrac{\sqrt{3}}{4}i$

SECTION 8.4 Roots of a Complex Number

LEARNING OBJECTIVES

1 Find both square roots of a complex number.
2 Find all nth roots of a complex number.
3 Graph the nth roots of a complex number.
4 Use roots to solve an equation.

What is it about mathematics that draws some people toward it? In many cases, it is the fact that we can describe the world around us with mathematics; for some people, mathematics gives a clearer picture of the world in which we live. In other cases, the attraction is within mathematics itself. That is, for some people, mathematics itself is attractive, regardless of its connection to the real world. What you will find in this section is something in this second category. It is a property that real and complex numbers contain that is, by itself, surprising and attractive for people who enjoy mathematics. It has to do with roots of real and complex numbers.

If we solve the equation $x^2 = 25$, our solutions will be square roots of 25. Likewise, if we solve $x^3 = 8$, our solutions will be cube roots of 8. Further, the solutions to $x^4 = 81$ will be fourth roots of 81.

Without showing the work involved, here are the solutions to these three equations:

Equation	Solutions
$x^2 = 25$	-5 and 5
$x^3 = 8$	$2, -1 + i\sqrt{3}$, and $-1 - i\sqrt{3}$
$x^4 = 81$	$-3, 3, -3i$, and $3i$

The number 25 has two square roots, 8 has three cube roots, and 81 has four fourth roots. As we progress through this section, you will see that the numbers we used in the equations given are unimportant; every real (or complex) number has exactly two square roots, three cube roots, and four fourth roots. In fact, every real (or complex) number has exactly n distinct nth roots, a surprising and attractive fact about numbers. The key to finding these roots is trigonometric form for complex numbers.

Suppose that z and w are complex numbers such that w is an nth root of z. That is,

$$w = \sqrt[n]{z}$$

If we raise both sides of this last equation to the nth power, we have

$$w^n = z$$

Now suppose $z = r(\cos \theta + i \sin \theta)$ and $w = s(\cos \alpha + i \sin \alpha)$. Substituting these expressions into the equation $w^n = z$, we have

$$[s(\cos \alpha + i \sin \alpha)]^n = r(\cos \theta + i \sin \theta)$$

We can rewrite the left side of this last equation using de Moivre's Theorem.

$$s^n(\cos n\alpha + i \sin n\alpha) = r(\cos \theta + i \sin \theta)$$

The only way these two expressions can be equal is if their moduli are equal and their arguments are coterminal.

Moduli Equal	*Arguments Coterminal*	
$s^n = r$	$n\alpha = \theta + 360°k$	k is any integer

Solving for s and α, we have

$$s = r^{1/n} \qquad \alpha = \frac{\theta + 360°k}{n} = \frac{\theta}{n} + \frac{360°}{n}k$$

To summarize, we find the nth roots of a complex number by first finding the real nth root of the modulus. The angle for the first root will be θ/n, and the angles for the remaining roots are found by adding multiples of $360°/n$. The values for k will range from 0 to $n - 1$, because when $k = n$ we would be adding

$$\frac{360°}{n} \cdot n = 360°$$

which would be coterminal with the first angle.

NOTE If using radians, replace 360° with 2π.

THEOREM (ROOTS)

The nth roots of the complex number

$$z = r(\cos\theta + i\sin\theta) = r\operatorname{cis}\theta$$

are given by

$$w_k = r^{1/n}\left[\cos\left(\frac{\theta}{n} + \frac{360°}{n}k\right) + i\sin\left(\frac{\theta}{n} + \frac{360°}{n}k\right)\right]$$

$$= r^{1/n}\operatorname{cis}\left(\frac{\theta}{n} + \frac{360°}{n}k\right)$$

where $k = 0, 1, 2, \ldots, n-1$. The root w_0, corresponding to $k = 0$, is called the *principal nth root* of z.

PROBLEM 1
Find the four fourth roots of $z = 81(\cos 30° + i\sin 30°)$.

EXAMPLE 1 Find the four fourth roots of $z = 16(\cos 60° + i\sin 60°)$.

SOLUTION According to the formula given in our theorem on roots, the four fourth roots will be

$$w_k = 16^{1/4}\left[\cos\left(\frac{60°}{4} + \frac{360°}{4}k\right) + i\sin\left(\frac{60°}{4} + \frac{360°}{4}k\right)\right] \qquad k = 0, 1, 2, 3$$

$$= 2[\cos(15° + 90°k) + i\sin(15° + 90°k)]$$

Replacing k with 0, 1, 2, and 3, we have

$$w_0 = 2(\cos 15° + i\sin 15°) = 2\operatorname{cis} 15° \qquad \text{when } k = 0$$
$$w_1 = 2(\cos 105° + i\sin 105°) = 2\operatorname{cis} 105° \qquad \text{when } k = 1$$
$$w_2 = 2(\cos 195° + i\sin 195°) = 2\operatorname{cis} 195° \qquad \text{when } k = 2$$
$$w_3 = 2(\cos 285° + i\sin 285°) = 2\operatorname{cis} 285° \qquad \text{when } k = 3$$

It is interesting to note the graphical relationship among these four roots, as illustrated in Figure 1. Because the modulus for each root is 2, the terminal point of the vector used to represent each root lies on a circle of radius 2. The vector representing the first root makes an angle of $\theta/n = 15°$ with the positive x-axis, then each following root has a vector that is rotated an additional $360°/n = 90°$ in the counterclockwise direction from the previous root. Because the fraction $360°/n$ represents one complete revolution divided into n equal parts, the nth roots of a complex number will always be evenly distributed around a circle of radius $r^{1/n}$.

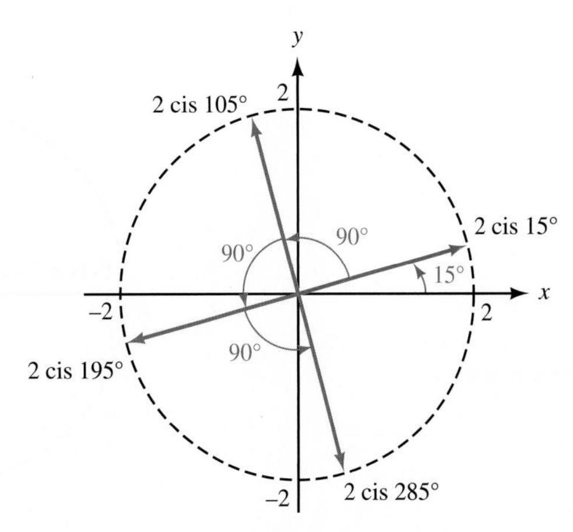

Figure 1

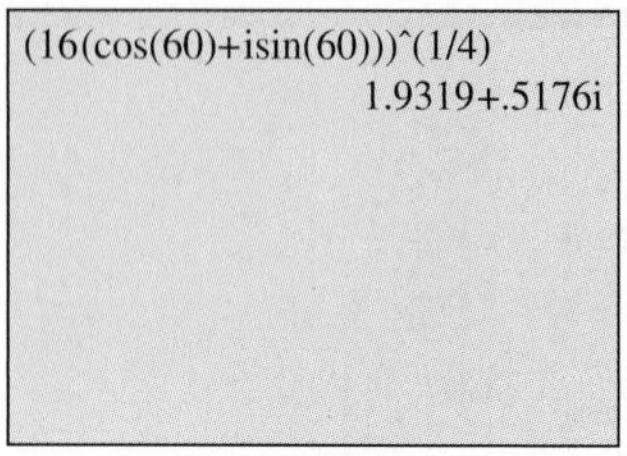

(16(cos(60)+isin(60)))^(1/4)
1.9319+.5176i

Figure 2

CALCULATOR NOTE If we use a graphing calculator to solve Example 1, we will only get one of the four fourth roots. Figure 2 shows the result with the values rounded to four decimal places. The complex number given by the calculator is an approximation of 2 cis 15°.

One application of finding roots is in solving certain equations, as our next examples illustrate.

PROBLEM 2
Solve $x^3 - 8 = 0$.

EXAMPLE 2 Solve $x^3 + 1 = 0$.

SOLUTION Adding -1 to each side of the equation, we have

$$x^3 = -1$$

The solutions to this equation are the cube roots of -1. We already know that one of the cube roots of -1 is -1. There are two other complex cube roots as well. To find them, we write -1 in trigonometric form and then apply the formula from our theorem on roots. Writing -1 in trigonometric form, we have

$$-1 = 1(\cos \pi + i \sin \pi)$$

The 3 cube roots are given by

$$w_k = 1^{1/3}\left[\cos\left(\frac{\pi}{3} + \frac{2\pi}{3}k\right) + i \sin\left(\frac{\pi}{3} + \frac{2\pi}{3}k\right)\right]$$

where $k = 0$, 1, and 2. Replacing k with 0, 1, and 2 and then simplifying each result, we have

$$w_0 = \cos\frac{\pi}{3} + i\sin\frac{\pi}{3} = \frac{1}{2} + \frac{\sqrt{3}}{2}i \qquad \text{when } k = 0$$

$$w_1 = \cos\pi + i\sin\pi = -1 \qquad \text{when } k = 1$$

$$w_2 = \cos\frac{5\pi}{3} + i\sin\frac{5\pi}{3} = \frac{1}{2} - \frac{\sqrt{3}}{2}i \qquad \text{when } k = 2$$

The graphs of these three roots, which are evenly spaced around the unit circle, are shown in Figure 3.

y
1
1 cis π/3 = 1/2 + √3/2 i
1 cis π = −1
−1
x
1
1 cis 5π/3 = 1/2 − √3/2 i
−1

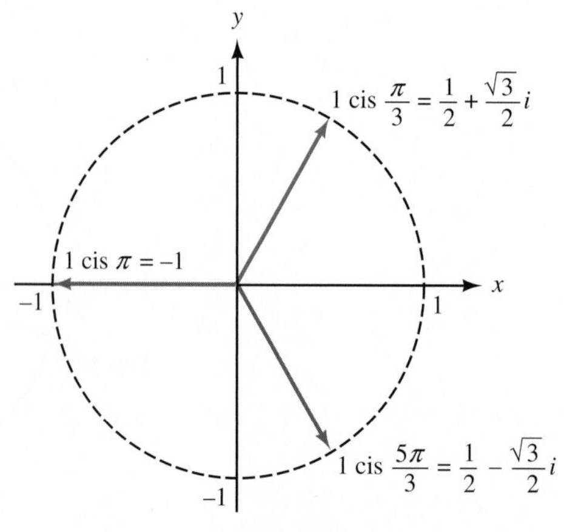

Figure 3

Note that the two complex roots are conjugates. Let's check root w_0 by cubing it.

$$\begin{aligned} w_0{}^3 &= \left(\cos \frac{\pi}{3} + i \sin \frac{\pi}{3}\right)^3 \\ &= \cos 3 \cdot \frac{\pi}{3} + i \sin 3 \cdot \frac{\pi}{3} \\ &= \cos \pi + i \sin \pi \\ &= -1 \end{aligned}$$

You may have noticed that the equation in Example 2 can be solved by algebraic methods. Let's solve $x^3 + 1 = 0$ by factoring and compare our results to those in Example 2. [Recall the formula for factoring the sum of two cubes as follows: $a^3 + b^3 = (a + b)(a^2 - ab + b^2)$.]

$$x^3 + 1 = 0$$

$$(x + 1)(x^2 - x + 1) = 0 \qquad \text{Factor the left side}$$

$$x + 1 = 0 \qquad x^2 - x + 1 = 0 \qquad \text{Set factors equal to 0}$$

The first equation gives us $x = -1$ for a solution. To solve the second equation, we use the quadratic formula.

$$\begin{aligned} x &= \frac{-(-1) \pm \sqrt{(-1)^2 - 4(1)(1)}}{2(1)} \\ &= \frac{1 \pm \sqrt{-3}}{2} \\ &= \frac{1 \pm i\sqrt{3}}{2} \end{aligned}$$

This last expression gives us two solutions; they are

$$x = \frac{1}{2} + \frac{\sqrt{3}}{2}i \qquad \text{and} \qquad x = \frac{1}{2} - \frac{\sqrt{3}}{2}i$$

As you can see, the three solutions we found in Example 2 using trigonometry match the three solutions found using algebra. Notice that the algebraic method of solving the equation $x^3 + 1 = 0$ depends on our ability to factor the left side of the equation. The advantage to the trigonometric method is that it is independent of our ability to factor.

PROBLEM 3
Solve $x^4 + 2x^2 + 2 = 0$.

EXAMPLE 3 Solve the equation $x^4 - 2\sqrt{3}x^2 + 4 = 0$.

SOLUTION The equation is quadratic in x^2. We can solve for x^2 by applying the quadratic formula.

$$\begin{aligned} x^2 &= \frac{2\sqrt{3} \pm \sqrt{12 - 4(1)(4)}}{2} \\ &= \frac{2\sqrt{3} \pm \sqrt{-4}}{2} \\ &= \frac{2\sqrt{3} \pm 2i}{2} \\ &= \sqrt{3} \pm i \end{aligned}$$

The two solutions for x^2 are $\sqrt{3} + i$ and $\sqrt{3} - i$, which we write in trigonometric form as follows:

$$x^2 = \sqrt{3} + i \qquad \text{or} \qquad x^2 = \sqrt{3} - i$$
$$= 2(\cos 30° + i \sin 30°) \qquad\qquad = 2(\cos 330° + i \sin 330°)$$

Now each of these expressions has two square roots, each of which is a solution to our original equation.

When $x^2 = 2(\cos 30° + i \sin 30°)$

$$x = 2^{1/2}\left[\cos\left(\frac{30°}{2} + \frac{360°}{2}k\right) + i \sin\left(\frac{30°}{2} + \frac{360°}{2}k\right)\right] \quad \text{for } k = 0 \text{ and } 1$$

$$= \sqrt{2} \text{ cis } 15° \qquad \text{when } k = 0$$

$$= \sqrt{2} \text{ cis } 195° \qquad \text{when } k = 1$$

When $x^2 = 2(\cos 330° + i \sin 330°)$

$$x = 2^{1/2}\left[\cos\left(\frac{330°}{2} + \frac{360°}{2}k\right) + i \sin\left(\frac{330°}{2} + \frac{360°}{2}k\right)\right] \quad \text{for } k = 0 \text{ and } 1$$

$$= \sqrt{2} \text{ cis } 165° \qquad \text{when } k = 0$$

$$= \sqrt{2} \text{ cis } 345° \qquad \text{when } k = 1$$

Using a calculator and rounding to the nearest hundredth, we can write decimal approximations to each of these four solutions.

SOLUTIONS

Trigonometric Form		Decimal Approximation
$\sqrt{2}$ cis 15°	=	$1.37 + 0.37i$
$\sqrt{2}$ cis 165°	=	$-1.37 + 0.37i$
$\sqrt{2}$ cis 195°	=	$-1.37 - 0.37i$
$\sqrt{2}$ cis 345°	=	$1.37 - 0.37i$

Getting Ready for Class

After reading through the preceding section, respond in your own words and in complete sentences.

a. How many nth roots does a complex number have?

b. How many degrees are there between the arguments of the four fourth roots of a number?

c. How many nonreal complex solutions are there to the equation $x^3 + 1 = 0$?

d. What is the first step in finding the three cube roots of -1?

8.4 PROBLEM SET

➤ CONCEPTS AND VOCABULARY

For Questions 1 through 4, fill in the blank with an appropriate word, number, or expression.

1. Every complex number has ___ distinct nth roots.

2. To find the principle nth root of a complex number, find the nth root of the _________ and divide the ___________ by ___. To find the remaining nth roots, add ________ to the argument repeatedly until you have them all.

3. The 5th roots of a complex number would have arguments that differ from each other by _______ degrees.

4. When graphed, all of the nth roots of a complex number will be evenly distributed around a _______ of ______ $r^{1/n}$.

➤ EXERCISES

Find two square roots for each of the following complex numbers. Leave your answers in trigonometric form. In each case, graph the two roots.

5. $4(\cos 30° + i \sin 30°)$

6. $16(\cos 30° + i \sin 30°)$

7. $25(\cos 210° + i \sin 210°)$

8. $9(\cos 310° + i \sin 310°)$

9. $49 \text{ cis } \pi$

10. $81 \text{ cis } \dfrac{5\pi}{6}$

Find the two square roots for each of the following complex numbers. Write your answers in standard form.

11. $2 + 2i\sqrt{3}$

12. $-2 + 2i\sqrt{3}$

13. $4i$

14. $-4i$

15. -25

16. 25

17. $1 + i\sqrt{3}$

18. $1 - i\sqrt{3}$

Find three cube roots for each of the following complex numbers. Leave your answers in trigonometric form.

19. $8(\cos 210° + i \sin 210°)$

20. $27(\cos 303° + i \sin 303°)$

21. $4\sqrt{3} + 4i$

22. $-4\sqrt{3} + 4i$

23. -27

24. 8

25. $64i$

26. $-64i$

Solve each equation.

27. $x^3 + 8 = 0$

28. $x^3 - 27 = 0$

29. $x^4 + 81 = 0$

30. $x^4 - 16 = 0$

31. Find the 4 fourth roots of $z = 16(\cos \frac{2\pi}{3} + i \sin \frac{2\pi}{3})$. Write each root in standard form.

32. Find the 4 fourth roots of $z = \cos \frac{4\pi}{3} + i \sin \frac{4\pi}{3}$. Leave your answers in trigonometric form.

33. Find the 5 fifth roots of $z = 10^5 \text{ cis } 15°$. Write each root in trigonometric form and then give a decimal approximation, accurate to the nearest hundredth, for each one.

34. Find the 5 fifth roots of $z = 10^{10} \text{ cis } 75°$. Write each root in trigonometric form and then give a decimal approximation, accurate to the nearest hundredth, for each one.

35. Find the 6 sixth roots of $z = -1$. Leave your answers in trigonometric form. Graph all six roots on the same coordinate system.

36. Find the 6 sixth roots of $z = 1$. Leave your answers in trigonometric form. Graph all six roots on the same coordinate system.

Solve each of the following equations. Leave your solutions in trigonometric form.

37. $x^4 - 2x^2 + 4 = 0$

38. $x^4 + 2x^2 + 4 = 0$

39. $x^4 + 2x^2 + 2 = 0$

40. $x^4 - 2x^2 + 2 = 0$

➤ REVIEW PROBLEMS

The problems that follow review material we covered in Sections 4.2, 4.3, and 7.4. Graph each equation on the given interval.

41. $y = -2 \sin(-3x), 0 \le x \le 2\pi$

42. $y = -2 \cos(-3x), 0 \le x \le 2\pi$

Graph one complete cycle of each of the following:

43. $y = -\cos\left(2x + \frac{\pi}{2}\right)$

44. $y = 3 \sin\left(\frac{\pi}{3}x - \frac{\pi}{3}\right)$

Find the area of triangle ABC given the following information.

45. $A = 56.2°$, $b = 2.65$ cm, and $c = 3.84$ cm

46. $B = 21.8°$, $a = 44.4$ cm, and $c = 22.2$ cm

47. $a = 2.3$ ft, $b = 3.4$ ft, and $c = 4.5$ ft

48. $a = 5.4$ ft, $b = 4.3$ ft, and $c = 3.2$ ft

➤ EXTENDING THE CONCEPTS

Cardan

49. Recall from the introduction to Section 8.2 that Jerome Cardan's solutions to the equation $x^3 = 15x + 4$ could be written as

$$x = \sqrt[3]{2 + 11i} + \sqrt[3]{2 - 11i}$$

Let's assume that the two cube roots are complex conjugates. If they are, then we can simplify our work by noticing that

$$x = \sqrt[3]{2 + 11i} + \sqrt[3]{2 - 11i} = a + bi + a - bi = 2a$$

which means that we simply double the real part of each cube root of $2 + 11i$ to find the solutions to $x^3 = 15x + 4$. Now, to end our work with Cardan, find the 3 cube roots of $2 + 11i$. Then, noting the discussion above, use the 3 cube roots to solve the equation $x^3 = 15x + 4$. Write your answers accurate to the nearest thousandth.

➤ LEARNING OBJECTIVES ASSESSMENT

These questions are available for instructors to help assess if you have successfully met the learning objectives for this section.

50. Find the two square roots of $36i$.

a. $3\sqrt{2} + 3i\sqrt{2}, -3\sqrt{2} - 3i\sqrt{2}$

b. $-3\sqrt{2} + 3i\sqrt{2}, 3\sqrt{2} - 3i\sqrt{2}$

c. $6i, -6i$

d. $6, -6$

51. Which of the following is one of the fifth roots of 32 cis 40°?

a. 2 cis 152° **b.** 2 cis 200°

c. 2 cis 112° **d.** 6.4 cis 8°

52. Which graph could represent the fourth roots of some complex number?

a.

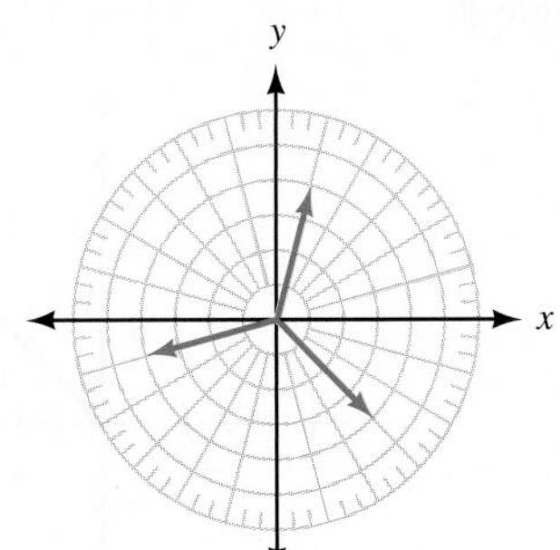

b.

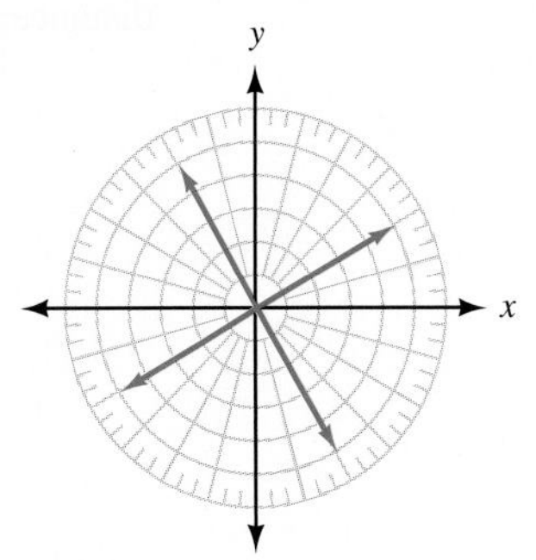

c.

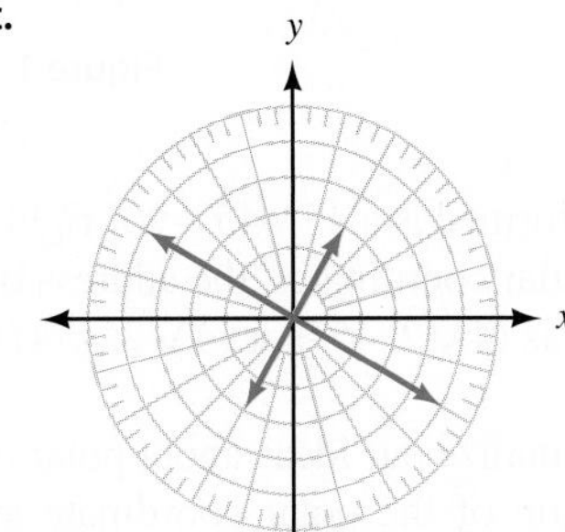

d.

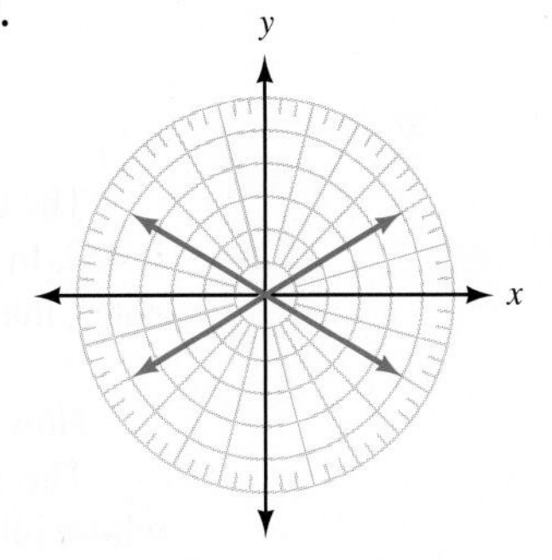

53. Solve $x^3 + 64 = 0$. Which of the following is a solution?

a. $2 - 2i\sqrt{3}$ **b.** $2\sqrt{3} + 2i$

c. $-4\sqrt{2} - 4i\sqrt{2}$ **d.** $-4\sqrt{2} + 4i\sqrt{2}$

SECTION 8.5 Polar Coordinates

LEARNING OBJECTIVES

1. Graph an ordered pair in polar coordinates.
2. Convert an ordered pair from polar to rectangular coordinates or vice-versa.
3. Express a polar equation in rectangular coordinates.
4. Express a rectangular equation in polar coordinates.

Up to this point in our study of trigonometry, whenever we have given the position of a point in the plane, we have used rectangular coordinates. That is, we give the position of points in the plane relative to a set of perpendicular axes. In a way, the rectangular coordinate system is a map that tells us how to get anywhere in the plane by traveling so far horizontally and so far vertically relative to the origin of our coordinate system. For example, to reach the point whose address is (2, 1), we start at the origin and travel 2 units right and then 1 unit up. In this section, we will see that there are other ways to get to the point (2, 1). In particular, we can travel $\sqrt{5}$ units on the terminal side of an angle in standard position. This type of map is the basis of what we call *polar coordinates:* The address of each point in the plane is given by an ordered pair (r, θ), where r is a directed distance on the terminal side of standard position angle θ.

PROBLEM 1
Write $(4, 4\sqrt{3})$ in polar coordinates.

EXAMPLE 1 A point lies at (4, 4) on a rectangular coordinate system. Give its address in polar coordinates (r, θ).

SOLUTION We want to reach the same point by traveling r units on the terminal side of a standard position angle θ. Figure 1 shows the point (4, 4), along with the distance r and angle θ.

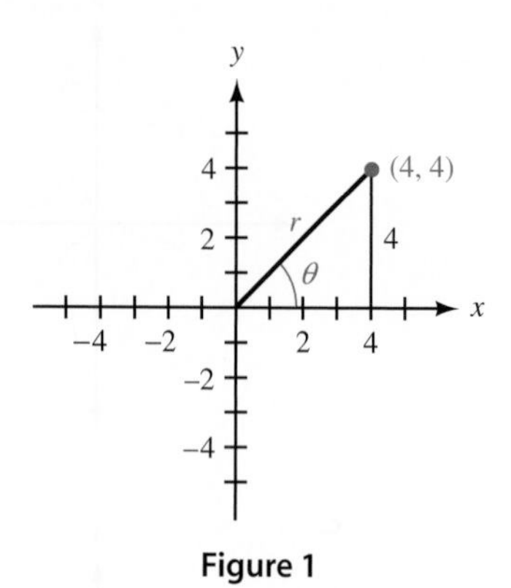

Figure 1

The triangle formed is a 45°–45°–90° right triangle. Therefore, r is $4\sqrt{2}$, and θ is 45°. In rectangular coordinates, the address of our point is (4, 4). In polar coordinates, the address is $(4\sqrt{2}, 45°)$ or $(4\sqrt{2}, \pi/4)$ using radians.

Now let's formalize our ideas about polar coordinates.

The foundation of the polar coordinate system is a ray called the *polar axis,* whose initial point is called the *pole* (Figure 2). In the polar coordinate system, points are named by ordered pairs (r, θ) in which r is the directed distance from the pole on the terminal side of an angle θ, the initial side of which is the polar axis. If the terminal side of θ has been rotated in a counterclockwise direction from the polar axis, then θ is a positive angle. Otherwise, θ is a negative angle. For example, the point (5, 30°) is 5 units from the pole along a ray that has been rotated 30° from the polar axis, as shown in Figure 3.

Figure 2 **Figure 3**

To simplify matters, in this book we place the pole at the origin of our rectangular coordinate system and take the polar axis to be the positive x-axis.

The circular grids used in Example 2 are helpful when graphing points given in polar coordinates. The lines on the grid are multiples of 15° or $\pi/12$ radians, and the circles have radii of 1, 2, 3, 4, 5, and 6.

PROBLEM 2
Graph the points (5, 240°), (4, −45°), $(-3, 7\pi/6)$, and $(-2, -\pi/3)$ on a polar coordinate system.

EXAMPLE 2 Graph the points (3, 45°), $(2, -4\pi/3)$, $(-4, \pi/3)$, and (−5, −210°) on a polar coordinate system.

SOLUTION To graph (3, 45°), we locate the point that is 3 units from the origin along the terminal side of 45°, as shown in Figure 4. The point $(2, -4\pi/3)$ is 2 units along the terminal side of $-4\pi/3$, as Figure 5 indicates.

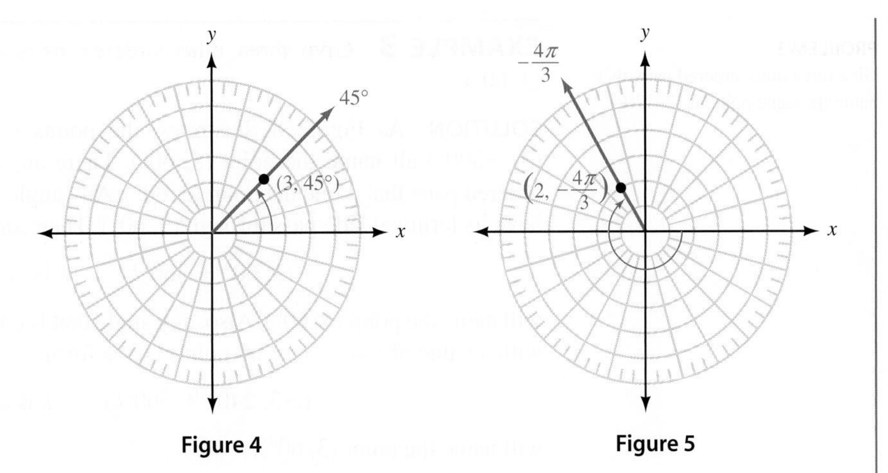

Figure 4

Figure 5

As you can see from Figures 4 and 5, if r is positive, we locate the point (r, θ) along the terminal side of θ. The next two points we will graph have negative values of r.

To graph a point (r, θ) in which r is negative, we look for the point that is r units in the *opposite* direction indicated by the terminal side of θ. That is, we extend a ray from the pole that is directly opposite the terminal side of θ and locate the point r units along this ray.

To graph $(-4, \pi/3)$, we first extend the terminal side of θ through the origin to create a ray in the opposite direction. Then we locate the point that is 4 units from the origin along this ray, as shown in Figure 6. The terminal side of θ has been drawn in blue, and the ray pointing in the opposite direction is shown in red.

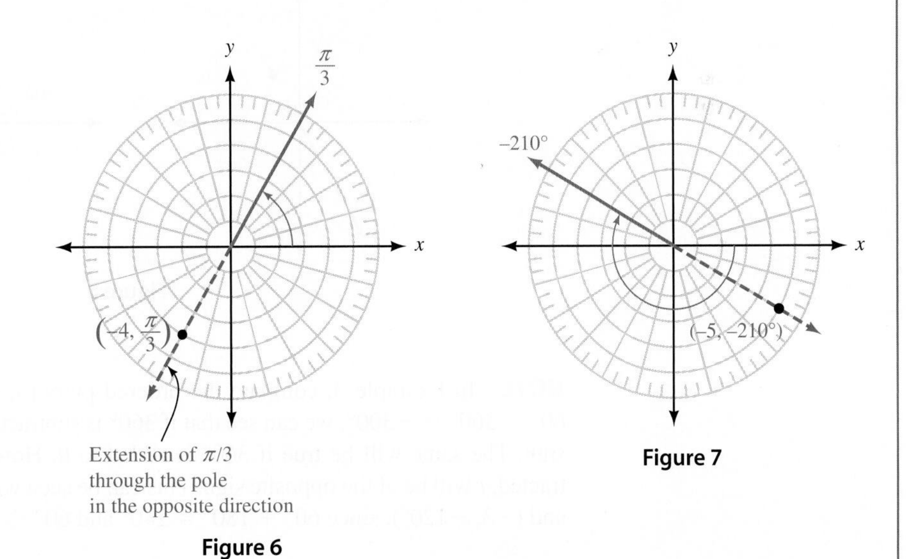

Figure 6

Figure 7

To graph $(-5, -210°)$, we look for the point that is 5 units from the origin along the ray in the opposite direction of $-210°$ (Figure 7).

In rectangular coordinates, each point in the plane is named by a unique ordered pair (x, y). That is, no point can be named by two different ordered pairs. The same is not true of points named by polar coordinates. As Example 3 illustrates, the polar coordinates of a point are not unique.

PROBLEM 3
Give three other ordered pairs that name the same point as $(2, 120°)$.

EXAMPLE 3 Give three other ordered pairs that name the same point as $(3, 60°)$.

SOLUTION As Figure 8 illustrates, the points $(-3, 240°)$, $(-3, -120°)$, and $(3, -300°)$ all name the point $(3, 60°)$. There are actually an infinite number of ordered pairs that name the point $(3, 60°)$. Any angle that is coterminal with 60° will have its terminal side pass through $(3, 60°)$. Therefore, all points of the form

$$(3, 60° + 360°k) \qquad k \text{ is any integer}$$

will name the point $(3, 60°)$. Also, any angle that is coterminal with 240° can be used with a value of $r = -3$, so all points of the form

$$(-3, 240° + 360°k) \qquad k \text{ is any integer}$$

will name the point $(3, 60°)$.

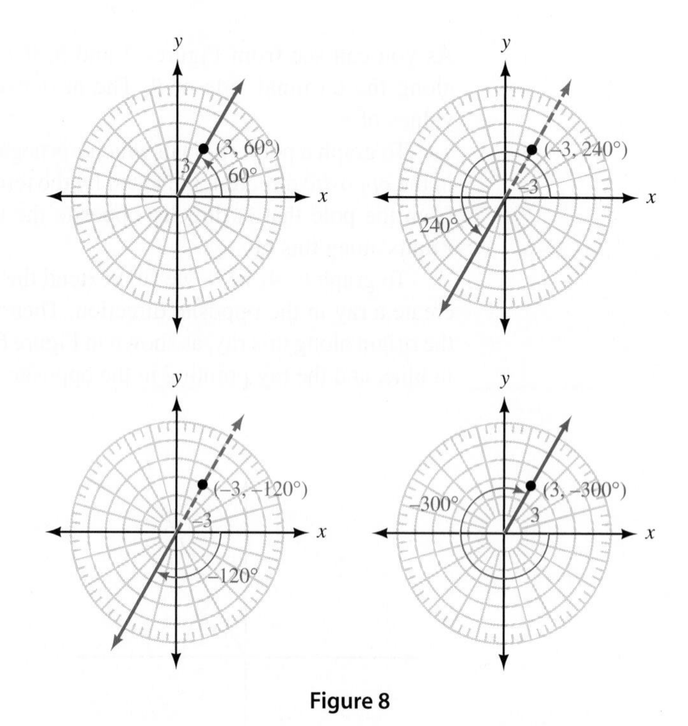

Figure 8

NOTE In Example 3, compare the ordered pairs $(3, 60°)$ and $(3, -300°)$. Because $60° - 360° = -300°$, we can see that if 360° is subtracted from θ, r will be of the same sign. The same will be true if 360° is added to θ. However, if 180° is added or subtracted, r will be of the opposite sign. This can be seen with the ordered pairs $(-3, 240°)$ and $(-3, -120°)$, since $60° + 180° = 240°$ and $60° - 180° = -120°$.

Polar Coordinates and Rectangular Coordinates

To derive the relationship between polar coordinates and rectangular coordinates, we consider a point P with rectangular coordinates (x, y) and polar coordinates (r, θ).

To convert back and forth between polar and rectangular coordinates, we simply use the relationships that exist among x, y, r, and θ in Figure 9.

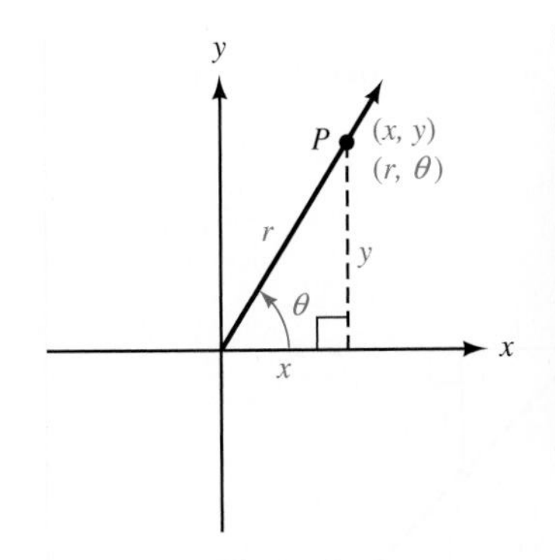

Figure 9

TO CONVERT RECTANGULAR COORDINATES TO POLAR COORDINATES

Let

$$r = \pm\sqrt{x^2 + y^2} \qquad \text{and} \qquad \tan\theta = \frac{y}{x}$$

where the sign of r and the choice of θ place the point (r, θ) in the same quadrant as (x, y).

TO CONVERT POLAR COORDINATES TO RECTANGULAR COORDINATES

Let

$$x = r\cos\theta \qquad \text{and} \qquad y = r\sin\theta$$

The process of converting to rectangular coordinates is simply a matter of substituting r and θ into the preceding equations. To convert to polar coordinates, we have to choose θ and the sign of r so the point (r, θ) is in the same quadrant as the point (x, y).

PROBLEM 4

Convert to rectangular coordinates.

a. $(2, 60°)$

b. $\left(4, \frac{\pi}{2}\right)$

c. $\left(-1, \frac{5\pi}{4}\right)$

EXAMPLE 4 Convert to rectangular coordinates.

a. $(4, 30°)$ **b.** $\left(-\sqrt{2}, \frac{3\pi}{4}\right)$ **c.** $(3, 270°)$

SOLUTION To convert from polar coordinates to rectangular coordinates, we substitute the given values of r and θ into the equations

$$x = r\cos\theta \qquad \text{and} \qquad y = r\sin\theta$$

Here are the conversions for each point along with the graphs in both rectangular and polar coordinates.

a. $x = 4\cos 30°$

$= 4\left(\frac{\sqrt{3}}{2}\right)$

$= 2\sqrt{3}$

$y = 4\sin 30°$

$= 4\left(\frac{1}{2}\right)$

$= 2$

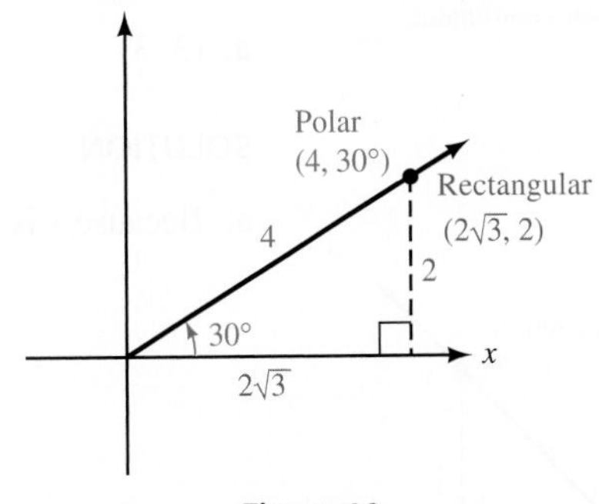

Figure 10

The point $(2\sqrt{3}, 2)$ in rectangular coordinates is equivalent to $(4, 30°)$ in polar coordinates. Figure 10 illustrates.

b. $x = -\sqrt{2}\cos\dfrac{3\pi}{4}$

$= -\sqrt{2}\left(-\dfrac{\sqrt{2}}{2}\right)$

$= 1$

$y = -\sqrt{2}\sin\dfrac{3\pi}{4}$

$= -\sqrt{2}\left(\dfrac{\sqrt{2}}{2}\right)$

$= -1$

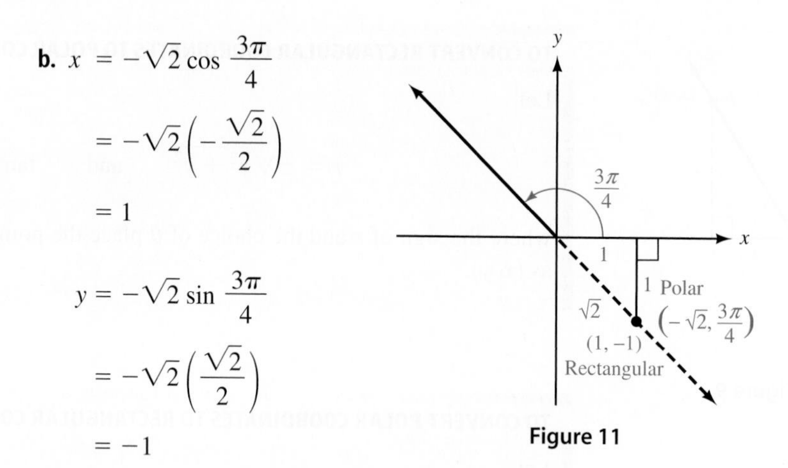

Figure 11

The point $(1, -1)$ in rectangular coordinates is equivalent to $(-\sqrt{2}, 3\pi/4)$ in polar coordinates. Figure 11 illustrates.

c. $x = 3\cos 270°$

$= 3(0)$

$= 0$

$y = 3\sin 270°$

$= 3(-1)$

$= -3$

Figure 12

The point $(0, -3)$ in rectangular coordinates is equivalent to $(3, 270°)$ in polar coordinates. Figure 12 illustrates.

PROBLEM 5
Convert to polar coordinates.
a. $(-3, 3)$
b. $(0, -4)$
c. $(-\sqrt{3}, -1)$

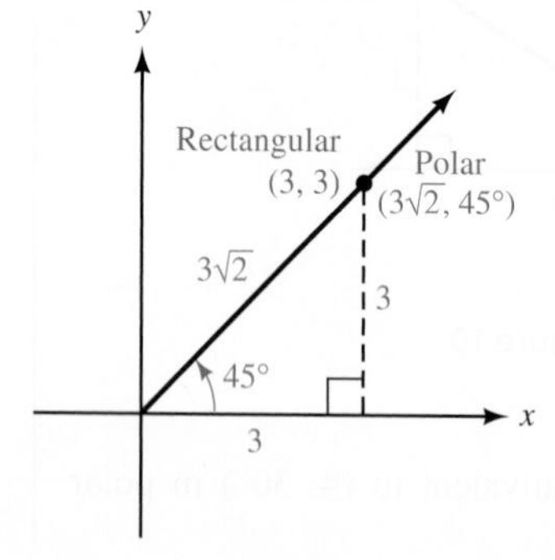

Figure 13

EXAMPLE 5 Convert to polar coordinates.

a. $(3, 3)$ **b.** $(-2, 0)$ **c.** $(-1, \sqrt{3})$

SOLUTION

a. Because x is 3 and y is 3, we have

$$r = \pm\sqrt{9 + 9} = \pm 3\sqrt{2}$$

$$\text{and} \quad \tan\theta = \frac{3}{3} = 1$$

Because $(3, 3)$ is in QI, we can choose $r = 3\sqrt{2}$ and $\theta = 45°$ (or $\pi/4$), as shown in Figure 13. Remember, there are an infinite number of ordered pairs in polar coordinates that name the point $(3, 3)$. The point $(3\sqrt{2}, 45°)$ is just one of them. Generally, we choose r and θ so that r is positive and θ is between 0° and 360°.

b. We have $x = -2$ and $y = 0$, so

$$r = \pm\sqrt{4 + 0} = \pm 2 \quad \text{and} \quad \tan\theta = \frac{0}{-2} = 0$$

Because $(-2, 0)$ is on the negative x-axis, we can choose $r = 2$ and $\theta = 180°$ (or π) to get the point $(2, 180°)$. Figure 14 illustrates.

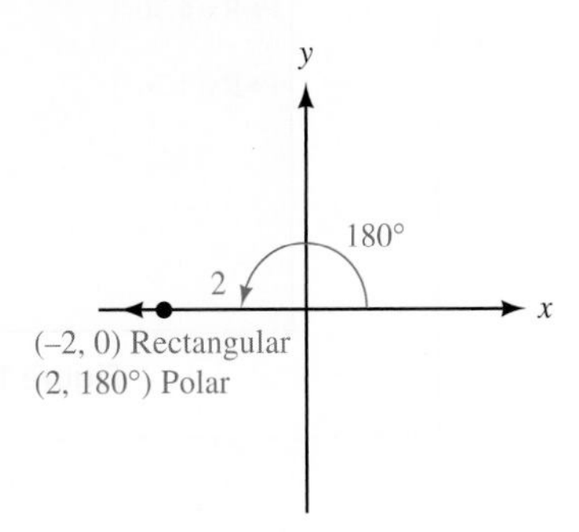

Figure 14

c. Because $x = -1$ and $y = \sqrt{3}$, we have

$$r = \pm\sqrt{1 + 3} = \pm 2 \quad \text{and} \quad \tan\theta = \frac{\sqrt{3}}{-1}$$

Because $(-1, \sqrt{3})$ is in QII, we can let $r = 2$ and $\theta = 120°$ (or $2\pi/3$). In polar coordinates, the point is $(2, 120°)$. Figure 15 illustrates.

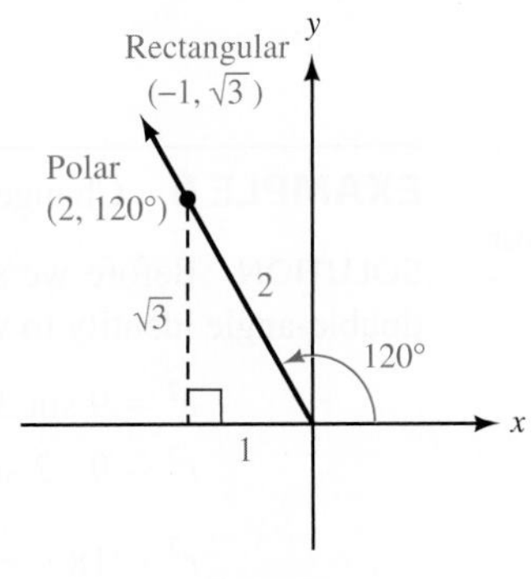

Figure 15

CALCULATOR NOTE If we use the $\boxed{\tan^{-1}}$ key on a calculator (set to degree mode) to find the angle θ in Example 5c, we will get

$$\theta = \tan^{-1}(-\sqrt{3}) = -60°$$

This would place the terminal side of θ in QIV. To locate our point in QII, we could use a negative value of r. The polar coordinates would then be $(-2, -60°)$.

Using Technology — Converting Between Polar and Rectangular Coordinates

In Section 8.2 we mentioned that most graphing calculators are able to convert a complex number between standard form and trigonometric form. The same is true for converting ordered pairs between rectangular coordinates and polar coordinates. In fact, sometimes the very same commands are used.

In Figure 16 we used a TI-84 set to degree mode to solve Example 4a. The value 3.4641 is an approximation of $2\sqrt{3}$. Example 5a is shown in Figure 17, where 4.2426 is an approximation of $3\sqrt{2}$. For a TI-84, the conversion commands are located in the ANGLE menu.

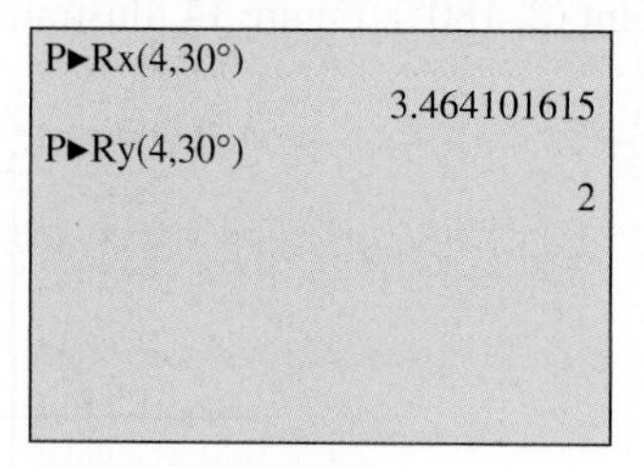

Figure 16

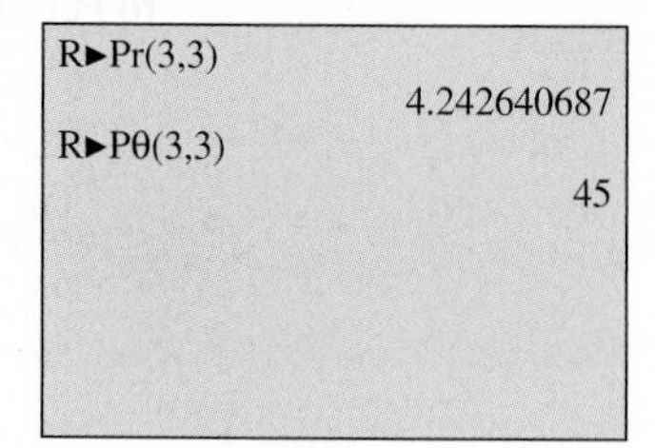

Figure 17

Equations in Polar Coordinates

Equations in polar coordinates have variables r and θ instead of x and y. The conversions we used to change ordered pairs from polar coordinates to rectangular coordinates and from rectangular coordinates to polar coordinates are the same ones we use to convert back and forth between equations given in polar coordinates and those in rectangular coordinates.

PROBLEM 6
Change $r^2 = 4 \cos 2\theta$ to rectangular coordinates.

EXAMPLE 6 Change $r^2 = 9 \sin 2\theta$ to rectangular coordinates.

SOLUTION Before we substitute to clear the equation of r and θ, we must use a double-angle identity to write $\sin 2\theta$ in terms of $\sin \theta$ and $\cos \theta$.

$$r^2 = 9 \sin 2\theta$$

$$r^2 = 9 \cdot 2 \sin \theta \cos \theta \qquad \text{Double-angle identity}$$

$$r^2 = 18 \cdot \frac{y}{r} \cdot \frac{x}{r} \qquad \text{Substitute } y/r \text{ for } \sin\theta \text{ and } x/r \text{ for } \cos\theta$$

$$r^2 = \frac{18xy}{r^2} \qquad \text{Multiply}$$

$$r^4 = 18xy \qquad \text{Multiply both sides by } r^2$$

$$(x^2 + y^2)^2 = 18xy \qquad \text{Substitute } x^2 + y^2 \text{ for } r^2$$

PROBLEM 7
Change $2x + y = 9$ to polar coordinates.

EXAMPLE 7 Change $x + y = 4$ to polar coordinates.

SOLUTION Because $x = r \cos \theta$ and $y = r \sin \theta$, we have

$$r \cos \theta + r \sin \theta = 4$$

$$r(\cos \theta + \sin \theta) = 4 \qquad \text{Factor out } r$$

$$r = \frac{4}{\cos \theta + \sin \theta} \qquad \text{Divide both sides by } \cos\theta + \sin\theta$$

The last equation gives us r in terms of θ.

Getting Ready for Class

After reading through the preceding section, respond in your own words and in complete sentences.

a. How do you plot the point whose polar coordinates are (3, 45°)?
b. Why do points in the coordinate plane have more than one representation in polar coordinates?
c. If you convert (4, 30°) to rectangular coordinates, how do you find the x-coordinate?
d. What are the rectangular coordinates of the point (3, 270°)?

8.5 PROBLEM SET

➤ CONCEPTS AND VOCABULARY

For Questions 1 through 6, fill in the blank with an appropriate word, expression, or equation.

1. The polar coordinate system consists of a point, called the _____, and a ray extending out from it, called the _______ _____.

2. In polar coordinates, r is the ________ ________ on the terminal side of an angle θ whose vertex is at the _____ and whose initial side lies along the ________ ______.

3. An angle θ is considered positive if its terminal side has been rotated ____________ from the ______ axis.

4. To graph a point (r, θ) with negative r, plot a point ___ units along a ray in the _______ direction from the terminal side of θ.

5. When converting between rectangular and polar coordinates, we assume the pole lies on the _______ and the polar axis lies along the ________ ___-axis.

6. To convert an ordered pair from polar to rectangular coordinates, use the relationships $x =$ _______ and $y =$ _______. To convert an ordered pair from rectangular to polar coordinates, use the relationships $r =$ _______ and _______, keeping in mind the quadrant the point lies in.

For Questions 7 and 8, determine if the statement is true or false.

7. In rectangular coordinates, each point is represented by a unique ordered pair.

8. In polar coordinates, each point is represented by a unique ordered pair.

➤ EXERCISES

Graph each ordered pair on a polar coordinate system.

9. (2, 45°) **10.** (3, 60°) **11.** (3, 150°) **12.** (4, 135°)

13. (1, −225°) **14.** (2, −240°) **15.** (−3, 45°) **16.** (−4, 60°)

17. $\left(-4, -\frac{7\pi}{6}\right)$ **18.** $\left(-5, -\frac{5\pi}{4}\right)$ **19.** (−2, 0) **20.** $\left(-2, \frac{3\pi}{2}\right)$

For each ordered pair, give three other ordered pairs with θ between $-360°$ and $360°$ that name the same point.

21. $(2, 60°)$ **22.** $(1, 30°)$ **23.** $(5, 135°)$ **24.** $(3, 120°)$

25. $(-3, 30°)$ **26.** $(-2, 45°)$

Convert to rectangular coordinates. Use exact values.

27. $(2, 60°)$ **28.** $(-2, 60°)$ **29.** $\left(3, \dfrac{3\pi}{2}\right)$ **30.** $(1, \pi)$

31. $(\sqrt{2}, -135°)$ **32.** $(\sqrt{2}, -225°)$ **33.** $\left(-4\sqrt{3}, \dfrac{\pi}{6}\right)$ **34.** $\left(4\sqrt{3}, -\dfrac{\pi}{6}\right)$

Use your graphing calculator to convert to rectangular coordinates. Round all values to four significant digits.

35. $(2, 19°)$ **36.** $(3, 124°)$ **37.** $(-3, 293°)$ **38.** $(-4, 261°)$

Convert to polar coordinates with $r \geq 0$ and $0° \leq \theta < 360°$.

39. $(-3, 3)$ **40.** $(-3, -3)$ **41.** $(2, -2\sqrt{3})$ **42.** $(-2\sqrt{3}, 2)$

Convert to polar coordinates with $r \geq 0$ and $0 \leq \theta < 2\pi$.

43. $(2, 0)$ **44.** $(-2, 0)$ **45.** $(-\sqrt{3}, -1)$ **46.** $(-1, -\sqrt{3})$

Convert to polar coordinates. Use a calculator to find θ to the nearest tenth of a degree. Keep r positive and θ between $0°$ and $360°$.

47. $(3, 4)$ **48.** $(4, 3)$ **49.** $(-1, 2)$ **50.** $(1, -2)$

51. $(-2, -3)$ **52.** $(-3, -2)$

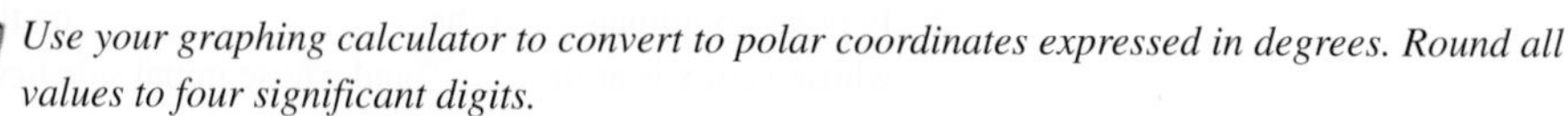

Use your graphing calculator to convert to polar coordinates expressed in degrees. Round all values to four significant digits.

53. $(5, 8)$ **54.** $(-2, 9)$ **55.** $(-1, -6)$ **56.** $(7, -3)$

Write each equation with rectangular coordinates.

57. $r^2 = 9$ **58.** $r^2 = 4$

59. $r = 6 \cos \theta$ **60.** $r = 6 \sin \theta$

61. $r^2 = 4 \cos 2\theta$ **62.** $r^2 = 4 \sin 2\theta$

63. $r(\cos \theta + \sin \theta) = 3$ **64.** $r(\cos \theta - \sin \theta) = 2$

Write each equation in polar coordinates. Then isolate the variable r when possible.

65. $x + y = 5$ **66.** $x - y = 5$

67. $x^2 + y^2 = 4$ **68.** $x^2 + y^2 = 9$

69. $x^2 + y^2 = 6x$ **70.** $x^2 + y^2 = 4x$

71. $y = x$ **72.** $y = -x$

➤ REVIEW PROBLEMS

The problems that follow review material we covered in Sections 4.2 and 4.3. Reviewing these problems will help you with the next section.

Graph one complete cycle of each equation.

73. $y = 6 \sin x$ **74.** $y = 6 \cos x$ **75.** $y = 4 \sin 2x$

76. $y = 2 \sin 4x$ **77.** $y = 4 + 2 \sin x$ **78.** $y = 4 + 2 \cos x$

➤ LEARNING OBJECTIVES ASSESSMENT

These questions are available for instructors to help assess if you have successfully met the learning objectives for this section.

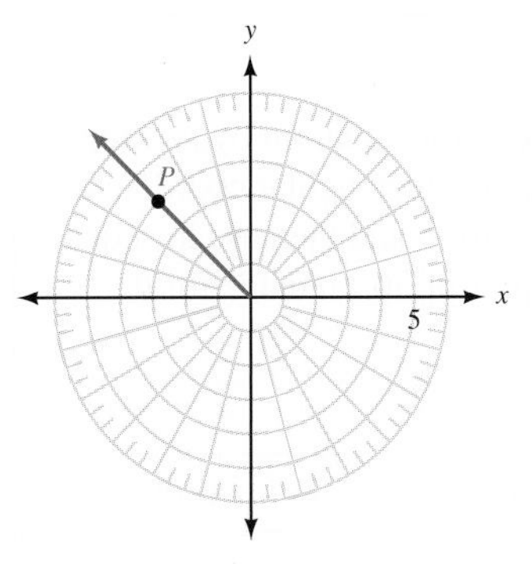

Figure 18

79. Which of the following ordered pairs is *not* a valid representation for point P shown in Figure 18?

a. $(4, 135°)$ **b.** $(-4, -45°)$ **c.** $(4, -405°)$ **d.** $(-4, 315°)$

80. Convert $\left(-2, \dfrac{5\pi}{6}\right)$ to rectangular coordinates.

a. $(1, -\sqrt{3})$ **b.** $(-\sqrt{3}, 1)$ **c.** $(-1, \sqrt{3})$ **d.** $(\sqrt{3}, -1)$

81. Write $r = 6(\cos\theta - \sin\theta)$ in rectangular coordinates.

a. $x^2 + y^2 = 6(x - y)$ **b.** $\sqrt{x^2 + y^2} = 6(x - y)$

c. $\sqrt{x^2 + y^2} = 6(y - x)$ **d.** $x^2 + y^2 = 6(y - x)$

82. Write $x^2 + xy + y^2 = 1$ in polar coordinates, and isolate r if possible.

a. $r = \dfrac{1}{\cos\theta + \sin\theta}$ **b.** $r = \sqrt{1 - \cos\theta\sin\theta}$

c. $r = \dfrac{1}{\sqrt{1 + \cos\theta\sin\theta}}$ **d.** $r = \dfrac{1}{1 - \cos\theta\sin\theta}$

SECTION 8.6 Equations in Polar Coordinates and Their Graphs

LEARNING OBJECTIVES

1. Graph an equation in polar coordinates by plotting points.
2. Graph an equation in polar coordinates by analyzing a graph in rectangular coordinates.
3. Graph an equation in polar coordinates using a graphing calculator.
4. Identify the graph of a polar equation.

© Hulton-Deutsch Collection/CORBIS

Archimedes

More than 2,000 years ago, Archimedes described a curve as starting at a point and moving out along a half-line at a constant rate. The end point of the half-line was anchored at the initial point and was rotating about it at a constant rate also. The curve, called the spiral of Archimedes, is shown in Figure 1, with a rectangular coordinate system superimposed on it so that the origin of the coordinate system coincides with the initial point on the curve.

In rectangular coordinates, the equation that describes this curve is

$$\sqrt{x^2 + y^2} = 3\tan^{-1}\frac{y}{x}$$

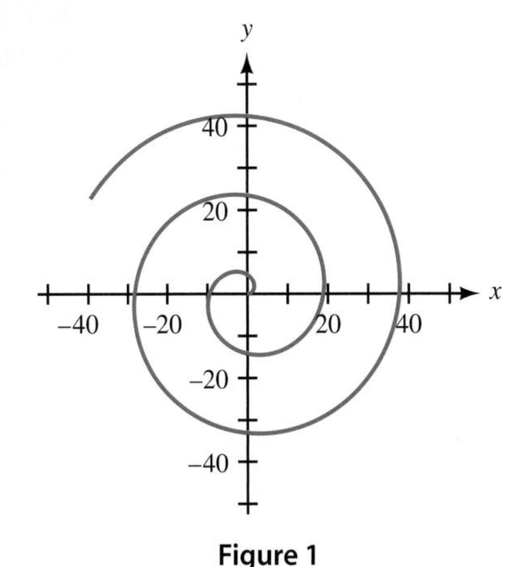

Figure 1

If we were to superimpose a polar coordinate system on the curve in Figure 1, instead of the rectangular coordinate system, then the equation in polar coordinates that describes the curve would be simply

$$r = 3\theta$$

As you can see, the equations for some curves are best given in terms of polar coordinates.

In this section, we will consider the graphs of polar equations. The solutions to these equations are ordered pairs (r, θ), where r and θ are the polar coordinates we defined in Section 8.5.

PROBLEM 1
Sketch the graph of $r = 4 \cos \theta$.

EXAMPLE 1 Sketch the graph of $r = 6 \sin \theta$.

SOLUTION We can find ordered pairs (r, θ) that satisfy the equation by making a table. Table 1 is a little different from the ones we made for rectangular coordinates. With polar coordinates, we substitute convenient values for θ and then use the equation to find corresponding values of r. Let's use multiples of 30° and 45° for θ.

TABLE 1

θ	$r = 6 \sin \theta$	r	(r, θ)
0°	$r = 6 \sin 0° = 0$	0	(0, 0°)
30°	$r = 6 \sin 30° = 3$	3	(3, 30°)
45°	$r = 6 \sin 45° = 4.2$	4.2	(4.2, 45°)
60°	$r = 6 \sin 60° = 5.2$	5.2	(5.2, 60°)
90°	$r = 6 \sin 90° = 6$	6	(6, 90°)
120°	$r = 6 \sin 120° = 5.2$	5.2	(5.2, 120°)
135°	$r = 6 \sin 135° = 4.2$	4.2	(4.2, 135°)
150°	$r = 6 \sin 150° = 3$	3	(3, 150°)
180°	$r = 6 \sin 180° = 0$	0	(0, 180°)
210°	$r = 6 \sin 210° = -3$	−3	(−3, 210°)
225°	$r = 6 \sin 225° = -4.2$	−4.2	(−4.2, 225°)
240°	$r = 6 \sin 240° = -5.2$	−5.2	(−5.2, 240°)
270°	$r = 6 \sin 270° = -6$	−6	(−6, 270°)
300°	$r = 6 \sin 300° = -5.2$	−5.2	(−5.2, 300°)
315°	$r = 6 \sin 315° = -4.2$	−4.2	(−4.2, 315°)
330°	$r = 6 \sin 330° = -3$	−3	(−3, 330°)
360°	$r = 6 \sin 360° = 0$	0	(0, 360°)

NOTE If we were to continue past 360° with values of θ, we would simply start to repeat the values of r we have already obtained, because $\sin \theta$ is periodic with period 360°.

Plotting each point on a polar coordinate system and then drawing a smooth curve through them, we have the graph in Figure 2.

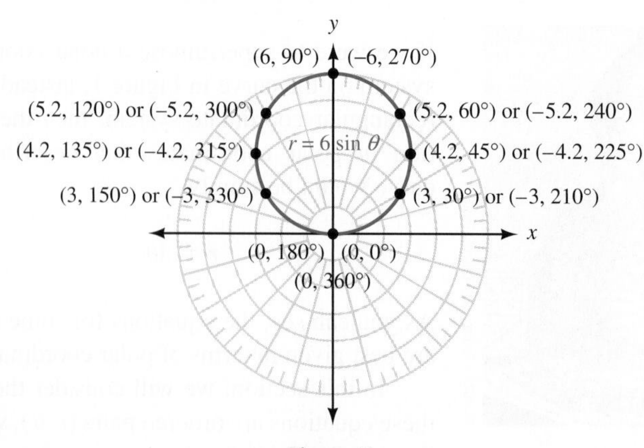

Figure 2

Using Technology — Creating Tables for Polar Equations

A graphing calculator may be used to create a table of values for a polar equation. For example, to make a table for the equation in Example 1, first set the calculator to degree mode and polar mode. Define the function $r_1 = 6 \sin \theta$. Set up your table to automatically generate values starting at $\theta = 0$ and using an increment of 15° (Figure 3). On some calculators, this is done by setting the independent variable to Auto.

```
Plot1  Plot2  Plot3
\r1=6sin(θ)
\r2=
\r3=
\r4=
\r5=
\r6=
```

```
TABLE SETUP
 TblStart=0
 ΔTbl=15
Indpnt: Auto Ask
Depend: Auto Ask
```

Figure 3

Now display the table and you should see values similar to those shown in Figure 4. Scroll down the table to see additional pairs of values (Figure 5).

θ	r1	
0	0	
15	1.5529	
30	3	
45	4.2426	
60	5.1962	
75	5.7956	
90	6	
θ=0		

Figure 4

θ	r1	
105	5.7956	
120	5.1962	
135	4.2426	
150	3	
165	1.5529	
180	0	
195	−1.553	
θ=195		

Figure 5

Could we have found the graph of $r = 6 \sin \theta$ in Example 1 without making a table? The answer is yes, there are a couple of other ways to do so. One way is to convert to rectangular coordinates and see if we recognize the graph from the rectangular equation. We begin by replacing $\sin \theta$ with y/r.

$$r = 6 \sin \theta$$

$$r = 6\frac{y}{r} \qquad \sin \theta = \frac{y}{r}$$

$$r^2 = 6y \qquad \text{Multiply both sides by } r$$

$$x^2 + y^2 = 6y \qquad r^2 = x^2 + y^2$$

The equation is now written in terms of rectangular coordinates. If we add $-6y$ to both sides and then complete the square on y, we will obtain the rectangular equation of a circle with center at (0, 3) and a radius of 3.

$$x^2 + y^2 - 6y = 0 \qquad \text{Add } -6y \text{ to both sides}$$

$$x^2 + y^2 - 6y + 9 = 9 \qquad \text{Complete the square on } y \text{ by adding 9 to both sides}$$

$$x^2 + (y - 3)^2 = 3^2 \qquad \text{Standard form for the equation of a circle}$$

This method of graphing, by changing to rectangular coordinates, works well only in some cases. Many of the equations we will encounter in polar coordinates do not have graphs that are recognizable in rectangular form.

In Example 2, we will look at another method of graphing polar equations that does not depend on the use of a table.

PROBLEM 2
Sketch the graph of $r = 2 \cos 2\theta$.

EXAMPLE 2 Sketch the graph of $r = 4 \sin 2\theta$.

SOLUTION One way to visualize the relationship between r and θ as given by the equation $r = 4 \sin 2\theta$ is to sketch the graph of $y = 4 \sin 2x$ on a rectangular coordinate system. (We have been using degree measure for our angles in polar coordinates, so we will label the x-axis for the graph of $y = 4 \sin 2x$ in degrees rather than radians as we usually do.) The graph of $y = 4 \sin 2x$ will have an amplitude of 4 and a period of $360°/2 = 180°$. Figure 6 shows the graph of $y = 4 \sin 2x$ between 0° and 360°.

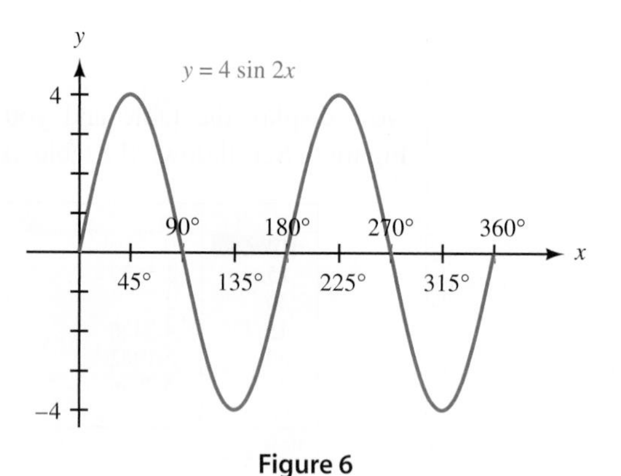

Figure 6

As you can see in Figure 6, as x goes from 0° to 45°, y goes from 0 to 4. This means that, for the equation $r = 4 \sin 2\theta$, as θ goes from 0° to 45°, r will go from 0 out to 4. A diagram of this is shown in Figure 7.

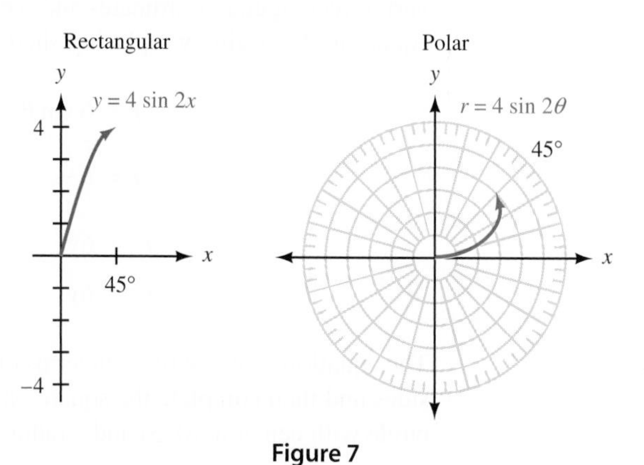

Figure 7

As x continues from 45° to 90°, y decreases from 4 down to 0. Likewise, as θ rotates through 45° to 90°, r will decrease from 4 down to 0. A diagram of this is shown in Figure 8. The numbers 1 and 2 in Figure 8 indicate the order in which those sections of the graph are drawn.

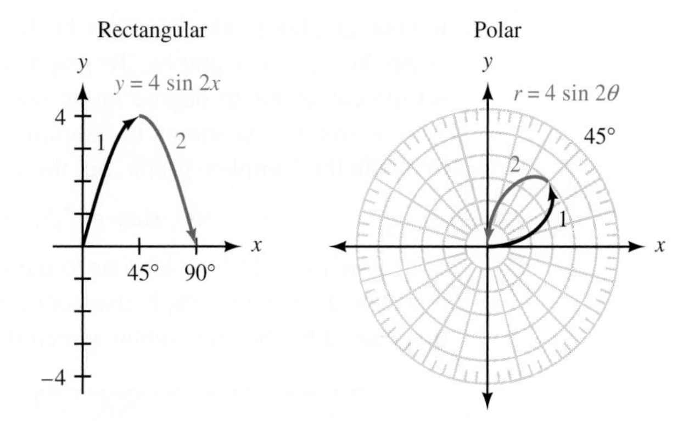

Figure 8

If we continue to reason in this manner, we will obtain a good sketch of the graph of $r = 4 \sin 2\theta$ by watching how y is affected by changes in x on the graph of $y = 4 \sin 2x$. Table 2 summarizes this information, and Figure 9 contains the graphs of both $y = 4 \sin 2x$ and $r = 4 \sin 2\theta$.

TABLE 2

Reference Number on Graphs	Variations in x (or θ)	Corresponding Variations in y (or r)
1	0° to 45°	0 to 4
2	45° to 90°	4 to 0
3	90° to 135°	0 to −4
4	135° to 180°	−4 to 0
5	180° to 225°	0 to 4
6	225° to 270°	4 to 0
7	270° to 315°	0 to −4
8	315° to 360°	−4 to 0

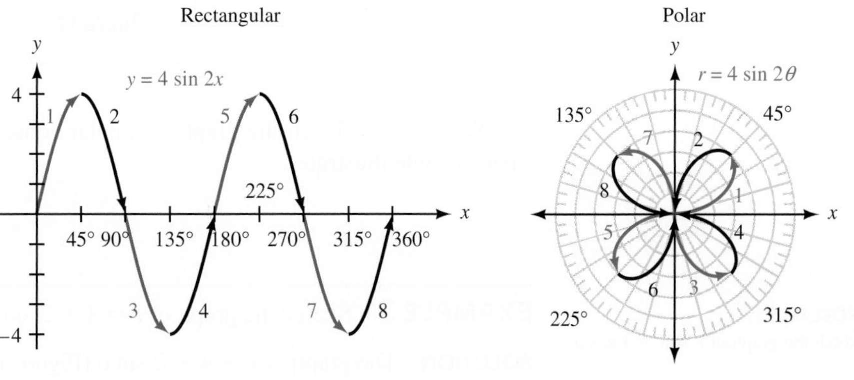

Figure 9

Using Technology Polar Graphs

If your graphing calculator can be set to polar mode, then it should be capable of producing polar graphs. To graph the equation $r = 4 \sin 2\theta$ from Example 2, set the calculator to degree mode and polar mode, and then define the function $r_1 = 4 \sin (2\theta)$. As shown in Example 2, we must allow θ to vary from 0° to 360° to obtain the complete graph. Set the window variables so that

$$0 \le \theta \le 360,\ \text{step} = 7.5;\ -4.5 \le x \le 4.5;\ -4.5 \le y \le 4.5$$

Your graph should look similar to the one shown in Figure 10. For a more "true" representation of the graph, use your zoom-square command to remove the distortion caused by the rectangular screen (Figure 11).

Figure 10

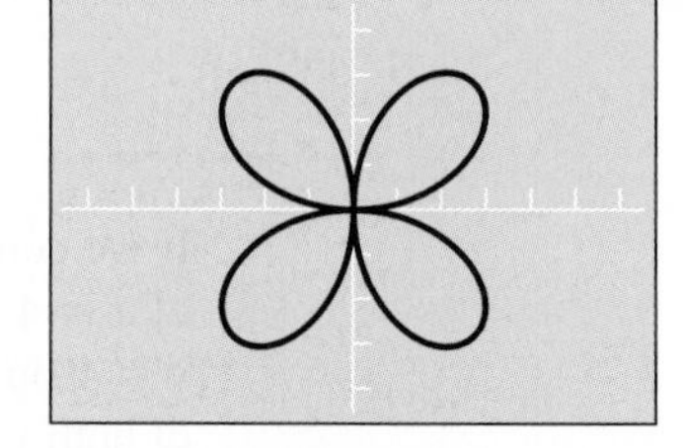

Figure 11

To find points on the graph, you can use the TRACE key. Depending on your model of calculator, you may be able to obtain points in rectangular coordinates, polar coordinates, or both. (On the TI-84, select PolarGC in the FORMAT menu to see polar coordinates.) An added benefit of tracing the graph is that it reinforces the formation of the graph using the method explained in Example 2.

You may also have a command that allows you to find a point (x, y) on the graph directly for a given value of θ. Figure 12 shows how this command was used to find the point corresponding to $\theta = 112°$ in rectangular coordinates and then in polar coordinates.

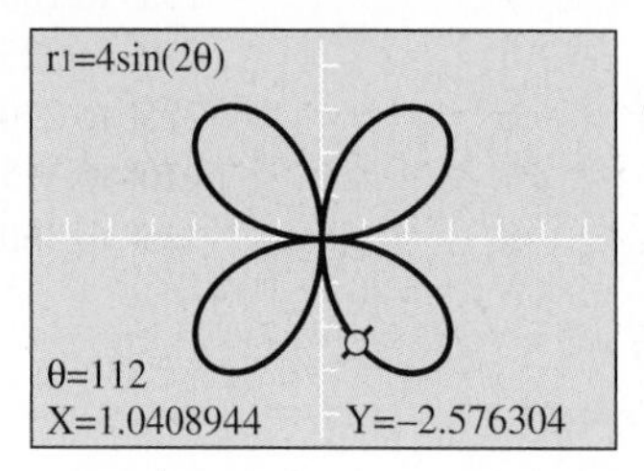

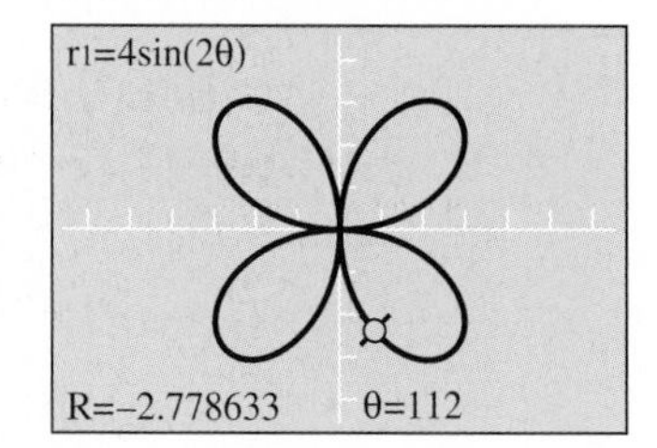

Figure 12

We can also sketch the graph of a polar equation by working in radians, as our final example illustrates.

PROBLEM 3
Sketch the graph of $r = 8 + 4 \cos \theta$.

EXAMPLE 3 Sketch the graph of $r = 4 + 2 \sin \theta$.

SOLUTION The graph of $r = 4 + 2 \sin \theta$ (Figure 14) is obtained by first graphing $y = 4 + 2 \sin x$ (Figure 13) and then noticing the relationship between variations in

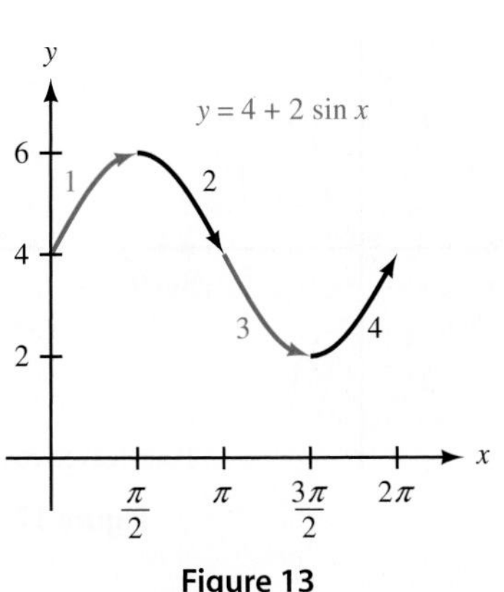

Figure 13

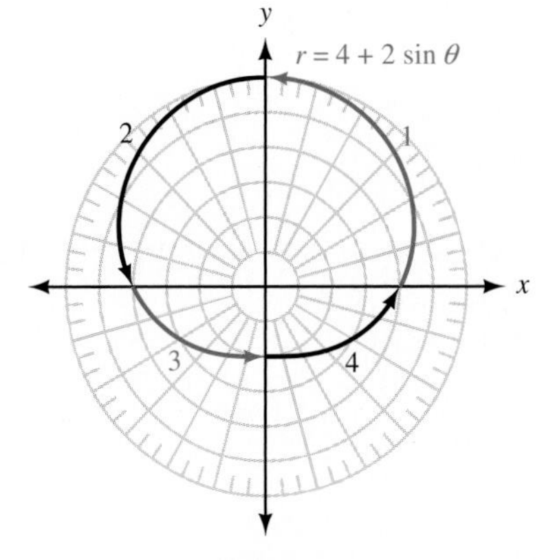

Figure 14

TABLE 3

Reference Number on Graphs	Variations in x (or θ)	Corresponding Variations in y (or r)
1	0 to $\frac{\pi}{2}$	4 to 6
2	$\frac{\pi}{2}$ to π	6 to 4
3	π to $\frac{3\pi}{2}$	4 to 2
4	$\frac{3\pi}{2}$ to 2π	2 to 4

x and the corresponding variations in y (see Table 3). These variations are equivalent to those that exist between θ and r.

Although the method of graphing presented in Examples 2 and 3 is sometimes difficult to comprehend at first, with a little practice it becomes much easier. In any case, the usual alternative is to make a table and plot points until the shape of the curve can be recognized. Probably the best way to graph these equations is to use a combination of both methods.

Here are some other common graphs in polar coordinates along with the equations that produce them (Figures 15–21). When you start graphing some of the equations in Problem Set 8.6, you may want to keep these graphs handy for reference. It is sometimes easier to get started when you can anticipate the general shape of the curve.

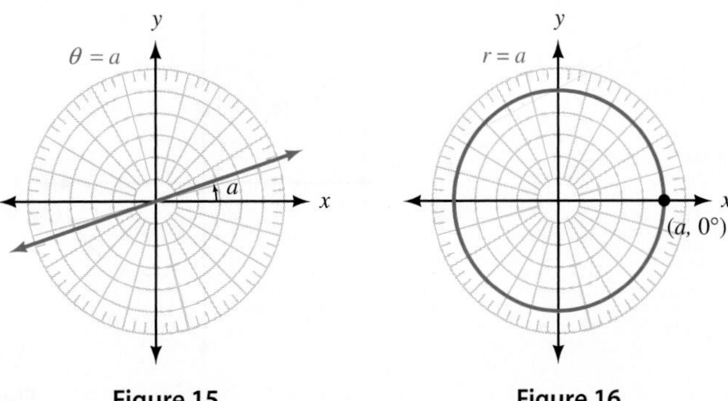

Figure 15

Figure 16

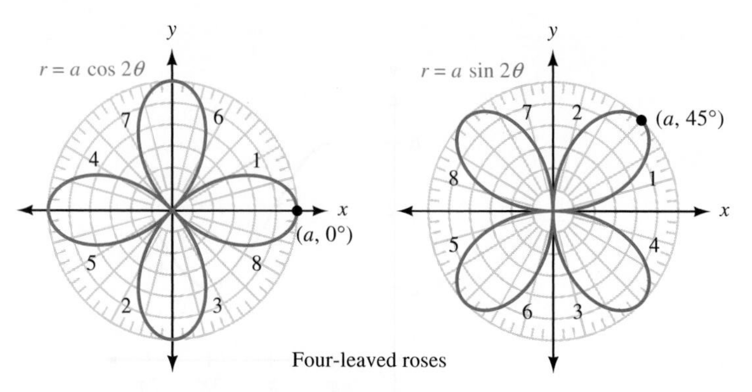

Four-leaved roses

Figure 17

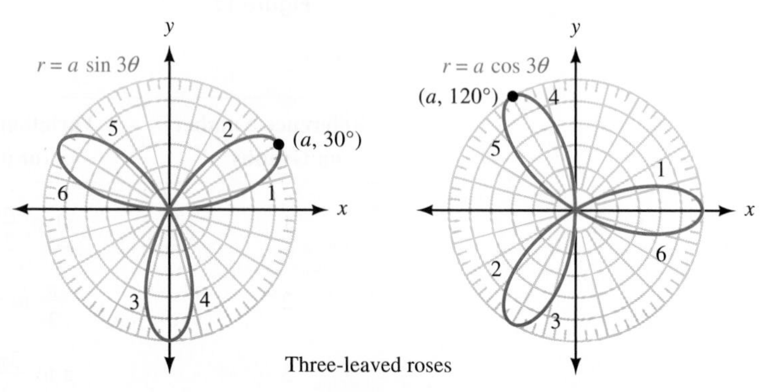

Three-leaved roses

Figure 18

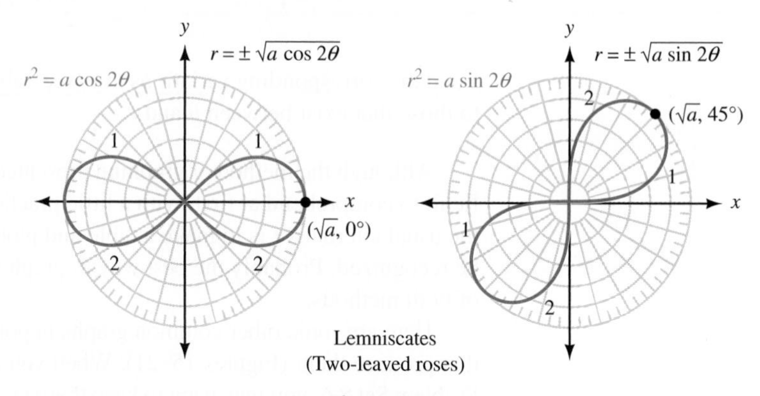

Lemniscates
(Two-leaved roses)

Figure 19

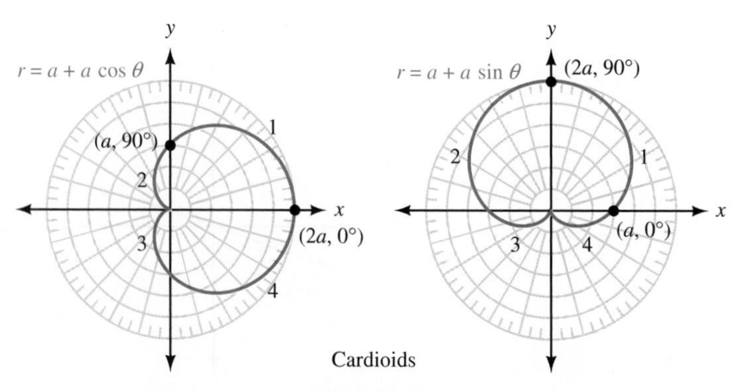

Cardioids

Figure 20

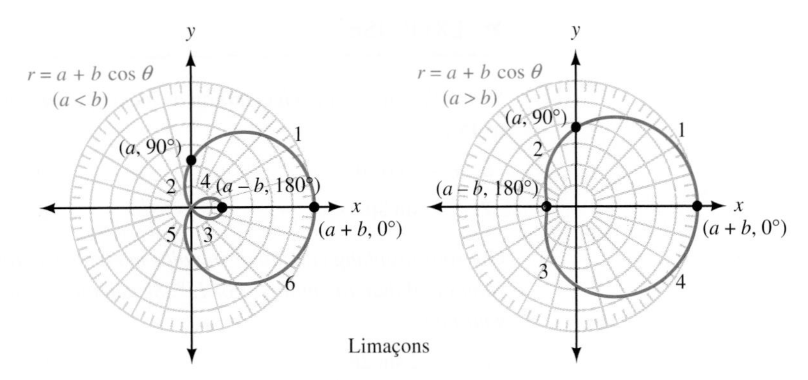

Figure 21

Getting Ready for Class

After reading through the preceding section, respond in your own words and in complete sentences.

a. What is one way to graph an equation written in polar coordinates?
b. What type of geometric figure is the graph of $r = 6 \sin\theta$?
c. What is the largest value of r that can be obtained from the equation $r = 6 \sin\theta$?
d. What type of curve is the graph of $r = 4 + 2 \sin\theta$?

8.6 PROBLEM SET

➤ CONCEPTS AND VOCABULARY

For Questions 1 through 3, fill in the blank with an appropriate word or symbol.

1. The simplest method of graphing a polar equation is to plot points by substituting convenient values of ___ into the equation and solving for ___.

2. Another way to graph a polar equation is to convert it to __________ __________ and see if the resulting equation has a recognizable graph.

3. A more advanced method of graphing a polar equation is to first graph a similar equation on a __________ coordinate system, and then analyze the behavior to determine how the polar graph will look.

4. Match each equation with its appropriate graph. Assume a and b are constants, and that n is a natural number.

a. $r = a$	**i.** Cardioid
b. $\theta = a$	**ii.** Circle
c. $r = a \cos 5\theta$	**iii.** Lemniscate
d. $r = a \sin 6\theta$	**iv.** Limaçon
e. $r^2 = a \cos 2\theta$	**v.** Line
f. $r = a + a \cos\theta$	**vi.** Rose (even number of leaves)
g. $r = a + b \cos\theta$	**vii.** Rose (odd number of leaves)

➤ EXERCISES

Sketch the graph of each equation by making a table using values of θ that are multiples of 45°.

5. $r = 6 \cos \theta$

6. $r = 4 \sin \theta$

7. $r = \sin 2\theta$

8. $r = \cos 2\theta$

Use your graphing calculator in polar mode to generate a table for each equation using values of θ that are multiples of 15°. Sketch the graph of the equation using the values from your table.

9. $r = 2 \sin 2\theta$

10. $r = 2 \cos 2\theta$

11. $r = 3 + 3 \sin \theta$

12. $r = 3 + 3 \cos \theta$

Determine whether the graph of the given equation will be a line, circle, rose curve, lemniscate, cardioid, or limaçon. Use Figures 15–21 to help you identify each equation, but do not actually sketch the graph.

13. $r = 5 \sin 3\theta$

14. $r^2 = 6 \sin 2\theta$

15. $r = 4 + 4 \cos \theta$

16. $r = 6$

17. $r = 4$

18. $r = 5 + 3 \cos \theta$

19. $r^2 = 3 \cos 2\theta$

20. $r = 3 \cos 2\theta$

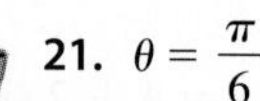

21. $\theta = \dfrac{\pi}{6}$

22. $r = 5 + 5 \sin \theta$

23. $r = 3 + 5 \sin \theta$

24. $\theta = \dfrac{2\pi}{3}$

Graph each equation.

25. $r = 3$

26. $r = 2$

27. $\theta = \dfrac{3\pi}{4}$

28. $\theta = \dfrac{\pi}{4}$

29. $r = 3 \sin \theta$

30. $r = 3 \cos \theta$

31. $r = 4 + 2 \sin \theta$

32. $r = 4 + 2 \cos \theta$

33. $r = 2 + 4 \cos \theta$

34. $r = 2 + 4 \sin \theta$

35. $r = 2 + 2 \sin \theta$

36. $r = 2 + 2 \cos \theta$

37. $r^2 = 4 \cos 2\theta$

38. $r^2 = 9 \sin 2\theta$

39. $r = 2 \sin 2\theta$

40. $r = 2 \cos 2\theta$

41. $r = 4 \cos 3\theta$

42. $r = 4 \sin 3\theta$

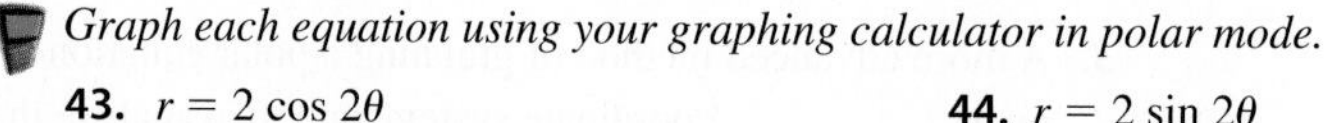

Graph each equation using your graphing calculator in polar mode.

43. $r = 2 \cos 2\theta$

44. $r = 2 \sin 2\theta$

45. $r = 4 \sin 5\theta$

46. $r = 6 \cos 6\theta$

47. $r = 3 + 3 \cos \theta$

48. $r = 3 + 3 \sin \theta$

49. $r = 1 - 4 \cos \theta$

50. $r = 4 - 5 \sin \theta$

51. $r = 2 \cos 2\theta - 3 \sin \theta$

52. $r = 3 \sin 2\theta + 2 \cos \theta$

53. $r = 3 \sin 2\theta + \sin \theta$

54. $r = 2 \cos 2\theta - \cos \theta$

Convert each equation to polar coordinates and then sketch the graph.

55. $x^2 + y^2 = 16$

56. $x^2 + y^2 = 25$

57. $x^2 + y^2 = 6x$

58. $x^2 + y^2 = 6y$

59. $(x^2 + y^2)^2 = 2xy$

60. $(x^2 + y^2)^2 = x^2 - y^2$

Change each equation to rectangular coordinates and then graph.

61. $r(2 \cos \theta + 3 \sin \theta) = 6$

62. $r(3 \cos \theta - 2 \sin \theta) = 6$

63. $r(1 - \cos \theta) = 1$

64. $r(1 - \sin \theta) = 1$

65. $r = 4 \sin \theta$

66. $r = 6 \cos \theta$

67. Graph $r_1 = 2 \sin \theta$ and $r_2 = 2 \cos \theta$ and then name two points they have in common.

68. Graph $r_1 = 2 + 2 \cos \theta$ and $r_2 = 2 - 2 \cos \theta$ and name three points they have in common.

➤ REVIEW PROBLEMS

The problems that follow review material we covered in Section 4.6.

Graph each equation.

69. $y = \sin x - \cos x, 0 \le x \le 4\pi$

70. $y = \cos x - \sin x, 0 \le x \le 4\pi$

71. $y = x + \sin \pi x, 0 \le x \le 8$

72. $y = x + \cos \pi x, 0 \le x \le 8$

73. $y = 3 \sin x + \cos 2x, 0 \le x \le 4\pi$

74. $y = \sin x + \frac{1}{2} \cos 2x, 0 \le x \le 4\pi$

➤ LEARNING OBJECTIVES ASSESSMENT

These questions are available for instructors to help assess if you have successfully met the learning objectives for this section.

TABLE 4

θ	r
0°	4
45°	2.8
90°	0
135°	−2.8
180°	−4
225°	−2.8
270°	0
315°	2.8
360°	4

75. Table 4 shows ordered pairs for a polar equation. Use the data in the table to sketch the graph of the equation.

a.

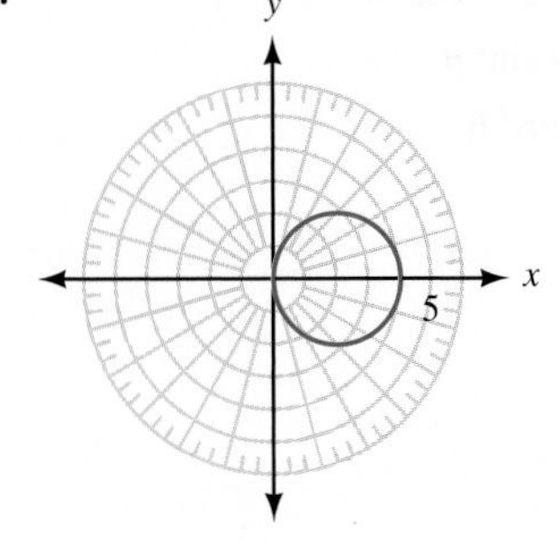

b.

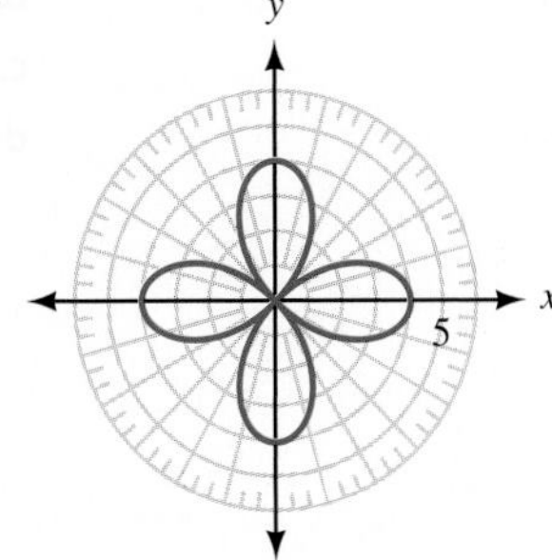

c.

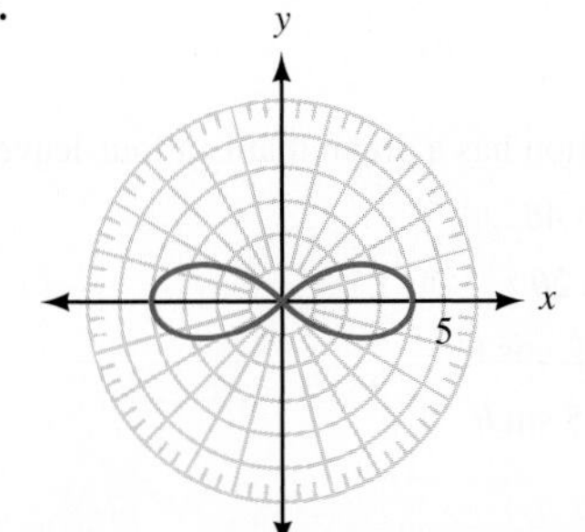

d.

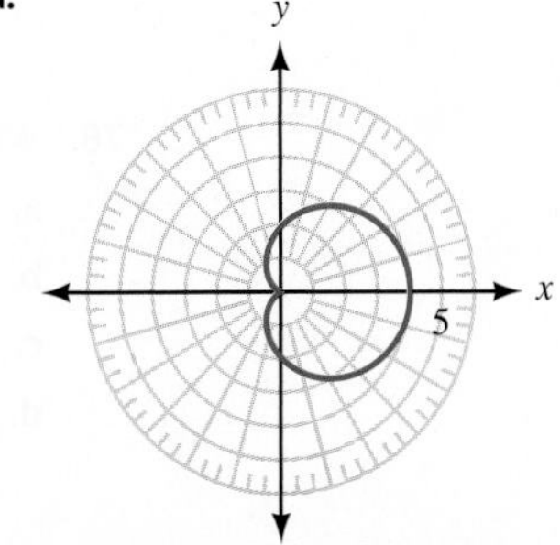

Figure 22

76. Use the rectangular graph of $r = f(\theta)$ shown in Figure 22 to sketch the polar graph for the equation.

a.

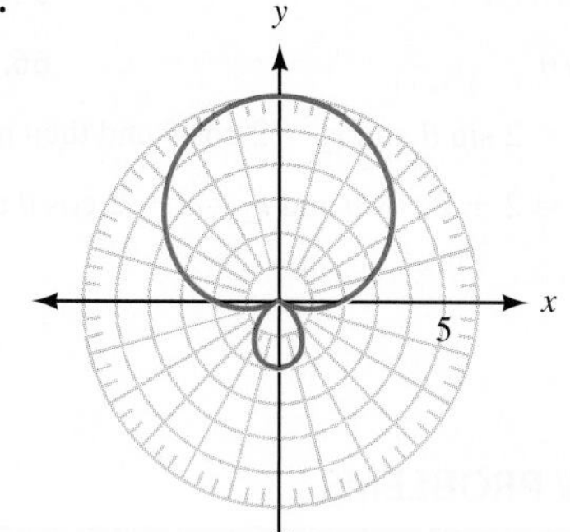

b.

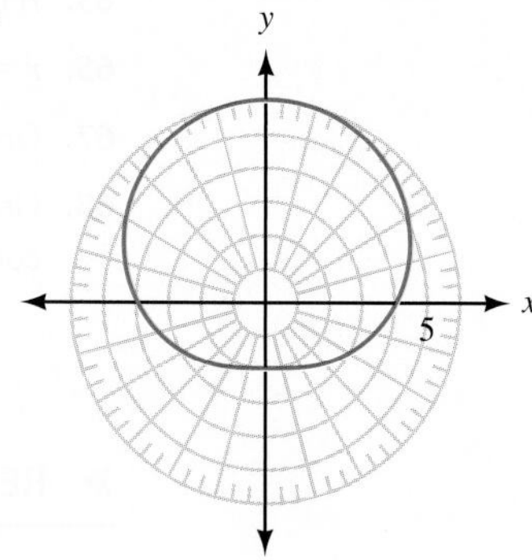

c.

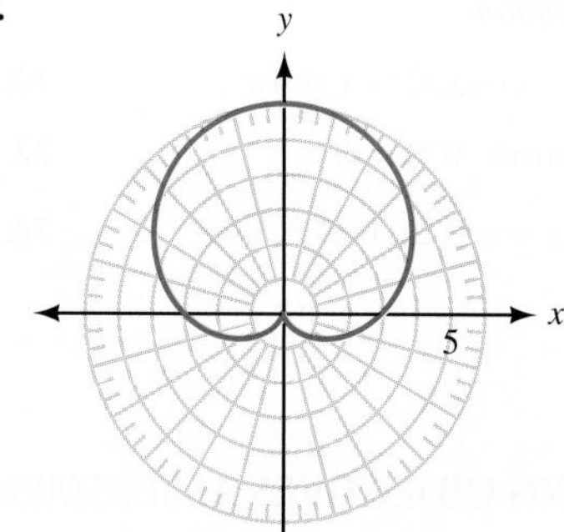

d.

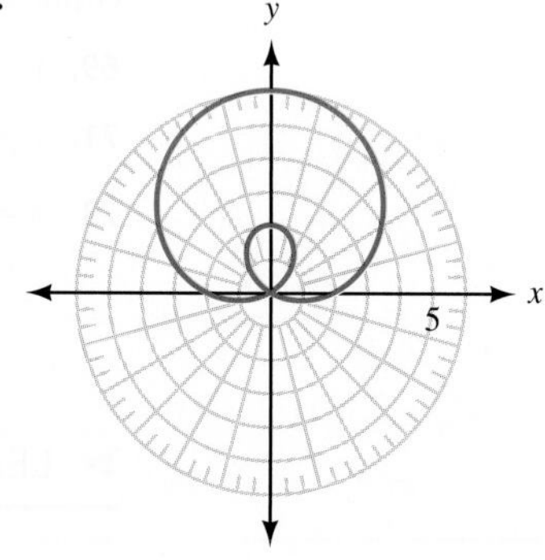

77. Use your graphing calculator to determine which of the following equations has the graph shown in Figure 23.

a. $r = 2(\sin\theta + \cos\theta)^2$

b. $r = 4\sin^2\theta - \cos^2\theta$

c. $r = 1 + 4\sin^2\theta$

d. $r = 3 + \cos^2\theta$

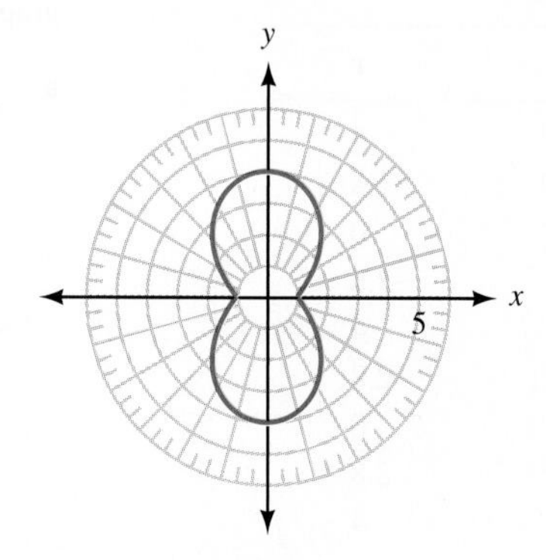

Figure 23

78. Which equation has a graph that is a four-leaved rose?

a. $r = 3\cos 4\theta$

b. $r = 5\sin 2\theta$

c. $r = 2 + 2\cos\theta$

d. $r = 3 + 5\sin\theta$

CHAPTER 8 SUMMARY

EXAMPLES

DEFINITIONS [8.1]

1 Each of the following is a complex number.

$$5 + 4i$$
$$-\sqrt{3} + i$$
$$7i$$
$$-8$$

The number $7i$ is complex because

$$7i = 0 + 7i$$

The number -8 is complex because

$$-8 = -8 + 0i$$

The number i is such that $i^2 = -1$. If $a > 0$, the expression $\sqrt{-a}$ can be written as $\sqrt{ai^2} = i\sqrt{a}$.

A *complex number* is any number that can be written in the form

$$a + bi$$

where a and b are real numbers and $i^2 = -1$. The number a is called the *real part* of the complex number, and b is called the *imaginary part.* The form $a + bi$ is called *standard form.*

All real numbers are also complex numbers since they can be put in the form $a + bi$, where $b = 0$. If $a = 0$ and $b \neq 0$, then $a + bi$ is called an *imaginary number.*

EQUALITY FOR COMPLEX NUMBERS [8.1]

2 If $3x + 2i = 12 - 4yi$, then

$$3x = 12 \quad \text{and} \quad 2 = -4y$$
$$x = 4 \qquad y = -1/2$$

Two complex numbers are equal if and only if their real parts are equal and their imaginary parts are equal. That is,

$$a + bi = c + di \quad \text{if and only if} \quad a = c \quad \text{and} \quad b = d$$

OPERATIONS ON COMPLEX NUMBERS IN STANDARD FORM [8.1]

3 If $z_1 = 2 - i$ and $z_2 = 4 + 3i$, then

If $z_1 = a_1 + b_1 i$ and $z_2 = a_2 + b_2 i$ are two complex numbers in standard form, then the following definitions and operations apply.

$$z_1 + z_2 = 6 + 2i$$

Addition

$$z_1 + z_2 = (a_1 + a_2) + (b_1 + b_2)i$$

Add real parts; add imaginary parts.

$$z_1 - z_2 = -2 - 4i$$

Subtraction

$$z_1 - z_2 = (a_1 - a_2) + (b_1 - b_2)i$$

Subtract real parts; subtract imaginary parts.

$$\begin{aligned} z_1 z_2 &= (2 - i)(4 + 3i) \\ &= 8 + 6i - 4i - 3i^2 \\ &= 11 + 2i \end{aligned}$$

Multiplication

$$z_1 z_2 = (a_1 a_2 - b_1 b_2) + (a_1 b_2 + a_2 b_1)i$$

In actual practice, simply multiply as you would multiply two binomials.

The conjugate of $4 + 3i$ is $4 - 3i$ and

$$\begin{aligned} (4 + 3i)(4 - 3i) &= 16 + 9 \\ &= 25 \end{aligned}$$

Conjugates

The conjugate of $a + bi$ is $a - bi$. Their product is the real number $a^2 + b^2$.

$$\begin{aligned} \frac{z_1}{z_2} &= \frac{2 - i}{4 + 3i} \cdot \frac{\mathbf{4 - 3i}}{\mathbf{4 - 3i}} \\ &= \frac{5 - 10i}{25} \\ &= \frac{1}{5} - \frac{2}{5}i \end{aligned}$$

Division

Multiply the numerator and denominator of the quotient by the conjugate of the denominator.

4 $i^{20} = (i^4)^5 = 1$

$i^{21} = (i^4)^5 \cdot i = i$

$i^{22} = (i^4)^5 \cdot i^2 = -1$

$i^{23} = (i^4)^5 \cdot i^3 = -i$

POWERS OF i [8.1]

If n is an integer, then i^n can always be simplified to i, -1, $-i$, or 1.

5 The graph of $4 + 3i$ is

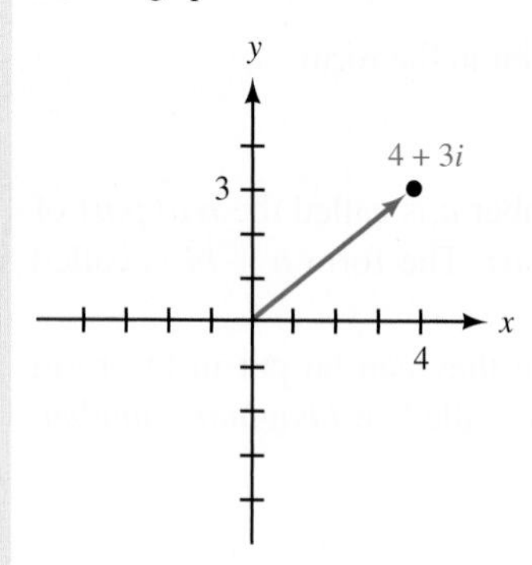

GRAPHING COMPLEX NUMBERS [8.2]

The graph of the complex number $z = x + yi$ is the arrow (vector) that extends from the origin to the point (x, y).

6 If $z = \sqrt{3} + i$, then $|z| = |\sqrt{3} + i| = \sqrt{3 + 1} = 2$

ABSOLUTE VALUE OF A COMPLEX NUMBER [8.2]

The *absolute value* (or *modulus*) of the complex number $z = x + yi$ is the distance from the origin to the point (x, y). If this distance is denoted by r, then

$$r = |z| = |x + yi| = \sqrt{x^2 + y^2}$$

7 For $z = \sqrt{3} + i$, θ is the smallest positive angle for which

$$\sin\theta = \frac{1}{2} \text{ and } \cos\theta = \frac{\sqrt{3}}{2}$$

which means $\theta = 30°$ or $\frac{\pi}{6}$.

ARGUMENT OF A COMPLEX NUMBER [8.2]

The *argument* of the complex number $z = x + yi$ is the smallest positive angle from the positive x-axis to the graph of z. If the argument of z is denoted by θ, then

$$\sin\theta = \frac{y}{r}, \qquad \cos\theta = \frac{x}{r}, \qquad \text{and} \qquad \tan\theta = \frac{y}{x}$$

8 If $z = \sqrt{3} + i$, then in trigonometric form

$$z = 2(\cos 30° + i\sin 30°)$$
$$= 2 \text{ cis } 30°$$

TRIGONOMETRIC FORM OF A COMPLEX NUMBER [8.2]

The complex number $z = x + yi$ is written in trigonometric form when it is written as

$$z = r(\cos\theta + i\sin\theta) = r \text{ cis } \theta$$

where r is the absolute value of z and θ is the argument of z.

9 If

$z_1 = 8(\cos 40° + i\sin 40°)$ and

$z_2 = 4(\cos 10° + i\sin 10°)$,

then

$z_1z_2 = 32(\cos 50° + i\sin 50°)$

$\frac{z_1}{z_2} = 2(\cos 30° + i\sin 30°)$

PRODUCTS AND QUOTIENTS IN TRIGONOMETRIC FORM [8.3]

If $z_1 = r_1(\cos\theta_1 + i\sin\theta_1)$ and $z_2 = r_2(\cos\theta_2 + i\sin\theta_2)$ are two complex numbers in trigonometric form, then their product is

$$z_1z_2 = r_1r_2[\cos(\theta_1 + \theta_2) + i\sin(\theta_1 + \theta_2)] = r_1r_2 \text{ cis } (\theta_1 + \theta_2)$$

and their quotient is

$$\frac{z_1}{z_2} = \frac{r_1}{r_2}[\cos(\theta_1 - \theta_2) + i\sin(\theta_1 - \theta_2)] = \frac{r_1}{r_2} \text{ cis } (\theta_1 - \theta_2)$$

10 If $z = \sqrt{2}$ cis 30°, then

$$z^{10} = (\sqrt{2})^{10} \text{ cis } (10 \cdot 30°) = 32 \text{ cis } 300°$$

DE MOIVRE'S THEOREM [8.3]

If $z = r(\cos\theta + i\sin\theta)$ is a complex number in trigonometric form and n is an integer, then

$$z^n = r^n(\cos n\theta + i\sin n\theta) = r^n \text{ cis } (n\theta)$$

11 The 3 cube roots of $z = 8(\cos 60° + i\sin 60°)$ are given by

$$w_k = 8^{1/3} \text{ cis } \left(\frac{60°}{3} + \frac{360°}{3}k\right) = 2 \text{ cis } (20° + 120°k)$$

where $k = 0, 1, 2$. That is,

$$w_0 = 2(\cos 20° + i\sin 20°)$$
$$w_1 = 2(\cos 140° + i\sin 140°)$$
$$w_2 = 2(\cos 260° + i\sin 260°)$$

ROOTS OF A COMPLEX NUMBER [8.4]

The n nth roots of the complex number $z = r(\cos\theta + i\sin\theta)$ are given by the formula

$$w_k = r^{1/n}\left[\cos\left(\frac{\theta}{n} + \frac{360°}{n}k\right) + i\sin\left(\frac{\theta}{n} + \frac{360°}{n}k\right)\right] = r^{1/n} \text{ cis } \left(\frac{\theta}{n} + \frac{360°}{n}k\right)$$

where $k = 0, 1, 2, \ldots, n - 1$. If using radians, replace 360° with 2π.

12

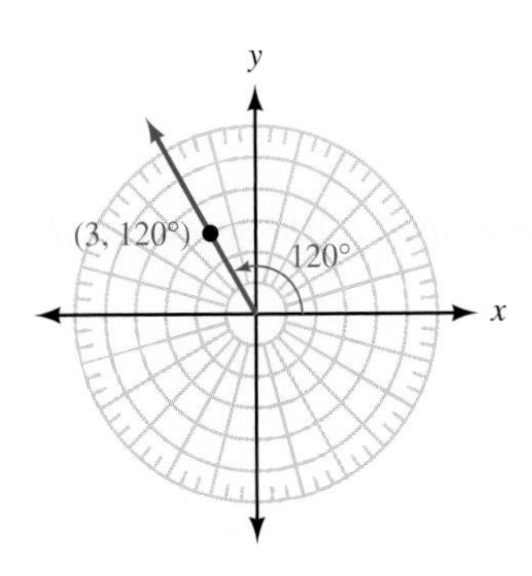

POLAR COORDINATES [8.5]

The ordered pair (r, θ) names the point that is r units from the origin along the terminal side of angle θ in standard position. The coordinates r and θ are said to be the *polar coordinates* of the point they name.

13 Convert $(-\sqrt{2}, 135°)$ to rectangular coordinates.

$$x = r\cos\theta = -\sqrt{2}\cos 135° = 1$$
$$y = r\sin\theta = -\sqrt{2}\sin 135° = -1$$

POLAR COORDINATES AND RECTANGULAR COORDINATES [8.5]

To derive the relationship between polar coordinates and rectangular coordinates, we consider a point P with rectangular coordinates (x, y) and polar coordinates (r, θ).

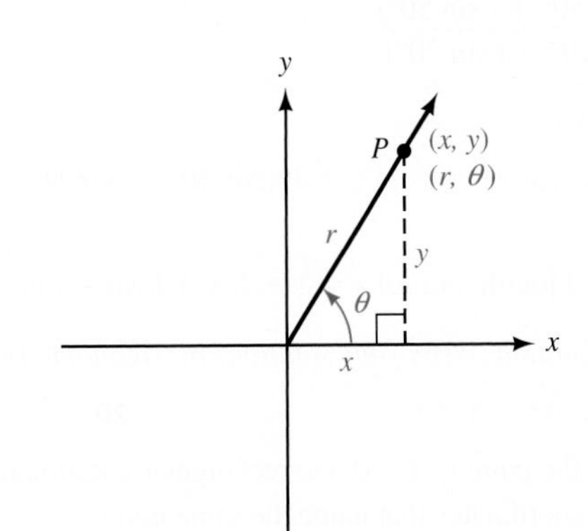

$$r = \pm\sqrt{x^2 + y^2}$$
$$\tan\theta = \frac{y}{x}$$
$$x = r\cos\theta$$
$$y = r\sin\theta$$

EQUATIONS IN POLAR COORDINATES [8.5, 8.6]

Equations in polar coordinates have variables r and θ instead of x and y. The conversions we use to change ordered pairs from polar coordinates to rectangular coordinates and from rectangular coordinates to polar coordinates are the same ones we use to convert back and forth between equations given in polar coordinates and those in rectangular coordinates.

14 Change $x + y = 4$ to polar coordinates.

Because $x = r\cos\theta$ and $y = r\sin\theta$, we have

$$r\cos\theta + r\sin\theta = 4$$
$$r(\cos\theta + \sin\theta) = 4$$
$$r = \frac{4}{\cos\theta + \sin\theta}$$

The last equation gives us r in terms of θ.

CHAPTER 8 TEST

1. Write $\sqrt{-12}$ in terms of i.
2. Find x and y so that the equation $(x^2 - 3x) + 16i = 10 + 8yi$ is true.
3. Simplify $(6 - 3i) + [(4 - 2i) - (3 + i)]$.
4. Simplify i^{17}.

Multiply. Leave your answer in standard form.

5. $(8 + 5i)(8 - 5i)$
6. $(3 + 5i)^2$
7. Divide $\dfrac{6 + 5i}{6 - 5i}$. Write your answer in standard form.

For each of the following complex numbers give: (a) the absolute value; (b) the opposite; and (c) the conjugate.

8. $3 - 4i$
9. $8i$

Write each complex number in standard form.

10. $8(\cos 330° + i\sin 330°)$
11. $2\text{ cis }\dfrac{3\pi}{4}$

Write each complex number in trigonometric form.

12. $-\sqrt{3} + i$
13. $5i$

Multiply or divide as indicated. Leave your answers in trigonometric form.

14. $5(\cos 25° + i\sin 25°) \cdot 3(\cos 40° + i\sin 40°)$
15. $\dfrac{10(\cos 50° + i\sin 50°)}{2(\cos 20° + i\sin 20°)}$
16. $[3\text{ cis }20°]^4$
17. Find two square roots of $z = 49(\cos 50° + i\sin 50°)$. Leave your answer in trigonometric form.
18. Find the 4 fourth roots of $z = 2 + 2i\sqrt{3}$. Leave your answer in trigonometric form.

Solve each equation. Write your solutions in trigonometric form.

19. $x^4 - 2\sqrt{3}x^2 + 4 = 0$
20. $x^3 = -1$
21. Convert the point $(-6, 60°)$ to rectangular coordinates, and state two other ordered pairs in polar coordinates that name the same point.
22. Convert $(-3, 3)$ to polar coordinates with r positive and θ between 0 and 2π.

23. Convert the equation $r = 6 \sin \theta$ to rectangular coordinates.

24. Convert the equation $x^2 + y^2 = 8y$ to polar coordinates.

Graph each equation.

25. $r = 4$ **26.** $r = 4 + 2 \cos \theta$ **27.** $r = \sin 2\theta$

 *Graph each equation using your graphing calculator in polar mode.*

28. $r = 4 - 4 \cos \theta$ **29.** $r = 6 \cos 3\theta$ **30.** $r = 3 \sin 2\theta - \sin \theta$

Intersections of Polar Graphs

OBJECTIVE: To find the points of intersection of two polar graphs.

Figure 1

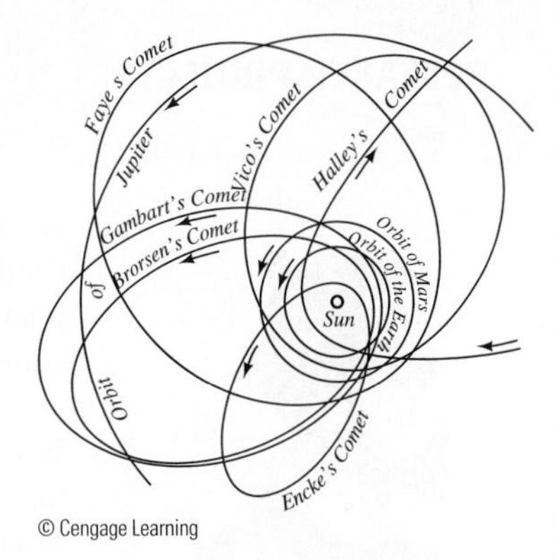

On July 16, 1994, Comet P/Shoemaker-Levy 9 collided with Jupiter (Figure 1). Potential collisions between planets and comets are found by identifying any intersection points in their orbits. These orbits are typically described using polar equations.

In Sections 8.5 and 8.6 we learned about the polar coordinate system and how to graph equations in polar coordinates. One of the things we might need to know about the graphs is if they have any points of intersection. In this project you will learn how to find these points.

First, we will find the points of intersection of the graphs of the polar equations $r = 1$ and $r = 1 + \cos \theta$.

1 We can approach the problem as a system of two equations in two variables. Using the substitution method, we can replace r in the second equation with the value 1 (from the first equation) to obtain an equation in just one variable:

$$1 = 1 + \cos \theta$$

Solve this equation for θ, if $0 \le \theta < 2\pi$. Then use your values of θ to find the corresponding values of r. Write your results as ordered pairs (r, θ).

2 Verify your results from Question 1 graphically. Sketch the graph of both equations on the same polar coordinate system. How many points of intersection are there? Did you find them all using the substitution method in Question 1?

Next, we will find the points of intersection of the graphs of the polar equations $r = 3 \sin \theta$ and $r = 1 + \sin \theta$.

3 Use the substitution method to find any points of intersection for the two equations, if $0 \le \theta < 2\pi$. Write your results as ordered pairs (r, θ).

4 Verify your results graphically. Sketch the graph of both equations on the same polar coordinate system. How many points of intersection are there? Did you find them all using the substitution method in Question 3?

5 Show that the pole $r = 0$ does satisfy each equation for some value of θ.

6 Use your graphing calculator to graph both equations. Make sure your calculator is set to polar mode, radians, and to graph equations *simultaneously.* Watch carefully as the graphs are drawn (at the same time) by your calculator. Did you notice any significant differences in how the intersection points occur? Write a small paragraph describing what you observed.

7 Based on your observations of the graphs, try to explain why you were not able to find one of the intersection points using the substitution method. You may find it helpful to keep in mind that each point in the plane has more than one different representation in polar coordinates. Write your explanation in a paragraph or two.

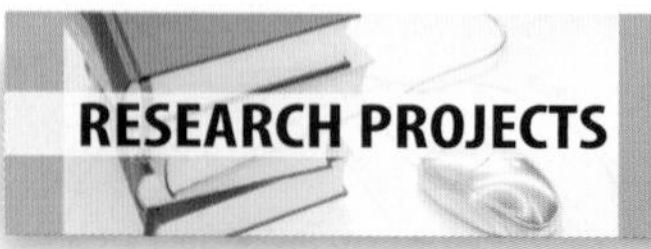

RESEARCH PROJECTS

Complex Numbers

Girolamo Cardano

As was mentioned in the introduction to this chapter, Jerome Cardan (Girolamo Cardano) was unable to solve certain equations because he did not know how to interpret the square root of a negative number. His work set the stage for the arrival of complex numbers.

Research the history of complex numbers. How were the works of Rafael Bombelli, Jean Robert Argand, Leonhard Euler, and Abraham de Moivre significant in the development of complex numbers? Write a paragraph or two about your findings.

A Bitter Dispute

Tartaglia

With the publication of *Ars Magna,* a dispute intensified between Jerome Cardan (Girolamo Cardano) and another mathematician named Niccolo Fontana, otherwise known as Tartaglia. What was the dispute, and how did it arise? In your opinion, who was at fault? Write a paragraph or two about your findings.

CUMULATIVE TEST 1–8

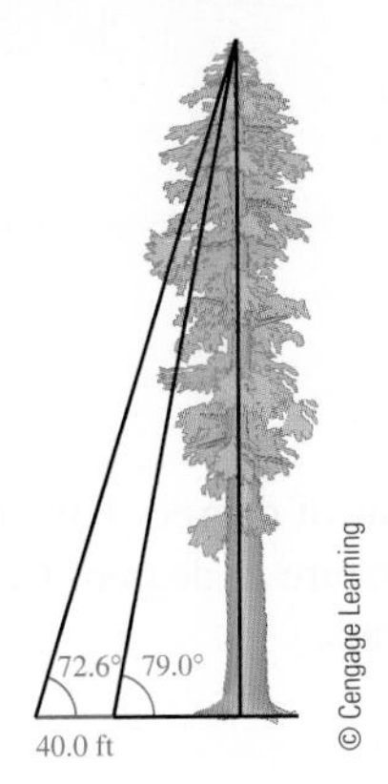

Figure 1

1. If $\cos\theta = 3/5$ with θ in QIV, find the values of the remaining five trigonometric functions for θ.
2. Use a calculator to find θ to the nearest tenth of a degree if θ is an acute angle and $\cos\theta = 0.9730$.
3. Use a calculator to find θ, to the nearest tenth of a degree, if $0° \le \theta < 360°$ and $\sec\theta = 1.5450$ with θ in QIV.
4. **Height of a Tree** An arborist needs to measure the height of a giant sequoia tree. He moves some distance along the ground from the base of the tree and measures the angle of elevation to the top of the tree, which is 79.0°. He then backs up another 40.0 feet and measures the angle of elevation to be 72.6°. Find the height of the sequoia tree (Figure 1).
5. If triangle ABC is a right triangle with $a = 20.5$, $b = 31.4$, and $C = 90°$, solve the triangle by finding the remaining sides and angles.
6. Convert $4\pi/3$ to degree measure.
7. If $\theta = 2\pi/3$ is a central angle that cuts off an arc length of $\pi/4$ centimeters, find the radius of the circle.
8. If a point is rotating with uniform circular motion on a circle of radius 1 foot, find v if the point rotates at 10 revolutions per minute. Give your answer in exact form.
9. Use the graph of $y = \sec x$ to find all values of x between -4π and 4π for which $\sec x$ is undefined.
10. Graph one complete cycle of $y = \frac{1}{2}\sin\frac{\pi}{2}x$.
11. Write an equivalent algebraic expression for $\tan(\sin^{-1}x)$ that involves x only.
12. Prove the equation $\sec^2 x\csc^2 x = \sec^2 x + \csc^2 x$ is an identity.

Find exact values for each of the following:

13. $\cos 15°$
14. $\cot\frac{\pi}{12}$
15. Solve $3\cos 2\theta + 5\sin\theta = 0$ for θ if $0° \le \theta < 360°$. Round your answers to the nearest tenth of a degree.

Problems 16 through 18 refer to triangle ABC, which is not necessarily a right triangle.

16. If $B = 118°$, $C = 37°$, and $c = 2.9$ inches, use the law of sines to find b.
17. Find two triangles for which $A = 26°$, $a = 4.8$ ft, and $b = 9.4$ ft.
18. If $a = 10$ km, $b = 12$ km, and $c = 11$ km, use the law of cosines to find B to the nearest tenth of a degree.
19. Find the area of the triangle in Problem 18.

For Problems 20 and 21 use the vectors $\mathbf{U} = 5\mathbf{i} + 12\mathbf{j}$ *and* $\mathbf{V} = -4\mathbf{i} + \mathbf{j}$.

20. Find $3\mathbf{U} - 5\mathbf{V}$.
21. Use the dot product to find the angle between $\mathbf{U}$ and $\mathbf{V}$ to the nearest tenth of a degree.
22. Simplify $(7 + 3i) - [(2 + i) - (3 - 4i)]$.
23. Simplify i^{16}.
24. For the complex number $3 + 4i$, give (a) the absolute value; (b) the opposite; and (c) the conjugate.
25. Write $2 + 2i$ in trigonometric form.
26. Simplify $[2(\cos 10° + i\sin 10°)]^5$. Give your answer in trigonometric form.
27. Find the two square roots of $4 - 4i\sqrt{3}$. Write your answers in standard form.
28. Convert the point $(4, 225°)$ to rectangular coordinates, and state two other ordered pairs in polar coordinates that name the same point.
29. Convert the equation $x + y = 2$ to polar coordinates.
30. Graph the equation $r = 2\cos 2\theta + 3\sin\theta$ using your graphing calculator in polar mode.

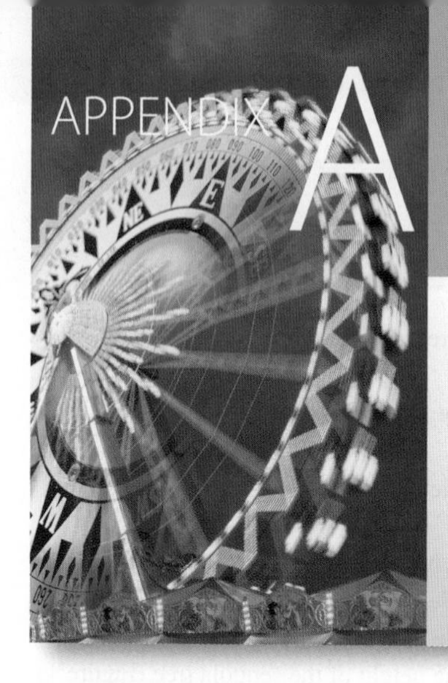

Review of Functions

Mathematics is not a careful march down a well-cleared highway, but a journey into a strange wilderness, where the explorers often get lost.

➤ *W. S. Anglin*

© Robert Brenner/PhotoEdit

INTRODUCTION

In the United States, temperature is usually measured in terms of degrees Fahrenheit; however, most European countries prefer to measure temperature in degrees Celsius. Table 1 shows some temperatures measured using both systems.

TABLE 1

Degrees Celsius (°C)	Degrees Fahrenheit (°F)
0	32
25	77
40	104
100	212

The relationship between these two systems of temperature measurement can be written in the form of an equation:

$$F = \frac{9}{5}C + 32$$

Every temperature measured in °C is associated with a unique measure in °F. We call this kind of relationship a function. In this appendix we will review the concept of a function, and see how in many cases a function may be "reversed."

SECTION A.1 Introduction to Functions

LEARNING OBJECTIVES

1. Determine the domain and range of a relation.
2. Use the vertical line test to identify a function.
3. Evaluate a function.
4. Graph a function or relation.

An Informal Look at Functions

To begin with, suppose you have a job that pays $7.50 per hour and that you work anywhere from 0 to 40 hours per week. The amount of money you make in one week depends on the number of hours you work that week. In mathematics we say that your weekly earnings are a *function* of the number of hours you work. If we let the variable x represent hours and the variable y represent the money you make, then the relationship between x and y can be written as

$$y = 7.5x \quad \text{for} \quad 0 \le x \le 40$$

EXAMPLE 1 Construct a table and graph for the function

$$y = 7.5x \quad \text{for} \quad 0 \le x \le 40$$

SOLUTION Table 1 gives some of the paired data that satisfy the equation $y = 7.5x$. Figure 1 is the graph of the equation with the restriction $0 \le x \le 40$.

TABLE 1
Weekly Wages

Hours Worked x	Rule $y = 7.5x$	Pay y
0	$y = 7.5(0)$	0
10	$y = 7.5(10)$	75
20	$y = 7.5(20)$	150
30	$y = 7.5(30)$	225
40	$y = 7.5(40)$	300

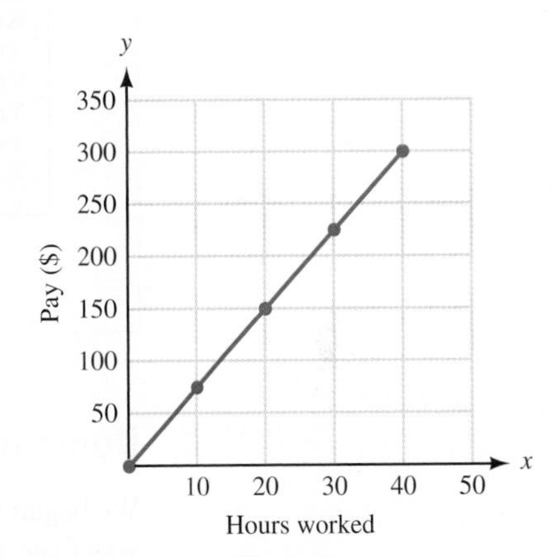

Figure 1

The equation $y = 7.5x$ with the restriction $0 \le x \le 40$, Table 1, and Figure 1 are three ways to describe the same relationship between the number of hours you work in one week and your gross pay for that week. In all three, we *input* values of x and then use the function rule to *output* values of y.

Using Technology Creating Tables and Graphs

To create the table and graph for the function $y = 7.5x$ in Example 1 using a graphing calculator, first define the function as $Y_1 = 7.5x$ in the equation editor (Figure 2). Set up your table to automatically generate values starting at $x = 0$ and using an increment of 10 (Figure 3). On some calculators, this is done by setting the independent variable to Auto. Now display the table and you should see values similar to those shown in Figure 4.

```
Plot1  Plot2  Plot3
\Y1=7.5X
\Y2=
\Y3=
\Y4=
\Y5=
\Y6=
\Y7=
```

Figure 2

```
TABLE SETUP
  TblStart=0
  ΔTbl=10
Indpnt: Auto Ask
Depend: Auto Ask
```

Figure 3

X	Y1	
0	0	
10	75	
20	150	
30	225	
40	300	
50	375	
60	450	

X=0

Figure 4

To create the graph of the function, set the window variables in agreement with the graph in Figure 1:

$$0 \le x \le 50, \text{ scale} = 10; 0 \le y \le 350, \text{ scale} = 50$$

By this, we mean that Xmin = 0, Xmax = 50, Xscl = 10, Ymin = 0, Ymax = 350, and Yscl = 50, as shown in Figure 5. Then graph the function. Your graph should be similar to the one shown in Figure 6.

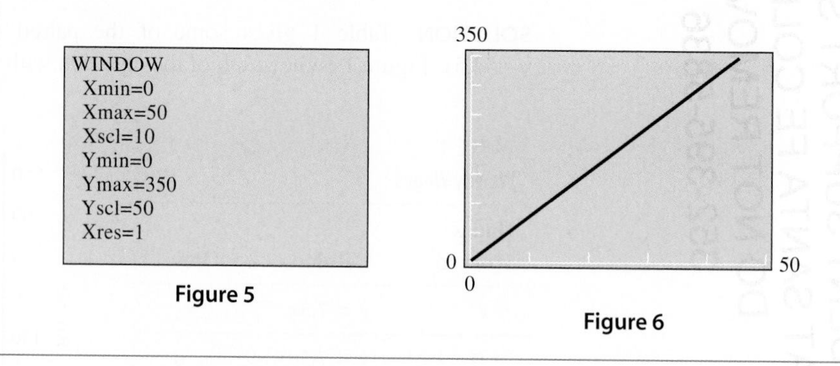

Figure 5

Figure 6

Domain and Range of a Function

We began this discussion by saying that the number of hours worked during the week was from 0 to 40, so these are the values that x can assume. From the line graph in Figure 1, we see that the values of y range from 0 to 300. We call the complete set of values that x can assume the *domain* of the function. The values that are assigned to y are called the *range* of the function.

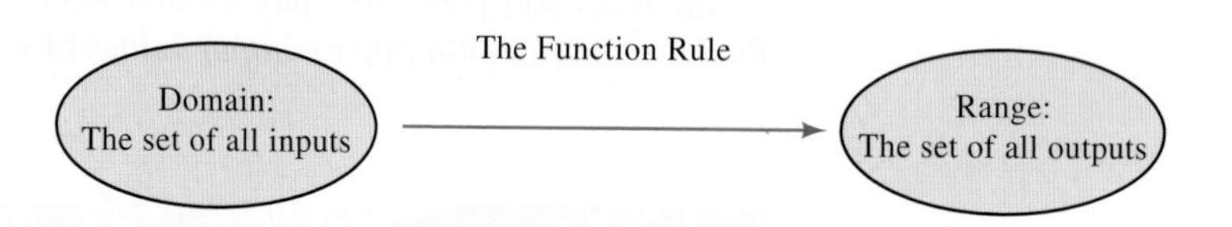

EXAMPLE 2 State the domain and range for the function

$$y = 7.5x, \quad 0 \le x \le 40$$

SOLUTION From the previous discussion we have

$$\text{Domain} = \{x \mid 0 \le x \le 40\}$$
$$\text{Range} = \{y \mid 0 \le y \le 300\}$$

■

A Formal Look at Functions

What is apparent from the preceding discussion is that we are working with paired data. The solutions to the equation $y = 7.5x$ are pairs of numbers, and the points on the line graph in Figure 1 come from paired data. We are now ready for the formal definition of a function.

DEFINITION

A *function* is a rule that pairs each element in one set, called the *domain,* with exactly one element from a second set, called the *range.*

In other words, a function is a rule for which each input is paired with exactly one output.

Functions as Ordered Pairs

The function rule $y = 7.5x$ from Example 1 produces ordered pairs of numbers (x, y). The same thing happens with all functions: The function rule produces ordered pairs of numbers. We use this result to write an alternative definition for a function.

NOTE The restriction on first coordinates in the alternative definition keeps us from assigning a number in the domain to more than one number in the range.

ALTERNATIVE DEFINITION

A *function* is a set of ordered pairs in which no two different ordered pairs have the same first coordinate. The set of all first coordinates is called the *domain* of the function. The set of all second coordinates is called the *range* of the function.

A Relationship That Is Not a Function

You may be wondering if any sets of paired data fail to qualify as functions. The answer is yes, as the next example reveals.

EXAMPLE 3 Sketch the graph of $x = y^2$ and determine if this relationship is a function.

SOLUTION Without going into much detail, we graph the equation $x = y^2$ by finding a number of ordered pairs that satisfy the equation, plotting these points, then drawing a smooth curve that connects them. Some values for x and y that satisfy the equation are given in Table 2, and the graph of $x = y^2$ is shown in Figure 7.

TABLE 2

x	y
0	0
1	1
1	−1
4	2
4	−2
9	3
9	−3

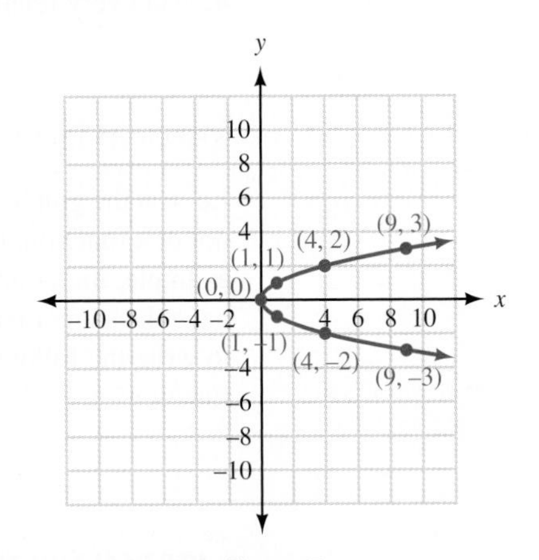

Figure 7

As you can see, several ordered pairs whose graphs lie on the curve have repeated first coordinates, such as (1, 1) and (1, −1), (4, 2) and (4, −2), and (9, 3) and (9, −3). The relationship is therefore not a function. ■

To classify all relationships specified by ordered pairs, whether they are functions or not, we include the following two definitions.

DEFINITION

A *relation* is a rule that pairs each element in one set, called the *domain,* with one or more elements from a second set, called the *range.*

Notice that the definition of a relation does not impose any condition on how the pairs of elements are assigned.

We can also think of a relation as the set of ordered pairs themselves.

ALTERNATIVE DEFINITION

A *relation* is a set of ordered pairs. The set of all first coordinates is the *domain* of the relation. The set of all second coordinates is the *range* of the relation.

Here are some facts that will help clarify the distinction between relations and functions.

1. Any rule that assigns numbers from one set to numbers in another set is a relation. If that rule makes the assignment so that no input has more than one output, then it is also a function.
2. Any set of ordered pairs is a relation. If none of the first coordinates of those ordered pairs is repeated, the set of ordered pairs is also a function.
3. Every function is a relation.
4. Not every relation is a function.

Vertical Line Test

Look at the graph shown in Figure 7. The reason this graph is the graph of a relation, but not of a function, is that some points on the graph have the same first coordinates—for example, the points (4, 2) and (4, −2). Furthermore, any time two points on a graph have the same first coordinates, those points must lie on a vertical line. This allows us to write the following test that uses the graph to determine whether a relation is also a function.

VERTICAL LINE TEST

If a vertical line crosses the graph of a relation in more than one place, the relation cannot be a function. If no vertical line can be found that crosses a graph in more than one place, then the graph is the graph of a function.

EXAMPLE 4 Graph $y = |x|$. Use the graph to determine whether we have the graph of a function. State the domain and range.

SOLUTION We let x take on values of $-4, -3, -2, -1, 0, 1, 2, 3,$ and 4. The corresponding values of y are shown in Table 3. The graph is shown in Figure 8.

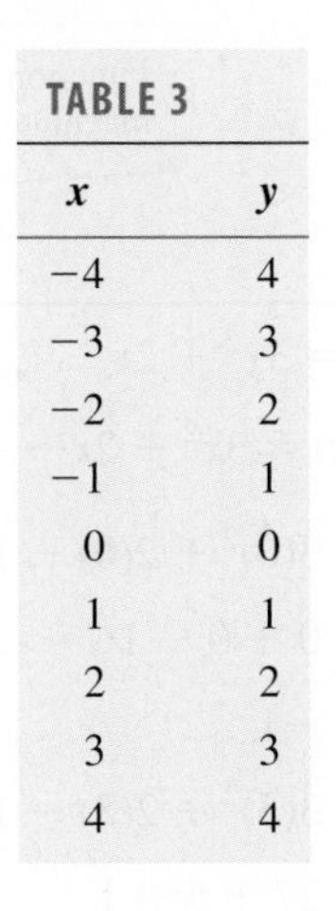

TABLE 3

x	y
−4	4
−3	3
−2	2
−1	1
0	0
1	1
2	2
3	3
4	4

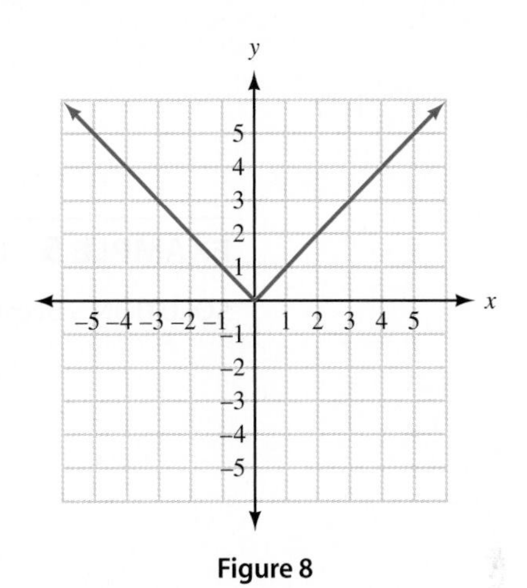

Figure 8

Because no vertical line can be found that crosses the graph in more than one place, $y = |x|$ is a function. The domain is all real numbers. The range is $\{y \mid y \geq 0\}$.

Function Notation

Let's return to the discussion that introduced us to functions. If a job pays \$7.50 per hour for working from 0 to 40 hours a week, then the amount of money y earned in one week is a function of the number of hours worked, x. The exact relationship between x and y is written

$$y = 7.5x \quad \text{for} \quad 0 \leq x \leq 40$$

Because the amount of money earned y depends on the number of hours worked x, we call y the *dependent variable* and x the *independent variable*. Furthermore, if we let f represent all the ordered pairs produced by the equation, then we can write

$$f = \{(x, y) \mid y = 7.5x \quad \text{and} \quad 0 \leq x \leq 40\}$$

Once we have named a function with a letter, we can use an alternative notation to represent the dependent variable y. The alternative notation for y is $f(x)$. It is read "f of x" and can be used instead of the variable y when working with functions. The notation y and the notation $f(x)$ are equivalent—that is,

$$y = 7.5x \Longleftrightarrow f(x) = 7.5x$$

When we use the notation $f(x)$ we are using *function notation.* The benefit of using function notation is that we can write more information with fewer symbols than we can by using just the variable y. For example, asking how much money a person will make for working 20 hours is simply a matter of asking for $f(20)$. Without function notation, we would have to say "find the value of y that corresponds to a value of $x = 20$." To illustrate further, using the variable y we can say "y is 150 when x is 20." Using the notation $f(x)$, we simply say "$f(20) = 150$." Each expression indicates that you will earn \$150 for working 20 hours.

NOTE Some students like to think of functions as machines. Values of x are put into the machine, which transforms them into values of $f(x)$, which then are output by the machine.

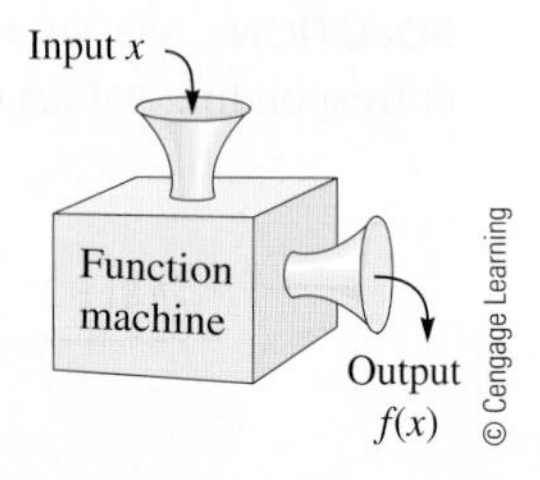

EXAMPLE 5 If $f(x) = 3x^2 + 2x - 1$, find $f(0)$, $f(3)$, and $f(-2)$.

SOLUTION Because $f(x) = 3x^2 + 2x - 1$, we have

$$\begin{aligned} f(\mathbf{0}) &= 3(\mathbf{0})^2 + 2(\mathbf{0}) - 1 \\ &= 0 + 0 - 1 \\ &= -1 \\ f(\mathbf{3}) &= 3(\mathbf{3})^2 + 2(\mathbf{3}) - 1 \\ &= 27 + 6 - 1 \\ &= 32 \\ f(\mathbf{-2}) &= 3(\mathbf{-2})^2 + 2(\mathbf{-2}) - 1 \\ &= 12 - 4 - 1 \\ &= 7 \end{aligned}$$

In Example 5, the function f is defined by the equation $f(x) = 3x^2 + 2x - 1$. We could just as easily have said $y = 3x^2 + 2x - 1$; that is, $y = f(x)$. Saying $f(-2) = 7$ is exactly the same as saying y is 7 when x is -2.

Function Notation and Graphs

We can visualize the relationship between x and $f(x)$ on the graph of the function. Figure 9 shows the graph of $f(x) = 7.5x$ along with two additional line segments. The horizontal line segment corresponds to $x = 20$, and the vertical line segment corresponds to $f(20)$. (Note that the domain is restricted to $0 \le x \le 40$.)

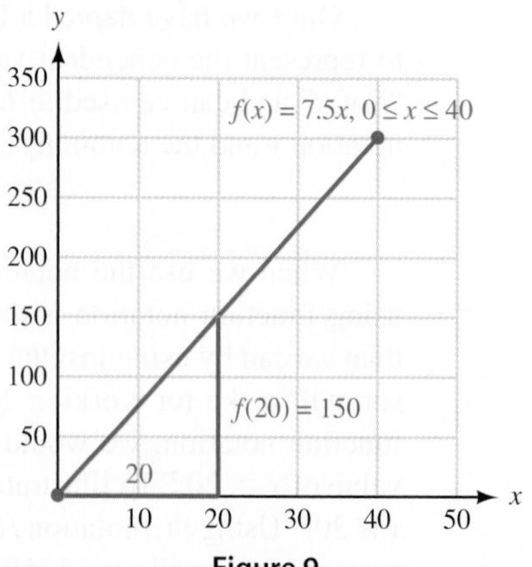

Figure 9

Using Function Notation

The final example in this section shows how to use and interpret function notation.

EXAMPLE 6 A painting is purchased as an investment for $125. If its value increases continuously so that it doubles every 5 years, then its value is given by the function

$$V(t) = 125 \cdot 2^{t/5} \quad \text{for} \quad t \geq 0$$

where t is the number of years since the painting was purchased, and $V(t)$ is its value (in dollars) at time t. Find $V(5)$ and $V(10)$, and explain what they mean.

SOLUTION The expression $V(5)$ is the value of the painting when $t = 5$ (5 years after it is purchased). We calculate $V(5)$ by substituting 5 for t in the equation $V(t) = 125 \cdot 2^{t/5}$. Here is our work:

$$V(\mathbf{5}) = 125 \cdot 2^{\mathbf{5}/5} = 125 \cdot 2^1 = 125 \cdot 2 = 250$$

In words: After 5 years, the painting is worth $250.

The expression $V(10)$ is the value of the painting after 10 years. To find this number, we substitute 10 for t in the equation:

$$V(\mathbf{10}) = 125 \cdot 2^{\mathbf{10}/5} = 125 \cdot 2^2 = 125 \cdot 4 = 500$$

In words: The value of the painting 10 years after it is purchased is $500.

The fact that $V(5) = 250$ means that the ordered pair (5, 250) belongs to the function V. Likewise, the fact that $V(10) = 500$ tells us that the ordered pair (10, 500) is a member of function V. ■

We can generalize the discussion at the end of Example 6 this way:

$$(a, b) \in f \quad \text{if and only if} \quad f(a) = b$$

Getting Ready for Class

After reading through the preceding section, respond in your own words and in complete sentences.

a. What is a function?
b. What is the vertical line test?
c. Explain what you are calculating when you find $f(2)$ for a given function f.
d. If $f(2) = 3$ for a function f, what is the relationship between the numbers 2 and 3 and the graph of f?

A.1 PROBLEM SET

➤ CONCEPTS AND VOCABULARY

For Questions 1 through 6, fill in the blank with an appropriate word.

1. A relation is any set of ________ ________.

2. A function is a rule for which each input is paired with ________ ____ output.

3. The set of all permissible inputs for a function is called the ________, and the set of all corresponding outputs is called the ________.

4. A function can also be defined as a set of ordered pairs in which no two different pairs have the same ________ ____________. The domain is the set of all ________ coordinates, and the range is the set of all _________ coordinates.

5. The ________ ______ test can be used to determine if a graph is the graph of a function.

6. If $y = f(x)$, then x is formally called the ___________ variable and y is called the ___________ variable.

For Questions 7 and 8, determine if the statement is true or false.

7. Every function is also a relation.

8. Every relation is also a function.

➤ EXERCISES

For each of the following relations, give the domain and range, and indicate which are also functions.

9. $\{(1, 3), (2, 5), (4, 1)\}$

10. $\{(3, 1), (5, 7), (2, 3)\}$

11. $\{(-1, 3), (1, 3), (2, -5)\}$

12. $\{(3, -4), (-1, 5), (3, 2)\}$

13. $\{(7, -1), (3, -1), (7, 4)\}$

14. $\{(5, -2), (3, -2), (5, -1)\}$

State whether each of the following graphs represents a function.

15.

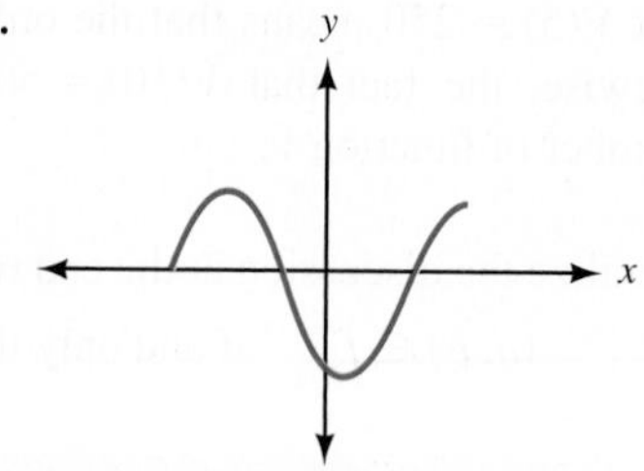

16.

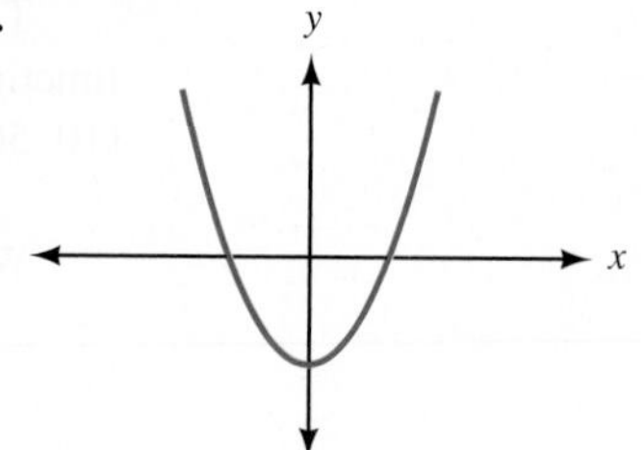

17.

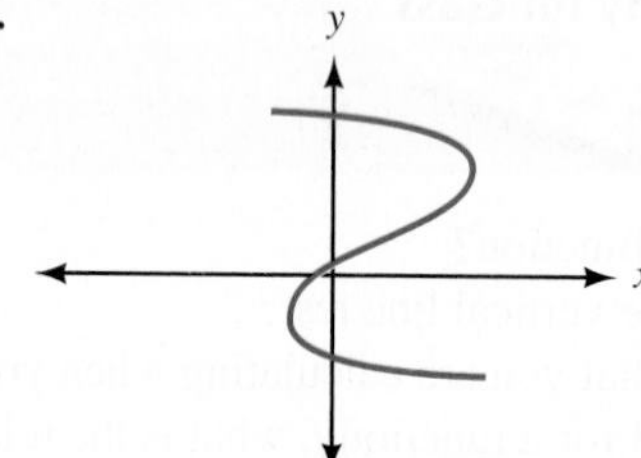

18.

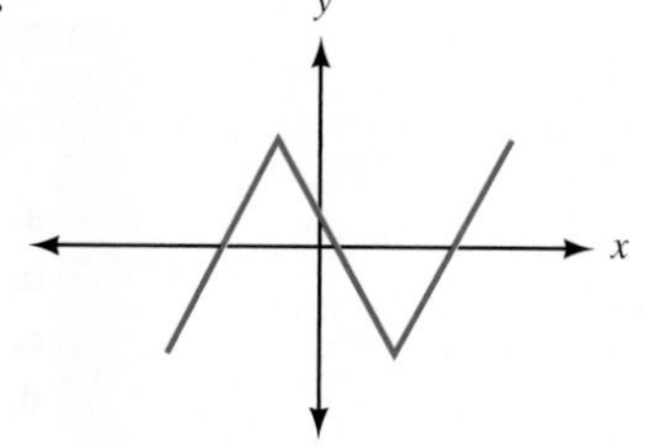

19.

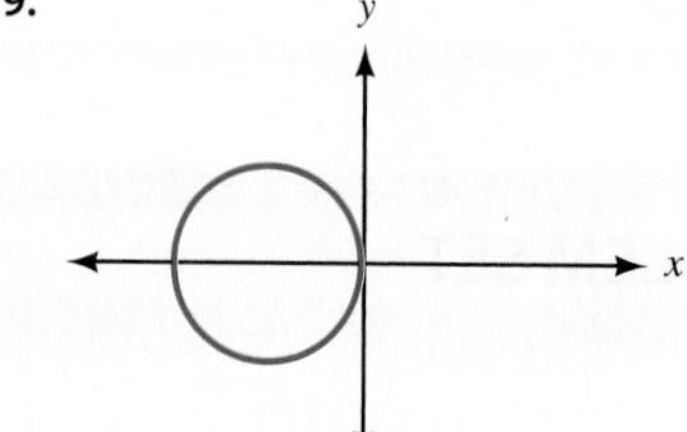

20.

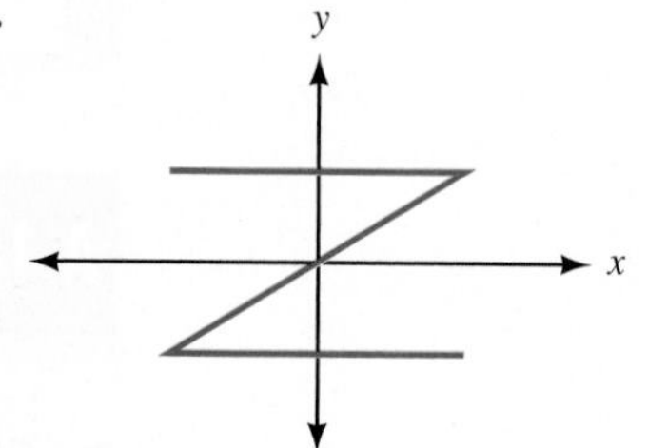

21.

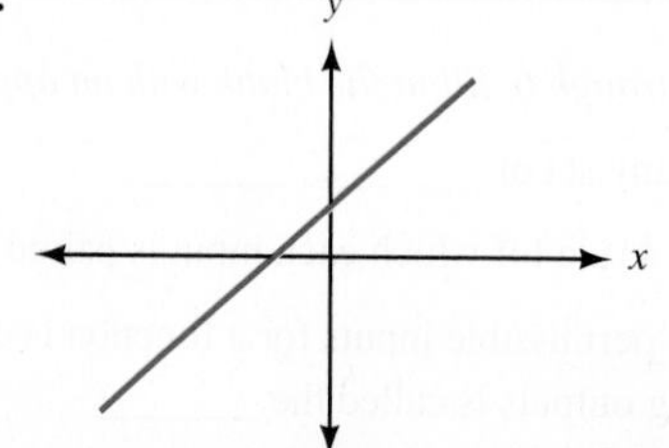

22.

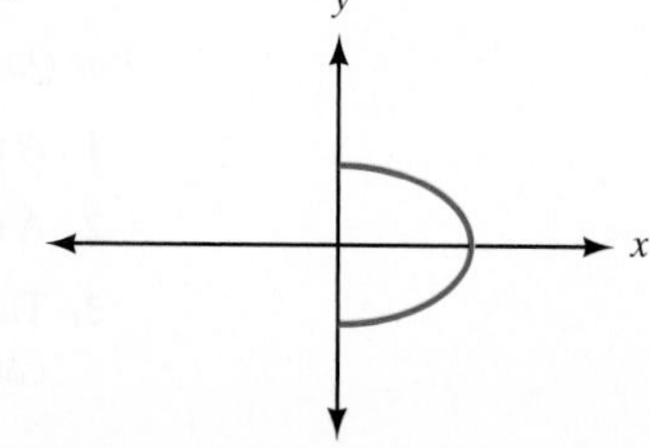

Graph each of the following relations. In each case, use the graph to find the domain and range, and indicate whether the graph is the graph of a function.

23. $y = x^2 - 1$

24. $y = x^2 + 4$

25. $x = y^2 + 4$

26. $x = y^2 - 9$

27. $y = |x - 2|$

28. $y = |x| + 2$

29. Suppose you have a job that pays $8.50 per hour and you work anywhere from 10 to 40 hours per week.

a. Write an equation, with a restriction on the variable x, that gives the amount of money, y, you will earn for working x hours in one week.

b. Use the function rule you have written in part (a) to complete Table 4.

TABLE 4

Hours Worked x	Gross Pay ($) y
10	
20	
30	
40	

c. Construct a line graph from the information in Table 4.

d. State the domain and range of this function.

30. The ad shown at left was in the local newspaper. Suppose you are hired for the job described in the ad.

312 Help Wanted

ESPRESSO BAR OPERATOR
Must be dependable, honest, service-oriented. Coffee exp desired. 15–30 hrs per wk. $5.25/hr. Start 5/31. Apply in person: Espresso Yourself, Central Coast Mall. Deadline 5/23.

a. If x is the number of hours you work per week and y is your weekly gross pay, write the equation for y. (Be sure to include any restrictions on the variable x that are given in the ad.)

b. Use the function rule you have written in part (a) to complete Table 5.

TABLE 5

Hours Worked x	Gross Pay ($) y
15	
20	
25	
30	

c. Construct a line graph from the information in Table 5.

d. State the domain and range of this function.

e. What is the minimum amount you can earn in a week with this job? What is the maximum amount?

31. Tossing a Coin Hali is tossing a quarter into the air with an underhand motion. The distance the quarter is above her hand at any time is given by the function

$$h = 16t - 16t^2 \quad \text{for} \quad 0 \le t \le 1$$

where h is the height of the quarter in feet, and t is the time in seconds.

a. Use the table feature of your graphing calculator to find the value of h every tenth of a second between $t = 0$ and $t = 1$.

b. State the domain and range of this function.

c. Graph the function on your calculator using an appropriate window.

32. **Intensity of Light** The following formula gives the intensity of light that falls on a surface at various distances from a 100-watt light bulb:

$$I = \frac{120}{d^2} \quad \text{for} \quad d > 0$$

where I is the intensity of light (in lumens per square foot), and d is the distance (in feet) from the light bulb to the surface.

   a. Use the table feature of your graphing calculator to find the value of I at every foot between $d = 1$ and $d = 6$.

   b. Graph the function on your calculator using an appropriate window.

33. **Area of a Circle** The formula for the area A of a circle with radius r is given by $A = \pi r^2$. The formula shows that A is a function of r.

   a. Graph the function $A = \pi r^2$ for $0 \leq r \leq 3$. (On the graph, let the horizontal axis be the r-axis, and let the vertical axis be the A-axis.)

   b. State the domain and range of the function $A = \pi r^2$, $0 \leq r \leq 3$.

34. **Area and Perimeter of a Rectangle** A rectangle is 2 inches longer than it is wide. Let x be the width and P be the perimeter.

   a. Write an equation that will give the perimeter P in terms of the width x of the rectangle. Are there any restrictions on the values that x can assume?

   b. Graph the relationship between P and x.

Let $f(x) = 2x - 5$ and $g(x) = x^2 + 3x + 4$. Evaluate the following.

35. $f(2)$
36. $f(3)$
37. $f(-3)$
38. $g(-2)$
39. $g(-1)$
40. $f(-4)$
41. $g(-3)$
42. $g(2)$
43. $f\left(\frac{1}{2}\right)$
44. $f\left(\frac{1}{4}\right)$
45. $f(a)$
46. $g(b)$
47. $g(4) + f(4)$
48. $f(2) - g(3)$
49. $f(3) - g(2)$
50. $g(-1) + f(-1)$

If $f = \{(1, 4), (-2, 0), (3, 0.5), (\pi, 0)\}$ and $g = \{(1, 1), (-2, 2), (0.5, 0)\}$, find each of the following values of f and g.

51. $f(1)$
52. $g(1)$
53. $g(0.5)$
54. $f(3)$
55. $g(-2)$
56. $f(\pi)$

57. Graph the function $f(x) = \frac{1}{2}x + 2$. Then draw and label the line segments that represent $x = 4$ and $f(4)$.

58. Graph the function $f(x) = -\frac{1}{2}x + 6$. Then draw and label the line segments that represent $x = 4$ and $f(4)$.

59. **Investing in Art** A painting is purchased as an investment for \$150. If its value increases continuously so that it doubles every 3 years, then its value is given by the function

$$V(t) = 150 \cdot 2^{t/3} \quad \text{for} \quad t \geq 0$$

where t is the number of years since the painting was purchased, and $V(t)$ is its value (in dollars) at time t. Find $V(3)$ and $V(6)$, and then explain what they mean.

60. **Average Speed** If it takes Minke t minutes to run a mile, then her average speed $s(t)$, in miles per hour, is given by the formula

$$s(t) = \frac{60}{t} \quad \text{for} \quad t > 0$$

Find $s(4)$ and $s(5)$, and then explain what they mean.

Area of a Circle *The formula for the area A of a circle with radius r can be written with function notation as $A(r) = \pi r^2$.*

61. Find $A(2)$, $A(5)$, and $A(10)$. (Use $\pi \approx 3.14$.)

62. Why doesn't it make sense to ask for $A(-10)$?

➤ EXTENDING THE CONCEPTS

The graphs of two functions are shown in Figures 10 and 11. Use the graphs to find the following.

63. a. $f(2)$ **b.** $f(-4)$ **c.** $g(0)$ **d.** $g(3)$

64. a. $g(2) - f(2)$ **b.** $f(1) + g(1)$ **c.** $f[g(3)]$ **d.** $g[f(3)]$

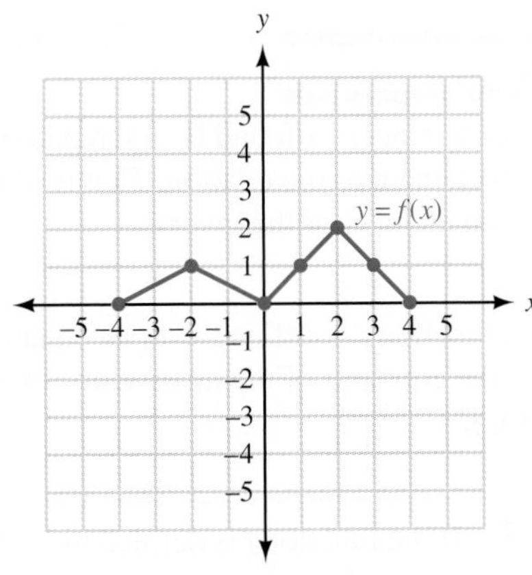

Figure 10

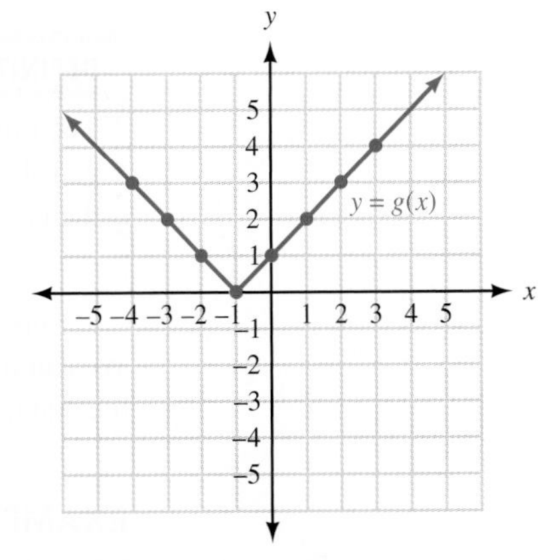

Figure 11

SECTION A.2 The Inverse of a Function

LEARNING OBJECTIVES

1. Find the equation of the inverse of a function.
2. Graph the inverse of a relation.
3. Use the horizontal line test to determine if a function is one-to-one.
4. Evaluate an inverse function.

The diagram in Figure 1 shows the route Justin takes to school. He leaves his home and drives 3 miles east and then turns left and drives 2 miles north. When he leaves school to drive home, he drives the same two segments but in the reverse order and the opposite direction; that is, he drives 2 miles south, turns right, and drives 3 miles west. When he arrives home from school, he is right where he started. His route home "undoes" his route to school, leaving him where he began.

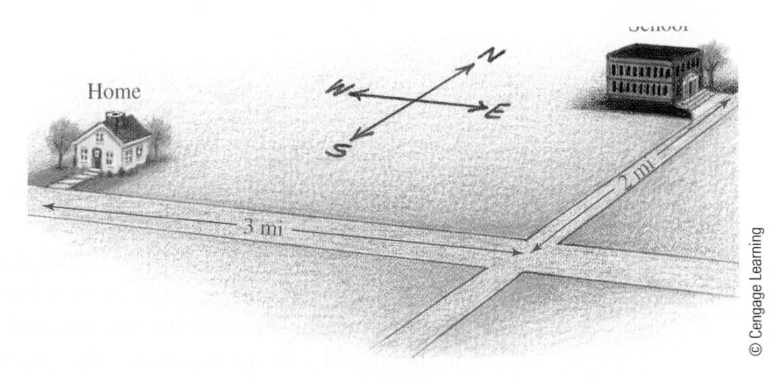

Figure 1

As you will see, the relationship between a function and its inverse function is similar to the relationship between Justin's route from home to school and his route from school to home.

Suppose the function f is given by

$$f = \{(1, 4), (2, 5), (3, 6), (4, 7)\}$$

The inverse of f is obtained by reversing the order of the coordinates in each ordered pair in f. The inverse of f is the relation given by

$$g = \{(4, 1), (5, 2), (6, 3), (7, 4)\}$$

It is obvious that the domain of f is now the range of g, and the range of f is now the domain of g. Every function (or relation) has an inverse that is obtained from the original function by interchanging the components of each ordered pair.

DEFINITION

The *inverse* of a relation is found by interchanging the coordinates in each ordered pair that is an element of the relation. That is, if (a, b) is an element of the relation, then (b, a) is an element of the inverse.

Suppose a function f is defined with an equation instead of a list of ordered pairs. We can obtain the equation of the inverse of f by interchanging the role of x and y in the equation for f.

EXAMPLE 1 If the function f is defined by $f(x) = 2x - 3$, find the equation that represents the inverse of f.

SOLUTION Because the inverse of f is obtained by interchanging the components of all the ordered pairs belonging to f, and each ordered pair in f satisfies the equation $y = 2x - 3$, we simply exchange x and y in the equation $y = 2x - 3$ to get the formula for the inverse of f:

$$x = 2y - 3$$

We now solve this equation for y in terms of x:

$$x + 3 = 2y$$

$$\frac{x+3}{2} = y$$

$$y = \frac{x+3}{2}$$

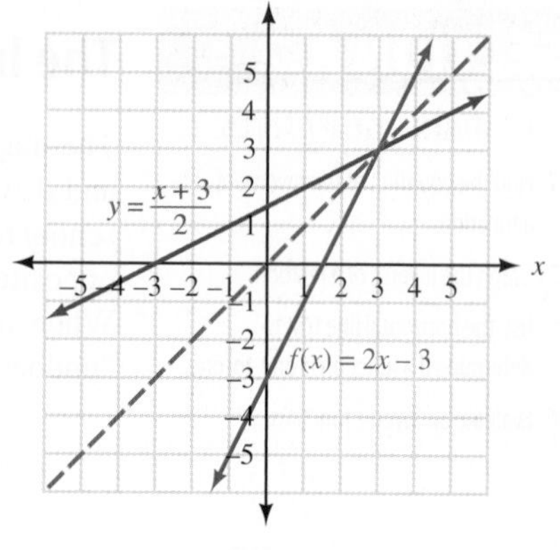

Figure 2

The last line gives the equation that defines the inverse of f. Let's compare the graphs of f and its inverse (see Figure 2).

The graphs of f and its inverse have symmetry about the line $y = x$. We say that the graph of the inverse is a *reflection* of the graph of f about the line $y = x$. This is a reasonable result because the one function was obtained from the other by interchanging x and y in the equation. The ordered pairs (a, b) and (b, a) always have symmetry about the line $y = x$.

SYMMETRY PROPERTY OF INVERSES

The graph of the inverse of a relation (or function) will be a reflection of the graph of the original relation (or function) about the line $y = x$.

EXAMPLE 2 Graph the function $y = x^2 - 2$ and its inverse. Give the equation for the inverse.

SOLUTION We can obtain the graph of the inverse of $y = x^2 - 2$ by graphing $y = x^2 - 2$ by the usual methods and then reflecting the graph about the line $y = x$ (Figure 3).

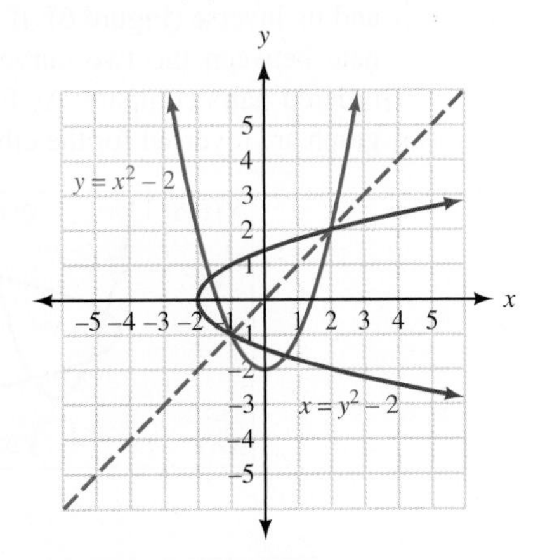

Figure 3

The equation that corresponds to the inverse of $y = x^2 - 2$ is obtained by interchanging x and y to get $x = y^2 - 2$.

We can solve the equation $x = y^2 - 2$ for y in terms of x as follows:

$$x = y^2 - 2$$
$$x + 2 = y^2$$
$$y = \pm\sqrt{x + 2}$$

Using Technology Graphing an Inverse

One way to graph a function and its inverse is to use parametric equations, which we cover in more detail in Section 6.4. To graph the function $y = x^2 - 2$ and its inverse from Example 2, first set your graphing calculator to parametric mode. Then define the following set of parametric equations (Figure 4).

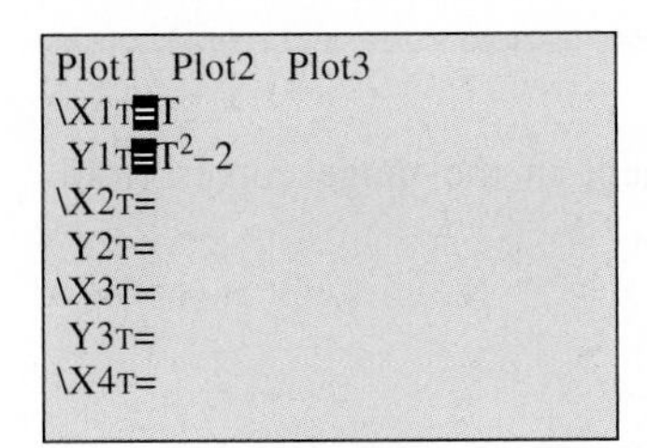

Figure 4

$$X_1 = t;\ Y_1 = t^2 - 2$$

Set the window variables so that

$$-3 \le t \le 3,\ \text{step} = 0.05;\ -4 \le x \le 4;\ -4 \le y \le 4$$

Graph the function using the zoom-square command. Your graph should look similar to Figure 5.

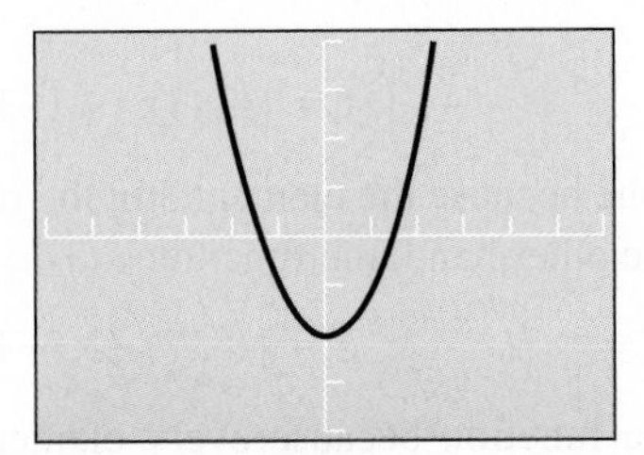

Figure 5

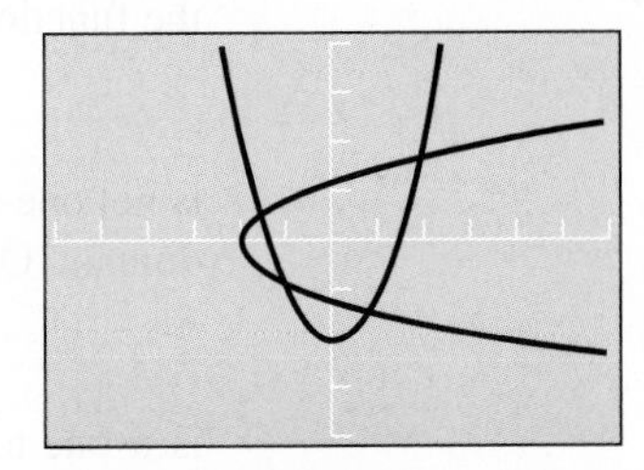

Figure 6

To graph the inverse, we need to interchange the roles of x and y for the original function. This is easily done by defining a new set of parametric equations that is just the reverse of the pair given above:

$$X_2 = t^2 - 2, Y_2 = t$$

Press GRAPH again, and you should now see the graphs of the original function and its inverse (Figure 6). If you trace to any point on either graph, you can alternate between the two curves to see how the coordinates of the corresponding ordered pairs compare. As Figure 7 illustrates, the coordinates of a point on one graph are reversed for the other graph.

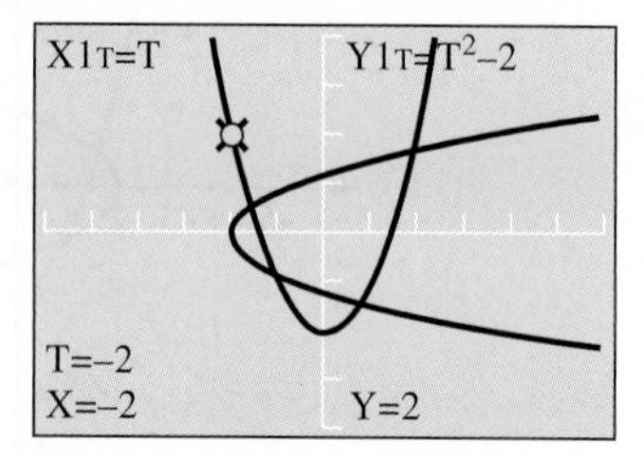

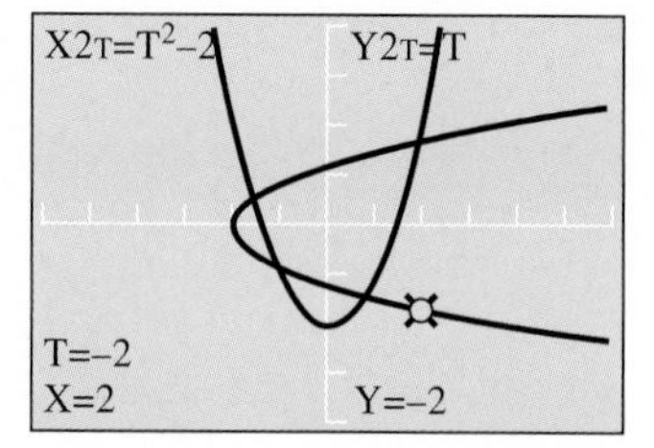

Figure 7

Comparing the graphs from Examples 1 and 2, we observe that the inverse of a function is not always a function. In Example 1, both f and its inverse have graphs that are nonvertical straight lines and therefore both represent functions. In Example 2, the inverse of function f is not a function because a vertical line crosses it in more than one place.

One-to-One Functions

We can distinguish between those functions with inverses that are also functions and those functions with inverses that are not functions with the following definition.

DEFINITION

A function is a *one-to-one function* if every element in the range comes from exactly one element in the domain.

This definition indicates that a one-to-one function will yield a set of ordered pairs in which no two different ordered pairs have the same second coordinates. For example, the function

$$f = \{(2, 3), (-1, 3), (5, 8)\}$$

is not one-to-one because the element 3 in the range comes from both 2 and -1 in the domain. On the other hand, the function

$$g = \{(5, 7), (3, -1), (4, 2)\}$$

is a one-to-one function because every element in the range comes from only one element in the domain.

Horizontal Line Test

If we have the graph of a function, we can determine if the function is one-to-one with the following test. If a horizontal line crosses the graph of a function in more than one place, then the function is not a one-to-one function because the points at which the horizontal line crosses the graph will be points with the same y-coordinates but different x-coordinates. Therefore, the function will have an element in the range that comes from more than one element in the domain.

Functions Whose Inverses Are Functions

Because one-to-one functions do not repeat second coordinates, when we reverse the order of the ordered pairs in a one-to-one function, we obtain a relation in which no two ordered pairs have the same first coordinate—by definition, this relation must be a function. In other words, every one-to-one function has an inverse that is itself a function. Because of this, we can use function notation to represent that inverse.

INVERSE FUNCTION NOTATION

If $y = f(x)$ is a one-to-one function, then the inverse of f is also a function and can be denoted by $y = f^{-1}(x)$.

NOTE The notation f^{-1} does not represent the reciprocal of f; that is, the -1 in this notation is not an exponent. The notation f^{-1} is defined as representing the inverse function for a one-to-one function.

To illustrate, in Example 1 we found the inverse of $f(x) = 2x - 3$ was the function $y = \dfrac{x+3}{2}$. We can write this inverse function with inverse function notation as

$$f^{-1}(x) = \frac{x+3}{2}$$

However, the inverse of the function in Example 2 is not itself a function, so we do not use the notation $f^{-1}(x)$ to represent it.

EXAMPLE 3 Find the inverse of $g(x) = \dfrac{x-4}{x-2}$.

SOLUTION To find the inverse for g, we begin by replacing $g(x)$ with y to obtain

$$y = \frac{x-4}{x-2} \qquad \text{The original function}$$

To find an equation for the inverse, we exchange x and y.

$$x = \frac{y-4}{y-2} \qquad \text{The inverse of the original function}$$

Now we multiply each side by $y - 2$ and then solve for y.

$$x(y-2) = y - 4$$

$$xy - 2x = y - 4 \qquad \text{Distributive property}$$

$$xy - y = 2x - 4 \qquad \text{Collect all terms containing } y \text{ on the left side}$$

$$y(x-1) = 2x - 4 \qquad \text{Factor } y \text{ from each term on the left side}$$

$$y = \frac{2x-4}{x-1} \qquad \text{Divide each side by } x - 1$$

Because our original function is one-to-one, as verified by the graph in Figure 8, its inverse is also a function. Therefore, we can use inverse function notation to write

$$g^{-1}(x) = \frac{2x - 4}{x - 1}$$

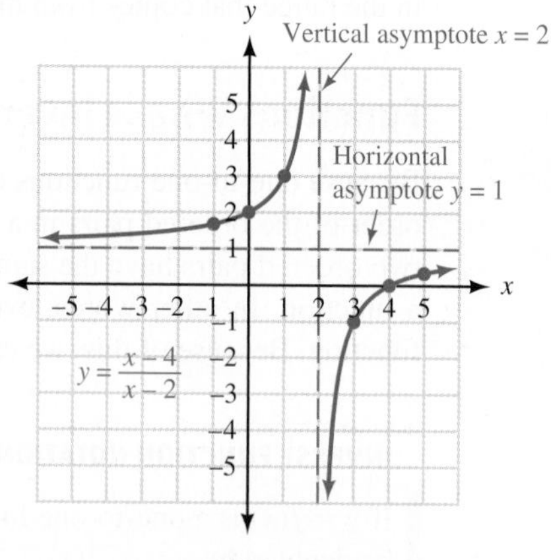

Figure 8

As we mentioned in the introduction to this section, one of the important relationships between a function and its inverse is that an inverse "undoes," or reverses, whatever actions were performed by the function. The next example illustrates this property of inverses.

EXAMPLE 4 Use the function $f(x) = \sqrt[3]{x - 2}$ and its inverse function $f^{-1}(x) = x^3 + 2$ to find the following:

a. $f(10)$ **b.** $f^{-1}[f(10)]$ **c.** $f^{-1}[f(x)]$

SOLUTION

a.
$$\begin{aligned} f(\mathbf{10}) &= \sqrt[3]{\mathbf{10} - 2} \\ &= \sqrt[3]{8} \\ &= 2 \end{aligned}$$

The function f takes an input of 10 and gives us a result of 2.

b.
$$\begin{aligned} f^{-1}[f(10)] &= f^{-1}[\mathbf{2}] && \text{Since } f(10) = 2 \\ &= (\mathbf{2})^3 + 2 && \text{Replace } x \text{ with 2 in } x^3 + 2 \\ &= 8 + 2 \\ &= 10 \end{aligned}$$

The inverse f^{-1} takes the output of 2 from f and "undoes" the steps performed by f, with the end result that we return to our original input value of 10.

c.
$$\begin{aligned} f^{-1}[f(x)] &= f^{-1}[\sqrt[3]{x - 2}] \\ &= (\sqrt[3]{x - 2})^3 + 2 \\ &= (x - 2) + 2 \\ &= x \end{aligned}$$

We see that when f^{-1} is applied to the output of f, the actions of f will be reversed, returning us to our original input x.

Example 4 illustrates another important property between a function and its inverse.

> If $y = f(x)$ is a one-to-one function with inverse $y = f^{-1}(x)$, then for all x in the domain of f,
>
> $$f^{-1}[f(x)] = x$$
>
> and for all x in the domain of f^{-1},
>
> $$f[f^{-1}(x)] = x$$

Functions, Relations, and Inverses—A Summary

Here is a summary of some of the things we know about functions, relations, and their inverses:

1. Every function is a relation, but not every relation is a function.
2. Every function has an inverse, but only one-to-one functions have inverses that are also functions.
3. The domain of a function is the range of its inverse, and the range of a function is the domain of its inverse.
4. If $y = f(x)$ is a one-to-one function, then we can use the notation $y = f^{-1}(x)$ to represent its inverse function.
5. The graph of a function and its inverse have symmetry about the line $y = x$.
6. If (a, b) belongs to the function f, then the point (b, a) belongs to its inverse.
7. The inverse of f will "undo," or reverse, the actions performed by f.

Getting Ready for Class *After reading through the preceding section, respond in your own words and in complete sentences.*

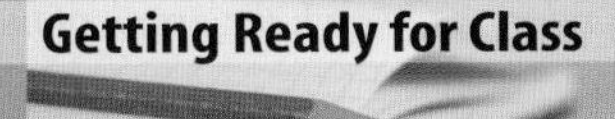

a. What is the inverse of a function?
b. What is the relationship between the graph of a function and the graph of its inverse?
c. Explain why only one-to-one functions have inverses that are also functions.
d. Describe the vertical line test, and explain the difference between the vertical line test and the horizontal line test.

A.2 PROBLEM SET

➤ CONCEPTS AND VOCABULARY

For Questions 1 through 8, fill in each blank with the appropriate word, symbol, or equation.

1. To find the inverse of a relation, ____________ the coordinates in each ordered pair for the relation.
2. The domain of the inverse is the same as the ________ of the relation, and the range of the inverse is the same as the _______ of the relation.

3. The graph of the inverse of a relation is a __________ of the graph of the relation about the line _______.
4. A function is one-to-one if no two different ordered pairs have the same __________.
5. Only functions whose graphs pass the __________ _____ test are one-to-one.
6. Only a ____ ___ _____ function will have an inverse that is also a function.
7. To find the formula for the inverse function, first ________ x and y, and then solve for _____.
8. Under composition, an inverse function _______ the steps performed by the function, so that $f^{-1}(f(x)) =$ ____.

For Questions 9 and 10, determine if the statement is true or false.

9. The notation $f^{-1}(x)$ means the reciprocal of $f(x)$.
10. Every function has an inverse function.

➤ EXERCISES

For each of the following one-to-one functions, find the equation of the inverse. Write the inverse using the notation $f^{-1}(x)$.

11. $f(x) = 3x - 1$
12. $f(x) = 2x - 5$
13. $f(x) = x^3$
14. $f(x) = x^3 - 2$
15. $f(x) = \dfrac{x-3}{x-1}$
16. $f(x) = \dfrac{x-2}{x-3}$
17. $f(x) = \dfrac{x-3}{4}$
18. $f(x) = \dfrac{x+7}{2}$
19. $f(x) = \dfrac{1}{2}x - 3$
20. $f(x) = \dfrac{1}{3}x + 1$
21. $f(x) = \dfrac{2x+1}{3x+1}$
22. $f(x) = \dfrac{3x+2}{5x+1}$

For each of the following relations, sketch the graph of the relation and its inverse, and write an equation for the inverse.

23. $y = 2x - 1$
24. $y = 3x + 1$
25. $y = x^2 - 3$
26. $y = x^2 + 1$
27. $y = x^2 - 2x - 3$
28. $y = x^2 + 2x - 3$
29. $y = 4$
30. $y = -2$
31. $y = \dfrac{1}{2}x + 2$
32. $y = \dfrac{1}{3}x - 1$

For each of the following functions, use your graphing calculator in parametric mode to graph the function and its inverse.

33. $y = \dfrac{1}{2}x^3$
34. $y = x^3 - 2$
35. $y = \sqrt{x+2}$
36. $y = \sqrt{x} + 2$

37. Determine if the following functions are one-to-one.

a.

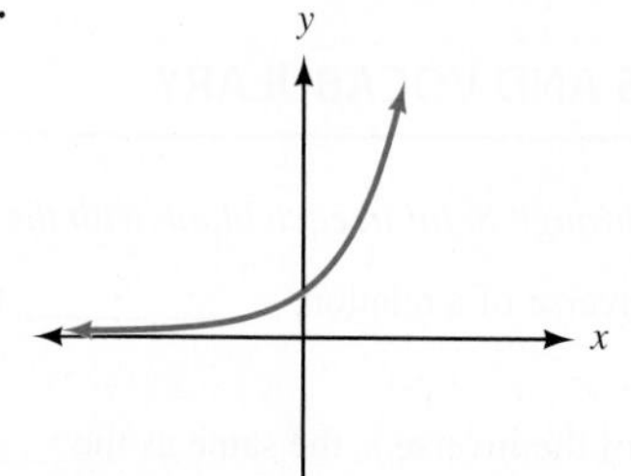

b.

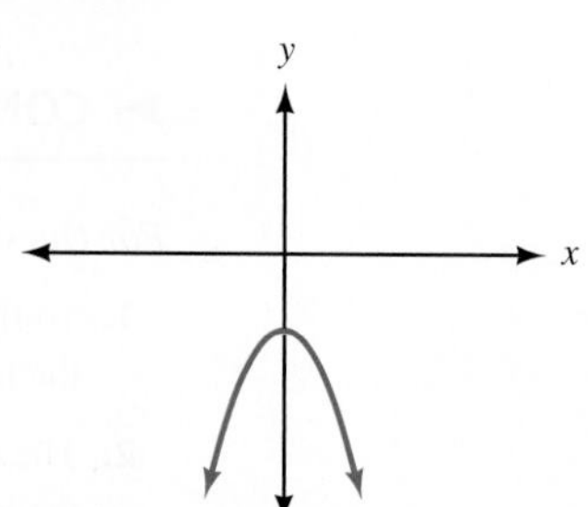

38. Could the following tables of values represent ordered pairs from one-to-one functions? Explain your answer.

a.

x	y
−2	5
−1	4
0	3
1	4
2	5

b.

x	y
1.5	0.1
2.0	0.2
2.5	0.3
3.0	0.4
3.5	0.5

39. If $f(x) = 3x - 2$, then $f^{-1}(x) = \dfrac{x+2}{3}$. Use these two functions to find

a. $f(2)$ **b.** $f^{-1}(2)$ **c.** $f[f^{-1}(2)]$ **d.** $f^{-1}[f(2)]$

40. If $f(x) = \dfrac{1}{2}x + 5$, then $f^{-1}(x) = 2x - 10$. Use these two functions to find

a. $f(-4)$ **b.** $f^{-1}(-4)$ **c.** $f[f^{-1}(-4)]$ **d.** $f^{-1}[f(-4)]$

41. Let $f(x) = \dfrac{1}{x}$, and find $f^{-1}(x)$.

42. Let $f(x) = \dfrac{a}{x}$, and find $f^{-1}(x)$. (a is a real number constant.)

43. Reading Tables Evaluate each of the following functions using the functions defined by Tables 1 and 2.

TABLE 1

x	$f(x)$
−6	3
2	−3
3	−2
6	4

TABLE 2

x	$g(x)$
−3	2
−2	3
3	−6
4	6

a. $f[g(-3)]$ **b.** $g[f(-6)]$ **c.** $g[f(2)]$

d. $f[g(3)]$ **e.** $f[g(-2)]$ **f.** $g[f(3)]$

g. What can you conclude about the relationship between functions f and g?

44. Reading Tables Use the functions defined in Tables 1 and 2 in Problem 43 to answer the following questions.

a. What are the domain and range of f?

b. What are the domain and range of g?

c. How are the domain and range of f related to the domain and range of g?

d. Is f a one-to-one function?

e. Is g a one-to-one function?

For each of the following functions, find $f^{-1}(x)$. Then show that $f[f^{-1}(x)] = x$.

45. $f(x) = 3x + 5$ **46.** $f(x) = 6 - 8x$

47. $f(x) = x^3 + 1$ **48.** $f(x) = x^3 - 8$

➤ EXTENDING THE CONCEPTS

49. Inverse Functions in Words Inverses may also be found by *inverse reasoning*. For example, to find the inverse of $f(x) = 3x + 2$, first list, in order, the operations done to variable x:

a. Multiply by 3.

b. Add 2.

Then, to find the inverse, simply apply the inverse operations, in reverse order, to the variable x:

a. Subtract 2.

b. Divide by 3.

The inverse function then becomes $f^{-1}(x) = \frac{x-2}{3}$. Use this method of "inverse reasoning" to find the inverse of the function $f(x) = \frac{x}{7} - 2$.

50. Inverse Functions in Words Refer to the method of inverse reasoning explained in Problem 49. Use *inverse reasoning* to find the following inverses:

a. $f(x) = 2x + 7$

b. $f(x) = \sqrt{x} - 9$

c. $f(x) = x^3 - 4$

d. $f(x) = \sqrt{x^3 - 4}$

51. The graphs of a function and its inverse are shown in Figure 9. Use the graphs to find the following:

a. $f(0)$

b. $f(1)$

c. $f(2)$

d. $f^{-1}(1)$

e. $f^{-1}(2)$

f. $f^{-1}(5)$

g. $f^{-1}[f(2)]$

h. $f[f^{-1}(5)]$

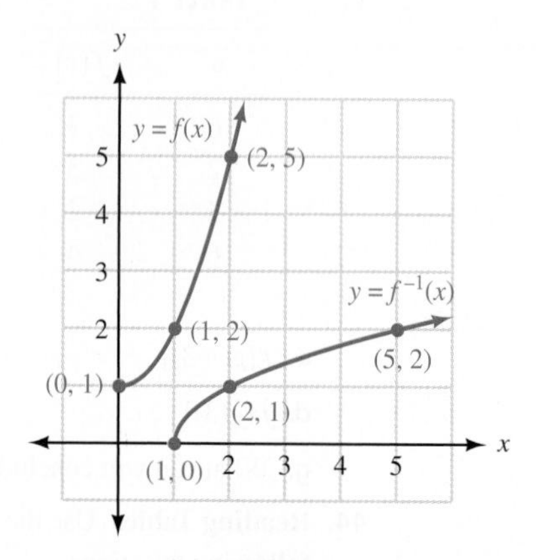

Figure 9

ANSWERS TO SELECTED EXERCISES

CHAPTER 1

MATCHED PRACTICE PROBLEMS 1.1

1. a. complement is 65°, supplement is 155° **b.** complement is $-28°$, supplement is 62° **c.** complement is $90° - \beta$, supplement is $180° - \beta$
2. $x = 12$ **3.** 5,862 ft **4.** $7, 7\sqrt{3}$ **5.** 6 ft, $3\sqrt{3} \approx 5.2$ ft **6.** $4\sqrt{2} \approx 5.7$ ft

PROBLEM SET 1.1

1. counterclockwise, clockwise **3.** 180° **5.** hypotenuse, sum, legs **7.** equal, $\sqrt{2}$ **9.** acute, complement is 80°, supplement is 170° **11.** acute, complement is 45°, supplement is 135° **13.** obtuse, complement is $-30°$, supplement is 60° **15.** we can't tell if x is acute or obtuse (or neither), complement is $90° - x$, supplement is $180° - x$ **17.** 60° **19.** 30° **21.** 50° (Look at it in terms of the big triangle *ABC*.) **23.** complementary **25.** 38° **27.** 1 sec **29.** 70° **31.** 5 (This triangle is called a 3–4–5 right triangle. You will see it again.) **33.** 15 **35.** 5 **37.** 5 (Note that this must be a 45°–45°–90° triangle.) **39.** 4 (This is a 30°–60°–90° triangle.)

41. 1 **43.** $\sqrt{41}$ **45.** 6 **47.** 22.5 ft **49.** $2, \sqrt{3}$ **51.** $4, 4\sqrt{3}$ **53.** $\dfrac{6}{\sqrt{3}} = 2\sqrt{3}, 4\sqrt{3}$ **55.** 40 ft **57.** 101.6 ft^2

59. $\dfrac{4\sqrt{2}}{5}$ **61.** 8 **63.** $\dfrac{4}{\sqrt{2}} = 2\sqrt{2}$ **65.** 1,414 ft **67.** $a = 2\sqrt{3}, b = \sqrt{3}, d = \dfrac{3\sqrt{2}}{2}$ **69. a.** $\sqrt{2}$ inches **b.** $\sqrt{3}$ inches

71. a. $x\sqrt{2}$ **b.** $x\sqrt{3}$ **73.** See Figure 1 on page 1.

MATCHED PRACTICE PROBLEMS 1.2

1.

2. $y = -\dfrac{1}{70}(x - 70)^2 + 70$ for $0 \le x \le 140$ **3.** 5 **4.** 13

5. $\left(\dfrac{\sqrt{2}}{3}\right)^2 + \left(\dfrac{\sqrt{7}}{3}\right)^2 = \dfrac{2}{9} + \dfrac{7}{9} = 1$

6. **7.**

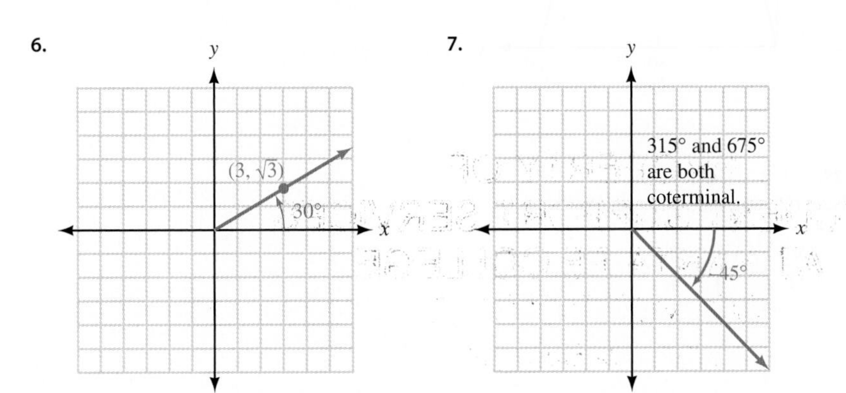

8. $\theta = 90° + 360°k$ for any integer k

PROBLEM SET 1.2

1. quadrants, I, IV, counterclockwise **3.** origin, positive x-axis **5.** quadrantal

7. $d = \sqrt{(x_2 - x_1)^2 + (y_2 - y_1)^2}$ **9.** QIV **11.** QII

13.

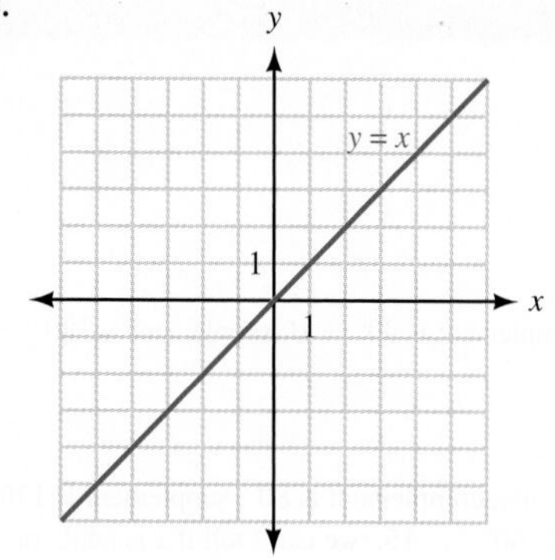

15.

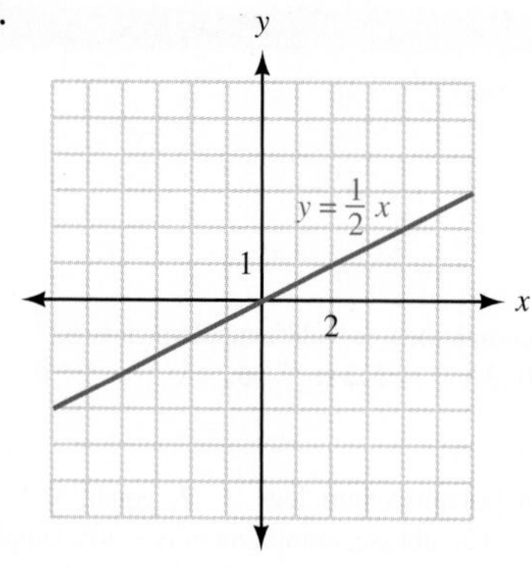

17. QII and QIII **19.** QIII

21.

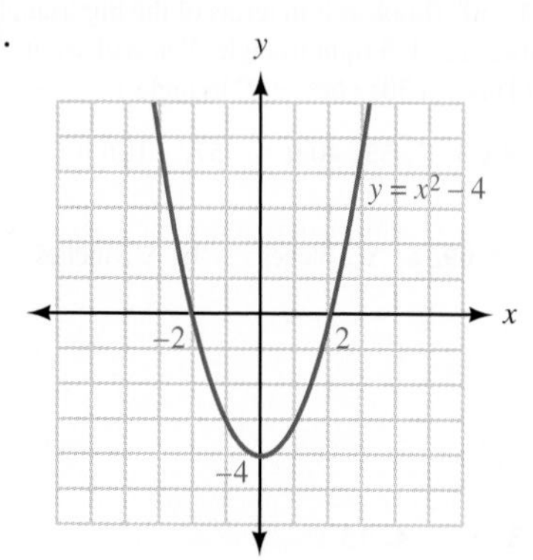

23.

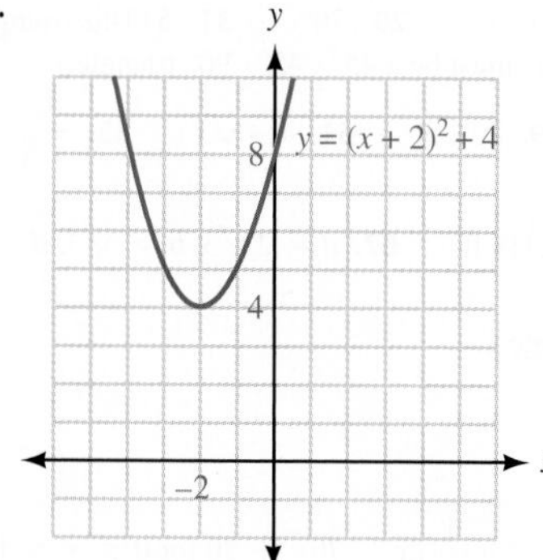

25.

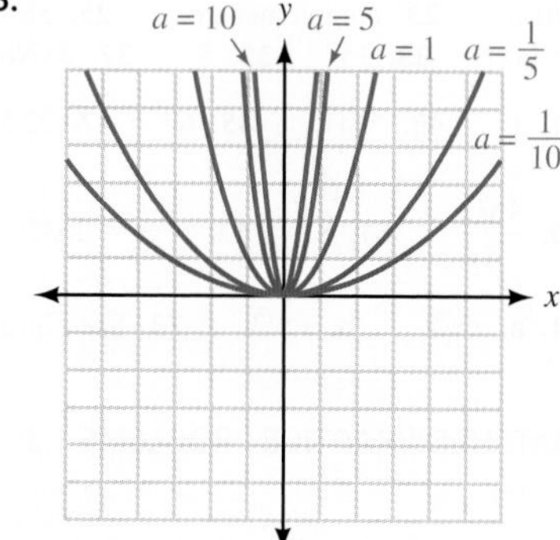

The parabola gets wider; the parabola gets narrower.

27.

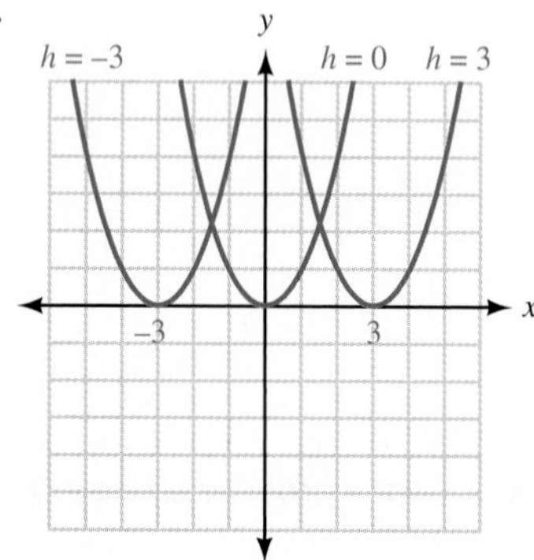

The parabola is shifted to the left; the parabola is shifted to the right.

29.

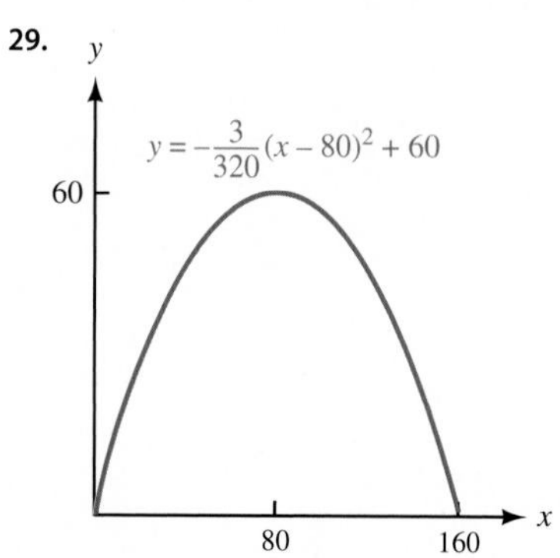

31. 5 **33.** 13 **35.** $\sqrt{130}$ **37.** 5 **39.** $-1, 3$ **41.** 1.3 mi

43. homeplate: (0, 0); first base: (60, 0); second base: (60, 60); third base: (0, 60)

45. $(0)^2 + (-1)^2 = 0 + 1 = 1$ **47.** $\left(\frac{1}{2}\right)^2 + \left(\frac{\sqrt{3}}{2}\right)^2 = \frac{1}{4} + \frac{3}{4} = 1$

49.

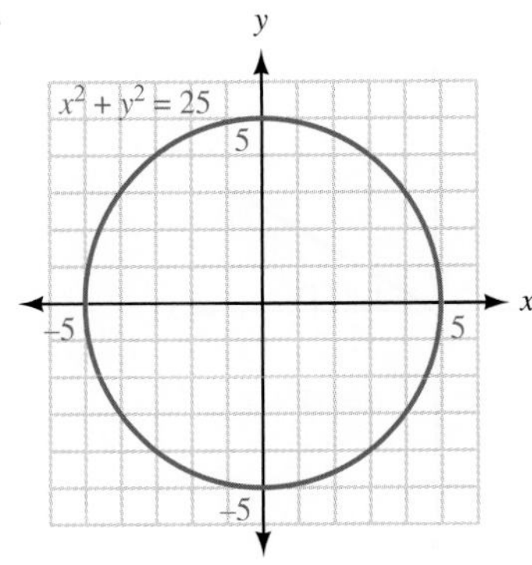

51. (0.5, 0.8660), (0.5, −0.8660) **53.** (0.7071, 0.7071), (0.7071, −0.7071)

55. (−0.8660, 0.5), (−0.8660, −0.5) **57.** (5, 0) and (0, 5)

59. $\left(-\frac{\sqrt{2}}{2}, -\frac{\sqrt{2}}{2}\right)$ and $\left(\frac{\sqrt{2}}{2}, \frac{\sqrt{2}}{2}\right)$ **61.** 45° **63.** 30° **65.** 60° **67.** 90°

69. 225° **71.** 150°

73.

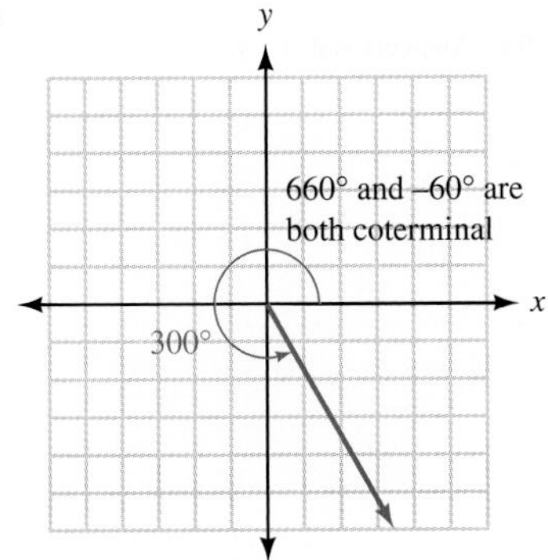

75.

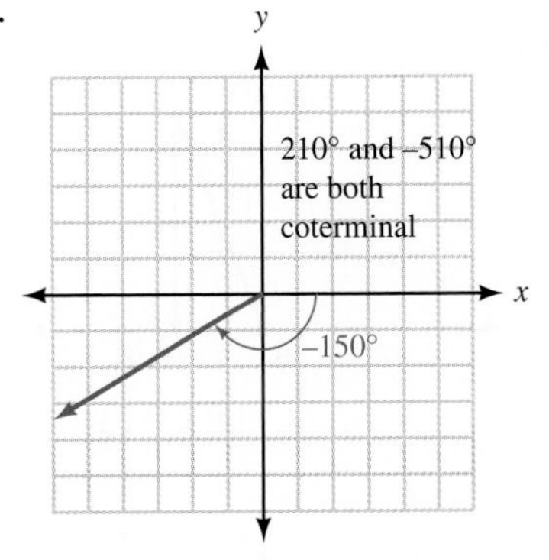

Note For Problems 77–84, other answers are possible.

77.

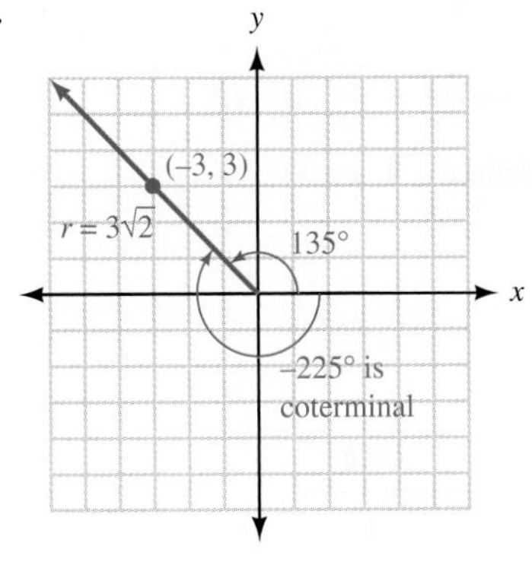

79.

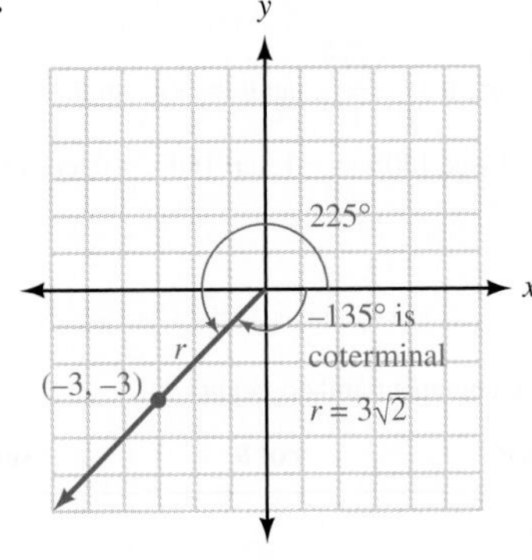

81.

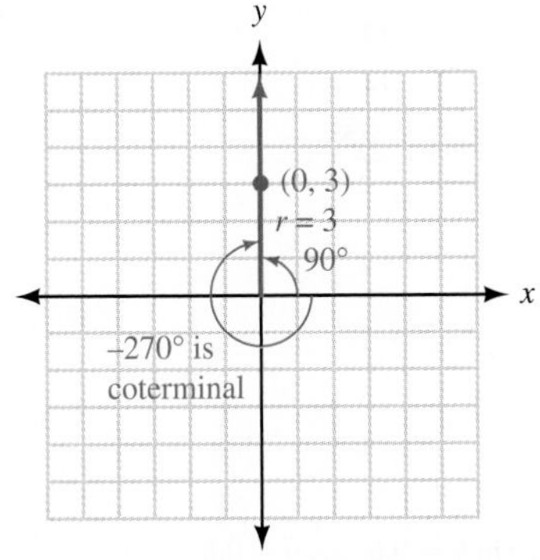

83.

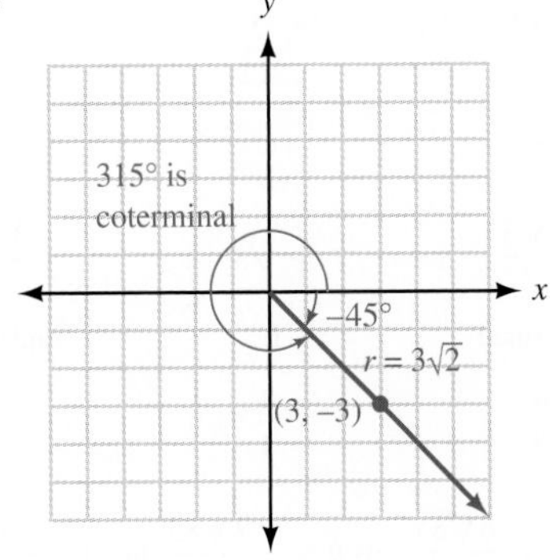

85. $30° + 360°k$ for any integer k **87.** $-135° + 360°k$ for any integer k **89.**

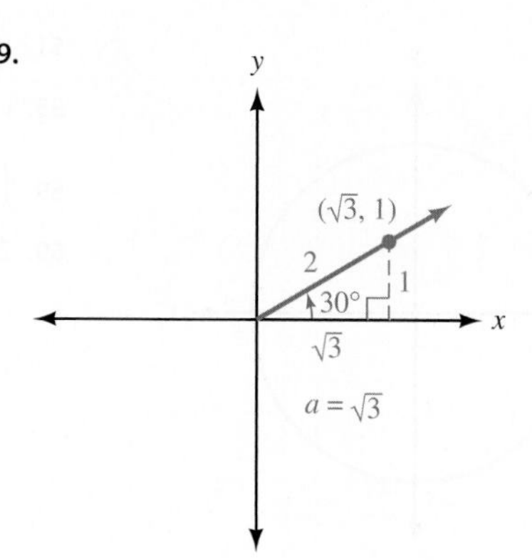

Note For Problems 91 and 92, one possible angle is shown in standard position.

91.

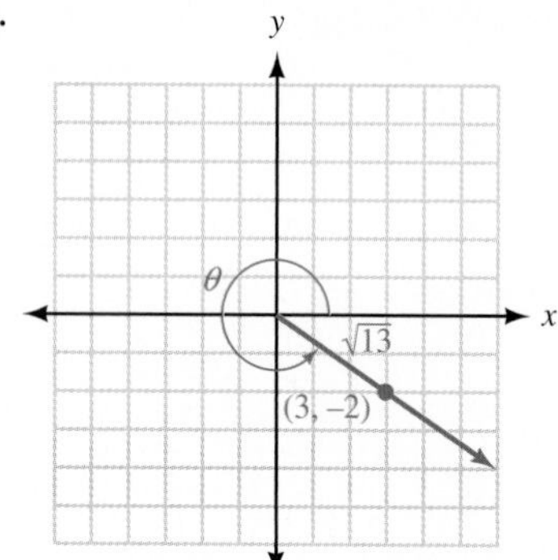

93.

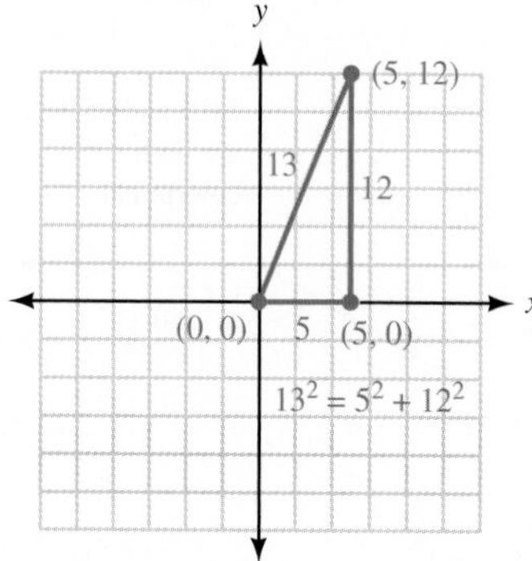

95. Answers will vary.

MATCHED PRACTICE PROBLEMS 1.3

1. $\sin\theta = -\frac{4\sqrt{17}}{17}$, $\csc\theta = -\frac{\sqrt{17}}{4}$, $\cos\theta = \frac{\sqrt{17}}{17}$, $\sec\theta = \frac{\sqrt{17}}{1}$, $\tan\theta = -4$, $\cot\theta = -\frac{1}{4}$ **2.** $\sin 60° = \frac{\sqrt{3}}{2}$, $\cos 60° = \frac{1}{2}$

3. $\sin 180° = 0$, $\csc 180°$ is undefined, $\cos 180° = -1$, $\sec 180° = -1$, $\tan 180° = 0$, $\cot 180°$ is undefined **4.** $\cot 30°$

5. $\sin\theta = -\frac{4}{5}$, $\cos\theta = -\frac{3}{5}$

PROBLEM SET 1.3

1. terminal, distance, origin **3.** tangent and secant, cotangent and cosecant

	$\sin\theta$	$\cos\theta$	$\tan\theta$	$\cot\theta$	$\sec\theta$	$\csc\theta$
5.	$\frac{4}{5}$	$\frac{3}{5}$	$\frac{4}{3}$	$\frac{3}{4}$	$\frac{5}{3}$	$\frac{5}{4}$
7.	$\frac{12}{13}$	$-\frac{5}{13}$	$-\frac{12}{5}$	$-\frac{5}{12}$	$-\frac{13}{5}$	$\frac{13}{12}$
9.	$-\frac{2\sqrt{5}}{5}$	$-\frac{\sqrt{5}}{5}$	2	$\frac{1}{2}$	$-\sqrt{5}$	$-\frac{\sqrt{5}}{2}$
11.	$-\frac{1}{2}$	$\frac{\sqrt{3}}{2}$	$-\frac{\sqrt{3}}{3}$	$-\sqrt{3}$	$\frac{2\sqrt{3}}{3}$	-2
13.	-1	0	undefined	0	undefined	-1
15.	$-\frac{4}{5}$	$-\frac{3}{5}$	$\frac{4}{3}$	$\frac{3}{4}$	$-\frac{5}{3}$	$-\frac{5}{4}$

17. $\sin\theta = \frac{3}{5}$, $\cos\theta = \frac{4}{5}$, $\tan\theta = \frac{3}{4}$ **19.** $\sin\theta = \frac{6\sqrt{85}}{85}$, $\cos\theta = -\frac{7\sqrt{85}}{85}$, $\tan\theta = -\frac{6}{7}$ **21.** $\sin\theta = 0.6$, $\cos\theta = 0.8$

23.

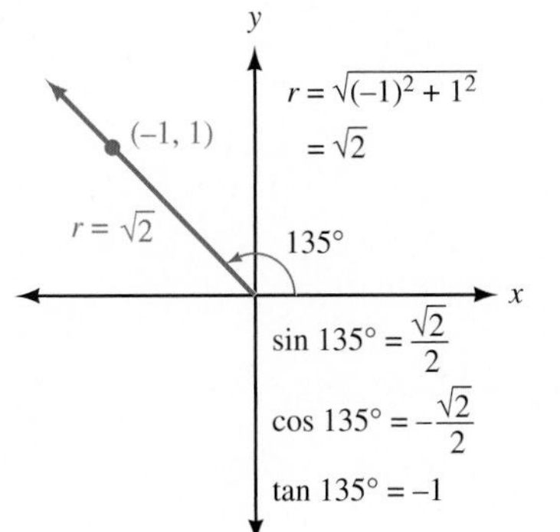

25.

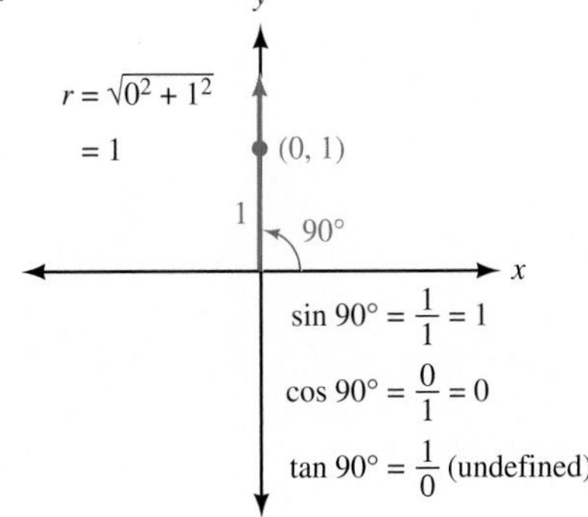

27. **29.**

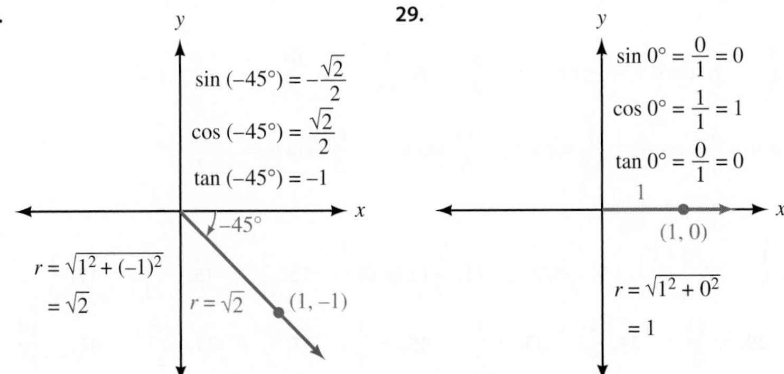

31. false **33.** true **35.** If $\sin\theta = \dfrac{y}{r} = 2$, then y would have to be greater than r, which is not possible.

37. Because $|\csc\theta| = \dfrac{r}{|y|}$ and $r \geq |y|$, this ratio will be equal to 1 or something greater.

39. 1 **41.** ∞ **43.** QI, QII **45.** QII, QIII **47.** QI, QIII **49.** QIII, QIV **51.** QIII

53. QI (both positive), QIV (both negative)

	sin θ	cos θ	tan θ	cot θ	sec θ	csc θ
55.	$\frac{12}{13}$	$\frac{5}{13}$	$\frac{12}{5}$	$\frac{5}{12}$	$\frac{13}{5}$	$\frac{13}{12}$
57.	$\frac{21}{29}$	$-\frac{20}{29}$	$-\frac{21}{20}$	$-\frac{20}{21}$	$-\frac{29}{20}$	$\frac{29}{21}$
59.	$-\frac{1}{2}$	$\frac{\sqrt{3}}{2}$	$-\frac{\sqrt{3}}{3}$	$-\sqrt{3}$	$\frac{2\sqrt{3}}{3}$	-2
61.	$-\frac{3}{5}$	$-\frac{4}{5}$	$\frac{3}{4}$	$\frac{4}{3}$	$-\frac{5}{4}$	$-\frac{5}{3}$
63.	$\frac{5}{13}$	$-\frac{12}{13}$	$-\frac{5}{12}$	$-\frac{12}{5}$	$-\frac{13}{12}$	$\frac{13}{5}$
65.	$\frac{2\sqrt{5}}{5}$	$\frac{\sqrt{5}}{5}$	2	$\frac{1}{2}$	$\sqrt{5}$	$\frac{\sqrt{5}}{2}$
67.	$\frac{a}{\sqrt{a^2+b^2}}$	$\frac{b}{\sqrt{a^2+b^2}}$	$\frac{a}{b}$	$\frac{b}{a}$	$\frac{\sqrt{a^2+b^2}}{b}$	$\frac{\sqrt{a^2+b^2}}{a}$

69. $\sin\theta = \dfrac{2\sqrt{5}}{5}$, $\cos\theta = \dfrac{\sqrt{5}}{5}$ **71.** $\sin\theta = \dfrac{3\sqrt{10}}{10}$, $\tan\theta = -3$

73.

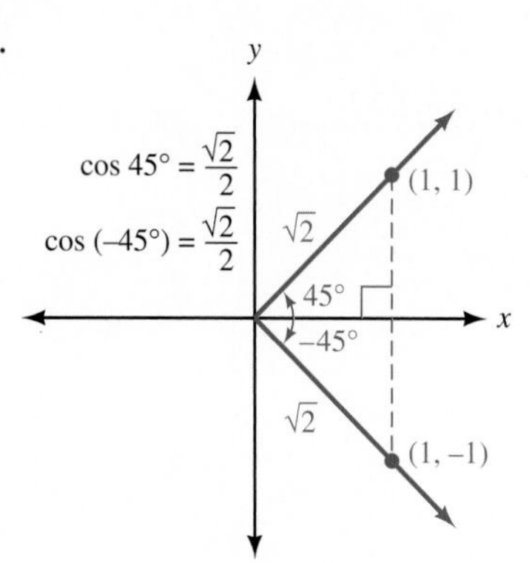

75. ± 12

MATCHED PRACTICE PROBLEMS 1.4

1. $-\frac{3}{2}$ **2.** $\frac{5}{3}$ **3.** 8 **4.** $\frac{3}{4}$ **5.** $\frac{3}{7}$ **6.** $-\frac{5}{7}$ **7.** $\tan\theta = -\frac{4}{3}$, $\cot\theta = -\frac{3}{4}$ **8.** $\frac{1}{16}$ **9.** $\frac{64}{27}$

10. $\cos\theta = -\frac{2\sqrt{2}}{3}$, $\tan\theta = \frac{1}{2\sqrt{2}} = \frac{\sqrt{2}}{4}$ **11.** $\sin\theta = \frac{12}{13}$, $\csc\theta = \frac{13}{12}$, $\sec\theta = -\frac{13}{5}$, $\tan\theta = -\frac{12}{5}$, $\cot\theta = -\frac{5}{12}$

PROBLEM SET 1.4

1. true, defined **3. a.** ii **b.** iii **c.** i **5.** $\frac{1}{7}$ **7.** $-\frac{3}{2}$ **9.** $-\sqrt{2}$ **11.** $\frac{1}{x}$ $(x \neq 0)$ **13.** $\frac{5}{4}$ **15.** $-\frac{1}{2}$ **17.** $\frac{1}{a}$

19. 2 **21.** $\frac{12}{5}$ **23.** $\frac{1}{2}$ **25.** 8 **27.** $\frac{12}{5}$ **29.** $-\frac{13}{5}$ **31.** $\frac{4}{5}$ **33.** $-\frac{1}{2}$ **35.** $-\frac{3}{5}$ **37.** $\frac{1}{2}$ **39.** $\frac{\sqrt{2}}{4}$ **41.** $-\frac{17}{15}$

43. $\frac{29}{20}$

	sin θ	cos θ	tan θ	cot θ	sec θ	csc θ
45.	$\frac{5}{13}$	$\frac{12}{13}$	$\frac{5}{12}$	$\frac{12}{5}$	$\frac{13}{12}$	$\frac{13}{5}$
47.	$-\frac{1}{2}$	$\frac{\sqrt{3}}{2}$	$-\frac{\sqrt{3}}{3}$	$-\sqrt{3}$	$\frac{2\sqrt{3}}{3}$	-2
49.	$\frac{1}{2}$	$-\frac{\sqrt{3}}{2}$	$-\frac{\sqrt{3}}{3}$	$-\sqrt{3}$	$-\frac{2\sqrt{3}}{3}$	2
51.	$-\frac{3\sqrt{13}}{13}$	$\frac{2\sqrt{13}}{13}$	$-\frac{3}{2}$	$-\frac{2}{3}$	$\frac{\sqrt{13}}{2}$	$-\frac{\sqrt{13}}{3}$
53.	$\frac{1}{a}$	$\frac{\sqrt{a^2-1}}{a}$	$\frac{1}{\sqrt{a^2-1}}$	$\sqrt{a^2-1}$	$\frac{a}{\sqrt{a^2-1}}$	a
55.	0.23	0.97	0.24	4.23	1.03	4.35
57.	0.59	−0.81	−0.73	−1.36	−1.24	1.69

Your answers for Problems 55–58 may differ from the answers here in the hundredths column if you found the reciprocal of a rounded number.

Note As Problems 59–62 indicate, the slope of a line through the origin is the same as the tangent of the angle the line makes with the positive x-axis.

59. 3 **61.** 3

MATCHED PRACTICE PROBLEMS 1.5

1. $\pm\dfrac{1}{\sqrt{1-\sin^2\theta}}$ **2.** $\dfrac{1}{\cos\theta}$ **3.** $\dfrac{\sin\theta\cos\theta+1}{\cos\theta}$ **4.** $6\cos^2\theta-\cos\theta-2$ **5.** $4|\sin\theta|$

6. $$\begin{aligned}\sin\theta+\cos\theta\cot\theta &= \sin\theta+\cos\theta\cdot\frac{\cos\theta}{\sin\theta}\\ &= \frac{\sin\theta}{1}\cdot\frac{\sin\theta}{\sin\theta}+\frac{\cos^2\theta}{\sin\theta}\\ &= \frac{\sin^2\theta+\cos^2\theta}{\sin\theta}\\ &= \frac{1}{\sin\theta}\\ &= \csc\theta\end{aligned}$$

7. $$\begin{aligned}(\cos\theta+\sin\theta)^2 &= \cos^2\theta+2\sin\theta\cos\theta+\sin^2\theta\\ &= 2\sin\theta\cos\theta+(\cos^2\theta+\sin^2\theta)\\ &= 2\sin\theta\cos\theta+1\end{aligned}$$

PROBLEM SET 1.5

1. sines, cosines

3. $\pm\sqrt{1-\sin^2\theta}$ **5.** $\pm\dfrac{\sqrt{1-\sin^2\theta}}{\sin\theta}$ **7.** $\dfrac{1}{\cos\theta}$ **9.** $\pm\dfrac{1}{\sqrt{1-\cos^2\theta}}$ **11. a.** $\dfrac{b}{a}$ **b.** $\dfrac{\sin\theta}{\cos\theta}=\tan\theta$ **13.** $\dfrac{\cos\theta}{\sin^2\theta}$

15. $\dfrac{1}{\cos\theta}$ **17.** $\dfrac{\sin\theta}{\cos\theta}$ **19.** $\dfrac{1}{\sin\theta}$ **21.** $\dfrac{\sin^2\theta}{\cos^2\theta}$ **23.** $\sin^2\theta$ **25.** $\dfrac{\sin\theta+1}{\cos\theta}$ **27.** $2\cos\theta$ **29.** $\cos\theta$

31. a. $\dfrac{b-a}{ab}$ **b.** $\dfrac{\cos\theta-\sin\theta}{\sin\theta\cos\theta}$ **33. a.** $\dfrac{b^2+a}{ab}$ **b.** $\dfrac{\cos^2\theta+\sin\theta}{\sin\theta\cos\theta}$ **35.** $\dfrac{\sin\theta\cos\theta+1}{\cos\theta}$ **37.** $\dfrac{\sin^2\theta}{\cos\theta}$

39. $\dfrac{\sin^2\theta+\cos\theta}{\sin\theta\cos\theta}$ **41. a.** $a^2-2ab+b^2$ **b.** $\cos^2\theta-2\cos\theta\sin\theta+\sin^2\theta=1-2\cos\theta\sin\theta$

43. a. a^2-5a+6 **b.** $\sin^2\theta-5\sin\theta+6$ **45.** $\sin^2\theta+7\sin\theta+12$ **47.** $8\cos^2\theta+2\cos\theta-15$ **49.** $\cos^2\theta$

51. $1-\tan^2\theta$ **53.** $1-2\sin\theta\cos\theta$ **55.** $\sin^2\theta-8\sin\theta+16$ **57.** $2|\sec\theta|$ **59.** $3|\cos\theta|$ **61.** $6|\tan\theta|$ **63.** $8|\cos\theta|$

For Problems 65–96, a few selected solutions are given here. See the Solutions Manual for solutions to problems not shown.

65. $\cos\theta\tan\theta=\cos\theta\cdot\dfrac{\sin\theta}{\cos\theta}=\sin\theta$

69. $$\begin{aligned}\frac{\sin\theta}{\csc\theta} &= \frac{\sin\theta}{\frac{1}{\sin\theta}}\\ &= \sin\theta\cdot\frac{\sin\theta}{1}\\ &= \sin^2\theta\end{aligned}$$

77. $$\begin{aligned}\sin\theta\tan\theta+\cos\theta &= \sin\theta\cdot\frac{\sin\theta}{\cos\theta}+\cos\theta\\ &= \frac{\sin^2\theta}{\cos\theta}+\cos\theta\\ &= \frac{\sin^2\theta+\cos^2\theta}{\cos\theta}\\ &= \frac{1}{\cos\theta}\\ &= \sec\theta\end{aligned}$$

81. $$\begin{aligned}\csc\theta-\sin\theta &= \frac{1}{\sin\theta}-\sin\theta\\ &= \frac{1}{\sin\theta}-\frac{\sin^2\theta}{\sin\theta}\\ &= \frac{1-\sin^2\theta}{\sin\theta}\\ &= \frac{\cos^2\theta}{\sin\theta}\end{aligned}$$

89. $$\begin{aligned}\frac{\cos\theta}{\sec\theta}+\frac{\sin\theta}{\csc\theta} &= \frac{\cos\theta}{\frac{1}{\cos\theta}}+\frac{\sin\theta}{\frac{1}{\sin\theta}}\\ &= \cos^2\theta+\sin^2\theta\\ &= 1\end{aligned}$$

93. $$\begin{aligned}\sin\theta(\sec\theta+\csc\theta) &= \sin\theta\cdot\sec\theta+\sin\theta\cdot\csc\theta\\ &= \sin\theta\cdot\frac{1}{\cos\theta}+\sin\theta\cdot\frac{1}{\sin\theta}\\ &= \frac{\sin\theta}{\cos\theta}+\frac{\sin\theta}{\sin\theta}\\ &= \tan\theta+1\end{aligned}$$

CHAPTER 1 TEST

1. $20°, 110°$ **2.** $3\sqrt{3}$ **3.** $h = 5\sqrt{3}, r = 5\sqrt{6}, y = 5, x = 10$ **4.** $2\sqrt{13}$ **5.** $90°$ **6.** $\frac{5}{2}$ and $\frac{5\sqrt{3}}{2}$ **7.** $15\sqrt{2}$ ft

8. $108°$ **9.** 13 **10.** $-5, 1$ **11.** $\left(\frac{1}{2}\right)^2 + \left(-\frac{\sqrt{3}}{2}\right)^2 = \frac{1}{4} + \frac{3}{4} = 1$ **12.** $225° + 360°k$ for any integer k

13.

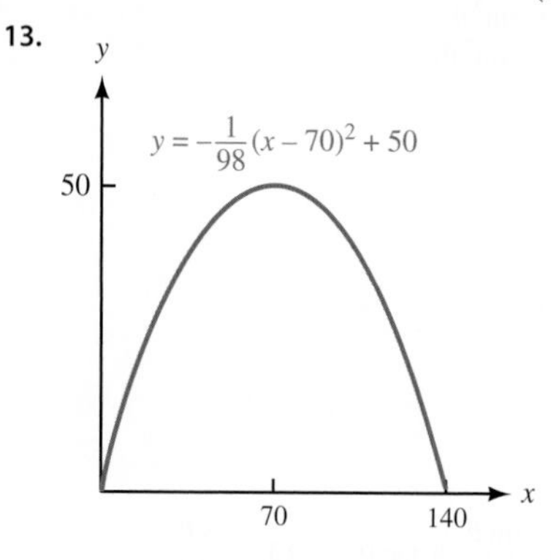

14. $\sin 90° = 1, \cos 90° = 0$, $\tan 90°$ is undefined **15.** $\sin(-45°) = -\frac{\sqrt{2}}{2}, \cos(-45°) = \frac{\sqrt{2}}{2}, \tan(-45°) = -1$ **16.** QII, QIII

17. QII **18.** $\sin\theta = -\frac{\sqrt{10}}{10}, \cos\theta = -\frac{3\sqrt{10}}{10}, \tan\theta = \frac{1}{3}, \cot\theta = 3, \sec\theta = -\frac{\sqrt{10}}{3}, \csc\theta = -\sqrt{10}$

19. Because $\sin\theta = \frac{y}{r}$ and $|y| \le r$, this ratio will be no larger than 1.

20. $\sin\theta = \frac{1}{2}, \cos\theta = -\frac{\sqrt{3}}{2}, \tan\theta = -\frac{\sqrt{3}}{3}, \cot\theta = -\sqrt{3}, \sec\theta = -\frac{2\sqrt{3}}{3}, \csc\theta = 2$ **21.** $\sin\theta = -\frac{2\sqrt{5}}{5}, \cos\theta = \frac{\sqrt{5}}{5}$

22. $-\frac{4}{3}$ **23.** $\frac{1}{27}$ **24.** $\cos\theta = \frac{1}{3}, \sin\theta = -\frac{2\sqrt{2}}{3}, \tan\theta = -2\sqrt{2}$ **25.** $1 - 2\sin\theta\cos\theta$ **26.** $\frac{\cos^2\theta}{\sin\theta}$ **27.** $2|\cos\theta|$

For Problems 28–30, see the Solutions Manual.

CHAPTER 2

MATCHED PRACTICE PROBLEMS 2.1

1. $\sin A = \frac{4}{5}, \csc A = \frac{5}{4}, \cos A = \frac{3}{5}, \sec A = \frac{5}{3}, \tan A = \frac{4}{3}, \cot A = \frac{3}{4}$

2. It is impossible for the ratio to be 3/2 because $\cos\theta = \frac{\text{adj}}{\text{hyp}}$ and the hypotenuse has to be the longest side of the triangle.

3. a. $30°$ **b.** $25°$ **c.** $(90° - x)$ **4. a.** $\left(\frac{\sqrt{3}}{2}\right)^2 + \left(\frac{1}{2}\right)^2 = \frac{3}{4} + \frac{1}{4} = 1$ **b.** $\left(\frac{\sqrt{2}}{2}\right)^2 - \left(\frac{\sqrt{2}}{2}\right)^2 = \frac{1}{2} - \frac{1}{2} = 0$

5. a. $\frac{3\sqrt{3}}{2}$ **b.** 0 **c.** 5

PROBLEM SET 2.1

1. triangle measure **3.** complement

	sin *A*	cos *A*	tan *A*	cot *A*	sec *A*	csc *A*
5.	$\frac{4}{5}$	$\frac{3}{5}$	$\frac{4}{3}$	$\frac{3}{4}$	$\frac{5}{3}$	$\frac{5}{4}$
7.	$\frac{2\sqrt{5}}{5}$	$\frac{\sqrt{5}}{5}$	2	$\frac{1}{2}$	$\sqrt{5}$	$\frac{\sqrt{5}}{2}$
9.	$\frac{2}{3}$	$\frac{\sqrt{5}}{3}$	$\frac{2\sqrt{5}}{5}$	$\frac{\sqrt{5}}{2}$	$\frac{3\sqrt{5}}{5}$	$\frac{3}{2}$
11.	$\frac{5}{6}$	$\frac{\sqrt{11}}{6}$	$\frac{5\sqrt{11}}{11}$	$\frac{\sqrt{11}}{6}$	$\frac{5}{6}$	$\frac{\sqrt{11}}{5}$

	sin *A*	cos *A*	tan *A*	cot *A*	sec *A*	csc *A*
13.	$\frac{\sqrt{2}}{2}$	$\frac{\sqrt{2}}{2}$	1	$\frac{\sqrt{2}}{2}$	$\frac{\sqrt{2}}{2}$	1
15.	$\frac{3}{5}$	$\frac{4}{5}$	$\frac{3}{4}$	$\frac{4}{5}$	$\frac{3}{5}$	$\frac{4}{3}$
17.	$\frac{\sqrt{3}}{2}$	$\frac{1}{2}$	$\sqrt{3}$	$\frac{1}{2}$	$\frac{\sqrt{3}}{2}$	$\frac{\sqrt{3}}{3}$

19. $(4, 3)$, $\sin A = \frac{3}{5}$, $\cos A = \frac{4}{5}$, $\tan A = \frac{3}{4}$

21. If $\cos\theta = \frac{\text{side adjacent } \theta}{\text{hypotenuse}} = 3$, then the side adjacent to θ would have to be longer than the hypotenuse, which is not possible.

23. If we let the side adjacent to θ have length 1, then because $\tan\theta = \frac{\text{side opposite } \theta}{\text{side adjacent } \theta} = \frac{\text{side opposite } \theta}{1}$, we can make this ratio as large as we want by making the opposite side long enough.

25. 80° **27.** 82° **29.** $90° - x$ **31.** x

33.

x	sin x	csc x
0°	0	undefined
30°	$\frac{1}{2}$	2
45°	$\frac{\sqrt{2}}{2}$	$\sqrt{2}$
60°	$\frac{\sqrt{3}}{2}$	$\frac{2\sqrt{3}}{3}$
90°	1	1

35. 2 **37.** 3 **39.** 1 **41.** 0 **43.** $4 + 2\sqrt{3}$ **45.** 1 **47.** $2\sqrt{3}$ **49.** $-\frac{3\sqrt{3}}{2}$ **51.** $\sqrt{2}$ **53.** $\frac{2\sqrt{3}}{3}$ **55.** $\frac{2\sqrt{3}}{3}$

57. 1 **59.** $\sqrt{2}$ **61.** $\frac{\sqrt{3}}{3}$ **63.** 1

	sin *A*	cos *A*	sin *B*	cos *B*
65.	0.38	0.92	0.92	0.38
67.	0.96	0.28	0.28	0.96

69. $\frac{\sqrt{3}}{3}, \frac{\sqrt{6}}{3}$ **71.** $\frac{\sqrt{3}}{3}, \frac{\sqrt{6}}{3}$ **73.** $2\sqrt{5}$

75.

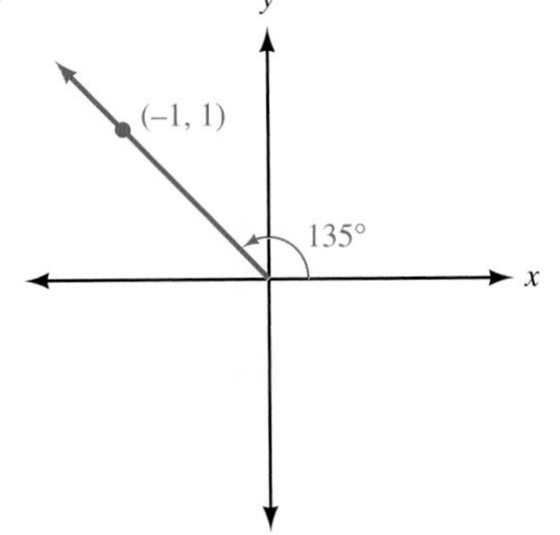

77. 225°

MATCHED PRACTICE PROBLEMS 2.2

1. 105° 3′ **2.** 47° 46′ **3.** 18°45′ **4.** 46.25° **5.** 0.4909 **6.** 0.5452 **7.** 0.8214 **8.** 1.1412
9. $\sin 52° = \cos 38° = 0.7880$ **10.** 68.3° **11.** 37.1° **12.** 75.31° **13.** 33°

PROBLEM SET 2.2

1. minutes, seconds **3.** value, angle **5.** 64° 9′ **7.** 89° 40′ **9.** 106° 49′ **11.** 55° 48′ **13.** 59° 43′ **15.** 53° 50′ **17.** 35° 24′ **19.** 16° 15′ **21.** 92° 33′ **23.** 19° 54′ **25.** 45.2° **27.** 62.6° **29.** 17.33° **31.** 48.45° **33.** 0.4571 **35.** 0.9511 **37.** 21.3634 **39.** 1.6643 **41.** 1.5003 **43.** 4.0906 **45.** 0.9100 **47.** 0.9083 **49.** 0.8355 **51.** 1.4370

53.

x	sin x	csc x
0°	0	Error
30°	0.5	2
45°	0.7071	1.4142
60°	0.8660	1.1547
90°	1	1

55. 12.3° **57.** 34.5° **59.** 78.9° **61.** 11.1° **63.** 33.3° **65.** 55.5° **67.** 65° 43′ **69.** 10° 10′ **71.** 8° 8′ **73.** 0.3907 **75.** 1.2134 **77.** 0.0787 **79.** 1 **81.** You get an error message. The sine of an angle can never exceed 1. **83.** You get an error message; tan 90° is undefined.

85. a.

x	tan x
87°	19.1
87.5°	22.9
88°	28.6
88.5°	38.2
89°	57.3
89.5°	114.6
90°	undefined

b.

x	tan x
89.4°	95.5
89.5°	114.6
89.6°	143.2
89.7°	191.0
89.8°	286.5
89.9°	573.0
90°	undefined

87. 18.4° **89.** $\sin\theta = -\frac{2\sqrt{13}}{13}$, $\cos\theta = \frac{3\sqrt{13}}{13}$, $\tan\theta = -\frac{2}{3}$ **91.** $\sin 90° = 1$, $\cos 90° = 0$, $\tan 90°$ is undefined

93. $\sin\theta = -\frac{12}{13}$, $\tan\theta = \frac{12}{5}$, $\cot\theta = \frac{5}{12}$, $\sec\theta = -\frac{13}{5}$, $\csc\theta = -\frac{13}{12}$ **95.** QII

MATCHED PRACTICE PROBLEMS 2.3

1. $B = 40°$, $a = 11$ cm, $b = 9.0$ cm **2.** $A = 34.7°$, $B = 55.3°$, $c = 6.22$ **3.** 19 inches **4.** 8.6 **5.** 110 ft

PROBLEM SET 2.3

1. left, right, first nonzero, not **3.** sides, angles **5. a.** 2 **b.** 3 **c.** 2 **d.** 2 **7. a.** 4 **b.** 6 **c.** 4 **d.** 4 **9.** 66 cm **11.** 39 m **13.** 2.19 cm **15.** 8.535 yd **17.** 37.5° **19.** 55° **21.** 59.20° **23.** $B = 65°$, $a = 10$ m, $b = 22$ m **25.** $B = 57.4°$, $b = 67.9$ in., $c = 80.6$ in. **27.** $B = 79° 18'$, $a = 1.121$ cm, $c = 6.037$ cm **29.** $A = 14°$, $a = 1.4$ ft, $b = 5.6$ ft **31.** $A = 63° 30'$, $a = 650$ mm, $c = 726$ mm **33.** $A = 66.55°$, $b = 2.356$ mi, $c = 5.921$ mi **35.** $A = 23°$, $B = 67°$, $c = 95$ ft **37.** $A = 50.12°$, $B = 39.88°$, $a = 451.6$ in. **39.** 49° **41.** 11 **43.** 16 **45.** $x = 79$, $h = 40$ **47.** 42° **49.** $x = 7.4$, $y = 6.2$ **51.** $h = 18$, $x = 11$ **53.** 17 **55.** 35.3° **57.** 35.3° **59.** 5.2° **61.** To the side, so that the shooter has a smaller angle in which to make a scoring kick. **63.** 200 ft

65. a. 160 ft **b.** 196 ft **c.** 40.8 ft **67.** 70.1° **69.** $\frac{1}{4}$ **71.** $-\frac{\sqrt{5}}{3}$

	sin θ	cos θ	tan θ	cot θ	sec θ	csc θ
73.	$\frac{\sqrt{3}}{2}$	$-\frac{1}{2}$	$-\sqrt{3}$	$-\frac{\sqrt{3}}{3}$	-2	$\frac{2\sqrt{3}}{3}$
75.	$-\frac{\sqrt{3}}{2}$	$-\frac{1}{2}$	$\sqrt{3}$	$\frac{\sqrt{3}}{3}$	-2	$-\frac{2\sqrt{3}}{3}$

77. The angle is approximately 60°.

MATCHED PRACTICE PROBLEMS 2.4

1. altitude is 10 cm, base is 19 cm **2.** 60° 20′ **3.** 195 m **4.** 14.0° **5.** S 14° E **6.** 171 mi north and 180 mi east **7.** 75 ft **8.** 4.3 mi

PROBLEM SET 2.4

1. elevation, depression **3.** north-south

5.

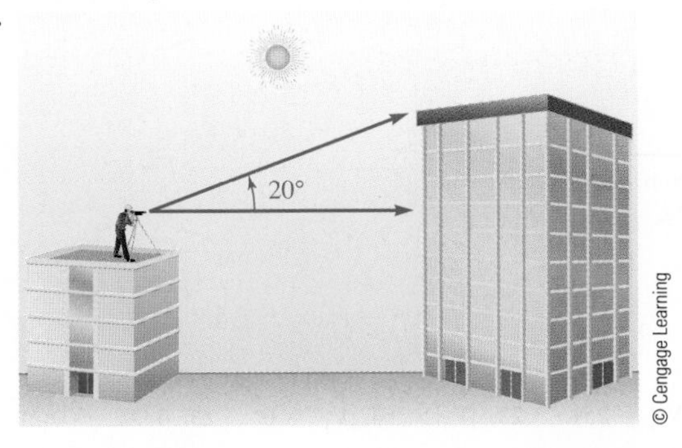

7.

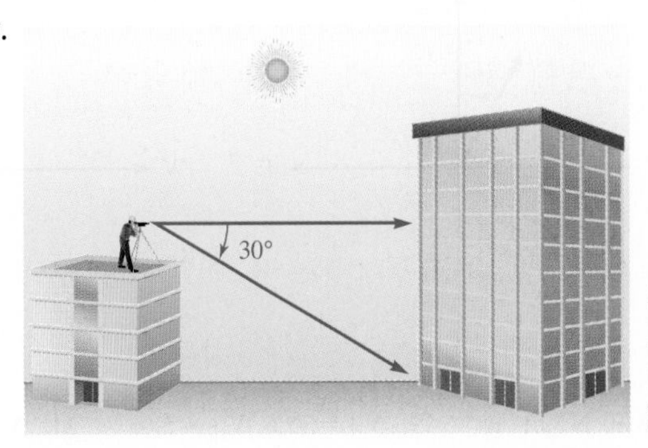

9.

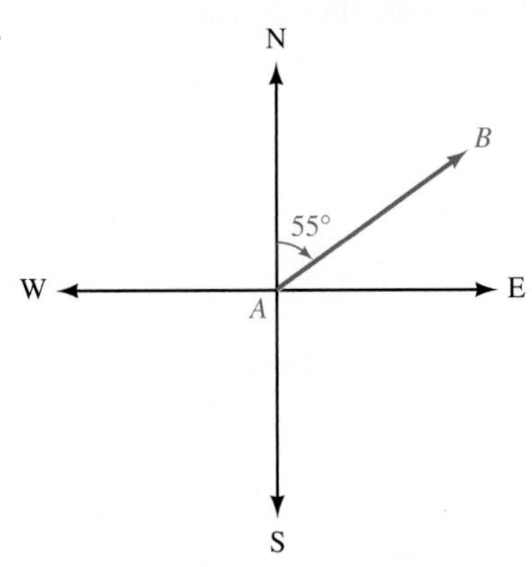

11.

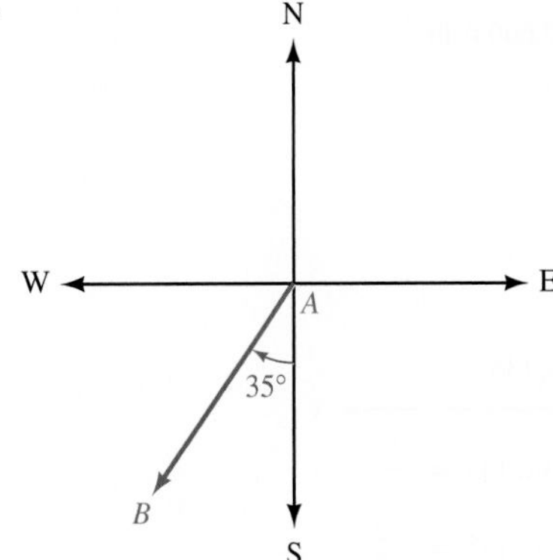

13. 39 cm, 68° **15.** 78.4° **17.** 39 ft **19.** 36.6° **21.** 55.1° **23.** 61 cm **25. a.** 800 ft **b.** 200 ft **c.** 14° **27.** 6.3 ft **29.** 31 mi, N 78° E **31.** 39 mi **33.** 63.4 mi north, 48.0 mi west **35.** 161 ft **37.** 78.9 ft **39.** 26 ft **41.** 4,000 mi **43.** 6.2 mi **45.** $\theta_1 = 45.00°$, $\theta_2 = 35.26°$, $\theta_3 = 30.00°$ **47.** $1 - 2 \sin\theta \cos\theta$

For Problems 49–54, see the Solutions Manual.

55. Answers will vary.

MATCHED PRACTICE PROBLEMS 2.5

1. 16 mph at N 72.3° E **2.** $|\mathbf{V}_x| = 27$ mph, $|\mathbf{V}_y| = 40$ mph **3.** 28 ft/sec at an angle of elevation of 41° **4.** 78 mi north and 56 mi east **5.** Tension is 39.1 lb, force exerted by Aaron is 22.4 lb **6.** 950 ft-lb

PROBLEM SET 2.5

1. scalar, vector **3.** resultant, diagonal **5.** horizontal, component, vertical, component
7. zero, static equilibrium

9.

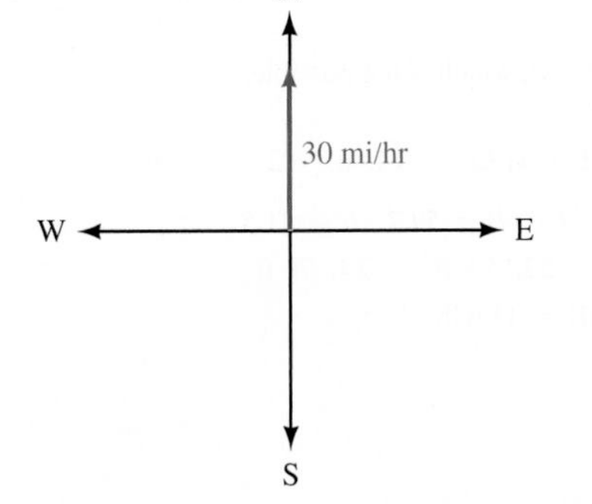

11.

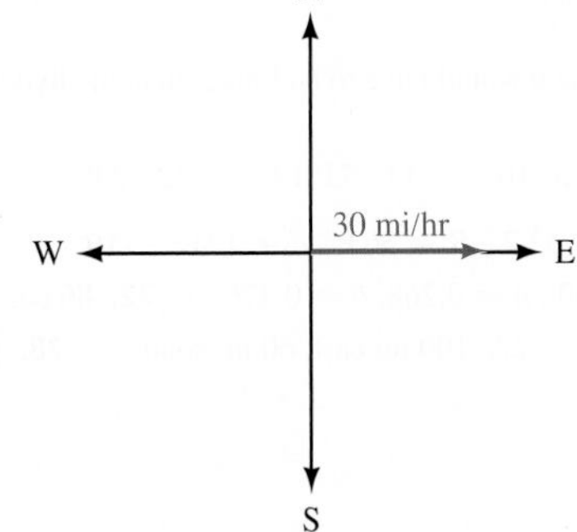

13. **15.**

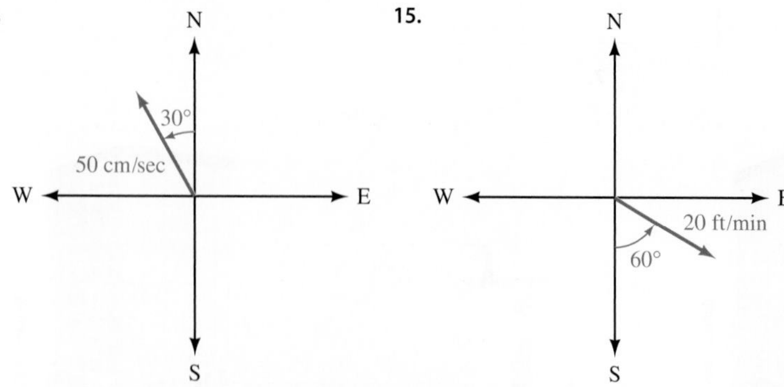

17. 15.3 mi, N 89.1° E **19.** $|\mathbf{V}_x| = 12.6$, $|\mathbf{V}_y| = 5.66$ **21.** $|\mathbf{V}_x| = 343$, $|\mathbf{V}_y| = 251$ **23.** $|\mathbf{V}_x| = 64$, $|\mathbf{V}_y| = 0$ **25.** 43.6
27. 5.9 **29.** 1.37 mi **31.** Both are 850 ft/sec. **33.** 2,550 ft **35.** 97 km south, 87 km east **37.** 38.1 ft/sec at an elevation of 23.2°
39. 240 mi north, 140 mi east **41.** $|\mathbf{H}| = 42.0$ lb, $|\mathbf{T}| = 59.4$ lb **43.** $|\mathbf{N}| = 7.7$ lb, $|\mathbf{F}| = 2.1$ lb **45.** $|\mathbf{F}| = 33.1$ lb
47. 2,900 ft-lb **49.** 7,600 ft-lb

51.

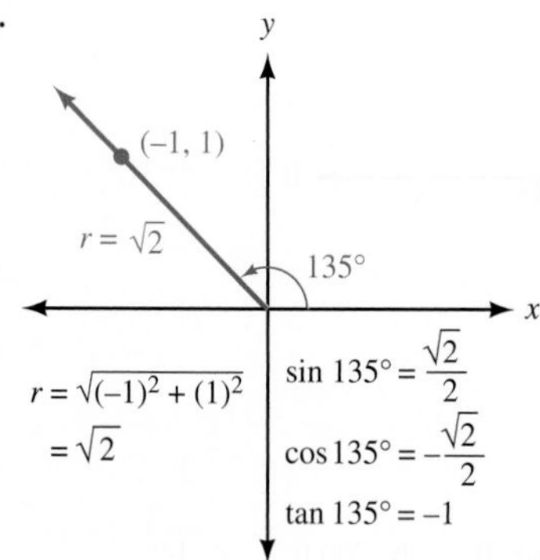

53. $\sin\theta = \dfrac{2\sqrt{5}}{5}$, $\cos\theta = \dfrac{\sqrt{5}}{5}$ **55.** $x = \pm 6$

CHAPTER 2 TEST

	sin A	cos A	tan A	sin B	cos B	tan B
1.	$\frac{\sqrt{5}}{5}$	$\frac{2\sqrt{5}}{5}$	$\frac{1}{2}$	$\frac{2\sqrt{5}}{5}$	$\frac{\sqrt{5}}{5}$	2
2.	$\frac{\sqrt{3}}{2}$	$\frac{1}{2}$	$\sqrt{3}$	$\frac{1}{2}$	$\frac{\sqrt{3}}{2}$	$\frac{\sqrt{3}}{3}$
3.	$\frac{3}{5}$	$\frac{4}{5}$	$\frac{3}{4}$	$\frac{4}{5}$	$\frac{3}{5}$	$\frac{4}{3}$

4. If $\sin\theta = \dfrac{\text{side opposite } \theta}{\text{hypotenuse}} = 2$, then the side opposite θ would have to be longer than the hypotenuse, which is not possible.

5. 76° **6.** $\frac{5}{4}$ **7.** 2 **8.** 0 **9.** $\frac{1}{2}$ **10.** 73° 10′ **11.** 73° 12′ **12.** 2.8° **13.** 0.4120 **14.** 0.7902
15. 0.3378 **16.** 71.2° **17.** 58.7° **18.** $A = 33.2°$, $B = 56.8°$, $c = 124$ **19.** $A = 30.3°$, $B = 59.7°$, $b = 41.5$
20. $A = 65.1°$, $a = 657$, $c = 724$ **21.** $B = 54° 30'$, $a = 0.268$, $b = 0.376$ **22.** 86 cm **23.** 5.8 ft **24.** 70 ft
25. 70° **26.** $|\mathbf{V}_x| = 380$ ft/sec, $|\mathbf{V}_y| = 710$ ft/sec **27.** 100 mi east, 60 mi south **28.** $|\mathbf{H}| = 45.6$ lb
29. $|\mathbf{F}| = 8.57$ lb **30.** 2,300 ft-lb

CHAPTER 3

MATCHED PRACTICE PROBLEMS 3.1

1. a. 40° **b.** 20° **c.** 35° **d.** 35° **e.** 70° **f.** 50° **2.** $-\frac{1}{2}$ **3.** 1 **4.** 2 **5.** $\frac{\sqrt{2}}{2}$ **6.** 153° **7.** 308.8° **8.** 240° **9.** 320° **10.** 151°

PROBLEM SET 3.1

1. reference, x **3.** positive

5.

y
x
150°
$\hat{\theta} = 30°$

7.

y
x
253.8°
$\hat{\theta} = 73.8°$

9.

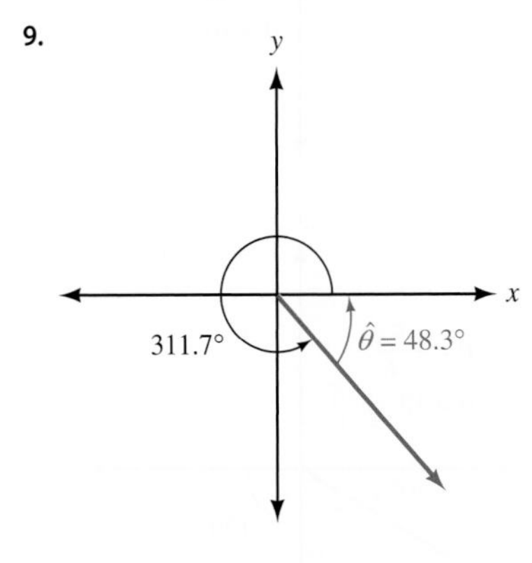

11.

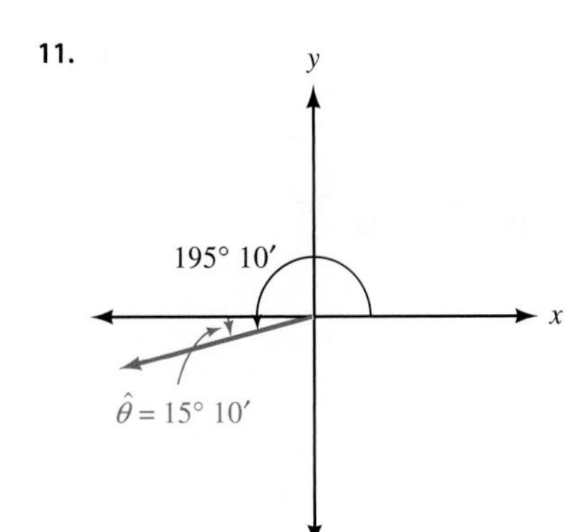

13.

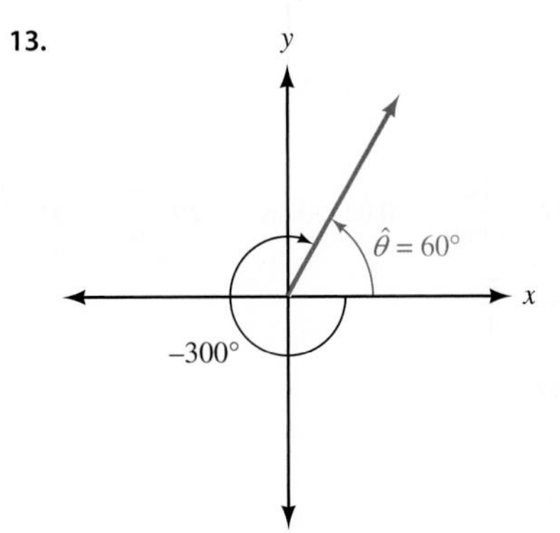

15.

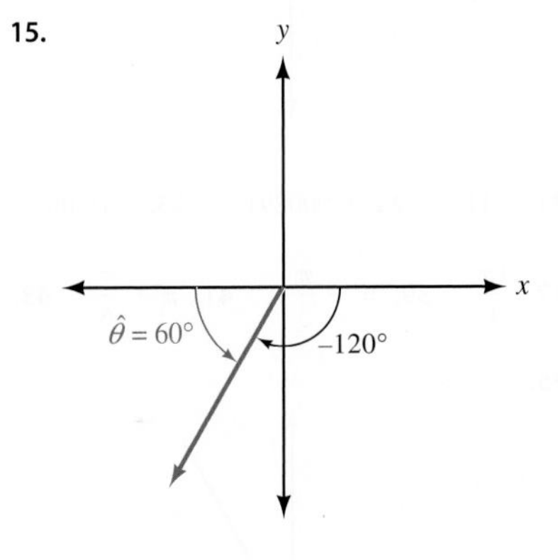

17. $-\frac{\sqrt{2}}{2}$ **19.** $-\frac{1}{2}$ **21.** −1 **23.** $-\frac{1}{2}$ **25.** $\frac{2\sqrt{3}}{3}$ **27.** $-\frac{2\sqrt{3}}{3}$ **29.** $\frac{1}{2}$ **31.** $-\frac{\sqrt{3}}{3}$ **33.** 0.9744 **35.** −4.8901 **37.** −0.7427 **39.** 1.5032 **41.** 1.7321 **43.** −1.7263 **45.** 0.7071 **47.** 0.2711 **49.** −1.3118 **51.** −0.8660 **53.** 198.0° **55.** 140.0° **57.** 210.5° **59.** 74.7° **61.** 105.2° **63.** 314.3° **65.** 156.4° **67.** 126.4° **69.** 236.0° **71.** 240° **73.** 135° **75.** 300° **77.** 240° **79.** 120° **81.** 135° **83.** 315° **85.** Complement is 20°, supplement is 110° **87.** Complement is $90° - x$, supplement is $180° - x$ **89.** Side opposite 30° is 5, side opposite 60° is $5\sqrt{3}$ **91.** $\frac{1}{4}$ **93.** 1

MATCHED PRACTICE PROBLEMS 3.2

1. 3 radians **2.** $\frac{5\pi}{18} \approx 0.873$ **3.** $\frac{43\pi}{18} \approx 7.50$ **4.** −0.829 **5.** 60° **6.** 420° **7.** −194.8° **8.** $\frac{\sqrt{3}}{2}$ **9.** $\frac{5\sqrt{2}}{2}$ **10.** $-\frac{3}{2}$ **11.** 1.0000

PROBLEM SET 3.2

1. radius **3. a.** θ **b.** $\pi - \theta$ **c.** $\theta - \pi$ **d.** $2\pi - \theta$ **5.** 3 **7.** $\frac{\pi}{4}$ **9.** 2 **11.** 0.1125 radian

13.

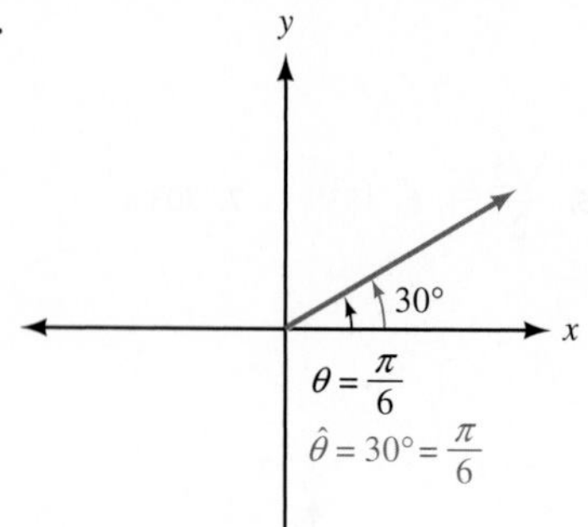

15.

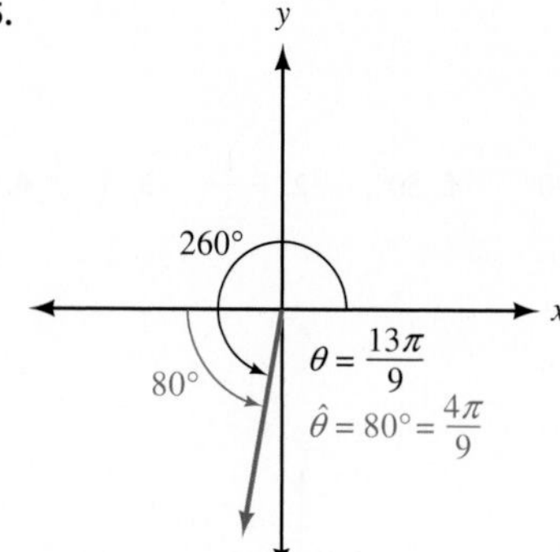

17.

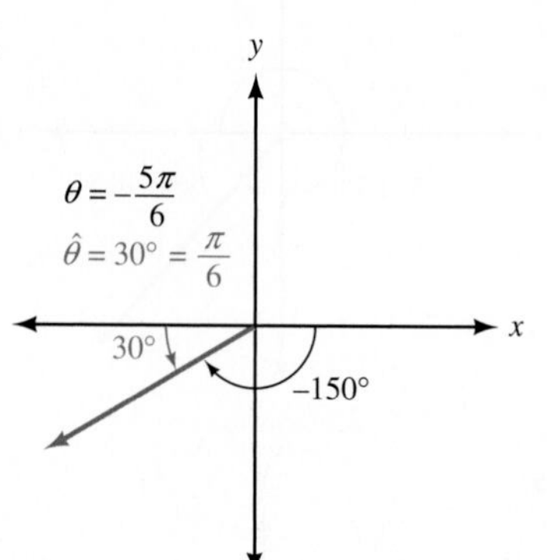

19.

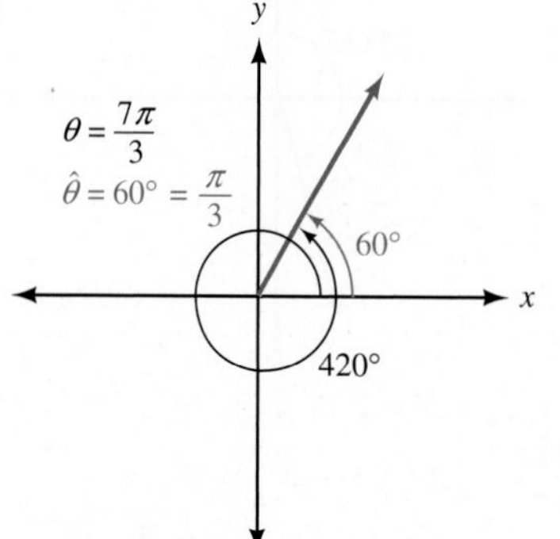

21. 2.11 **23.** 0.000291 **25.** 1.16 mi **27.** $\frac{7\pi}{1{,}080} \approx 0.02$ radian **29.** $\frac{\pi}{6}$ **31.** $\frac{2\pi}{3}$ **33.** $\frac{7\pi}{4}$ **35.** $\frac{2\pi}{3}$

37. $\frac{3\pi}{4}$ **39.** $\pi - \frac{\pi}{3}$ **41.** $\pi + \frac{\pi}{6}$ **43.** $2\pi - \frac{\pi}{4}$

45.

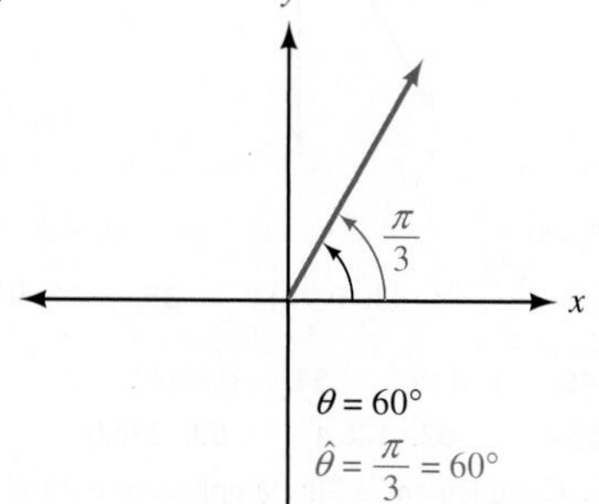

47.

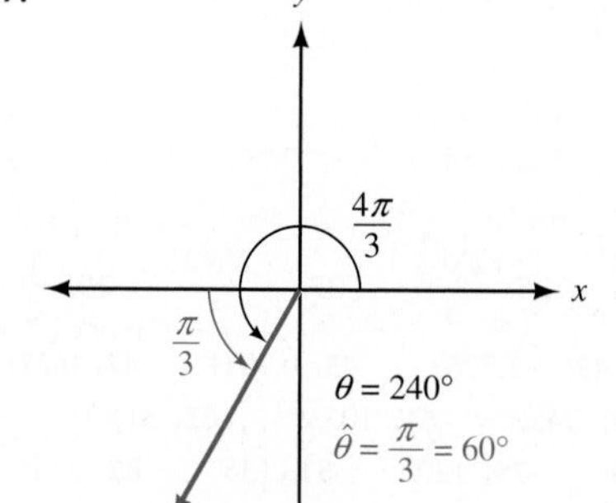

49.

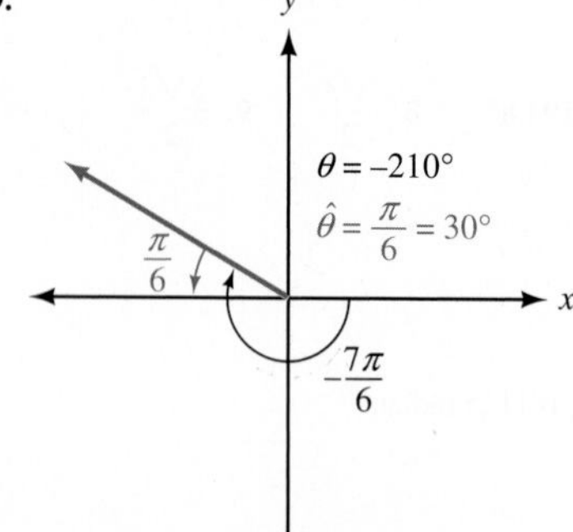

51.

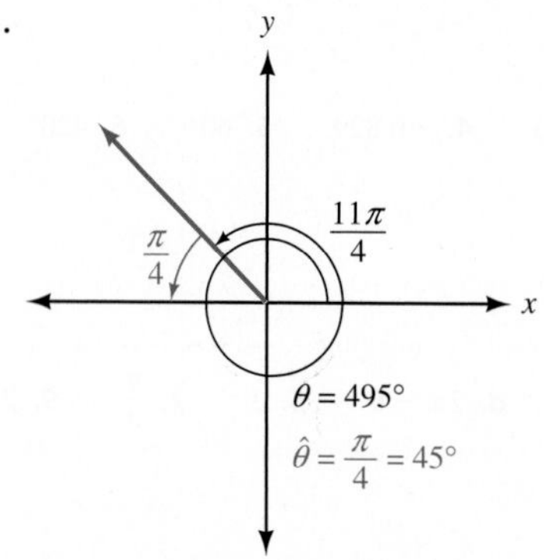

53. 57.3° **55.** 14.3° **57.** $-\frac{\sqrt{3}}{2}$ **59.** $\frac{\sqrt{3}}{3}$ **61.** -2 **63.** 2 **65.** $-\frac{1}{2}$ **67.** $2\sqrt{2}$ **69.** $\frac{\sqrt{3}}{2}$ **71.** $\frac{1}{4}$ **73.** $\frac{5}{2}$

75. $\frac{\sqrt{3}}{2}$ **77.** -2 **79.** $\frac{14}{3}$ **81.** $(0, 0), \left(\frac{\pi}{4}, \frac{\sqrt{2}}{2}\right), \left(\frac{\pi}{2}, 1\right), \left(\frac{3\pi}{4}, \frac{\sqrt{2}}{2}\right), (\pi, 0)$ **83.** $(0, -1), \left(\frac{\pi}{2}, 0\right), (\pi, 1), \left(\frac{3\pi}{2}, 0\right), (2\pi, -1)$

85. $\left(0, \frac{1}{2}\right), \left(\frac{\pi}{2}, 0\right), \left(\pi, -\frac{1}{2}\right), \left(\frac{3\pi}{2}, 0\right), \left(2\pi, \frac{1}{2}\right)$ **87.** $(0, 0), \left(\frac{\pi}{4}, 1\right), \left(\frac{\pi}{2}, 0\right), \left(\frac{3\pi}{4}, -1\right), (\pi, 0)$

89. $\left(\frac{\pi}{6}, 1\right), \left(\frac{\pi}{3}, \frac{\sqrt{3}}{2}\right), \left(\frac{2\pi}{3}, 0\right), \left(\pi, -\frac{\sqrt{3}}{2}\right), \left(\frac{7\pi}{6}, -1\right)$ **91.** $(0, 3), \left(\frac{\pi}{2}, 2\right), (\pi, 1), \left(\frac{3\pi}{2}, 2\right), (2\pi, 3)$

93. $\left(-\frac{\pi}{4}, 0\right), (0, 3), \left(\frac{\pi}{4}, 0\right), \left(\frac{\pi}{2}, -3\right), \left(\frac{3\pi}{4}, 0\right)$ **95.** $\frac{\pi}{11}$ **97.** 0.9965

	sin θ	cos θ	tan θ	cot θ	sec θ	csc θ
99.	$-\frac{3\sqrt{10}}{10}$	$\frac{\sqrt{10}}{10}$	-3	$-\frac{1}{3}$	$\sqrt{10}$	$-\frac{\sqrt{10}}{3}$
101.	$\frac{1}{2}$	$-\frac{\sqrt{3}}{2}$	$-\frac{\sqrt{3}}{3}$	$-\sqrt{3}$	$-\frac{2\sqrt{3}}{3}$	2
103.	$\frac{2\sqrt{5}}{5}$	$\frac{\sqrt{5}}{5}$	2	$\frac{1}{2}$	$\sqrt{5}$	$\frac{\sqrt{5}}{2}$

MATCHED PRACTICE PROBLEMS 3.3

1. $\sin\frac{2\pi}{3} = \frac{\sqrt{3}}{2}, \cos\frac{2\pi}{3} = -\frac{1}{2}, \tan\frac{2\pi}{3} = -\sqrt{3}$ **2.** $\frac{\pi}{6}, \frac{5\pi}{6}$ **3.** 0.6745

4. $\cos\frac{7\pi}{3} = \frac{1}{2}$, cosine is the function, $\frac{7\pi}{3}$ is the argument, and $\frac{1}{2}$ is the value.

5. -1.0565 **6. a.** Not possible **b.** Possible **c.** Not possible

7. When $t = 0$, $\tan t = 0$. When $t = \frac{\pi}{2}$, $\tan t$ is undefined. As t increases from 0 to $\frac{\pi}{2}$, $\tan t$ increases from 0 toward ∞.

PROBLEM SET 3.3

1. cosine, sine **3.** argument **5.** sine, cosine, tangent, cotangent **7.** $\frac{1}{2}$ **9.** 0 **11.** -2 **13.** $\frac{1}{2}$ **15.** $-\frac{\sqrt{3}}{3}$ **17.** $-\sqrt{2}$

	sin θ	cos θ	tan θ	cot θ	sec θ	csc θ
19.	$\frac{1}{2}$	$-\frac{\sqrt{3}}{2}$	$-\frac{\sqrt{3}}{3}$	$-\sqrt{3}$	$-\frac{2\sqrt{3}}{3}$	2
21.	$-\frac{\sqrt{3}}{2}$	$\frac{1}{2}$	$-\sqrt{3}$	$-\frac{\sqrt{3}}{3}$	2	$-\frac{2\sqrt{3}}{3}$
23.	0	-1	0	undefined	-1	undefined
25.	$\frac{\sqrt{2}}{2}$	$-\frac{\sqrt{2}}{2}$	-1	-1	$-\sqrt{2}$	$\sqrt{2}$

27. $\frac{\pi}{6}, \frac{5\pi}{6}$ **29.** $\frac{5\pi}{6}, \frac{7\pi}{6}$ **31.** $\frac{2\pi}{3}, \frac{5\pi}{3}$ **33.** $\sin\frac{4\pi}{3} \approx -0.8660, \cos\frac{4\pi}{3} = -0.5$ **35.** $\sin\frac{7\pi}{6} = -0.5, \cos\frac{7\pi}{6} \approx -0.8660$

37. 2.0944, 4.1888 **39.** 1.5708 **41.** $\sin 120° \approx 0.8660, \cos 120° = -0.5$ **43.** $\sin 225° \approx -0.7071, \cos 225° \approx -0.7071$

45. 210°, 330° **47.** 45°, 225° **49.** $\sin\theta = -\frac{2\sqrt{5}}{5}, \cos\theta = \frac{\sqrt{5}}{5}, \tan\theta = -2$ **51.** $\sin t = 0.8415, \cos t = 0.5403, \tan t = 1.5575$

53. $(-0.6536, 0.7568)$ **55.** 0.4 **57.** -10 **59.** 1 **61.** 2.5, 3.75 **63.** 5 **65.** $2A$ **67.** $x + \frac{\pi}{2}$

69. $-\frac{1}{2}$; function is cosine, argument is $\frac{2\pi}{3}$, value is $-\frac{1}{2}$ **71.** -0.7568 **73.** -1.6198 **75.** 1.4353 **77.** 1 **79.** $\frac{1}{2}$

81. cosecant and cotangent **83.** No **85.** Yes **87.** No **89.** No **91.** Yes **93.** No

95. The value of csc t is undefined at $t = 0$. For values of t near 0, csc t will be a large positive number. As t increases to $\pi/2$, csc t will decrease to 1.

97. The value of sin t will decrease from 1 to 0. **99.** See the Solutions Manual.

101. a. Close to 0 **b.** Close to 1 **c.** Close to 0 **d.** Becoming infinitely large **e.** Close to 1 **f.** Becoming infinitely large

103. The shortest length of OE occurs when point A is at either $\theta = 0$ or $\theta = \pi$, when this distance is the radius of the circle, or 1. For all other positions, OE is greater in length.

105. a. $\frac{\pi}{3}$ **b.** $\frac{\pi}{6}$ **c.** $\frac{\pi}{3}$ **107.** $B = 48°, a = 24, b = 27$ **109.** $A = 68°, a = 790, c = 850$ **111.** $A = 44.7°, B = 45.3°, b = 4.41$

MATCHED PRACTICE PROBLEMS 3.4

1. 17 inches **2.** 200 ft and 390 ft **3.** 3.8 cm **4.** 24,000 ft **5.** 9.8 m^2 **6.** $2\sqrt{2}$ cm **7.** 340 ft^2

PROBLEM SET 3.4

1. radius, angle **3.** sector **5.** 6 in. **7.** 2.25 ft **9.** 2π cm $\approx$ 6.28 cm **11.** $\frac{2\pi}{3}$ mm $\approx$ 2.09 mm **13.** $\frac{35\pi}{4}$ in. $\approx$ 27.5 in. **15.** 5.03 cm **17.** 4,400 mi **19.** $\frac{4\pi}{9}$ ft $\approx$ 1.40 ft **21.** 33.0 feet **23.** 1.92 radians, 110° **25.** 2,100 mi **27.** 65.4 ft **29.** 480 ft **31. a.** 103 ft **b.** 361 ft **c.** 490 ft **33.** 0.5 ft **35.** 3 in. **37.** 4 cm **39.** 1 m **41.** $\frac{6}{\pi}$ km $\approx$ 1.91 km **43.** 9 cm^2 **45.** 19.2 $in.^2$ **47.** $\frac{9\pi}{5}$ $m^2 \approx$ 5.65 m^2 **49.** $\frac{25\pi}{24}$ $m^2 \approx$ 3.27 m^2 **51.** $\frac{18}{\pi}$ $ft^2 \approx$ 5.73 ft^2 **53.** 2 cm **55.** $\frac{4\sqrt{3}}{3}$ in. $\approx$ 2.31 in. **57.** 900π $ft^2 \approx$ 2,830 ft^2 **59.** $\frac{350\pi}{11}$ mm $\approx$ 100 mm **61. a.** 29.3 mm **b.** 42.6 mm^2 **63.** 60.2° **65.** 2.31 ft **67.** 74.0° **69.** 0.009 radian $\approx$ 0.518° **71.** The sun is also about 400 times farther away from the earth.

MATCHED PRACTICE PROBLEMS 3.5

1. 1.6 cm/sec **2.** $\frac{\pi}{6}$ rad/sec **3.** 4,500 inches **4.** $d = 12 \tan 4\pi t, l = 12 \sec 4\pi t$ **5.** $\omega = 66\pi$ rad/min and $v = 69.1$ ft/min **6. a.** 0.71 mi/hr **b.** $H = 114 - 100 \cos\left(\frac{\pi}{5}t\right)$

PROBLEM SET 3.5

1. uniform circular **3.** proportional, twice **5.** 1.5 ft/min **7.** 3 cm/sec **9.** 15 mi/hr **11.** 80 ft **13.** 22.5 mi **15.** 7 mi **17.** 4 rad/min **19.** $\frac{8}{3}$ rad/sec $\approx$ 2.67 rad/sec **21.** 37.5π rad/hr $\approx$ 118 rad/hr **23.** $d = 100 \tan\left(\frac{\pi}{2}t\right)$; when $t = \frac{1}{2}$, $d = 100$ ft; when $t = \frac{3}{2}$, $d = -100$ ft; when $t = 1$, d is undefined because the light rays are parallel to the wall. **25.** 40 in. **27.** 180π m $\approx$ 565 m **29.** 7,200 ft **31.** 20π rad/min $\approx$ 62.8 rad/min **33.** $\frac{200\pi}{3}$ rad/min $\approx$ 209 rad/min **35.** 11.6π rad/min $\approx$ 36.4 rad/min **37.** 10 in./sec **39.** 0.5 rad/sec **41.** 80π ft/min $\approx$ 251 ft/min **43.** $\frac{\pi}{12}$ rad/hr $\approx$ 0.262 rad/hr **45.** 300π ft/min $\approx$ 942 ft/min **47.** 9.50 mi/hr **49.** 24.0 rpm **51.** 5.65 ft/sec **53.** 0.47 mi/hr **55.** $h = 110.5 - 98.5 \cos\left(\frac{2\pi}{15}t\right)$ **57.** 889 rad/min (53,300 rad/hr) **59.** 12 rps **61.** See the Solutions Manual. **63.** 18.8 km/hr **65.** 52.3 mm **67.** 80.8 rpm **69.** $|V_x| = 54.3$ ft/sec, $|V_y| = 40.9$ ft/sec **71.** 71.9 mi west, 46.2 mi south

CHAPTER 3 TEST

1.

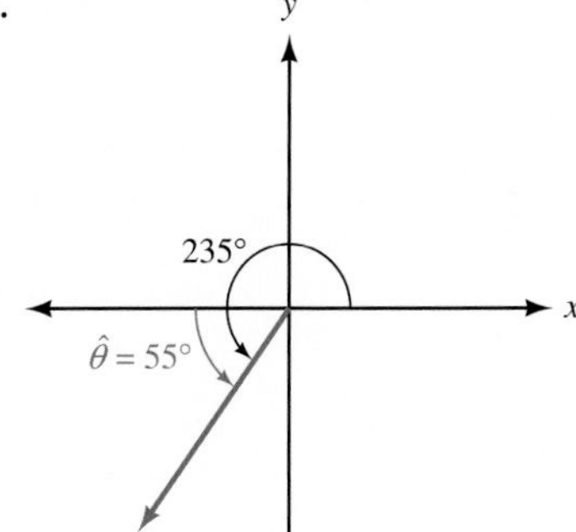

2.

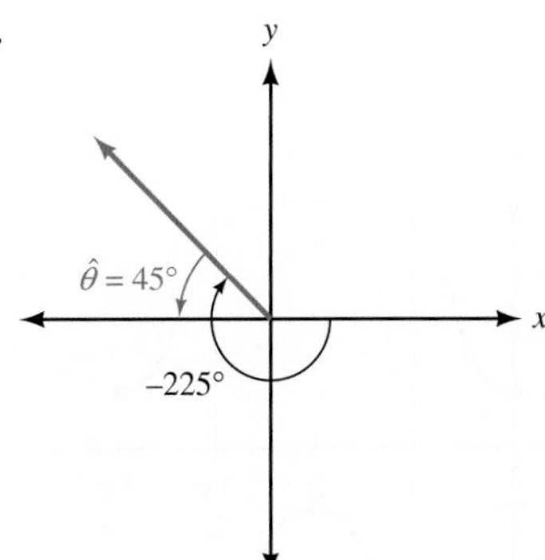

3. -1.1918 **4.** 1.1964 **5.** -1.2991 **6.** $174.0°$ **7.** $226.0°$ **8.** $-\frac{\sqrt{2}}{2}$ **9.** $-\frac{\sqrt{3}}{3}$ **10.** $\frac{25\pi}{18}$ **11.** $105°$

12. $-\frac{\sqrt{2}}{2}$ **13.** $-\frac{2\sqrt{3}}{3}$ **14.** $\sin t = -\frac{3\sqrt{13}}{13}, \cos t = \frac{2\sqrt{13}}{13}, \tan t = -\frac{3}{2}$ **15.** $4x$ **16.** 0.2837 **17.** No

18. 0 **19.** 2π ft ≈ 6.28 ft **20.** 4 cm **21.** 10.8 cm^2 **22.** 10π ft ≈ 31.4 ft **23.** 8 in.^2 **24.** 72 in.

25. 0.5 rad/sec **26.** 80π ft/min **27.** 4 rad/sec for the 6-cm pulley and 3 rad/sec for the 8-cm pulley

28. $2{,}700\pi$ ft/min $\approx 8{,}480$ ft/min **29.** 22.0 rpm **30.** 41.6 km/hr

CUMULATIVE TEST CHAPTERS 1–3

1. $x = 6, h = 3\sqrt{3}, s = 3\sqrt{3}, r = 3\sqrt{6}$ **2.** $120°$ **3.** $\sqrt{a^2 + b^2}$ **4.** $120° + 360°k$ for any integer k **5.** QIV

6. $\sin\theta = \frac{4}{5}, \cos\theta = -\frac{3}{5}, \tan\theta = -\frac{4}{3}, \cot\theta = -\frac{3}{4}, \sec\theta = -\frac{5}{3}, \csc\theta = \frac{5}{4}$

7. $\sin\theta = -\frac{12}{13}, \cos\theta = -\frac{5}{13}, \cot\theta = \frac{5}{12}, \sec\theta = -\frac{13}{5}, \csc\theta = -\frac{13}{12}$ **8.** $-\frac{1}{2}$ **9.** $\sin^2\theta - 4\sin\theta - 21$

10. A solution is given in the Solutions Manual. **11.** $\sin A = \frac{5}{13}, \cos A = \frac{12}{13}, \tan A = \frac{5}{12}, \sin B = \frac{12}{13}, \cos B = \frac{5}{13}, \tan B = \frac{12}{5}$

12. $17°$ **13.** $9°43'$ **14.** $4.7°$ **15. a.** 2 **b.** 2 **c.** 4 **16.** $A = 42°, a = 240, c = 360$ **17.** 42.4 mi, N 76.4° E

18. 62.3 ft **19.** $|\mathbf{V}_x| = 4.3, |\mathbf{V}_y| = 2.5$ **20.** $|\mathbf{H}| = 43.7$ lb, $|\mathbf{T}| = 91.6$ lb

21.

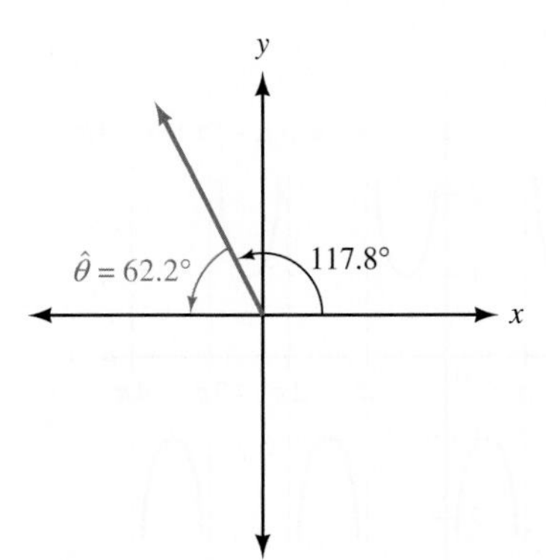

22. $241.5°$ **23.** $-\frac{13\pi}{6}$ **24.** $\frac{\sqrt{3}}{2}$ **25.** $\frac{2\pi}{3}, \frac{4\pi}{3}$

26. When $t = \pi/2$, $\tan t$ is undefined. For values of t slightly larger than $\pi/2$, $\tan t$ will be a large negative number. As t increases to π, $\tan t$ will increase to 0. **27.** 6.28 m **28.** $4\pi \text{ inches}^2$ **29.** 120π ft **30.** 26.2 mi/hr

CHAPTER 4

MATCHED PRACTICE PROBLEMS 4.1

1.

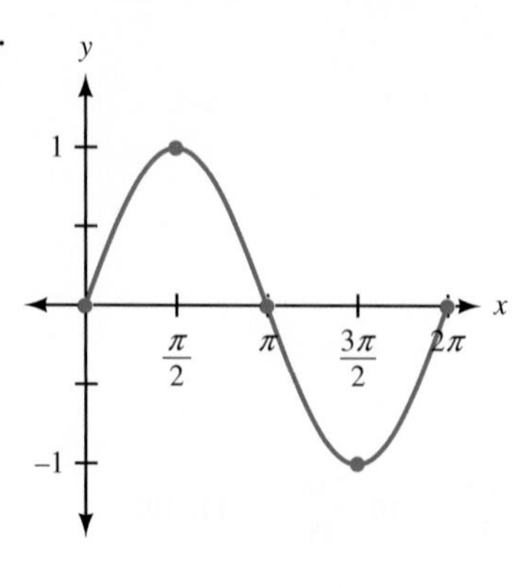

2.

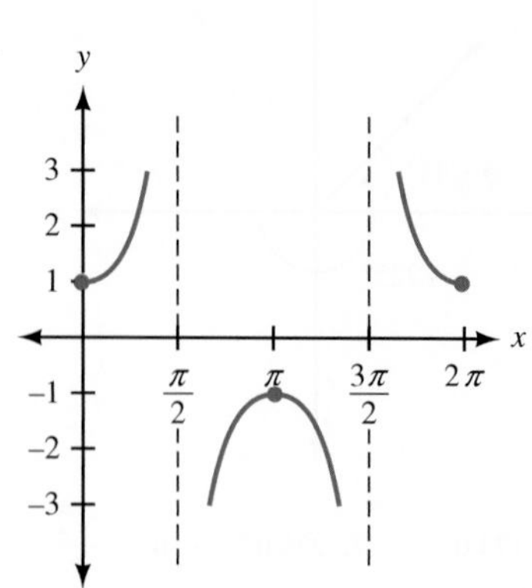

3. $\sec(-x) = \dfrac{1}{\cos(-x)} = \dfrac{1}{\cos x} = \sec x$ **4. a.** $-\dfrac{1}{2}$ **b.** $-\dfrac{2\sqrt{3}}{3}$

PROBLEM SET 4.1

1. y-coordinate, arc length **3.** difference, greatest, least **5.** equal, opposite **7.** sine and cosine **9.** secant and cosecant **11.** sine, cosine, secant, and cosecant

13.

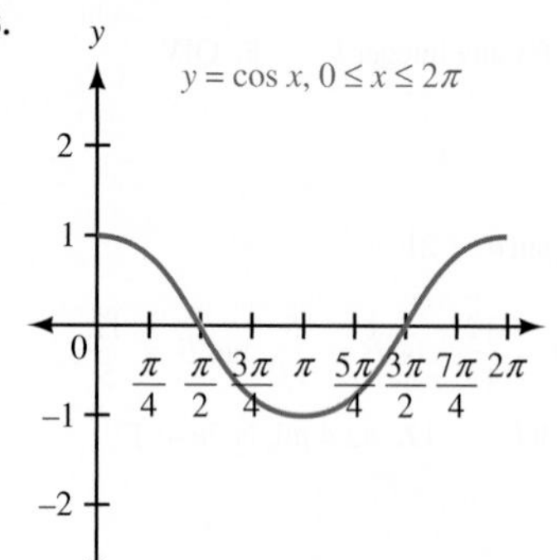

15.

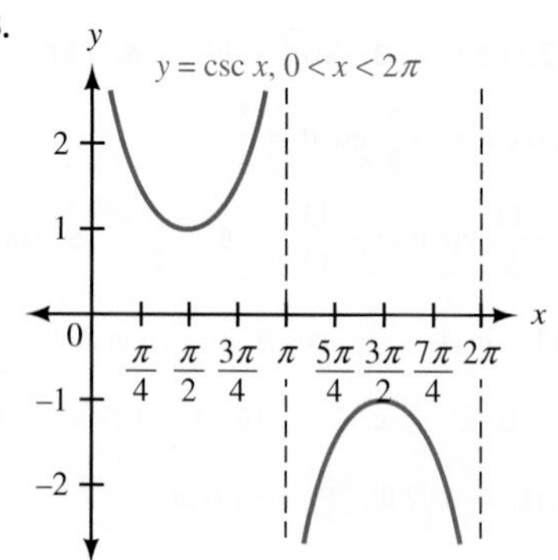

17.

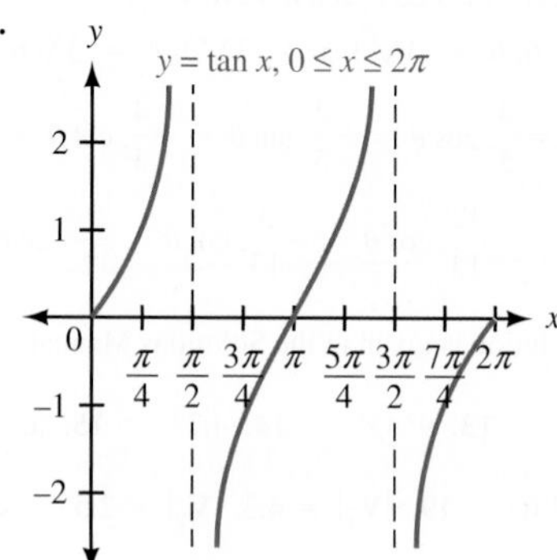

19.

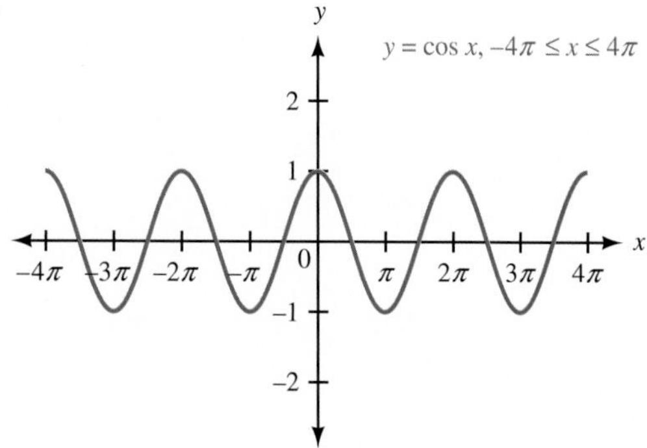

21.

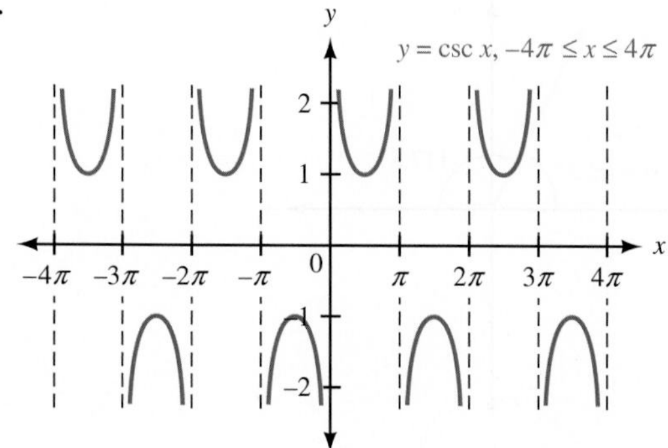

23.

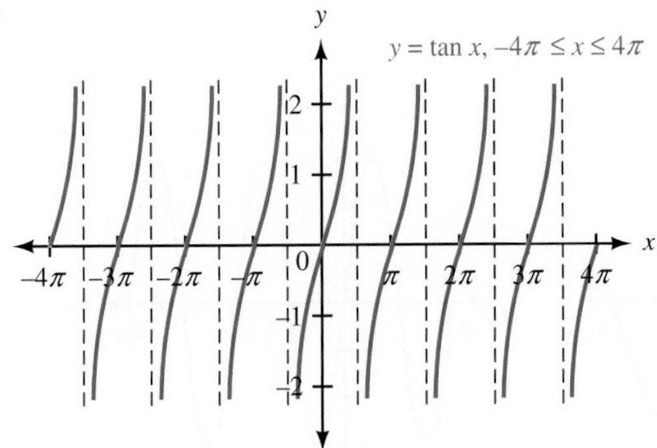

25. $0, \pi, 2\pi$ **27.** $\frac{\pi}{2}$ **29.** $0, \pi, 2\pi$ **31.** $0, 2\pi$

33. $\frac{\pi}{2}, \frac{3\pi}{2}$ **35.** $0, \pi, 2\pi$ **37.** $\frac{1}{2}$ **39.** $-\frac{\sqrt{3}}{2}$

41. $-\frac{1}{2}$ **43.** $-\frac{\sqrt{2}}{2}$ **45.** $\frac{1}{3}$

47.

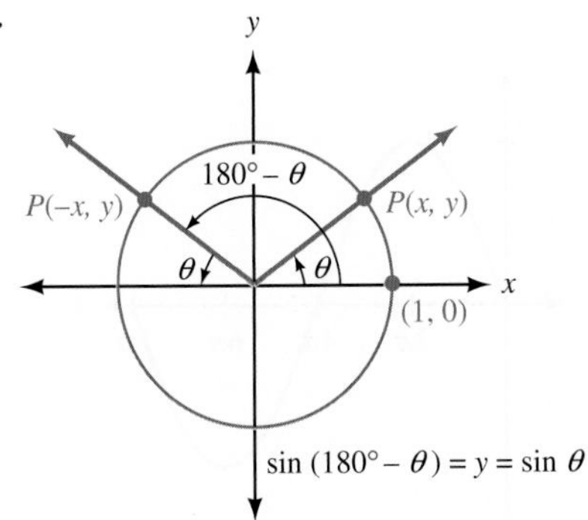

For Problems 48–60, see the Solutions Manual.

61. 120° **63.** 330°

65.

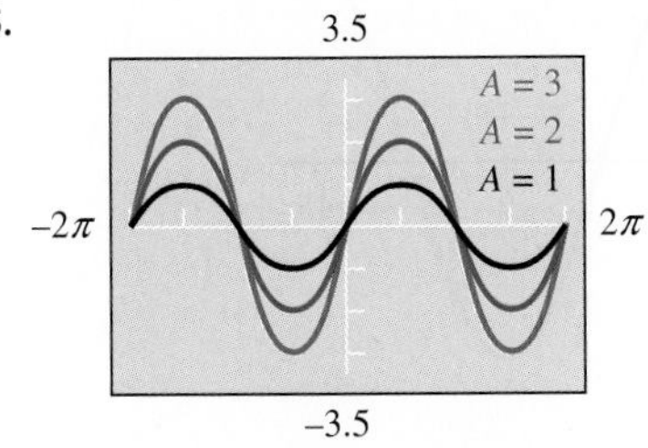

The amplitude is increased.

67.

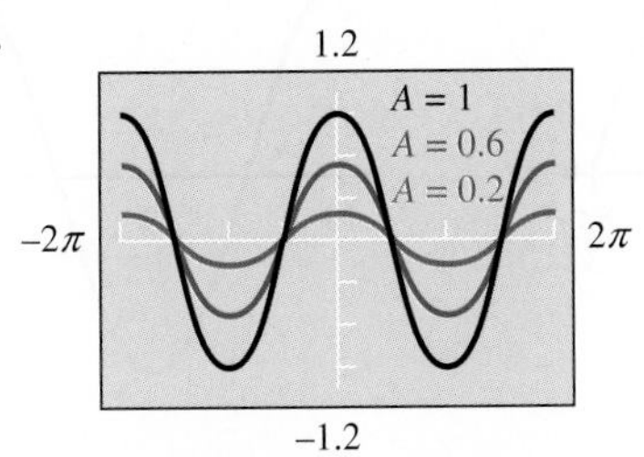

The amplitude is decreased.

69.

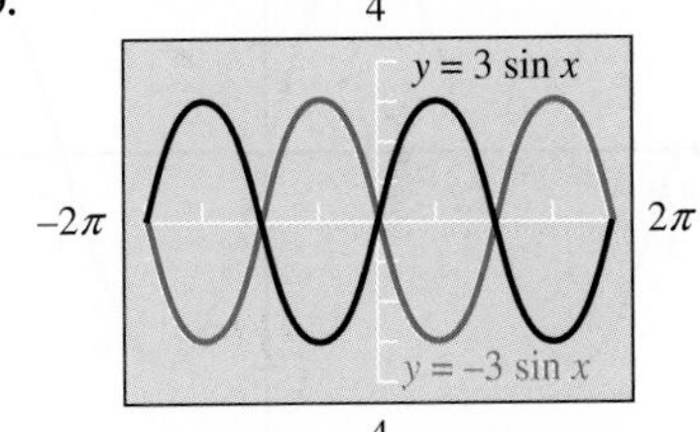

The graph is reflected about the x-axis.

71.

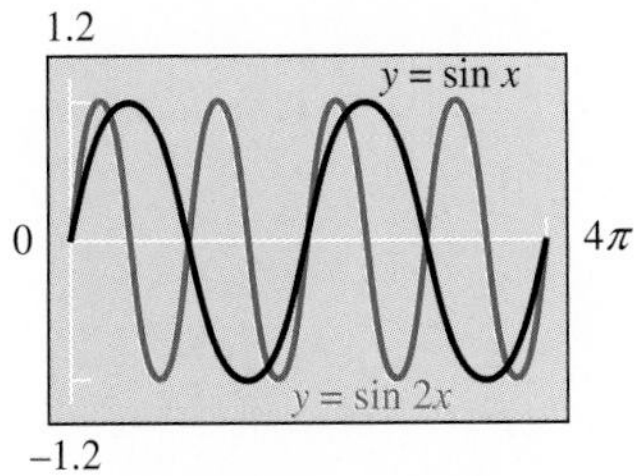

The period is halved.

73.

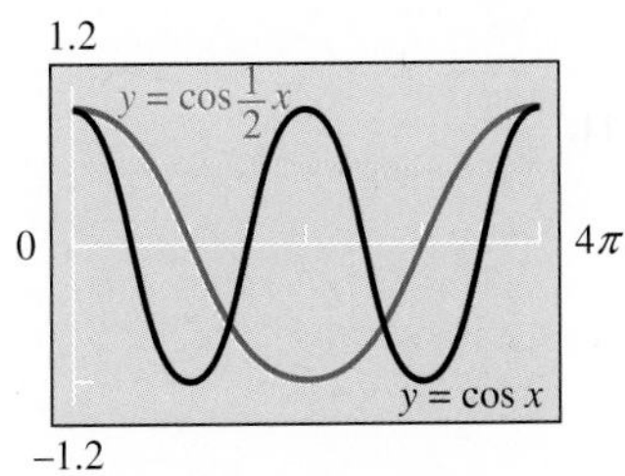

The period is doubled.

MATCHED PRACTICE PROBLEMS 4.2

1.

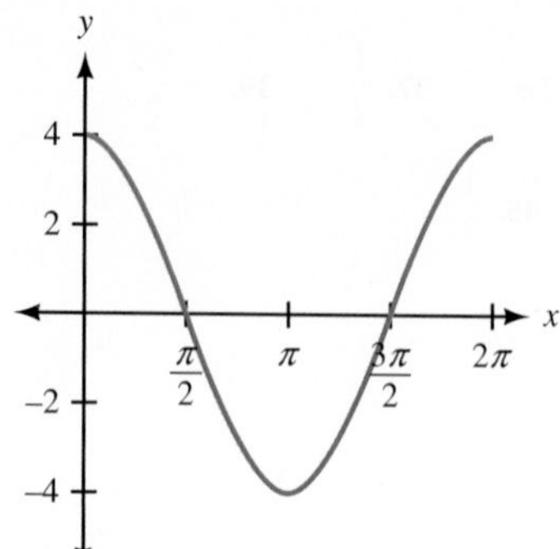

2.

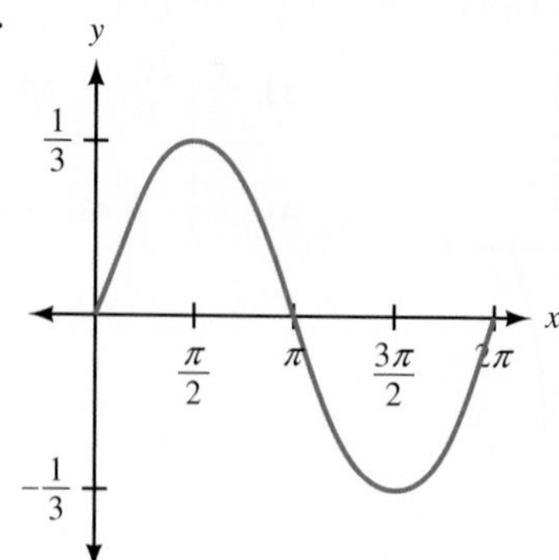

3.

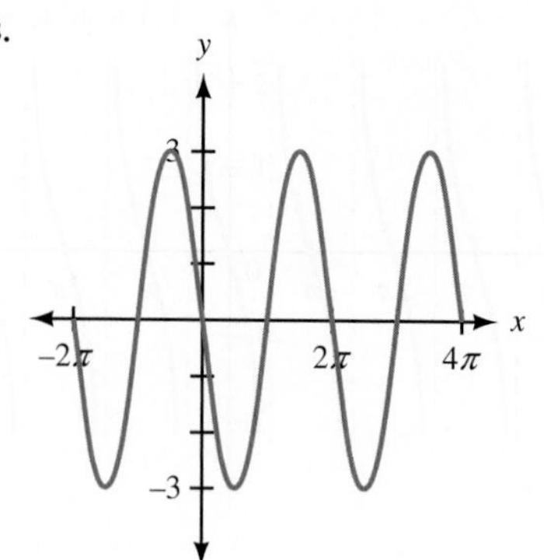

4.

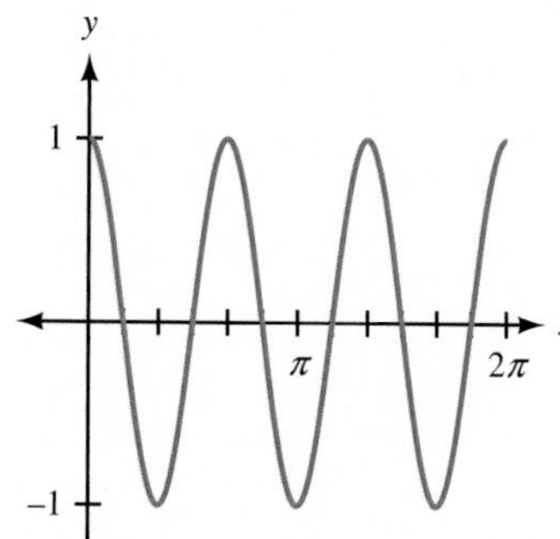

5.

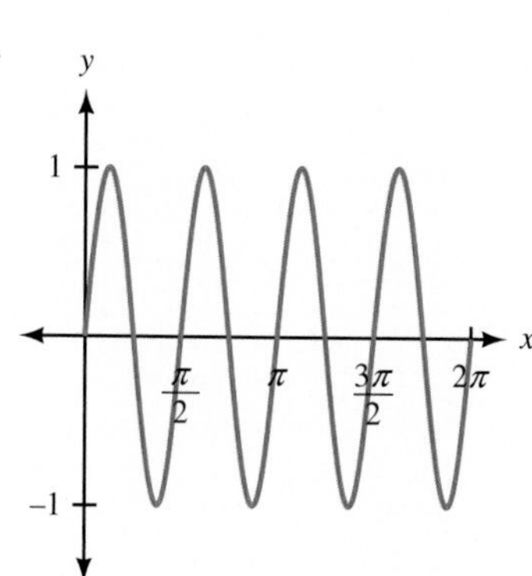

6.

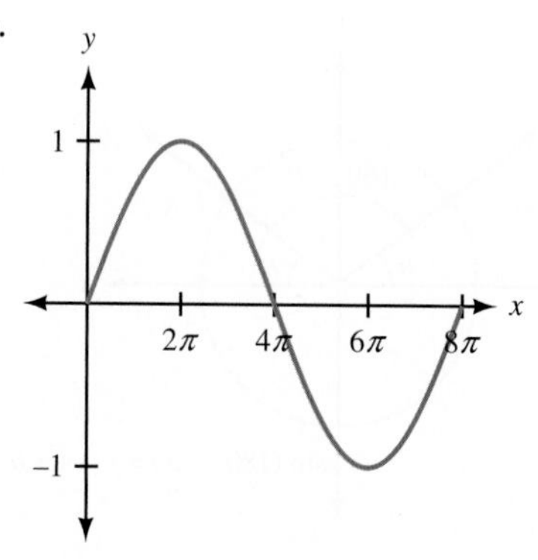

7.

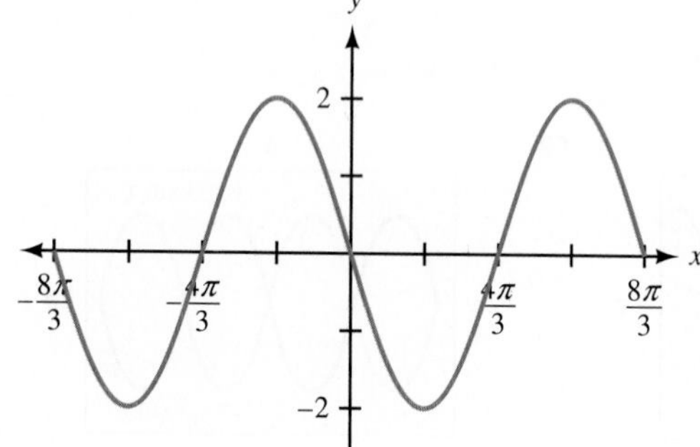

8.

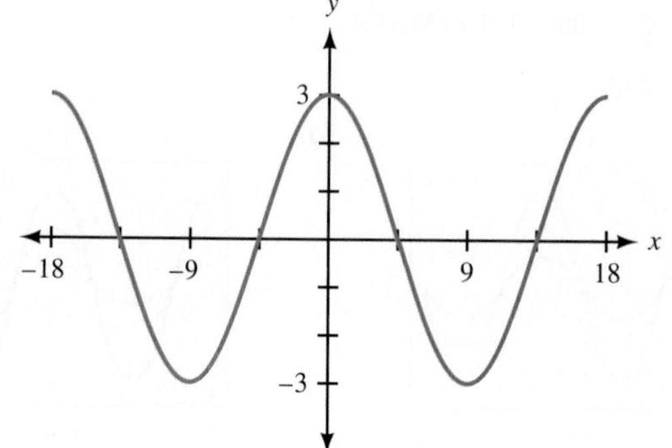

PROBLEM SET 4.2

1. $|A|$, $[-|A|, |A|]$ **3.** 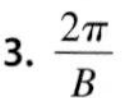 $\frac{2\pi}{B}$ **5.** $-\sin Bx$, odd

7.

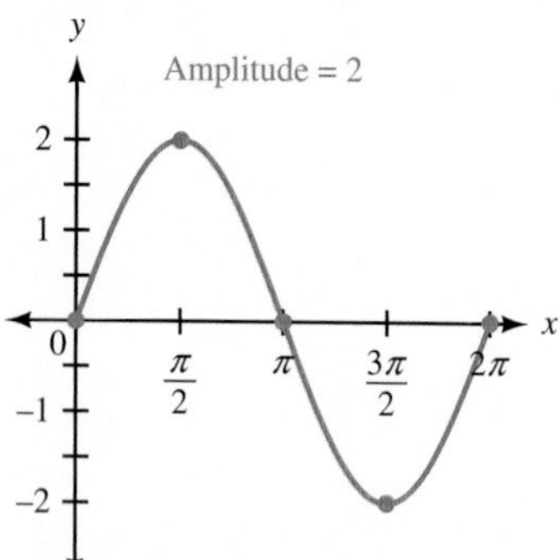

9. 5 **11.** $\frac{1}{4}$

13.

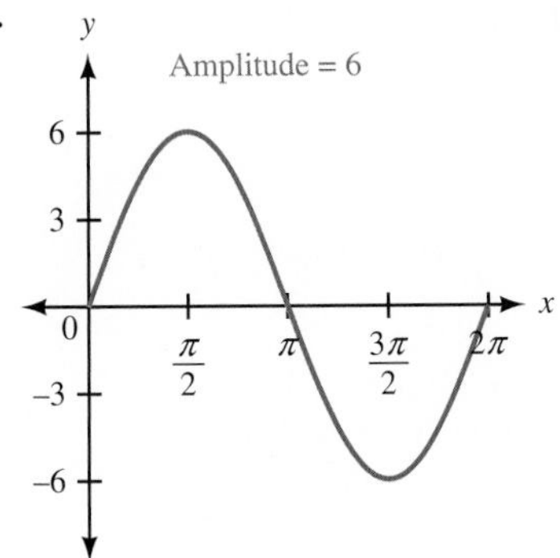

15.

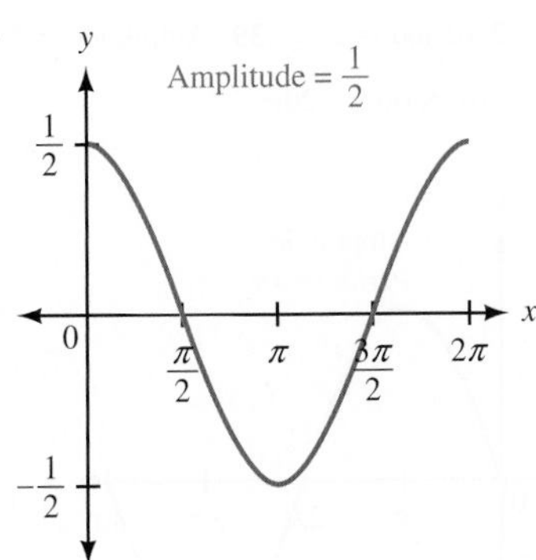

17.

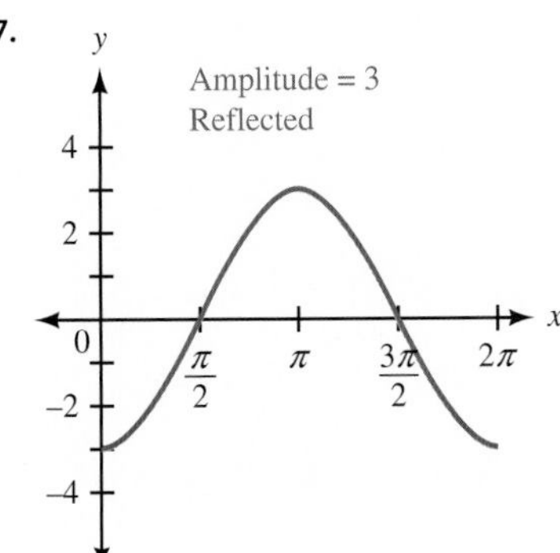

19.

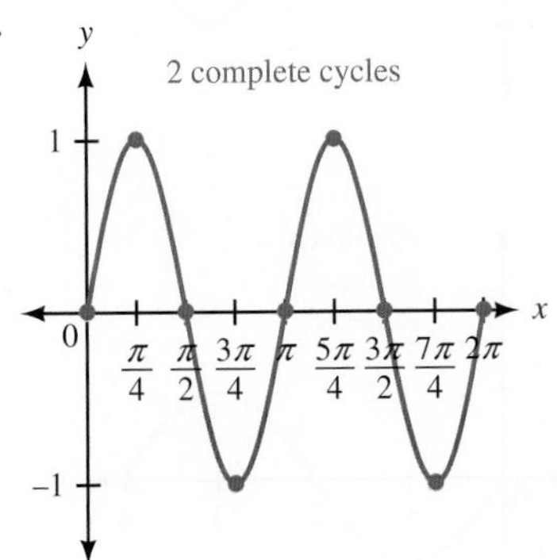

21. $\frac{\pi}{2}$ **23.** 12π **25.** 1

27.

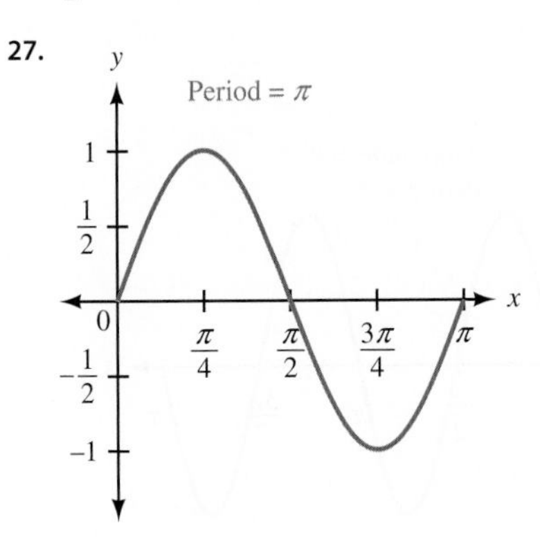

29.

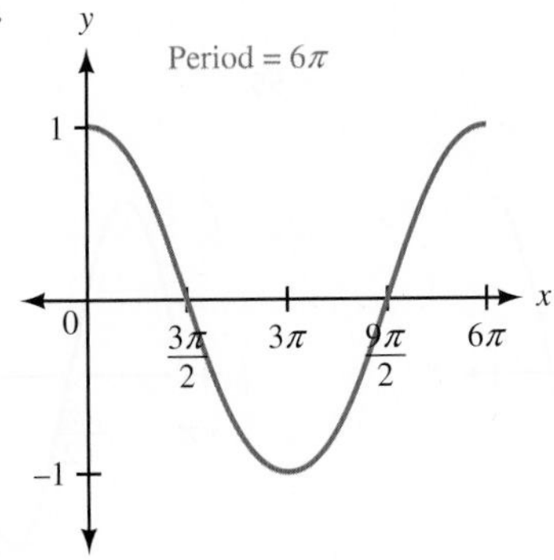

31.

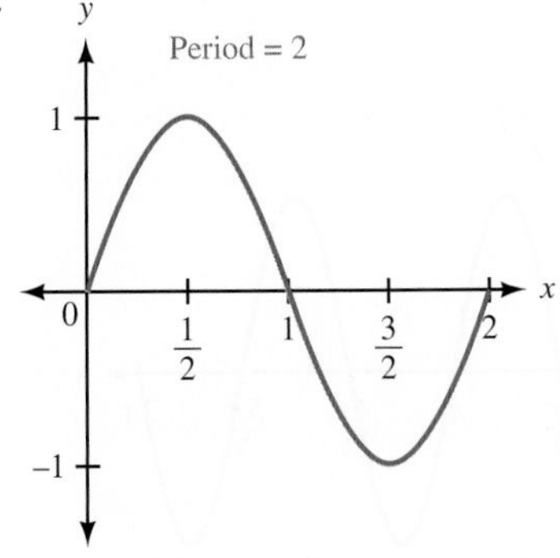

33.

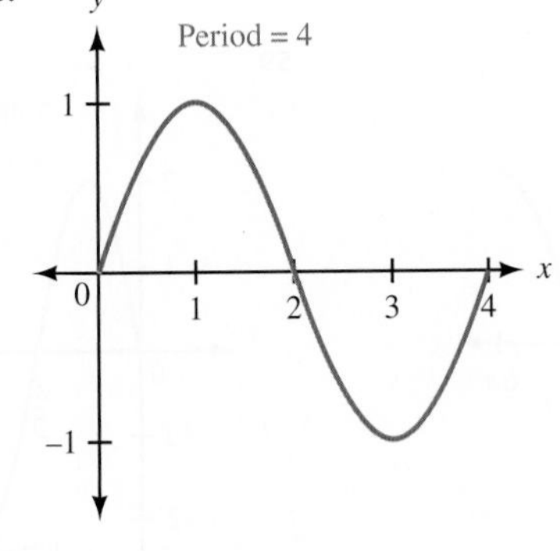

35. Amplitude = 4, period = 4π **37.** Amplitude = 2, period = 2 **39.** Amplitude = 5, period = 3

41. Amplitude = $\frac{1}{3}$, period = $\frac{2\pi}{3}$ **43.** Amplitude = 10, period = 20π

45.

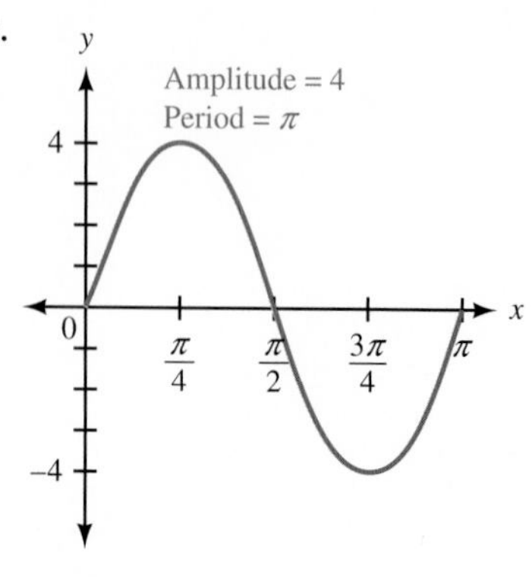

47.

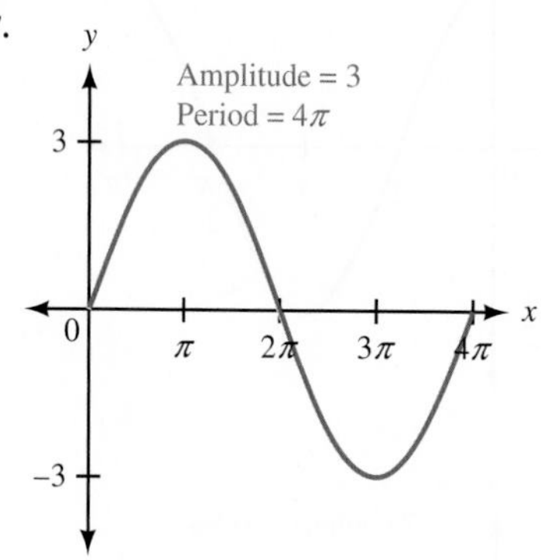

49.

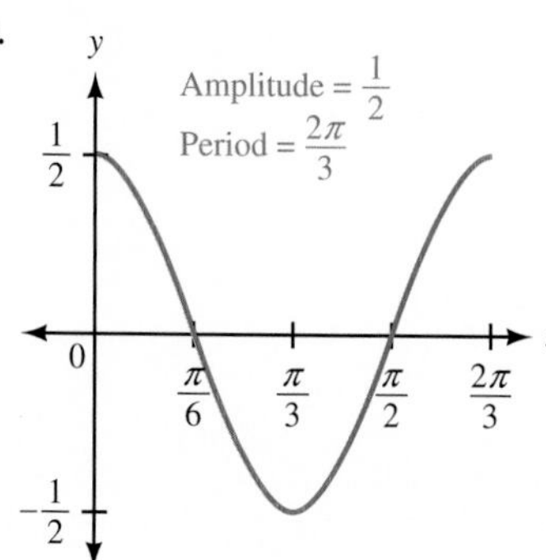

51.

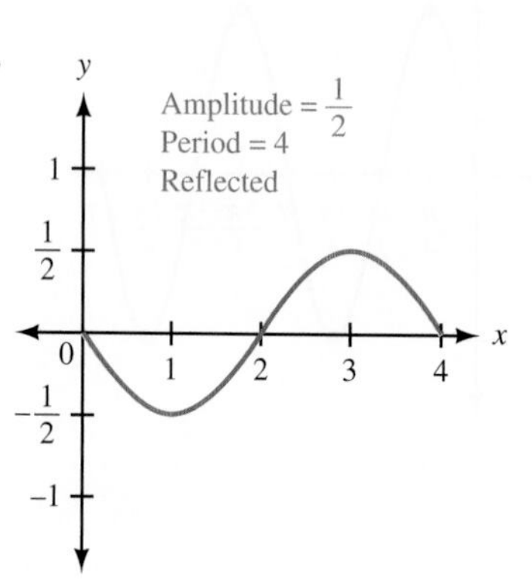

53.

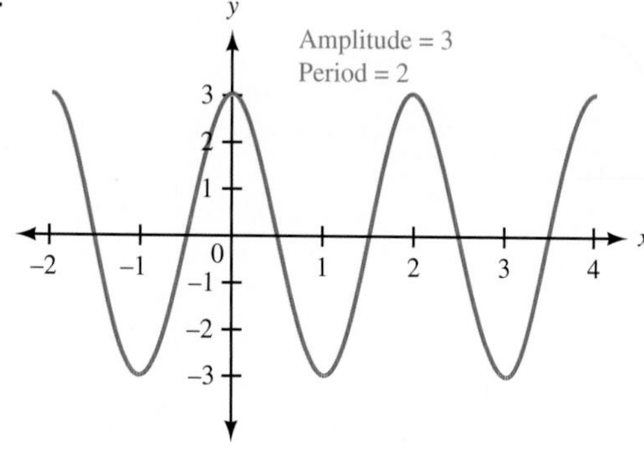

55.

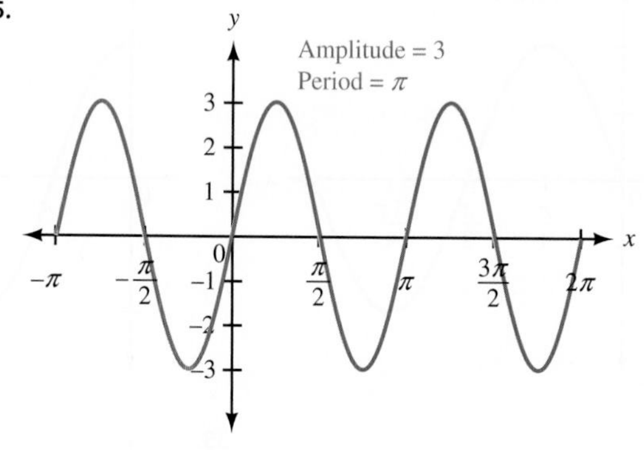

57.

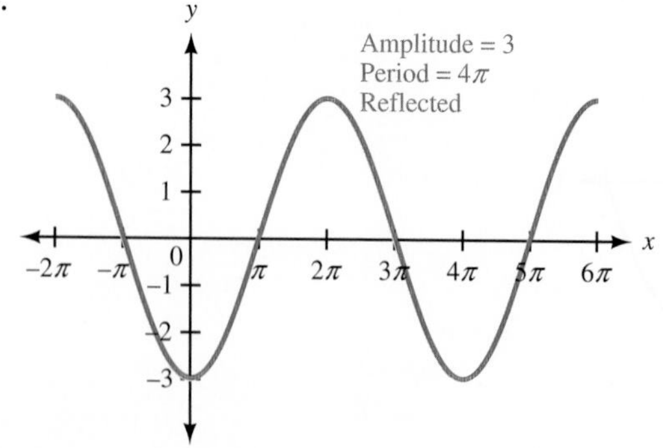

59.

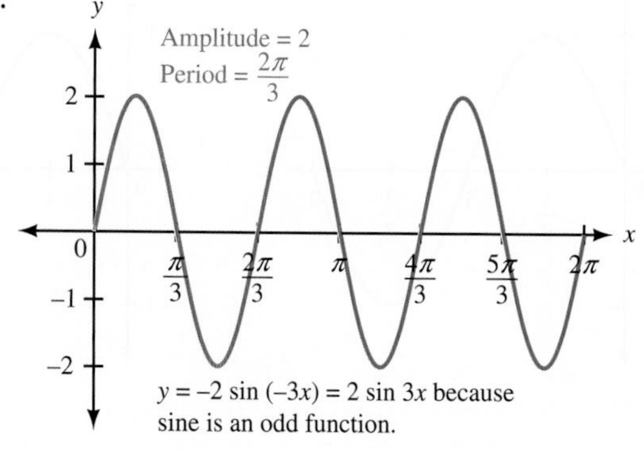

61. Maximum value of I is 20 amperes; one complete cycle takes 1/60 second.

63. a.

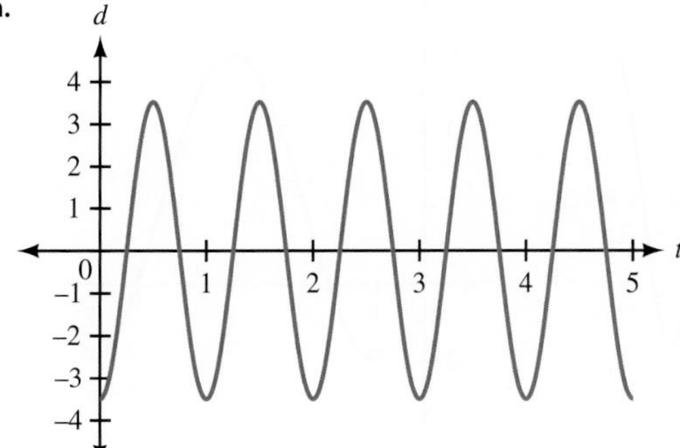

65.

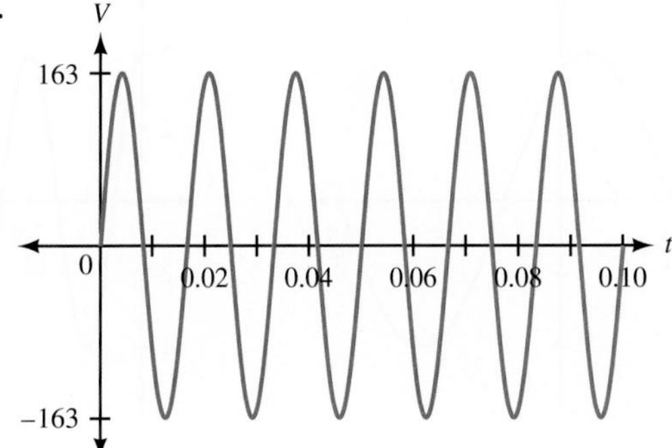

b. 3.5 cm **c.** 1 sec

67. 50 Hz **69.** 0 **71.** 1 **73.** $\frac{\sqrt{3}}{2}$ **75.** $\frac{3}{2}$ **77.** $\frac{5\pi}{6}$ **79.** $\frac{5\pi}{4}$

81.

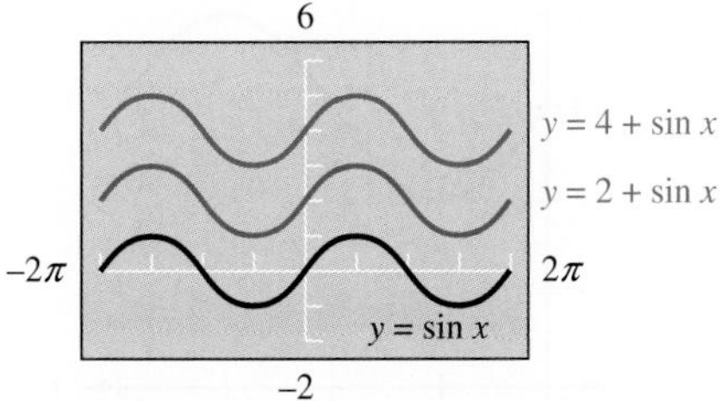

The graph is shifted k units upward.

83.

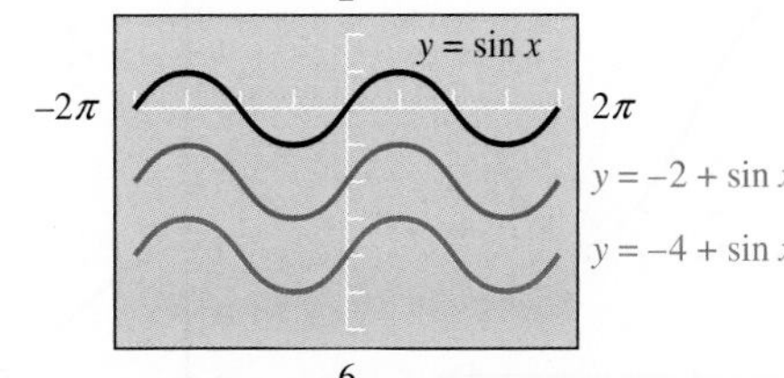

The graph is shifted k units downward.

85.

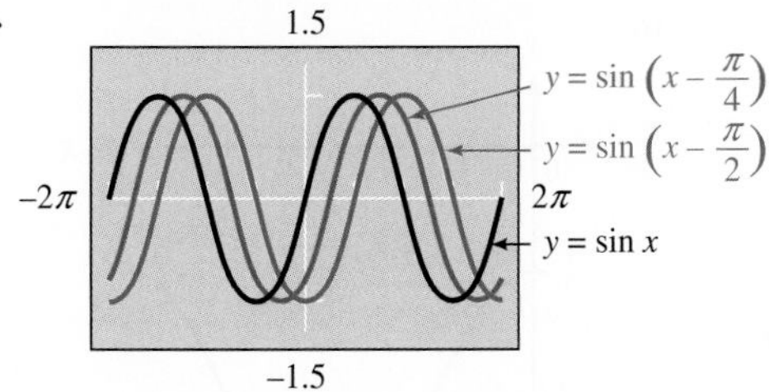

The graph is shifted h units to the right.

87.

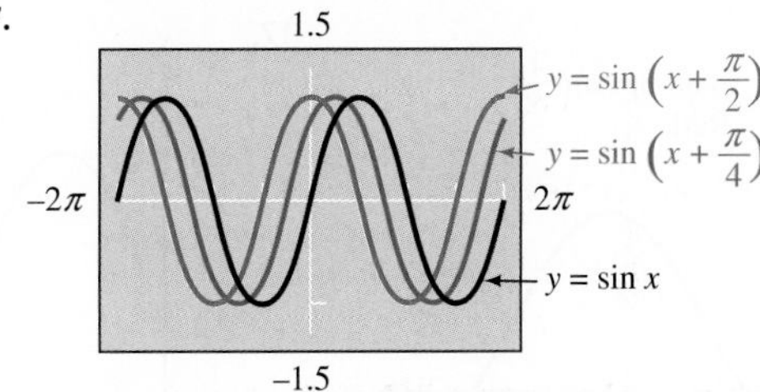

The graph is shifted h units to the left.

MATCHED PRACTICE PROBLEMS 4.3

1.

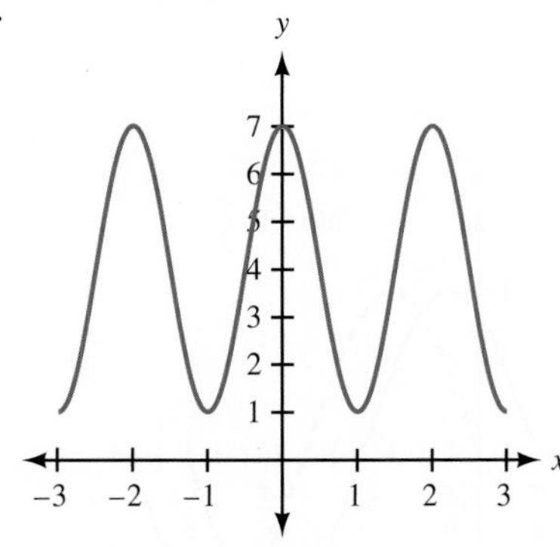

2.

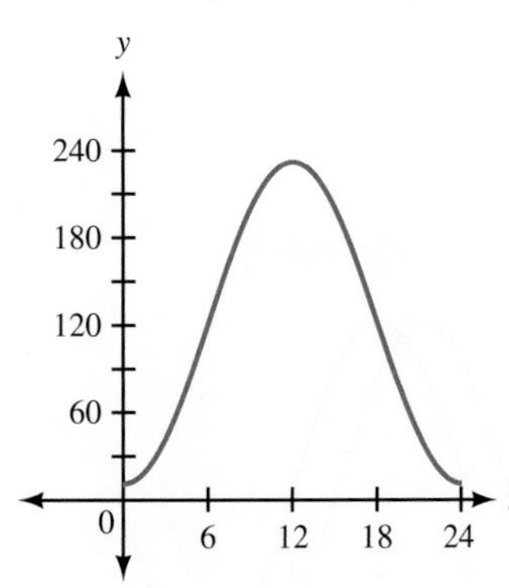

3.

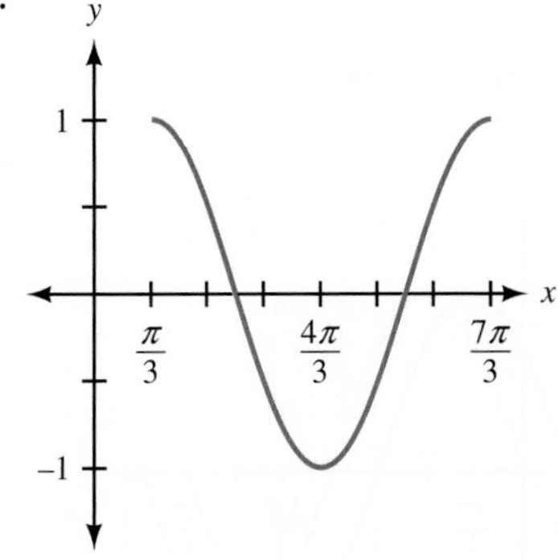

4.

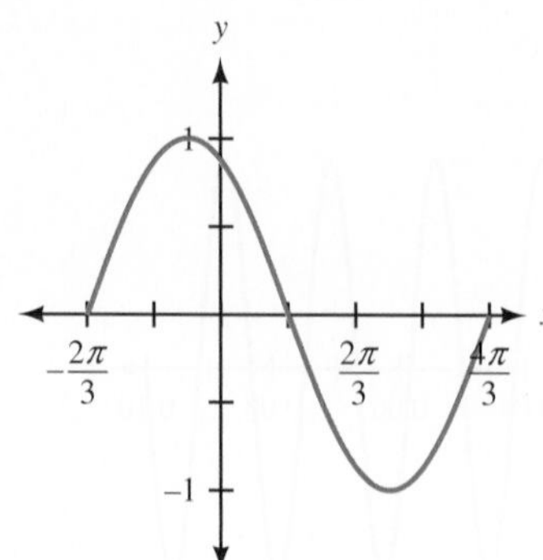

5.

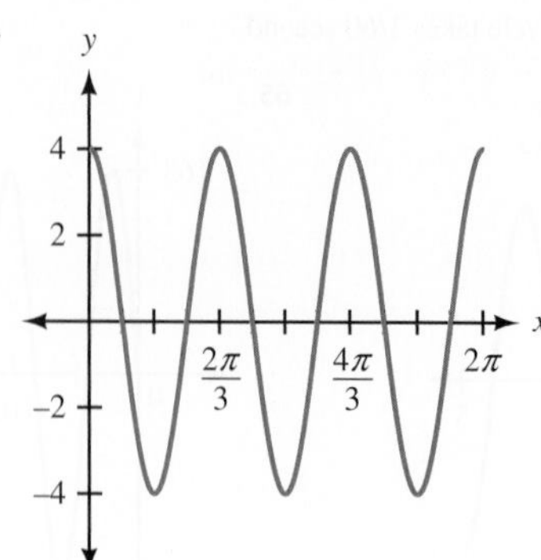

6.

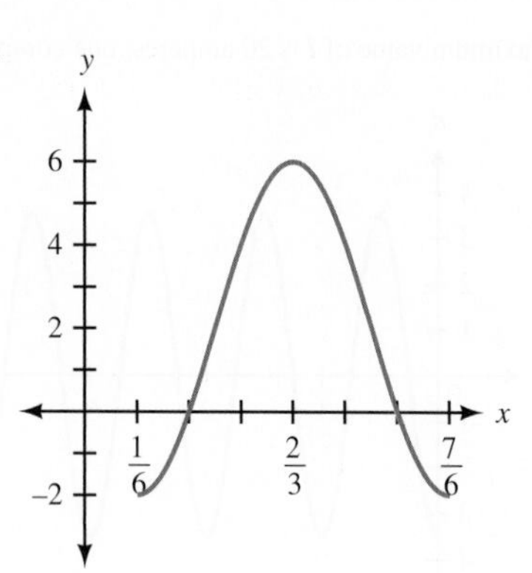

PROBLEM SET 4.3

1. vertical, upward, downward 3. phase 5. Up 5 units 7. Down $\frac{1}{4}$ unit

9.

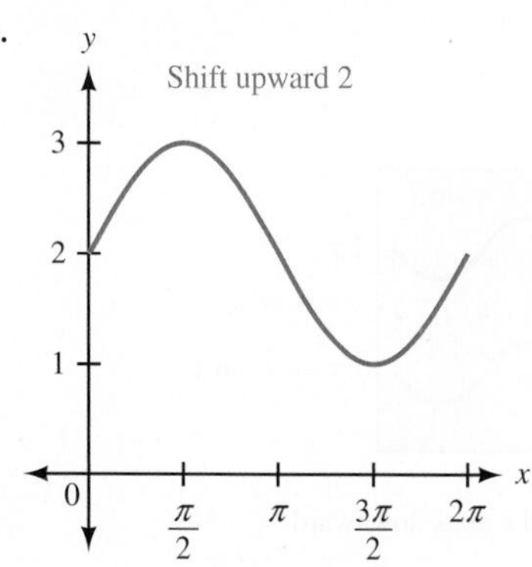

11.

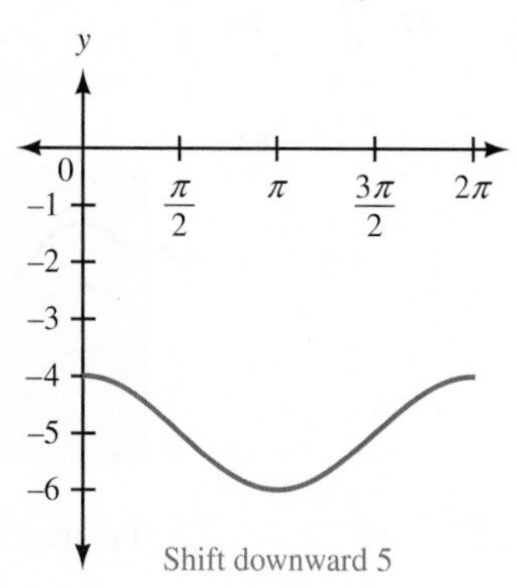

13.

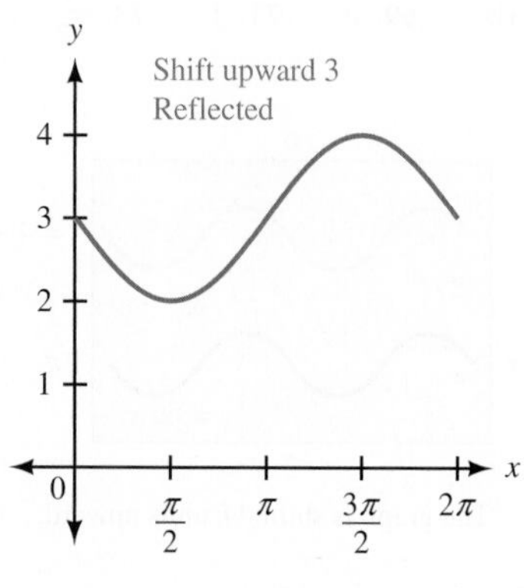

15.

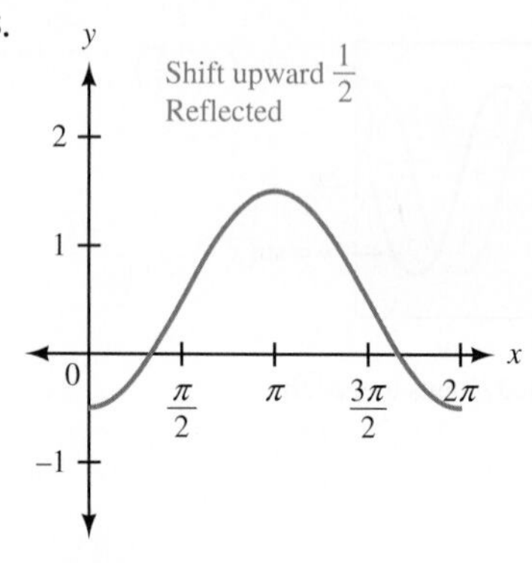

17.

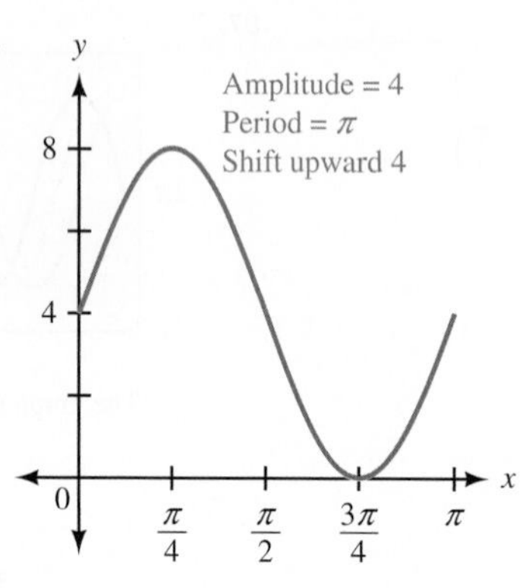

19.

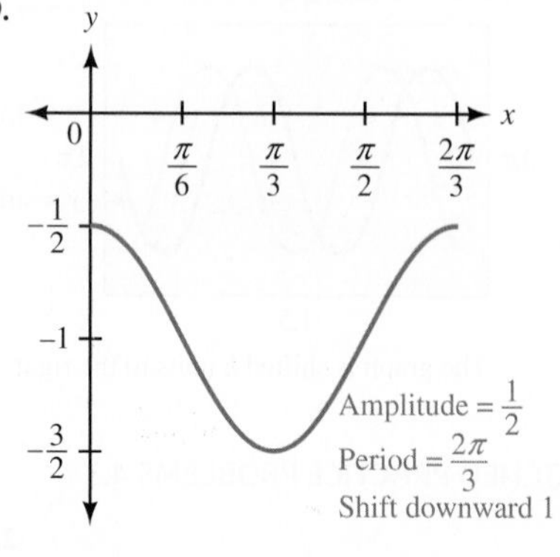

21. Left π units 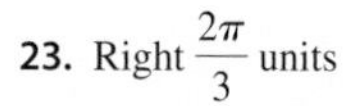 23. Right $\frac{2\pi}{3}$ units

25.

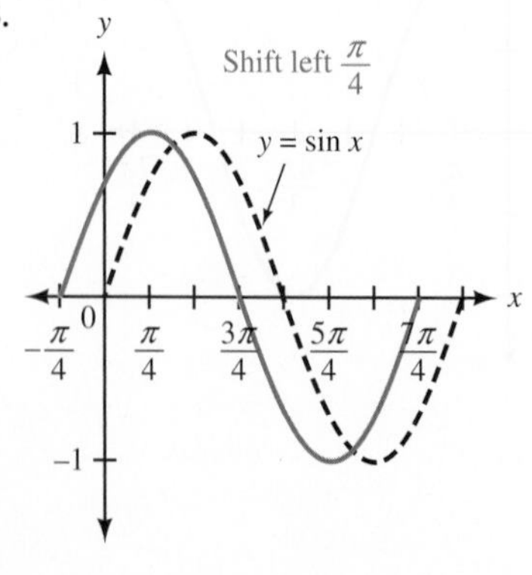

27.

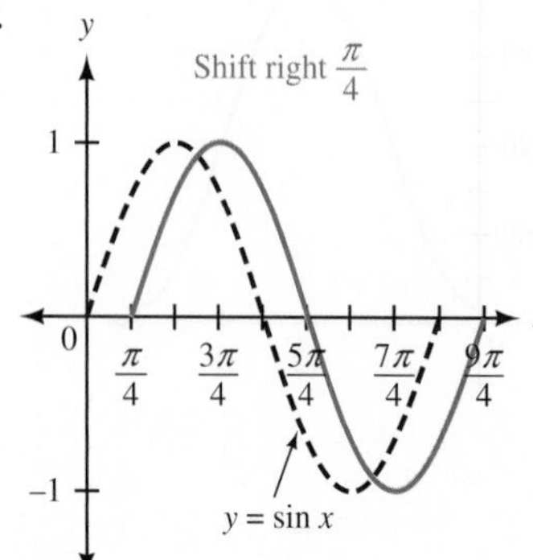

29.

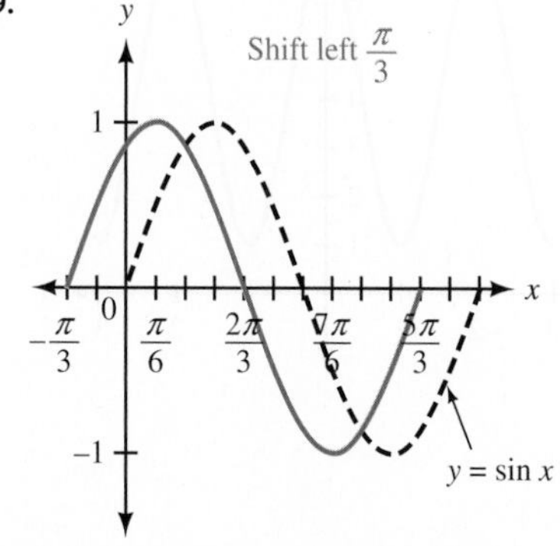

31.

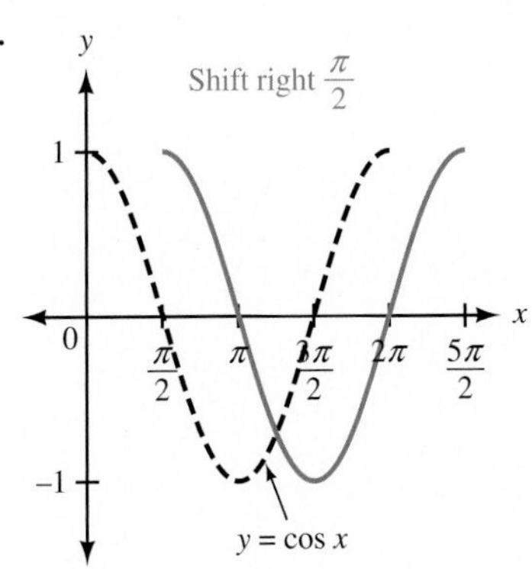

33.

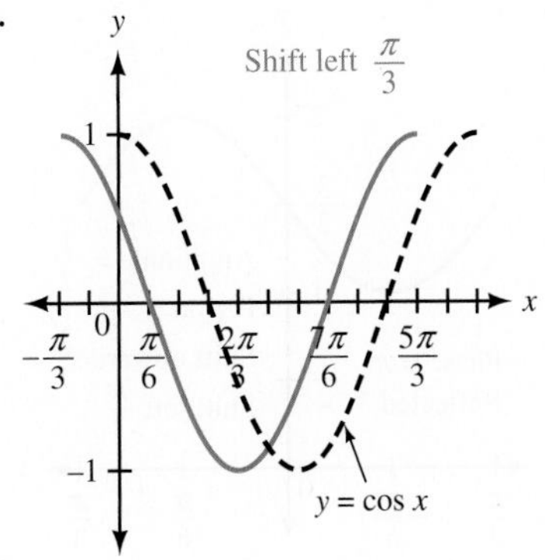

35. Period = 6, horizontal shift = −1, phase = $\frac{\pi}{3}$

37. Period = $\frac{\pi}{3}$, horizontal shift = $\frac{\pi}{6}$, phase = $-\pi$

39. Period = 4π, horizontal shift = $-\frac{\pi}{3}$, phase = $\frac{\pi}{6}$

41.

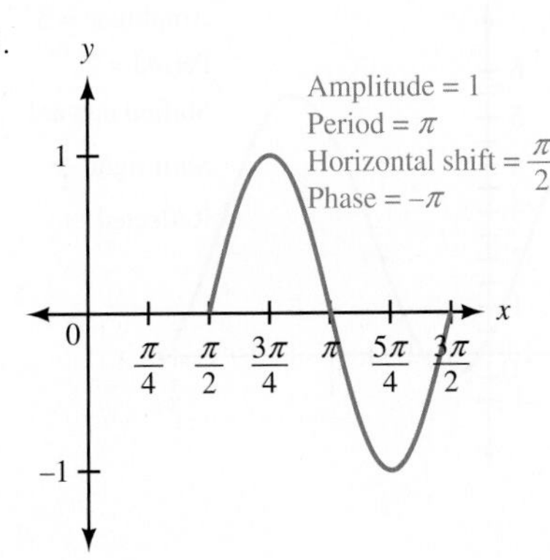

43.

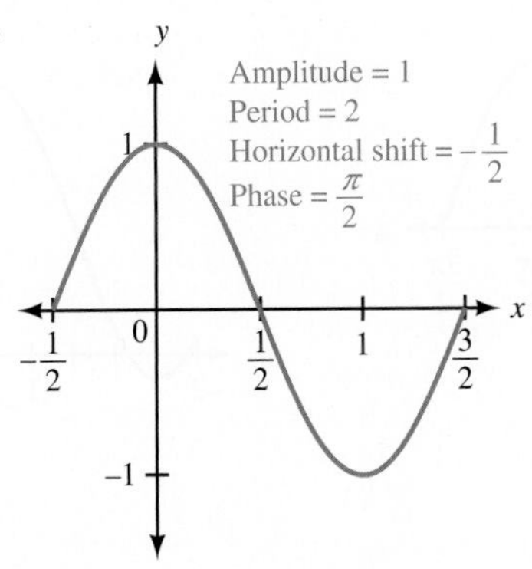

45.

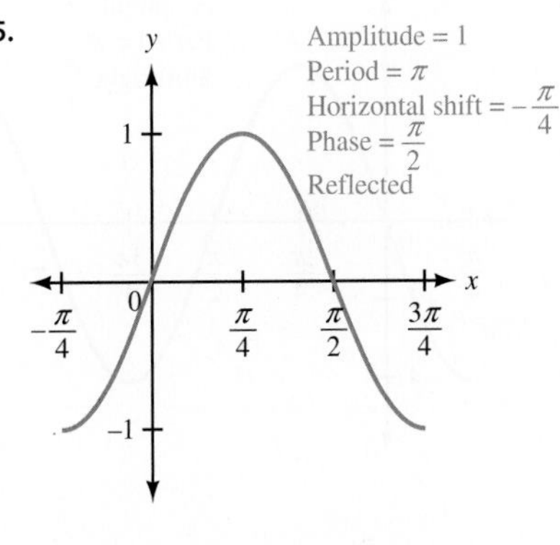

47.

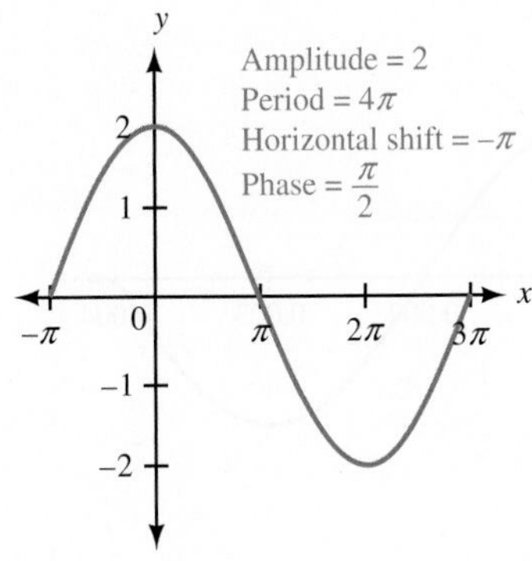

49.

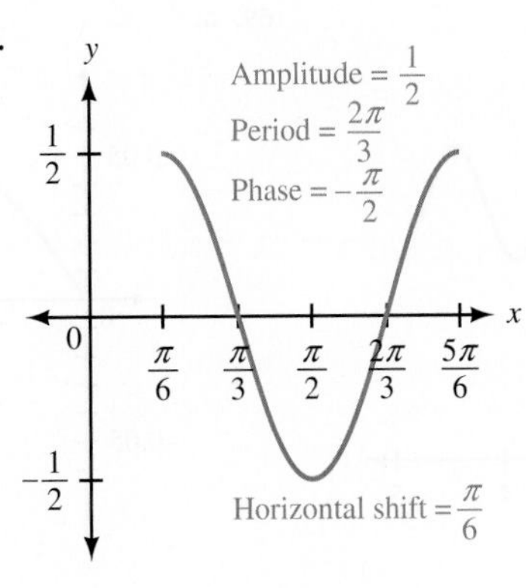

51.

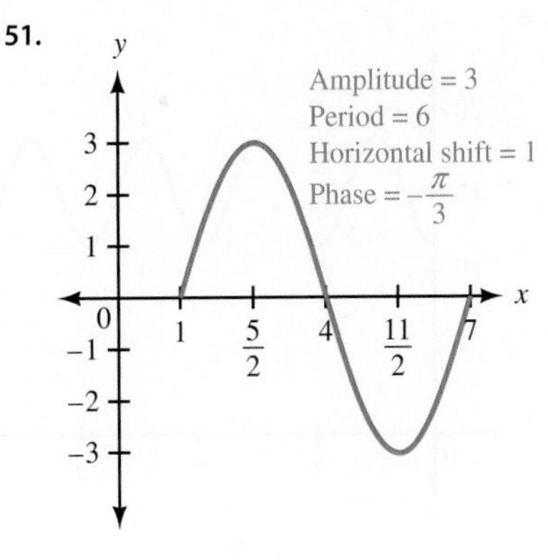

53.

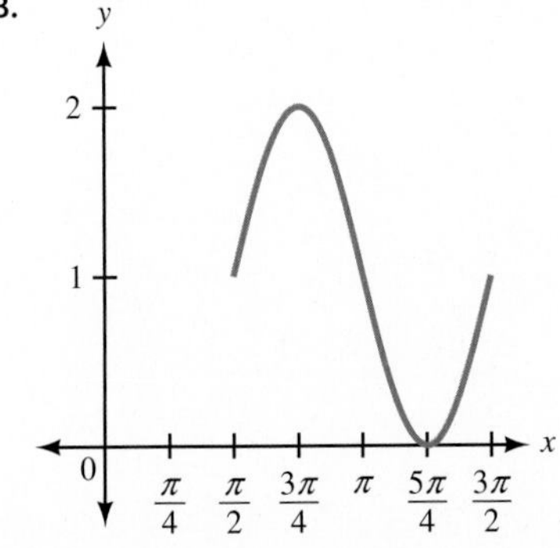

55.

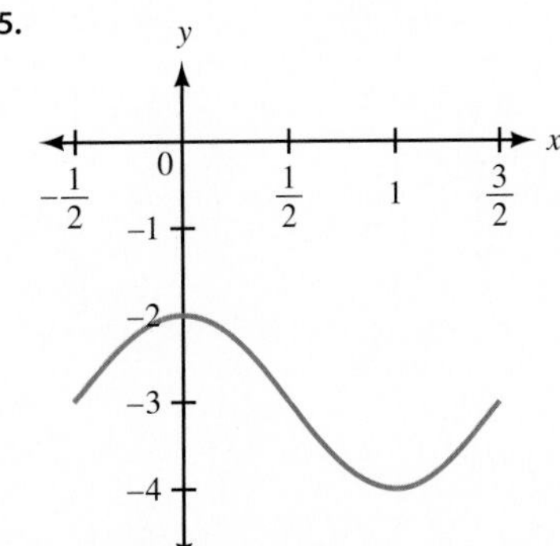

57.

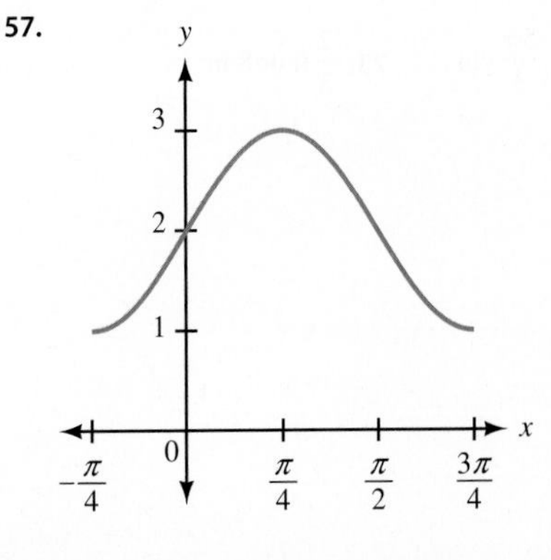

59.

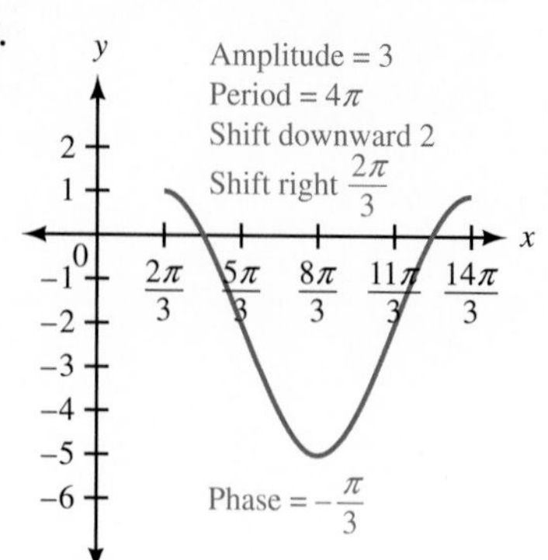

61.

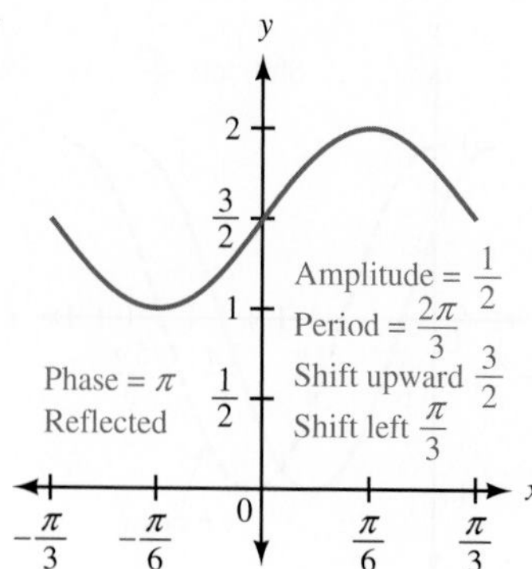

63.

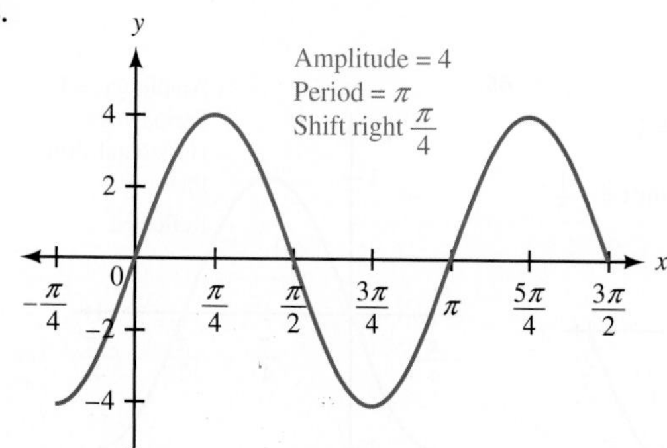

65.

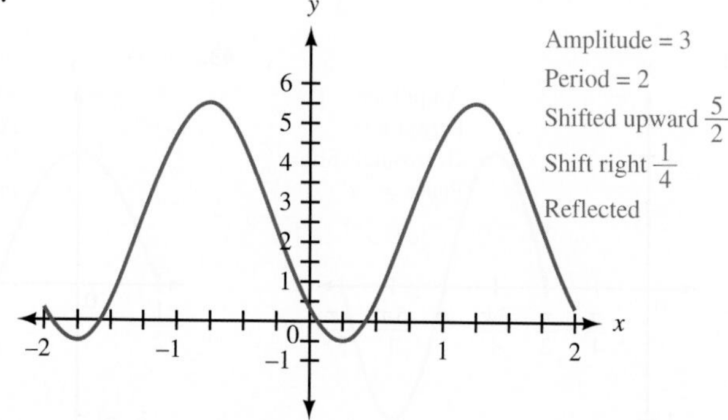

67. a.

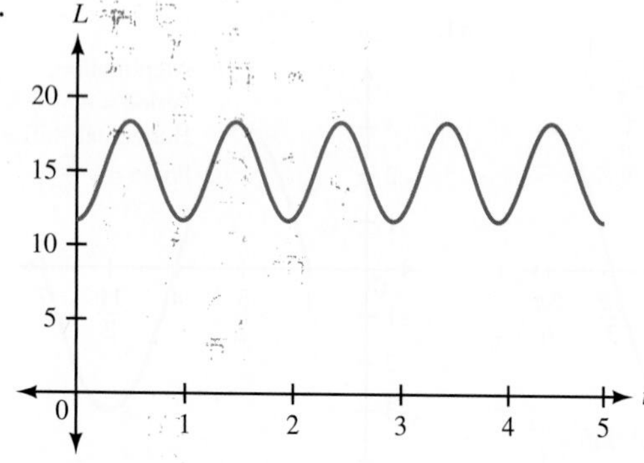

b. 15 cm **c.** 11.5 cm **d.** 18.5 cm

69. a.

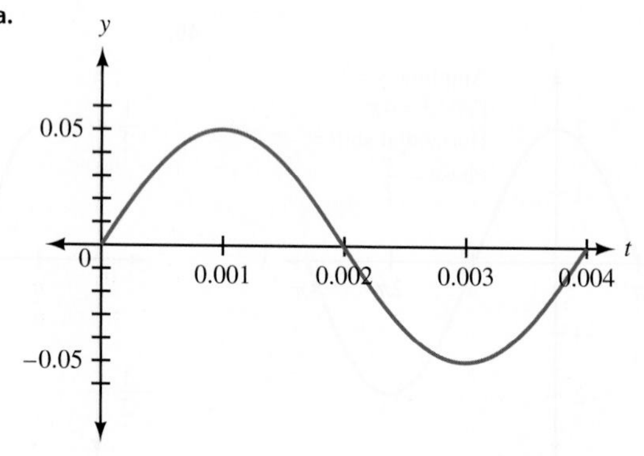

b. 10π

71. $\frac{5\pi}{3}$ cm **73.** $\frac{2}{3}$ ft or 8 in.

MATCHED PRACTICE PROBLEMS 4.4

1.

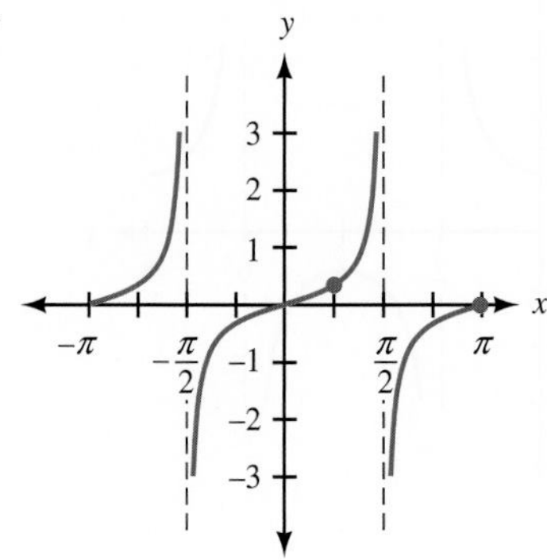

2.

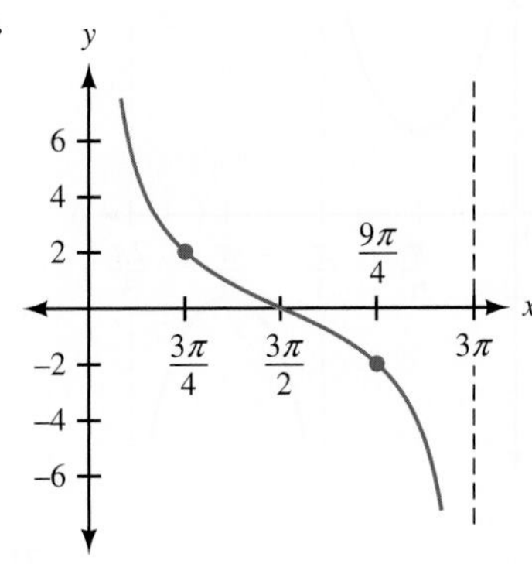

3.

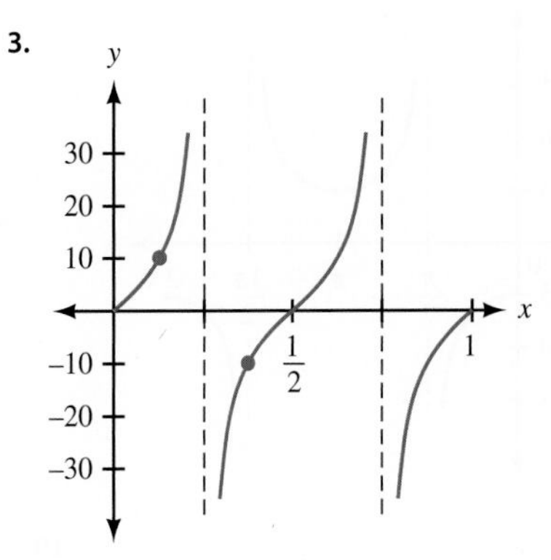

4.

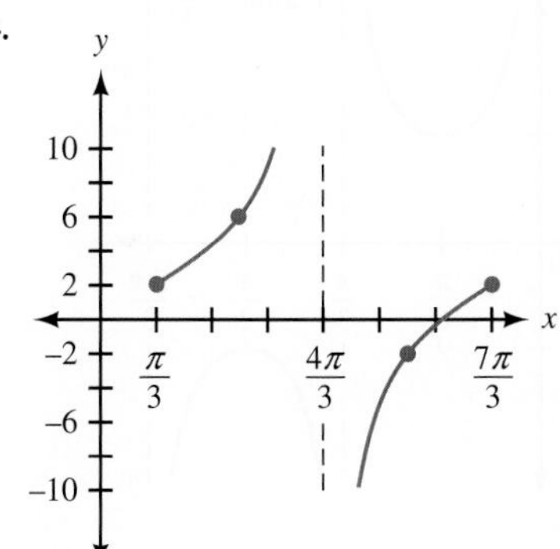

5.

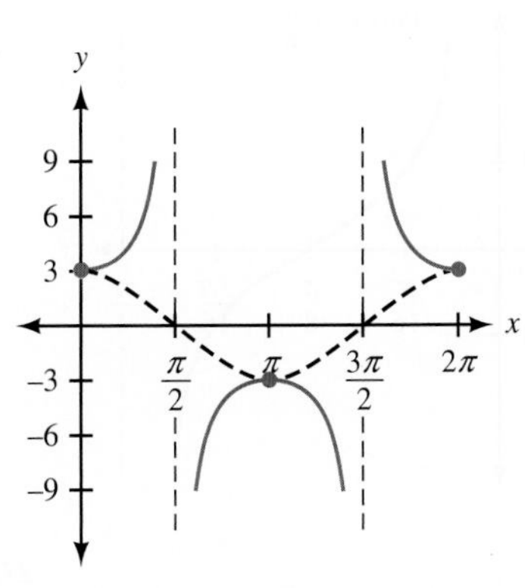

6.

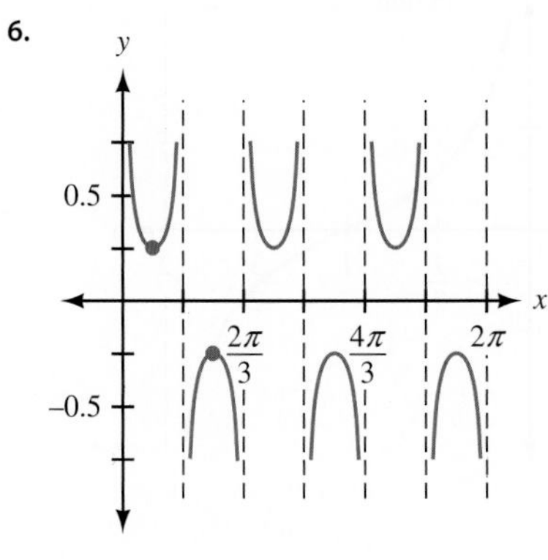

7.

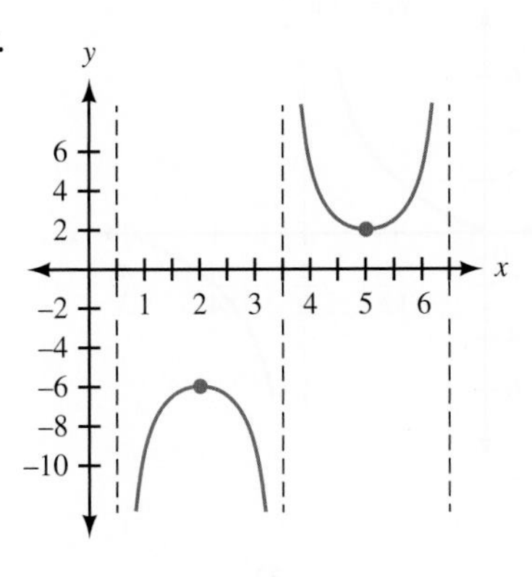

PROBLEM SET 4.4

1. expanded, contracted, multiply **3.** $\frac{\pi}{B}$

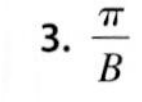

5.

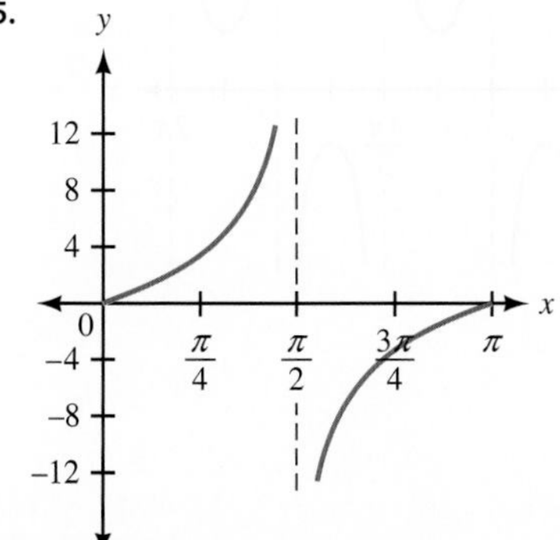

7.

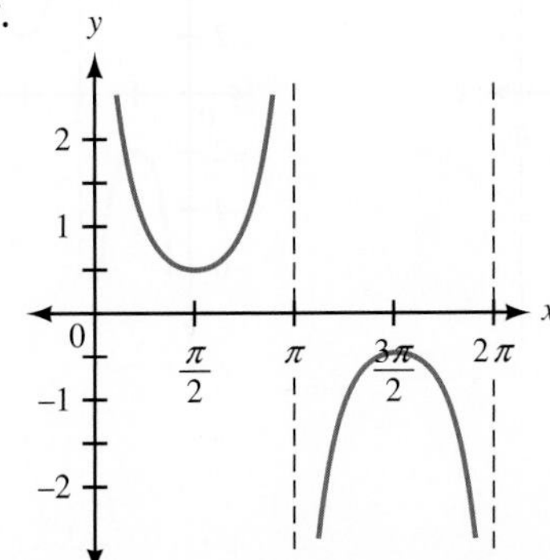

9.

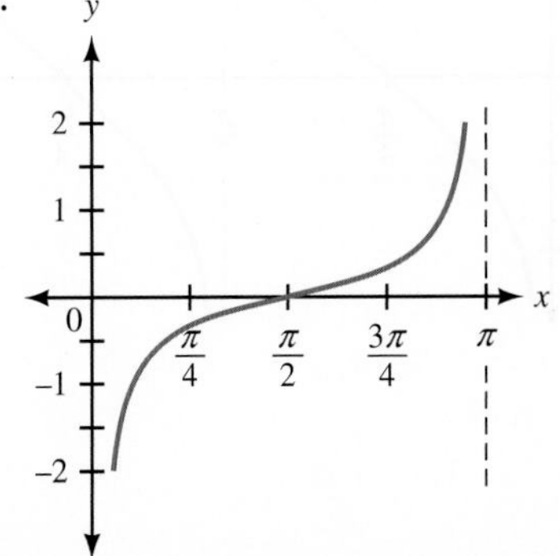

11.

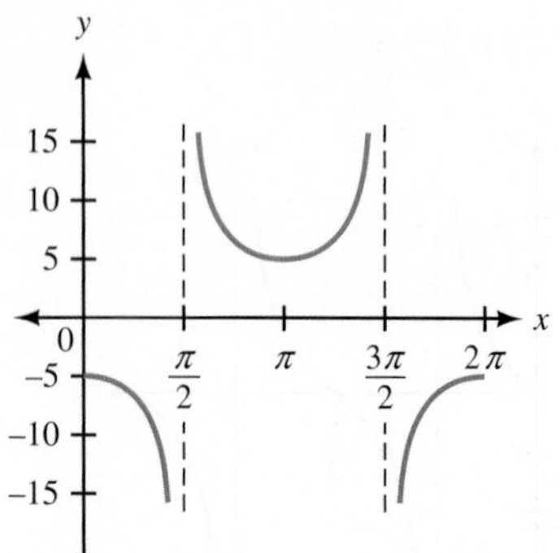

13.

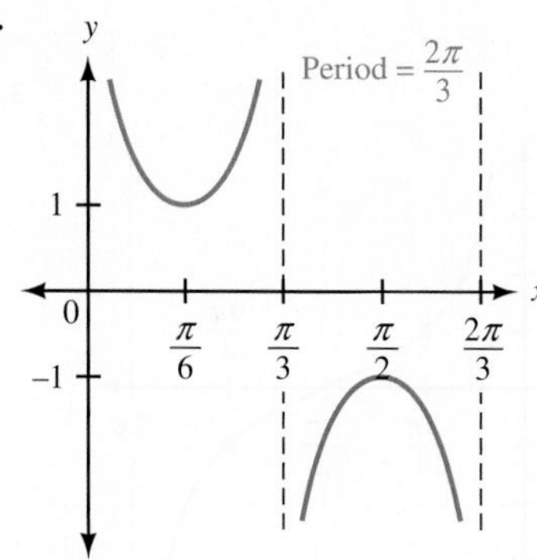

15.

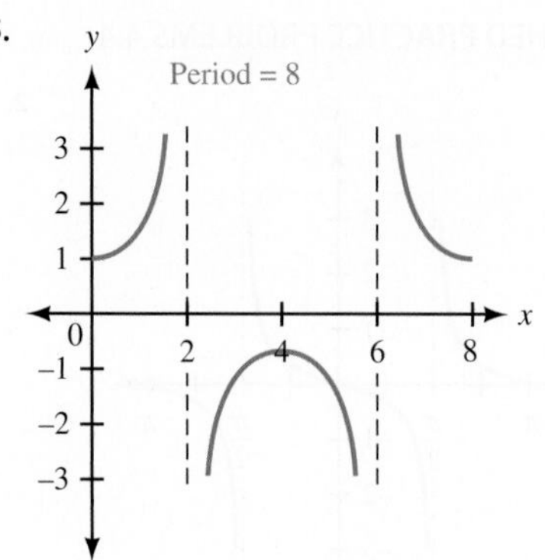

17.

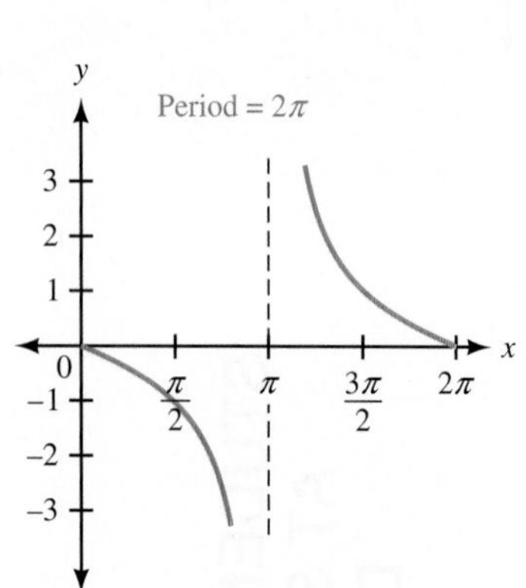

19.

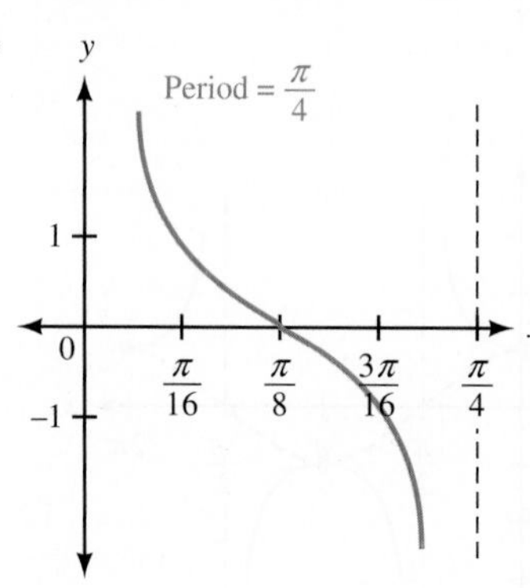

21.

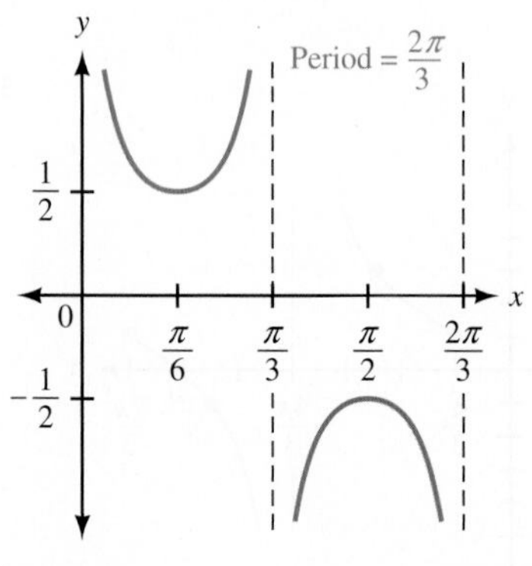

23.

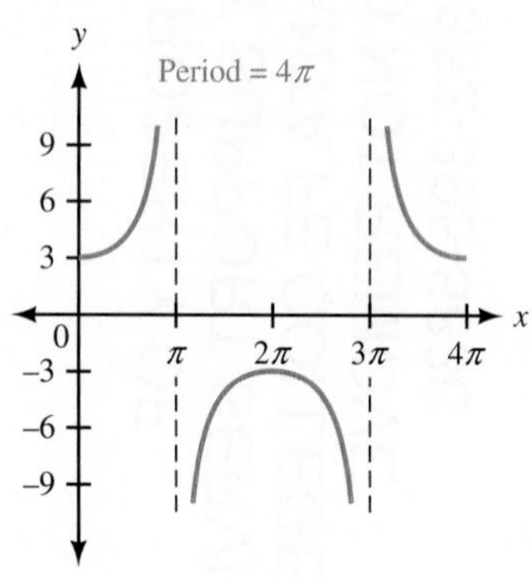

25.

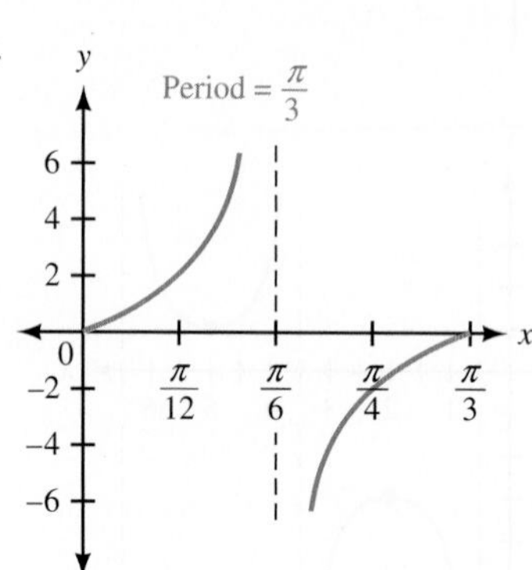

27.

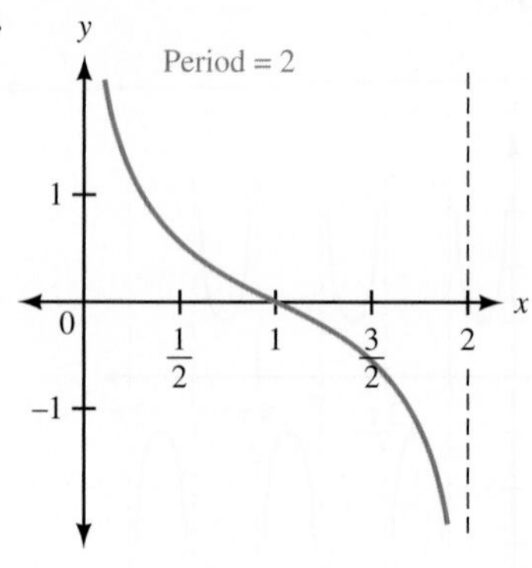

29.

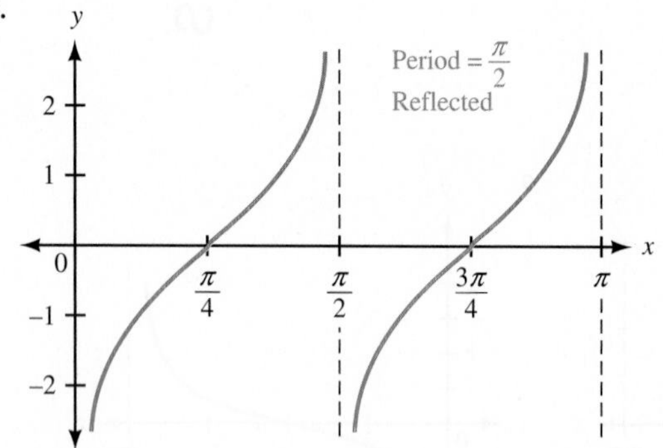

31.

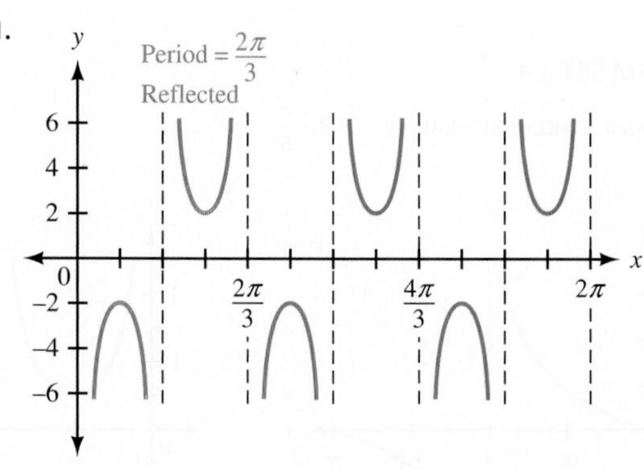

33. a.

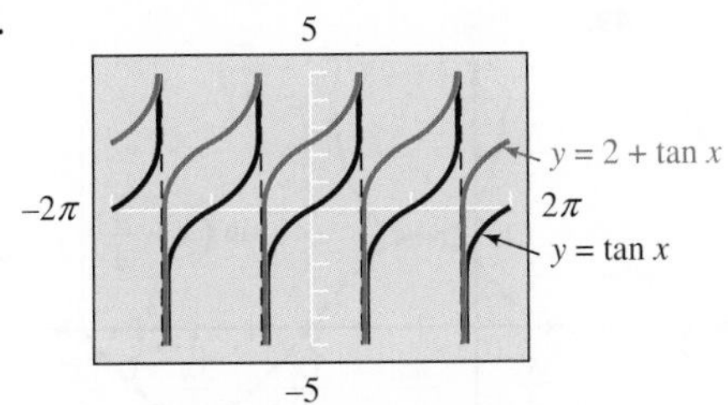

b.

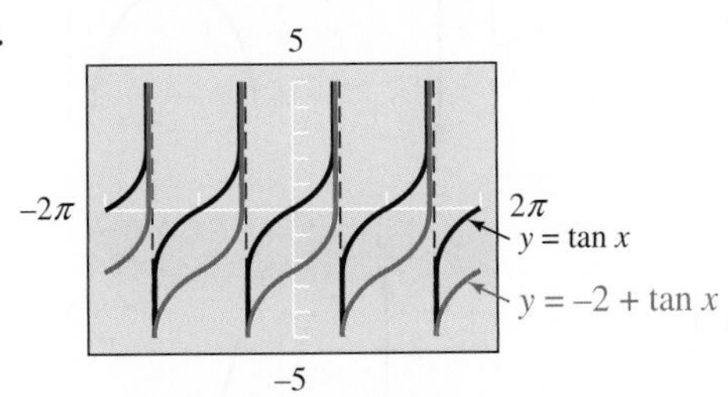

c.

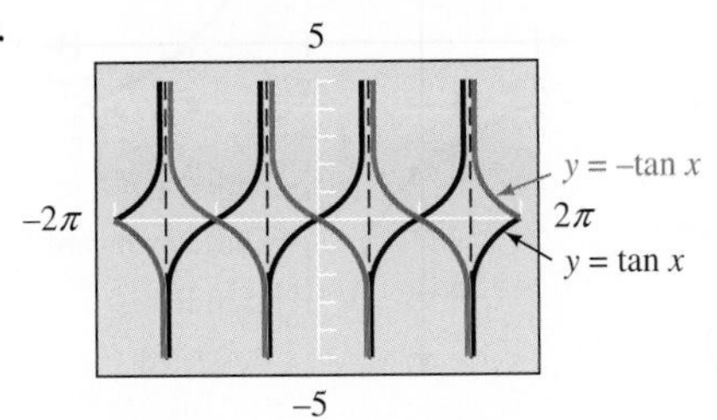

35. a.

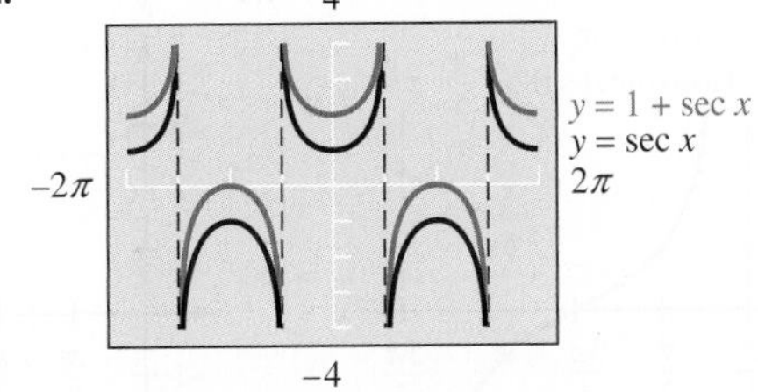

b.

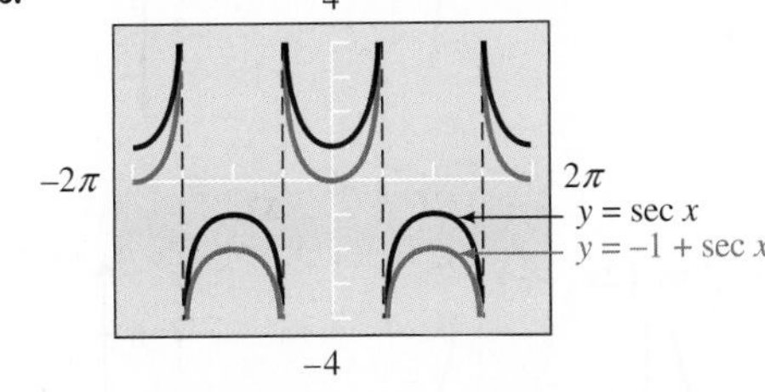

c.

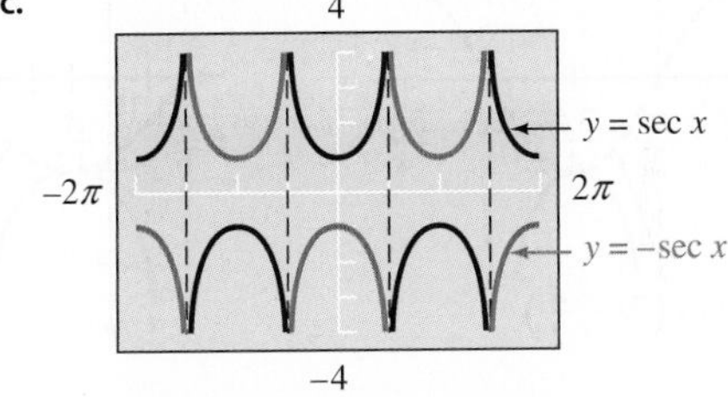

37. a.

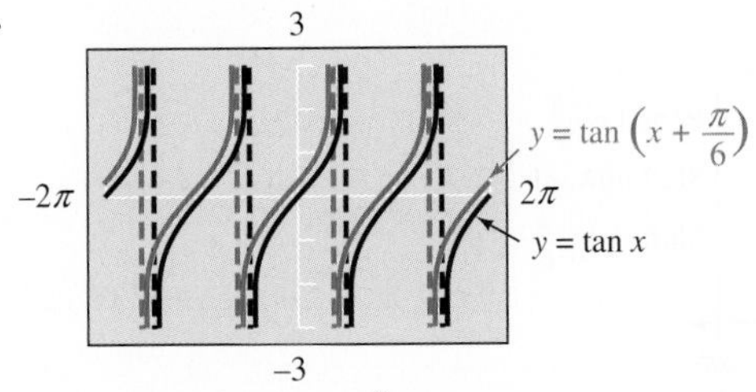

The graph is shifted $\frac{\pi}{6}$ to the left.

b.

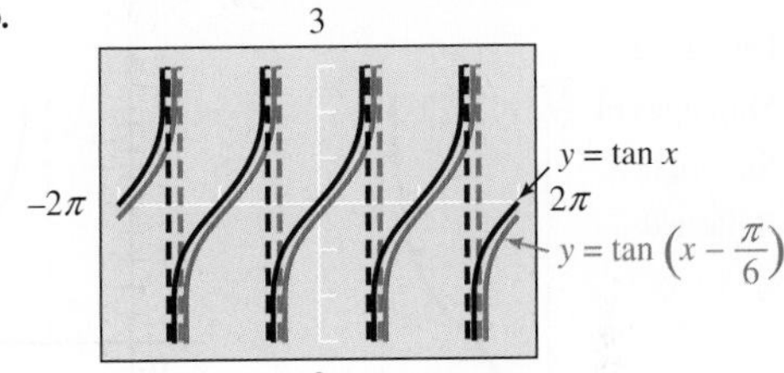

The graph is shifted $\frac{\pi}{6}$ to the right.

39.

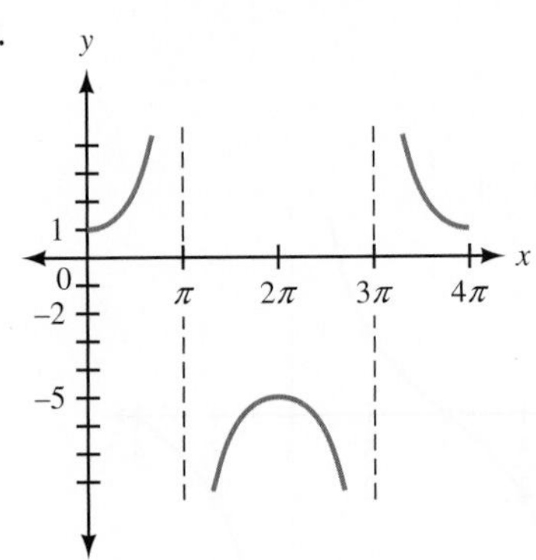

41.

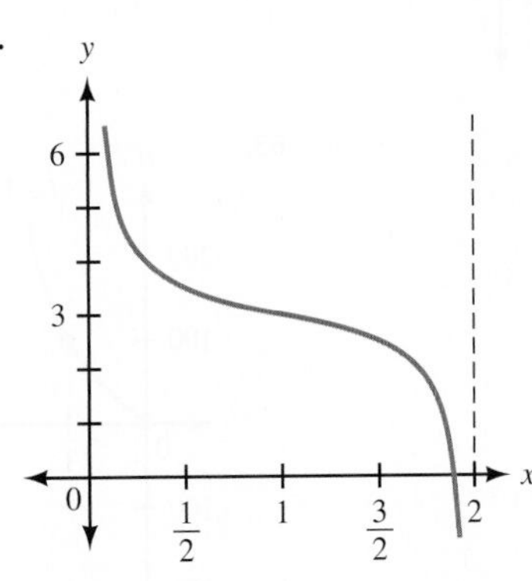

43.

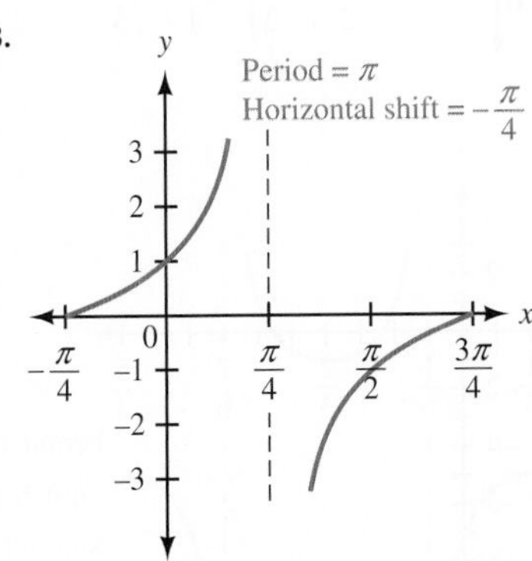

45.

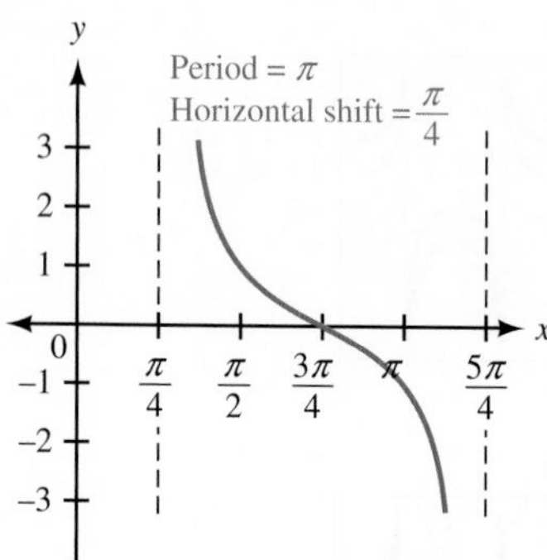

47.

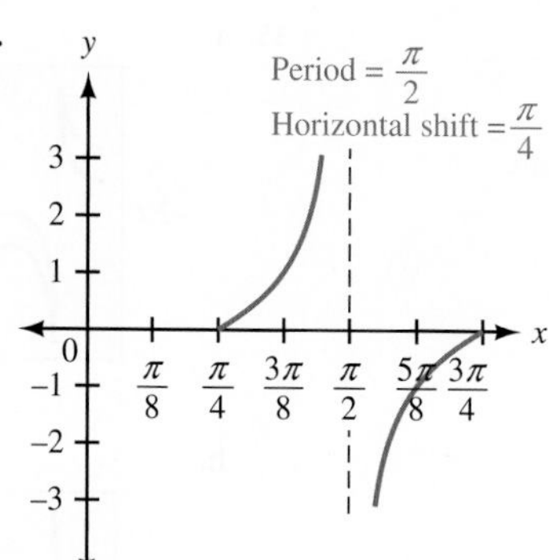

49.

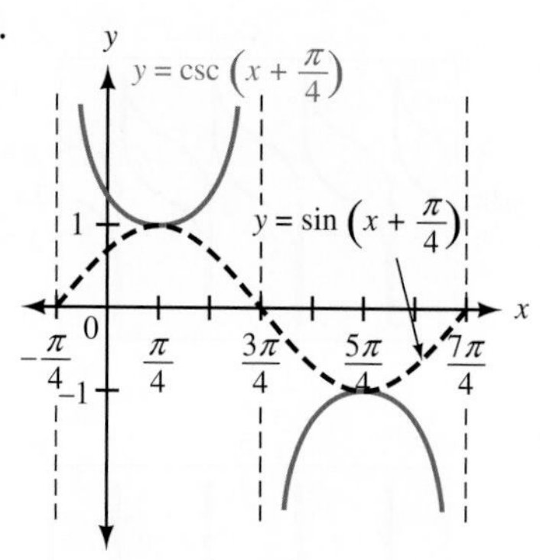

51.

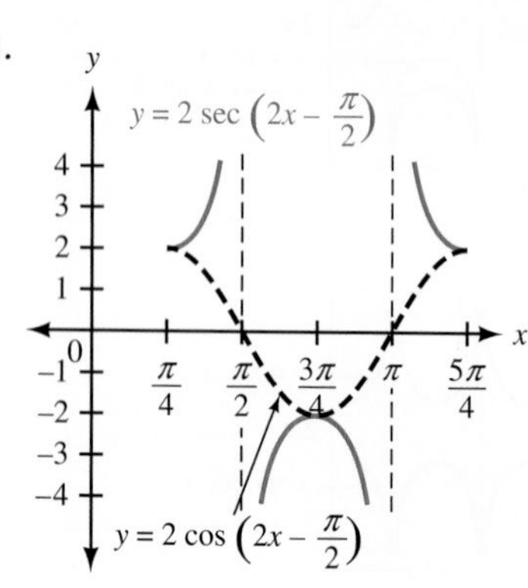

53.

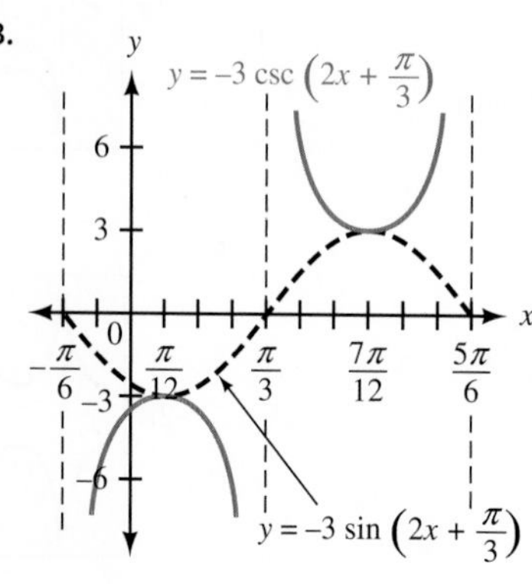

55.

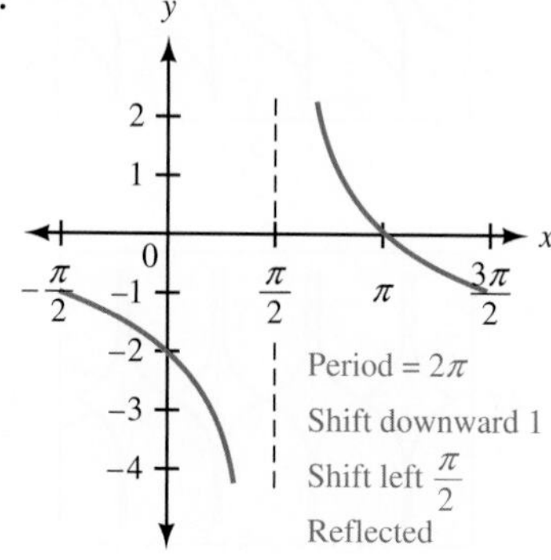

57.

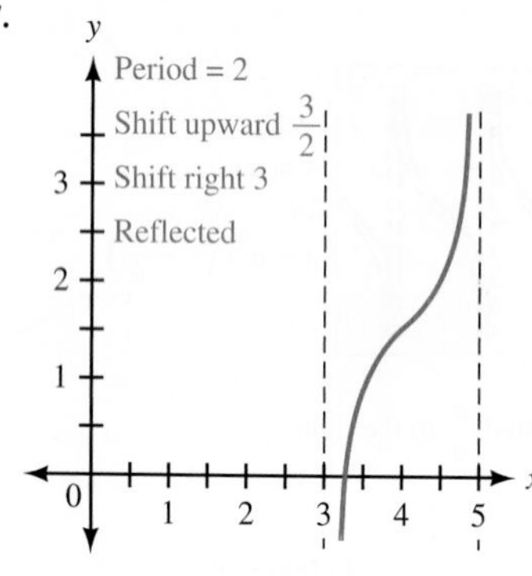

59.

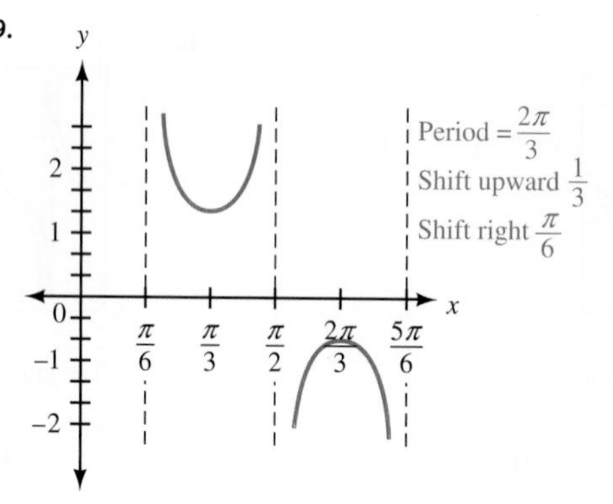

61.

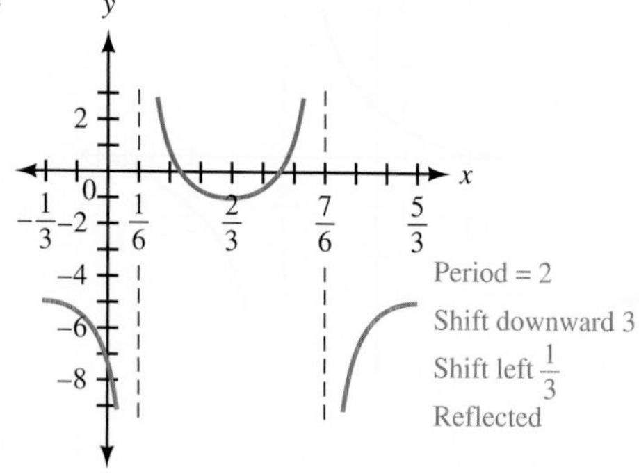

63.

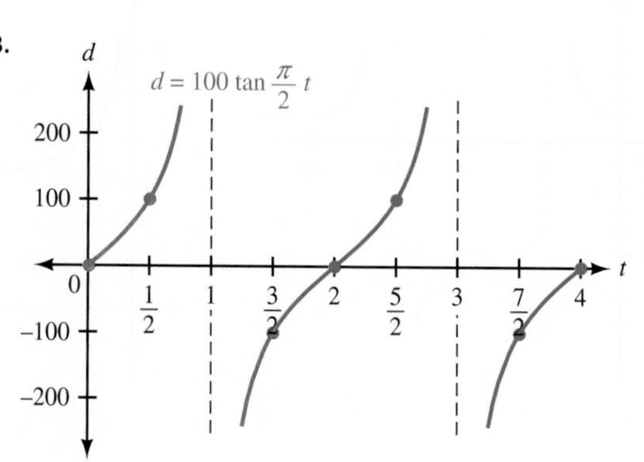

65. 4θ **67.** $2x + \pi$ **69.** -0.8391 **71.** -0.4663 **73.** -1.7919

MATCHED PRACTICE PROBLEMS 4.5

1. $y = 2x - 4$ **2.** $y = -5 \sin \frac{x}{4}$ **3.** $y = -4 \cos \frac{\pi x}{3}$ **4.** $y = 2 - 5 \sin 3x$ **5.** $y = 1 + 3 \sin\left(\frac{\pi x}{2} + \frac{\pi}{8}\right)$
6. $H = 118 - 110 \cos\left(\frac{\pi}{12}t\right)$ **7.** $y = 74 - 20 \cos\left(\frac{\pi x}{6} - \frac{\pi}{6}\right)$

PROBLEM SET 4.5

1. half, period **3.** average **5.** $y = -\frac{1}{2}x + 1$ **7.** $y = 2x - 3$

For Problems 9–34, we give one possible equation. Other answers are possible.

9. $y = \sin x$ **11.** $y = 3 \cos x$ **13.** $y = -3 \cos x$ **15.** $y = \sin 3x$ **17.** $y = \sin \frac{1}{3}x$

19. $y = 2 \cos 3x$ **21.** $y = 4 \sin \pi x$ **23.** $y = -4 \sin \pi x$ **25.** $y = 2 - 4 \sin \pi x$ **27.** $y = 3 \cos\left(2x + \frac{3\pi}{4}\right)$

29. $y = -2 \cos\left(3x - \frac{\pi}{4}\right)$ **31.** $y = -3 - 3 \sin(2x)$ **33.** $y = 3 - 3 \sin\left(2x - \frac{\pi}{2}\right)$ **35.** $d = 5 \cos\left(\frac{8\pi}{5}t\right)$

37.

t (min)	h (ft)
0	12
1.875	40.8
3.75	110.5
5.625	180.2
7.5	209
9.375	180.2
11.25	110.5
13.125	40.8
15	12

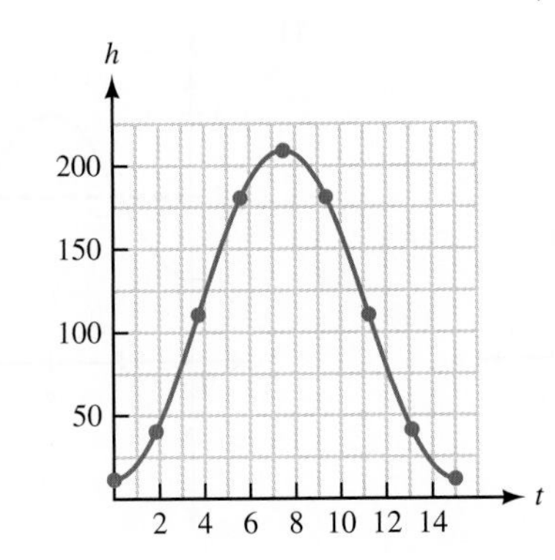

$h = 110.5 - 98.5 \cos\left(\frac{2\pi}{15}t\right)$

39. Answers will vary. **41.** Answers will vary. **43.** 39° **45.** 56° **47.** 84° **49.** 131.8° **51.** 205.5° **53.** 281.8°

MATCHED PRACTICE PROBLEMS 4.6

1.

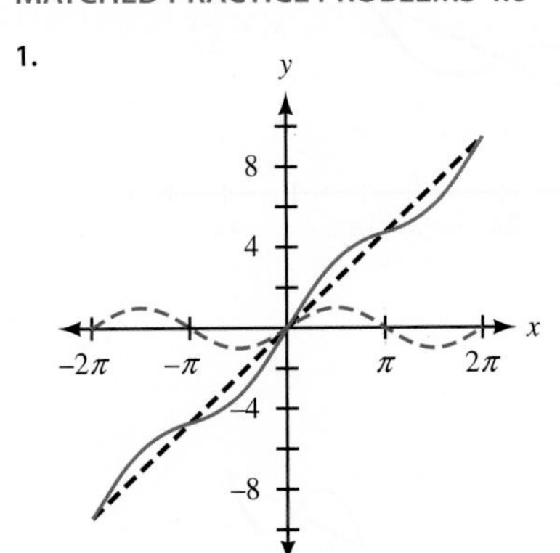

2.

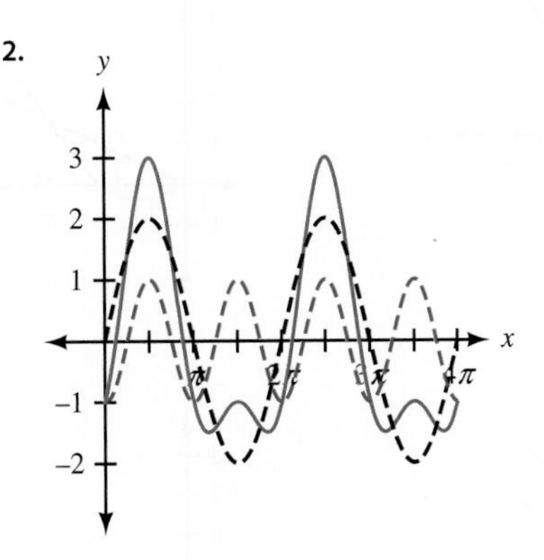

3.

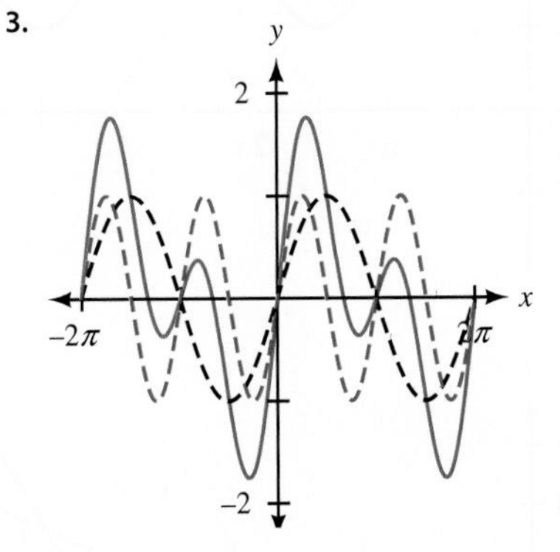

4.

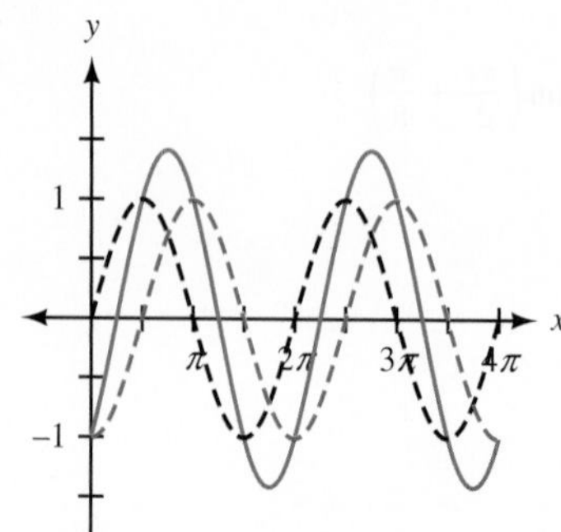

5.

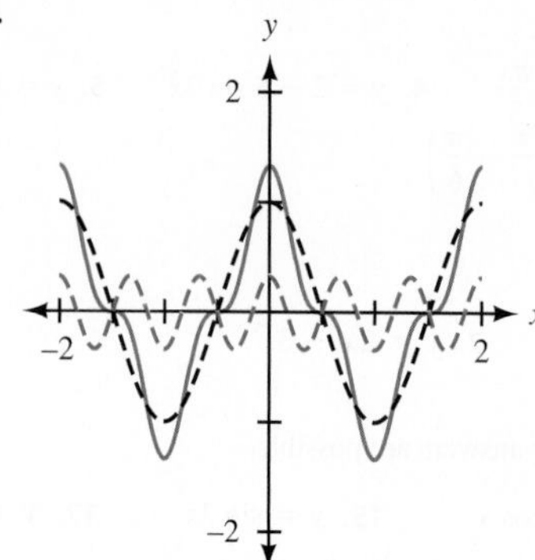

PROBLEM SET 4.6

1. y, y **3.** below, y_2 **5.** $(1, 3)$ **7.** $(\pi, -1.5)$

9.

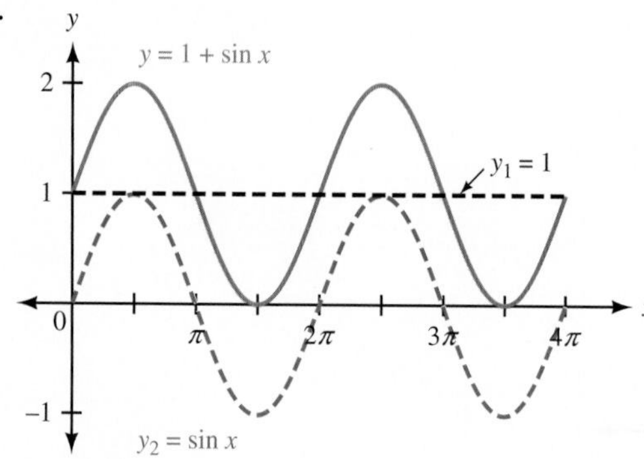

11.

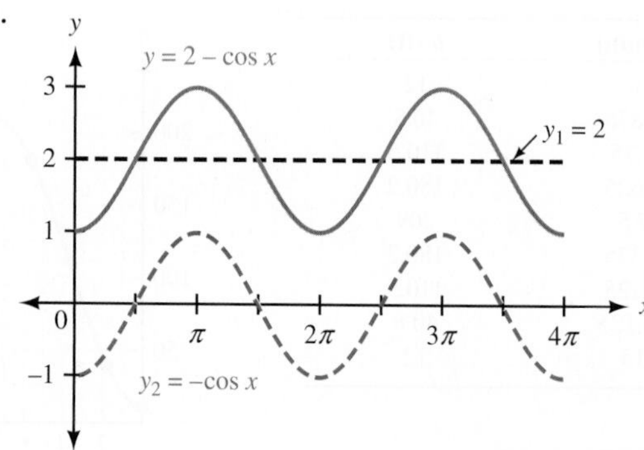

13.

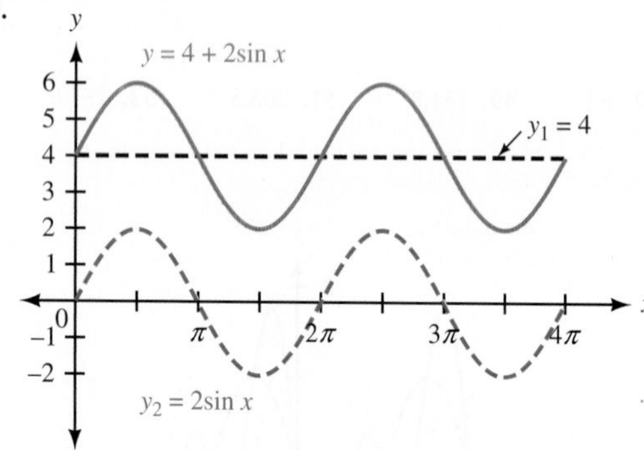

15.

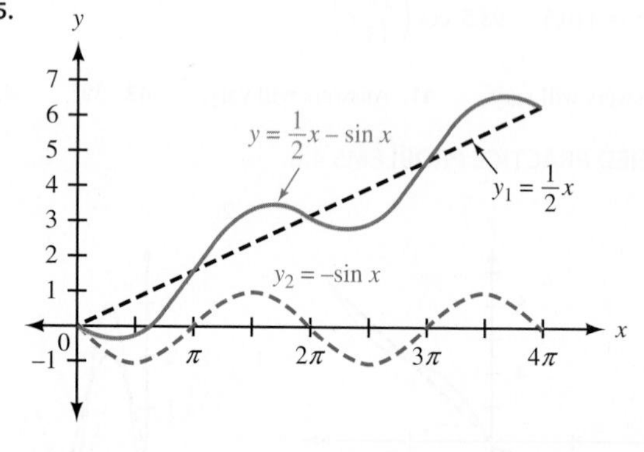

17.

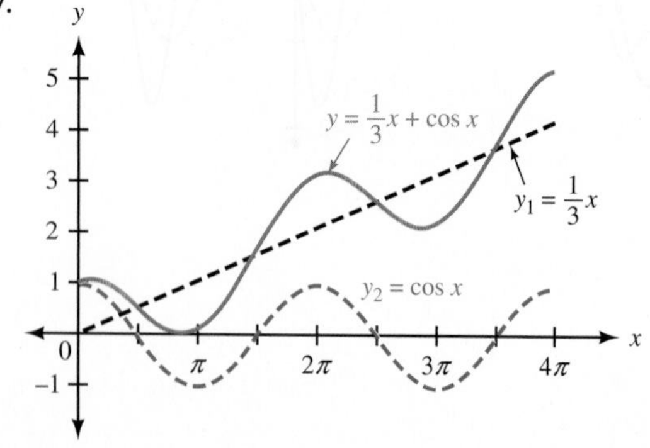

19.

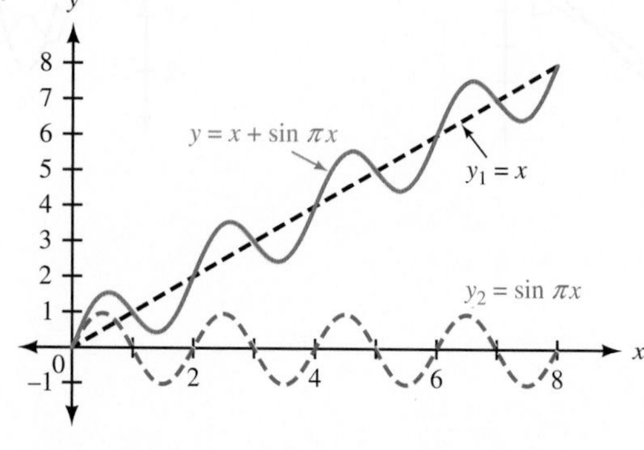

21.

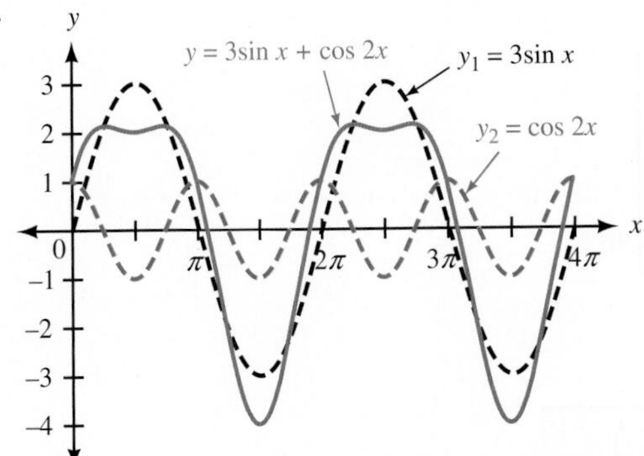

23.

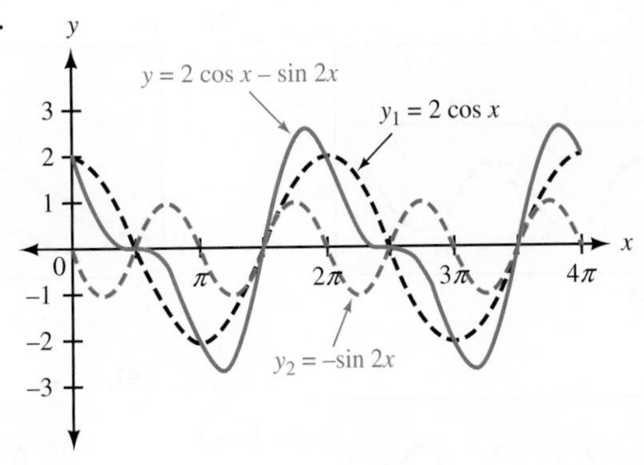

25.

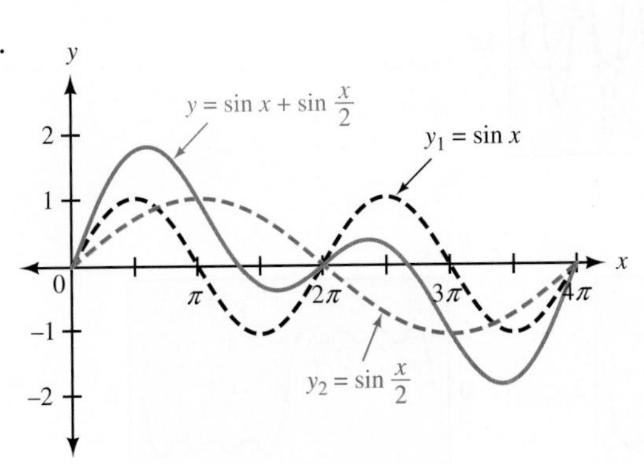

27.

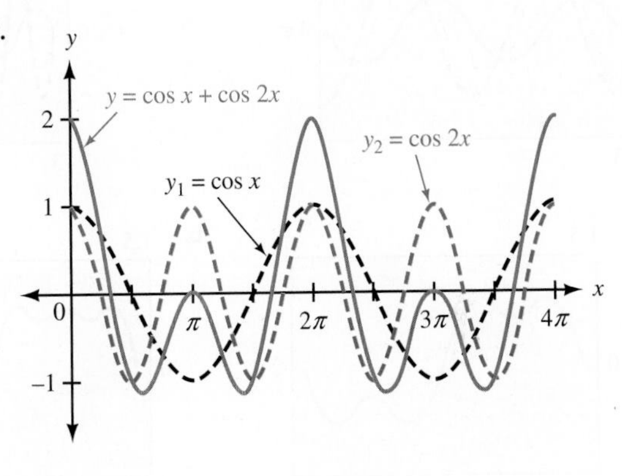

29.

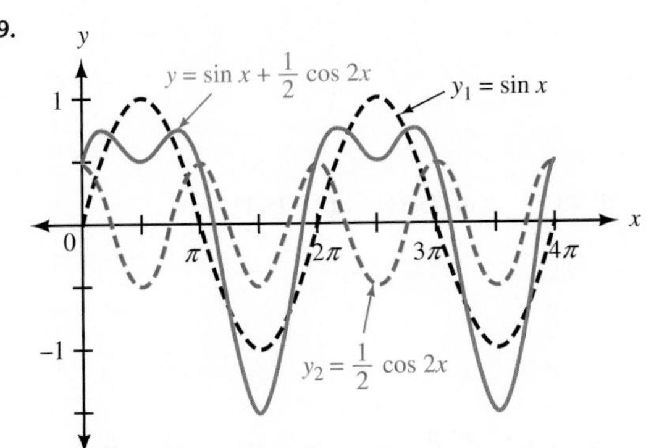

31.

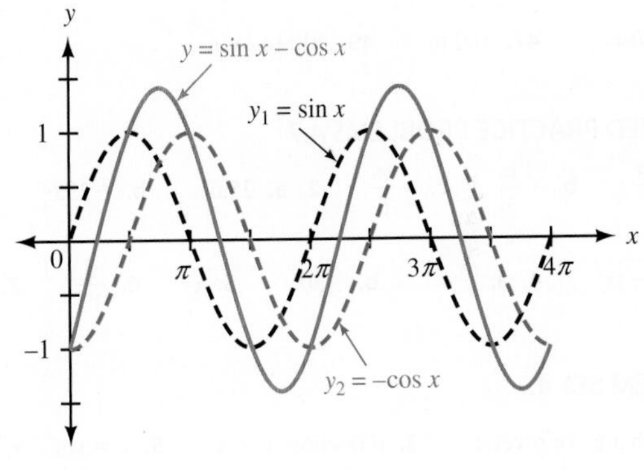

33.

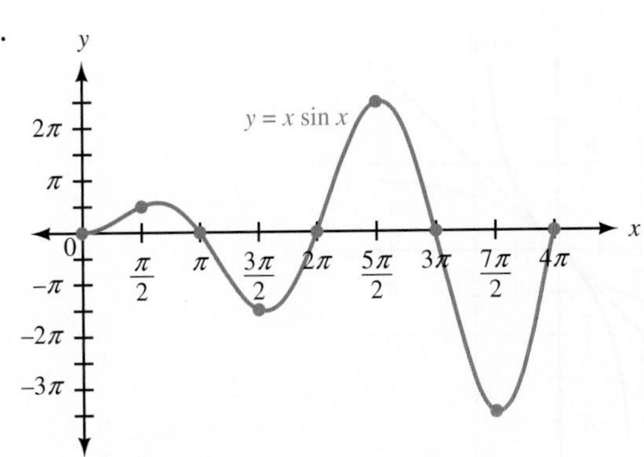

35.

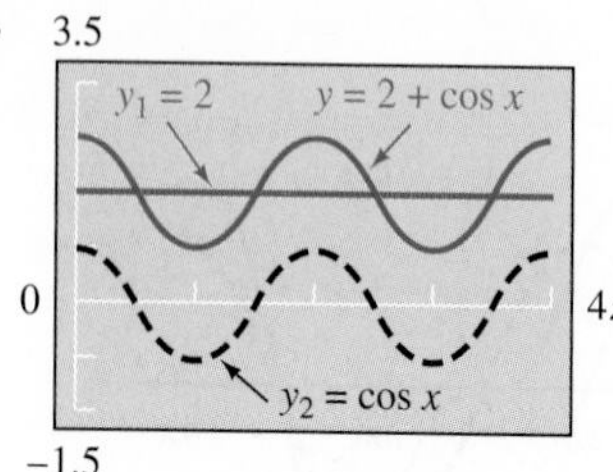

37.

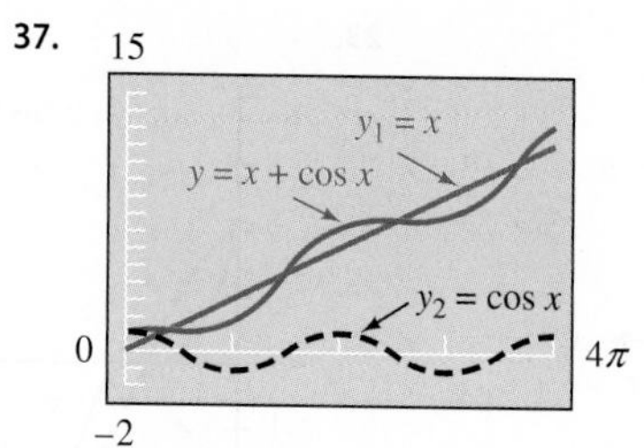

39.

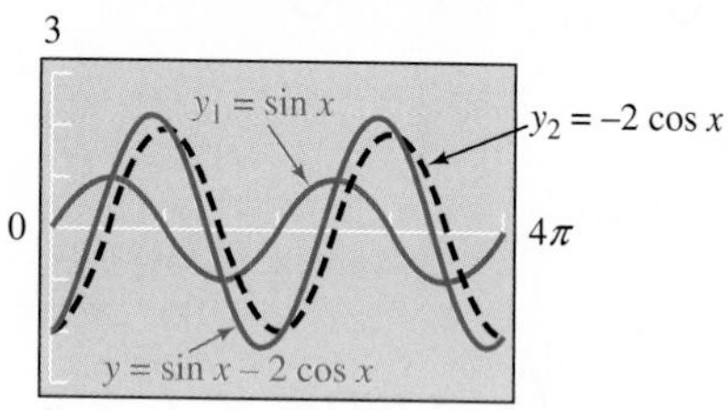

41.

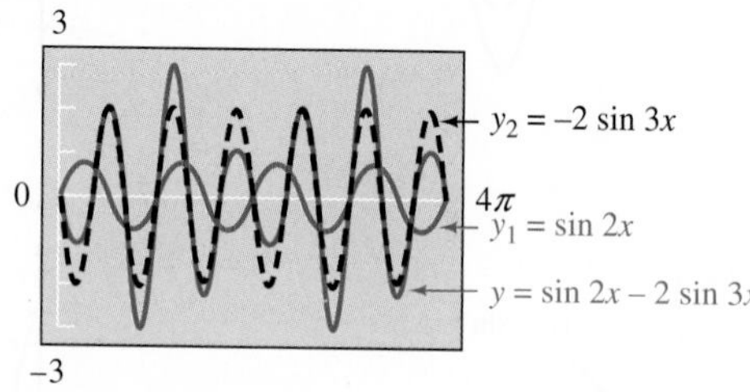

43. a.

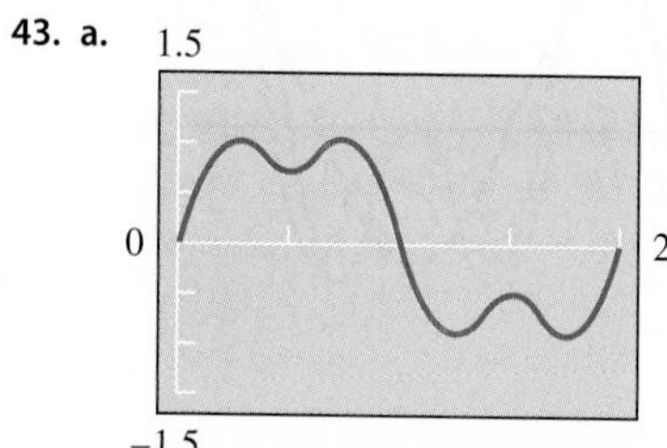

b.

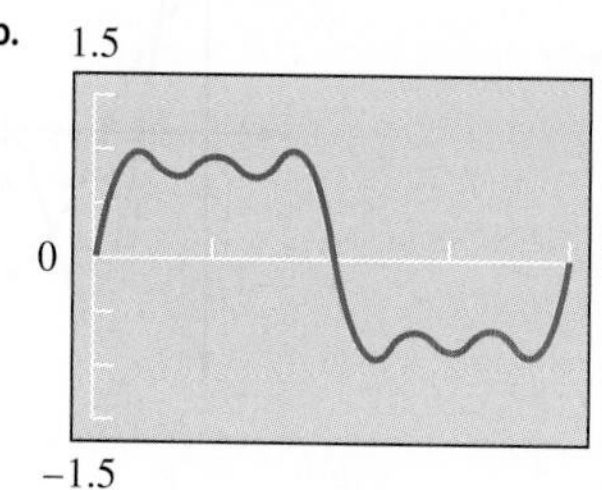

c.

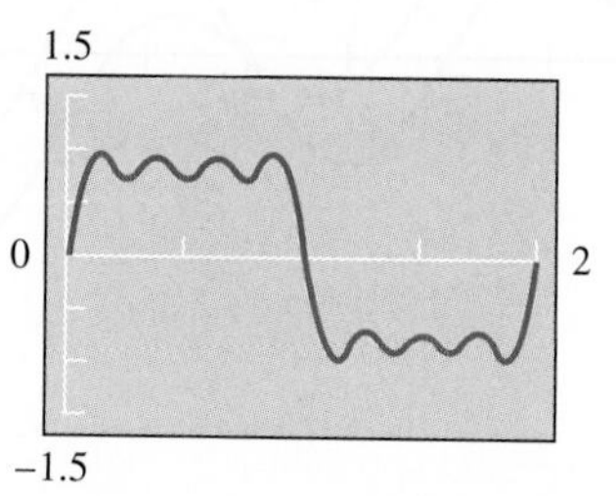

45. $\frac{1}{4}$ ft/sec **47.** 180 m **49.** 60π in./min

MATCHED PRACTICE PROBLEMS 4.7

1. a. $\frac{2\pi}{3}$ **b.** $-\frac{\pi}{3}$ **c.** $-\frac{\pi}{6}$ **2. a.** 25.6° **b.** −25.6° **c.** 136.2° **d.** 43.8° **e.** 75.3° **f.** −75.3°

3. $\sqrt{4 - x^2}$ **4. a.** $\frac{\sqrt{3}}{2}$ **b.** −30° **5.** x **6.** $\frac{5}{13}$ **7.** $\frac{\sqrt{4 - x^2}}{2}$

PROBLEM SET 4.7

1. interchange or reverse **3.** reflection, $y = x$ **5.** $y = \sin^{-1} x$, $y = \arcsin x$ **7.** angle, $-\frac{\pi}{2}, \frac{\pi}{2}$, sine **9.** I, IV **11.** I, IV

13.

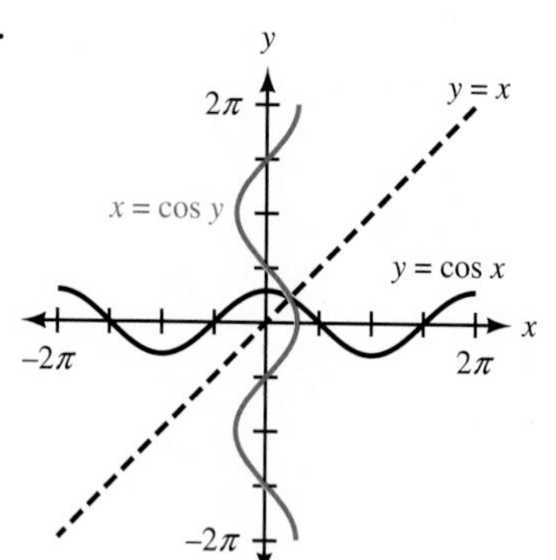

15.

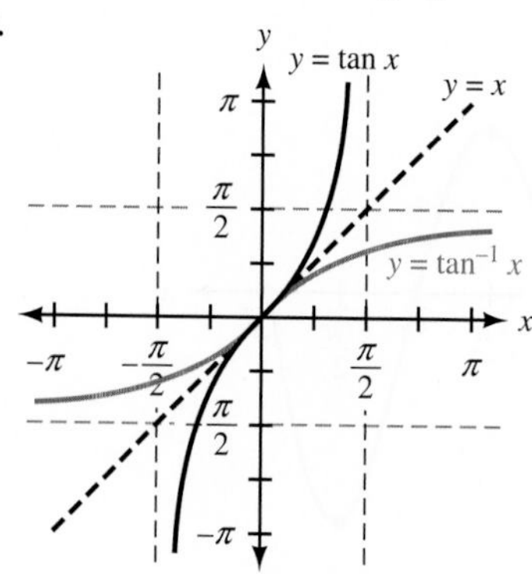

17. $\frac{\pi}{3}$ **19.** π **21.** $\frac{\pi}{4}$ **23.** $\frac{3\pi}{4}$ **25.** $-\frac{\pi}{6}$ **27.** $\frac{\pi}{3}$ **29.** 0 **31.** $-\frac{\pi}{6}$ **33.** $\frac{\pi}{2}$ **35.** $\frac{\pi}{6}$ **37.** 9.8°

39. 32.6° **41.** 20.8° **43.** 117.8° **45.** −70.0° **47.** −50.0° **49. a.** 60° **b.** 150° **c.** 45°

51. a. −45° **b.** 60° **c.** −30° **53.** $4 \cos \theta$ **55.** $\frac{3}{5}$ **57.** $\frac{1}{2}$ **59.** $\frac{1}{2}$ **61.** −45° **63.** $\frac{\pi}{3}$ **65.** 120° **67.** $\frac{\pi}{4}$

69. 45° **71.** $-\frac{\pi}{6}$ **73.** $\frac{3}{4}$ **75.** $\sqrt{5}$ **77.** $\frac{\sqrt{3}}{2}$ **79.** 2 **81.** x **83.** x **85.** $\sqrt{1 - x^2}$ **87.** $\frac{x}{\sqrt{x^2 + 1}}$

89. $\frac{\sqrt{x^2 - 1}}{x}$ **91.** x **93.** 1,370 miles

95.

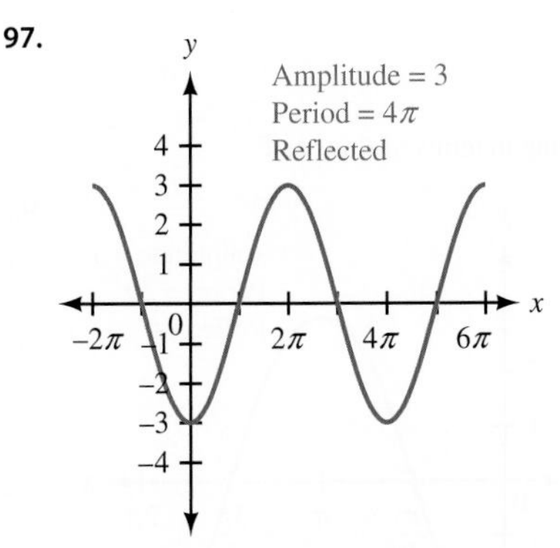

97.

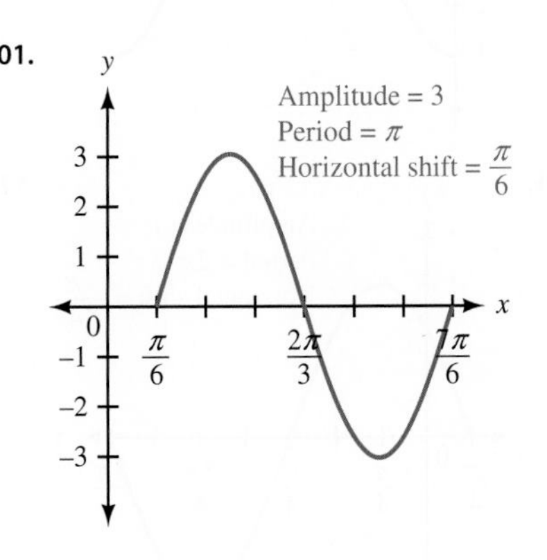

99.

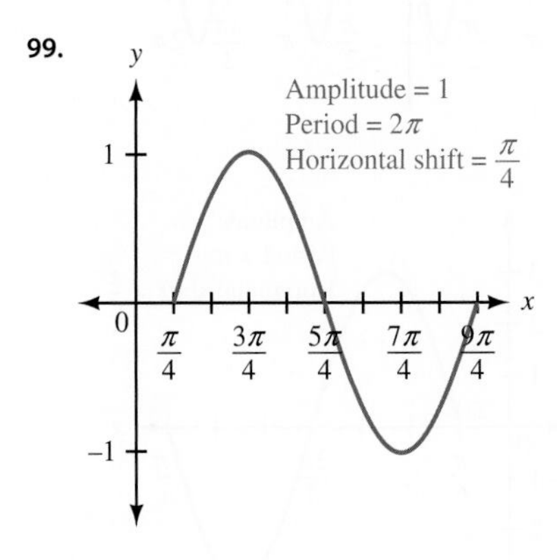

101.

CHAPTER 4 TEST

1.

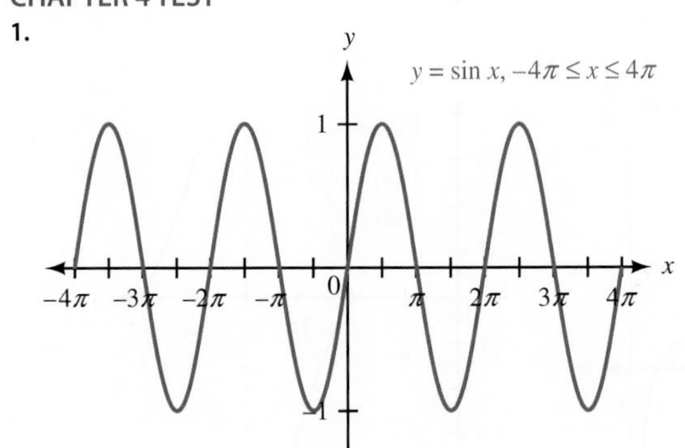

2.

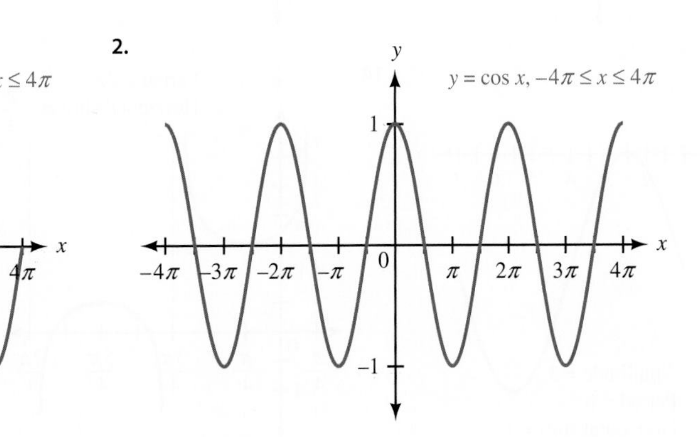

3.

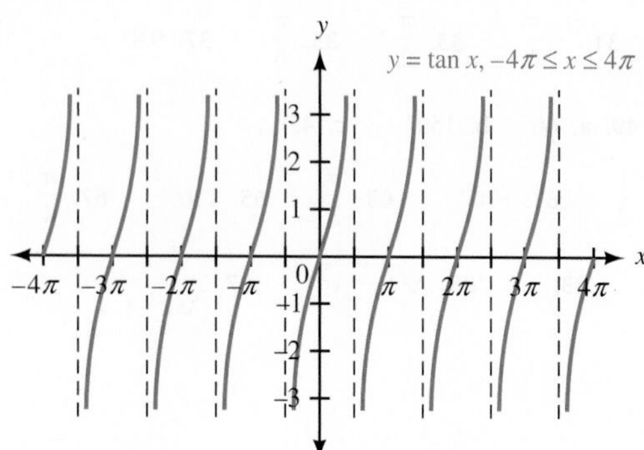

4.

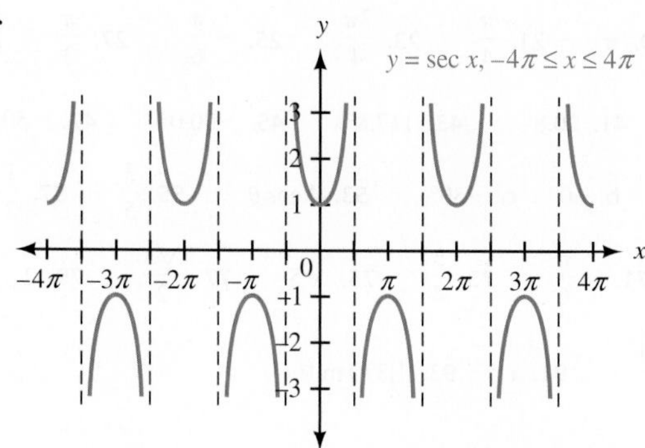

5. Begin by writing $\cot(-\theta) = \dfrac{\cos(-\theta)}{\sin(-\theta)}$

6. First use odd and even functions to write everything in terms of θ instead.

7.

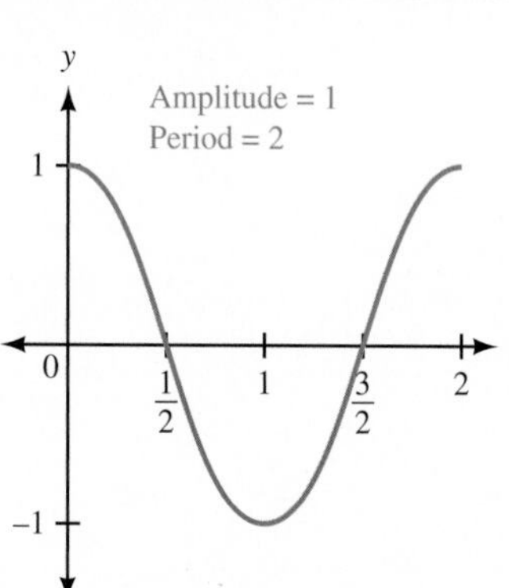

8.

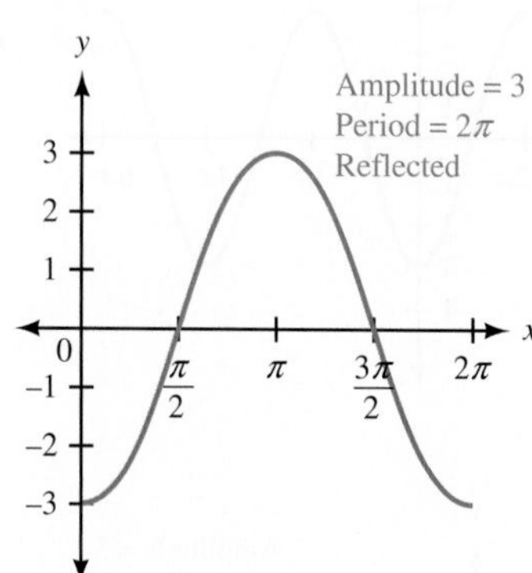

9.

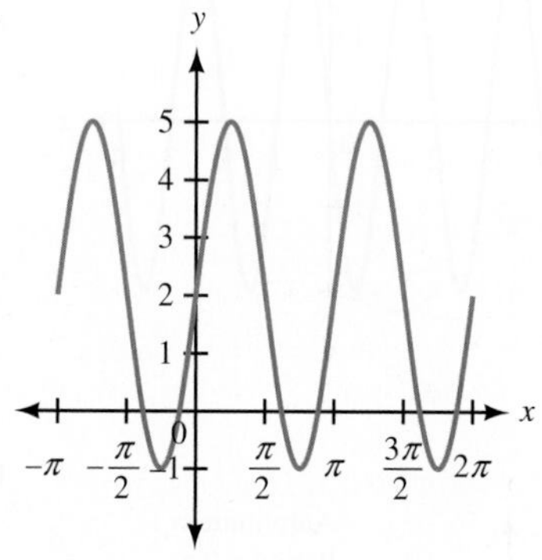

10.

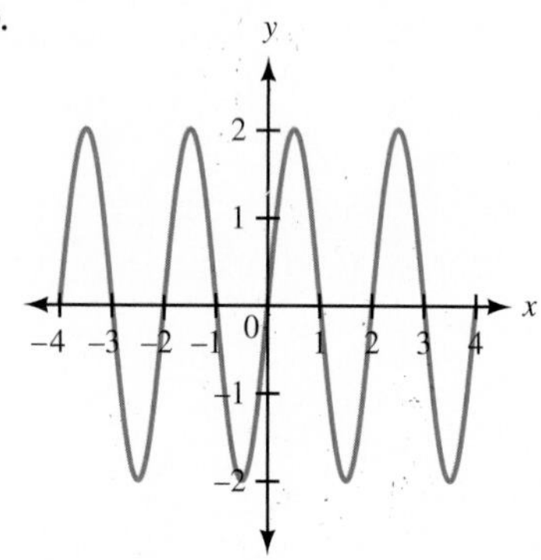

11.

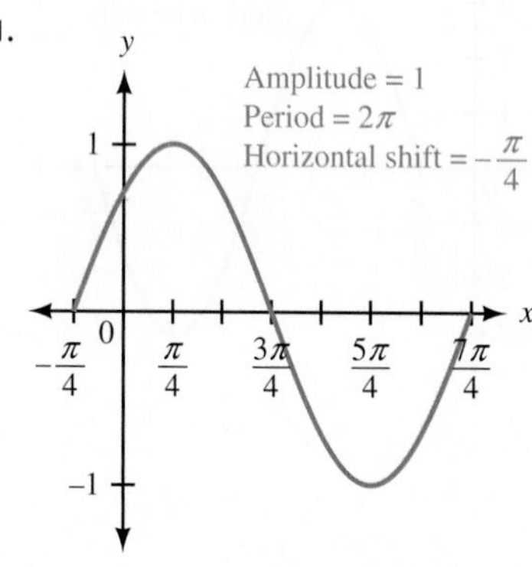

12.

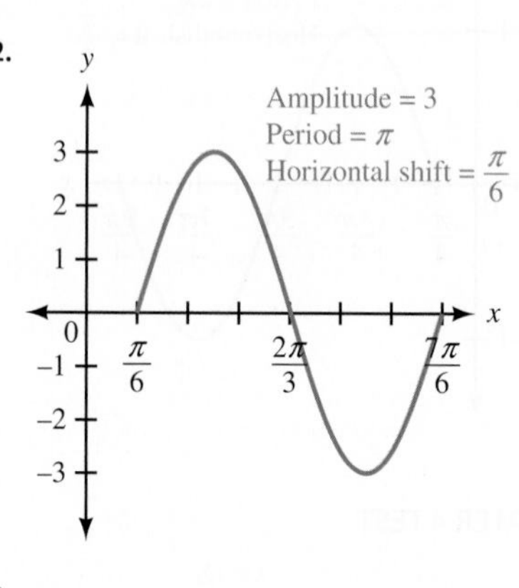

13.

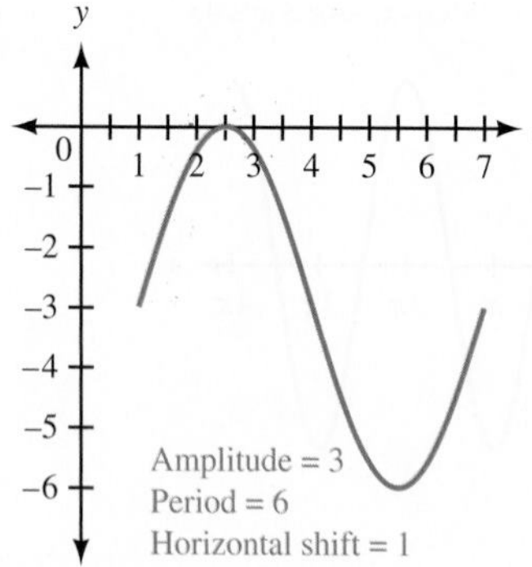

14.

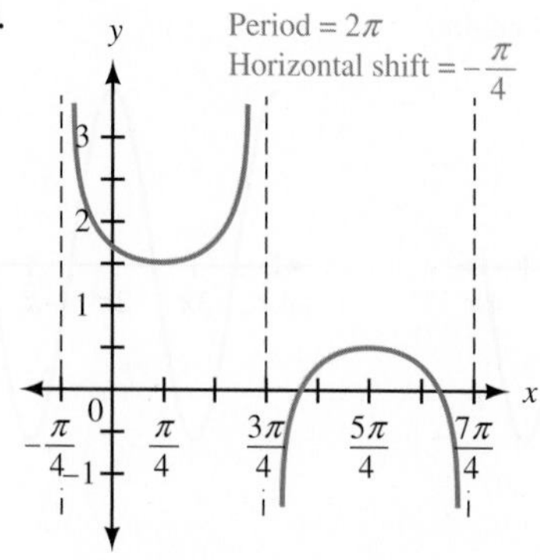

15.

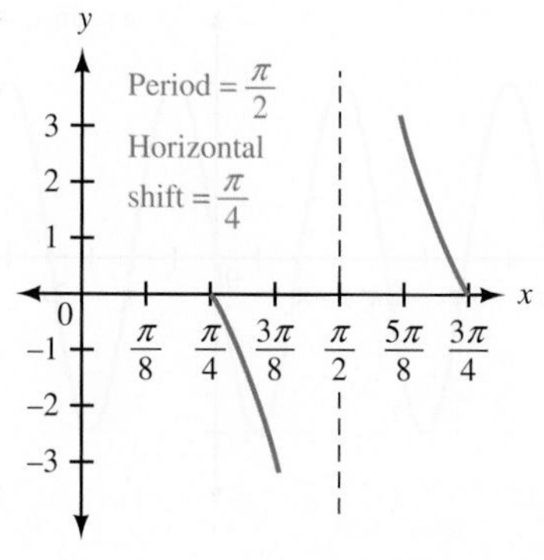

16.

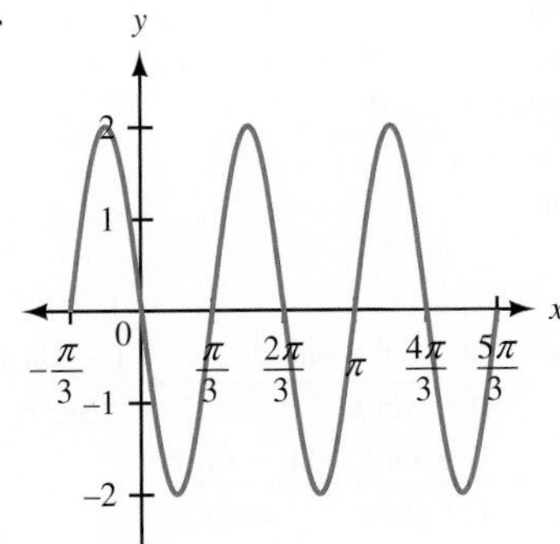

17.

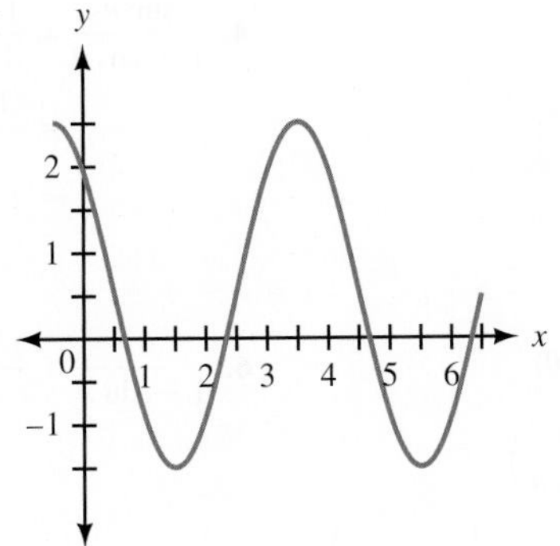

18. $y = 2\sin\left(\frac{1}{2}x + \frac{\pi}{2}\right)$ **19.** $y = \frac{1}{2} + \frac{1}{2}\sin\frac{\pi}{2}x$

20.

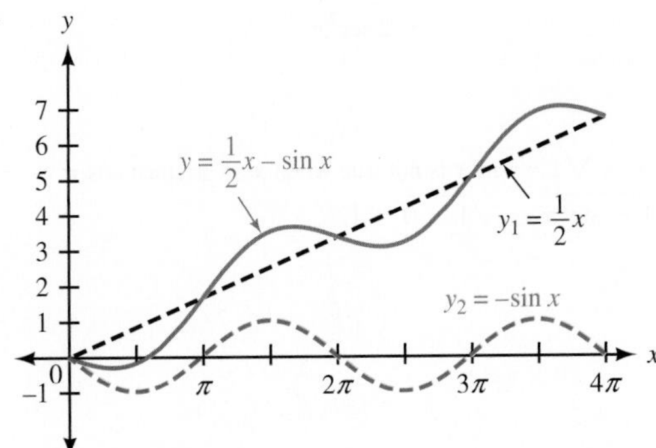

21.

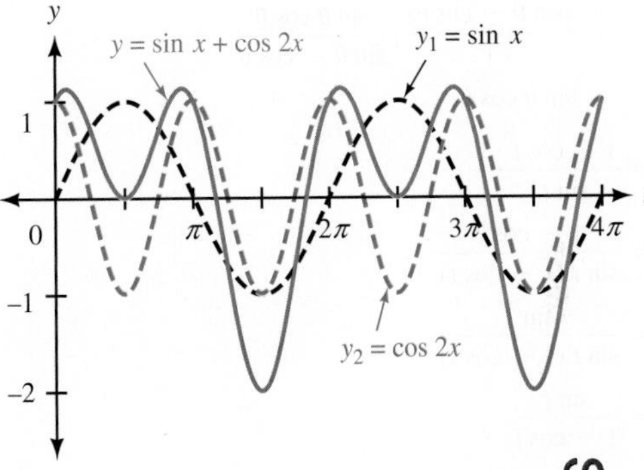

22.

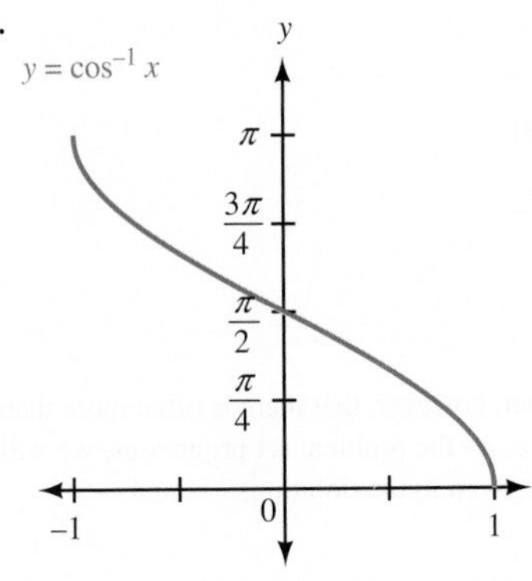

23.

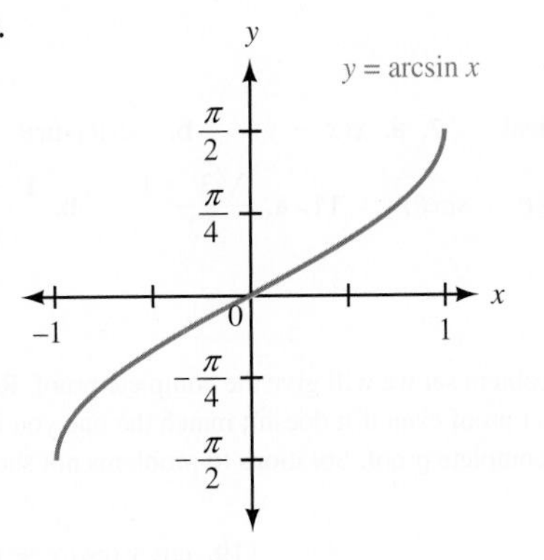

24. $\frac{\pi}{6}$ **25.** $-\frac{\pi}{4}$ **26.** 36.4° **27.** −39.7° **28.** $\frac{\sqrt{5}}{2}$ **29.** $\frac{\pi}{6}$ **30.** $\sqrt{1 - x^2}$

CHAPTER 5

MATCHED PRACTICE PROBLEMS 5.1

1.
$$\frac{\tan\theta}{\sec\theta} = \tan\theta \div \sec\theta$$
$$= \frac{\sin\theta}{\cos\theta} \div \frac{1}{\cos\theta}$$
$$= \frac{\sin\theta}{\cos\theta}\cdot\frac{\cos\theta}{1}$$
$$= \sin\theta$$

2.
$$\cos x(\csc x + \sec x) = \cos x\left(\frac{1}{\sin x} + \frac{1}{\cos x}\right)$$
$$= \frac{\cos x}{\sin x} + 1$$
$$= \cot x + 1$$

3. $\dfrac{\csc^2 x - 1}{\csc^2 x - \csc x} = \dfrac{(\csc x - 1)(\csc x + 1)}{\csc x(\csc x - 1)}$

$= \dfrac{\csc x + 1}{\csc x}$

$= \dfrac{\csc x}{\csc x} + \dfrac{1}{\csc x}$

$= 1 + \sin x$

4. $\dfrac{\sin^2\theta}{1+\cos\theta} = \dfrac{1-\cos^2\theta}{1+\cos\theta}$

$= \dfrac{(1-\cos\theta)(1+\cos\theta)}{1+\cos\theta}$

$= 1 - \cos\theta$

5. $\dfrac{\sin\theta - \cos\theta}{\sec\theta - \csc\theta} = (\sin\theta - \cos\theta) \div (\sec\theta - \csc\theta)$

$= (\sin\theta - \cos\theta) \div \left(\dfrac{1}{\cos\theta} - \dfrac{1}{\sin\theta}\right)$

$= (\sin\theta - \cos\theta) \div \left(\dfrac{1}{\cos\theta}\cdot\dfrac{\sin\theta}{\sin\theta} - \dfrac{1}{\sin\theta}\cdot\dfrac{\cos\theta}{\cos\theta}\right)$

$= (\sin\theta - \cos\theta) \div \dfrac{\sin\theta - \cos\theta}{\sin\theta\cos\theta}$

$= \dfrac{(\sin\theta - \cos\theta)}{1}\cdot\dfrac{\sin\theta\cos\theta}{\sin\theta - \cos\theta}$

$= \sin\theta\cos\theta$

6. $\dfrac{1}{1-\sin x} + \dfrac{1-\sin x}{\cos^2 x} = \dfrac{1(1+\sin x)}{(1-\sin x)(1+\sin x)} + \dfrac{1-\sin x}{\cos^2 x}$

$= \dfrac{1+\sin x}{1-\sin^2 x} + \dfrac{1-\sin x}{\cos^2 x}$

$= \dfrac{1+\sin x}{\cos^2 x} + \dfrac{1-\sin x}{\cos^2 x}$

$= \dfrac{2}{\cos^2 x}$

$= 2\sec^2 x$

7. $\dfrac{1+\cos t}{\sin t} = \dfrac{1+\cos t}{\sin t}\cdot\dfrac{1-\cos t}{1-\cos t}$

$= \dfrac{1-\cos^2 t}{\sin t(1-\cos t)}$

$= \dfrac{\sin^2 t}{\sin t(1-\cos t)}$

$= \dfrac{\sin t}{1-\cos t}$

8. $\cos x = \sqrt{1-\sin^2 x}$ is not true when $x = \pi$ since $\cos\pi = -1$ but $\sqrt{1-\sin^2\pi} = \sqrt{1-0} = 1$.

PROBLEM SET 5.1

1. equal, defined **3.** complicated **5.** identical **7. a.** $x(x - y)$ **b.** $\sin\theta(\sin\theta - \cos\theta)$

9. a. $(x + y)(x - y)$ **b.** $(\cos\theta + \sin\theta)(\cos\theta - \sin\theta)$ **11. a.** $\dfrac{\sqrt{3}-1}{2}$ **b.** $\dfrac{1-\cos x}{\sin^2 x}$

13. a. $-3 + 2\sqrt{2}$ **b.** $\dfrac{(1-\sin x)^2}{\cos^2 x}$

For some of the problems in the early part of this problem set we will give the complete proof. Remember, however, that there is often more than one way to prove an identity. You may have a correct proof even if it doesn't match the one you find here. As the problem set progresses, we will give hints on how to begin the proof instead of the complete proof. Solutions to problems not shown are given in the Solutions Manual.

15. $\csc\theta\tan\theta = \dfrac{1}{\sin\theta}\cdot\dfrac{\sin\theta}{\cos\theta}$

$= \dfrac{1}{\cos\theta}$

$= \sec\theta$

19. $\cos x(\csc x + \tan x) = \cos x\csc x + \cos x\tan x$

$= \cos x\cdot\dfrac{1}{\sin x} + \cos x\cdot\dfrac{\sin x}{\cos x}$

$= \dfrac{\cos x}{\sin x} + \sin x$

$= \cot x + \sin x$

25. $\dfrac{\cos^4 t - \sin^4 t}{\sin^2 t} = \dfrac{(\cos^2 t + \sin^2 t)(\cos^2 t - \sin^2 t)}{\sin^2 t}$

$= \dfrac{\cos^2 t - \sin^2 t}{\sin^2 t}$

$= \dfrac{\cos^2 t}{\sin^2 t} - \dfrac{\sin^2 t}{\sin^2 t}$

$= \cot^2 t - 1$

27. Write the numerator on the right side as $1 - \sin^2\theta$ and then factor it. **33.** Factor the left side and then write it in terms of sines and cosines. **35.** Change the left side to sines and cosines and then add the resulting fractions. **41.** See Example 6 in this section.

45. Rewrite the left side in terms of cosine and then simplify. **69.** $\theta = -\frac{\pi}{3}$ is one possible answer. **71.** $\theta = 0$ is one possible answer.

73. $\theta = \frac{\pi}{4}$ is one possible answer.

Note For Problems 75–82, when the equation is an identity, the proof is given in the Solutions Manual.

75. Is an identity. **77.** Not an identity; $x = \pi/3$ is one possible counterexample. **79.** Not an identity; $A = \pi/6$ is one possible counterexample. **81.** Is an identity. **83.** See the Solutions Manual.

85. $\cos A = \frac{4}{5}$, $\tan A = \frac{3}{4}$ **87.** $\frac{\sqrt{3}}{2}$ **89.** $\frac{\sqrt{3}}{2}$ **91.** $15°$ **93.** See the Solutions Manual.

MATCHED PRACTICE PROBLEMS 5.2

1. $\frac{\sqrt{6}+\sqrt{2}}{4}$

2.
$$\begin{aligned}\sin\left(x+\frac{\pi}{2}\right) &= \sin x \cos\frac{\pi}{2} + \cos x \sin\frac{\pi}{2}\\ &= \sin x\,(0) + \cos x\,(1)\\ &= \cos x\end{aligned}$$

3. $\sin 6x$

4.
$$\begin{aligned}\sin(90° - B) &= \sin 90° \cos B - \cos 90° \sin B\\ &= (1)\cos B - (0)\sin B\\ &= \cos B\end{aligned}$$

5.

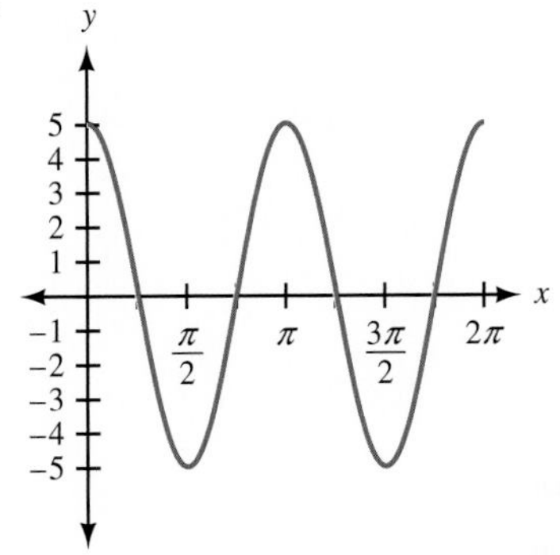

6. $\frac{\sqrt{6}-\sqrt{2}}{4}$

7. $\sin(A-B) = \frac{64}{1025}$, $\cos(A-B) = -\frac{1023}{1025}$, $\tan(A-B) = -\frac{64}{1023}$

8. $-\frac{304}{297}$

PROBLEM SET 5.2

1. $\sin x \cos y + \cos x \sin y$ **3.** $\cos\theta\cos\phi - \sin\theta\sin\phi$ **5.** $\frac{\tan C + \tan D}{1 - \tan C \tan D}$ **7.** False **9.** $\frac{\sqrt{6}-\sqrt{2}}{4}$

11. $\frac{\sqrt{6}-\sqrt{2}}{\sqrt{6}+\sqrt{2}} = 2 - \sqrt{3}$ **13.** $\frac{\sqrt{2}-\sqrt{6}}{4}$ **15.** $\frac{\sqrt{2}-\sqrt{6}}{4}$

17.
$$\begin{aligned}\sin(x + 2\pi) &= \sin x \cos 2\pi + \cos x \sin 2\pi\\ &= \sin x\,(1) + \cos x\,(0)\\ &= \sin x\end{aligned}$$

For Problems 18–28, proceed as in Problem 17. Expand the left side and simplify. Solutions are given in the Solutions Manual.

29. $\sin 5x$ **31.** $\cos 6x$ **33.** $\cos 90° = 0$

35.

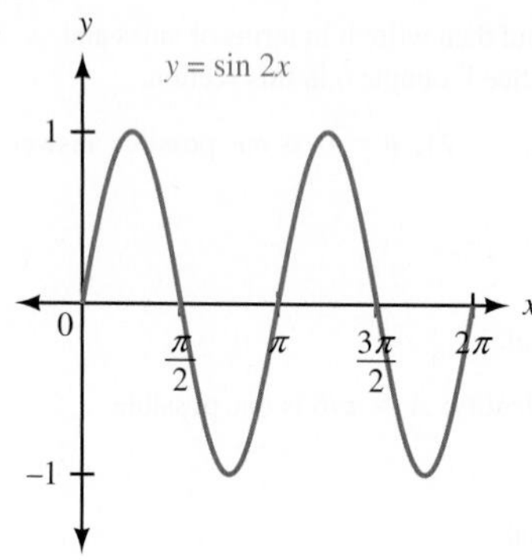

37.

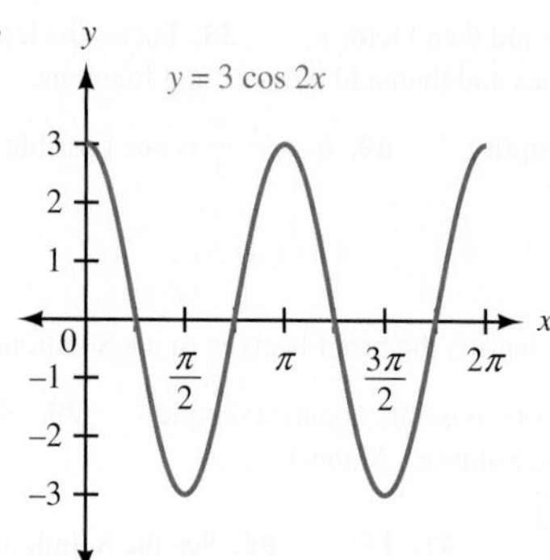

39.

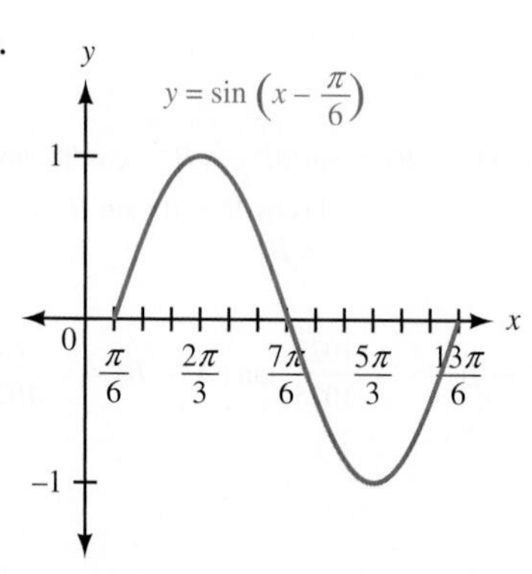

41.

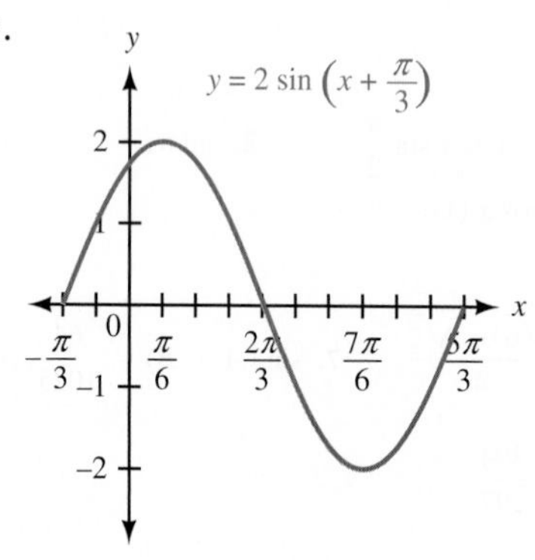

43. $-\frac{16}{65}, \frac{63}{65}, -\frac{16}{63}$, QIV **45.** $2, \frac{1}{2}$, QI **47.** 1 **49.** $\cos 2x = \cos^2 x - \sin^2 x$

For Problems 51–66, see the Solutions Manual.

Note For Problems 67–72, when the equation is an identity, the proof is given in the Solutions Manual.

67. Is an identity. **69.** Not an identity; $x = 0$ is one possible counterexample. **71.** Is an identity.

73.

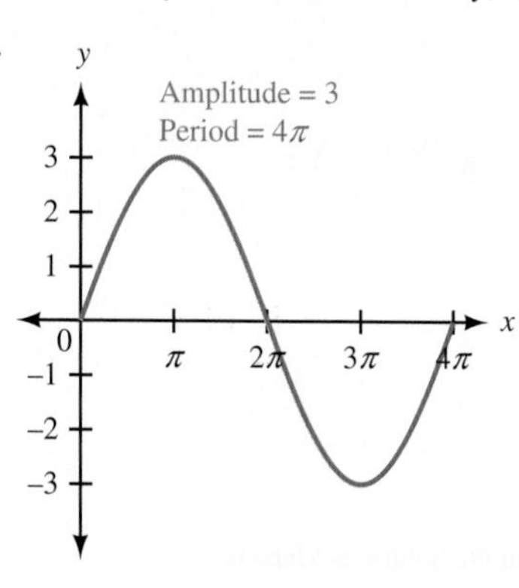

75.

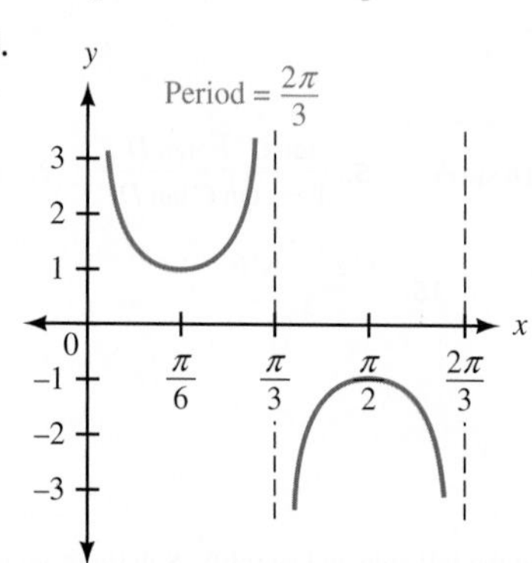

77.

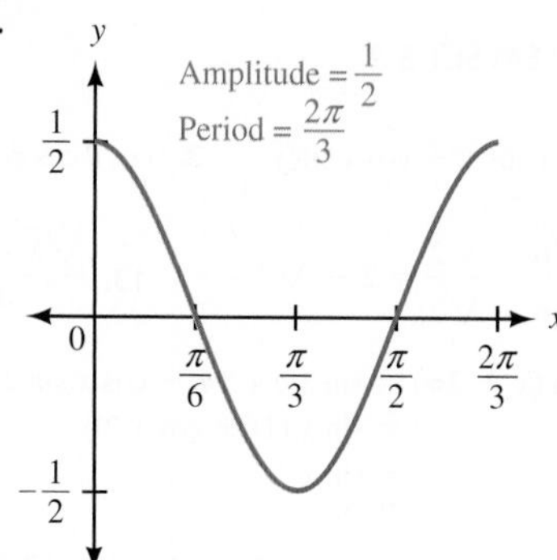

MATCHED PRACTICE PROBLEMS 5.3

1. $\dfrac{3\sqrt{55}}{32}$

2. $\dfrac{4}{\tan x + \cot x} = 4 \div (\tan x + \cot x)$
$= 4 \div \left(\dfrac{\sin x}{\cos x} + \dfrac{\cos x}{\sin x}\right)$
$= 4 \div \left(\dfrac{\sin^2 x + \cos^2 x}{\sin x \cos x}\right)$
$= (2 \cdot 2) \div \left(\dfrac{1}{\sin x \cos x}\right)$
$= 2 \cdot 2 \sin x \cos x$
$= 2 \sin 2x$

3. $\dfrac{2 \tan \theta}{1 + \tan^2\theta} = \dfrac{2 \tan \theta}{\sec^2\theta}$
$= 2\dfrac{\sin \theta}{\cos \theta} \cdot \cos^2 \theta$
$= 2 \sin \theta \cos \theta$
$= \sin 2\theta$

4. $\dfrac{1}{8}$

5. $\sin 2x \cos 2x = 2 \sin x \cos x \cdot (1 - 2 \sin^2 x)$
$= 2 \sin x \cos x - 4 \cos x \sin^3 x$
$= \sin 2x - 4 \cos x \sin^3 x$

6.

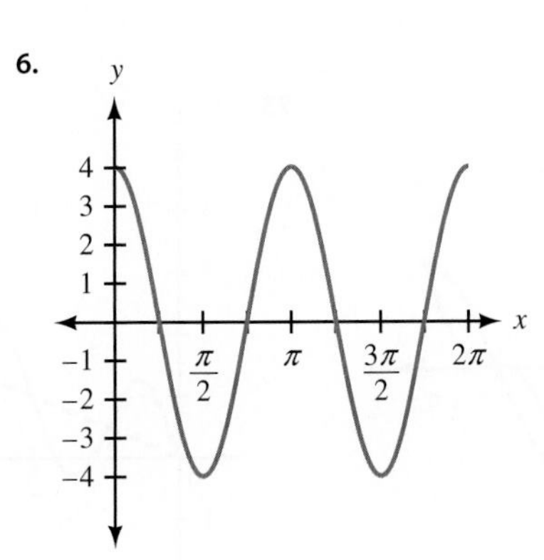

7. $\dfrac{1 + \cos 2\theta}{\sin 2\theta} = \dfrac{1 + 2\cos^2\theta - 1}{2 \sin \theta \cos \theta}$
$= \dfrac{2 \cos^2\theta}{2 \sin \theta \cos \theta}$
$= \dfrac{\cos \theta}{\sin \theta}$
$= \cot \theta$

8. $\dfrac{\sqrt{2}}{2}$ **9.** $\cos^{-1}\dfrac{x}{2} - \dfrac{x\sqrt{4 - x^2}}{4}$

PROBLEM SET 5.3

1. $2 \sin x \cos x$ **3.** $\dfrac{2 \tan y}{1 - \tan^2 y}$ **5.** False **7.** $\dfrac{24}{25}$ **9.** $\dfrac{24}{7}$ **11.** $-\dfrac{4}{5}$ **13.** $\dfrac{4}{3}$ **15.** $\dfrac{120}{169}$ **17.** $\dfrac{169}{119}$ **19.** $\dfrac{3}{5}$ **21.** $\dfrac{5}{3}$

23.

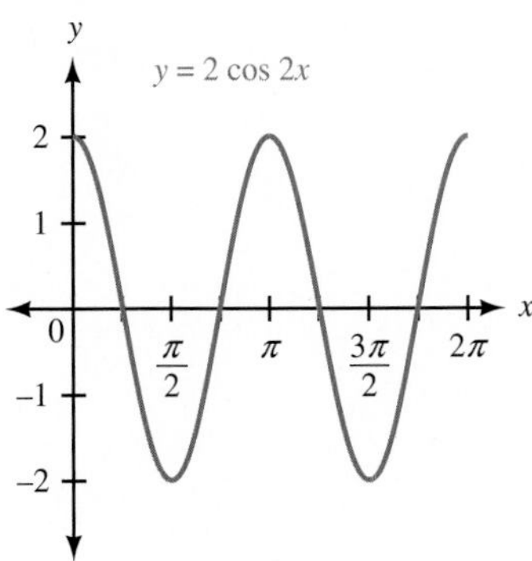

25.

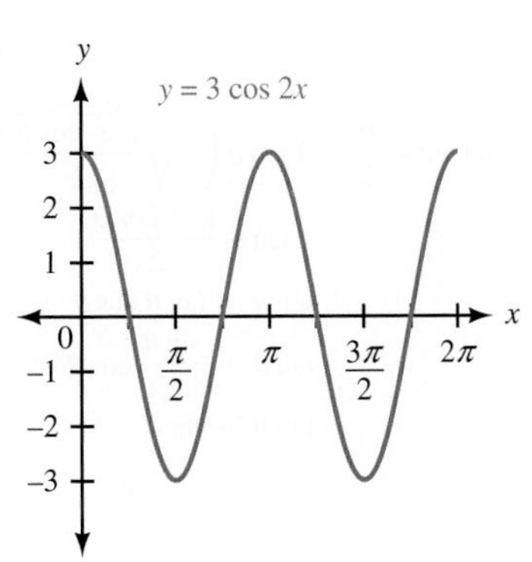

27.

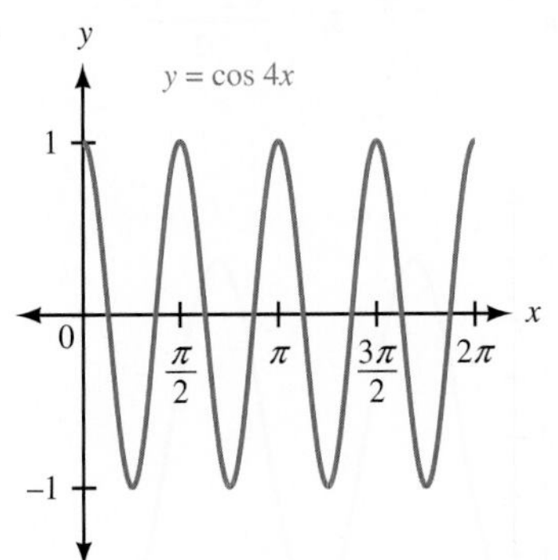

For Problems 29–32, see the Solutions Manual.

33. $\sqrt{3}$ **35.** $\dfrac{1}{2}$ **37.** $-\dfrac{\sqrt{3}}{2}$ **39.** $\dfrac{1}{4}$ **41.** $\dfrac{1}{2}$

For Problems 43–60, see the Solutions Manual.

Note For Problems 61–64, when the equation is an identity, the proof is given in the Solutions Manual.

61. Is an identity **63.** Not an identity; $x = \pi/3$ is one possible counterexample.

65. $\frac{1}{2}\left(\tan^{-1}\frac{x}{5} - \frac{5x}{x^2 + 25}\right)$ **67.** $\frac{1}{2}\left(\sin^{-1}\frac{x}{3} - \frac{x\sqrt{9 - x^2}}{9}\right)$ **69.**

$y = 2 - 2\cos x$

71.

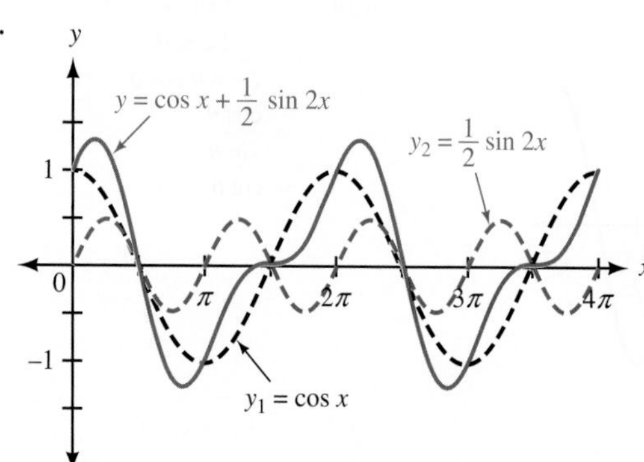

73.

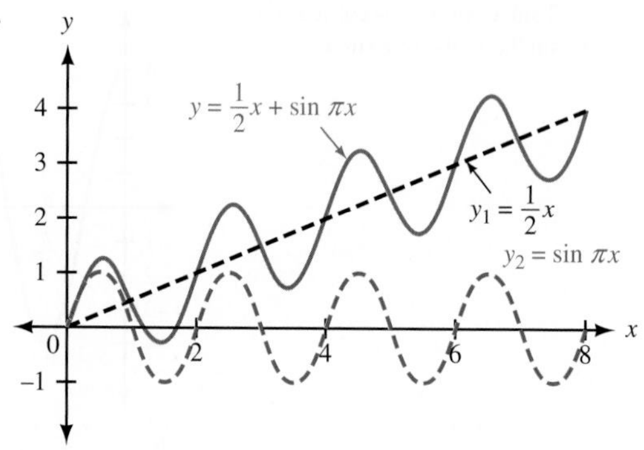

MATCHED PRACTICE PROBLEMS 5.4

1. $\sin\frac{A}{2} = \frac{4\sqrt{17}}{17}$, $\cos\frac{A}{2} = -\frac{\sqrt{17}}{17}$, $\tan\frac{A}{2} = -4$ **2.** $\sin\frac{A}{2} = \frac{\sqrt{2}}{10}$, $\cos\frac{A}{2} = -\frac{7\sqrt{2}}{10}$, $\tan\frac{A}{2} = -\frac{1}{7}$, $\csc\frac{A}{2} = 5\sqrt{2}$, $\sec\frac{A}{2} = -\frac{5\sqrt{2}}{7}$, $\cot\frac{A}{2} = -7$ **3.** $-\frac{\sqrt{2 - \sqrt{2}}}{2}$

4.

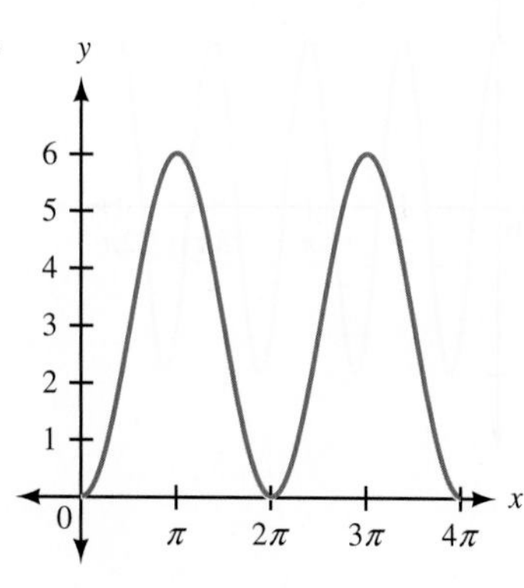

5.
$$
\begin{aligned}
2\tan\theta\cos^2\frac{\theta}{2} &= 2\tan\theta\left(\pm\sqrt{\frac{1+\cos\theta}{2}}\right)^2 \\
&= 2\tan\theta\,\frac{1+\cos\theta}{2} \\
&= \tan\theta + \tan\theta\cos\theta \\
&= \tan\theta + \frac{\sin\theta}{\cos\theta}\cos\theta \\
&= \tan\theta + \sin\theta
\end{aligned}
$$

PROBLEM SET 5.4

1. $\frac{A}{2}$ **3.** $\pm\sqrt{\frac{1 - \cos x}{2}}$ **5.** $\frac{1 - \cos y}{\sin y}$ or $\frac{\sin y}{1 + \cos y}$ **7.** False **9.** QI **11.** QII **13.** Negative **15.** False

17. $\frac{1}{2}$ **19.** $-\frac{2\sqrt{3}}{3}$ **21.** $-\frac{\sqrt{10}}{10}$ **23.** $-\sqrt{10}$ **25.** $\sqrt{\frac{3 + 2\sqrt{2}}{6}}$ **27.** $-\sqrt{\frac{3 - 2\sqrt{2}}{6}}$ **29.** $-3 + 2\sqrt{2}$

31. $\frac{\sqrt{5}}{5}$ **33.** $\frac{\sqrt{10}}{10}$

35.

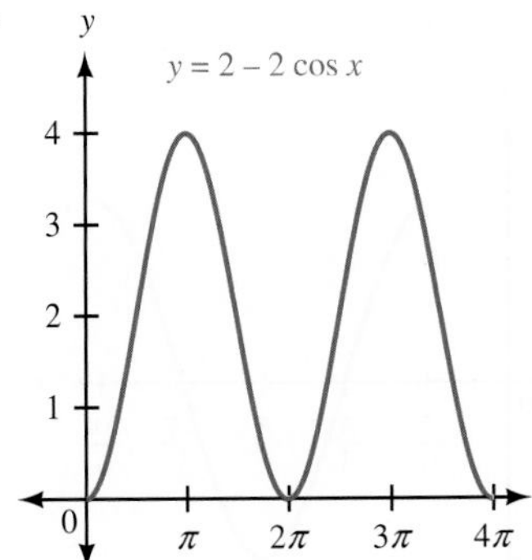

37.

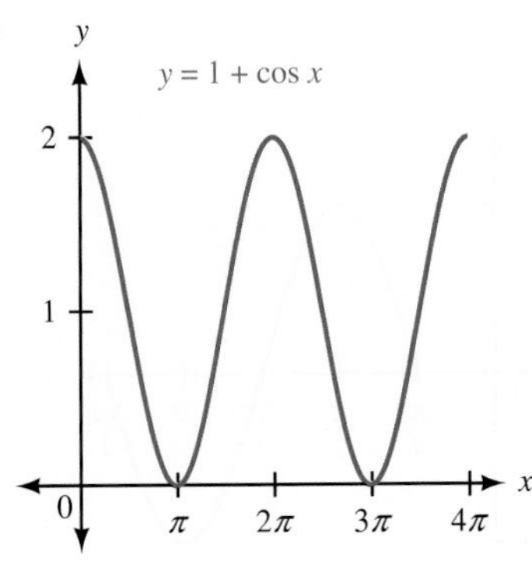

39. $\dfrac{\sqrt{2+\sqrt{3}}}{2}$ **41.** $\dfrac{\sqrt{2+\sqrt{3}}}{2}$ **43.** $-\dfrac{\sqrt{2-\sqrt{3}}}{2}$

For Problems 45–56, see the Solutions Manual.

57. $\dfrac{3}{5}$ **59.** $\dfrac{\sqrt{5}}{5}$ **61.** $\dfrac{x}{\sqrt{x^2+1}}$ **63.** $\dfrac{x}{\sqrt{1-x^2}}$

MATCHED PRACTICE PROBLEMS 5.5

1. $\dfrac{-3-4\sqrt{3}}{10}$ **2.** $\dfrac{1-x^2}{1+x^2}$

3.
$$\begin{aligned}\cos 240° \sin 30° &= \frac{1}{2}[\sin(240° + 30°) - \sin(240° - 30°)]\\ -\frac{1}{2}\cdot\frac{1}{2} &= \frac{1}{2}[\sin 270° - \sin 210°]\\ -\frac{1}{4} &= \frac{1}{2}\left[-1 - \left(-\frac{1}{2}\right)\right]\\ -\frac{1}{4} &= \frac{1}{2}\left[-\frac{1}{2}\right]\\ -\frac{1}{4} &= -\frac{1}{4}\end{aligned}$$

4. $3\cos 10x + 3\cos 4x$

5.
$$\begin{aligned}\sin 240° + \sin 60° &= 2\sin\frac{240° + 60°}{2}\cos\frac{240° - 60°}{2}\\ \sin 240° + \sin 60° &= 2\sin 150° \cos 90°\\ -\frac{\sqrt{3}}{2} + \frac{\sqrt{3}}{2} &= 2\left(\frac{1}{2}\right)(0)\\ 0 &= 0\end{aligned}$$

6.
$$\begin{aligned}\frac{\sin 5x + \sin 7x}{\cos 5x - \cos 7x} &= \frac{2\sin\left(\frac{5x+7x}{2}\right)\cos\left(\frac{5x-7x}{2}\right)}{-2\sin\left(\frac{5x+7x}{2}\right)\sin\left(\frac{5x-7x}{2}\right)}\\ &= \frac{2\sin(6x)\cos(-x)}{-2\sin(6x)\sin(-x)}\\ &= \frac{2\sin(6x)\cos x}{-2\sin(6x)(-\sin x)}\\ &= \frac{\cos x}{\sin x}\\ &= \cot x\end{aligned}$$

PROBLEM SET 5.5

1. angle **3.** $\frac{1}{2}[\sin(x+y) - \sin(x-y)]$ **5.** $2\cos\frac{x+y}{2}\sin\frac{x-y}{2}$ **7.** $\frac{2\sqrt{5}}{5}$ **9.** $\frac{2\sqrt{3}-1}{2\sqrt{5}} = \frac{2\sqrt{15}-\sqrt{5}}{10}$ **11.** $\frac{4}{5}$

13. $-\frac{5}{13}$ **15.** $\frac{x}{\sqrt{1-x^2}}$ **17.** $\frac{\sqrt{9-x^2}}{x}$ **19.** $2x\sqrt{1-x^2}$ **21.** $2x^2 - 1$ **23.** $\frac{\sqrt{x^2+2x-8}}{x+1}$

25. See the Solutions Manual. **27.** $5(\sin 8x + \sin 2x)$ **29.** $\frac{1}{2}(\cos 10x + \cos 6x)$ **31.** $\frac{1}{2}[\cos 270° + \cos(-90°)] = 0$

33. $\frac{1}{2}(\sin 4\pi - \sin 2\pi) = 0$ **35.** See the Solutions Manual. **37.** $2\sin 5x \cos 2x$ **39.** $2\cos 30° \cos 15° = \sqrt{3}\cos 15°$

41. $2\cos\frac{\pi}{3}\sin\frac{\pi}{4} = \frac{\sqrt{2}}{2}$

For Problems 43–48, see the Solutions Manual.

49.

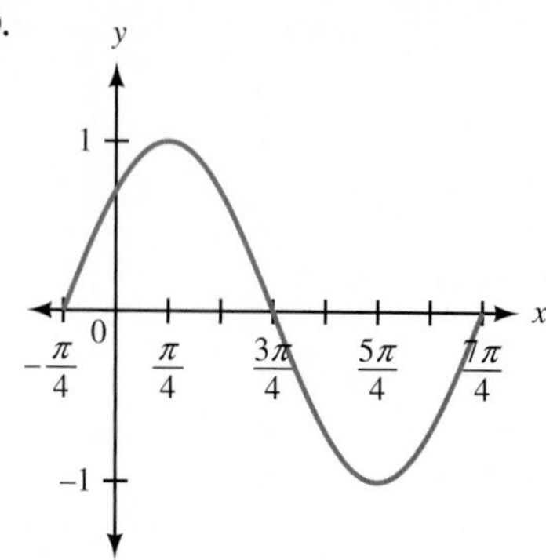

51.

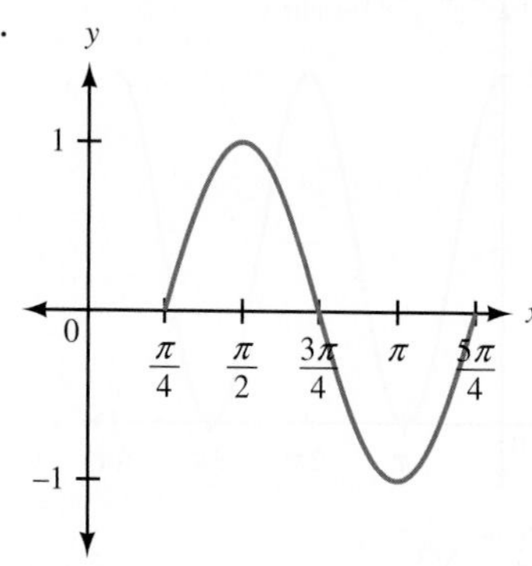

53.

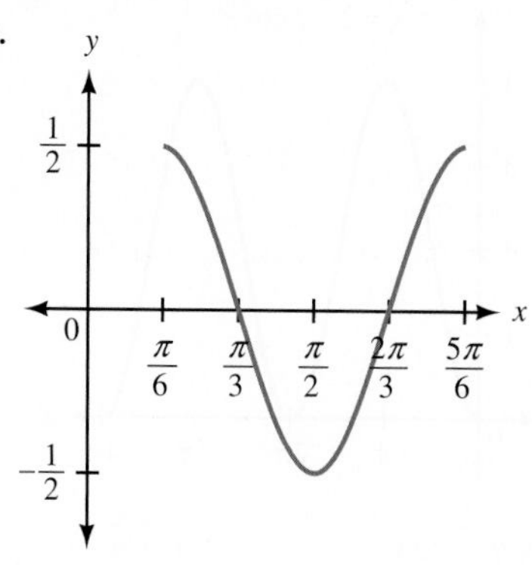

CHAPTER 5 TEST

For Problems 1–10, see the Solutions Manual.

Note For Problems 11 and 12, when the equation is an identity, the proof is given in the Solutions Manual.

11. Is an identity. **12.** Not an identity; $\theta = \frac{\pi}{4}$ is one possible counterexample. **13.** $\frac{63}{65}$ **14.** $-\frac{56}{65}$

15. $-\frac{119}{169}$ **16.** $\frac{\sqrt{10}}{10}$

Note For Problems 17–18, other answers are possible depending on the identity used.

17. $\frac{\sqrt{6} + \sqrt{2}}{4}$ **18.** $\frac{\sqrt{3} - 1}{\sqrt{3} + 1} = 2 - \sqrt{3}$ **19.** $\cos 9x$ **20.** $\sin 90° = 1$ **21.** $\frac{3}{5}, -\sqrt{\frac{5 - 2\sqrt{5}}{10}}$

22. $\frac{3}{5}, \sqrt{\frac{10 - \sqrt{10}}{20}}$ **23.** 1 **24.** $\pm\frac{\sqrt{3}}{2}$ **25.** $\frac{11\sqrt{5}}{25}$ **26.** $\frac{11\sqrt{5}}{25}$ **27.** $1 - 2x^2$ **28.** $2x\sqrt{1 - x^2}$

29. $\frac{1}{2}(\cos 2x - \cos 10x)$ **30.** $2 \cos 45° \cos(-30°) = \frac{\sqrt{6}}{2}$

CHAPTER 6

MATCHED PRACTICE PROBLEMS 6.1

1. 1 **2.** 60°, 300° **3.** No solution **4.** $180° + 360°k, 300° + 360°k$ **5.** 109.5°, 250.5° **6.** $-\frac{1}{2}, -1$ **7.** $\frac{2\pi}{3}, \pi, \frac{4\pi}{3}$

8. 4.04, 5.39

PROBLEM SET 6.1

1. angle, true **3.** reference, 0°, 360°, 360° **5. a.** $30° + 360°k, 150° + 360°k$ **b.** 30°, 150°

7. a. $30° + 360°k, 330° + 360°k$ **b.** 30°, 330° **9. a.** $60° + 180°k$ **b.** 60°, 240° **11. a.** $\frac{\pi}{3} + 2k\pi, \frac{2\pi}{3} + 2k\pi$ **b.** $\frac{\pi}{3}, \frac{2\pi}{3}$

13. a. $\frac{\pi}{6} + 2k\pi, \frac{11\pi}{6} + 2k\pi$ **b.** $\frac{\pi}{6}, \frac{11\pi}{6}$ **15. a.** $\frac{3\pi}{2} + 2k\pi$ **b.** $\frac{3\pi}{2}$ **17. a.** $48.6° + 360°k, 131.4° + 360°k$

b. 48.6°, 131.4° **19. a.** ∅ **b.** ∅ **21. a.** $228.6° + 360°k, 311.4° + 360°k$ **b.** 228.6°, 311.4° **23. a.** $x(1 - 2y)$

b. $\sin\theta(1 - 2\cos\theta)$ **25. a.** $(2x - 1)(x - 3)$ **b.** $(2\cos\theta - 1)(\cos\theta - 3)$

27. a. $\frac{\pi}{6} + 2k\pi, \frac{\pi}{2} + 2k\pi, \frac{5\pi}{6} + 2k\pi$ **b.** $\frac{\pi}{6}, \frac{\pi}{2}, \frac{5\pi}{6}$ **29. a.** $k\pi, \frac{2\pi}{3} + 2k\pi, \frac{4\pi}{3} + 2k\pi$ **b.** $0, \frac{2\pi}{3}, \pi, \frac{4\pi}{3}$

31. a. $\frac{\pi}{2} + 2k\pi, \frac{7\pi}{6} + 2k\pi, \frac{11\pi}{6} + 2k\pi$ **b.** $\frac{\pi}{2}, \frac{7\pi}{6}, \frac{11\pi}{6}$

33. a. $120° + 360°k, 150° + 360°k, 210° + 360°k, 240° + 360°k$ **b.** $120°, 150°, 210°, 240°$
35. a. $0° + 180°k, 60° + 360°k, 120° + 360°k$ **b.** $0°, 60°, 120°, 180°$
37. a. $120° + 360°k, 240° + 360°k$ **b.** $120°, 240°$ **39. a.** $201.5° + 360°k, 338.5° + 360°k$ **b.** $201.5°, 338.5°$
41. a. $51.8° + 360°k, 308.2° + 360°k$ **b.** $51.8°, 308.2°$ **43. a.** $17.0° + 360°k, 163.0° + 360°k$ **b.** $17.0°, 163.0°$
45. $20° + 360°k, 80° + 360°k$ **47.** $360°k, 120° + 360°k$ **49.** $\frac{7\pi}{9} + 2k\pi, \frac{13\pi}{9} + 2k\pi$ **51.** $\frac{7\pi}{6} + 2k\pi, \frac{5\pi}{3} + 2k\pi$

For Problems 53–70, see the answer for the corresponding problem.

71. $h = -16t^2 + 750t$ **73.** 1,436 ft **75.** 15.7° **77.** $\cos 2A = 1 - 2\sin^2 A$ **79.** $\frac{\sqrt{2}}{2}\sin\theta + \frac{\sqrt{2}}{2}\cos\theta$ **81.** $\frac{\sqrt{6} + \sqrt{2}}{4}$
83. See the Solutions Manual.

MATCHED PRACTICE PROBLEMS 6.2

1. 0 **2.** 0°, 60°, 180°, 300° **3.** 60°, 180°, 300° **4.** $\frac{\pi}{6}, \frac{5\pi}{6}$ **5.** 30°, 210°

PROBLEM SET 6.2

1. reciprocal, clear, multiplying **3.** square, extraneous **5.** 30°, 330° **7.** 225°, 315° **9.** 45°, 135°, 225°, 315° **11.** 30°, 150°
13. 30°, 90°, 150°, 270° **15.** 90°, 210°, 330° **17.** $\frac{7\pi}{6}, \frac{3\pi}{2}, \frac{11\pi}{6}$ **19.** $0, \frac{2\pi}{3}, \frac{4\pi}{3}$ **21.** $\frac{\pi}{2}, \frac{7\pi}{6}, \frac{11\pi}{6}$ **23.** $\frac{\pi}{3}, \frac{5\pi}{3}$ **25.** $\frac{2\pi}{3}, \frac{4\pi}{3}$
27. $\frac{\pi}{4}$ **29.** 30°, 90° **31.** 60°, 180° **33.** 60°, 300° **35.** 120°, 180° **37.** 210°, 330° **39.** 41.8°, 48.6°, 131.4°, 138.2°
41. 36.9°, 143.1°, 216.9°, 323.1° **43.** $225° + 360°k, 315° + 360°k$ **45.** $\frac{\pi}{4} + 2k\pi$ **47.** $120° + 360°k, 180° + 360°k$
49. See the Solutions Manual. **51.** 68.5°, 291.5° **53.** 218.2°, 321.8° **55.** 73.0°, 287.0° **57.** 0.3630, 2.1351
59. 3.4492, 5.9756 **61.** 0.3166, 1.9917 **63.** $\sqrt{\frac{3 - \sqrt{5}}{6}}$ **65.** $\sqrt{\frac{3 - \sqrt{5}}{3 + \sqrt{5}}}$ or $\frac{3 - \sqrt{5}}{2}$

67.

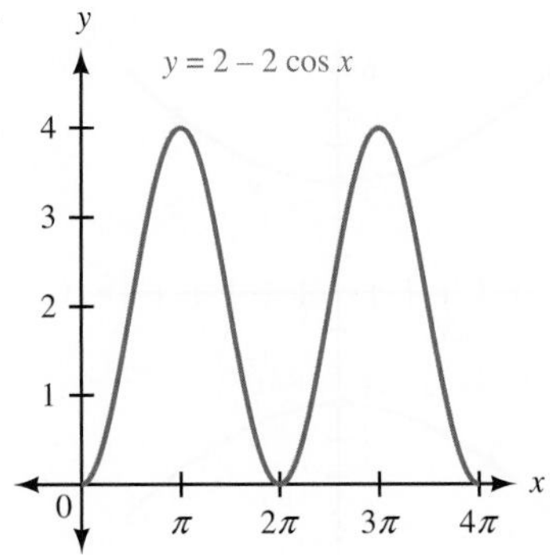

y = 2 − 2 cos x

y; 4, 3, 2, 1, 0; x; π, 2π, 3π, 4π

69. $\frac{\sqrt{2 - \sqrt{2}}}{2}$

MATCHED PRACTICE PROBLEMS 6.3

1. 20°, 40°, 140°, 160°, 260°, 280° **2.** $\frac{3\pi}{8} + \frac{k\pi}{2}$ **3.** $\frac{\pi}{9}, \frac{5\pi}{9}, \frac{7\pi}{9}, \frac{11\pi}{9}, \frac{13\pi}{9}, \frac{17\pi}{9}$ **4.** 30°, 60°, 120°, 150°, 210°, 240°, 300°, 330°
5. $\frac{\pi}{6} + \frac{k\pi}{2}, \frac{\pi}{3} + \frac{k\pi}{2}$ **6.** 135° **7.** 57.7%

PROBLEM SET 6.3

1. $n\theta, n$ **3.** sum **5.** 30°, 60°, 210°, 240° **7.** 67.5°, 157.5°, 247.5°, 337.5° **9.** 60°, 180°, 300° **11.** $\frac{\pi}{8}, \frac{7\pi}{8}, \frac{9\pi}{8}, \frac{15\pi}{8}$
13. $\frac{\pi}{3}, \pi, \frac{5\pi}{3}$ **15.** $\frac{\pi}{6}, \frac{2\pi}{3}, \frac{7\pi}{6}, \frac{5\pi}{3}$ **17.** $15° + 180°k, 75° + 180°k$ **19.** $30° + 60°k$ **21.** $6° + 36°k, 12° + 36°k$
23. 112.5°, 157.5°, 292.5°, 337.5° **25.** 20°, 100°, 140°, 220°, 260°, 340° **27.** 15°, 105°, 195°, 285°
29. $\frac{\pi}{18}, \frac{5\pi}{18}, \frac{13\pi}{18}, \frac{17\pi}{18}, \frac{25\pi}{18}, \frac{29\pi}{18}$ **31.** $\frac{5\pi}{18}, \frac{7\pi}{18}, \frac{17\pi}{18}, \frac{19\pi}{18}, \frac{29\pi}{18}, \frac{31\pi}{18}$ **33.** $\frac{\pi}{10} + \frac{2k\pi}{5}$ **35.** $\frac{\pi}{8} + \frac{k\pi}{4}$ **37.** $\frac{\pi}{5} + \frac{2k\pi}{5}$
39. $10° + 120°k, 50° + 120°k, 90° + 120°k$ **41.** $60° + 180°k, 90° + 180°k, 120° + 180°k$

43. $20° + 60°k, 40° + 60°k$ **45.** 0°, 270° **47.** 180°, 270° **49.** 96.8°, 173.2°, 276.8°, 353.2°

51. 27.4°, 92.6°, 147.4°, 212.6°, 267.4°, 332.6° **53.** 50.4°, 84.6°, 140.4°, 174.6°, 230.4°, 264.6°, 320.4°, 354.6°

55. 4.0 min and 16.0 min **57.** 6 **59.** $\frac{1}{4}$ second (and every second after that)

For Problems 61–64, see the Solutions Manual.

65. $-\frac{4\sqrt{2}}{9}$ **67.** $\sqrt{\frac{3-2\sqrt{2}}{6}}$

MATCHED PRACTICE PROBLEMS 6.4

1.

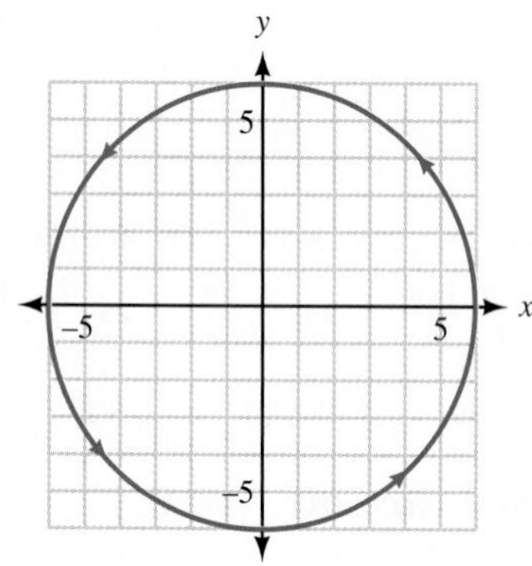

2. $\frac{x^2}{25} + \frac{y^2}{4} = 1$ **3.** $(x - 5)^2 + (y + 3)^2 = 1$ **4.** $\frac{(x + 4)^2}{9} - \frac{(y + 1)^2}{25} = 1$

5.

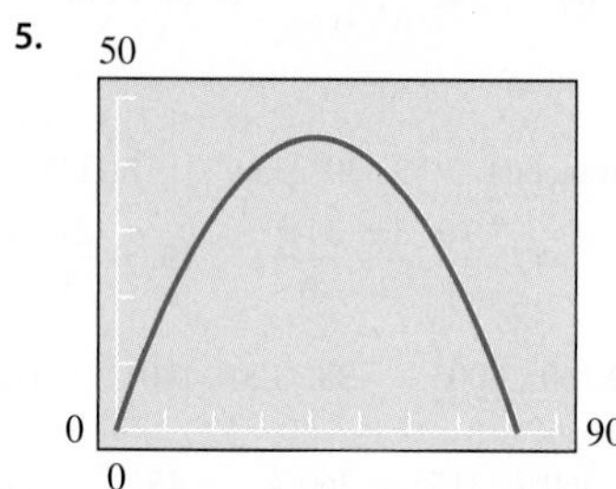

PROBLEM SET 6.4

1. parametric, parameter **3.** orientation, direction, increases **5.** $\sin t$, $\cos t$, Pythagorean

7.

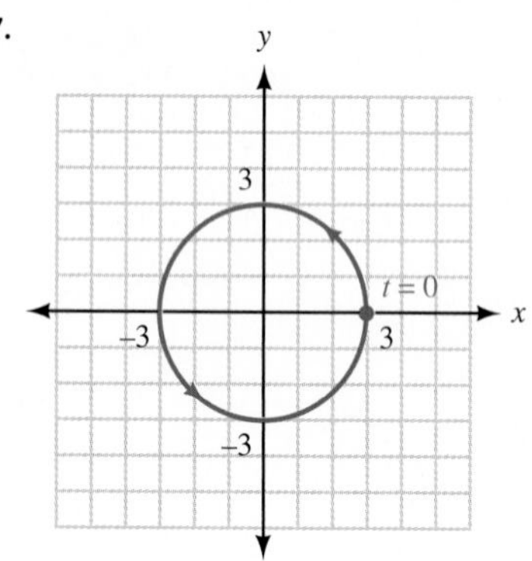

9.

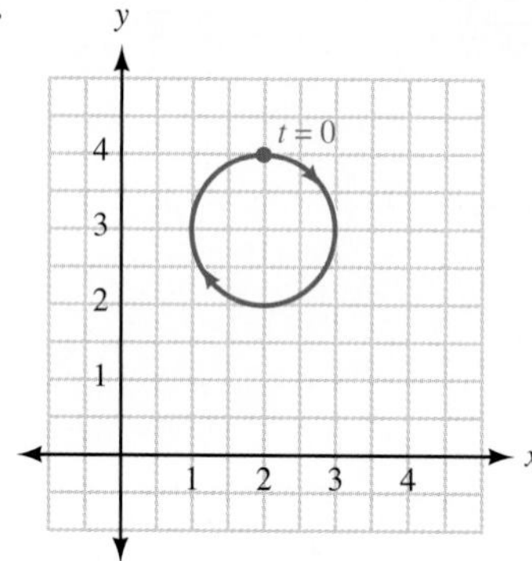

11.

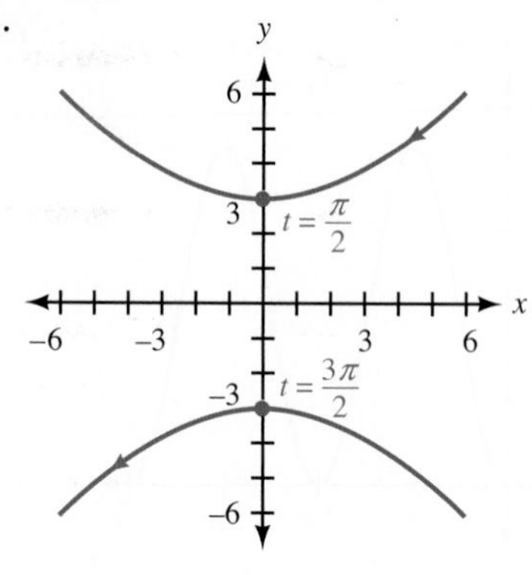

13.

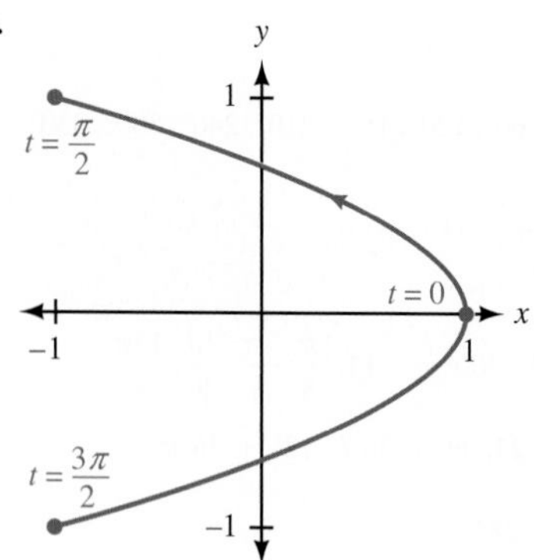

15.

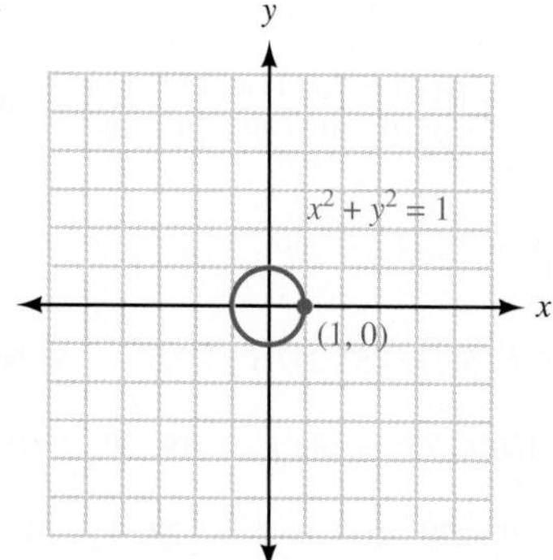

17.

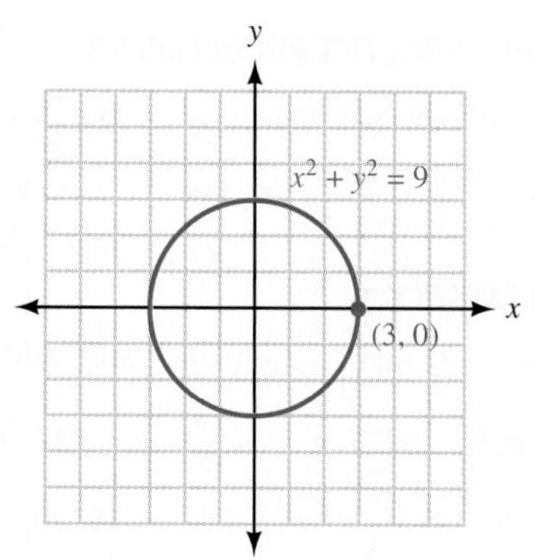

19.

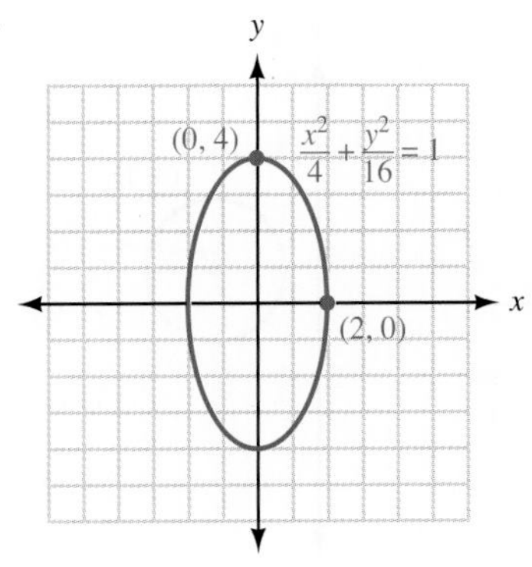

21.

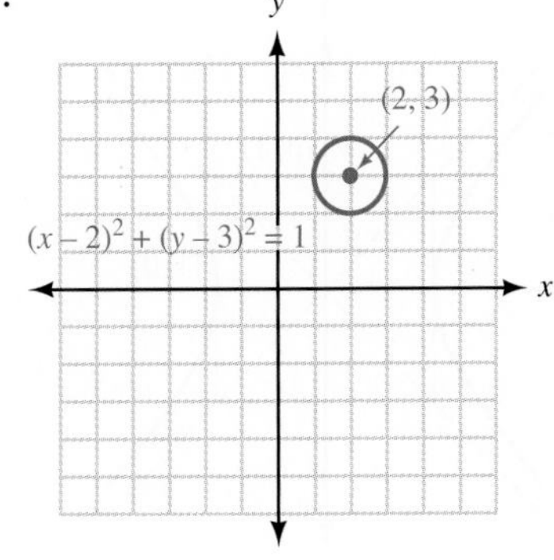

23.

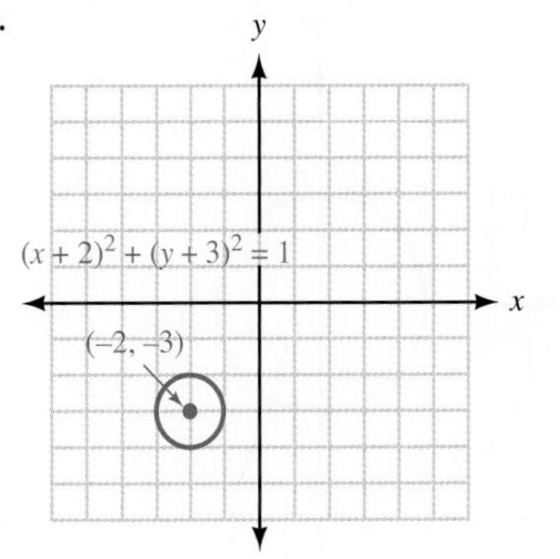

25.

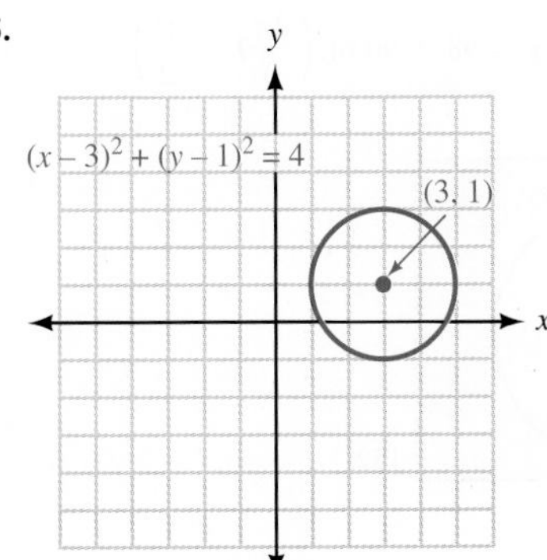

27.

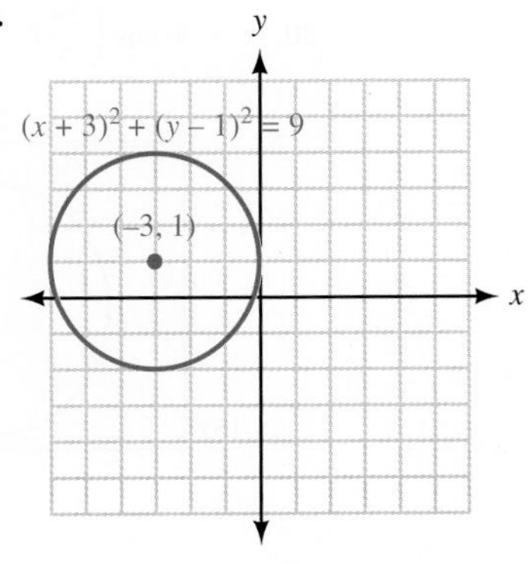

29. $x^2 - y^2 = 1$ **31.** $\frac{x^2}{9} - \frac{y^2}{9} = 1$ **33.** $\frac{(y-4)^2}{9} - \frac{(x-2)^2}{9} = 1$ **35.** $x = 1 - 2y^2$ **37.** $y = x$ **39.** $2x = 3y$

41. $x = 98.5 \cos\left(\frac{2\pi}{15}T - \frac{\pi}{2}\right)$

$y = 110.5 + 98.5 \sin\left(\frac{2\pi}{15}T - \frac{\pi}{2}\right)$

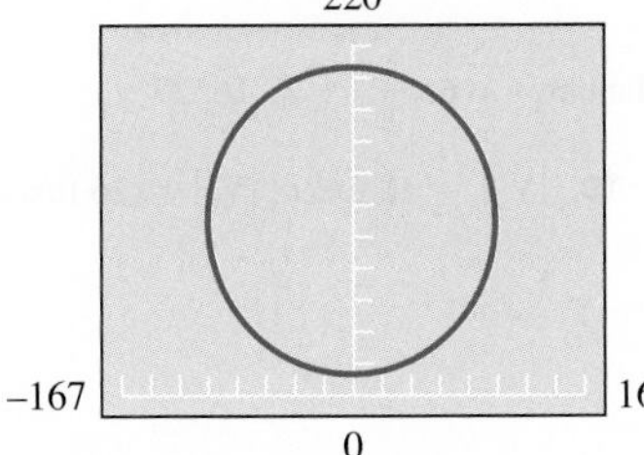

43. $x = 125 + 75 \cos\left(\frac{\pi}{16}t - \frac{\pi}{2}\right)$, $y = 90 + 75 \sin\left(\frac{\pi}{16}t - \frac{\pi}{2}\right)$

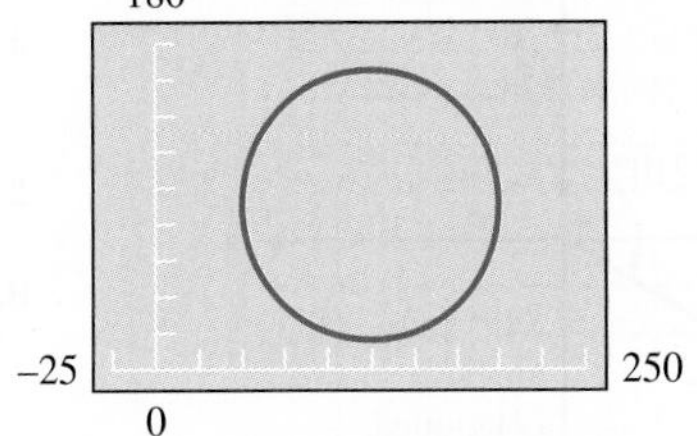

45. To three significant digits:
maximum height = 63.0 ft
maximum distance = 145 ft
time he hits the net = 3.97 sec
The graph is shown in Figure 19 of this section.

47. $\sqrt{1 - x^2}$ **49.** $\frac{1 - x^2}{1 + x^2}$ **51.** $4(\sin 5x + \sin x)$

53. $|\mathbf{V}_0| = 22.2$ ft/sec, $\theta = 50.8°$

CHAPTER 6 TEST

1. 30°, 150° **2.** 150°, 330° **3.** 30°, 90°, 150°, 270° **4.** 0°, 60°, 180°, 300° **5.** 45°, 135°, 225°, 315° **6.** 90°, 210°, 330° **7.** 180° **8.** 0°, 240° **9.** 48.6°, 131.4°, 210°, 330° **10.** 95°, 115°, 215°, 235°, 335°, 355° **11.** 0°, 90° **12.** 90°, 180° **13.** 40°, 80°, 160°, 200°, 280°, 320° **14.** 22.5°, 112.5°, 202.5°, 292.5°

15. $2k\pi, \frac{\pi}{3} + 2k\pi, \frac{5\pi}{3} + 2k\pi$ **16.** $\frac{\pi}{6} + k\pi$ **17.** $\frac{\pi}{2} + \frac{2k\pi}{3}$ **18.** $\frac{\pi}{8} + \frac{k\pi}{2}$ **19.** 90°, 203.6°, 336.4°

20. 111.5°, 248.5° **21.** 0.7297, 2.4119 **22.** 1.0598, 2.5717 **23.** 0.3076, 2.8340 **24.** 0.3218, 1.2490, 3.4633, 4.3906 **25.** 5.3 min and 14.7 min

26.

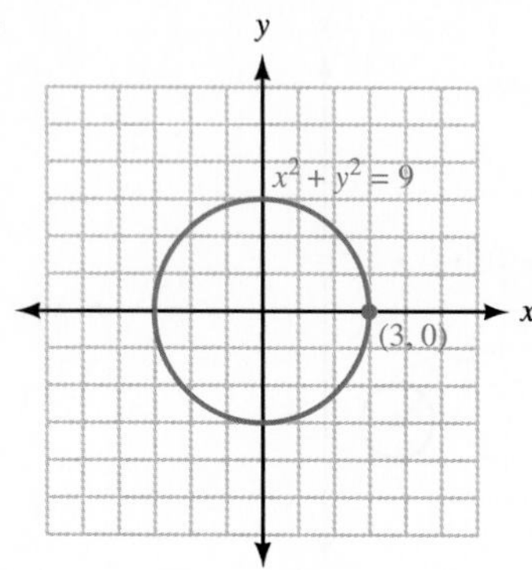

27.

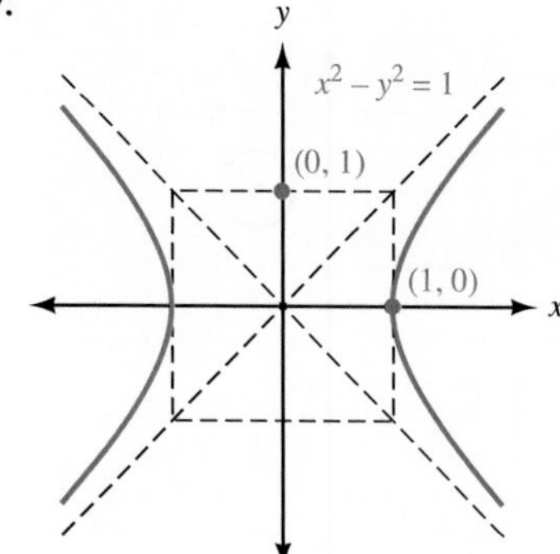

28.

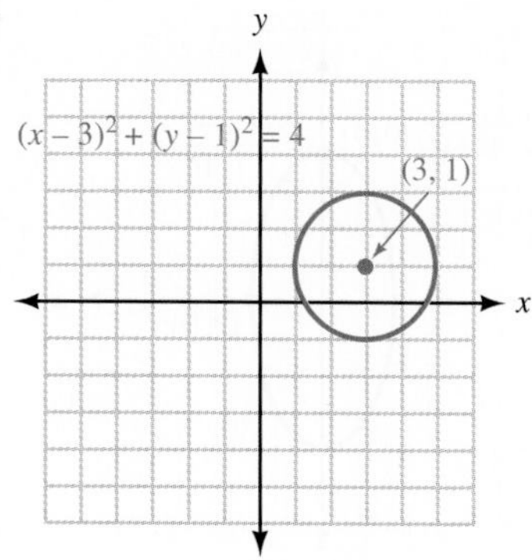

29.

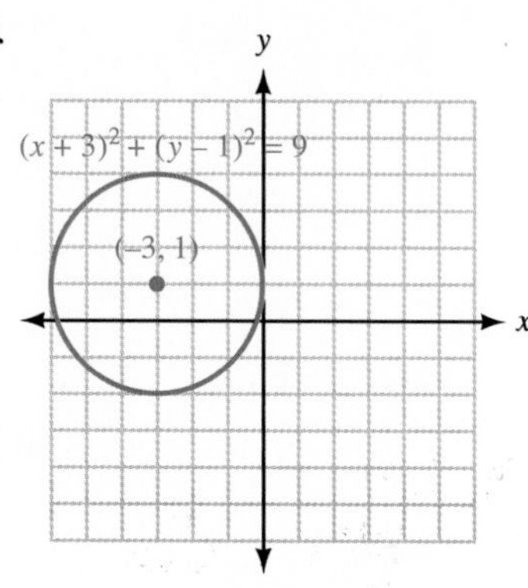

30. $x = 90\cos\left(\frac{2\pi}{3}T - \frac{\pi}{2}\right),\ y = 98 + 90\sin\left(\frac{2\pi}{3}T - \frac{\pi}{2}\right)$

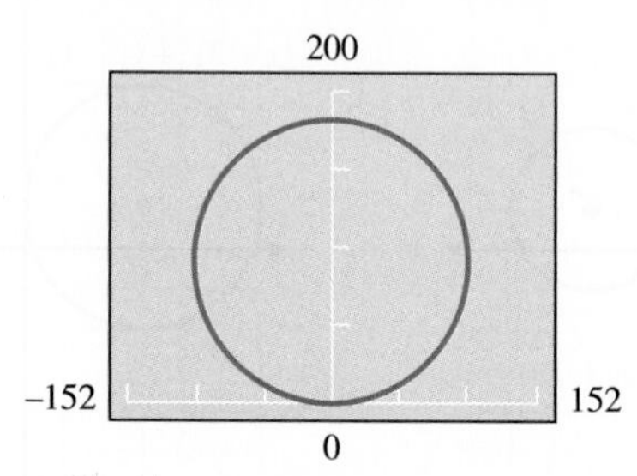

CUMULATIVE TEST CHAPTERS 1-6

1. 3 **2.**

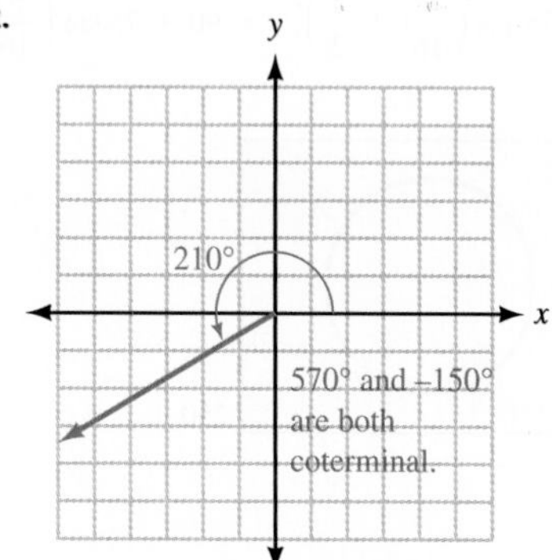

3. $\cos\theta = -\frac{7}{25},\ \tan\theta = -\frac{24}{7},\ \cot\theta = -\frac{7}{24},\ \sec\theta = -\frac{25}{7},\ \csc\theta = \frac{25}{24}$

4. $\cos\theta = \frac{\sqrt{a^2 - 1}}{a},\ \csc\theta = a,\ \cot\theta = \sqrt{a^2 - 1}$

5. A proof is given in the Solutions Manual. **6.** $\frac{1}{2}$ **7.** $16°27'$

8. $x = 17,\ y = 16$ **9.** 2.3 mi **10.** $|\mathbf{V}_x| = 40$ ft/sec, $|\mathbf{V}_y| = 26$ ft/sec

11.

y
x
$\hat{\theta} = 50°\ 20'$
$410°\ 20'$

12. $2\sqrt{2}$ **13.** $\cos\theta = -0.2537,\ \sin\theta = 0.9673,\ \tan\theta = -3.8128$

14. 2π cm **15.** $\frac{5}{3}$ rad/sec **16.** $-3\pi, -\pi, \pi, 3\pi$

17.

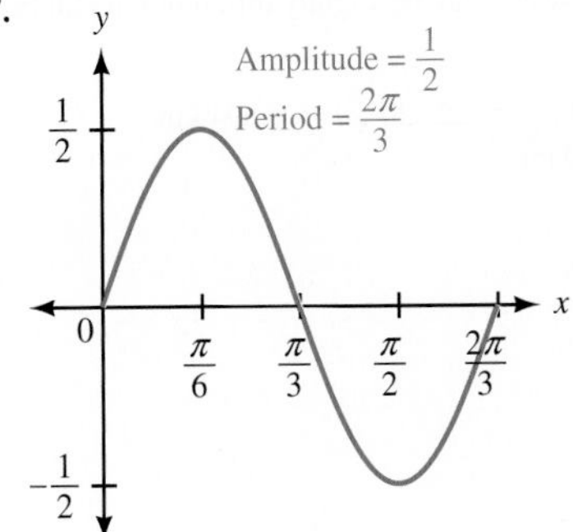

18.

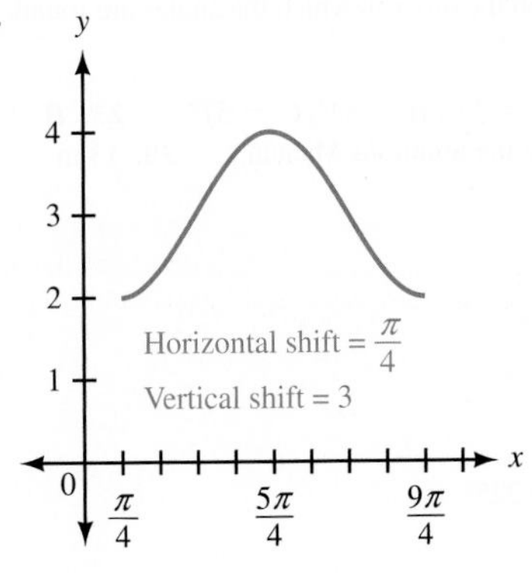

19.

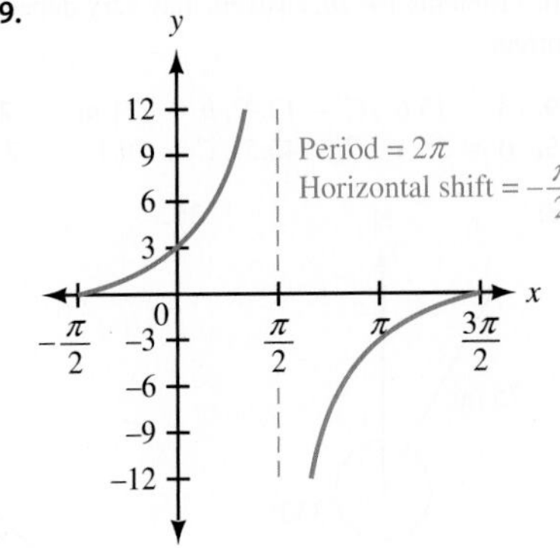

20. $y = 3 - 2\cos\left(\pi x - \frac{\pi}{2}\right)$ **21.** $\frac{3\sqrt{13}}{13}$

For Problems 22–23, see the Solutions Manual.

24. $-\frac{120}{169}$ **25.** $-\frac{3\sqrt{10}}{10}$ **26.** $2(\cos 4x - \cos 10x)$ **27.** $0°, 120°, 240°$ **28.** $0, \frac{\pi}{3}, \pi, \frac{5\pi}{3}$ **29.** $40° + 120°k, 80° + 120°k$

30.

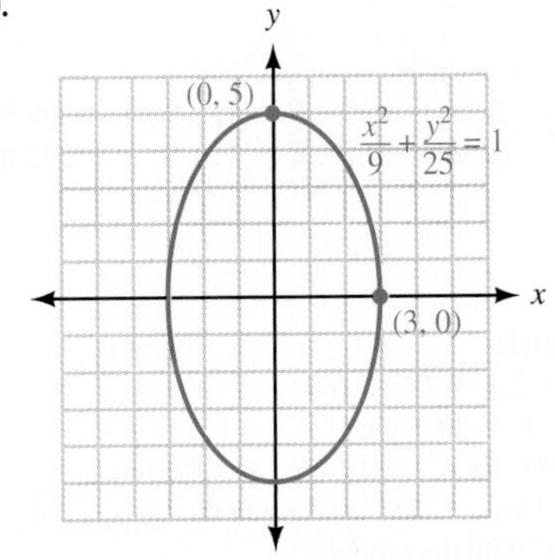

CHAPTER 7

MATCHED PRACTICE PROBLEMS 7.1

1. 11 in. **2.** $B = 56°, a = 6.5$ cm, $c = 8.8$ cm **3.** 82.7 ft **4.** 59.5 mi **5.** 17° **6.** $|\mathbf{T}_1| = 13$ lb, $|\mathbf{T}_2| = 17$ lb

PROBLEM SET 7.1

1. oblique **3.** sine, opposite, constant **5.** 28 cm **7.** 71 in. **9.** 54 yd **11.** $C = 70°, c = 44$ km **13.** $C = 80°, a = 11$ cm **15.** $C = 66.1°, b = 302$ in., $c = 291$ in. **17.** $A = 92°, b = 6.1$ m, $c = 3.8$ m **19.** $B = 16°, b = 1.39$ ft, $c = 4.36$ ft **21.** $A = 141.8°, b = 118$ cm, $c = 214$ cm **23.** $\sin B = 5$, which is impossible **25.** 11 **27.** 20 **29.** 209 ft **31.** 273 ft **33.** 137 ft **35.** 42 ft **37.** 14 mi, 9.3 mi **39.** $|\mathbf{CB}| = 341$ lb, $|\mathbf{CA}| = 345$ lb **41.** $|\mathbf{CA}| = 3{,}240$ lb, $|\mathbf{CB}| = 3{,}190$ lb **43.** 45°, 135° **45.** 90°, 270° **47.** 30°, 90°, 150° **49.** 47.6°, 132.4°

MATCHED PRACTICE PROBLEMS 7.2

1. $B = 47°, C = 85°, a = 19$ in. **2.** 16.6 cm **3.** $A = 99°, B = 46°, C = 35°$ **4.** 229 mph at 112° from due north

PROBLEM SET 7.2

1. sides, angle, sides, unique **3.** side, cosines, sines, smaller **5.** clockwise, north, intended **7.** air, heading **9.** $b^2 + c^2 - 2bc\cos A$ **11.** 100 in. **13.** $C = 67°$ **15.** 9.4 m **17.** $A = 128°$

For Problems 19–26, answers may vary depending on the order in which the angles are found. Your answers may be slightly different but still be correct.

19. $A = 15.6°, C = 12.9°, b = 731$ m **21.** $A = 39°, B = 84°, C = 57°$ **23.** $B = 114°\ 10', C = 22°\ 30', a = 0.694$ km
25. $A = 55.4°, B = 45.5°, C = 79.1°$ **27.** See the Solutions Manual. **29.** 13 m **31.** 130 mi

33. **35.**

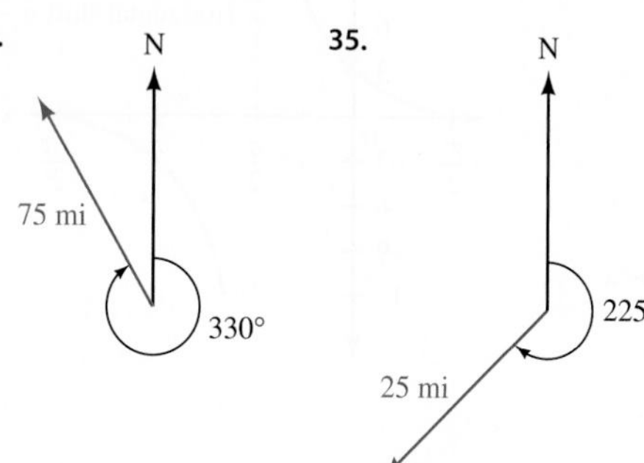

37. 462 mi **39.** 190 mi/hr (two significant digits) with heading 153° **41.** 18.5 mi/hr at 81.3° from due north
43. 130 lb **45.** 27 knots, 161° **47.** 61.2 cm, 55.6° **49. a.** 45 cm **b.** 51° **51.** 13.4 mi

53. $\frac{\pi}{18} + \frac{2k\pi}{3}, \frac{5\pi}{18} + \frac{2k\pi}{3}$ **55.** $\frac{\pi}{12} + \frac{k\pi}{6}$ **57.** $20° + 120°k, 100° + 120°k$ **59.** $45° + 60°k$ **61.** 0°, 90°

MATCHED PRACTICE PROBLEMS 7.3

1. No triangle is possible. **2.** $B = 55°, C = 83°, c = 77$ cm; $B' = 125°, C' = 13°, c' = 18$ cm **3.** $A = 17.2°, B = 125.6°, b = 289$ ft
4. No triangle is possible. **5.** $B = 56°, C = 80°, c = 78$ cm; $B' = 124°, C' = 12°, c' = 17$ cm **6.** $A = 17.8°, B = 129.5°, b = 292$ ft
7. 116 mph or 209 mph

PROBLEM SET 7.3

1. two, opposite **3.** angle, reference, I, II **5.** $\sin B = 1.5$; because $\sin B > 1$, there is no triangle
7. $B = 35°$; $B' = 145°$, but $A + B' = 120° + 145° = 265° > 180°$ **9.** $B = 77°$ or $B' = 103°$; $A + B' < 180°$
11. $B = 54°, C = 88°, c = 67$ ft; $B' = 126°, C' = 16°, c' = 18$ ft **13.** $B = 28.1°, C = 39.7°, c = 30.2$ cm
15. $A = 117°\ 20', B = 34°\ 50', a = 660$ m; $A' = 7°, B' = 145°\ 10', a' = 90.6$ m **17.** $A = 90°\ 10', C = 27°\ 10', a = 7.63$ in.
19. No triangle is possible **21.** No triangle is possible **23.** $A = 126.4°, B = 26.8°, a = 65.7$ km **25.** 19 ft or 35 ft
27. 32° or 178° from due north **29.** $6\sqrt{3}$ mi/hr ≈ 10 mi/hr **31.** Yes, it makes an angle of 86° with the ground.
33. S 55° E or S 75° E **35.** 30°, 150°, 210°, 330° **37.** 90°, 270° **39.** 41.8°, 48.6°, 131.4°, 138.2°

41. $\frac{7\pi}{6} + 2k\pi, \frac{11\pi}{6} + 2k\pi$ **43.** $\frac{3\pi}{4} + k\pi$

MATCHED PRACTICE PROBLEMS 7.4

1. 5.42 cm^2 **2.** 10.8 ft^2 **3.** 95 m^2

PROBLEM SET 7.4

1. half, sides, sine **3.** sum, two **5.** $\frac{1}{2}(a + b + c)$ **7.** 6 ft **9.** 4.15 m **11.** 1,520 cm^2 **13.** 342 m^2 **15.** 0.123 km^2

17. 26.3 m^2 **19.** 28,300 in.^{2} **21.** 2.09 ft^2 **23.** 1,410 in.^{2} **25.** 15.0 yd^2 **27.** 8.15 ft^2 **29.** 156 in.^{2} **31.** 14.3 cm

33.

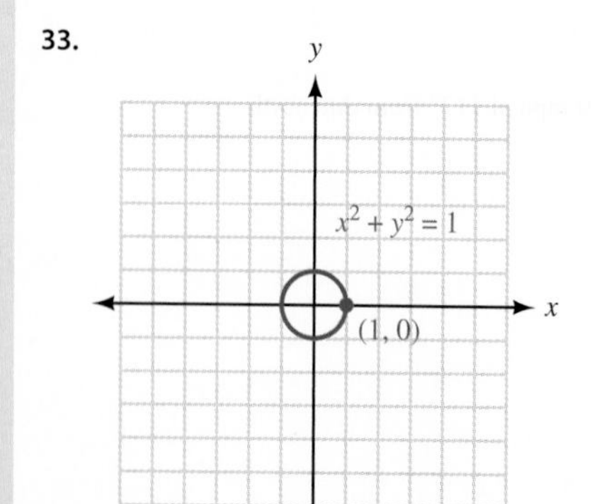

35.

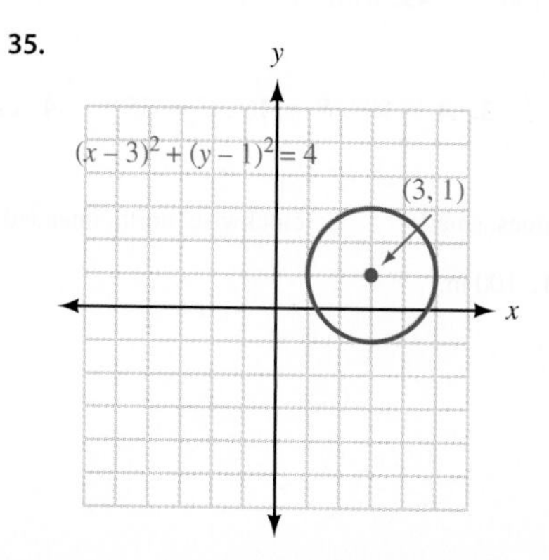

37. $\frac{y^2}{9} - \frac{x^2}{4} = 1$ **39.** $y = 1 - 2x^2$

MATCHED PRACTICE PROBLEMS 7.5

1.

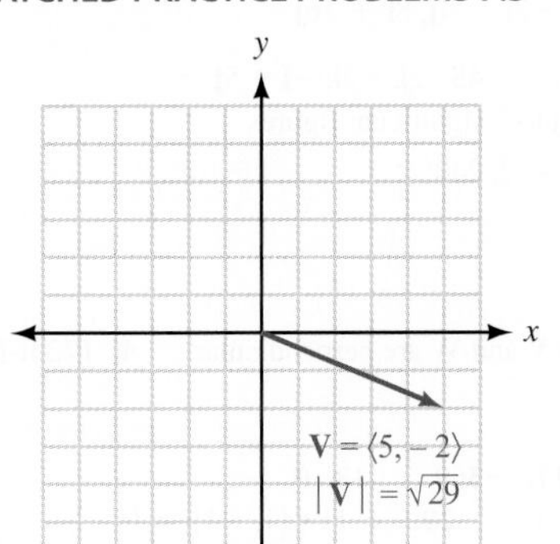

2. a. $\langle -5, 1\rangle$ **b.** $\langle 34, -32\rangle$ **3.** $5\mathbf{i} + 12\mathbf{j}$ **4.** $8.6\mathbf{i} + 12.3\mathbf{j}$
5. a. $\mathbf{i} + 2\mathbf{j}$ **b.** $24\mathbf{i} - 43\mathbf{j}$ **6.** $|\mathbf{T}| = 74$ lb, $|\mathbf{H}| = 43$ lb

PROBLEM SET 7.5

1. tail, origin **3.** components **5.** length, direction **7.** unit **9.** vector component, horizontal vector component, vertical vector component

11.

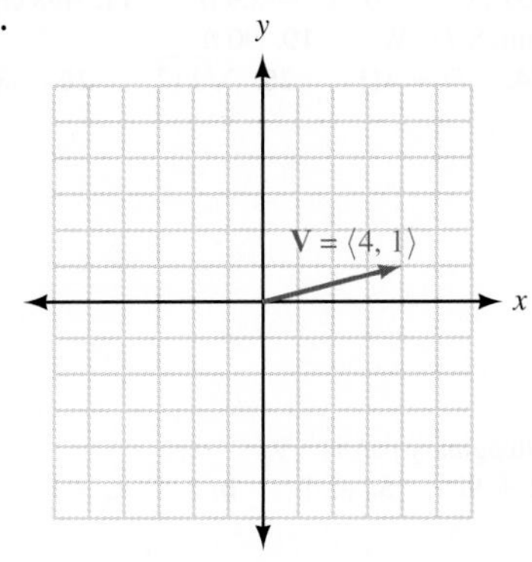

13.

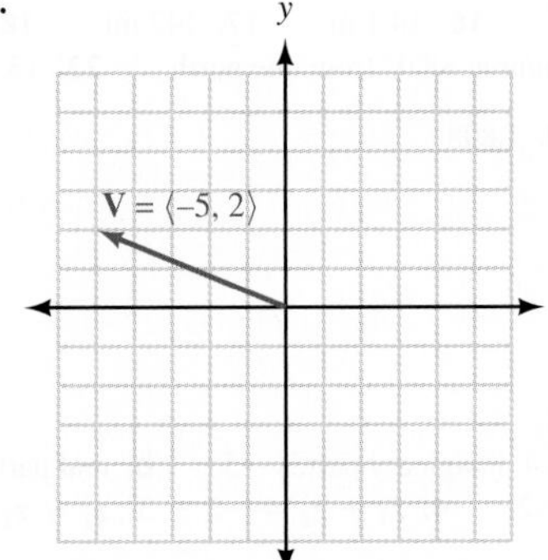

15.

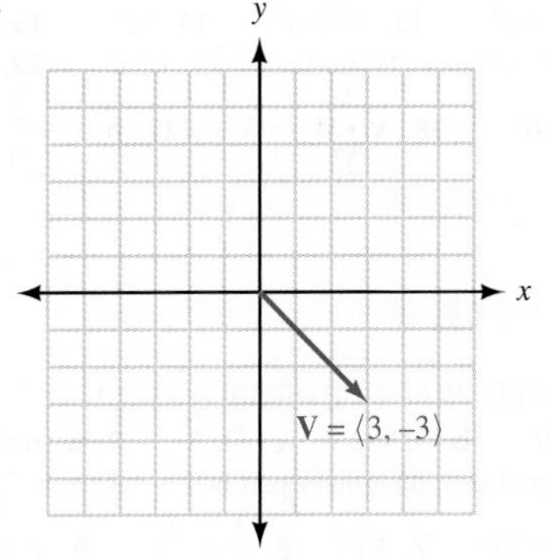

17.

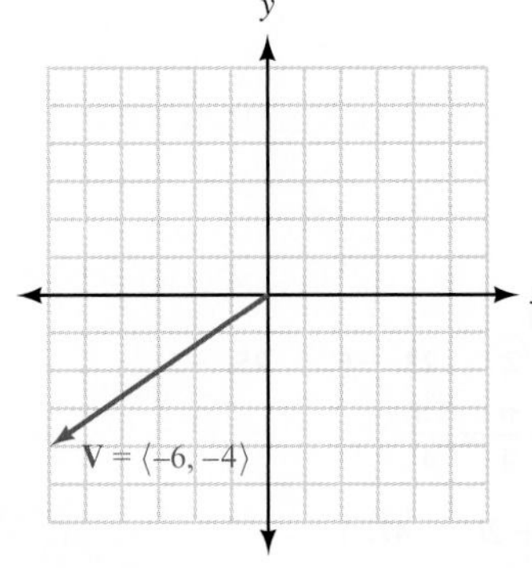

19.

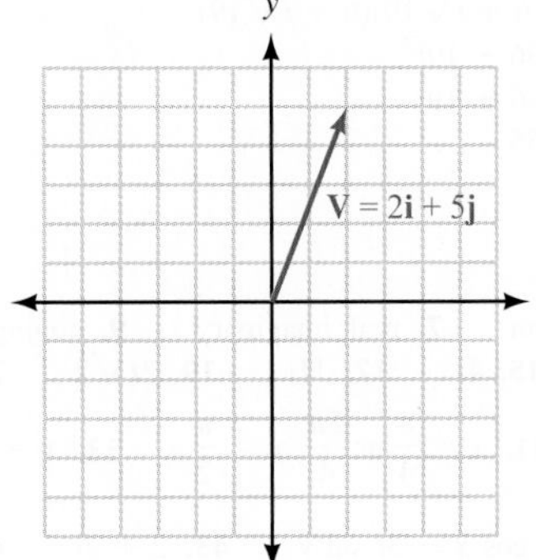

21.

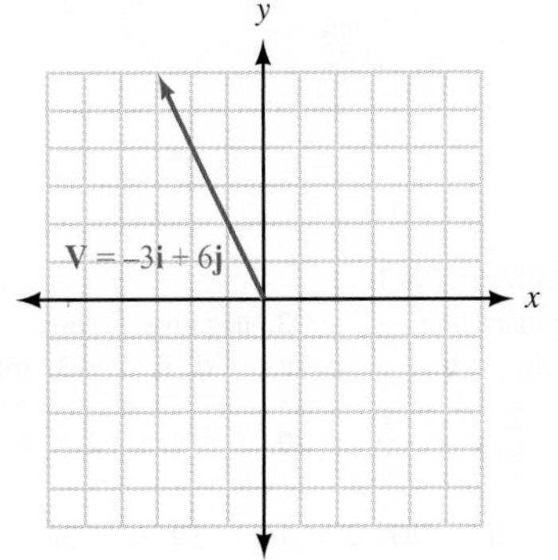

23.

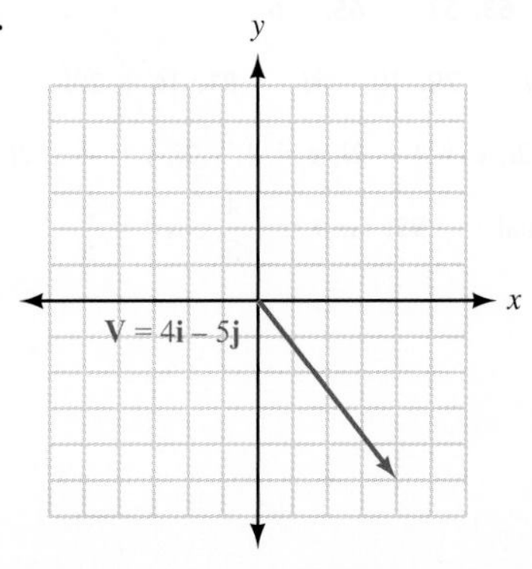

25.

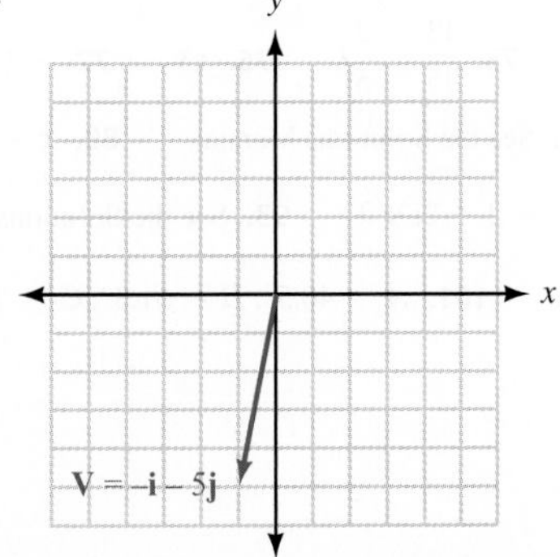

27. $\sqrt{61}$ **29.** 5 **31.** 13 **33.** $\sqrt{5}$ **35.** $\left\langle -\frac{3}{2}, \frac{7}{2} \right\rangle$, $\langle 3, -7 \rangle$, $\langle -12, 28 \rangle$ **37.** $\mathbf{i} + 2\mathbf{j}$, $-2\mathbf{i} - 4\mathbf{j}$, $8\mathbf{i} + 16\mathbf{j}$
39. $\langle 8, 0 \rangle$, $\langle 0, 8 \rangle$, $\langle -4, 20 \rangle$ **41.** $\langle -5, 1 \rangle$, $\langle -5, -1 \rangle$, $\langle -10, -3 \rangle$ **43.** $\langle -1, 3 \rangle$, $\langle 9, -1 \rangle$, $\langle 23, -4 \rangle$ **45.** $2\mathbf{j}$, $-2\mathbf{i}$, $-\mathbf{i} + 5\mathbf{j}$
47. $6\mathbf{i} - 8\mathbf{j}$, $6\mathbf{i} + 8\mathbf{j}$, $18\mathbf{i} - 16\mathbf{j}$ **49.** $7\mathbf{i} + 7\mathbf{j}$, $-3\mathbf{i} + 3\mathbf{j}$, $16\mathbf{i} + 19\mathbf{j}$ **51.** $\mathbf{V} = \langle 14, 12 \rangle = 14\mathbf{i} + 12\mathbf{j}$ to 2 significant figures
53. $\mathbf{W} = \langle -5.1, -6.1 \rangle = -5.1\mathbf{i} - 6.1\mathbf{j}$ to 2 significant figures **55.** $|\mathbf{U}| = 3\sqrt{2}, 45°$ **57.** $|\mathbf{W}| = 2, 240°$
59. $|\mathbf{N}| = 7.7$ lb, $|\mathbf{F}| = 2.1$ lb **61.** $|\mathbf{H}| = 45.6$ lb **63.** $|\mathbf{T}_1| = 16$ lb, $|\mathbf{T}_2| = 20$ lb

MATCHED PRACTICE PROBLEMS 7.6

1. a. 39 **b.** −31 **c.** −42 **2. a.** 63.4° **b.** 127.2° **3.** **U** and **W** are perpendicular, and **V** and **W** are perpendicular. **4.** 172 ft-lb

PROBLEM SET 7.6

1. multiply, add **3.** magnitudes, cosine **5.** orthogonal **7.** projection, onto **9.** 35 **11.** −369 **13.** 0
15. 0 **17.** 20 **19.** 10 **21.** −134 **23.** $\theta = 90.0°$ **25.** $\theta = 94.4°$ **27.** $\theta = 111.3°$ **29.** $\langle 1, 1 \rangle \cdot \langle 1, -1 \rangle = 0$
31. $\langle -1, 0 \rangle \cdot \langle 0, 1 \rangle = 0$ **33.** $-4(6) + (-3)(-8) = 0$ **35.** −8 **37.** 696 ft-lb **39.** 997 ft-lb **41.** 510 ft-lb **43.** 0 ft-lb
45. See the Solutions Manual. **47.** 2,900 ft-lb **49.** 9,000 ft-lb

CHAPTER 7 TEST

1. 6.7 in. **2.** $C = 78.4°$, $a = 26.5$ cm, $b = 38.3$ cm **3.** $B = 49.2°$, $a = 18.8$ cm, $c = 43.2$ cm **4.** 11 cm **5.** 95.7°
6. $A = 43°$, $B = 18°$, $c = 8.1$ m **7.** $B = 34°$, $C = 111°$, $a = 3.8$ m **8.** $\sin B = 3.0311$, which is impossible
9. $B = 29°$; $B' = 151°$, but $A + B' = 193° > 180°$ **10.** $B = 71°$, $C = 58°$, $c = 7.1$ ft; $B' = 109°$, $C' = 20°$, $c' = 2.9$ ft **11.** 498 cm^2
12. 52 cm^2 **13.** 17 km^2 **14.** 51° **15.** 410 ft **16.** 14.1 m **17.** 142 mi **18.** 4.2 mi, S 75° W **19.** 90 ft
20. 300 mi/hr or 388 mi/hr **21.** 65 ft **22.** 260 mi/hr at 88.9° from due north **23.** 13 **24.** $-5\mathbf{i} + 41\mathbf{j}$ **25.** $\sqrt{117}$ **26.** −8
27. 98.6° **28.** $\mathbf{V} \cdot \mathbf{W} = 0$ **29.** $b = -\frac{5}{3}$ **30.** 1,808

CHAPTER 8

MATCHED PRACTICE PROBLEMS 8.1

1. a. $4i$ **b.** $3i\sqrt{2}$ **c.** $i\sqrt{7}$ **2. a.** real part is 4, imaginary part is −5 **b.** real part is 0, imaginary part is −9
c. real part is 3, imaginary part is 0 **3.** $x = 7, y = -2$ **4.** $z_1 + z_2 = -2 - 3i$, $z_1 - z_2 = 12 + 9i$ **5. a.** 1 **b.** i **c.** −1
6. $14 - 2i$ **7.** 53 **8.** $\frac{3}{5} + \frac{6}{5}i$ **9.** $x + y = (6 + i\sqrt{19}) + (6 - i\sqrt{19}) = 12$

$$\begin{aligned} xy &= (6 + i\sqrt{19})(6 - i\sqrt{19}) \\ &= 36 - 19i^2 \\ &= 36 + 19 \\ &= 55 \end{aligned}$$

PROBLEM SET 8.1

1. imaginary unit, −1 **3.** negative, before **5.** sum **7.** real, imaginary **9.** conjugate, real
11. a. iii **b.** i **c.** iv **d.** ii **13.** true **15.** $4i$ **17.** $11i$ **19.** $3i\sqrt{2}$ **21.** $2i\sqrt{2}$ **23.** −6 **25.** −3
27. $x = \frac{2}{3}, y = -\frac{1}{2}$ **29.** $x = 4$ or -2, $y = 1$ **31.** $x = \frac{\pi}{4}$ or $\frac{5\pi}{4}$, $y = \frac{\pi}{2}$ **33.** $x = \frac{\pi}{2}$, $y = \frac{\pi}{4}$ or $\frac{5\pi}{4}$
35. $10 - 2i$ **37.** $2 - 4i$ **39.** $9 - 2i$ **41.** $5 \cos x - 3i \sin y$ **43.** $2 + 2i$ **45.** $12 + 2i$ **47.** 1
49. $-i$ **51.** −1 **53.** i **55.** $-48 - 18i$ **57.** $10 + 10i$ **59.** $5 - 12i$ **61.** 41 **63.** 53 **65.** −6
67. $\frac{1}{5} + \frac{3}{5}i$ **69.** $-\frac{5}{13} + \frac{12}{13}i$ **71.** $-2 - 5i$ **73.** $\frac{13}{15} - \frac{2}{5}i$ **75.** 13 **77.** $-7 + 22i$ **79.** $10 - 3i$ **81.** $16 + 20i$
83. $x^2 + 9$ **85.** See the Solutions Manual. **87.** See the Solutions Manual. **89.** $x = 4 + 2i, y = 4 - 2i$; $x = 4 - 2i, y = 4 + 2i$
91. $x = 1 + i\sqrt{3}, y = 2 - 2i\sqrt{3}$; $x = 1 - i\sqrt{3}, y = 2 + 2i\sqrt{3}$ **93.** See the Solutions Manual. **95.** $\sin\theta = -\frac{4}{5}$, $\cos\theta = \frac{3}{5}$
97. 135° **99.** $B = 69.6°$, $C = 37.3°$, $a = 248$ cm **101.** $A = 40.5°$, $B = 61.3°$, $C = 78.2°$

MATCHED PRACTICE PROBLEMS 8.2

1.

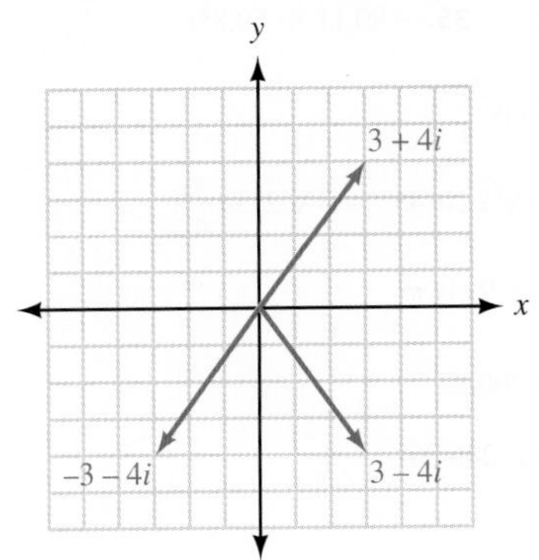

2.

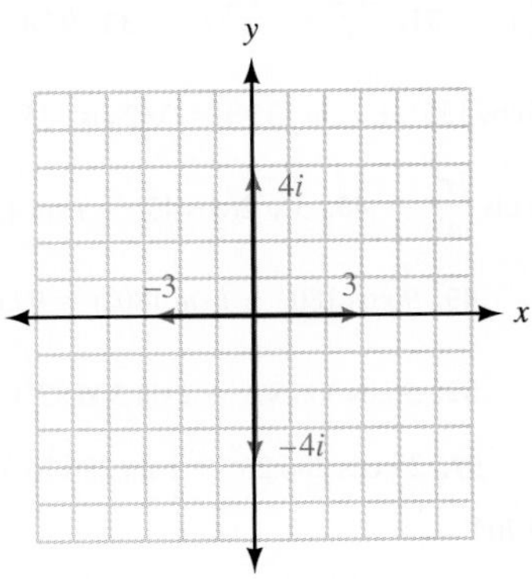

3. 4, 6, $\sqrt{10}$ **4.** 2 cis 60° **5.** $\frac{3\sqrt{2}}{2} + \frac{3\sqrt{2}}{2}i$ **6.** $2\sqrt{13}$ cis 56.31°, $2\sqrt{13}$ cis 303.69°

PROBLEM SET 8.2

1. Argand, real, imaginary **3.** modulus, origin, (a, b) **5.** standard, trigonometric **7.** evaluate, distribute

9.

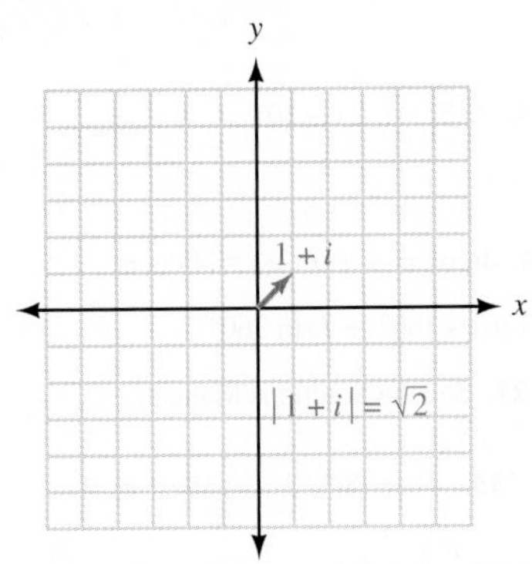

11.

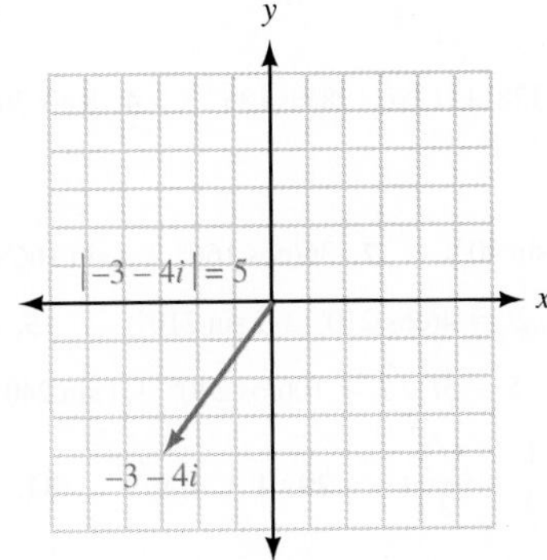

13.

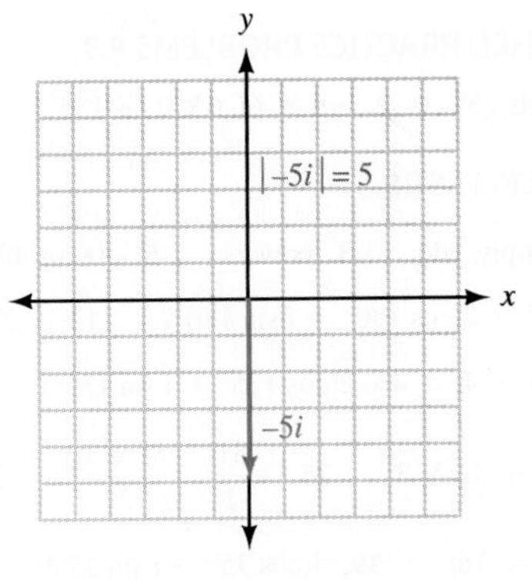

15.

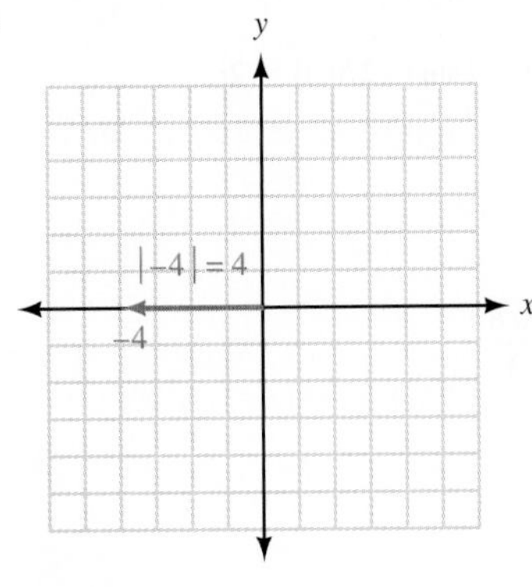

17.

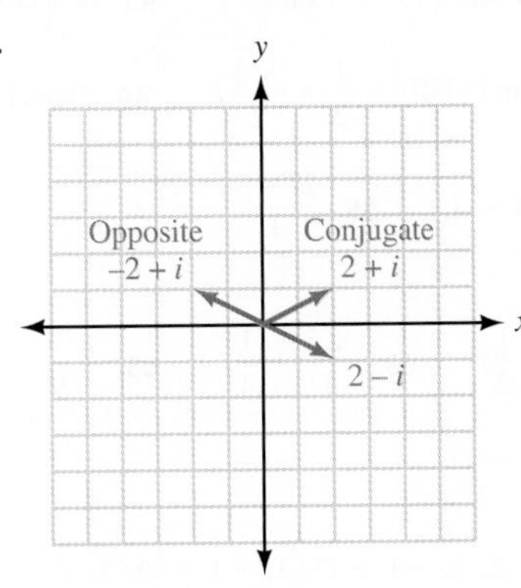

19.

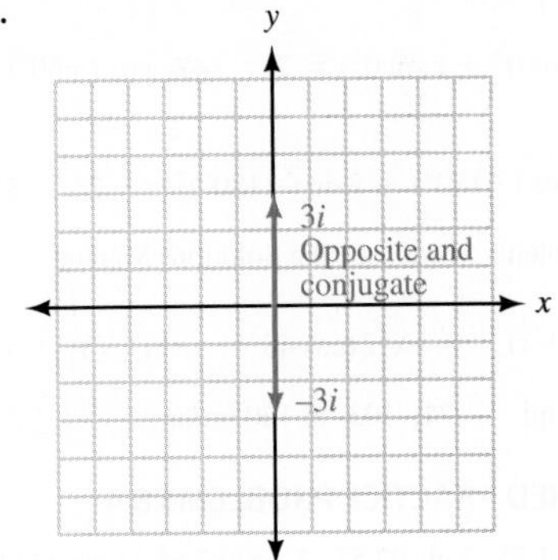

21.

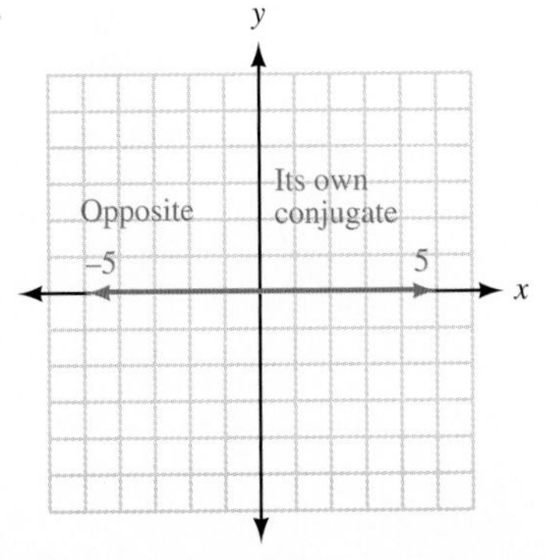

23.

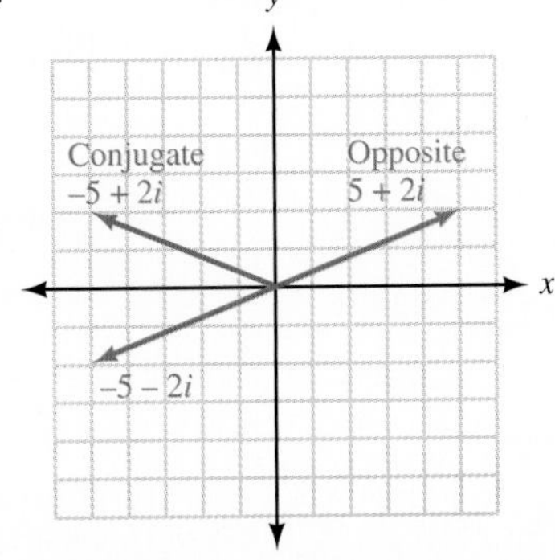

25. $\sqrt{3} + i$ **27.** $-2 + 2i\sqrt{3}$ **29.** $-\frac{\sqrt{3}}{2} - \frac{1}{2}i$ **31.** $\frac{\sqrt{2}}{2} - \frac{\sqrt{2}}{2}i$ **33.** $9.78 + 2.08i$ **35.** $-80.11 + 59.85i$

37. $-0.91 - 0.42i$ **39.** $9.60 - 2.79i$ **41.** $\sqrt{2}(\cos 135° + i \sin 135°) = \sqrt{2} \text{ cis } 135° = \sqrt{2} \text{ cis} \frac{3\pi}{4}$

43. $\sqrt{2}(\cos 315° + i \sin 315°) = \sqrt{2} \text{ cis } 315° = \sqrt{2} \text{ cis} \frac{7\pi}{4}$ **45.** $3\sqrt{2}(\cos 45° + i \sin 45°) = 3\sqrt{2} \text{ cis } 45° = 3\sqrt{2} \text{ cis} \frac{\pi}{4}$

47. $8(\cos 270° + i \sin 270°) = 8 \text{ cis } 270° = 8 \text{ cis} \frac{3\pi}{2}$ **49.** $9(\cos 180° + i \sin 180°) = 9 \text{ cis } 180° = 9 \text{ cis } \pi$

51. $4(\cos 120° + i \sin 120°) = 4 \text{ cis } 120° = 4 \text{ cis} \frac{2\pi}{3}$ **53.** $5(\cos 306.87° + i \sin 306.87°) = 5 \text{ cis } 306.87°$

55. $29(\cos 133.60° + i \sin 133.60°) = 29 \text{ cis } 133.60°$ **57.** $25(\cos 286.26° + i \sin 286.26°) = 25 \text{ cis } 286.26°$

59. $5\sqrt{5}(\cos 169.70° + i \sin 169.70°) = 5\sqrt{5} \text{ cis } 169.70°$

For Problems 61–68, see the answer to the corresponding problem.

For Problems 69–72, see the Solutions Manual.

73. $\frac{\sqrt{6} - \sqrt{2}}{4}$ **75.** $\frac{56}{65}$ **77.** $\sin 120° = \frac{\sqrt{3}}{2}$ **79.** No triangle is possible. **81.** $B = 62.7°$ or $B = 117.3°$

MATCHED PRACTICE PROBLEMS 8.3

1. $12 \text{ cis } 75°$ **2.** $-6 + 6i, 6\sqrt{2} \text{ cis } 135°$ **3.** $-128 + 128i, 128 \text{ cis } 135°$ **4.** $8 \text{ cis } 30°$ **5.** $\sqrt{3} + i, 2 \text{ cis } 30°$

PROBLEM SET 8.3

1. multiply, add **3.** powers **5.** $10(\cos 40° + i \sin 40°)$ **7.** $36(\cos 166° + i \sin 166°)$ **9.** $4(\cos \pi + i \sin \pi) = 4 \text{ cis } \pi$

11. $-2 = 2(\cos 180° + i \sin 180°)$ **13.** $-2\sqrt{3} - 2i = 4(\cos 210° + i \sin 210°)$ **15.** $12 = 12(\cos 360° + i \sin 360°)$

17. $-4 + 4i = 4\sqrt{2}(\cos 135° + i \sin 135°)$ **19.** $-5 - 5i\sqrt{3} = 10(\cos 240° + i \sin 240°)$ **21.** See the Solutions Manual.

23. $32 + 32i\sqrt{3}$ **25.** $-\frac{81}{2} - \frac{81\sqrt{3}}{2}i$ **27.** $-\frac{1}{2} + \frac{\sqrt{3}}{2}i$ **29.** $4 + 4i\sqrt{3}$ **31.** -4 **33.** $-8 - 8i\sqrt{3}$ **35.** $-8i$

37. $16 + 16i$ **39.** $4(\cos 35° + i \sin 35°)$ **41.** $1.5(\cos 19° + i \sin 19°)$ **43.** $0.5\left(\cos \frac{\pi}{3} + i \sin \frac{\pi}{3}\right) = 0.5 \text{ cis } \frac{\pi}{3}$

45. $2(\cos 0° + i \sin 0°) = 2$ **47.** $\cos(-60°) + i \sin(-60°) = \frac{1}{2} - \frac{\sqrt{3}}{2}i$ **49.** $2[\cos(-270°) + i \sin(-270°)] = 2i$

51. $2[\cos(-180°) + i \sin(-180°)] = -2$ **53.** $-4 - 4i$ **55.** $\frac{27\sqrt{3}}{4} - \frac{27}{4}i$

For Problems 57–60, see the Solutions Manual.

61. $(1 + i)^{-1} = [\sqrt{2}(\cos 45° + i \sin 45°)]^{-1} = \frac{1}{2} - \frac{1}{2}i$ **63.** $\frac{\sqrt{3}}{4} + \frac{1}{4}i$ **65.** $-\frac{7}{9}$ **67.** $\frac{\sqrt{6}}{3}$

69. 6.0 mi **71.** 103° at 160 mi/hr

MATCHED PRACTICE PROBLEMS 8.4

1. $3 \text{ cis } 7.5°, 3 \text{ cis } 97.5°, 3 \text{ cis } 187.5°, 3 \text{ cis } 277.5°$ **2.** $2, -1 + i\sqrt{3}, -1 - i\sqrt{3}$

3. $\sqrt[4]{2} \text{ cis } 67.5°, \sqrt[4]{2} \text{ cis } 112.5°, \sqrt[4]{2} \text{ cis } 247.5°, \sqrt[4]{2} \text{ cis } 292.5°$

PROBLEM SET 8.4

1. n **3.** 72

5.

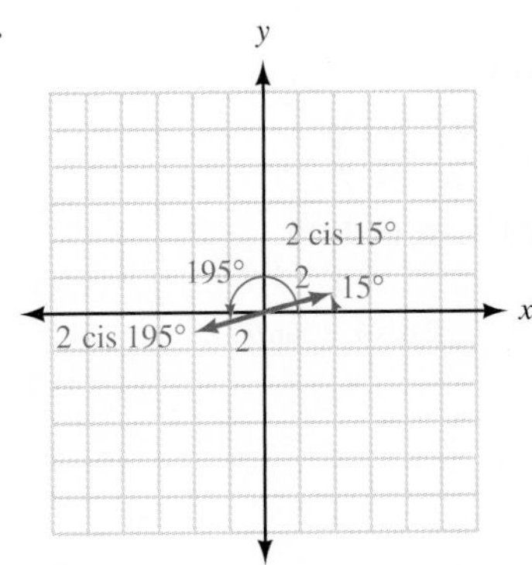

7.

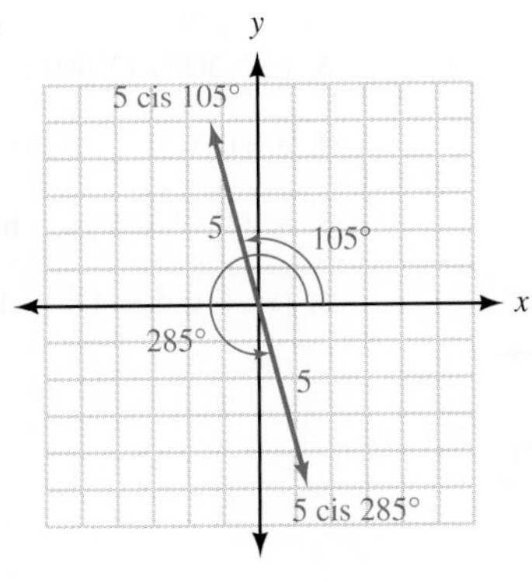

9.

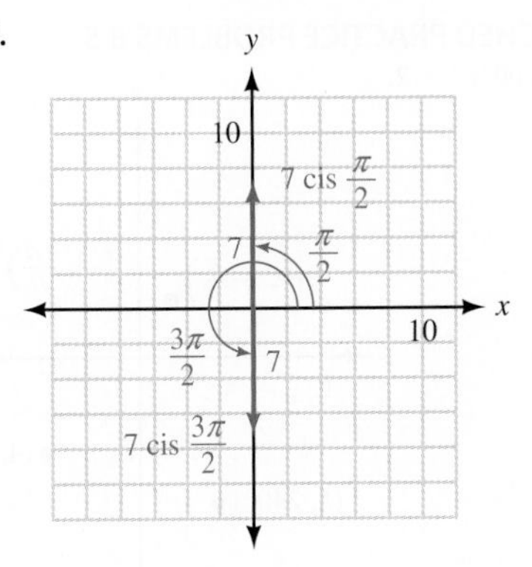

11. $\sqrt{3} + i, -\sqrt{3} - i$ **13.** $\sqrt{2} + i\sqrt{2}, -\sqrt{2} - i\sqrt{2}$ **15.** $5i, -5i$ **17.** $\frac{\sqrt{6}}{2} + \frac{\sqrt{2}}{2}i, -\frac{\sqrt{6}}{2} - \frac{\sqrt{2}}{2}i$

19. $2(\cos 70° + i \sin 70°), 2(\cos 190° + i \sin 190°), 2(\cos 310° + i \sin 310°)$

21. $2(\cos 10° + i \sin 10°), 2(\cos 130° + i \sin 130°), 2(\cos 250° + i \sin 250°)$

23. $3(\cos 60° + i \sin 60°), 3(\cos 180° + i \sin 180°), 3(\cos 300° + i \sin 300°)$

25. $4(\cos 30° + i \sin 30°), 4(\cos 150° + i \sin 150°), 4(\cos 270° + i \sin 270°)$

27. $-2, 1 + i\sqrt{3}, 1 - i\sqrt{3}$ **29.** $\pm\frac{3\sqrt{2}}{2} \pm \frac{3\sqrt{2}}{2}i$ **31.** $\sqrt{3} + i, -1 + i\sqrt{3}, -\sqrt{3} - i, 1 - i\sqrt{3}$

33.
$10(\cos 3° + i \sin 3°) \approx 9.99 + 0.52i$
$10(\cos 75° + i \sin 75°) \approx 2.59 + 9.66i$
$10(\cos 147° + i \sin 147°) \approx -8.39 + 5.45i$
$10(\cos 219° + i \sin 219°) \approx -7.77 - 6.29i$
$10(\cos 291° + i \sin 291°) \approx 3.58 - 9.34i$

35.

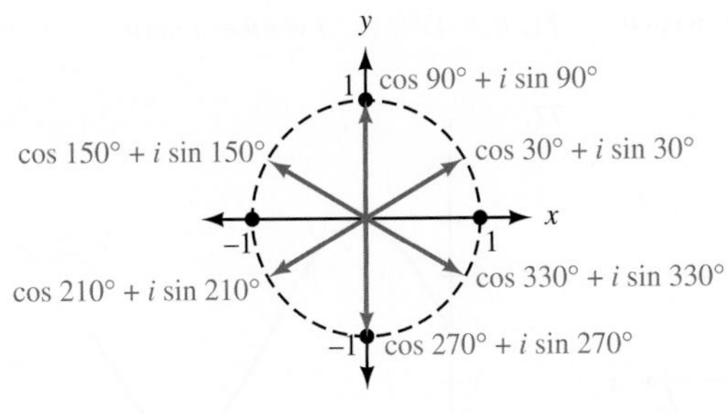

37. $\sqrt{2}(\cos\theta + i\sin\theta)$ where $\theta = 30°, 150°, 210°, 330°$

39. $\sqrt[4]{2}(\cos\theta + i\sin\theta)$ where $\theta = 67.5°, 112.5°, 247.5°, 292.5°$

41.

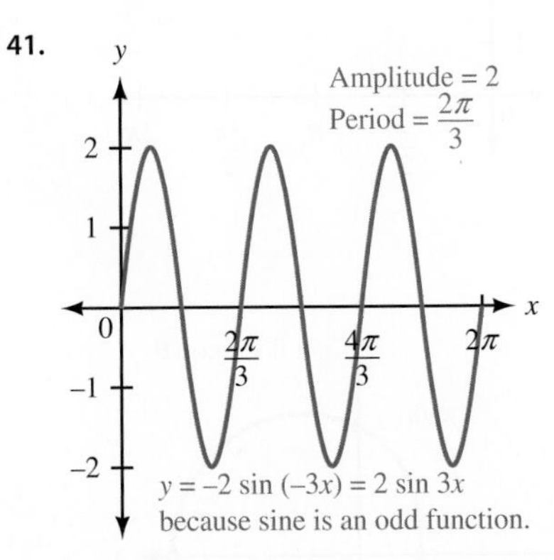

43.

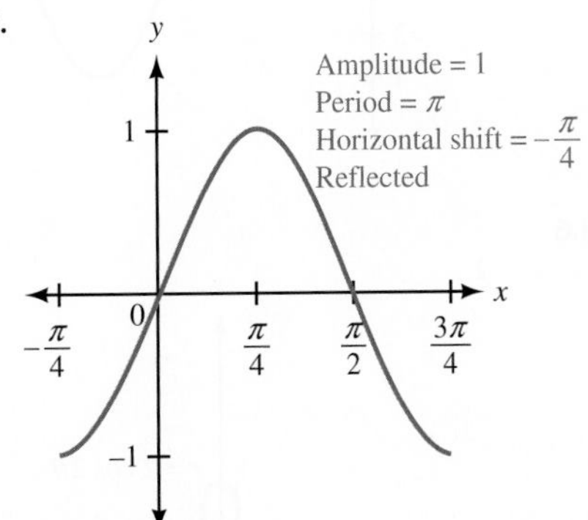

45. 4.23 cm^2 **47.** 3.8 ft^2 **49.** −3.732, −0.268, 4.000

MATCHED PRACTICE PROBLEMS 8.5

1. $(8, 60°)$ **2.**

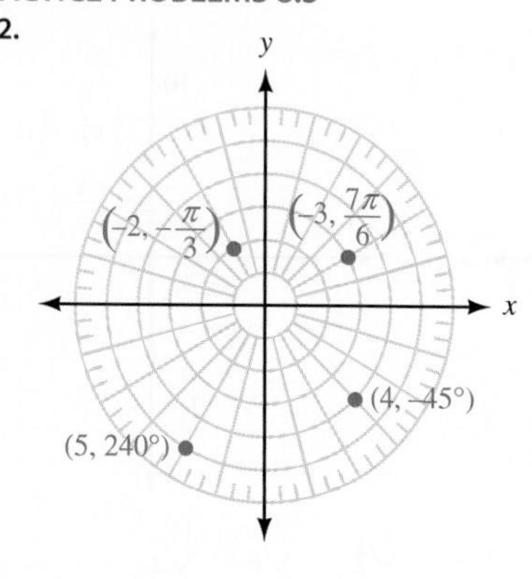

3. $(-2, 300°), (2, 480°), (-2, -60°)$

4. a. $(1, \sqrt{3})$ **b.** $(0, 4)$ **c.** $\left(\frac{\sqrt{2}}{2}, \frac{\sqrt{2}}{2}\right)$

5. a. $(3\sqrt{2}, 135°)$ **b.** $(4, 270°)$ **c.** $(2, 210°)$

6. $(x^2 + y^2)^2 = 4x^2 - 4y^2$ **7.** $r = \dfrac{9}{2\cos\theta + \sin\theta}$

PROBLEM SET 8.5

1. pole, polar axis **3.** counterclockwise, polar **5.** origin, positive x **7.** true

9–19. (odd)

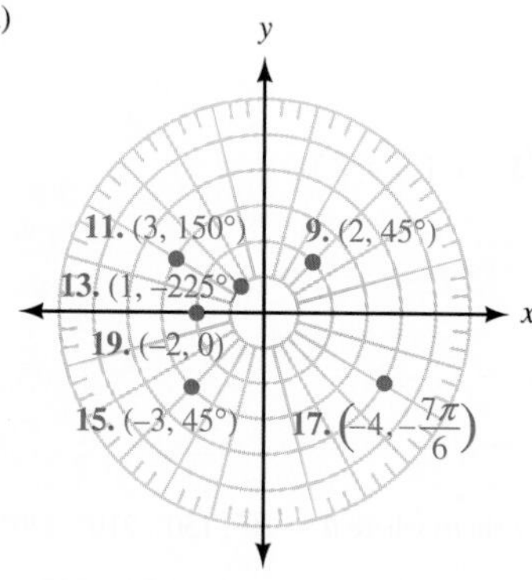

21. $(2, -300°), (-2, 240°), (-2, -120°)$ **23.** $(5, -225°), (-5, 315°), (-5, -45°)$ **25.** $(-3, -330°), (3, -150°), (3, 210°)$ **27.** $(1, \sqrt{3})$ **29.** $(0, -3)$ **31.** $(-1, -1)$ **33.** $(-6, -2\sqrt{3})$ **35.** $(1.891, 0.6511)$ **37.** $(-1.172, 2.762)$ **39.** $(3\sqrt{2}, 135°)$ **41.** $(4, 300°)$ **43.** $(2, 0)$ **45.** $\left(2, \frac{7\pi}{6}\right)$ **47.** $(5, 53.1°)$ **49.** $(\sqrt{5}, 116.6°)$ **51.** $(\sqrt{13}, 236.3°)$ **53.** $(9.434, 57.99°)$ **55.** $(6.083, -99.46°)$ **57.** $x^2 + y^2 = 9$ **59.** $x^2 + y^2 = 6x$ **61.** $(x^2 + y^2)^2 = 4(x^2 - y^2)$ **63.** $x + y = 3$ **65.** $r = \dfrac{5}{\cos\theta + \sin\theta}$ **67.** $r = 2$ **69.** $r = 6\cos\theta$ **71.** $\theta = 45°$ or $r\cos\theta = r\sin\theta$

73.

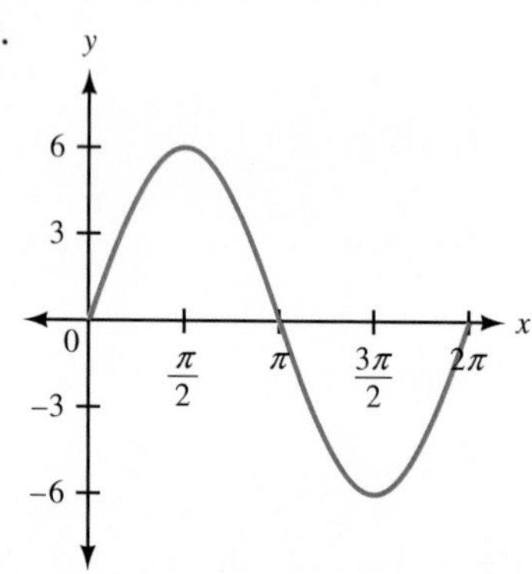

75.

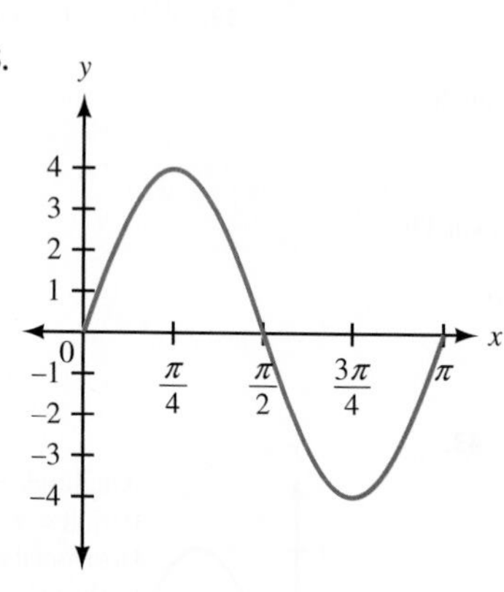

77.

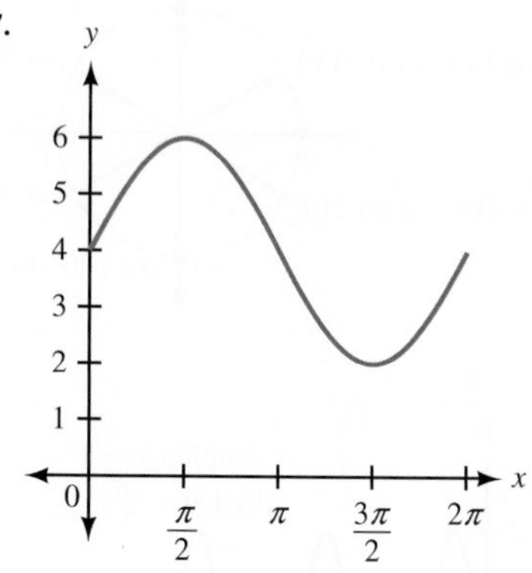

MATCHED PRACTICE PROBLEMS 8.6

1.

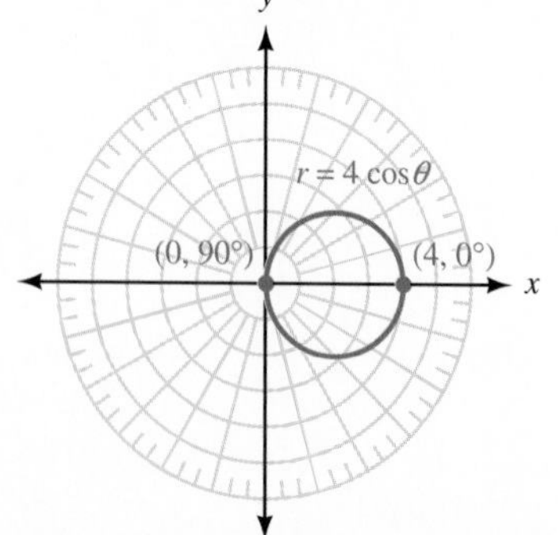

2.

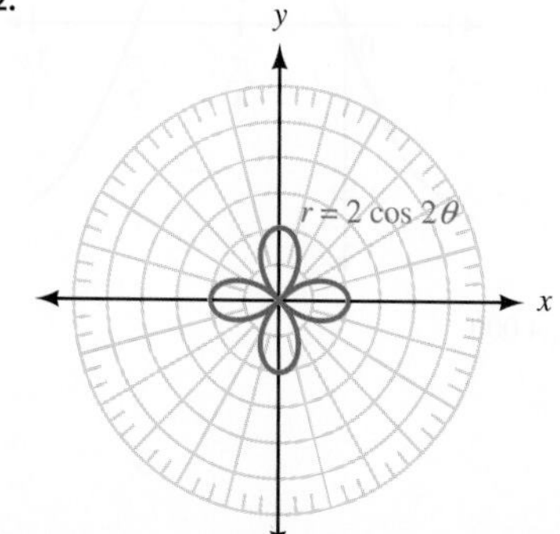

3.

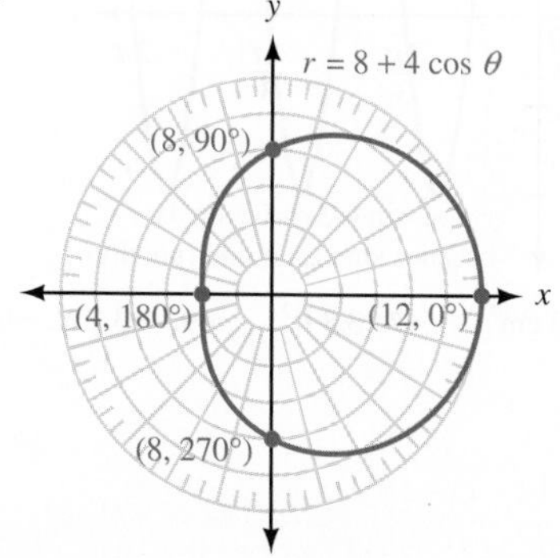

PROBLEM SET 8.6

1. θ, r **3.** rectangular

5.

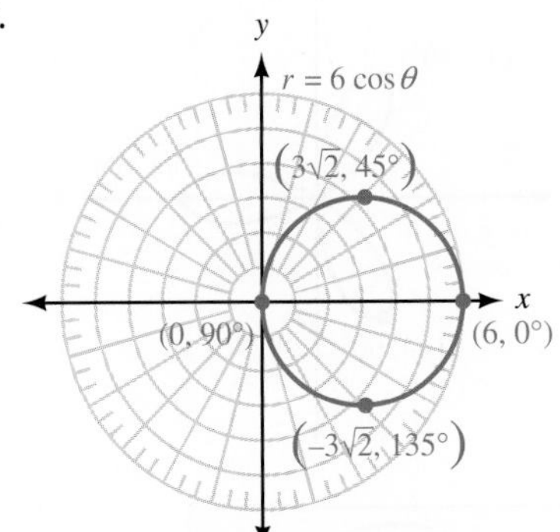

7.

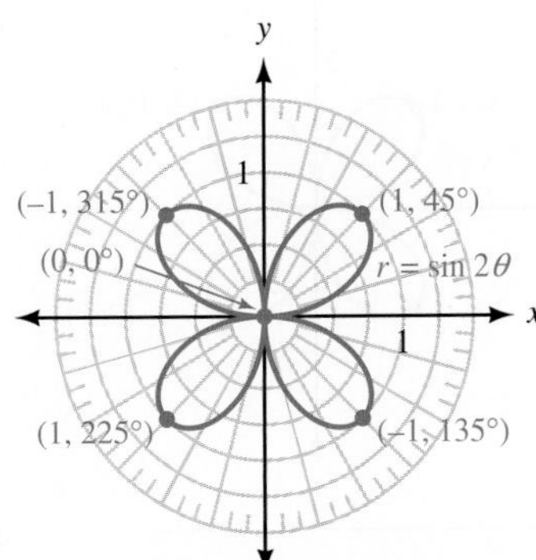

9.

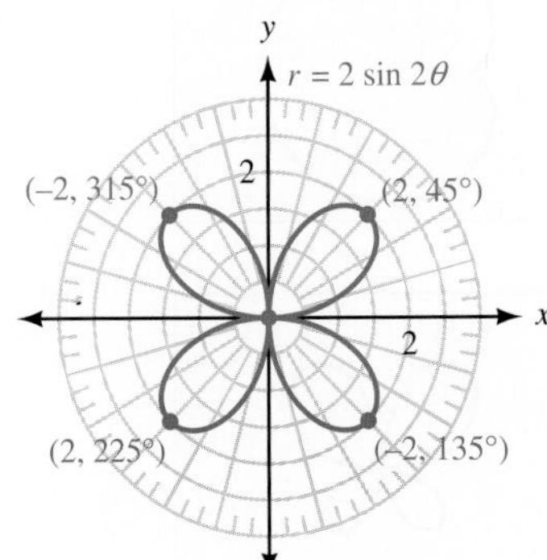

11.

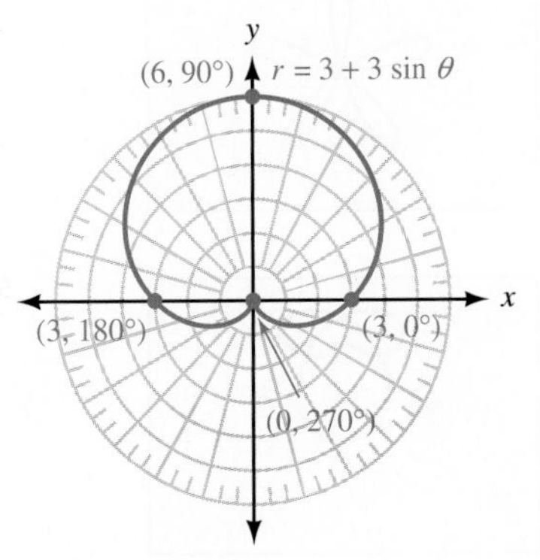

13. Rose curve **15.** Cardioid **17.** Circle **19.** Lemniscate **21.** Line **23.** Limaçon

25.

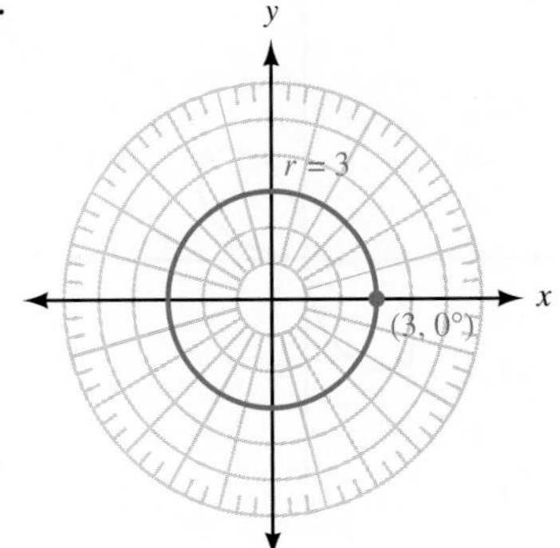

27.

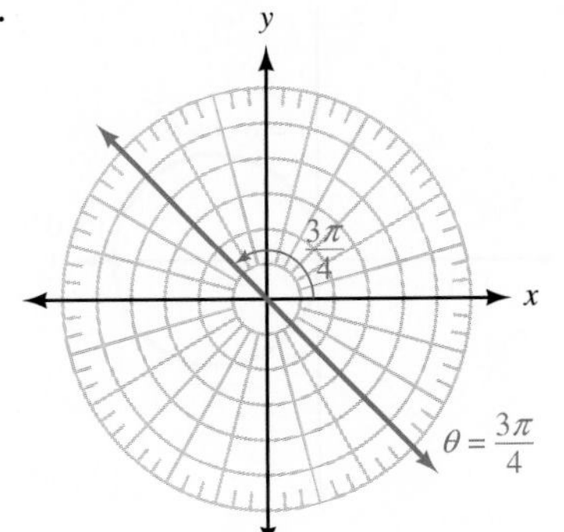

29.

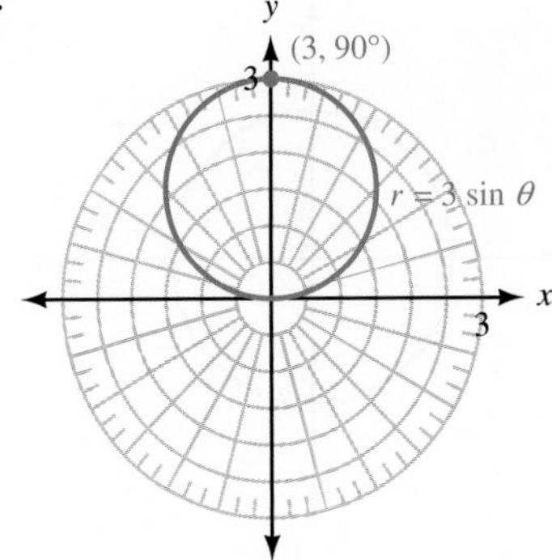

31.

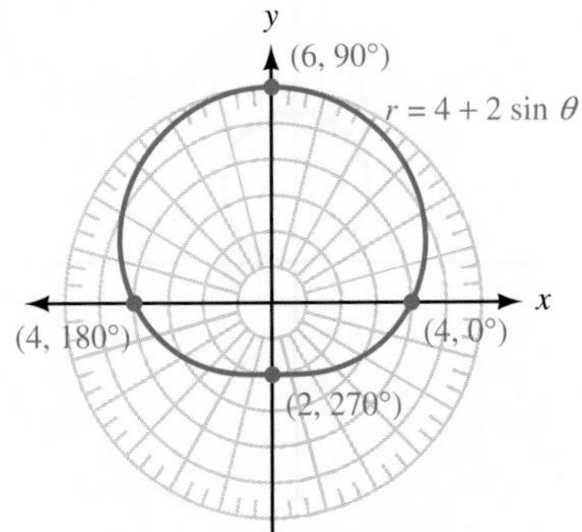

33.

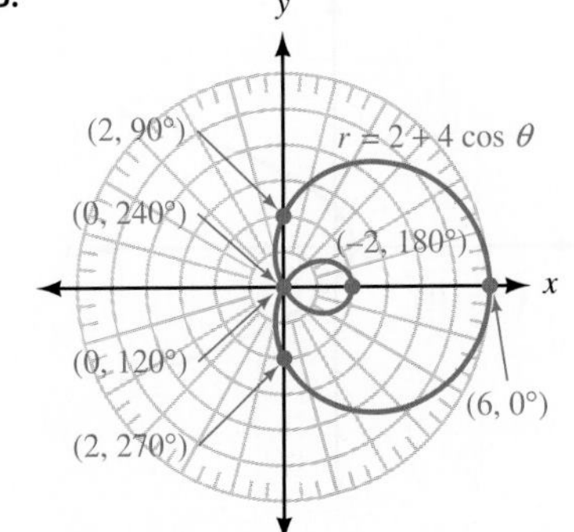

35.

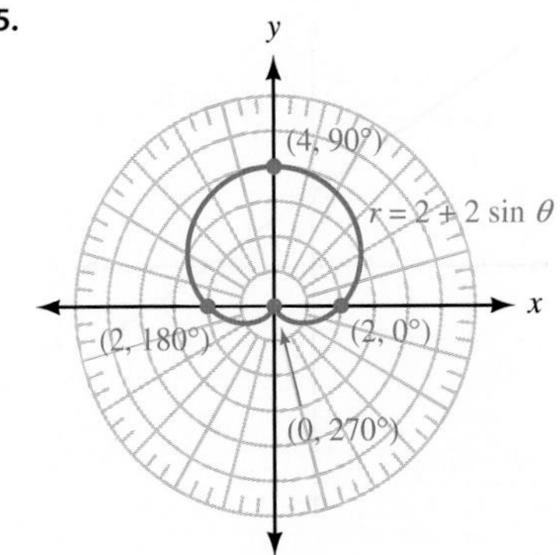

37.

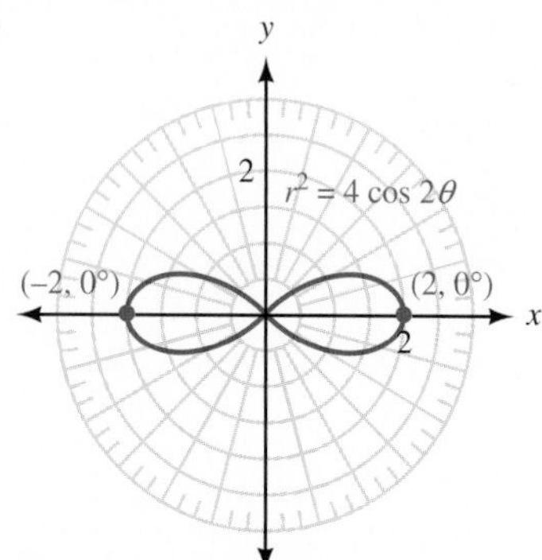

39.

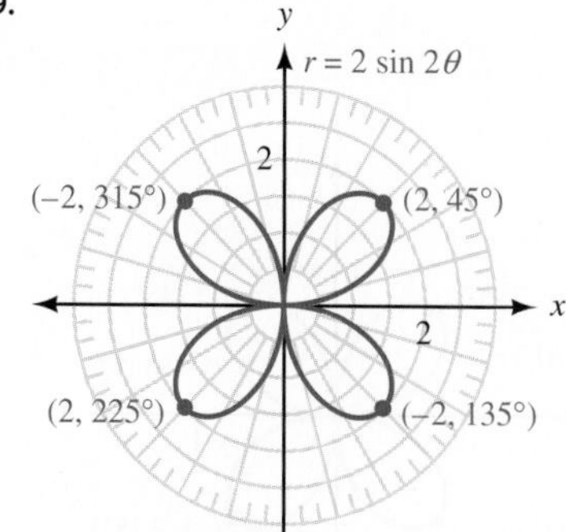

41.

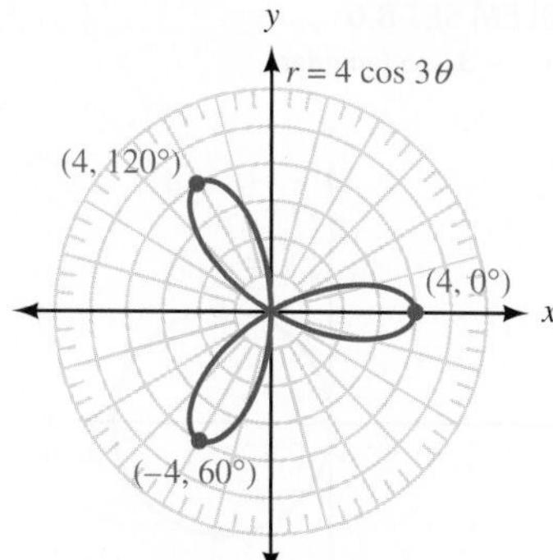

43.

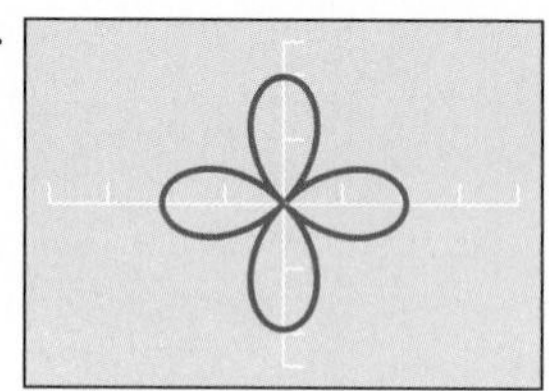

45.

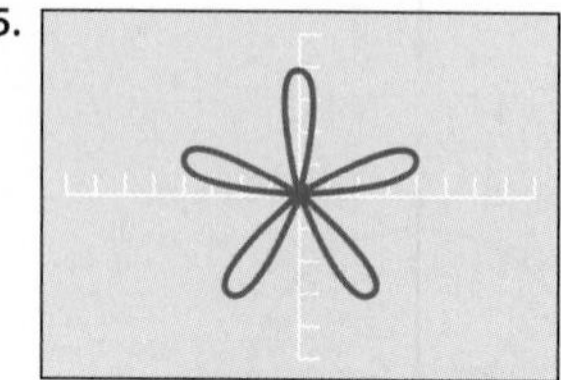

47.

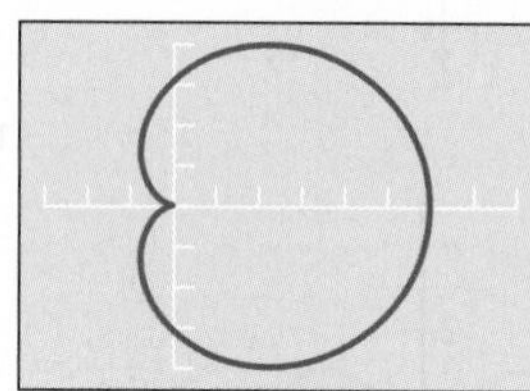

49.

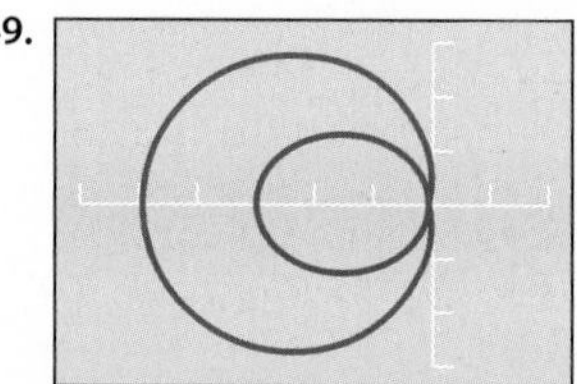

51.

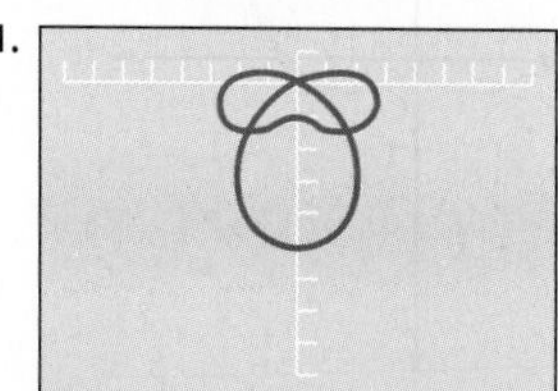

53.

55.

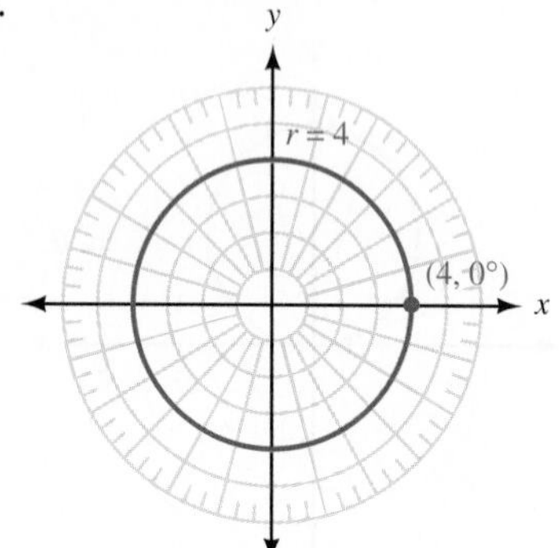

57.

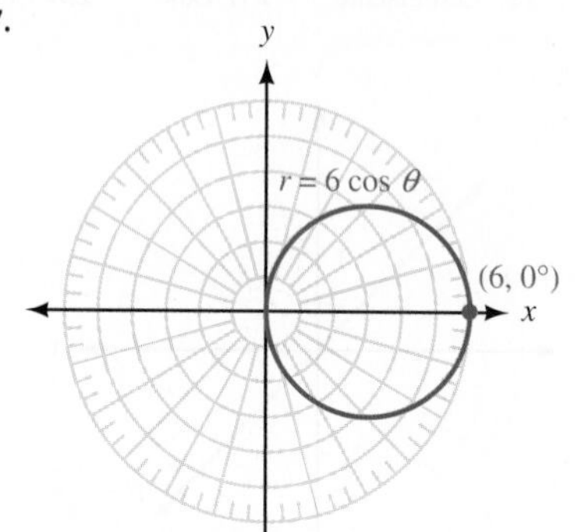

59.

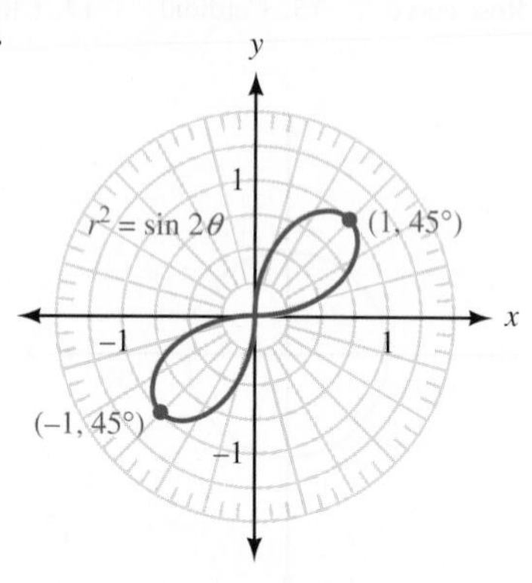

61.

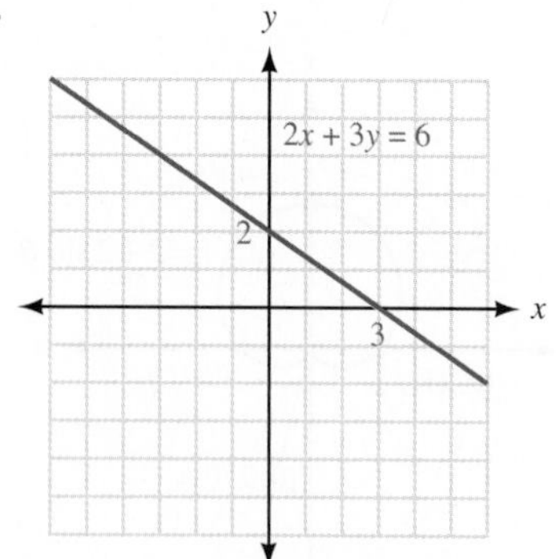

63.

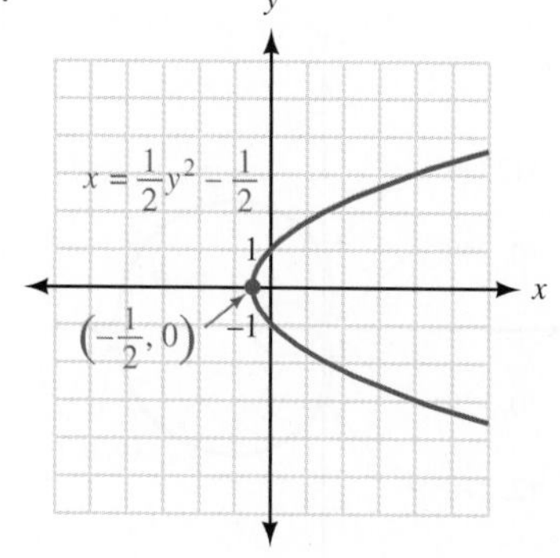

65.

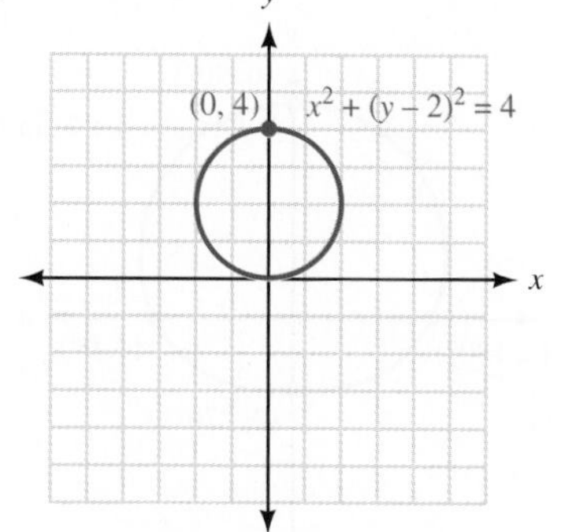

67.

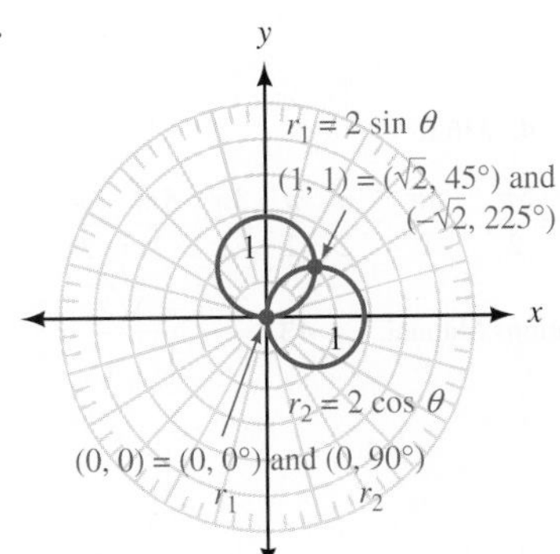

69.

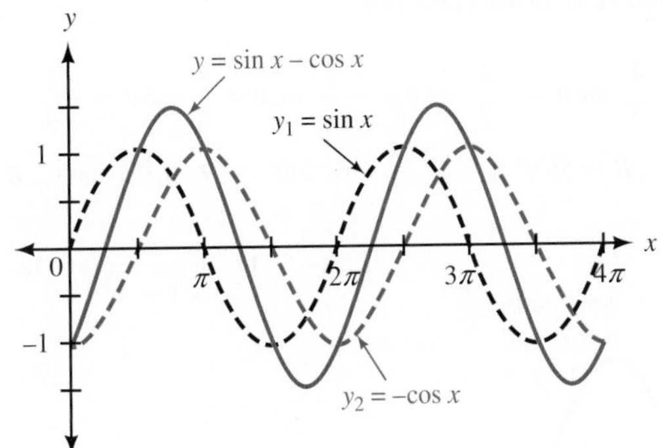

71.

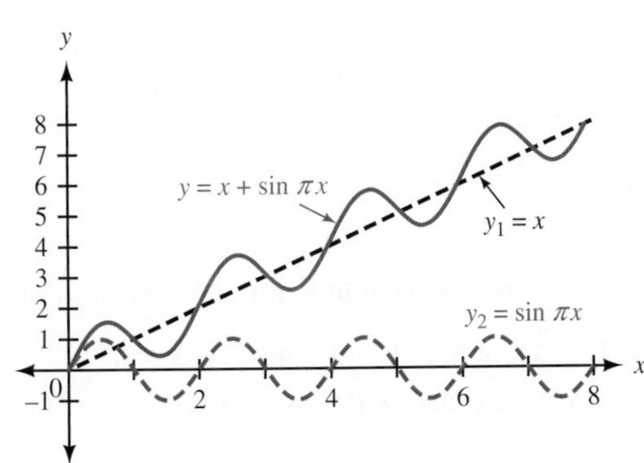

73.

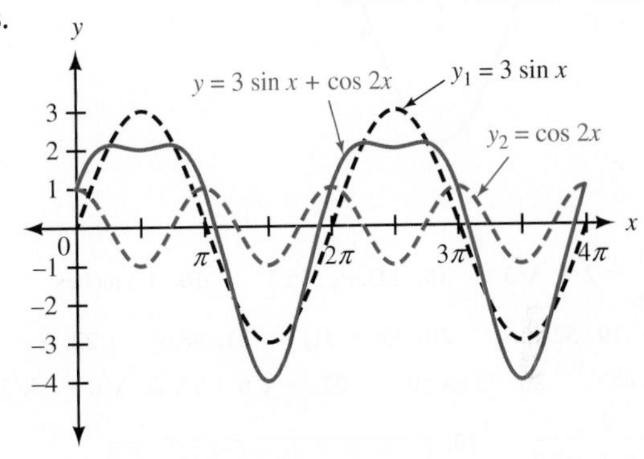

CHAPTER 8 TEST

1. $2i\sqrt{3}$ **2.** $x = -2$ or 5, $y = 2$ **3.** $7 - 6i$ **4.** i **5.** 89 **6.** $-16 + 30i$ **7.** $\frac{11}{61} + \frac{60}{61}i$

8. a. 5 **b.** $-3 + 4i$ **c.** $3 + 4i$ **9. a.** 8 **b.** $-8i$ **c.** $-8i$ **10.** $4\sqrt{3} - 4i$ **11.** $-\sqrt{2} + i\sqrt{2}$
12. $2(\cos 150° + i \sin 150°)$ **13.** $5(\cos 90° + i \sin 90°)$ **14.** $15(\cos 65° + i \sin 65°)$ **15.** $5(\cos 30° + i \sin 30°)$
16. $81(\cos 80° + i \sin 80°) = 81 \text{ cis } 80°$ **17.** $7(\cos 25° + i \sin 25°)$, $7(\cos 205° + i \sin 205°)$
18. $\sqrt{2}(\cos \theta + i \sin \theta)$ where $\theta = 15°, 105°, 195°, 285°$ **19.** $x = \sqrt{2}(\cos \theta + i \sin \theta)$ where $\theta = 15°, 165°, 195°, 345°$
20. $x = \cos \theta + i \sin \theta$ where $\theta = 60°, 180°, 300°$ **21.** $(6, 240°)$, $(6, -120°)$; $(-3, -3\sqrt{3})$
22. $\left(3\sqrt{2}, \frac{3\pi}{4}\right)$ **23.** $x^2 + y^2 = 6y$ **24.** $r = 8 \sin \theta$

25.

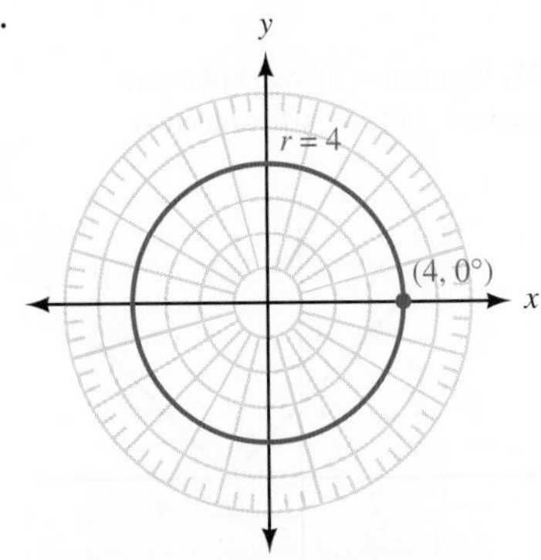

26.

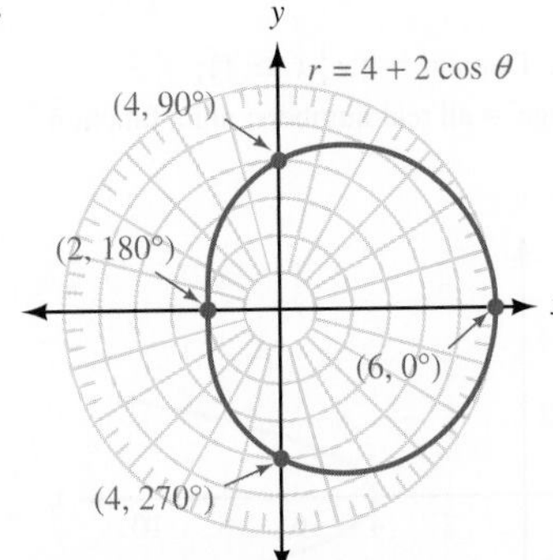

27.

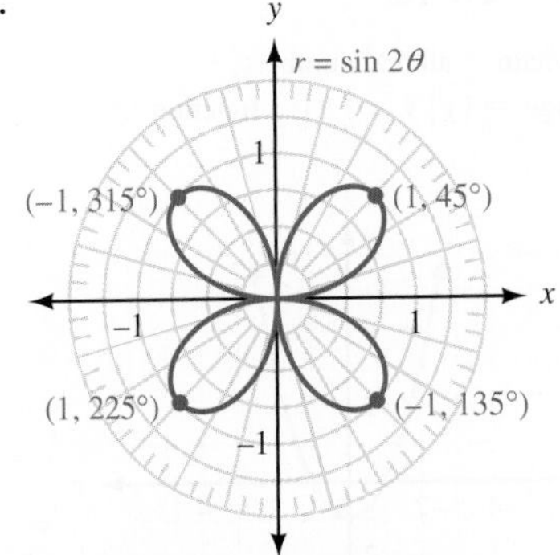

28.

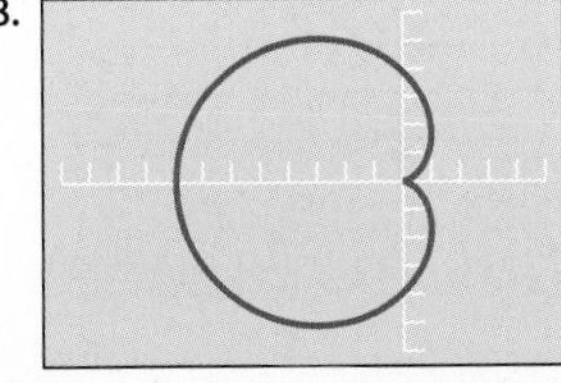

29.

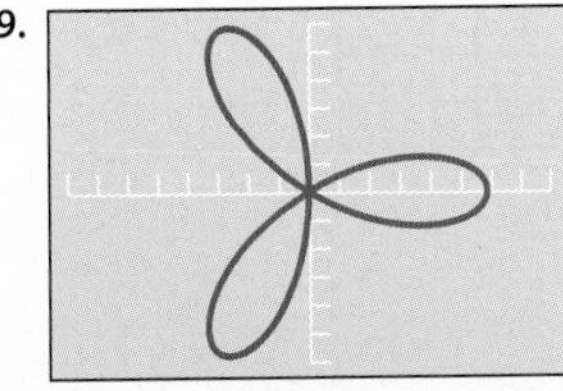

30.

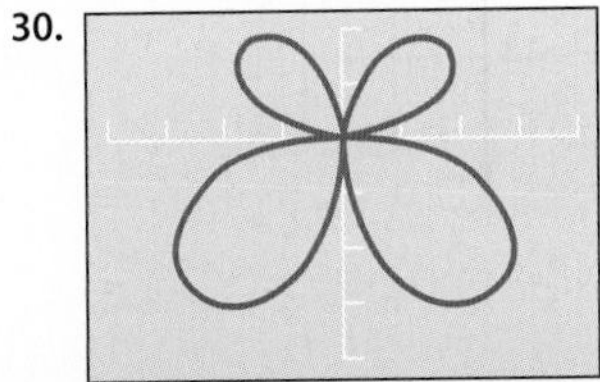

CUMULATIVE TEST CHAPTERS 1–8

1. $\sin\theta = -\frac{4}{5}$, $\tan\theta = -\frac{4}{3}$, $\cot\theta = -\frac{3}{4}$, $\sec\theta = \frac{5}{3}$, $\csc\theta = -\frac{5}{4}$ **2.** 13.3° **3.** 310.3° **4.** 336 ft

5. $A = 33.1°$, $B = 56.9°$, $c = 37.5$ **6.** 240° **7.** 0.375 cm **8.** 20π ft/min **9.** $-\frac{7\pi}{2}, -\frac{5\pi}{2}, -\frac{3\pi}{2}, -\frac{\pi}{2}, \frac{\pi}{2}, \frac{3\pi}{2}, \frac{5\pi}{2}, \frac{7\pi}{2}$

10. Amplitude $= \frac{1}{2}$, Period $= 4$

11. $\frac{x}{\sqrt{1-x^2}}$ **12.** A proof is given in the Solutions Manual. **13.** $\frac{\sqrt{6}+\sqrt{2}}{4}$

14. $\frac{\sqrt{3}+1}{\sqrt{3}-1} = 2 + \sqrt{3}$ **15.** 203.8°, 336.2° **16.** 4.3 inches **17.** $B = 59°$, $C = 95°$, $c = 11$ ft; $B' = 121°$, $C' = 33°$, $c' = 6.0$ ft

18. 69.5° **19.** 52 km^2 **20.** $35\mathbf{i} + 31\mathbf{j}$ **21.** 98.6° **22.** $8 - 2i$ **23.** 1 **24. a.** 5 **b.** $-3 - 4i$ **c.** $3 - 4i$

25. $2\sqrt{2}$ cis 45° **26.** 32 cis 50° **27.** $-\sqrt{6} + i\sqrt{2},\ \sqrt{6} - i\sqrt{2}$ **28.** $(-2\sqrt{2}, -2\sqrt{2})$; $(-4, 45°)$, $(4, -135°)$

29. $r = \frac{2}{\cos\theta + \sin\theta}$ **30.**

APPENDIX A

PROBLEM SET A.1

1. ordered pairs **3.** domain, range **5.** vertical line **7.** True **9.** Domain = {1, 2, 4}; range = {1, 3, 5}; a function

11. Domain = {−1, 1, 2}; range = {−5, 3}; a function **13.** Domain = {3, 7}; range = {−1, 4}; not a function **15.** Yes **17.** No

19. No **21.** Yes

23. Domain = all real numbers; range = $\{y \mid y \geq -1\}$; a function

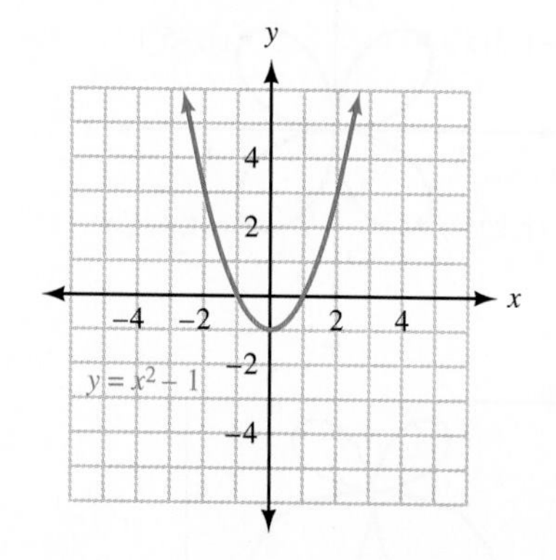

25. Domain = $\{x \mid x \geq 4\}$; range = all real numbers; not a function

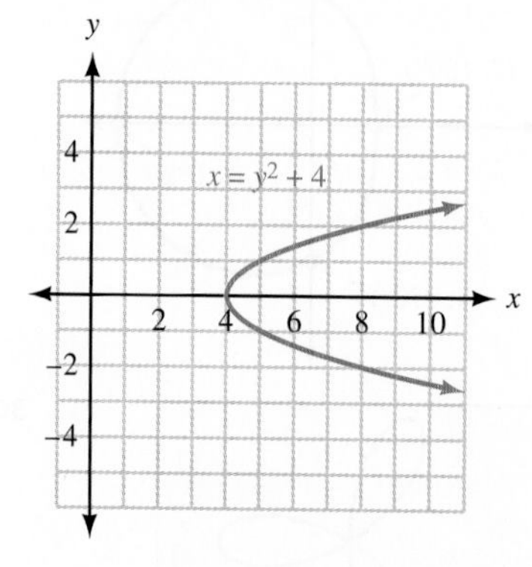

27. Domain = all real numbers; range = $\{y \mid y \geq 0\}$; a function

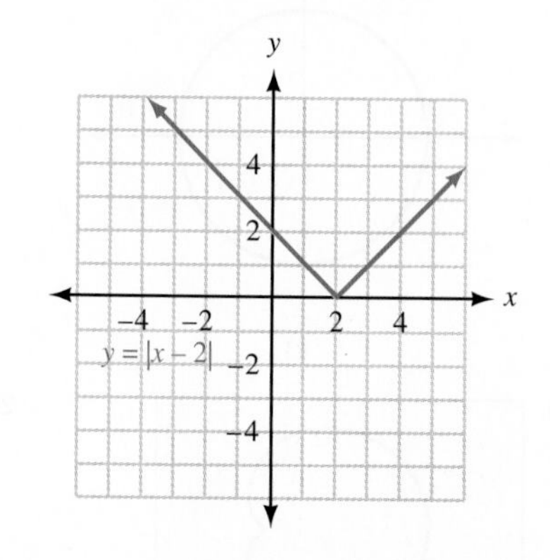

29. a. $y = 8.5x \quad \text{for} \quad 10 \le x \le 40$

b.

Hours Worked x	Gross Pay ($) y
10	85
20	170
30	255
40	340

c.

Gross pay ($)
350
300
250
200
150
100
50
0
0 10 20 30 40
Hours worked

d. Domain $= \{x \mid 10 \le x \le 40\}$; range $= \{y \mid 85 \le y \le 340\}$

31. a.

Time (sec) t	Distance (ft) h
0	0
0.1	1.44
0.2	2.56
0.3	3.36
0.4	3.84
0.5	4
0.6	3.84
0.7	3.36
0.8	2.56
0.9	1.44
1	0

b. Domain $= \{t \mid 0 \le t \le 1\}$; range $= \{h \mid 0 \le h \le 4\}$

c.

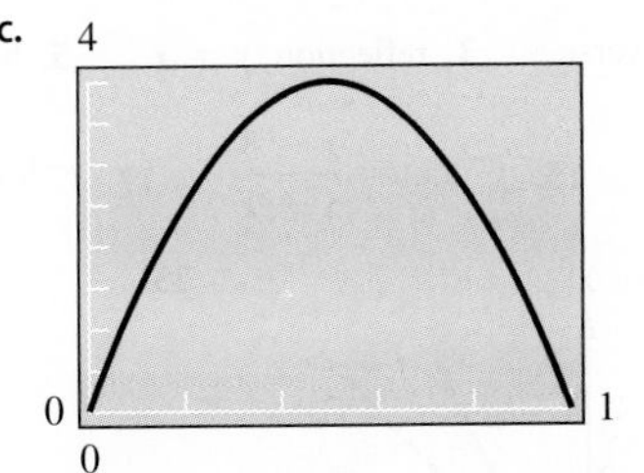

33. a.

A
25
20
15
10
5
$A = \pi r^2, 0 \le r \le 3$
1 2 3 4
r

b. Domain $= \{r \mid 0 \le r \le 3\}$; range $= \{A \mid 0 \le A \le 9\pi\}$

35. -1 **37.** -11 **39.** 2 **41.** 4 **43.** -4 **45.** $2a - 5$ **47.** 35 **49.** -13 **51.** 4 **53.** 0 **55.** 2

57.

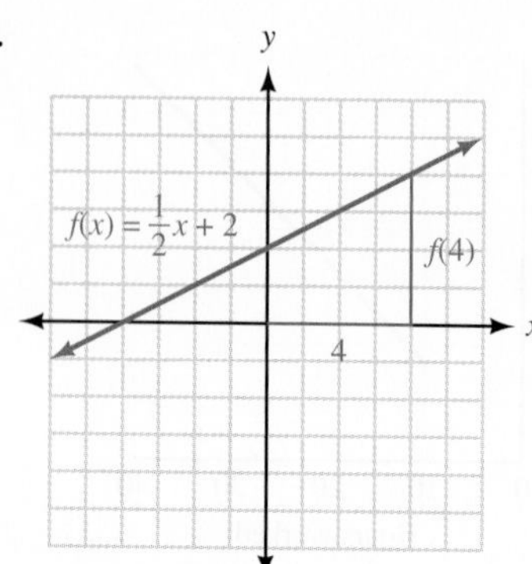

59. $V(3) = 300$, the painting is worth $300 in 3 years; $V(6) = 600$, the painting is worth $600 in 6 years.
61. $A(2) = 12.56$; $A(5) = 78.5$; $A(10) = 314$
63. **a.** 2 **b.** 0 **c.** 1 **d.** 4

PROBLEM SET A.2

1. interchange or reverse **3.** reflection, $y = x$ **5.** horizontal line **7.** exchange, y **9.** False **11.** $f^{-1}(x) = \frac{x+1}{3}$

13. $f^{-1}(x) = \sqrt[3]{x}$ **15.** $f^{-1}(x) = \frac{x-3}{x-1}$ **17.** $f^{-1}(x) = 4x + 3$ **19.** $f^{-1}(x) = 2x + 6$ **21.** $f^{-1}(x) = \frac{1-x}{3x-2}$

23.

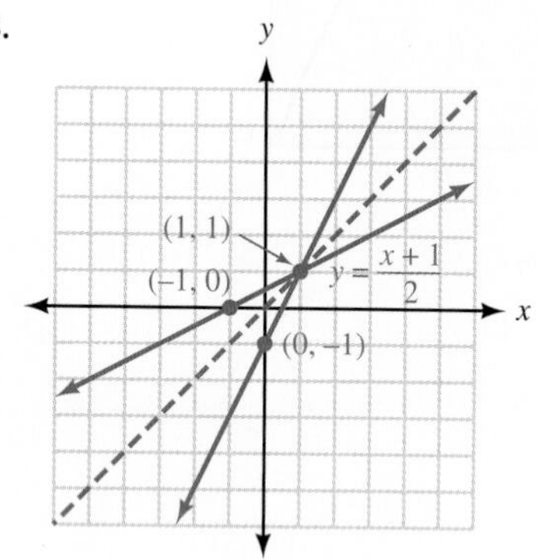

25.

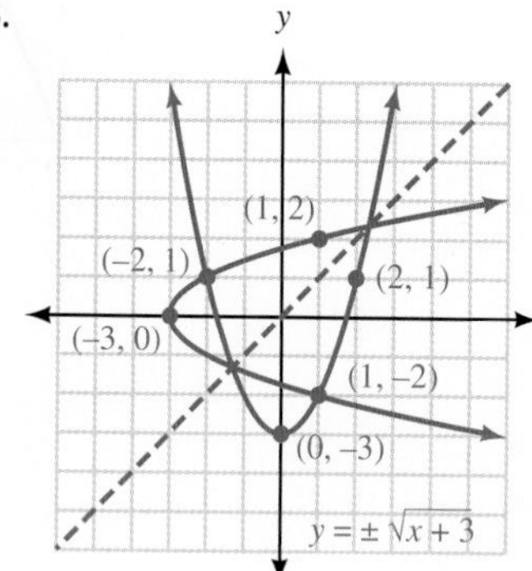

27.

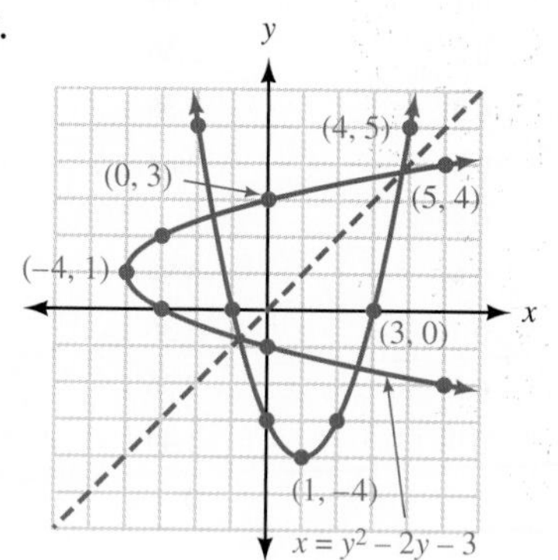

29.

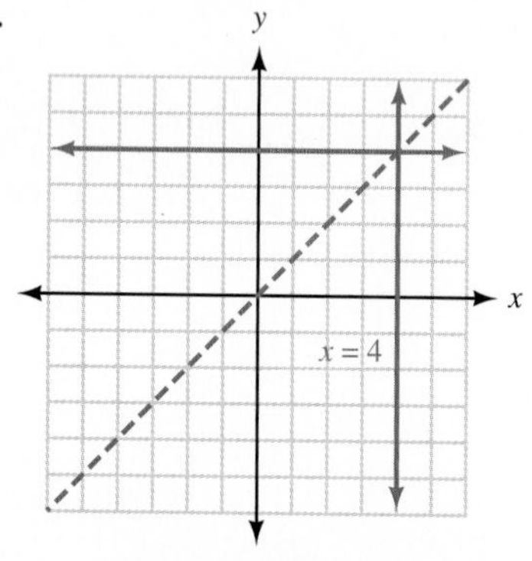

31.

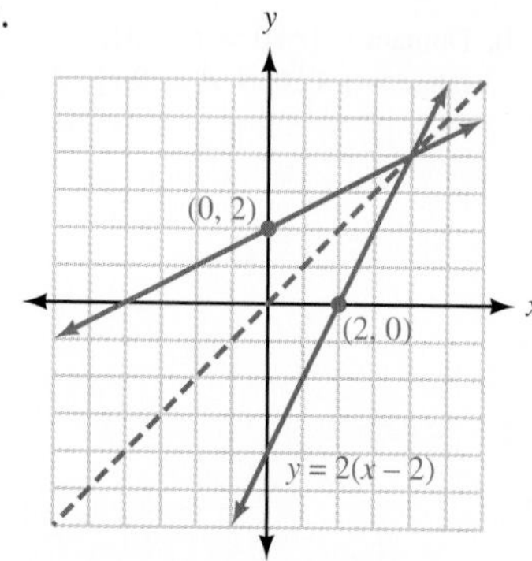

33.

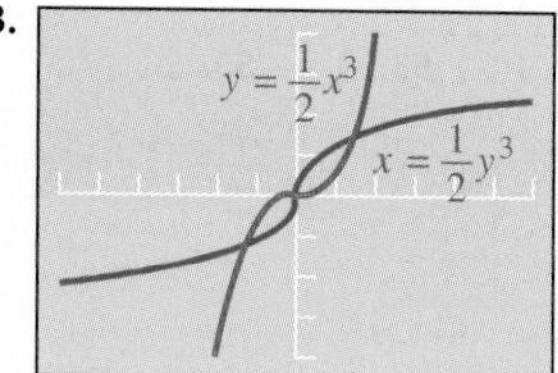

35.

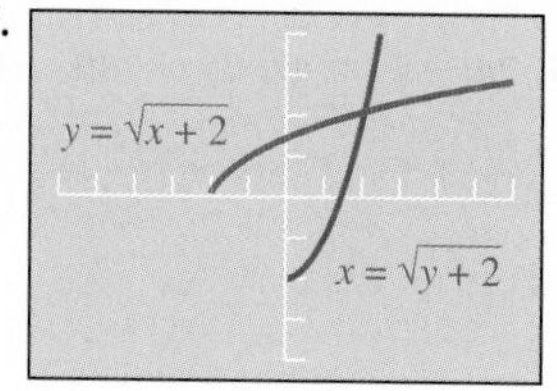

37. **a.** Yes **b.** No **39.** **a.** 4 **b.** $\frac{4}{3}$ **c.** 2 **d.** 2 **41.** $f^{-1}(x) = \frac{1}{x}$ **43.** **a.** −3 **b.** −6 **c.** 2 **d.** 3 **e.** −2 **f.** 3 **g.** They are inverses of each other. **45.** $f^{-1}(x) = \frac{x-5}{3}$ **47.** $f^{-1}(x) = \sqrt[3]{x-1}$ **49.** $f^{-1}(x) = 7(x + 2)$

51. **a.** 1 **b.** 2 **c.** 5 **d.** 0 **e.** 1 **f.** 2 **g.** 2 **h.** 5

INDEX

B

C

D

L

M

N

O

P

Q

R

MISCELLANEOUS FACTS

Exact Values on the Unit Circle [3.3]

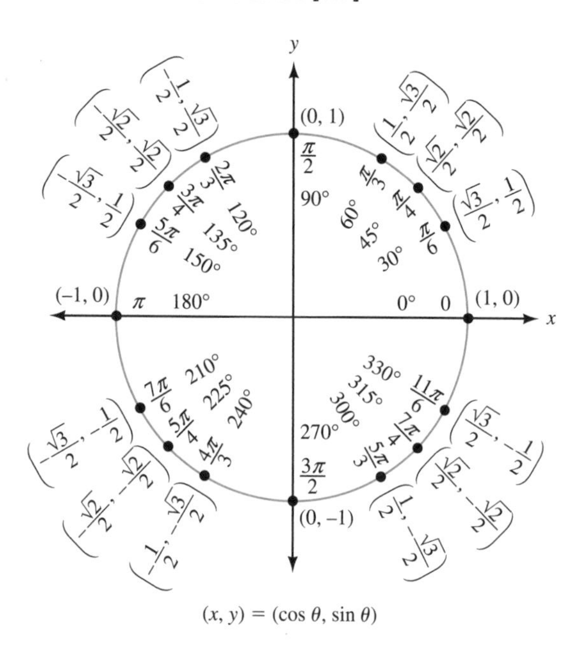

$(x, y) = (\cos\theta, \sin\theta)$

Radian Measure [3.2]

$$\theta \text{ (in radians)} = \frac{s}{r}$$

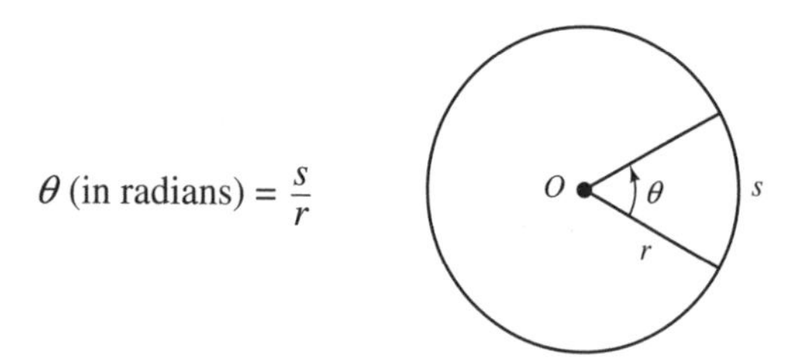

Converting Between Radians and Degrees [3.2]

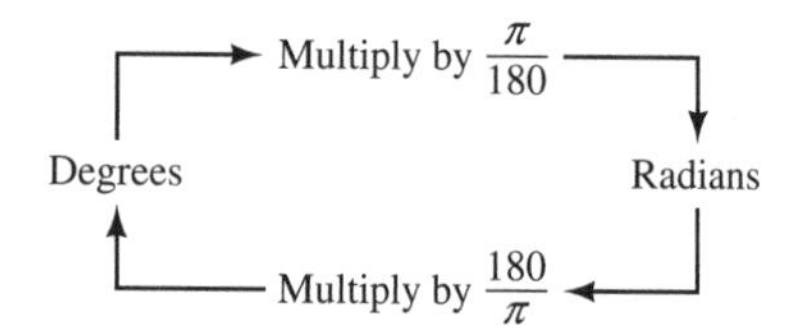

Reference Angles [3.1]

The *reference angle* $\hat{\theta}$ for any angle θ in standard position is the positive acute angle between the terminal side of θ and the x-axis.

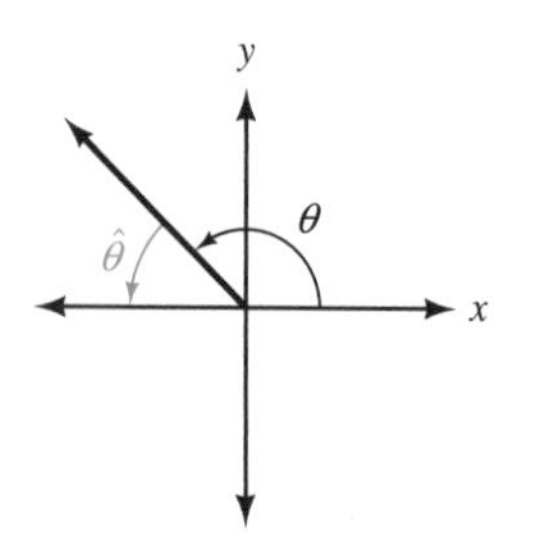

Inverse Trigonometric Functions [4.7]

Inverse Function	Meaning
$y = \sin^{-1} x$ or $y = \arcsin x$	$x = \sin y$ and $-\frac{\pi}{2} \le y \le \frac{\pi}{2}$
In words: y is the angle between $-\pi/2$ and $\pi/2$, inclusive, whose sine is x.	
$y = \cos^{-1} x$ or $y = \arccos x$	$x = \cos y$ and $0 \le y \le \pi$
In words: y is the angle between 0 and π, inclusive, whose cosine is x.	
$y = \tan^{-1} x$ or $y = \arctan x$	$x = \tan y$ and $-\frac{\pi}{2} < y < \frac{\pi}{2}$
In words: y is the angle between $-\pi/2$ and $\pi/2$ whose tangent is x.	

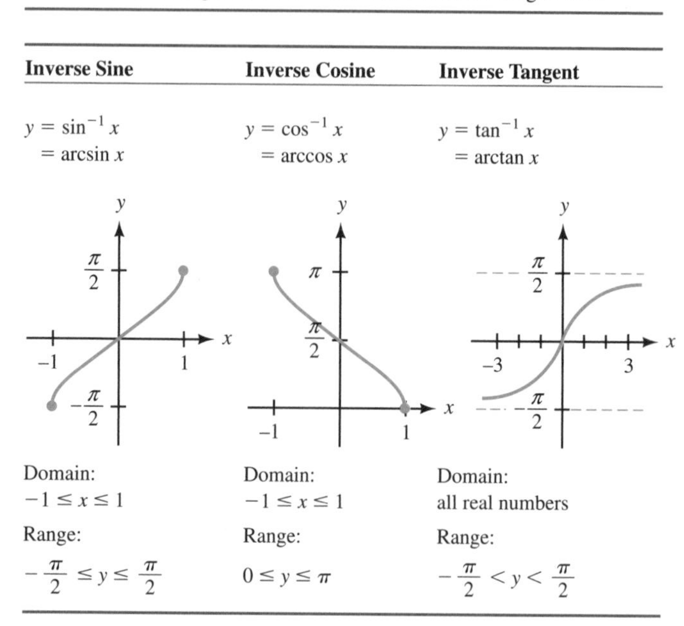

Inverse Sine	Inverse Cosine	Inverse Tangent
$y = \sin^{-1} x = \arcsin x$	$y = \cos^{-1} x = \arccos x$	$y = \tan^{-1} x = \arctan x$
Domain: $-1 \le x \le 1$	Domain: $-1 \le x \le 1$	Domain: all real numbers
Range: $-\frac{\pi}{2} \le y \le \frac{\pi}{2}$	Range: $0 \le y \le \pi$	Range: $-\frac{\pi}{2} < y < \frac{\pi}{2}$

Uniform Circular Motion [3.4, 3.5]

A point on a circle of radius r moves a distance s on the circumference of the circle, in an amount of time t.

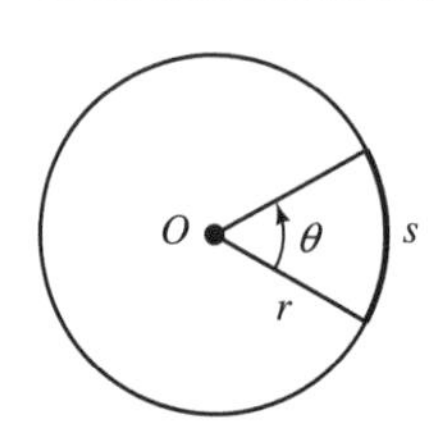

θ is measured in radians

Angular Velocity

$$\omega = \frac{\theta}{t}$$

Linear Velocity

$$v = \frac{s}{t}$$

$$v = r\omega$$

Arc Length

$$s = r\theta$$

Area of a Sector

$$A = \frac{1}{2}r^2\theta$$

The Area of a Triangle [7.4]

$$S = \frac{1}{2}ab\sin C$$

$$S = \sqrt{s(s-a)(s-b)(s-c)}$$

where $s = \frac{1}{2}(a + b + c)$

$$S = \frac{a^2 \sin B \sin C}{2 \sin A}$$

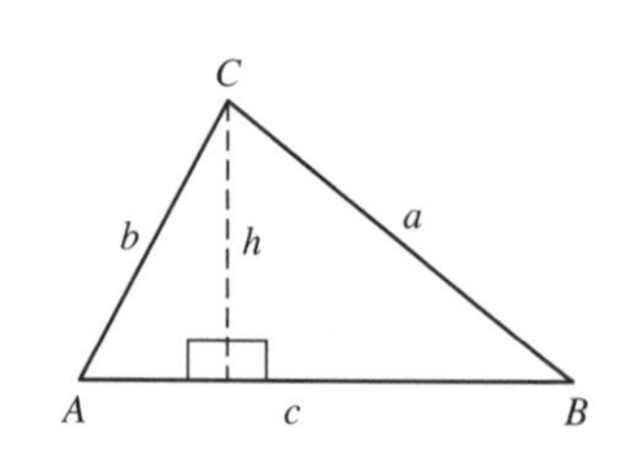